Auto Engine Repair

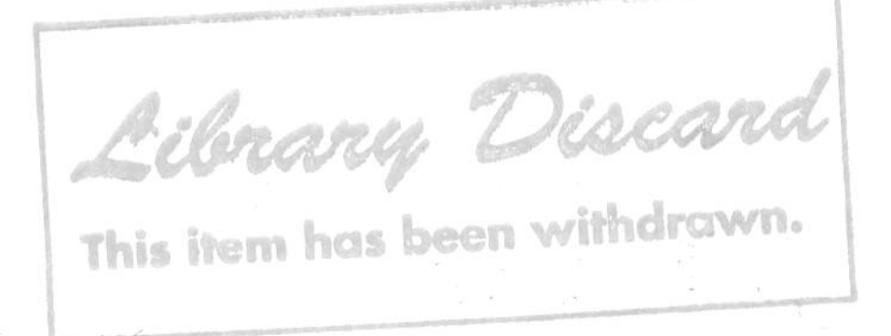

by

James E. Duffy
Automotive Writer

Publisher
The Goodheart-Willcox Company, Inc.
Tinley Park, Illinois
www.g-w.com

Manufactured in the United States of America.

Library of Congress Catalog Card Number 2004052354
International Standard Book Number 1-59070-400-2

1 2 3 4 5 6 7 8 9 – 05 – 09 08 07 06 05

Library of Congress Cataloging-in-Publication Data

Duffy, James E.
Auto Engine Repair / by James E. Duffy.

p. cm.
Includes index.
ISBN 1-59070-400-2
1. Automobiles--Motors--Maintenance and repair. I. Title.

TL210.D7797 2004
629.25'04'0288--dc22 2004052354

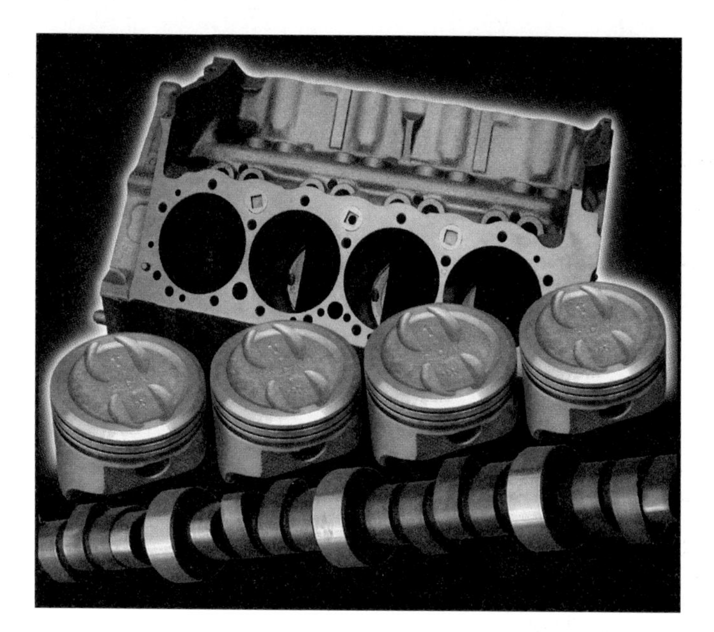

Introduction

Auto Engine Repair will help you learn to diagnose and rebuild all makes and models of gasoline engines. The text will prepare you to use factory service manuals and specifications to complete competent engine service and repair work on any gasoline engine found in today's cars, vans, SUVs, and light trucks.

This book has been revised and reorganized to help you pass the ASE certification test on engine repair. The engine repair test primarily includes questions on how to troubleshoot, repair, and rebuild passenger vehicle engines. This edition also has new information on engine machining and rebuilding.

Automobile manufacturers are now using advanced engine designs to improve power, efficiency, and dependability. Today's cars and light trucks are equipped with computer-controlled; high-performance; 4-, 6-, 8-, 10-, and 12-cylinder engines. Many engines have three or four valves per cylinder, overhead camshafts, aluminum heads, aluminum blocks, turbochargers, superchargers, plastic parts, and many other engineering advances. This textbook was written to help you understand and service engines with these challenging innovations.

Auto Engine Repair has 26 chapters that are grouped into six sections. Section 1 reviews fundamentals of engine operation, equipment, and safety. Section 2 details engine construction and design so you will know how engines are supposed to operate and how they are manufactured. Section 3 summarizes the theory and service of engine-related systems. To be a competent engine technician, you must understand and be able to service starting systems, ignition systems, fuel systems, emission control systems, diesel injection systems, turbochargers, and more. The ASE engine repair test includes questions on engine support systems. Section 4 explains how to diagnose both mechanical and performance problems. It also summarizes engine tune-up procedures. Section 5 covers the overhauling or rebuilding of engines. This section has been expanded to include more information on automotive machining. Section 6 summarizes ASE certification tests and career success.

Auto Engine Repair is written so you may study the chapters in the sequence presented or concentrate on certain topics. The chapters are small and cover specific engine assemblies or related subjects. Cross-references help you find additional information elsewhere in the book.

Each chapter begins with learning objectives, which emphasize the important topics you will study in that chapter. A summary is provided at the end of the chapter to help you review the most important concepts. The review questions at the end of each chapter allow you to measure your progress. ASE-type questions are also provided at the end of the chapter to help you prepare for certification.

Auto Engine Repair was developed to help you become a professional engine technician as quickly as possible. Since engine repair is very technical and precise, the book was written using short, easy-to-understand sentences accompanied by carefully selected illustrations. Color photos and line drawings have been used to illustrate as much as possible about engine repair. Technical terms are printed in ***bold italic*** and defined so you can quickly learn the "language" of an engine technician.

Chapter 5 provides general information about shop safety. Additional instruction regarding safety is stressed throughout the book. You can never know too much about safety. There are three types of special notes you will encounter in this book:

Note: A note may highlight important technical information, present information related to the topic, or cross-reference the material.

Caution: A caution identifies a situation that may result in damage to the vehicle, tools, or equipment if the proper procedure is not followed.

Warning: A warning identifies operations that may potentially result in injury.

Specific repair procedures are presented in a sequential, step-by-step list. This allows you to easily follow the procedure as you perform the operation.

Auto Engine Repair is a valuable guide to anyone wanting to understand or work on today's engines. It will help a student enter the trade more easily, but it will also help the vehicle owner know more about the operation and servicing of engines. An experienced technician can also benefit from this text by using it to become better informed about the latest engine technology and repair methods.

—James E. Duffy

Brief Contents

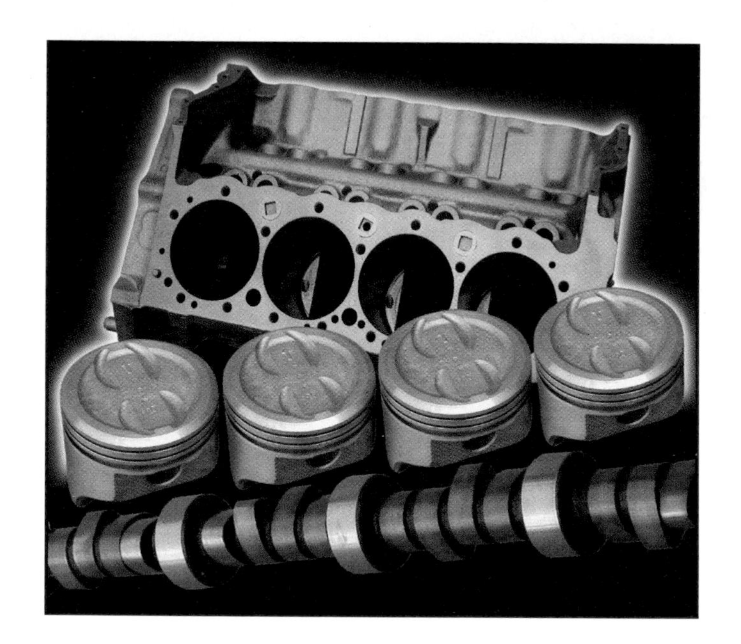

Expanded Contents

Section 3 Engine Systems

Section 4 Engine Diagnosis and Tune-Up

Section 5 Engine Overhaul

Section 6 ASE Certification and Careers

The engine in a modern vehicle is a complex piece of machinery. In order to properly service and maintain the engine, you must completely understand all of its components, systems, and subsystems. (Ford)

Chapter 1

Review of Engine Operation

After studying this chapter, you will be able to:

- ❑ List the major parts of an automotive engine.
- ❑ Explain the purpose of major engine parts and assemblies.
- ❑ Describe the relationship between the major parts of an engine.
- ❑ Summarize the four-stroke cycle.
- ❑ List and describe the related systems of an engine.

Know These Terms

BDC
Camshaft
Charging system
Combustion
Computer system
Connecting rod
Control module
Cooling system
Crankcase
Crankshaft
Cylinder
Cylinder block
Cylinder head
Drive train
Emission control systems
Engine block
Engine valves
Exhaust gas recirculation system
Exhaust manifold
Four-stroke cycle
Fuel system
Gasoline injection
Ignition system
Intake manifold
Internal combustion engine
Lubrication system
Main bore
Oil galleries
Piston
Piston pin
Piston rings
Ports
Positive crankcase ventilation system
Spark plug
Starting system
TDC
Throttle valve
Valve guides
Valve seats
Valve springs
Valve train
Water jackets
Water pump

This chapter provides a quick review of the operating principles of a four-stroke-cycle, piston engine. The interaction of basic engine components are discussed. Related systems—cooling, lubrication, fuel, computer control, and other systems—are explained. This review will prepare you for later text chapters that discuss these topics in much more detail.

This chapter uses words and illustrations to construct a basic, one-cylinder engine. You will see how each part is installed in the basic engine and learn how that part performs an important function. Then, near the end of the chapter, the systems that supplement engine operation and protect the engine from damage are reviewed.

If you have completed an introductory course that covered engine operation, you should still read through this chapter to refresh your memory. If you are not familiar with the operation of an engine, study this chapter carefully. This will let you catch up with the students that have already had some training in engines.

Automotive Engine

An ***engine*** is the source of power for moving the vehicle and operating the other systems. Sometimes termed the power plant, it burns a fuel (usually gasoline or diesel fuel) to produce heat, expansion of gasses, pressure, and resulting part movement.

Since a vehicle's engine burns fuel inside of itself, it is termed an ***internal combustion engine.*** As you will learn, the arrangement of an engine's parts allows it to harness the energy of the burning fuel.

Figure 1-1 illustrates the major parts of a modern, multi-cylinder engine. Study them as they are introduced:

- The block is the supporting structure for the engine.
- The piston slides up and down in the block.
- The piston rings seal the space between the block and sides of the piston.

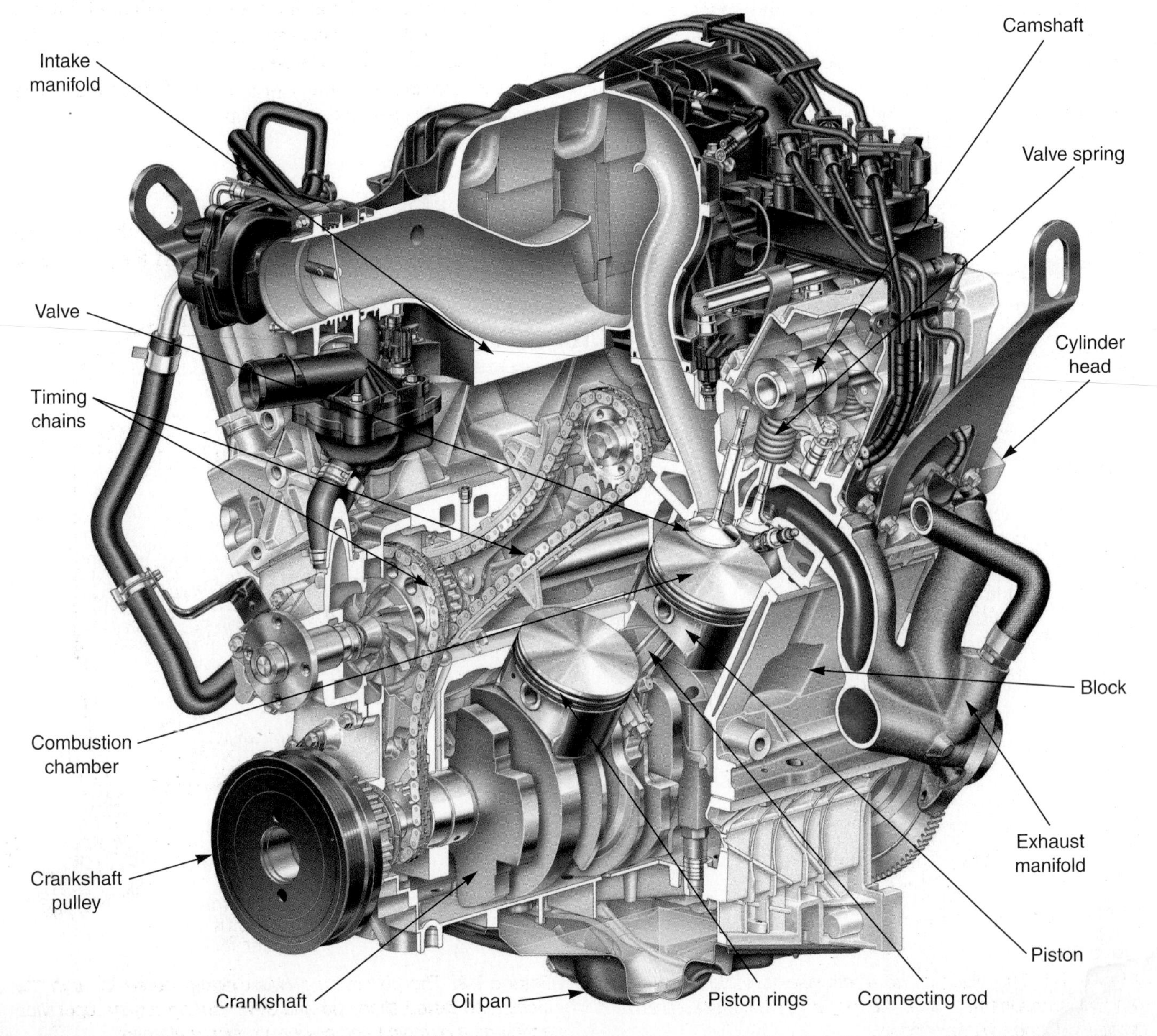

Figure 1-1. *Study the general location of parts in an engine. This will help you while reviewing the operation of an engine in this chapter. (Ford)*

- The connecting rod connects the piston to the crankshaft.
- The crankshaft converts the up and down action of the piston into rotary motion.
- The cylinder head fits over the top of the block and holds the valves.
- The valves open and close to control fuel entry into and exhaust exit from the combustion chamber.
- The combustion chamber is a cavity formed above the piston and below the cylinder head for containing the burning fuel.
- The camshaft opens the valves at the proper time.
- The valve springs close the valves.
- The timing belt or chain turns the camshaft at one-half of the engine speed.

Engine Block

The ***engine block,*** also called the ***cylinder block,*** forms the framework or "backbone" of an engine. This is because many of the other components of an engine fasten to the block. Cast from iron or aluminum, the block is the largest part of an engine. **Figure 1-2** shows a cutaway view of a basic engine block. Note the part names.

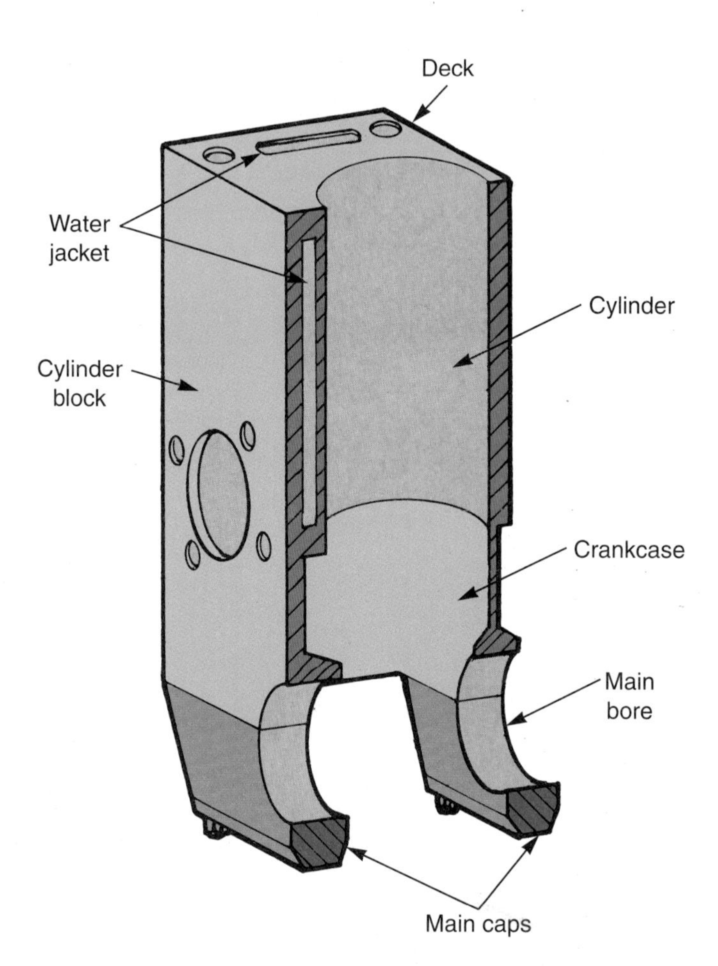

Figure 1-2. *The block is the framework for holding the other engine components. Note the parts of the block. This is a simplified, one-cylinder engine.*

A ***cylinder*** is a large hole machined through the top of the engine block. The piston fits into the cylinder. During engine operation, the cylinder, also called the cylinder bore, guides the piston as it slides up and down. The cylinder is slightly larger than the piston to produce a clearance (space) between the two parts.

Main caps are bolted to the bottom of the block. They hold the crankshaft in place and form the bottom half of the main bore.

The ***main bore*** is a series of holes machined from the front to the rear of the block. The crankshaft fits into these holes. With the engine running, the crankshaft spins or rotates in the main bore.

The ***deck*** is a flat surface machined on the top of the block for the cylinder head. The head is bolted to the deck. Coolant and oil passages in the deck align with openings in the cylinder head.

Coolant passages, or ***water jackets,*** surround the cylinders and combustion chamber. They are hollow areas inside the block and head for coolant. Coolant circulated through the water jackets removes the heat generated by the fuel burning in the cylinders.

The ***crankcase*** is the lower area of the block. The crankshaft spins inside the crankcase.

Piston

The ***piston*** converts the pressure of combustion into movement. See **Figure 1-3.** ***Combustion*** is the burning of

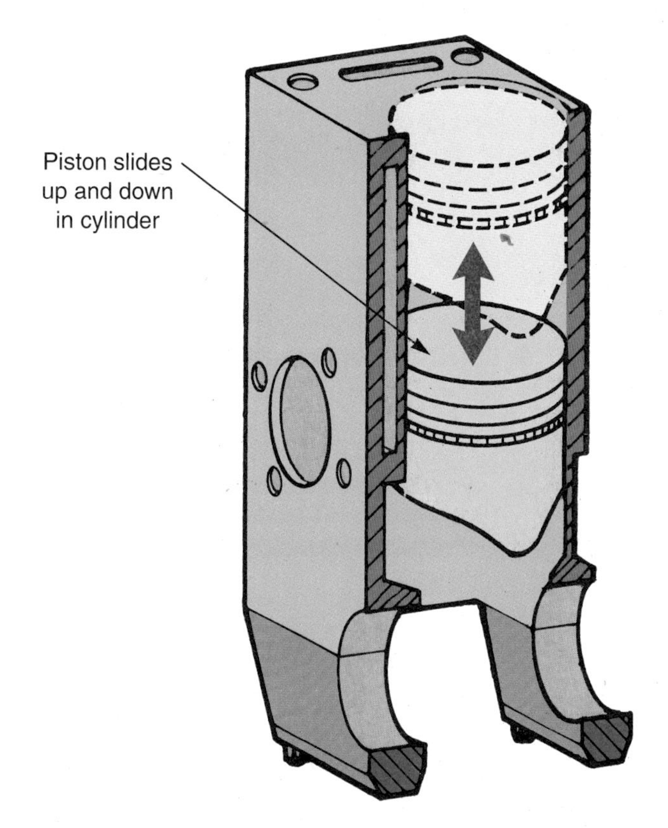

Figure 1-3. *The piston fits into a cylinder bored through the block. The piston slides up and down during engine operation acting as a pumping and power-producing element.*

fuel, which results in expanding gas inside the cylinder. The piston transfers the pressure of combustion to the piston pin, connecting rod, and crankshaft. It also holds the piston rings and piston pin.

During engine operation, the piston slides up and down in the cylinder at tremendous speeds. At a vehicle speed of about 55 mph (88 km/h), the piston can accelerate from zero to 60 miles an hour and then back to zero in one movement from top to bottom in the cylinder. This places tremendous stress on the piston and its related parts.

Piston Rings

The ***piston rings*** fit into grooves machined into the sides of the piston. These rings keep combustion pressure from entering the crankcase and engine oil from entering the combustion chamber. Look at **Figure 1-4.**

The ***compression rings*** seal the clearance between the block and piston. They are normally the two upper piston rings. Their job is to contain the pressure formed in the combustion chamber, **Figure 1-5.** Without compression rings, pressure would blow past the outside diameter of the piston and into the lower area of the engine block.

The ***oil ring*** fits into the lowest groove in the piston. It is designed to scrape excess oil from the cylinder wall to keep it from being burned in the combustion chamber, **Figure 1-6.** If oil enters the area above the piston and burns, blue smoke blows out of the tailpipe.

Piston Pin

A ***piston pin,*** also called a wrist pin, allows the connecting rod to swing back and forth inside the piston. The pin fits through a hole machined in the piston and through a hole in the upper end of the connecting rod. Refer to **Figure 1-7.**

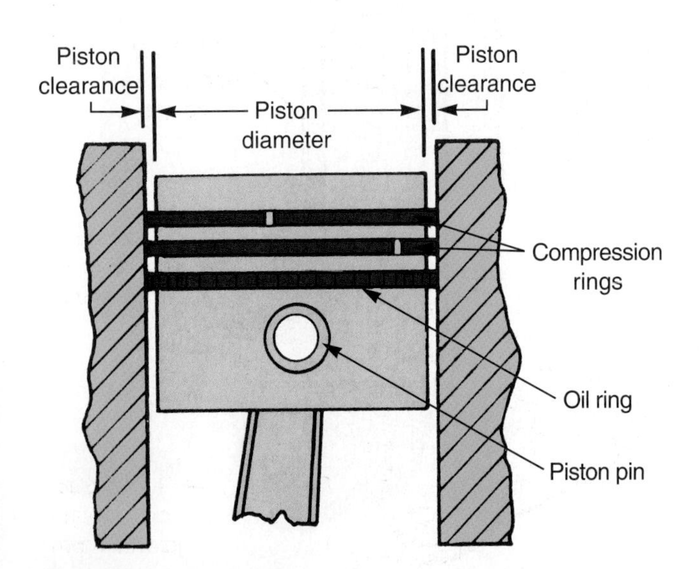

Figure 1-4. *The clearance between the piston and cylinder allows the piston to move freely in the cylinder. Rings seal the clearance.*

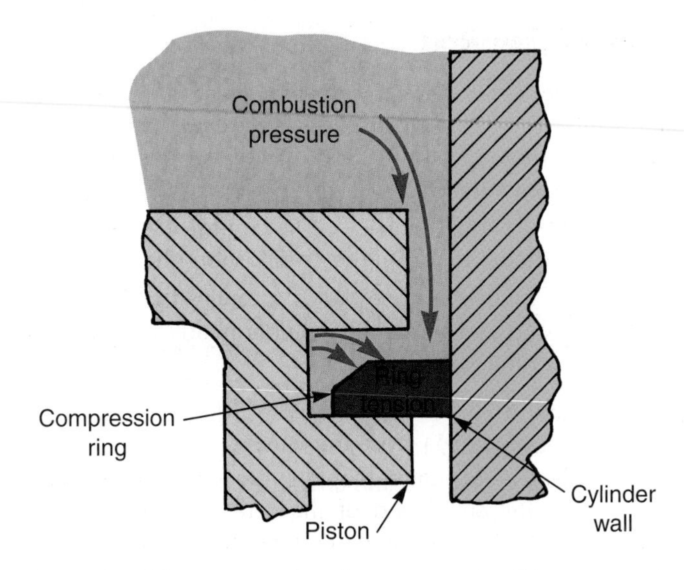

Figure 1-5. *The compression rings use combustion pressure to help seal against the cylinder wall. This keeps pressure in the combustion chamber and out of the crankcase.*

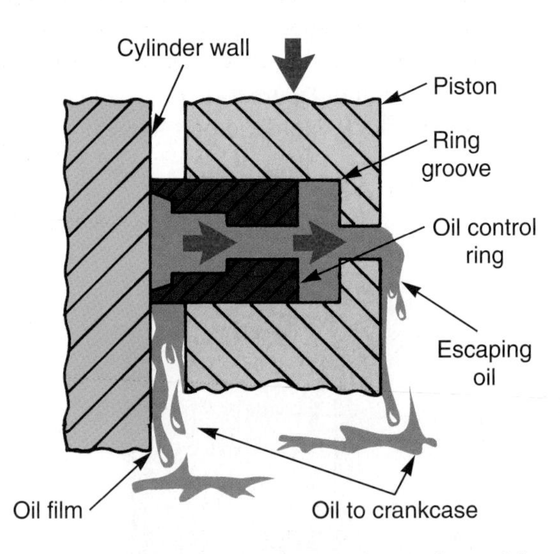

Figure 1-6. *Oil rings act as a scraper to keep oil out of the combustion chamber. (Deere & Co.)*

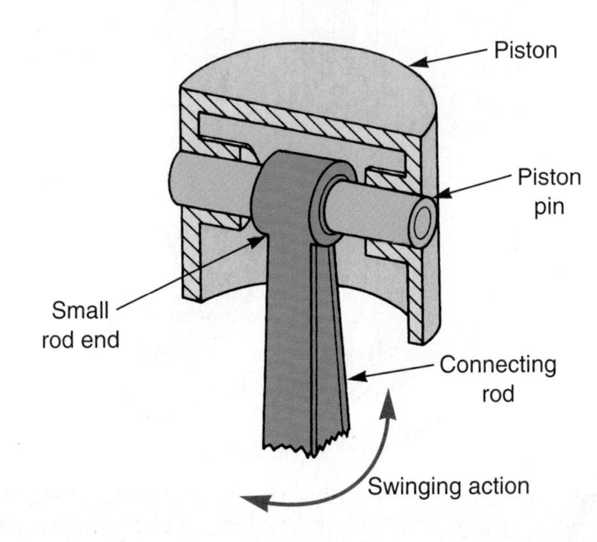

Figure 1-7. *The piston pin fits into a hole bored in the piston. The pin attaches the piston to the connecting rod.*

Connecting Rod

The ***connecting rod*** transfers the force of the piston to the crankshaft. It is fastened to the piston pin at the top and the crankshaft at the bottom. It also causes piston movement on nonpower-producing events (up and down piston movements). See **Figure 1-8.**

The small, top end of the connecting rod has a hole machined in it for the piston pin. The top of the rod extends inside of the piston.

The big, bottom end of the connecting rod fits around the crankshaft journal. It has a removable cap that allows the installation and removal of the rod-piston assembly. Special rod bolts and nuts hold the cap in place.

As discussed in later chapters, bushings are normally installed in the small end of the rod. Rod bearings are installed in the big end of the connecting rod.

Crankshaft

The ***crankshaft*** converts the up and down (reciprocating) movement of the connecting rod and piston into rotary motion. The rotary motion is used to power gears, chains, belts, and the drive train.

The crankshaft fits into the main bore of the engine block, as shown in **Figure 1-9.** It mounts on the main bearings and is free to spin inside the block. The connecting rods are attached to the crankshaft journals. **Figure 1-10**

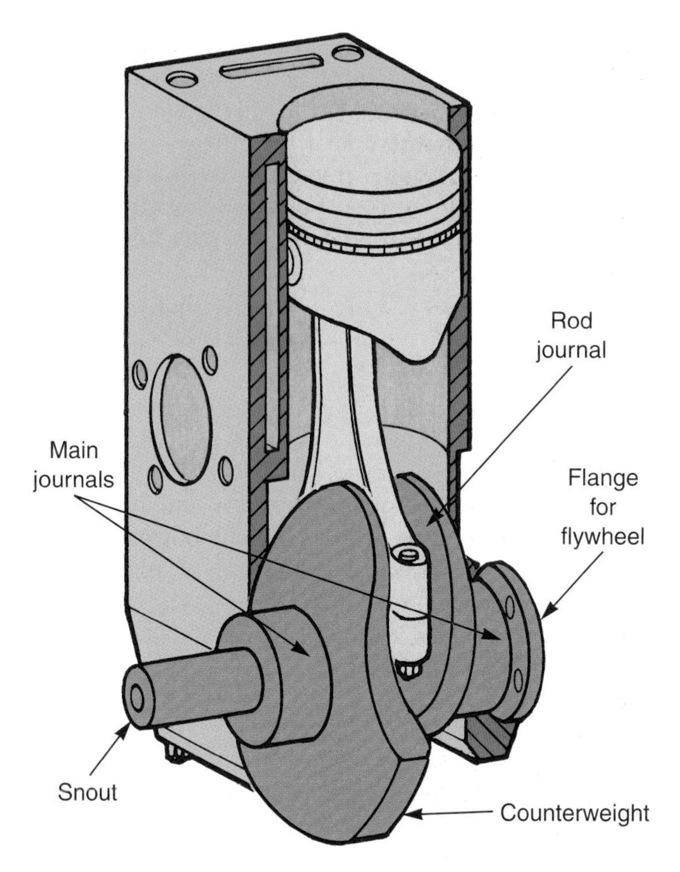

Figure 1-9. *The crankshaft takes the reciprocating motion of the piston and produces rotary motion for vehicle's drive train and accessory system.*

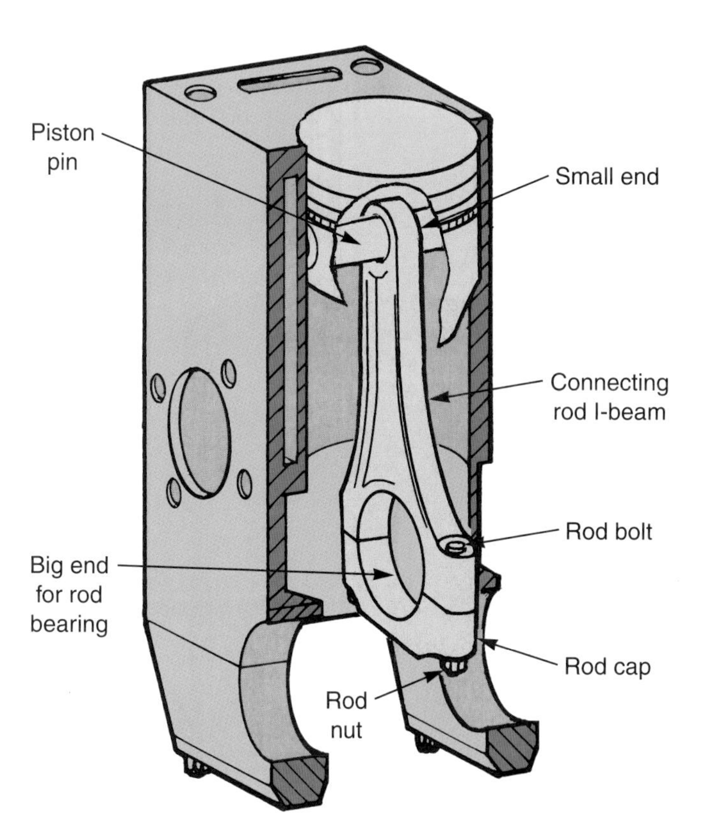

Figure 1-8. *The connecting rod links the piston and crankshaft together. The large end has a removable cap that allows the rod to be bolted around the crankshaft journal. The small end has a hole for the piston pin.*

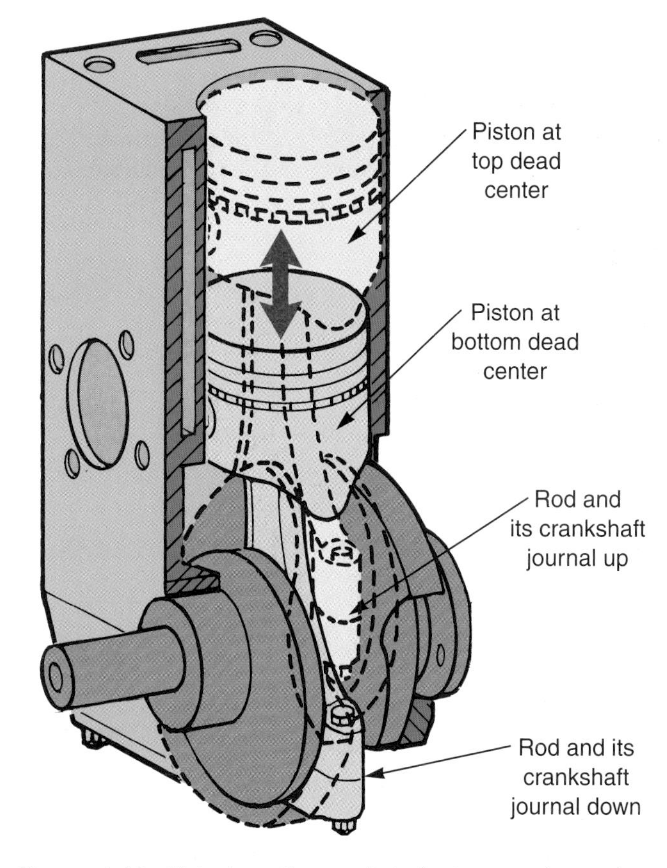

Figure 1-10. *Note how the crankshaft changes the reciprocating motion of the piston into rotary motion.*

shows how the crankshaft changes the reciprocating (up and down) motion of the piston and connecting rod into a rotary motion.

The ***engine flywheel*** is a very heavy, round disk mounted to the back of the crankshaft. It helps to keep the crankshaft spinning between power strokes and smooth engine operation. It also holds a large gear used by the starter.

TDC and BDC

The abbreviation ***TDC*** stands for top dead center. This is the point of travel where the piston is at its highest point in the cylinder. The abbreviation ***BDC*** stands for bottom dead center. This is the point of travel where the piston is at its lowest point in the cylinder. Refer to **Figure 1-11.**

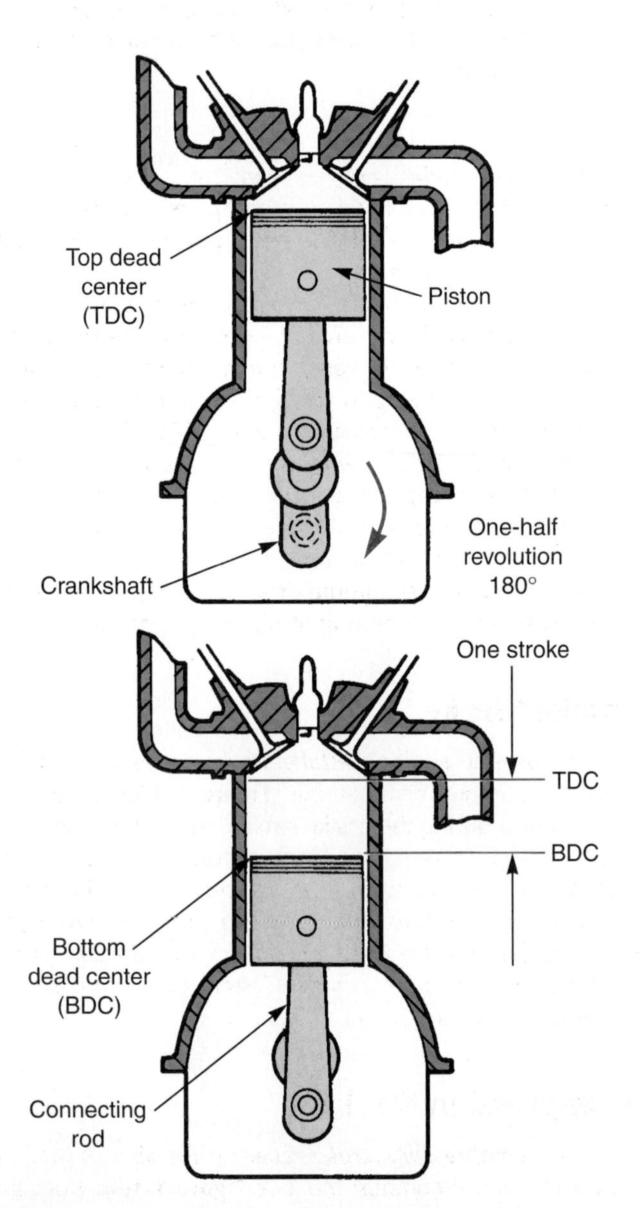

Figure 1-11. *TDC is when the piston is at the top of its stroke. BDC is when the piston is at the bottom of its stroke. One stroke is the piston movement from TDC to BDC or vice versa. (Ford)*

Cylinder Head

The ***cylinder head*** is bolted to the top of the block deck to enclose the top of the cylinders and form the top of the combustion chamber, **Figure 1-12.** Like the block, the cylinder head contains water jackets for cooling and oil passages for lubricating moving parts on or in the cylinder head.

Valve guides are machined through the top of the head for the valves. The valves slide up and down in these guides.

Cylinder head ***ports*** are passages for the air-fuel mixture to enter the combustion chamber and for exhaust gasses to flow out of the engine. These are located in the cylinder head.

Valve seats are machined in the opening where the ports enter the combustion chamber. The valves close against the seats to make a leakproof seal at high temperatures.

Engine Valves

Engine valves control the flow into and out of the engine cylinder or combustion chamber. They fit into the cylinder head, operate inside the valve guides, and close on the valve seats. ***Valve springs*** fit over the top end of the valves to keep the valves in a normally closed position, **Figure 1-13.**

Figure 1-14 shows how a valve opens and closes the ports in the cylinder head. When the valve slides down, the valve head moves away from the valve seat and the port is opened. When the valve slides up, the valve head moves toward the valve seat until the valve face makes contact with the valve seat. This seals the combustion chamber from the port.

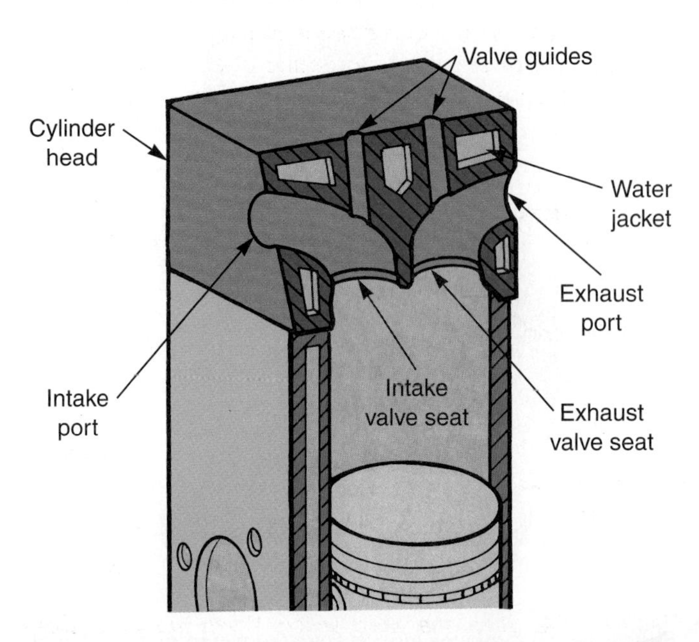

Figure 1-12. *The cylinder head is bolted to the top of the block. It forms a cover over the cylinder. The head also holds the valves that control flow into and out of the cylinder.*

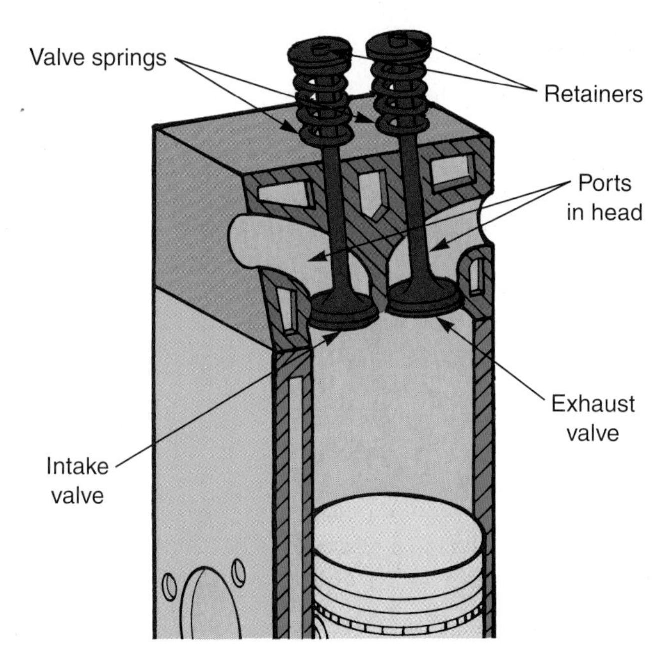

Figure 1-13. *Valves fit into guides in the cylinder head. Valve springs hold the valves closed. The valves seal against valve seats in the head to close off ports from the combustion chamber.*

There are two types of valves—intake and exhaust. The intake valve is the larger valve and it allows a fuel charge to flow into the cylinder. The exhaust valve is the smaller valve and it opens to let burned gasses (exhaust) out of the cylinder. **Figure 1-15** shows how the air-fuel mixture flows through the intake port, past the valve, and into the combustion chamber when the valve is open.

Four-Stroke Cycle

The ***four-stroke cycle*** needs four up or down piston movements, or strokes, to produce one complete cycle. Every two up and two down strokes of the piston results

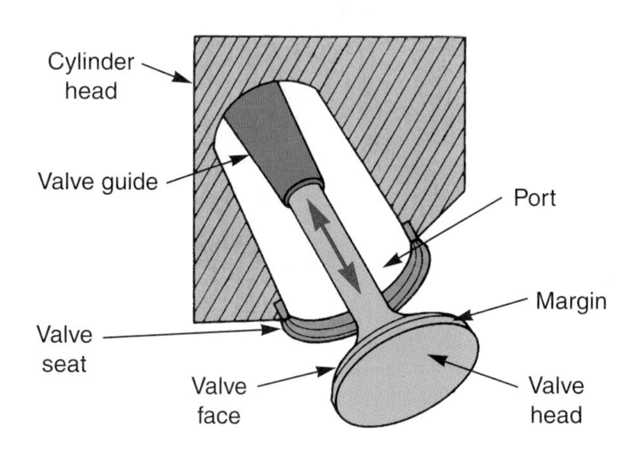

Figure 1-14. *Study the valve action. When the valve slides open, the valve face is lifted off of the valve seat. This opens the port to the combustion chamber and gasses are free to enter or exit the cylinder.*

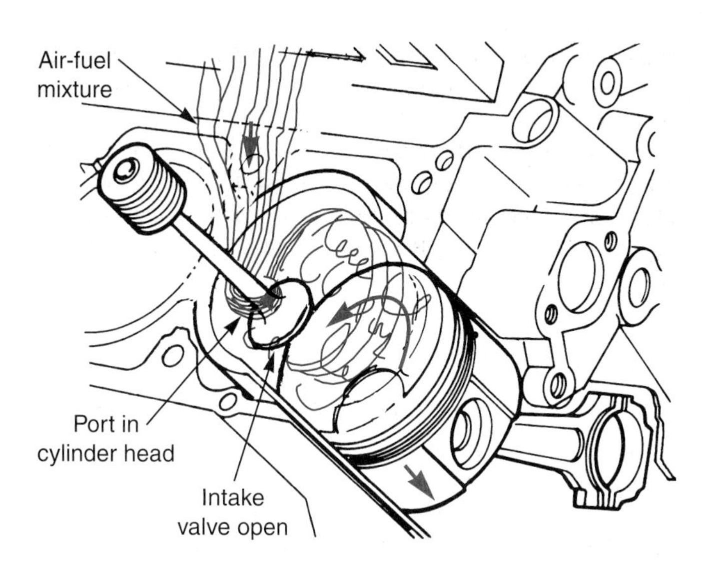

Figure 1-15. *Note the action as the intake valve opens. Downward movement of the piston forms a vacuum in the cylinder. Atmospheric pressure pushes the air-fuel charge into the cylinder. (Ford)*

in one power-producing cycle. Two complete revolutions of the crankshaft are needed to complete one four-stroke cycle. Automotive engines, both gas and diesel, are four-stroke-cycle engines.

The four strokes are intake, compression, power, and exhaust. With the engine operating, these strokes happen over and over very rapidly. At idle, an engine might be running at 800 revolutions per minute (rpm), which means the crankshaft rotates 800 times in one minute. Since it takes two complete revolutions of the crankshaft to complete a four-stroke cycle, an engine completes 400 four-stroke cycles per minute at idle. In other words, the piston must slide up 800 times and down 800 times per minute. You can imagine how fast these events are happening at highway speeds!

Intake Stroke

A gasoline engine's ***intake stroke*** draws air and fuel into the combustion chamber. **Figure 1-16A** shows the basic action during the intake stroke. Study the position of the valves and movement of the piston. The piston slides down to form a vacuum (low pressure area). The intake valve is open and the exhaust valve is closed. Atmospheric pressure (outside air pressure) pushes the air-fuel charge into the vacuum in the cylinder. This fills the cylinder with a burnable mixture of fuel and air.

Compression Stroke

The ***compression stroke*** squeezes the air-fuel mixture to make it more combustible. See **Figure 1-16B.** Both the intake and exhaust valves are closed. The piston slides up and compresses the mixture into the small area in the combustion chamber.

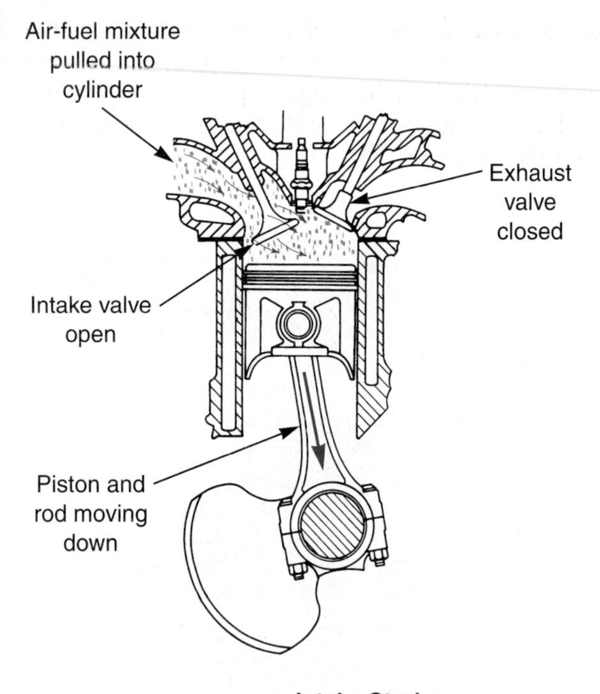

Intake Stroke

A—Intake stroke. The piston slides down with the intake valve open and the exhaust valve closed. The air-fuel charge is pulled into the cylinder.

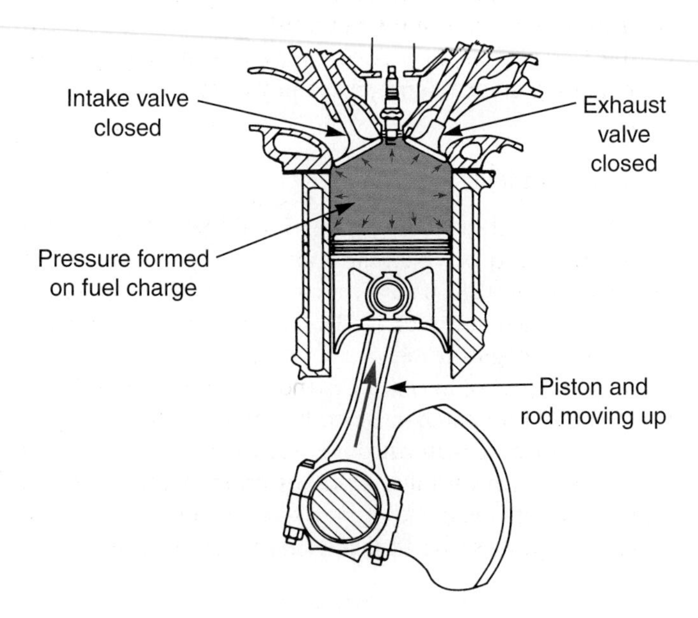

Compression Stroke

B—Compression stroke. Both valves are closed and the piston slides up. This compresses the air-fuel charge and prepares it for combustion.

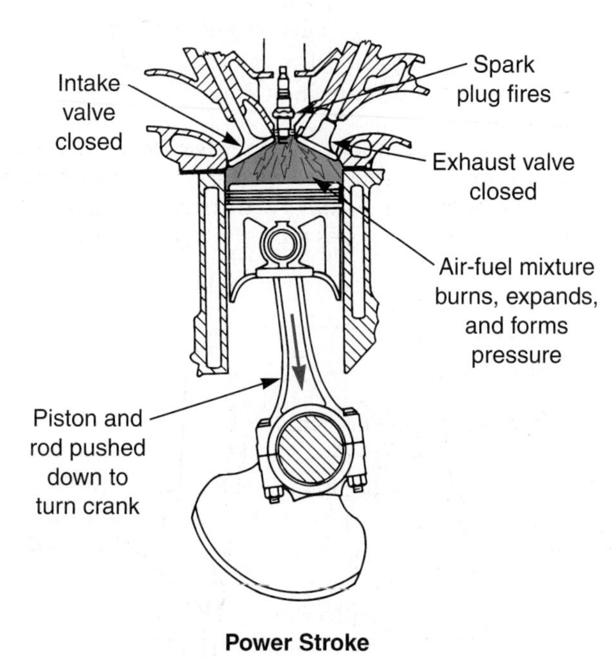

Power Stroke

C—Power stroke. The spark plug fires and the fuel begins to burn. The heat of combustion causes expansion of the gasses and creates pressure. This pushes the piston down with tremendous force to spin the crankshaft.

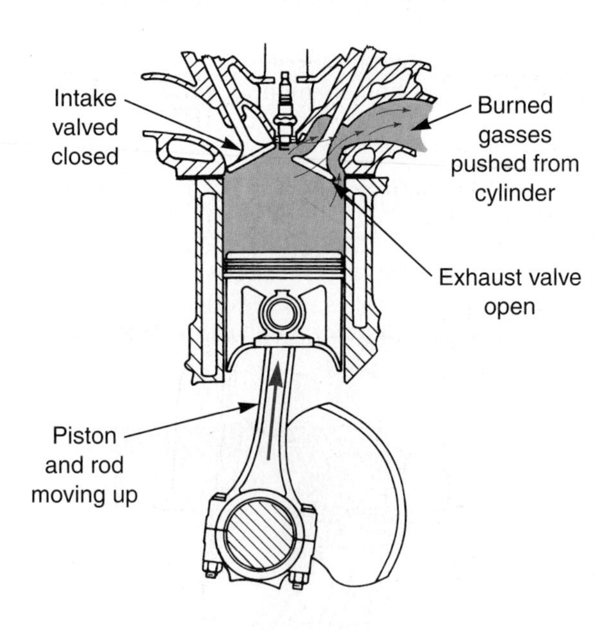

Exhaust Stroke

D—Exhaust stroke. The piston slides up with the intake valve closed and the exhaust valve open. This pushes the exhaust gasses out of the cylinder so a fresh charge can enter.

Figure 1-16. *Review the four-stroke cycle. You must be able to visualize these events to be a competent engine technician.*

For proper combustion (burning), it is very important that the valves, rings, and other components do *not* allow pressure leakage out of the combustion chamber. Leakage during the compression stroke may prevent the mixture from igniting and burning on the power stroke.

Power Stroke

The air-fuel mixture is ignited and burned during the ***power stroke*** to produce gas expansion, pressure, and a powerful downward piston movement. See **Figure 1-16C.** Both valves are still closed. The spark plug fires and the fuel mixture begins to burn.

As the mixture burns, it expands and builds pressure in the combustion chamber. Since the piston is the only part that can move, it is thrust downward with several tons of force. This downward thrust pushes on the connecting rod and crankshaft forcing the crankshaft to turn. The power stroke is the only stroke that does not consume (use) energy.

Exhaust Stroke

The ***exhaust stroke*** pushes the burned gasses out of the cylinder and into the vehicle's exhaust system. See **Figure 1-16D.** The intake valve remains closed, but the exhaust valve is open. Since the piston is now moving up, the burned gasses are pushed out of the exhaust port to ready the cylinder for another intake stroke.

Valve Train

The ***valve train*** operates the engine valves. It times valve opening and closing to produce the four-stroke cycle. Basic valve train parts are shown in **Figure 1-17.**

The ***camshaft*** opens the valves and allows the valve springs to close the valves at the proper times. The camshaft has a series of lobes (egg-shaped bumps) that act on the valves or valve train to slide the valve down in its guide. See **Figure 1-18.**

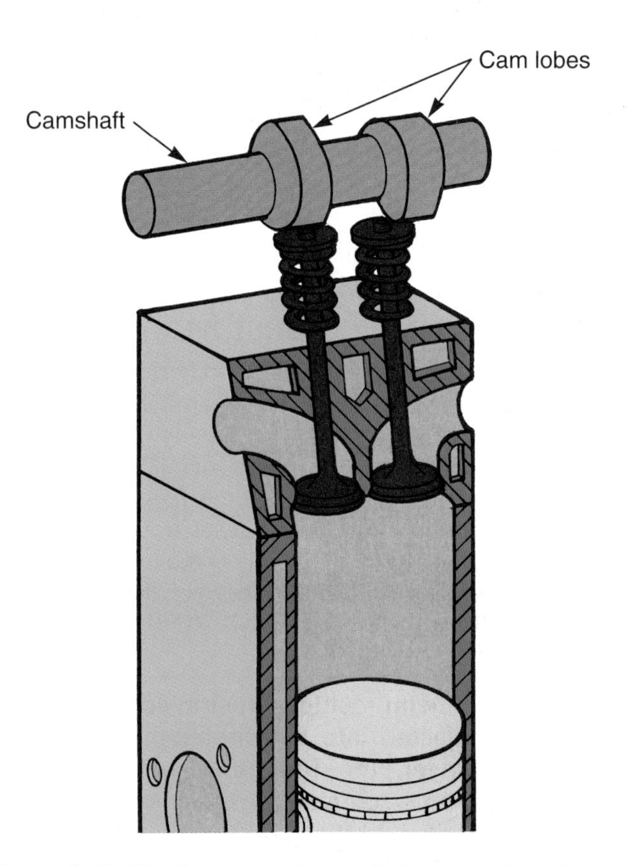

Figure 1-18. *The lobes on the camshaft act on the valves or valve train to open and close the valves. As shown here, the camshaft fits into the cylinder head on many engines. This lets it operate directly on valves without using push rods.*

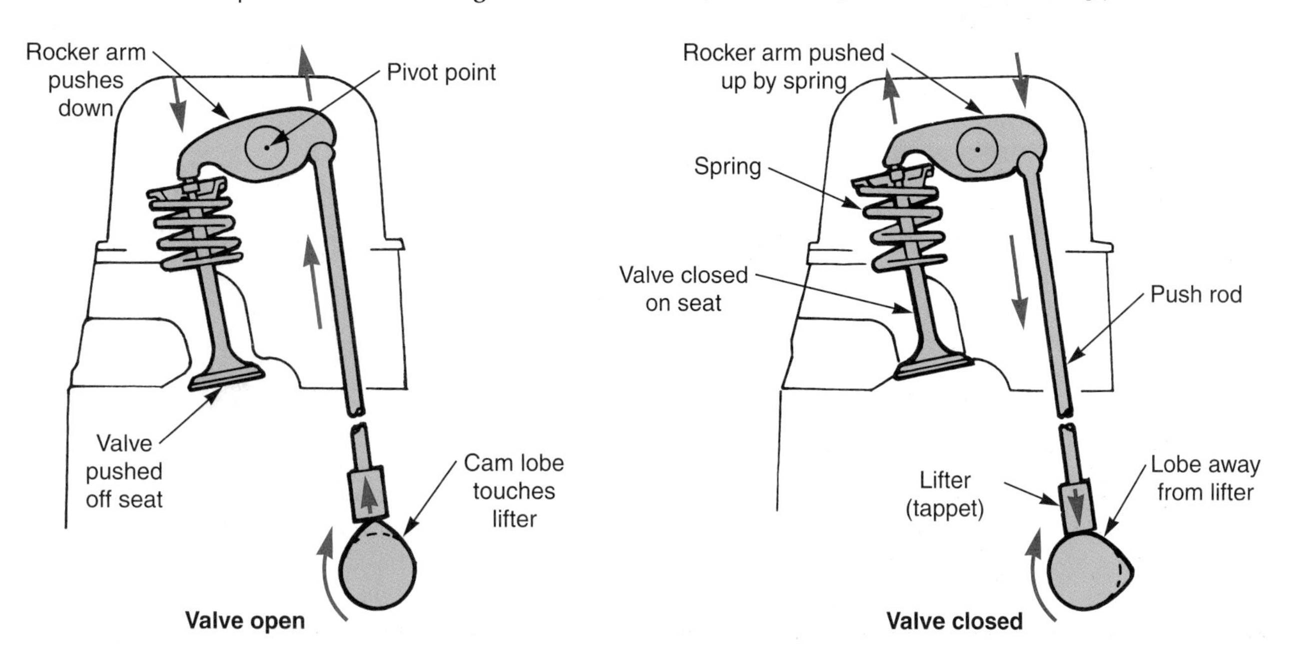

Figure 1-17. *The camshaft operates the valve train. When the cam lobe rotates into a lifter, the valve is opened. When a lobe moves out of a lifter, the valve spring closes the valve. (Ford)*

In **Figure 1-17,** note how the cam lobe acts on the valve train. When the lobe moves into the lifter, the lifter, pushrod, and one side of the rocker arm are pushed up. This opens the valve. When the lobe rotates away from the lifter, the valve spring pushes the valve and other parts into the closed position.

Camshaft timing is needed to ensure that the valves properly open and close in relation to the crankshaft. Either a belt, chain, or set of gears is used to turn the camshaft at one-half of the crankshaft speed and keep the camshaft in time with the crankshaft. **Figure 1-19** shows how a timing belt is used to operate the camshaft on our basic engine.

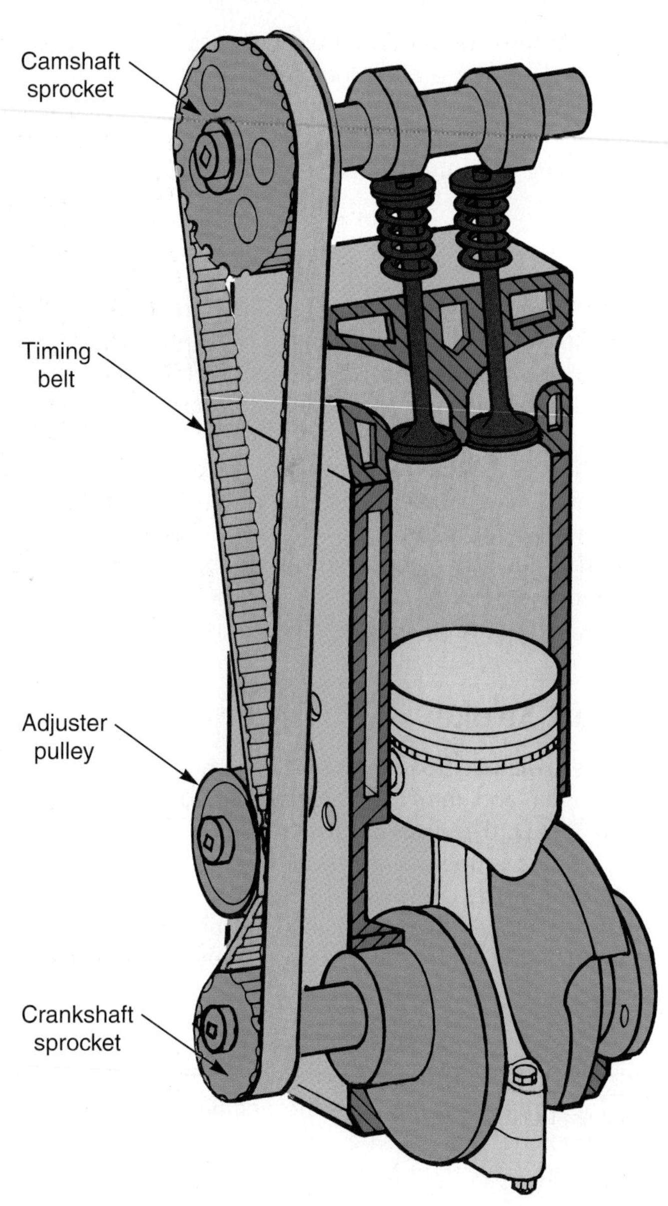

Figure 1-19. *The camshaft is turned at one-half of engine speed. A timing belt is the most common method of turning the camshaft in time with the crankshaft rotation. Note the part names.*

Intake and Exhaust Manifolds

The ***intake manifold*** carries the air-fuel mixture into the cylinder head intake ports. It normally is bolted to the cylinder head. Ports in the intake manifold match the intake ports in the cylinder head. The ***exhaust manifold,*** as its name implies, carries burned gasses from the cylinder head exhaust port to the other parts of the exhaust system. **Figure 1-20** shows the basic action of the intake and exhaust manifolds.

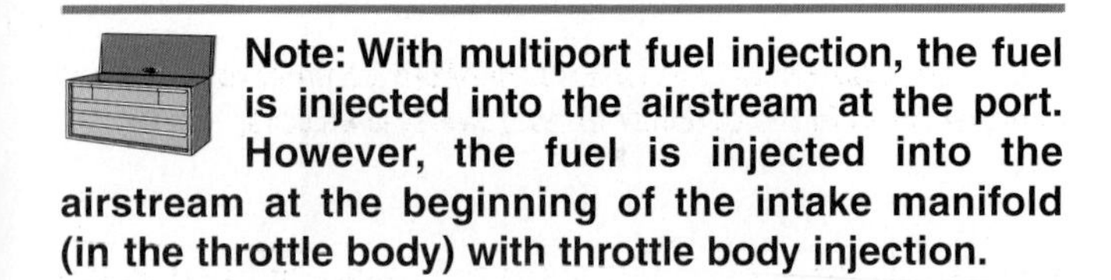

Note: With multiport fuel injection, the fuel is injected into the airstream at the port. However, the fuel is injected into the airstream at the beginning of the intake manifold (in the throttle body) with throttle body injection.

Cooling System

The ***cooling system*** is needed to carry the heat of combustion and friction away from the engine. Without a cooling system, the piston, valves, cylinder, and other parts could be ruined in a matter of minutes. The head and block could also crack from the tremendous heat. Basically, a cooling system consists of a radiator, water pump, fan, thermostat, water jackets, and connecting hoses, **Figure 1-21.**

The ***water pump*** circulates an antifreeze and water solution through the water jackets, hoses, and radiator. It is often driven by a fan belt running off of the crankshaft pulley, but may be driven by an electric motor. The coolant (antifreeze-water solution) picks up heat from the metal parts of the engine and carries it to the radiator.

The ***radiator*** transfers heat from the coolant to the outside air. A fan is used to pull air through the radiator. Large radiator hoses connect the radiator to the engine.

The ***thermostat*** is a temperature-sensing valve that controls the operating temperature of the engine. When the engine is cold, the thermostat blocks coolant flow through the radiator and speeds warm-up of the engine. When the engine is warm, the thermostat opens to allow coolant to circulate through the radiator, thus removing heat from the engine.

Lubrication System

The ***lubrication system*** circulates engine oil to high-friction points in the engine. Without oil, friction will result in wear, scoring, and damage to parts very quickly. The lubrication system basically consists of an oil pump, oil pickup, oil pan, and oil galleries. See **Figure 1-22.**

The ***oil pump*** is the "heart" of the lubrication system because it circulates oil through the oil galleries. The ***oil galleries*** are small passages that lead to the crankshaft

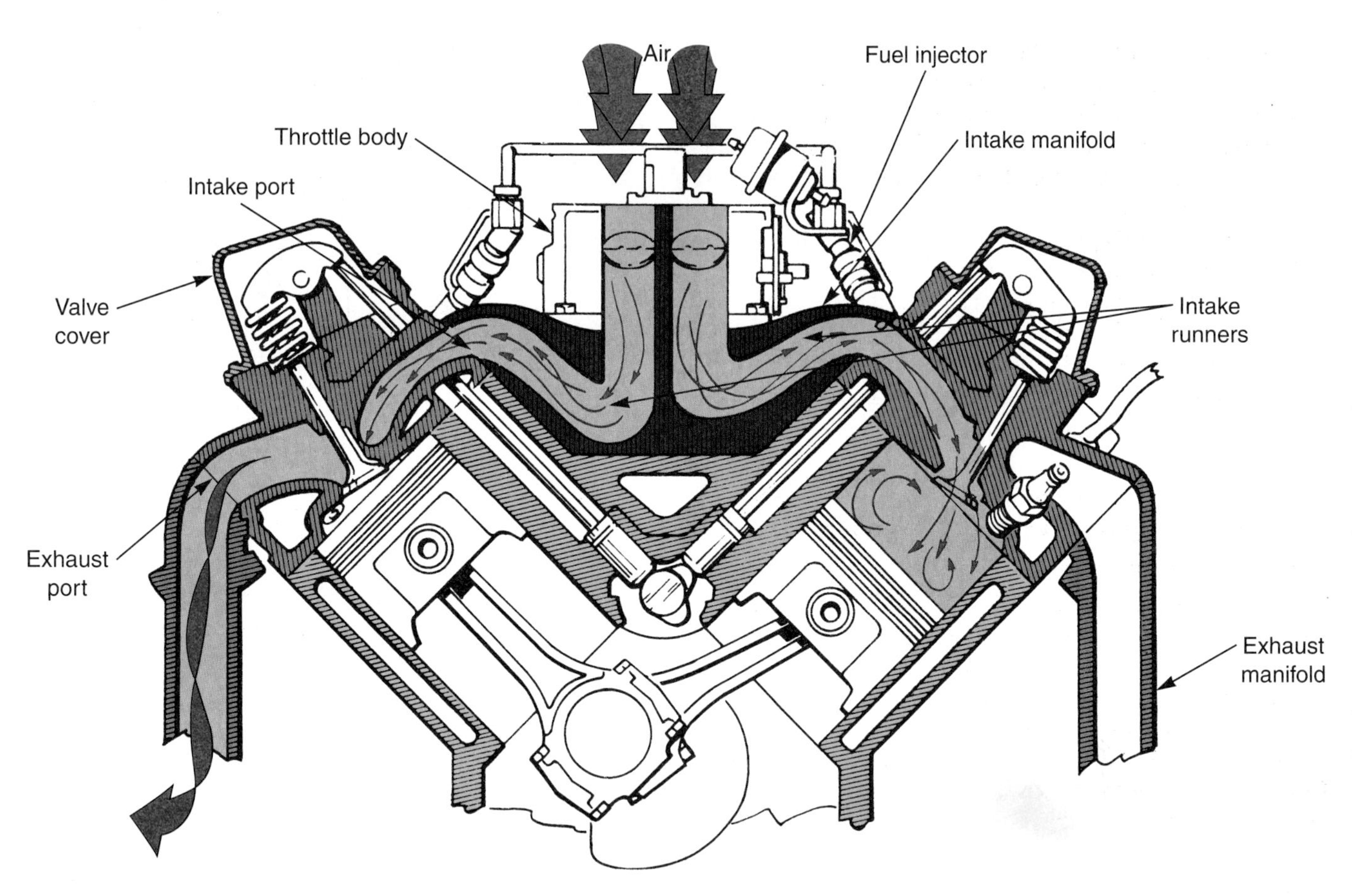

Figure 1-20. *This is a cutaway of an engine. Note the intake and exhaust manifolds. Ports in the intake manifold connect to the intake ports in the cylinder head. The exhaust manifold is bolted over the exhaust ports in the cylinder head. (General Motors)*

bearings, camshaft bearings, and valve train components. These are high-friction points in the engine and need oil for protection.

Ignition System

An ***ignition system*** is needed on a gasoline engine to ignite and burn the air-fuel mixture. It must increase battery voltage enough to produce a high-voltage electric arc, or "spark," at the tip of each spark plug in the combustion chamber. Refer to **Figure 1-23.**

A fundamental ignition system consists of a spark plug, plug wire, ignition coil, switching device, and power source. On many late-model vehicles, the ignition coil is mounted directly on top of the spark plug eliminating the need for a spark plug wire.

The ***switching device*** in the ignition system is an electronic control unit that makes and breaks electrical current flow to the ignition coil(s). Ignition coil operation is timed with crankshaft rotation so that the spark occurs in the combustion chamber at the end of the compression stroke.

The ***ignition coil*** is used to step up battery voltage to over 60,000 volts. This is enough voltage to make the electricity jump the spark plug gap. The ignition coil fires every time the switching device *stops* current flow from the battery, **Figure 1-23.** This causes the magnetic field in the coil to collapse and induce a higher voltage in the coil's output wire.

The ***spark plug*** is the "match" that starts the air-fuel mixture burning in the combustion chamber. When ignition coil fires and sends current through the spark plug wire, an electric arc (spark) forms at the tip of the spark plug. This makes the fuel and air start to burn, producing the power stroke.

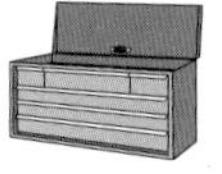

Note: Older ignition systems used a mechanical switching device called points. Ignition systems with points have not been used on production vehicles since the mid 1970s. If you encounter this type of ignition system, refer to the appropriate service manual for service procedures.

Starting System

The ***starting system*** turns the engine crankshaft until the engine can begin running on its own power. It uses a battery, ignition switch, high-current relay, and electric motor to rotate the crankshaft, **Figure 1-24.**

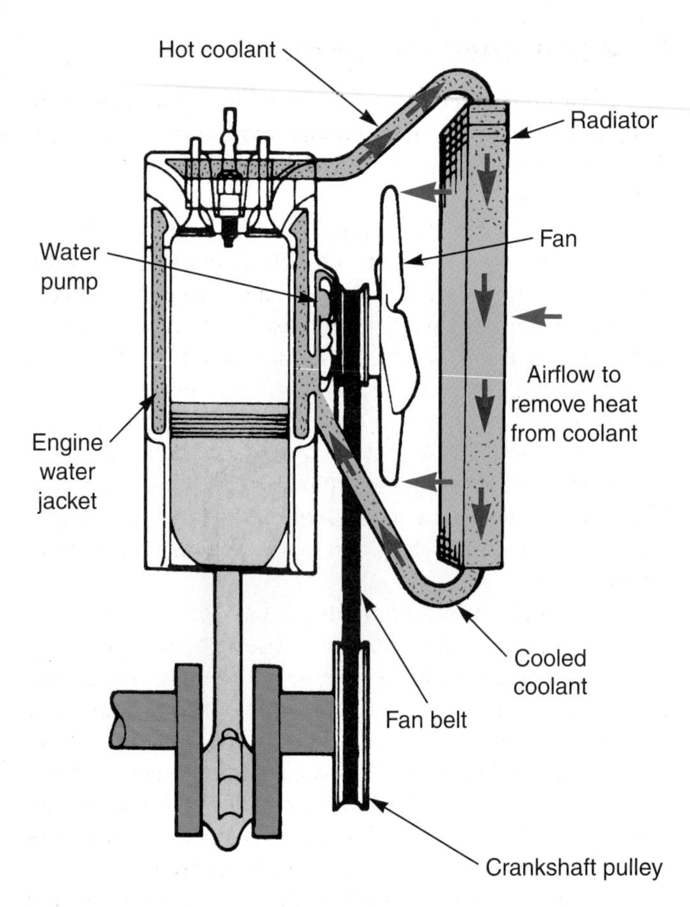

Figure 1-21. *A cooling system is needed to remove heat from the engine and prevent severe engine damage. Water jackets allow coolant to flow around the cylinders and through the cylinder head. A water pump circulates coolant through the system. The radiator dissipates heat into the outside air. A fan pulls air through the radiator. (DaimlerChrysler)*

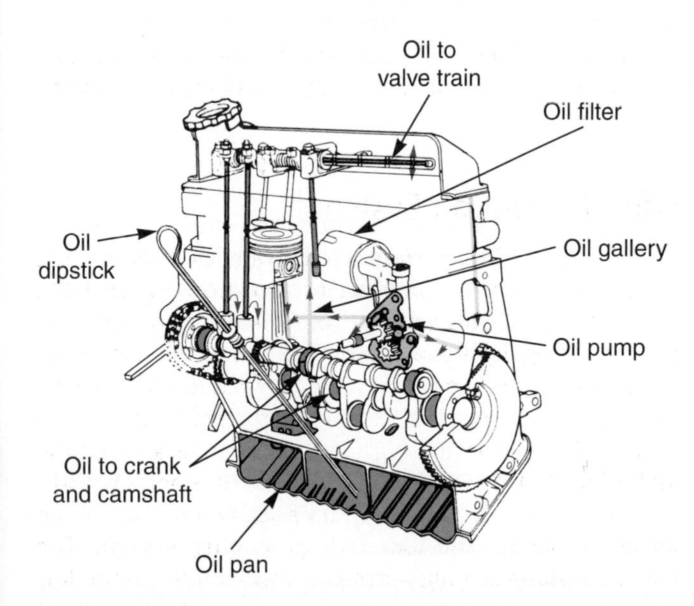

Figure 1-22. *The lubrication system prevents excess friction that may damage the engine. Note the part names. (DaimlerChrysler)*

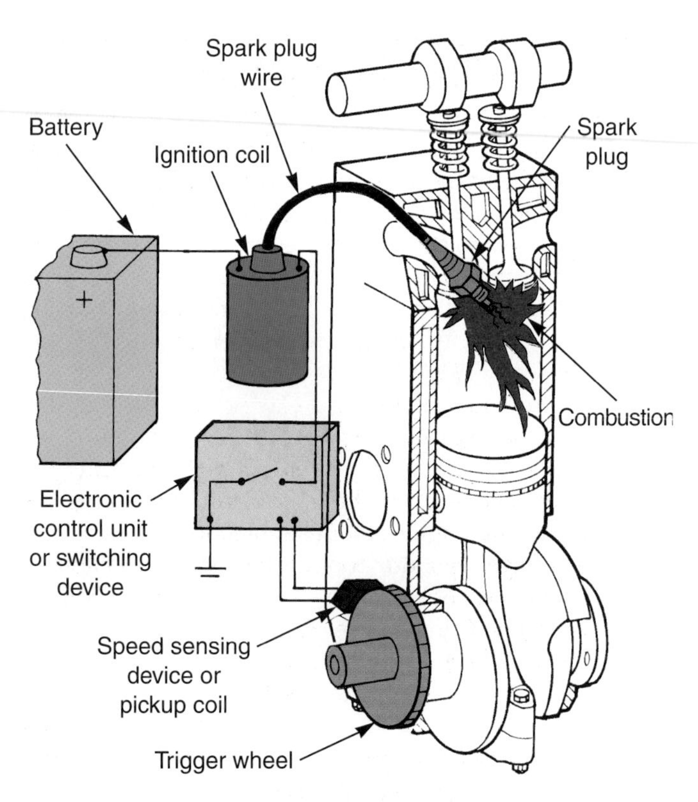

Figure 1-23. *The ignition system is used on a gasoline engine to ignite the fuel in the combustion chamber. A coil produces high voltage for the spark plug. When the switching device breaks the flow of current to the coil, the coil and spark plug fire to ignite the fuel.*

The ***battery*** stores chemical energy that can be changed into electrical energy. When the driver turns ignition switch (start switch), the solenoid (high-current relay) sends battery current to the starter motor. The ***starter motor*** has a small gear that engages a large gear on the crankshaft flywheel. The motor has enough torque (turning force) to spin the flywheel, and thus the crankshaft, until the engine starts and runs. Then, the driver releases the ignition key and deactivates the starting system.

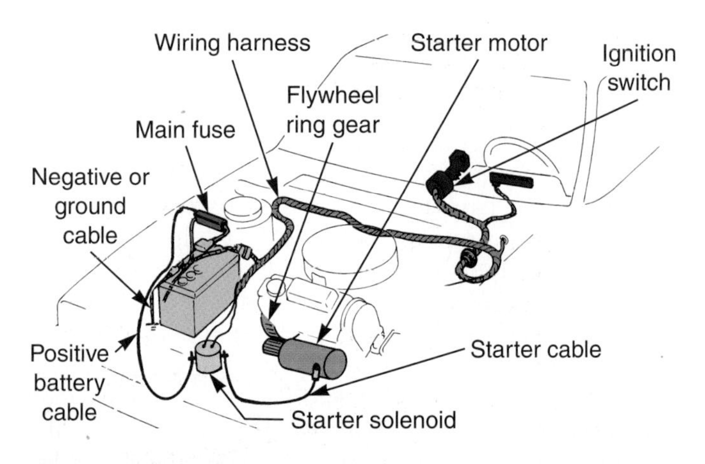

Figure 1-24. *The starting system rotates the crankshaft until the engine starts. A powerful electric starter motor has a gear that meshes with the gear on the engine flywheel. A solenoid makes the electrical connection between the battery and the starter motor when the ignition key is turned to the start position. (Honda)*

Charging System

The ***charging system*** is needed to recharge (re-energize) the battery after starting system or other electrical system operation. The battery can become discharged (run down) after only a few minutes of starter motor operation. The charging system also provides all of the vehicle's electrical needs while the engine is running. Basically, the charging system consists of the alternator and a voltage regulator. Look at **Figure 1-25.**

The ***alternator*** produces the electricity to recharge the battery. It is driven by a belt from the engine crankshaft pulley. The alternator sends current through the battery to reactivate the chemicals in the battery. This again prepares the battery for starting or other electrical loads.

The ***voltage regulator*** controls the electrical output of the alternator. It ensures that about 14.5 volts are produced by the alternator. Current then flows back into the battery, since battery voltage is only about 12.5 volts.

Fuel System

The ***fuel system*** must meter the right amount of fuel (usually gasoline or diesel oil) into the engine for efficient combustion under different conditions. At low speeds, it must meter a small amount of fuel into the airstream. As engine speed and load increase, the fuel system must meter more fuel into the airstream. The fuel system must also alter the fuel metering with changes in engine temperature and other variables.

There are two basic types of automotive fuel systems in current use—gasoline injection and diesel injection. A third type of fuel system—carburetion—has not been commonly used since the mid 1980s.

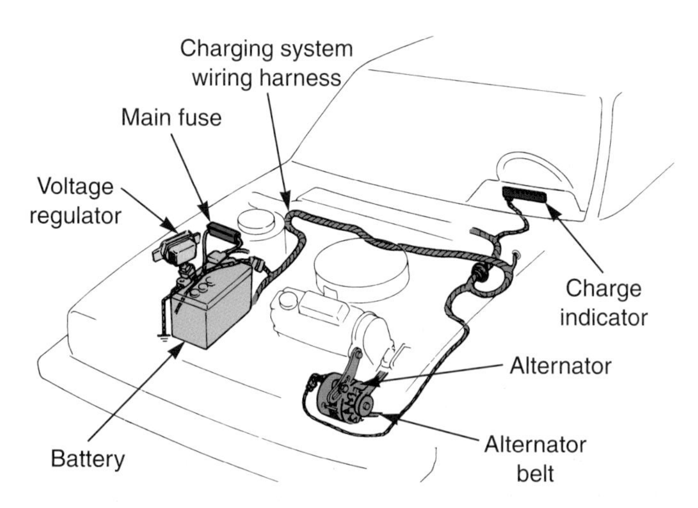

Figure 1-25. *The charging system recharges the battery and provides the vehicle's electrical needs while the engine is running. (Honda)*

Gasoline Injection System

A ***gasoline injection*** system uses fuel pump pressure to spray fuel into the engine intake manifold, usually near the cylinder head's intake port. A basic system is pictured in **Figure 1-26.**

An electric ***fuel pump*** forces fuel from the fuel tank to the fuel injector. A constant pressure is maintained at the injector. The ***fuel injector*** is simply an electrically-operated fuel valve. When energized by the control module, it opens and squirts fuel into the intake manifold. When not energized, it closes and prevents fuel entry into the engine.

Modern gasoline injection systems open the injector when the engine intake valve opens. Then, fuel is partially forced into the combustion chamber by pump pressure. This helps control how much fuel enters the cylinder and also increases combustion efficiency, as you will learn in later chapters.

A control module (computer) is used to regulate when and how long the injector opens. It uses electrical information from various sensors to analyze the needs and operating conditions of the engine. The engine sensors monitor various operating conditions, such as engine temperature, speed, load, and so on. In this way, the computer can determine whether more or less fuel is needed and whether the injector should be opened for a longer or shorter period of time based on the current operating conditions.

A ***throttle valve*** controls airflow, engine speed, and engine power. It is connected to the accelerator pedal. When the pedal is pressed, the throttle valve opens to allow more air into the combustion chambers. In turn, the control module holds the injectors open for a longer period of time, allowing more fuel into the combustion chamber. The increase in air and fuel results in an increase in engine power output.

As the accelerator pedal is released, the throttle valve closes, reducing the amount of air allowed into the combustion chamber. The control module, in turn, reduces the amount of time the injectors are open, thus reducing the amount of fuel released into the injection chamber. The decrease in air and fuel results in a decrease in engine power output.

Diesel Injection System

A ***diesel injection*** system forces fuel directly into the engine's combustion chamber, as shown in **Figure 1-27.** The heat resulting from highly compressed air, *not* an electric spark plug, ignites and burns the fuel. When the intake valve opens, a full charge of air is allowed to flow into the cylinder. Then, on the compression stroke, the air is squeezed until it is at a high temperature. As soon as the fuel is injected into the hot air, the fuel burns and expands.

A diesel injection system basically consists of an injection pump, injector, and glow plug system. The ***injection pump*** is a high-pressure, mechanical pump. It is powered by the engine and forces fuel to the diesel injector under very-high pressure. A conventional fuel pump feeds fuel from the tank to the injection pump.

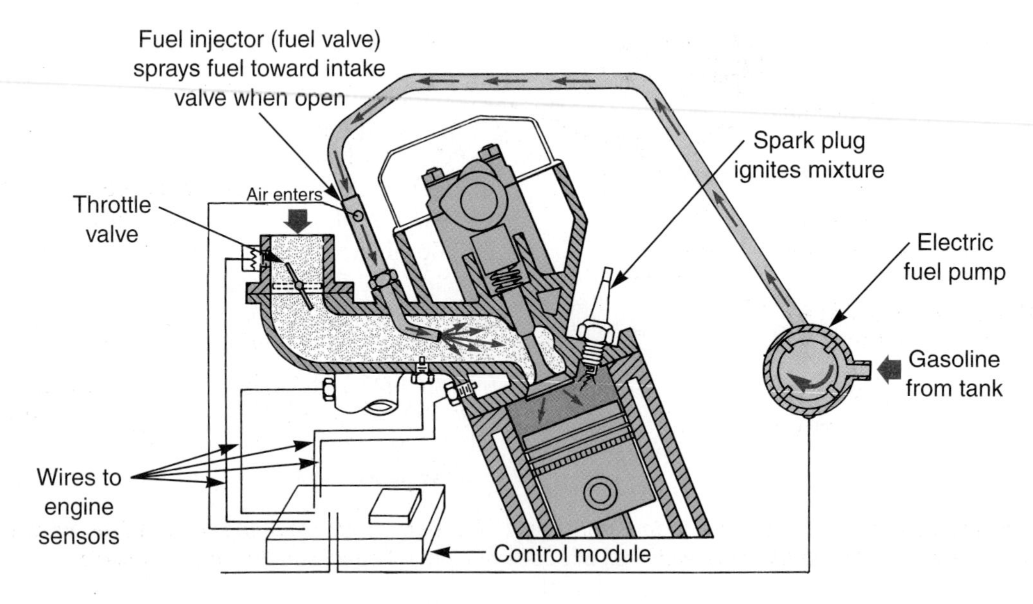

Figure 1-26. *Gasoline injection uses pressure from an electric fuel pump to spray fuel into engine intake manifold through fuel injectors. A throttle valve is used to control the airflow into the engine.*

The ***diesel injector*** is simply a spring-loaded valve. It is normally closed and blocks fuel flow. However, when the injection pump forces fuel into the injector under high pressure, the injector opens. The fuel is sprayed directly into the combustion chamber or a precombustion chamber. This is called a direct injection system.

Note in **Figure 1-27** that a diesel does *not* have a throttle valve or a spark plug. Engine power and speed are controlled by the injection pump, which is controlled by a control module. The more fuel injected into the combustion chamber, the more speed and power are produced by the engine. As less fuel is injected into the combustion chamber, engine speed and power decrease.

A glow plug system is used to aid cold starting of a diesel engine. The ***glow plug*** is an electric heating element that warms the air in the combustion chamber. This helps the air become hot enough to start combustion until the heat from engine operation can warm the incoming air.

Carburetion Fuel System

Instead of injecting fuel into the airstream or combustion chamber, a ***carburetion*** system uses engine vacuum (suction) to pull fuel into the engine, **Figure 1-28.** Airflow through the carburetor and the vacuum created by the engine's intake stroke draw fuel out of the carburetor's fuel bowl. As the throttle valve is opened, more air flows through the carburetor and, thus, more fuel flows into the airstream. A low-pressure, mechanical fuel pump draws fuel from the tank and delivers it to the carburetor's fuel bowl. The carburetion system is not currently used on late-model vehicles.

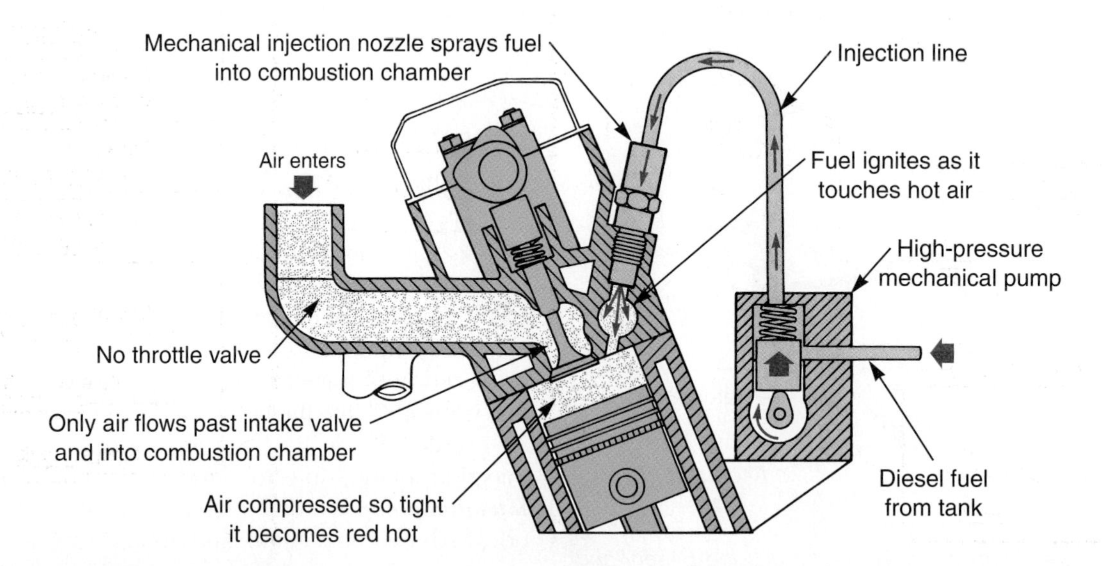

Figure 1-27. *Diesel injection is primarily mechanical. A mechanical injection pump forces fuel into a spring-loaded injector nozzle. The pressure opens the injector and fuel sprays directly into combustion chamber. A diesel does not use a throttle valve or spark plug.*

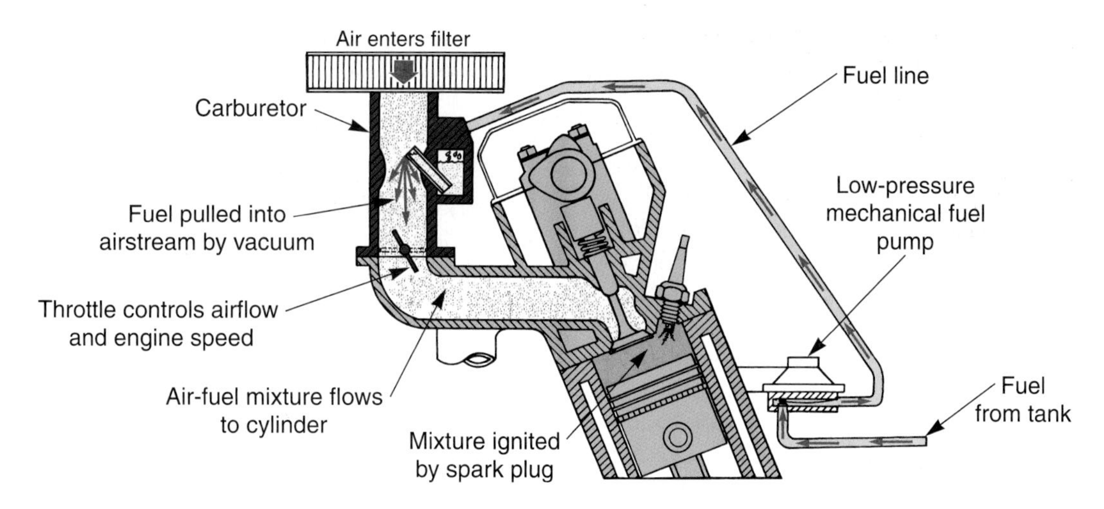

Figure 1-28. *Carburetion systems are not used today. In this type of system, airflow into engine pulls fuel out of carburetor's fuel bowl. A mechanical fuel pump fills the fuel bowl with fuel, but does not force the fuel into the intake manifold.*

Computer System

A ***computer system*** is used to increase the overall efficiency of a vehicle. The computer, or ***control module,*** can control the ignition system, fuel system, transmission or transaxle, emission control systems, and other systems. **Figure 1-29** shows a diagram of a simplified vehicle computer system. A modern vehicle may have several control modules.

To understand how a vehicle's computer control system works, think of the human body's central nervous system. For example, if your finger touches a hot stove, the nerves (sensors) in your hand send a signal to your brain. Your brain (control module) analyzes these signals and decides that you are in pain. Your brain (control module) quickly activates your muscles (actuators) to pull your hand away from the hot stove. A computer control system works in the same manner. It controls actions based on sensory inputs. For simplicity, a vehicle's computer control system can be divided into three subsystems: sensor, control, and actuator.

The ***sensor subsystem*** checks various operating conditions using sensors. A ***sensor*** is a device that can

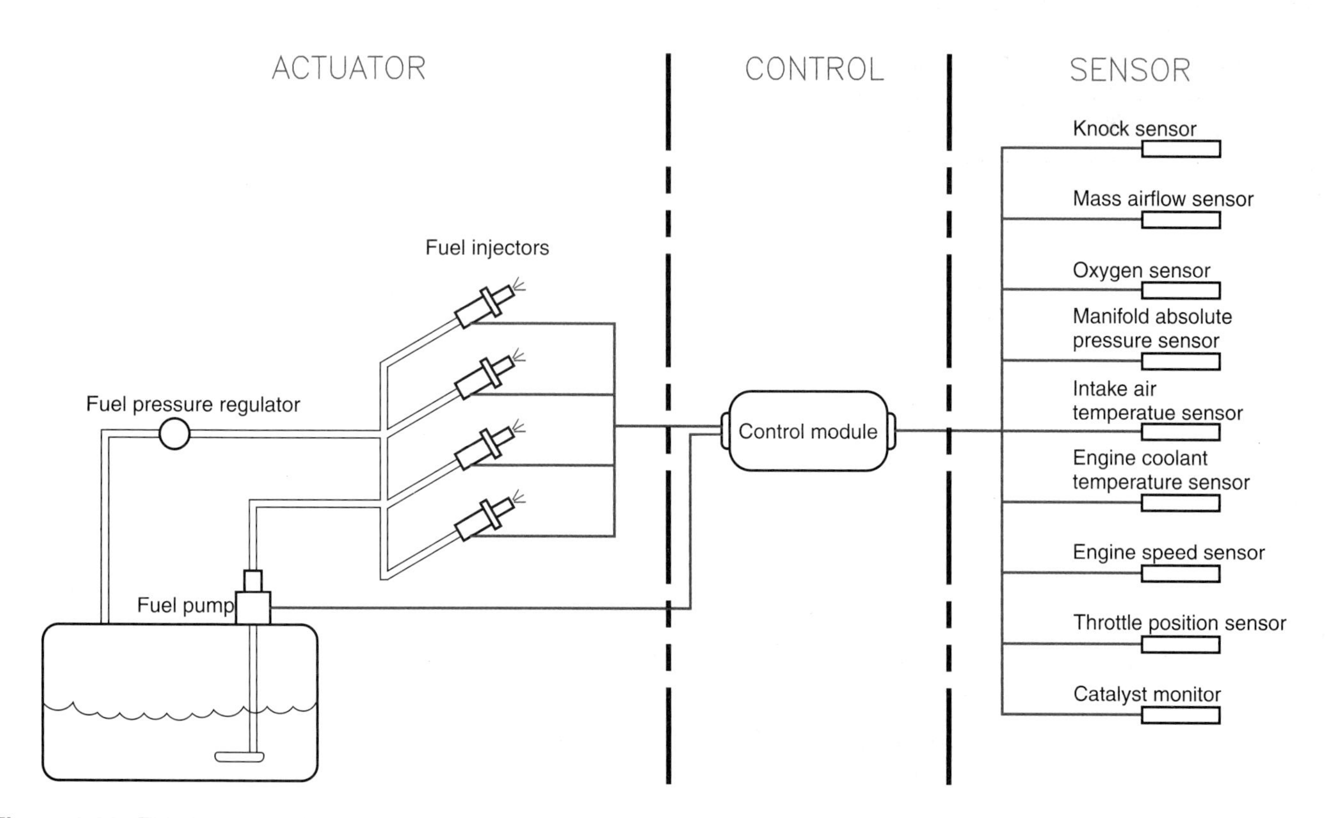

Figure 1-29. *This is a simplified diagram of the sensor, control, and actuator subsystems of a computer control system for fuel injector operation.*

change its electrical signal based on a change in a condition. Sensors might measure intake manifold vacuum, throttle opening, engine speed, transmission gear position, road speed, turbocharger boost pressure, and other conditions. The sensing system sends different electrical current values back to the control module.

The ***control subsystem*** looks at the inputs from the sensors and determines what actions need to take place. A control module, or computer, contains miniaturized electrical circuits that collect, store, and analyze information. The control module then provides signals to the actuator subsystem.

The ***actuator subsystem*** serves as the "hands" of the computer system. Based on signals from the control module, this system moves parts, opens injectors, closes the throttle, turns on the fuel pump, and performs other tasks needed to increase the overall efficiency of the vehicle. Electric motors, solenoids or relays, and switches are the actuators. The actuators turn on or off, open or close, or change position based on the signals from the control module. This is discussed in detail later in this textbook.

Emission Control Systems

Emission control systems are designed to reduce the amount of harmful chemicals and compounds (emissions) that enter the atmosphere from a vehicle. There are several types of emission control systems:

- **PCV.** The ***positive crankcase ventilation system*** pulls fumes from the engine crankcase into the intake manifold so they can be burned before entering the atmosphere.
- **Evaporative emissions control.** This system uses a charcoal-filled canister to collect and store gasoline fumes from the fuel tank when the engine is not running. Air is drawn through the canister and into the intake manifold while the engine is running so the collected fumes are burned.
- **EGR.** The ***exhaust gas recirculation system*** injects exhaust gasses into the engine to lower combustion temperatures and reduce one form of pollution in the engine exhaust.
- **Air injection.** This system forces air into the exhaust stream leaving the engine to help burn any unburned fuel that exits the combustion chamber.
- **Catalytic converter.** This device chemically converts byproducts of combustion into harmless substances.

Many of these systems work together, all reducing their share of harmful emissions. The computer also plays an important part in reducing pollution. It improves the efficiency of many systems.

Drive Train

The ***drive train*** uses power from the engine to turn the vehicle's drive wheels. Drive train configurations vary, but can generally be classified as rear-wheel drive or front-wheel drive. **Figure 1-30** shows simplified drive trains.

Clutch

The ***clutch*** allows the driver to engage or disengage the engine power from the drive train. It is mounted onto the engine flywheel between the engine and transmission or transaxle. A clutch is needed when the vehicle has a manual transmission. A vehicle with an automatic transmission does not have a clutch.

Transmission

The ***transmission*** uses a series of gears to allow the amount of torque going to the drive wheels to be varied. The driver can shift gears to change the ratio of crankshaft revolutions to drive wheel rotation. When first accelerating, more torque is needed to get the vehicle moving. Then, at higher road speeds, less torque is needed to maintain road speed. Engine speed also needs to be reduced at highway speeds.

A ***manual transmission*** is shifted by hand. Levers, cables, or rods connect the driver's shift lever to the internal parts of the transmission. An ***automatic transmission*** uses a hydraulic (fluid pressure) system to shift gears. A torque converter (fluid clutch) and special planetary gear sets provide automatic operation. However, the driver must manually change the shift lever position to change from forward to reverse.

A ***transaxle*** is a transmission and a differential (axle drive mechanism) combined into one housing. It is commonly used with front-wheel-drive vehicles, but may also be found on rear-engine, rear-wheel-drive vehicles. A transaxle may be manual or automatic.

Drive Shaft and Drive Axle

A ***drive shaft*** is used with a front-engine, rear-wheel-drive vehicle. It connects the transmission to the differential.

A ***drive axle*** connects the differential to the drive hubs or wheels. On most rear-wheel-drive vehicles, the drive axle is a solid steel shaft. On vehicles with a transaxle, the drive axle is a flexible shaft extending from the transaxle to the front wheel hubs.

Summary

This chapter reviewed the basic operation of a four-stroke-cycle, piston engine. The engine found in most vehicles is called an internal combustion engine because it burns fuel inside of combustion chambers.

The cylinder block holds the other parts. The piston and connecting rod transfer combustion pressure to the crankshaft. The crankshaft converts the up and down action of the piston into rotary motion.

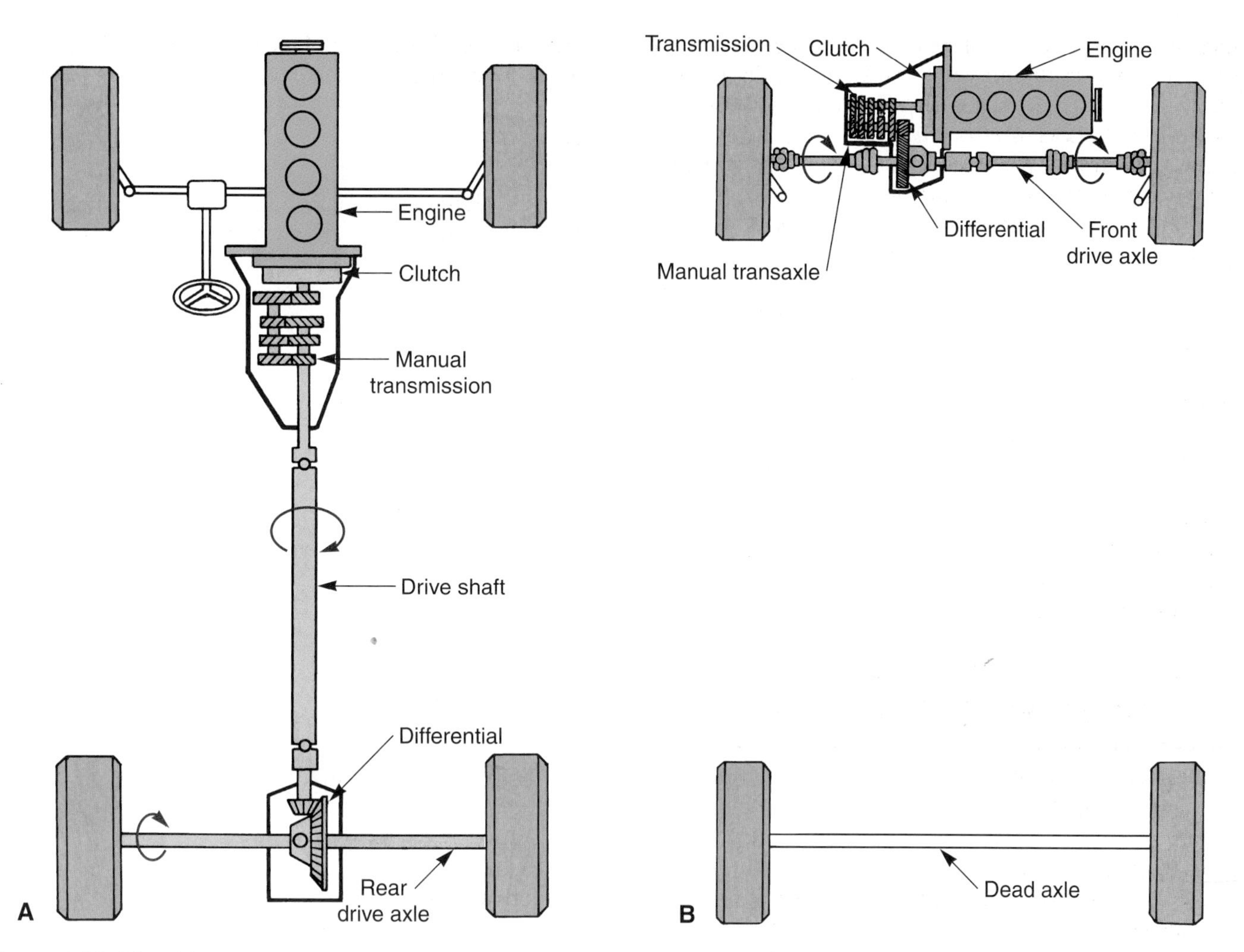

Figure 1-30. *The drive train uses engine power and crankshaft rotation to turn the vehicle's drive wheels. A—A front-engine, rear-wheel-drive vehicle. B—A front-engine, front-wheel-drive vehicle. Note the differential and transmission in one housing.*

The cylinder head is bolted on top of the cylinder block. It contains the valves that control flow into and out of the cylinder. The camshaft operates the valves so that they open and close to correspond to piston action, producing the four-stroke cycle.

The intake stroke draws air-fuel mixture into the engine. The compression stroke squeezes the mixture and readies it for burning. The power stroke ignites the mixture and the expanding gasses push the piston down with tremendous force. The exhaust stroke purges the burned gasses to prepare the cylinder for another intake stroke.

Every two crankshaft revolutions complete one four-stroke cycle. Thus, one power stroke is produced every two crankshaft revolutions. The flywheel helps keep the crankshaft spinning on the nonpower-producing strokes and smoothes out engine operation.

Various systems are needed to keep the engine running. The cooling system removes excess combustion heat and prevents engine damage. The lubrication system also prevents engine damage by reducing friction between moving engine parts.

The ignition system is needed on a gasoline engine to ignite the fuel and start it burning. A diesel engine uses high-compression-stroke pressure to heat the air in the cylinder enough to start combustion, instead of an electric spark. Also, a diesel does not use a throttle valve to control engine speed. The amount of fuel injected into the cylinder controls engine speed and power output.

The two types of fuel systems in use today are gasoline injection and diesel injection. Modern fuel injection increases fuel economy over older fuel systems by closer control of the fuel use by each cylinder. Carburetion is not used today.

The starting system uses a powerful electric motor to turn the engine flywheel until the engine can run on its own power. The charging system recharges the battery and supplies electricity to the vehicle while the engine is running.

A vehicle's computer system is similar to the human body's central nervous system. The computer system monitors various conditions by analyzing input signals from sensors. Then, the control module (computer) determines what action should be taken to maintain maximum efficiency. Based on this, the control module sends signals to actuators that control the operation of the fuel system, ignition system, transmission, and other devices.

Emission control systems reduce the amount of harmful pollutants that enter the atmosphere. Some emission control systems prevent fuel vapors from evaporating into the air. Others ensure complete combustion of fuel that leaves the combustion chamber and others just increase engine efficiency to reduce air pollution.

The drive train transfers engine power to the drive wheels. Both front-wheel and rear-wheel drive are found on today's vehicles. Front-wheel-drive vehicles have a transaxle, which is the transmission and differential combined in a single unit.

Review Questions—Chapter 1

Please do not write in this text. Write your answers on a separate sheet of paper.

1. The _____ is the main supporting structure of an engine.
2. These hold the crankshaft in the bottom of the engine block.
   (A) Cam bearings.
   (B) Main caps.
   (C) Decks.
   (D) Rod bearings.
3. _____ surround the cylinders in most engines to provide a way for coolant to remove heat from the cylinders.
4. What is the basic function of a piston?
5. _____ keep combustion pressure from blowing into the crankcase and they also keep _____ out of the combustion chamber.
6. Why is a piston pin or wrist pin needed?
7. The crankshaft converts the _____ motion of the piston into rotary motion.
8. Describe the basic parts of a cylinder head.
9. Which type of engine valve is larger and which type is smaller?
10. Summarize the four-stroke cycle.
11. The _____ opens the valves and allows them to close.
12. Why are the cooling and lubrication systems important?
13. An electric ignition system can be found on a(n) _____ engine but not on a(n) _____ engine.
14. What are the two types of fuel systems in use today?
15. Describe the three basic subsystems in a computer control system.

ASE-Type Questions—Chapter 1

1. Technician A says an engine's timing belt turns the camshaft at one-half of engine speed. Technician B says the camshaft opens the exhaust valve at the beginning of the power stroke. Who is correct?
   (A) A only.
   (B) B only.
   (C) Both A and B.
   (D) Neither A nor B.
2. Technician A says that the illustration is of a gasoline injection system. Technician B says that a diesel injection system injects fuel directly into the cylinder, as shown in the illustration. Who is correct?
   (A) A only.
   (B) B only.
   (C) Both A and B.
   (D) Neither A nor B.

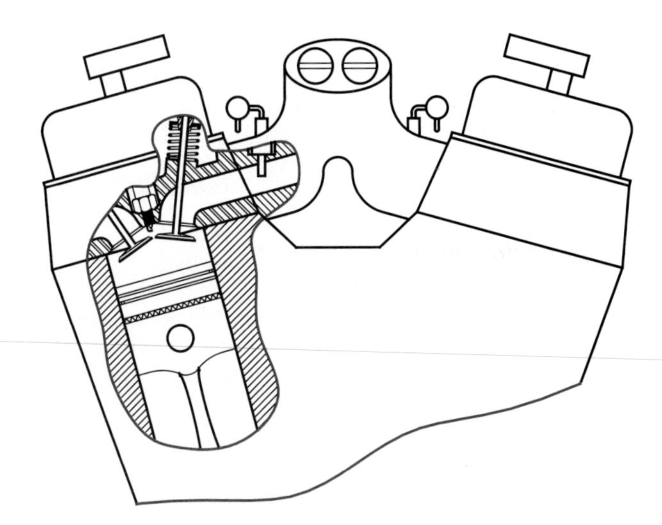

3. All of the following are major parts of a modern, multi-cylinder automotive engine *except:*
   (A) valve springs.
   (B) spool valves.
   (C) piston rings.
   (D) connecting rods.
4. A(n) _____ is the flat surface machined on the top of the cylinder block for the cylinder head.
   (A) cam housing
   (B) engine block deck
   (C) oil gallery
   (D) None of the above.

5. Technician A says the big end of a connecting rod fits around the piston pin. Technician B says the big end of a connecting rod fits around the crankshaft. Who is correct?
   (A) A only.
   (B) B only.
   (C) Both A and B.
   (D) Neither A nor B.
6. Technician A says the engine's oil rings always fit into the top groove on the pistons. Technician B says the engine's oil rings are normally located in the lowest groove on the pistons. Who is correct?
   (A) A only.
   (B) B only.
   (C) Both A and B.
   (D) Neither A nor B.
7. All of the following are basic parts of an automotive engine's piston assembly *except:*
   (A) connecting rod.
   (B) piston rings.
   (C) crankshaft journal.
   (D) piston pin.
8. The intake valve is _____ the exhaust valve.
   (A) smaller than
   (B) the same size as
   (C) larger than
   (D) All of the above.
9. While discussing the operation of a four-stroke cycle engine, Technician A says every two up and down strokes of the engine's piston results in one power-producing stroke. Technician B says every four up and down strokes of the engine's piston results in one power-producing stroke. Who is correct?
   (A) A only.
   (B) B only.
   (C) Both A and B.
   (D) Neither A nor B.
10. A(n) _____ is used to increase battery voltage to over 60,000 volts.
   (A) starter
   (B) magnet
   (C) ignition coil
   (D) alternator
11. All of the following are modern automotive emission control systems *except:*
   (A) air injection.
   (B) PCV.
   (C) intake valve.
   (D) evaporative emissions control.

A leveling device is a useful accessory for an engine lift. Here, a leveling device is attached to the engine lift chain. The two mounting brackets must be securely attached to the engine. By turning the handle on the leveling device, the engine can be shifted up in the front or back. (OTC Div. of SPX Corp.)

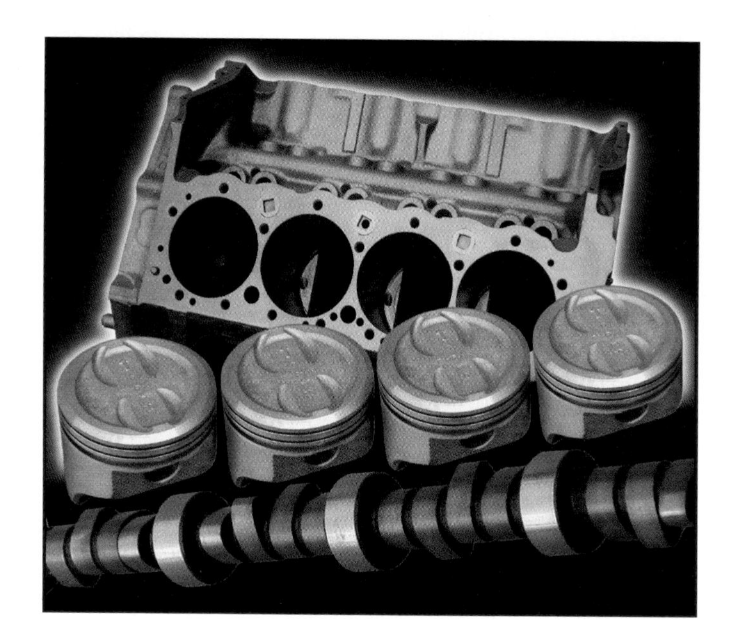

Chapter 2

Engine Service Tools and Equipment

After studying this chapter, you will be able to:

- ❑ Summarize the use of specialized engine service tools and equipment.
- ❑ Select the proper engine service tools when performing engine repairs.
- ❑ Explain the use of common engine cleaning tools.
- ❑ Use precision engine measuring tools accurately: micrometer, caliper, dial indicator, etc.
- ❑ List machine tools commonly used during engine rebuilding.
- ❑ Explain how to use typical electric-electronic and mechanical engine testing devices.
- ❑ Describe the function of commonly used engine test equipment.

Know These Terms

Boring bar
Brush hone
Cam bearing driver
Camshaft grinder
Compression gauge
Cooling system pressure tester
Crankshaft grinder
Crankshaft welder
Cylinder hone
Cylinder leakage tester
Degree wheel
Dial bore gauge
Dial indicator
Digital caliper
Dye penetrant testing
Electronic fuel injection tester
Engine balancer
Engine prelubricator
Exhaust gas analyzer
Feeler gauge
Flex hone
Flywheel lock
Flywheel rotating tool
Hole gauge
Hot tank
Hydraulic lifter tester
Hydraulic press
Line hone
Magnafluxing
Media blaster
Micrometer
Milling machine
Multimeter
Number punch set
Oil pressure gauge
Piston knurler
Piston ring compressor
Piston ring expander
Plastigage
Pressure gauge
Puller
Ridge reamer
Rigid hone
Ring groove cleaner
Rod rebuilding machine
Sensor sockets
Snap-ring pliers
Spark plug cleaner
Spark tester
Steel rule
Stethoscope
Straightedge
Telescoping gauge
Temperature probes
Torque wrench
Tubing equipment
Vacuum gauge
Valve adjusting wrench
Valve grind machine
Valve guide tools
Valve seat grinder
Valve spring compressor
Valve spring tester
VOM

Modern engines are complex, electro-mechanical assemblies. An engine has both mechanical parts and electronic components to monitor and control engine operation. The internal combustion engine provides a dependable source of power for our passenger vehicles. During manufacturing, engines are computer designed; precision cast, often in exotic metal alloys; machined; and assembled to extremely close tolerances. These processes result in even more complex repair methods and many more specialized tools than ever before.

A wide variety of tools and equipment are now required to repair worn or damaged engines. A professional engine technician will invest thousands of dollars on not just a set of basic hand tools, but also a broad array of test instruments, pressure-vacuum gauges, specialized service tools, and engine repair-rebuild equipment. To be a competent engine repair technician, you must know when and how to use the wide variety of service tools required on present-day vehicle engines.

Today's car engines are supported by very complex and often interacting support systems. To diagnose whether a problem source is mechanical or electrical frequently requires the use of very specialized testing equipment. Also, modern engine designs require specialized tools to disassemble, rebuild, and reassemble engines.

This chapter provides an overview of the selection and use of engine service tools, cleaning tools, measuring tools, electrical test instruments, and engine test equipment. This will give you a solid background when learning to diagnose, service, and repair today's "high tech" engines.

Note: Basic hand tools are not covered in this chapter. This text is written for students specializing in engine repair. If you need more information on basic automotive hand tools, refer to *Modern Automotive Technology* published by The Goodheart-Willcox Company, Inc.

Engine Service Tools and Equipment

There are many specialized engine service tools and pieces of equipment. This section describes the function of many of these tools and pieces of equipment.

Engine Stand

An ***engine stand*** holds an engine for easy disassembly and reassembly. It is a metal framework on small casters or wheels. The cylinder block bolts to the stand.

The engine stand holds the engine at approximately waist level. It also allows the engine to be rotated into different positions for access to the top or bottom of the cylinder block. A catch tray is sometimes mounted on the lower part of the engine stand to catch dripping coolant and oil, **Figure 2-1**.

Figure 2-1. *An engine stand is required when disassembling or reassembling an engine. The engine can be rotated into different positions for easy access to the crankcase, cylinders, deck surfaces, cam bore, and other features. The catch pan collects oil and fluid that may drip from the engine.*

Engine Crane

A portable ***engine crane*** is used to remove (raise) an engine from or install (lower) an engine into a vehicle, **Figure 2-2.** It has a hydraulic hand jack for raising and a pressure release valve for lowering. An engine crane is also handy for lifting heavy parts, such as intake manifolds, cylinder heads, transmissions, and transaxles.

Hydraulic Press

A ***hydraulic press*** is capable of producing tons of force for assembly or disassembly of pressed-together components, **Figure 2-3.** In engine mechanics, the hydraulic press is commonly used to force piston pins out of the pistons. A special fixture is needed to support the piston and prevent piston damage while driving.

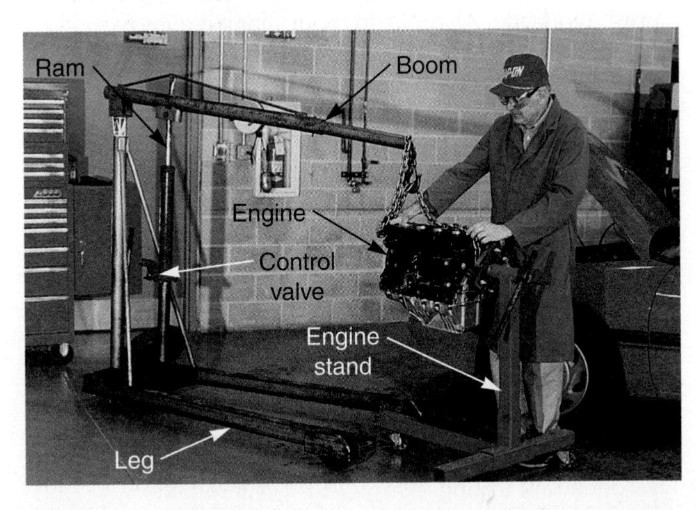

Figure 2-2. *An engine crane is for lifting engines and other heavy assemblies.*

Figure 2-3. *A hydraulic press produces a driving force powerful enough to push a piston pin out of its press-fit bore, for example. Always wear eye protection and stand to one side when using a press.*

Figure 2-4. *Various types of hydraulic jacks are used in engine repair. This is an extension-type transmission jack often needed for transmission or transaxle removal during engine service operations. (OTC Div. of SPX Corp.)*

Warning: Respect the force generated by a hydraulic press. Always wear eye protection and stand to one side during pressing operations. If a part breaks, it can fly out of the press with deadly power.

Hydraulic Jacks

Hydraulic jacks are used to lift the vehicle, engine, and transmission during engine removal or installation. They can also be used to lift the engine when replacing motor mounts, exhaust manifolds, oil pans, and so on. A special hydraulic transmission jack is shown in **Figure 2-4.**

Never work under a vehicle that is only supported by a floor jack. The vehicle must be mounted on jack stands before working. Also, keep your hands and feet from under an engine or transmission that is raised on a hydraulic jack. A hydraulic jack can fail, dropping the engine or transmission.

Ridge Reamer

A ***ridge reamer*** is used to remove the metal lip formed at the top of a worn cylinder. The top of the cylinder is not exposed to the rubbing, wearing action of the piston rings. It will remain unworn, while the lower section of the cylinder will wear larger in diameter. This lip is called a cylinder or ring ridge. If the cylinder ridge is *not* removed, the piston can be damaged when you are trying to push it out during engine teardown. As shown in **Figure 2-5,** the ridge reamer tool has small cutter blades for removing a cylinder ridge.

Ring Groove Cleaner

A ***ring groove cleaner*** removes carbon deposits from inside of piston ring grooves, **Figure 2-6.** It is turned or rotated around the piston by hand. Small cutters the same size as the ring grooves scrape out accumulated carbon.

If carbon is *not* removed from the piston ring grooves, the new rings could be forced out against the cylinder wall after engine operation and piston expansion due to heat. This may result in the rings and cylinder becoming scored and ruined.

Piston Ring Expander

A ***piston ring expander*** is used to remove and install piston rings. Compression rings are normally made of brittle, easy-to-break cast iron. When installing new piston rings, be very careful not to over expand and break the rings. A piston ring expander will help prevent ring breakage while speeding engine assembly, **Figure 2-7.**

Snap-Ring Pliers

Snap-ring pliers are used to expand or contract snap rings for removal and installation. These pliers have specially shaped, usually pointed tips. See **Figure 2-8.** Snap-ring pliers are used to grasp and remove or install small, spring-clip-type fasteners.

Sensor Sockets

Sensor sockets have a special shape or opening in the side to allow service of engine and other sensors. Refer to **Figure 2-9.** The opening in the side of the socket allows the socket to fit over sensors with wiring pigtails attached. Other sensor sockets may have a special shape to fit a particular sensor.

Piston Knurler

A ***piston knurler*** will form (not cut) grooves in the piston skirt to increase the outside diameter of the piston. Knurling is needed on slightly worn pistons to restore their

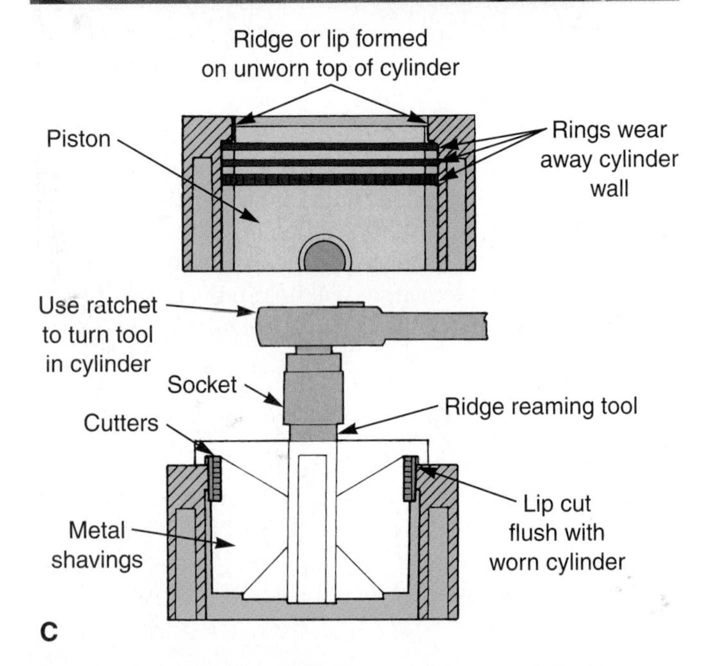

Figure 2-5. *A—This is a ridge reamer used to remove the metal lip formed at top of a cylinder where the rings do not rub and wear. B—Note the carbide steel cutter on the outer edge of the ridge reamer. C—This drawing details the action of a ridge reamer. (Snap-on Tool Corp.)*

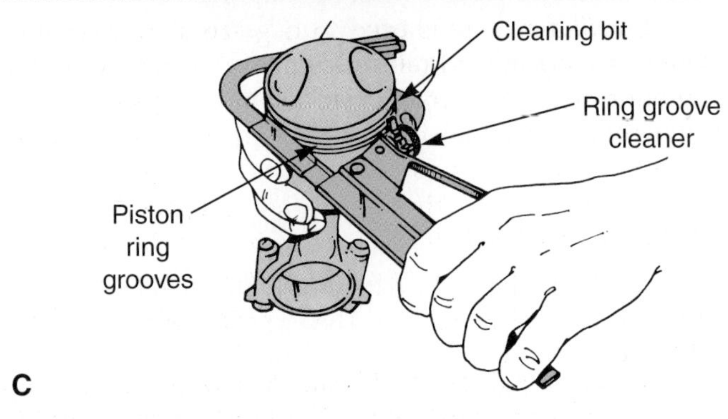

Figure 2-6. *A—A ring groove cleaner is needed to remove hard carbon deposits from inside of the piston ring grooves. B—Select the cutter bit that is almost as wide as, but fits into, the piston ring groove. C—Note how the ring groove cleaner is rotated around the piston to remove carbon deposits from inside the grooves. (Lisle Tools)*

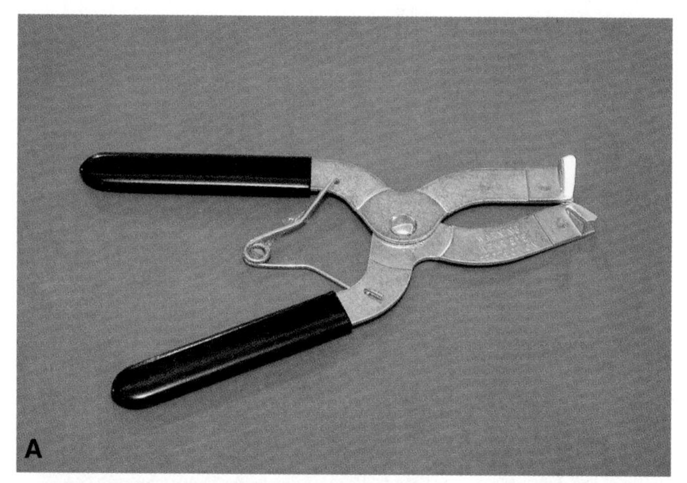

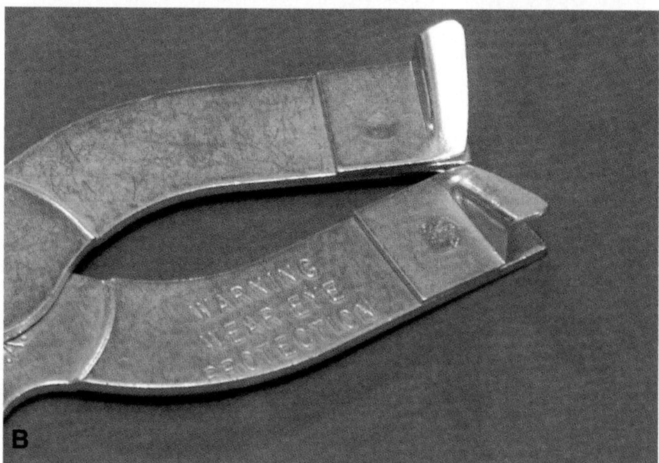

Figure 2-7. *A—A ring expander makes ring installation quicker and can help prevent ring breakage. B—The jaws on ring expander pliers will grasp the ends of compression rings.*

size to within specifications. The knurled grooves also retain oil to prevent further piston and cylinder wear. Look at **Figure 2-10.**

Cylinder Hone

A ***cylinder hone*** is used to deglaze a cylinder wall during an engine rebuild. Deglazing is removing the smooth surface and providing a lightly roughened or textured surface. The cylinder hone contains fine-grit stones that are spun inside the cylinder by an electric drill attached to the tool. The drill is pulled up and down to produce a crosshatched hone pattern on the cylinder wall. Then, when new piston rings are installed, the rings will properly wear into the cylinder wall surface and produce a good ring seal.

A ***brush hone*** has small beads of abrasive material on the ends of spring-steel wires. The brush hone is used to final deglaze cylinders before assembly. See **Figure 2-11A.**

A ***flex hone*** has hard, flat, abrasive stones affixed to spring-loaded, movable arms. It is used when cylinder wear is minor and only a small amount of cylinder honing is needed. Refer to **Figure 2-11B.**

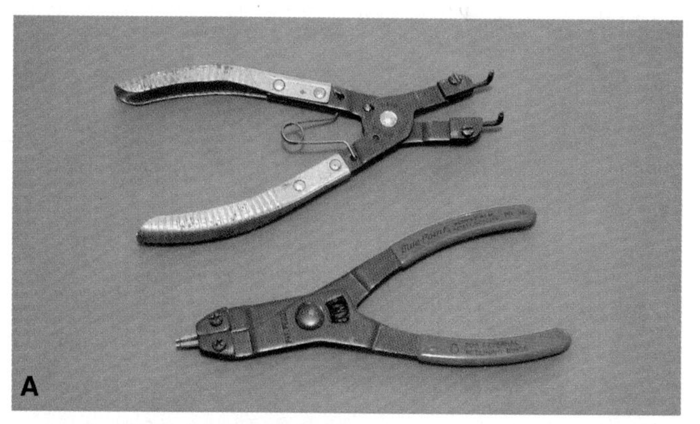

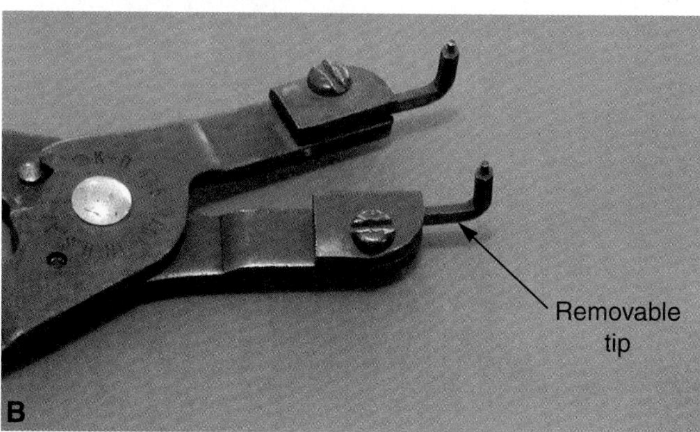

Figure 2-8. *A—Snap-ring pliers are often needed to remove spring-steel clips (snap rings). B—Snap-ring pliers have specially shaped, removable tips to fit different snap rings. (Snap-on Tool Corp.)*

A ***rigid hone*** has hard, flat, abrasive stones attached to stationary, but adjustable, arms. It is used when the worn cylinders do not require boring but a large amount of honing is needed to true and texture the cylinder walls in the block. Look at **Figure 2-11C.**

Figure 2-9. *Sensor sockets are designed to fit over large or odd-shaped engine sensors. The socket shown on the left is for removing an oil sending unit. The socket shown on the right is designed to fit over a wire pigtail that sticks out of a sensor. (Snap-on Tool Corp.)*

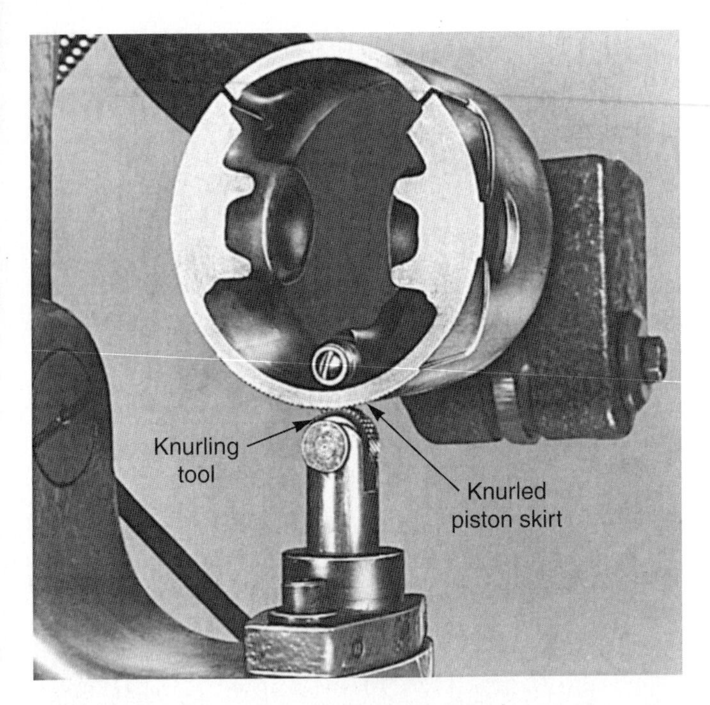

Figure 2-10. *A knurling tool forces a pattern of grooves into the piston side to increase the outside diameter of the piston. In this way, the diameter of used pistons can be restored to specifications. (Deere & Co.)*

Flywheel Rotating Tool and Flywheel Lock

A ***flywheel rotating tool*** is a special pry bar with fingers for grasping the teeth on the flywheel ring gear. It is used to rotate the engine crankshaft by hand, **Figure 2-12A.**

A ***flywheel lock*** will keep the engine crankshaft from turning during engine disassembly or reassembly. You may need to lock the flywheel, and thus the crankshaft, when torquing flywheel bolts or the front damper bolt, for example. Refer to **Figure 2-12B.**

Piston Ring Compressor

A ***piston ring compressor*** is used to squeeze the piston rings into their grooves so that the piston assembly can be installed into its cylinder. The tool is simply a spring-steel clamp that can be tightened around the piston and piston rings. See **Figure 2-13.** The wooden handle of a hammer is commonly used to force the piston assembly into the block as the rod is guided over the crankshaft.

Valve Adjusting Wrench

A ***valve adjusting wrench*** is a special tool for turning the rocker arm adjusting screws. Using this tool, you can set the clearance between the rocker arm and the valve stem, **Figure 2-14.**

Degree Wheel

A ***degree wheel*** is used to measure how much the crankshaft has been rotated when performing camshaft degreeing. One is shown in **Figure 2-15.** Some high-performance engines have an adjustable timing sprocket or gear. The camshaft timing can be changed in relation to the crankshaft to affect engine performance.

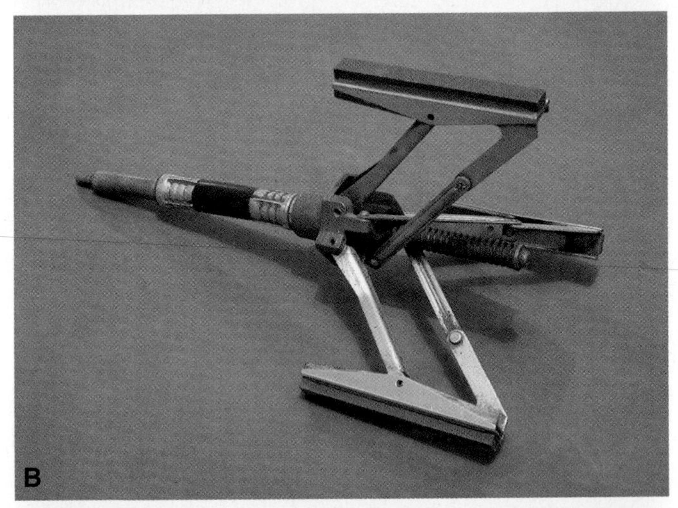

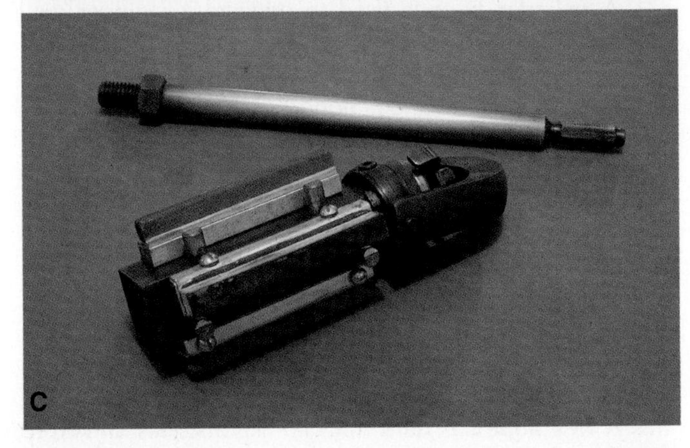

Figure 2-11. *Cylinder hones are needed to prepare the cylinder wall surface so new rings will seal and the engine will not consume oil or smoke. A—Brush hone. B—Flex hone. C—Rigid hone.*

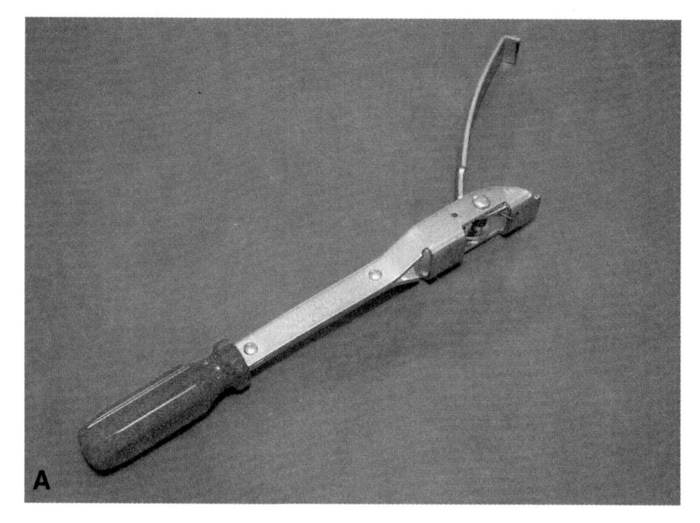

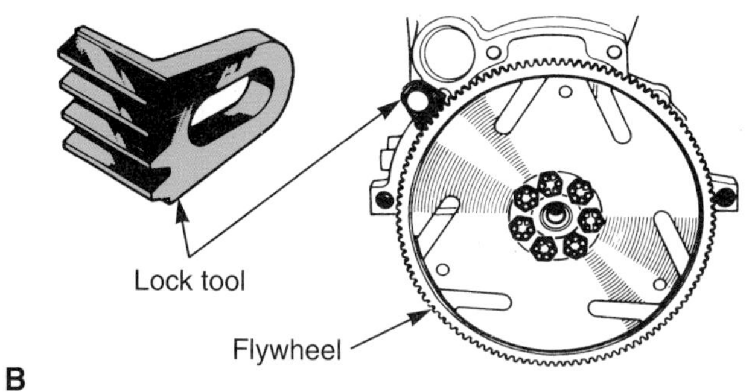

Figure 2-12. *A—This flywheel turning tool is used to rotate an engine crankshaft by hand. B—A flywheel lock keeps the crankshaft from turning while torquing flywheel bolts or the front damper. (Renault)*

Hydraulic Lifter Tester

A ***hydraulic lifter tester*** is used to measure the wear inside of a lifter. It preloads the lifter plunger and then produces a controlled collapse of the lifter. By timing how long it takes for the plunger to move to the bottom of its bore, you can find out if a lifter is good or bad. See **Figure 2-16.**

Engine Prelubricator

An ***engine prelubricator*** can be used to fill the oil passages in an engine with oil before starting the engine. It can also be used to find worn engine bearings and other lubrication system problems during pre-teardown diagnosis, **Figure 2-17.**

Pullers

A ***puller*** uses a screw action to force a hub, damper, pulley, gear, etc., off of its shaft. Several pullers are pictured in **Figure 2-18.** The puller is clamped or bolted onto the part. Then, when the puller screw is tightened, the component is pulled off its shaft. For example, a puller can be used to force a harmonic balancer off of the crankshaft.

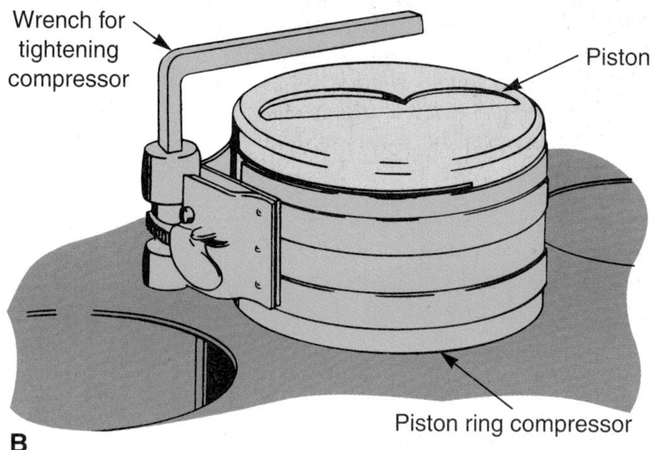

Figure 2-13. *A ring compressor holds the rings into their grooves as the piston assembly is forced into the cylinder block. A—This style of ring compressor has different-size sleeves to fit various diameter pistons and has an action similar to pliers. B—This style of ring compressor has a band that is tightened by a wrench. (Snap-on Tool Corp., General Motors)*

Warning: Pullers can exert tons of force. Wear eye protection and observe appropriate safety precautions.

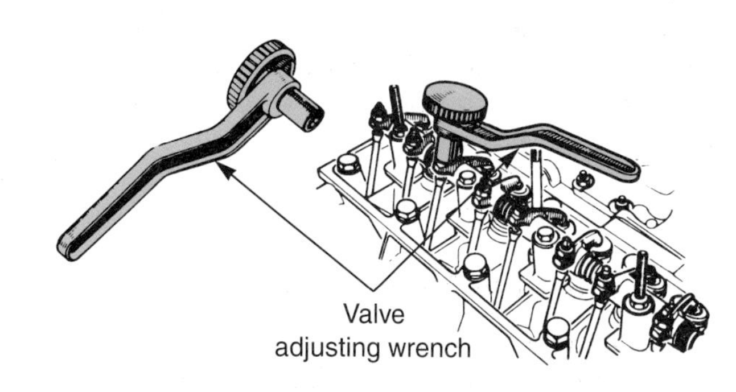

Figure 2-14. *This valve adjusting wrench is a specialized tool used to make valve clearance adjustment. (Renault)*

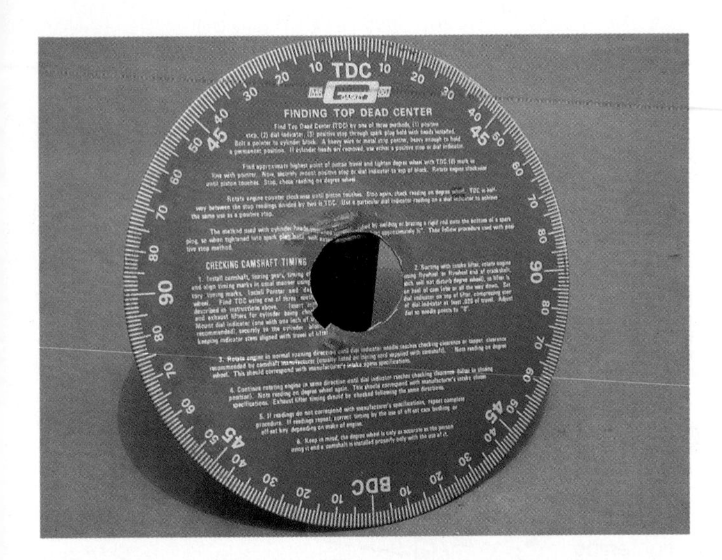

Figure 2-15. *A degree wheel is used when degreeing a cam for racing applications. It can also be used when adjusting valves to accurately position the crank and camshaft during engine blueprinting.*

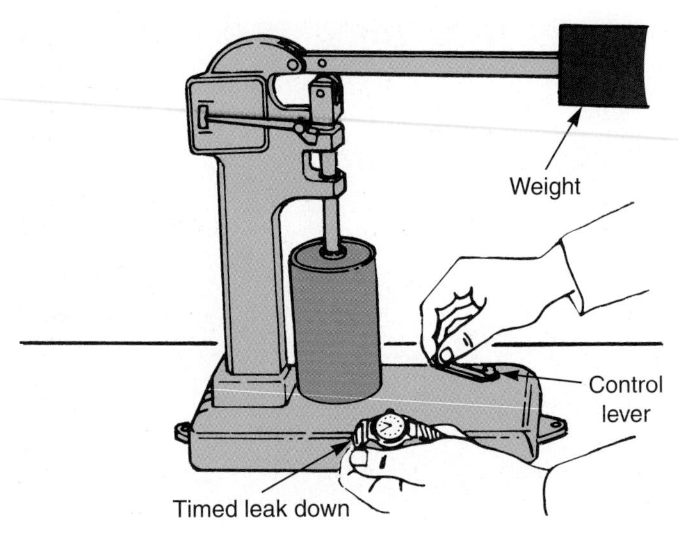

Figure 2-16. *A hydraulic lifter tester places a load on the lifter plunger. The amount of time it takes to push the plunger down can be used to determine the condition of the lifter. (General Motors)*

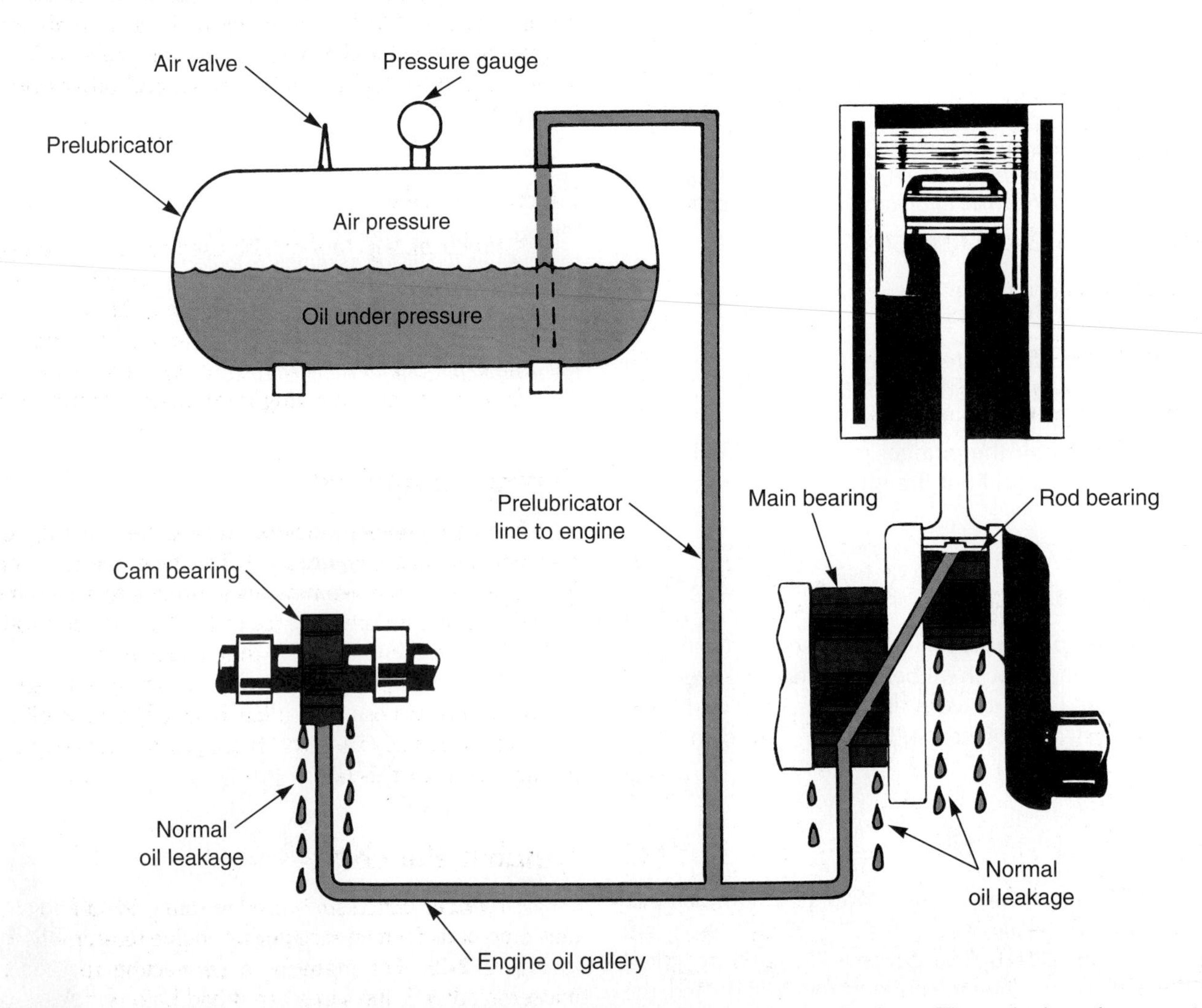

Figure 2-17. *An engine prelubricator consists of a steel tank with an air fitting and an outlet hose. The outlet hose is connected to the engine lubrication system. As air pressure is injected into the tank, engine oil is forced through the engine.*

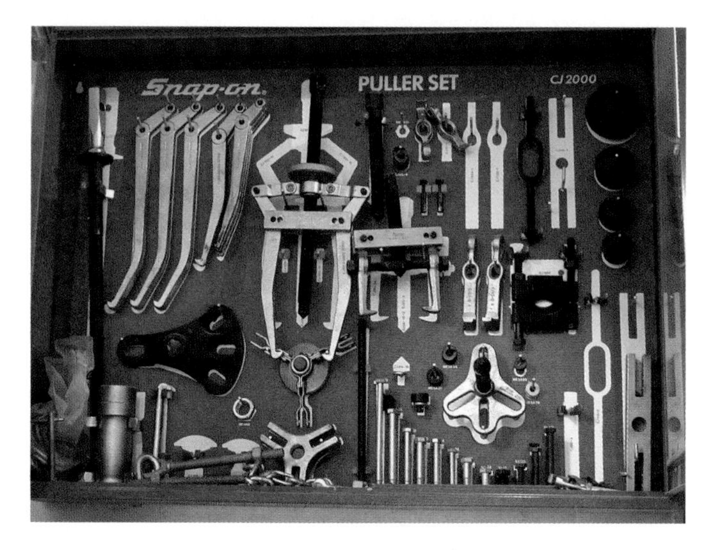

Figure 2-18. *Pullers are used to force gears, pulleys, dampers, sprockets, and other parts off of shafts. (Snap-on Tool Corp.)*

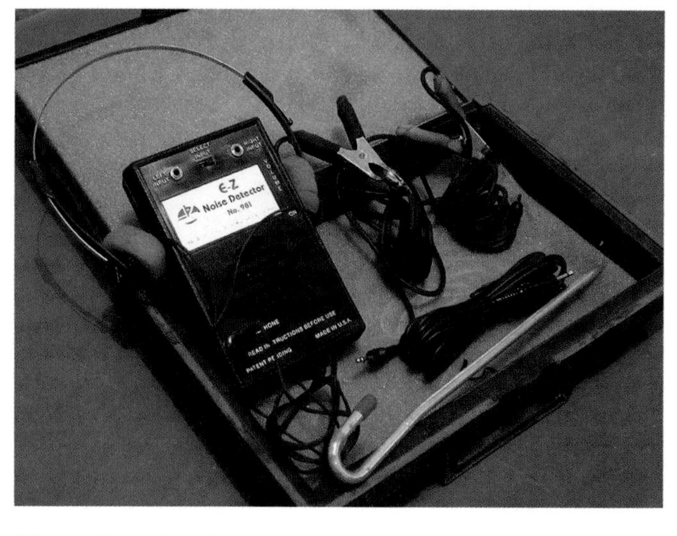

Figure 2-20. *An electronic noise detector will help you find the source of abnormal engine noise.*

Stethoscope

A ***stethoscope*** is a listening device that amplifies the sound generated by internal parts, **Figure 2-19.** It can help you find abnormal, internal noises. To locate a noise, touch the tip of the tool on different engine components. When the sound is the loudest, you have found the source of the noise.

Electronic Noise Detector

An ***electronic noise detector*** or stethoscope amplifies sounds emitted from engine parts. See **Figure 2-20.** Metal clips can be attached to the external parts of an engine. Noise from inside engine parts is electronically amplified to operate headphones or an earpiece. The clips can be moved around to find the loudest point and source of the engine noise.

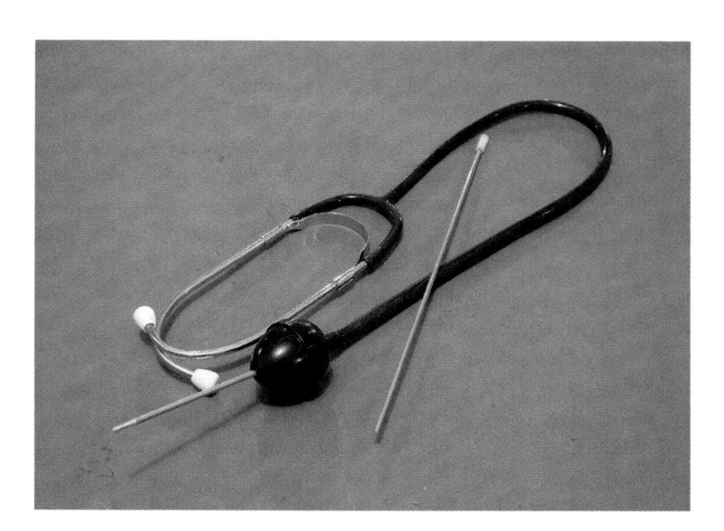

Figure 2-19. *A stethoscope can be used to find the source of abnormal engine sounds.*

Mirror Probe

A ***mirror probe*** can be used to inspect for problems in an area that is blocked from view. It is a small mirror hinged on a long handle. You can use the probe to find the source of hidden leaks, cracked parts, and other troubles, **Figure 2-21.**

Pickup Tools

A ***finger pickup tool*** can be used to retrieve dropped parts or tools. A special type will grasp engine lifters and pull them out of the engine. See **Figure 2-22A.**

A ***magnetic pickup tool*** serves a similar purpose as the finger pickup tool. However, it will attract and hold only ferrous (metal containing iron) objects, **Figure 2-22B.**

Tubing Equipment

Tubing equipment includes tube cutters, flaring tools, reamers, and so on, **Figure 2-23.** This type of equipment is frequently used to make new lines running to an engine. If an old fuel line is damaged, for instance, you can produce a new line from tubing stock using these tools.

Tube cutting pliers are used to cut soft tubing, such as plastic or rubber hose. The pliers have a sharp, steel-blade jaw that presses against a flat, blunt jaw to squarely cut the tubing. See **Figure 2-24.**

Number Punch Set

A ***number punch set*** is used to stamp identifying numbers onto parts for reference during engine reassembly. Refer to **Figure 2-25.** For example, a connecting rod must be reassembled with the same cap it had before. If these parts are not numbered in the casting, numbers must be stamped into the parts so they are matched during reassembly.

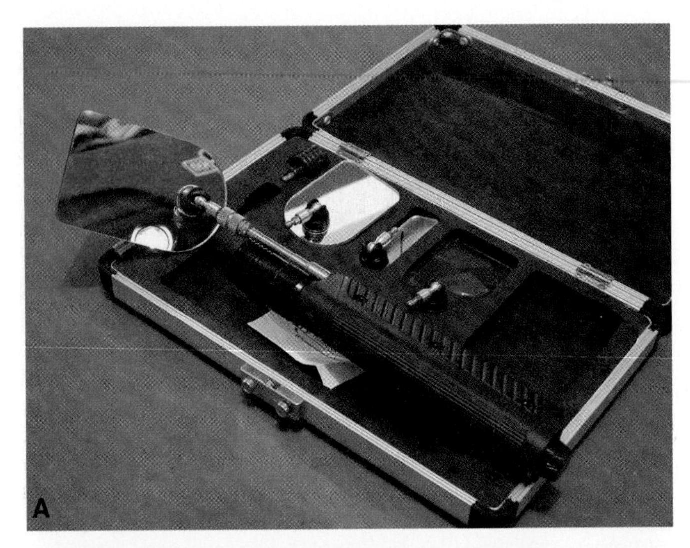

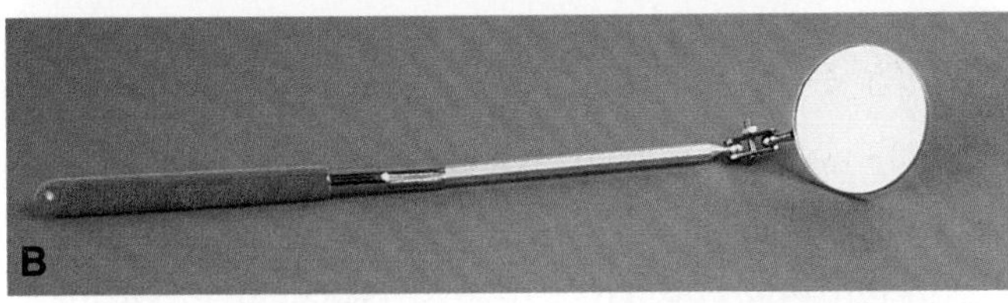

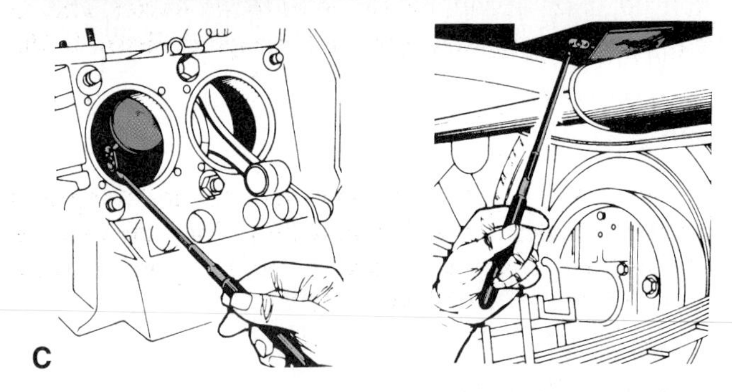

Figure 2-21. *A—This is a flashlight-mirror-probe tool that can be used to look or search in tight, dark areas of the engine compartment. B—An extension mirror will help you inspect visually obstructed locations. C—Mirror probes in use. (General Tool Corp.)*

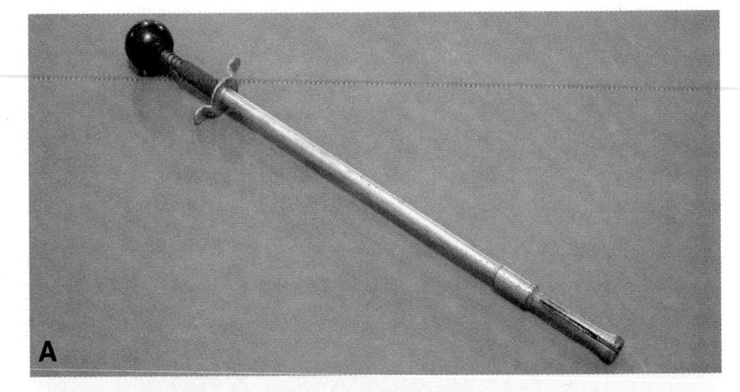

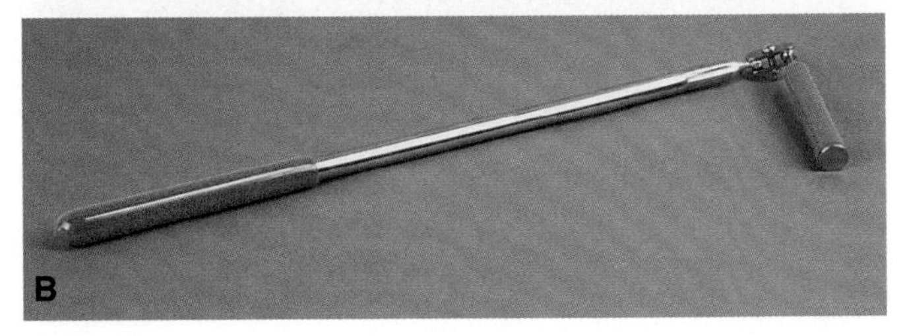

Figure 2-22. *A—This finger pickup tool can grasp and pick up small objects. The fingers are opened by pressing the handle knob. B—A magnetic pickup tool can be used to retrieve small iron or steel parts and tools. (Snap-on Tool Corp.)*

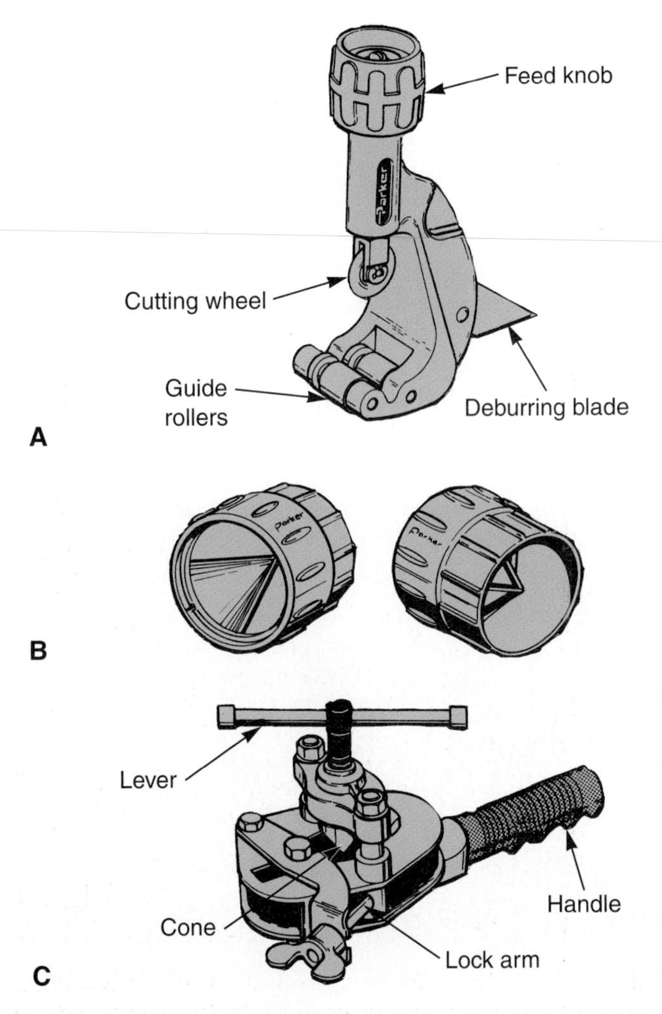

Figure 2-23. *Basic tubing tools. A—Tube cutter. B—Deburring tools. C—Flaring tool. (Parker)*

Clutch Alignment Tool

A ***clutch alignment tool*** is needed when installing a transmission/clutch assembly onto the rear of the engine. Look at **Figure 2-26.** The tool is used to hold the manual clutch disc onto the flywheel at the exact center of the flywheel. This allows the other clutch parts to be attached in perfect alignment with the centerline of the engine crankshaft. After the alignment tool is removed, the transmission input shaft can be installed through the clutch assembly and into the pilot bearing in the engine crankshaft.

Distributor Wrench

A ***distributor wrench*** is a long-handled, box-end wrench for removing fasteners that are under obstructions. See **Figure 2-27.** This type of wrench is handy when servicing distributors, sensors, and other parts with hard-to-reach bolts or nuts.

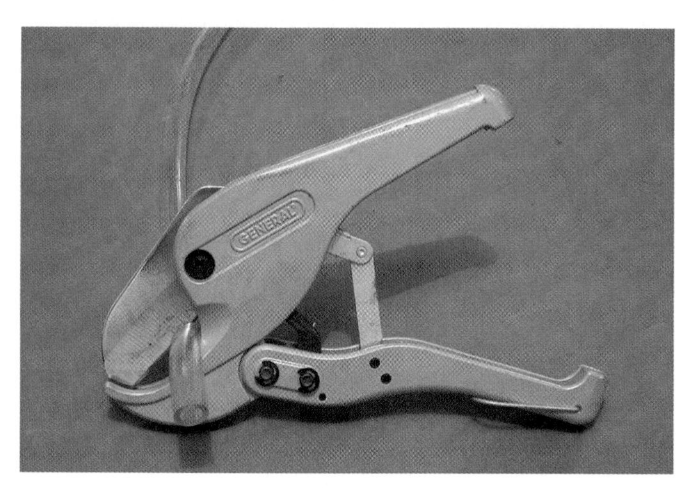

Figure 2-24. *A hose cutting tool has a sharp jaw that clamps against a flat jaw. This produces a shearing action for squarely cutting rubber and plastic tubing. (General Tool Corp.)*

Figure 2-25. *A number set has hardened tips that can easily imprint a number into steel, cast iron, and aluminum parts. (Snap-on Tool Corp.)*

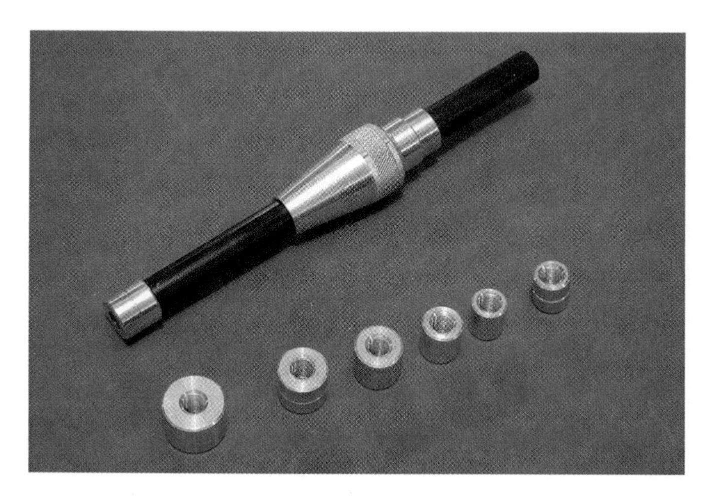

Figure 2-26. *A clutch alignment tool holds the clutch disc as the clutch pressure plate is bolted on. After the tool is removed, the transmission input shaft will slide through the clutch assembly and into the crankshaft pilot bearing. (Snap-on Tool Corp.)*

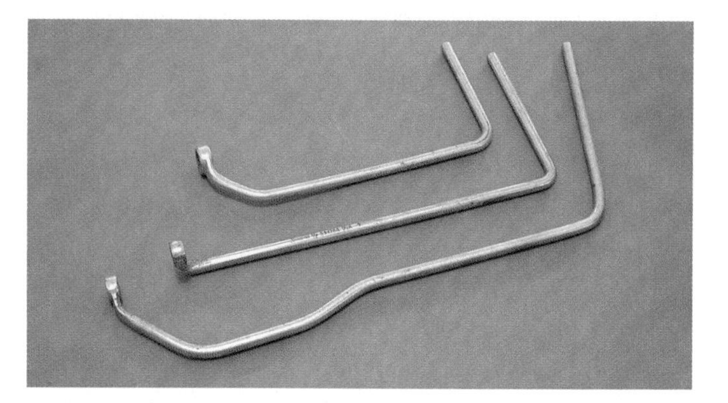

Figure 2-27. *Note how these distributor wrenches have long, odd-shaped handles that allow the wrench to reach under obstructions to remove or install fasteners. (Snap-on Tool Corp.)*

Oil Deflecting Rocker Arm Clips

Oil deflecting rocker arm clips are small metal clips used to prevent oil from squirting out of the engine rocker arms when operating the engine with the valve covers removed. They fit over the oil holes in the rocker arms to keep the oil from spraying out of the engine, **Figure 2-28.**

Drivers

Drivers are plastic or metal hubs for installing and removing press-fit parts, **Figure 2-29A.** For example, the front and rear crankshaft seals on some engines must be installed using a driver, **Figure 2-29B.**

Cam Bearing Driver

A ***cam bearing driver*** is used to force cam bearings into or out of the cylinder block, **Figure 2-30.** The driver is a long steel bar with a threaded end and special bushings. When the tool is screwed down, a powerful pulling action is produced that pulls cam bearings out of the block.

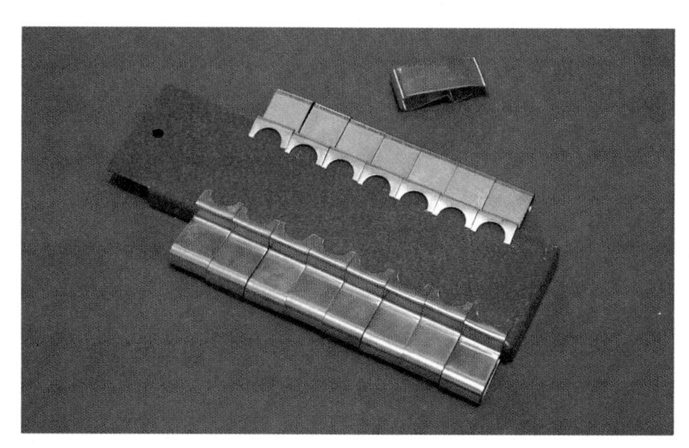

Figure 2-28. *Oil deflecting rocker arm clips can be placed over rocker arms to prevent oil from spraying out of the engine while adjusting lifters.*

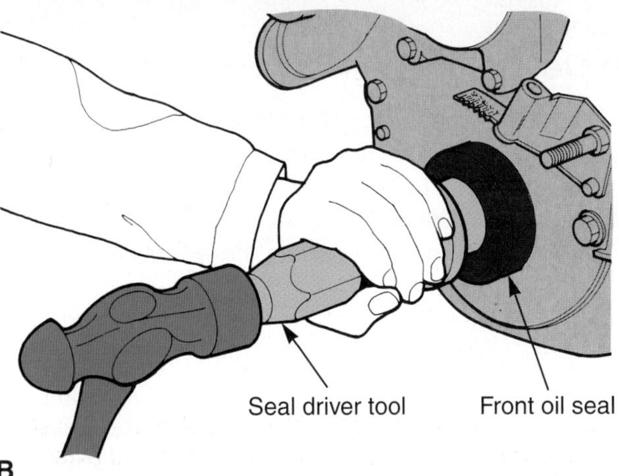

Figure 2-29. *A—This seal driver set has different-size collars to fit various bearing and seal diameters. B—The correct-size driver produces an equal force around the seal or bearing body to prevent damage during installation. (Snap-on Tool Corp., General Motors)*

Caution: When installing cam bearings, the oil hole(s) in each bearing must line up with the oil hole(s) in the block or head.

Oil Filter Wrench

An ***oil filter wrench*** is designed to grasp and hold the outside diameter of an engine oil filter. It allows the oil filter to be turned for removal. Several types are shown in **Figure 2-31.** An oil filter wrench is for removal only. Never use one to install an oil filter.

Jumper Cables

Jumper cables are used to start a vehicle with a dead (discharged) battery. The cables are connected between the dead battery and another vehicle's battery. This provides electricity to crank and start the engine of the vehicle with the dead battery.

Cylinder Head Rebuild Tools

Various special tools are needed to rebuild and repair cylinder heads. Machine shops often have a dedicated workstation or area for cylinder head rebuilding. See **Figure 2-32.**

Cylinder Head Stand

A ***cylinder head stand*** is used to hold a cylinder head while servicing the valves and valves seats, **Figure 2-33.** A cylinder head, especially one made of cast iron, can be heavy and clumsy. The stand makes service work much easier.

Valve Spring Compressor

A ***valve spring compressor*** is used when removing or installing a valve assembly, **Figure 2-34.** The tool is a clamp-like device that squeezes (compresses) the valve spring so that the valve keepers can be removed or installed. One end of the compressor fits onto the valve head. The other end fits over the spring retainer.

A valve spring compressor variation is available for servicing valves in overhead cam engines. The cylinder heads in some of these engines have a pocket in which the valve spring operates. The traditional valve spring compressor cannot be used for this design.

Valve Seat Grinder

A ***valve seat grinder,*** or a seat cutting tool, is used to resurface the valve seats in a cylinder head. One is shown in **Figure 2-35.** After prolonged use, valve seats can become pitted and irregular. The seat grinder or cutter smoothes the valve seat and makes it concentric (round). In this way, the ground, resurfaced valve face forms a leakproof seal when the valve is closed on the valve seat.

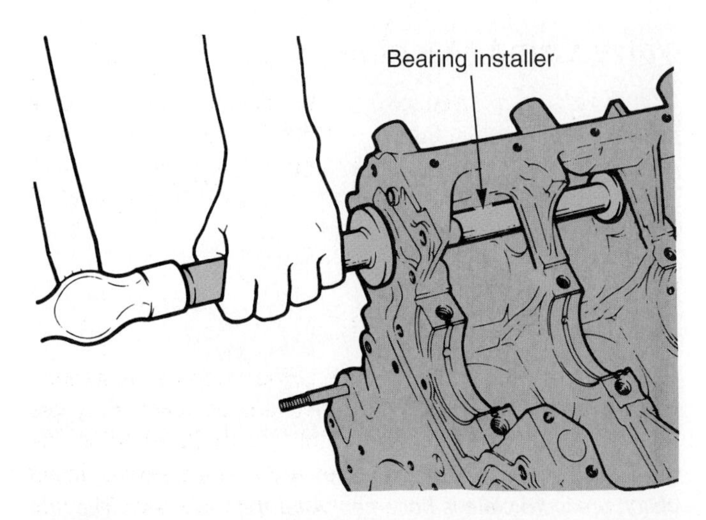

Figure 2-30. *A cam bearing driver is needed to force bearings into and out of the cylinder block or head. (DaimlerChrysler)*

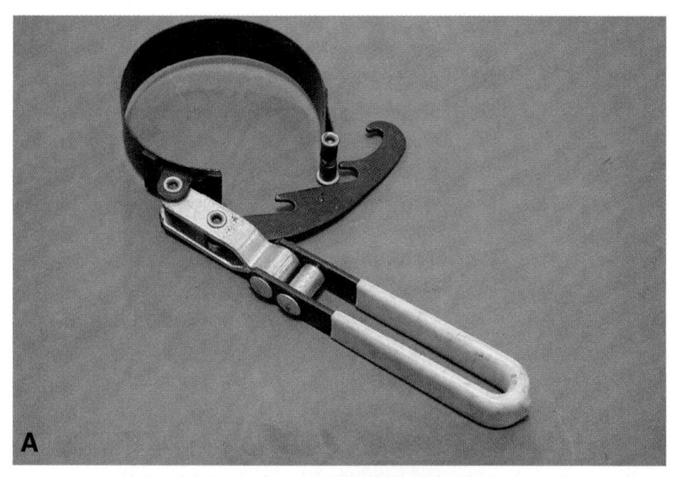

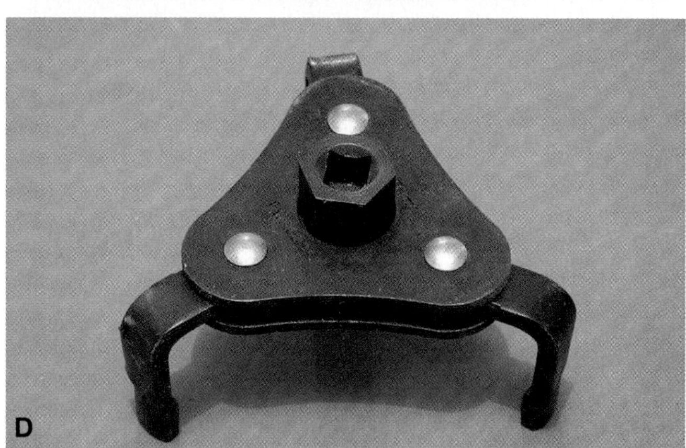

Figure 2-31. *An oil filter wrench is used to unscrew old, used oil filters. There are various styles of oil filter wrenches. A—Adjustable with metal strap. B—Metal strap for use with a ratchet and extension. C—Nylon strap for use with a ratchet and extension. D—Mechanical-finger-type for use with a ratchet and extension.*

A ***grinding stone dresser*** is a tool used to resurface the grinding stones in the valve seat grinder. The tool has a diamond-tipped screw that can be adjusted up to the surface of the grinding stone. When the stone is spun in the dressing tool, the diamond tip can be moved across the grinding stone to clean up and true its surface. Refer to **Figure 2-35.**

Valve Grind Machine

A ***valve grind machine*** can be used to resurface valve faces and valve stem tips. Refer to **Figure 2-36.** It uses grinding stones to produce smooth surfaces on used valves. The machine must be set to the correct angle when grinding valve faces.

Valve Guide Tools

Valve guide tools or a valve guide machine are needed to restore the cylinder head valve guides when they are worn. See **Figure 2-37.** With an integral (built-in) guide, the old guide can be machined larger and a new sleeve or insert pressed into the head. With a pressed-in guide, the old guide can be driven out and a new one driven into the head.

Valve Spring Tester

As springs are used, they tend to weaken and lose tension. A ***valve spring tester*** can be used to measure the

Figure 2-32. *A cylinder head rebuild station or work area has all of the tools and equipment needed to resurface valve seats, check for warpage, and perform other head-rebuild tasks.*

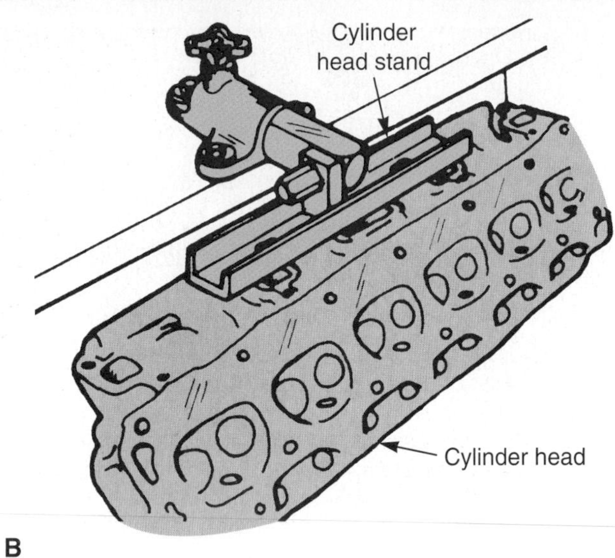

Figure 2-33. *A cylinder head stand makes valve service more convenient by securely holding the head during rebuild. A—The head rests on top of the stands in this style. B—The head is clamped to the work surface in this style of stand. (K-Line Tools)*

Figure 2-35. *A valve seat grinder is used to resurface valve seats. The grinding stone dresser is used to resurface the cutting stones. (Sioux Tools, Inc.)*

tension of valve springs, **Figure 2-38A.** The spring is compressed to a specific height in the tester. The spring pressure required to compress the spring is displayed by the tester in pounds or kilograms. If the pressure is below specifications, the valve spring must be replaced or valve spring shims must be used with the old spring, **Figure 2-38B.**

Machine Tools

There are several machine tools that may be required to complete engine rebuilding or repairs. These tools are usually

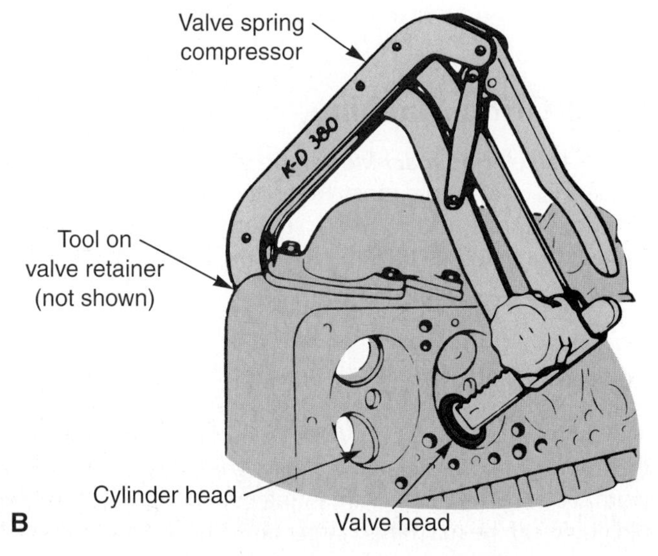

Figure 2-34. *A—A valve spring compressor is used to squeeze the spring together so the valve assembly can be removed from the cylinder head. B—Note how the valve spring compressor fits around the head to grasp the valve and spring. (K-D Tools)*

Figure 2-36. *A valve grind machine uses electric-motor-driven abrasive stones and a spinning chuck to resurface valve faces and valve stem tips. (Sioux Tools, Inc.)*

large and costly. Work requiring machine tools is usually performed by a machine shop, not an automotive technician.

Milling Machine

A ***milling machine*** is used to resurface cylinder heads and blocks. See **Figure 2-39.** This is done to true the surface, thus preventing head gasket leakage and other problems. The head or block is mounted in the machine. Then, a large, multi-tip cutter rotating at high speed is moved across the part to machine off a very thin layer of metal. This will correct any warpage in the head or block.

Boring Bar

A ***boring bar*** is a machine shop tool for making a worn engine cylinder larger in diameter, **Figure 2-40.** After prolonged service, the piston rings can wear a cylinder

Figure 2-38. *A—A valve spring tester is used to check the spring pressure. Compress the spring to a specified height and check the pressure. The pressure must be within specifications. B—Valve spring shims can be used to equalize valve spring pressure. They are placed under valve springs to increase spring tension.*

Figure 2-37. *This guide and seat machine is designed to machine the head for replacement of valve guides and valve seats.*

Figure 2-39. *A milling machine is being used to resurface a cylinder head gasket-sealing surface.*

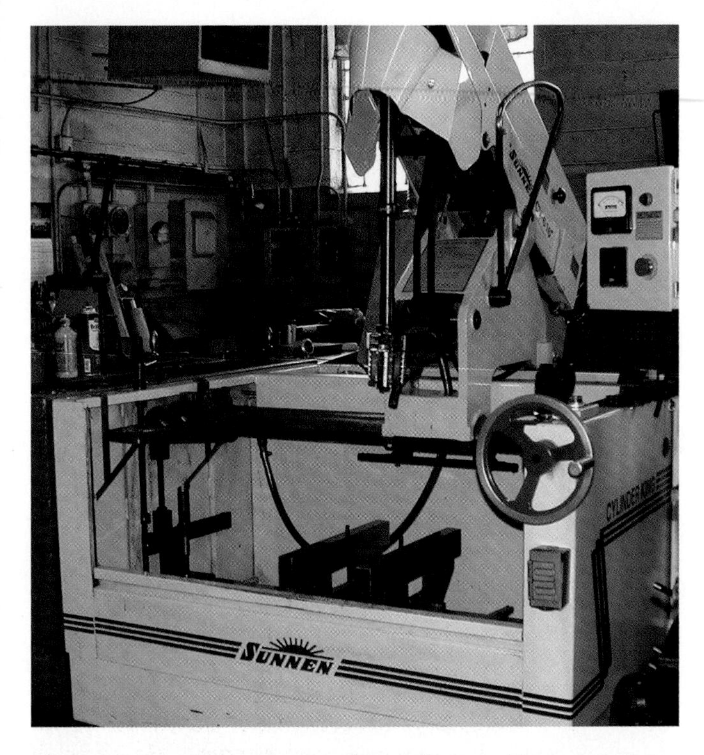

Figure 2-40. *A boring bar can be used to machine a cylinder to a larger diameter for oversize pistons or repair sleeves.*

out-of-round and cause taper. Part breakage can also damage a cylinder. If the cylinders are not too severely worn or damaged, they can be bored oversize so that new, oversize pistons can be installed during the engine overhaul. If the cylinders are badly damaged, the boring bar can be used to machine the block in preparation for inserting cylinder liners.

A boring bar can also be used to increase the diameter of bearing bores. For example, if the main bearing bores are badly damaged, they can be machined oversize. Then, oversize main bearings can be installed.

Line Hone

A ***line hone*** is used to repair minor damage and straighten the main bearing bores of a cylinder block. It is a rigid hone that is rotated in the main bearing bore of a block. See **Figure 2-41.** A cylinder block main bore can be damaged when an engine is run without oil and the main bearings seize or "burn."

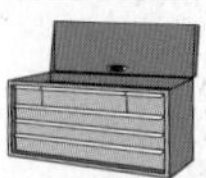

Note: Honing removes a smaller amount of material than boring. Boring removes a large amount of material as when installing sleeves in a cylinder block, for example.

Rod Rebuilding Machine

A ***rod rebuilding machine*** is used to re-machine engine connecting rod bores during repairs, **Figure 2-42.** The machine can be used to rebuild and salvage connecting rods with minor damage. Badly damaged rods should be replaced.

Figure 2-41. *A line hone is being used to straighten the main bearing bores of a cylinder block.*

Crankshaft Welder

A ***crankshaft welder*** is used to weld damaged journals during crankshaft rebuilding. For example, if a rod bearing spins or fails, it will quickly cut metal off of the crankshaft journal and badly damage the journal. Metal must be welded onto the journal and then the journal ground to specifications. A crankshaft welder is pictured in **Figure 2-43.**

Crankshaft Grinder

Worn crankshaft journals can be resurfaced using a ***crankshaft grinder***. This is a large machine tool that rotates

Figure 2-42. *A rod rebuilding machine can be used to rebuild and salvage connecting rods with minor wear.*

Figure 2-43. *This machine can be used to weld damaged crankshaft journals.*

the crankshaft against a spinning stone, **Figure 2-44.** The crank journals are ground to a smaller diameter. New, undersize rod and main bearings are installed during engine reassembly to bring bearing clearances within specifications.

Camshaft Grinder

Cam lobes are prone to wear because of the high friction between the lobes and lifters or rocker arms. A ***camshaft grinder*** is a machine shop tool that can be used to resurface the cam lobes and bearing journals on a worn camshaft, **Figure 2-45.** The camshaft is rotated against a grinding wheel. This will restore the shape of the cam lobes.

Figure 2-44. *A crankshaft grinder can be used to resurface worn or welded crankshaft journals. It rotates the crankshaft journals against a spinning stone to resurface journals for new bearings.*

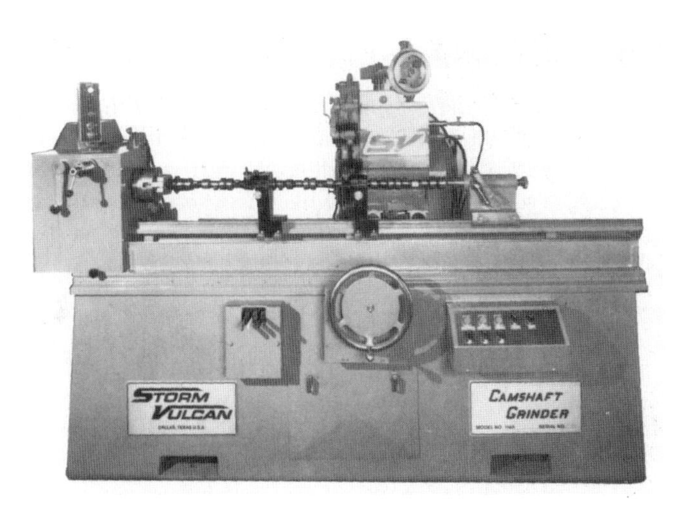

Figure 2-45. *Camshaft grinder is used to restore the shape of the cam lobes. (Storm Vulcan)*

Engine Balancer

Normally, an engine is balanced from the factory. If the same pistons, rods, and crankshaft are reused during an engine rebuild, balancing is *not* needed. However, if pistons, rods, or the crankshaft are changed, balancing may be required to prevent engine vibration. An ***engine balancer*** is used to spin the engine crankshaft to determine if weight must be added to or removed from its counterweights, **Figure 2-46.**

To balance a crankshaft, the pistons, rods, rings, bearings, and piston pins are weighed on an accurate scale. Then, a formula is used to determine how much weight must be bolted to the crankshaft journals to simulate the weight of the piston assemblies. With this weight attached, the crankshaft is spun on the balancing machine. The machine shows where and how much weight must be changed (added or subtracted) to make the crankshaft rotate without vibrating.

Figure 2-46. *An engine balancer spins the engine crankshaft to determine if weight must be added or removed from the crankshaft counterweights to avoid engine vibration.*

Bench Grinder

A ***bench grinder*** is frequently used by the engine technician. A wire wheel installed on the grinder can be used to remove hard carbon deposits from engine valves, for example. The grinding wheel is used to resurface chisels and perform other metal-removal tasks. A few bench grinder rules to follow are:

- ❑ Wear eye protection and keep your hands away from the wheel.
- ❑ Keep the tool rest adjusted close to the wheel. Otherwise, the part can catch in the grinder.
- ❑ Make sure the grinder shields are in place.

Caution: Do *not* use a wire wheel to remove carbon from soft aluminum parts—pistons, for instance. The abrasive action of the wheel can quickly remove the soft metal and ruin the part.

Compressed-Air System

The components of a compressed-air system include an air compressor, air lines, and air tools. In addition, a pressure regulator, filter, and lubricator may be attached to the system. Air-powered tools are driven by the compressed-air system. Air-powered tools can be found in nearly every automotive service facility.

Air Compressor

An ***air compressor*** is the source of compressed air for an automotive service facility. An air compressor normally has an electric motor that spins an air pump. The air pump forces air into a large, metal storage tank. See **Figure 2-47A.** The air compressor turns on and off automatically to maintain a preset pressure in the system.

The compressor air pump has an inlet filter that should be cleaned and oiled periodically to reduce

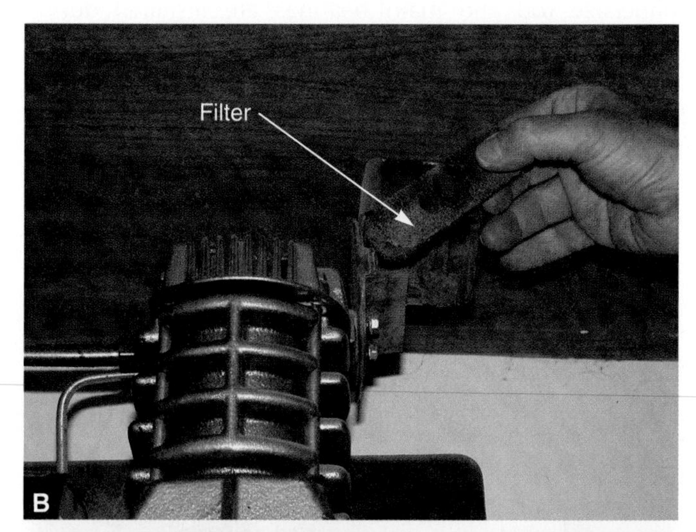

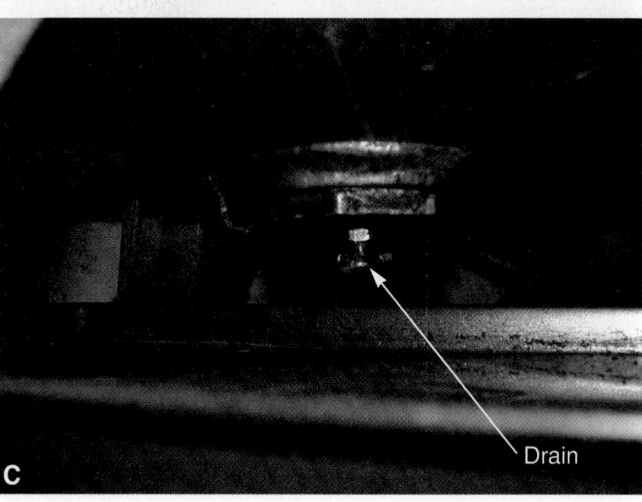

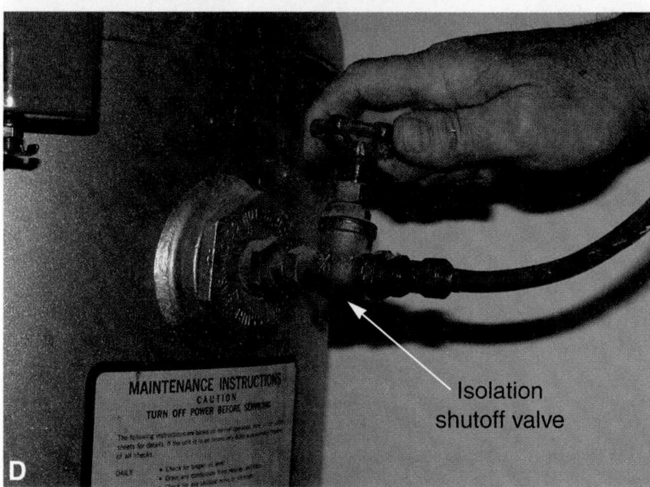

Figure 2-47. *A—The air compressor provides air for all air-powered tools and equipment in automotive service facility. Note the major parts. B—The compressor air inlet filter should be cleaned periodically. C—The drain valve on the air compressor tank is used to remove the water that condenses during compression. D—A shutoff valve allows the air tank to be isolated from the air lines and hoses.*

compressor pump wear, **Figure 2-47B.** A drain on the bottom of the tank must be opened periodically to remove condensed water from the tank, **Figure 2-47C.**

Metal air lines feed out from the tank to several locations in the shop. A main shutoff valve on the outlet line or fitting is used to isolate the pressure tank from the shop air hoses, **Figure 2-47D.**

A technician can connect flexible, high-pressure air hoses to the metal lines from the air compressor. These hoses allow the technician to take a source of air pressure to the vehicle being repaired. Quick-disconnect connectors are used on air hoses. These allow a technician to connect or disconnect hoses or tools without using a wrench. The connectors have an outer sleeve that slides back with finger pressure as the technician pushes or pulls on the hose.

Warning: Shop air pressure is high enough to kill or severely injure a person if air is injected through the skin and into the bloodstream. Respect shop air pressure!

Pressure Regulator

A ***pressure regulator*** is used to set a specific pressure in the compressed-air system. This pressure is often called "shop pressure" and may be lower than the pressure generated by the air compressor. In most cases, shop pressure is between 100 and 150 pounds per square inch (psi).

A filter-dryer may be connected to the system. The filter-dryer traps water so that it can be removed. This increases the life of air tools.

A lubricator may also be connected to the system. The lubricator introduces oil into the airstream. This also increases the life of air tools.

Air Tools

Air tools, also called pneumatic tools, use air pressure for operation. Air tools are labor-saving devices and well worth their cost. Always lubricate an air tool before use. Squirt a few drops of air tool oil into the tool's air inlet fitting. This protects the internal parts of the tool, increasing the life and power of the tool. See **Figure 2-48.**

Engine Cleaning Tools and Equipment

A great deal of time must be spent cleaning parts during an engine overhaul. One missed piece of gasket material can cause pressure or fluid leakage and, thus, catastrophic engine failure. It is important to understand the proper selection and use of cleaning tools and equipment.

Figure 2-48. *Air tool oil should be injected into the inlet fitting of air tools every evening to prevent rust formation in the air tool motor.*

Steam Cleaner and High-Pressure Washer

A ***steam cleaner*** or a ***high-pressure washer*** is used to remove heavy deposits of dirt, grease, and oil from the outside of engines, **Figure 2-49.** They provide an easy and rapid method of cleaning before disassembly.

Warning: A steam cleaner operates at relatively high pressure and temperature. Follow all safety rules and specific operating instructions. You can be seriously burned by a steam cleaner.

Cold Solvent Tank

A ***cold solvent tank*** is used to remove grease and oil from engine parts, **Figure 2-50.** After removing all old

Figure 2-49. *Pressure washers, or a steam cleaner like this one, can be used to clean oil and grease deposits from the external surfaces of an engine before teardown. (Snap-on Tool Corp.)*

Figure 2-50. *A cold solvent tank contains a mild cleaning agent for removing oil and grease from parts. Wear protective gloves, a solvent-proof apron, and eye protection. In a poorly ventilated area, also wear a respirator.*

gaskets and scraping off excess grease, the parts can be brushed clean in the solvent.

Hot Tank

A ***hot tank*** uses heat and a very powerful cleaning agent to clean engine parts. This is normally a piece of machine shop equipment. See **Figure 2-51.**

An engine block is frequently hot tanked during an overhaul. Hot tanking removes all paint, grease, oil, and mineral deposits in the water jackets. Before hot tank cleaning, you should remove all core plugs. Also, cam bearings will be eroded by hot tanking. New ones must be installed after hot tanking.

Caution: Aluminum parts cannot be hot tanked.

Figure 2-51. *Hot tanks use a very powerful cleaning agent to remove paint, mineral deposits, sludge, and so on. Wear eye protection and gloves when working around a hot tank.*

Figure 2-52. *When using a blow gun, do not aim the nozzle toward your body. If air enters your bloodstream, it could kill you.*

Blow Guns

An air powered ***blow gun*** is commonly used to dry and clean engine parts that have been washed in solvent. A blow gun is also used to blow dust and loose dirt off of an engine part before disassembly. When using a blow gun, wear eye protection. Also, direct the blast of air away from yourself and others, **Figure 2-52.**

Media Blaster

A ***media blaster,*** or "sand blaster," can be used to clean badly rusted parts, **Figure 2-53.** A blast cabinet fitted with long rubber gloves allows you to blast rusted surfaces while viewing through a glass window. Shop air pressure is used to direct a stream of air and media, such as sand, iron slag, or other abrasive material, into the part. This scrubs the surfaces clean.

Figure 2-53. *A media blaster can be used to clean badly rusted parts. Long rubber gloves allow you to blast rusted surfaces while viewing through a glass window.*

Figure 2-54. *A shot peen machine is used for cleaning and smoothing part surfaces by tumbling the part in a large metal cage while bombarding it with small beads of iron.*

A shot peen machine is used for cleaning part surfaces by tumbling the part in a large metal cage while bombarding the surface with metal shot (small beads of iron). This cleans and smoothes the surfaces of engine parts. See **Figure 2-54.**

Spark Plug Cleaner

A ***spark plug cleaner*** can remove deposits from spark plug tips and allow reuse of the plugs in the engine. The cleaner uses shop air pressure and blast media (sand-like particles) to remove carbon deposits from a spark plug. Some manufacturers do not recommend spark plug cleaning.

Scrapers, Brushes, and Scuff Wheels

Scrapers are commonly used by an engine technician to remove old gasket material and oil buildups. See **Figures 2-55A** and **2-55B.** ***Hand brushes*** are used with solvent to remove oil and grease. See **Figure 2-55C.** A rotary brush is also used in a drill to remove hard carbon deposits from cylinder head combustion chambers. Refer to **Figure 2-56.**

A ***scuff wheel*** is an abrasive pad used in a air tool for rapid removal of gasket material and sealer. See **Figure 2-57.** It is one of the most commonly used cleaning tools used in engine repair. The abrasive wheel is spun on the part surface to remove debris with minimum abrasion to the part surface. It can be used on iron and aluminum parts. However, do not press down as hard when cleaning soft metal or you could damage the part.

Some manufacturers do not recommend using metal scrapers, brushes, and scuff wheels on aluminum parts. Instead, plastic or wooden tools should be used. Be sure to check the service manual for exact recommendations.

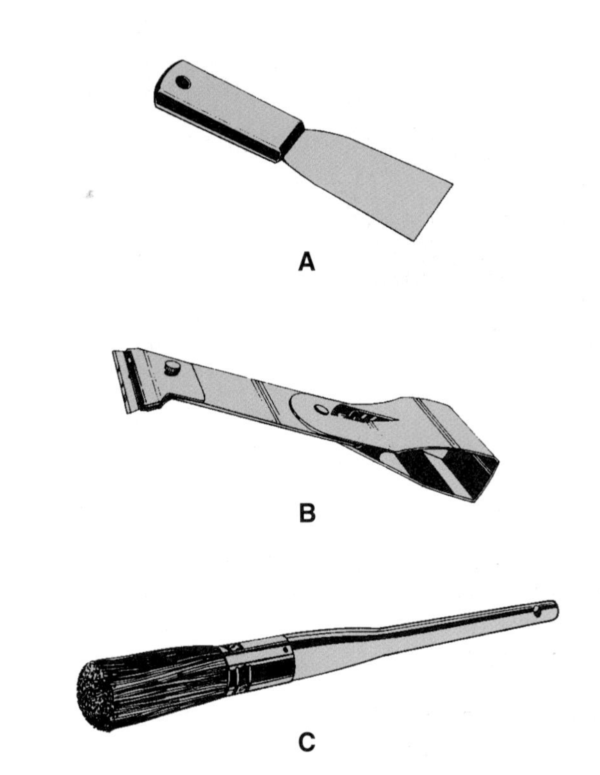

Figure 2-55. *A—A putty-knife-type scraper can be used to remove gaskets. B—A razor-blade-type scraper can be used to remove paint and other thin materials. C—A hand brush, or parts cleaning brush, for use in cold solvent tank. (Plew Tools)*

Caution: When using a rotary brush to clean a combustion chamber, be careful not to damage the valve seats.

Die Grinder

A ***die grinder*** is a hand-held tool with a shaft that spins at high speed. See **Figure 2-58.** The tool can be fitted with a grinding stone for use on a cast iron part or a carbide steel cutter for use on an aluminum part. Die grinders can rapidly remove metal, such as a weld bead. A die grinder may be used to port (enlarge) intake or exhaust passages or repair cylinder heads.

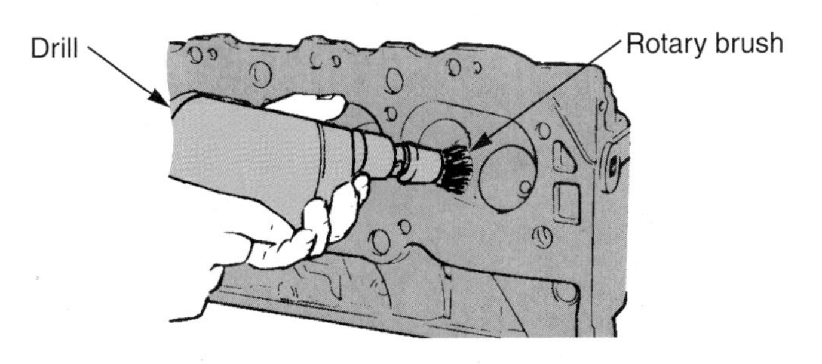

Figure 2-56. *A rotary brush is commonly used in a hand drill to remove hard carbon deposits from inside the combustion chambers in a cylinder head.*

Figure 2-57. *A—A scuff wheel mounted in an angle grinder can be used to remove old gasket material from engine surfaces. This method is much faster than hand scraping. B—The spinning abrasive wheel quickly removes hardened gasket material with minimum abrasion to part surfaces.*

Spray Gun

A ***spray gun*** is often used to repaint an engine block and related parts during a complete rebuild. See **Figure 2-59.** After cleaning and assembly of the long block (block, heads, and related parts), external surfaces should be painted with an epoxy primer and engine enamel. This will restore the outer appearance of the engine and prevent rusting of cast iron and steel surfaces.

Figure 2-58. *A die grinder uses wheels or stones spun at a high speed to remove metal from parts after welding or for modifications.*

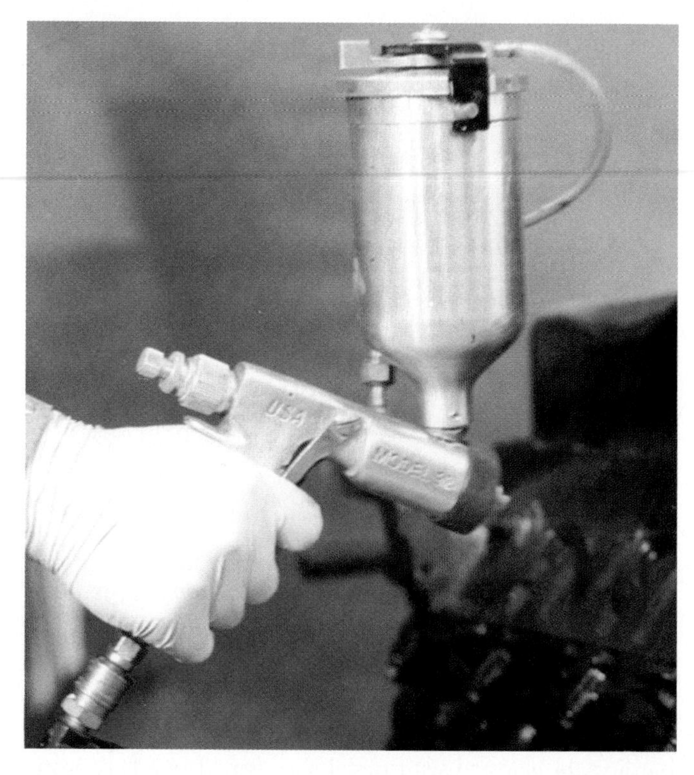

Figure 2-59. *A spray gun is used to prime and repaint engine parts after a major rebuild. A rebuilt engine is normally repainted in the factory-original color.*

Caution: Never paint any surface of the engine that is exposed to the oiling system. Over time, paint can flake off, enter the oiling system, and damage friction surfaces.

Pressure Testing

There are various pressure and vacuum measurements that must be made during engine service. To perform these measurements, several types of gauges and testers are available. This section discusses some of the common pressure and vacuum gauges and testers.

Pressure Gauge

A ***pressure gauge*** measures pounds per square inch (psi) and/or kilopascals (kPa), **Figure 2-60.** It is used to check various pressures, such as fuel pump pressure, pressure regulator pressure, engine oil pressure, or turbo boost pressure. An engine technician usually has several pressure gauges, each designed for a specific job.

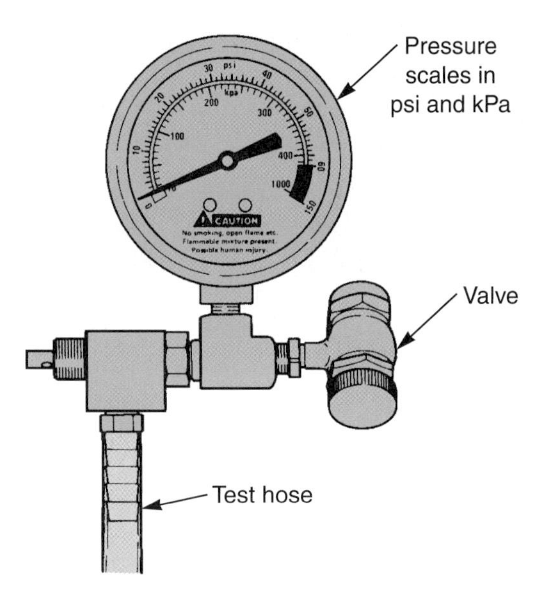

Figure 2-60. *Various types of pressure gauges are used in engine repair to diagnose potential problems. An engine technician will have pressure gauges for checking engine oil pressure, fuel pressure (like this one), negative air pressure or vacuum, and so on. (Ford)*

Compression Gauge

A ***compression gauge*** is used to measure the amount of pressure in a cylinder during the engine compression stroke, **Figure 2-61.** This test is called a compression test and it provides a means of testing the mechanical condition of the engine. The tester is threaded or held into the spark plug hole in a cylinder as the engine is turned over.

A compression test should be performed when symptoms point to major engine problems. Some symptoms that may indicate a major engine problem include engine missing, rough idle, and a puffing noise in induction system or exhaust. If the compression gauge readings obtained during the compression test are not within specifications, something is mechanically wrong in the engine.

Warning: Disable the ignition system when performing a compression test to prevent severe electrical shock and to prevent the engine from starting.

Oil Pressure Gauge

An ***oil pressure gauge*** is designed to screw into a oil plug hole for measuring engine oil pressure. Refer to **Figure 2-62.** It measures the actual pressure developed by the oil pump. If the vehicle's oil pressure gauge reads low or indicator light comes on, its accuracy can be checked by referring to the test gauge reading and factory specifications.

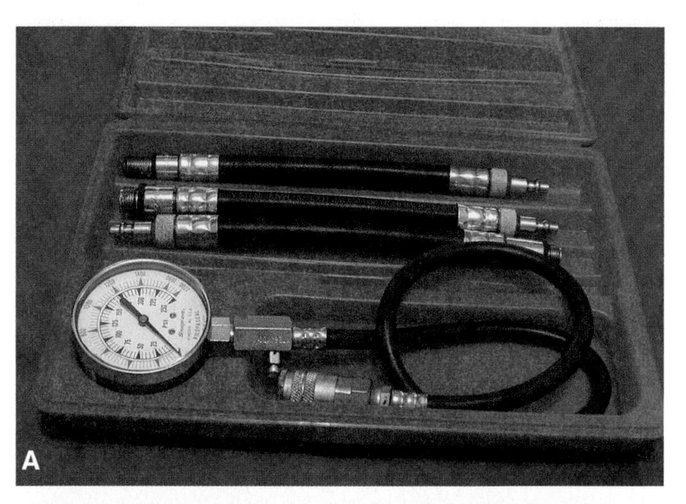

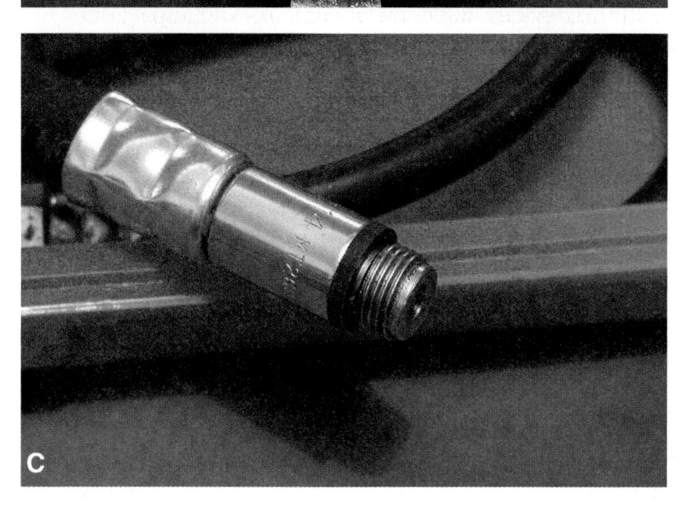

Figure 2-61. *A—An engine compression gauge is used to measure the compression in a cylinder. B—The gauge on the tool shows the maximum compression developed by the cylinder. C—The tool is threaded into the spark plug hole. (Snap-on Tool Corp.)*

Vacuum Gauge and Vacuum Pump

A ***vacuum gauge*** is commonly used to measure negative pressure or vacuum (suction), **Figure 2-63.** It is similar to a pressure gauge. However, the gauge reads in

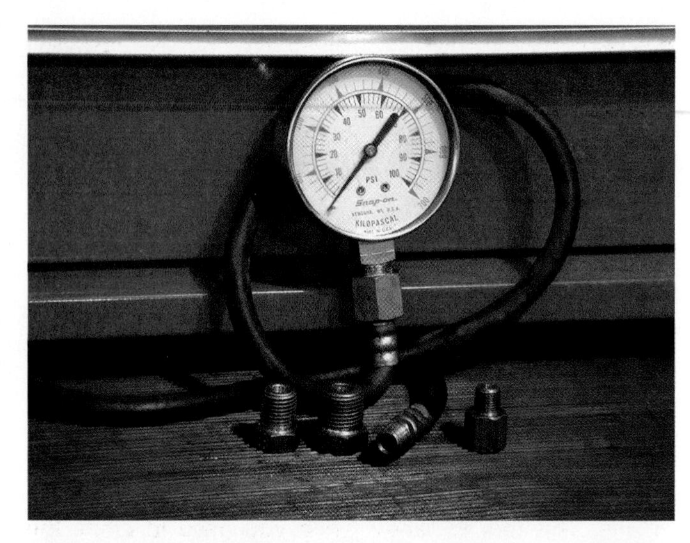

Figure 2-62. *An oil pressure gauge has small, interchangeable fittings for threading the gauge into the engine's oil pressure sensor hole or any hole that taps into an engine oil gallery. A low oil pressure reading on the test gauge indicates worn engine bearings or a bad oil pump. (Snap-on Tool Corp.)*

inches of mercury (in/Hg) or centimeters of mercury (cm/Hg). As one example, a vacuum gauge is used to measure the vacuum in an engine's intake manifold. If the reading is low or fluctuating, it may indicate an engine problem.

Figure 2-64 shows a hand-held vacuum pump and gauge assembly. It can be used to test many engine-related vacuum devices. If a component leaks vacuum or does not operate at its specified vacuum, it is faulty.

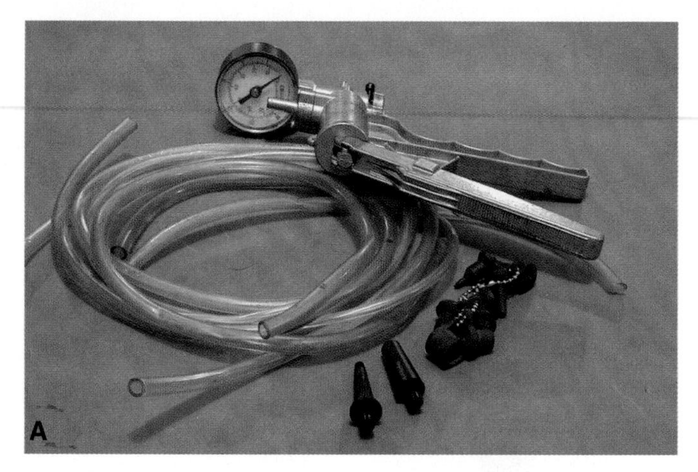

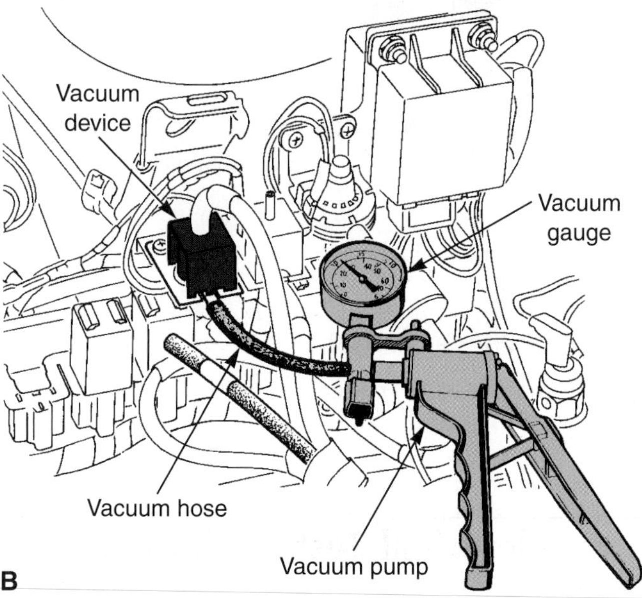

Figure 2-64. *A—A hand-held vacuum pump-gauge is used to test vacuum-operated devices. Note the rubber adapters for connecting to different diameter fittings and hoses. B—The device being tested should function properly when the pump is activated.*

Cylinder Leakage Tester

A ***cylinder leakage tester*** measures the amount of air leakage out of a combustion chamber, **Figure 2-65.** The tester forces external air pressure into the cylinder, which should be done with the piston at top dead center (TDC) on the compression stroke. The pressure gauge on the tester indicates the amount of pressure in the cylinder. As air leaks out of the cylinder, the pressure reading drops. A final reading is taken after a specified period of time. Using the two readings, you can determine the percentage of air leakage and compare the result to specifications.

Figure 2-63. *This is a general use vacuum gauge. A vacuum gauge, as you will learn in later chapters, can be used to diagnose engine problems.*

If leakage is severe enough, you will be able to hear and feel air blowing out of the engine. Air may blow out the intake manifold (bad intake valve), exhaust system (burned exhaust valve), breather (bad rings, piston, or cylinder), or into an adjacent cylinder (blown head gasket).

Cooling System Pressure Tester

A ***cooling system pressure tester*** is used to test for leaks in the cooling system, **Figure 2-66.** The tester is

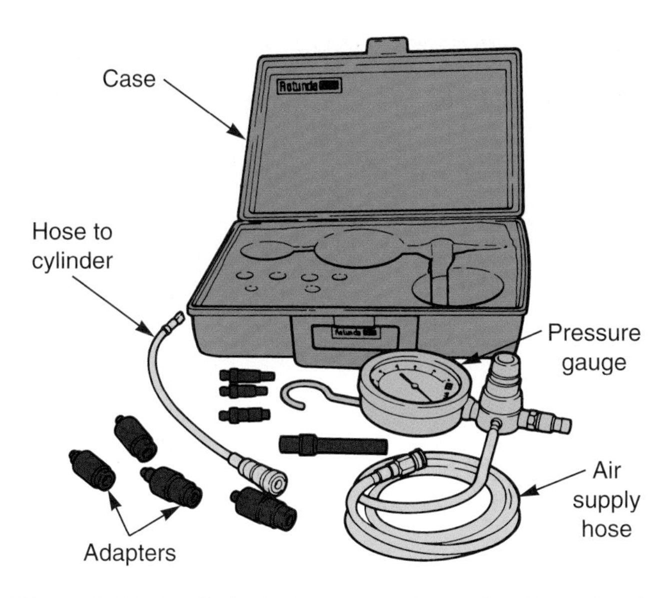

Figure 2-65. *A cylinder leakage tester is used to determine the amount of air leakage out of a combustion chamber. (Ford)*

connected to the radiator in place of the cap and used to pressurize the cooling system. If the pressure reading on the tester gauge drops, there is a leak in the cooling system, such as a cracked head or block, blown head gasket, leaking hose, or leaking radiator. Most cooling system pressure testers also have an adapter for testing the operation of the radiator cap.

Electrical Test Equipment

Today's engines are controlled by electronics. Sensors and a computer monitor various engine operating conditions. Then, electrical signals are used to control various engine functions, such as ignition timing, fuel injection, idle speed, turbocharger boost, and so on.

To be a competent engine technician in a modern automotive service facility, you *must* know how to make electrical tests. This section discusses common electrical test equipment and their use.

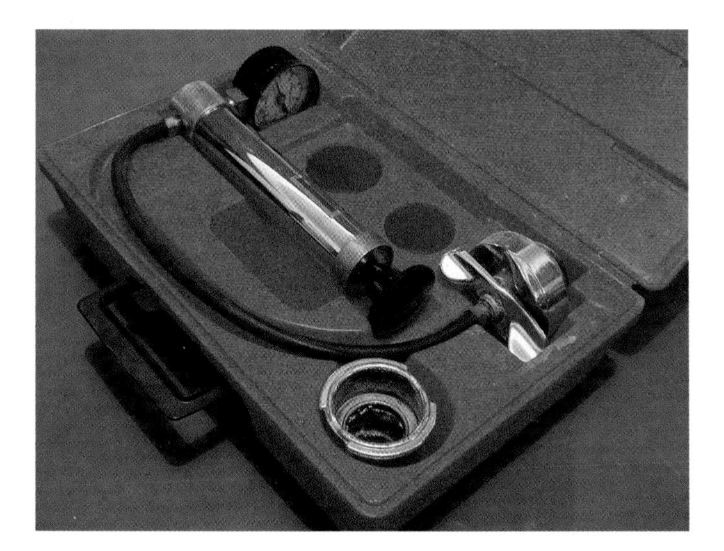

Figure 2-66. *A cooling system pressure tester pressurizes the cooling system to find leaks in the radiator, water hoses, water pump, and other parts. It can also be used to check the radiator cap seals and for proper pressure release. (Snap-on Tool Corp.)*

Remote Starter Switch

A ***remote starter switch*** is used to turn the engine over without turning the ignition key to the start position, **Figure 2-67.** The remote starter switch is connected to the starter solenoid and the battery. When the ignition key is in the run position and the button on the remote starter switch is pressed, the starter is energized to crank the engine. This tool is handy when you need to crank the engine to rotate the crankshaft. It is commonly used when adjusting engine valves, during engine removal for flywheel-torque converter bolt removal, and to make compression tests.

VOM (Multimeter)

A ***VOM*** (volt-ohm-milliammeter), also called ***multimeter,*** is used to measure voltage, current, and resistance. When you suspect a faulty electrical component on an engine, check it with your multimeter. If the current, voltage, or ohms test value is *not* within specifications, something is wrong with the circuit or component.

An analog multimeter displays the electrical value using a gauge with a needle, **Figure 2-68A.** This is an older design that can still be useful in engine repair. For example, an analog multimeter is handy for finding poor electrical connections. The multimeter can be set to measure ohms (resistance), connected to the circuit, and then the connection in question wiggled. If the gauge's needle moves as the connection is wiggled, the loose

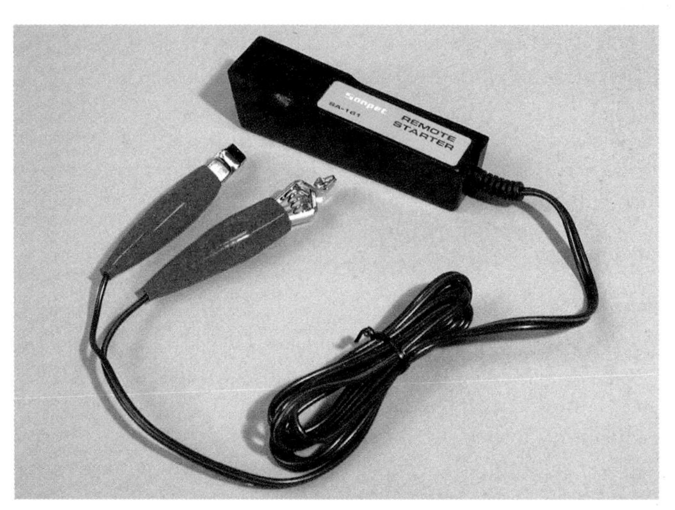

Figure 2-67. *A remote starter switch is connected to the starter solenoid so the engine can be cranked while working in the engine compartment or under the vehicle.*

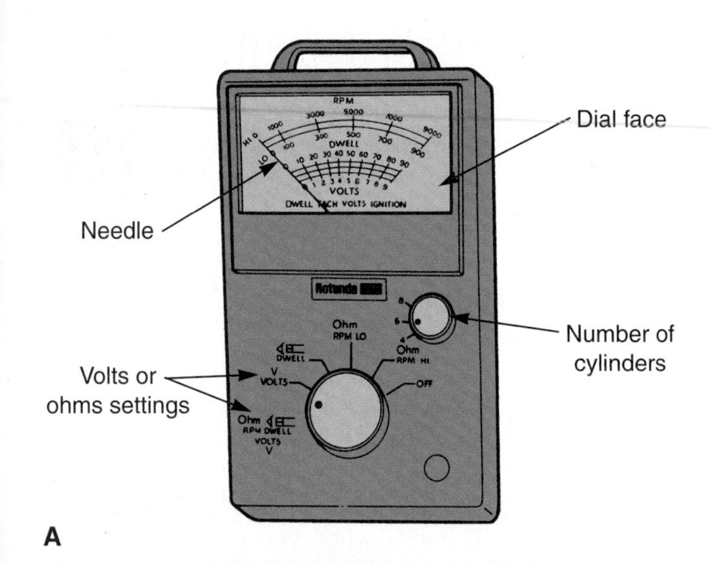

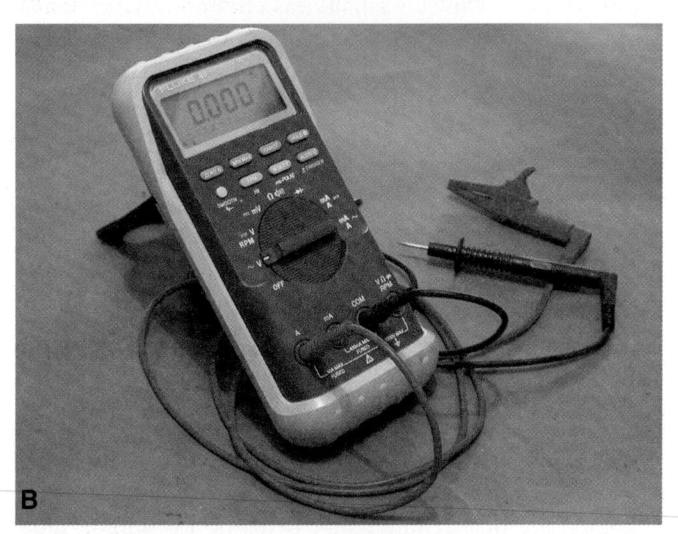

Figure 2-68. *A multimeter is needed for engine repair on vehicles with computer-controlled engines. A—An analog multimeter has a needle that swings across the meter face. B—A digital multimeter has an LED or LCD display. (Ford, Fluke)*

electrical connection is loose. An analog multimeter is also easier to read when the input, such as voltage, is fluctuating.

A digital multimeter displays the electrical value on an LED or LCD display, **Figure 2-68B.** This is a more modern type that is normally recommended when checking engine sensors. A digital multimeter normally draws less current than an analog multimeter. It is, therefore, less likely to damage delicate electronic components.

Tach-Dwell Meter

A ***tach-dwell meter*** is a tachometer and a dwell meter combined into one instrument. The tachometer is for measuring engine speed (rpm). The dwell meter measures in degrees. For example, the dwell meter is used when adjusting ignition contact points or calibrating a computer-controlled carburetor. A tach-dwell is sometimes used during engine tune-ups.

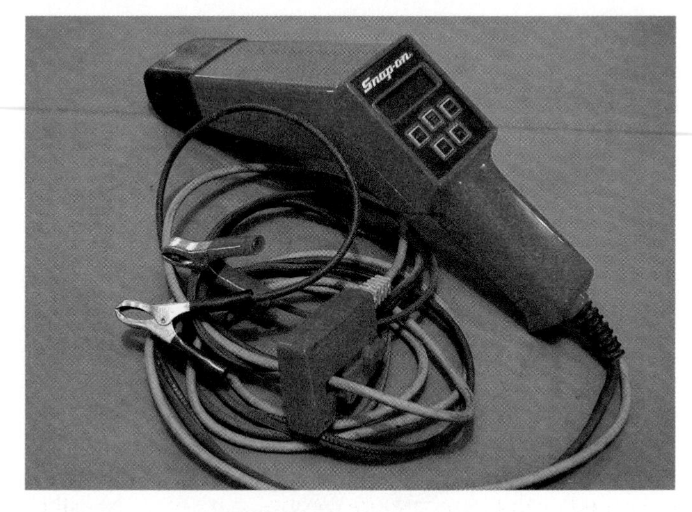

Figure 2-69. *A timing light is a strobe light that flashes when the number one cylinder fires. (Snap-on Tool Corp.)*

Note: A tach-dwell meter is only used on old cars that have points ignition systems. This type of ignition system has not been used on production vehicles since the mid 1970s.

Timing Light

A ***timing light*** is a strobe (flashing) light used to check and adjust ignition timing, **Figure 2-69.** Normally, the timing light has two small-diameter leads that are connected to the vehicle's battery. A third, larger-diameter lead is connected to the spark plug wire going to the number one cylinder. The timing light is then aimed on the engine timing marks. These marks are usually located on the crankshaft damper or flywheel.

The timing light, by flashing on and off, makes the spinning crankshaft pulley, balancer, or flywheel appear to stand still. This makes the timing mark, which is actually on a rotating part, visible. The timing can then be compared to specifications.

Mag-Tach Timing Meter

A ***mag-tach timing meter*** is a magnetically triggered tachometer for measuring engine speed (rpm) and for adjusting diesel injection timing. It is used on both diesel and gasoline engines. A diesel engine does *not* have an electrically-operated ignition system to power a conventional tachometer or timing light.

The mag-tach is operated by a magnet that senses a notch in the engine flywheel or damper. In this way, engine speed can be measured without a connection to the ignition system.

Some mag-tachs, also called timing testers, are capable of measuring ignition timing advance and injection timing advance. A luminosity probe is used to detect a combustion flame for setting injection timing on diesels.

Load Tester

A ***load tester*** can be used to check the condition of a vehicle's battery, charging system, and starting system. It can also perform other electrical tests. A load tester is commonly used because it is very fast and easy to use. One is shown in **Figure 2-70.**

Temperature Probe

Special ***temperature probes*** or gauges are frequently used to measure the temperature of various components; radiator temperature, for example. Exhaust manifold temperature, air inlet temperature, and catalytic converter temperature are other examples of temperatures that may be measured. Most temperature probes can read in either degrees Fahrenheit (°F) or Celsius (°C). The temperature reading can be compared to specifications. If the measured temperature is too low or too high, a repair or adjustment is needed. **Figure 2-71** shows one type of temperature probe.

Engine Analyzer

An ***engine analyzer*** is a group of different testing instruments mounted in one assembly, **Figure 2-72.** An engine analyzer may have a tachometer, vacuum-pressure gauge, multimeter (VOM), exhaust gas analyzer, and oscilloscope in a single, roll-around cabinet.

Scan Tool

A ***scan tool*** provides an interface with the vehicle's on-board computer. See **Figure 2-73.** The tool is normally attached to a special electrical connector under the dash. Then, a digital code is displayed on the tool, usually as a number or a series of flashing lights. This "engine trouble code" is used to determine the location of any faulty parts.

Figure 2-71. *Temperature probes accurately measure various engine related temperatures as a means of diagnosing problems.*

The service manual contains a chart indicating what each number code or value means.

Note: Many other electrical-electronic testers are used by an engine technician. These are detailed in later textbook chapters.

Black Light

A ***black light*** (ultraviolet light) can be used with a dye to locate engine leaks. A special dye solution can be added to the engine oil, for example. Then, if oil is leaking, the black light makes the fluid glow and show up distinctly. Leak solutions are available for engine oil and engine coolant, **Figure 2-74.** These solutions are different and should not be interchanged.

Figure 2-70. *A load tester will quickly check the condition of the battery, charging system, and starting system. (Snap-on Tool Corp.)*

Figure 2-72. *An engine analyzer contains various testing devices in one cabinet. (Snap-on Tool Corp.)*

Figure 2-73. *A scan tool connects to the diagnostic port under the dash of the vehicle. Then, any engine trouble codes are displayed to determine the location of any engine faults.*

Battery Charger

A ***battery charger*** is used to recharge (energize) a discharged (de-energized) car battery. It forces current back through the battery. Normally, the red charger lead is connected to the battery positive (+) terminal. The black charger lead is connected to the negative (–) battery terminal.

Warning: The gasses around the top of a battery can explode. Always connect the battery charger leads to the battery before turning the charger on. This will prevent sparks that could ignite any battery gas.

Spark Tester

A ***spark tester*** is used to check the basic operation of the engine ignition system. The tester has construction similar to a spark plug, but includes a clip for grounding the tester. The spark tester is connected between the end of a spark plug wire and ground. When the engine is cranked or started, a "hot" spark should jump across the tester gap. This shows ignition system operation.

Electronic Fuel Injection Tester

An ***electronic fuel injection tester*** is used to check the operation of electronic fuel injection systems. Most electronic fuel injection testers come with numerous accessories. The tester usually tests the system's sensors, circuit wiring, and computer. A pressure gauge may be provided for measuring fuel pump and pressure regulator outputs.

Electronic fuel injection test equipment varies. Make sure you follow the operation instructions carefully. Frequently, vehicle manufacturers recommend a particular brand of electronic fuel injection tester. The service manual should explain the use of the recommended tester.

Exhaust Gas Analyzer

An ***exhaust gas analyzer*** measures the chemical content of the engine's exhaust. This allows the technician to check combustion efficiency and the amount of pollutants produced by the engine. The results produced by an exhaust analyzer can be used to examine engine, fuel system, and emission control system operation.

Special Electrical Testers

Numerous types of specialized electrical test equipment is available. A charging system tester is an electronic device for quickly checking the voltage output of the alternator. When it is connected to the electrical system, indicator lights show whether the charging system is working normally. An ignition system tester checks the major components of an electronic ignition system. Other specialized electrical test equipment is also used.

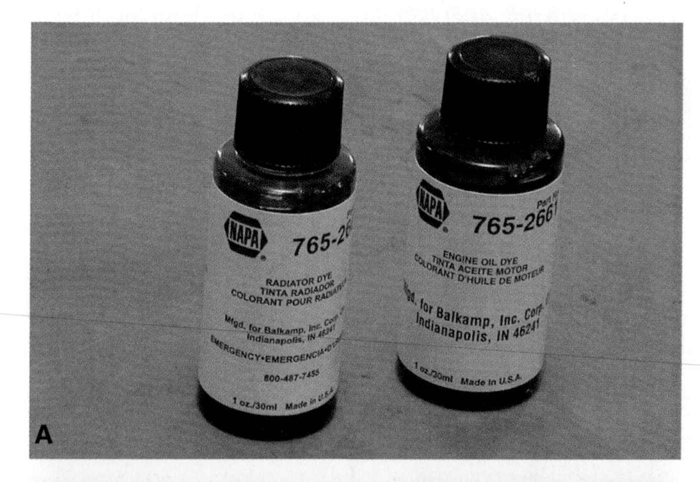

Figure 2-74. *A—A dye can be added to coolant or to engine oil. B—A black light can then be used to find fluid leaks; the dye glows under black light.*

Testing Safety Rules

When using test equipment, there are several rules to remember. Always observe appropriate safety precautions.

- ❑ Read the operating instructions for the test equipment. Failure to follow directions may cause injury and severe damage to the part or instrument.
- ❑ If the engine is to be running during your tests, set the parking brake and block the wheels. Place an automatic transmission in park or manual transmission in neutral. Connect an exhaust vent hose to the tailpipe if you are working in an enclosed auto shop.
- ❑ Keep test equipment leads (wires) or hoses away from the engine belts, fan, and hot engine parts. Leads and hoses can be easily damaged.
- ❑ When working around an engine fan, wear eye protection, remove jewelry, and secure long hair and loose clothing.
- ❑ Never look directly into a throttle body when cranking or running the engine. Do not cover the air inlet with your hand. Serious burns can result if the engine backfires.
- ❑ Refer to the vehicle manufacturer's service manual for specific testing procedures.

Warning: Failure to observe proper safety precautions may result in injury to you or others. Damage to the vehicle or test equipment may also result.

Part Crack Detection

Cracks in engine parts must be found so that the part can be repaired or replaced. Two common crack detection methods are magnafluxing and dye penetrant testing.

Magnafluxing involves using a magnet and iron particles to find cracks in cast iron or steel parts. The magnet is placed over areas that might be cracked, like combustion chambers in a cylinder head. Then, particles of iron are sprinkled over the part. The magnetic field jumping across any crack will make the particles collect over the crack. This highlights the crack and makes it visible to the naked eye. See **Figure 2-75.**

Figure 2-75. *In magnafluxing, a magnet is placed over the areas of cylinder head that might be cracked. Iron particles are then sprinkled over the part to highlight any crack.*

Dye penetrant testing is needed on nonmagnetic engine parts, such as those made of aluminum. A special dye is sprayed over the part. Look at **Figure 2-76.** The dye collects in any cracks that are in the part. A crack appears as a dark line on the part.

Figure 2-76. *An engine technician is spraying dye penetrant over the combustion chambers in an aluminum head to find cracks.*

Measuring Tools

Numerous types of measuring tools are used by an engine technician. As an engine operates, its parts rub together and slowly wear out. Precision measurement is needed to find out if the parts are too worn to be reused in the engine. Engine manufacturers give specifications for maximum wear limits for most engine parts. If the measurements are *not* within these specifications, the parts must be reconditioned or replaced. This section of the chapter provides a brief review of engine measuring tools and equipment.

Measuring Systems

The two measuring systems are the US Customary system, also called conventional system, and the SI system, also called the metric system. Both are commonly used when working on engines.

The US Customary system is mainly used in the United States. Almost all other countries use the metric system. The US is slowly replacing its system with the metric system. All foreign cars and many new, American-made engines use metric bolts, nuts, and other parts. Manufacturer specifications are also given in both conventional and metric values.

US Customary Measuring System

The US Customary measuring system originated from sizes taken from parts of the human body. For example, the width of the human thumb was used to standardize the inch. The length of the human foot helped standardize the foot (12 inches). The distance between the tip of a finger and nose was used to set the standard of the yard (3 feet). Obviously, these are not very scientific standards.

Metric (SI) Measuring System

The metric measuring system uses a power of 10 for all basic units. It is a simpler and more logical system than the US Customary system. Computation often requires nothing more than adding zeros or moving a decimal point. For instance one meter equals 10 decimeters, 100 centimeters, or 1000 millimeters.

The SI measuring system is based on the metric system, however there are some differences. SI stands for Système International d'Unités, which in English is the International System of Units. In practice, SI is referred to as the "metric" system, even though the two systems are slightly different.

Conversion Charts

A measuring system conversion chart is needed when changing from one measuring system to another: inches to centimeters and centimeters to inches, gallons to liters and liters to gallons. A conversion chart lets the technician quickly convert US Customary values into equivalent metric values or vice versa. A decimal conversion chart is commonly used by the technician to find equivalent decimal values for fractions of an inch. Refer to the back of this textbook for conversion charts.

Note: Instruments that take fractional measurements are only accurate to about 1/64 of an inch. For smaller measurements, instruments that measure in either decimal inches or millimeters should be used.

Rule (Scale)

A ***steel rule,*** also called scale, is frequently used to make linear (straight-line) measurements that are accurate to about 1/64″ (0.5 mm), **Figure 2-77.**

A US Customary rule has numbered graduations that equal full inches. The smaller, unnumbered lines or graduations represent fractions of an inch (1/2, 1/4, 1/8, 1/16, and so on). The shortest graduation lines equal the smallest fractions of an inch.

Figure 2-77. *A steel rule is handy for making rough measurements. This one has a US Customary scale that is accurate to 1/16 of an inch and a metric scale that is accurate to 1 mm.*

A metric rule normally has lines or divisions representing millimeters (mm). Each numbered line usually equals 10 mm (or one centimeter).

A pocket rule or pocket scale is short, typically 6″ long. It fits in your shirt pocket and is very handy.

Combination Square

A ***combination square*** has a sliding square mounted on a steel rule, **Figure 2-78.** The square is a frame with an edge that is precisely 90° to the long edge of the steel rule. A combination square is needed when the rule must be held perfectly square (straight) against the part being measured.

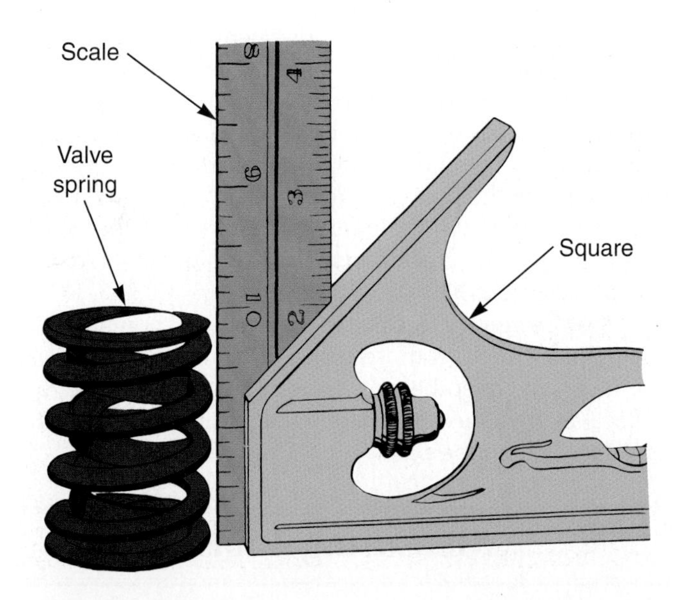

Figure 2-78. *A combination square can be used to check valve spring height. (General Motors)*

Micrometers

A ***micrometer,*** nicknamed a "mike," is commonly used when making very accurate engine measurements. It is accurate to one one-thousandth of an inch (.001″) or one one-hundredth of a millimeter (0.01 mm). There are several types of micrometers used in engine repair.

An ***outside micrometer*** is for measuring external dimensions, diameters, or thicknesses, **Figure 2-79**. It is fitted around the outside of the part. Large micrometers often have removable inserts for changing the measurement range of the tool. Then, the thimble is turned until the part is lightly touching both the spindle and anvil. Measurement is obtained by reading graduations on the hub and thimble.

An ***inside micrometer*** is for measuring internal measurements in large holes, cylinders, or other part openings. It is read in the same manner as an outside micrometer, **Figure 2-80.**

A ***depth micrometer*** is helpful when precisely measuring the depth of an opening. The base of the micrometer is positioned squarely on the part. Then, the thimble is turned until the spindle contacts the bottom of the opening. The depth micrometer is read in the same way as an outside micrometer. However, the hub markings are *reversed.* See **Figure 2-81.** As the thimble is turned in, the reading *increases.*

Note: Some micrometers have a digital readout. A micrometer with a digital readout can be quickly read.

Reading an Inch-Based Micrometer

To read an inch-based micrometer, follow the steps listed below. One complete turn of the thimble is equal to .025″. Refer to **Figure 2-82** as you follow the procedure.

1. Note the *largest* number visible on the micrometer sleeve (barrel). Each number equals .100″ (2 = .200″, 3 = .300″, 4 = .400″, and so on).

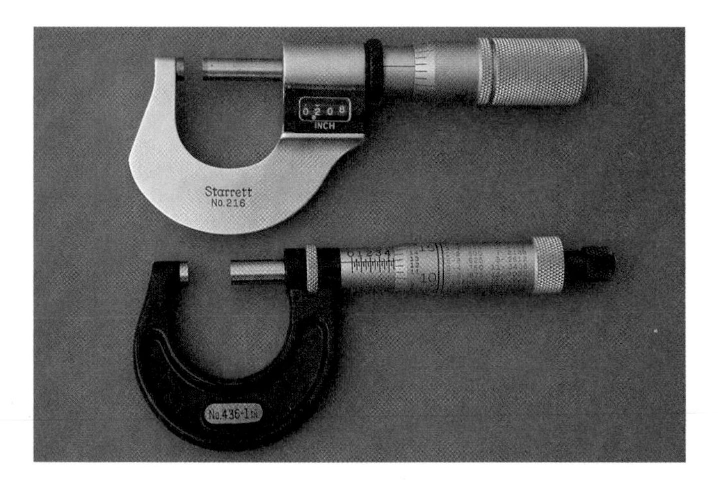

Figure 2-79. *An outside micrometer is used to make accurate outside measurements. (Starrett)*

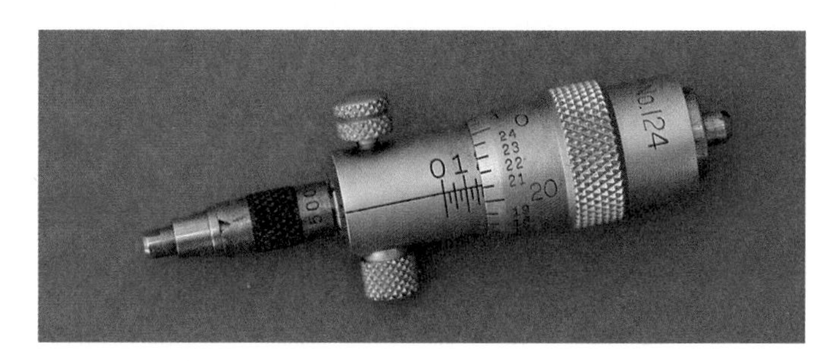

Figure 2-80. *An inside micrometer is used to make accurate inside measurements. (Starrett)*

2. Count the number of *full* graduations visible to the right of the sleeve number. Each full sleeve graduation equals .025″ (two full lines = .050″, three full lines = .075″).
3. Note the *thimble* graduation most closely aligned with the horizontal sleeve (index) line. Each thimble graduation equals .001″ (two thimble graduations = .002″, three thimble graduations = .003″). Round off when the sleeve line is not directly aligned with a thimble graduation.
4. Add the decimal values from Steps 1, 2, and 3. Also, add any full inches if the micrometer does not start reading at 0. This is the total micrometer reading.

Reading a Metric-Based Micrometer

To read an metric-based micrometer, follow the steps listed below. The procedure is similar to that for reading an inch-based micrometer. However, one complete revolution of the thimble equals 0.5 mm. Refer to **Figure 2-83** as you follow the procedure.

1. Note the *largest* number visible on the sleeve (barrel). Each number equals 1.00 mm (2 = 2.00 mm, 3 = 3.00 mm).
2. Count the number of *full* graduations visible to the right of the sleeve number. Each full sleeve graduation equals 0.50 mm (two sleeve lines = 1.00 mm, three = 1.50 mm).
3. Read the *thimble* graduation most closely aligned with the horizontal sleeve (index) line. Each thimble graduation equals 0.01 mm (two graduations = 0.02 mm, three = 0.03 mm).

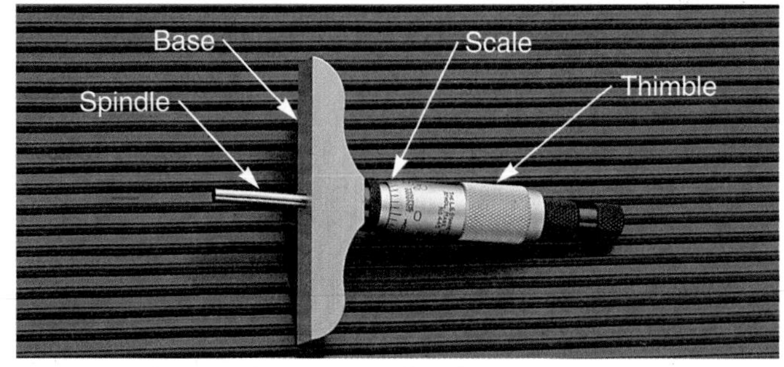

Figure 2-81. *A depth micrometer is used to make accurate depth measurements. (Starrett)*

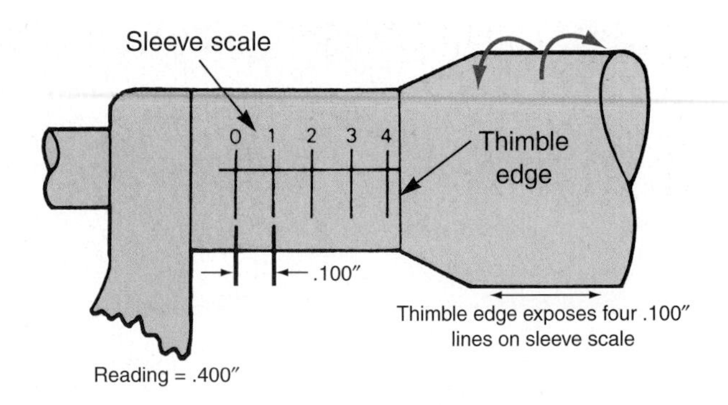

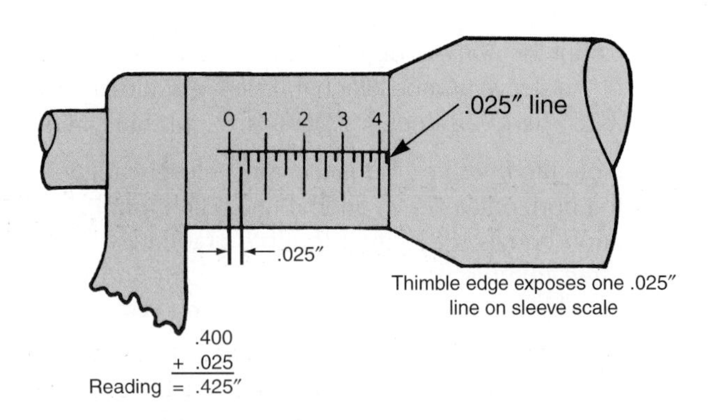

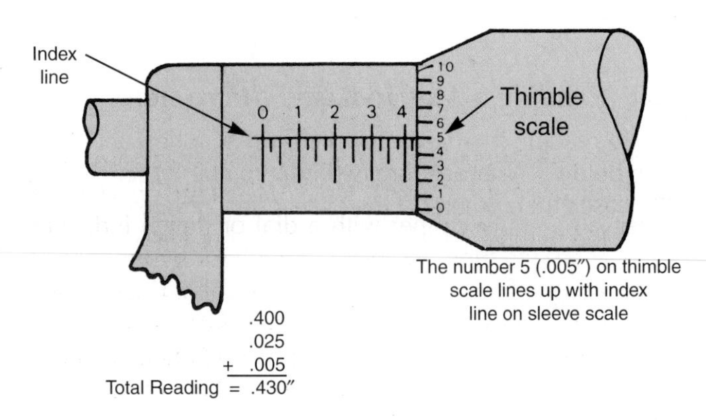

Figure 2-82. *These are the basic steps for reading a micrometer. Add the three steps to get total micrometer reading.*

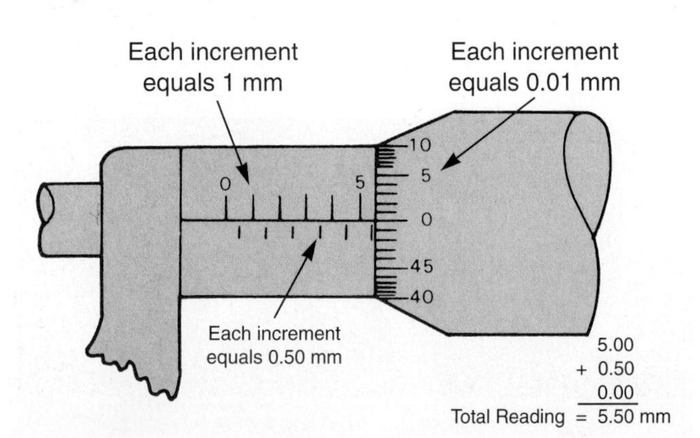

Figure 2-83. *Note scale divisions on metric micrometer. They are in millimeters. It is read like conventional micrometer.*

4. Add the values obtained in Steps 1, 2, and 3. Add any measurement if the micrometer does not begin reading at 0. This is the total micrometer reading.

Micrometer Rules

A few micrometer rules to remember are:

- Never drop or overtighten a micrometer. It is very delicate and its accuracy can easily be thrown off.
- Store micrometers where they cannot be damaged. Keep them in wooden or plastic storage boxes.
- Grasp the micrometer frame in your palm and turn the thimble with your thumb and forefinger. The micrometer should just "drag" on the part being measured.
- Hold the micrometer squarely with the work or false readings can result. Closely watch how the spindle is contacting the part.
- On round parts, rock or swivel the micrometer as it touches the part. This will ensure that the most accurate diameter measurement is obtained.
- Place a thin film of oil on the micrometer during storage. This will keep the tool from rusting.
- Always check the accuracy of a micrometer if it is dropped, struck, or after a long period of use. Tool salespeople sometimes have standardized gauge blocks for checking micrometer accuracy. See **Figure 2-84.**

Can you read the micrometers given in **Figure 2-85?**

Telescoping and Hole Gauges

A ***telescoping gauge*** measures internal part bores or openings, **Figure 2-86.** It is commonly used to measure block cylinder bores. The spring-loaded gauge expands to the size of the opening when the set screw is loosened. Then, the set screw is tightened to lock the size. Finally, an outside micrometer is used to measure the gauge, **Figure 2-87.**

A ***hole gauge*** is needed for measuring very small holes in parts where an internal micrometer cannot be

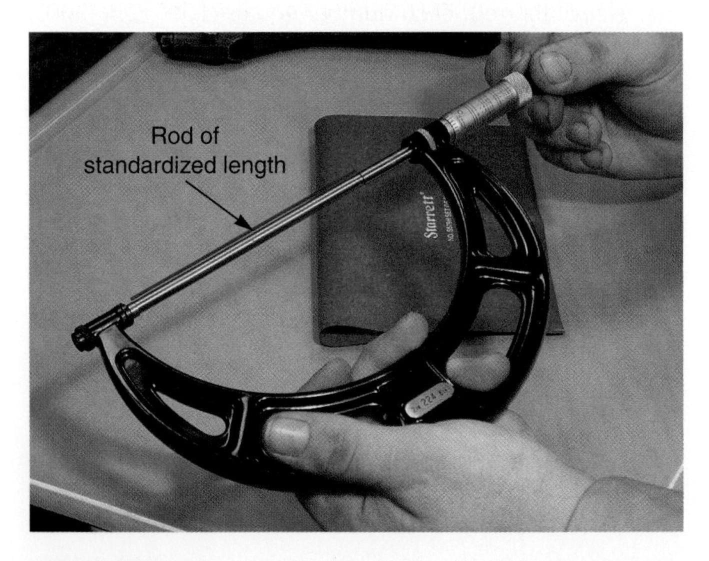

Figure 2-84. *Here, a technician is checking the accuracy of a micrometer by measuring a standard-length rod. (Starrett)*

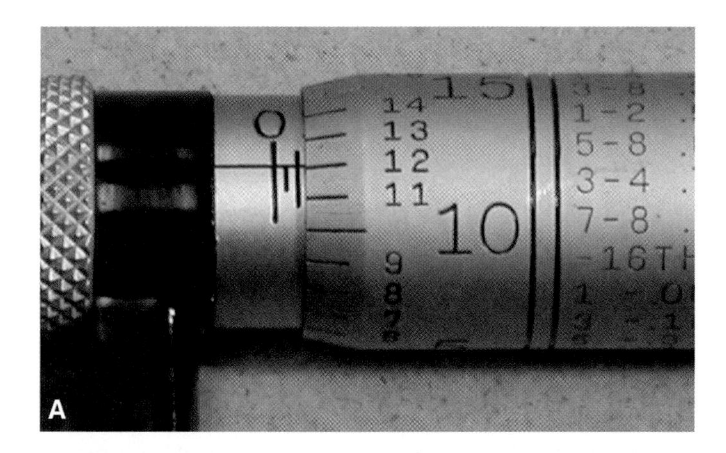

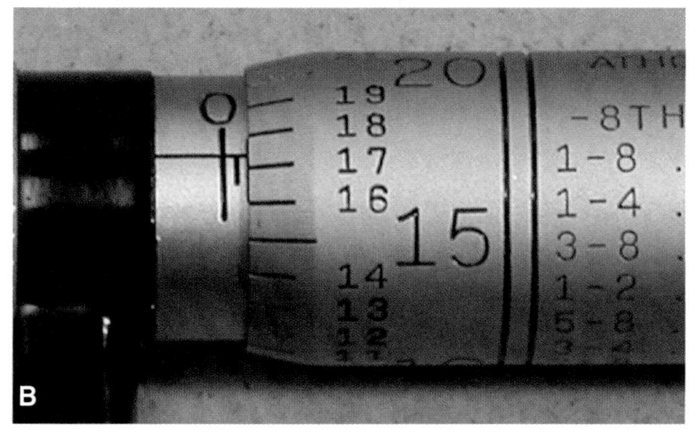

Figure 2-85. *Can you read these micrometers? A—This reading is .062" (.050 + .012). B—This reading is .042" (.025 + .017). (Starrett)*

used. See **Figure 2-88.** The hole gauge, like a telescoping gauge, is inserted and adjusted to fit the hole. Then, it is removed and measured with an outside micrometer.

Dividers

Dividers look similar to a compass used to draw circles but have straight, sharply pointed tips. They are commonly

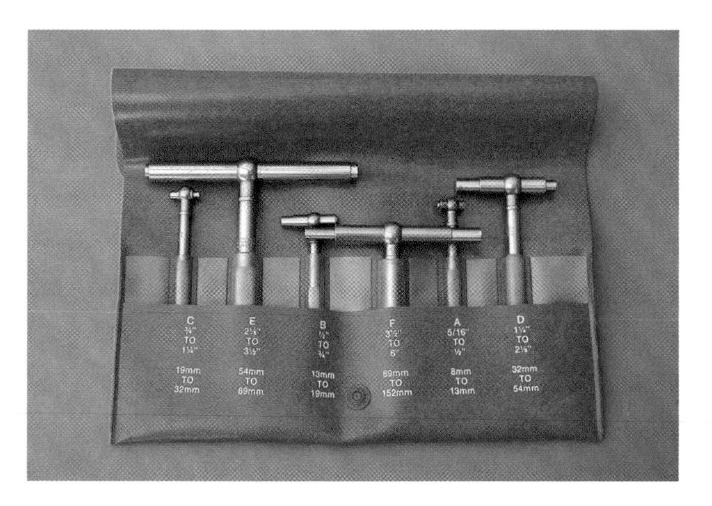

Figure 2-86. *Telescoping gauges can be used with a micrometer or vernier caliper to measure larger internal diameters.*

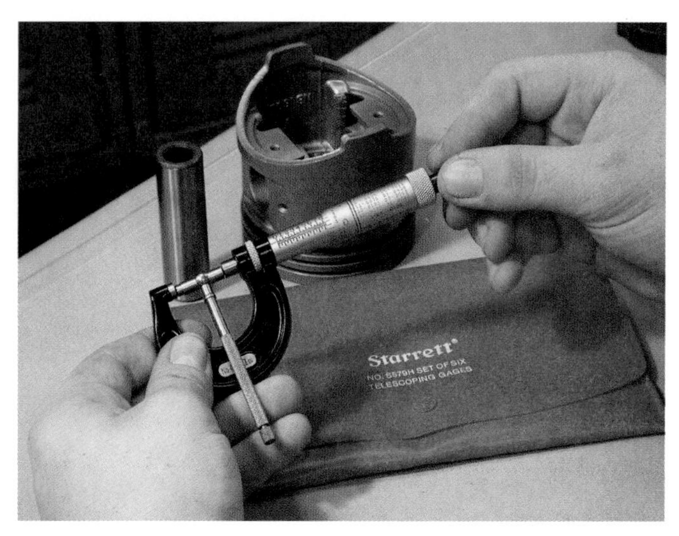

Figure 2-87. *After the telescoping gauge is adjusted to fit inside a bore, an outside micrometer is used to measure the gauge and determine the diameter of the bore. (Starrett)*

used for layout work on sheet metal parts. The sharp points can scribe circles and lines on sheet metal and plastic. Dividers can also be used to transfer and make surface measurements.

Dial or Digital Calipers

A dial or ***digital caliper*** is a very useful measuring tool with an accuracy of typically .001" (0.01 mm). This type of caliper is a vernier caliper with a dial or digital indicator that provides the thousandths of an inch (or hundredths of a millimeter). Unlike a caliper, a *vernier* caliper has a scale and can be used to make measurements directly. See **Figure 2-89.** Most types of vernier, dial, and digital calipers can be used to make inside, outside, and depth measurements quickly and precisely.

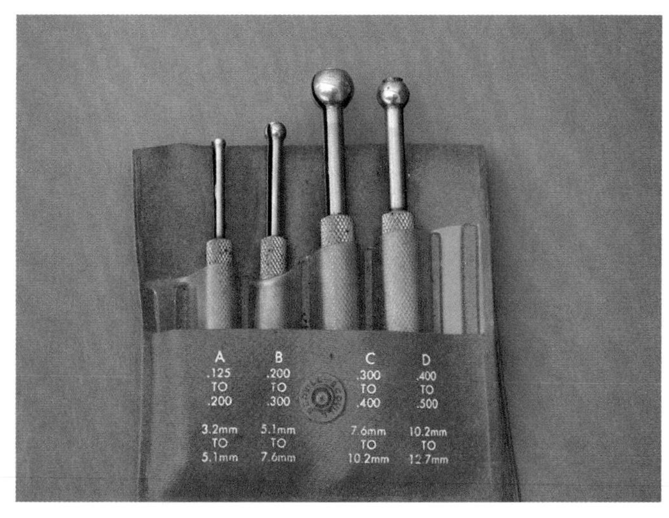

Figure 2-88. *Hole gauges are used to measure the diameter of very small openings or holes. A micrometer or vernier caliper is used to measure the gauge.*

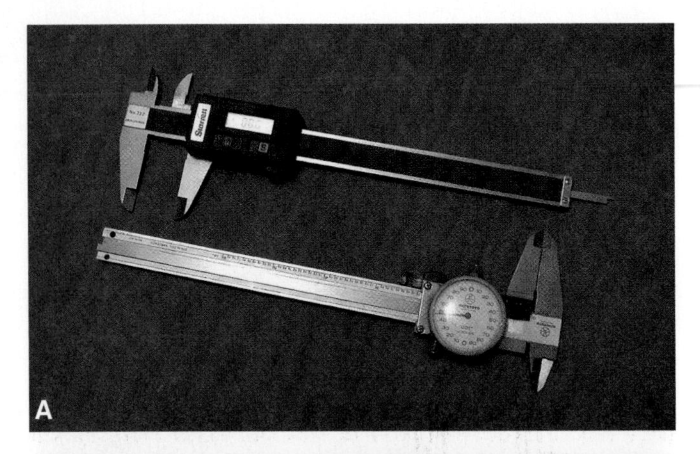

Figure 2-89. *A—Dial and digital calipers are types of vernier calipers. They can be used to measure accurately to .001″ or .01 mm. B—A digital caliper displays the total measurement. There is no scale on the beam that must be added to the reading to determine the total measurement. (Starrett)*

The readout on a digital caliper displays the *total* measurement. When using a dial caliper, you must add the parts of the measurement in a manner similar to reading a micrometer. The dial reading is from 0 to .010″ in increments of one one-thousandth of an inch, or from 0 to 0.10 mm in increments of one one-hundredth of a millimeter. This measurement is added to the reading on the sliding scale of the dial caliper.

Feeler Gauges

A ***feeler gauge*** is used to measure small clearances or gaps between parts. The two basic types of feeler gauges are flat and wire, **Figure 2-90.**

A ***flat feeler gauge*** has precision ground, steel blades of various thicknesses. Thickness is written on each blade in thousandths of an inch (.001″, .010″, .017″, for example) and/or in hundredths of a millimeter (0.01 mm, 0.06 mm, 0.20 mm, 0.23 mm, for example). A flat feeler gauge is normally used to measure small distances between *parallel surfaces.*

A ***wire feeler gauge*** has precision-sized wires that are also labeled by diameter or thickness. It is normally used to measure slightly larger spaces or gaps than a flat feeler gauge. The wire gauge's round shape also makes it more

Figure 2-90. *A wire feeler gauge (left) is for checking clearance between unparallel surfaces. A flat feeler gauge (right) is for checking clearances between parallel surfaces. (Snap-on Tool Corp.)*

accurate for measuring between *unparallel* or *curved surfaces.* For example, a wire feeler gauge should be used to check spark plug gaps because of the uneven wear of the electrodes.

To measure with either type of feeler gauge, find the gauge thickness that just fits between the two parts being measured. The wire or blade should drag slightly when pulled between the two surfaces. The size given on the side of the gauge is the clearance between the two components.

Straightedge

A ***straightedge*** is normally used with a flat feeler gauge to check or measure the warpage (flatness) of engine parts. The steel straightedge is held tightly on the part surface. Then, different sizes of flat feeler gauge blades are slid between the straightedge and part surface. The largest blade that fits under the straightedge equals the warpage.

Machining or resurfacing will usually straighten the part and allow its use on the engine. Cylinder heads, blocks, and manifolds are the most common engine parts checked with a straightedge, **Figure 2-91.**

Dial Indicator

A ***dial indicator*** is used to measure part movement in thousandths of an inch or hundredths of a millimeter. The needle on the indicator face registers the amount of plunger movement, **Figure 2-92A.**

A dial indicator is frequently used to check gear teeth backlash (clearance), shaft end play, cam lobe lift, and other similar kinds of part movements, **Figure 2-92B.** A magnetic mounting base or clamp mechanism is normally used to secure the dial indicator to or near the work. **Figure 2-93** shows a special fixture for using a dial indicator to measure valve spring installed height. Some rules to follow when using a dial indicator to measure are:

- ❑ Mount the indicator securely and position the dial plunger parallel with the movement to be measured.

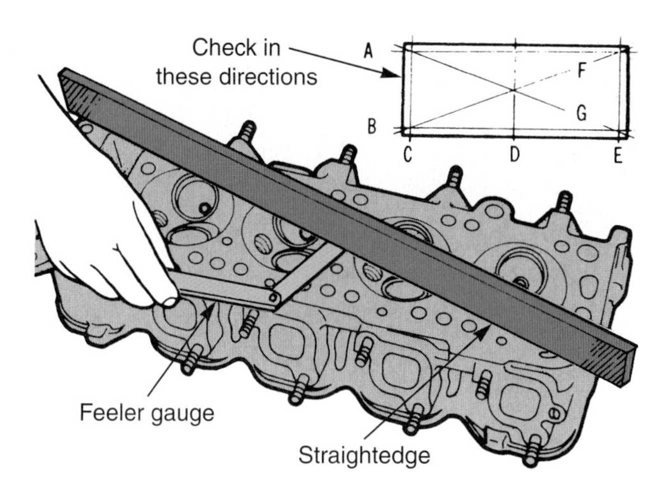

Figure 2-91. *A straightedge can be used with a feeler gauge to show if a part is warped. The feeler gauge that fits between part and straightedge equals the amount of warpage. Check at various angles. (DaimlerChrysler)*

- ❑ Preload or partially compress the indicator plunger before locking the indicator into place. Part movement in either direction should cause dial pointer movement.

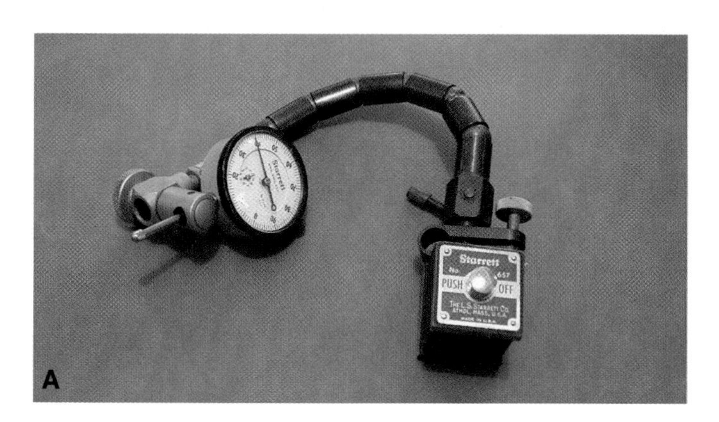

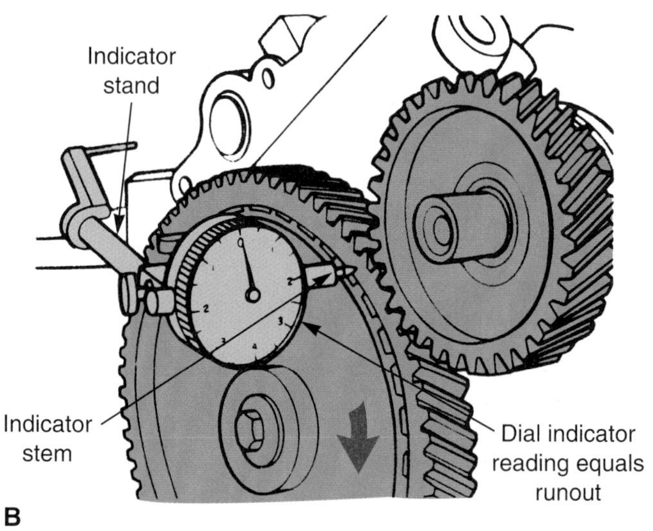

Figure 2-92. *A—A dial indicator is used to measure part movement. It has an accuracy of one one-thousandth of an inch or one one-hundredth of a millimeter. B—Here, a dial indicator is being used to measure gear runout or wobble. (Starrett, Ford)*

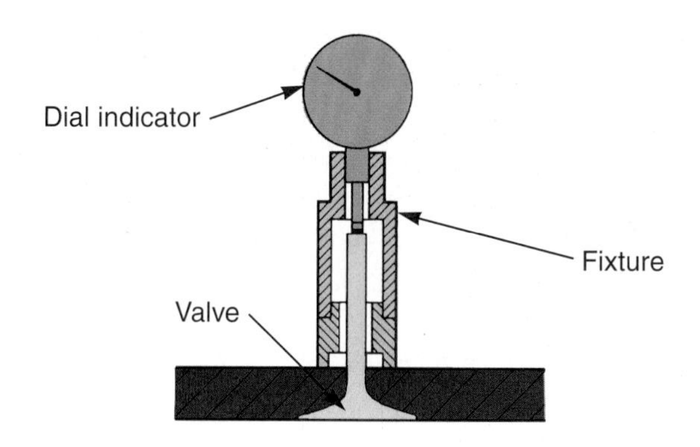

Figure 2-93. *This special fixture allows the installed spring height to be checked with a dial indicator. This fixture is provided by one automaker and will not work on other engines. (K-D Tools)*

- ❑ Move the part back and forth or rotate the part while reading the indicator. The movement of the pointer equals the part movement, clearance, or runout.
- ❑ Be careful not to damage a dial indicator. It is very delicate.

V-Blocks

V-blocks are metal stands for holding parts, usually a camshaft, for measuring straightness. See **Figure 2-94.** The camshaft journals can be placed in the v-blocks. A dial indicator can then be used to measure camshaft straightness when the cam is rotated in the v-blocks.

Dial Bore Gauge

A ***dial bore gauge*** is often used for measuring cylinder wear in an engine block. This to is a variation of a dial indicator and is slid up and down in the cylinder bore. Any needle deflection shows cylinder wear, taper, and out-of-round. Refer to **Figure 2-95.**

Burette

The size of cylinder head combustion chambers can be measure with a burette and plastic or glass cover. A

Figure 2-94. *V-blocks can be used to hold camshafts and balancer shafts while measuring them for runout or straightness. (Starrett)*

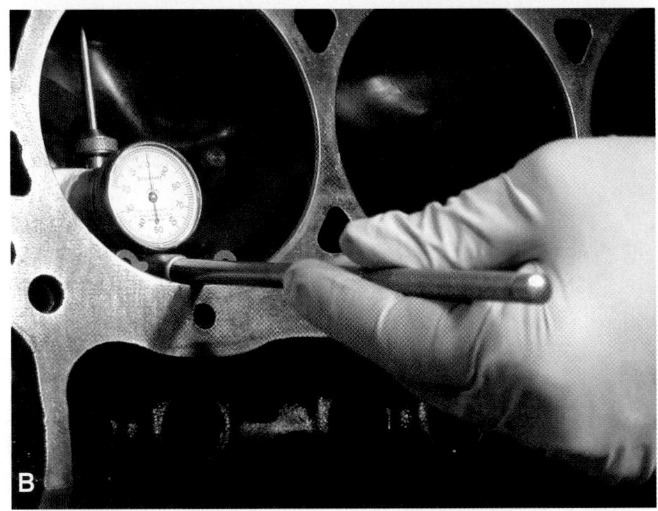

Figure 2-95. *A—A dial bore gauge is a very fast way of checking the condition of engine cylinders. B—A technician is using a dial bore gauge. The gauge's indicator movement shows amount of taper and out-of-roundness.*

burette is a graduated cylinder with a valve for metering out liquid. The plastic or glass cover is sealed over the top of the combustion chamber with the valves and springs installed. Then, use the burette to meter liquid into the combustion chamber through a small hole in the cover. Refer to **Figure 2-96.**

Graduations on the burette allow you to measure combustion chamber volume. On a racing engine, it is very important that all combustion chamber volumes are equal so that compression stroke and power stroke pressures are the same.

Plastigage™

Plastigage is a special measuring device normally used to check clearances between internal surfaces. Shown in **Figure 2-97,** it is commonly used to measure clearances in engine connecting rod bearings, crankshaft main bearings, and oil pumps.

To measure with Plastigage, place a small piece of the round, plastic-like thread on a clean, unoiled part surface.

Figure 2-96. *A technician is measuring the volume of a combustion chamber. The burette is suspended over the cover plate. Note the scale on the side of the burette.*

Bolt the parts together, torque to specifications, and disassemble the parts. Compare the smashed Plastigage strip to the scale provided with the Plastigage. The corresponding scale width indicates the clearance between the parts.

Torque Wrenches

A ***torque wrench*** measures the amount of turning force being applied to a fastener (bolt or nut). Torque wrench scales read in foot-pounds (ft-lb) for US Customary measurements and newton-meters (N·m) for metric measurements. The three general types of torque wrenches are flex bar, dial indicator, and sound indicating (clicker). These are shown in **Figure 2-98.**

Figure 2-97. *Plastigage is placed between two parts to measure clearance. When assembled, the parts will smash Plastigage wire. Then, the parts are disassembled and the width of the smashed Plastigage is compared to the scale to determine the clearance.*

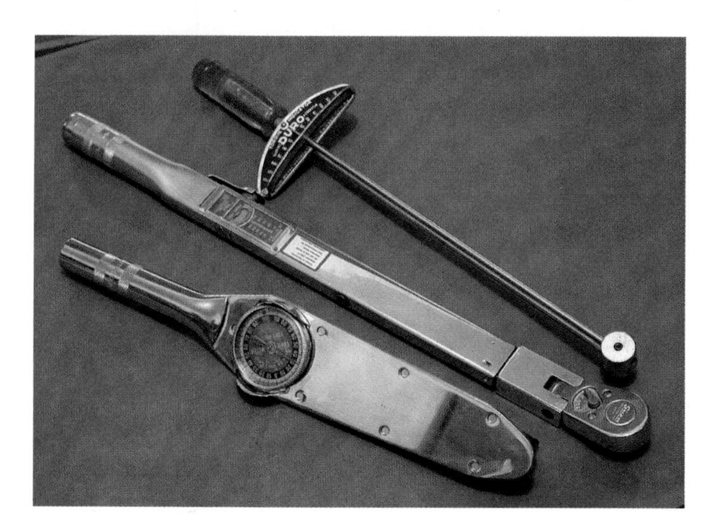

Figure 2-98. *These are three types of torque wrench. From left to right, dial indicator, sound indicating (clicker), and flex bar. (Snap-on Tool Corp.)*

The flex bar torque wrench is inexpensive and adequate. The dial indicator torque wrench is very accurate, but can be hard to use in tight quarters. The sound indicating type torque wrench is very fast and easy to use. It makes a "pop" or "click" sound when a preset torque value is reached. The advantage of this type is you do not have to watch an indicating needle while torquing.

Other Engine Tools and Equipment

This chapter has only covered the most commonly used engine tools and equipment. Throughout this text, you will learn about many other more specialized tools and equipment. Refer to the text index if you need specific information on a tool not covered in this chapter.

Summary

A good engine technician knows when and how to use the many tools and equipment needed to service today's engines. Selecting the right tool for the job will increase work efficiency and quality.

A crane is commonly used to remove and install an engine. Always use an engine stand when disassembling an engine. Remove the ring ridge before removing the piston from a cylinder. A cylinder hone is needed to deglaze cylinders to make new piston rings seal properly.

A ring expander speeds piston ring replacement and helps prevent ring breakage. A ring compressor is needed to push the rings into their grooves for piston installation into the block.

A cylinder head stand makes head work easier. A spring compressor is needed to remove and install valves in the head. A valve spring tester is used to check the springs to verify they are within specifications and reusable.

A valve grind machine is used to resurface the faces and stem tips of valves. A valve seat grinder is used to resurface the cylinder head valve seats so a good seal is formed between the valve and seat.

A boring bar is a machine tool for making cylinders or bearing bores oversized. This is done to repair wear or damage. Crankshaft grinders and camshaft grinders are also machine tools. They are used to restore crankshafts and camshafts to within specifications.

A cold solvent tank is useful for removing oil and grease deposits from parts. A hot tank is used to remove more stubborn deposits, such as paint, grease, and mineral deposits.

Numerous measuring tools are needed in engine repair. Some of these include micrometers, feeler gauges, dial indicators, Plastigage, and torque wrenches. You must know how to properly use all of these tools.

Modern engines have numerous sensors to feed information back to the on-board computer. As an engine technician, you must be able to test and replace these engine sensors. Voltmeters, ohmmeters, ammeters, and special testers are normally used to diagnose electrical-electronic troubles that could affect engine operation.

Tune-up equipment is also used by the engine technician. You must be able to use pressure gauges, compression gauges, cylinder leakage testers, vacuum gauges, timing lights, temperature probes, electronic fuel injection analyzers, and engine analyzers.

Always follow all safety rules and equipment operating instructions. Tools and equipment vary and so do the methods for using them. Failure to follow safety rules and operating instructions may result in injury or damage to equipment or the vehicle.

Review Questions—Chapter 2

Please do not write in this text. Write your answers on a separate sheet of paper.

1. A(n) _____ allows for easy engine service because the cylinder block can be rotated into different positions.
2. Which tool is used to remove the metal lip that forms at the top of a worn cylinder to allow piston removal?
   (A) Boring bar.
   (B) Ridge reamer.
   (C) Hone.
   (D) Ring groove cleaner.
3. A cylinder hone is used to _____ a cylinder wall to let the _____ seal properly.
4. What may happen if the carbon is not cleaned from inside of the piston ring grooves on a used piston?
5. Explain the difference between a piston ring expander and a piston ring compressor.
6. A valve spring compressor fits onto the _____ and onto the valve spring retainer.

7. Briefly describe how to determine if a valve spring should be replaced.
8. A(n) _____ is a machine tool for increasing the diameter of a cylinder.
9. When would a machine shop need to balance an engine?
10. You should wear _____ and stand to one side when using a hydraulic press because it can exert tons of force.
11. A(n) _____ is commonly used for installing and removing piston pins.
12. What is the difference between a cold solvent tank and a hot tank?
13. How do you use a flat feeler gauge and straightedge to check for warpage on a cylinder head?
14. A dial or digital caliper is accurate to approximately _____ of an inch or _____ of a millimeter.
15. Summarize the basic steps for using an inch-based outside micrometer.
16. A(n) _____ is used to measure part movement to within one one-thousandth of an inch (or one one-hundredth of a millimeter).
17. What is a VOM?
18. Define the term *engine trouble code.*

ASE-Type Questions—Chapter 2

1. A ridge reamer is used to remove the metal lip formed at the top of a worn _____.
   (A) valve guide
   (B) engine cylinder
   (C) piston
   (D) valve
2. Technician A says a cylinder hone is used to repair minor damage in an engine's cylinder wall. Technician B says a cylinder hone is used to deglaze an engine's cylinder wall. Who is correct?
   (A) A only.
   (B) B only.
   (C) Both A and B.
   (D) Neither A nor B.
3. All of the following are common types of cylinder hones *except:*
   (A) grinding hone.
   (B) brush hone.
   (C) flex hone.
   (D) rigid hone.
4. Technician A says piston knurling is used to restore slightly worn piston rings to within specifications. Technician B says piston knurling is used to restore slightly worn pistons to within specifications. Who is correct?
   (A) A only.
   (B) B only.
   (C) Both A and B.
   (D) Neither A nor B.
5. Technician A says a crankshaft grinder is used to resurface worn crankshaft journals. Technician B says a crankshaft grinder is used to resurface worn crankshaft counterweights. Who is correct?
   (A) A only.
   (B) B only.
   (C) Both A and B.
   (D) Neither A nor B.
6. A(n) _____ should be used to clean a cast iron engine block during an engine overhaul.
   (A) acid dip
   (B) cold solvent tank
   (C) steam cleaner
   (D) hot tank
7. All of the following cleaning equipment should be used when rebuilding an automotive engine *except:*
   (A) blow gun.
   (B) rotary brush.
   (C) Plastigage.
   (D) bench grinder.
8. Technician A says an outside micrometer is for measuring external dimensions, diameters, or thicknesses. Technician B says an outside micrometer is used to measure internal depths, lengths, and diameters. Who is correct?
   (A) A only.
   (B) B only.
   (C) Both A and B.
   (D) Neither A nor B.

9. The reading on this micrometer is:
   (A) .815″.
   (B) .865″.
   (C) .915″.
   (D) .965″.

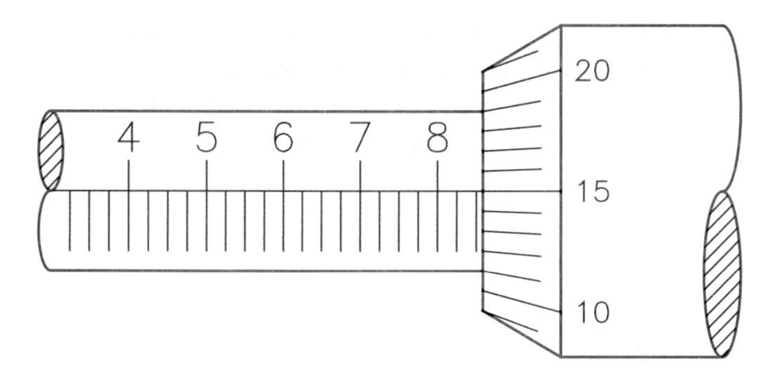

10. Technician A says a compression gauge can be used to detect compression leakage in an engine's combustion chamber. Technician B says a cylinder leakage tester can detect compression leakage in an engine's combustion chamber. Who is correct?
    (A) A only.
    (B) B only.
    (C) Both A and B.
    (D) Neither A nor B.

11. A car engine is missing and a puffing sound can be heard at the tailpipe. It was just tuned-up by another shop. Technician A says that a compression test should be done to check the mechanical condition of the engine. Technician B says that the carburetor should be rebuilt because that is a common cause of a miss. Who is correct?
    (A) A only.
    (B) B only.
    (C) Both A and B.
    (D) Neither A nor B.

Modern automotive engines are complex pieces of machinery that contain many different types of fasteners. Many more types of fasteners can be found on the body, interior, and trim of the vehicle. (Land Rover)

Chapter 3
Engine Hardware

After studying this chapter, you will be able to:

- ❑ Explain the purpose of the various fasteners used on automotive engines.
- ❑ Summarize thread specifications.
- ❑ Properly torque fasteners to specifications.
- ❑ Identify the various types of nuts and screw heads.
- ❑ Repair fastener and thread damage.
- ❑ Efficiently remove and install engine gaskets.
- ❑ Describe the use of sealers.
- ❑ Diagnose gasket and seal failures.
- ❑ Replace metal lines and rubber hoses properly.
- ❑ Form a double-lap flare on tubing.
- ❑ Service engine belts.

Know These Terms

Anaerobic sealer
Antifriction bearings
Bolts
Cap screws
Cogged belt
Cotter pin
Die
Fasteners
Flaring tool
Form-in-place gasket
Friction bearing
Gasket
Grade markings
Hardening sealers
Hose clamps
Hoses
Key
Keyway
Line bending tool
Machine screws
Metal lines
Nonhardening sealers
Nuts
O-ring seal
Pins
Ribbed belt
RTV sealer
Seals
Serpentine belt
Setscrew
Snap ring
Split pin
Tap
Thread chaser
Thread pitch gauge
Thread repair insert
Torque sequence
Torque-to-yield
Washers

This chapter provides an overview of the various types of hardware found on automotive engines. During engine repair, you must know how to properly service all of the parts of an engine. In doing so, you must work with bolts, nuts, snap rings, lock pins, machine screws, studs, washers, gaskets, seals, sealants, hoses, belts, tubing, and so on. Remember, if you install even one component incorrectly, the whole engine may fail in service. Look at all of the fasteners visible on the engine in **Figure 3-1.**

This chapter will help you develop some of the basic skills required of a professional engine technician. It provides vital information to prepare you for later, more advanced textbook chapters on engine overhaul, cooling system service, fuel system service, and so on. Study this chapter carefully and you will be taking another "big step" toward becoming an certified engine technician.

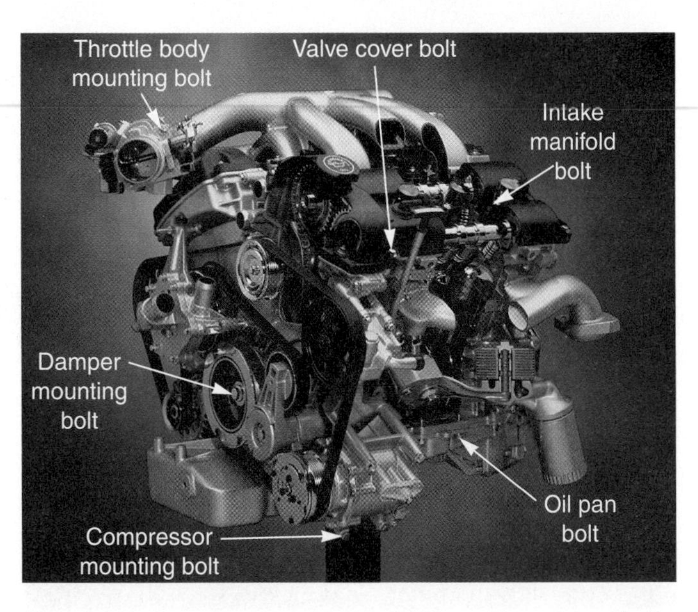

Figure 3-1. *An engine is assembled with a wide variety of bolts, nuts, belts, gaskets, seals, and sealants. (Ford)*

Fasteners

Fasteners include any type of holding device used to secure parts together. Hundreds are used in the construction of an engine. The most common ones are pictured in **Figure 3-2.**

Bolts are metal shafts with threads on one end and a head on the other end. When threaded into a part without the use of a nut, they are called ***cap screws. Nuts*** have internal threads and are threaded onto a bolt or stud. The most common nuts have a hex-shaped (six-sided) head. Various types of nuts can be found on an engine, **Figure 3-3.**

Washers are used under bolt heads and nuts. A flat washer, or plain washer, increases the clamping surface under the fastener. It prevents the bolt or nut head from gouging or digging into the part. A lock washer prevents the bolt or nut from loosening. Under stress and vibration, a fastener may unscrew without a lock washer. Lock tabs or plates perform the functions of both flat and lock washers. They increase clamping surface area and secure the fastener. They are found on engine exhaust manifolds.

Machine screws are similar to bolts, but they normally have screwdriver-type heads. They are threaded their full length and are relatively weak and small. Machine screws are used to secure parts when clamping loads are light. They come in various head shapes, as pictured in **Figure 3-4.** You might find machine screws on accessory units mounted on an engine.

Figure 3-2. *Study the names of these basic fasteners. (Deere & Co.)*

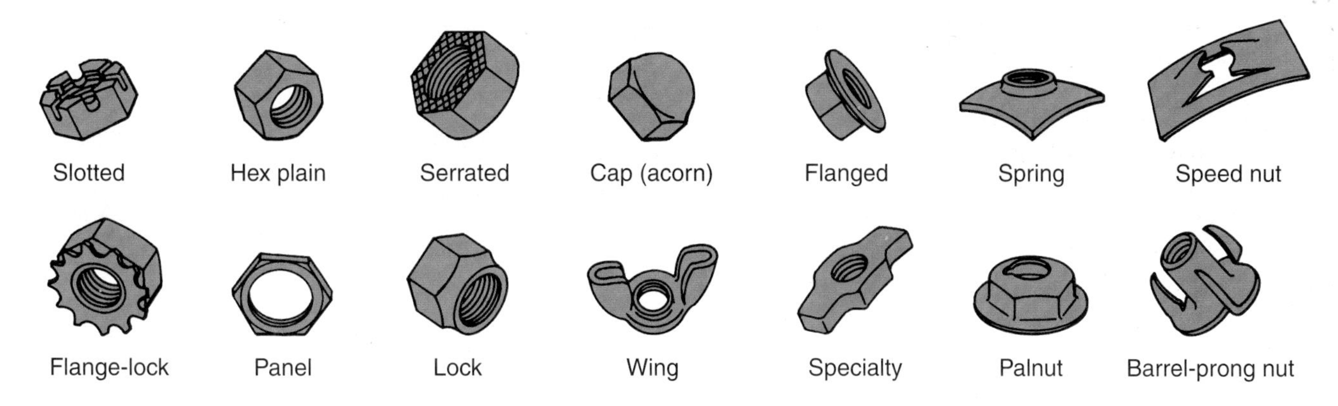

Figure 3-3. *There are various types of nuts found on engines and related hardware. (Deere & Co.)*

Bolt and Nut Dimensions

The basic dimensions of a bolt are bolt size, bolt head size, bolt length, and thread pitch. Refer to **Figure 3-5.**

- ❑ **Bolt size.** A measurement of the outside diameter of the bolt threads.
- ❑ **Bolt head size.** The distance across the flats of the bolt head. This size is the same as the wrench size. Remember that the bolt head size is not the size of the bolt, a common mistake.
- ❑ **Bolt length.** Measured from the bottom of the bolt head to the threaded end of the bolt.
- ❑ **Thread pitch.** This is the coarseness of the thread. The pitch on US fasteners is the number of threads per inch. The pitch on metric fasteners is the distance between each thread, measured in millimeters.

Thread Series

There are different standard thread series. However, as an engine technician, the most common thread series you will encounter are fine and coarse. Refer to **Figure 3-6.**

For US Customary fasteners, the Unified National thread series is used. Threads with a coarse thread series

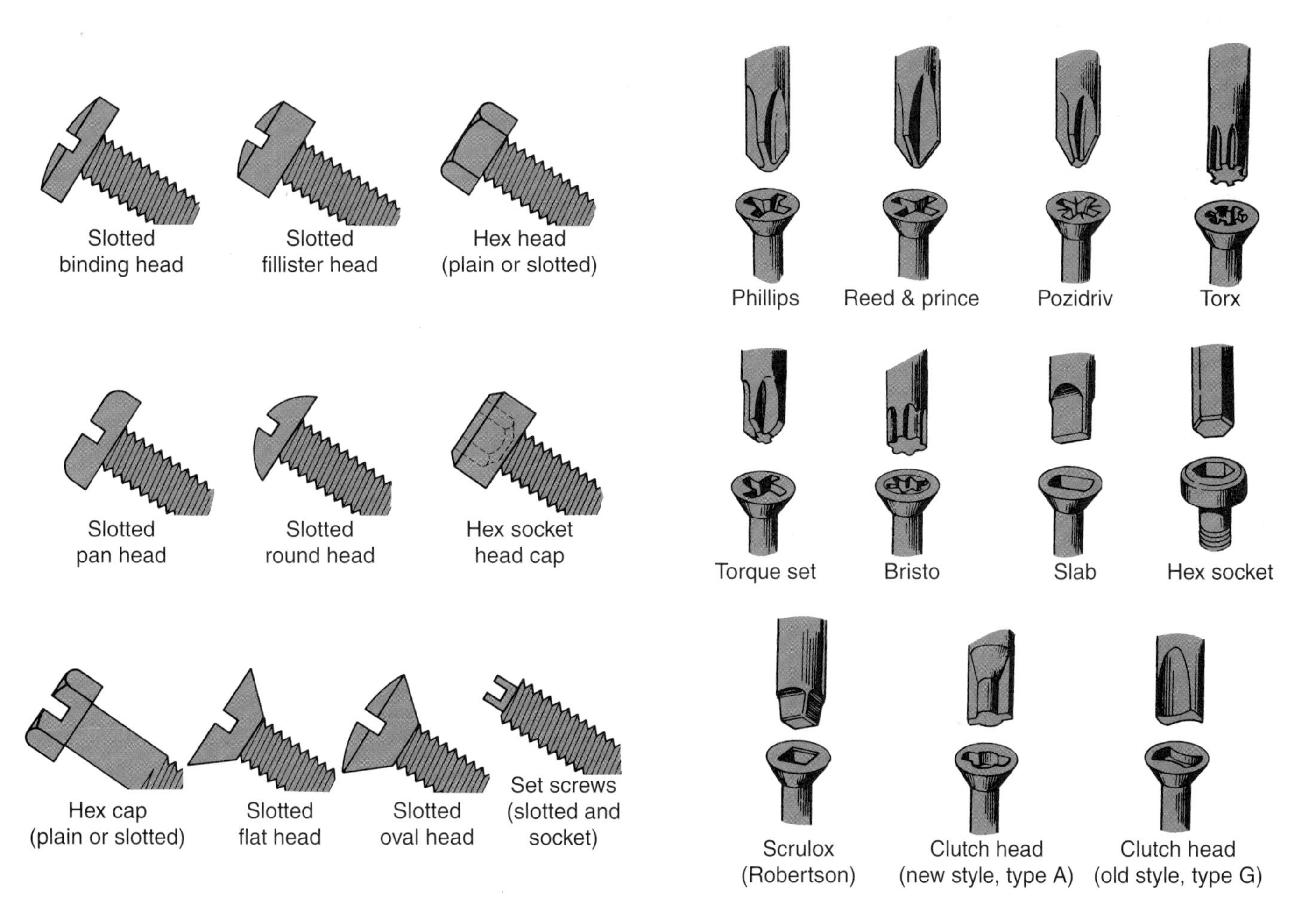

Figure 3-4. *These are basic machine screw and drive styles. (Heyco, Klein Tools)*

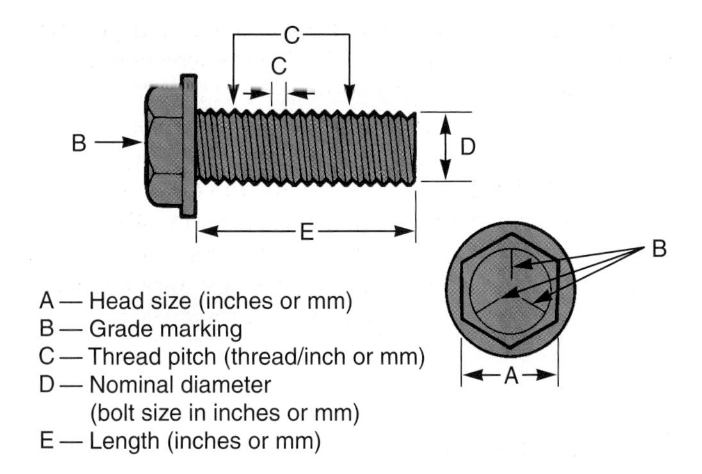

Figure 3-5. *These are the basic dimensions of a bolt.*

are called Unified National Coarse, abbreviated UNC. Threads with a fine thread series are called Unified National Fine, abbreviated UNF. For example, look at the thread specification:

3/4 - 10 UNC

This is a 3/4″ diameter bolt or nut having 10 threads per inch and a coarse thread form.

For metric fasteners, coarse threads with a pitch of 1.25 mm are assumed unless otherwise specified. For example, look at the metric thread specification:

M20 × 2.0

The M20 indicates this is a metric thread specification with a bolt or nut diameter of 20 mm. The 2.0 indicates a thread pitch of 2 mm, which is a coarse thread.

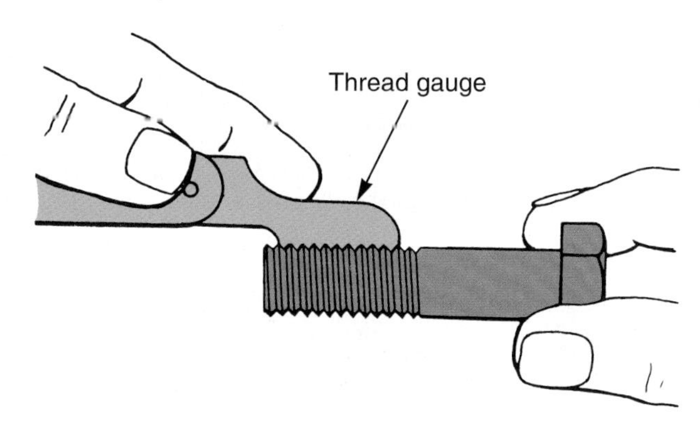

Figure 3-7. *A thread gauge can be used to check the thread type. (Deere & Co.)*

Never interchange thread series or pitches. If a bolt is forced into a hole with the wrong threads, either the bolt or part threads can be ruined. Metric threads can be easily mistaken for Unified National threads if not inspected carefully. A ***thread pitch gauge*** can be used to determine thread series, **Figure 3-7.** A bolt size gauge will also help you determine bolt thread diameters or sizes and thread pitch quickly, as shown in **Figure 3-8.**

Left- and Right-Hand Threads

Fasteners may have right- or left-hand threads. Right-hand threads are the most common. A fastener with right-hand threads must be turned clockwise to tighten. Left-hand threads are less common. A fastener

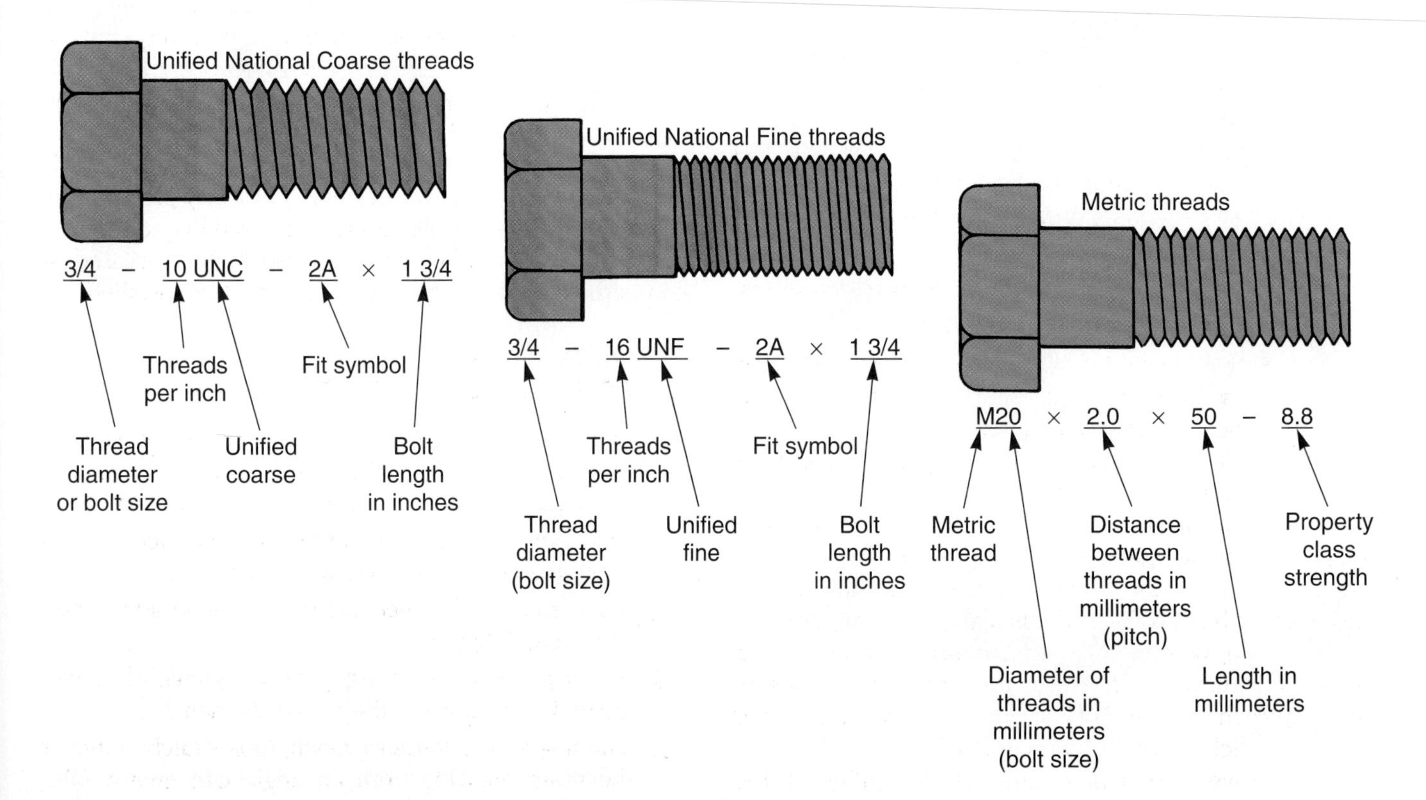

Figure 3-6. *Unified National and metric threads. Study the thread specifications.*

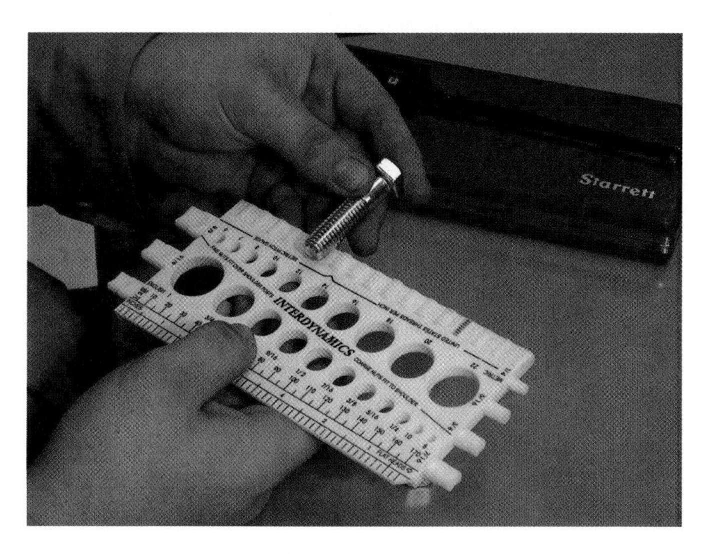

Figure 3-8. *This technician is using a bolt size gauge. By inserting the bolt into the correct-size hole you can determine the size and thread pitch of the bolt.*

with left-hand threads must be turned counterclockwise to tighten. The letter L may be stamped on fasteners with left-hand threads.

Grade Markings

The grade refers to the relative strength of a bolt. It indicates the amount of torque, compression, or tension a fastener can withstand before breaking. ***Grade markings*** specify the strength of the bolt. The markings are located on the bolt head. US Customary bolts are marked with lines or slash marks. The more lines in the grade marking, the stronger the bolt. A metric bolt is marked by a numbering system. The larger the number in the grade marking, the stronger the bolt. **Figure 3-9** shows a torque specification chart based on bolt diameter and grade.

Never replace a bolt with one that is a lower grade. The weaker bolt may fail under normal operating conditions. This may cause part failure and a dangerous condition.

Torquing Fasteners

It is critical that engine bolts and nuts are tightened correctly. The process of properly tightening a fastener is called ***torquing*** the fastener, **Figure 3-10.** ***Torque specifications*** are values given by the vehicle manufacturer to which fasteners should be tightened. These values are given in foot-pounds or newton-meters. Torque specifications are normally given for the fasteners on all precision assemblies, such as head bolts, rod bolts, and oil pump bolts.

A torque wrench is used to torque fasteners. As discussed in Chapter 2, there are three basic types of torque wrench. **Figure 3-11** shows how to set a sound indicating (clicker) type torque wrench.

The power, efficiency, and dependability of the rebuilt engine is greatly dependent on proper fastener torque. If overtightened, a bolt will stretch and possibly break. The threads may also fail. If undertightened, a bolt may loosen over time. The resulting part movement may shear the fastener or break a gasket, causing leakage.

Some fasteners found on an engine have a "right" and a "wrong" side. For example, connecting rod nuts are considered "critical" fasteners. These nuts must be installed in the correct orientation. When reused, they also need to be installed on the same bolt from which they were removed. The side of the used nut that goes down is shiny or shows marks of being tightened up against the part. See **Figure 3-12.**

Torque Sequence

A ***torque sequence*** or pattern ensures parts are clamped together evenly. An incorrect or uneven pattern may cause warpage, leakage, and part breakage. A criss-cross pattern is generally used, starting in the middle. The pattern is from side to side and end to end. Each bolt is tightened a little at a time, jumping back and forth from the middle outward. However, some manufactures may specify an "outside-in" pattern. See **Figure 3-13.**

A shop manual should be consulted to determine the proper torque sequence for any critical assembly. Assemblies that likely have a torque sequence specified include cylinder heads, intake manifolds, and exhaust manifolds. Proper procedures are detailed later in this text.

Torque-to-Yield

Torque-to-yield is a bolt tightening method in which the bolt is tightened until it is deformed (stretched). The specification for this method includes a bolt torque and a number of degrees. The bolt is first torqued to the specification. Then, a degree wheel adapter is placed between the socket and wrench. The fastener is then turned until the degree wheel reads as specified by the manufacturer. This stretches the bolt to its correct yield point and preloads the fastener for better part clamping under varying conditions. Since the bolt is deformed when this method is used, it cannot be reinstalled. New fasteners must be installed.

Using a Torque Wrench

Follow these basic rules when using a torque wrench:

- ❑ Place a steady pull on the torque wrench. Do *not* use short, jerky motions or inaccurate readings can result.
- ❑ Only pull on the handle of the torque wrench. Do *not* allow the wrench beam to touch anything.
- ❑ Make sure the fastener threads are clean and, when specified, lightly oiled.
- ❑ When possible, avoid using swivel joints. They can affect the accuracy of the torque wrench.
- ❑ When reading a torque wrench, look straight down at the scale. Viewing from an angle can give a false reading.

Caution
The torque specifications listed below are approximate guidelines only and may vary depending on conditions when used such as amount and type of lubricant, type of plating on bolt, etc.

SAE Standard / Foot-pounds

Grade of bolt	SAE 1 & 2	SAE 5	SAE 6	SAE 8		
Min. tensile strength	64,000 PSI	105,000 PSI	133,000 PSI	150,000 PSI		
Markings on head					Size of socket or wrench opening	
US standard	Foot-pounds				US regular	
Bolt diameter					Bolt head	Nut
1/4	5	7	10	10.5	3/8	7/16
5/16	9	14	19	22	1/2	9/16
3/8	15	25	34	37	9/16	5/8
7/16	24	40	55	60	5/8	3/4
1/2	37	60	85	92	3/4	13/16
9/16	53	88	120	132	7/8	7/8
5/8	74	120	167	180	15/16	1
3/4	120	200	280	296	1-1/8	1-1/8

Metric Standard

Grade of bolt		5D	8G	10K	12K	
Min. tensile strength		71,160 PSI	113,800 PSI	142,200 PSI	170,679 PSI	
Grade markings on head		5D	8G	10K	12K	Size of socket or wrench opening
Metric		Foot-pounds				Metric
Bolt dia.	US dec. equiv.					Bolt head
6 mm	.2362	5	6	8	10	10 mm
8 mm	.3150	10	16	22	27	14 mm
10 mm	.3937	19	31	40	49	17 mm
12 mm	.4720	34	54	70	86	19 mm
14 mm	.5512	55	89	117	137	22 mm
16 mm	.6299	83	132	175	208	24 mm
18 mm	.709	111	182	236	283	27 mm
22 mm	.8661	182	284	394	464	32 mm

Figure 3-9. *A basic bolt torque chart can be used when factory specifications are not available.*

- ❑ A general torque value chart should only be used when manufacturer's specifications are not available.
- ❑ When a torque pattern is not available from the manufacturer, use a general crisscross pattern.
- ❑ Tighten fasteners in at least four steps: to one-half of the recommended torque, to three-fourths torque, to full torque, and to full torque a second time.
- ❑ Retorque fasteners when required by the vehicle manufacturer. The fasteners on cylinder heads, intake manifolds, and exhaust manifolds may need to be retightened after a period of engine operation.
- ❑ Replace bolts when recommended by the vehicle manufacturer.

Thread Repairs

External and internal threads may become damaged. Minor thread damage includes nicks in threads, partial flattened threads, and other less serious problems. Major thread damage generally includes badly smashed threads, stripped threads, or threads that cannot be easily repaired.

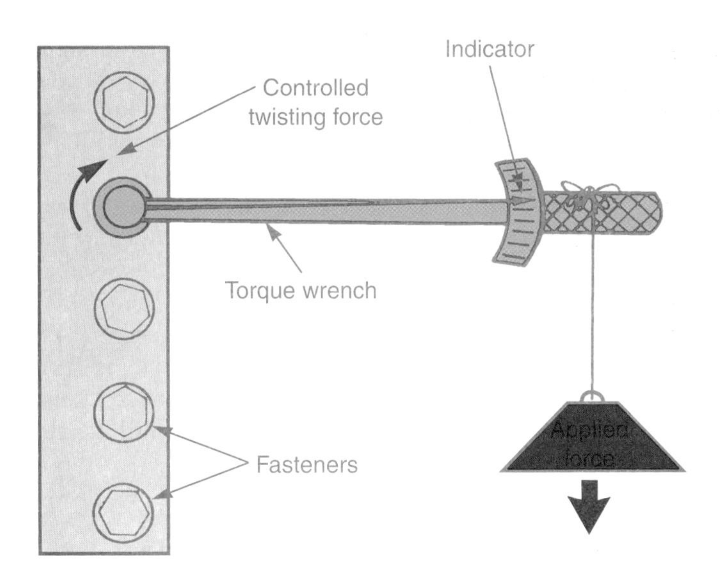

Figure 3-10. *Torque is measured in foot-pounds or newton-meters. One foot-pound of torque is produced when one pound of weight is suspended from a one foot long lever arm.*

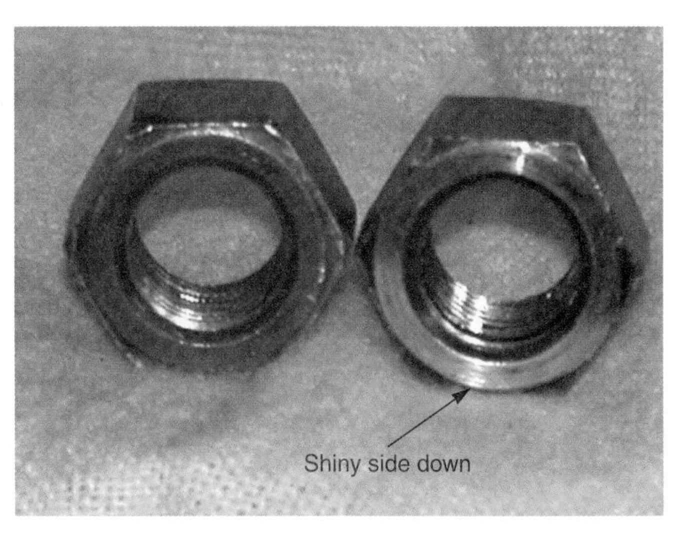

Figure 3-12. *A common mistake is to install rod nuts that are being reused upside down. The side of the nut that goes down or against the part is shiny and shows friction marks from contact with the rod or part.*

Many times, the parts can be salvaged by repairing the damaged threads. Therefore, you must be able to quickly and properly repair damaged threads.

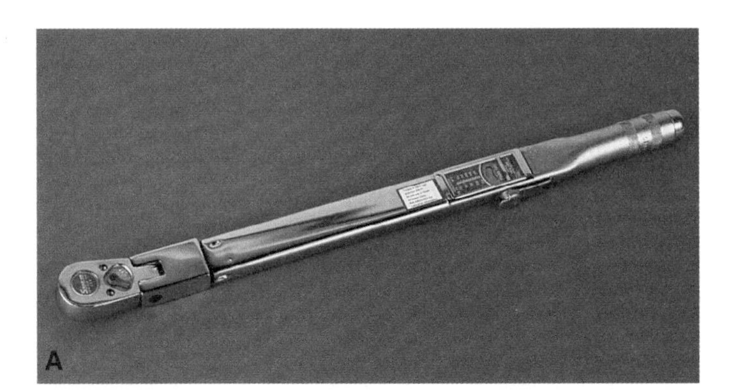

Figure 3-11. *A quality torque wrench is vital to proper repair. A—This is a sound indicating (clicker) torque wrench. B—By turning the knob on the side of the wrench, you can adjust the torque at which the wrench sounds (clicks).*

Minor Thread Repairs

Minor thread damage can usually be repaired with a thread chaser, **Figure 3-14.** A ***thread chaser*** is a rethreading tool for cleaning up damaged threads. This type of tool is available for fixing both internal and external threads. The chaser is run through or over the threads to restore them, **Figure 3-15.**

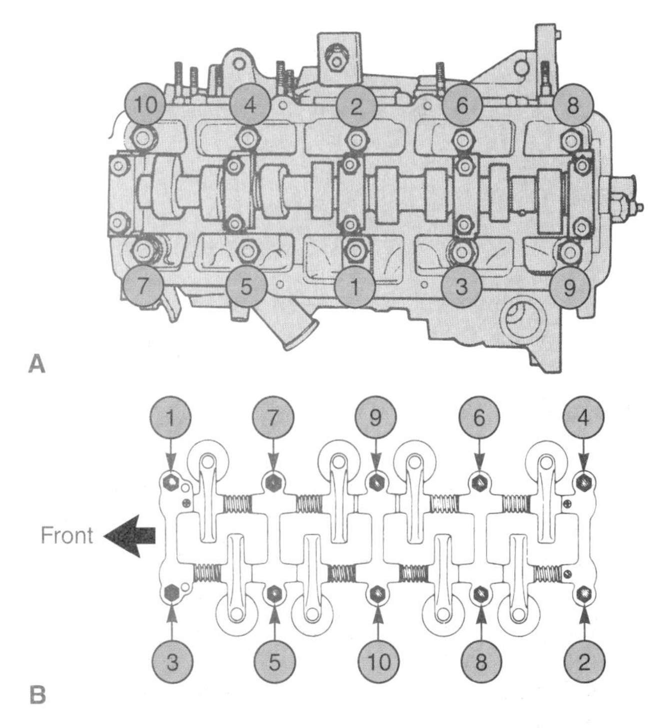

Figure 3-13. *A—This is a basic crisscross tightening sequence. Note that the pattern is "inside-out." B—Sometimes, an "outside-in" torque pattern is specified. Note that the pattern is still crisscross.*

Figure 3-14. *Thread chasers are similar to taps and dies, but have cutting surfaces that will not cut as deep into threaded fasteners. (Snap-on Tool Corp.)*

Taps and Dies

When a thread chaser cannot be used to clean up damaged threads, the hole can be drilled and tapped oversize. First, drill out the hole one diameter or size larger. Then, cut new threads in the drilled hole with the correct size tap. A larger bolt can then be installed.

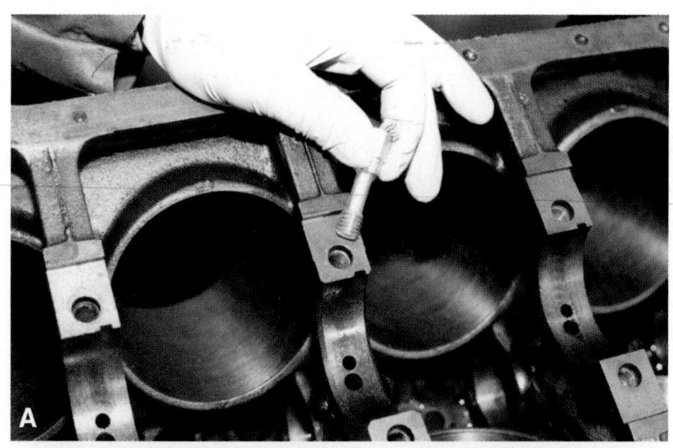

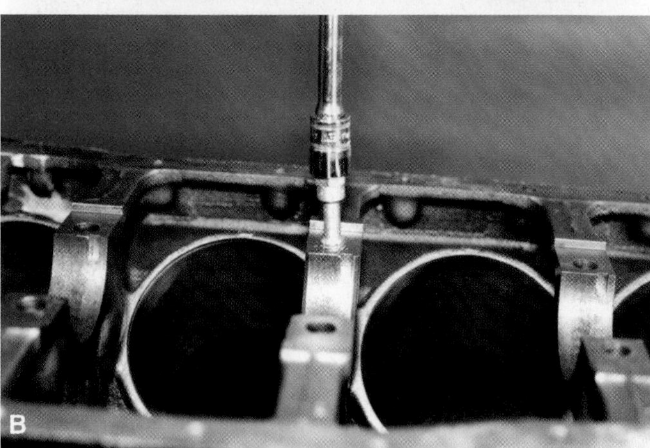

Figure 3-15. *Using a thread chaser. A—Select the correct thread chaser. Start the chaser in the hole by hand to prevent cross threading. B—Use a speed handle to spin the thread chaser to the bottom of the hole.*

A ***tap*** is a tool for cutting internal threads, **Figure 3-16.** A ***die*** is for cutting external threads, such as on rods, bolts, shafts, and pins. There are various shapes of taps. Some are for starting the threads, others are for cutting threads all the way to the bottom of a hole.

Taps are mounted in special handles called tap handles, and dies in die handles, **Figure 3-17.** Hold the handle squarely while rotating it into the work. As soon as the tap or die begins to bind, back the tool off about a quarter turn to prevent tool breakage. This cleans out the metal cuttings that have collected in the hole or tool. Then, the cut can be made another half turn deeper. Keep rotating a half turn in and a quarter turn out until the cut is complete.

There are a few rules for using taps and dies. These include:

- Never force a tap handle or the tap may break. Back off the handle as described to clean out metal shavings.
- Keep the tap and die well oiled to ease cutting.
- Always use the proper size tap in a correctly sized and drilled hole.

A drill and tap size chart is given in **Figure 3-18.** This chart will help you select the correct size of drill bit for a given tap size. For example, a 27/64″ hole must be drilled when using a 1/2″ coarse tap.

Caution: Be extremely careful not to break a tap or a screw extractor. They are case hardened and cannot be easily drilled out of a hole. You will compound your problems if you break one of these tools in a hole.

Thread Repair Insert

A ***thread repair insert*** should be used when using an oversize hole and fastener is *not* acceptable. An insert can

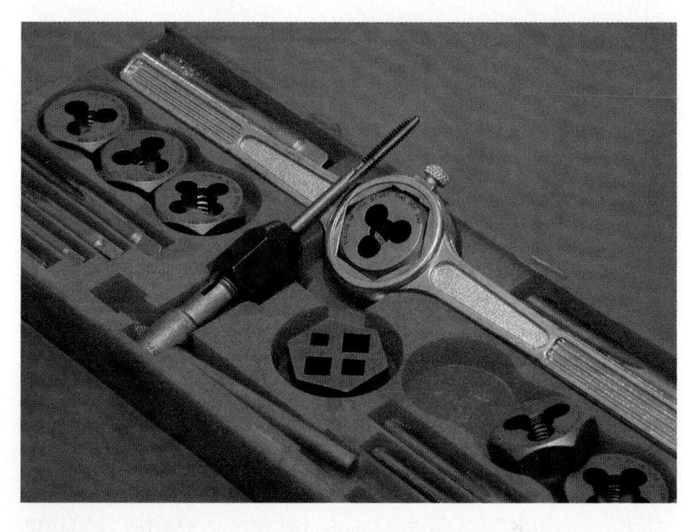

Figure 3-16. *Tap and die set is often needed for thread repair. A tap is for cutting threads in hole. A die is for cutting threads on shaft or bolt.*

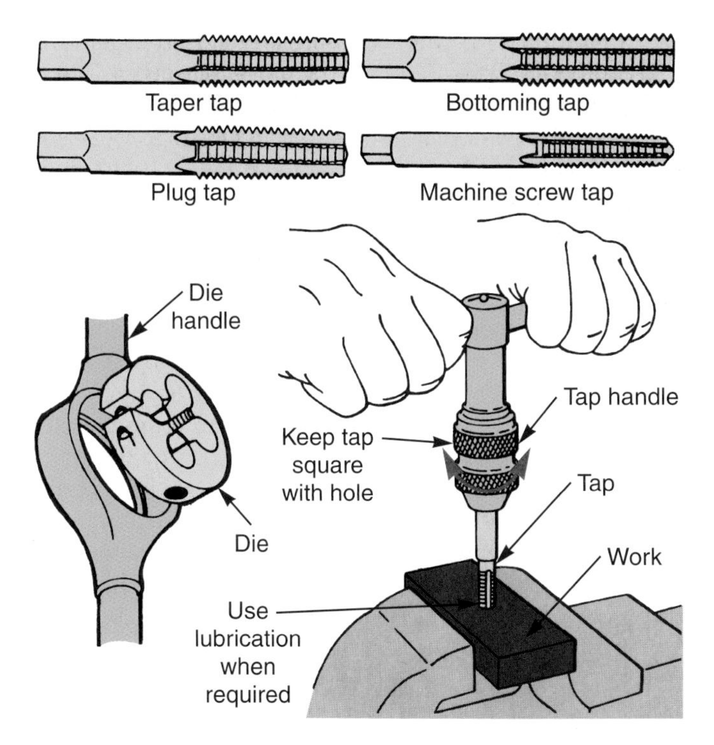

Figure 3-17. *Taps and dies are held in a special handle. Four types of taps are shown at the top of this figure.*

repair damaged internal threads and allow the use of the original size fastener.

To use a thread repair insert, drill the hole oversize as described in the instructions that come with the insert. Then, tap the hole and screw the insert into the threaded hole. The internal threads of the insert act as threads of the same size as the previously damaged threads. These steps are summarized in **Figure 3-19.**

It is very easy to overtighten spark plugs in an aluminum cylinder head during a tune-up. This can make the threads on the spark plug lock or stick into the threads in the spark plug hole. When the overtightened spark plugs are removed, the soft aluminum threads in the cylinder head can be stripped or pulled out.

To repair spark plug hole threads, you can install a special thread repair insert. **Figure 3-20** shows a thread repair kit for spark plug holes. To prevent metal shavings from falling down into the cylinder, the head should be removed. Then, the hole can be drilled out, threaded, and repaired with the insert.

Caution: If a customer requests spark plug hole thread repairs with the cylinder head on the engine, coat the drill bit and tap with grease. This will help hold the metal shavings onto the drill and tap. However, avoid repairing spark plug hole threads with the cylinder head installed. The metal shavings from drilling and taping could fall down into the cylinder. These shavings can score the cylinder wall or stick on the valve faces, causing engine damage. Warn the customer of the dangers of this procedure and get the customer's approval in writing.

American National Screw Thread Pitches

UNC				
Bolt or tap size	**Threads per inch**	**Outside diameter at screw**	**Drill sizes**	**Decimal equivalent of drill**
1	64	.073	53	0.0595
2	56	.086	50	0.0700
3	48	.099	47	0.0785
4	40	.112	43	0.0890
5	40	.125	38	0.1015
6	32	.138	36	0.1065
8	32	.164	29	0.1360
10	24	.190	25	0.1495
12	24	.216	16	0.1770
1/4	20	.250	7	0.2010
5/16	18	.3125	F	0.2570
3/8	16	.375	5/16	0.3125
7/16	14	.4375	U	0.3680
1/2	13	.500	27/64	0.4219
9/16	12	.5625	31/64	0.4843
5/8	11	.625	17/32	0.5312
3/4	10	.750	21/32	0.6562
7/8	9	.875	49/64	0.7656
1	8	1.000	7/8	0.875
1 1/8	7	1.125	63/64	0.9843
1 1/4	7	1.250	1 7/64	1.1093
UNF				
Bolt or tap size	**Threads per inch**	**Outside diameter at screw**	**Drill sizes**	**Decimal equivalent of drill**
0	80	.060	3/64	0.0469
1	72	.073	53	0.0595
2	64	.086	50	0.0700
3	56	.099	45	0.0820
4	48	.112	42	0.0935
5	44	.125	37	0.1040
6	40	.138	33	0.1130
8	36	.164	29	0.1360
10	32	.190	21	0.1590
12	28	.216	14	0.1820
1/4	28	.250	3	0.2130
5/16	24	.3125	I	0.2720
3/8	24	.375	Q	0.3320
7/16	20	.4375	25/64	0.3906
1/2	20	.500	29/64	0.4531
9/16	18	.5625	0.5062	0.5062
5/8	18	.625	0.5687	0.5687
3/4	16	.750	11/16	0.6875
7/8	14	.875	0.8020	0.8020
1	14	1.000	0.9274	0.9274
1 1/8	12	1.125	1 3/64	1.0468
1 1/4	12	1.250	1 11/64	1.1718

Figure 3-18. *A tap drill chart gives the correct size of hole to drill for each size of tap.*

Removing Damaged Fasteners

An engine technician must be able to remove nuts and bolts that are broken, rusted in place, or have rounded-off

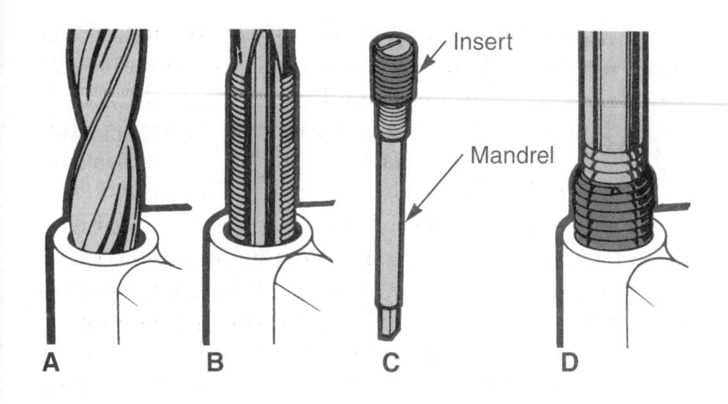

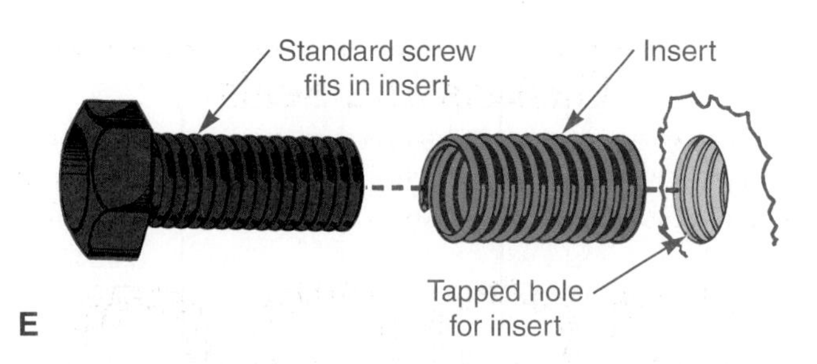

Figure 3-19. *A thread repair insert can be used to repair stripped threads. A—Drill the hole oversize. B—Tap the hole. C—Place the thread repair insert on the installation tool. D—Use the installation tool to screw the thread repair insert into the hole. E—The original size bolt will thread into the repaired threads. (General Motors, DaimlerChrysler)*

heads. This important skill is not covered in most service manuals. Certain tools and methods are needed for removing problem fasteners.

Locking pliers can sometimes be used to remove broken bolts or bolts with rounded heads, **Figure 3-21A.** Lock the pliers tightly on the bolt. Then, rotate the bolt to remove it.

Figure 3-20. *This is a kit for repairing damaged spark plug threads in an aluminum cylinder head. The spark plug hole is first drilled larger. Then, the hole is tapped with new threads and the insert installed. This kit is recommended for aluminum cylinder heads, but not for cast iron heads.*

A stud puller or stud wrench can be used to remove studs and bolts that are broken off above the surface of the part. This tool is also used to install studs.

In some cases, a bolt is broken with little of it above the surface of the part. You can cut a slot in the top of the bolt with a hacksaw. Then, a screwdriver can be used to unscrew the broken bolt. If this does not work, you can weld on another bolt head and use a wrench to remove the bolt.

When the fastener is broken flush with the part surface, a hammer and punch can be used for removal. Angle the punch so that blows from the hammer unscrew the broken bolt.

A screw extractor can also be used to remove bolts that are broken flush or below the surface of the part. See **Figure 3-21B.** To use a screw extractor, drill a hole in the exact center of the broken bolt. Then, lightly drive the extractor into the hole using a hammer. Unscrew the broken bolt using a wrench. **Figure 3-22** shows a complete set of screw extractors. You must drill the appropriate size hole for the extractor you are using.

On some broken bolts, you may have to drill a hole almost as large as the inside diameter of the threads. Then, use a tap or punch to remove the thin layer of threads remaining in the hole. When using this method, be careful not to damage the threads in the hole.

For nuts that are frozen onto a bolt, you can use a nut splitter. This tool breaks the nut into two pieces. The pieces can then be easily removed from the bolt.

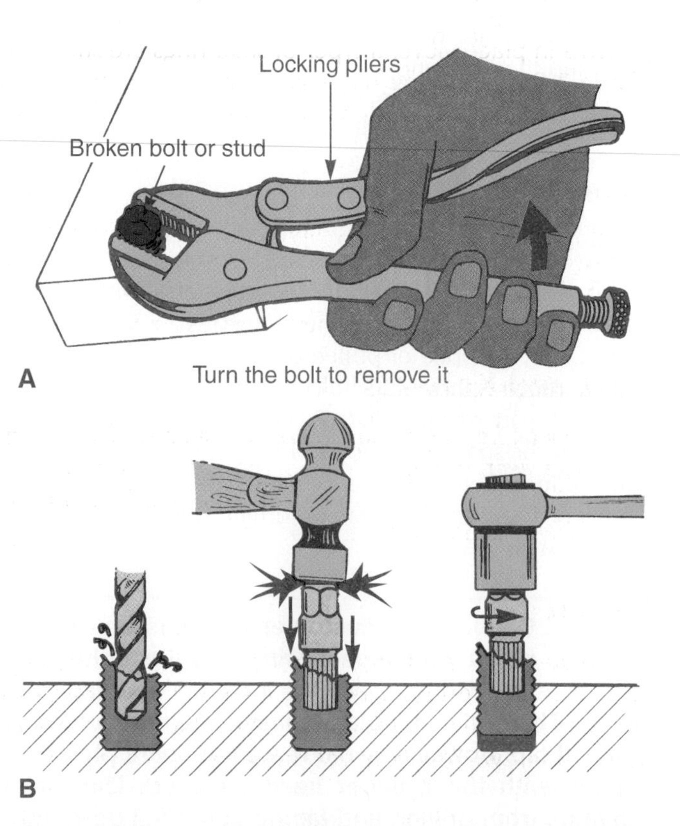

Figure 3-21. *A—Locking pliers can sometimes be used to remove a broken bolt or stud when part of the fastener is sticking up above the part. B—A screw extractor can be used to remove a broken bolt when it is below the surface of the part. (Florida Voc. Ed., Lisle Tools)*

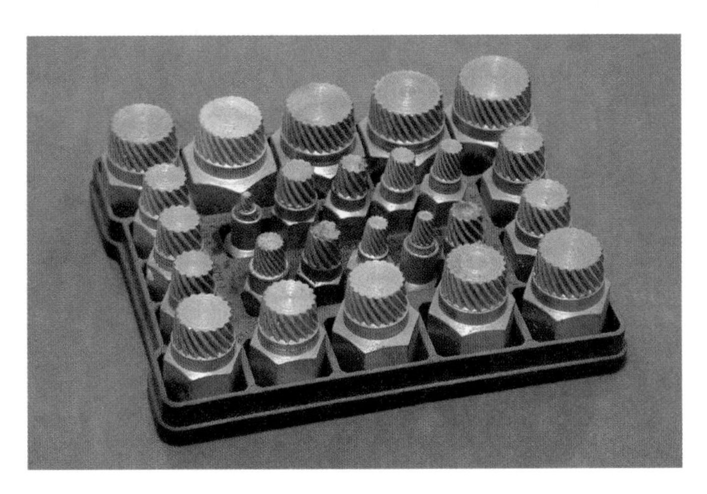

Figure 3-22. *This screw extractor set will fit most any size of broken bolt. (Snap-on Tool Corp.)*

Nonthreaded Fasteners

There are numerous types of nonthreaded fasteners found on a modern vehicle. It is essential to learn the most common types.

Snap Rings

A ***snap ring*** fits into a groove in a part and commonly holds shafts, bearings, gears, pins, and other similar components in place. Several types of snap rings are shown in **Figure 3-23.** Snap ring pliers are needed to remove and install snap rings. These tools have special jaws that grasp the snap ring.

Warning: Wear eye protection when working with snap rings. A flexed snap ring can fly off of pliers with considerable force.

Keys and Keyways

A metal ***key*** fits into a groove cut into a shaft and part, such as a gear, pulley, or collar. The groove is called the ***keyway.*** The key and keyway prevent the part from turning on its shaft. Refer to **Figure 3-24.**

Setscrews

A ***setscrew*** is a headless fastener designed to accept an Allen wrench (hex wrench) or screwdriver. Setscrews are normally used to lock a part onto a shaft. They may be used with or without keys and keyways.

Pins

Pins are frequently used to secure a gear or pulley to its shaft. A pin may be straight, tapered, split, or another variation. Several pins are shown in **Figure 3-25.**

A ***cotter pin*** is a safety device often used in conjunction with a slotted nut. It fits through a hole in the bolt or part. This keeps the slotted nut from turning and possibly coming off. Cotter pins are also used with shafts, linkages, and other pins.

A ***split pin,*** also known as a roll pin or spring pin, is commonly used to hold the drive gear on the distributor shaft. Since the oil pump may be driven by the distributor shaft, the pin serves as a safety device. If the oil pump locks up, the pin will shear off and stop the engine from running (no distributor rotation). This prevents severe engine damage from lack of lubrication. There are many other applications for split pins.

Gaskets and Seals

Gaskets and seals are used between engine parts to prevent fluid leakage. It is important to understand a few

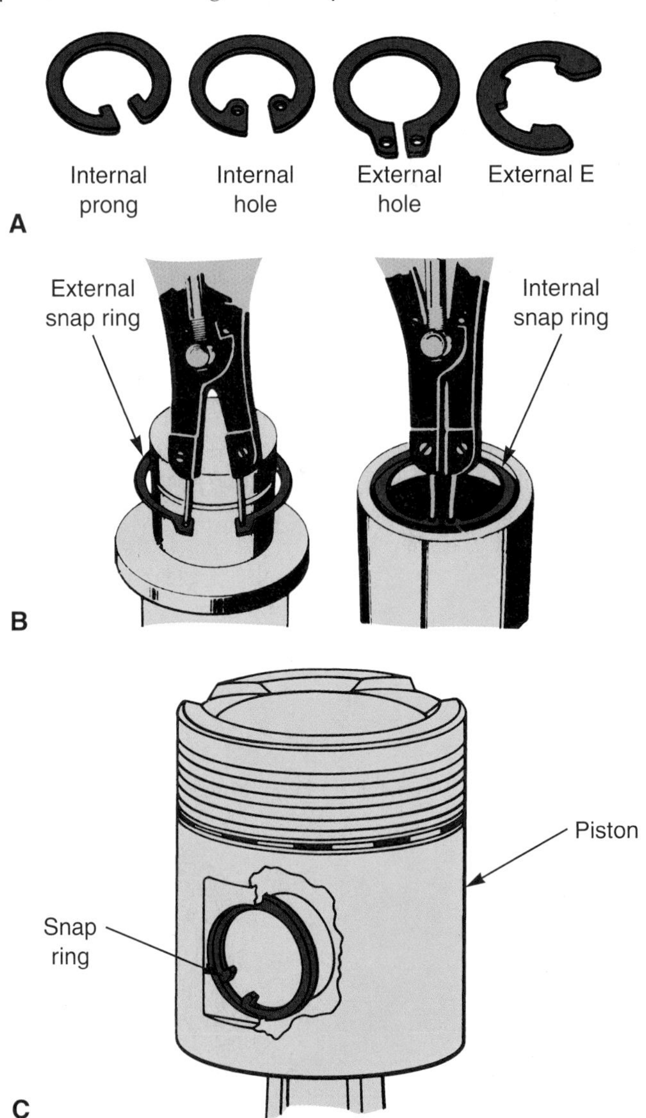

Figure 3-23. *A—These are four common types of snap ring. B—Using snap ring pliers. C—Snap rings are sometimes used to hold the piston pin inside of the piston. The snap ring fits into a small groove machined into the pin bore. (Deere & Co.)*

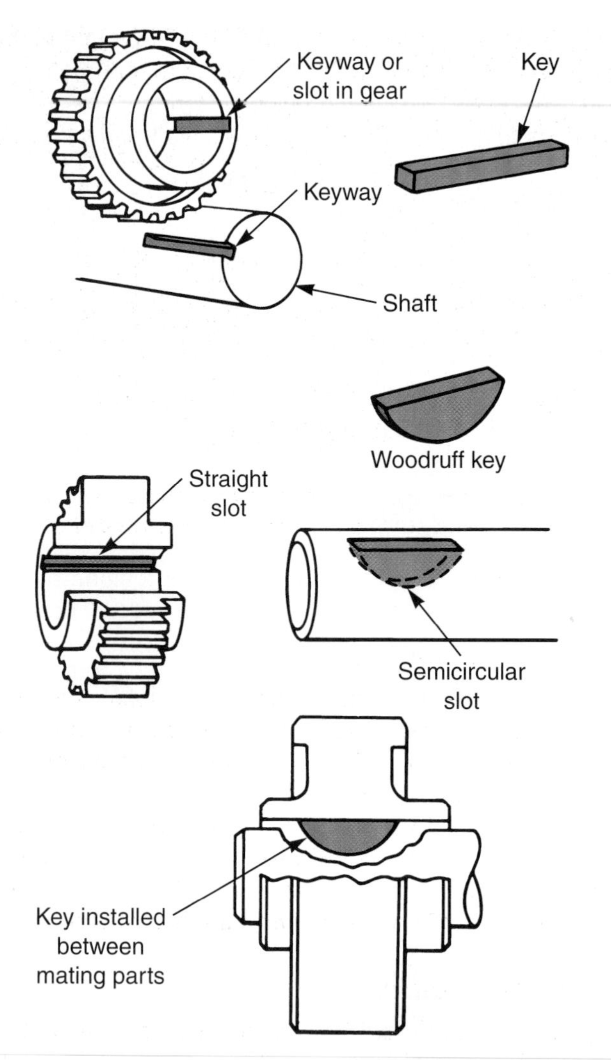

Figure 3-24. *A key and keyway lock a part onto its shaft. This is commonly done to hold the damper and pulley onto the front of the crankshaft. (Deere & Co.)*

principles about gaskets and seals. If they are serviced improperly, serious customer complaints and mechanical failures can result.

Gaskets

A ***gasket*** is a soft, flexible material placed between parts, **Figure 3-26.** It can be made of a fiber material, rubber,

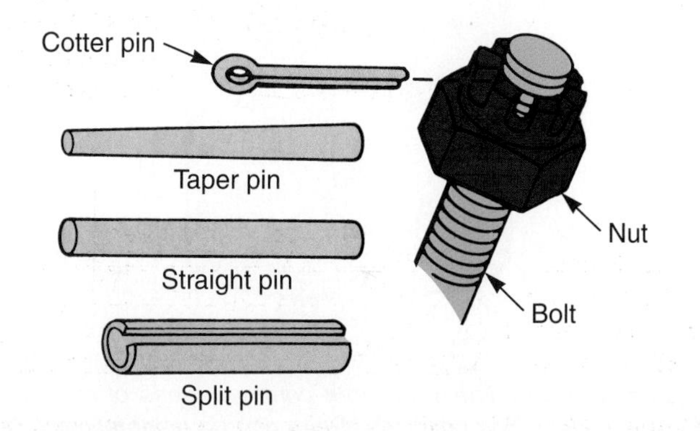

Figure 3-25. *These are four common types of pin. Note the common application of cotter pins. (Florida Voc. Ed., Deere & Co.)*

Figure 3-26. *Gaskets prevent leakage between parts. Various sealers and adhesives can be used with gaskets. Here, spray tack is being applied to hold the gaskets in place during assembly. (Fel-Pro Gaskets)*

neoprene (synthetic rubber), cork, treated paper, or thin metal. When the parts are tightly fastened together, the gasket is compressed and deformed. This forces the gasket material to fill small gaps, scratches, or other imperfections in the mating part surfaces. A leakproof seal is produced between the parts.

Gasket Rules

When working with gaskets, inspect for leaks before engine disassembly. If the two parts are leaking, the mating part surfaces should be inspected closely for problems.

Avoid part damage during disassembly. Be careful not to nick, gouge, or dent mating surfaces while removing parts. The slightest unevenness could cause leakage.

Clean off the old engine gasket carefully. All of the old gasket material must be scraped or wire brushed from the parts. Use care on aluminum and other soft metals, which are easily damaged. Use a dull scraper and wire brush with light pressure.

After gasket removal, wash the parts in solvent. Blow dry with compressed air. Then, wipe mating surfaces with a clean shop towel.

Compare the new gasket to the shape of the mating surface. Lay the new gasket into place. All holes and sealing surfaces must match perfectly.

Some gaskets require sealer, **Figure 3-27.** Sealer is normally used where two different gaskets come together. It will prevent leakage where gaskets overlap. Check the service manual for details. Use sealer sparingly. Too much sealer may clog internal passages in the assembly.

After fitting the gasket and parts in place, hand-start all fasteners before tightening. This will ensure proper part alignment and threading of fasteners. It also lets you check for proper bolt lengths.

When more than one bolt is used to hold a part, tighten each bolt a little at a time. Tighten one to about half of its torque specification, then the others. Use either a basic crisscross or factory-recommended torque pattern. This

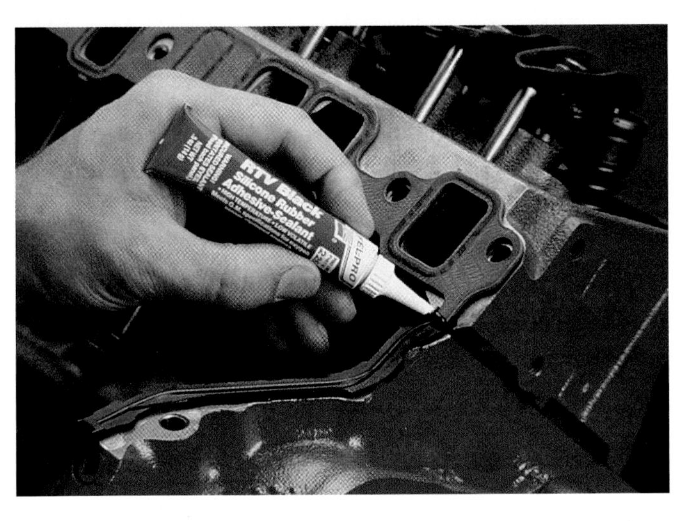

Figure 3-27. *Sealer is often used where two different gaskets meet.*

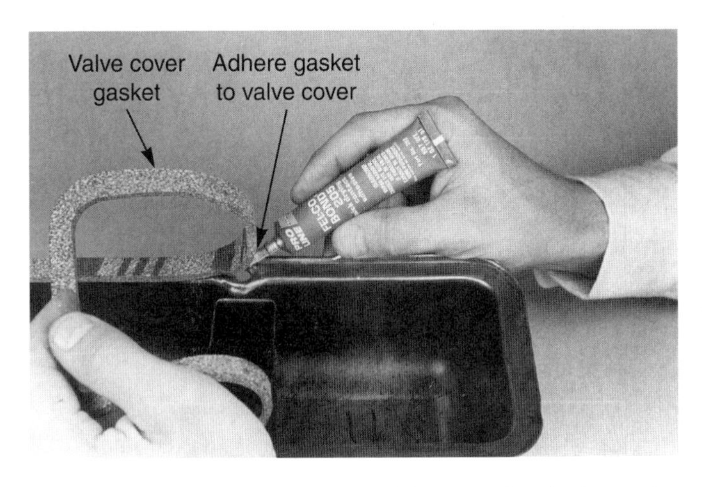

Figure 3-28. *This technician is applying a nonhardening sealer between the valve cover and gasket. (Fel-Pro Gaskets)*

will ensure even gasket compression and sealing. Tighten to three-fourth torque and to full torque. Then, retorque each fastener. Do not overtighten the fasteners. It is very easy to tighten the bolts enough to dent sheet metal parts and smash or break gaskets. Apply only the specified torque.

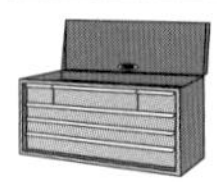

Note: Some engine parts, such as some valve covers, are plastic. They can be damaged easily during removal and cleaning.

Sealers

A sealer is commonly coated on a gasket to help prevent leakage and to hold the gasket in place during assembly. There are numerous kinds of sealers. They have different properties and are designed for different uses. Always read the sealer's label and the factory service manual before selecting a sealer.

Hardening sealers are used on permanent assemblies, such as fittings and threads. They are also used for filling uneven surfaces. Hardening sealers are usually resistant to heat and most chemicals.

Nonhardening sealers are for semipermanent assemblies, such as cover plates, flanges, threads, hose connections, and valve covers. See **Figures 3-28.** They are also resistant to most chemicals and moderate heat. Shellac is a nonhardening sealer. It is a gummy, sticky substance that remains pliable. It can be used on fiber gaskets as a sealer and to hold the gasket in place during assembly.

A ***form-in-place gasket*** is a special sealer used in place of a conventional fiber or rubber gasket. Two common types of form-in-place gaskets are RTV and anaerobic sealer.

Caution: Use only oxygen-sensor-safe sealers.

RTV or Silicone Sealer

RTV stands for room temperature vulcanizing and is also called silicone sealer. ***RTV sealer*** cures (dries) from moisture in the air. It is used to form a gasket on thin, flexible flanges.

RTV sealer normally comes in a tube. Depending on the brand, it can have a shelf life from one year to two years. Always inspect the package for the expiration date before use. If too old, RTV sealer will *not* cure or seal properly.

Using RTV Sealer

RTV sealer should be applied in a continuous bead approximately 1/8″ (3 mm) to 3/16″ (5 mm) in diameter, **Figure 3-29.** All mounting holes must be circled. Uncured RTV sealer may be removed with a rag.

Components should be torqued in place while the RTV sealer is still wet to the touch, within about 10 minutes. The use of locating dowels is often recommended to prevent the sealing bead from being smeared. If the continuous bead of sealer is broken, a leak may result.

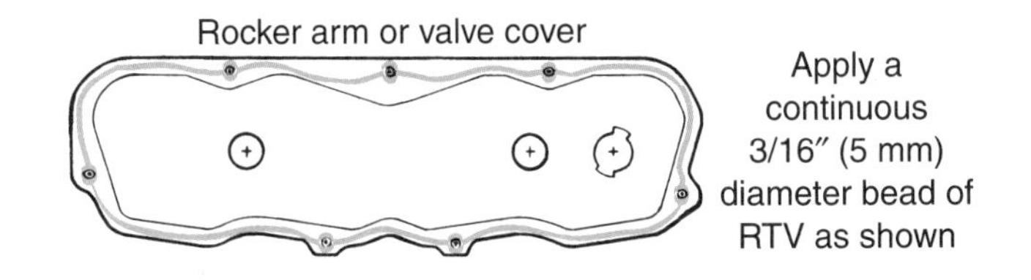

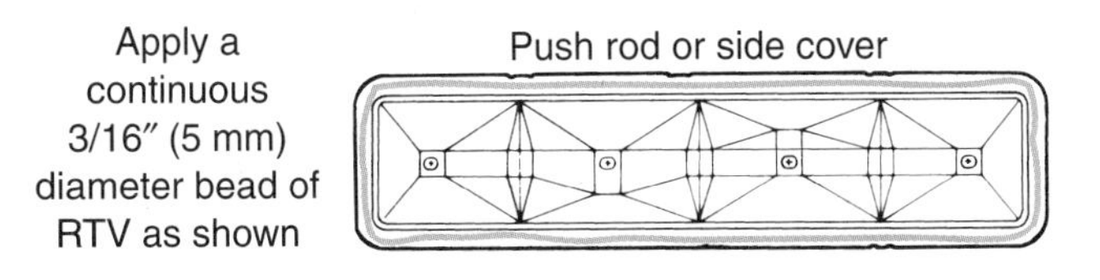

Figure 3-29. *RTV or silicone sealer can be used in place of a cork or rubber gasket. Note the size of bead run around the covers and that bolt holes are circled. (General Motors)*

Anaerobic Sealer

Anaerobic sealer cures in the absence of air and is designed for tightly fitting, thick parts. It is used between smooth, true surfaces *not* on thin, flexible flanges.

Anaerobic sealer should be applied sparingly. Use 1/16″ to 3/32″ (1.5 to 2 mm) diameter bead on one gasket surface. Be certain that the sealer surrounds each mounting hole. Typically, the bolts should be torqued within 15 minutes of applying the sealer.

Sealing Engine Parts

When selecting a form-in-place gasket, refer to the vehicle manufacturer's service manual for recommendations. Scrape or wire brush gasket surfaces to remove all loose material. Check that all gasket rails are flat. Using a shop towel and solvent, wipe off any oil and grease. The sealing surfaces must be clean and dry before using a form-in-place gasket.

A few gasket manufacturers sell pre-cut gaskets designed to replace form-in-place gaskets. When working on an engine installed in a vehicle, it can be difficult to properly clean the sealing surfaces. Or, it may be almost impossible to fit a part on the engine without hitting and breaking the bead of sealant. When this is the case, a pre-cut gasket might work better than sealer alone.

Sealer is sometimes recommended on bolts that extend into water jackets. **Figure 3-30** shows sealer on a front cover bolt. Use approved sealer on the threads of the bolts.

Seals

Seals prevent leakage between a stationary part and a moving part. For example, there is a seal between the engine front cover and the crankshaft. A seal allows the shaft to spin or slide inside the nonmoving part without fluid leakage, **Figure 3-31.** Seals are normally made of synthetic rubber molded onto a metal body.

Servicing Seals

When working with seals, inspect the seal for leakage before disassembly. If a seal is leaking, there may be other problems besides a defective seal. Look for a scored shaft, misaligned seal housing, or damaged parts. Leakage requires close inspection of the seal and parts after disassembly.

Remove the old seal carefully. Pry it out without scratching the seal housing. Sometimes, a special puller is required for seal removal. This is discussed in later chapters.

Inspect the shaft for wear and burrs. Look at the shaft closely where it contacts the seal, **Figure 3-32.** It should be smooth and flat. File off any burrs that could cut the new seal. A badly worn shaft will require polishing, a shaft sleeve repair kit, or replacement.

Compare the old seal to the new seal. The inside diameter (ID) and outside diameter (OD) on the new seal must be the same as on the old seal. To double-check the inside diameter, slip the seal over the shaft. It should fit snugly.

Coat the outside of the seal housing with approved sealer. Lubricate the inner lip of the seal with system fluid.

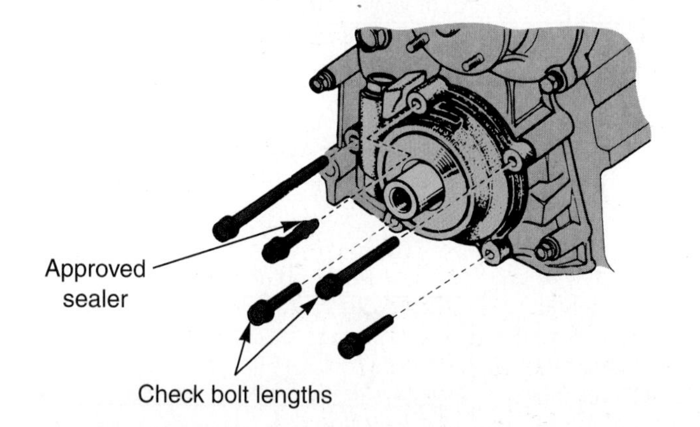

Figure 3-30. *Bolts that extend into water and oil passages often have an approved sealer applied to their threads. Check the service manual for details. (Toyota)*

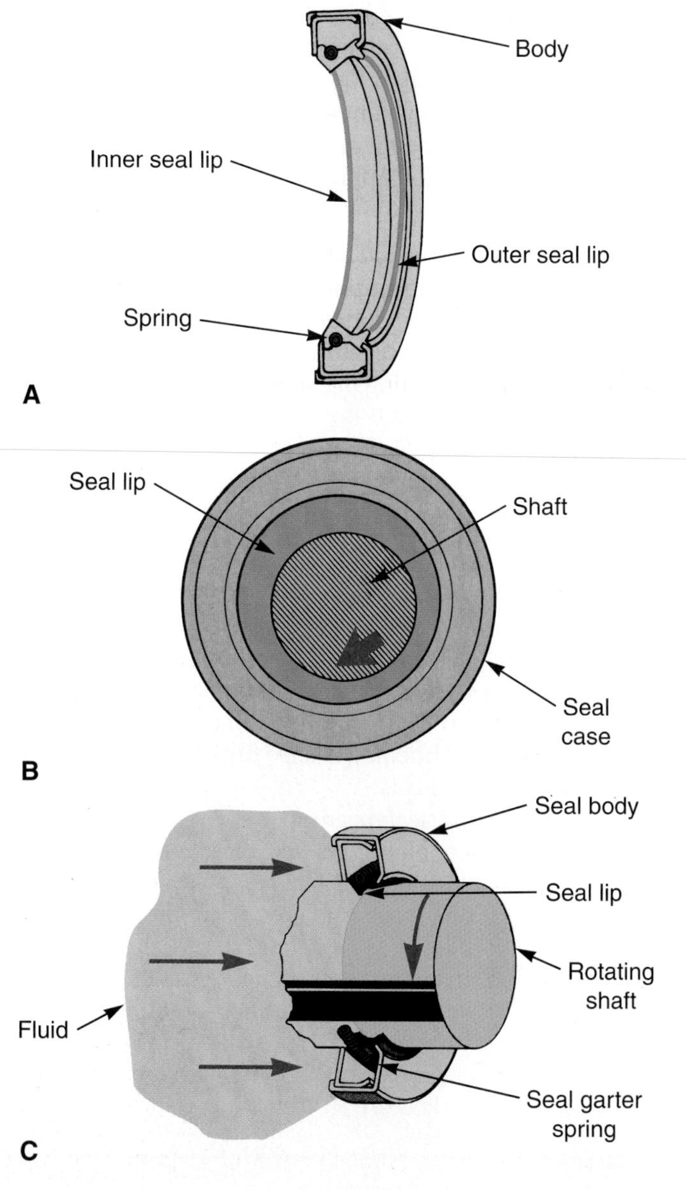

Figure 3-31. *A seal prevents leakage between a rotating shaft and a stationary housing or cover. A—Seal parts. B—Seal on a shaft. C—Seal action. (DaimlerChrysler, Caterpillar)*

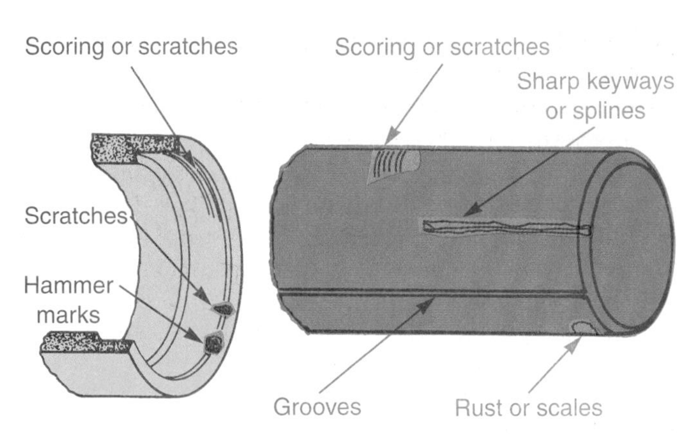

Figure 3-32. *When servicing a seal, check for these kinds of problems. (Federal Mogul)*

Install the seal with the sealing lip facing the inside of the part. If installed backwards, the seal will leak. Also, check that the seal is squarely and fully seated in its bore.

O-Ring Seals

An ***O-ring seal*** is a stationary type seal that fits into a groove between two parts, **Figure 3-33.** When the parts are assembled, the synthetic rubber seal is partially compressed and forms a leakproof joint.

Normally, O-ring seals should be coated with system fluid. The system fluid is engine oil, diesel fuel, or the fluid used in the component. This helps the parts slide together without scuffing or cutting the seal.

Usually, sealants are *not* used on O-ring type seals. When in doubt about any seal installation, refer to the vehicle's service manual.

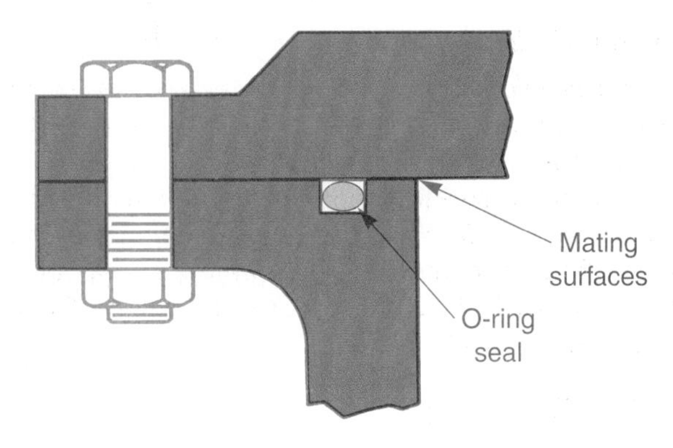

Figure 3-33. *An O-ring seal is usually a stationary seal between mating parts. (Deere & Co.)*

Note: Special gaskets and seals sometimes require other installation techniques. These special situations are discussed in later chapters where they apply.

Hoses and Lines

When working on engines, you will have to service numerous types of hoses and lines. **Figure 3-34** shows some of the types of lines and hoses found in an engine compartment. It is very important that you handle them properly. If not, dangerous or damaging leaks can result.

Figure 3-34. *There are many different hoses and lines in the engine compartment. (Honda)*

Figure 3-35. *Using a tubing cutter. A—Tighten the cutter down lightly around the metal line. Turn the cutter around the line, tighten, and turn again. Repeat until the line is cut. B—Use the pointed tool to remove burrs from the inside lip of the cut.*

Warning: Diesel injection lines can contain over 8000 psi (56,000 kPa) of pressure. Never attempt to remove a diesel injection component with the engine running. Fuel could squirt out, blinding you or puncturing your skin and causing blood poisoning or death! Wear eye protection when working around running diesel engines.

Metal Lines

Metal lines or tubing are used as fuel lines and brake lines. Metal fuel lines usually run from the fuel tank to the engine compartment. They are usually double-wall steel tubing.

A ***tubing cutter*** is used to cut a metal line, **Figure 3-35.** Secure the line in a vise with v-groove caps. These caps will hold the metal tubing without damaging it. Next, place the tubing cutter over the line so that both rollers contact the line. The cutting wheel should be centered over the line. Turn the knob on the cutter to move the cutting wheel down against the tubing. The wheel should just touch the metal line.

With the cutting wheel in contact with the line, rotate the cutter all the way around the tubing. Then, tighten the knob a little more and again rotate the cutter all the way around the line. Continue tightening the knob and rotating the cutter until the line is cut. Then, use a pointed tool to deburr the inside edge of the cut. Often, this tool is provided on the cutter, **Figure 3-35B.** You do not want metal shavings to fall into the line and contaminate the system.

You will often have to bend the new replacement line to match the bends or curves of the old line. You can use a ***line bending tool*** as shown in **Figure 3-36.** This tool allows you to create bends without pinching or collapsing the line. If a bending tool is not available, a bending spring can be used, **Figure 3-37.**

A ***flaring tool*** is used to form a flare on the end of the line, **Figure 3-38.** First, form the tubing inward with the flaring tool adapter. Refer to **Figure 3-39.** Then, force the flaring tool cone into the end of the line to produce a double-lap flare. When the line is installed, the fitting nut will draw the flare tight into the connector creating a leakproof seal. Be sure to blow the line clean before installing it into the vehicle.

Figure 3-36. *A—Using a line bending tool to bend a line. B—The completed bend.*

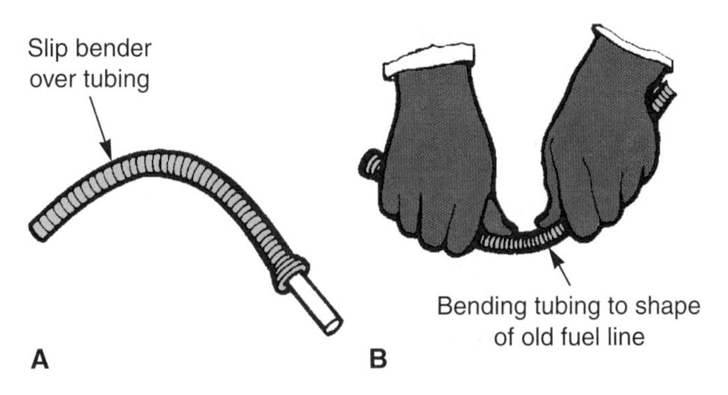

Figure 3-37. *If a line bending tool is not available, a bending spring can be used. A—The tubing should fit in spring without much clearance. B—Use both hands to bend tubing to desired shape.*

Note: Place the fitting nut on the line in the proper orientation before flaring the ends. If both ends of the line are flared, the fitting nut cannot be installed.

Hoses

Hoses are used on an engine to make a flexible connection between two fittings. **Figure 3-40** shows common hoses. Hoses may carry fuel, coolant, oil, air, and other fluids to the engine. Like metal lines, hoses must be serviced properly to prevent engine failure. ***Hose clamps*** secure hoses tightly to their fittings. The three basic types are illustrated in **Figure 3-41.**

To check a hose, visually inspect it as you squeeze the hose. Do this for the radiator hoses, heater hoses, fuel system hoses, vacuum hoses, and so on. You may see cracking of the outer rubber layer, the hose may have hardened with age, or the hose may have softened due to

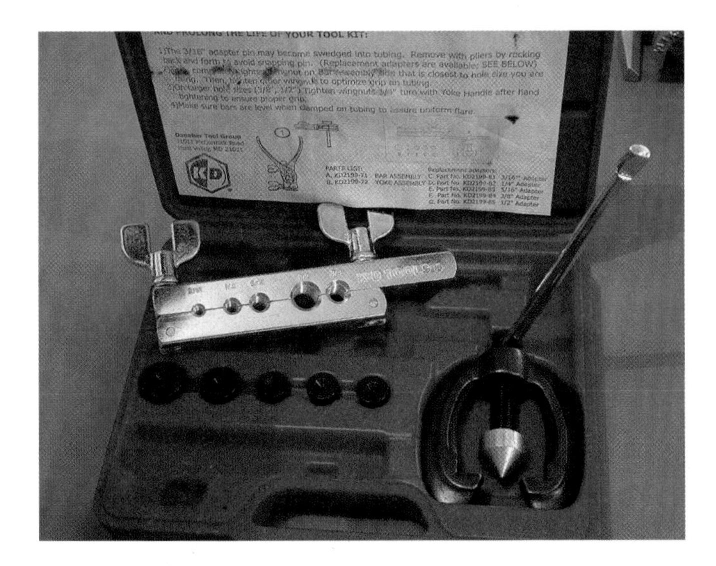

Figure 3-38. *A tube flaring set can be used to form flared ends on metal lines.*

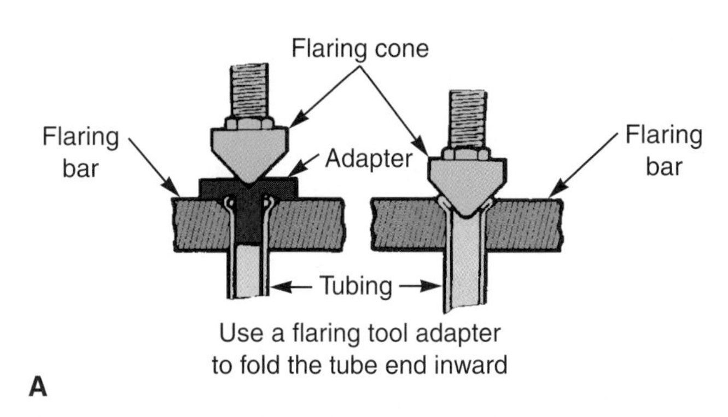

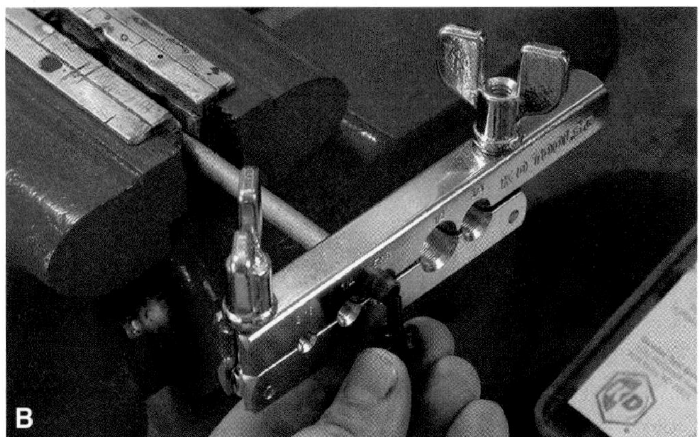

Figure 3-39. *A—A double-lap flare must be put on the end of the new fuel line. B—With the tubing clamped into place, fit the correct adapter into the line. C—To produce the second lap, remove adapter and use the cone. D—A smooth, uniform edge is formed on the end of the tubing.*

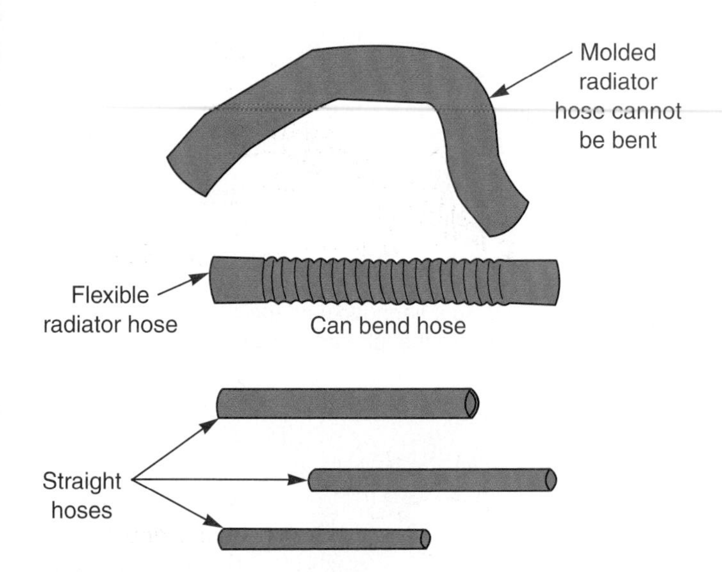

Figure 3-40. *These are four types of hose that may be found in an engine compartment.*

fluid contamination. See **Figure 3-42.** In any of these cases, install a new hose.

To remove a hose, loosen the hose clamps. If the hose is being replaced, you can cut a slit in the hose with a razor blade knife to ease removal. Twist and pull the hose off.

Compare the new hose to the old one. The bends and diameter should be the same. Clean the hose fittings. Then, install the new hose following recommended procedures for the type of hose being serviced. Finally, check the hose and connections for leaks.

Caution: Antifreeze must be disposed of properly. Drain the cooling system before removing the radiator or heater hoses.

Line and Hose Service Rules

Remember these rules when working with lines and hoses:

- Place a shop towel around the fuel line fitting during removal. This will keep fuel from spraying on you or the hot engine.

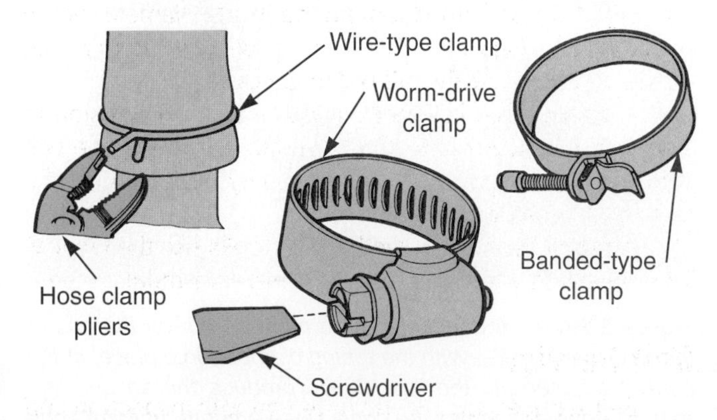

Figure 3-41. *These are three types of hose clamps. (Mopar)*

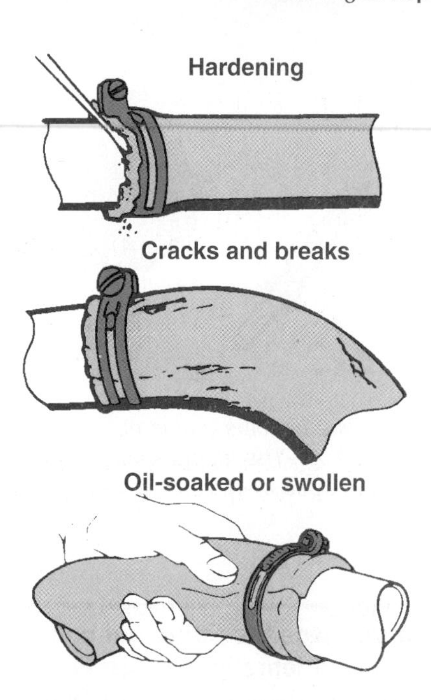

Figure 3-42. *Inspect hoses for these problems when working on an engine. (Gates Rubber Co., Ford)*

- Use a flare nut or tubing wrench. Two wrenches may be needed where two fittings join, **Figure 3-43.**
- Do not bend or kink plastic fuel lines.
- Use only approved double-wall steel tubing for metal fuel lines. Never use copper tubing.
- Make smooth bends when forming a new line. Use a bending tool or bending spring.
- Form double-lap flares on the ends of fuel lines. A single-lap flare is *not* approved for fuel lines.
- Reinstall all line hold-down clamps and brackets. If not properly supported, the line can vibrate and fail.
- Route all lines and hoses away from hot or moving engine parts. Double-check clearances after installation.

Figure 3-43. *Sometimes, two wrenches are needed to remove a fitting. This prevents the line from being twisting off.*

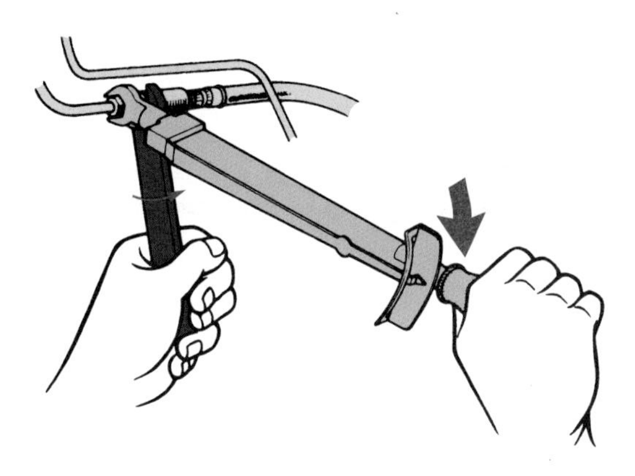

Figure 3-44. *When specified in the service manual, torque line fittings. This is especially important on high-pressure injection lines.*

- ❑ Torque fittings when recommended in the service manual, **Figure 3-44.**
- ❑ Use only hoses approved for the application. If the wrong type is used, the system fluid may chemically attack and rapidly ruin the hose. A dangerous or damaging leak could result.
- ❑ Check the condition of all hoses before reinstallation.
- ❑ Cut hoses off squarely, **Figure 3-45.**
- ❑ Make sure a hose fully covers its fitting or line before installing the clamps. If not installed properly, pressure in the system may force the hose off the fitting or line. Be careful not to break plastic fittings when removing vacuum hoses. See **Figure 3-46A.**
- ❑ Label hoses if their routing is confusing, **Figure 3-46B.**
- ❑ Double-check all fittings for leaks. Start the engine and inspect the connections closely.

Warning: Many gasoline injection systems retain pressure after the engine has been shut off. Relieve system pressure at the pressure test fitting before disconnecting fuel lines.

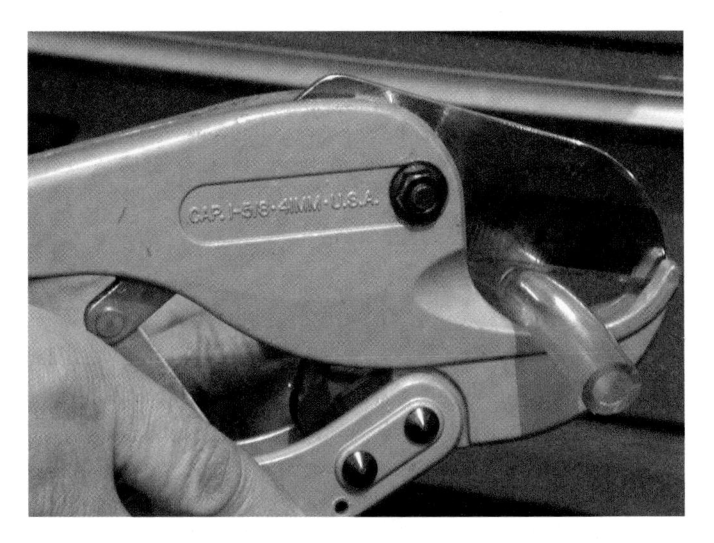

Figure 3-45. *Cut hoses squarely. These special cutters will help.*

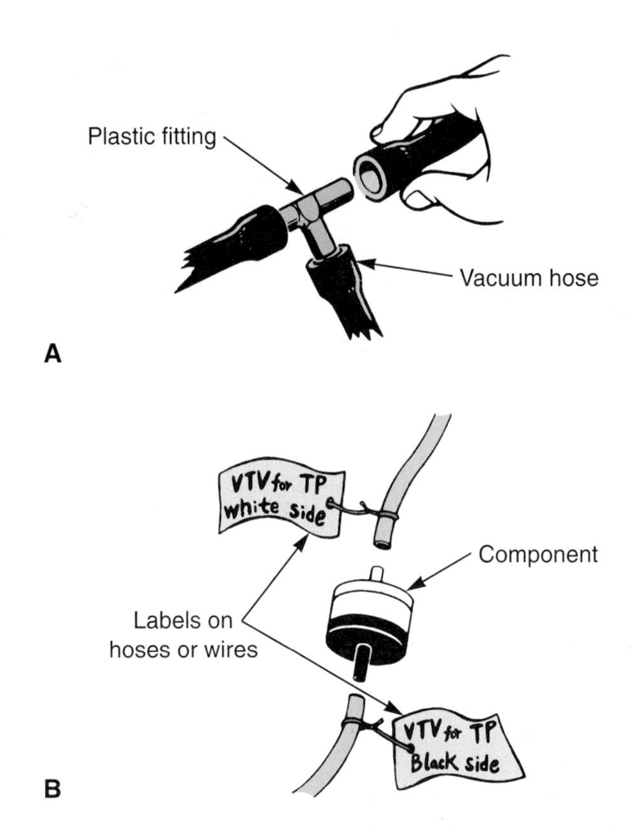

Figure 3-46. *A—Be careful not to break plastic fittings when disconnecting vacuum hoses. Grasp the fitting and twist the hose while pulling. B—Tag hoses and wires or label them with tape to avoid confusion during reassembly. (Toyota)*

Belts

Vehicles use several types of belts to draw power from the engine to operate the water pump, alternator, power steering pump, air conditioning compressor, and sometimes the air injection pump, and auxiliary shaft. Often, a single belt operates all of the accessory components. Since the belt "snakes" around many pulleys, this configuration is called a ***serpentine belt.*** However, some designs have multiple belts. See **Figure 3-47.**

Belt Types

A ***ribbed belt*** has a flat or rectangular cross section with small ridges formed around the inside diameter of the belt. Refer to **Figure 3-47.** A ribbed belt can handle the torque required to drive all of the accessories.

A ***cogged belt*** has teeth or ribs that run from side to side on the belt. These teeth fit in matching ribs in the belt sprockets. This type of belt is commonly used to drive the engine camshaft and auxiliary shaft.

A ***V-belt*** has a cross section that looks like the letter V. The pulleys in which the belt rides have a similar shape.

Belt Service

Belt service typically includes periodic inspection and replacement. To inspect a belt, look for cracking and

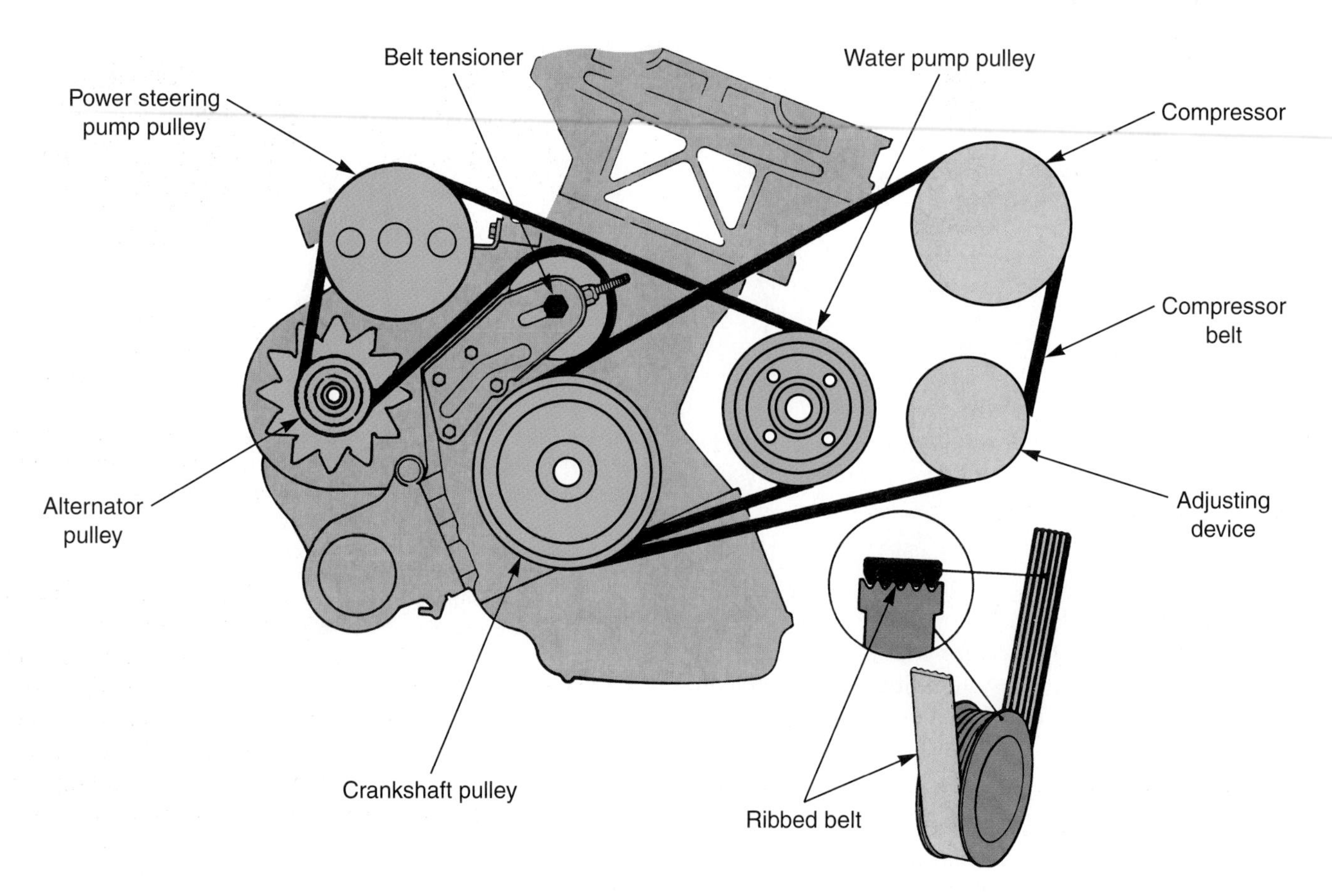

Figure 3-47. *Note the arrangement of drive belts on this engine. Two ribbed belts are used. (Saab)*

fraying. Twist the belt over and check its underside. Look for the troubles shown in **Figures 3-48** and **3-49.** If any belt problems are found, install a new belt. Follow service manual procedures and tighten the belt to specifications.

If the alternator is driven by its own belt, the belt should be tightened just enough to prevent slippage. Look at **Figure 3-50.** Overtightening could cause premature failure of the alternator bearings. Other accessories, such

A—Do not bend, twist, or turn the belt excessively. Oil and water will deteriorate the belt. Fix all engine leaks.

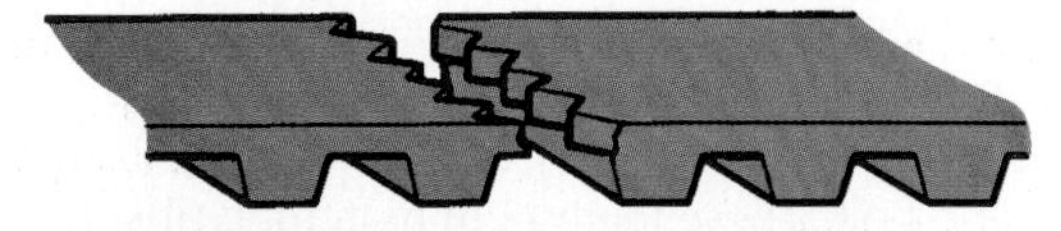

B—Belt breakage may be caused by a sprocket or tensioner problem. Check these parts before installing a new belt.

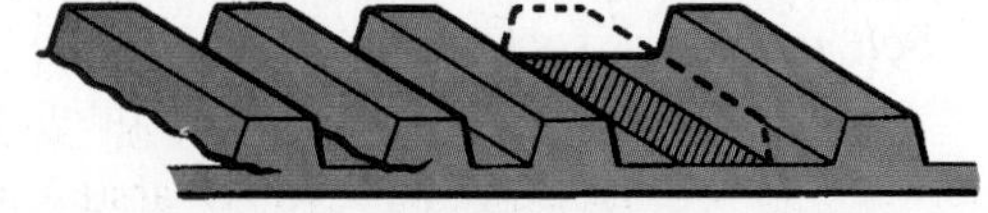

C—If timing belt teeth are missing, check for a locked component. The oil pump, injection pump, etc., could be frozen.

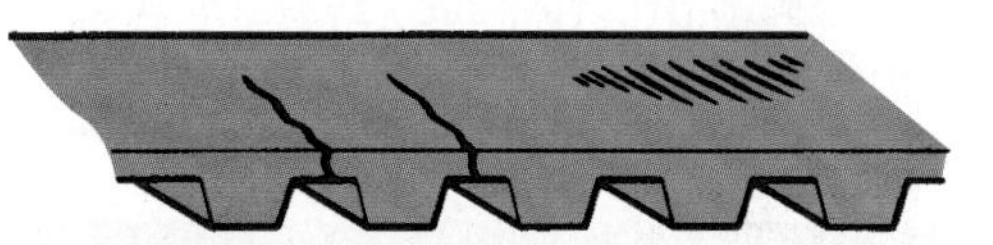

D—If there is wear or cracks on the smooth side of the belt, check the idler or tensioner pulley.

E—With damage or wear on one side of the belt, check the belt guide and pulley alignment.

F—Wear on timing belt teeth may be caused by a timing sprocket problem. Inspect sprockets carefully. Oil contamination will also cause this trouble.

Figure 3-48. *Check cogged belts for these problems. (Toyota)*

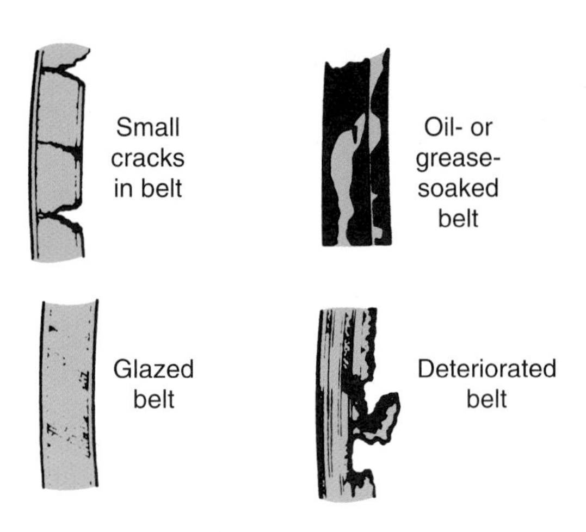

Figure 3-49. *Inspect belts closely for problems. (Sun Electric)*

as the power steering pump, air conditioning compressor, and auxiliary drive, have bearings that are exposed to oil. These belts can generally be tighter than an alternator belt. However, always follow the vehicle manufacturer's recommendations for belt tension (tightness).

Bearings

There are basically two types of bearings used in an engine—friction and antifriction. See **Figure 3-51.** A ***friction bearing*** is a plain bearing that has a smooth surface of soft metal, such as copper or babbit, on which a part slides or rotates. ***Antifriction bearings*** use balls, rollers, or needle bearings between two moving surfaces. A rolling,

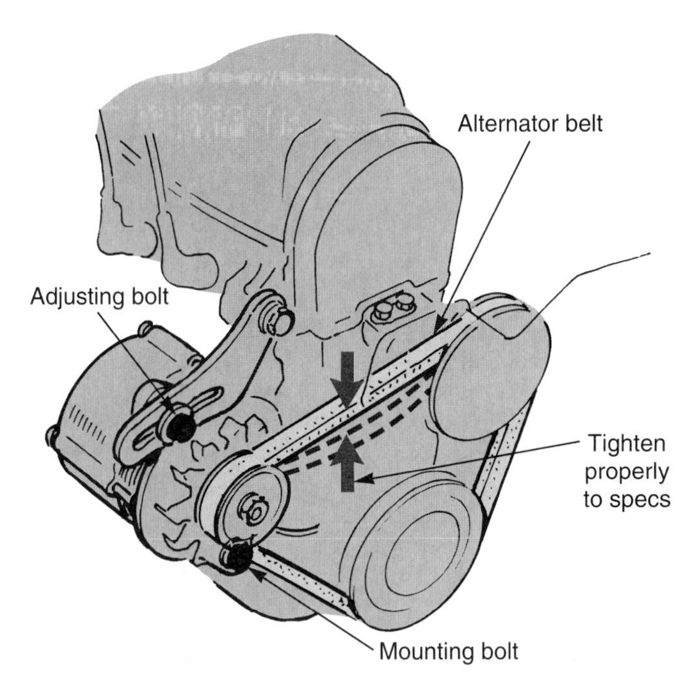

Figure 3-50. *When the alternator is driven by a separate belt, the belt should be just tight enough to prevent slippage. Always follow manufacturer recommendations for belt tension. (Honda)*

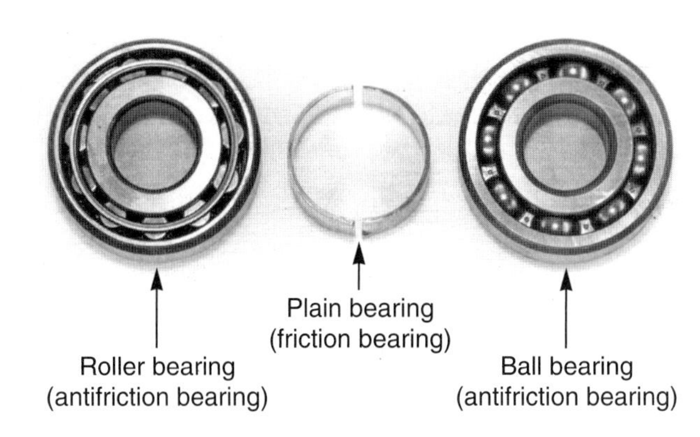

Figure 3-51. *Friction bearings are the most common type of bearing found in an engine. A few parts, however, use needle bearings, as you will learn in later textbook chapters.*

instead of sliding, action reduces friction. Oil or grease is used with both types of bearings to reduce friction.

Needle and friction bearings are commonly used in an engine. Needle bearings are used on roller lifters and sometimes the rocker arms. They are also found in engine accessories, such as the alternator and water. Friction bearings are commonly used as engine main, connecting rod, and camshaft bearings. Roller and ball bearings are used elsewhere in the vehicle.

Summary

To be a competent engine technician, you must know how to work with many types of fasteners. Fasteners include any type of holding device, from a cap screw to gasket sealer. Bolts are the most common fasteners used on engines.

Bolt size is measured across the threads. The bolt length is from the bottom of the head to the tip of the threads. Threads may be coarse or fine, either UNF/UNC or metric. Thread types must never be mixed or damage will result.

Flat washers increase the clamping surface for a bolt or nut. They also keep the head of the fastener from digging into the part. Lock washers keep the fastener from loosening during operation service.

Bolts and nuts should be torqued to specifications in a crisscross pattern when required. Many engine components must be torqued during assembly.

Many nonthreaded fasteners are found in engines. Snap rings can be used to secure piston pins, for example. A key and keyway are normally used on the front of the engine crankshaft.

Gaskets, seals, and sealants also play an important role in engine operation. Their failure could cause major engine damage. Proper service methods are essential.

Hoses and metal lines carry fuel, coolant, and other fluids. They must be well maintained to assure engine dependability. Remember the danger of fuel spray from both gasoline and diesel injection systems. Replace any hose that shows signs of deterioration. Also keep the engine accessory belt(s) in good condition.

Review Questions—Chapter 3

Please do not write in this text. Write your answers on a separate sheet of paper.

1. Define the term *fastener.*
2. When a bolt is threaded into a part, the bolt is called a(n) _____.
3. Bolt size is measured across the _____.
4. List the three types of bolt threads.
5. Turn a fastener with _____ threads clockwise to tighten.
6. This refers to the strength of a bolt.
   (A) Grade.
   (B) Compression.
   (C) Tension.
   (D) None of the above are correct.
   (E) All of the above are correct.
7. Explain the purpose of a flat washer and a lock washer.
8. Why is a crisscross pattern of tightening fasteners normally recommended for parts with multiple fasteners?
9. Briefly describe a *thread repair insert.*
10. Snap-ring pliers are needed to remove and install _____.
11. A(n) _____ fits in a slot and prevents a part from rotating on its shaft.
12. How is RTV sealer used?
13. A(n) _____ is a stationary type seal that usually fits into a groove in a component.
14. The two types of bearings are _____ and _____.
15. Which bearing is often used in roller lifters?
    (A) Roller bearing.
    (B) Friction bearing.
    (C) Needle bearing.
    (D) Ball bearing.

ASE-Type Questions—Chapter 3

1. All of the following are basic dimensions of a bolt *except:*
   (A) bolt length.
   (B) thread pitch.
   (C) thread shape.
   (D) bolt head size.
2. Technician A says that a Woodruff key is a common fastener used on an automotive engine. Technician B says that a snap ring is a common fastener used on an automotive engine. Who is correct?
   (A) A only.
   (B) B only.
   (C) Both A and B.
   (D) Neither A nor B.
3. Technician A says that metric threads can be easily confused for Unified National threads if not inspected carefully. Technician B says that a thread pitch gauge can determine if threads are metric or Unified National. Who is correct?
   (A) A only.
   (B) B only.
   (C) Both A and B.
   (D) Neither A nor B.
4. Technician A says that you should make sure that a fastener's threads are clean before torquing. Technician B says that you should view the torque wrench scale from an angle when torquing a fastener. Who is correct?
   (A) A only.
   (B) B only.
   (C) Both A and B.
   (D) Neither A nor B.
5. Technician A says that part A shown in the illustration is called a keyway. Technician B says that a key and keyway are used to prevent the gear from rotating on the shaft. Who is correct?
   (A) A only.
   (B) B only.
   (C) Both A and B.
   (D) Neither A nor B.

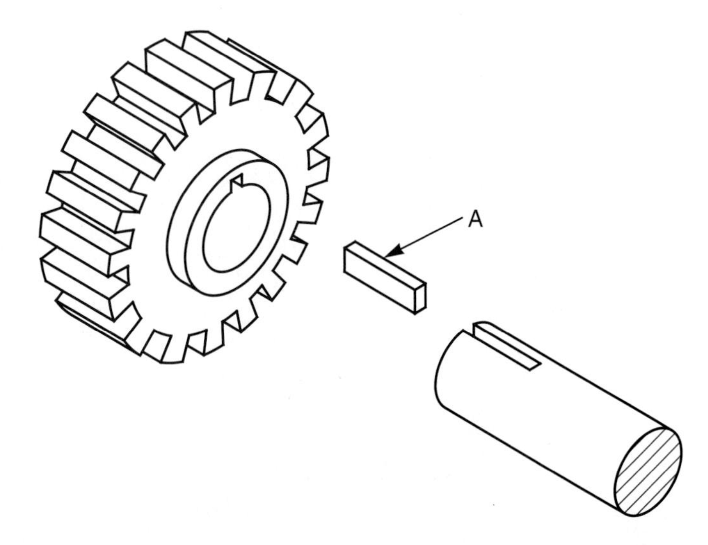

6. A die can be used to cut threads on all of the following *except:*
   (A) rods.
   (B) holes.
   (C) bolts.
   (D) shafts.
7. Technician A says that a thread repair insert will repair damaged external threads. Technician B says that a thread repair insert will repair damaged internal threads. Who is correct?
   (A) A only.
   (B) B only.
   (C) Both A and B.
   (D) Neither A nor B.
8. When installing a new gasket, you should _____.
   (A) tighten fasteners in a crisscross pattern
   (B) never overtighten the engine component fasteners
   (C) coat the new gasket with grease
   (D) Both A and B.
9. All of the following are basic rules to follow when working with automotive seals *except:*
   (A) check the size of the new seal.
   (B) use a special puller when needed to remove the old seal.
   (C) check the shaft or seal housing for damage.
   (D) coat the outside of the housing with sealer before installing a new seal.
10. Metal fuel lines are made of _____ tubing.
   (A) single-wall steel
   (B) double-wall copper
   (C) double-wall steel
   (D) single-wall copper
11. Technician A says that antifriction bearings are commonly used as engine main bearings. Technician B says that friction bearings are commonly used on roller lifters. Who is correct?
   (A) A only.
   (B) B only.
   (C) Both A and B.
   (D) Neither A nor B.

Fuel cell vehicles have complex electrical systems, as do most modern vehicles. A thorough understanding of the principles of electricity is needed to service these vehicles. (Ford)

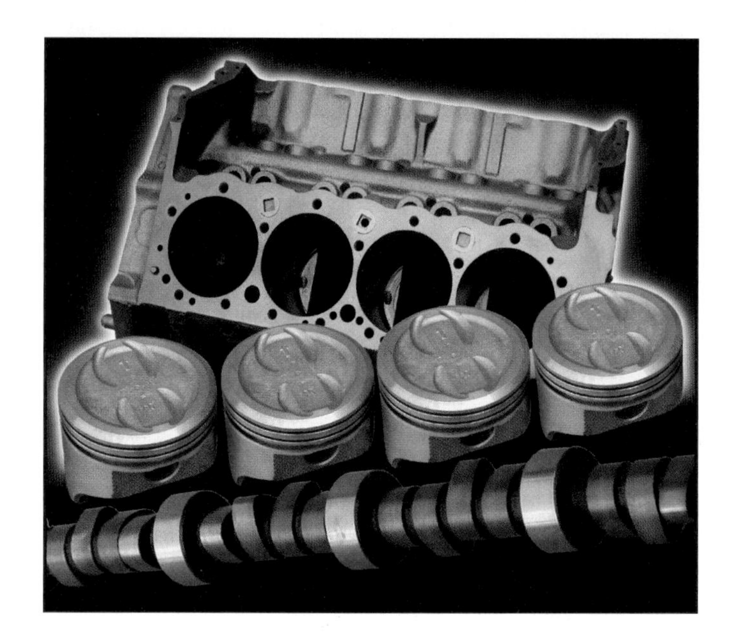

Chapter 4

Electricity and Electronics for Engines

After studying this chapter, you will be able to:

- ❑ Describe a simple electrical circuit.
- ❑ Explain the types of automotive circuits.
- ❑ Describe the function of basic electrical components.
- ❑ Explain the operation of electronic components.
- ❑ Use Ohm's law.
- ❑ Properly handle and repair automotive wiring.
- ❑ Identify electrical symbols and use wiring diagrams.
- ❑ Use electrical and electronic test instruments.

Know These Terms

Alternating current
Ammeter
Amplifier
Capacitor
Circuit
Circuit breaker
Conductor
Current
Diode
Direct current
Electricity
Electromagnet
Electronic control module (ECM)
Electrons
Flux
Frame ground circuit
Fuse
Ground wires
Induction
Insulator
Integrated circuit (IC)
Load
Ohm's law
Ohmmeter
Parallel circuit
Power source
Primary wire
Printed circuits
Relay
Resistance
Secondary wire
Series circuit
Series-parallel circuit
Short circuit
Solenoid
Switch
Test light
Transistor
Voltage
Voltmeter
Windings

Today's engines have a wide variety of electrical and electronic devices. Some of these devices monitor engine conditions. Others control engine-related systems. In general, electrical and electronic devices help make an engine more efficient and powerful because they are more precise and faster acting than mechanical devices. Due to the number of electrical and electronic devices, it is impossible to service almost any part of an engine without working with some type of wiring or electrical component.

This chapter provides a review of general, basic information related to engine electrical and electronic components. If you have not studied this topic before, it is an important chapter that will help you understand test equipment, ignition systems, charging and starting systems, computer control systems, and other subjects covered in this book.

What Is Electricity?

Atoms are the building blocks that make up everything in our universe. Your body, this book, the air, the Sun—everything is made of atoms! An atom is made up of tiny particles called protons, neutrons, and electrons.

An atom works something like the planets circling the Sun in our solar system. ***Electrons*** are negatively charged particles that circle around the center core of the atom. This center core is called the ***nucleus*** and it contains protons and neutrons. ***Protons*** are positively charged particles. ***Neutrons*** do not have an electrical charge, but contribute to the mass of the atom.

Different substances have different atomic structures, which is the arrangement of particles in the atom. Some atoms have extra electrons, called ***free electrons,*** that are not bound tightly to the atom. They can move from atom to atom. This movement of free electrons is what we call ***electricity.***

Insulators and Conductors

An ***insulator*** is a substance without free electrons; it blocks the flow of electricity. Plastic, rubber, ceramic, and air are examples of insulators. Wires are covered with insulation to prevent unwanted electrical flow.

A ***conductor*** is a substance having free electrons; it allows the flow of electricity. Metal is the most common conductor and it is used in wiring of electrical devices. Electrons can easily move through a conductor.

Current, Voltage, and Resistance

The three basic elements of electricity are current, voltage, and resistance. These components are interrelated. It is important to understand each component and how it affects the other two.

Current (abbreviated I) is the flow of electrons through a conductor. The unit for current is amperage (A). Just as water molecules flow through a garden hose, electrons flow through a wire in a circuit.

Voltage (abbreviated E) is the force or electrical pressure that causes current flow. The unit for voltage is volts (V). Just as water pressure causes water to squirt out of the end of a garden hose, voltage causes current to flow through a wire. An increase in voltage (pressure) causes an increase in current. A decrease in voltage causes a decrease in current. Most vehicles have a 12 V (actually 12.5 V) electrical system.

Resistance (abbreviated R) is the opposition to current flow. The unit for resistance is ohms (Ω). As an example, when current flows through a lightbulb, the electrons rub against the atoms in the bulb filament, which is a resistance wire inside the bulb. This "electrical friction," or resistance, heats the filament and makes it glow.

Resistance is needed to control the flow of current in a circuit. Just as the on-off valve on a garden hose can be opened or closed to control water flow, circuit resistance can be increased or decreased to control the flow of electricity. High resistance reduces current when voltage remains constant. Low resistance increases current when voltage remains constant.

Circuits

A ***circuit*** is a complete pathway through which electricity can travel. There are various types of fundamental circuits. It is important for you to understand the configuration of each when working with engine-related electrical devices.

Simple Circuit

A simple circuit is made up of a power source, conductors, and a load. These are the minimum requirements for a circuit, **Figure 4-1.** For example, the vehicle's battery or alternator is a power source, the wiring and metal frame of the vehicle are conductors, and a lightbulb is a load.

The ***power source*** provides the force that moves electrons through the circuit. See **Figure 4-2.** A circuit power source may be the battery or the alternator (charging system).

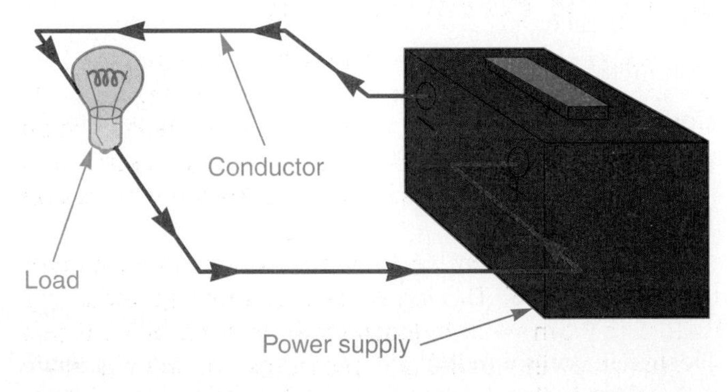

Figure 4-1. *To make a basic electric circuit, you need a power source, conductors or wiring, and a load.*

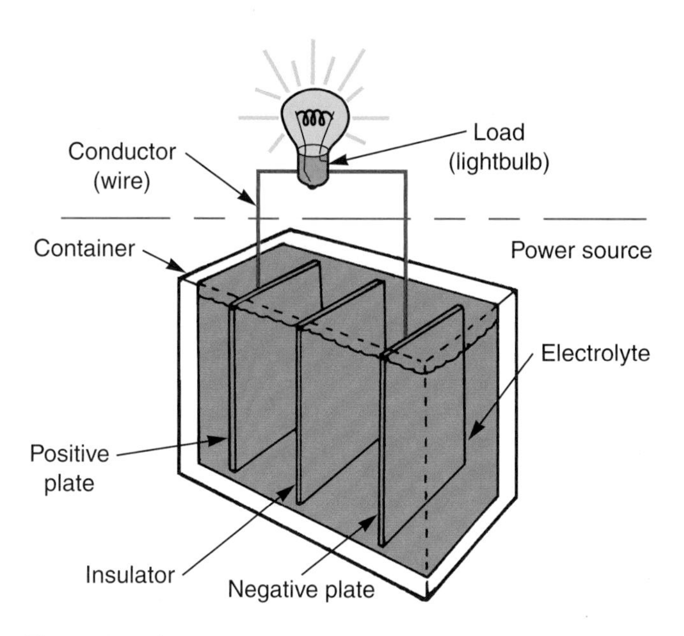

Figure 4-2. *A battery can be used as a power source for a circuit. The electrolyte (battery acid) produces a chemical reaction that makes electrons flow between the plates.*

The ***load*** can be any resistance element in the circuit, such as a lightbulb, electric motor, radio, and so on. The load is what performs the function of the circuit. For example, the load may be an electric cooling fan. The function of this circuit is to draw air through the radiator and protect the engine from overheating.

The ***conductors*** in the simple circuit are metal wires or the metal parts of the vehicle. They carry current from the power source to the load and back to the power source.

Frame Ground Circuit

A ***frame ground circuit,*** also called one wire circuit, uses the vehicle's metal unibody or frame as a conductor to complete an electrical circuit. This practice is common because it saves wiring. Wires do need to be run as return conductors to the battery. A frame ground circuit is shown in **Figure 4-3.**

Series and Parallel Circuits

A ***series circuit*** only has one path for current flow. Two or more loads are wired into this single path, **Figure 4-4A.** When two loads are wired in series, they are dependent on each other. For example, if two lightbulbs are in series and one burns out, the circuit is broken and the other bulbs stop working.

A ***parallel circuit*** has more than one path for current flow, **Figure 4-4B.** The loads are wired into separate legs (paths) and can work independently of each other. If two lightbulbs are in parallel and one burns out, the circuit for the second bulb is not broken and that bulb will continue to work.

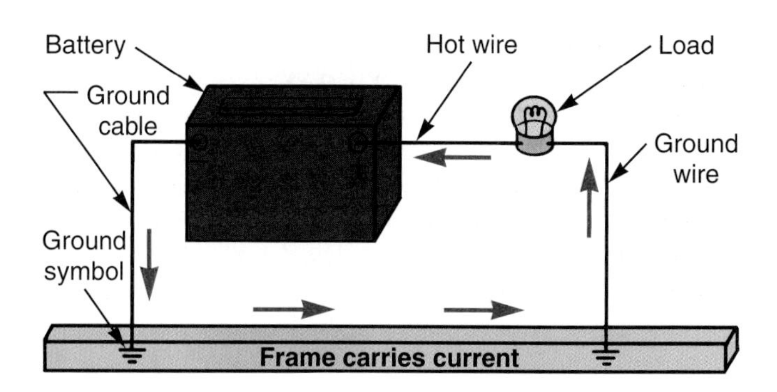

Figure 4-3. *A frame ground circuit is commonly used in vehicles. The metal frame or unibody structure is used as a conductor to complete the circuit.*

A ***series-parallel circuit*** is a circuit consisting of at least one series circuit and one parallel circuit. Refer to **Figure 4-4C.** This type of circuit has at least three loads, two in parallel and two in series.

All of these circuit types can be found in engine-related wiring. You must remember how series circuit loads affect each other and how a parallel circuit has independent legs that have little effect on each other.

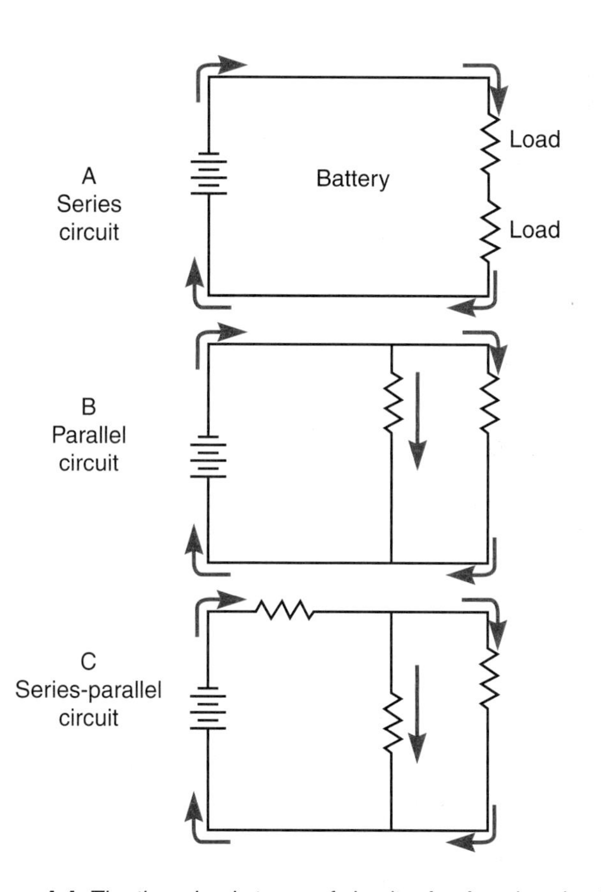

Figure 4-4. *The three basic types of circuits. A—A series circuit has only one path in which current can flow. B—A parallel circuit has more than one path in which current can flow. C—A series-parallel circuit is a combination of a series and a parallel circuit.*

Ohm's Law

Ohm's law is a basic formula for calculating volts, amps, and ohms in electrical circuits. If you know two values, you can use Ohm's law and basic math to find the missing value.

Figure 4-5 shows an Ohm's law pie chart that represents the use of the formula. Note the location of E (voltage), I (current), and R (resistance) in the pie chart. Voltage (volts) is above current (amps) and resistance (ohms). The basic formula in Ohm's law is:

Resistance = Voltage ÷ Current

or

Current = Voltage ÷ Resistance

or

Voltage = Current × Resistance

For example, suppose you know that a circuit has 12 volts applied and its load is 10 ohms. To find the current (amps) being drawn by the circuit, divide 12 by 10 to get 1.2 amps. If a 5 ohm load is placed in series with the 10 ohm load, divide 12 by 15 to get .8 amps. When in series, ohms are added to get the total load. See **Figure 4-6.**

Magnetic Field

You are probably familiar with magnetism from using a simple permanent magnet, such as a refrigerator magnet.

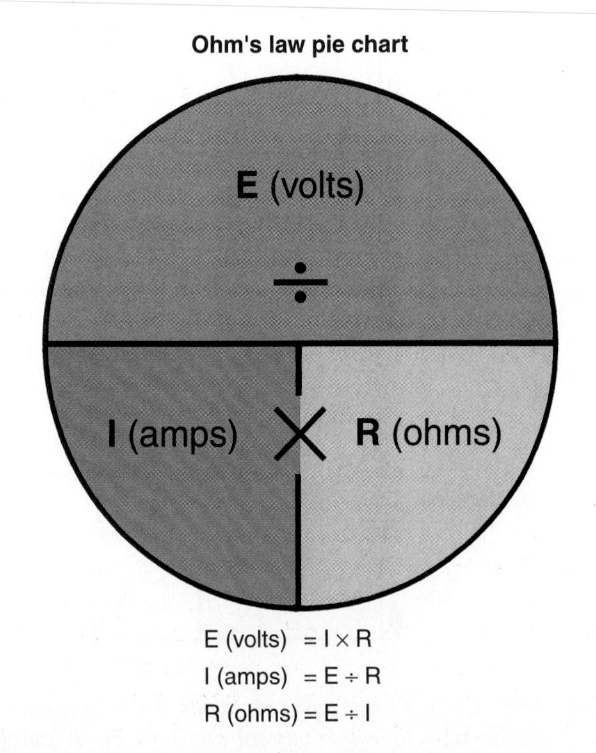

Figure 4-5. *This is an Ohm's law pie chart that represents the formulas for calculating voltage, current, or ohms in a circuit when two of the three values are known.*

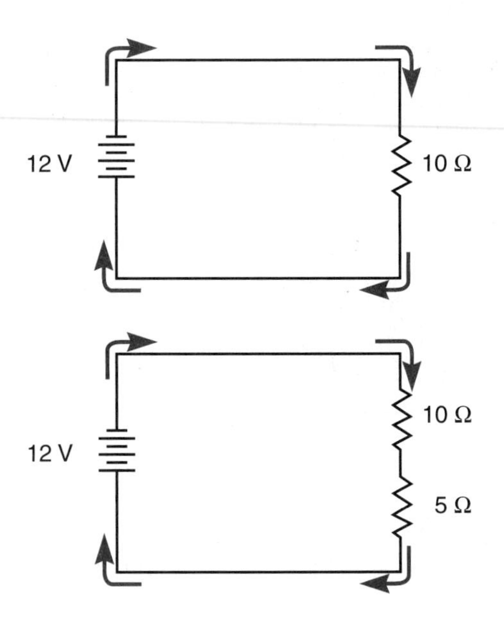

Figure 4-6. *These two simple circuits show the relationship of resistance, current, and voltage. Use Ohm's law to calculate the amperage in each circuit. As resistance increases, current drops.*

A magnet produces an invisible magnetic field that attracts iron-containing (ferrous) objects. The magnetic field, or ***flux,*** is formed from invisible lines of force.

A magnetic field can also be created using electricity. A long piece of wire can be wound into a coil. When the ends of the wire are connected to a power source and current passes through the wire, a magnetic field is produced. To make the field stronger, a soft iron bar or core can be inserted into the center of the coil. The iron core is magnetized when current flows through the wire, making an ***electromagnet.***

A magnetic field can be used to create electricity. If a magnetic field is passed over a wire, an electric current is generated, or induced, in the wire. As the wire cuts through the lines of force (flux), electricity flows through the wire. This action is called ***induction.***

Many engine-related components use magnetism. Electronic fuel injection, electric motors, relays, ignition systems, and on-board computers are just a few examples.

Electrical Components

Electrical components are devices that use electricity and moving parts to accomplish a task. Mechanical switches, fuses, and circuit breakers are examples of electrical components. These devices are commonly used and you must understand them fully.

Switch

A ***switch*** allows an electric circuit to be manually turned on or off, **Figure 4-7.** When the switch is closed (on), the circuit is complete (fully connected) and will

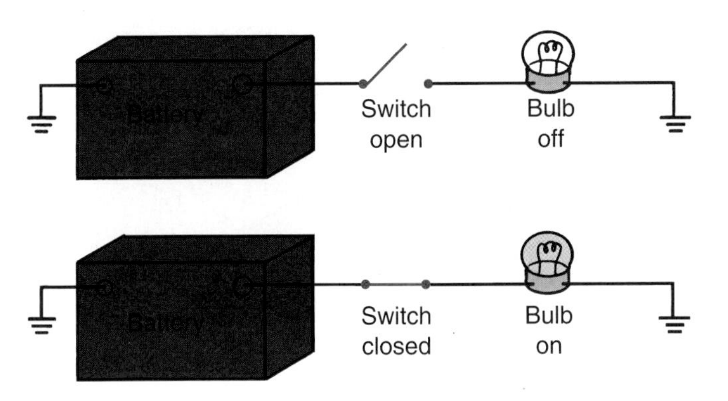

Figure 4-7. *A switch provides means of breaking the path for current in a circuit.*

operate. When the switch is open (off), the circuit is broken (disconnected) and does not function.

Fuse

A ***fuse*** protects a circuit against damage caused by excessive current draw (overcurrent), such as that created by a short circuit. A ***short circuit*** or "short" is caused when a defective wire or component touches ground and causes excess current flow. As shown in **Figure 4-8,** the fuse link will melt to stop current flow and prevent any further circuit damage.

A ***fuse box*** is often located under the vehicle's dashboard. It contains fuses for the various circuits. One is pictured in **Figure 4-9.**

Warning: If a short to ground exists in a circuit, unlimited current flow will result. This can cause an electrical fire. Therefore, never install a circuit, such as for aftermarket lights, without overcurrent protection.

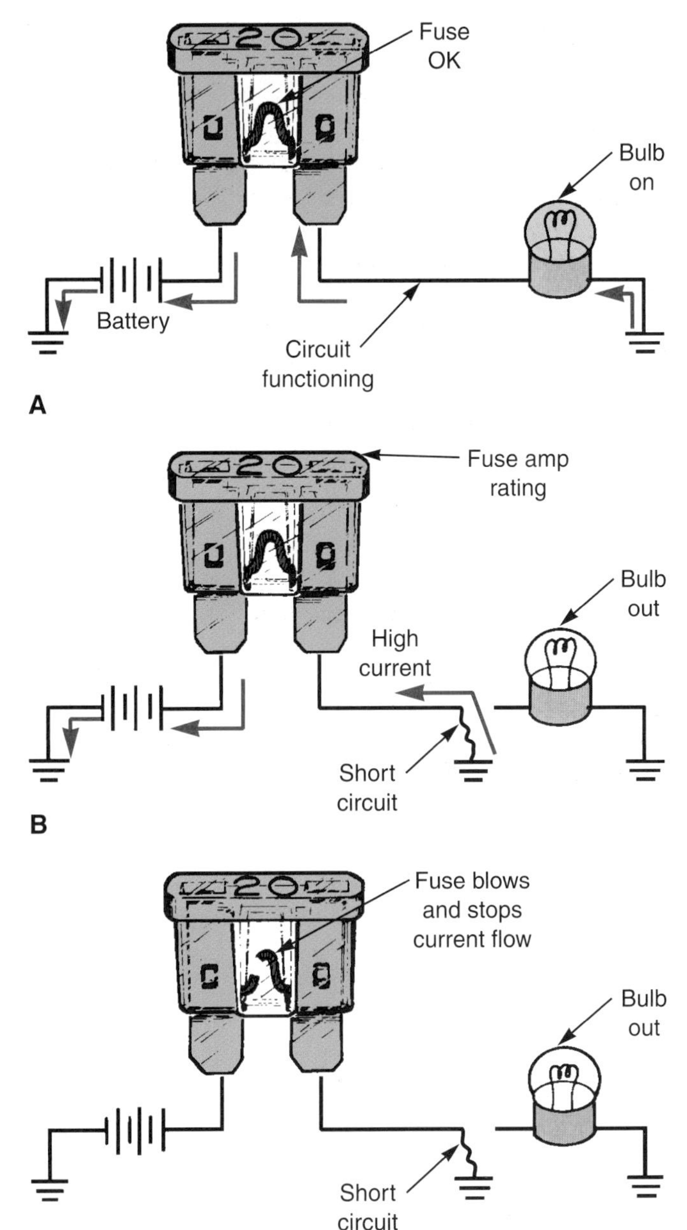

Figure 4-8. *A fuse provides overcurrent protection. A—The circuit is functioning normally. B—A short to ground occurs, resulting in high current in the circuit. C—When the current in the circuit exceeds the rating of the fuse, the filament melts and current stops flowing.*

Circuit Breaker

A ***circuit breaker*** performs the same function as a fuse. It disconnects the power source from the circuit when current becomes too high. Excess current heats a bimetal strip and causes it to bend, opening a set of breakers (contact points). Normally, a circuit breaker will automatically reset itself when current returns to normal levels and the bimetal strip cools. Circuit breaker action is shown in **Figure 4-10.**

Electric Motor

A basic electric motor can be created from a loop of wire and a permanent magnet. This is illustrated in **Figure 4-11.** When current is passed through the wire loop, a magnetic field forms around the wire. In other words, an electromagnet is formed. The north pole of the electromagnet is attracted to the south pole of the permanent magnet. This forces the loop of wire to move. By reversing the current flow through the wire loop, the poles of the electromagnet are reversed. The north pole of the electromagnet, which is now on the opposite side of the electromagnet, is attracted to the south pole of the permanent magnet. The loop of wire again moves so the poles align. In this way, rotary motion is produced.

In an electric motor, many loops of wire are formed from one continuous wire. These loops are called the ***windings.*** The poles of the permanent magnet are called the ***pole pieces*** or pole shoes. Electricity is transferred into

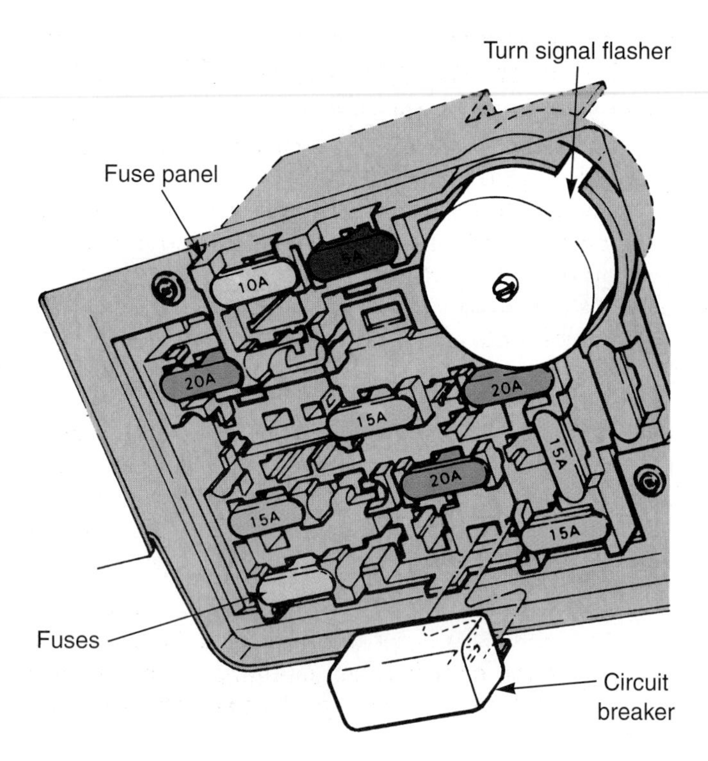

Figure 4-9. *The fuse box is normally installed under the dash. (Ford)*

the windings through brushes and a commutator. The ***commutator*** is a conducting ring that spins with the windings. It is connected to the two ends of the windings.

Relay

A ***relay*** is an electrically operated switch, **Figure 4-12.** It is often used to provide on-off control in a high current or voltage circuit using a lower current or voltage circuit. A relay is basically an electromagnet and a set of contact points. When an electric current passes through the electromagnet's coil, the contact points close. When the current stops flowing through the electromagnet, the contact points open.

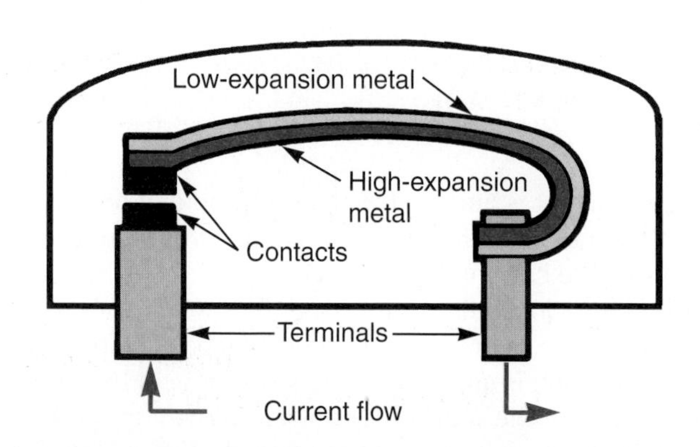

Figure 4-10. *A circuit breaker provides overcurrent protection for a circuit. Excess current heats the dissimilar metals at different rates, which makes the arm bend to disconnect the contacts. Unlike a fuse, however, a circuit breaker resets itself when current returns to a normal level. (Ford)*

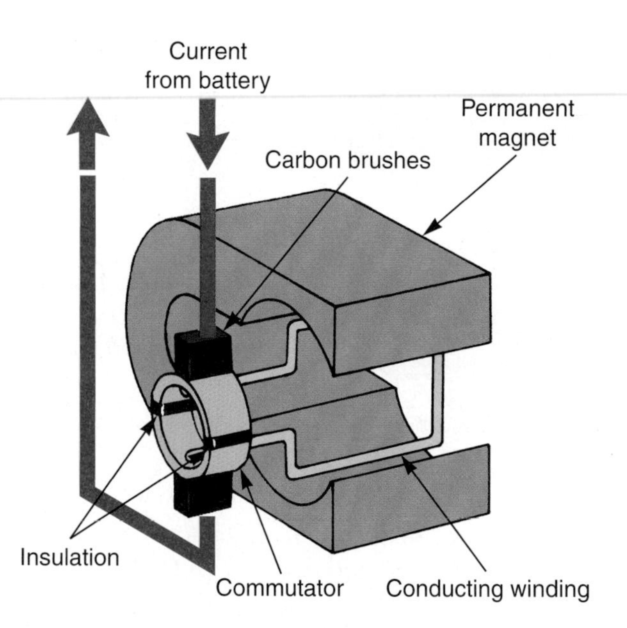

Figure 4-11. *Study the principle of an electric motor. The magnetic field of the permanent magnet acts on the magnetic field produced when current flows through the loop of wire. (Robert Bosch)*

A relay allows a small dash switch to control another circuit by remote control. Small, low-current wires can be used behind the dash while larger, high-current wires are used in the relay-operated circuit. See **Figure 4-13.**

Solenoid

A ***solenoid*** is an electromagnet with a sliding core, called a ***plunger.*** See **Figure 4-14.** When current flows through the electromagnet, the plunger is pulled into the solenoid. When current stops flowing through the electromagnet, the plunger is free to move out of the solenoid. Usually, a spring forces the plunger out of the solenoid.

As you will learn in later chapters, there are many applications for solenoids. A solenoid is commonly used on an engine starting motor. Also, a fuel injector is basically a solenoid used to meter fuel.

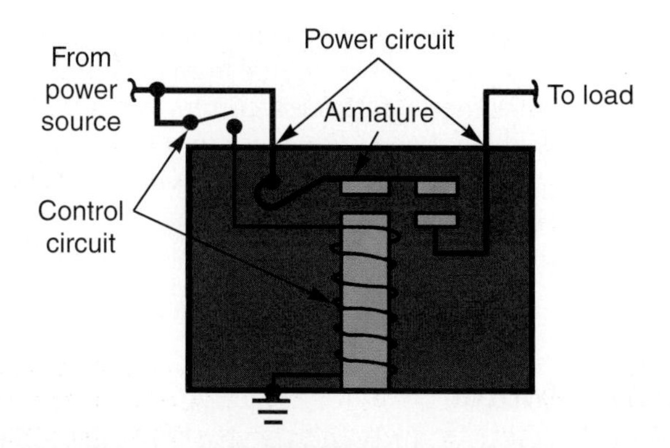

Figure 4-12. *A relay allows a low-voltage circuit to control a high-voltage circuit. As current flows through the low-voltage coil, a magnetic field is produced that pulls the armature down. This completes the high-voltage circuit and current flows through it. (Ford)*

Figure 4-13. *Relays can be located under the dash, a back seat, or in the engine compartment. The function of each relay is described in the service manual.*

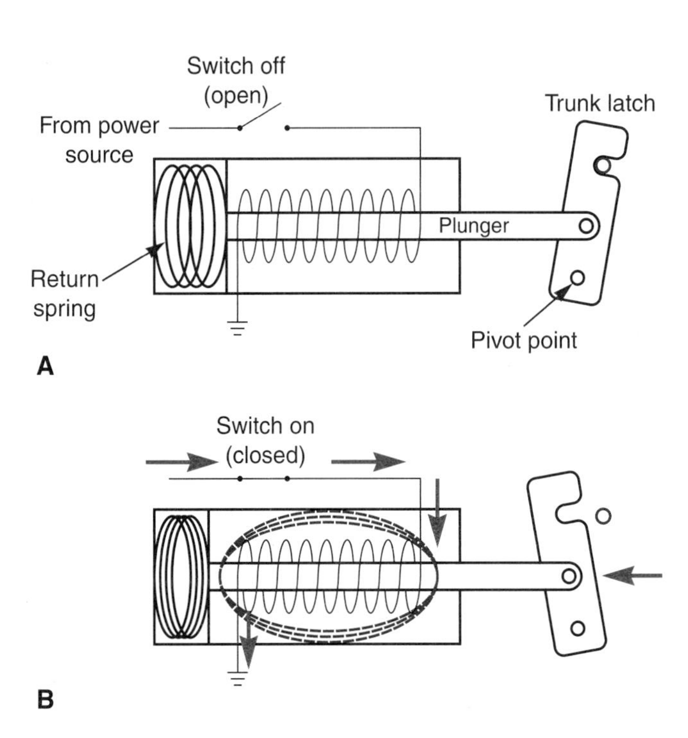

Figure 4-14. *Solenoid is an electromagnet with a movable core (plunger). Solenoids are often used to move other components, such as the trunk latch shown here. A—When the circuit is open, the return spring forces the plunger out of the solenoid and the latch is engaged. B—When the circuit is closed (the dash switch is pressed), current flows through the coil and a magnetic field is formed around the plunger. This field pulls the plunger into the coil and the latch is released.*

Alternating and Direct Current

Alternating current (ac) changes from positive polarity to negative polarity and the current flows back and forth through the circuit. ***Direct current*** (dc) is always the same polarity, either positive or negative, and the current only flows in one direction. Direct current is commonly used throughout an automobile.

A comparison of ac and dc is given in **Figure 4-15.** This illustration represents what you would see on an oscilloscope. When voltage is zero or off, the line (trace) is on the base or zero level. When voltage has positive polarity, the line moves up on the screen. The line moves down from zero for a negative polarity. Notice how the dc waveform is either zero or a positive polarity. The ac waveform varies from a positive polarity, to zero, to a negative polarity.

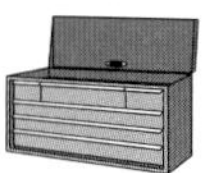

Note: The subject of ac and dc is covered in later chapters on charging systems and testing for engine performance problems.

Electronic Devices

Electronic devices, or solid state devices, do not use moving parts. They use special substances called ***semiconductors*** that can change from an insulator to a

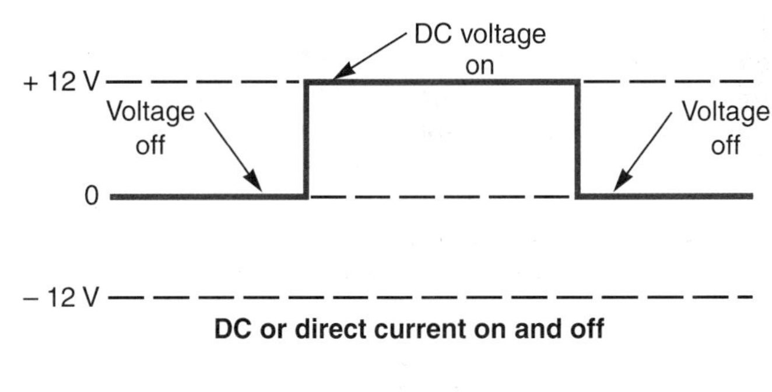

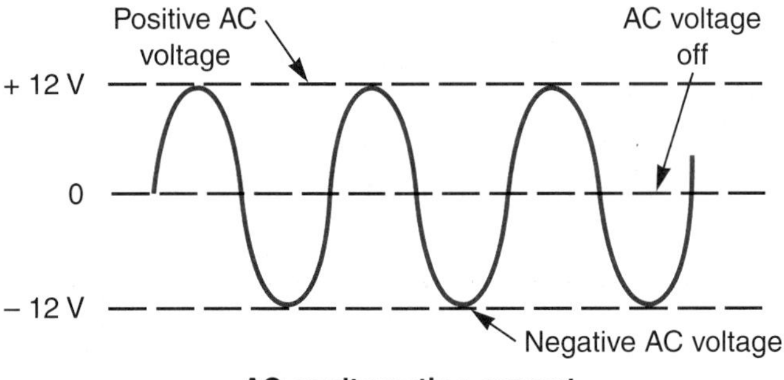

Figure 4-15. *A—Direct current always has either positive or negative polarity, but not both. B—Alternating current changes from positive to negative polarity. Note: Direct current is shown here as a square waveform, but it may occur as a sine waveform, as alternating current is shown here.*

conductor with electrical stimulation. To be a competent engine technician, you must be familiar with the operation of common electronic devices.

Diode

A ***diode*** allows current to flow in only one direction. This device can be thought of as an electronic check valve. When forward biased, current is entering the diode in the correct direction and the diode acts as a conductor. Current flows through the diode. When reverse biased, current is trying to enter in the wrong direction and the diode acts as an insulator. It stops current from passing through the circuit. A diode is commonly used to change alternating current into direct current. See **Figure 4-16.**

Transistor

A ***transistor*** performs the same basic function as a relay; it acts as a remote control switch. It is much more efficient than a relay, however. A transistor can sometimes turn on and off several times a second. It does this without using moving parts that can wear and deteriorate.

Look at **Figure 4-17.** A transistor amplifies (increases) a small control or base current. The small base current energizes the semiconductor material, changing it from an insulator to a conductor. This allows the much larger circuit current to pass through the transistor.

Capacitor

A ***capacitor,*** or condenser, is a device used to absorb unwanted electrical pulses in a circuit, such as voltage fluctuations. Capacitors are used in various types of electrical and electronic circuits.

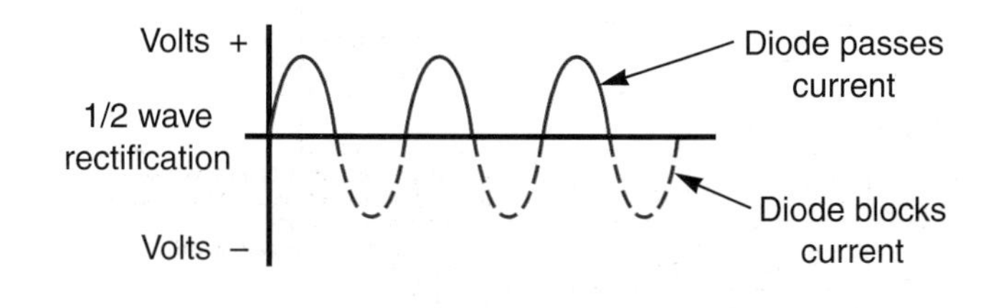

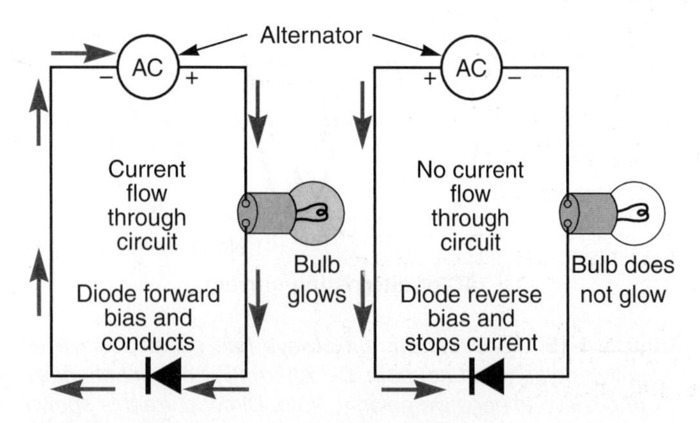

Figure 4-16. *A diode allows current pass in only one direction. A diode can be used to change ac into dc.*

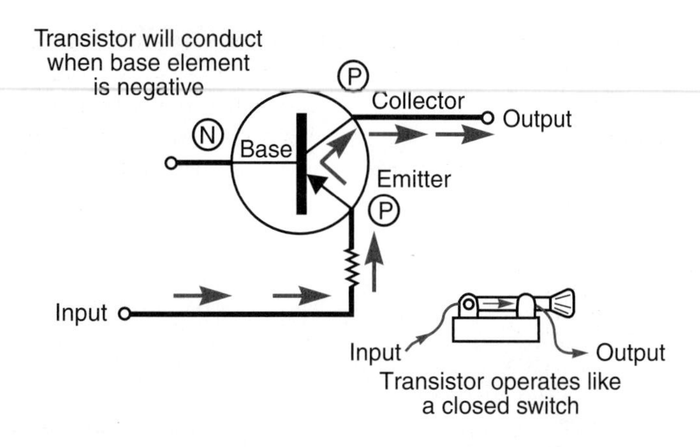

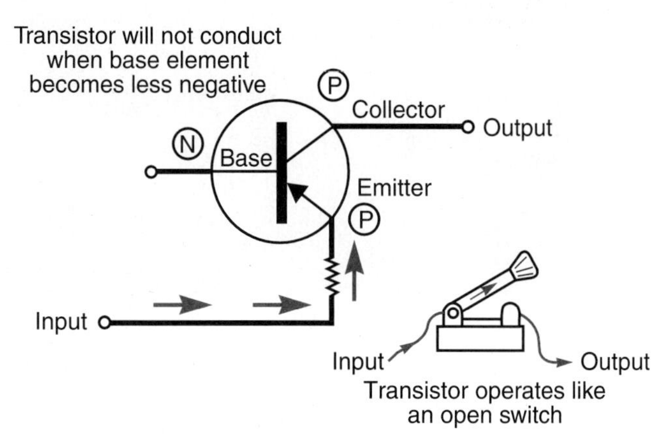

Figure 4-17. *A transistor is a solid state device that performs the same function as a relay. A low base current can be used to turn the transistor on and allow a higher current to flow. (Motorola)*

Integrated Circuit

An ***integrated circuit (IC)*** contains miniaturized diodes, transistors, resistors, and capacitors in a single chip. The chip is a small plastic housing with metal terminals. Integrated circuits are used in very complex electronic circuits; computers for example.

Printed Circuit

Printed circuits do not use conventional, round wires. They have flat conductor strips mounted on an insulating board. Printed circuits are normally used in place of wires on the back of the instrument panel. This eliminates the need for a bundle of wires going to the engine indicators, gauges, and instrument bulbs.

Amplifier

An ***amplifier*** is an electronic circuit designed to use a very small current to control a very large current. Its function is much the same as a transistor. However, higher output currents and smaller control currents are possible.

A good example of an amplifier is the ignition control module. This is an amplifier that uses small electrical pulses from the computer to produce strong on-off cycles to operate the ignition coil(s).

Computer

A computer or ***electronic control module (ECM)*** uses very complex electronic circuits to perform various functions, **Figure 4-18.** It uses input from engine sensors to gather information about engine efficiency. It is programmed

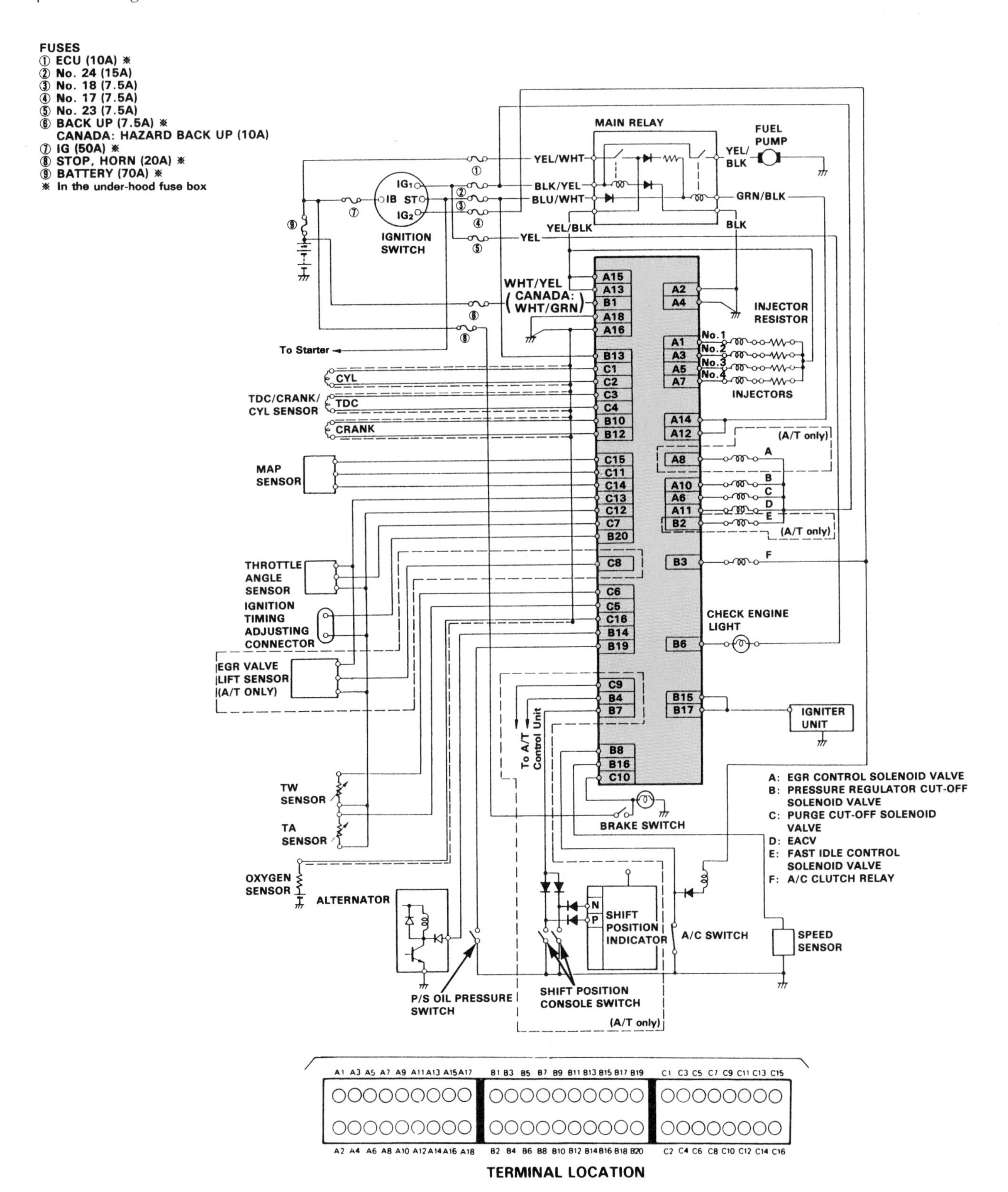

Figure 4-18. *Computer and engine sensors are used to increase engine efficiency. Various sensors provide data to the ECM. Study the location of major components and how they connect to the computer and each other. (Honda)*

to send out electrical signals to relays, solenoids, and other components to affect engine operation.

An integrated computer system uses feedback data from several subsystems, such as the engine, brakes, suspension, and traction control systems, to better control each individual system. All systems are controlled from a central processor instead of each system being controlled by an individual processor. In this way, the computer monitors all functions of the vehicle and can better decide how to control the engine, drive train, braking, traction control, ride stiffness, and so on.

Sensors

Sensors are electronic devices that change their voltage output or resistance with a change in a condition. They are mounted in various locations on the engine and other systems. Look at how the sensors are connected to the ECM in the schematic shown in **Figure 4-18.** A few of the most common engine sensors are:

- **Oxygen (O_2) sensor.** Measures the amount of oxygen in the engine exhaust. It changes voltage output with a change in oxygen content. The amount of oxygen in the exhaust indicates combustion efficiency. The computer can adjust the air-fuel mixture to increase combustion efficiency.
- **Manifold absolute pressure (MAP) sensor.** Monitors the vacuum in the engine intake manifold in relation to outside air pressure. Engine manifold vacuum changes with engine load.
- **Throttle position sensor (TPS).** Changes resistance as the engine throttle valve is opened and closed. The computer can alter ignition timing, fuel mixture, and other variables as needed based on how much power is requested by the driver.
- **Engine coolant temperature (ECT) sensor.** Changes resistance as engine coolant temperature changes. This allows the computer to richen the air-fuel mixture when the engine is cold.
- **Airflow sensor.** Measures the amount of air flowing into the engine. This lets the computer process data on engine speed, air density, and air temperature.
- **Crankshaft position sensor.** Produces an electrical signal corresponding to engine speed.

Other engine sensors are also used: knock sensors (KS), oil pressure sensors, intake air temperature (IAT) sensors, etc. They all use similar principles and are explained in more detail later in this book.

Automotive Wiring

A vehicle has various types of wiring in its many electrical systems. To be an engine technician, it is important to learn the different types, how they are used, and how to service them.

A ***wiring harness*** is a group of wires enclosed in a plastic or tape covering. This covering helps protect and organize wires. See **Figure 4-19.**

Wire Size

Wire size is determined by the diameter of the wire's metal conductor. The diameter is stated in gauge size, which is a number system. The larger the gauge number, the smaller the diameter of the wire conductor.

Wire Types

Primary wire is small and carries battery or alternator voltage. Primary wire normally has plastic insulation to prevent shorting. The insulation is usually color coded so that different wires are marked with different colors, **Figure 4-20.** This lets you trace (follow) wires that are partially hidden.

Secondary wire, also called high-tension cable, is only used in a vehicle's ignition system. It has extra thick insulation and carries high voltage from the ignition coil to the spark plug, **Figure 4-21.** The conductor, however, is designed to carry a very low current.

Battery cable is extremely large-gauge wire capable of carrying high current from the battery to the starting

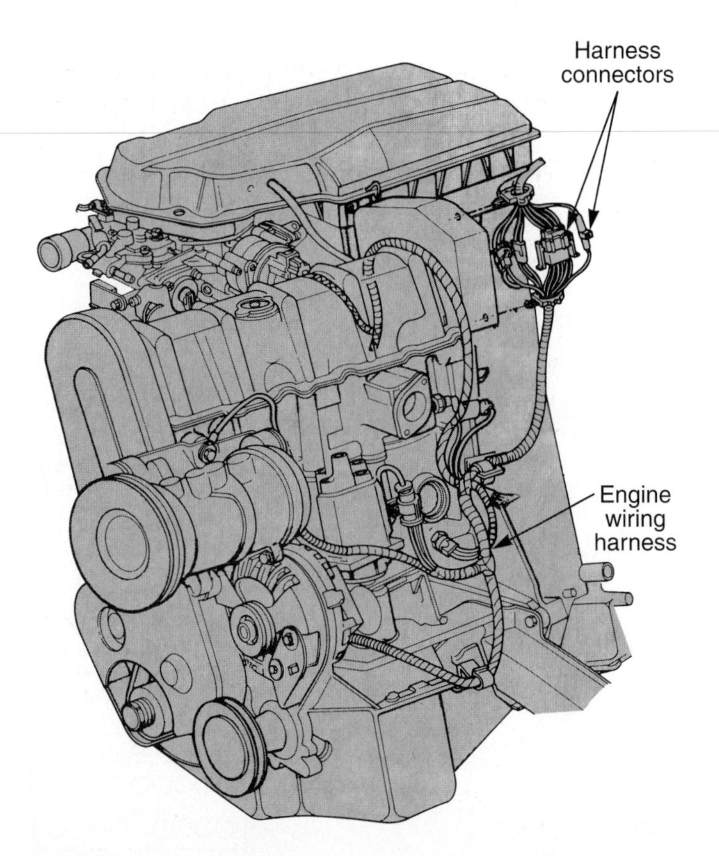

Figure 4-19. *The engine wiring harness surrounds and protects wires for engine sensors, ignition system, and other components. Plastic connectors are used to join sections of wires. (DaimlerChrysler)*

Code	Color
B	Black
Br	Brown
G	Green
Gy	Gray
L	Blue
Lb	Light blue
Lg	Light green
O	Orange
R	Red
W	White
Y	Yellow

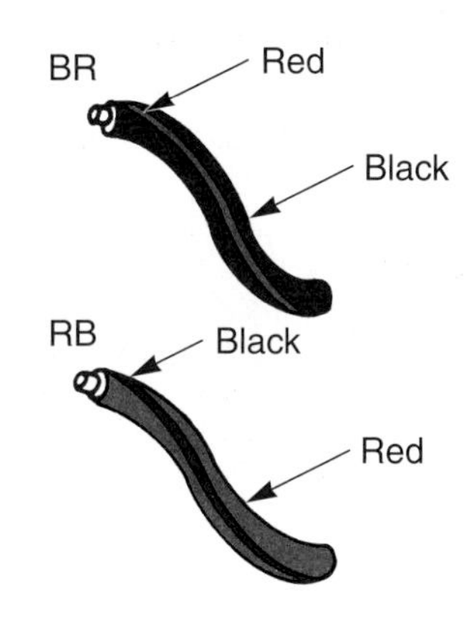

Figure 4-20. *Primary wires are normally color coded. This allows you to trace a wire through the vehicle. Wiring diagrams give abbreviations for the color coding.*

motor. Usually, a starting motor draws more current than all of the other electrical components combined, normally over 100 amps. Battery cable carries battery and charging system voltage, about 12.5 to 14.5 volts.

Ground wires or ground straps connect electrical components to the chassis or ground of the vehicle. These wires are not usually insulated since they connect circuits or parts to ground.

Note: When disconnecting wiring harness connectors, make sure you use approved methods. Most modern electrical connectors lock together with plastic or metal clips. You must release the clips to disconnect the wiring. *Figure 4-22* shows examples.

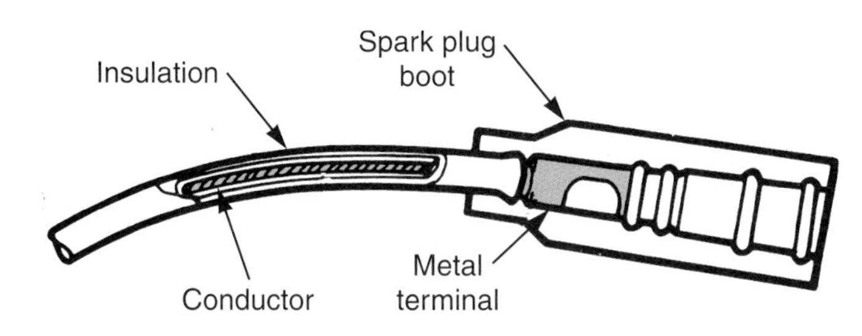

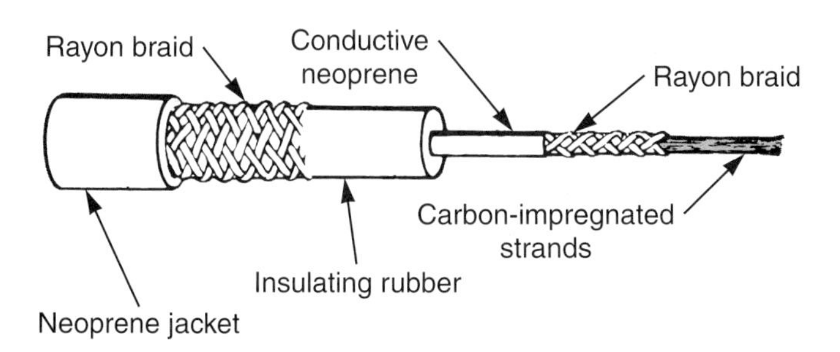

Figure 4-21. *Secondary wire has a large diameter to carry high voltage. Most secondary wires have carbon-impregnated strands that provide some internal resistance to prevent radio frequency (RF) interference. (Champion Spark Plugs)*

Wiring Repairs

When replacing a section of wire, always use wire of equal size. If a smaller wire is used, the circuit could malfunction due to high resistance. Undersize wire may heat up and melt resulting in an electrical fire.

Crimp connectors and ***crimp terminals*** can be used to quickly repair wiring. Several different types are shown in **Figure 4-23.** Terminals allow a wire to be connected to an electrical component. Connectors or splicers allow a wire to be connected to another wire. ***Crimping pliers*** are used to deform the crimp connector or terminal onto the wire. **Figure 4-24A** shows a technician installing a crimp terminal.

A ***soldering gun*** or iron can also be used to permanently fasten wires to terminals or to other wires, **Figure 4-24B.** The soldering gun produces enough heat to melt solder

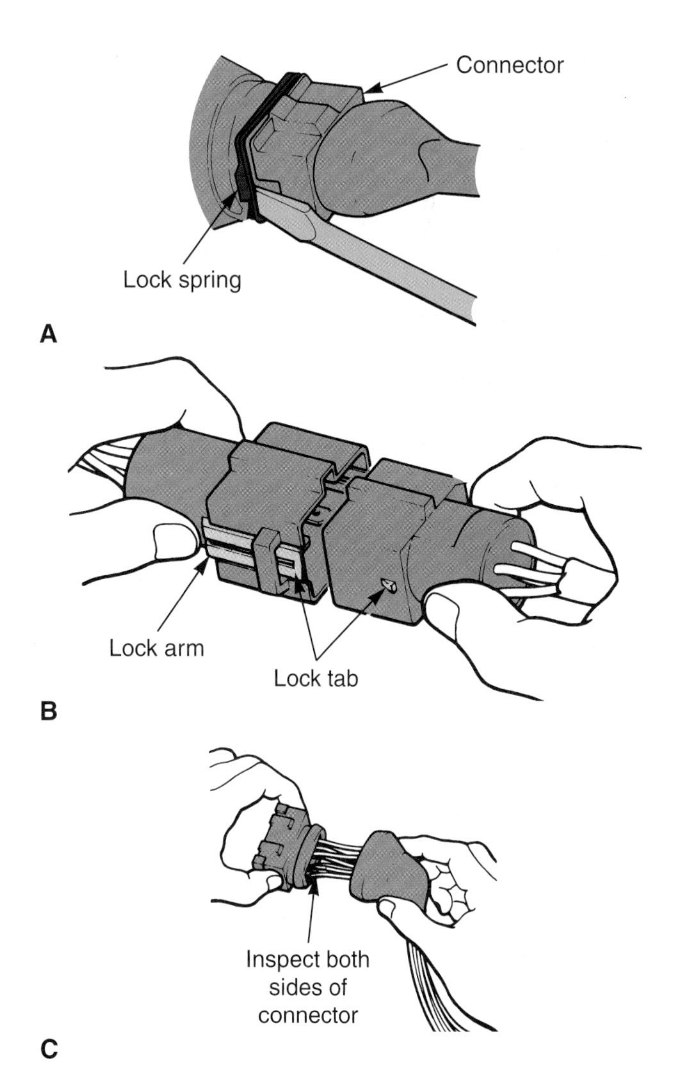

Figure 4-22. *Always disconnect the wiring harness connectors properly. A—This connector has a metal lock spring that keeps the connector from pulling apart. It must be released before disconnecting the connector. B—This connector has plastic arms that lock around tabs. You must squeeze arms to free the connector. C—When inspecting a connector, check for corrosion on the terminals and for moisture in either side of the connector. (Honda)*

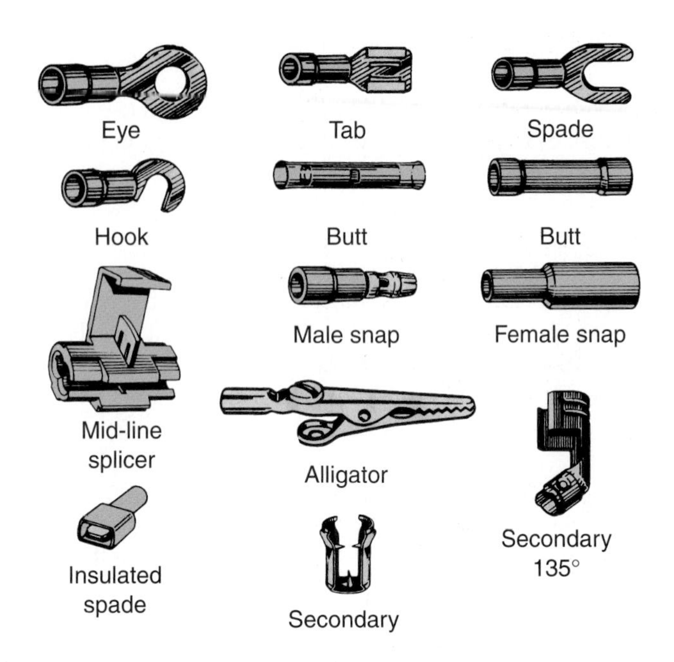

Figure 4-23. *There are various types of wire terminals and connectors. (Belden)*

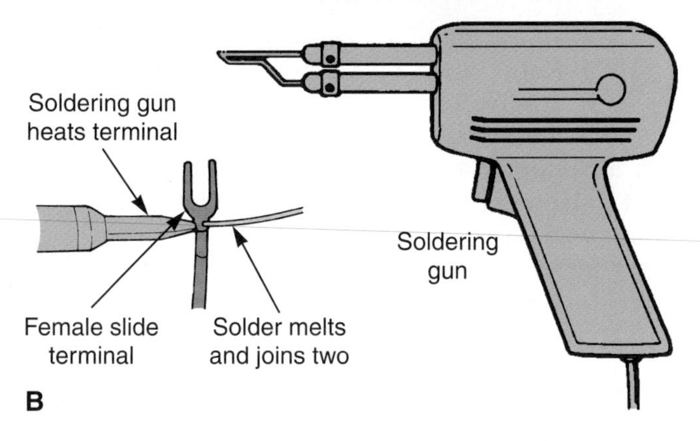

Figure 4-24. *Wire repairs can be done with crimp connectors or soldering gun. A—Strip the insulation from the wire so that the connector can make contact with the wire. Use crimping pliers to form the connector around the wire. B—A soldering gun can be used to heat the wire and terminal. Then, solder is melted onto the preheated connection. (Klein Tools, Florida Voc. Ed.)*

(a lead-tin alloy). The soldering gun is touched to the wire and other component to preheat them. Then, the solder is touched to the heated joint; the solder melts. When cooled, the solder makes a solid connection between the electrical components.

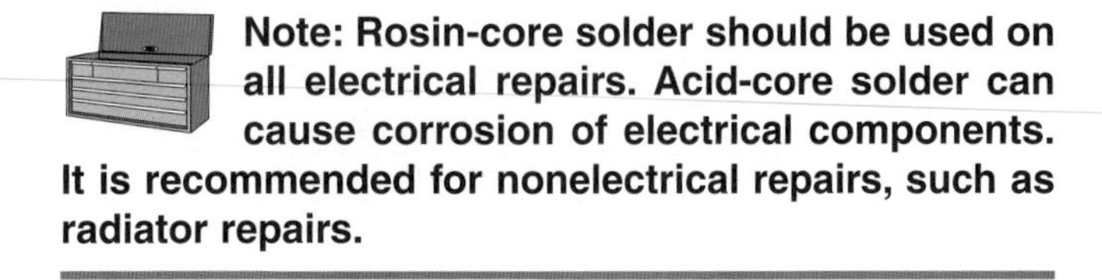

Note: Rosin-core solder should be used on all electrical repairs. Acid-core solder can cause corrosion of electrical components. It is recommended for nonelectrical repairs, such as radiator repairs.

Electrical Component Location Chart

Today's cars have dozens of electrical and electronic components in the engine compartment. A ***component location diagram*** illustrates where each electrical or electronic part is positioned in the engine compartment. One is shown in **Figure 4-25.** Using this chart, you can quickly find a component that is suspected of causing trouble.

Wiring Diagram

A ***wiring diagram*** is a schematic that shows how electrical components are connected by wires, **Figure 4-26.** A wiring diagram serves as an "electrical road map" that helps the technician with difficult repairs.

Wiring diagrams use ***electrical symbols*** to represent the electrical components in a circuit. The lines on the diagram represent the wires and connect to the symbols. In this way, you can trace each wire and see how it connects to each component. **Figure 4-27** shows several electrical symbols.

Basic Electrical Tests

Various electrical tests and testing devices are used by an engine technician. You should have a general understanding of these tools and how to use them.

Jumper Wire

A ***jumper wire*** is handy for testing switches, relays, solenoids, wires, and other nonresistive components. The

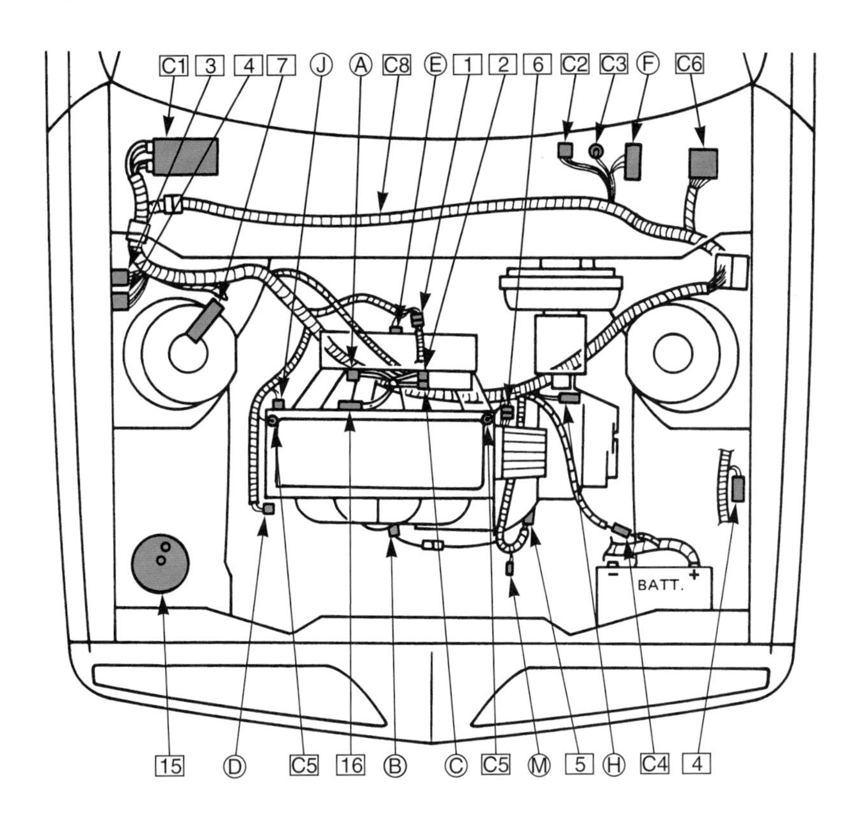

Computer System
- C1 Electronic control module (ECM)
- C2 ALDL connector
- C3 "Check engine" light
- C4 System power
- C5 System ground
- C6 Fuse panel
- C8 Computer control harness

Air/Fuel System
- 1 Fuel injectors jumper
- 2 Idle air motor
- 3 Fuel pump relay
- 4 Cooling fan relay

Transmission Converter Clutch Control System
- 5 Trans. conv. clutch connector

Ignition System
- 6 Electronic spark timing connector
- 7 Electronic spark control (ESC)

Fuel Vapor Control System
- 15 Vapor canister

Turbo System
- 16 Wastegate solenoid

Sensors/Switches
- A Manifold pressure sensor
- B Exhaust oxygen sensor
- C Throttle position sensor
- D Coolant sensor connector
- E Manifold air temperature sensor
- F Vehicle speed sensor
- H P/N switch
- J ESC sensor (knock)
- M Fuel pump test conector

Figure 4-25. *This is an example of a service manual electrical component location chart for the engine compartment. Can you find the fuse panel, oxygen sensor, and fuel pump test connector? (General Motors)*

jumper can be substituted for the component, as shown in **Figure 4-28.** If the circuit functions with the jumper in place, then the component being bypassed is defective.

Test Light

A ***test light*** can be used to check a circuit for power or voltage, **Figure 4-29.** For example, if there is no spark at the spark plugs when turning the engine over, a test light can verify that voltage is reaching the ignition coil. A test light can also be used to quickly check for circuitry continuity (completeness of circuit). Connect the alligator clip to ground. Then, touch the pointed tip to the circuit to check for power. The point on the test light can also be used to pierce a wire's insulation. If there is voltage, the light glows. If it does not glow, there is an open or break between the power source and the test point.

A ***self-powered test light*** contains a battery. It is used to check for circuit continuity in a circuit with its power source removed. To use a self-powered test light, the normal source of power, such as the battery or feed wire, must be disconnected. Then, connect the test light across the circuit or component. If the light glows, the circuit or part has continuity. If it does not glow, there is an open or break between the two test points.

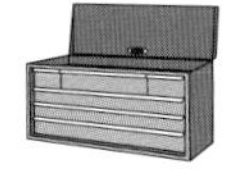

Note: Be sure an unpowered test light has a good ground. A common way to check this is by touching the tip of the test light to the positive terminal on the battery. If the test light has a good ground, the light will glow. If you do not first verify a good ground, you may incorrectly conclude that a circuit is not powered.

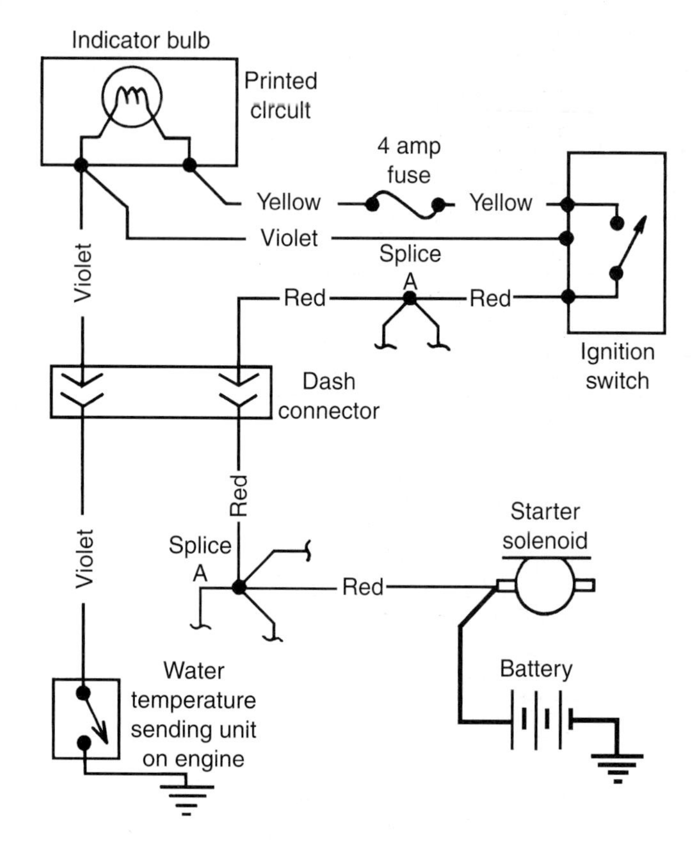

Figure 4-26. *Wiring diagram uses symbols and lines to represent electrical components and wires. This allows you to quickly see how wires connect to components and where the connections are in the wiring harness.*

Testing with a Multimeter

A ***multimeter,*** also called a VOM, is a device that consists of a voltmeter, ohmmeter, and ammeter. As pictured in **Figure 4-30,** a function or control knob is used to select the function to be used. Performing tests using the three different functions of a multimeter are described in the next sections.

A multimeter may be analog or digital. An analog multimeter displays the reading using a needle on a scale. A digital multimeter has an LED or LCD readout instead of a needle, **Figure 4-31.** This type is recommended for testing computer system components.

Note: An ohmmeter, ammeter, or voltmeter can be purchased as a separate piece of equipment. However, most technicians prefer the convenience of having the three functions in a single unit (the multimeter or VOM).

Voltmeter Function

The ***voltmeter*** function is used to measure the amount of voltage (number of volts) in a circuit. The multimeter is connected in parallel with the circuit. See **Figure 4-32A.** The voltmeter reading can be compared to specifications to determine whether an electrical problem exists.

Ammeter Function

The ***ammeter*** function measures the amount of current (number of amps) in a circuit. Multimeters with a conventional ammeter are connected in series with the circuit, as shown in **Figure 4-32B.** However, some multimeters have an inductive ammeter. A clip-on probe is slipped over the outside of the insulation of the wire being tested. The ammeter then uses the magnetic field around the outside of the wire to determine the amount of current in the wire. An inductive ammeter is very fast and easy to use.

Ohmmeter Function

The ***ohmmeter*** function measures the amount of resistance (number of ohms) in a circuit or component. As shown in **Figure 4-32C,** the multimeter is connected across the wire or component being tested with the circuit's power source disconnected. Then, the ohmmeter reading can be compared to specifications. If the resistance is not within specifications, the part is defective.

Multimeter Rules

There are a few rules to remember when using a multimeter. These are:

- ❑ When the ohmmeter function is selected, do not connect the multimeter to a source of voltage. This could damage the meter or blow its fuse. Disconnect the battery from the circuit before measuring resistance.
- ❑ Use a high-impedance meter when checking electronic components, especially in an engine computer control system. Some meters with low impedance will draw too much current through the electronic device, ruining it.
- ❑ Disconnect electric terminals carefully or you could damage the terminal and cause high circuit resistance and an engine malfunction.
- ❑ When using a pointed test probe, be careful not to stick the sharp probe into your hand. It is very easy for the tip to slip off of a wire and stab through the skin.
- ❑ Do not stick holes in wires going to some engine sensors, such as the oxygen sensor. These sensors produce such a small voltage that moisture entering through the hole in the insulation could affect the sensor signals to the computer.
- ❑ When measuring an unknown electrical quantity, start with the multimeter on a high setting. Then, reduce the setting until you obtain a proper scale. For example, when measuring current, set the multimeter to a high amperage range. This will prevent excess needle deflection or current flow through the meter that could damage the meter.
- ❑ Follow factory service manual procedures when making electrical tests. The wrong meter connection could cause serious damage to today's computer systems.
- ❑ Position the meter so that it cannot be damaged from a fall or hot or spinning engine parts. Keep the leads away from the hot exhaust manifolds and spinning parts.

+	Positive
−	Negative
	Ground
	Fuse
	Circuit breaker
	Capacitor
Ω	Ohms
	Resistor
	Variable resistor
	Series resistor
	Coil
	Step up coil
	Connector
	Male connector
	Female connector
	Multiple connector
	Denotes wire continues elsewhere
	Splice
J2 2	Splice identification
	Optional wiring with / wiring without
	Thermal element (bimetal strip)
	Y windings
88:88	Digital readout
	Single filament lamp

	Open contact
	Closed contact
	Closed switch
	Open switch
	Closed ganged switch
	Open ganged switch
	Two pole single throw switch
	Pressure switch
	Solenoid switch
	Mercury switch
	Diode or rectifier
	Bi-directional zener diode
	Dual filament lamp
	Light-emitting diode (LED)
	Thermistor
	Gauge
TIMER	Timer
	Motor
	Armature and brushes
	Denotes wire goes through grommet
#36	Denotes wire goes through 40 way disconnect
#19 STRG COLUMN	Denotes wire goes through 25 way steering column connector
INST PANEL #14	Denotes wire goes through 25 way instrument panel connector

Figure 4-27. *These are symbols commonly found on automotive wiring diagrams. (DaimlerChrysler)*

- ❑ Always read the meter's operating instructions before use. Meter ranges, specifications, and capabilities vary.

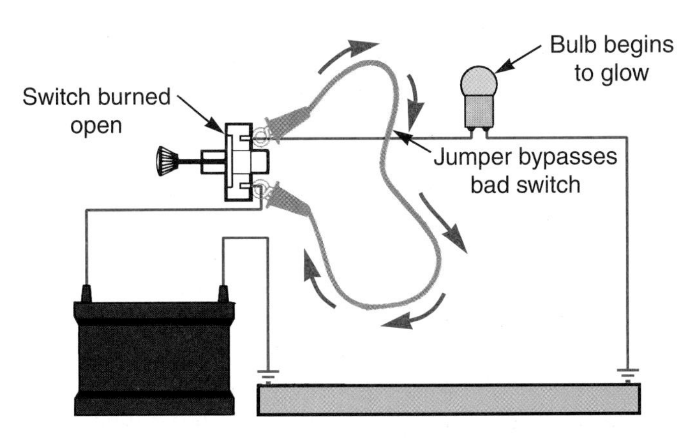

Figure 4-28. *Jumper wire can be used to bypass a component. For example, if you think a switch is bad, jumper around the switch. If the circuit functions, the switch is bad. (Ford)*

Summary

An engine technician working in a modern service facility must have a sound understanding of electrical and electronic components and how they are serviced. The engine compartment on a late-model vehicle is full of engine sensors, solenoids, relays, and other devices. As an engine technician, you must be trained in diagnosing and replacing these parts.

Electricity is the movement of free electrons through a conductor or wire. Insulation is used to prevent electron flow. Voltage is the electrical pressure that causes current flow. Resistance is needed to limit and control current. Ohm's law describes the relationship between resistance, current, and voltage, and can be used to find an unknown electrical value when two values are known.

A simple circuit consists of a power source, conductors, and a load. Vehicles commonly use a frame ground circuit to save wiring.

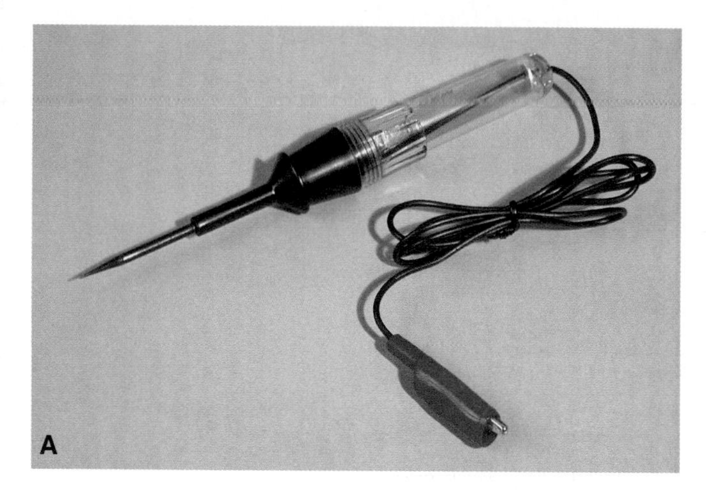

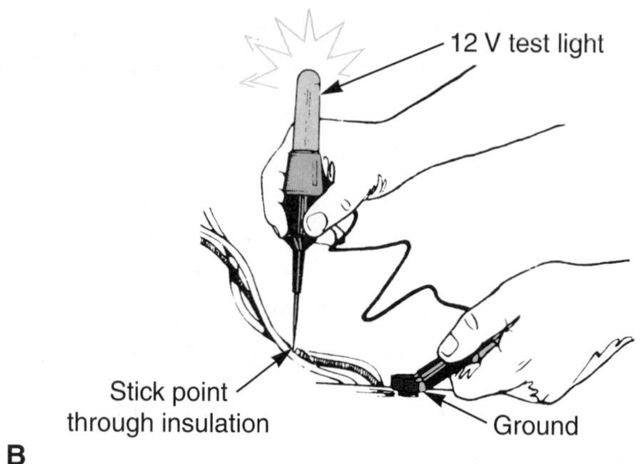

Figure 4-29. *A—This is a common test light. B—A test light can be used to quickly check for power in a circuit. (Lisle Tools)*

A fuse or circuit breaker provides overcurrent protection in a circuit, such as occurs with a short. Either will break the circuit and protect the circuit from excess current and heat.

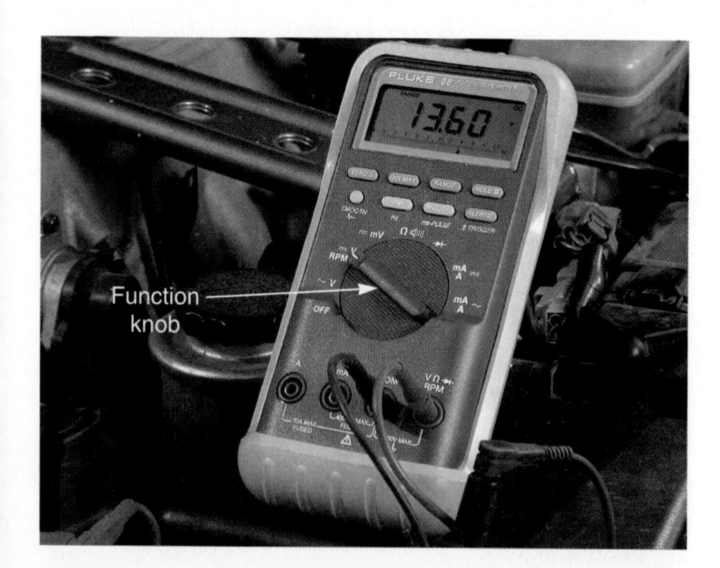

Figure 4-30. *A multimeter is being used to check for voltage. Notice the control knob. (Fluke)*

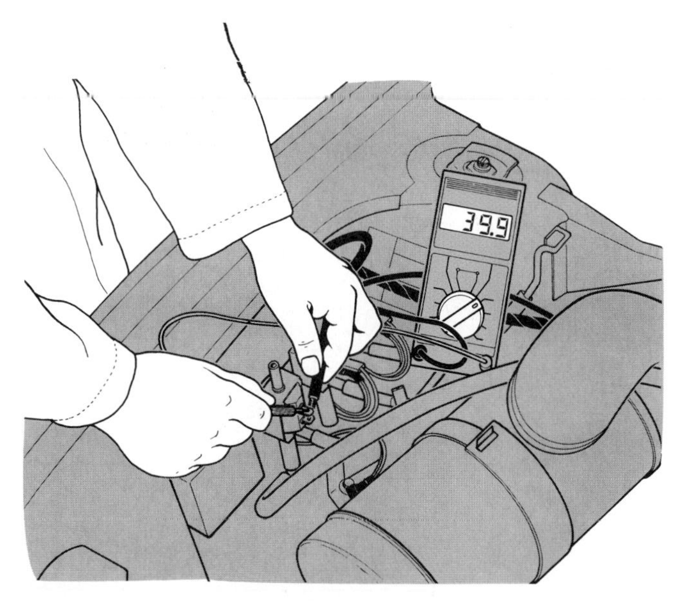

Figure 4-31. *A digital meter has an LED or LCD readout instead of needle. In this example, the technician is measuring the resistance of a vacuum valve's windings to check for an open or short. (Peugeot)*

A relay is an electrically operated switch. A solenoid is an electromagnet with a sliding core, or plunger.

Alternating current (ac) changes direction while direct current (dc) only flows in one direction. A vehicle primarily uses dc.

There are several solid state or semiconductor devices in the electrical system of a modern vehicle. A diode is an electrical check valve that only allows current flow in one direction. A transistor is a solid state switch. A small current can stimulate the semiconductor material in the transistor, changing it from an insulator to a conductor. An integrated circuit contains miniaturized components in a small chip. ICs are commonly used in computers.

Sensors monitor engine operation. The sensors convert a condition, such as temperature or movement, into an electrical signal. The computer uses these signals to determine the needs of the engine. It can then use relays, solenoids, and small dc motors to affect engine operation and improve efficiency.

Primary wire is the small-diameter wire that carries battery voltage to electrical components. Secondary wire has a larger diameter for carrying high voltage in the ignition system.

Wiring can be repaired with crimp connectors or soldering. Wiring diagrams are "electrical road maps" that show how all of the components in a circuit connect to each other.

Test lights, jumper wires, and a multimeter are commonly used to make electrical tests. A digital, high-impedance multimeter is normally recommended for tests on computer control systems.

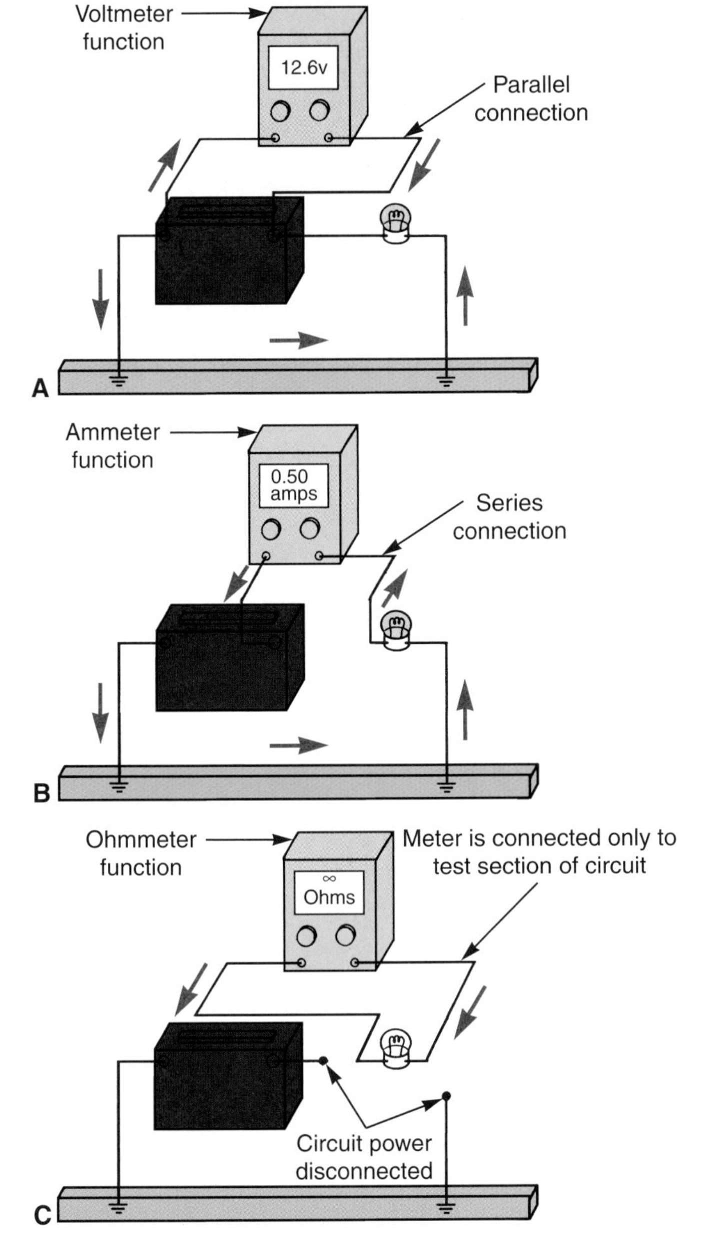

Figure 4-32. *Study basic multimeter connections. A—When using the voltmeter function, the multimeter is connected in parallel with the circuit. B—When using the ammeter function, the multimeter is connected in series. Inductive ammeters, however, have a clip that fits over or around insulation to measure current. C—When using the ohmmeter function, the multimeter is connected in series with the component being tested. The power source must be disconnected or meter damage may result.*

Review Questions—Chapter 4

Please do not write in this text. Write your answers on a separate sheet of paper.

1. Explain the difference between an *insulator* and a *conductor.*
2. _____ is the flow of electrons through a circuit.
3. Why is resistance needed in a circuit?
4. A(n) _____ uses the vehicle's body or metal structure as a return wire to the battery negative.
5. If a circuit has 12.5 volts applied and 25 ohms of resistance, how much current will flow in the circuit?
6. What is the purpose of a *fuse* or *circuit breaker?*
7. Magnetism is used in the following components:
   (A) Relay.
   (B) Solenoid.
   (C) Motor.
   (D) All of the above.
   (E) None of the above.
8. A(n) _____ has no moving parts and performs the same function as a relay by acting as a remote control switch.
9. How does a computer control system increase engine efficiency?
10. _____ wire is small-diameter wire that carries battery voltage and _____ wire has a larger diameter for carrying high voltage to the spark plugs.
11. _____ connectors are often used in electrical repairs.
12. _____ solder should be used for electrical repairs.
13. Describe a *wiring diagram.*

ASE-Type Questions—Chapter 4

1. Technician A says voltage controls the flow of current in a circuit. Technician B says that voltage is the electrical pressure that causes current flow. Who is correct?
   (A) A only.
   (B) B only.
   (C) Both A and B.
   (D) Neither A nor B.
2. All of the following are basic types of electrical circuits *except:*
   (A) frame ground circuit.
   (B) series circuit.
   (C) series-ground circuit.
   (D) parallel circuit.
3. Technician A says that if a circuit has 12 volts with a 2 ohm load, the circuit has 6 amps flowing through it. Technician B says that if a circuit has 12 volts with a 2 ohm load, the circuit has 24 amps flowing through it. Who is correct?
   (A) A only.
   (B) B only.
   (C) Both A and B.
   (D) Neither A nor B.

4. A circuit breaker performs the same function as a _____.
   (A) relay
   (B) solenoid
   (C) fuse
   (D) None of the above.

5. Technician A says that alternating current flows in one direction and is commonly used throughout an automobile. Technician B says that direct current flows back and forth through a circuit and is not commonly used throughout an automobile. Who is correct?
   (A) A only.
   (B) B only.
   (C) Both A and B.
   (D) Neither A nor B.

6. Technician A says that a relay is an electrically operated switch. Technician B says that a solenoid has a plunger that moves to induce a voltage in the coil's windings. Who is correct?
   (A) A only.
   (B) B only.
   (C) Both A and B.
   (D) Neither A nor B.

7. Technician A says that electrical devices use moving parts to conduct electricity through a circuit. Technician B says electronic devices do not use moving parts to perform their function in a circuit. Who is correct?
   (A) A only.
   (B) B only.
   (C) Both A and B.
   (D) Neither A nor B.

8. All of the following are solid state devices *except:*
   (A) condenser.
   (B) relay.
   (C) diode.
   (D) transistor.

9. Technician A says that primary wire has a small diameter and carries battery or alternator voltage. Technician B says that primary wire runs from the ignition coil to the spark plugs. Who is correct?
   (A) A only.
   (B) B only.
   (C) Both A and B.
   (D) Neither A nor B.

10. Technician A says that the circuit shown in the illustration is a series circuit. Technician B says that the circuit shown in the illustration is a series-parallel circuit. Who is correct?
    (A) A only.
    (B) B only.
    (C) Both A and B.
    (D) Neither A nor B.

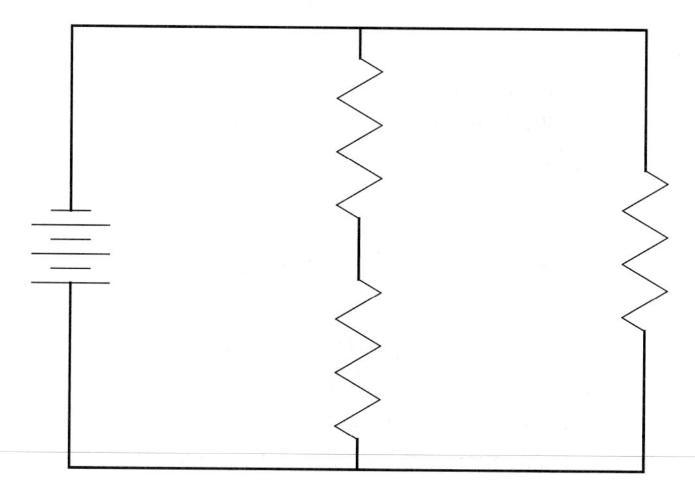

11. An ammeter measures the amount of _____ in a circuit.
    (A) current
    (B) resistance
    (C) voltage
    (D) wattage

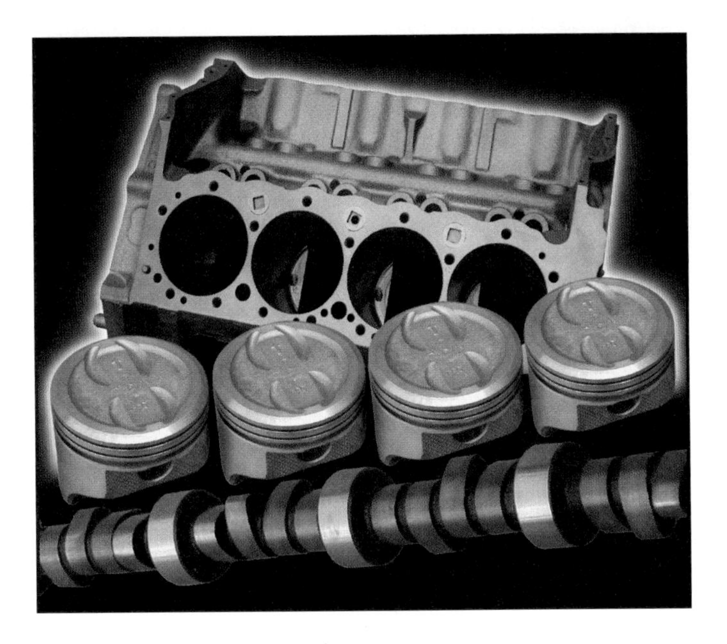

Chapter 5

Shop Safety

After studying this chapter, you will be able to:

- ❑ List common types of accidents.
- ❑ Identify fire hazards.
- ❑ Explain how an electrical fire can start.
- ❑ Identify asphyxiation hazards.
- ❑ List safety rules.
- ❑ Explain what qualifies a material as a hazardous waste.
- ❑ Describe the regulations concerning the disposal of hazardous wastes.
- ❑ Explain the importance of material safety data sheets.

Know These Terms

Asphyxiation

Corrosive hazard

Electrical fires

Explosion

Fire

Gasoline

Hazardous waste

Ignitable hazard

Material safety data sheet (MSDS)

Reactivity hazard

Respirators

Toxicity hazard

Hundreds of engine technicians are injured or killed each year while on the job. A majority of these incidents resulted from broken safety rules, **Figure 5-1.** As an engine technician, you will be exposed to many potentially dangerous situations. You will be working around running engines, engines suspended in the air on a hoist, gasoline, and numerous other possible hazards. Moving vehicles in the close quarters of a service facility can also be hazardous. You must learn to control these dangerous situations and keep your job safe and enjoyable. This chapter reviews basic shop safety and will help you gain basic safety skills.

Types of Accidents

There are six common types of accidents—fires, explosions, asphyxiation, chemical burns, electric shock, and physical injuries. To create a safe working environment, you must prevent these. Each could cripple or kill you or someone else.

Fires

A fire can cause horrible destruction, injury, and death. ***Fire*** is the result of heat, fuel, and oxygen in the correct proportions to start combustion. When working on engines, gasoline, diesel fuel, cleaning solvents, and oily rags are just a few of the many possible sources of fuel in a service facility. Fire extinguishers are rated in terms of the type of fire for which they should be used. See **Figure 5-2.**

Gasoline is the most dangerous flammable in the shop, **Figure 5-3.** Just a cup of ignited gasoline can engulf a whole engine compartment in flames. The fire may then consume the rest of the vehicle and maybe the entire shop. A few rules for handling gasoline include:

Figure 5-1. *A running engine can be dangerous. Hot parts can cause burns. Spinning fans can cause deep cuts. Exhaust fumes can cause asphyxiation.*

- ❑ *Never* use gasoline as a cleaning agent. Cleaning solvents are flammable, but not as flammable as gasoline.
- ❑ Keep sources of heat, such as welding and cutting equipment, away from the engine's fuel system.
- ❑ Wipe up gasoline spills right away. Do not spread oil absorbent (oil-dry) on a gasoline spill because the absorbent will become flammable.
- ❑ Disconnect the car battery before working on the fuel system.
- ❑ Wrap a shop towel around any fitting when disconnecting a fuel line. This will collect fuel that leaks or sprays out.
- ❑ Store gasoline and other flammables in approved, sealed containers.

Electrical fires can occur when a current-carrying wire shorts to ground. This causes unlimited current flow, which in turn causes the wire to heat up, melt its insulation, and burn. See **Figure 5-4.** Then, other wires may short and burn. In this repeating pattern, the wiring through much of the car can quickly begin to burn.

To prevent electrical fires, disconnect the car battery before working on any wiring. Since there is no power source connected, you cannot accidentally create a short.

Explosions

An ***explosion*** is a violent expansion of gasses due to rapid combustion. There are several possible sources of explosion when servicing an engine or its systems. For example, a vehicle's battery can explode. The chemical reaction that takes place inside of the battery to produce electricity also generates hydrogen gas. As shown in **Figure 5-5,** this highly explosive gas can surround the top of the battery. The slightest spark or flame can ignite the gas and cause the battery to explode. Fragments of battery case and the battery acid can travel through the air as the result of the explosion. Blindness, cuts, acid burns, and scars can result.

Various other sources can result in explosions. For example, sodium-filled engine valves, welding tanks, propane bottles, and fuel tanks can all explode if mishandled. These hazards are discussed in later chapters where appropriate.

Asphyxiation

Asphyxiation is a condition in which the body has too little oxygen or too much carbon dioxide. It may be caused by breathing toxic or poisonous substances in the air. Mild cases of asphyxiation will cause dizziness, headaches, and vomiting. Severe asphyxiation can cause death.

The most likely source of asphyxiation in a service facility is the exhaust gasses from the vehicle's engine. Exhaust contains very little oxygen. Coupled with the fact that the engine is consuming oxygen for combustion, the oxygen in an enclosed shop can quickly be depleted. In

Fire Extinguishers and Fire Classifications

Fires
Class A Fires Ordinary Combustibles (Materials such as wood, paper, textiles.) *Requires... cooling-quenching* Old: A New
Class B Fires Flammable Liquids (Liquids such as grease, gasoline, oils, and paints.) *Requires...blanketing or smothering.* Old: B New
Class C Fires Electrical Equipment (Motors, switches, and so forth.) *Requires... a nonconducting agent.* Old: C New
Class D Fires Combustible Metals (Flammable metals such as magnesium and lithium.) *Requires...blanketing or smothering.* D

Type	Use	Operation
Soda-acid Bicarbonate of soda solution and sulfuric acid	Okay for use on: A Not for use on: B C D	Direct stream at base of flame.
Pressurized Water Water under pressure	Okay for use on: A Not for use on: B C D	Direct stream at base of flame.
Carbon Dioxide (CO_2) Carbon dioxide (CO_2) gas under pressure	Okay for use on: B C Not for use on: A D	Direct discharge as close to fire as possible, first at edge of flames and gradually forward and upward.
Foam Solution of aluminum sulfate and bicarbonate of soda	Okay for use on: A B Not for use on: C D	Direct stream into the burning material or liquid. Allow foam to fall lightly on fire.
Dry Chemical	Multi-purpose type: Okay for A B C; Not okay for D Ordinary BC type: Okay for B C; Not okay for A D	Direct stream at base of flames. Use rapid left-to-right motion toward flames.
Dry Chemical Granular type material	Okay for use on: D Not for use on: A B C	Smother flames by scooping granular material from bucket onto burning metal.

Figure 5-2. *There are different types of fires. Fire extinguishers are rated by the type of fire on which they should be used.*

Figure 5-3. *Gasoline is the most dangerous and underestimated flammable. A cup of gasoline, when ignited, can engulf a vehicle in flames.*

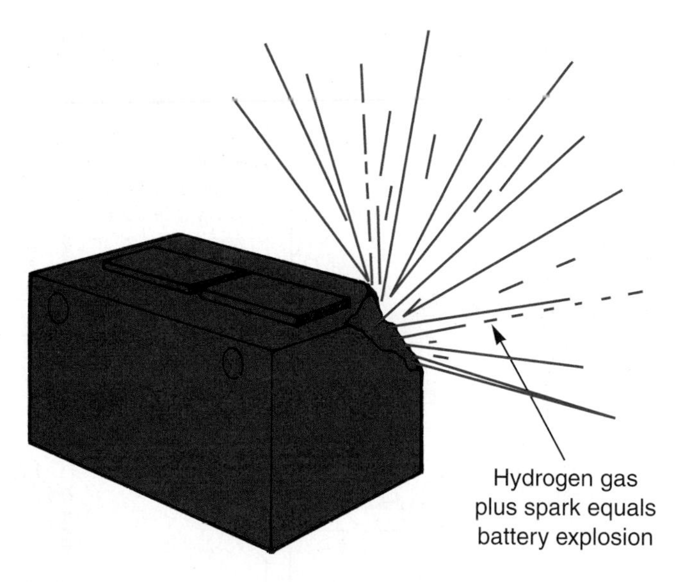

Figure 5-5. *A battery can explode, caused by a spark igniting the hydrogen gas that can collect around the top of the battery.*

addition, an engine's exhaust gasses contain carbon monoxide, which is deadly poison. Using an exhaust hose, connect the vehicle's tailpipe to the shop's exhaust ventilation system, **Figure 5-6.** Also, make sure the exhaust ventilation system is turned on whenever the engine is running.

There are other substances in the shop that are harmful if inhaled. One of these harmful substances is asbestos, which may be found in the dust from a clutch disc or older brake pads. ***Respirators*** (filter masks) should be worn when working around any kind of airborne impurities, **Figure 5-7.** If welding during major engine repair or in a machine shop, a special welding respirator should be worn. It will trap toxic welding fumes and block them from entering your nose, mouth, throat, and lungs. See **Figure 5-8.**

Warning: A filter mask or respirator will only protect you from the substances for which it is rated.

Chemical Burns

Various solvents, battery acid, and a few other substances found in the shop can cause chemical burns to the skin. Decarbonizing cleaner ("carb cleaner"), for example, is powerful and can severely burn your skin in a matter of seconds. Always read the directions and warnings on chemicals. See **Figure 5-9.** Wear rubber gloves and eye protection. If a skin burn occurs, follow the treatment directions on the product label.

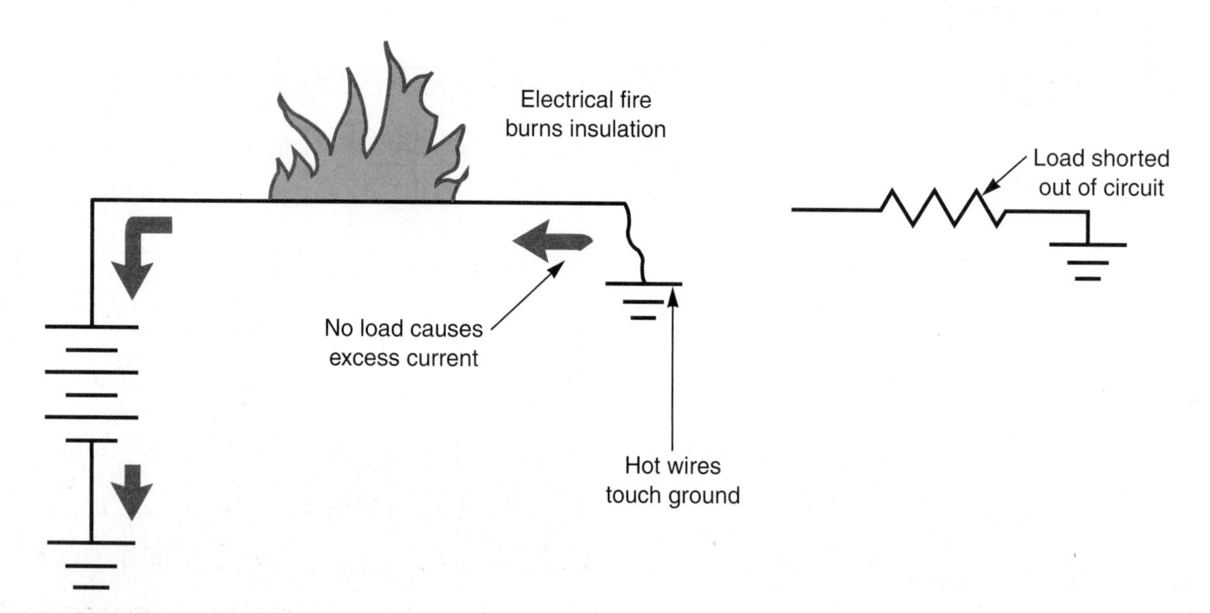

Figure 5-4. *Electrical fire can occur when a wire shorts to ground. The resulting high current causes enough heat to make the wire insulation burn. Always disconnect the battery before working on wiring.*

Figure 5-6. *Use an exhaust ventilation system when running an engine in an enclosed area.*

Electric Shock

Electric shock can occur when using improperly grounded electric power tools. Never use an electric tool unless it has a functional ground prong. This is the third, round prong on the plug. The ground prong prevents current from accidentally passing through your body. Also, never use an electric tool on a wet shop floor.

Physical Injury

Physical injuries, such as cuts, broken bones, and strained backs, can result from hundreds of different accidents. As an engine technician, you must constantly think and evaluate every repair technique. Decide whether a particular operation is safe or dangerous and take action as required. For example, why move an engine block by hand when a crane is available? You and a friend may be strong enough to lift the engine block, but why risk back injury? Once your back is injured, it will never be the same!

Figure 5-7. *When working around airborne toxins, wear an approved respirator. It will help keep harmful chemicals out of your lungs.*

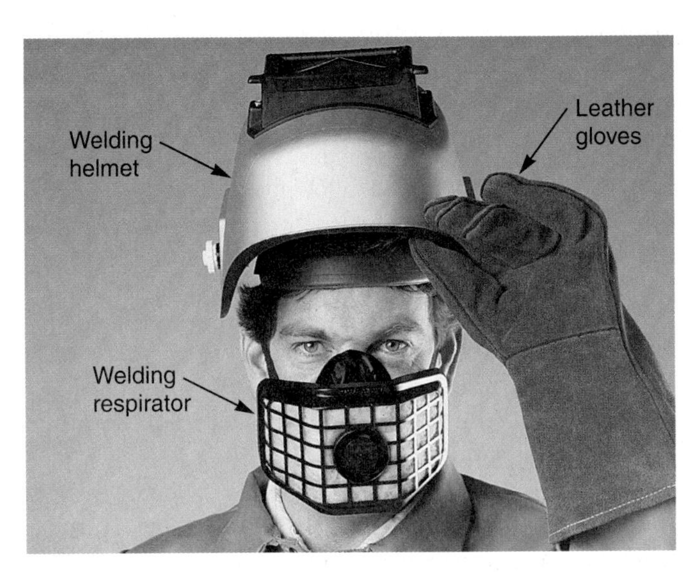

Figure 5-8. *This is a special welding respirator. It is designed to prevent inhalation of toxic welding fumes.*

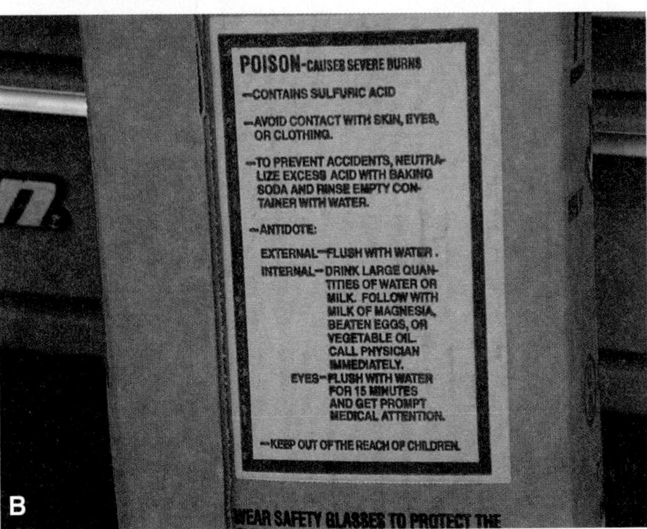

Figure 5-9. *A—Battery acid can cause severe chemical burns. B—Make sure you read and follow label directions on caustic products. Note the treatments (antidotes).*

General Safety Rules

The following list provides several general safety rules. Remember these and follow them at all times.

- ❑ Wear appropriate eye protection during any operation that could endanger your eyes! See **Figure 5-10.** This includes operating power tools, working around a spinning engine fan, carrying batteries, using a cutting torch, and welding.
- ❑ Keep your shop organized. Return all tools and equipment to their proper storage areas. Never lay tools or parts on the floor, **Figure 5-11.**
- ❑ Dress like a professional technician. Remove rings, bracelets, necklaces, watches, and other jewelry. They can get caught in engine fans, belts, etc., causing injury. Also, roll up long sleeves and secure long hair. They, too, can get caught in spinning parts.
- ❑ Work like a professional. When learning to be an engine technician, it is easy to get excited about your work. However, avoid working too fast. You could overlook a repair procedure or safety rule and cause an accident.
- ❑ Use the right tool for the job. There is usually a tool that is best suited for each repair task. Always ask yourself this question: Is there another tool that will work better?
- ❑ Never carry sharp tools or parts in your pockets. They can easily puncture your skin.

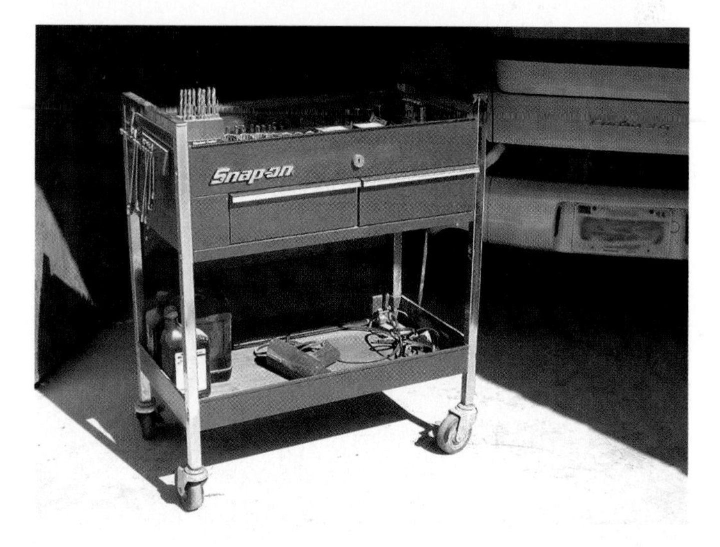

Figure 5-11. *Always keep tools organized while working. Do not lay tools on the floor. This roll around cart can be used to hold all tools needed for the job. (Snap-on Tool Corp.)*

- ❑ Keep equipment guards or shields in place. If a power tool has a safety guard, use it. Refer to **Figure 5-12.**
- ❑ Lift heavy parts with your legs, *not* with your back. When lifting, bend at your knees while keeping your back as straight as possible. On extremely heavy assemblies, such as transmissions, engine blocks, and transaxles, use a portable crane.
- ❑ Use adequate lighting. A portable shop light increases safety, work speed, and precision.

Figure 5-10. *Wear appropriate eye protection. There are several types available for different situations.*

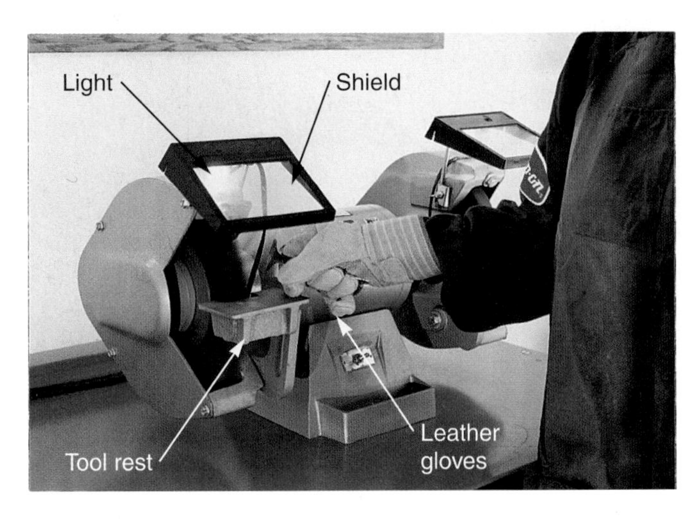

Figure 5-12. *Always use a protection device when needed. Here, the tool rest is being used to hold and secure a chisel while sharpening.*

- ❑ Ventilate your work area when needed. Turn on the shop ventilation fan or open the shop doors anytime fumes are present in the shop.
- ❑ Never stir up asbestos dust. Asbestos is a cancer-causing agent. Do not use compressed air to blow the dust off brake parts or clutch assemblies.
- ❑ Jack up or raise a vehicle slowly and safely, **Figure 5-13.** A car or light truck may weigh as much as two tons.
- ❑ Never work under a vehicle unless it is supported by jack stands, **Figure 5-14.** It is *not* safe to work under a vehicle held up by only a floor jack. Also, chock (block) the vehicle's wheels when the car is on jack stands.
- ❑ Drive slowly when in the shop area. With students and vehicles in the shop, it is very easy to have an accident.
- ❑ Keep away from spinning engine fans. The engine fan is like a spinning knife. It can inflict serious injuries. Also, if a part or tool is dropped into the fan, it can fly out and hit someone or damage the radiator.
- ❑ Respect running engines. When an engine is running, make sure that the transmission or transaxle is in park. Check that the emergency brake is set and that the wheels are blocked.
- ❑ Do not smoke in the shop. Smoking is a serious fire hazard considering fuel, cleaning solvents, and other flammables that are in the shop.

Figure 5-13. *A—A floor jack and jack stands. Never get under a vehicle that is supported by only the floor jack. B—Stay out from under vehicle as it is being raised or lowered. Also, raise and lower the vehicle slowly.*

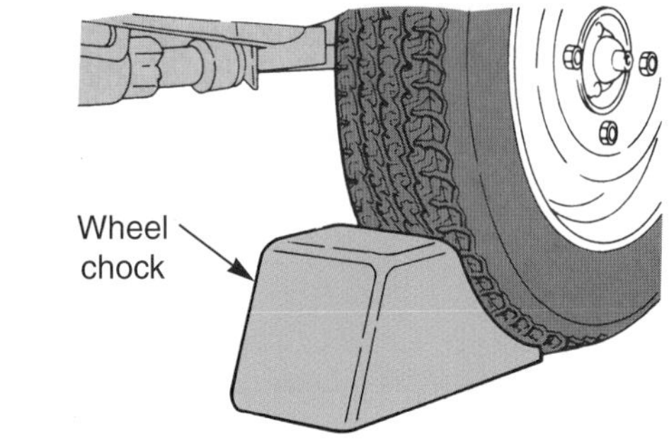

Figure 5-14. *A—Jack stands must be used while working under vehicle. It is not safe to work under vehicle held only by a floor jack. B—Chock (block) the wheels when the vehicle is lifted to keep the vehicle from rolling off the jack stands. (Subaru)*

- ❑ Chemical pneumonia can result from inhaling oil mist. Dermatitis can result from skin contact with oil.
- ❑ Obtain permission before using any new or unfamiliar power tool, lift, or shop equipment. Your instructor will need to provide training on proper use. See **Figure 5-15.**
- ❑ Wear ear protection when using loud power tools. For example, if you are using an air chisel to cut off an old exhaust system, the sound generated by the hammer action can be above a safe limit. To avoid partial loss of your hearing, wear ear plugs or ear muffs, as shown in **Figure 5-16.**
- ❑ Wear gloves when needed while working. Leather gloves will protect your hands from cuts and abrasions, **Figure 5-17.** Latex or rubber gloves can be used to protect your skin from oil and grease.
- ❑ Use compressed air with caution, **Figure 5-18.** Air lines can contain up to 150 psi (1000 kPa) of air

Figure 5-16. *Wear ear protection in loud or noisy situations. (Snap-on Tool Corp.)*

Figure 5-15. *A—Never use equipment for which you are not trained. This technician has been instructed in the proper locations for the lift arms. B—Since the technician that placed the vehicle on the lift was properly trained, the vehicle is secure and work can be safely completed under the vehicle.*

Figure 5-17. *Wear gloves to protect your hands when needed. Leather gloves are good when using a grinder or when drilling. Latex or rubber gloves should be worn when working with caustic chemicals.*

Figure 5-18. *Compressed air can be very dangerous. Use blow nozzles and other air tools with care.*

pressure. If air is forced through your skin and into the bloodstream, death can result. Never direct a blow nozzle at you or anyone else.

- ❑ Many air tools have a pressure regulating valve to limit tool speed or power, **Figure 5-19.** Always adjust the pressure control to the lowest-possible setting that will do the job. This will help avoid part damage and possible injury.
- ❑ A clean shop is a safe shop. Always do your part to keep the service facility clean and organized. See **Figure 5-20.** A cluttered, disorganized shop indicates a very disorganized technician.
- ❑ Report unsafe conditions to your instructor. If you notice a hazard, inform your instructor immediately.
- ❑ Avoid anyone who does not take shop work seriously.
- ❑ If an accident or injury ever occurs in the shop, notify your instructor immediately.

Figure 5-20. *These technicians are cleaning up outside their shop. Doing so increases shop safety, but it also makes the shop appear more professional to customers.*

Lift Safety

Safety is always important, but especially so when using floor jacks and vehicle lifts. A vehicle can weigh 3000 lbs (1360 kg) or more. The careless use of lift equipment results in numerous injuries and deaths each year. In addition to bodily harm, equipment and customer vehicles are damaged.

Before using any type of shop equipment, lifts and jacks included, learn the proper procedure for using the equipment. Respect the dangers that the equipment presents. Follow all safety rules and learn to "think before you act."

In addition to following safe operating procedures, the equipment must be in good operating condition to perform safely. Do not use equipment that has cracked or bent parts, faulty safety locks, leaking cylinders, or other problems.

Study the following safety precautions carefully. They apply to all types of lifting equipment.

Figure 5-19. *Many air tools, like this air impact wrench, have a power or speed adjustment knob. Set the tool for the lowest possible pressure needed to complete the task. (Snap-on Tool Corp.)*

- ❑ Before lifting any vehicle, determine how the weight is distributed on the vehicle. For example, a front-wheel drive vehicle is heavier at the front of the vehicle than a rear-wheel drive vehicle because of the transaxle. Therefore, the lift saddles at the front of the vehicle must be positioned to account for this.
- ❑ Position the lift saddles so they securely contact the vehicle's chassis. Most vehicles have certain lift points designed into the frame or unitized body.
- ❑ With saddles properly positioned, raise the vehicle until its weight is fully supported by the lift (wheels are just off the ground). Then, carefully push down on one end of the vehicle. If the vehicle moves on the saddles or feels unstable, lower the vehicle and reposition the saddles before proceeding.
- ❑ When raising a vehicle, watch for side or overhead obstructions.
- ❑ Securely engage the lift safety lock before working under the vehicle.
- ❑ Do not change the working height of the lift unless all personnel are out from under the vehicle.
- ❑ When using a floor jack, always place jack stands under the vehicle before working on the vehicle.
- ❑ Always check for equipment, parts, and personnel under the vehicle before lowering it.
- ❑ Lower a vehicle slowly. Watch the vehicle and lift closely until the lift is fully lowered and the vehicle is firmly on the ground.

Disposing of Shop Wastes

Automotive service and maintenance facilities frequently generate hazardous wastes. These wastes are regulated by the Resource Conservation and Recovery Act.

This federal act applies to businesses that generate, transport, or manage hazardous wastes. Any business that maintains or repairs vehicles, heavy equipment, or farm equipment must comply with the regulations of the act.

Hazardous Waste

Hazardous waste is a solid, liquid, or gas that can harm people or the environment. There are several criteria for determining if a substance is hazardous.

A material is considered an ***ignitable hazard*** if it will easily ignite and burn. Gasoline, diesel oil, solvents, and other chemicals are considered ignitable hazards.

A material or waste is a ***corrosive hazard*** if it dissolves metals and other materials or burns human skin. Battery acid and many part-cleaning solvents are considered corrosive hazards.

Anything that reacts violently or releases poisonous gasses when in contact with other materials is considered a ***reactivity hazard.*** Materials that generate toxic mists, fumes, vapors, and flammable gasses are also reactive hazards.

Materials like lead, cadmium, chromium, arsenic, and other heavy metals that can pollute and make water and soil harmful are considered a ***toxicity hazard.*** Used motor oil, solvents, and other chemicals in the auto shop are toxic hazards that must be disposed of properly.

Hazardous Automotive Waste

Draining automotive fluids and replacing nonrepairable components are the most common automotive repair activities that produce hazardous wastes. Some automotive fluids and solid wastes that are considered hazardous include:

- ❑ Used motor oil. It is combustible and may contain toxic chemicals.
- ❑ Other discarded lubricants, such as transmission and differential fluids. These may contain toxic chemicals.
- ❑ Used parts cleaners and degreasers. These are combustible and may contain toxic chemicals.
- ❑ Decarbonizing cleaners. These contain flammable or combustible liquids.
- ❑ Old batteries. These contain lead and toxic chemicals.
- ❑ Old tires and catalytic converters.
- ❑ Antifreeze.
- ❑ Refrigerant.

Recycling Motor Oil

Used motor oil should be recycled. One gallon of used motor oil can be recycled into two and one-half quarts of high-quality motor oil. Recycling old oil not only saves our environment from pollution, it helps conserve natural resources.

Always send used oil to a recycling center. The old oil should be stored in an approved container and kept separate from other fluids. Some recycling companies provide a pickup service, while others require you to take the old oil to their facility.

Note: As a facility that works with motor oil, you are required to accept used oil from the public, even if they are not a customer.

Antifreeze

Antifreeze is classified as a hazardous waste due in part to the heavy metal and chlorinated solvents that it picks up when circulating through an engine's cooling system. In addition, antifreeze presents several health hazards. See **Figure 5-21.** Used antifreeze should never be mixed with used oil. In addition, it must be collected and disposed of by a registered hazardous waste recycling/disposal company.

Refrigerant

The refrigerant in the air conditioning systems must not be vented to the atmosphere. Regulations require that they be recovered and recycled. Several types of refrigerant-recovery systems are available.

Disposal

One of the best ways to deal with hazardous wastes is to minimize the quantity produced. This can be accomplished by practicing good housekeeping, improving inventory control, and following proper spill-containment techniques.

When hazardous wastes are produced, they must be disposed of properly. Regulations require that these wastes be collected by a registered hazardous waste hauler. Several major companies offer pick up and recycling services. Repair or maintenance facilities that generate

Figure 5-21. *Antifreeze and other shop substances are poisonous. These substances should be stored in a location where children and animals cannot get to them. The warning label will provide treatment recommendations.*

220 lb (100 kg) of hazardous waste monthly must fill out a Uniform Hazardous Waste Manifest before shipping the wastes to a disposal or recycling site. The manifest is simply a tracking document that must accompany hazardous wastes when they are shipped from the work facility. It contains detailed information about the origin, character, and destination of the wastes. When shipping certain wastes, the proper Department of Transportation (DOT) shipping descriptions must be listed on the manifest. Tables listing these descriptions are available from each state's hazardous waste management agency or from a regional EPA office.

EPA regulations state that no manifest is needed for used oil or lead-acid batteries sent off for recycling. In such cases, the material is not regarded as hazardous. Your state might have its own requirements. Check with your state hazardous waste management agency.

Used oil filters are considered hazardous waste unless they are to be recycled for scrap metal. If not recycled, they must be listed on the manifest as hazardous. Before disposal, oil filters should be gravity drained so that they do not contain free-flowing oil. Then, store them in a closed, labeled container for pickup by a recycler.

Material Safety Data Sheets

Always read label directions when using chemicals. Rust penetrant, lubricants, part cleaners, and other substances can be dangerous if not used properly. The product label will give general precautions for using the product, **Figure 5-22.** Further information can be obtained from the product's material data safety sheet.

Chemical manufacturers are required to provide a ***material safety data sheet (MSDS)*** for each chemical that they produce. This sheet is important because it lists all of the known dangers and treatment procedures for a specific chemical. Employers are required to have an MSDS for each chemical or substance used in their facility. Be sure to read the MSDS for any chemical or substance that you are not familiar with.

Figure 5-22. *Always read label directions before using any shop chemical.*

Summary

A service facility can be a safe and enjoyable place to work if safety rules are followed. However, if safety regulations are not followed, the shop can be a very dangerous place to work. Always make sure you are using approved, safe practices when working on a vehicle.

You must prevent fires, explosions, chemical burns, electric shocks, and other physical injuries. Constantly think about what you are doing and take corrective action when needed.

Hazardous wastes are often generated in the service facility. These wastes are regulated by the Resource Conservation and Recovery Act and must be disposed of properly. A material data safety sheet provides information about a chemical or substance, such as common reactions and treatments for overexposure.

Review Questions—Chapter 5

Please do not write in this text. Write your answers on a separate sheet of paper.

1. List six common types of accidents.
2. _____ is the most dangerous flammable in the shop.
3. Which of the following is not an acceptable cleaning agent?
   (A) Decarbonizing cleaner.
   (B) Gasoline.
   (C) Soap.
   (D) Degreasing solvent.
4. What should you do when disconnecting an engine fuel line to prevent fuel from spaying out?
5. How can a short cause an electrical fire?
6. Batteries generate _____, which can explode.
7. Why is engine exhaust dangerous?
8. This can cause severe chemical burns.
   (A) Battery acid.
   (B) Decarbonizing cleaner.
   (C) Both A and B.
   (D) Neither A nor B.
9. Why is a ground prong provided on electrical equipment?
10. Why should you remove jewelry when on the job?

ASE-Type Questions—Chapter 5

1. Technician A says that you should disconnect the battery before removing a fuel line from an engine. Technician B says that you should wrap a shop towel around the fitting before disconnecting a fuel line. Who is correct?
   (A) A only.
   (B) B only.
   (C) Both A and B.
   (D) Neither A nor B.
2. Technician A says that a fire extinguisher bearing this symbol can be used on burning oil or gasoline. Technician B says that a fire extinguisher bearing this symbol can be used on an electrical fire. Who is correct?
   (A) A only.
   (B) B only.
   (C) Both A and B.
   (D) Neither A nor B.

3. The most likely source of asphyxiation in an auto shop is _____.
   (A) gasoline
   (B) oily shop rags
   (C) engine exhaust
   (D) None of the above.
4. Technician A says that a battery can produce gasses that can cause an explosion. Technician B says that sodium-filled engine valves, if mishandled, can cause an explosion. Who is correct?
   (A) A only.
   (B) B only.
   (C) Both A and B.
   (D) Neither A nor B.
5. Technician A says that a respirator will block all poisonous fumes that can be produced in a shop. Technician B says that a respirator will block only those fumes for which it is rated. Who is correct?
   (A) A only.
   (B) B only.
   (C) Both A and B.
   (D) Neither A nor B.
6. Technician A says that battery acid can cause chemical burns to the skin. Technician B says that some decarbonizing cleaners can cause chemical burns to the skin. Who is correct?
   (A) A only.
   (B) B only.
   (C) Both A and B.
   (D) Neither A nor B.
7. Technician A says that an electric power tool with a faulty ground prong can cause electrocution. Technician B says that if a power tool's faulty ground prong is removed, the power tool is safe to use. Who is correct?
   (A) A only.
   (B) B only.
   (C) Both A and B.
   (D) Neither A nor B.
8. Technician A says that eye protection should be worn when carrying a battery. Technician B says that eye protection should be worn when working around an engine fan. Who is correct?
   (A) A only.
   (B) B only.
   (C) Both A and B.
   (D) Neither A nor B.
9. Technician A says that used batteries are considered a hazardous waste. Technician B says that used decarbonizing cleaner is considered a hazardous waste. Who is correct?
   (A) A only.
   (B) B only.
   (C) Both A and B.
   (D) Neither A nor B.
10. All of the following are safety rules to follow when working in an auto shop *except:*
    (A) never carry sharp parts or tools in your pocket.
    (B) always lift heavy engine parts with your back, not with your legs.
    (C) keep the auto shop organized.
    (D) always keep equipment guards or shields in place.
11. Technician A says that refrigerants can be vented into the atmosphere as long as proper filtering systems are utilized. Technician B says that used motor oil should be recycled. Who is correct?
    (A) A only.
    (B) B only.
    (C) Both A and B.
    (D) Neither A nor B.

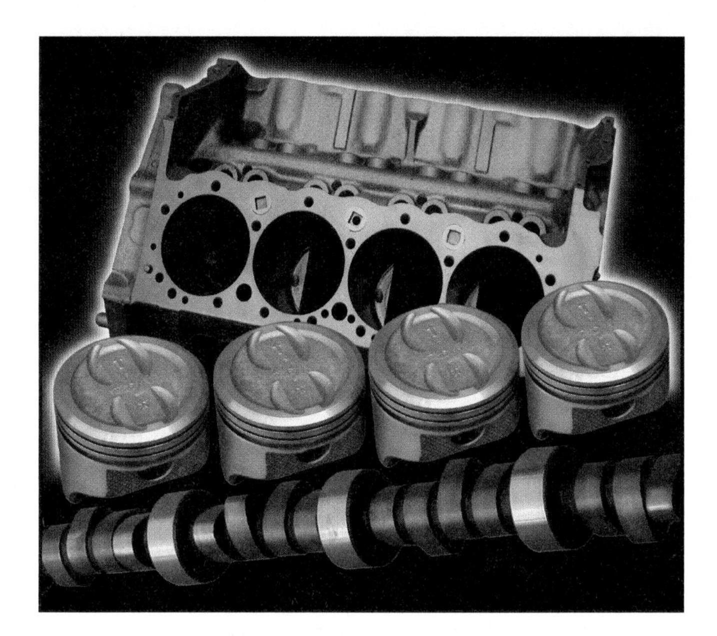

Chapter 6

Engine Types and Classifications

After studying this chapter, you will be able to:

- ❑ Describe the different engine cylinder arrangements.
- ❑ Explain cylinder numbering and firing order.
- ❑ Identify different crankshaft configurations.
- ❑ List the two types of cooling systems.
- ❑ Describe different fuel metering systems.
- ❑ Explain different types of ignition.
- ❑ Describe combustion chamber designs.
- ❑ Classify an engine by its valve location.
- ❑ Compare cam-in-block and overhead camshaft engine designs.
- ❑ Explain different methods of engine aspiration.
- ❑ Describe alternative engine designs.

Know These Terms

Backflow combustion chamber
Balancer shaft
Cam-in-block engine
Cast iron crankshaft
Crossflow combustion chamber
Cylinder arrangement
Dual overhead camshafts (DOHC)
Externally balanced
Firing order
Forged steel crankshaft
Four-valve combustion chamber
Fuel injected gasoline engine
Hemispherical combustion chamber
I-head engine
Inline engine
Internally balanced
L-head engine
Normally aspirated
Opposed engine
Overhead camshaft engine (OHC)
Pancake combustion chamber
Pent-roof combustion chamber
Single overhead camshaft (SOHC)
Spark ignition
Stratified charge combustion chamber
Supercharged engine
Swirl combustion chamber
Turbocharged engine
Two-valve combustion chamber
V-type engine
Wedge combustion chamber
W-type engine

To become a professional engine technician, you must be able to differentiate between the various engine types. Understanding how an engine is designed and constructed will help you when troubleshooting problems. An experienced technician can usually glance into an engine compartment and instantly describe numerous facts about how the engine is constructed and operates. For example, you might hear a technician say, "This is a dual overhead cam, four-cylinder engine with a direct ignition system and an intercooled turbocharger." This information would help the technician if he or she had to work on the engine.

This chapter introduces the many classifications and designs of modern engines. This information will prepare you for later chapters that discuss engine construction and service in more detail. In a sense, this chapter will help you develop the "language" of an engine technician; so study carefully!

Classifying Engines

There are many ways to classify an engine, even though the fundamental engine parts are basically the same. These small design differences, however, can greatly affect engine performance and service. Modern automotive engines are normally classified by one or more of the following:

- ❑ Arrangement of cylinders.
- ❑ Number of cylinders.
- ❑ Crankshaft design.
- ❑ Cooling system type.
- ❑ Type of fuel burned.
- ❑ Type of ignition.
- ❑ Fuel metering system.
- ❑ Combustion chamber shape.
- ❑ Cylinder head port design.
- ❑ Number of valves per cylinder.
- ❑ Valve location.
- ❑ Camshaft location and driving mechanism.
- ❑ Engine aspiration.
- ❑ Method of balancing the engine.

Cylinder Arrangement

Cylinder arrangement refers to the position of the cylinders in the engine block in relation to the crankshaft. There are four common, basic cylinder arrangements found in cars and light trucks: inline, V-type, W-type, and opposed.

Inline Engine

An ***inline engine*** has cylinders positioned one after the other in a straight line. The cylinders are located vertically in a line parallel with the crankshaft centerline. This is shown in **Figure 6-1.** The cylinders are usually vertical, but they may be at an angle to vertical to reduce the height of the engine. Inline engines with cylinders that are angled to vertical are called ***slant engines.***

Inline engines are very common today. They are well suited to small-displacement (size) engines. Three-, four-, five-, and six-cylinder engines are frequently an inline design. Because of their small size and good fuel economy, inline four-cylinder engines are one of the most common types found in today's cars. This design is found on domestic and import vehicles.

An inline engine is a very durable and powerful design. Each crankshaft rod journal carries only one connecting rod. Therefore, inline 4- and 6-cylinder engines can produce very high horsepower-to-displacement power ratios without part failure. For this reason, many high performance turbocharged gas and diesel engines are inline configurations.

V-Type Engine

A ***V-type engine*** looks like the letter V when viewed from the front or rear, **Figure 6-1.** The two banks (sets) of cylinders lay at an angle from vertical on each side of the crankshaft. A V-type design reduces the length and height of the engine. This can allow an engine to fit into a small

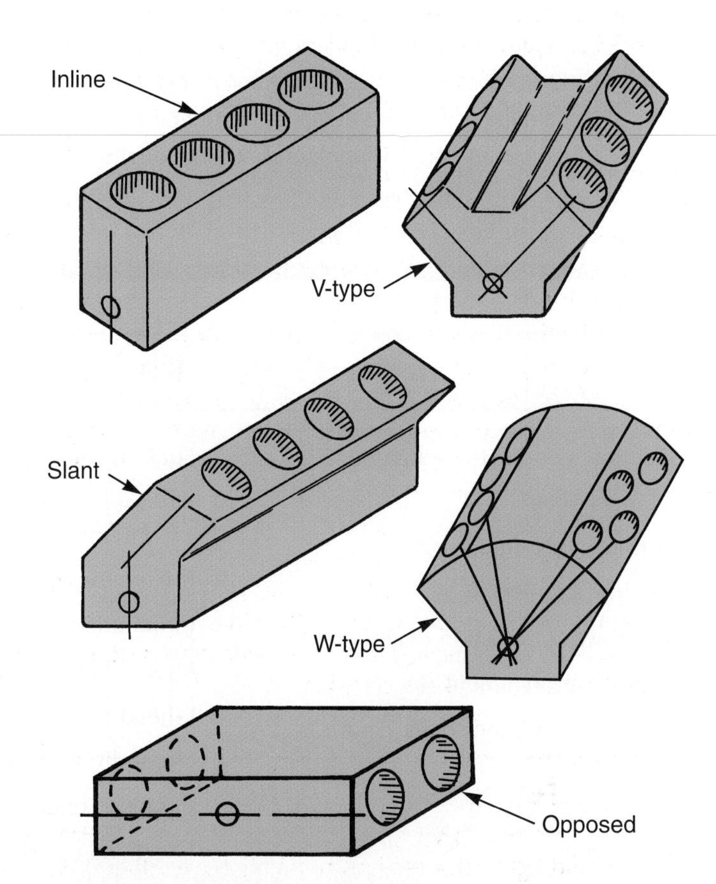

Figure 6-1. *These are common cylinder arrangements used in automobiles. Inline and V-type are more common than opposed and W-type configurations.*

engine compartment. An equal number of cylinders is needed in each bank. A V-type engine may have six, eight, ten, or twelve cylinders; V-6 and V-8 are the most common.

V-8 engines usually have a large displacement and can be found in performance cars, luxury cars, and light trucks. They are very powerful and idle very smoothly because there are four power strokes per crankshaft revolution.

The V-6 engine is now one of the most popular configurations. It is small, compact, produces good power, and provides good fuel economy. V-6 engines are lightweight and offer a good balance between power and fuel economy. They can be found as standard or optional equipment on most mid-size vehicles.

W-Type Engine

A ***W-type engine*** is similar to a V-type engine. It has two banks of cylinders, but the cylinder bores in each bank are staggered, **Figure 6-1.** This allows for a more compact engine.

Opposed Engine

An ***opposed engine,*** also termed "pancake engine," has two banks of cylinders that lay flat or horizontal on each side of the crankshaft, **Figure 6-1.** This configuration is also called a "boxer engine" because the pistons in opposite banks move toward each other, like the fists of two boxers.

An opposed cylinder arrangement is found in several makes of cars. An opposed 12-cylinder engine can be found in top-of-the-line Ferrari performance cars. Porsche also uses an opposed 6-cylinder engine in many of their high-performance sports cars. Subaru uses an opposed 4-cylinder engine in some of their four-wheel drive vehicles.

The opposed design allows for a vehicle with a very low center of gravity because most of the weight of the engine is near ground. Lowering the center of gravity improves the cornering of the vehicle. An opposed engine has the durability of an inline engine with the compact packaging of a V-type engine. In addition, the power strokes of the opposed cylinders tend to balance each other and reduce the thrust on the main bearings. Reducing this thrust also reduces the horsepower lost due to friction in the main bearings.

One disadvantage of an opposed engine is that oil leaks are more difficult to prevent, especially around the valve covers. The valve covers lay sideways and oil tends to seep out of the bottom of the valve cover gaskets. With inline and V-type engine designs, the valve covers sit more upright and are less likely to leak oil.

Number of Cylinders

Car and light truck engines normally have either 4, 6, or 8 cylinders. A few engines, however, have 3, 5, 10, 12, or 16 cylinders. A greater number of cylinders generally increases engine smoothness and power. For instance, an 8-cylinder engine produces twice as many power strokes per crankshaft revolution as a 4-cylinder engine. This reduces power pulsations and roughness (vibration), especially at idle. In turn, a 12-cylinder engine runs even more smoothly.

Four-cylinder engines usually have inline or opposed configurations. Six-cylinder engines can have inline, opposed, or V-type configurations. Five-cylinder engines normally have an inline configuration. Eight-, ten-, and twelve-cylinder engines have a V-type configuration.

Cylinder Numbering and Firing Order

Cylinder numbers identify the cylinders, pistons, and connecting rods of the engine. Cylinder numbers can be cast into the intake manifold; corresponding numbers are normally stamped into the sides of the connecting rods. Cylinder numbering varies with the engine design and cylinder arrangement.

Look at **Figure 6-2.** It shows typical cylinder numbers for V-8, V-6, inline 6-cylinder, and inline 4-cylinder engines. Four-cylinder engines are usually numbered 1-2-3-4 from the front to the rear. On V-type engines, you can normally tell the number one cylinder because it is located slightly ahead or in front of the front cylinder on the other side of the block.

Engine manufacturers use cylinder numbers so the engine technician can make repairs and do tune-up operations. For example, when rebuilding an engine, pistons and rods must be returned to the cylinder from which they were removed. Also, you will need to know which is the number one cylinder so a timing light can be connected during a tune-up.

The firing order is given in terms of the cylinder numbers. ***Firing order*** refers to the sequence in which combustion occurs in the engine. In other words, it is the order in which cylinders fire. The position of the crankshaft rod journals in relation to each other determines engine firing order. Two similar engines can have completely different firing orders.

Always refer to a service manual to verify cylinder numbers and firing order. Both will vary by manufacturer. For example, some V-type engines have one bank that contains all odd-numbered cylinders (such as 1, 3, 5) and the other bank all even-numbered cylinders (such as 2, 4, 6). Or, the cylinders may be numbered in sequence (1, 2, 3) on one bank and then in sequence (4, 5, 6) on the other bank.

Crankshaft Classification

There are several different ways to classify an engine crankshaft. The most common are:

- ❑ An ***inline crankshaft*** only has one connecting rod fastened to each rod journal, **Figure 6-3.**

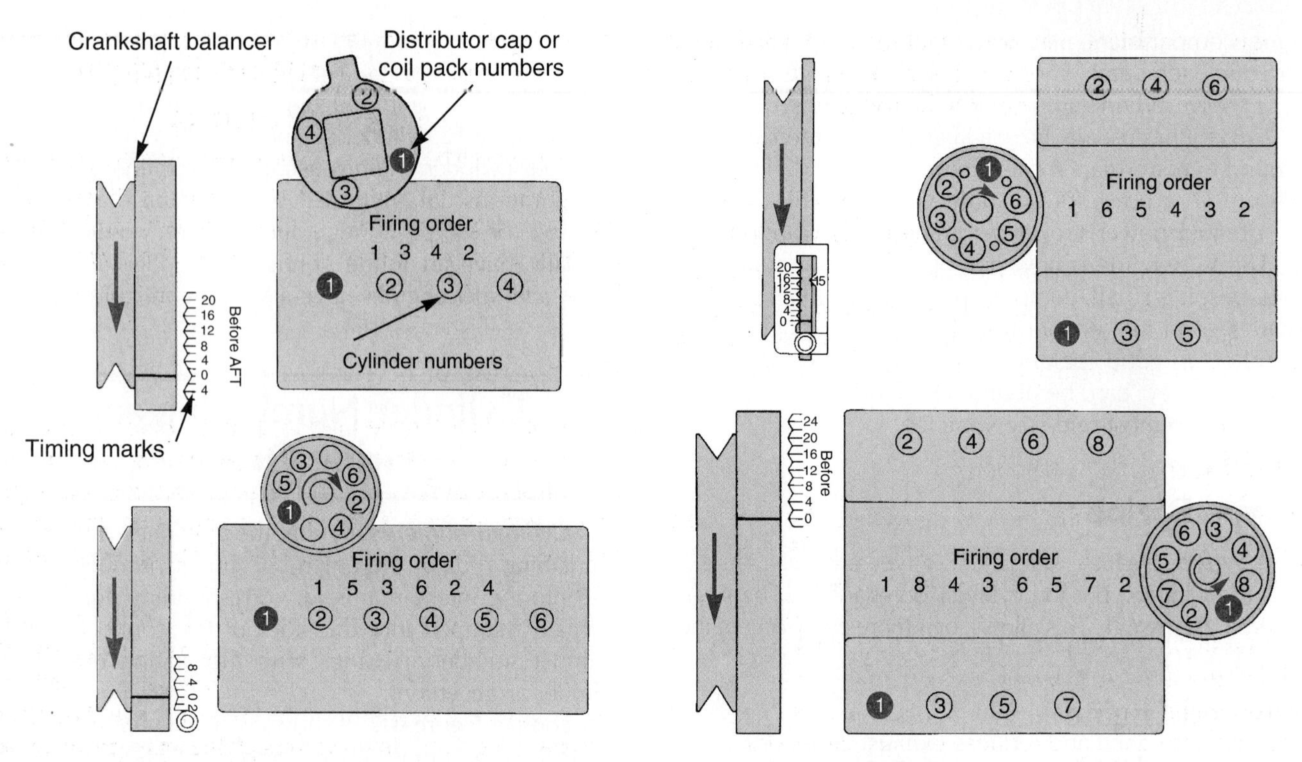

Figure 6-2. *Numbers usually run in sequence, from front to rear, on inline engines. Cylinder numbers vary with V-type engines. Note how you can determine the number one cylinder on V-type engines. It is slightly ahead of the front cylinder on the opposite bank. (Mitchell Manuals)*

- A ***V-type crankshaft*** has two connecting rods bolted to the same journal, **Figure 6-3.** The journal may be splayed or, more commonly, not splayed. A splayed crankshaft has the individual rod journals on the same crankshaft journal machined off-center. The splayed rod journal has an inherent weakness because the crankshaft pins can break under high load or engine speed conditions.
- A ***cast iron crankshaft*** is made by pouring molten iron into a mold. It is a common type for light- and medium-duty applications.
- A ***forged steel crankshaft*** is made by hammering hot steel into a mold using tons of force. It is much stronger and more rigid than a cast iron crankshaft and is used in high-performance applications.
- An ***internally balanced crankshaft*** uses the counterweights on the crankshaft to offset the weight of the rod and piston assemblies. Metal is added to or removed from the counterweights to balance the engine and prevent vibration.
- An ***externally balanced crankshaft*** uses internal counterweights, but also has weights on the flywheel and a

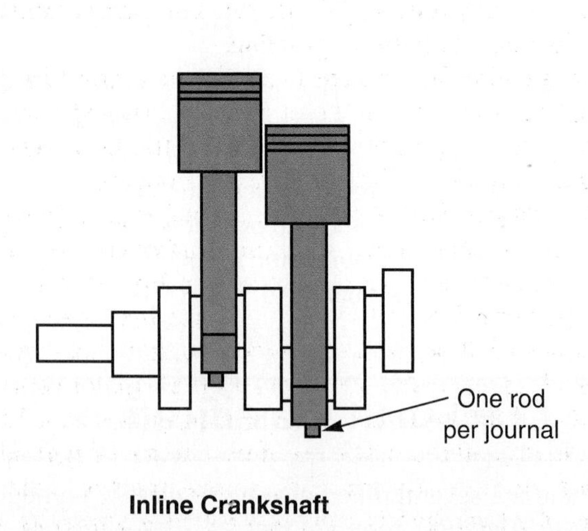

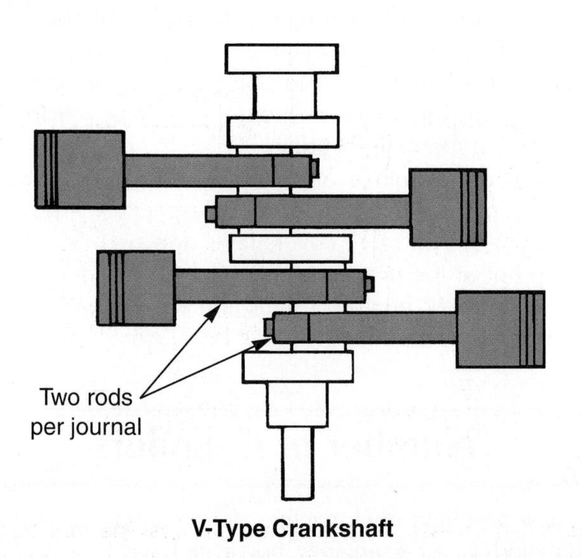

Figure 6-3. *A V-type crankshaft has two connecting rods per rod journal. This is why the number one cylinder on a V-type engine is ahead of the front cylinder in the opposite bank. Note how an inline engine only has one rod per journal.*

harmonic balancer to prevent vibration. A small metal pad is added to the flywheel or balancer to counteract a longer stroke machined on the rod journals.

Cooling System Classification

There are two types of systems providing cooling for the engine—liquid cooling and air cooling. See **Figure 6-4.** Almost all vehicles now have a liquid-cooled engine.

Liquid Cooling System

The ***liquid cooling system*** surrounds the cylinders with coolant (a water-antifreeze solution). The coolant carries combustion heat out of the cylinder head and engine block to prevent engine damage.

The liquid cooling system is very efficient because it will let the engine warm up quickly and can closely control engine operating temperature. This increases engine performance and reduces exhaust emissions.

Air Cooling System

An ***air cooling system*** circulates air over cooling fins on the cylinders and cylinder heads. This removes heat from the cylinders and heads to prevent overheating.

Air-cooled engines are not commonly used in modern cars and light trucks. The air cooling system cannot maintain as constant of an engine temperature as a liquid cooling system. This reduces engine efficiency and increases exhaust emissions. In order to comply with strict exhaust emission regulations, most vehicle manufacturers have phased out air-cooled engines.

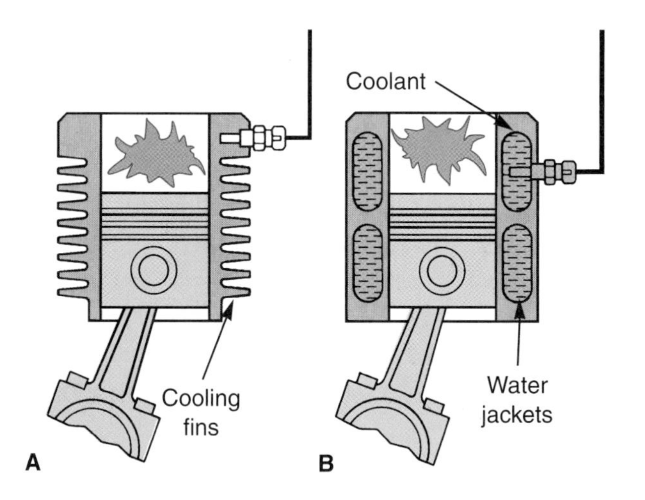

Figure 6-4. *A—An air cooling system has fins that transfer engine heat to the surrounding air. B—A liquid cooling system has pockets around the cylinders to hold coolant, which collects heat.*

Fuel and Ignition System Classifications

An automotive engine can also be classified by the type of fuel it burns, how it ignites the fuel, and how fuel is metered into the engine. These classifications are discussed in this section.

Fuel Type

A ***gasoline engine*** burns gasoline, which is normally metered into the intake manifold. A throttle valve is used to control the airflow into the engine. This controls engine speed and power. The piston compresses the air-fuel mixture. A spark plug ignites the compressed air-fuel mixture, **Figure 6-5A.**

A ***diesel engine*** burns diesel oil (fuel), which is a thicker fraction of crude oil. Diesel fuel is injected directly into the engine combustion chamber or a precombustion chamber. A precombustion chamber may be used to house the injector and glow plug. Only air flows through the intake manifold and a throttle valve is *not* used to control airflow and engine speed, **Figure 6-5B.** The amount of fuel injected into the combustion chambers controls engine speed. The compression ratio on a diesel is very high, generally between 17:1 and 20:1, which produces a very high pressure in the cylinder. When fuel is injected into the cylinder, it ignites and burns from the heat of high compression. A spark plug is not needed.

A ***liquefied petroleum gas (LPG) fuel system*** burns a very light fraction of crude oil. The LPG is stored in a high-pressure tank. At high pressure, the petroleum is in liquid form, not gas. A special converter is used to change the liquid into a gas and a metering system meters the gas into the engine intake manifold.

LPG is mainly propane and butane, but contains small amounts of other gasses. It has combustion qualities equal to or better than high-octane gasoline. LPG is a very good fuel. It produces good power, economy, and low exhaust pollution levels. LPG has operating characteristics almost identical to those of gasoline.

An ***alcohol fuel system*** is similar to a gasoline fuel system, but twice as much fuel must be metered into the engine. Also, for maximum efficiency, the compression ratio is usually higher than in a gasoline engine.

A ***hydrogen fuel system*** uses hydrogen gas. This is a very promising fuel system because it produces no pollution. Most systems in development use hydrogen as the fuel to power a fuel cell. However, since hydrogen gas is very explosive, it poses danger when stored in large quantities, such as at a distribution center or in a motor vehicle.

One of the best sources of hydrogen is water. Each molecule of water is made up of two atoms of hydrogen and one atom of oxygen. One simple method used to separate the hydrogen is a process called ***electrolysis.*** An electric current is sent through water to release the oxygen and hydrogen.

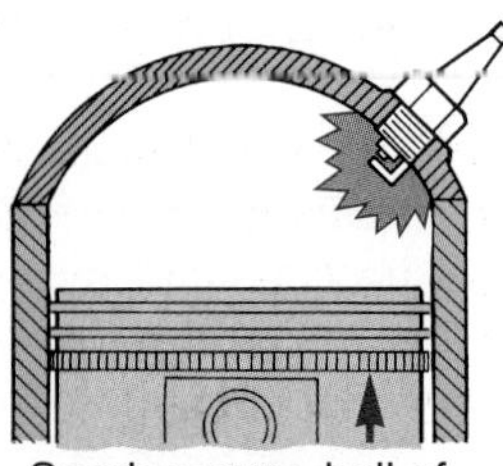

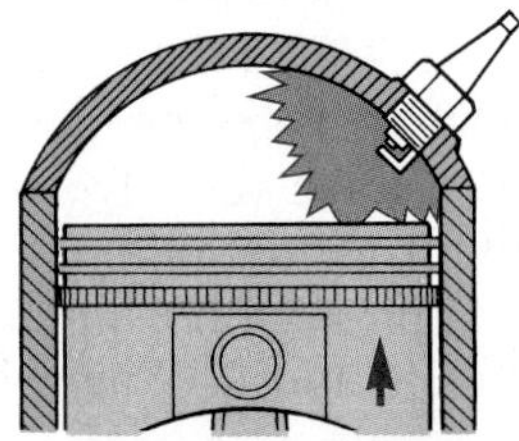

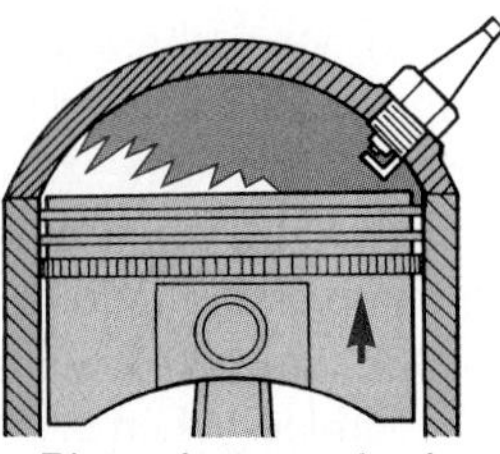

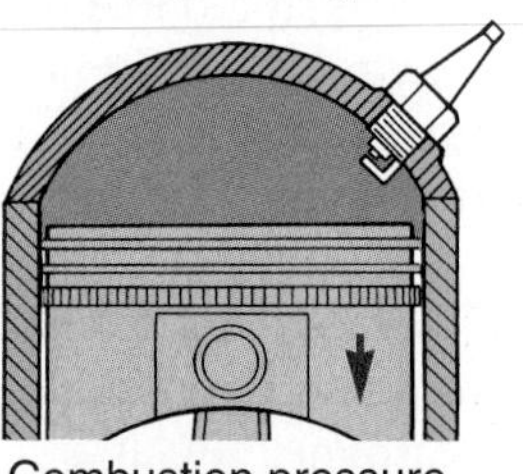

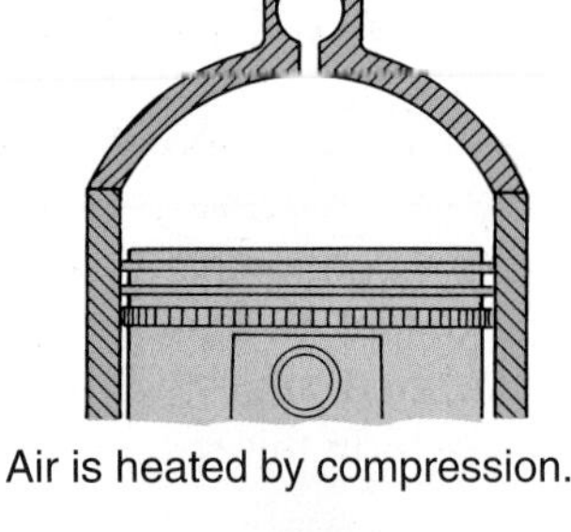

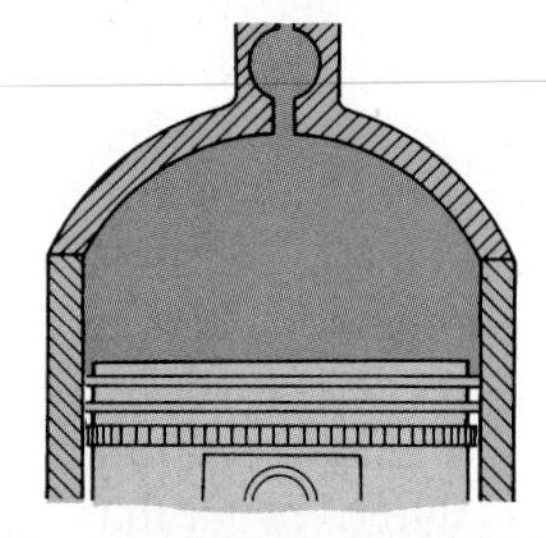

Figure 6-5. *A—The operation of a four-stroke cycle gasoline engine. Note that a spark plug provides the heat needed to ignite the fuel. B—The operation of a diesel engine. Note that the heat of compression ignites the fuel as the fuel is sprayed into the combustion chamber (or precombustion chamber).*

When burned with pure oxygen, hydrogen produces heat and water. When burned in an engine using air instead of oxygen, water and a very small amount of pollution (NO_X) are released. However, this is only a fraction of the pollutants formed by the combustion of most other fuels.

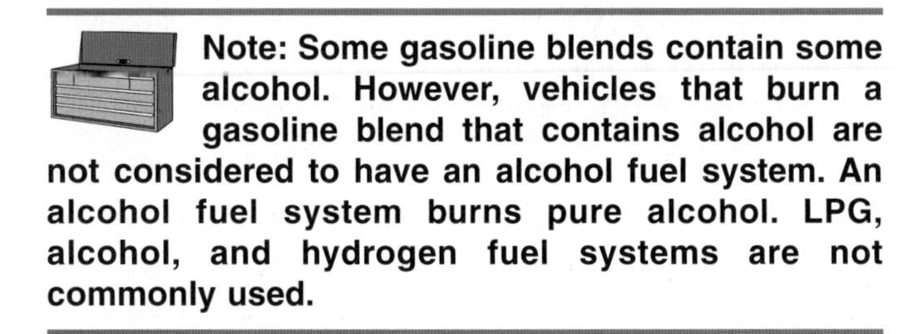

Note: Some gasoline blends contain some alcohol. However, vehicles that burn a gasoline blend that contains alcohol are not considered to have an alcohol fuel system. An alcohol fuel system burns pure alcohol. LPG, alcohol, and hydrogen fuel systems are not commonly used.

Ignition Classification

Two methods are commonly used to ignite (light) the air-fuel mixture in the engine combustion chamber—electric spark and compression. A ***spark ignition*** engine, **Figure 6-5A,** uses an electric ignition and a spark plug to start the combustion of the fuel. Gasoline-, LPG-, and alcohol-fueled engines use this method of ignition. A ***compression ignition*** engine, **Figure 6-5B,** uses the heat generated by the high compression pressure to heat the air and ignite the fuel. Diesel engines are compression ignition engines.

Fuel Metering

A gasoline engine can be classified by how fuel is metered into the engine. There are two basic types of fuel metering for gasoline engines—fuel injection and carburetion. Diesel engines are fuel injected.

A ***fuel injected gasoline engine*** sprays fuel into airstream in the intake manifold, either at the port or in the throttle body. An engine with this type of fuel metering is more properly termed a *gasoline* injected engine because a diesel engine also injects fuel into the engine. Computer-controlled gasoline injection is the most common type found on a modern car or light truck because it can closely match the fuel fed into the engine with engine load. Gasoline injection may be throttle body injection or multiport injection, **Figure 6-6.** In addition, the injection may be indirect, where fuel is injected into the airstream, or direct, where fuel is injected directly into the cylinder.

A ***carbureted engine*** uses engine vacuum to draw fuel out of the carburetor and into the engine intake manifold. Carbureted engines have been phased out for the more efficient computer-controlled gasoline injection.

Combustion Chamber Classifications

The design of the engine's combustion chamber can also be used to classify an engine. Its shape, number of valves per cylinder, port configuration, number of spark plugs, etc., can all be used to indicate how an engine is constructed. You should understand these differences if you are going to be a "top notch" engine technician.

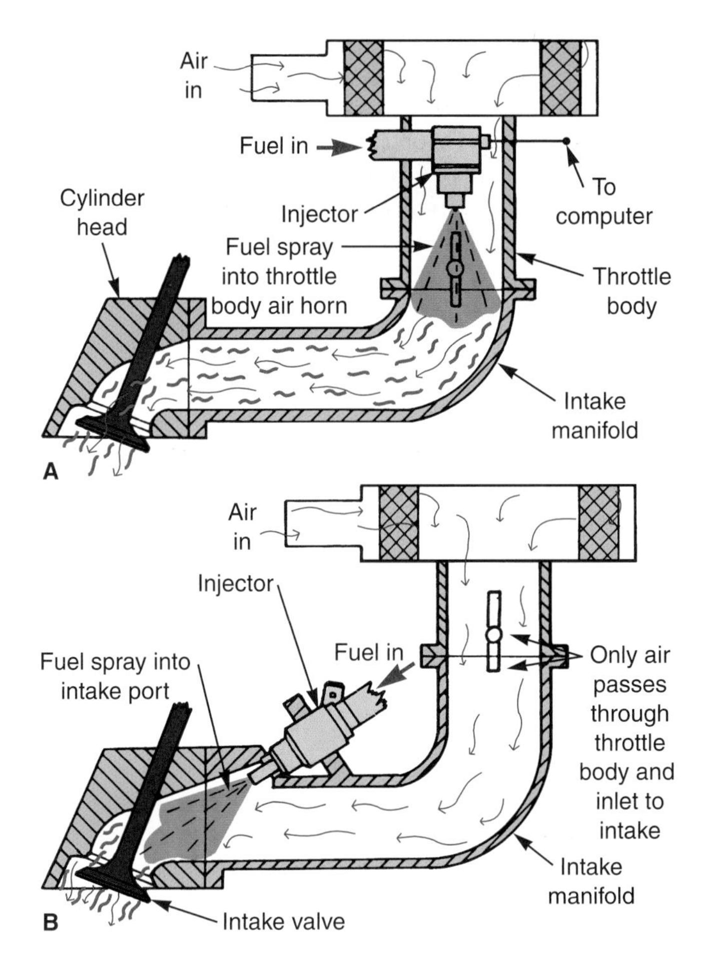

Figure 6-6. *A—In a throttle body injection system, the fuel injector is located in the throttle body. B—In a multiport fuel injection system, an injector is located right before each intake valve. Some multiport fuel injection systems inject the fuel directly into the combustion chamber. An indirect injection system is shown here.*

Combustion Chamber Shape

There are four basic combustion chamber shapes for gasoline engines. These are pancake, wedge, hemispherical, and pent-roof, **Figure 6-7.**

The ***pancake combustion chamber,*** also called bathtub chamber, has valve heads almost parallel with the top of the piston. The chamber forms a flat pocket over the piston head, **Figure 6-7A.** This is an older, less-common combustion chamber design.

A ***wedge combustion chamber,*** also called a wedge head, is shaped like a triangle or wedge when viewed from the side, as in **Figure 6-7B.** Valves are placed side-by-side with the spark plug next to the valves.

A ***squish area*** is commonly formed inside a wedge combustion chamber. When the piston reaches top dead center (TDC), it comes very close to the bottom of the cylinder head. This squeezes the air-fuel mixture in that area and forces it ("squishes" it) out into the main part of the chamber. Squish can be used to improve air-fuel mixing and burning at low engine speeds. A wedge combustion chamber normally uses a flat top piston.

A ***hemispherical combustion chamber,*** nicknamed a hemi head, is shaped like a dome when viewed from the side, **Figure 6-7C.** The valves are canted (tilted) on each side of the chamber. The spark plug is located near the center. This design is extremely efficient. There are no hidden pockets for incomplete combustion. The surface area is very small, reducing heat loss from the chamber. The centrally located spark plug produces a very short flame path for fast and complete combustion. The canted valves help increase the breathing ability of the engine.

The hemi head was first used in high-horsepower, racing engines. It is now used in many passenger car and light truck engines. It allows the engine to operate at high rpms and makes it very fuel efficient. It also produces complete burning of the fuel to reduce emissions.

One disadvantage of the hemi head is that it often requires a domed piston to obtain the needed compression

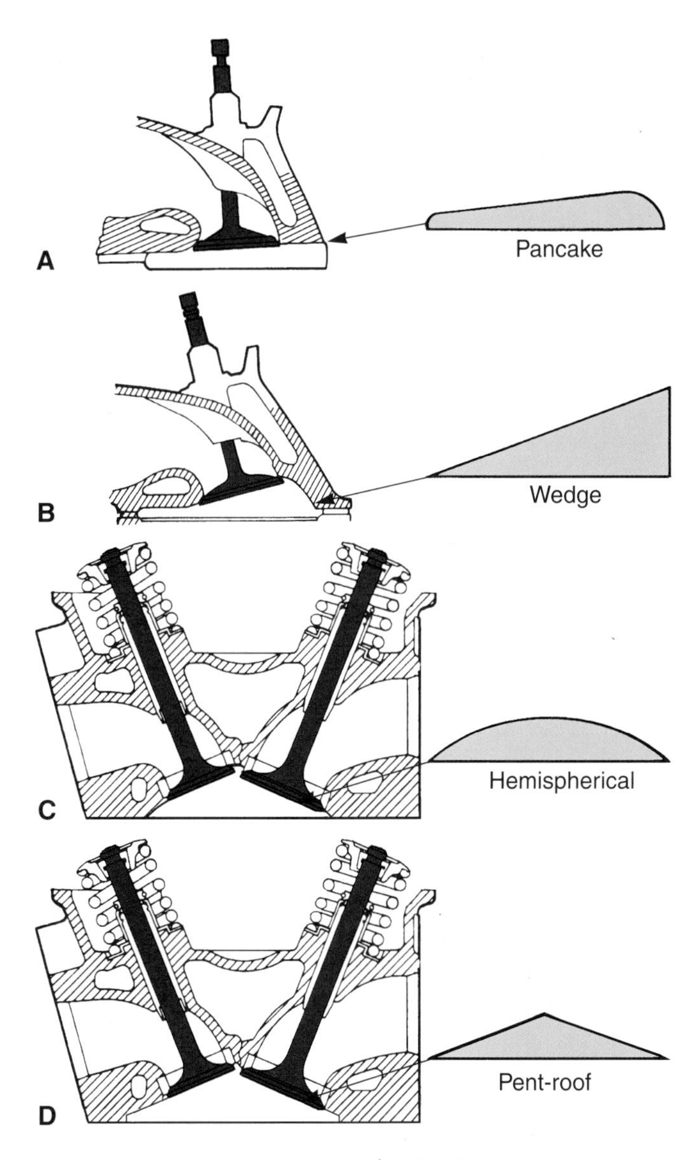

Figure 6-7. *There are four basic combustion chamber shapes. A—Pancake combustion chamber. B—Wedge combustion chamber. C—Hemispherical combustion chamber. D—Pent-roof combustion chamber. This type is common in four-valves-per-cylinder cylinder heads. (DaimlerChrysler)*

ratio or compression stroke pressure. The dome adds weight when compared to a flat top piston. The increased piston weight reduces mechanical efficiency during high-engine-speed operation.

The ***pent-roof combustion chamber*** is similar to the hemispherical combustion chamber, but it has flat, angled surfaces rather than a domed surface. See **Figure 6-7D.** This design improves volumetric efficiency and reduces emissions.

Combustion Chamber Type

There are many different types of combustion chamber. A combustion chamber may be classified by one or more of the characteristics described in this section.

A ***swirl combustion chamber*** uses the shape of the intake and exhaust ports and the shape of the combustion chamber roof to help mix the air-fuel mixture. As shown in **Figure 6-8,** a curve is provided in the intake port right before the intake valve and seat. Sometimes, a mask area is used to also control the movement of the mixture through the port and into the combustion chamber. Swirling the air-fuel charge improves combustion efficiency.

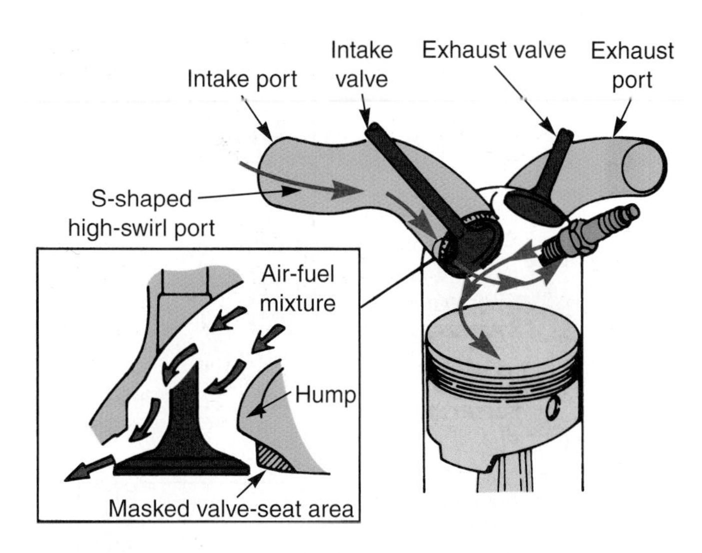

Figure 6-8. *A swirl combustion chamber uses the shape of the intake port and the entry of the air-fuel mixture into combustion chamber to mix the charge for more-efficient combustion.*

A ***crossflow combustion chamber*** has the intake ports on one side of the head and the exhaust ports on the other side. This is pictured in **Figure 6-9.** During engine operation, the air-fuel charge enters the combustion chamber on the intake stroke from one side of the head. Then, on the exhaust stroke, the burned gasses leave on the opposite side. The exiting exhaust gasses help pull more air-fuel charge into the combustion chamber to increase power.

A ***noncrossflow*** or ***backflow combustion chamber*** is illustrated in **Figure 6-10** as compared to a crossflow design. A noncrossflow combustion chamber is an older design that has been phased out for the more efficient crossflow combustion chamber.

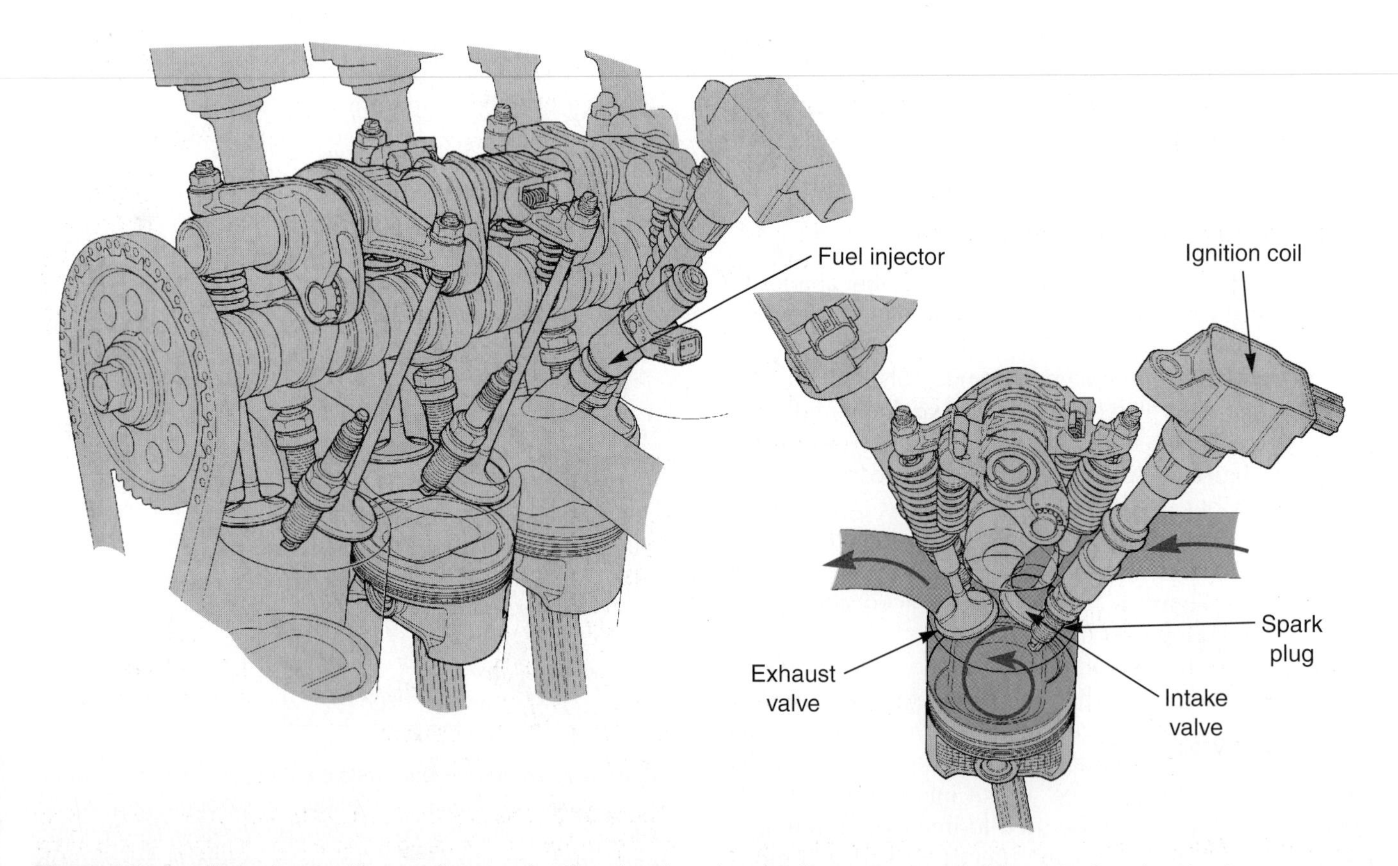

Figure 6-9. *A cylinder head with crossflow combustion chambers has intake ports on one side of the head and exhaust ports on the other. (Honda)*

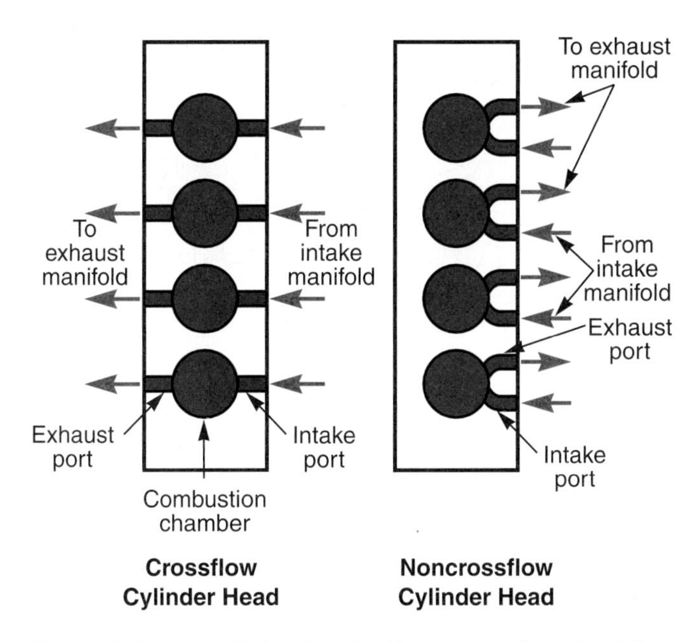

Figure 6-10. *A cylinder head with noncrossflow (backflow) combustion chambers has intake and exhaust ports on the same side of the head.*

A ***two-valve combustion chamber*** has one intake valve and one exhaust valve per cylinder. One is pictured in **Figure 6-9.** This is a conventional design that has been used for years. The intake valve is slightly larger than the exhaust valve.

A ***four-valve combustion chamber*** has two intake valves and two exhaust valves for each engine cylinder. This is a newer design used on several high-performance vehicles. A four-valve chamber is in **Figure 6-11.** A four-valve combustion chamber normally has ***siamese ports*** where one large port divides into two smaller ports right before the engine valve.

The valves in a four-valve chamber are generally smaller than the valves in a two-valve chamber. However, they expose a larger port area when open and will increase flow into and out of the chamber. This increases engine speed capability and the resulting power. A higher compression ratio can also be obtained with a four-valve combustion chamber. The valves do not have to be opened as wide to obtain the same flow. Therefore, the piston can be designed with less valve relief to compress the air-fuel charger tighter. This also increases combustion efficiency.

Twin (dual) camshafts are normally used with a four-valve chamber. Too many rocker arms and push rods would be needed to operate so many valves with a single camshaft. A view of a four-valve, dual camshaft cylinder head is given in **Figure 6-12.**

A ***mixture jet combustion chamber*** uses an extra passage running to the intake valve port. The mixture jet is used to aid swirl in the port and combustion chamber to help burning. The mixture jet mainly helps engine efficiency at low speeds. Refer to **Figure 6-13.**

An ***air jet combustion chamber*** has a small, extra valve that allows a stream of air to enter the combustion chamber to aid swirl and combustion efficiency. Shown in **Figure 6-14,** two conventional valves are provided. A third, smaller valve is also used. It opens to admit a gush of air that mixes the air-fuel charge to speed burning. The jet valve only works at idle and low engine speeds. At higher engine speeds, normal air-fuel mixing is adequate for efficient combustion.

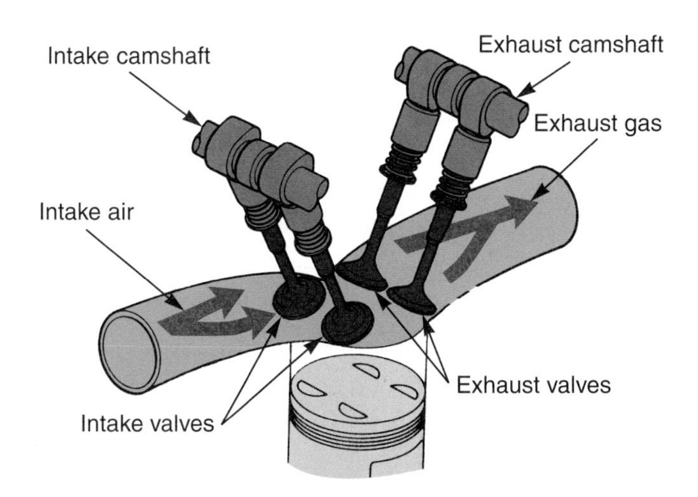

Figure 6-11. *Many modern high-performance engines have four valves per cylinder to allow better airflow and increase engine power. (Toyota)*

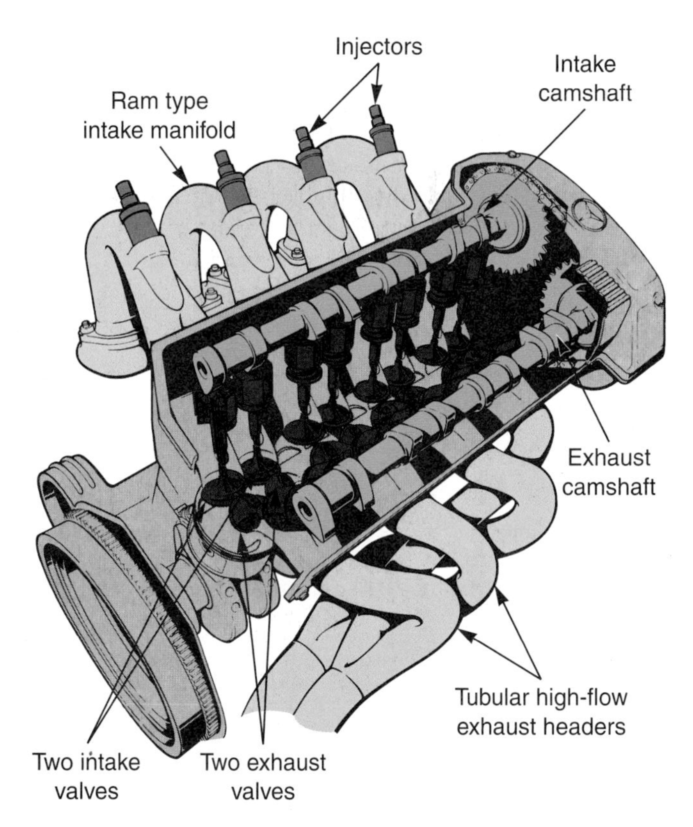

Figure 6-12. *This is a dual camshaft, four-valve cylinder head configuration. In this design, the intake runners and exhaust system are tubular to increase airflow through cylinder head at high speeds. (Mercedes-Benz)*

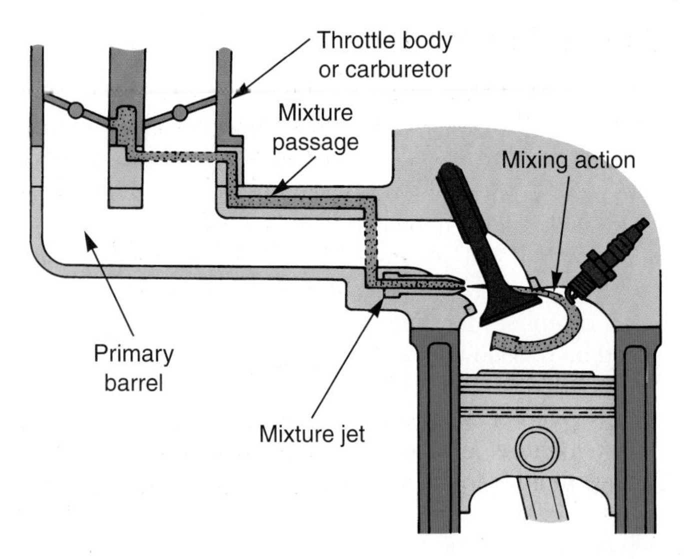

Figure 6-13. *A mixture jet injects a charge into the intake port right before the intake valve. This increases the mixing action and helps burning. (General Motors)*

A passage runs from the carburetor to the combustion chamber and jet valve. During the intake stroke at lower engine speeds, the engine camshaft opens both the conventional intake valve and the air jet valve. This allows fuel mixture to flow into the cylinder past the conventional intake valve. At the same time, a stream of air flows into the cylinder through the jet valve.

A ***stratified charge combustion chamber*** uses a flame in a small combustion prechamber to ignite and burn the fuel in the main, large combustion chamber, **Figure 6-15.** A stratified charge chamber allows the engine to operate on a lean, high-efficiency air-fuel ratio. Fuel economy is increased and exhaust emissions are reduced.

A very lean mixture (high ratio of air-to-fuel) is admitted into the main combustion chamber. The mixture is so lean that it will not easily ignite and burn. A richer mixture (high ratio of fuel-to-air) is admitted into the small prechamber by an extra valve. When the fuel mixture in the small prechamber is ignited, flames blow into and ignite the hard-to-burn lean fuel mixture in the main combustion chamber.

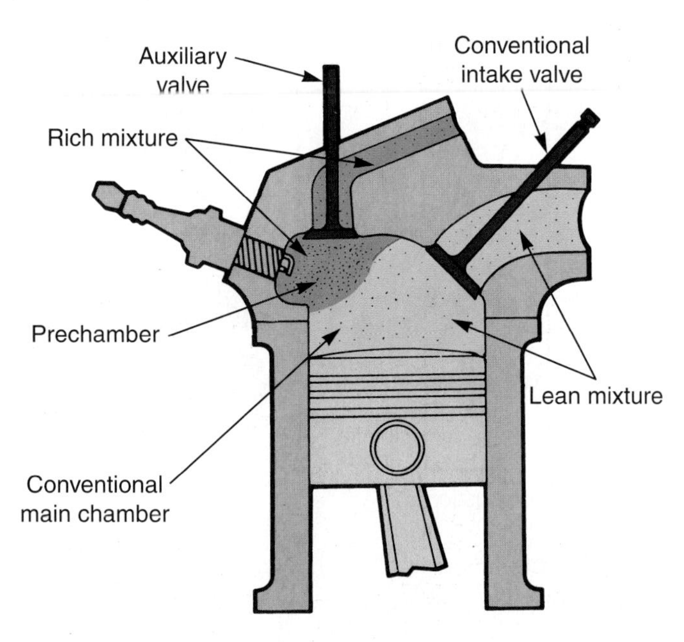

Figure 6-15. *A stratified charge combustion chamber has a small chamber to hold a rich air-fuel mixture. A lean air-fuel mixture enters the main combustion chamber. Then, the rich mixture is ignited by the spark plug; the flame blows into the lean mixture and ignites it. (Ford)*

Spark Plugs Per Cylinder

The number of spark plugs per cylinder is another way in which a combustion chamber may be classified. Most engines use one spark plug per cylinder. However, some high-performance, high-efficiency engines use two spark plugs per cylinder. Dual electric arcs (sparks) igniting the air-fuel mixture ensure more complete combustion of the fuel charge. Combustion is actually started in two

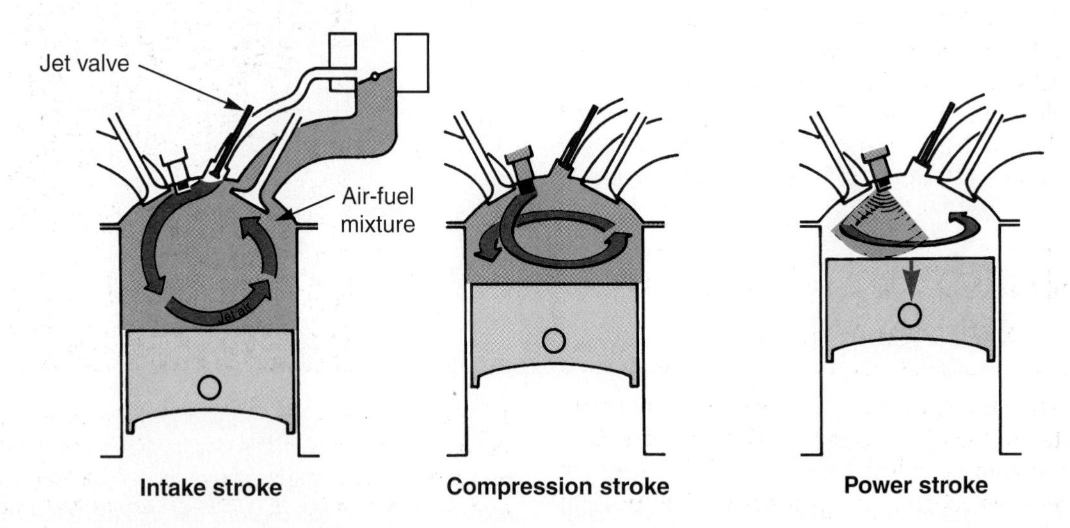

Figure 6-14. *An air jet chamber uses a small, third valve to inject an airstream directly into the combustion chamber. This helps mixing and burning at low speeds. (DaimlerChrysler)*

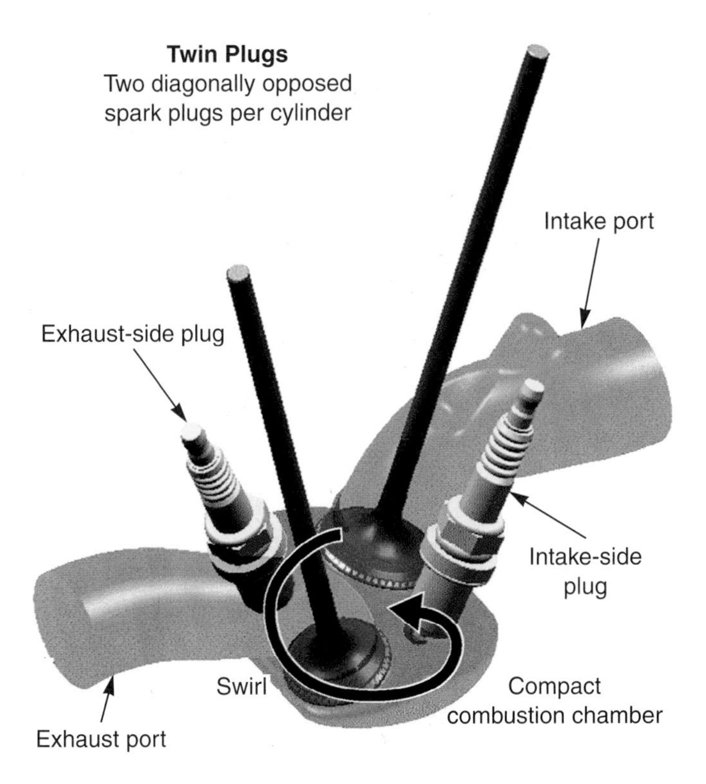

Figure 6-16. *By using two spark plugs per cylinder, horsepower can be increased and emissions decreased.*

locations in each combustion chamber. Two spark plugs per cylinder increases engine horsepower slightly while also decreasing exhaust emissions. Refer to **Figure 6-16.**

Valve Location

The location of the engine valves is still another way to describe the type of automotive engine. This classification is described in terms of the cylinder head.

In an ***I-head engine***, intake and exhaust valves are in the cylinder head. Another name for this design is overhead valve engine (OHV), **Figure 6-17.** The OHV engine is the most common design. Numerous variations are now in use.

An ***L-head engine*** has both the intake and exhaust valves in the block. See **Figure 6-18.** Also called a flat head engine, its cylinder head simply forms a cover over the cylinders and valves. This type of head is no longer used.

Camshaft Drive and Location Classifications

There are two basic locations for the engine camshaft. It can be located in the block or in the cylinder head. Both locations are common. Also, there are three types of camshaft drive—belt, chain, and gear drive. The location of the camshaft and the type of drive mechanism are two additional ways in which an engine can be classified.

Cam-in-Block Engine

A ***cam-in-block engine*** uses push rods to transfer camshaft motion to the rocker arms and valves, **Figure 6-19A.** The term *overhead valve (OHV)* is sometimes used to refer to this design instead of the term cam-in-block. A cam-in-block valve train requires a short timing chain and has less mass to rotate than an OHC valve train. This dependable design is used in many larger-displacement V-8 engines.

Overhead Camshaft Engine

In an ***overhead camshaft engine (OHC),*** the camshaft is located in the cylinder head. Refer to **Figure 6-19B.** Push rods are not needed to operate the rockers and valves. This

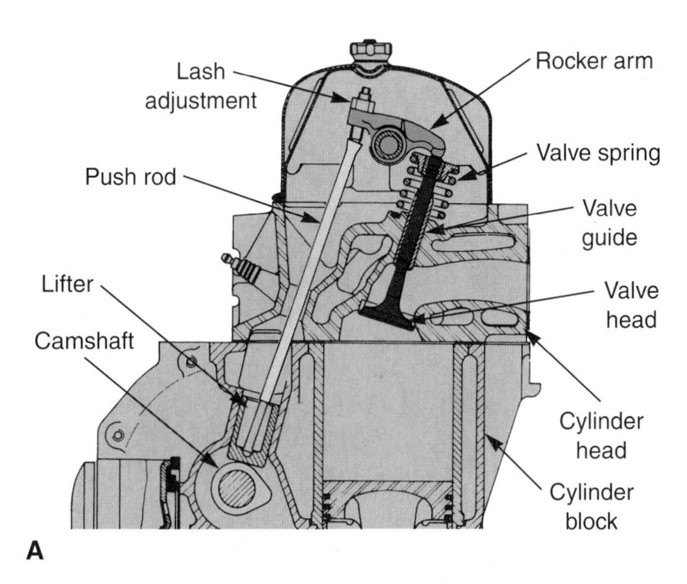

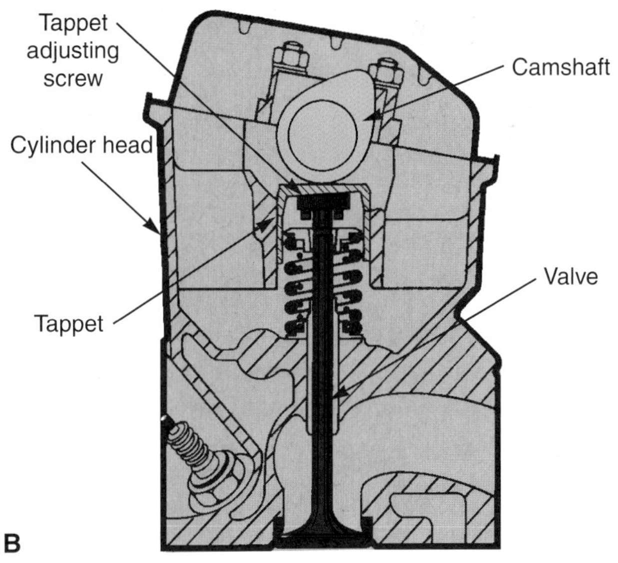

Figure 6-17. *An I-head engine has the valves in the cylinder head. A—Note how in this design the camshaft is in the block and a push rod transfers the camshaft motion to a rocker arm and then the valve. B—This is an I-head engine with an overhead camshaft. Notice how the camshaft acts directly on the valve. (Renault, DaimlerChrysler)*

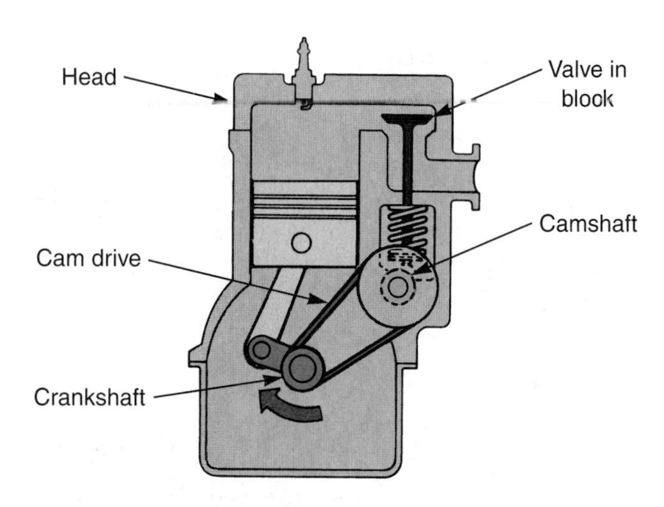

Figure 6-18. *The L-head engine is no longer used as an automotive engine. The valves are contained in the block in this design. (Ford)*

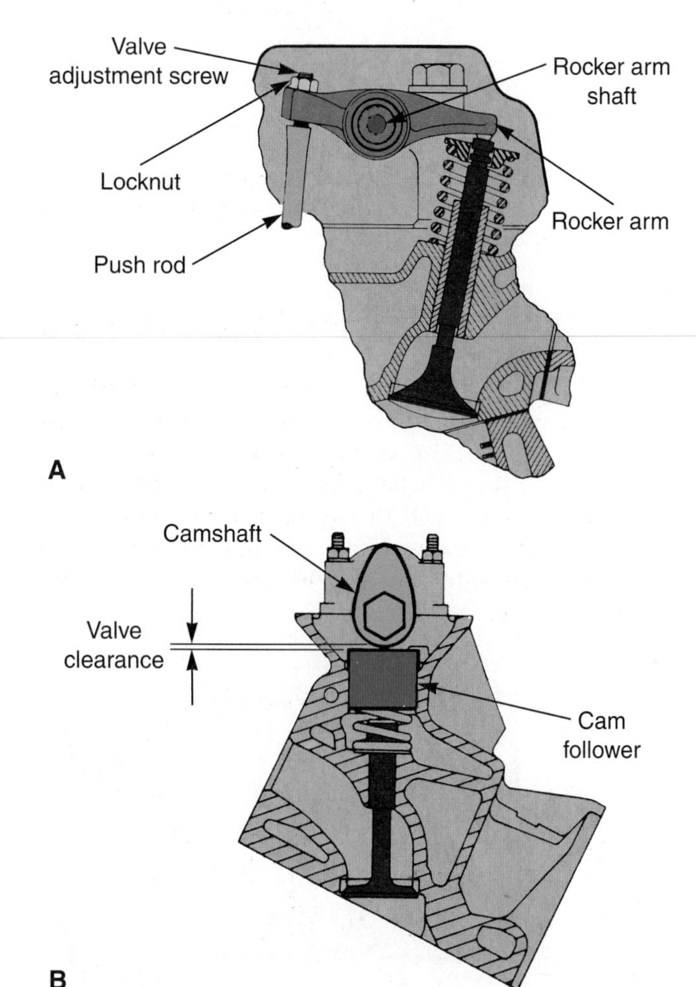

Figure 6-19. *Compare cam-in-block (overhead valve) and overhead camshaft configurations. A—When the camshaft is in the block, push rods and rocker arms transfer the camshaft motion to the valves. B—An overhead camshaft acts directly on the valve. (Federal Mogul)*

type engine is a refinement of the overhead valve engine. The industry trend is toward the OHC engine since it offers performance advantages over the cam-in-block engine.

With the camshaft in the head, the number of valve train parts is reduced. This cuts the weight of the valve train. Also, the valves can be placed at more of an angle to improve airflow through cylinder head ports.

OHC engines were first used in racing cars because of their efficiency at high engine speeds. Now, they are common in nearly all makes and models of vehicle. Without push rods to flex coupled with less valve train weight and improved valve positioning, the OHC increases high-speed efficiency and power output.

Single Overhead Camshaft Engine

A ***single overhead camshaft (SOHC)*** engine has one camshaft per cylinder head. The camshaft lobe may act directly on the valves or rocker arms may transfer motion to the valves. This is a very common design found on today's cars. See **Figure 6-20.** The single camshaft in a SOHC head operates both exhaust and intake valves. Rocker arms are needed to transfer camshaft lobe action to the valve stems.

Note: A V-type, W-type, or opposed engine may have two camshafts, but one *per cylinder head.* These are classified as SOHC engines.

Dual Overhead Camshaft Engine

A ***dual overhead camshaft (DOHC)*** engine has two camshafts per cylinder head. See **Figure 6-21.** One camshaft operates the intake valves and the other camshaft operates the exhaust valves. This design can be used with a two-valve, three-valve, or four-valve combustion chamber and is common in many late-model engines. A DOHC engine, although normally more powerful and efficient than an OHV or SOHC engine, is more complex and costly to repair.

Camshaft Drive

A camshaft ***belt drive*** uses a cogged rubber belt to turn the camshaft, **Figure 6-22.** A gear is mounted to the crankshaft. A larger gear is mounted to the camshaft. The timing belt connects the two gears. This type of drive is normally used when the camshaft is located in the cylinder head because it can span the long distance from the crankshaft to the camshaft. A belt drive provides very smooth, quiet, and trouble-free operation.

A camshaft ***chain drive*** is another common method of driving the camshaft. A gear is attached to the crankshaft. A larger gear is attached to the camshaft. A heavy-duty timing chain connects the two gears. This type of drive is normally used when the camshaft is located in the engine

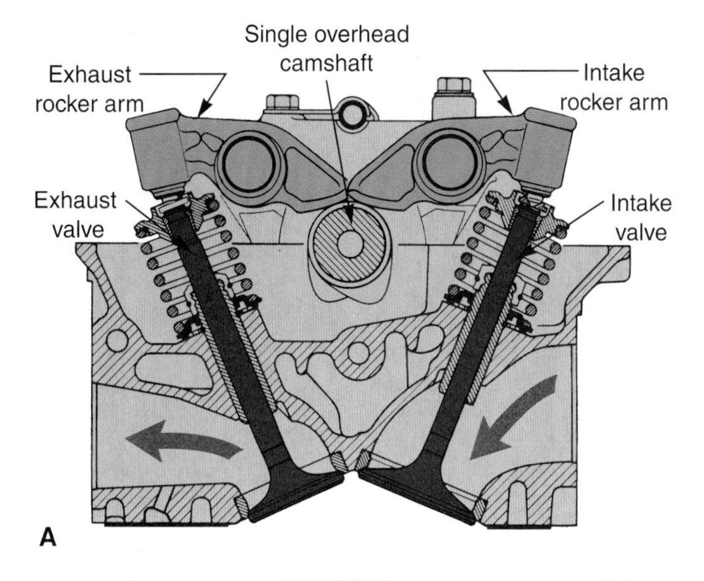

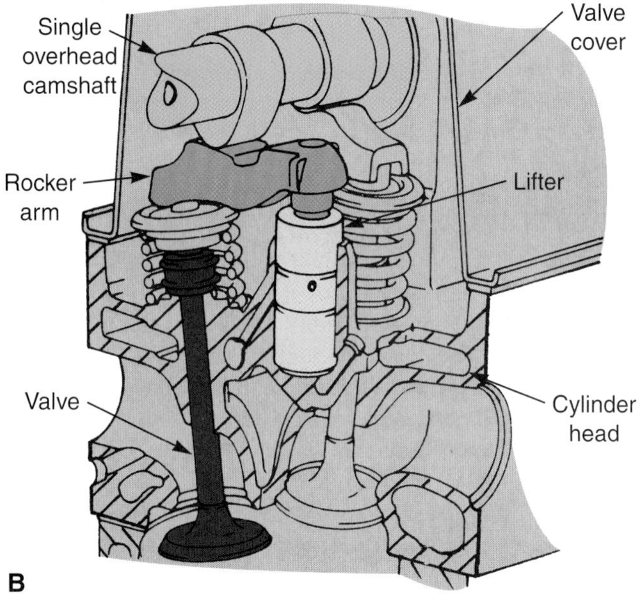

Figure 6-20. *A—A single overhead camshaft engine has one camshaft in the cylinder head. Camshaft lobes can act directly on the valves or rocker arms can be used. B—This SOHC engine uses hydraulic lifters to quiet engine operation. (Fiat, Ford)*

block. However, a timing chain can also be found on a few overhead camshaft engines.

A camshaft ***gear drive*** uses two or more meshed gears to turn the camshaft. The timing gears mesh directly; a belt or chain is not used. This type drive is a heavy-duty arrangement found on a few diesel engines and severe-service gasoline engines. The gears are extremely dependable, but heavy and noisy. They are only used with push rod–type engines.

Engine Aspiration

Engine aspiration refers to how air enters the engine, or how the engine "breathes." An engine can be classified

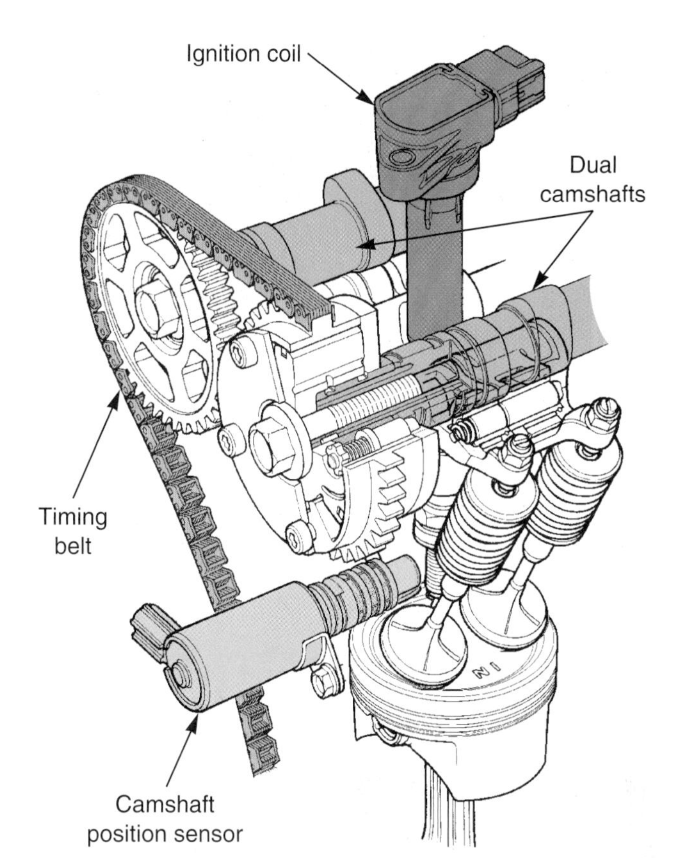

Figure 6-21. *A dual overhead camshaft engine has two camshafts in each cylinder head. The engine shown here has four valves per cylinder. (Honda)*

as normally aspirated, turbocharged, or supercharged, **Figure 6-23.**

A ***normally aspirated*** engine uses atmospheric pressure to force air into the engine. Atmospheric pressure is 14.7 psi (101.4 kPa) at sea level. This is the maximum pressure at which air is forced into the combustion chamber on each intake stroke. As a result, only a certain amount of air and fuel is forced into the combustion chamber. However, most vehicles have normally aspirated engines.

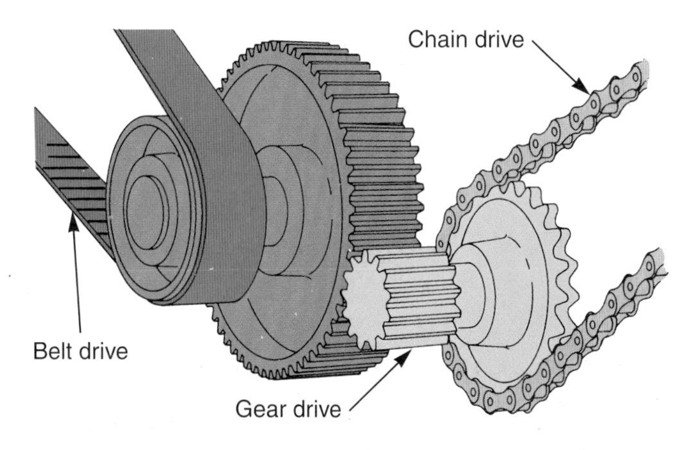

Figure 6-22. *Three methods of driving an engine camshaft: belt, gears, or chain. (Deere & Co.)*

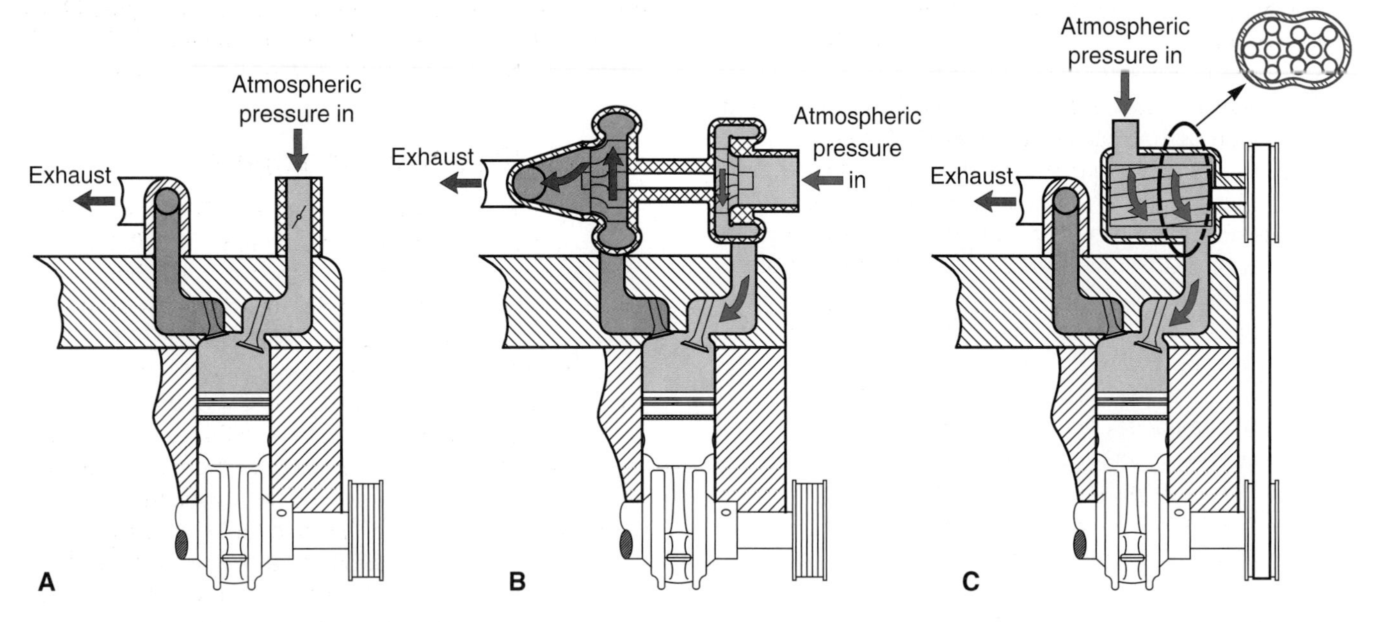

Figure 6-23. *A—A normally aspirated engine only uses atmospheric pressure to force the air-fuel charge into cylinders. B—A turbocharged engine uses exhaust gasses to spin a turbine. The turbine pressurizes the intake manifold and forces the air-fuel charge into the cylinders. C—A supercharged engine uses a belt, chain, or gears to spin a compressor (blower). The blower pressurizes the intake manifold and forces the air-fuel charge into the cylinders.*

At higher elevations above sea level, such as in the mountains, atmospheric pressure is lower than 14.7 psi. Since there is less pressure forcing air into the engine, the air-fuel charge is smaller. The power output from a normally aspirated engine is lower at higher elevations than at sea level.

A ***turbocharged engine*** forces air into the engine. As a result, the air-fuel mixture is forced into the cylinders under pressure, which allows a denser air-fuel charge. In other words, more air and fuel are forced into the cylinder on each intake stroke. The denser charge allows for more power output from the engine, up to 50% more.

A turbocharger, or "turbo," is driven by the flow of exhaust gasses from the engine. The time it takes for exhaust gasses to travel from the combustion chamber to the turbocharger can result in turbo lag. This is when the power increase from the turbocharger is slightly behind the increase in engine speed.

A ***supercharged engine*** also forces air into the engine. However, the supercharger, or "blower," is driven from the crankshaft rather than by exhaust flow. See **Figure 6-24.** This eliminates the turbo lag of a turbocharged engine.

This is a simplified explanation of turbocharging and supercharging. Several additional components are needed on turbocharged and supercharged engines compared to a normally aspirated engine. Refer to Chapter 16 for complete details of engine turbocharging and supercharging.

Engine Balancing Classification

Just as you balance a wheel and tire to keep the assembly from vibrating, an engine crankshaft and its related parts must be balanced to make the engine run smoothly. An engine can be classified by its method of balancing.

Most engines are ***internally balanced.*** The crankshaft counterweights are heavy enough to counteract the piston and rod weight. This prevents engine vibration. The flywheel and front damper are neutrally balanced. This means that they are equal in weight around their circumference.

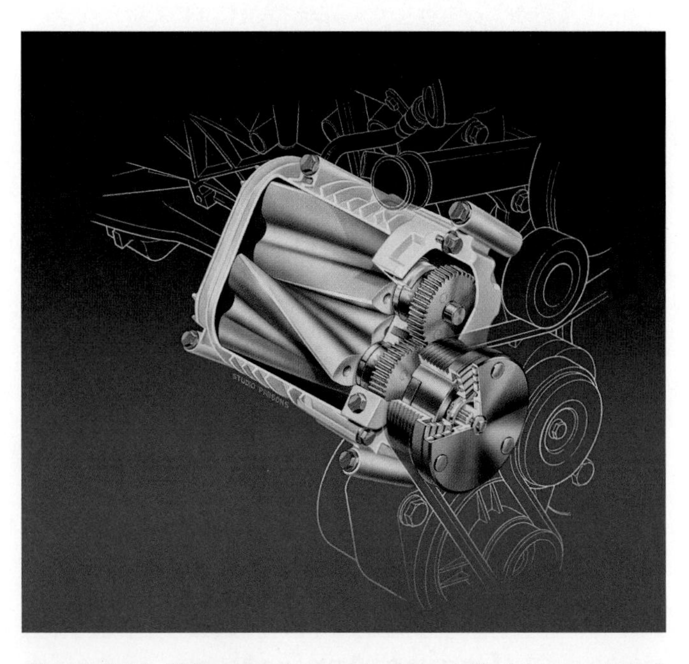

Figure 6-24. *This supercharger is driven off of the crankshaft by means of a belt. The supercharger contains two rotors that turn in opposite directions to compress the incoming air, which is discharged into the intake manifold. (Mercedes-Benz)*

A few engines are ***externally balanced.*** Extra weight is added to certain points on the flywheel and balancer to prevent engine vibration.

A ***balancer shaft*** is sometimes used to help smooth the operation of the engine, **Figure 6-25.** A balancer shaft, also called a silent shaft, is an extra shaft with counter-weights. The weights on the balancer shaft are positioned to counteract the non-power-producing strokes. A chain, belt, or set of gears connected to the crankshaft is used to spin the balancer shaft, usually at twice the engine speed. Two balancing shafts are frequently used in one engine.

A four-cylinder engine normally runs very rough at idle. Vibration can be felt at idle in the passenger compartment. The balancer shafts help keep the crankshaft spinning between power strokes. With the balancing shafts, low speed smoothness is increased tremendously. Only a small amount of energy is needed to spin the shafts so there is little reduction of fuel economy.

Alternative Engine Designs

Cars and light trucks generally have a four-stroke cycle, piston engine that operates on one of two fuels—gasoline or diesel fuel. However, there are several alternative engine designs in production or development. Some of these designs are variations of the four-stroke cycle, piston engine. Other designs have a completely different operational design. This section describes several alternative engine designs.

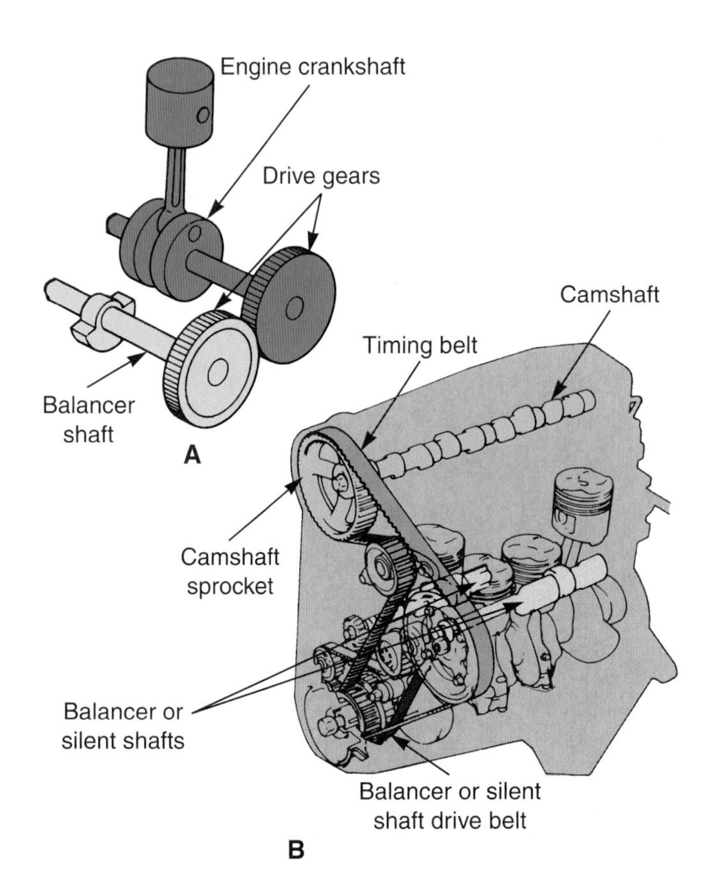

Figure 6-25. *Balancer shafts can be used to make the engine run smoother at idle. Counterweights on the shafts help keep the crankshaft spinning on non-power-producing strokes. A—Simple balancer shaft. B—Turbo diesel balancer shafts. (Ford)*

Cylinder Deactivation Engine

A ***cylinder deactivation engine (CDE),*** sometimes termed a variable displacement engine (VDE), uses solenoid-operated rocker arms to alter the number of engine cylinders that function during engine operation. By disabling rocker arms, valves, and fuel injectors, the number of cylinders firing and using fuel can be reduced. This, in effect, reduces the displacement of the engine.

When accelerating, all of the cylinders and rocker arms function normally. However, when cruising speed is reached and there is a less demand for power, the computer can deactivate cylinders to improve fuel economy.

Figure 6-26 shows the basic parts of a cylinder deactivation engine. This is not a very common design. Even with the valves deactivated, the rings, bearings, and other moving parts in the "dead" cylinders still contribute to frictional losses and reduce efficiency.

Variable Compression Ratio Engine

A ***variable compression ratio engine (VCRE)*** can alter the volume of its combustion chambers, and thus its compression ratio, for improved operating efficiency. In this design, the cylinder sleeves are attached to the cylinder head, but are free to move up and down in the cylinder block. The engine is also supercharged.

As shown in **Figure 6-27,** the cylinder head is movable or tiltable. One side of the cylinder head is mounted on a pivot bar. The other side of the head is positioned by an

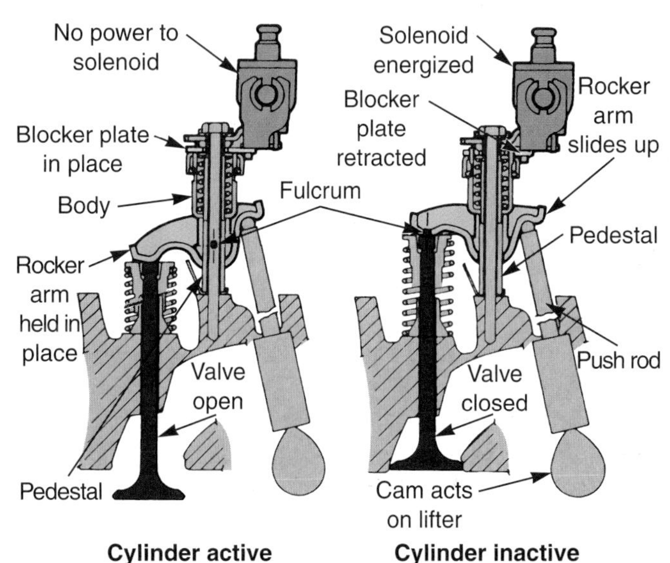

Figure 6-26. *Cylinder deactivation engines simply disable rocker arms to keep certain cylinders from working and consuming fuel. This design is old and these engines are rare. (General Motors)*

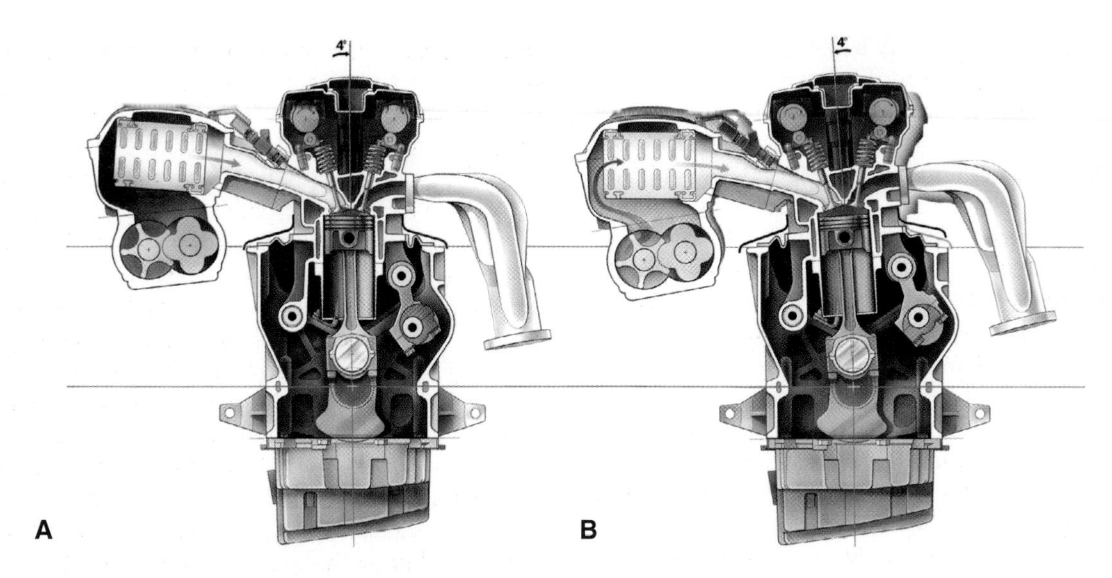

Figure 6-27. *This is a variable compression ratio engine. The cylinder head-cylinder assembly is mounted on pivot point. A membrane seals the gap between the block and cylinder head. A—In low-load operation, the compression ratio is high. B—During high-load operation, the cylinder head tilts (in this design), decreasing the compression ratio and allowing more supercharger boost. (Saab)*

electro-hydraulically actuated arm. The arm can be raised or lowered to tilt the cylinder walls, cylinder head, and related top end parts in relation to the crankshaft centerline. A flexible membrane between the block and the cylinder head seals the crankcase. The crankshaft is mounted in the cylinder block in a conventional manner.

During operation, the head and cylinders are held in the down position to increase the engine's compression ratio for low-load, low-emission, high-fuel-economy operation. The supercharger is also turned off. During high-load operation, the assembly is lifted up to move the head away from the crankshaft. This increases the area of the combustion chamber and reduces the compression ratio. The supercharger is turned on, which boosts the pressure in the cylinder and increases the horsepower output. However, fuel economy decreases and emissions increase.

An on-board computer matches the compression ratio to the load on the engine. In this way, a small-displacement, variable compression ratio engine can generate more power and better fuel economy than a larger, conventional engine.

Variable Valve Timing Engine

A ***variable valve timing engine (VVTE)*** can alter valve opening and closing independent of crankshaft rotation. These engines can match valve lift and duration to engine speed and load for improved power and lower exhaust emissions. Valve lift is how far the valve opens. Valve duration is how long the valve stays open. This improves engine efficiency compared to conventional engines that fix valve action with each degree of crankshaft rotation and single camshaft lobe profile. Most variable valve timing designs have electro-hydraulically operated rocker arms to alter valve operation. See **Figure 6-28.** A small hydraulic piston in the rocker arm shaft is used to engage and disengage the assembly. A rotating camshaft gear can also be used to change valve timing, but not lift and duration.

At low engine speeds, only one intake valve is fully opened per cylinder. The other intake rocker arm for that cylinder only slightly opens its intake valve. This slight opening is needed to avoid fuel accumulation behind the valve face. By only opening one of the two intake valves, air velocity through the intake port is increased. This, in turn, improves atomization and mixing of the fuel charge for improved combustion efficiency at low engine speeds.

At high engine speeds, the two intake rocker arms are locked to a mid-intake rocker arm that provides very high valve lift and longer duration. The mid-rocker arm rides on a high-speed camshaft lobe. This allows a larger, more-powerful air-fuel charge to enter the combustion chambers.

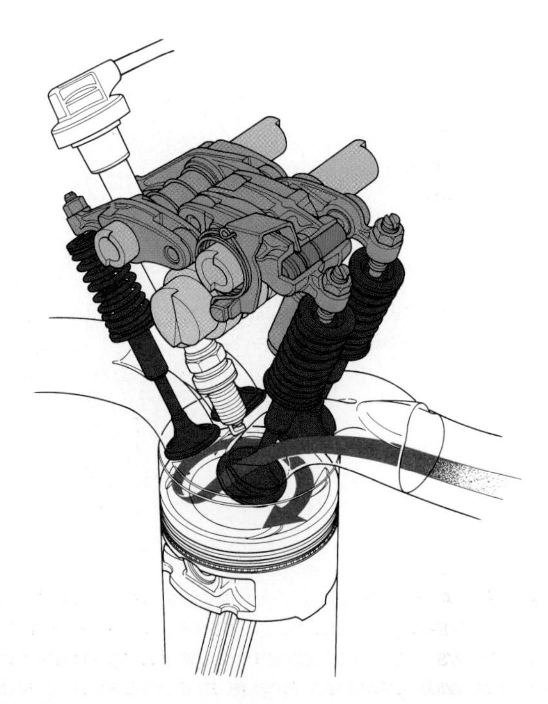

Figure 6-28. *This engine uses small hydraulic pistons acting on the rocker arm to alter valve lift and duration. (Honda)*

Variable Intake Engine

A ***variable intake engine (VIE)*** can adjust its intake manifold runner length to match engine speeds. At low engine speeds, long, small-diameter intake manifold runners increase the speed of the air-fuel charge, creating a "ram effect" that allows a denser charge into the combustion chamber. However, at high engine speeds, short, large-diameter intake manifold runners are needed to produce the same effect. Conventional, fixed-runner intake manifolds are designed as a compromise between low-speed and high-speed efficiency.

Various methods are used by engine designers to provide variable length intake manifolds. Some of these include dual intake runners (one long and one short), different routing of intake manifold runners using butterfly valves, or by a rotating runner device that lengthens or shortens the airflow path.

Figure 6-29 shows one type of VIE design. This engine simply has two intake manifold runner lengths. A butterfly valve in the intake manifold plenum can open and close to select either route for incoming air.

Figure 6-30 shows a variable runner, also called variable resonance, intake manifold design. A flap or butterfly valve in the intake manifold can be opened or closed to alter the manifold runner length. The long intake runner length helps low-speed operation. The short intake runner is for high-speed operation.

Another variation uses a stack of hollow plastic runner wheels mounted inside the manifold plenum to vary the intake. By rotating the runner wheels, each manifold runner can be shortened or lengthened as needed. During low-engine-speed operation, the runner wheels are turned so that incoming air has a very long path to flow through before entering the combustion chambers. Then, as engine speed increases, the runner wheels are turned to shorten the effective runner length.

Variable Backpressure Exhaust Engine

A ***variable backpressure exhaust engine (VBEE)*** uses different paths through the engine exhaust manifolds to improve engine efficiency. Similar to a variable intake manifold, the system can increase exhaust runner length for low-engine-speed operation and decrease exhaust runner length for high-engine-speed performance. A VBEE is rare and presently only used on exotic, high-performance engines from manufacturers such as Aston Martin, Ferrari, Honda, Lamborghini, Nissan, and Mitsubishi.

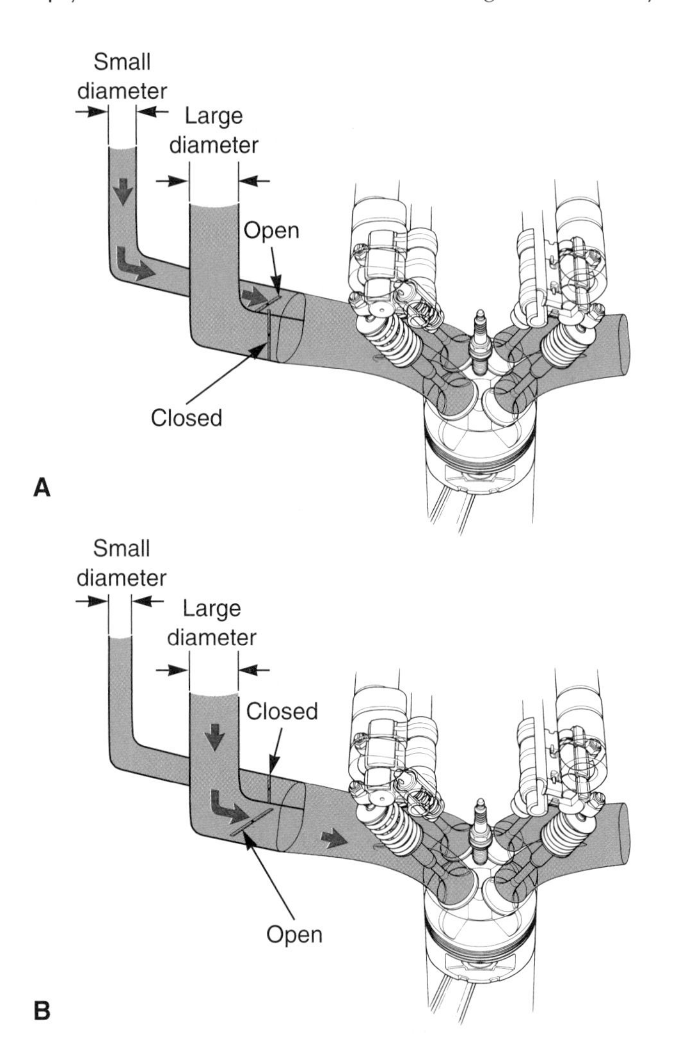

Figure 6-29. *A variable intake engine alters the path that air takes when entering combustion chambers. A—At low engine speeds, the airstream is directed through a long, small-diameter runner to provide good air velocity and mixing of fuel. B—At high engine speeds, the airstream is directed through a short, large-diameter runner to allow for higher power output.*

Figure 6-30. *This is a variable intake system that has a dual-resonance, variable-length intake manifold. At low speeds, intake air takes the long route through the curved intake runners. At high speeds, a butterfly valve opens to shorten the path for the incoming air. (Mercedes-Benz)*

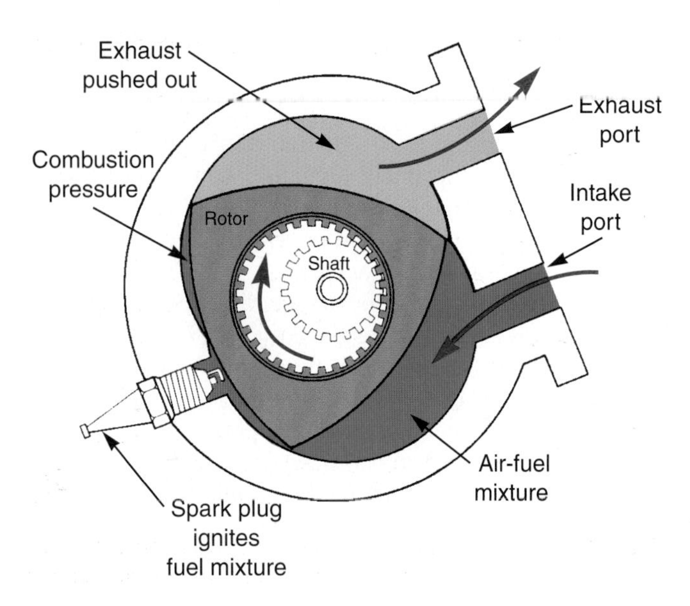

Figure 6-31. *Study the parts of a rotary automotive engine. The rotor spins inside the housing for smooth operation. There are no reciprocating pistons in this design.*

Wankel (Rotary) Engine

A ***Wankel engine,*** also termed a ***rotary engine,*** uses a triangular rotor instead of conventional pistons, **Figure 6-31.** The rotor turns or spins inside of a specially shaped housing. A rotary engine is one of the few alternative engine designs to be mass produced and installed in production vehicles.

Figure 6-32 illustrates the basic operation of a rotary engine. While the rotor spins on its own axis, it also orbits around a main shaft. This eliminates the normal reciprocating motion found in piston engines. One complete cycle (all four strokes) takes place every time a rotor face completes one revolution. Since there are three rotor faces, three power strokes are produced per rotor revolution.

A rotary engine is very powerful for its size. Also, because there is no up and down motion, engine operation is very smooth and vibration free.

In the past, a complicated emission control system was needed to make the rotary engine pass emission standards. This has limited use of the traditional rotary engine. The newest rotary engine, called the Renesis design, has intake and exhaust ports in the engine endplates and to the side of the rotor housing. See **Figure 6-33.** The ports are no longer in the periphery of the rotor housing. This helps reduce exhaust emissions because it eliminates the intake/exhaust port overlap of the older design, which contributed to higher emissions.

Miller-Cycle Engine

A ***Miller-cycle engine*** uses a modified four-stroke cycle. This engine is designed with a shorter compression stroke and a longer power stroke to increase efficiency. The intake valve remains open longer to delay compression. Because the intake valve remains open for a relatively long time, the air-fuel charge tends to travel back out the intake port. To compensate for this reverse flow, a supercharger is normally used to pressurize the intake manifold and block this flow, **Figure 6-34.**

Theoretically, a Miller-cycle engine can produce more power and is more economical than a conventional four-stroke cycle engine of equal size. However, the engine is more complex to produce and maintain, partly due to the supercharger. Miller-cycle engines are currently used in limited applications in some passenger cars.

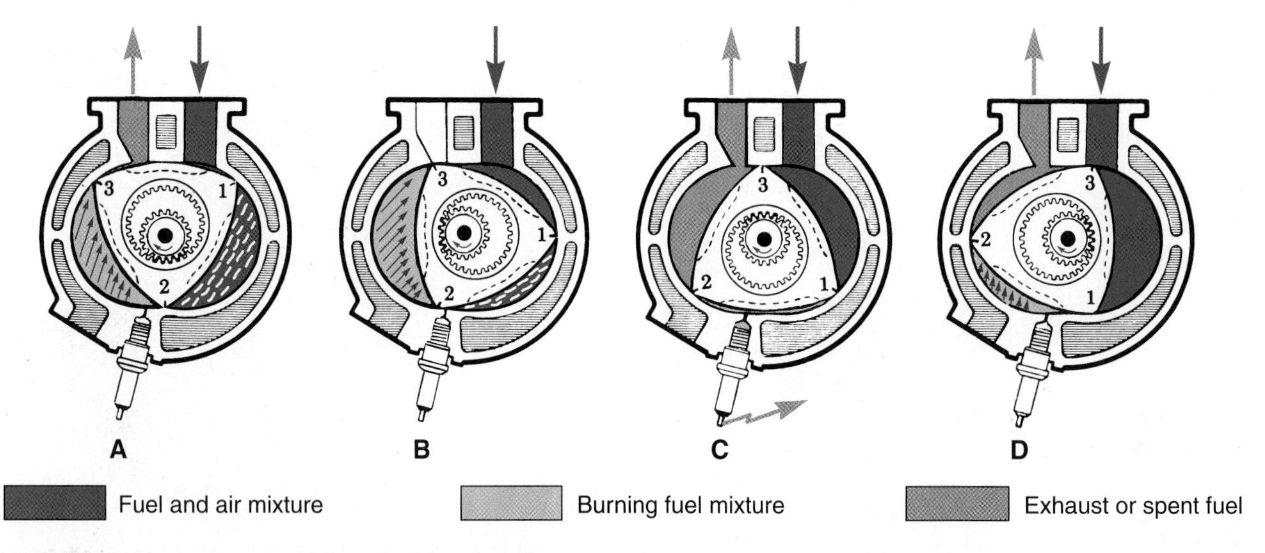

Figure 6-32. *The sequence of events in a rotary engine. A—Intake is starting between Points 1 and 3; compression is occurring between Points 1 and 2; power is being produced between Points 2 and 3; exhaust is finishing between Points 3 and 1. B—Intake continues between Points 1 and 3; compression continues between Points 1 and 2; power is finishing between Points 2 and 3. C—Intake continues between Points 1 and 3; spark occurs between Points 1 and 2 and power begins; exhaust begins between Points 2 and 3. D—Intake is finished between Points 1 and 3; power is being produced between Points 1 and 2; exhaust is continuing between Points 2 and 3.*

Figure 6-33. *In this new rotary engine design, the intake and exhaust ports are located in the endplates. This reduces emissions considerably.*

Figure 6-34. *Study this cutaway view of a Miller-cycle engine. Note the small supercharger to prevent blow-back into intake manifold. (Mazda)*

Hybrid Power Source

In response to emission and fuel economy standards, some vehicle manufacturers experimented in the past with using an electric motor and large storage batteries to power an automobile. However, overall their production was limited. These vehicles saw some success as a means of transportation for short trips. However, speed and driving distance was limited.

As an alternative to electric or gasoline-powered vehicles, several manufacturers have developed hybrid vehicles, or vehicles that use a hybrid power source. A ***hybrid power source*** uses two different methods of propulsion, usually a gasoline engine and a large electric motor. See **Figure 6-35.** In a hybrid vehicle, a high-efficiency gasoline engine, powerful electric motor-generator, and battery pack work together to propel the vehicle. Hybrid vehicles offer superior gas mileage and reduced emissions.

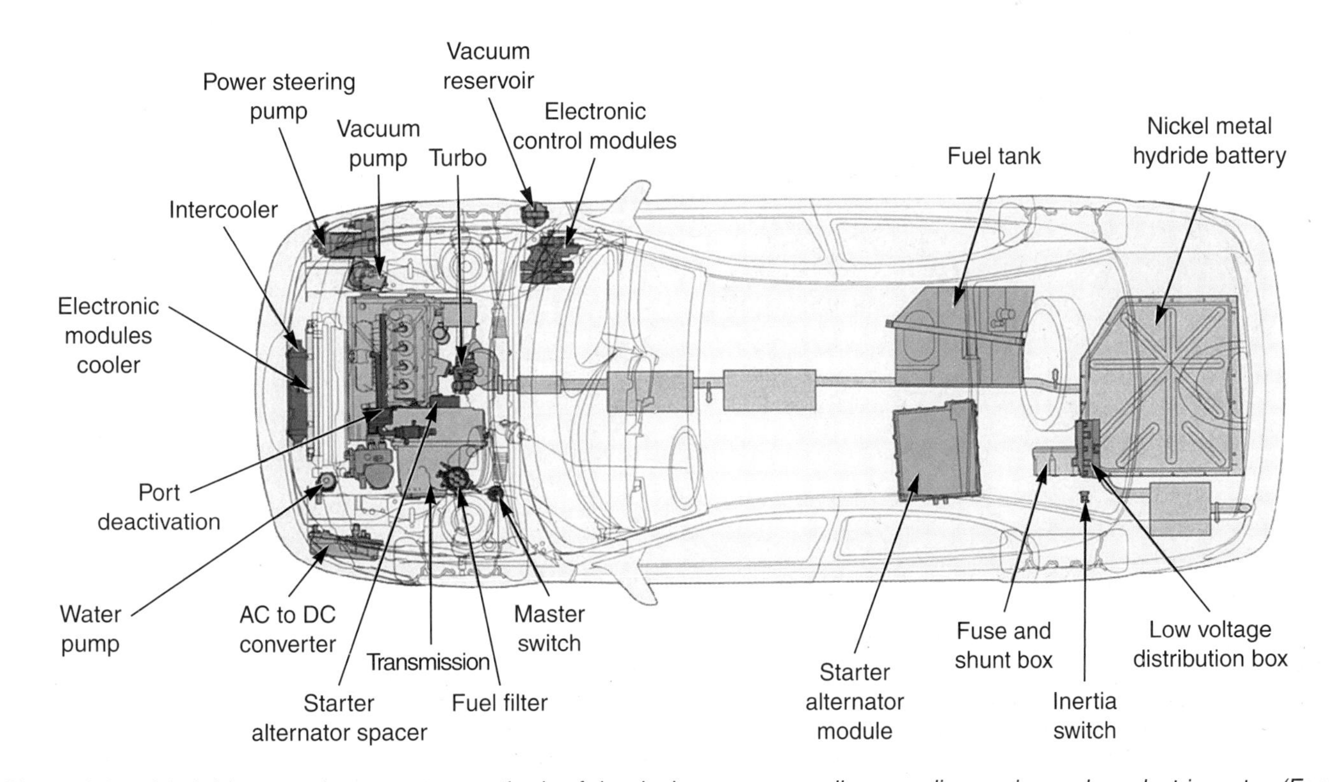

Figure 6-35. *A hybrid power plant uses two methods of developing power, usually a gasoline engine and an electric motor. (Ford)*

A power control unit is used to operate the motor-generator and gasoline engine. It is usually located between the passenger compartment and the trunk. The power control unit can internally switch the electrical connections on the motor-generator from the generator mode (charge batteries) to the motor mode (assist engine).

The batteries and electric motor supply power assist to the gasoline engine when the car accelerates. This greatly reduces the consumption of gasoline since the stored electrical energy and not engine power alone are used to propel the vehicle. Once cruising speeds are reached, the gasoline engine takes over and the motor is used as a generator to recharge the batteries.

Most hybrids use a continuous variable transmission (CVT). This type transmission allows the engine to run at an almost constant speed. The engine does not have to increase or decrease its speed (rpm) as much during vehicle speed changes. This allows the engine to be run at its most efficient speed, thus eliminating the need for varying other engine operating parameters.

The motor-generator is the key to modern hybrid technology. The motor-generator is an ultra-thin, brushless motor-generator. Refer to **Figure 6-36.** It can produce electricity (generator mode) and convert electricity into rotary motion (motor assist mode). The thin, but large-diameter motor-generator is normally located between the engine and the transmission. Large electrical leads connect the motor generator with its array of batteries.

For high electrical efficiency, a series of electrical coils are placed around the motor armature. The coil windings can produce a very powerful magnetic field to spin the motor armature or they can have current induced into them to recharge the battery array. The motor-generator can also replace the conventional engine starting motor. When you turn the key to start the engine, the large motor can easily spin the engine crankshaft until the engine fires and runs on its own power.

The battery array or battery pack is a series of high-power, nickel-metal hydride cells. The batteries are initially recharged through regenerative braking and then fully charged by engine and generator operation. The energy stored in the batteries is discharged into the motor-generator on acceleration or when added power is needed, as when climbing a hill for example. Since the motor armature is fastened to the engine crankshaft, the armature helps rotate the transmission input shaft and driveline. **Figure 6-37** shows the three modes of operation for a hybrid vehicle.

Regenerative braking uses the energy generated by slowing the vehicle to generate electricity, which is stored in the batteries. When the brakes are applied, the

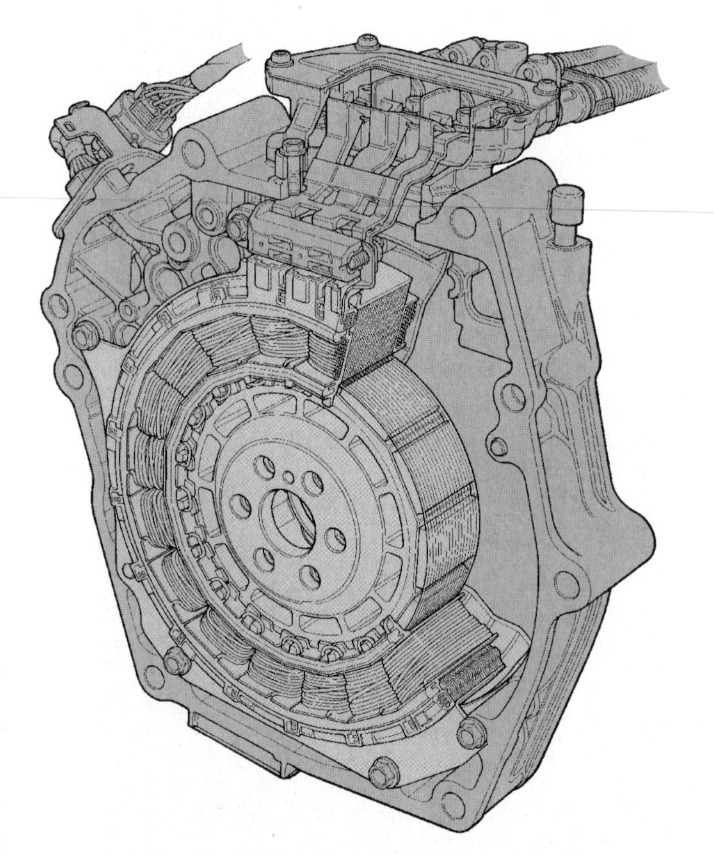

Figure 6-36. *The motor-generator is the key to modern hybrid power sources. It can be used as a motor to assist the engine in propelling the vehicle. It can also be used as a starting motor to spin the engine crankshaft for engine start-up or as an electric generator to help slow the vehicle while recharging the battery pack. (Honda)*

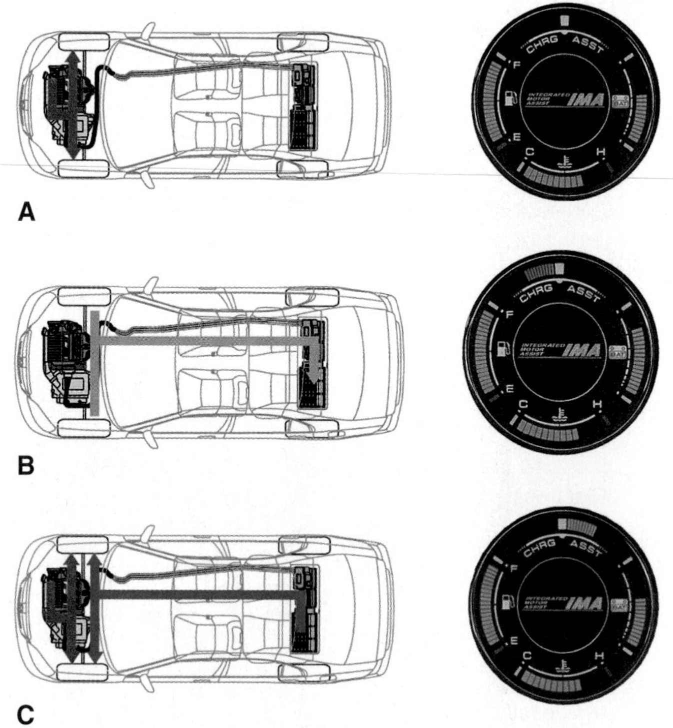

Figure 6-37. *Study the operation of a hybrid power plant. A—When cruising at a steady speed, the gasoline engine powers the vehicle normally. B—In the charging mode, the motor-generator acts as an electrical generator to recharge the battery pack. The load of turning the generator also helps slow the vehicle during braking. C—In assist mode, a large flow of current is directed to the motor-generator to help spin the crankshaft. (Honda)*

motor-generator converts the movement of its armature into electricity. Since it takes considerable energy to turn the armature, the car is slowed, assisting the braking of the vehicle.

Fuel Cell

A ***fuel cell*** can serve as a power plant by converting chemicals into electricity. See **Figure 6-38.** In a fuel

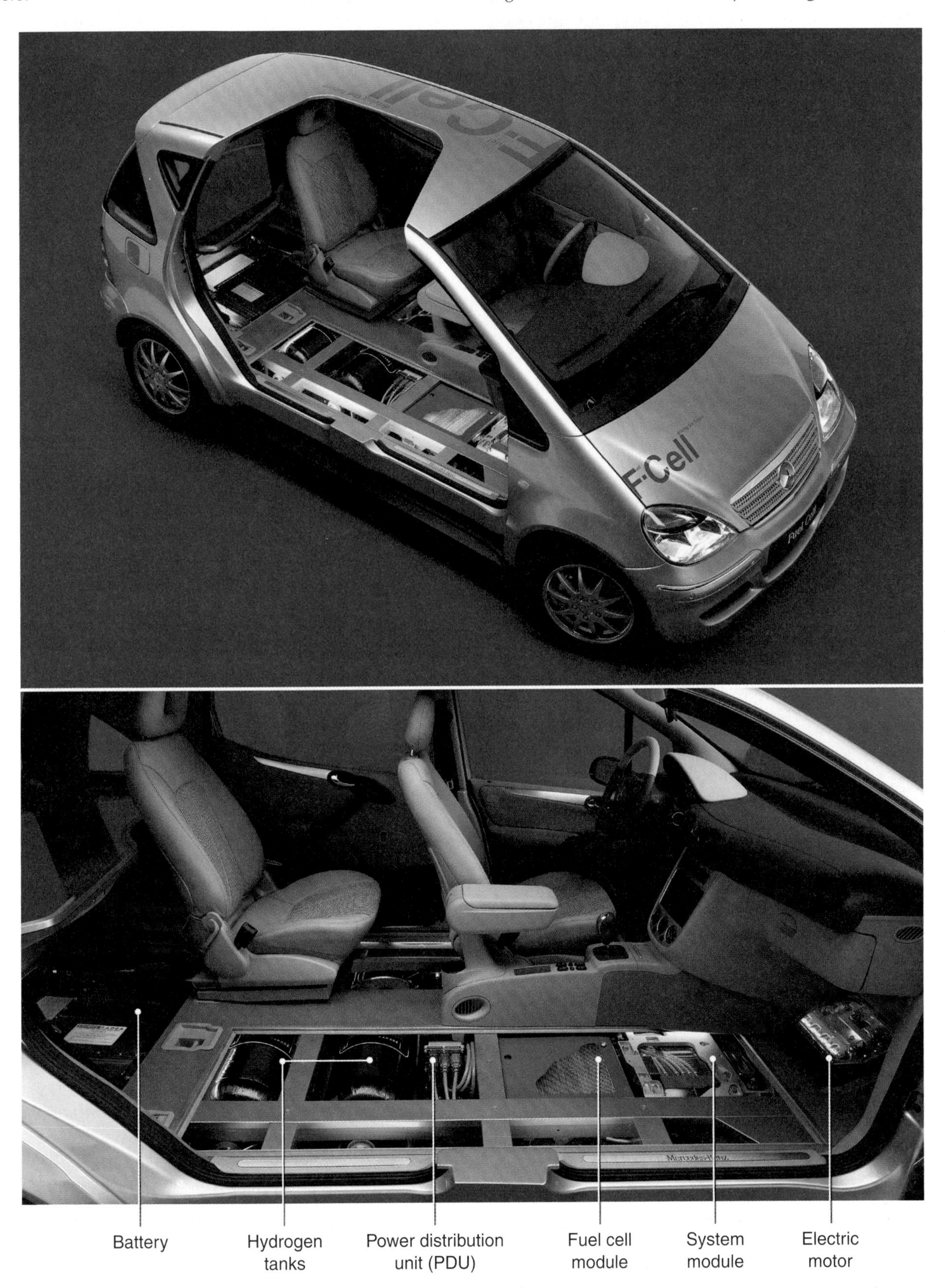

Figure 6-38. *This is a modern design for a fuel cell–powered vehicle. Note the parts of the system. (Mercedes-Benz)*

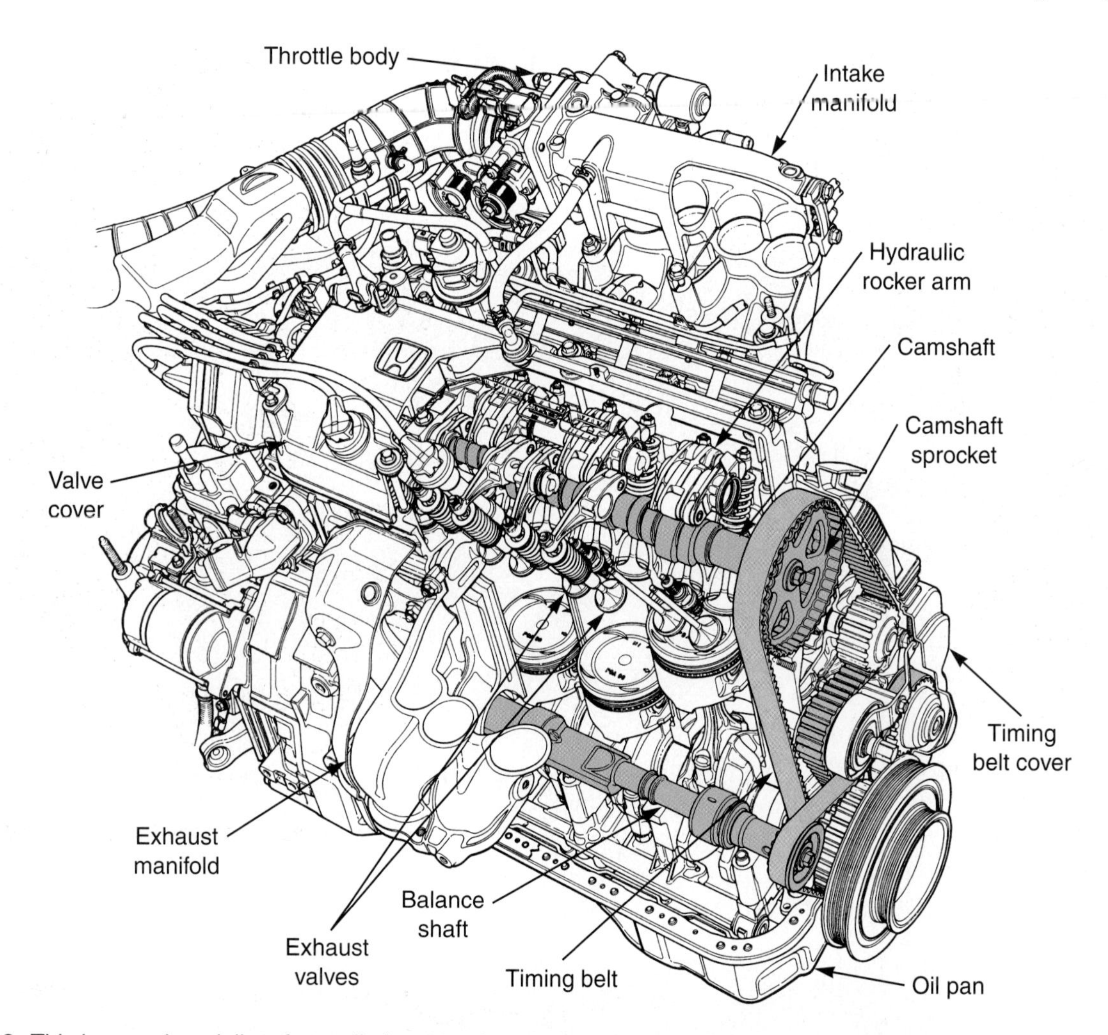

Figure 6-39. *This is a modern, inline, four-cylinder, 16-valve engine. It has hydraulically controlled rocker arms that can shift to use different camshaft lobes on the intake camshaft. One set of camshaft lobes has profiles for low-speed efficiency, the other set has profiles for high-speed power and fuel economy. Note the use of a balancer shaft in this engine. (Honda)*

cell, chemicals are combined or altered to produce water and electrons (electrical energy). The electrical energy from the fuel cell can then be used to operate large electric motors that propel the vehicle. Theoretically, fuel cells can convert hydrogen into electricity with very high efficiency.

A fuel cell is made of a plastic, gas-permeable membrane coated with a catalytic foil, usually made of platinum. This membrane is sandwiched between two electrode plates. Hydrogen gas flows through the anode plate (negative electrode) and the membrane while compressed air flows the other way. The catalyst causes the hydrogen gas to separate into positive hydrogen ions and electrons. This causes a potential difference of electrical energy across the anode and cathode (positive electrode). When current is drawn out of the fuel cell, oxygen molecules separate and combine with hydrogen ions to form water and heat.

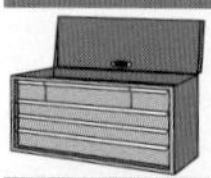

Note: Some fuel cell designs derive the needed hydrogen from gasoline.

Engines

Figures 6-39 through **6-44** show several typical automotive engines. Study each carefully. Note the design variations between the different types. Also, study the names of all of the parts. This will help you in later chapters as you continue to learn more about engines.

Summary

Cylinder arrangement refers to the position of the cylinders in the engine block in relation to the crankshaft. The four common cylinder arrangements are inline, V-type, W-type, and opposed.

An inline engine has cylinders positioned one after the other in a straight line. A V-type engine looks like the letter V when viewed from the front or rear. A W-type engine is similar to a V-type engine, but with staggered cylinders. An opposed engine has two banks of cylinders that lay flat or horizontal on each side of the crankshaft.

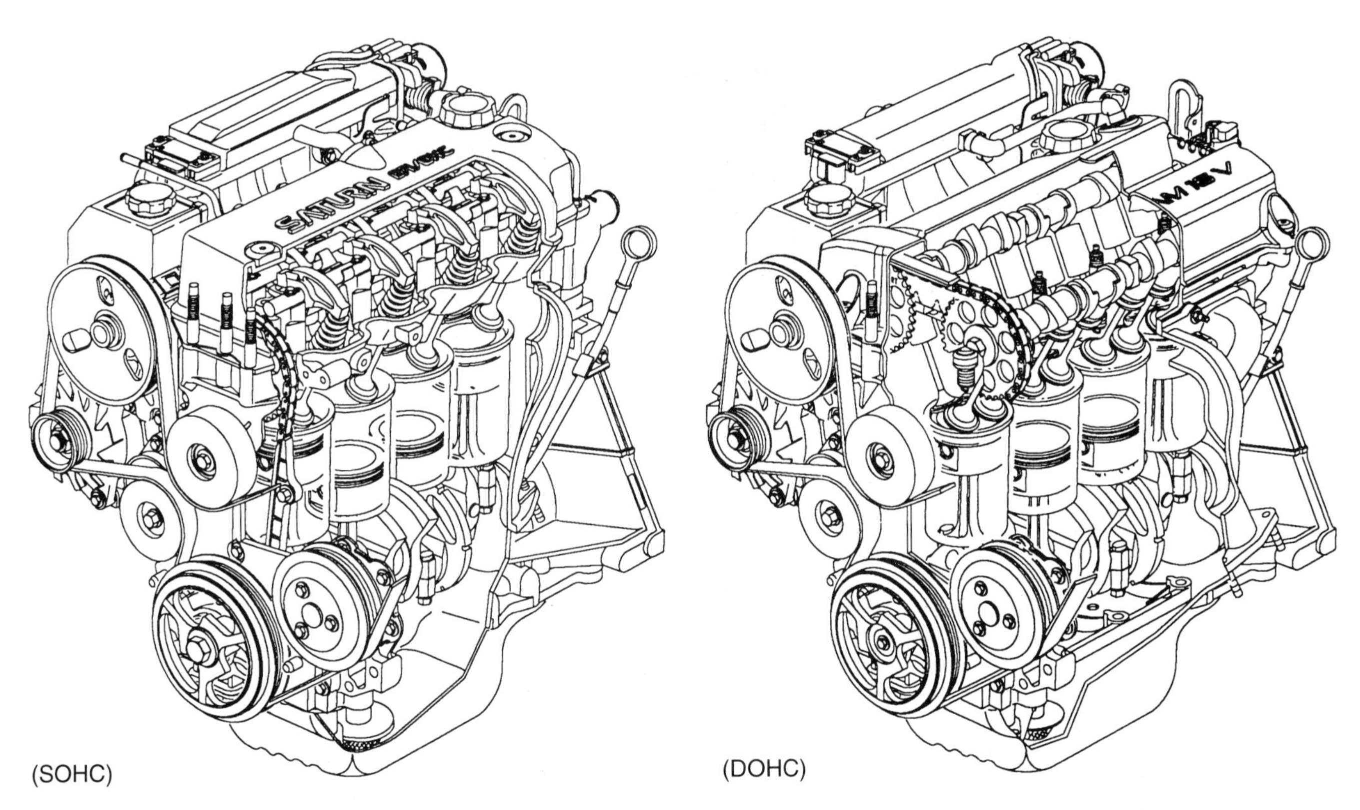

Figure 6-40. *Compare the single overhead camshaft engine on the left with the dual overhead camshaft engine on the right. How many differences can you find? (Saturn)*

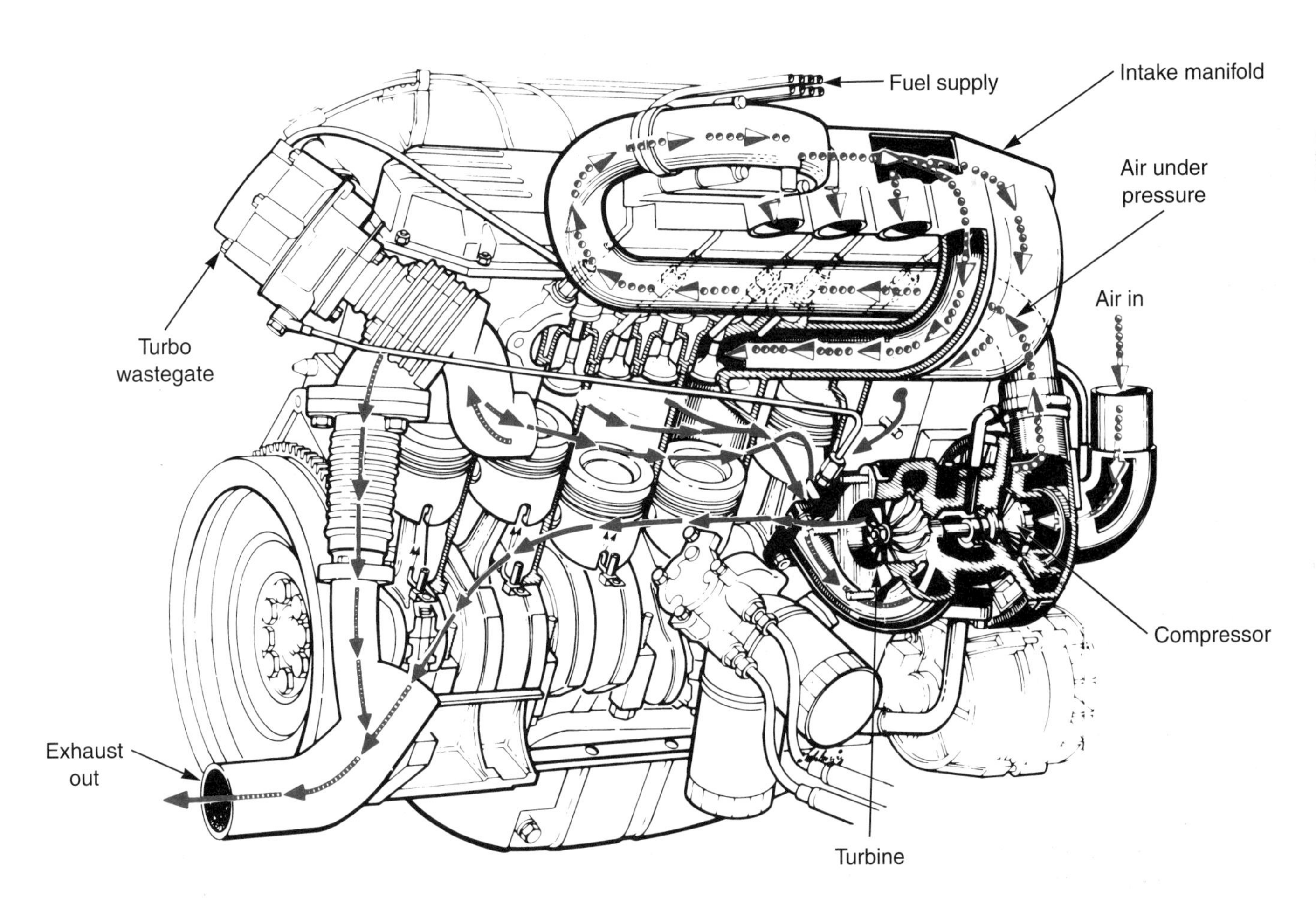

Figure 6-41. *This is a turbocharged, inline engine that uses a wastegate to limit boost pressure. Note the airflow. (Audi)*

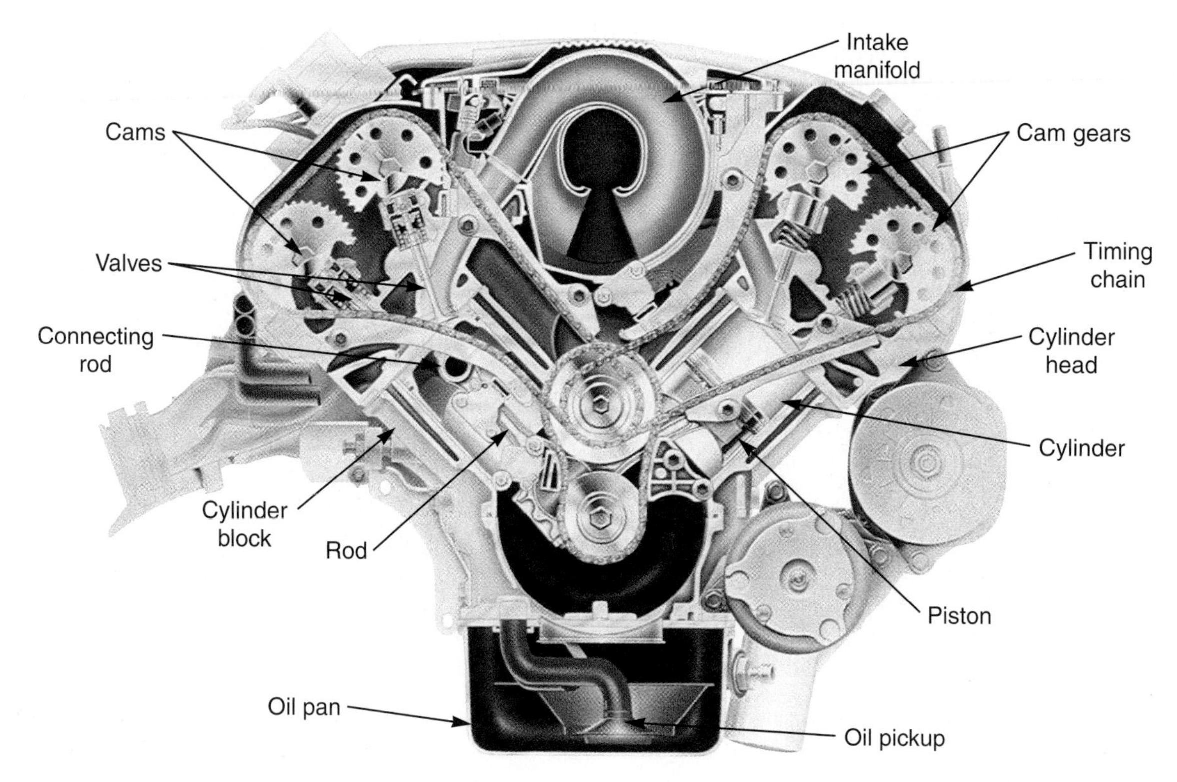

Figure 6-42. *This is a front view of a dual overhead camshaft engine. Notice the long timing chains used to spin the camshaft sprockets at one-half of the crankshaft speed. (Cadillac)*

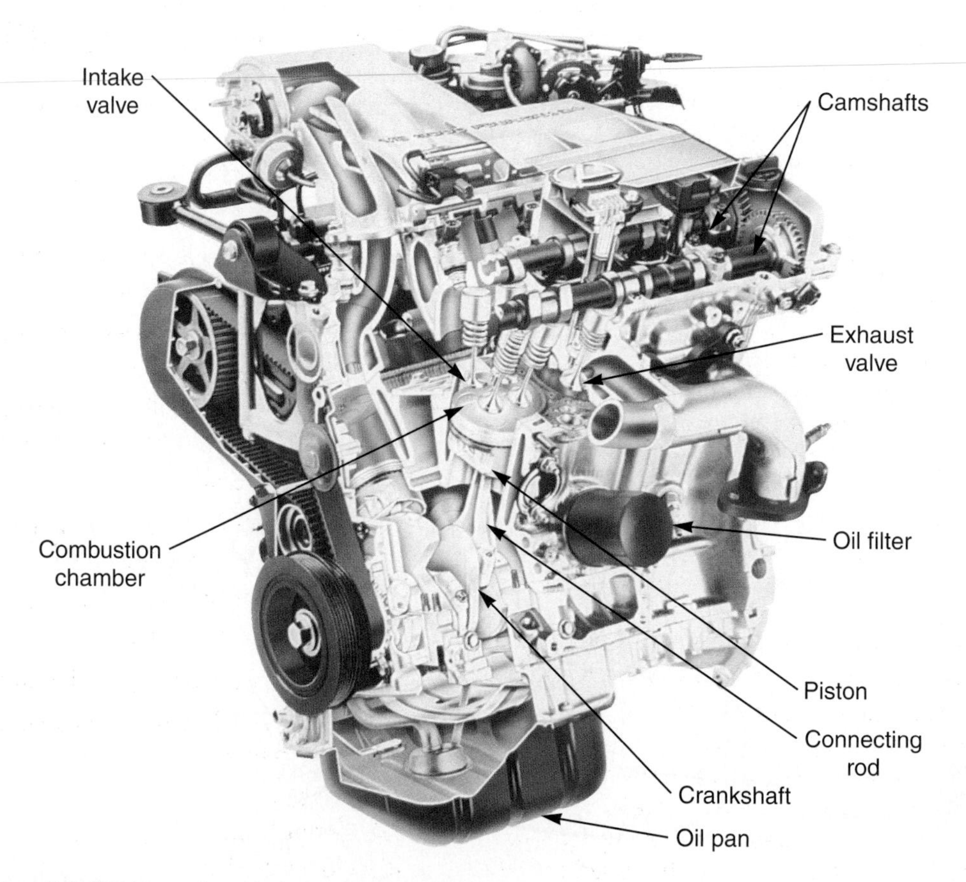

Figure 6-43. *This is a DOHC V-6 engine. Note the parts. (Lexus)*

Figure 6-44. *This is a horizontally opposed engine. It provides for a very low center of gravity and the lower main bearing friction of the boxer design. It is a 24-valve, 6-cylinder engine equipped with dual overhead cams and fuel injection. (Subaru)*

Cylinder numbers identify the cylinders, pistons, and connecting rods of the engine. Firing order refers to the sequence in which combustion occurs in the engine.

There are several different ways to classify an engine crankshaft. Some classifications include inline or V-type, cast iron or forged steel, and externally or internally balanced.

There are two types of systems providing cooling for the engine—liquid cooling and air cooling. Almost all vehicles now have a liquid-cooled engine.

The two common types of fuel are gasoline and diesel. Fuel metering is achieved through fuel injection. Diesel engines are multiport fuel injected. A gasoline engine may be multiport or throttle body fuel injected. Gasoline engines have spark ignition. Diesel engines have compression ignition.

There are four basic combustion chamber shapes for gasoline engines. These are pancake, wedge, hemispherical, and pent-roof. In addition, there are many different types of combustion chamber. The number of valves or spark plugs per chamber often define the type of combustion chamber

Engines may also be classified by the location of the valves. In addition, the location and number of camshafts are important. There are three basic camshaft drive systems—belt, chain, and gears.

Engine aspiration refers to how air enters the engine, or how the engine "breathes." There are three types of engine aspiration. An engine can be classified as normally aspirated, turbocharged, or supercharged.

An engine may be internally or externally balanced. An internally balanced engine uses the counterweights on the crankshaft to counteract the piston and rod weight. An externally balanced engine has extra weight added to the flywheel and balancer.

There are several alternative engine designs in production or development. These include rotary and Miller cycle engines. Other alternative designs include hybrid and fuel cell vehicles.

Review Questions—Chapter 6

Please do not write in this text. Write your answers on a separate sheet of paper.

1. Explain the four common types of automotive engine cylinder arrangements.
2. Car and light truck engines normally have _____, _____, or _____ cylinders.
3. _____ identify the cylinders, pistons, and connecting rods of an engine.
4. How can you normally tell which cylinder is the number one cylinder in a V-type engine?
5. Define *firing order.*
6. The most common type of cooling system on automobiles is the _____ cooling system.
7. Describe the major differences between a gasoline engine and a diesel engine.
8. A diesel engine is a(n) _____ ignition engine.
9. Which of the following is *not* a common combustion chamber shape?
   (A) Wedge.
   (B) Symmetrical.
   (C) Hemispherical.
   (D) Pancake.

10. How does swirl help combustion efficiency?
11. Explain the advantages of a four-valve combustion chamber.
12. What is a *stratified charge combustion chamber?*
13. In a(n) _____ or _____ valve engine, both valves are located in the cylinder head.
14. A(n) _____ engine has one camshaft in each cylinder head and a(n) _____ engine has two camshafts per cylinder head.
15. Describe the operation of a *rotary engine.*

ASE-Type Questions—Chapter 6

1. The term *cylinder arrangement* refers to the position of the cylinders in the engine block in relation to the _____.
   (A) camshaft
   (B) crankshaft
   (C) pistons
   (D) valves
2. All of the following are basic cylinder arrangements found in modern automotive engines *except:*
   (A) horizontally opposed.
   (B) V-type.
   (C) vertically opposed.
   (D) inline.
3. Technician A says that modern automotive engines can be classified by firing order. Technician B says that modern automotive engines can be classified by intake valve size. Who is correct?
   (A) A only.
   (B) B only.
   (C) Both A and B.
   (D) Neither A nor B.
4. Technician A says that some automotive engine camshafts are gear driven. Technician B says that chain- and belt-driven camshafts are the most common on automotive engines. Who is correct?
   (A) A only.
   (B) B only.
   (C) Both A and B.
   (D) Neither A nor B.
5. Technician A says that an air cooling system can maintain a constant engine temperature better than a liquid cooling system. Technician B says that a liquid cooling system is used on most modern automotive engines. Who is correct?
   (A) A only.
   (B) B only.
   (C) Both A and B.
   (D) Neither A nor B.
6. All of the following are examples of modern fuel systems *except:*
   (A) multiport gasoline injection.
   (B) throttle body gasoline injection.
   (C) multiport diesel injection.
   (D) throttle body diesel injection.
7. Technician A says that some gasoline engines use compression ignition to ignite the fuel in their combustion chambers. Technician B says that there are some gasoline engines that inject fuel directly into the combustion chamber. Who is correct?
   (A) A only.
   (B) B only.
   (C) Both A and B.
   (D) Neither A nor B.
8. All of the following are crankshaft classifications *except:*
   (A) cast iron.
   (B) aluminum.
   (C) forged steel.
   (D) internally balanced.
9. Technician A says that the crankshaft in the illustration is for a V-6 engine. Technician B says that the crankshaft in the illustration is for a V-8 engine. Who is correct?
   (A) A only.
   (B) B only.
   (C) Both A and B.
   (D) Neither A nor B.

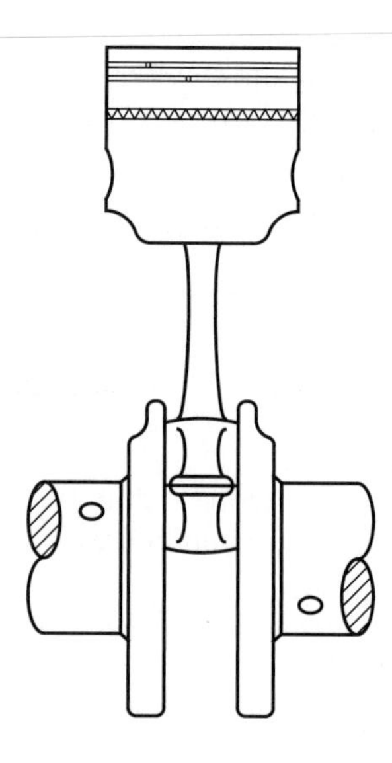

10. Technician A says that push rods are not needed to operate the rocker arms and valves in an overhead camshaft engine. Technician B says that push rods are used to operate the rocker arms and valves in an overhead valve engine. Who is correct?
    (A) A only.
    (B) B only.
    (C) Both A and B.
    (D) Neither A nor B.

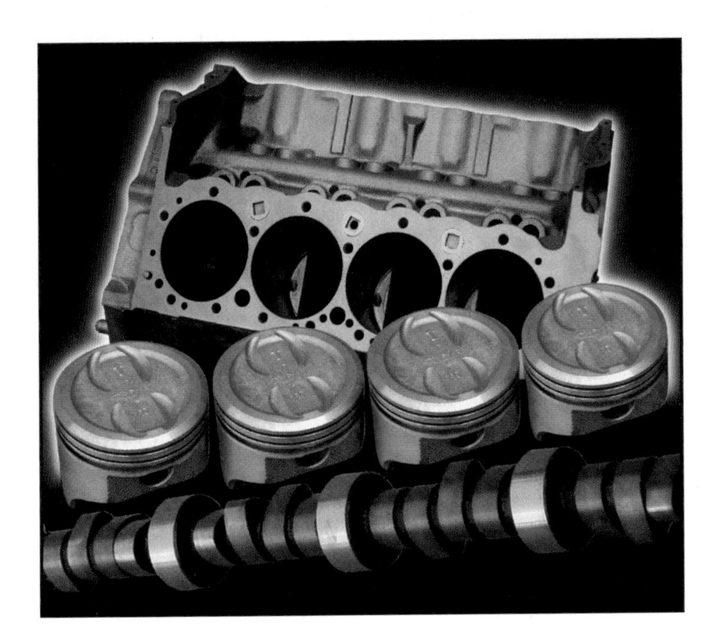

Chapter 7

Engine Size and Performance Ratings

After studying this chapter, you will be able to:

- ❑ Explain engine bore, stroke, and displacement measurements.
- ❑ Calculate engine displacement.
- ❑ Explain bore/stroke ratio.
- ❑ Define oversquare engine and undersquare engine.
- ❑ Describe piston protrusion.
- ❑ Explain compression ratio.
- ❑ Summarize rotational and angle measurement in degrees.
- ❑ Explain how atmospheric pressure affects engine performance.
- ❑ Describe engine horsepower and torque.
- ❑ Explain various horsepower ratings.
- ❑ Define volumetric, mechanical, and thermal efficiency.

Know These Terms

Atmospheric pressure
Bore/stroke ratio
Brake horsepower (bhp)
Chassis dynamometer
Compression pressure
Cylinder bore
Displacement
Drive wheel horsepower
Engine compression ratio
Engine displacement
Engine dynamometer
Engine efficiency
Engine torque
Force
Frictional horsepower (fhp)
Gross horsepower (ghp)
Horsepower
Indicated horsepower (ihp)
Mechanical efficiency
Net horsepower
Oversquare engine
Piston displacement
Piston protrusion
Piston stroke
Power
Prony brake
Square engine
Taxable horsepower (thp)
Thermal efficiency
Undersquare engine
Vacuum
Volumetric efficiency
Work

This chapter discusses general engine specifications. Specifications are the many measurements or ratings given by the engine manufacturer. Engine specifications include the displacement of the engine, number of cylinders, and horsepower and torque output of the engine.

To be a specialized engine technician, you should understand general engine specifications. You might use this information when explaining something to a customer or when trying to analyze a problem. You could also continue your training into automotive engineering and find this subject matter even more essential.

Engine Size Measurements

The size of an engine is referred to as its ***displacement.*** Displacement is determined by cylinder diameter, the amount of piston travel per stroke, and the number of cylinders. Any of these three variables can be changed to alter engine displacement. Engine displacement is commonly used when ordering parts. Also, measurement specifications are often grouped in service manuals by engine displacement.

Bore and Stroke

The ***cylinder bore*** is the diameter of the engine cylinder wall. It is measured across the cylinder wall, parallel with the top of the block. Refer to **Figure 7-1.** The cylinder bore in automotive engines is usually between 3″ and 4″ (75 mm and 100 mm).

The ***piston stroke*** is the distance that the piston moves from top dead center (TDC) to bottom dead center (BDC). See **Figure 7-1.** The amount of offset (throw) built into the crankshaft connecting rod journals determines piston stroke. Piston stroke is usually between 3″ and 4″ (75 mm and 100 mm).

A service manual normally gives bore and stroke specifications together, with bore listed first. For instance, a specification for bore and stroke may be given as 4.00″ × 3.00″. This means that the engine cylinder is 4 inches in diameter and the piston stroke is 3 inches.

Generally, a larger bore and stroke makes the engine more powerful. It can pull in more fuel and air on each intake stroke. Then, more pressure is exerted on the head of the piston during the power stroke. The trend by auto manufacturers, however, is to make their engines smaller, but more powerful and fuel efficient.

Note: Cylinder bore is the diameter of the cylinder, not the diameter of the piston. The diameter of the piston is slightly less than the diameter of the cylinder.

Piston Displacement

Piston displacement is the volume the piston displaces (moves) from BDC to TDC. It is determined by the cylinder diameter and the piston stroke. A large cylinder diameter and large piston stroke result in a larger piston displacement than a small cylinder diameter and small piston stroke. The formula for finding piston displacement is:

$$\text{piston displacement} = (\text{bore squared} \times 3.14 \times \text{stroke}) \div 4$$

For example, if an engine has a bore of 3″ and a stroke of 3″, what is its piston displacement? Plug the bore

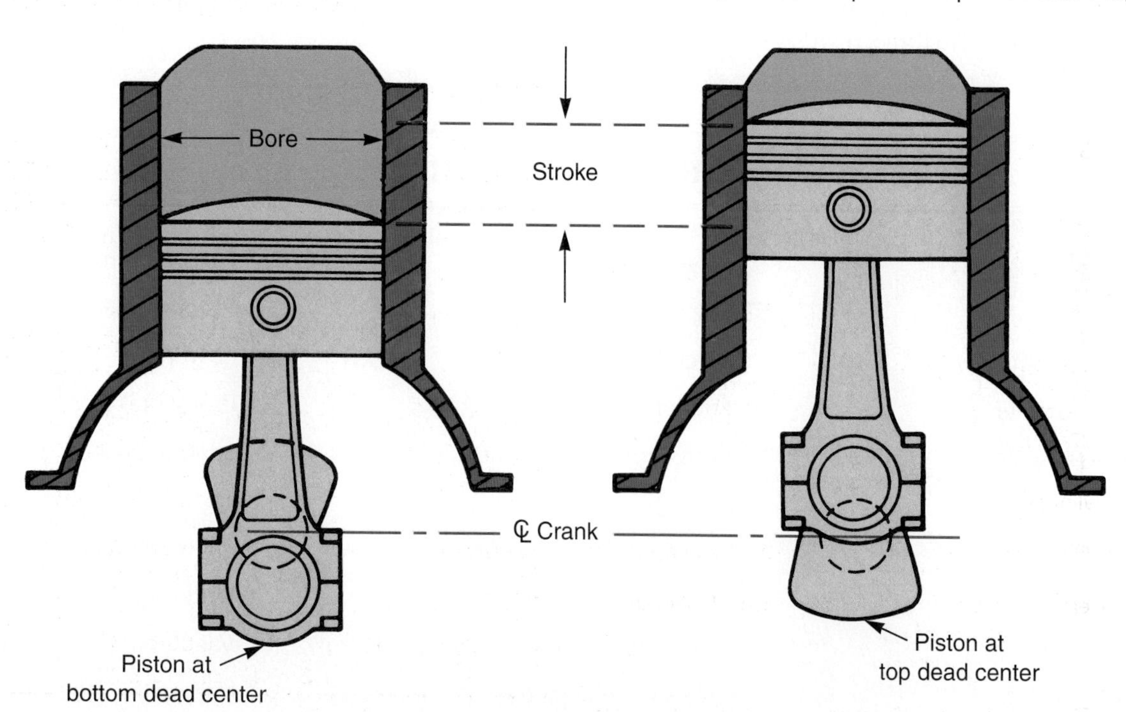

Figure 7-1. *Bore is the diameter of the cylinder. Stroke is the distance that the piston travels between top dead center (TDC) and bottom dead center (BDC). (Ford)*

and stroke values into the above formula to get piston displacement.

piston displacement = (3 inches^2 × 3.14 × 3 inches) ÷ 4
= 84.78 in^3/4
= 21.195 in^3 (cubic inches)

If you were to enlarge (bore over) the above cylinder by .050″ (bore = 3.050″), what would the new piston displacement be? Plug the new bore into the above equation and calculate the displacement. The correct answer is 21.907 cubic inches.

Engine Displacement

Engine size, or ***engine displacement,*** is the piston displacement times the number of cylinders in the engine, **Figure 7-2.** For example, if one piston in an engine displaces 21.195 cubic inches and the engine has six cylinders, what would the engine displacement be?

engine displacement = piston displacement × number of cylinders
= 21.195 cubic inches × 6
= 127.17 cubic inches

The actual displacement is often different from the displacement by which the engine is referred to. For example, this engine may likely be called a 130 cubic inch engine, even though its actual displacement is 127.17 cubic inches.

Cubic inches (cu. in., ci, or in^3), cubic centimeters (cc), and liters (L) are common ways to state engine displacement. For example, a V-8 engine might have a 350 ci engine. A V-6 may have a 3.3 L engine. A four-cylinder engine might have a displacement of 2300 cc. Since one liter equals 1000 cc, this engine may also be called a 2.3 L engine.

Engine displacement is usually matched to the weight of the car. A heavier car, truck, or van needs a larger engine that produces more power. A light, economy car only needs a small, low-power engine for adequate acceleration.

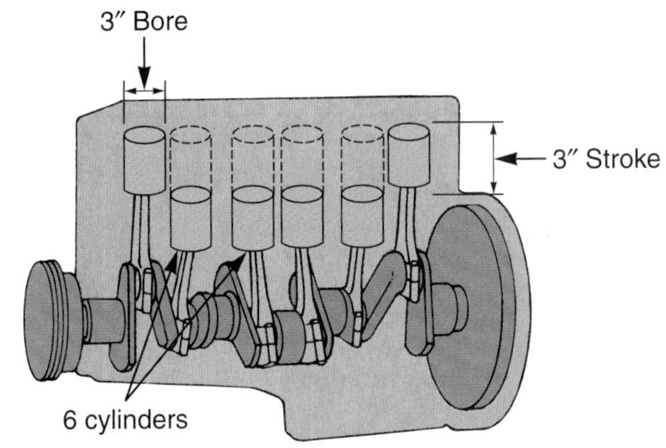

Figure 7-2. *To find engine displacement, find the piston displacement for one cylinder. Then, multiply that result by the number of cylinders in the engine. (Deere & Co.)*

Bore/Stroke Ratio

An engine's ***bore/stroke ratio*** is the relationship between the bore and the stroke. It is calculated by simply dividing the bore by the stroke. Read through the chart in **Figure 7-3.**

An ***oversquare engine*** has a bore dimension that is larger than the stroke dimension. The bore/stroke ratio is larger than one. All of the engines given in **Figure 7-3** are oversquare. This is the most common engine design. A larger bore and shorter stroke allows for higher engine speeds.

Most racing engines are oversquare. A short stroke allows the engine to operate at high engine speeds. For example, a racing engine might operate at 18,000 rpm. The large bore and very short stroke allows the engine to operate at higher speeds without piston and connecting rod breakage. At these speeds, each connecting rod must withstand about 7000 pounds of force. A longer connecting rod is weaker than a shorter connecting rod. This force applied to a longer connecting rod would likely result in rod and engine failure.

An ***undersquare engine*** has a stroke dimension that is larger than the bore dimension. The bore/stroke ratio is smaller than one, such as .93:1. See **Figure 7-4.** Undersquare engines are less common, but some diesel engines are undersquare. These engines have a long stroke for low-engine-speed power. Large industrial and tractor engines are sometimes undersquare because they operate at low engine speeds.

A ***square engine*** has equal bore and stroke dimensions. In other words, the bore/stroke ratio is 1:1.

Piston Protrusion

Frequently, the engine is designed so the piston protrudes or sticks out the top of the block at TDC. ***Piston protrusion*** is a comparison of the piston head location at TDC in relation to the block deck, **Figure 7-5.**

Cubic Inches	Liters	Bore × Stroke	Bore/Stroke Ratio
140	2.3	3.78 × 3.13	1.20:1
153	2.5	3.68 × 3.30	1.11:1
177	2.9	3.66 × 2.83	1.29:1
183	3.0	3.50 × 3.14	1.11:1
232	3.8	3.81 × 3.39	1.12:1
245	4.0	3.94 × 3.31	1.19:1
281	4.6	3.55 × 3.54	1.01:1
302	5.0	4.00 × 3.00	1.33:1
351	5.8	4.00 × 3.50	1.14:1
460	7.5	4.36 × 3.85	1.13:1

Figure 7-3. *This chart shows the displacement, bore, stroke, and bore/stroke ratio for several engines. (Ford)*

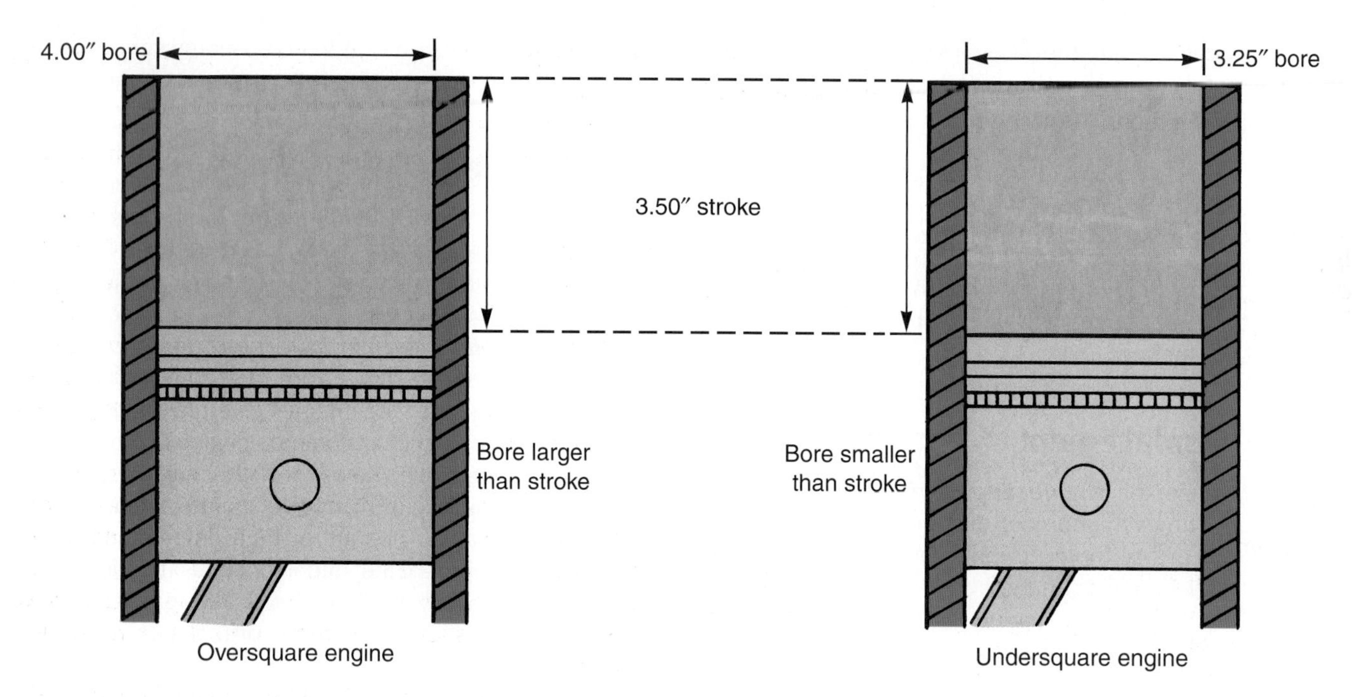

Figure 7-4. *An oversquare engine has a bore dimension that is larger than its stroke dimension. This is the most common design. An undersquare engine has a stroke dimension that is larger than its bore dimension.*

Piston protrusion is especially critical with diesel engines. As you will learn in later chapters, you must commonly measure piston protrusion to select the correct diesel head gasket thickness. Also, if you have to mill (machine) the block deck surface or the cylinder head to correct warpage, piston protrusion can be critical.

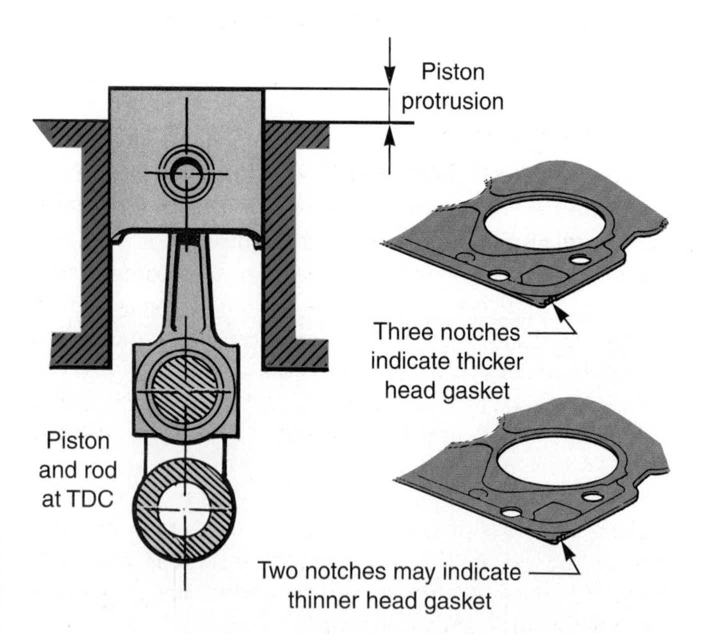

Figure 7-5. *Piston protrusion is a comparison of piston head location at TDC in relation to the block deck. This is critical with a diesel engine. The correct head gasket thickness must be used because of the small clearance between the cylinder head and the piston head. (Peugeot)*

Force, Work, and Power

Force is a pushing or pulling action. When a compressed spring is released, an outward movement is produced due to the force from the spring. Force is measured in pounds or newtons.

Work is done when force causes movement. If the compressed spring moves another engine part, *useful* work has been done. If the spring does not cause movement, no *useful* work has been done. Work is measured in foot-pounds (ft-lb) or newton-meters (N·m). The formula for work is:

$$\text{work} = \text{distance moved} \times \text{force applied}$$

For example, if you use a hoist to lift a 400 pound engine 3 feet in the air, how much work has been done? Plug these values into the formula.

$$\text{work} = 3' \times 400 \text{ pounds}$$
$$= 1200 \text{ foot-pounds (ft-lb)}$$

Power is work done over a period of time. It is measured in foot-pounds per unit of time (ft-lb/sec or ft-lb/min) or horsepower (hp). The metric unit for power is the watt or kilowatt (kW).

In a given amount of time, high power output can do more work than low power output. The formula for power is:

$$\text{power} = (\text{distance} \times \text{force}) \div \text{time}$$

If an engine moves a 3000 pound car 1000 feet in one minute, how much power is needed? Plug these values into the formula.

$$\text{power} = (1000 \text{ feet} \times 3000 \text{ pounds}) \div 1 \text{ minute}$$
$$= 3{,}000{,}000 \text{ ft-lb/min}$$

One horsepower (hp) is equal to 33,000 ft-lb/min. Therefore, the horsepower required to move the 3000 pound car 1000 feet in one minute is:

3,000,000 ft-lb/min ÷ 33,000 ft-lb/min/hp = 90.9 horsepower

This may seem high, but keep in mind that the drive train multiplies the power output from the engine. Also, the wheels and tires reduce the friction between the vehicle and the ground. The above formula does not take these factors into account. Therefore, the actual horsepower required is much less.

Compression Ratio

Engine compression ratio compares cylinder volumes with the piston at the extreme top and bottom of its travel. At bottom dead center (BDC), a cylinder has maximum volume. At top dead center (TDC), a cylinder has minimum volume. An engine's compression ratio controls how tightly the air-fuel mixture is pressurized or squeezed on the compression stroke. See **Figure 7-6.**

A compression ratio is expressed in relation to 1. For example, an engine may have a compression ratio of 9:1 (9 to 1). This means the maximum cylinder volume is nine times greater than the minimum cylinder volume.

When a gasoline engine piston is at BDC, the cylinder volume might be 40 cubic inches (0.66 L). When the piston slides to TDC, the volume may reduce to 5 cubic inches (0.08 L). Dividing 40 by 5, the compression ratio for this engine is 8:1, **Figure 7-7.**

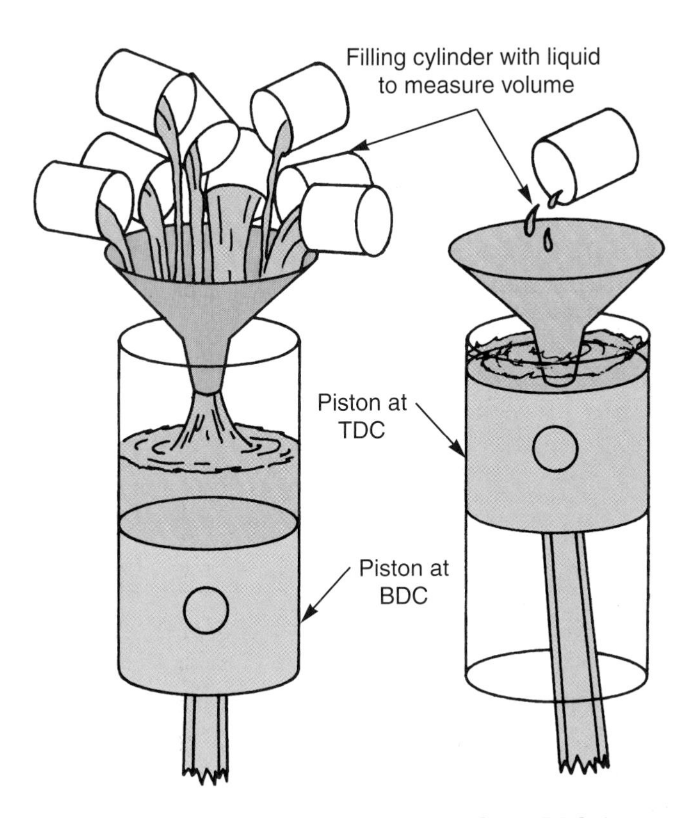

Figure 7-6. *Filling a cylinder with fluid at TDC and BDC demonstrates the compression ratio of the cylinder. This cylinder has an 8:1 compression ratio.*

In response to emission standards, gasoline engines began to be designed in the 1970s with low compression ratios, about 8:1. This allowed the use of clean-burning, unleaded fuel, but also resulted in a slight reduction in engine power and efficiency. However, compression ratios on modern, normally aspirated engines may be as high as 10:1 while still maintaining emissions compliance. This is due to computer control and advances in ignition systems. Turbocharged or supercharged engines usually have lower compression ratios to allow for the boost generated by the turbo or blower.

A high-compression-ratio gasoline engine requires the use of high-octane fuel. Octane is a relative measure of a fuel's ability to resist preignition. If a low-octane fuel is used in a high-compression engine, the fuel can ignite and burn prematurely, resulting in engine damage. Automotive fuels and combustion are explained in detail in the next chapter.

With a compression ignition engine (diesel), BDC cylinder volume is between 17 and 20 times as large as TDC cylinder volume. See **Figure 7-7.** The compression ratio is between 17:1 and 20:1. This very high compression ratio results in the air being compressed so tightly that it heats up enough to ignite the diesel fuel.

Note: Many gasoline engines in the muscle car era of the late 1960s and early 1970s had high compression ratios, between 11:1 and 12:1. A high compression ratio helped some of these engines produce over 400 horsepower. However, these engines burned leaded gasoline and produced high levels of exhaust emissions.

Compression Pressure

Compression pressure is the amount of pressure produced in the engine cylinder on the compression stroke. Compression pressure is normally measured in pounds per square inch (psi) or kilopascals (kPa). A gasoline engine may have compression pressure from 130 psi to 180 psi (895 kPa to 1240 kPa). A diesel engine has a much higher compression pressure of about 250 psi to 400 psi (1725 kPa to 2755 kPa).

A compression gauge is used to measure compression stroke pressure. It is screwed into the spark plug or glow plug hole. The ignition or injection system is disabled. Then, the engine is turned with the starter. The gauge will read compression stroke pressure.

Discussed in later chapters, compression stroke pressure is an indicator of engine condition. If low, there is a problem allowing air to leak out of the cylinder. The engine might have bad rings, burned valves, or a blown head gasket.

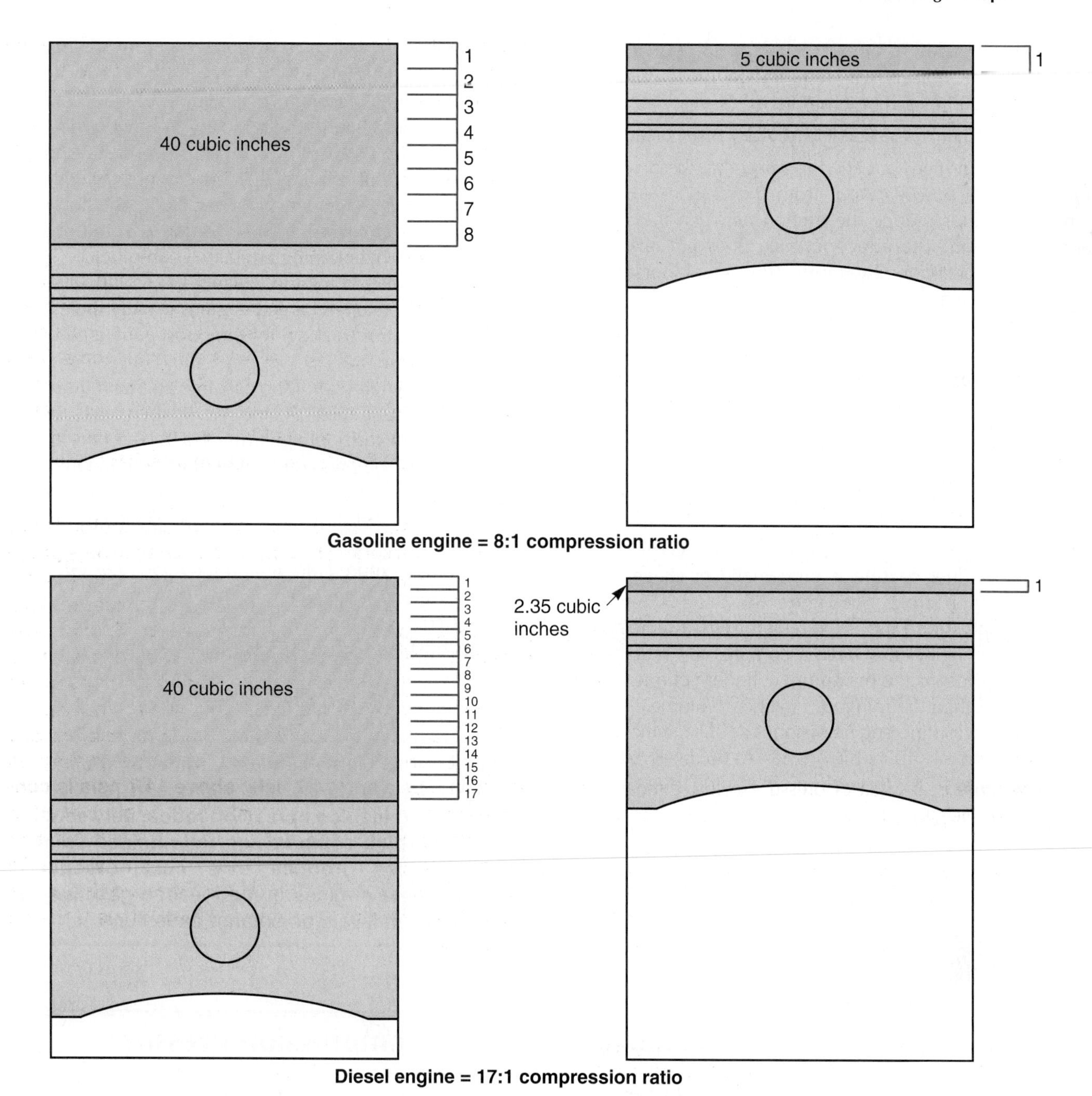

Figure 7-7. *A comparison of compression ratios of gasoline and diesel engines. A diesel engine has a much higher compression ratio because it must squeeze the air until it is hot enough to ignite the fuel.*

Rotation or Angle Measurement

Part rotation or angle measurements are made in degrees. A circle is divided into 360 equal parts called degrees (°). This is shown in **Figure 7-8.** One-half of a complete revolution is equal to 180°. One-fourth of a revolution is equal to 90°; one-eighth revolution is 45°.

Degree measurements are common in engine repair. For example, crankshaft rotation for ignition or injection timing adjustment is given in degrees. Throttle plate angle adjustment can also be given in degrees. When "dialing in" a camshaft in a racing engine, you will have to measure crankshaft rotation in degrees while adjusting the mounting locations of the camshaft sprockets.

Atmospheric Pressure

Atmospheric pressure is air pressure produced by the weight of the air above the surface of the earth. It is measured in pounds per square inch (psi) or kilopascals (kPa). At sea level, atmospheric pressure is 14.7 psi (101.4 kPa). See **Figure 7-9.** As altitude above sea level increases, such as in the mountains, atmospheric pressure decreases because there is less air above pushing down.

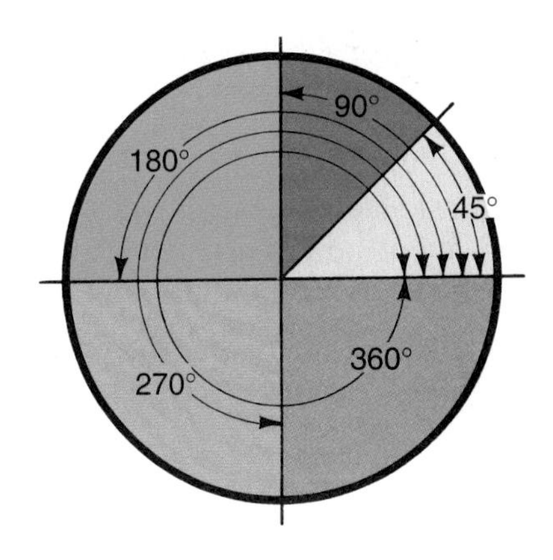

Figure 7-8. *Note how angles and rotation are measured in degrees. Crankshaft rotation is one example of a degree measurement.*

Atmospheric pressure pushes equally in all directions, **Figure 7-10.** As a result, normally aspirated engines use atmospheric pressure to force the air-fuel mixture into the cylinders. Supercharged or turbocharged engines assist atmospheric pressure by compressing the air using a large air pump (turbocharger or supercharger) and "blowing" the air into the cylinders.

Vacuum

A ***vacuum*** is an area of pressure below atmospheric pressure, or negative pressure. For example, when the engine piston slides down, it forms a vacuum. Since the outside air pressure is greater, air can rush in to fill the cylinder on the intake stroke.

Vacuum is measured in inches of mercury (in/Hg) or millimeters of mercury (mm/Hg). A vacuum gauge, discussed in Chapter 2, is commonly used to measure intake manifold vacuum. Manifold vacuum can be used to help diagnose several engine problems, as will be explained in later chapters.

A diesel engine does not produce intake manifold vacuum. This is because a diesel engine does not have a throttle valve to meter air into the engine. If you connect a vacuum gauge to a port on the intake manifold on a diesel engine, the reading would be almost zero. For this reason, a vacuum pump is commonly used on diesel engines to produce the vacuum needed to operate devices such as the power brake booster and vacuum switches.

Note: There are two basic methods or systems of measuring pressure—gauge and absolute. In gauge pressure, normal atmospheric pressure is read as zero pressure. This is 0 pounds per square inch gauge (psig). Any pressure below atmospheric pressure is read as a vacuum in inches of mercury (in/Hg). In absolute pressure, normal atmospheric pressure is read as 14.7 pounds per square inch absolute (psia) at sea level, not zero. Vacuum in the absolute system is between 0 psia and 14.7 psia; above 14.7 psia is considered pressure. A reading of 0 psia is equivalent to 30 in/Hg and considered a perfect vacuum. Most of the pressure and vacuum measurements you make as an engine technician will be done using the gauge system, where atmospheric pressure reads 0 psi on the pressure gauge.

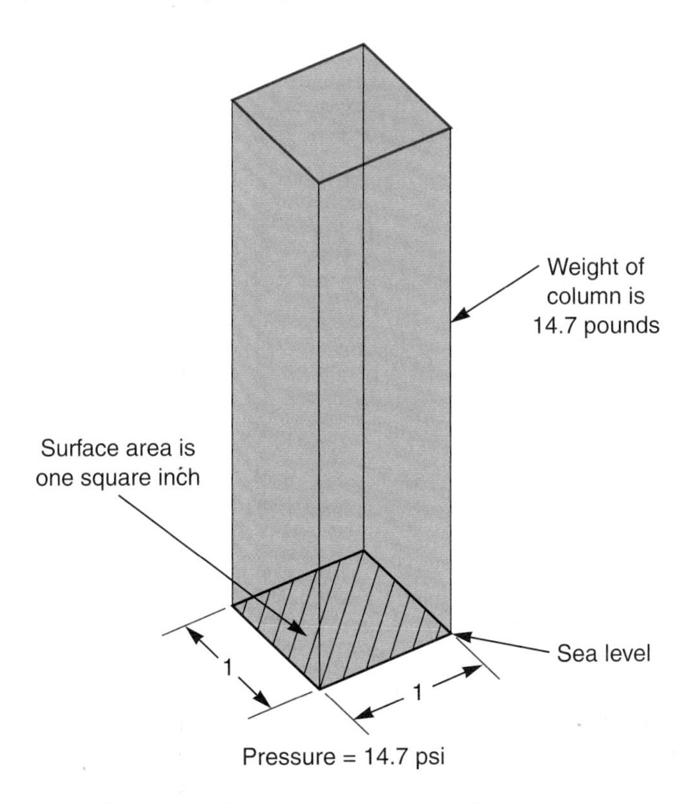

Figure 7-9. *The column of air over a one-inch-square area at sea level weighs 14.7 pounds and produces an atmospheric pressure of 14.7 pounds per square inch.*

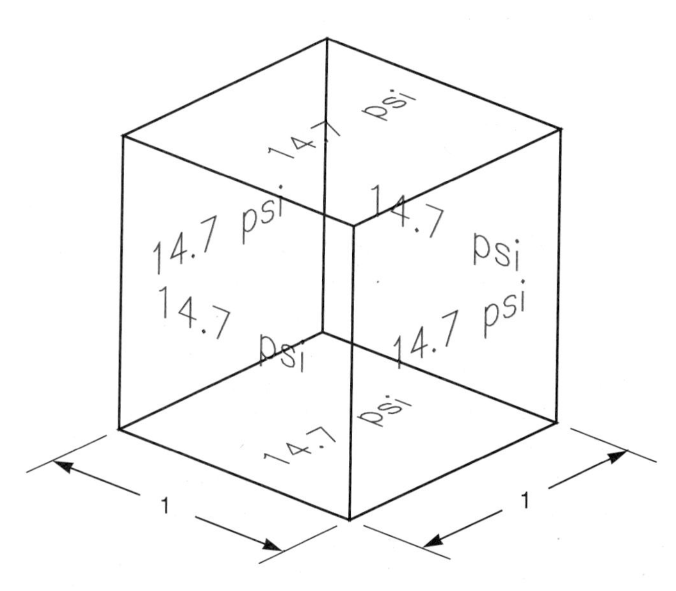

Figure 7-10. *Atmospheric pressure pushes equally in all directions.*

Engine Torque

Engine torque is a rating of the turning force at the engine crankshaft. When combustion pressure pushes the piston down, a strong rotating force is applied to the crank. This turning force is sent to the transmission or transaxle, drive train or drive axles, and drive wheels to move the vehicle. Torque is measured in foot-pounds (ft-lb) or newton-meters (N·m).

Engine torque specifications are given in a service manual. For example, 278 ft-lb at 3000 rpm may be given as the torque specification for a particular engine. This engine would be capable of producing a maximum of 278 foot-pounds of torque when operating at an engine speed of 3000 revolutions per minute.

Engine Power

Engine power is a measure of an engine's ability to perform work over time. Engine power is measured in horsepower (hp) or kilowatts (kW); horsepower is used in the US. The unit ***horsepower*** originated as the average strength of one horse. Therefore, theoretically, a 300 hp engine can do the work of 300 horses.

One horsepower equals 550 ft-lb of work per second, or 33,000 ft-lb of work per minute, **Figure 7-11.** To find engine horsepower, use the formula:

$$\text{hp} = (\text{distance (ft)} \times \text{weight (lb)}) \div (33{,}000 \times \text{time in minutes})$$

or

$$\text{hp} = (\text{work (ft-lb)}) \div (33{,}000 \times \text{time in minutes})$$

For an engine to lift 500 pounds a distance of 700 feet in one minute, about how much horsepower would be needed? To calculate, apply these values to the formula:

$$\begin{aligned}\text{hp} &= (500 \text{ lb} \times 700 \text{ ft}) \div (33{,}000 \times 1 \text{ minute})\\ &= 10.6 \text{ hp}\end{aligned}$$

Engine Horsepower Ratings

Vehicle manufacturers rate engine horsepower output at a specific engine speed. This engine power rating is normally stated in a service manual. For instance, a high-performance, turbocharged engine might be rated at 400 hp at 5000 rpm. There are several different methods of calculating engine horsepower.

Brake Horsepower

Brake horsepower (bhp) is the usable horsepower at the engine crankshaft. Shown in **Figure 7-12,** a ***prony brake*** was originally used to measure brake horsepower. The engine crankshaft spun the prony brake while the braking mechanism was applied. The resulting amount of pointer deflection was then used to find brake horsepower.

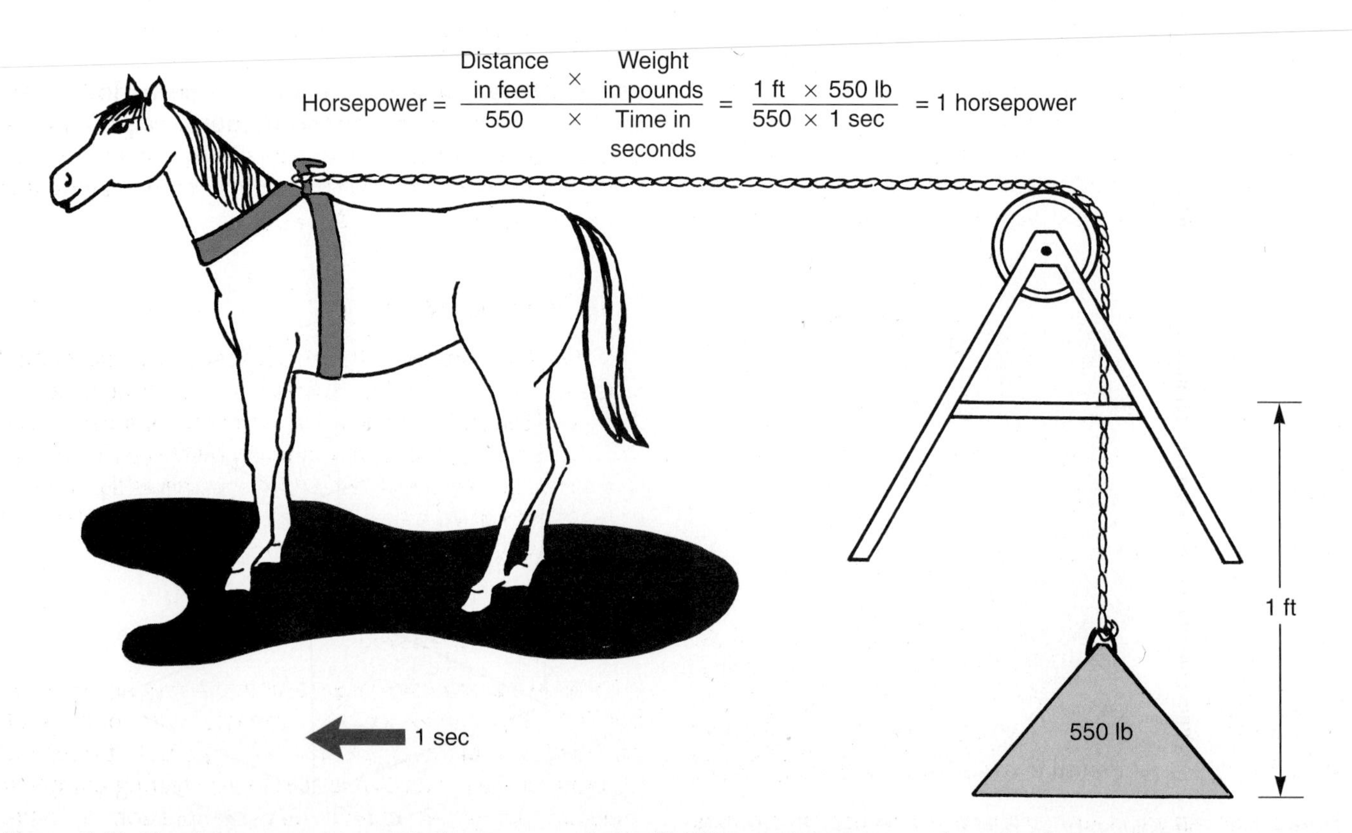

Figure 7-11. *One horsepower is equal to the power needed to lift 550 pounds a distance of one foot in one second.*

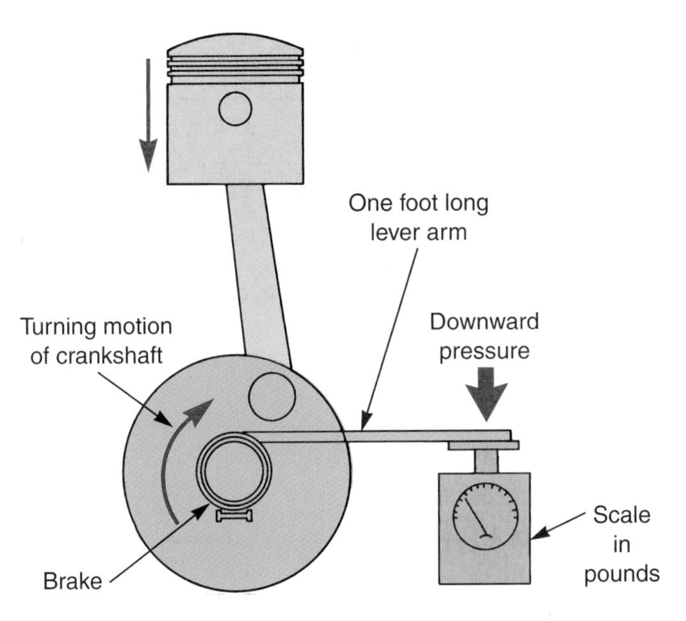

Figure 7-12. *Engine brake horsepower was originally measured using a prony brake.*

An ***engine dynamometer*** (dyno) is now used to measure brake horsepower, **Figure 7-13.** It functions in much the same way as a prony brake. Either an electric motor or fluid coupling is used to place a drag on the engine crankshaft. Then, power output can be determined.

Drive Wheel Horsepower

Drive wheel horsepower is an engine horsepower rating that takes into account the friction losses required to operate the transmission/transaxle, drive axles or drive shaft joints, differential, axle bearings, and other drive train parts that consume energy. Drive wheel horsepower indicates the amount of horsepower available to propel the car. A ***chassis dynamometer*** measures the horsepower delivered to the vehicle's drive wheels, or the drive wheel horsepower. See **Figure 7-14.**

Figure 7-13. *An engine dynamometer is now used to measure engine brake horsepower. This engine is being tested before installation into a car. (MSD Ignition Systems)*

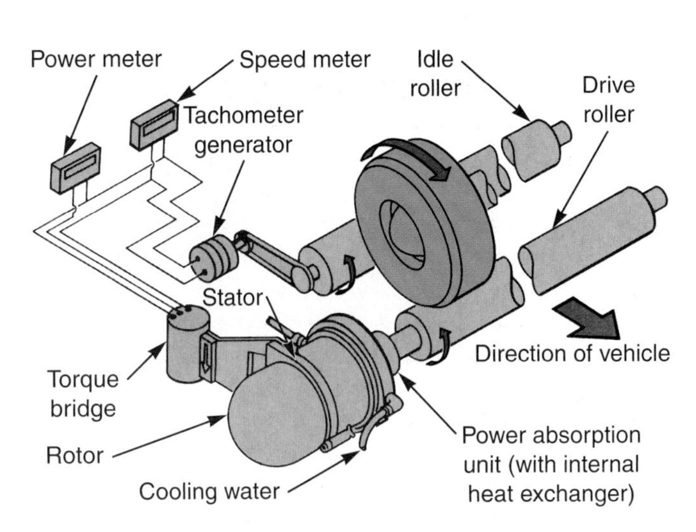

Figure 7-14. *A chassis dynamometer measures the horsepower at the drive wheels. (Clayton)*

Indicated Horsepower

Indicated horsepower (ihp) refers to the amount of power formed in the engine combustion chambers. A special pressure sensing device is placed in the cylinder. The pressure readings are used to find indicated horsepower.

Frictional Horsepower

Frictional horsepower (fhp) is the power lost to friction. It is the power needed to overcome engine friction and a measure of the resistance to movement between engine parts. Frictional horsepower reduces the amount of power available to propel the vehicle.

Net Horsepower

Net horsepower is the maximum power developed when an engine is loaded down with all of its accessories. Accessories include the alternator, water pump, supercharger, air injection pump, air conditioning, power steering pump, and so on. See **Figure 7-15.** Net horsepower is the amount of useful power with the engine installed in the vehicle. It is the amount of power available to the drive train.

Gross Horsepower

Gross horsepower (ghp) is similar to net horsepower, but it is the engine power available with only basic accessories installed (alternator, water pump). It does *not* account for the power lost to the power steering pump, air injection pump, air conditioning compressor, or other extra units. For a given engine, gross horsepower is higher than net horsepower.

Figure 7-15. *Net horsepower is a rating that takes into account all of the accessories on the engine. Notice how many accessories and pulleys are found on this engine. Yet, this is the high-performance engine in the Ford GT. (Ford)*

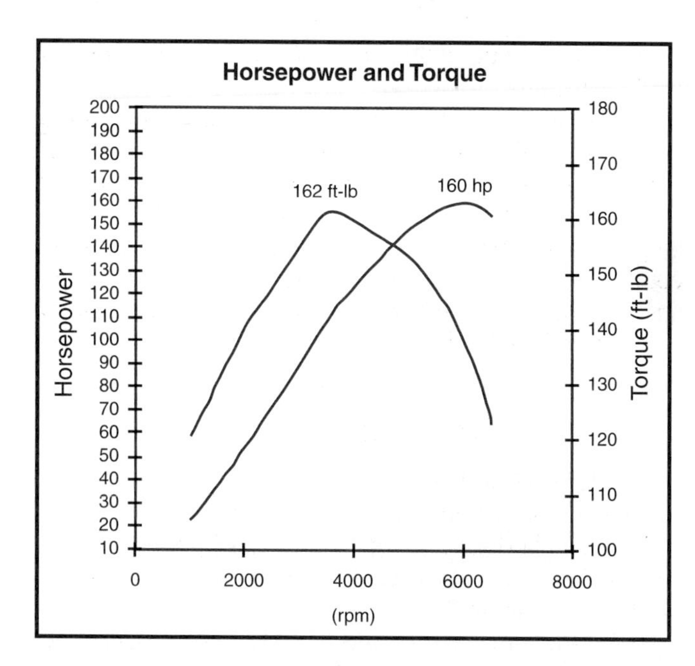

Figure 7-16. *Torque and horsepower curves show operating characteristics of an engine. How much torque and horsepower does this engine produce at 2000 rpm? At 4000 rpm? (Honda)*

Taxable Horsepower

Taxable horsepower (thp) is simply a general rating of engine size. In many states, it is used to find the tax placed on a vehicle. The formula for taxable horsepower is:

$$\text{thp} = \text{cylinder bore} \times \text{cylinder bore} \times \text{number of cylinders} \times .4$$

Horsepower and Torque Curves

Horsepower and torque curves are used to show how engine horsepower and torque change with engine speed. Refer to **Figure 7-16.** These curves show the operating characteristics of an engine.

In a passenger vehicle, it is important that an engine makes adequate torque and horsepower over a wide range of engine speeds. Then, the vehicle will accelerate properly in all transmission gears. A passenger vehicle engine has a level, gradual curve for both horsepower and torque.

A racing or performance engine is usually designed to make maximum power at high engine speeds. A performance engine does not operate well at lower engine speeds. However, low-speed power is not usually important in performance applications. A performance engine generally has a steep curve that peaks at a high engine speed for both horsepower and torque.

Figure 7-17 shows the power curves of a hybrid vehicle. Note how the hybrid's electric motor assist greatly increases low speed torque for improved acceleration. Also, the maximum horsepower is increased by the use of the electric motor.

Engine Efficiency

Engine efficiency is the ratio of power produced by the engine and the power supplied to the engine (energy content of the fuel). By comparing fuel consumption to

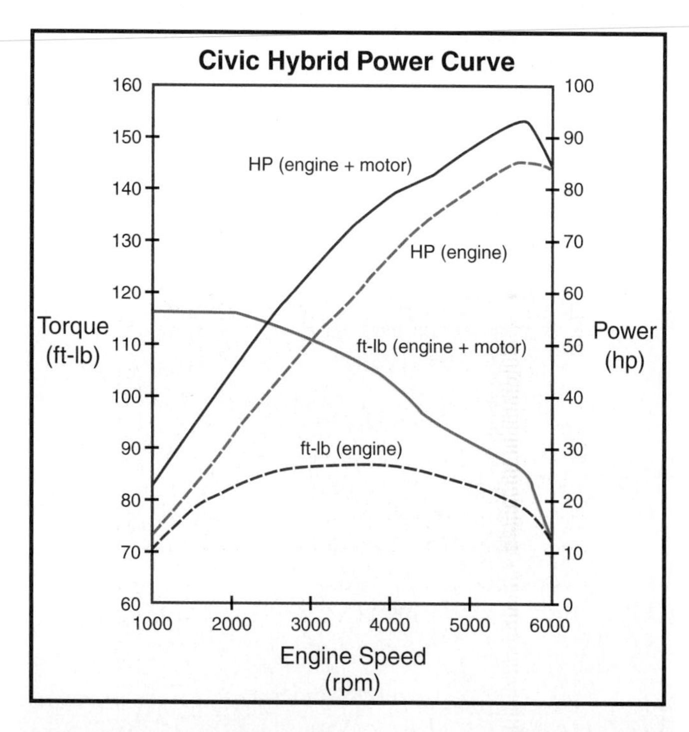

Figure 7-17. *Compare the power curves for this hybrid vehicle with and without the electric motor assist. At 4000 rpm, what are the horsepower and torque outputs of just the engine? What are the outputs when the electric motor assists at the same engine speed? (Honda)*

engine power output (brake horsepower), you can find engine efficiency.

If all of the energy in the fuel is converted into useful work, an engine is 100% efficient. However, modern piston engines are only about 20% efficient. **Figure 7-18** illustrates how the energy of the fuel is used by a piston engine. About 70% of the fuel's energy is lost to heat to the cooling and exhaust systems. About 5% is lost to friction and another 5% is lost to airflow. This leaves very little energy to move the engine pistons.

Volumetric Efficiency

Volumetric efficiency is the ratio of air drawn into the cylinder to the maximum possible amount of air that can enter the cylinder. It refers to how well an engine can "breathe" on its intake stroke. Volumetric efficiency is illustrated in **Figure 7-19.**

If volumetric efficiency is 100%, the cylinder completely fills with air on the intake stroke. However, engines are only about 80% to 90% volumetrically efficient. Airflow restriction in the ports and around the valves limit airflow.

High-volumetric efficiency increases engine power because more fuel and air can be burned in the combustion chambers. Turbocharging and supercharging increase volumetric efficiency because air is forced into the engine under pressure. The formula for volumetric efficiency is:

volumetric efficiency = volume of air taken into cylinder ÷ maximum possible volume in cylinder

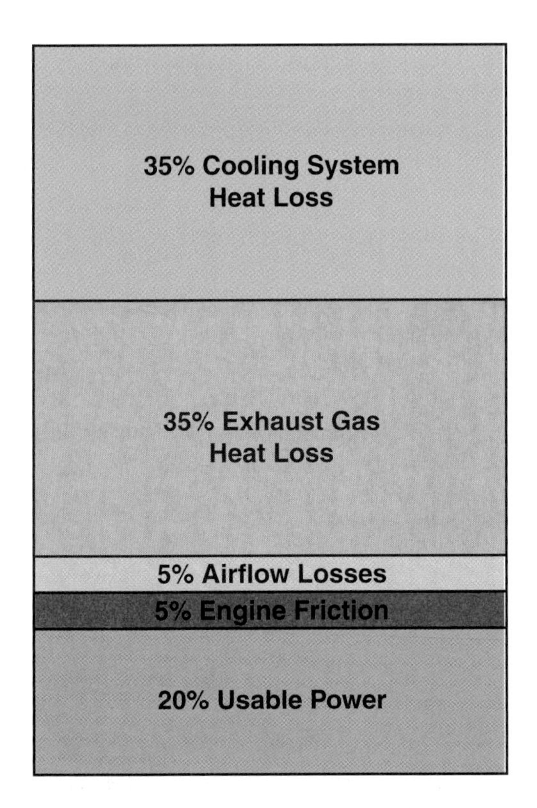

Figure 7-18. *A piston engine is not very efficient. Note that only about 20% of the fuel's energy is used to power the vehicle. The other 80% is lost to heat and friction.*

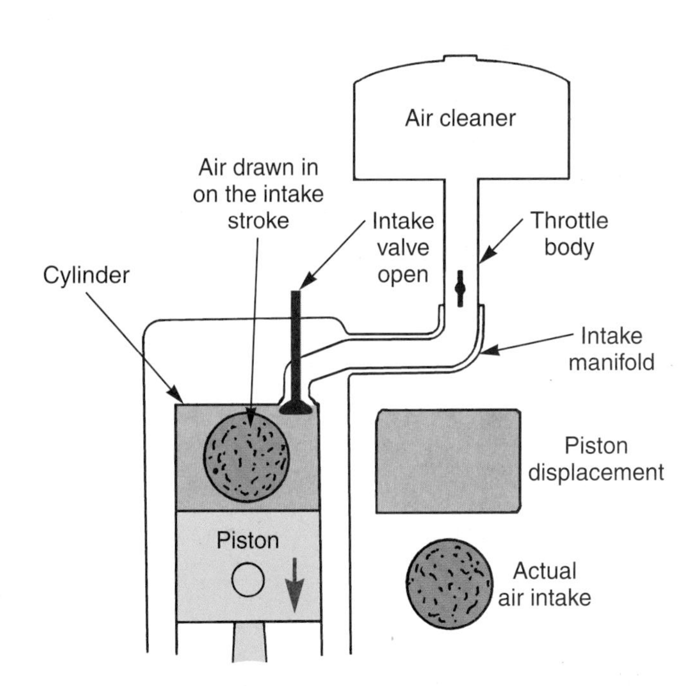

Figure 7-19. *Volumetric efficiency is a comparison of the actual air intake to the theoretically maximum air intake. (Deere & Co.)*

Mechanical Efficiency

Mechanical efficiency compares brake horsepower (bhp) to indicated horsepower (ihp). It is a measurement of mechanical friction. Remember that ihp is the theoretical power produced by combustion. On the other hand, bhp is the actual power at the engine crankshaft. The difference between the two is due to frictional losses.

Mechanical efficiency of around 75% to 95% is normal. This means that about 5% to 25% of the engine's power is lost to friction. The friction between the piston rings and cylinder walls accounts for most of this loss.

Thermal Efficiency

Thermal efficiency is a comparison of the amount of energy contained in the fuel used to the amount of energy produced by the engine. Thermal efficiency is found by comparing the energy stored in the fuel that is burned to the horsepower output of the engine. Thermal efficiency is another indicator of how well an engine can use the fuel's energy.

For example, one gallon of gasoline contains about 19,000 British thermal units (Btu) of energy. One horsepower equals about 42.4 Btu of heat energy released per minute. With this information, along with the brake horsepower and amount of fuel burned, you can find engine thermal efficiency.

thermal efficiency = energy output ÷ energy input
= (brake horsepower × 42.4 Btu/min/hp)
÷ (19,000 Btu/gallon × gallons/min)

In order to use this formula, you must convert brake horsepower to Btus by multiplying by 42.4 Btu/minute.

Generally, engine thermal efficiency is around 20% to 30%. The rest of the energy is absorbed as heat into the metal parts of the engine, blown out the exhaust as heat, or lost to friction.

Society of Automotive Engineers

The Society of Automotive Engineers (SAE) is a group of professionals that sets standards and improves technical communication in the automotive trade. SAE membership includes over 84,000 engineers, technical professionals, academics, and government professionals in almost 100 counties around the globe.

SAE helps standardize industry measurement procedures, standardize abbreviations, and define technical terms. SAE also publishes magazines and technical papers, and test various products used in the industry. To further your knowledge, visit the SAE website at www.sae.org.

Summary

Engine size is determined by cylinder bore, piston stroke, and number of cylinders. Cylinder bore is the diameter of the cylinder. Stroke is piston movement from TDC to BDC and is determined by the crankshaft throw. An oversquare engine has a bore dimension that is larger than the stroke dimension. An undersquare engine is just the opposite.

Piston displacement is the volume the piston displaces as it moves from BDC to TDC. Engine displacement is the sum of all piston displacements. Engine displacement is measured in cubic inches, cubic centimeters, or liters and is usually matched to the weight of the car.

Piston protrusion is a comparison of the piston head location in relation to the block deck when the piston is at TDC. It is a critical measurement on diesel engines.

Compression ratio is a comparison of cylinder volumes with the piston at TDC and BDC. A gasoline engine typically has a compression ratio of about 8:1 or 9:1. A diesel engine has a higher compression ratio, about 17:1 or 20:1, because it is a compression ignition engine.

Compression pressure is a result of the upward piston movement on the compression stroke. The higher the compression ratio, the higher the compression pressure. A gasoline engine may have compression pressure from 130 psi to 180 psi.

Atmospheric pressure is the air pressure produced by the weight of air above the earth. In a normally aspirated engine, it is used to fill the engine cylinders on the intake stroke. Vacuum is pressure below atmospheric pressure, or negative pressure.

Engine torque is a rating of the turning force at the engine crankshaft. Engine torque specifications are given at an engine speed. This is the engine speed at which the maximum torque is produced.

Engine power is a measure of the engine's ability to perform work over time. Engine power is measured in horsepower or kilowatts. In the US, engine power is rated in horsepower.

There are several horsepower ratings. Brake horsepower is the usable power at the engine crankshaft and is measured on an engine dynamometer. A chassis dynamometer measures drive wheel horsepower at the drive wheels. Indicated horsepower refers to the amount of power formed in the combustion chambers. Frictional horsepower is the power lost to friction in the engine. Net horsepower is the maximum power left when an engine is loaded with all of its accessory units. Taxable horsepower is a rating based on engine size. Curves can be used to show how horsepower and torque change with engine speed.

Engine efficiency is the ratio of power produced by the engine to the power supplied to the engine in the form of energy in the fuel. Volumetric efficiency is the ratio of actual air intake and potential air intake. It rates the engine's "breathing ability." Mechanical efficiency compares brake horsepower to indicated horsepower and is a measure of power loss to friction. Thermal efficiency is a comparison of the amount of energy contained in fuel used to the amount of energy produced by the engine, as indicated by horsepower output.

Review Questions—Chapter 7

Please do not write in this text. Write your answers on a separate sheet of paper.

1. Engine size (displacement) is determined by which of the following?
   (A) Main journal offset, cylinder bore, and number of cylinders.
   (B) Rod journal offset, cylinder bore, and piston stroke.
   (C) Cylinder bore, piston stroke, and number of cylinders.
   (D) Piston stroke, rod journal offset, and main journal offset.
2. How do you measure cylinder bore?
3. Which of the following is an undersquare engine?
   (A) 4.00 × 3.00
   (B) 4.23 × 3.25
   (C) 3.25 × 3.50
   (D) 3.50 × 3.00
4. If an engine has a bore of 4″ and a stroke of 3″, what is its piston displacement?
5. If the engine in Question #4 is a V-6, what is the displacement of the engine?
6. If the engine in Question #5 has its cylinders bored .030″ oversize, what is the new displacement?
7. The bore/stroke ratio for an engine with a bore of 3.5″ and a stroke of 3.0″ is _____.

8. Define *piston protrusion.*
9. If an engine has a compression ratio of 8.5:1, what does that mean?
10. Define *compression pressure.*
11. Atmospheric pressure at sea level is _____ psi.
12. Define *vacuum.*
13. What does an engine dynamometer measure? What does a chassis dynamometer measure?
14. List and explain the seven engine horsepower ratings.
15. _____ efficiency is a measurement of frictional losses in an engine.

ASE-Type Questions—Chapter 7

1. Technician A says one of the factors that determines an engine's size is the amount of piston travel per stroke. Technician B says one of the factors that determines an engine's size is cylinder diameter. Who is correct?
   (A) A only.
   (B) B only.
   (C) Both A and B.
   (D) Neither A nor B.
2. If one piston in a V-8 engine displaces 30 cubic inches, the engine displacement is _____ cubic inches.
   (A) 180
   (B) 240
   (C) 248
   (D) 3.75
3. Technician A says the term compression ratio refers to the total pressure developed in a cylinder. Technician B says the term piston protrusion refers to a comparison of the location of the piston head at TDC in relation to the block deck. Who is correct?
   (A) A only.
   (B) B only.
   (C) Both A and B.
   (D) Neither A nor B.
4. How much power is needed to lift 2000 lb a distance of 500 ft in one minute?
   (A) 1,000,000 foot-pounds per minute.
   (B) 2,000,000 foot-pounds per minute.
   (C) 2500 foot-pounds per minute.
   (D) None of the above.
5. A cylinder's compression pressure is normally measured in _____.
   (A) foot-pounds
   (B) newton-meters
   (C) pounds per square inch
   (D) None of the above.
6. When the piston in a given engine is at BDC, the cylinder volume is 30 cubic inches. When this piston reaches TDC, the cylinder volume is 5 cubic inches. What is the compression ratio of this engine?
   (A) 6:1.
   (B) 5:1.
   (C) 150:1.
   (D) None of the above.
7. Technician A says that as altitude above sea level increases, more air is forced into a normally aspirated engine. Technician B says that as altitude above sea level increases, atmospheric pressure decreases. Who is correct?
   (A) A only.
   (B) B only.
   (C) Both A and B.
   (D) Neither A nor B.
8. Engine torque is a rating of the turning force at the engine's _____.
   (A) camshaft
   (B) connecting rods
   (C) crankshaft
   (D) drive wheel(s)
9. Approximately how much power is needed to lift 600 lb a distance of 1200 ft in one minute?
   (A) 6.5 horsepower.
   (B) 21.8 horsepower.
   (C) 218 horsepower.
   (D) 65 horsepower.

10. Technician A says that the cylinder shown in the illustration is oversquare. Technician B says that the bore of the cylinder shown in the Illustration is 3.750″. Who is correct?
    (A) A only.
    (B) B only.
    (C) Both A and B.
    (D) Neither A nor B.

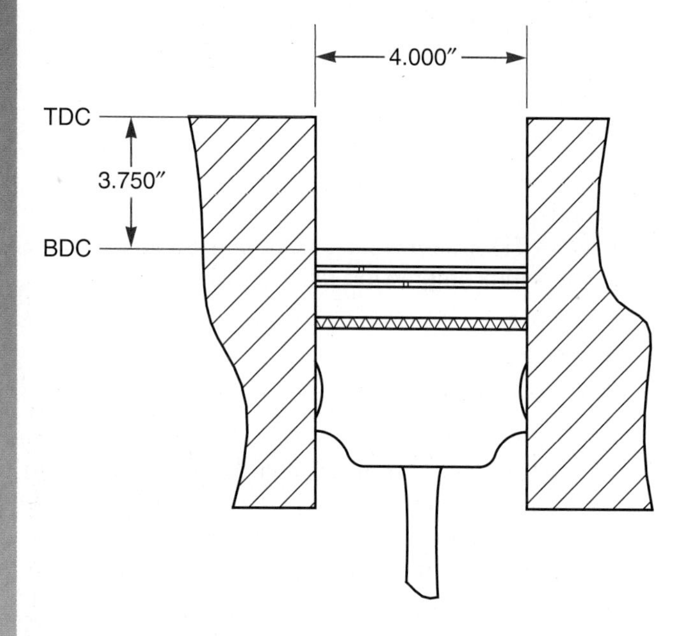

11. Technician A says that turbocharging increases an engine's volumetric efficiency. Technician B says that supercharging decreases an engine's volumetric efficiency. Who is correct?
    (A) A only.
    (B) B only.
    (C) Both A and B.
    (D) Neither A nor B.

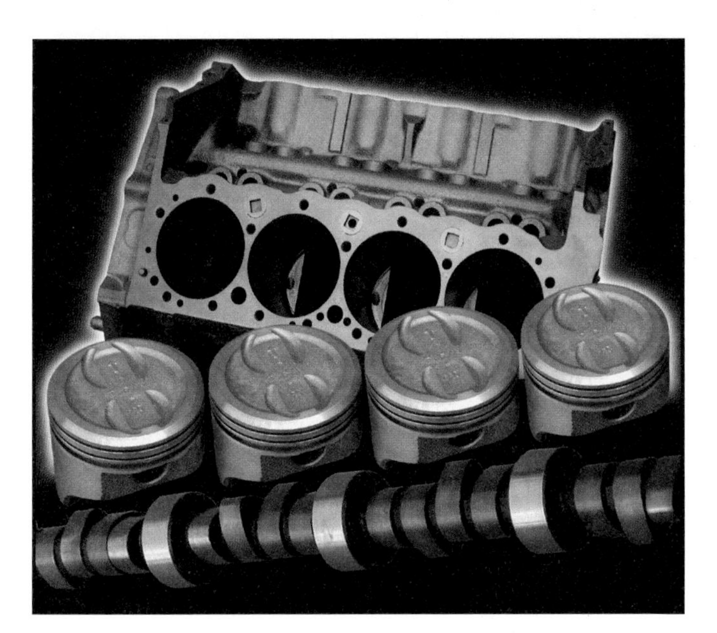

Chapter 8

Engine Combustion and Fuels

After studying this chapter, you will be able to:

- ❑ Explain how crude oil is converted into gasoline, diesel fuel, and other products.
- ❑ Discuss gasoline octane ratings.
- ❑ Describe normal and abnormal combustion.
- ❑ Explain preignition and detonation.
- ❑ List the factors contributing to preignition and detonation.
- ❑ Identify the damage that can be caused by detonation.
- ❑ Describe the operation of a knock sensing system.
- ❑ Explain dieseling and spark knock.
- ❑ Discuss the factors that affect engine combustion.

Know These Terms

Air-fuel ratio
Atomization
Combustion
Crude oil
Detonation
Dieseling
Duration
Fractionating tower
Ignition timing
Intercooler
Knock sensing system
Knock sensor
Lean fuel mixture
Normal combustion
Octane rating
Petroleum
Ping
Pollutant
Postignition
Preignition
Refining
Rich fuel mixture
Scavenging
Spark knock
Surface ignition
Turbulence
Valve timing

Today's engines operate over 90% cleaner than the engines used in the 1960s. This is due to higher compression ratios, fuel injection, turbocharging or supercharging, intercoolers, computer control, and many other design improvements that help extract as much energy as possible out of "each drop of fuel." However, these improvements also leave little room for error. The slightest engine problem can cause abnormal combustion. As you will learn, improper combustion or a fuel-related problem can reduce engine efficiency or cause severe engine damage.

This chapter explains where fuel comes from and what is happening inside an engine when fuel burns. Once you understand the theory, it can help you when trying to diagnose dozens of engine performance problems. If you can understand what is occurring inside an engine, you will be better prepared to troubleshoot internal engine problems. As an engine technician, you must be able to analyze all types of troubles. Study this chapter carefully!

Petroleum

Crude oil, or ***petroleum,*** is oil taken directly from the ground. It is a mixture of various elements, including carbon, hydrogen, sulfur, nitrogen, and oxygen. Crude oil varies in consistency from very thin, like water, to a very thick, semi-solid. It is largely made up of hydrocarbons, which are chemical mixtures of hydrogen and carbon. Hydrocarbons contain a lot of energy and are what make fuel flammable.

Traditionally, petroleum is obtained from wells. However, there are alternative methods of obtaining petroleum from oil shale and tar sands. In addition, automotive fuel can be synthesized from coal.

Oil Shale

Oil shale is a natural resource that can be altered and made into petroleum. Oil shale is a sedimentary rock that contains a tar-like substance called kerogen. However, oil shale contains a relatively large amount of sulfur and other impurities that must be removed through refinement.

Tar Sand

Tar sands are heavy hydrocarbons mixed with sand and dirt. There are several methods of removing these tars from the sand. One is to submerge the tar sand in hot water and steam. This forms a hot slurry that melts and liquefies the tar. Since oil is lighter than water, the petroleum floats to the top of the slurry where it can be easily removed.

Usually, tar sands contain about 12-1/2% petroleum compounds by weight. Around four to five tons of tar sand will produce one barrel of oil.

Coal Synthesized Fuel

Through various processes, coal can be converted into either a hydrocarbon gas or a liquid suitable for use in internal combustion engines. Changing coal into a gas is called gasification. Changing coal into a liquid is known as liquefaction. One ton of coal normally yields around two and one half barrels of synthetic oil suitable for refinement into gasoline.

Making Engine Fuels

Crude oil must be heated and broken down into different compounds, such as gasoline, diesel fuel, and liquefied petroleum gas. This process is known as ***refining.*** Refining begins with distillation. The crude oil is heated under pressure. The heated petroleum enters the base of a device called a ***fractionating tower.*** Since the different hydrocarbons, or fractions, contained in crude oil have different boiling points, they can be separated by weight, **Figure 8-1.** This is how crude oil is separated into natural gas, gasoline, motor oil, lubricating oils, and other products.

The temperature in the fractionating tower, or "pipe still," is hottest near the bottom and becomes cooler near the top. When the hot oil enters the bottom of the tower, it instantly "flashes" or vaporizes and begins to rise in the tower. As the oil vapor rises, the different fractions condense and settle out by weight on the trays in the tower. The heavier fractions in the crude oil settles on the lower trays. The lighter fractions rise to the higher trays in the tower. The lightest fractions flow out of the top of the tower. Each fraction of the crude oil is then collected in storage tanks near the fractionating tower.

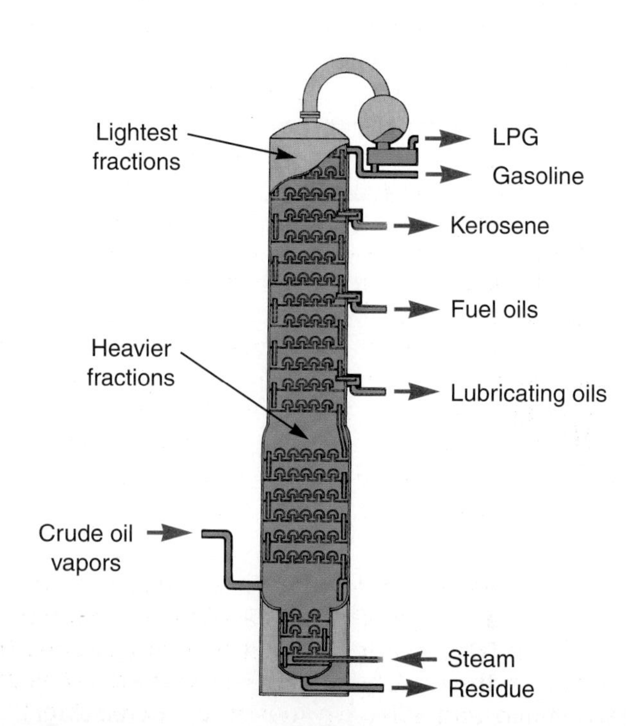

Figure 8-1. *A fractionating tower is needed to break crude oil into its parts. Heavier fractions, like motor oil, settle on the lower trays. Lighter fractions, like gasoline, settle on the upper trays. (Ford)*

Octane Ratings

The ***octane rating*** of a gasoline indicates its ability to resist ping, or knock. ***Ping*** is a sound produced by abnormal, rapid combustion. For example, if gasoline with a low octane rating is used in a turbocharged engine, the high compression stroke pressure can heat the air-fuel charge to a temperature that is high enough to ignite the fuel before the spark plug fires. This results in the violent action of the piston slamming against the expanding gasses, which causes a loud knocking noise. This can also damage engine parts.

In general, the higher the fuel's octane rating, or number, the greater the amount of heat and compression the gasoline can withstand before abnormal combustion occurs. For example, if an engine makes a knocking or pinging sound when burning fuel with an octane rating of 87, a fuel with a higher octane rating may be needed. The octane rating of the gasoline must be matched to the compression pressure in the engine cylinders—high compression pressure requires high octane. The manufacturer recommends the correct octane rating for a particular engine.

Gasoline is generally sold in two or three generic grades based on the octane rating—low, medium, and high octane. These are general categories; the octane rating of each grade may vary from one gas station to another. The high-octane grade, also called premium, high-test, or super, has the most resistance to knock or ping. It is generally required in high-compression, turbocharged, or supercharged engines. The generic grade may be listed on the gas pump, but the octane rating will always be listed. Follow the octane rating, not the generic grade.

It is very important to use fuel with the manufacturer's specified octane rating. There are many engine design variables that affect engine knock and octane ratings, such as compression ratio, combustion chamber shape, ignition timing, operating temperature, and boost pressure. For this reason, every automobile engine has been factory tested in order to determine its minimum and most-efficient octane requirement.

If you use fuel with a lower-than-recommended octane number, it will hurt overall engine performance. Power may be reduced, fuel economy may be lowered, run-on may occur, and the engine may experience knocking or pinging. Under extreme conditions, a low-octane fuel can actually cause physical damage to the engine. Detonation, the most severe and harmful type of engine knock, can burn or knock holes in the top of pistons, bend connecting rods, burn valves, and finally destroy the engine.

Note: There are additives used to increase the octane rating of gasoline. For example, MTBE, ethanol, or toluene is often added to gasoline. In the past, tetraethyl lead was added as an inexpensive and effective means of increasing the octane rating. However, this compound is *not* used as an octane booster in modern gasoline blends due to health concerns.

Increasing Octane with Alcohol

Alcohol is excellent as an octane-boosting additive for gasoline. It is clean burning and can be manufactured from renewable resources. The alcohol acts like an antiknock additive. It can make low-octane fuel perform like high-octane fuel. For example, 10% ethanol added to 87 octane gasoline can increase the fuel's octane rating to more than 90 octane. This is equal to the octane rating of premium gasoline. While the percentage of alcohol in the blend can vary from one supplier to another, most blends are 10% alcohol. This mixture can be burned in gasoline engines without modifying the engine.

The alcohol used as an additive is usually ethyl alcohol, also called ethanol or "grain alcohol." It is colorless, harsh tasting, and highly flammable. Ethanol can be made from numerous farm crops, such as wheat, corn, sugar cane, potatoes, fruit, oats, soybeans, or any crop rich in carbohydrates. Crop wastes can also be used to produce ethanol.

Gasoline Combustion

Combustion is a chemical reaction that releases the energy stored in the fuel as heat. In a gasoline engine, it is a rapid oxidation (burning) of carbon and hydrogen. The release of energy occurs when the hydrogen in gasoline combines with oxygen to form water (H_2O) and the carbon in gasoline combines with oxygen to form carbon dioxide gas (CO_2).

However, actual combustion in an engine is not perfect or complete. Air is not pure oxygen. In fact, it is about 78% nitrogen. As a consequence, combustion produces chemical compounds other than water and carbon dioxide. These other compounds are undesirable and are called ***pollutants.*** As you can see in **Figure 8-2,** several harmful compounds are created because of the other substances in air and because some of the gasoline does not burn.

Combustion occurs inside the engine's cylinders above the piston. The released heat energy causes the rapid expansion of the gasses, **Figure 8-3.** This expansion of gasses generates force inside the combustion chamber and on top of the piston. As a result, the piston is forced down. Proper combustion of any automotive fuel requires the fuel to be mixed with the correct amount of air and compressed in the cylinder before ignition.

Atomizing Fuel

Gasoline is very volatile; in other words, it easily evapo rates. As a result, there is a layer of gasoline vapor on the sur face of liquid gasoline. While this vapor is easily ignited, th rate of evaporation and combustion is too slow to produc adequate engine power, **Figure 8-4A.** In order to increase th

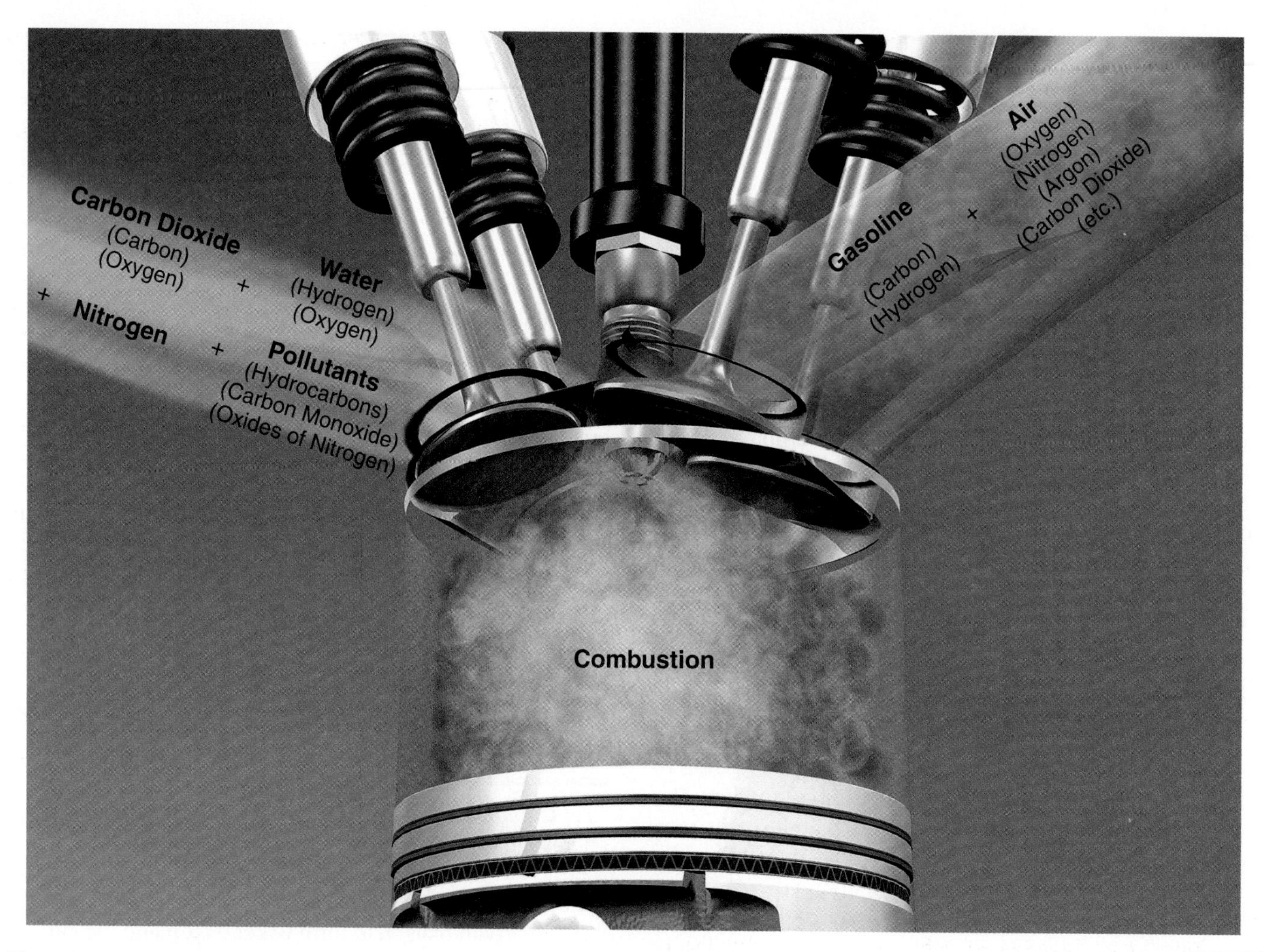

Figure 8-2. *Normal combustion in an engine produces several different types of pollution. Proper control of combustion can help reduce these emissions. (Ford)*

rate of combustion, the amount of vapor must be increased by increasing the surface area of the liquid gasoline. This is done by breaking the fuel into tiny droplets in a process called ***atomization.*** A vehicle's fuel injectors atomize the fuel for better combustion, **Figure 8-4B.** In the past, carburetors were used to atomize fuel. Atomization allows gasoline vapor and oxygen to be mixed in the proper proportion to obtain the best release of energy (power output).

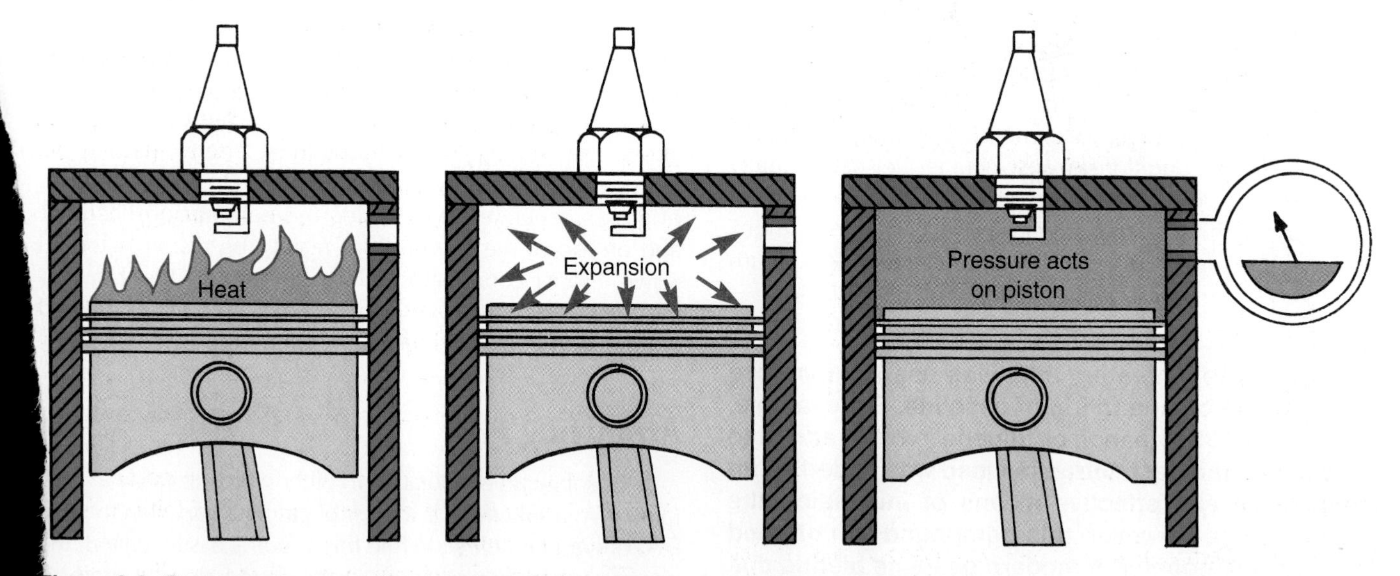

Figure 8-3. *During engine operation, burning fuel first produces heat. Heat then causes expansion of gasses. Since the gasses are confined in the cylinder, they produce pressure on the head of the piston.*

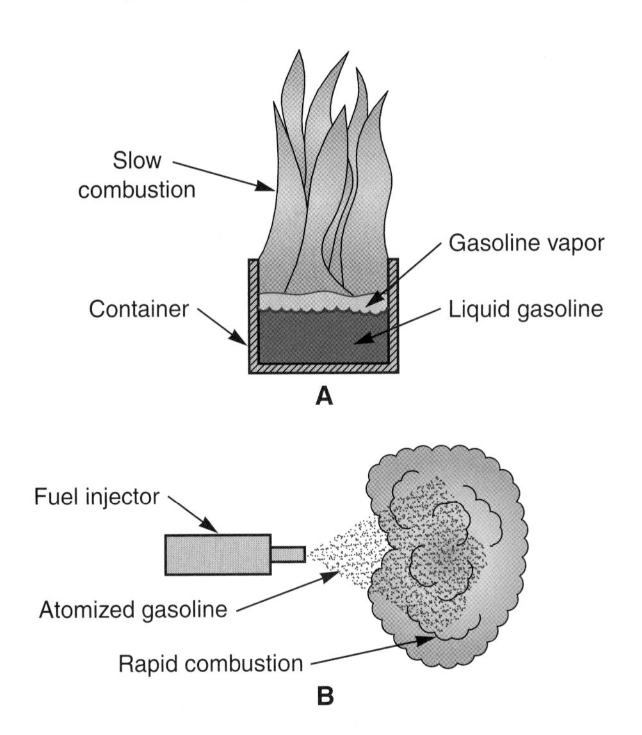

Figure 8-4. *A—Gasoline burns too slowly when in liquid form. B—If it is atomized into a mist, it will burn much better.*

Combustion Temperature and Pressure

At the start of the compression stroke, the combustion chamber temperature may be around 575°F (300°C) in a warm engine. After combustion, however, the temperature may rise to as high as 4000°F or 5000°F (2200°C or 2760°C). See **Figure 8-5.** Some of the heat is transferred to surrounding components. The temperature increase also results in expansion of the gasses in the cylinder, which increases pressure inside the cylinder.

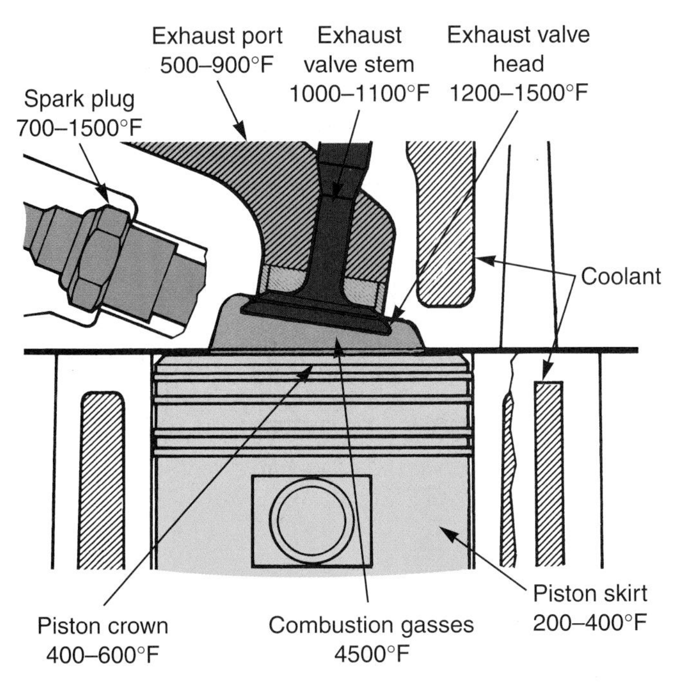

Figure 8-5. *Combustion produces tremendous heat in the engine. The combustion flame can reach 4500°F. Also, note the temperatures of various engine parts.*

The cylinder pressure on the compression stroke may be around 150 psi (1000 kPa) in an engine with an 8.5:1 compression ratio. In an engine with a compression ratio of 11:1, the pressure may be about 230 psi (1585 kPa) on the compression stroke. After combustion, the cylinder pressure can increase to well over 1000 psi (6900 kPa). This pressure, as you will learn in later chapters, can be much higher in supercharged or turbocharged engines. This is a tremendous amount of force pushing on the top of the piston, which forces the piston down.

For example, say the pressure increases to 800 psi. This means that every square inch of the piston top has 800 pounds of pressure pushing on it. If the piston has an area of 12.6 in^2 (a cylinder bore of about 4″), the combustion pressure would be like having a 5 ton (4525 kg) weight resting on the piston (12.6 in^2 × 800 psi = 10,080 pounds or about 5 tons). This tremendous force pushes the piston down in the cylinder, which in turn spins the crankshaft.

Normal Combustion

There are several factors that affect combustion, which are discussed later in this chapter. These include:

- ❑ Air/fuel ratio.
- ❑ Temperature.
- ❑ Compression.
- ❑ Spark timing.
- ❑ Fuel characteristics.

If all of the factors affecting combustion are favorable, the fuel burning process will be normal. ***Normal combustion*** occurs in a gasoline engine when a single flame spreads evenly and uniformly away from the spark plug electrodes. **Figure 8-6** illustrates normal combustion.

Normal combustion takes around 3/1000 of a second, which is comparatively slow. Normal combustion is *not* an explosion. An explosion, like that of dynamite, takes only around 1/50,000 of a second. This is a fraction of the time it takes for the normal combustion in a gasoline engine. However, under some conditions gasoline can burn too fast inside the engine. This results in combustion of the gasoline having properties more like an explosion than normal combustion, which is an undesirable situation.

Preignition

Preignition, also known as ping, occurs when a hot spot in the combustion chamber ignites the air-fuel mixture before the spark plug fires. This is abnormal combustion. **Figure 8-7** shows what happens during preignition. Study each phase carefully.

The effects of preignition are usually a slight loss c power and fuel economy. You can hear preignition as a mil

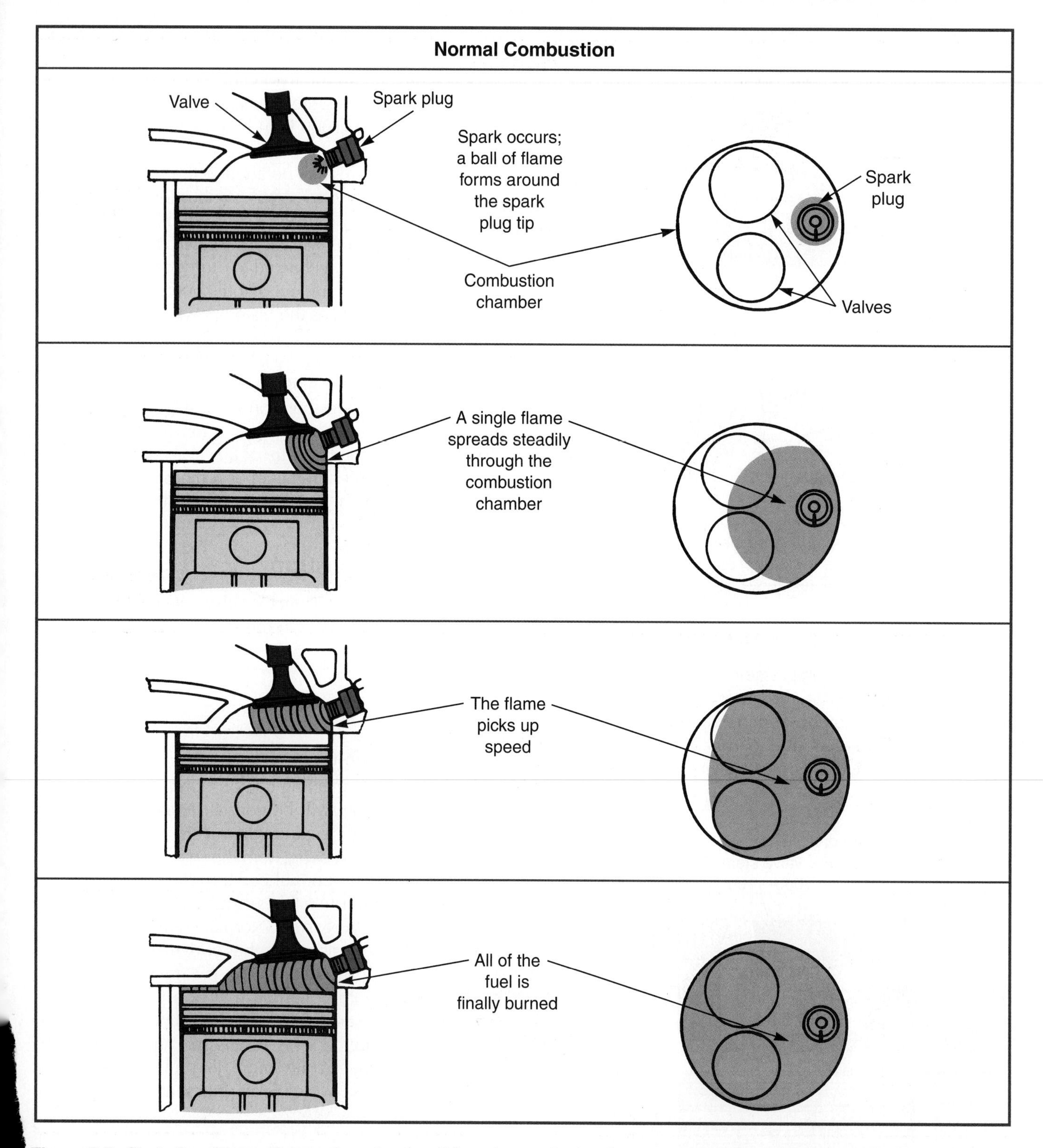

Figure 8-6. *Study the phases of normal combustion. When the spark plug fires, a smooth, even flame front spreads through the combustion chamber. Peak cylinder pressure is developed a few degrees past TDC on the power stroke.*

pinging, knocking, or rattling. However, severe or prolonged cases of preignition can cause more serious problems.

Surface Ignition

To fully understand preignition, you must first understand its cause—surface ignition. ***Surface ignition*** is when a surface in the combustion chamber is heated to a temperature that is high enough to ignite the air-fuel mixture. It can be caused by any of the following:

- ❑ Incandescent (glowing hot) piece of carbon in the combustion chamber.
- ❑ Overheated engine from improper cooling system operation.

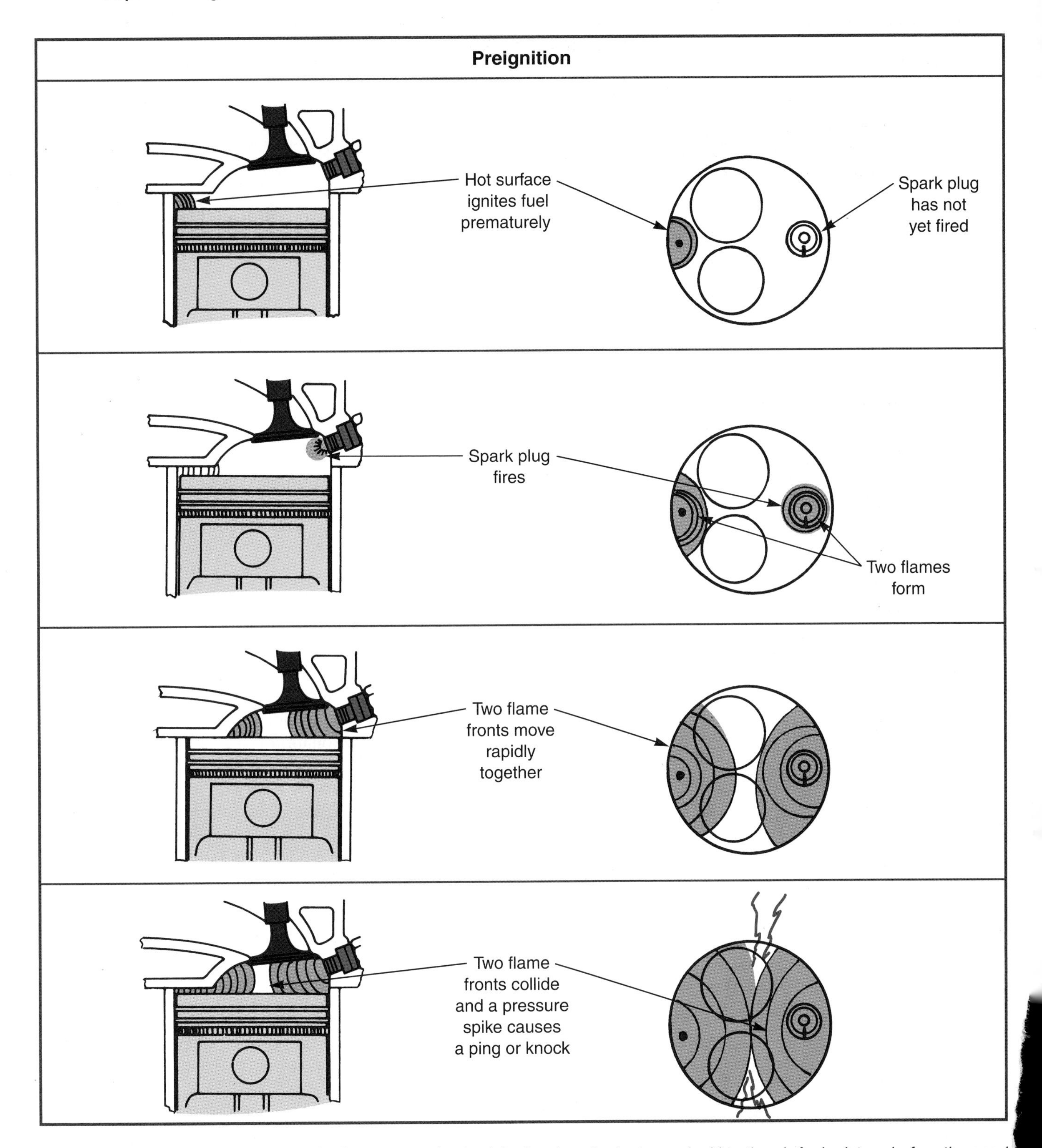

Figure 8-7. *Preignition is the result of a hot spot, a glowing bit of carbon for instance, igniting the air-fuel mixture before the spark plug fires. When the plug fires, a second flame front is created. When the two flame fronts collide, a pressure spike rattles parts and produces a pinging noise.*

- ❑ Exhaust valve overheated by lean fuel mixture. A lean fuel mixture can result from a clogged injector strainer, vacuum leak, stuck EGR valve, computer system malfunction, or other problem.
- ❑ Overheated spark plug. This may be due to the heat range of the spark plug being too high.
- ❑ Exhaust valve overheated by leakage. This may be due to an insufficient tappet clearance, weak valve spring, or sticking valve.

- ❑ Sharp edges in the combustion chamber. Overheated threads on the spark plug, the edge of the head gasket, and machined surfaces can all result in sharp edges in the combustion chamber.
- ❑ Excessively dry and hot atmospheric conditions.
- ❑ Air filter door stuck shut.

Postignition

Surface ignition can also happen after the spark plug fires. Since it occurs after normal spark ignition, it is termed ***postignition.*** The effects of postignition are minimal.

Dieseling

Preignition or postignition can also cause an engine to run after the ignition key switch is turned off. The term ***dieseling*** is used to identify this condition because, like a diesel engine, the gasoline engine runs without a spark igniting the fuel. The ignition switch is off and no voltage is present in the ignition system. Usually, an engine diesels because hot spots in the combustion chambers are igniting the fuel. For example, a leaky fuel injector may allow fuel into the intake even after the engine is shut off. A hot spot in the combustion chamber may then ignite the fuel. Dieseling places undue strain on pistons, cylinders, valves, throttle body gaskets, and the air cleaner.

A knocking, coughing, or fluttering noise can be heard when an engine diesels. In some cases, an unpleasant sulfur (rotten egg) odor may also be evident. The fluttering or coughing sound usually occurs right before the engine stops. This coughing sound can be a result of the engine actually "kicking" backwards. Air can blow back through the intake ports.

A dieseling engine can be stopped by, with the brake applied, leaving the automatic transmission in drive or dragging a manual transmission clutch to "lug" the engine. However, for a permanent fix, the source of the problem must be found and corrected. The reasons an engine may diesel include:

- ❑ Engine idle speed is set too high.
- ❑ Carbon deposits are increasing the engine's compression ratio.
- ❑ Using gasoline with the incorrect octane rating.
- ❑ Engine overheating because of low coolant level, a bad thermostat, or other problem.
- ❑ Spark plug heat range too high.
- ❑ Sticking throttle or gas pedal linkage.
- ❑ Engine needs a tune-up.
- ❑ The ignition switch turned off before the engine returns to curb idle speed.
- ❑ Oil entry into cylinders from an engine mechanical problem.

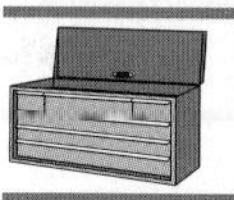

Note: The terms *run-on* and *after running* are also used to describe dieseling.

Spark Knock

Spark knock is a form of preignition, but it is not caused by surface ignition. Instead, the electric arc at the spark plug happens too early in the compression stroke. Spark knock is an ignition timing problem, not an engine mechanical problem. The timing is advanced too far and is causing combustion pressures to slam into the upward-moving piston. In other words, the maximum cylinder pressure is generated before TDC. If retarding the ignition timing cures or corrects a pinging problem in an engine, the problem was primarily spark knock. Ignition timing is discussed later in this chapter.

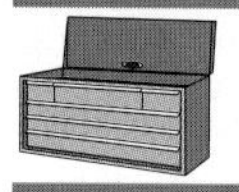

Note: Sometimes, early spark can cause detonation.

Detonation

Detonation is abnormal combustion where the last unburned portion of the fuel mixture almost explodes in the combustion chamber. A part of detonation is, in fact, almost as fast as an explosion of dynamite. The total process of engine detonation is about six times faster than normal combustion. See **Figure 8-8.**

Detonation can cause a very rapid pressure rise in the engine cylinder. It can be so fast and extreme that parts of the engine, such as the cylinders and cylinder heads, actually flex and vibrate. This vibration can be heard as a loud knock or ping. A detonation knock will sound something like a ball peen hammer rapidly striking the pistons or engine block. It is much louder than preignition ping.

Detonation is a very serious, abnormal combustion problem. Severe detonation can cause connecting rods to bend, cylinder heads to crack, pistons to melt, spark plugs to shatter, and bearings can be ruined. Also, the extreme heat can soften the top of a piston. Then, the pressure can force a hole in the softened aluminum, **Figure 8-9.**

Detonation-caused knocking will usually be heard when you push the gas pedal all the way to the floorboard and the engine is accelerating rapidly at moderate engine speeds. Normally, at higher engine speeds, there is not enough time for the end gas to heat up, ignite without a spark, and detonate.

Detonation Factors

You can think of detonation as a "race" between the normal flame front and the autoignition of the unburned air-fuel mixture. This unburned portion of the air-fuel

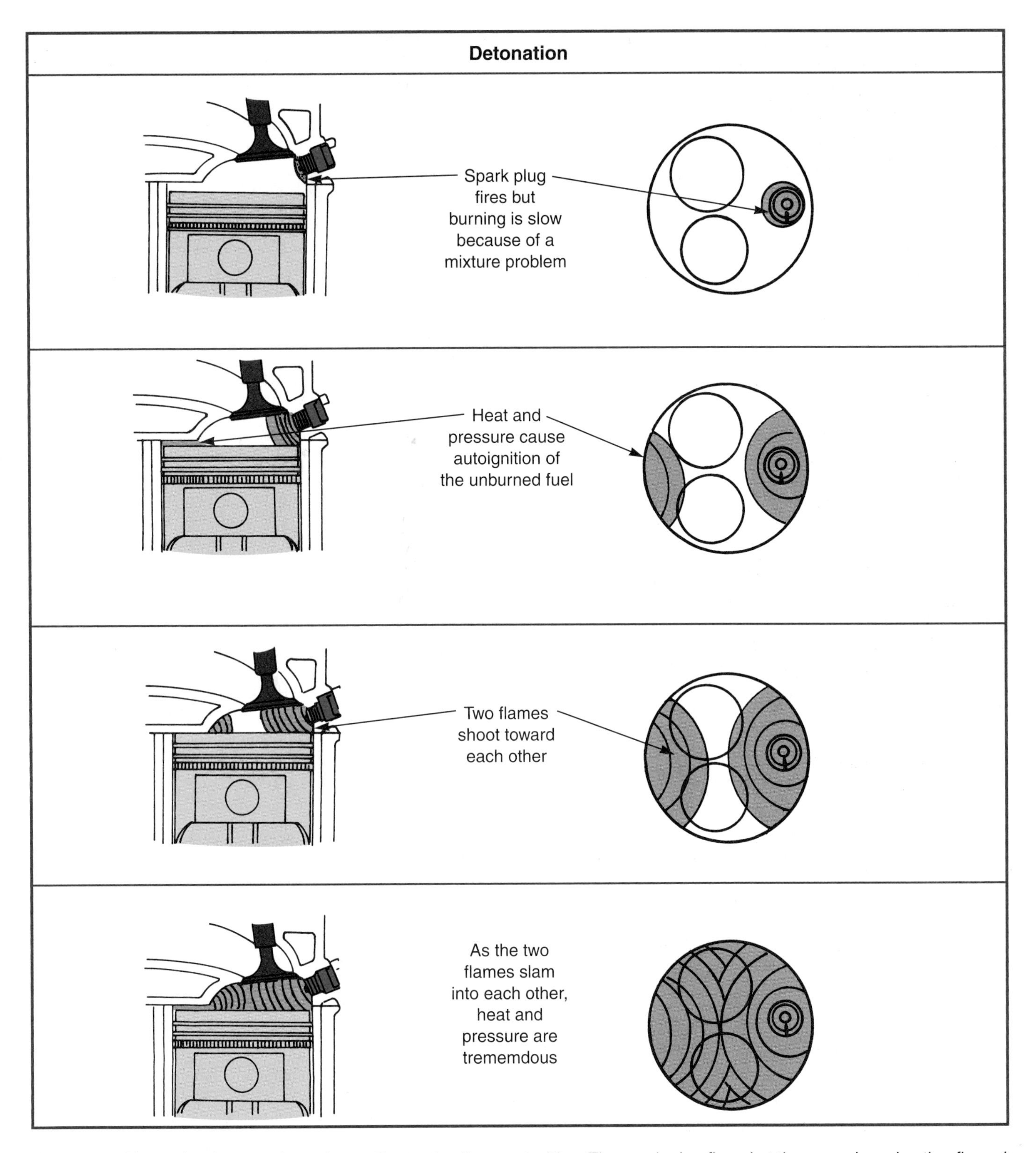

Figure 8-8. *Detonation is more damaging to the engine than preignition. The spark plug fires, but the normal combustion flame is too slow because of a mixture problem. The air-fuel mixture on the other side of the combustion chamber is heated, pressurized, and ignited. A violent explosion occurs that can damage the piston, head, and other components.*

mixture is called the ***end gas.*** If normal combustion is faster than the heating and squeezing on the unburned fuel, normal combustion "wins" and detonation is prevented. Conditions causing detonation include:

- Slow-burning, lean air-fuel mixture. This may be caused by a faulty fuel injector or fuel pump, blocked fuel filter or line, vacuum leak at higher engine speeds (bad PCV valve or EGR valve), or computer troubles.
- Gasoline with too low of an octane rating.
- Carbon deposits increasing the compression ratio. This is the result of oil entering cylinders or poor detergent action of the gasoline.

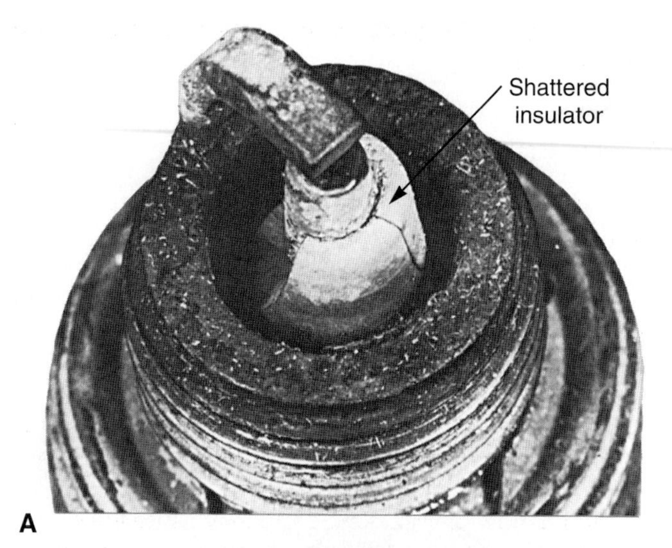

Figure 8-9. *A—Detonation has shattered this spark plug insulator. B—Prolonged detonation has overheated the head of this piston and blown a hole through it. (Champion Spark Plugs)*

- Engine operating at above-normal temperature due to low coolant level, water jacket blockage, or other trouble.
- Ignition timing too advanced.
- Bad oil rings and/or valve seals allowing oil to be burned in the cylinders.
- Air cleaner door stuck allowing too much hot exhaust manifold air to enter the engine.
- Excessive turbocharger or supercharger boost pressure from a bad pressure-limiting valve or faulty knock sensor.

Detonation vs Preignition

Detonation is similar to, yet quite different from, preignition. The processes of each are reversed. Detonation occurs *after* the start of normal combustion. Preignition begins *before* normal combustion. However, one condition can cause the other, and both are capable of causing engine damage.

Wild Knock

When carbon breaks loose from the combustion chamber wall, it can cause ***wild knock.*** When this happens, the carbon particle bounces around in the cylinder. While suspended in the burning fuel, it will heat up and ignite the fuel mixture each time fuel enters that cylinder. This can cause a pinging sound that can last for a few moments or come and go for no apparent reason.

Knock Sensing System

Many late-model engines are equipped with a ***knock sensing system.*** High-compression engines, turbocharged engines, and supercharged engines often incorporate a knock sensor to avoid detonation damage. The system uses a knock sensor and the control module (computer) to retard ignition timing or limit boost pressure to prevent detonation and preignition. When turbocharger boost is high, heat and pressure in the combustion chambers are also very high. This can cause ping or knock.

Figure 8-10 shows a simple diagram of a knock sensing system. The ***knock sensor*** acts as a microphone to "listen" for the sound of engine knock. When the sensor

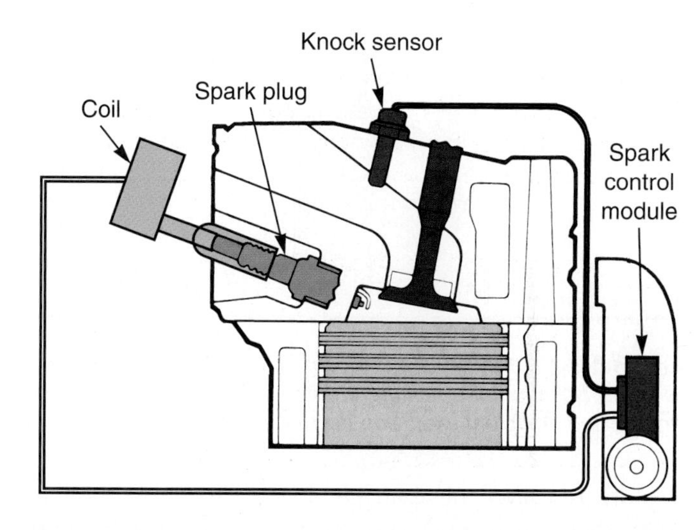

Figure 8-10. *This is a schematic of a knock sensor system. The knock sensor sends an electric pulse to the control module if it "hears" engine knock. Then, the control module can retard the ignition timing or open the turbo wastegate to stop the knocking. (DaimlerChrysler)*

"hears" engine knock, an electrical pulse is sent to the control module. The control module is programmed to retard the timing or open the wastegate (pressure limiting valve) until the knock sound is eliminated.

To check the action of a knock sensor, tap on it with a wrench with the engine running. This should make the computer retard the ignition timing or activate the wastegate. Engine speed should drop slightly. Follow service manual instructions for testing details.

Factors Affecting Combustion

Ideally, combustion should burn the fuel as fast as possible without producing detonation, utilize as much of the fuel's energy as possible, and produce as little exhaust emissions as possible. As you will learn in this section, there are many interacting factors affecting combustion. These factors determine whether combustion takes place normally or abnormally.

Ignition Timing

Ignition timing, or spark timing, is an important condition controlling combustion efficiency. Spark timing should be set to factory specifications. The specifications are usually given in degrees before TDC. On late-model vehicles, the timing is controlled by the computer and may not be adjustable.

Advancing the ignition timing causes the spark to happen earlier in the compression stroke. In other words, the piston is farther from TDC as ignition timing is advanced. When ignition timing is advanced without harmful knock, engine power is generally increased.

Retarding the ignition timing causes the spark to happen later in the compression stroke. In other words, the piston is closer to TDC as ignition timing is retarded. If the ignition timing is retarded too much, energy is wasted because combustion heat leaves the engine, as is shown in **Figure 8-11.** Much of the heat is passed into the engine block, cylinder head, and exhaust system instead of being used to power the engine. In fact, retarded ignition timing can cause an engine to overheat so much that the exhaust manifolds crack.

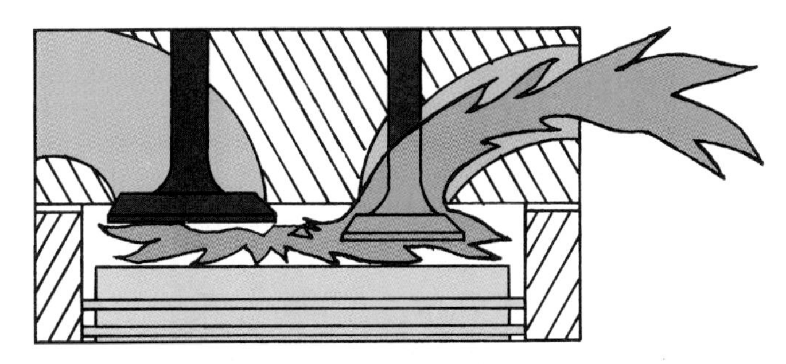

Figure 8-11. *Retarded ignition timing lowers engine efficiency because less pressure is developed and more combustion heat can blow out of the open exhaust valve. This can even overheat and crack the exhaust manifold.*

Spark Intensity and Duration

The size and duration of the spark at the spark plug electrodes is very important to combustion. This is especially true with today's lean fuel mixtures. Such mixtures are hard to ignite and keep burning. This is the main reason for wider spark plug gaps and high ignition voltages.

The wider the plug gap, the more secondary ignition voltage (40,000V or more) is required to sustain an arc across the spark plug electrodes. The resulting spark has more energy or ***intensity*** to start and maintain the combustion of a lean fuel mixture.

Too narrow of a spark plug gap may cause the engine to have a "lean misfire." This is when the spark is not hot enough, or has too low of an intensity, to ignite the lean mixture. A lean misfire generally occurs at moderate to high engine speeds.

For complete fuel burning, it is important to "hold" or sustain a spark for a certain period of time. This is called the ***duration*** of the spark. The proper duration is about 1/3 of the time it takes for combustion. Normally, a spark should occur at a spark plug for at least one millisecond (1/1000 of a second).

Note: Some engine analyzers are capable of measuring the duration of the sparks at the spark plugs. If spark duration is too long or short, there are several possible problems including a faulty ECM, wiring, or improper spark plug gap.

Air-Fuel Ratio

The ***air-fuel ratio*** is a comparison of the amount of fuel and air entering an engine, **Figure 8-12.** The

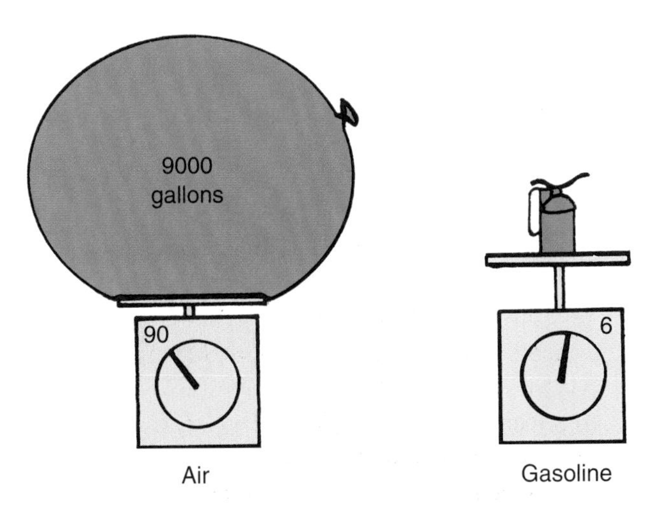

Figure 8-12. *Proper combustion requires a tremendous amount of air. (General Motors)*

proportions can be adjusted to meet the engine's changing needs.

A chemically correct air-fuel ratio is called a ***stoichiometric fuel mixture.*** It is a theoretically ideal ratio of around 14.7:1 (14.7 parts air to 1 part fuel). When engine conditions remain steady (unchanged engine load, temperature, fuel distribution, etc.), this ratio of fuel to air ensures that all of the fuel blends with all of the air and is burned. The air-fuel ratio affects the speed of flame travel during combustion. Flame travel is quite slow whenever the mixture is extremely lean or rich.

A ***lean fuel mixture*** contains more than 14.7 parts of air for every one part of fuel. A lean mixture provides better fuel economy and fewer exhaust emissions, **Figure 8-13.** However, power is reduced.

A ***rich fuel mixture*** contains less than 14.7 parts of air for every one part of fuel. This improves engine power and cold engine starting. However, it also increases emissions and fuel consumption, **Figure 8-13.**

Figure 8-14 shows a graph comparing engine power and fuel consumption to the air-fuel ratio. Note how for the best fuel economy, the mixture must be lean. Also, for the maximum power, the mixture must be rich.

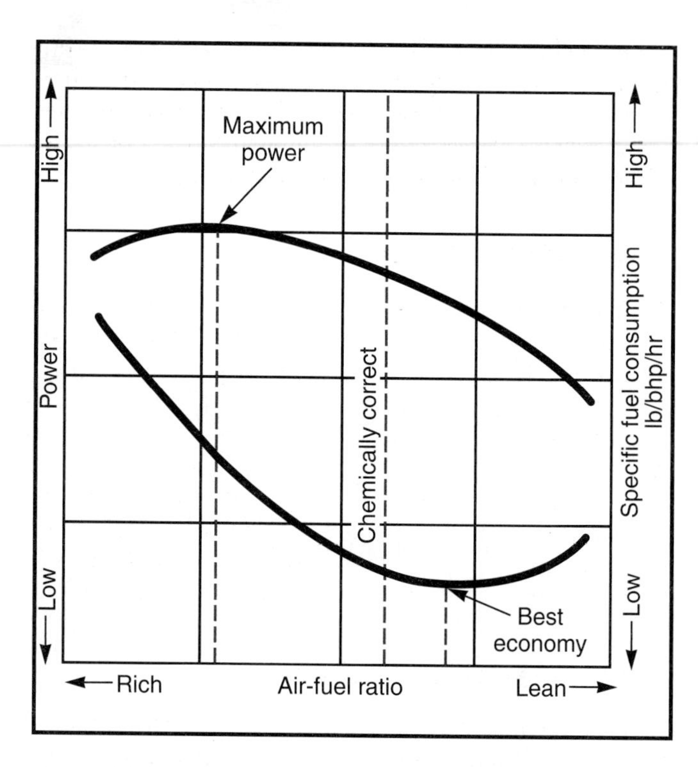

Figure 8-14. *This graph shows the relationship of engine power and fuel consumption to the air-fuel ratio. Note how the stoichiometric (chemically correct) mixture is between maximum power and best economy. (Ethyl Corporation)*

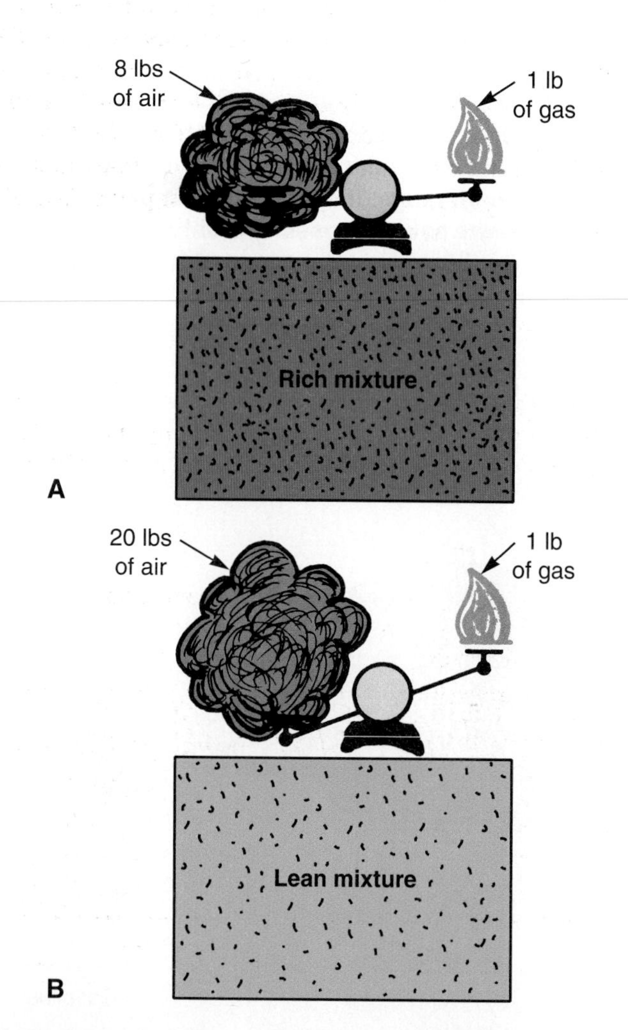

Figure 8-13. *A—A rich fuel mixture has more fuel mixed with the air. B—A lean fuel mixture has less fuel in the air.*

As mentioned earlier, a stoichiometric fuel mixture is about 14.7:1 in theoretically ideal conditions. However, multicylinder automobile engines do not operate under ideal conditions. They suffer from variations in fuel delivery from cylinder to cylinder, as well as differences in compression, temperature, and timing. For this reason, a stoichiometric fuel mixture is not always desirable, as is explained later.

A diesel engine uses a much leaner mixture than that used by a gasoline engine. The air-fuel ratio on a diesel engine is usually between 20:1 and 100:1. This leaner mixture is partially why a diesel engine is more fuel efficient than a gasoline engine. However, power is sacrificed.

Compression Ratio

The compression ratio or compression pressure of an engine is very important to combustion. Generally, increasing the compression ratio increases engine power and fuel economy, but also increases harmful exhaust emission levels (oxides of nitrogen [NO_X]). See the text index to find more information on compression ratios.

Combustion Chamber Turbulence

Air and fuel ***turbulence*** (swirl) in the combustion chamber normally speeds up and improves combustion efficiency. Turbulence mixes or "stirs" the fuel mixture and exposes more of the atomized fuel to the combustion flame. With more fuel area touching the hot flame, combustion is faster and produces less exhaust pollution. Refer to **Figure 8-15.** The shapes of ports, valves, pistons, and combustion chambers all affect combustion. They affect flow and turbulence of the incoming fuel charge.

Piston Head Shape

Since the piston head serves as one surface of the combustion chamber, it is very important that it be shaped to work with the cylinder head. A critical period of combustion occurs while the piston is at TDC. If the piston head gets too close to the cylinder head, it can restrict combustion by either blocking off areas or by quenching (cooling) a portion of the fuel mixture. Any area not easily reached by the flame can lower combustion efficiency. Also, these areas can get an uneven mixing of fuel. Thus, the fuel in these areas may ignite and burn poorly, **Figure 8-16.**

Intake Manifold Pressure

The amount of negative (vacuum) or positive (boost) pressure inside the intake manifold of an engine also affects combustion. It regulates how much air and fuel enters the engine. A high-intake-manifold pressure forces a large air-fuel charge into the combustion chamber. This increases power output. On the other hand, low-intake-manifold pressure reduces the power developed by decreasing the quantity of fuel and air entering the cylinders.

A high-intake-manifold pressure occurs when the gas pedal is pushed to the floor. This opens the throttle or air valves completely, permitting full atmospheric pressure (14.7 psi at sea level) to push the air and fuel into the engine cylinders.

As is discussed later, supercharging and turbocharging raise the maximum intake manifold pressure higher than atmospheric pressure. This increases the air-fuel charge entering the engine and raises the compression pressure before combustion. As a result, supercharging or turbocharging can increase engine power by as much as 50%.

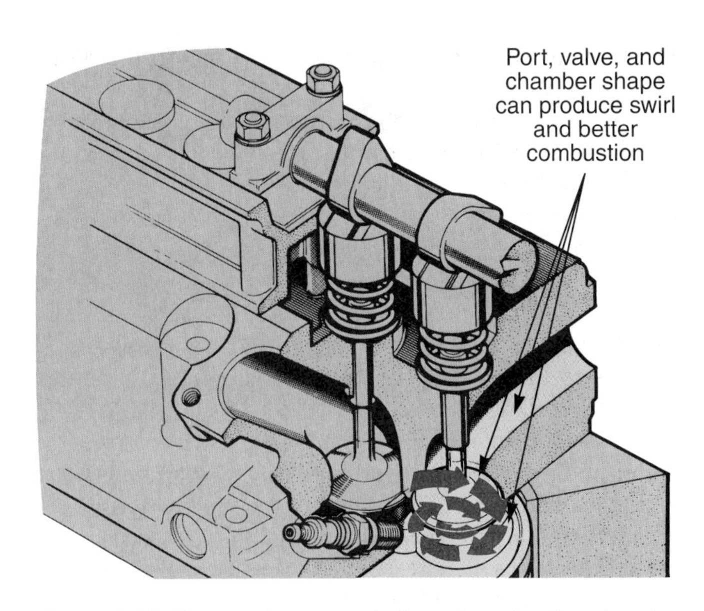

Figure 8-15. *Ports, valves, camshaft, and combustion chamber shape all interact and affect the turbulence in the combustion chamber. (Jaguar)*

Cylinder Bore and Stroke

To a certain degree, reducing the cylinder bore and stroke improves combustion efficiency. With a smaller bore, the distance the combustion flame must travel is reduced. As a result, combustion speed increases, heat loss is reduced, and efficiency is improved. With a short stroke, less energy is lost to the friction of piston rings rubbing against the cylinder walls. Therefore, more power is available to spin the crankshaft.

Fuel Metering System

The type of fuel metering system has a slight effect on combustion. Fuel injection improves combustion over carburetion for two basic reasons. First, injection systems allow more air to enter the cylinders. An air-fuel mixture is

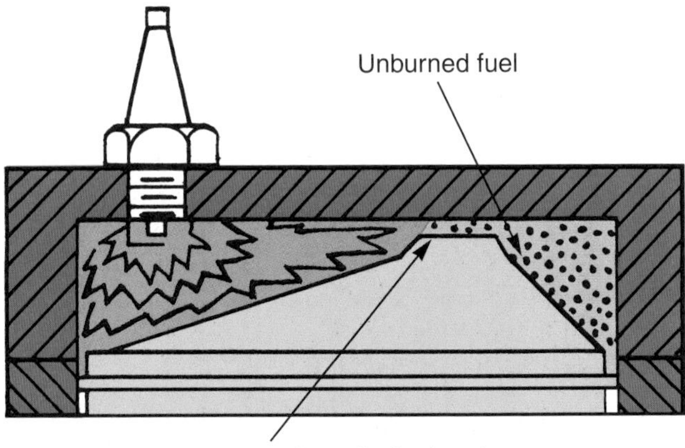

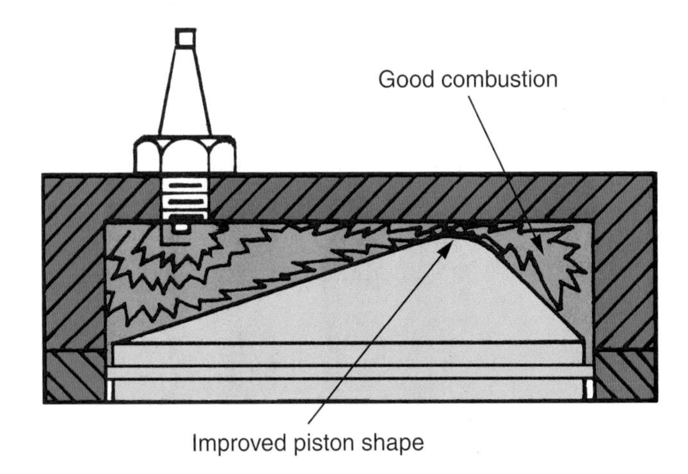

Figure 8-16. *The piston must be designed to work with the shape of the cylinder head. Note how the piston on the left is blocking the flame from reaching some of the mixture. The result is reduced power and increased emissions.*

much heavier than air alone. In a carbureted system, efficiency is lower because the heavier air-fuel mixture must be pulled through the intake manifold. Second, injection systems have more precise control over the amount of fuel entering each cylinder. Gasoline injection is covered in later chapters.

Valve Timing

Valve timing controls when the valves open in relation to the piston position. This is another important factor controlling combustion.

Both the intake and exhaust valves are open at a certain point in the cycle, based on valve timing. This is called ***valve overlap.*** The inertia of the fuel mixture in the intake manifold port helps to force more fuel and air into the cylinder. This increases combustion power, **Figure 8-17.**

This principle is also used to draw out a little more of the burned gasses through the exhaust port. This is called ***scavenging.*** It allows more room for the fresh fuel mixture that is entering the cylinder.

Valve timing can also be used to "clean up" combustion by retaining some of the exhaust gasses in the engine cylinders where they can limit peak combustion temperature and NO_X pollution.

Valve Lift and Duration

Valve lift is how far the engine valves open. ***Valve duration*** is how long the valves stay open. Valve lift and duration can have a pronounced effect on combustion, engine power, fuel economy, and drivability. Generally, lift and duration determine at what engine speed the best performance is developed.

Figure 8-18 shows some basic camshaft lobe profiles (shapes). Valve lift is determined by the height of the camshaft lobe from the centerline of camshaft to the lobe tip. A tall lobe produces a high valve lift. Valve duration is controlled by the width of the camshaft lobe. The wider the lobe, the longer the valve remains open. Camshafts are covered in Chapter 10.

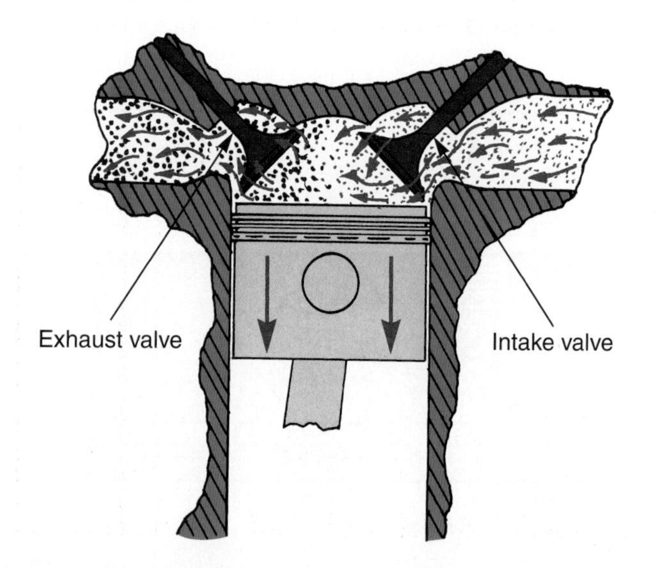

Figure 8-17. *Valve overlap is when intake and exhaust valves open at same time.*

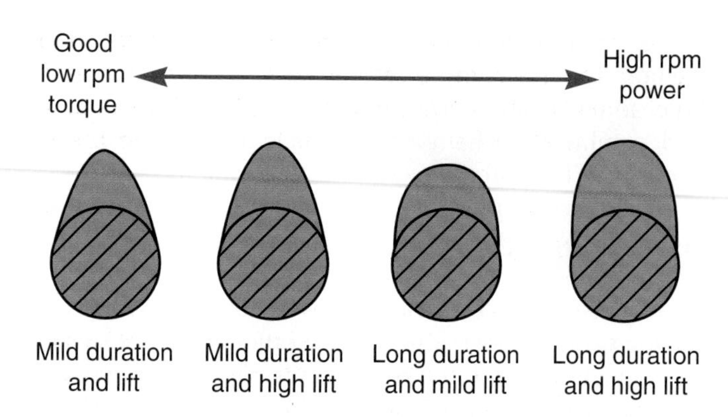

Figure 8-18. *Study these basic camshaft profiles and how they affect engine operation.*

In most stock, unmodified engines, lift and duration are designed to give maximum combustion efficiency between idle speed (around 800 rpm) and highway speed (around 3000 rpm). This provides the best combustion power, fuel economy, and cleanliness over the range of engine speeds at which most vehicles are operated.

In specialized applications, the lift and duration are matched to the application. For example, a race engine is designed for maximum power at maximum engine speeds. As a result, the valve lift and duration are increased. The valves are opened more and for a longer period of time than on a stock engine to help the engine breathe at high engine speeds (6000 to 10,000 rpm). This increases high-speed horsepower, but at the cost of fuel economy and emissions. On the other hand, recreational vehicles (RVs) are heavy and need a lot of torque at low engine speeds. The camshaft in an RV is matched to these requirements.

There is always a trade-off when changing valve lift and duration. For example, a camshaft designed for a racing application usually results in a *decrease* in engine power, fuel economy, and exhaust cleanliness at normal engine speeds. When the torque is increased at low engine speeds, such as in an RV or off-road application, there is usually a loss of horsepower at higher engine speeds. Vehicle manufacturers spend a lot of time and money developing the best camshaft design for various applications.

Intake Air Temperature

The temperature of the air entering the air cleaner and intake manifold has some effect on combustion. Generally, cool air increases engine performance and inhibits detonation. Cool air is more dense (compact) than hot air. Therefore, it carries more oxygen into the cylinders.

Some vehicles have a thermostatically controlled air cleaner that maintains a constant inlet air temperature of between 80°F and 110°F (27°C and 43°C). This intake air temperature provides a happy medium between engine power and low exhaust emissions.

An ***intercooler*** is used on supercharged and turbocharged engines to lower air inlet temperature. Superchargers and turbochargers, by compressing intake air, heat the air flowing into the engine. The intercooler helps cool the air charge before it enters the combustion chambers. This allows the engine to run with more boost without suffering detonation or abnormal combustion. Intercoolers can be cooled by outside airflow or engine coolant.

Spark Plug Location

Ideally, a spark plug should be centrally located in the combustion chamber. With the plug in the center of the chamber, the distance from the plug electrodes (start of combustion) to the farthest point in the chamber is at a minimum. This speeds up combustion because the flame has the shortest possible distance to travel.

Figure 8-19 shows that in an engine with a 4″ cylinder bore, a centrally located spark plug results in a maximum flame travel distance of around 2-1/4″. If the plug is located on one side of the combustion chamber, the distance is increased to 3″ or more. As a consequence, combustion takes longer and more heat energy is absorbed into the metal walls of the cylinder, combustion chamber, and piston head.

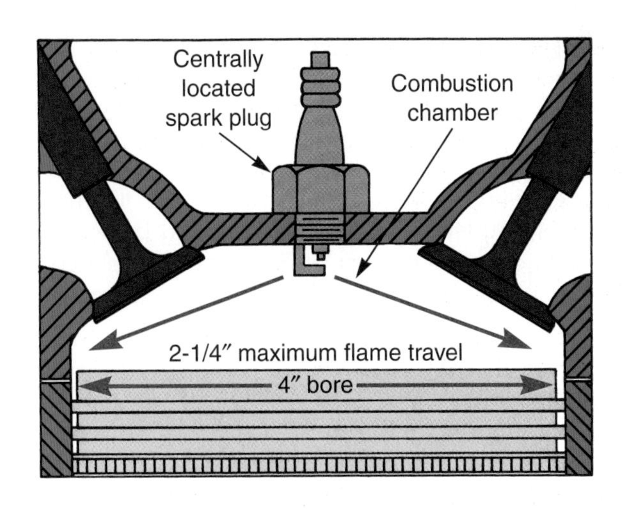

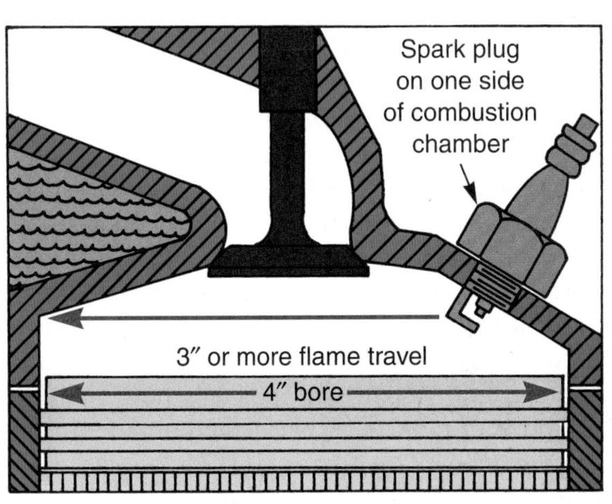

Figure 8-19. *The spark plug location in the combustion chamber shown on top results in a shorter path for the flame to travel. This design is more efficient than the combustion chamber shown on the bottom, which results in a longer path for the flame to travel.*

Combustion Chamber Surface Area

The smaller the surface area in the combustion chamber, the more efficient the combustion. By having less area to absorb heat, there is more heat remaining in the burning fuel to cause expansion, pressure, and piston movement. Notice in **Figure 8-20** that the irregularly shaped combustion chamber has more surface area than the smoother, domed combustion chamber.

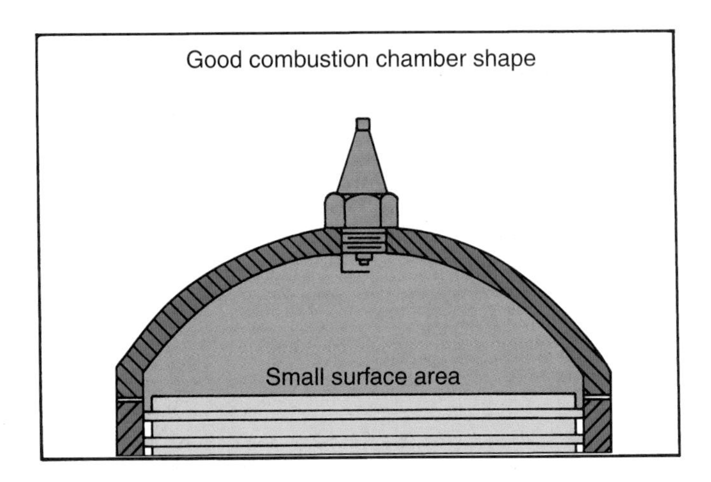

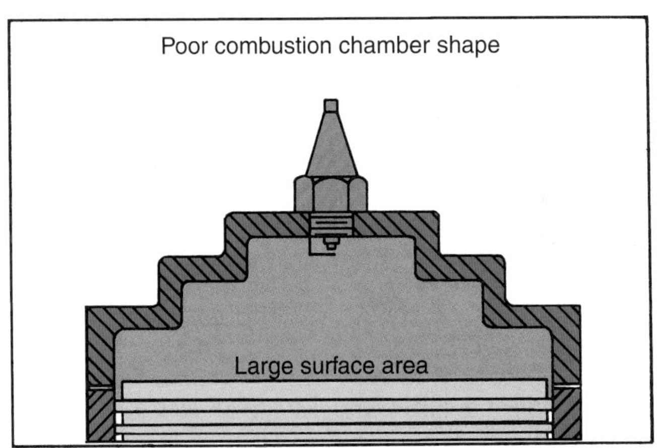

Figure 8-20. *The combustion chamber on the left has less surface area and is more efficient. The irregular shape shown on the right has more surface area and can also block the flame from burning some of the fuel.*

Combustion Chamber Temperature

In general, combustion is improved as the temperature of the combustion chamber is increased. This is true so long as the temperature is not high enough to cause surface ignition. If the chamber wall is cool, fuel near the cold metal surface does not vaporize and burn properly. This unburned fuel can leave the engine as wasted energy and pollution. Also, a cool combustion chamber wall absorbs more combustion heat and results in less cylinder pressure produced.

Combustion chamber temperatures are affected by compression ratio, cooling system design, air-fuel ratios, and ignition timing. Normally, a combustion chamber will average a little under 1000°F (538°C) during moderate engine operation.

Summary

Automotive fuels and combustion play an important role in determining the performance and service life of an engine. Combustion must be a controlled burning of the air-fuel mixture in the cylinders so that maximum pressure is developed on the head of the piston a few degrees past TDC.

Crude oil is converted into gasoline, diesel fuel, and other products through refinement. Gasoline is graded by an octane rating system that indicates how well the fuel resists knock or ping. A higher octane number means the fuel burns more slowly and steadily under heat and pressure. It is ideal for a high-compression, supercharged, or turbocharged engine.

Ignition timing, fuel mixing, fuel distribution, valve action, and other factors must be perfect for combustion to be efficient. Normal combustion occurs in a gasoline engine when the spark plug fires and a single flame burns through the mixture. Normal combustion is not an explosion.

Preignition results when a hot spot in the combustion chamber ignites the fuel prematurely. This is called surface ignition. Then, the spark plug fires. The two flame fronts slam into each other producing a pressure spike and a ping noise. Dieseling is a surface ignition condition in a gasoline engine where the engine keeps running when the driver turns the key and engine off.

Detonation occurs when normal combustion is too slow. It is more damaging than preignition. The unburned fuel autoignites and almost explodes in the combustion chamber. Pistons, connecting rods, cylinder heads, etc., can be ruined by prolonged detonation.

Knock sensing systems are used to detect and correct for ping or knock. The knock sensor acts like a microphone "listening" for knock. Any knocking or pinging sound will make the sensor produce an electrical signal for the control module. The control module can then retard the ignition timing or open the turbo wastegate to prevent knock.

Review Questions—Chapter 8

Please do not write in this text. Write your answers on a separate sheet of paper.

1. Crude oil contains:
   (A) Carbon.
   (B) Hydrogen.
   (C) Sulfur.
   (D) All of the above.
   (E) None of the above.
2. How is crude oil changed into gasoline, diesel fuel, and other products?
3. High-octane gasoline ignites and burns _____ than a low-octane fuel.
4. _____ refers to how the fuel injector breaks fuel into tiny droplets.
5. Normal gasoline combustion occurs when a single _____ spreads evenly away from the spark plug electrodes.
6. Normal combustion takes about _____ second.
7. Define *preignition.*
8. Define *detonation.*
9. List eight factors contributing to detonation.
10. How does a knock sensing system work?
11. List two possible causes for engine dieseling in a gasoline engine.
12. List five factors that affect engine combustion.
13. How does air-fuel ratio affect the operation of a gasoline engine?
14. Define *scavenging.*

ASE-Type Questions—Chapter 8

1. Technician A says that natural gas can be obtained from crude oil. Technician B says that gasoline can be obtained from crude oil. Who is correct?
   (A) A only.
   (B) B only.
   (C) Both A and B.
   (D) Neither A nor B.

2. Technician A says that gasoline with a low octane rating can reduce an engine's power efficiency. Technician B says that gasoline with a low octane rating can lower an automotive engine's fuel economy. Who is correct?
   (A) A only.
   (B) B only.
   (C) Both A and B.
   (D) Neither A nor B.

3. Technician A says that the camshaft lobe shown on the left will provide better torque at lower engine speeds than the lobe shown on the right. Technician B says that the camshaft lobe shown on the right will provide better power at higher engine speeds than the lobe shown on the left. Who is correct?
   (A) A only.
   (B) B only.
   (C) Both A and B.
   (D) Neither A nor B.

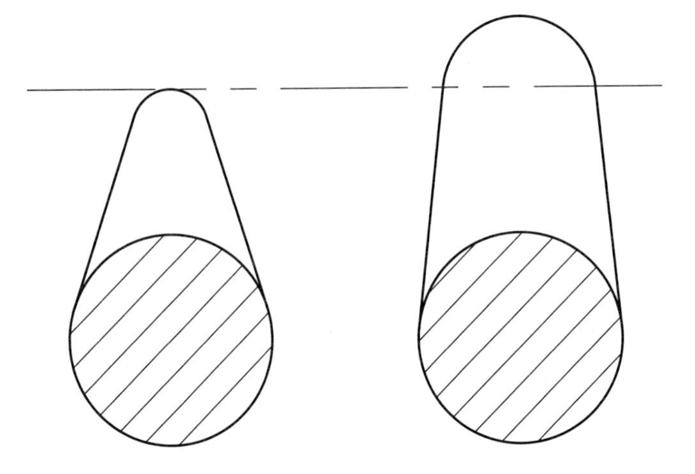

4. All of the following are general grades of gasoline sold in the US *except:*
   (A) medium.
   (B) leaded.
   (C) premium.
   (D) regular.

5. Technician A says that normal combustion occurs in a gasoline engine when a single flame spreads evenly and uniformly away from the spark plug electrodes. Technician B says that normal combustion occurs in a gasoline engine when dual flames spread evenly and uniformly away from the spark plug electrodes. Who is correct?
   (A) A only.
   (B) B only.
   (C) Both A and B.
   (D) Neither A nor B.

6. Technician A says that preignition occurs when the last unburned portion of the fuel mixture almost explodes in the combustion chamber. Technician B says that preignition occurs when a hot spot ignites the fuel mixture in the combustion chamber before the spark plug fires. Who is correct?
   (A) A only.
   (B) B only.
   (C) Both A and B.
   (D) Neither A nor B.

7. Surface ignition can be caused by a(n) _____.
   (A) defected exhaust valve
   (B) overheated spark plug
   (C) glowing piece of carbon in the combustion chamber
   (D) All of the above.

8. Technician A says that under certain circumstances engine detonation can crack an engine's cylinder head. Technician B says that under certain circumstances engine detonation can actually melt an engine's piston. Who is correct?
   (A) A only.
   (B) B only.
   (C) Both A and B.
   (D) Neither A nor B.

9. Technician A says that ignition timing can affect the quality of engine combustion. Technician B says that intake manifold pressure can affect the quality of engine combustion. Who is correct?
   (A) A only.
   (B) B only.
   (C) Both A and B.
   (D) Neither A nor B.

This is a bare block for a four-cylinder engine. Once all of the internal parts are installed, such as the bearings, crankshaft, and pistons, the assembly is called a short block. (DaimlerChrysler)

Chapter 9

Short Block Construction

After studying this chapter, you will be able to:

- ❑ Identify the different stages of cylinder block assembly.
- ❑ Describe the construction of a cylinder block.
- ❑ Summarize the function of the crankshaft.
- ❑ Describe the construction of a crankshaft.
- ❑ Identify the components of a crankshaft.
- ❑ Describe the function of connecting rods.
- ❑ Explain the construction of connecting rods.
- ❑ Summarize the construction and operation of piston pins.
- ❑ Identify different types of pistons.
- ❑ Describe the function of piston rings.
- ❑ Explain the function of engine bearings.
- ❑ Describe the construction of engine bearings.

Know These Terms

Bare block
Bearings
Camshaft bore
Compression rings
Connecting rod
Core plugs
Counterweights
Crankshaft
Cylinder block
Cylinders
Dry sleeve
Engine piston
Floating piston pin
Flywheel
Integral cylinder
Lifter bores
Long block
Main bearings
Main bore
Main caps
Main journals
Oil ring
Oil spurt hole
Piston head
Piston pin
Press-fit piston pin
Rear main oil seal
Short block
Sleeved cylinder
Thrust main bearing
Thrust washers
Wet sleeve

Late-model engines are made using "high tech" materials and machining techniques. To work on today's engines, you must have a full understanding of how engine components are constructed and designed. This knowledge will help you solve problems and make important decisions concerning how to service engines. If you do not fully understand engine construction, it can be very easy to overlook a repair task and make it very difficult to diagnose many engine troubles.

Study the information in this chapter carefully! It provides a base for many later chapters on engine troubleshooting and repair.

Cylinder Block Assembly

There are three basic states or stages of assembly for a cylinder block. These are bare block, short block, and long block. It is important to understand the differences between each stage of assembly. **Figure 9-1** shows a complete engine assembly. As you read through this section, consider which parts of this assembly are included in each stage of assembly for the cylinder block.

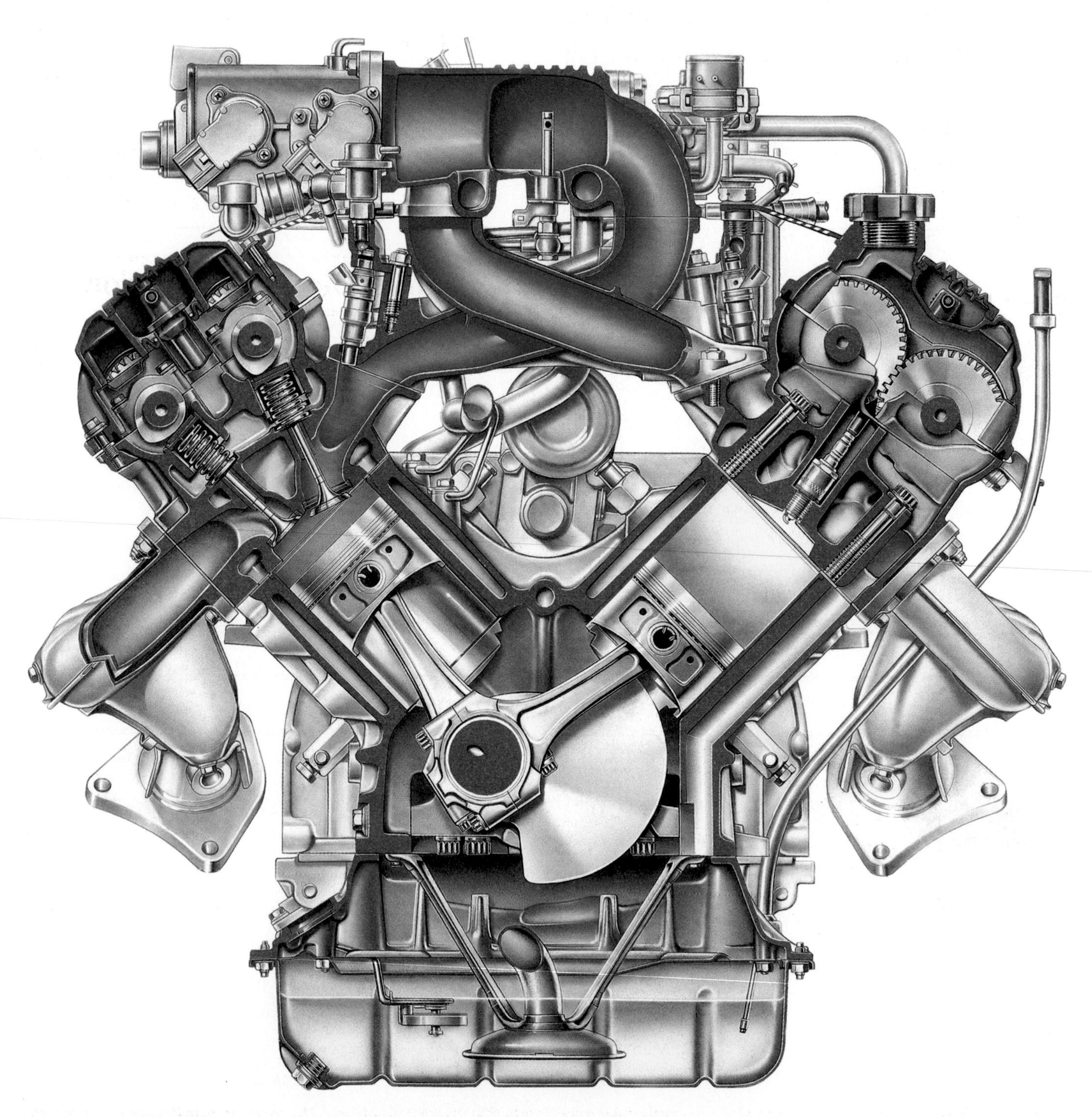

Figure 9-1. *Can you identify the components of the short block and the long block? (Lexus)*

Figure 9-2. *This is a bare block. No parts are installed.*

Bare Block

A ***bare block*** is the cylinder block without any parts installed. This is the finished, machined casting. The pistons, rods, a crankshaft, camshaft, balancer shaft, and other parts are not installed in the block. See **Figure 9-2.**

Short Block

A ***short block*** is the cylinder block with all of its internal parts installed. The pistons, connecting rods, crankshaft, and bearings are installed in the block. The term ***bottom end*** is often used to mean the same thing as short block. **Figure 9-3** shows the parts of a typical short block. Study which parts are *not* considered in the short block. The heads, oil pan, and front-end parts are not installed.

A technician may replace the short block with a new or rebuilt short block. The front cover, water pump, motor mounts, flywheel, reconditioned old heads, etc., are cleaned and reused on the new short block. The engine technician may also rebuild or overhaul the short block by replacing or reconditioning worn parts. This is explained in detail in later chapters.

Long Block

A ***long block*** is the short block with the completed heads installed, **Figure 9-3.** Parts like the valve covers, front cover, flywheel, and mounts are *not* included in the long block. Long blocks can be purchased from major auto manufacturers and specialized engine rebuilders to replace a high-mileage, badly worn engine.

Cylinder Block Construction

The ***cylinder block*** is the main foundation for an engine. It supports or holds all of the other engine components.

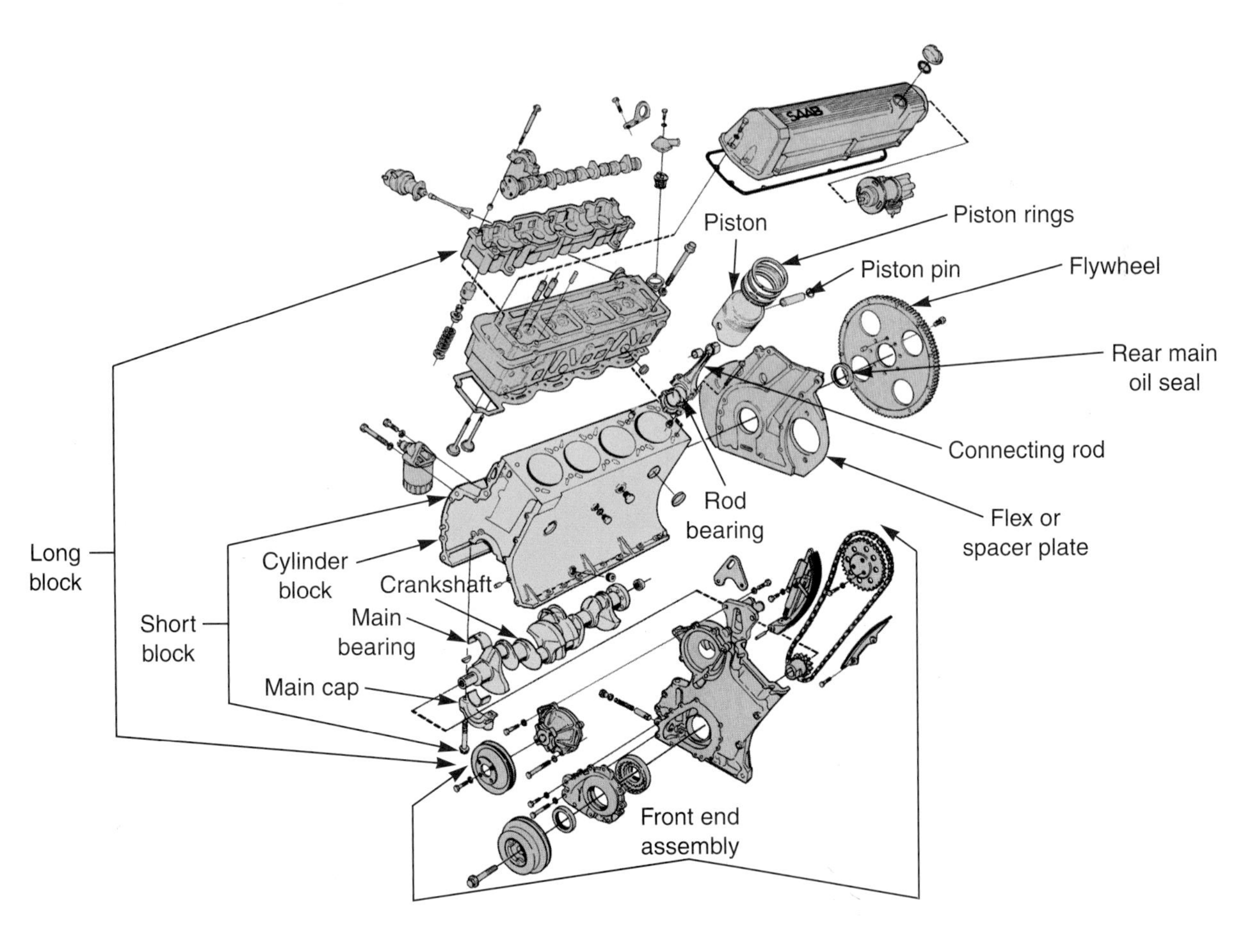

Figure 9-3. *The short block is the bare block plus all of the major parts that install in the block. The long block is the short block with the complete cylinder heads installed. (Saab)*

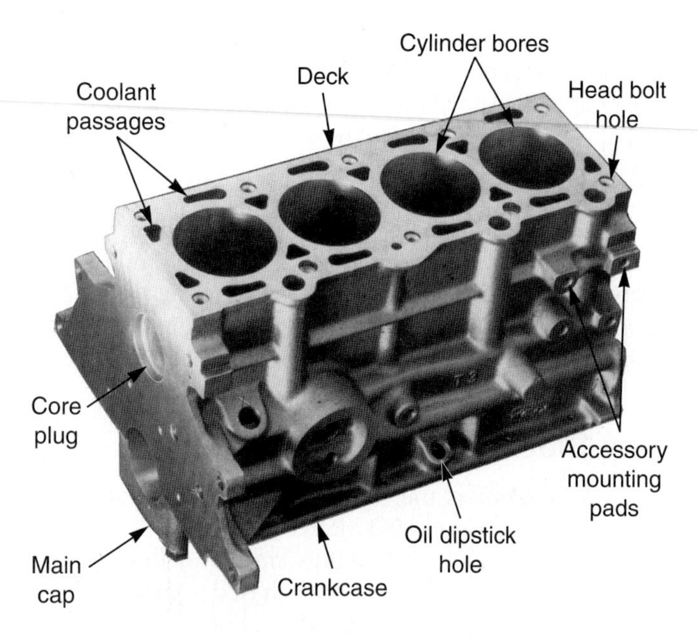

Figure 9-4. *A cylinder block is made by pouring molten metal into a mold. The internal sand mold allows hollow areas for the cylinders, water jackets, and so on. (Ford)*

Various surfaces and holes are machined in the block for attaching these other parts. **Figure 9-4** shows a typical cylinder block for a four-cylinder, inline, overhead camshaft engine.

Cylinder Block Materials

Cylinder blocks are normally made of cast iron or aluminum alloy. A cast iron block is heavy and strong. Older engines, diesel engines, and extreme-duty engines normally use cast iron because of its strength and dependability. Nickel, chromium, and other metals can be added to iron to form an iron alloy that is more wear resistant and stronger than plain gray iron.

An aluminum block is much lighter and almost as strong as a cast iron. However, the aluminum casting must be much thicker than an iron casting to achieve the same strength. Aluminum blocks are becoming more popular on late-model vehicles because the reduced weight increases fuel economy. Aluminum also dissipates heat much faster than cast iron.

Casting a Cylinder Block

A cylinder block is manufactured by casting, which is pouring molten metal (iron or aluminum) into a mold. The outside of the mold forms the external shape of the block. The inside of the mold is made of sand and forms the coolant passages (water jackets) and main oil passages in the block. After the molten metal cools and solidifies, the mold is removed and the sand is shaken out of the block to make the block hollow, **Figure 9-5.**

There are holes in the block that allow the internal sand mold to be removed from inside the block during the casting process. ***Core plugs,*** also called casting plugs, are small metal plugs pressed or screwed into these holes. See **Figure 9-6.**

Various oil passages are also machined in the block so motor oil can reach high friction areas. This is discussed in the chapter on lubrication systems.

The cylinder block for a diesel engine is cast in basically the same way as a gasoline engine block. However, the casting is thicker in high-stress areas. The cylinder walls are usually thicker. Also, the block is usually a four-bolt-main configuration. This is discussed later in the chapter.

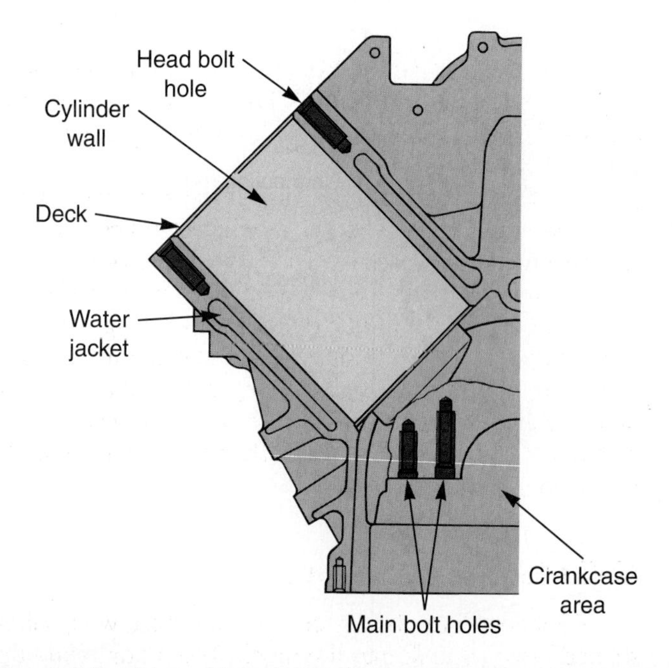

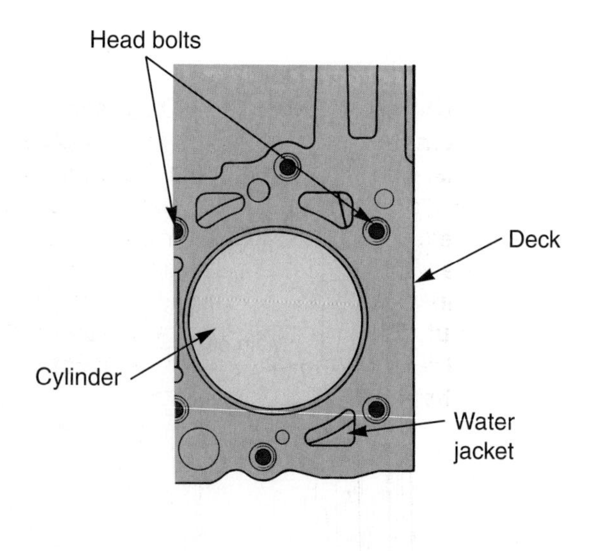

Figure 9-5. *Side and top views of a block show bolt holes, openings for coolant, the crankcase area, and other parts of the block (Mercedes-Benz)*

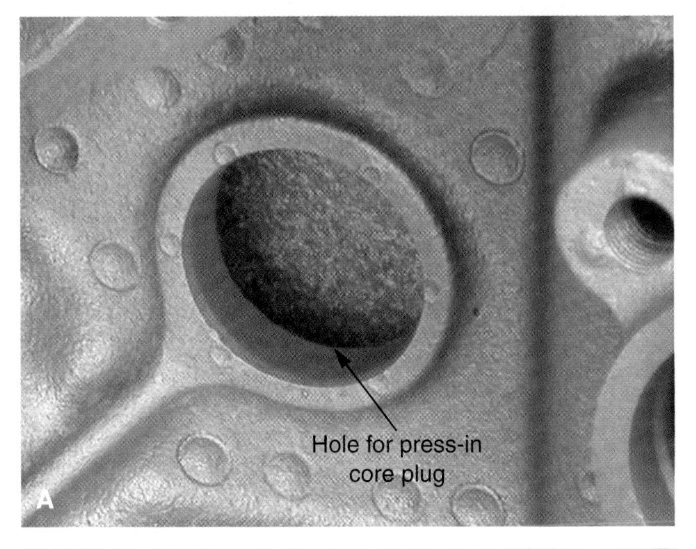

Figure 9-6. *Core plugs fill the holes that are needed during the block manufacturing so the sand core can be removed to produce water jackets. A—This core hole uses a press-in core plug. B—This core hole uses a screw-in core plug.*

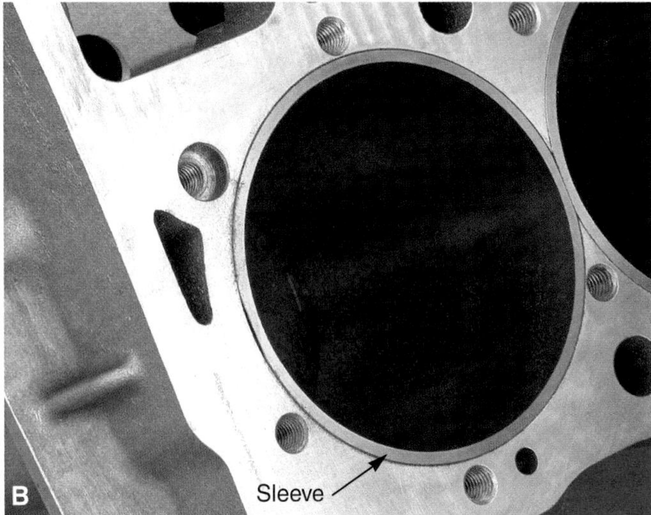

Figure 9-7. *A—This cylinder is integral to the block. B—This aluminum block has press-in cast iron sleeves.*

Caution: The term *freeze plug* is commonly, but incorrectly, used to refer to a core plug. This is based on the misconception that core plugs are designed to pop out if the water inside the block freezes. Core plugs do *not* protect the block from damage when a weak coolant solution or water in the coolant passages freezes. Even if the core plugs pop out, the cylinder block will frequently be cracked by the tremendous pressure of ice expansion inside the block.

Cylinders

Cylinders, also called cylinder walls, are large holes machined into the cylinder block for the pistons. An ***integral cylinder*** is part of the block. A ***sleeved cylinder*** is a separate part pressed into the block. Sleeves are discussed in the next section. **Figure 9-7** compares blocks with and without sleeves.

A cast iron block normally has integral, machined cylinders. Cast iron has excellent wear characteristics and can withstand the rubbing action of the piston rings very well.

An aluminum block commonly has cast iron sleeves as the cylinder walls. Aluminum does not have good wear resistance, so iron liners are press fit into the lightweight aluminum block. Cast iron is very durable and the weight of the cylinder sleeves is minor. Sometimes, iron is electroplated as a coating on machined aluminum cylinder walls instead of pressing in a sleeve.

Cylinder Block Sleeves

Cylinder block sleeves or liners are thick-wall, tube-shaped inserts that fit into the block. The pistons slide up and down in them. New sleeves can be installed to repair badly worn, cracked, or scored cylinder walls. There are two basic types of cylinder sleeves—dry sleeves and wet sleeves. See **Figure 9-8.**

Dry Sleeves

A ***dry sleeve*** is pressed into a cylinder that has been bored (machined) oversize. A dry sleeve is relatively thin and not exposed to engine coolant. See **Figure 9-9A.** The outside of a dry sleeve touches the existing walls of the block. The block provides support for the thin, dry sleeve.

When an integral cylinder becomes badly worn or damaged, a dry sleeve can be installed to repair the cylinder block. The original cylinder must be bored almost as large as the outside diameter of the sleeve. Then, the sleeve is pressed into the oversize hole. Next, the inside of the sleeve is machined to the original bore diameter. This allows the reuse of the standard piston.

Wet Sleeves

A ***wet sleeve*** is exposed to the engine coolant, **Figure 9-9B.** It must withstand combustion pressure and heat without the added support of the cylinder block. Therefore, it must be made thicker than a dry sleeve.

A wet sleeve normally has a flange at the top. When the head is installed, the clamping action pushes down on the flange and holds the sleeve in position. The cylinder head gasket keeps the top of the sleeve from leaking.

A rubber or copper O-ring is used at the bottom, and sometimes the top, of a wet sleeve to prevent coolant leakage into the crankcase. The O-ring seal is squeezed between the block and liner to form a leakproof joint.

The trend is to use light aluminum cylinder blocks with cast iron, wet sleeves. The thickness of the wet sleeves helps maintain a uniform temperature during engine operation. The cast iron sleeves also wear very well. Both characteristics help increase engine service life.

Figure 9-8. *The two types of sleeves are wet and dry. (Dana)*

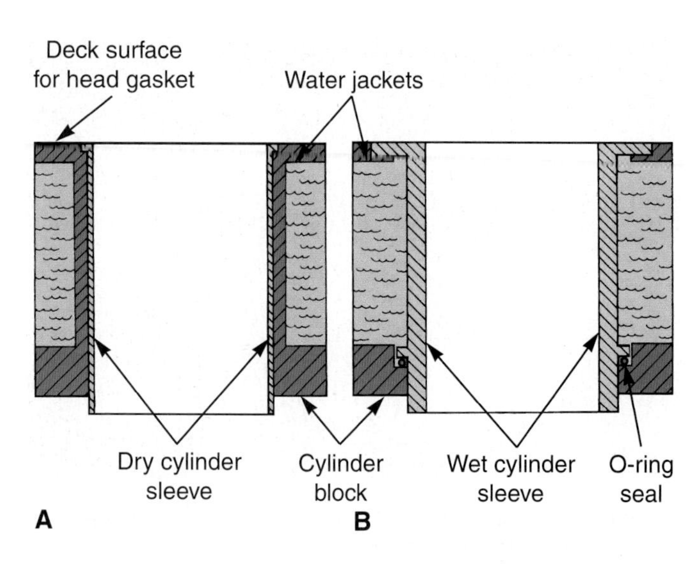

Figure 9-9. *A—Dry sleeves are very thin and fit against the existing cylinder wall. B—Wet sleeves are thicker and not supported by wall of block.*

Cylinder Wall Coatings

A few exotic or high-performance aluminum blocks do not have cast iron sleeves. Termed a ***sleeveless aluminum block,*** a special alloy coating is bonded onto the cylinder walls to improve piston ring and cylinder wall sealing and wear resistance.

In the past, the cylinder walls in these exotic aluminum blocks were chrome plated. This provided a very hard surface on which the piston rings could operate. However, this coating suffered from wear problems under minimum oil supply conditions. For example, if the engine was raced at high engine speeds for prolonged periods, the high friction and heat resulted in a lack of lubrication that caused piston ring and cylinder wall wear. The hard chrome coating could actually be pulled off of the aluminum block.

Now, a nickel-ceramic coating is being applied to improve cylinder wall performance under high friction conditions. One type of nickel-ceramic coating is a nickel-silicone carbide matrix that is about .0025″ to .003″ (0.06 mm to 0.08 mm) thick. A mixture of nickel and silicone carbide is electroplated onto the aluminum cylinder walls. This nickel-ceramic coating provides better wear and is more ductile (flexible) than chrome plating.

Cylinder Block Deck

The ***cylinder block deck*** is a flat, machined surface for the cylinder head. See **Figure 9-10.** The head gasket and head fit onto the deck surface. Bolt holes are drilled and tapped in the deck for head bolts. Coolant and oil passages allow fluids to circulate through the block, head gasket, and cylinder heads.

Figure 9-10. *The cylinder block deck is the flat surface on which the cylinder head is installed.*

Siamese Cylinders

A block with ***siamese cylinders*** has the cylinders connected with cast metal ribs. Conventional blocks have water jackets between the cylinders. Siamese cylinders do not have this water jacket because the extra metal is needed to strengthen the block and hold the cylinders and deck surface in place. Refer to **Figure 9-11.**

Siamese cylinders are used when the cylinders are extremely close together or when the block is aluminum and needs reinforcement. Sometimes, grooves are machined in the top of the siamese section to allow limited coolant circulation.

Lifter Bores

Lifter bores are precision-machined and honed holes in the cylinder block on overhead valve (OHV) engines to accept tappets or lifters. The tappets or lifters ride on the camshaft. Lifter bores are located in the lifter valley. The ***lifter valley*** is the area in the block for the lifters and push rods. Lifter bores and the lifter valley area are shown in **Figure 9-12.** Some engine designs place the starter motor in the lifter valley.

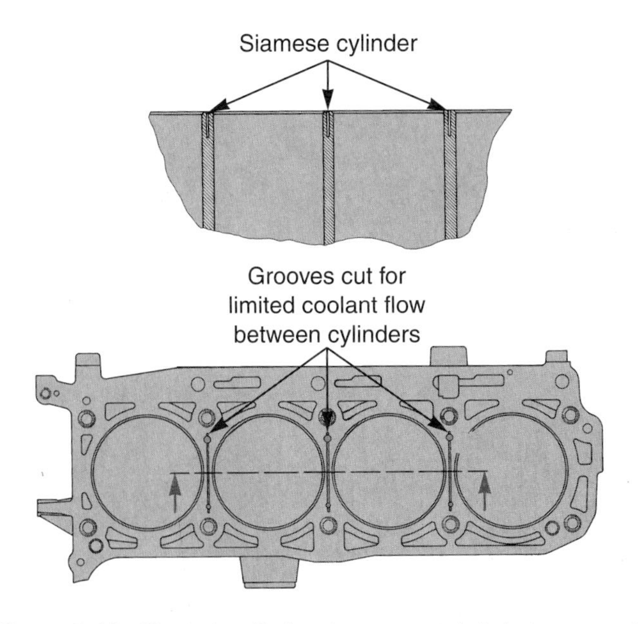

Figure 9-11. *Siamese cylinders have a metal rib between each cylinder for added strength. Siamese cylinders were also shown in **Figure 9-4.** (Mercedes-Benz)*

Camshaft and Main Bores

The ***camshaft bore*** is a series of holes machined through the center of the cylinder block on cam-in-block engines for the camshaft bearings. See **Figure 9-13.** One-piece, bushing-type camshaft bearings are press-fit into the camshaft bore holes. The oil holes in the block must line up with oil holes in the camshaft bearings to provide proper lubrication. On overhead camshaft engines, the camshaft bore is in the cylinder head(s).

The ***main bore*** of a cylinder block is a series of holes for the crankshaft main bearings. The main bore is machined through the lower area of the block below the cylinders, **Figure 9-13.** Split-shell main bearings fit into these holes.

Main Caps

Main caps fasten to the bottom of the cylinder block and form one-half of the main bore. See **Figure 9-14.** Large, case-hardened bolts screw into holes in the block to secure the main caps to the block. The crankshaft and main bearings are held in place by the main caps. Therefore, they must be very strong. There are several types of main cap designs. The most common ones are:

- **Two-bolt mains.** Two large cap screws (main bolts) hold each main cap onto the cylinder block. See **Figure 9-15A.** The cap screws thread straight into the block from the bottom. This is a very common design.
- **Four-bolt mains.** Two large and two small cap screws are used per main cap, **Figure 9-15B.** Like a two-bolt main, the cap screws thread straight into the block from the bottom. This design is very common in applications where a two-bolt main does not provide enough strength, such as high-horsepower engines. Diesel engines are often four-bolt-main engines.
- **Cross-bolt mains.** Two conventional cap screws thread in through the bottom of the caps. Two additional cap screws fit horizontally through the sides of the block and thread into the sides of the main caps, **Figure 9-15C.**
- **Girdle cap, bedplate, or unit cap.** One large main cap is used to hold all of the lower main bearings. This is a very strong design because more surface area of the block is used to hold the crankshaft, **Figure 9-16.**

Line boring is a term that refers to the machining of the main bore and the camshaft bore. A long boring bar is rotated through the bore to cut a perfectly straight hole of equal diameter for each bearing insert. Line boring is explained fully in Chapter 21, *Short Block Rebuilding and Machining.*

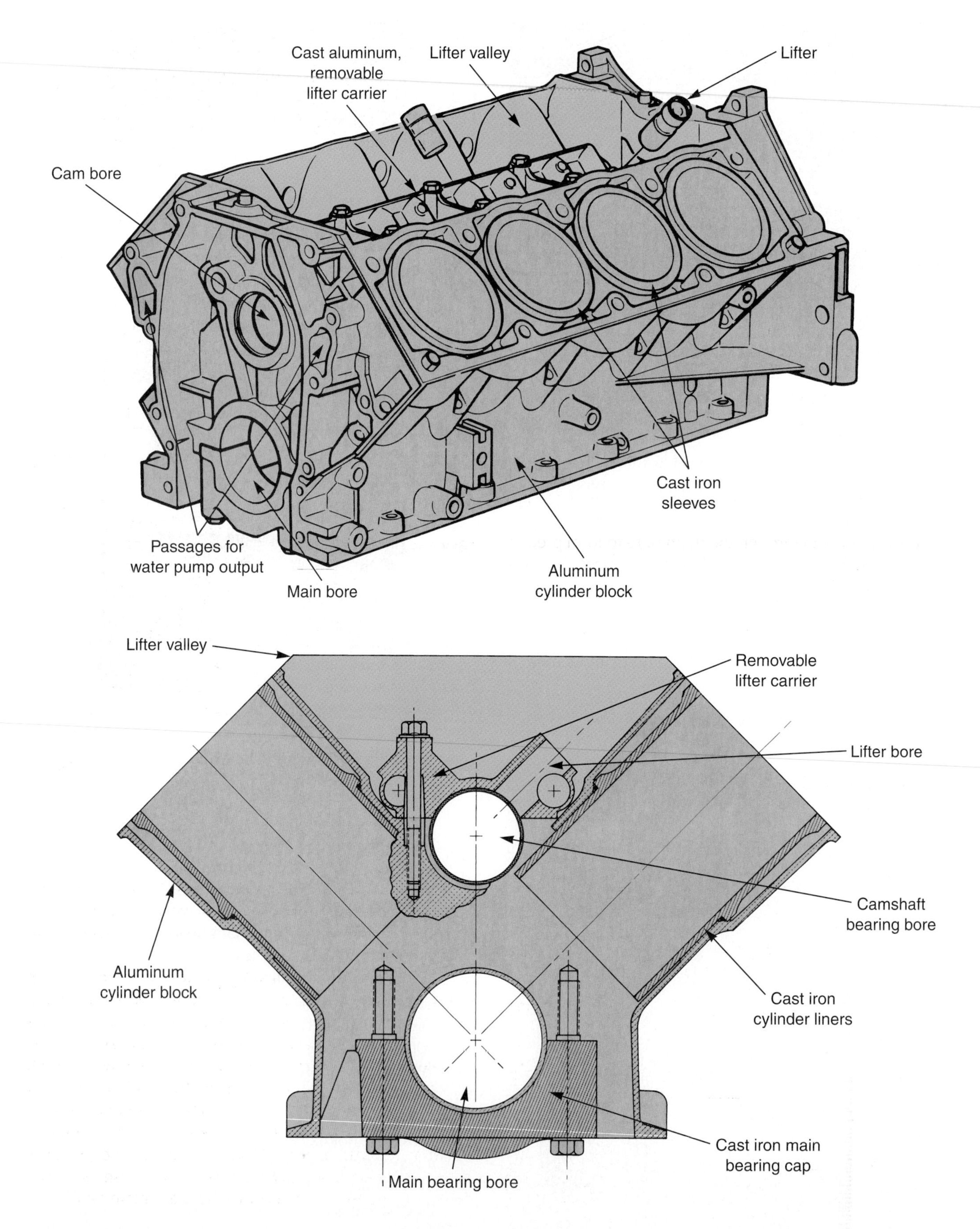

Figure 9-12. *This engine has a removable lifter carrier that contains the lifter bores. The carrier is installed in the lifter valley. (General Motors)*

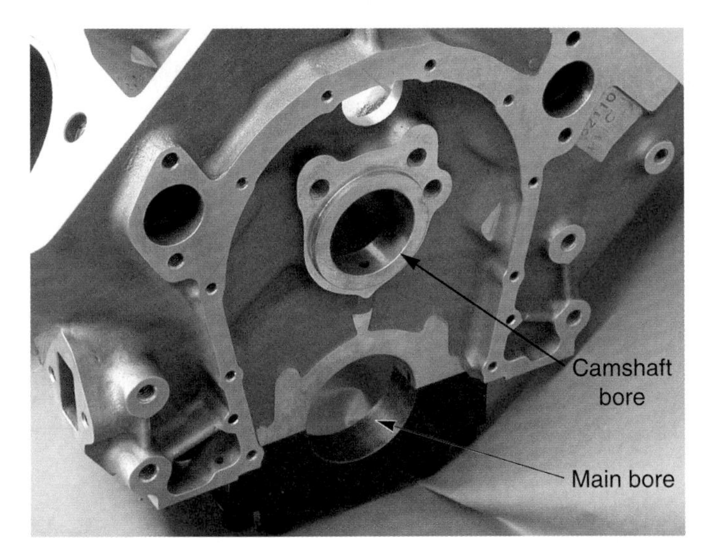

Figure 9-13. *The camshaft bore holds the camshaft and camshaft bearings. The main bore holds the crankshaft and main (crankshaft) bearings.*

When line boring the main bore, the main caps are installed and the bolts are torqued to specifications. Then, the boring bar is run through the main bore to form precise holes where the caps and block surfaces meet. For this reason, *never* mix up the main caps. They cannot be interchanged or the bore diameters will become out-of-round. This could lock the main bearings against the crank and damage the bearings and crankshaft. Main caps are usually numbered 1, 2, 3, 4, etc., from the front to the rear of the block so they will not be mixed up. Arrows are sometimes cast on the caps to show the front of the engine.

Main Bearings

Main bearings provide an operating surface for the crankshaft. The crankshaft main journals rotate on the lubricated main bearings as the crankshaft spins. The bearings snap-fit into the cylinder block and main caps.

Figure 9-14. *The main caps hold the crankshaft in the block and form one-half of the main bore.*

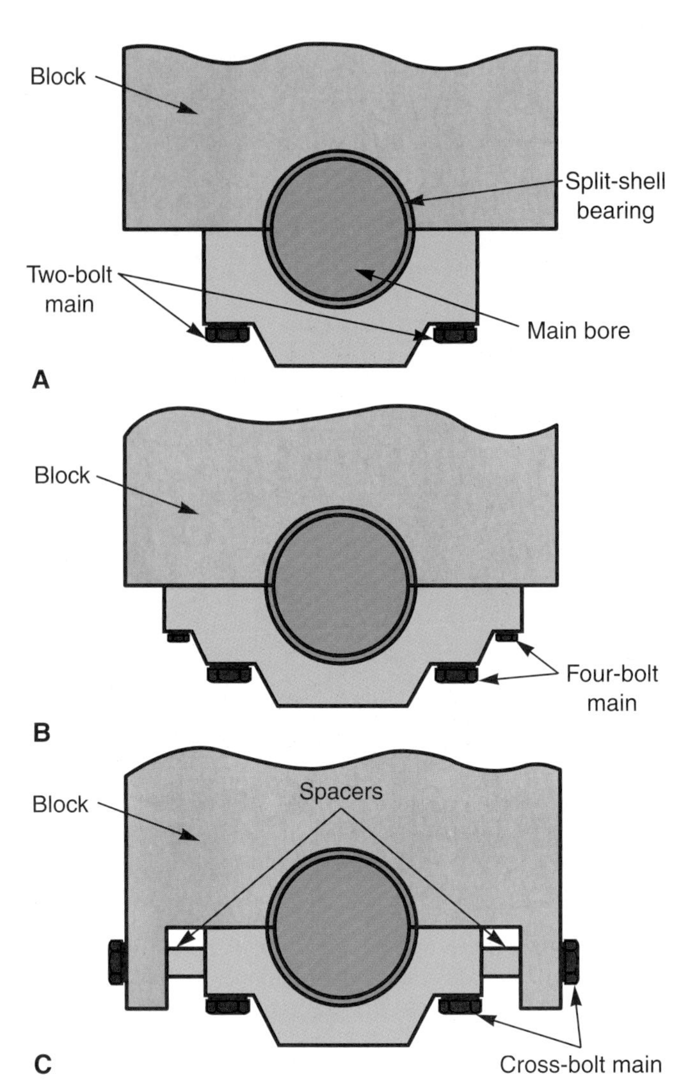

Figure 9-15. *Study basic main cap designs. A—Two-bolt main. B—Four-bolt main. C—Cross-bolt main.*

A ***plain main bearing*** is a split-type bearing. This type of bearing simply limits up and down movement of the crankshaft in the block. Plain main bearings are used on all but one of the main bore holes.

One ***thrust main bearing*** is also needed to limit crankshaft endplay, or crankshaft front-to-rear movement in the block. The bearing has thrust surfaces on its sides that have a small clearance between the cheeks, or thrust surfaces, machined on the crankshaft. In this way, clutch action, for example, will only make the crank move a few thousandths of an inch forward and rearward in the block.

Thrust washers serve the same purpose as a conventional main thrust bearing. However, they are not made as part of the main bearing, **Figure 9-17.** The washers fit into the block and main cap, next to the plain main bearing.

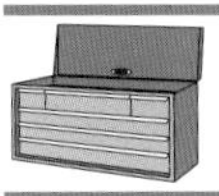

Note: Bearing construction is discussed in more detail at the end of this chapter.

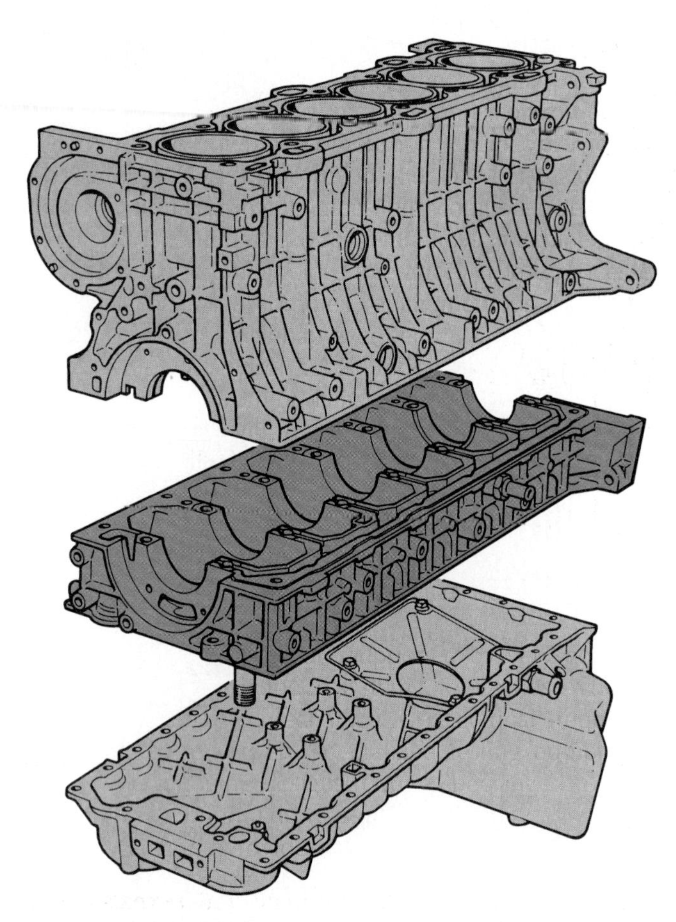

Figure 9-16. *A girdle or unit cap is a high-performance main cap design. It provides tremendous strength to hold the crankshaft and main bearings in place under high-engine-speed, high-power conditions. (Volvo)*

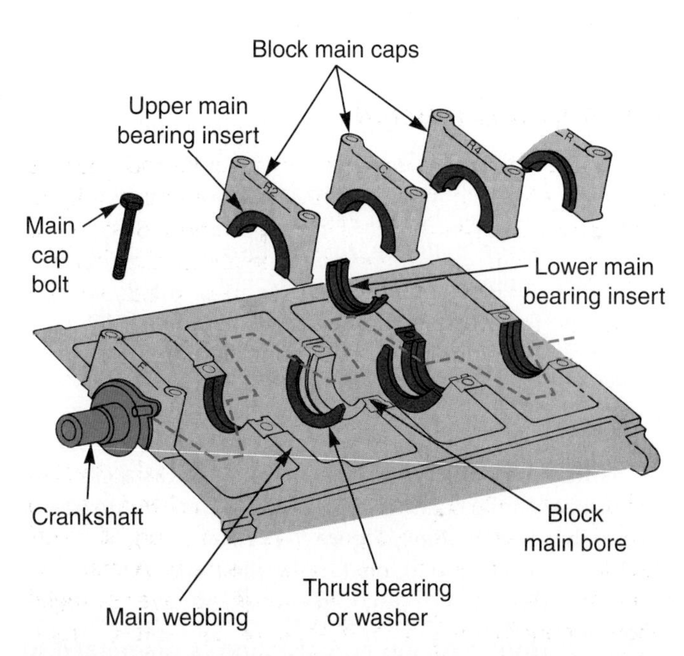

Figure 9-17. *A thrust washer or bearing limit crankshaft endplay. (Ford)*

Crankshaft Construction

The ***crankshaft*** harnesses the force produced by the downward thrust of the pistons on their power strokes, **Figure 9-18.** It converts the reciprocating (up and down) motion of the pistons into a rotary (spinning) motion. The crankshaft fits into the main bore of the block.

Engine crankshafts are usually made of cast iron or forged steel. Forged steel crankshafts are needed for heavy duty applications, such as turbocharged, supercharged, or diesel engines. A steel crankshaft is stiffer and stronger than a cast iron crankshaft. It can withstand greater forces than a cast iron crankshaft without flexing, twisting, or breaking.

Crank Main Journals

Crankshaft ***main journals*** are precision-machined and polished surfaces that ride on the main bearings. They are machined along the centerline of the crankshaft. When the engine is running, the main journals spin about the crankshaft centerline, but remain stationary in relation to the rods and pistons, **Figure 9-19.**

Grooved main journals have a small notch cut around their surface. The main groove is used to help distribute oil evenly around the journal and main bearing.

Crankshaft Oil Passages

Crankshaft oil passages are cast or machined holes through the crankshaft for lubricating oil. The oil pump forces oil through the block, main bearings, and into the crankshaft. Then, the oil can flow through the crankshaft oil passages and to the engine bearings.

Figure 9-20 shows the oil passages in one crankshaft. Note how the holes have been drilled. Screw-in or press-in plugs are installed on the ends of the drilled oil passages.

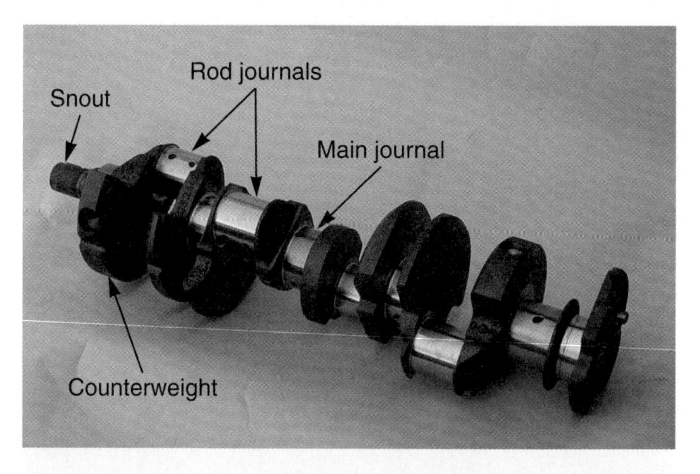

Figure 9-18. *The crankshaft changes reciprocating motion into rotary motion. A crankshaft can be made from cast iron or forged steel. Note the basic parts identified here.*

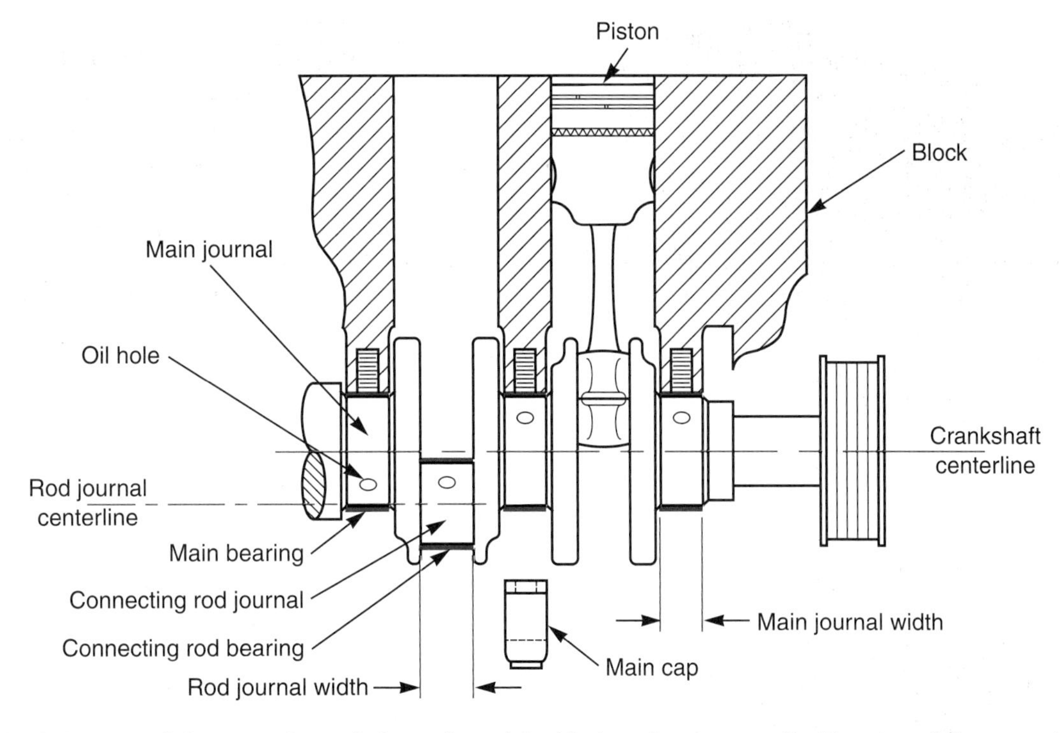

Figure 9-19. *Main bearings fit between the main journals and the block and main caps. Rod bearings fit between the rod journals on the crankshaft and the connecting rod.*

Main Bearing Clearance

Main bearing clearance is the space between the crankshaft main journal and the main bearing insert. The clearance allows lubricating oil to enter and separate the journal and bearing. With oil between the two parts, they rotate without rubbing on each other and wearing. A typical bearing clearance is .001″ (0.03 mm).

Crankshaft Rod Journals

Crankshaft rod journals, also termed crankpins, are precision-machined and polished surfaces for the connecting rod bearings. The connecting rods bolt around these journals. The rod journals are offset from the main journals so that the rods and pistons slide up and down in the block when the crankshaft rotates. Normally, rod journals are not grooved like main journals.

Since automobile engines have multiple cylinders, the crankshaft rod journals are arranged so that there is always at least one cylinder on a power stroke. In this way, force is always being transmitted to the crankshaft, which smoothes engine operation.

A current trend in engine design is to reduce the diameter of rod journals. This reduces bearing surface velocity and friction in the bearings.

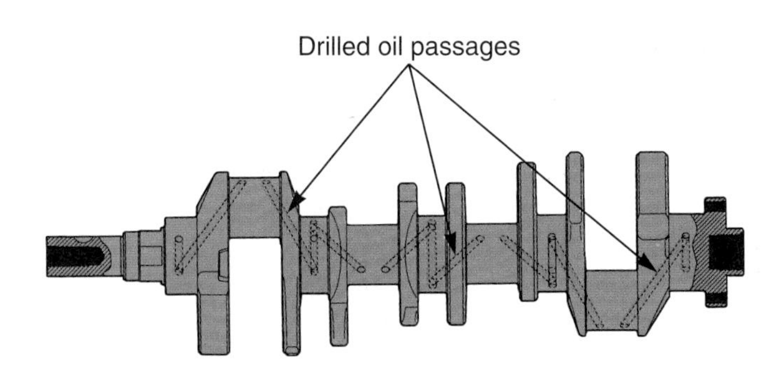

Figure 9-20. *Oil passages drilled through the crankshaft allow oil to pass through the block, main bearings, and crankshaft into the rod bearing clearance. (Lexus)*

Splayed Rod Journal

A ***splayed rod journal*** has two separate rod journals machined offset from each other on the same crankpin. See **Figure 9-21.** This design is used on some V-6 and V-10 engines to help smooth engine operation. Combustion strokes are staggered away from each other so the cylinders fire more evenly.

Roll Hardened Journal

A ***roll hardened journal*** has been specially treated during manufacturing. Basically, a large roller is driven around the journal to compress the freshly forged metal and make it harder and more wear resistant. A machine shop must not grind through this hardened layer of metal when reconditioning a crankshaft or the softer, more-wear-prone metal will be exposed. Crankshaft service life can then be reduced.

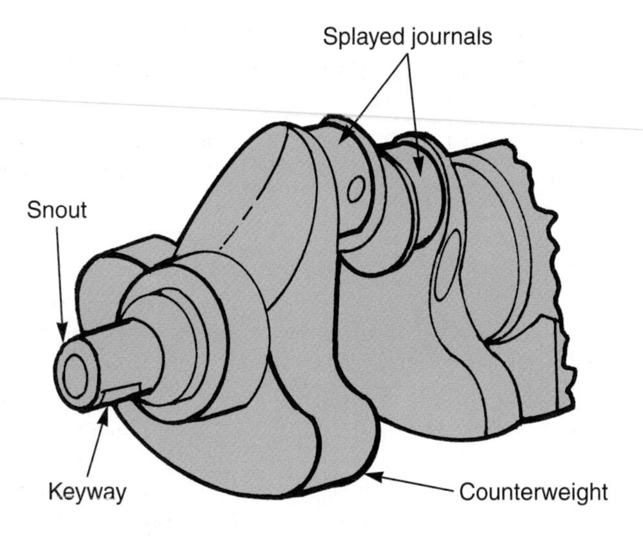

Figure 9-21. *This is a splayed rod journal. It is sometimes used in V-6 engines to help smooth operation.*

Crankshaft Thrust Surfaces

The ***crankshaft thrust surfaces,*** or cheeks, are located on each side of the main and rod journals. This is the area where the sides of the grinding stone rub on the crankshaft during its initial machining. The distance between cheeks determines journal width.

Crankshaft Counterweights

Counterweights are heavy lobes on the crankshaft to help prevent engine vibration. Refer to **Figure 9-18**. They are used to counteract the weight of the pistons, piston rings, connecting rods, and rod bearings. Counterweights are normally formed as part of the crankshaft. However, on a few older engines, they are bolted to the crankshaft. A fully counterweighted crankshaft has weights formed opposite every crankpin (rod journal). A partially counterweighted crankshaft only has weights formed on the center areas of the crankshaft. A fully counterweighted crankshaft will operate with less vibration than a partially counterweighted crankshaft.

Crankshaft Snout and Flange

The ***crankshaft snout*** is a shaft machined on the front of the crankshaft to hold the damper, camshaft drive sprocket or gear, and crank pulleys. It sticks through the front of the block and front cover. A keyway is cut into the snout to keep the sprocket, gear, damper, or pulley from spinning or slipping.

The ***crankshaft flange*** is machined on the rear of the crank to hold the engine flywheel. It is a round disk with threaded holes for the flywheel bolts.

Flywheel

A ***flywheel*** is a large, steel disk mounted on the rear flange of the crankshaft, **Figure 9-22.** It can have several functions:

- On a manual transmission vehicle, it is heavy and helps smooth the engine operation.
- It connects the engine crankshaft to the transmission or transaxle. Either the manual clutch or the automatic transmission torque converter is bolted to the flywheel.
- A large ring gear on the flywheel allows engine starting. A small gear on the starting motor engages the flywheel ring gear and turns it.

Hardened flywheel bolts secure the flywheel to the crankshaft. The holes in the crankshaft and flywheel are usually staggered unevenly so that the flywheel can only be installed in one position. This ensures that the engine stays in balance when the flywheel is removed and then reinstalled. The flywheel is centered on the crankshaft by a hub or lip machined on the crankshaft flange.

Pilot Bearing

A ***pilot bearing*** is mounted in the rear of the crankshaft to support the manual transmission input shaft, **Figure 9-22.** It can be a bushing-, roller-, or ball-type bearing. The pilot bearing is press-fit into a pocket machined in the crankshaft.

An engine coupled to an automatic transmission does not need a pilot bearing. The large hub on the torque converter normally fits directly into the crankshaft.

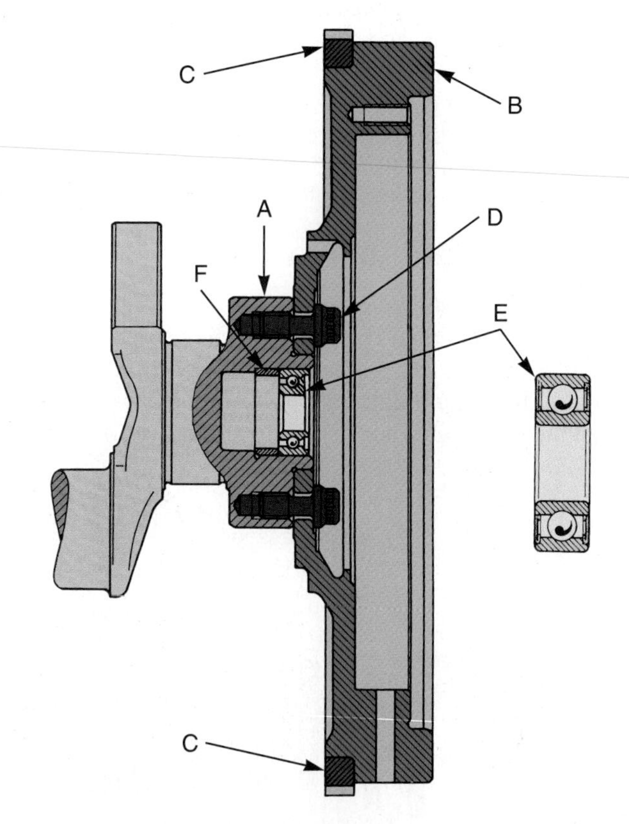

Figure 9-22. *The flywheel is bolted to the crankshaft flange. Note the components. A—Crankshaft flange. B—Flywheel. C—Ring gear. D—Flywheel bolt. E—Ball bearing–type pilot bearing. F—Spacer ring. (Mercedes-Benz)*

Crankshaft Oil Seals

Crankshaft oil seals keep oil from leaking out of the front and rear of the engine. The oil pump forces oil into the main and rod bearings. This causes oil to spray out of the bearings. Seals are placed around the front and rear of the crank to contain this oil, **Figure 9-23.**

The front seal prevents oil leakage at the front of the crankshaft. It is covered later in this book. The ***rear main oil seal*** fits around the rear of the crankshaft to prevent oil leakage. The seal lip rides on a smooth, machined and polished surface on the crankshaft. It can be a one- or two-piece seal. There are different types or styles of rear main oil seals.

A two-piece neoprene rear main oil seal usually is installed in a groove cut into the block and rear main cap. The seal has two lips. One lip traps oil and the other lip keeps dust and dirt out of the engine. The sealing lips ride on a machined surface of the crankshaft. Spiral grooves may be cut into this surface to help throw oil inward and prevent leakage.

A one-piece neoprene rear main oil seal fits around the rear flange on the crankshaft. It has sealing lips similar to a two-piece neoprene seal. This is the most common type of rear main oil seal on modern engines.

A rope or wick rear main oil seal was commonly used in the past. It is simply a woven rope impregnated (filled) with graphite. One piece of the rope seal fits into a groove in the block. Another piece fits into a groove in the main cap. This type seal has been replaced by one- and two-piece neoprene seals.

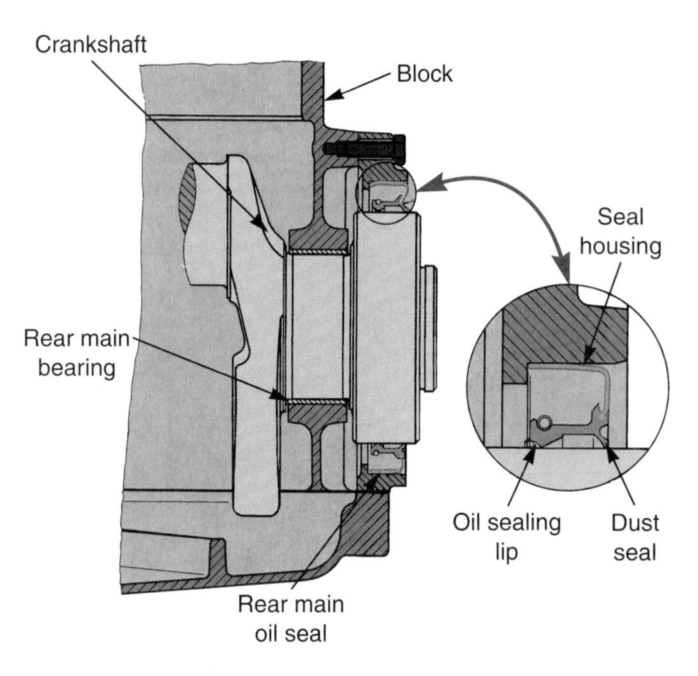

Figure 9-23. *The rear main oil seal keeps oil from spraying out of the rear main bearing. Oil pressure tends to push the sealing lip tight against the crankshaft. (Mercedes-Benz)*

Connecting Rod Construction

The ***connecting rod*** fastens the piston to the crankshaft. See **Figure 9-24.** It transfers piston movement to the crankshaft rod journals. The connecting rod also causes piston movement during the non-power-producing strokes (intake, compression, and exhaust).

A rod and piston are stationary at TDC. When the vehicle is at highway speeds, the piston and rod accelerate to about 60 mph halfway down in the cylinder and decelerate to zero at BDC, all in about 3″ of travel. This produces tremendous force that can tear or shear metal, and is repeated millions of times during the life of an engine. As you can see, the connecting rod must withstand some of the highest loads in an engine assembly.

Connecting rods normally have an I-beam shape because of this contour's high strength-to-weight ratio. H-beam shaped rods are also available in the aftermarket. Connecting rods can be made from various materials. Most passenger vehicles have steel connecting rods.

- **Cast steel.** Cast steel connecting rods are inexpensive. They can be found in low-horsepower, low-engine-speed engines.

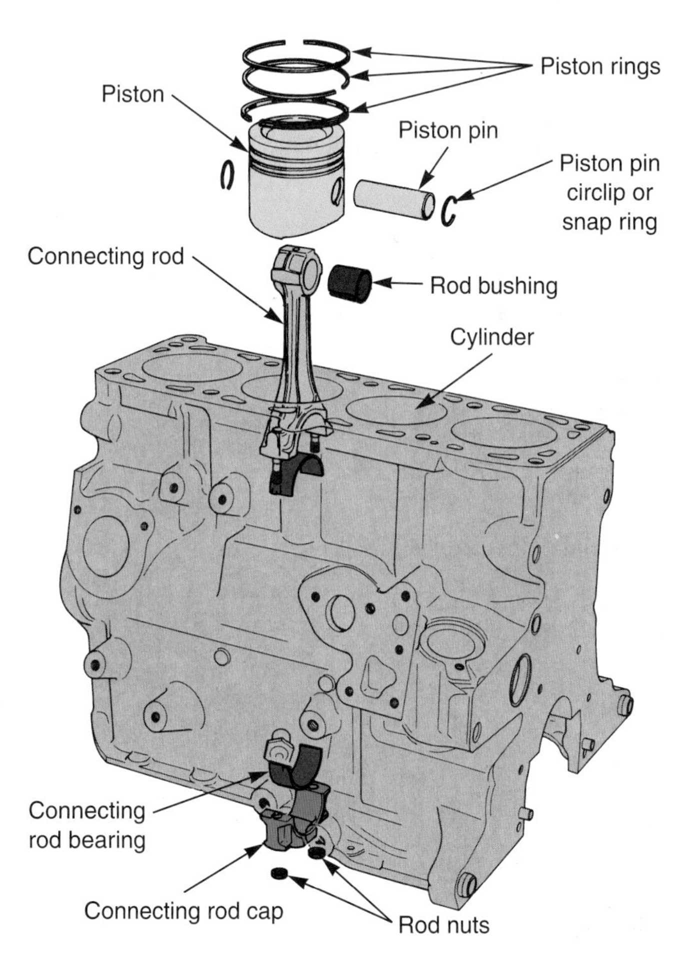

Figure 9-24. *The connecting rod transfers the piston motion to the crankshaft. (DaimlerChrysler)*

- **Forged steel.** Forged steel connecting rods are much stronger than cast steel rods. They are used in many modern high-performance engines because of their higher strength-to-weight ratio.
- **Titanium.** Titanium connecting rods are used in a few exotic, high-performance engines. Titanium connecting rods are almost as light as aluminum, but as strong as steel. However, they are very expensive to manufacture. Titanium connecting rods are primarily limited to racing applications.
- **Aluminum.** Aluminum connecting rods may be installed in some types of racing engines. For example, drag racing engines often use aluminum rods because the light weight allows extremely high engine speed for a short burst of high horsepower output.

The connecting rods in a diesel engine are thicker and heavier than those in a gasoline engine. This is needed to withstand the higher combustion pressures. A diesel engine operates a much lower engine speeds than a gasoline engine. Therefore, the large reciprocating mass of the pistons and connecting rods does not damage the connecting rod bearings.

Rod Oil Holes

Some connecting rods have an ***oil spurt hole*** that provides added lubrication for the cylinder walls, piston pin, and other surrounding parts, **Figure 9-25A.** Oil from inside the crankshaft will spray out when the holes in the crankshaft journal and bearing align with the spurt hole.

A drilled connecting rod has a hole machined through its entire length. This passage allows oil to lubricate the piston pin. Refer to **Figure 9-25B.**

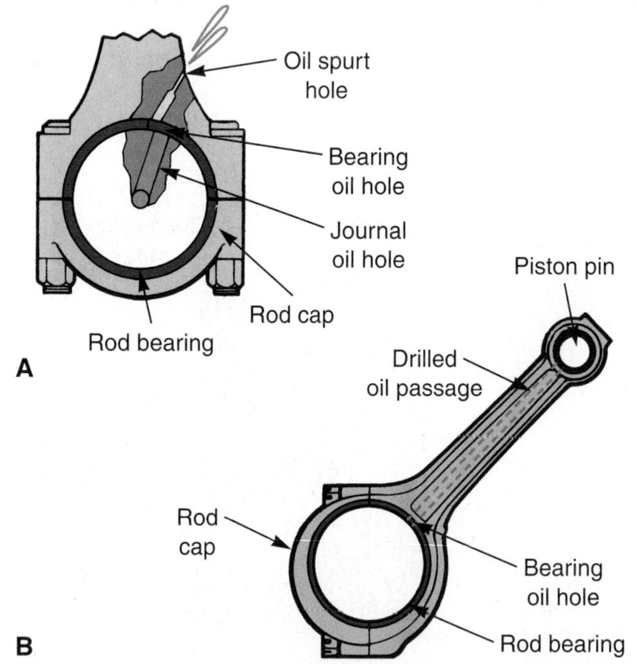

Figure 9-25. *A—When the holes line up, oil can squirt out of the spurt hole to provide extra lubrication on the cylinder walls. B—A drilled connecting rod provides positive lubrication for the piston pin. (Federal-Mogul)*

Connecting Rod Cap

The ***connecting rod cap*** is bolted to the bottom of the connecting rod, as shown in **Figure 9-24.** It can be removed for disassembly of the engine.

The connecting rod and rod cap are generally produced as a single unit. In most cases, the cap is cut (machined) from the rod during manufacture. This creates a smooth mating surface, **Figure 9-26A.**

However, the cap on a broken-surface connecting rod is scribed and cracked off to produce a rough, irregular mating surface, **Figure 9-26B.** This surface helps lock the rod and cap in perfect alignment. It also prevents the components from shifting during engine operation. Although a broken-surface connecting rod cannot be rebuilt, oversize rod bearings can be installed during an engine overhaul. Oversize and undersize bearings are discussed at the end of this chapter.

The connecting rod big end or lower end is the hole machined in the rod body and cap. The connecting rod bearing fits into the big end. The big end fits over the rod journal on the crankshaft. The piston pin fits into the other end of the connecting rod.

Connecting rod bolts and nuts clamp the rod cap and rod together. They are special, high-tensile strength fasteners. Some rods use cap screws without a nut. The cap screw threads into the rod itself. See **Figure 9-27.**

Connecting Rod Numbers

Connecting rods are numbered to ensure proper location of each connecting rod in the engine. They also ensure that the rod cap is installed on the rod body correctly. See **Figure 9-28.**

During manufacture, the connecting rod caps are bolted to the connecting rods. Then, the big end holes are machined. Since the holes may not be perfectly centered,

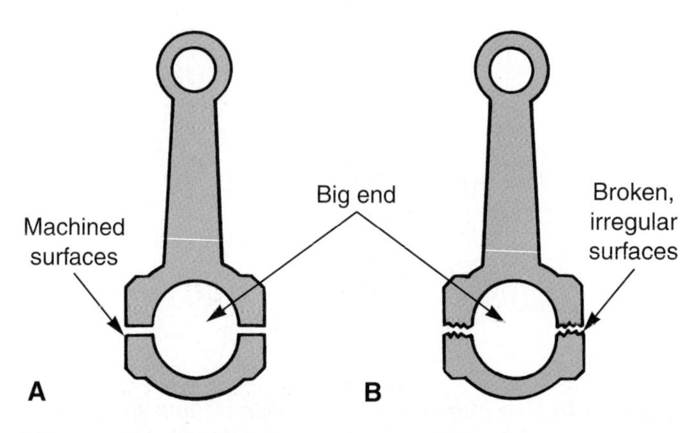

Figure 9-26. *Connecting rod assemblies. A—Conventional connecting rod. B—Broken-surface connecting rod.*

Figure 9-27. *Most connecting rods in production vehicles have a bolt and nut to secure the rod cap (left). Some high-performance production engines and most race engines use a cap screw that threads into the rod body (right).*

rod caps must not be interchanged or reversed. If the cap is installed without the rod numbers in alignment, the bore will not be perfectly round. Severe rod, crankshaft, and bearing damage will result.

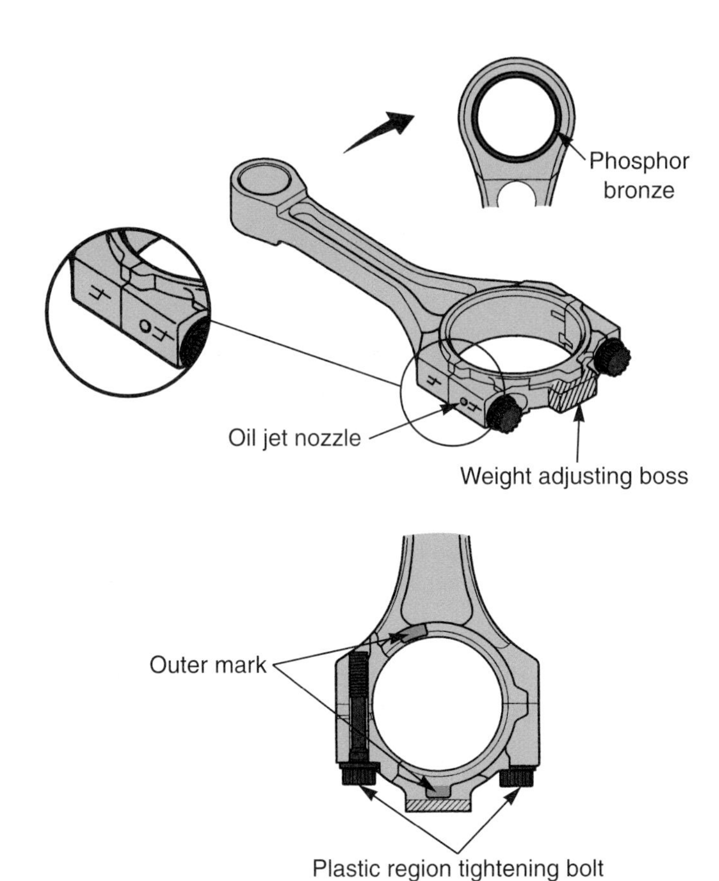

Figure 9-28. *Connecting rods and caps are numbered to ensure they are matched. (Lexus)*

Connecting Rod Bearings

The connecting rod bearings fit between the connecting rods and the rod journals on the crankshaft, as was shown in **Figure 9-19.** The rod bearings are removable inserts, as are main bearings.

Rod bearing clearance is the small space between the rod bearing and the crankshaft rod journals. As with the main bearings, this clearance allows oil to enter the bearing. The oil prevents metal-to-metal contact that would wear out the crankshaft and bearings.

A rod bushing is normally used in the small end bore of the connecting rod. It is usually a one-piece, phosphor bronze bushing pressed into the rod.

Connecting Rod Dimensions

Connecting rod length is measured from the center of the piston pin hole to the center of the big end bore. Special machine shop fixtures are available for checking rod length. This measurement is important when doing engine service because two connecting rods may look the same, but may in fact have slightly different lengths.

Small end diameter is the distance across the bushing in the top of the rod. Big end diameter is the distance across the big end bore with the bearing removed. As you will learn, both of these are checked during an engine overhaul to make sure the rod is in good condition.

Connecting rod width is the distance across the sides of the big end. Rod width partially determines rod side play, which is the distance the rod can move sideways when installed on the crankshaft. Look at **Figure 9-29.**

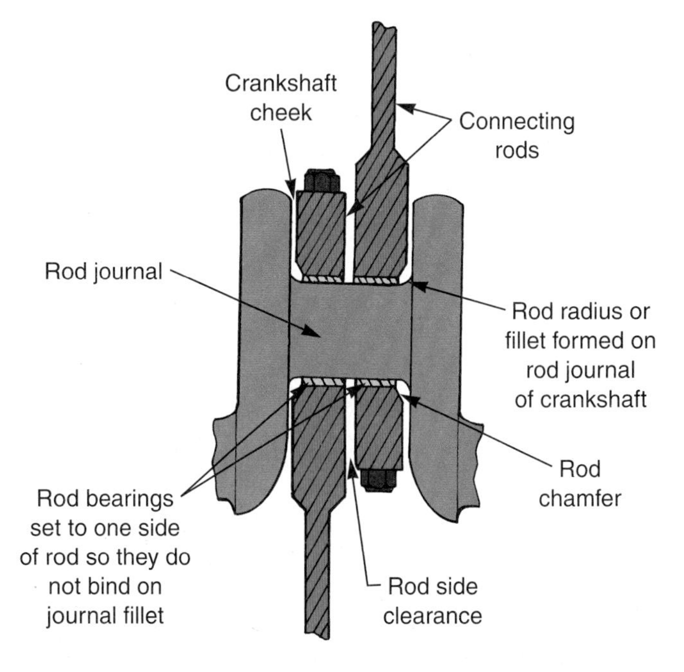

Figure 9-29. *Rod width partially determines rod side play.*

Piston Pin Construction

The ***piston pin,*** also called a ***wrist pin,*** allows the piston to swing on the connecting rod. The pin fits through the hole in the piston and the small end of the connecting rod. The piston pin is hollow and machined and polished to a very precise finish.

Piston pins are normally made of case-hardened steel. Case-hardening is a heating and cooling process that increases the resistance to wear of the piston pin. The outer layer, or "case," of metal on the pin is hardened. The inner metal remains unhardened so the pin is not too brittle.

Piston pins in modern engines are normally classified by one of two designs—floating and press fit. This classification refers to how the piston pin is held in the piston.

Floating Piston Pin

A full ***floating piston pin*** is secured by snap rings. The pin is free to "float" or rotate in both the piston pin bore and the connecting rod small end. See **Figure 9-30A.**

A bronze bushing is usually used in the connecting rod. The piston pin hole serves as the other bearing surface for the pin. The snap rings fit into grooves machined inside the piston pin hole, **Figure 9-31.** Full floating piston pins are better than press-fit pins since they reduce friction and wear.

Press-Fit Piston Pin

A ***press-fit piston pin*** is forced tightly into the connecting rod's small end. See **Figure 9-30B.** The pin can rotate freely in the piston. However, the pin is not free to move in the connecting rod. This holds the pin inside the piston and prevents it from sliding out and rubbing on the cylinder.

The press-fit piston pin is a very dependable design. It is also inexpensive to manufacture. The trend is toward press-fit piston pins in today's engines.

Piston-Guided Connecting Rod

A ***piston-guided connecting rod*** has thrust surfaces formed next to the piston pin to limit axial movement of the rod small end. Axial movement is movement along the axis of the hole; in other words, side-to-side movement. Most engines use only the rod big end and crankshaft to limit rod side-to-side movement. As shown in **Figure 9-32,** a few high-performance engines have two thrust surfaces at the inner faces of the piston pin bore. This keeps the rod from cocking sideways and sliding to one side of the pin under a heavy load.

Piston Pin Offset

If the pin hole is centered in the piston, the piston could slap (knock) in the cylinder. As the piston moves up in the cylinder, it may be positioned opposite of the major thrust surface. Then, during combustion, the piston can be rapidly pushed to the opposite side of the cylinder, producing a knock sound. This is undesirable.

Piston pin offset locates the piston pin hole slightly to one side of the piston centerline. See **Figure 9-33.** This is designed to quiet the piston operation. The pin hole is moved toward the piston's major thrust surface. This results in the piston surface being pushed tightly against the cylinder during power strokes. The tendency to slap sideways in the cylinder is reduced.

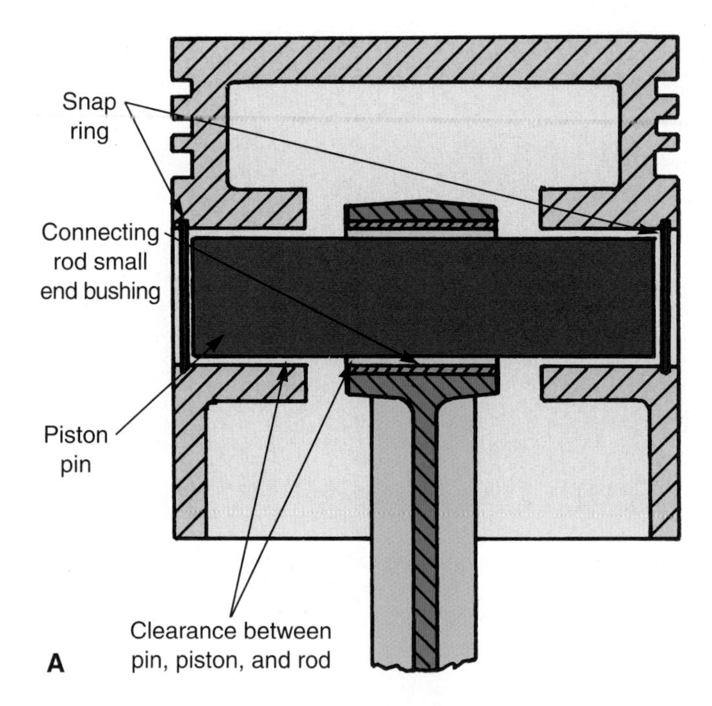

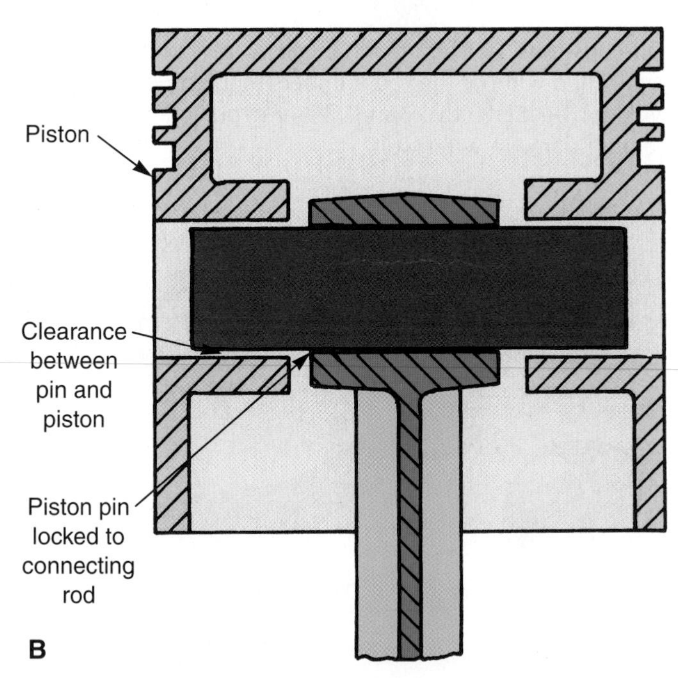

Figure 9-30. *Two methods of holding the piston pin in the piston. A—A press-fit pin is locked in the connecting rod small end. B—A floating piston pin is free to rotate in both the piston and rod. Snap rings keep the pin inside the piston.*

Figure 9-31. *A full floating piston pin is held in place by a snap ring that fits into a groove in the piston.*

A piston notch, arrow, or other mark on the head of the piston is used to indicate piston pin offset and the front of the piston. The piston may also have the word *Front* stamped on it. This information lets you know how to position the piston in the block for correct location of piston pin offset.

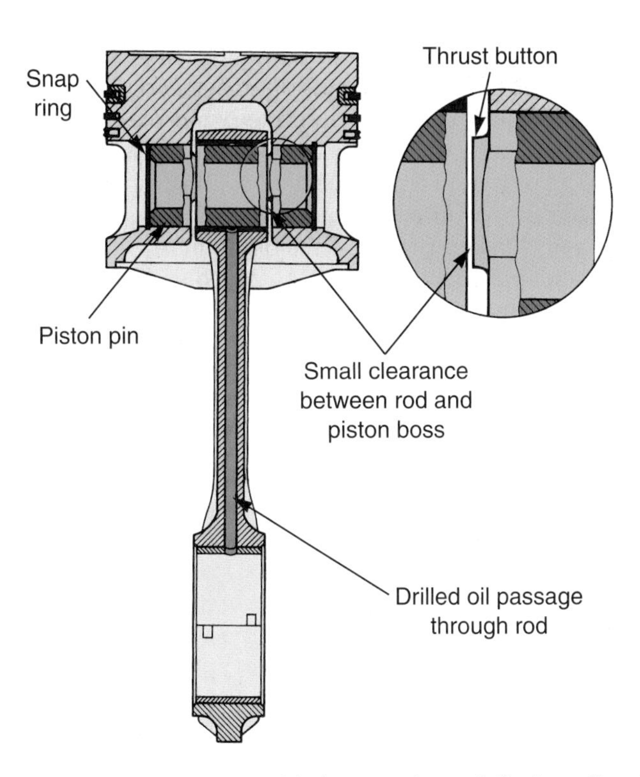

Figure 9-32. *In a piston-guided connecting rod, the boss thrust surface only allows slight side-to-side movement under heavy load. (Mercedes-Benz)*

Piston Construction

The ***engine piston*** transfers the pressure of combustion (expanding gas) to the connecting rod and crankshaft. It must also hold the piston rings and piston pin while operating in the cylinder. **Figure 9-33** shows a cutaway of a piston. Study this illustration as you read through this section.

Piston Material

Pistons are made of aluminum. Since aluminum is very light and relatively strong, it is an excellent material for engine pistons. When an engine is running, the piston must withstand tremendous heat and pressure as well as severe forces due to the almost instantaneous change of direction at TDC and BDC.

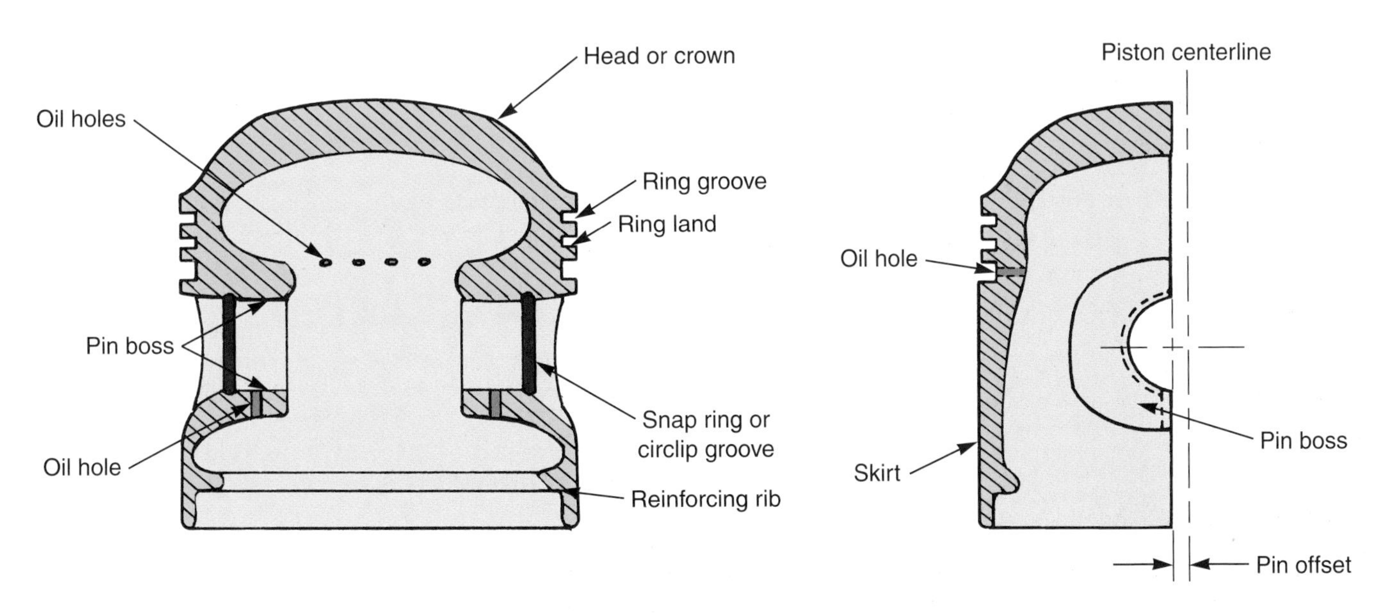

Figure 9-33. *This cutaway shows the basic parts of a piston.*

An engine piston is normally *cast* aluminum. However, *forged* aluminum pistons may be found in turbocharged, supercharged, fuel-injected, and diesel engines. In these engines, the pistons are exposed to severe stress.

The pistons in a diesel engine are thicker and heavier than the pistons in a gasoline engine. The piston head is especially thicker. The top of the piston must withstand the heat and pressure generated as the diesel fuel rapidly detonates and burns.

Parts of a Piston

The ***piston head*** or ***crown*** is the top of the piston. It is exposed to the heat and pressure of combustion and must be thick enough to withstand these forces. The head must also be shaped to match and work with the shape of the combustion chamber for complete combustion.

Piston ring grooves are slots machined in the piston for the piston rings, **Figure 9-34.** The upper two grooves hold the compression rings. The lower piston groove holds the oil ring. Oil holes in the bottom groove allow the oil to pass through the piston. The oil then drains back into the crankcase.

The ***ring lands*** are the areas between and above the ring grooves. They separate and support the piston rings as they slide on the cylinder.

A ***piston skirt*** is the side of the piston below the last ring. It keeps the piston from tipping in its cylinder. Without a skirt, the piston could cock and jam in the cylinder.

The ***piston boss*** is a reinforced area around the piston pin hole. It must be strong enough to support the piston pin under severe loads.

A ***pin hole*** is machined through the pin boss for the piston pin. It is slightly larger than the pin.

Piston Shapes

Piston shape generally refers to the shape of the piston head or crown. Usually, a piston head is shaped to match and work with the shape of the cylinder head combustion chamber.

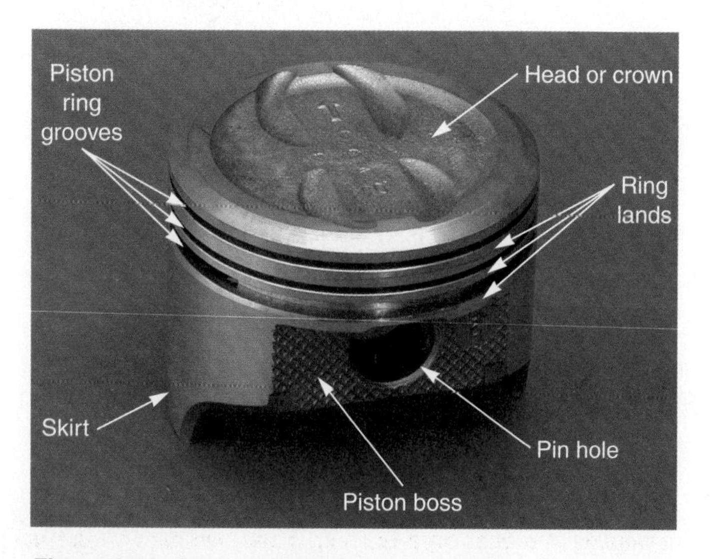

Figure 9-34. *Study the major parts of a piston.*

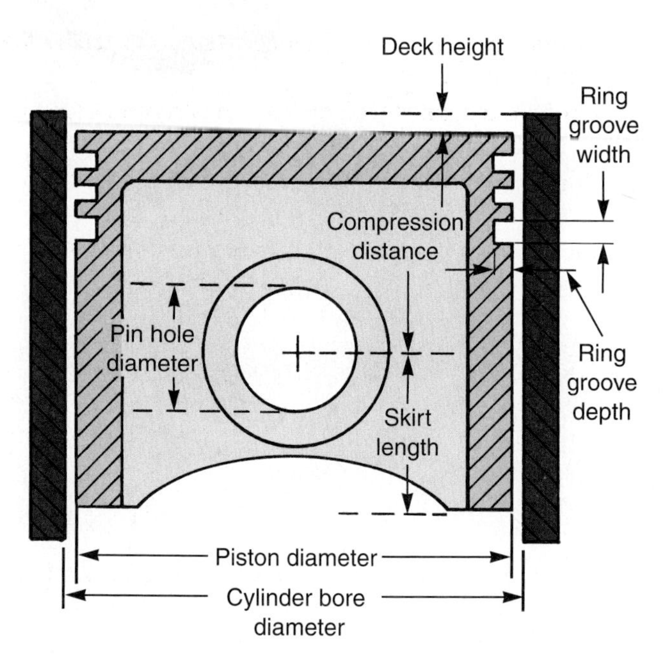

Figure 9-35. *This is a flat top piston. Also, study the basic piston dimensions.*

A ***flat top piston*** has a crown that is almost flat and is parallel to the block deck, **Figure 9-35.** A flat top piston is commonly used with a wedge or pancake type cylinder head.

A ***dished piston*** has most of the piston head recessed to lower compression, **Figure 9-36A.** The crown is concave. It might be found in a turbocharged engine.

A ***valve relief piston*** has small indentations either cast or machined into the piston crown, **Figure 9-36B.** They provide ample piston-to-valve clearance. Without valve reliefs, the valve heads could strike the pistons.

A ***domed*** or ***pop-up piston*** has a head that is curved upward, **Figure 9-36C.** This type is normally used with a hemispherical cylinder head. In these engines, the piston crown must be enlarged to fill the domed combustion chamber and produce enough compression pressure.

Skirt Design

A ***slipper skirt*** is produced when the portion of the piston skirt below the piston pin ends are removed. A slipper skirt provides clearance between the piston and crankshaft counterweights. The piston can slide farther down in the cylinder without hitting the crankshaft.

A ***straight skirt*** is flat across the bottom. This style is no longer common in auto engines that are oversquare.

Piston Dimensions

There are several dimensions associated with pistons. Refer to **Figure 9-35.** As you will learn, these dimensions affect how the piston functions in the cylinder. Many are also important when working on an engine.

- **Piston diameter.** The distance measured across the sides of the piston.
- **Pin hole diameter.** The distance measured across the inside of the hole for the piston pin.
- **Ring groove width.** The distance measured from the top to the bottom of a ring groove.
- **Ring groove depth.** The distance measured from the ring land to the back of a ring groove.

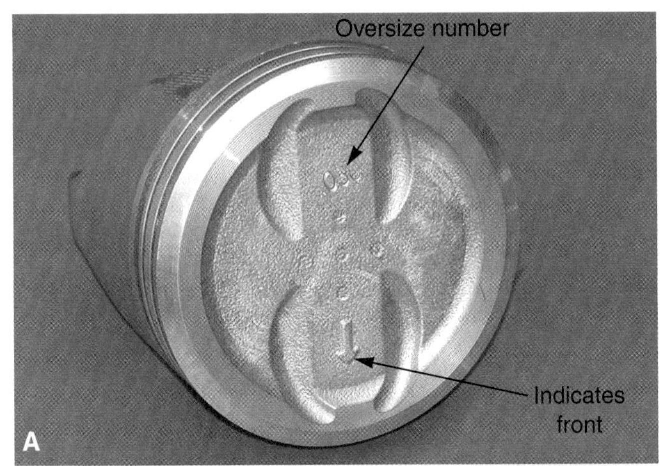

Figure 9-36. *Note common piston designs. A—Dished piston lowers the compression ratio. B—Valve relief piston. C—Domed piston raises the compression ratio.*

- **Skirt length.** The distance from the bottom of the skirt to the centerline of the pin hole.
- **Compression distance.** The distance from the centerline of the pin hole to the top of the piston.

Piston Clearance

Piston clearance is the amount of space between the sides of the piston and the cylinder wall. Clearance is needed for a lubricating film of oil and to allow for expansion when the piston heats up. The piston must always be free to slide up and down in the cylinder block.

Specialized Designs

Manufacturers are constantly trying to get more power and fuel efficiency from engines. As a result, several designs and manufacturing procedures have been developed. Some of these remain in the experimental stage while others, such as cam grinding, are common in today's production engines.

Cam Ground Piston

A ***cam ground piston*** is machined slightly out-of-round. The piston is a few thousandths of an inch (or hundredths of a millimeter) larger in diameter perpendicular to the piston pin centerline. See **Figure 9-37.** Cam grinding is done to compensate for different rates of piston expansion (enlargement) due to differences in the material thickness. As the piston is heated by combustion, the thicker material around the pin boss expands more in a line parallel with the piston pin. The cam ground piston then becomes round when hot.

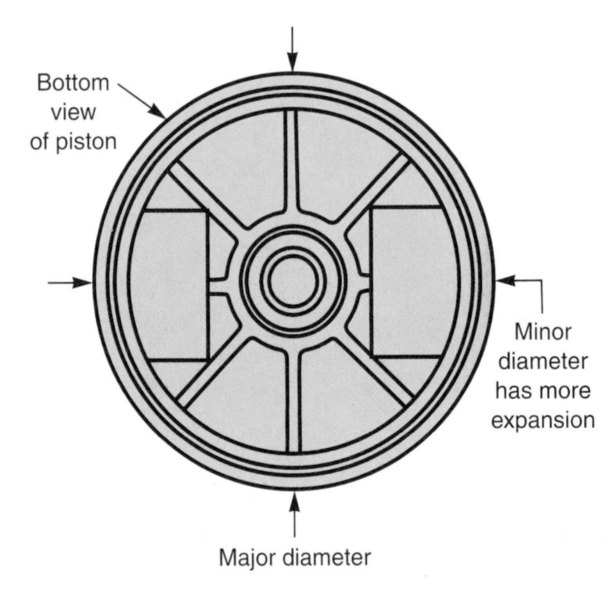

Figure 9-37. *A cam ground piston is not perfectly round when cold. However, when the piston heats up during engine operation, the boss area expands more than the rest of the piston and the piston becomes round. (Ford)*

A cold cam ground piston has the correct piston-to-cylinder clearance. The unexpanded piston will not slap, flop sideways, or knock in the cylinder because of too much clearance. Yet, the cam ground piston will not become too tight in the cylinder when heated to full operating temperature.

Piston Taper

Like cam grinding, ***piston taper*** is normally used to maintain the correct piston-to-cylinder clearance, **Figure 9-38.** The top of the piston is machined slightly smaller than the bottom. Since the piston head gets hotter than the skirt and expands more, piston taper makes the piston almost equal in size at the top and bottom when in operation (hot).

Fiber-Reinforced Piston

A ***fiber-reinforced piston*** is made from a special alloy reinforced with fibers to increase piston dependability under severe operating conditions. As shown in **Figure 9-39,** the area above the top compression ring may be reinforced with fibers to increase wear resistance, heat resistance, and cooling ability. This area of the piston is exposed to extreme loads and is prone to failure.

Piston Head Coatings

A ***piston head coating*** refers to the aluminum piston head being covered with another substance to improve durability and power production, **Figure 9-40.** The coating may be ceramic, stainless steel, or other material. For example, one piston manufacturer bonds a stainless steel cap to the aluminum piston. A steel mesh is used between the aluminum and stainless steel to form a good bond. The stainless steel can resist heat, carbon buildup, and pressure better than aluminum.

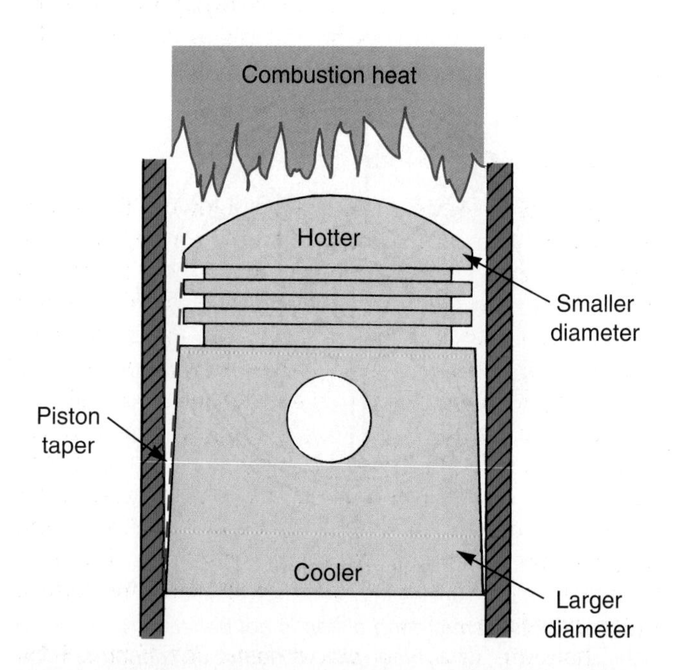

Figure 9-38. *Piston taper is commonly used because the head of the piston gets hotter and expands more than the skirt.*

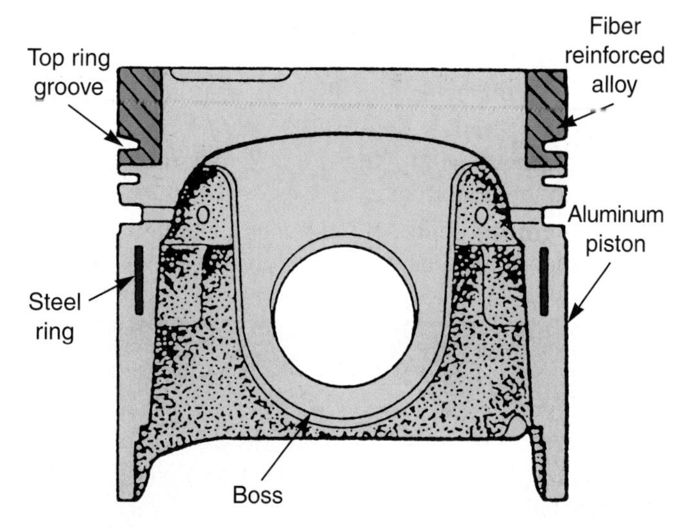

Figure 9-39. *This fiber-reinforced piston uses fibers in a special alloy matrix to increase strength and heat resistance on the critical top land of the piston. (Toyota)*

A ceramic coating can also be bonded to the piston head as a thermal barrier. Aluminum conducts heat very quickly. As a result, an uncoated aluminum piston conducts considerable combustion heat through the piston. This leaves less heat in the combustion chamber to produce gas expansion and pressure. By coating the piston head with a thin layer of heat-insulating ceramic, heat conduction out of the combustion chamber is reduced. This allows the engine to generate more power from the same amount of fuel.

Teflon-Coated Piston Skirts

Teflon-coated piston skirts reduce friction by reducing drag. This is especially true during hot restart conditions. When an engine is shut down, heat can soak into aluminum pistons. This can cause slow cranking and hard starting problems. Teflon's nonstick surface is ideal for this application.

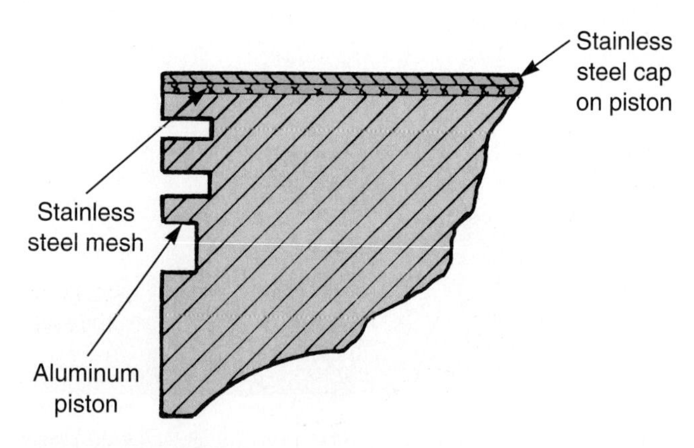

Figure 9-40. *A few high-performance pistons have a steel cap bonded to the aluminum piston.*

Two-Piece Piston

A ***two-piece piston*** is constructed in two parts—the crown and the skirt. As pictured in **Figure 9-41,** the piston pin holds the two together. This is an experimental design that has proven very successful in racing engines. The design also allows the use of different materials in the crown and skirt. For example, the crown can be made of fiber-reinforced aluminum and stainless steel and the skirt from lightweight plastic.

Oil-Cooled Piston

An ***oil-cooled piston*** uses oil pressure to direct a stream of oil through and onto the piston to help cool the piston head. A simple drawing of this type piston is shown in **Figure 9-42.**

Some diesel engines have a different form of cooling the piston with oil. Small tubes or jets direct a spray of engine oil onto the bottom of the pistons. This helps cool the pistons and increases engine dependability.

Variable Compression Piston

A ***variable compression piston*** is an experimental, two-piece design controlled by engine oil pressure. The piston head fits over the piston main body and can slide up and down. Engine oil pressure fed between the halves forms a hydraulic cushion. Under normal driving conditions, the piston top extends for maximum compression ratio and power. When engine speed increases, combustion pressure pushes the piston head down to lower compression and prevent engine knocking.

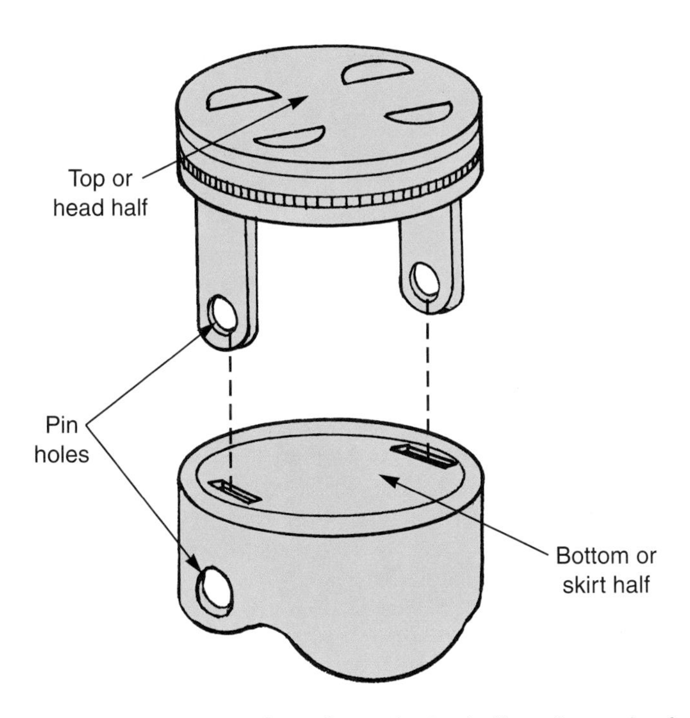

Figure 9-41. *In a two-piece piston, the top half can be made of one metal alloy and the bottom from another.*

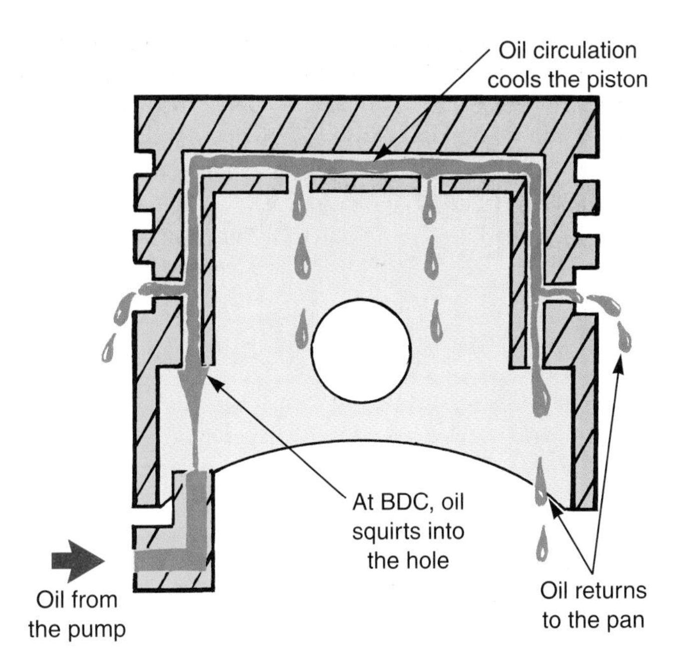

Figure 9-42. *An oil-cooled piston has oil circulated through it to help cool the piston.*

Piston Ring Construction

The pistons in an automotive engine normally have three piston rings—two compression rings and one oil ring. See **Figure 9-43.** It is important for you to understand how variations in ring construction provide different operating characteristics.

The piston rings seal the clearance between the outside of the piston and the cylinder wall , **Figure 9-44.** They must keep combustion pressure from entering the crankcase and oil from entering the combustion chambers. This must be done in very hot condition and when the rings are traveling a high speed.

Compression Rings

The ***compression rings*** prevent blowby. Blowby is combustion pressure leaking past the piston rings into the engine crankcase. On the compression stroke, pressure is trapped between the cylinder and piston grooves by the compression rings. Combustion pressure pushes the compression rings down in their grooves and out against the cylinder wall. This produces an almost leakproof seal.

Compression rings are usually made of cast iron. An outer layer of chrome, molybdenum, or other metal may be coated on the face of the ring to increase wear resistance and oil absorption. The face of the compression ring may also be grooved to speed ring seating. Seating is the initial ring wear that makes the ring match the surface of the cylinder perfectly.

There are several special designs. See **Figure 9-45.** Dyke or L-shaped ring also uses combustion pressure to force ring out against cylinder wall. The cross section of

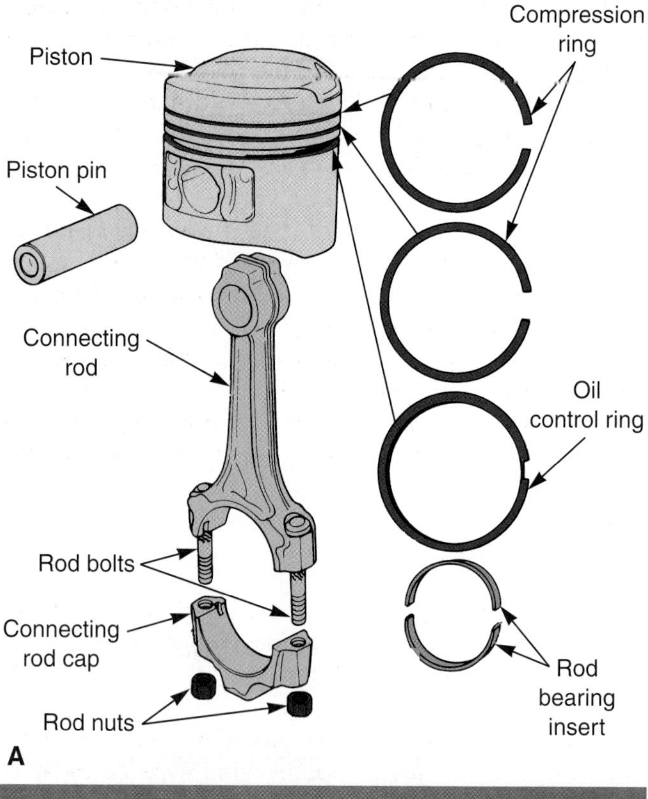

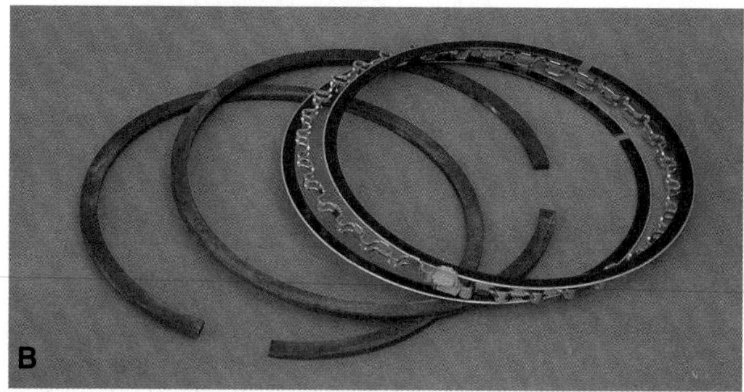

Figure 9-43. *A—Two compression rings and one oil ring are normally used on passenger car pistons. B—A set of piston rings. (Ford)*

these rings looks like the letter L. A head land ring also has an L-shape, but the ring is partially outside of the ring groove. A spring expander can be used to increase tension forcing compression ring out against cylinder. It increases friction, however. There is also a special ring design that uses small holes around head of piston to allow combustion pressure to act on the back of the ring. This design is called gas ported and allows the ring to be held tightly against the cylinder when combustion pressure is high. However, outward force is reduced when pressure in the combustion chamber is low to reduce ring drag or friction.

Piston ring shape refers to the cross-section of a piston ring. See **Figure 9-46.** The shape controls how the piston ring operates in its groove and on the cylinder wall. Based on the shape, there is usually a top and a bottom to the ring. Make sure the ring is properly installed in the piston.

Oil Rings

Primarily, an ***oil ring*** must prevent engine oil from entering the combustion chamber. It scrapes excess oil off the cylinder wall. Refer to **Figure 9-44B.** If too much oil gets into the combustion chamber and is burned, blue smoke will come out of the vehicle's exhaust pipe.

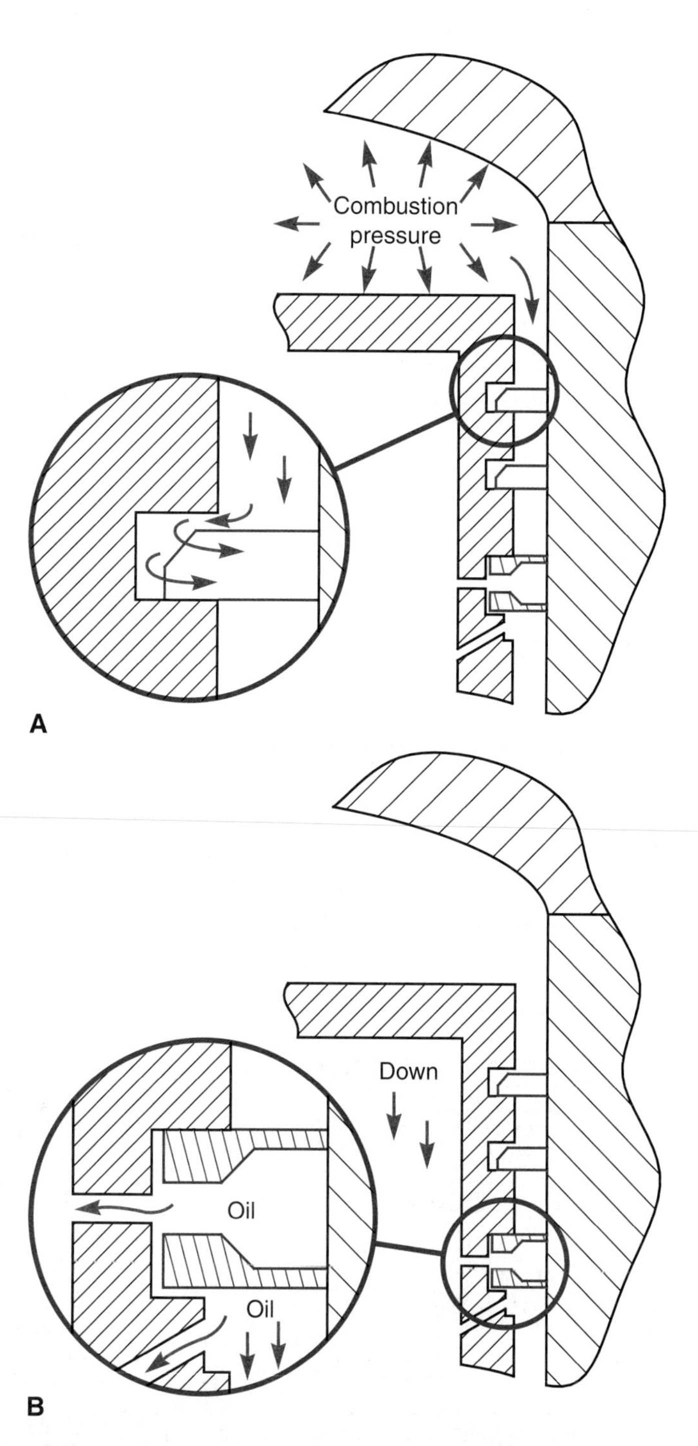

Figure 9-44. *Note the basic action of compression and oil rings. A—Combustion pressure flows down between the piston and cylinder. It actually helps push the compression rings out against the cylinder. B—The oil ring scrapes excess oil from the cylinder. Oil flows through the ring and piston and back into the crankcase.*

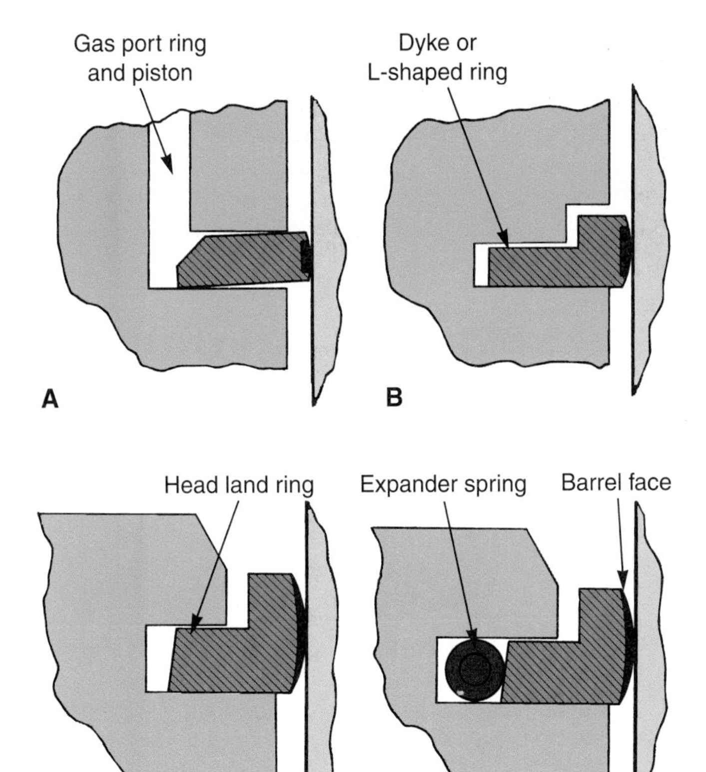

Figure 9-45. *Special compression ring designs. A—Gas ported. B—Dyke or L-shaped. C—Head land ring. D—Spring expander. (Sealed Power)*

Oil rings are available in two basic designs—rail-spacer type and one-piece type. An oil ring consisting of two rails and an expander-spacer (rail-spacer type) is the most common, **Figure 9-47.**

Piston Ring Dimensions

Basic piston ring dimensions include the ring width, ring wall thickness, and ring gap. The ***ring width*** is the distance from the top to the bottom of the ring. The difference between the ring width and the piston ring groove width determines ring side clearance.

Ring radial wall thickness is the distance from the face of the ring to its inner wall. The difference between the ring wall thickness and ring groove depth determines ring back clearance.

The ***ring gap*** is the split or space between the ends of a piston ring when installed in the cylinder. The ring gap allows the ring to be spread open and installed on the piston. There are different designs for the joint at the gap. A butt joint is the most common. Refer to **Figure 9-48.** The ring gap also allows the ring to be made a slightly larger diameter than the cylinder. When installed, the ring spreads outward and presses on the cylinder wall to aid ring sealing.

Ring Expanders

A ring expander can be used behind the second compression ring, as shown in **Figure 9-45D.** The expander helps push the ring out against the cylinder wall, increasing the ring's sealing action.

A ring expander can also be placed behind a one-piece oil ring to increase ring tension. A ring expander-spacer is part of a three-piece oil ring. It holds the two steel oil ring scrapers apart and helps push them outward.

Piston Ring Coatings

The face of piston rings can be coated with chrome or other metal. Soft ring coatings, usually iron or molybdenum, help the ring wear-in quickly and form a good seal. The soft, grooved, outer surface wears away rapidly so the ring conforms to the shape of the cylinder. Also called ***quick seal rings,*** they are commonly recommended for used cylinders that are slightly worn.

Hard ring coatings, such as chrome, are used to increase ring life and reduce friction. They are used in new

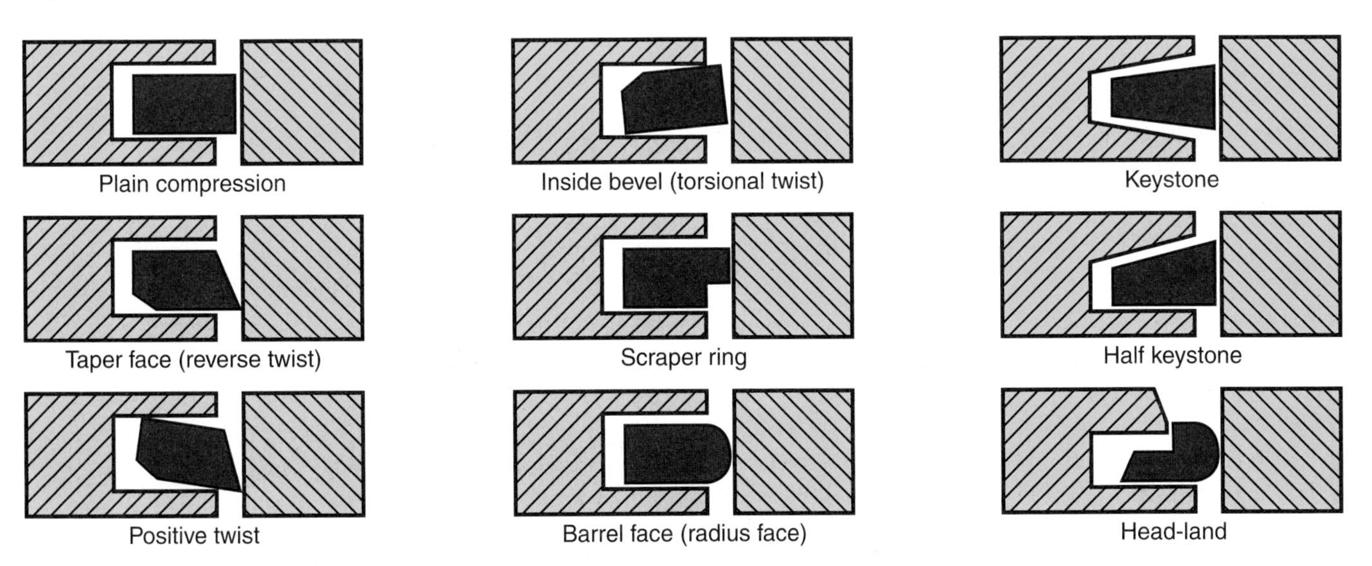

Figure 9-46. *Study the different compression ring shapes.*

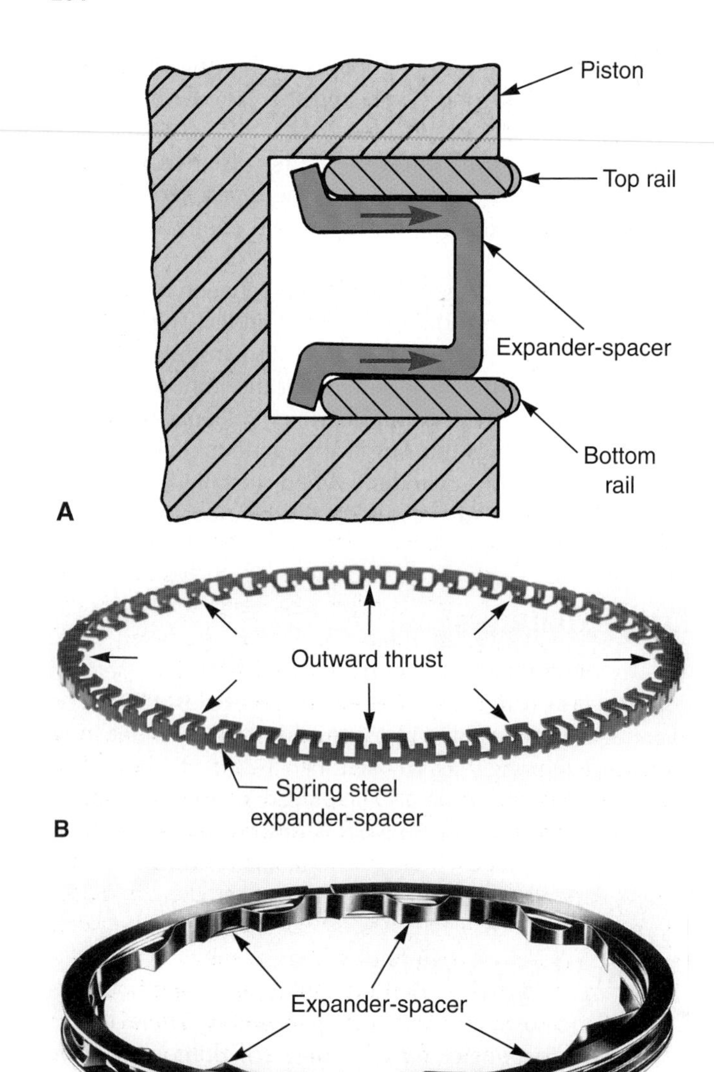

Figure 9-47. *A—The expander-spacer holds and pushes the scraper rails outward. B—The expander-spacer is made of spring steel. C—A complete rail-spacer type oil ring. (Dana, Perfect Circle)*

or bored cylinders that are perfectly round and not worn. To aid break-in, chrome plated rings usually have ribbed faces. The ribs hold oil and wear quickly to produce a good seal.

Porous ring coatings help break-in and service life. These coatings hold oil and reduce friction. Molybdenum is a porous, as well as soft, metal sometimes used on compression rings.

Piston Rings for Coated Cylinder Walls

When rebuilding an engine that has nickel-ceramic cylinder walls, you must use piston rings designed to operate on that coating. The top compression ring is usually a plasma moly or positive vapor deposit (PVD) type. The second compression ring is usually ductile iron.

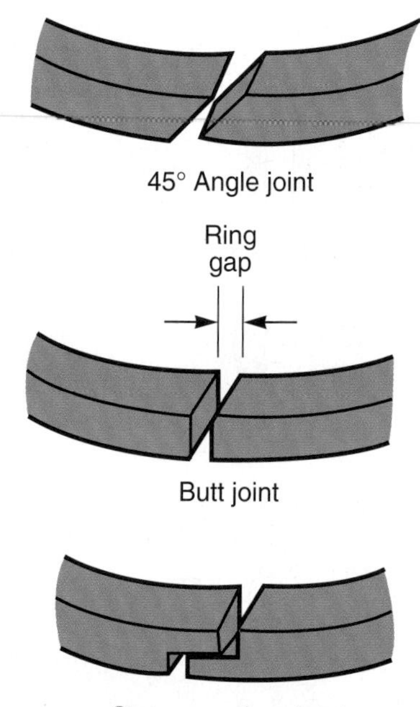

Figure 9-48. *Note the basic ring gap designs. A butt joint is most common, but a step or gapless ring provides improved sealing.*

The oil control rings must not be chrome plated. Never use chrome piston rings, not even chrome oil rings, in coated cylinders because wall scuffing and scoring will result.

Engine Bearing Construction

There are three basic types of engine ***bearings***—connecting rod, crankshaft main, and camshaft bearings. This is illustrated in **Figure 9-49.**

Steel is normally used for the portion of the bearing that contacts the stationary part. This is called the backing material. Softer metal is bonded over the backing to form the bearing surface. See **Figure 9-50.** Any one of three metals is commonly plated over the top of the steel backing:

- ❑ Babbitt (a lead-tin alloy).
- ❑ Copper.
- ❑ Aluminum.

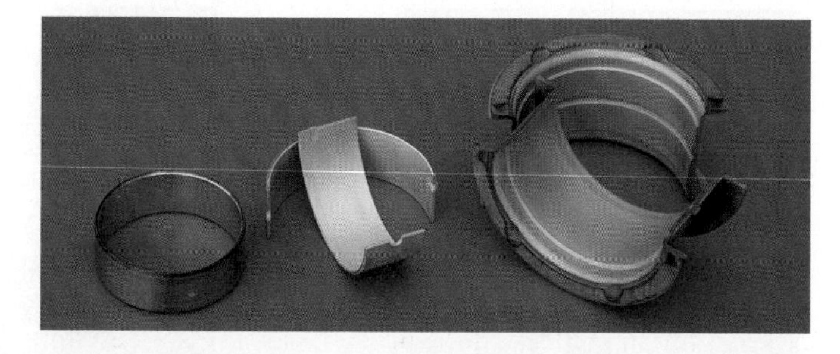

Figure 9-49. *An engine contains three basic types of bearings. From left to right, camshaft, connecting rod, and main bearing. Note the thrust surfaces on the main bearing.*

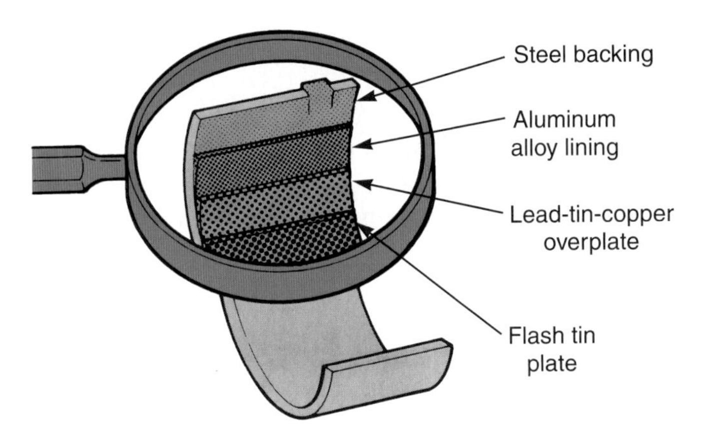

Figure 9-50. *Engine bearings are made with several different layers of metal. The steel backing is hard; the top layer is very thin and soft. (Federal Mogul)*

These three metals may be used in different combinations to design bearings for light-, medium-, heavy-, or extra-heavy-duty applications.

Some exotic engines have special materials bonded to the steel backing to reduce friction, heat, wear. Some of these high-performance bearing coatings are as thick as .001″ (0.03 mm). This is thick enough that you must allow for it when machining an engine to accept these high-performance bearings.

Bearing Characteristics

Engine bearings must operate under tremendous loads, severe temperature variations, abrasive action, and corrosive surroundings. Essential bearing characteristics include:

- ❑ Load strength.
- ❑ Conformability.
- ❑ Embedability.
- ❑ Corrosion resistance.
- ❑ Crush.
- ❑ Spread.

The ***bearing load strength*** is the bearing's ability to withstand pounding and crushing during engine operation. The piston and rod transfer several tons of downward force through the bearing to the crankshaft. The bearing must not fatigue, flatten, or split under this load. If the bearing load strength is too low, the bearing can smash, fail, or spin in its bore. This can ruin the bore or journal.

Bearing conformability is the bearing's ability to move, shift, or adjust to imperfections in the journal surface. See **Figure 9-51.** The soft metal of the bearing surface can conform to any defects in the journal.

Dirt or metal is sometimes carried into the bearings. ***Bearing embedability*** refers to the bearing's ability to absorb dirt, metal, or other hard particles. The bearing should allow the particles to sink beneath the surface and into the bearing material. This will prevent the particles from scratching, wearing, and damaging the journal.

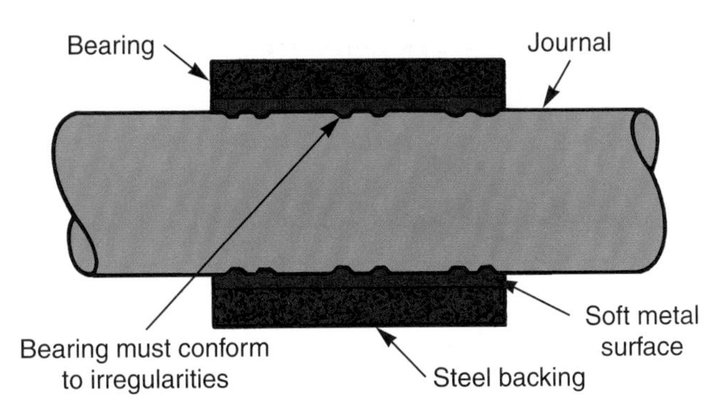

Figure 9-51. *Bearings must conform to irregularities in the journal.*

Bearing corrosion resistance is the bearing's ability to resist corrosion from acids, water, and other impurities in the engine oil. Combustion blowby gasses cause oil contamination that may corrode the engine bearings. Aluminum-lead and other alloys are now commonly used because of their excellent corrosion resistance.

Bearing crush is used to help prevent the bearing from spinning inside its bore during engine operation. The bearing is made slightly larger in diameter than the bearing bore. The ends of each part of the bearing are slightly above the mating surfaces, **Figure 9-52.** When the rod or main cap is tightened, the bearing ends hit each other. This jams the backside of the bearing inserts tightly against the bore, locking them in place. This is one reason why the back of the bearing must be clean and dry (not lubricated) during engine assembly.

Bearing spread is used on split-type engine bearings to hold the bearing in place during assembly. The distance across the parting line of the bearing is slightly wider than the bearing bore, **Figure 9-53.** This causes the bearing insert to stick in its bore when pushed into place with your fingers. Tension from bearing spread keeps the bearing from falling out of its bore as you assemble the engine.

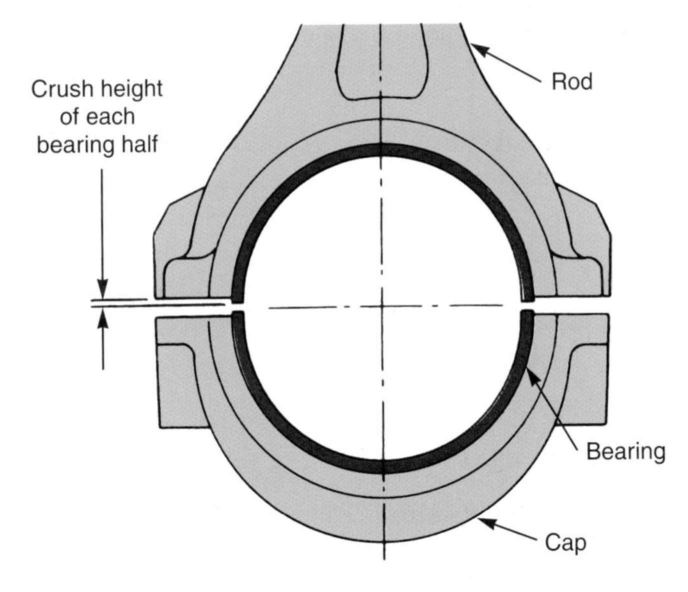

Figure 9-52. *Bearing crush ensures the bearing does not spin inside its bore. (Deere & Co)*

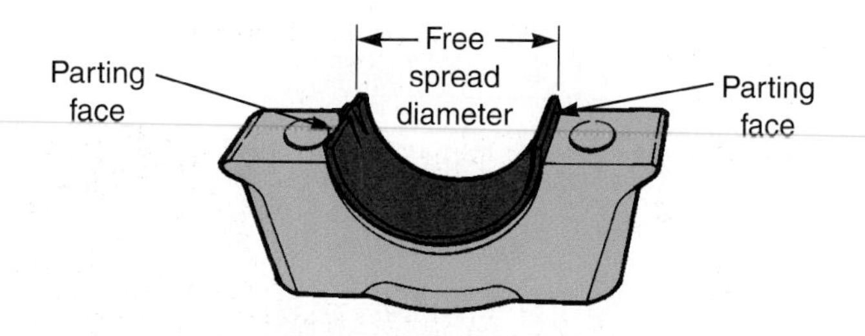

Figure 9-53. *Bearing spread holds the bearing in place during engine assembly. The bearing diameter is slightly larger than the bore diameter. (Federal Mogul)*

Standard, Undersize, and Oversize Bearings

A ***standard-size bearing*** has the original dimensions specified by the engine manufacturer. A standard bearing may have the abbreviation STD stamped on the back. However, parts are sometimes machined to correct for wear. For example, the journals on a crankshaft may be machined to a smaller diameter to correct for scoring. Or, the bore in a connecting rod may be machined to a larger diameter to correct for an alignment problem.

An ***undersize bearing*** is designed to be used on a journal that has been machined to a smaller diameter. If the journal has been worn or damaged, it can be ground undersize by a machine shop. Then, undersize bearings are needed to make up for the increased clearance, **Figure 9-54.** Connecting rod and main bearings are available in undersizes of .010″, .020″, .030″, and sometimes .001″ and .040″. The amount of undersize is normally stamped on the back of the bearing. The crankshaft may also have an undersize number stamped on it by the machine shop.

An ***oversize bearing*** is designed to be used in a bore that has been machined to a larger diameter. For example, the main bore in an engine block may be bored to a larger diameter to correct for a badly misaligned bore. Oversize main bearings must then be installed to make up for the increased clearance.

Caution: Before machining a crankshaft or camshaft bores oversize or undersize, be sure that bearing insert sizes are available to match the amount of machining.

Bearing Locating Lugs and Dowels

Lugs or dowels are used to locate split bearings in their bores, **Figure 9-55.** The bearing usually has a lug that fits into a recess machined in the bearing bore or cap. Sometimes, however, a dowel in the cap or bore fits in a hole in the bearing insert. Either method helps keep the insert from shifting or turning during crankshaft rotation.

Bearing Oil Holes and Grooves

Oil holes and grooves in the engine bearings allow oil to flow through the block and into the clearance between the bearing and crankshaft journal. The grooves provide a channel so oil can completely circle the bearing before flowing over and out of the bearing. See **Figure 9-56.**

Select Fit Parts

Select fit means that some engine parts are selected and installed in a certain position to improve the fit or clearance between parts. For example, pistons are commonly select fit into their cylinders, **Figure 9-57.** The engine manufacturer measures the diameter of the cylinders. If one cylinder is machined slightly larger than another, a slightly larger piston is installed (select fit) in that cylinder. Because of select fit parts, it is important that you reinstall parts in their original locations when possible.

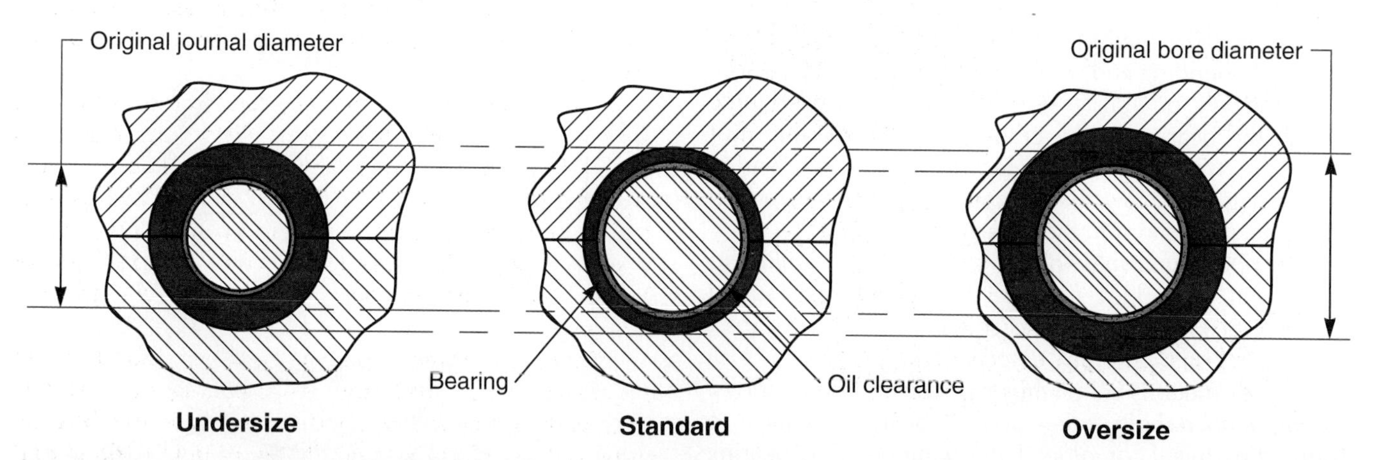

Figure 9-54. *An undersize bearing is thicker than standard bearing to make up for the increased clearance when the journal has been machined smaller. An oversize bearing is also thicker, but makes up for the increased clearance when the bore has been machined larger. Notice how the oil clearance is the same in all three situations.*

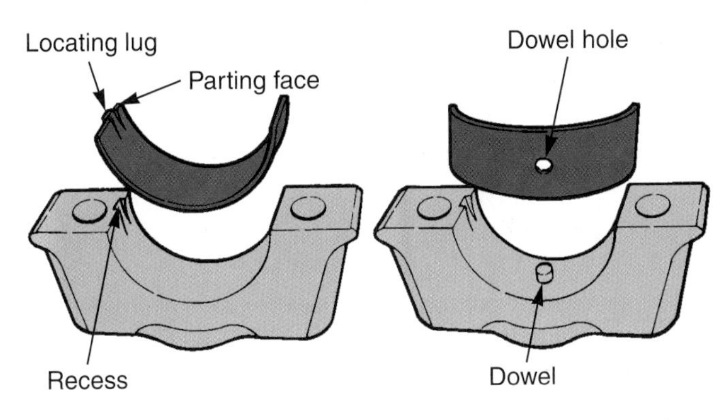

Figure 9-55. *Lugs and dowels are two methods of securing a bearing in its bore. Lugs are more common. (Federal Mogul)*

Summary

A bare block is the cylinder block without any parts installed. A short block includes everything installed in the cylinder block. A long block also includes the cylinder heads.

Blocks can be made of cast iron or aluminum. Aluminum blocks normally have cast iron sleeves. A wet sleeve is thick and serves as the cylinder wall. A dry sleeve is normally installed in a bored cylinder. Core plugs are in the sides of the block to seal holes that were needed during manufacturing.

The crankshaft is installed in the main bore on main bearing inserts. With an OHV engine, the camshaft bore is also in the block. Main caps secure the crankshaft and main bearings in the block. There are several main cap designs—two-bolt, four-bolt, cross-bolt, and girdle cap. A thrust bearing limits crankshaft endplay.

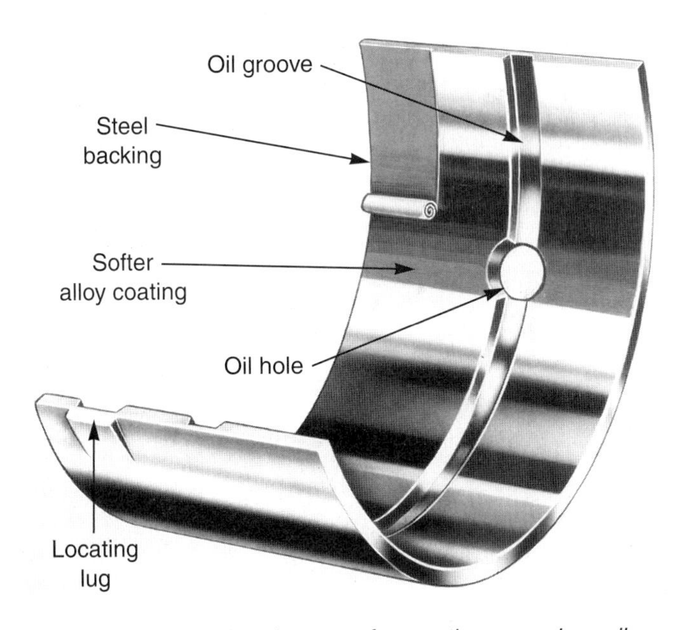

Figure 9-56. *Main bearings are frequently grooved so oil can circle the bearing and be evenly distributed. Grooves also let a constant supply of oil enter the crankshaft for the rod bearings. (Clevite)*

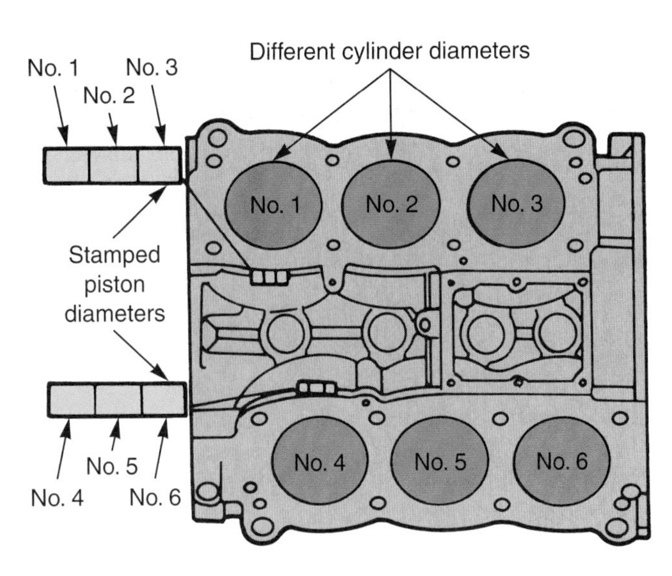

Figure 9-57. *Pistons and other parts are often select fit. Note that the required piston diameter for each cylinder is stamped on the engine block. (Ford)*

Journals are machined and polished surfaces on the crankshaft for the main and rod bearings. A steel crankshaft is stronger than a cast iron crankshaft and can be found in high-performance or diesel engines. A splayed rod journal has the journals offset from each other to smooth V-6 engine operation. A rear main oil seal keeps engine oil from leaking out the rear of the engine from the rear main bearing.

Connecting rods are normally made of steel. Bolts or cap screws hold the cap on the rod body. Rod numbers are used so the caps can be reinstalled correctly on the same rods. Connecting rod length is measured from the center of the small end to the center of the big end.

Piston pins are case-hardened steel. A floating piston pin is free to rotate in both the rod and piston. It is held in place with snap rings. A press-fit pin is locked in the connecting rod. A piston-guided connecting rod fits closely in the piston boss and the piston limits side movement of the rod small end. Piston pin offset is used to prevent piston slap. Several piston head shapes are used in today's engines.

Automotive pistons normally have three rings—two compression rings and one oil ring. Piston ring shape can be used to help increase sealing efficiency. A ring expander is used to force the ring out against the cylinder wall with more pressure. Soft ring coatings are recommended for worn cylinders. Harder coatings are for new or freshly bored cylinders that are perfectly round and not tapered.

There are three types of bearings found in an engine—camshaft, rod, and main. Bearings are usually steel with a soft metal bonded to create the bearing surface. A standard bearing fits on an unmachined part (journal or bore). Undersize or oversize bearings are needed when a journal or bore is machined to a different size.

Review Questions—Chapter 9

Please do not write in this text. Write your answers on a separate sheet of paper.

1. An aluminum cylinder block normally has:
   (A) Aluminum cylinder walls.
   (B) Cast iron sleeves.
   (C) Silicone-impregnated cylinders.
   (D) Chrome sleeves.
2. _____ are used to seal holes in the block from manufacturing, but do not protect the engine block from damage if a weak coolant solution freezes inside the block.
3. Explain the difference between a *wet sleeve* and a *dry sleeve.*
4. What are *siamese cylinders*?
5. Why must you never mix up main caps?
6. Describe four main cap designs.
7. A(n) _____ is used to limit crankshaft endplay.
8. Main bearing clearance is the space between the crankshaft main _____ and the main _____.
9. Why is a *splayed journal* sometimes used in V-6 engines?
10. This part is usually installed in the rear of the crankshaft when the vehicle has a manual transmission.
    (A) Snout.
    (B) Torque converter.
    (C) Pilot bearing.
    (D) Pulley.
11. Describe three types of rear main seals.
12. Engine connecting rods in passenger vehicles are usually made of _____.
13. Connecting rod length is measured from the center of the _____ to the center of the _____.
14. A(n) _____ piston pin uses snap rings to keep the pin inside the piston.
15. Why is piston pin offset used?
16. Why are valve reliefs formed in some piston heads?
17. Why are pistons cam ground?
18. _____ accounts for the fact that the head of the piston runs hotter and expands more than the skirt.
19. In what application would a soft ring coating be used?
20. This makes it important for many engine parts to be reinstalled in the same location during repairs.
    (A) Oversize parts.
    (B) Undersize parts.
    (C) Worn parts.
    (D) Select fit parts.

ASE-Type Questions—Chapter 9

1. Technician A says that the piston assemblies are included in an engine's short block. Technician B says that the valve train components are included in an automotive engine's short block. Who is correct?
   (A) A only.
   (B) B only.
   (C) Both A and B.
   (D) Neither A nor B.
2. Another term used to refer to an engine's short block is _____.
   (A) bare block
   (B) cylinder block
   (C) bottom end
   (D) None of the above.
3. Technician A says that an aluminum cylinder block dissipates heat much faster than a cast iron cylinder block. Technician B says that an aluminum cylinder block has aluminum sleeves. Who is correct?
   (A) A only.
   (B) B only.
   (C) Both A and B.
   (D) Neither A nor B.
4. Technician A says that an engine's lifter bores are located in the cylinder block deck. Technician B says that an engine's lifter bores are where the lifters are installed. Who is correct?
   (A) A only.
   (B) B only.
   (C) Both A and B.
   (D) Neither A nor B.
5. All of the following are main cap designs *except:*
   (A) cross-bolt mains.
   (B) two-bolt mains.
   (C) three-bolt mains.
   (D) four-bolt mains.

6. Technician A says that forged steel crankshafts are normally used in turbocharged automotive gasoline engines. Technician B says that forged steel crankshafts are normally used in diesel engines. Who is correct?
   (A) A only.
   (B) B only.
   (C) Both A and B.
   (D) Neither A nor B.
7. The flywheel is used to _____.
   (A) smooth engine operation
   (B) connect the crankshaft to the transmission or transaxle
   (C) hold the starter motor
   (D) Both A and B.
8. Technician A says that a floating piston pin is press fit into the connecting rod's small end. Technician B says that a floating piston pin is held in place by snap rings. Who is correct?
   (A) A only.
   (B) B only.
   (C) Both A and B.
   (D) Neither A nor B.
9. Technician A says that the term *piston clearance* refers to the amount of space between the piston head and cylinder head. Technician B says that the term *piston clearance* refers to the amount of space between the sides of the piston and the engine's cylinder wall. Who is correct?
   (A) A only.
   (B) B only.
   (C) Both A and B.
   (D) Neither A nor B.
10. Technician A says that the cylinder shown on the left has a sleeve. Technician B says that the cylinder shown on the right has a wet sleeve. Who is correct?
   (A) A only.
   (B) B only.
   (C) Both A and B.
   (D) Neither A nor B.

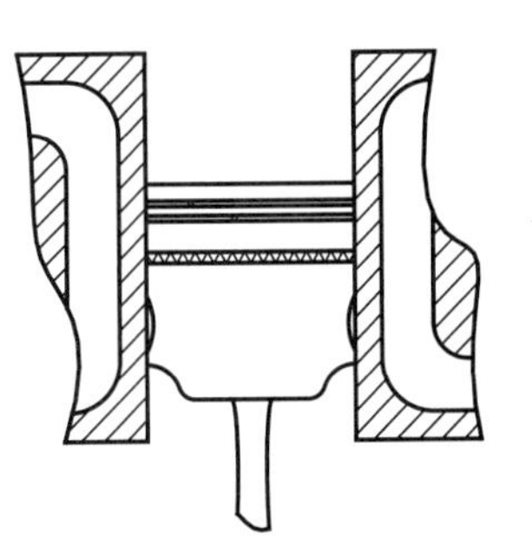

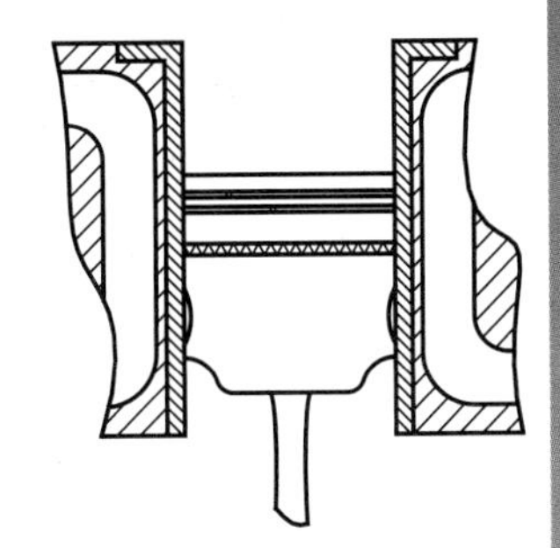

11. Technician A says that bearing crush is used to help keep the bearing from spinning inside its bore during engine operation. Technician B says that bearing crush is used to hold the bearing in place during engine assembly. Who is correct?
   (A) A only.
   (B) B only.
   (C) Both A and B.
   (D) Neither A nor B.

Cylinder heads come in many different designs. This cylinder head has three valves per cylinder. (Ford)

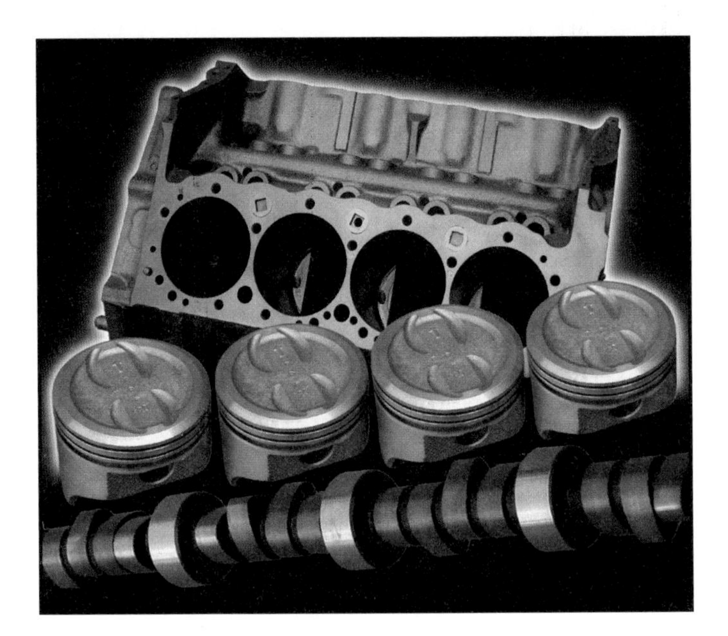

Chapter 10

Top End Construction

After studying this chapter, you will be able to:

- ❑ List the parts of a top end assembly.
- ❑ Compare overhead valve and overhead camshaft valve trains.
- ❑ Describe the parts of a cylinder head.
- ❑ Compare cylinder head design variations.
- ❑ Summarize the construction of engine valves.
- ❑ Explain the construction of valve seats and valve guides.
- ❑ Describe the purpose of valve stem seals.
- ❑ Explain the construction of valve keepers, retainers, and valve springs.
- ❑ Identify the parts of a camshaft.
- ❑ Define valve lift, duration, and overlap.
- ❑ Explain lifters and cam followers.
- ❑ Compare rocker arm design variations.

Know These Terms

Adjustable rocker arms
Bare cylinder head
Cam follower
Camshaft
Camshaft lobes
Combustion chambers
Coolant passages
Cylinder head
Duration
Engine top end
Exhaust ports
Hydraulic camshaft
Induction-hardened valve seats
Intake ports
Integral valve guides
Integral valve seat
Interference angle
Lift
Mechanical camshaft
Nonadjustable rocker arms
Oil passages
Pressed-in valve guides
Pressed-in valve seat
Push rods
Rocker arms
Valve face angle
Valve guides
Valve lash
Valve overlap
Valve seals
Valve seat angle
Valve stem-to-guide clearance
Valve timing
Valve train

Engine top end refers to the parts that fasten above the engine's short block. These parts include the cylinder heads, valves, camshaft, rocker arms and other related valve train components. These parts work together to control the flow of the air/fuel charge into the engine cylinders, as well as the flow of exhaust into the exhaust system. Refer to **Figure 10-1.**

There are many variations in the designs of today's engines. The top end is one area where considerable variations can exist from one manufacturer to another. For example, one engine may have two valves per cylinder and use push rods. On the other hand, an engine found in an exotic sports car may have four valves per cylinder, dual overhead camshafts, and variable valve timing. Today's top-end designs can be much more complex than in the past.

This chapter will help you learn more about how the parts of any engine top end are constructed and designed. As a result, you will be better prepared to diagnose, test, and repair any make or model of engine.

Cylinder Head Construction

The ***cylinder head*** is fastened to the block deck and covers the top of the cylinders. Large ***head bolts*** secure the head to the block. A ***head gasket*** creates a seal between the block and head surfaces to prevent oil, coolant, and combustion-chamber pressure leakage. Refer to **Figure 10-2.** Gaskets are discussed in detail later in this chapter.

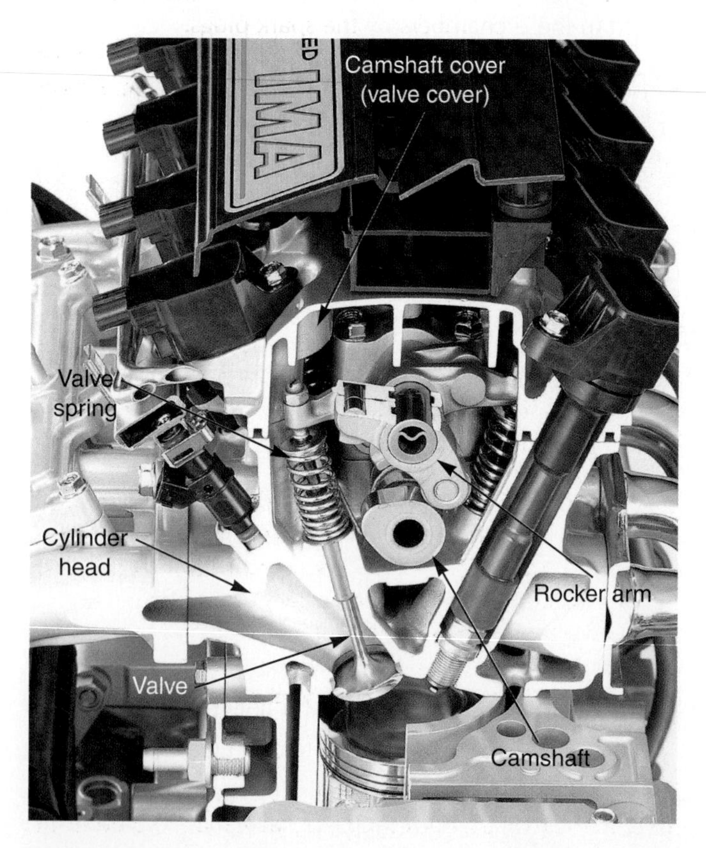

Figure 10-1. *This cutaway view shows the major parts of the engine top end. (Honda)*

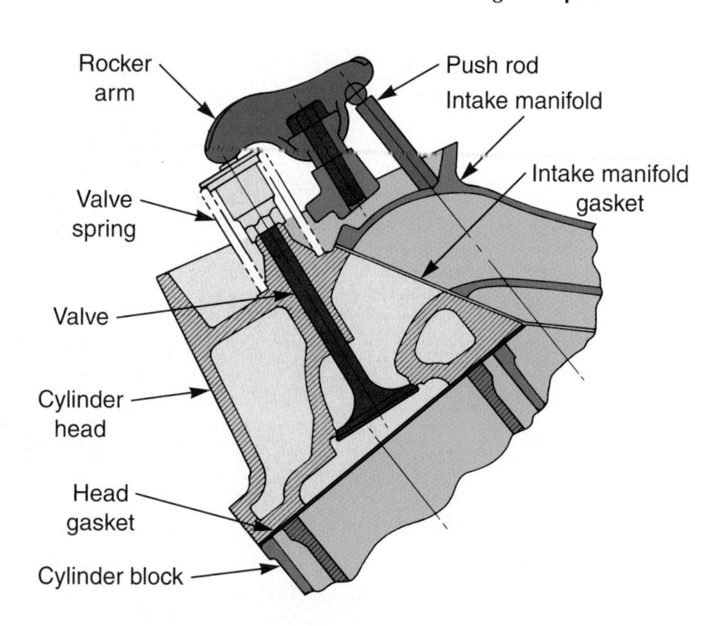

Figure 10-2. *The cylinder head is bolted to the block deck. A head gasket seals the surface between the head and block. (General Motors)*

Cylinder head construction can vary considerably, depending on whether the engine is a cam-in-block (OHV) or overhead camshaft (OHC) configuration. With an OHV engine, the head contains the valves and rocker arms. However, with an OHC engine, it must also hold the camshaft(s), **Figure 10-3.**

Figure 10-4 pictures a cylinder head for a dual overhead cam (DOHC) engine. Note how camshaft housings bolted to the top of the heads contain the camshafts in this design. Camshaft housings are used in some overhead camshaft engine designs.

Bare Head

A ***bare cylinder head*** is a machined casting without any parts installed. It is commonly made of cast iron or aluminum, **Figure 10-5.** A ***cylinder head assembly*** is the bare head with the valves, springs, retainers, and related parts installed. If a cylinder head becomes badly damaged, the technician may need to replace the bare head and install all of the reusable parts in the new head.

Parts of Cylinder Head

The major parts of a bare cylinder head include the following. Refer to **Figure 10-5.**

- ❑ Combustion chambers.
- ❑ Intake ports.
- ❑ Exhaust ports.
- ❑ Oil passages.
- ❑ Coolant passages.
- ❑ Intake deck.
- ❑ Exhaust deck.
- ❑ Dowel holes.

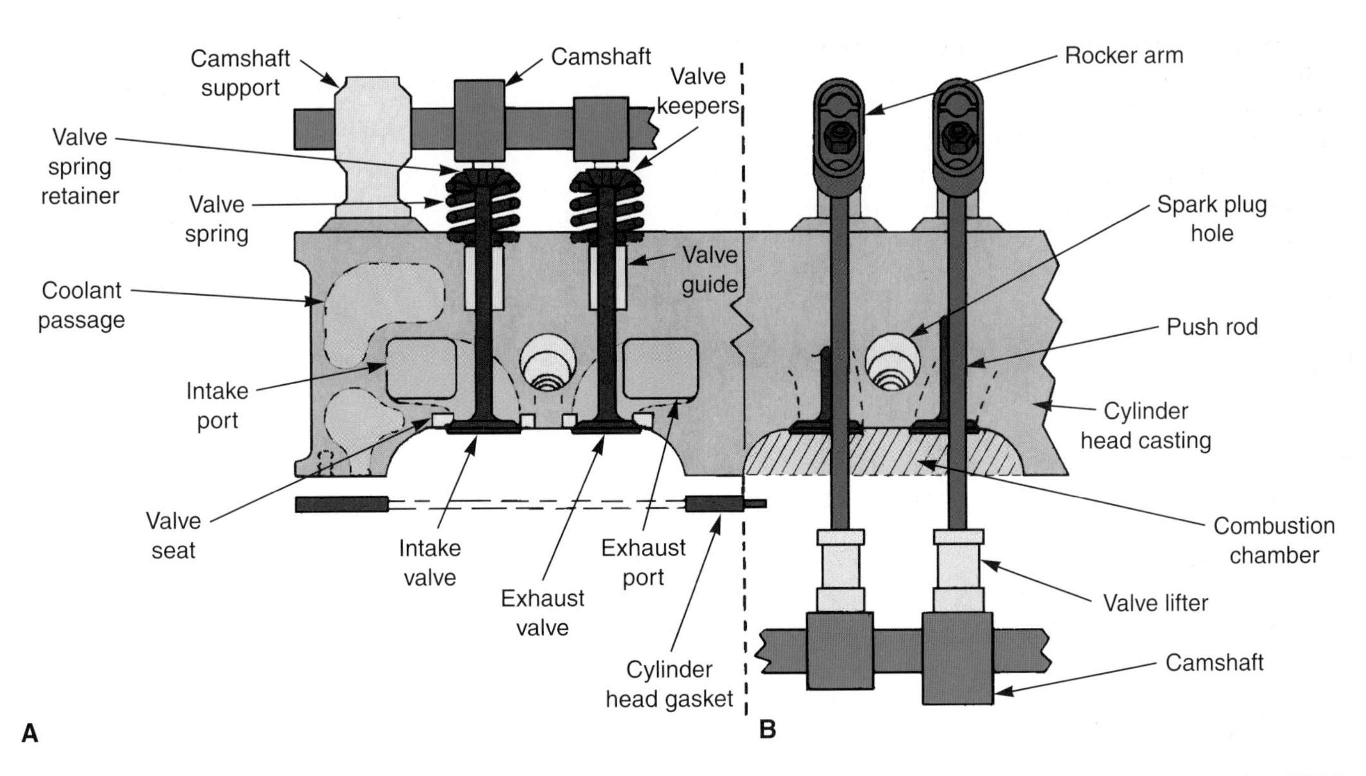

Figure 10-3. *These are the two major variations of engine top end. A—Overhead camshaft (OHC). B—Overhead valve (OHV). (Ford)*

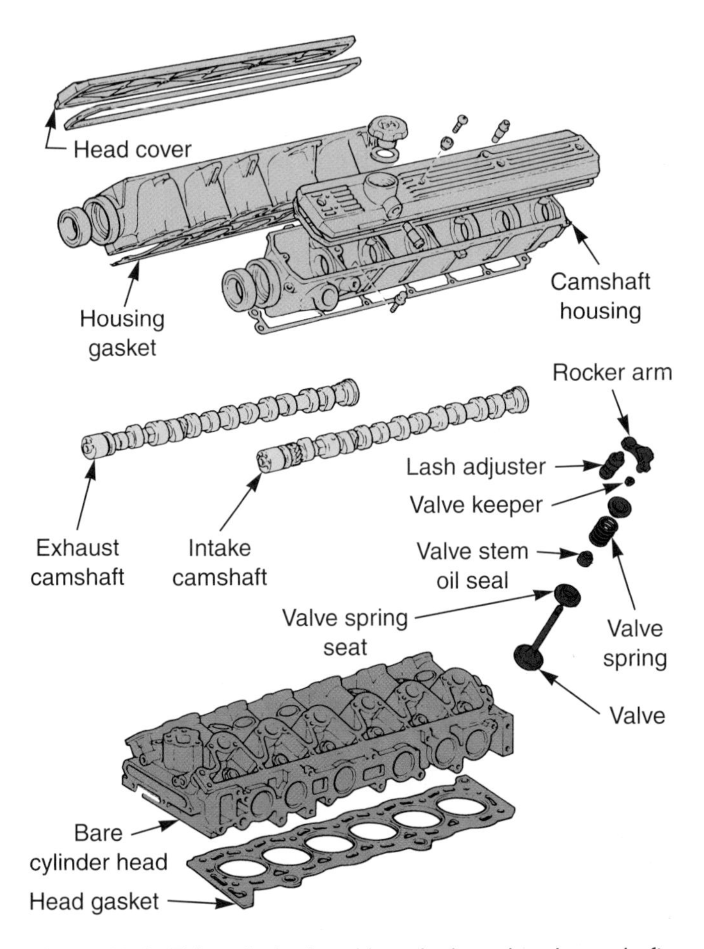

Figure 10-4. *This cylinder head has dual overhead camshafts. Note how camshaft housings are bolted to the top of head to hold the camshafts. (Toyota)*

Combustion chambers are cavities or pockets formed in the head deck surface above the cylinders. They are located directly over the pistons. The air-fuel mixture is ignited in these chambers by the spark plugs.

Intake ports are large passages through the cylinder head from the intake manifold runners to the combustion chambers, **Figure 10-6.** These ports carry the air-fuel mixture to the cylinders.

Exhaust ports are smaller passages leading from the combustion chambers to the exhaust manifold. They are usually on the opposite side of the head from the intake port. See **Figure 10-6.** These ports carry burned gasses out of the cylinders.

Oil is needed in the head to lubricate the moving parts of the valve train. ***Oil passages*** are holes in the head so oil can enter from the block. See **Figure 10-7.** These holes align with matching holes in the head gasket and block deck.

As in the cylinder block, ***coolant passages*** (water jackets) are formed in the cylinder head. They allow coolant to circulate around the combustion chambers, which removes excess heat from the burning fuel.

The ***intake deck*** is a machined surface on the head for bolting on the intake manifold. The intake ports are in the intake deck. Also, the intake deck has threaded holes for the intake manifold bolts. The ***exhaust deck*** is a machined surface on the head for bolting on the exhaust manifold. The exhaust ports are in this surface. There are threaded holes in the exhaust deck for the exhaust manifold bolts. In some older designs, both the intake and exhaust ports are on the same surface.

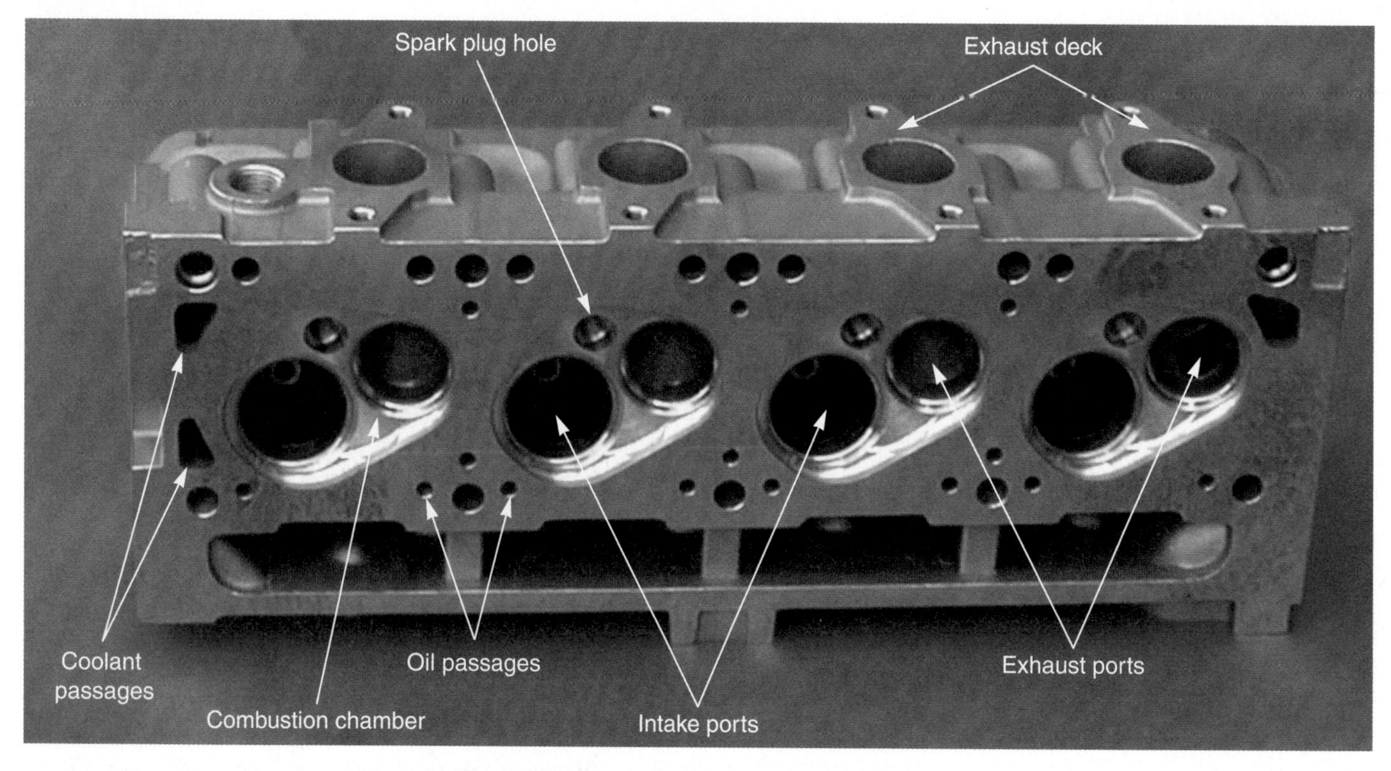

Figure 10-5. *A bare cylinder head is the machined casting without any parts installed.*

In some designs, ***dowel holes*** are provided in the head. Dowels in the block fit into these holes. This holds the head gasket and cylinder head in alignment when installing the head on the block.

Other machined surfaces and holes are provided on the ends of the cylinder head for accessory mounting brackets. For example, the brackets on which the alternator, air conditioning compressor, power steering pump, etc., often have at least one mounting bolt threaded into the head.

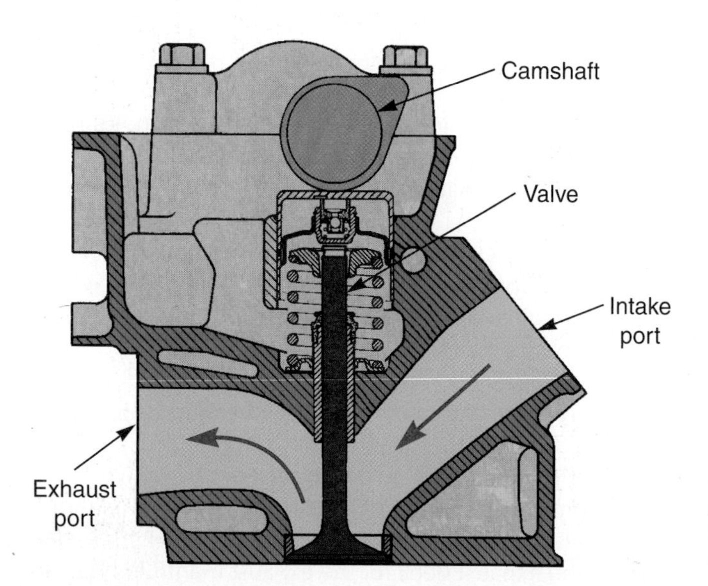

Figure 10-6. *This cutaway shows a side view of a cylinder head. Note the intake and exhaust ports. (Mercedes-Benz)*

Figure 10-8 shows a cutaway view of a typical OHC cylinder head. OHC cylinder heads also require camshaft bearing caps to secure the camshaft and camshaft bearings. Two bolts normally fasten each camshaft cap to the top of the cylinder head.

Diesel Cylinder Heads

The construction of a diesel cylinder head is basically the same as for a gasoline cylinder head. However, there are a couple of significant differences. First, the fuel injectors in a diesel engine are normally installed in the head. On a gasoline engine, the fuel injectors are normally installed in the intake manifold, or sometimes in the throttle body. See **Figure 10-9.**

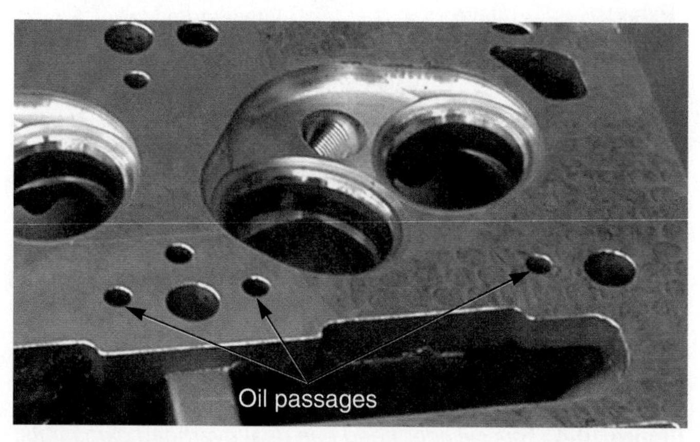

Figure 10-7. *This close-up of a cylinder head deck shows the oil passages in the cylinder head.*

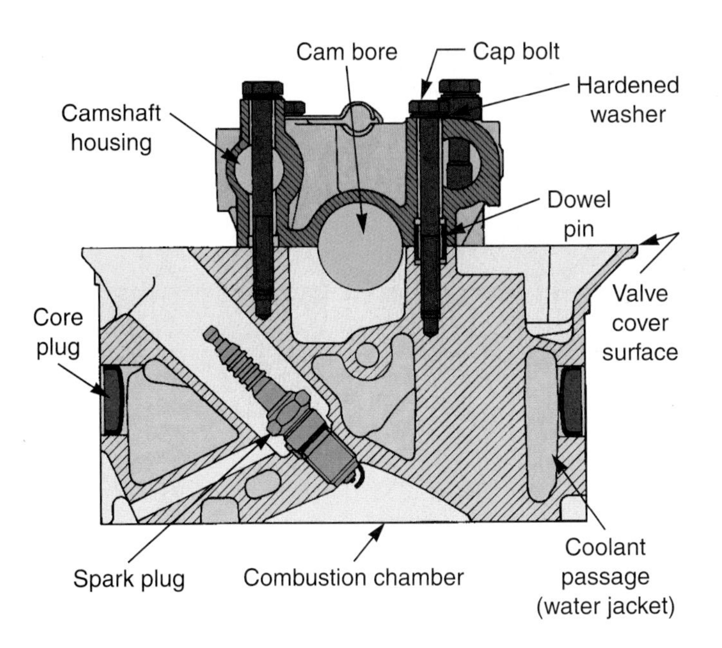

Figure 10-8. *This cutaway of a cylinder head shows how the spark plug screws into the head and how a camshaft housing holds the camshaft and cam bearings. (Mercedes-Benz)*

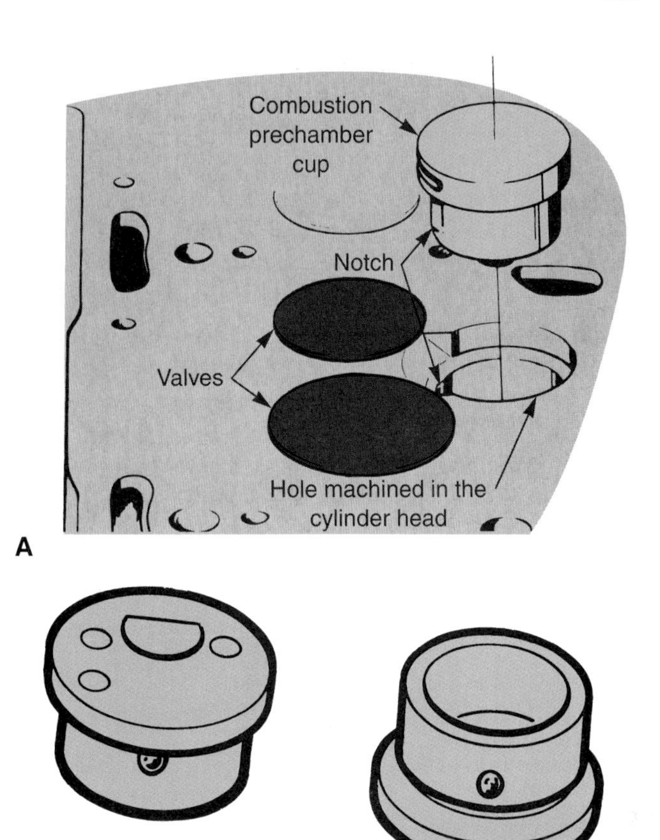

Figure 10-10. *A—A precombustion chamber is formed by a cup press-fit into the cylinder head in a diesel engine. B—The deck-side of the cup has a small hole that allows burning fuel to blow out into the main combustion chamber. C—The head-side of the cup forms the precombustion chamber into which the injector and glow plug extend. (General Motors, Peugeot)*

Second, most diesel engines have a precombustion chamber in the cylinder head for each cylinder. The fuel is injected into this chamber instead of the combustion chamber. The precombustion chamber is formed by pressing a small cup into a machined hole in the cylinder head. See **Figure 10-10.** Since this is a press fit, the cup stays in place during operation.

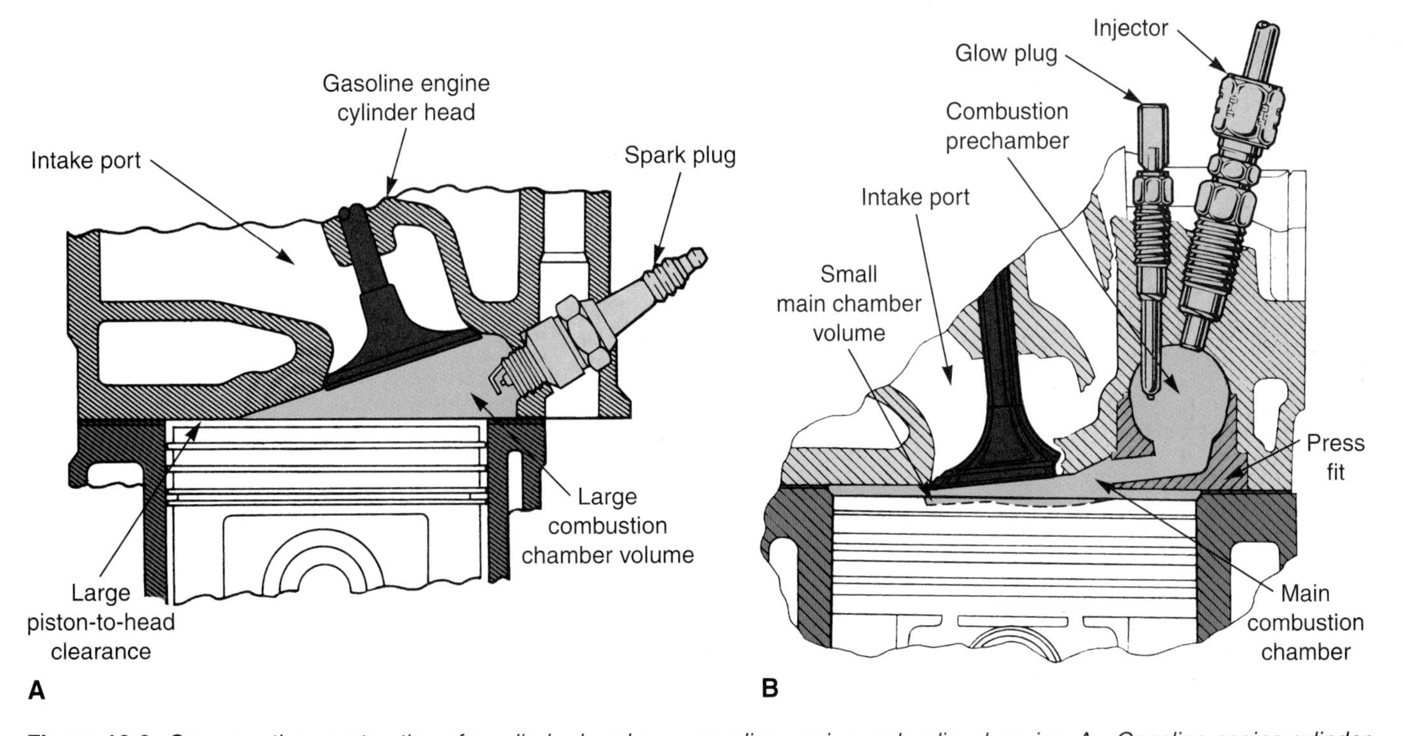

Figure 10-9. *Compare the construction of a cylinder head on a gasoline engine and a diesel engine. A—Gasoline engine cylinder head. B—Diesel engine cylinder head. (Ford, General Motors)*

Valve Guides

Valve guides are small holes machined through the top of the head into the intake and exhaust ports, **Figure 10-11.** The engine valve stems slide up and down in these guides. The two basic types of valve guides are integral and pressed-in. Both are commonly used in modern passenger car engines.

Integral Valve Guides

Integral valve guides are made as part of the cylinder head casting, **Figure 10-12A.** The guide is simply drilled and reamed in the head itself. Integral valve guides are inexpensive to produce and very common in cast iron cylinder heads. They are not used in aluminum heads because the soft aluminum would wear too quickly.

Pressed-In Valve Guides

Pressed-in valve guides or ***valve guide inserts*** are sleeves made of iron, steel, or bronze. They are force-fitted into the cylinder head, **Figure 10-12B.** A hole, slightly smaller than the outside diameter of the guide, is machined in the head. Then, a press is used to drive the guide into the smaller hole, locking the guide in place.

Pressed-in guides can be found in some cast iron and all aluminum cylinder heads. When in cast iron heads, the pressed-in guide may have been installed by a machine shop to recondition worn guides. This is discussed in later chapters.

Knurled Valve Guides

Knurled valve guides have spiral grooves pressed into the inside diameter of the guide. Knurling is a machine shop process that decreases the inside diameter of the guide. It is often used to restore slightly worn guides to specifications. This process is discussed in later chapters.

Figure 10-11. *This is a close-up of the top of the cylinder head showing the valve guides.*

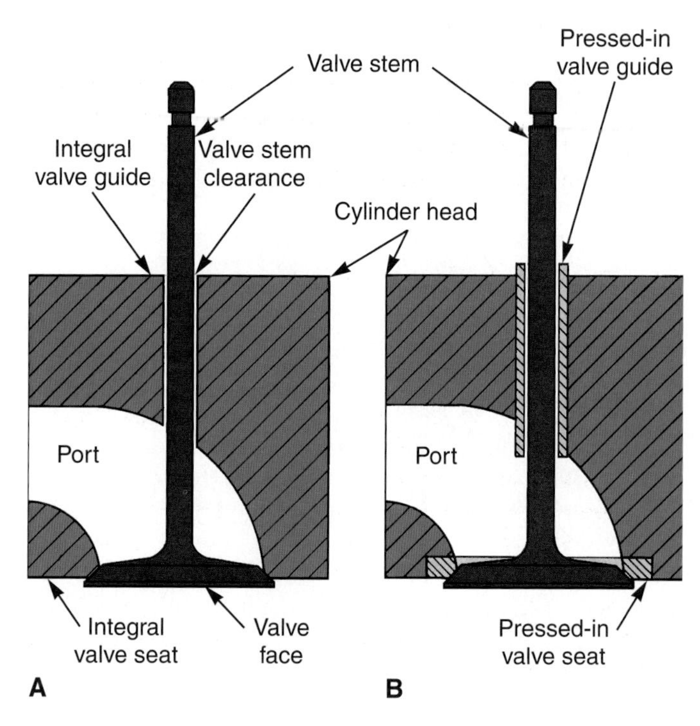

Figure 10-12. *A—Integral valve guides and seats. B—Pressed-in valve guides and seats.*

Valve Seats

Valve seats are round, machined surfaces in the port openings to the combustion chambers. When the engine valve closes, the valve touches the seat to close off the port and seal the combustion chamber. Refer to **Figure 10-13.** As with valve guides, valve seats can be part of the head (integral) or a separate, pressed-in components. Both types can be found in today's engines.

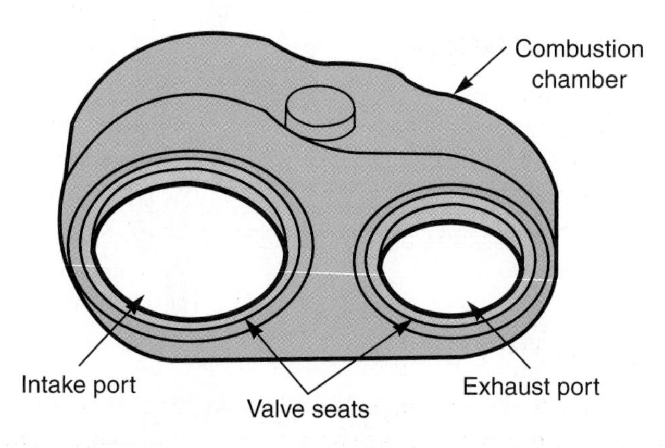

Figure 10-13. *Valve seats are located at the openings of ports into combustion chamber. They can be integral or pressed-in. (Ford)*

Integral Valve Seat

An ***integral valve seat*** is made from the head casting. Cutters are used to machine a precise face on the port opening. The seat is centered on the valve guide so the valve aligns with the seat. Look at **Figure 10-12B.**

Pressed-In Valve Seat

A ***pressed-in valve seat*** or ***seat insert*** is a separate, machined part force-fit into the cylinder head, **Figure 10-12A.** A recess is cut into the combustion chamber that is slightly smaller than the outside diameter of the insert. Then, a press is used to drive the insert into the head. Friction keeps the seat from falling out.

Steel valve seat inserts are used primarily in aluminum cylinder heads. When inserts are used in cast iron heads, they have normally been installed to repair worn or damaged seats. A seat insert is not commonly used in cast iron heads because it does not dissipate heat as quickly as an integral seat. As a result, the valves run slightly hotter.

Induction-Hardened Valve Seats

Induction-hardened valve seats are commonly used on late-model engines to increase service life. Induction hardening is an electric-heating operation that makes the surface of the metal much harder and more resistant to wear. This is shown in **Figure 10-14A.**

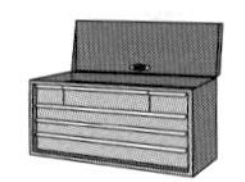

Note: In the past, gasoline contained tetraethyl lead to increase octane. The lead also acted as a high-temperature lubricant between the valves and valve seats. However, leaded gasoline is not available today. Also, engine operating temperatures are higher than in the past. Hardened valve faces and seats are needed to withstand these severe conditions, especially on the exhaust side.

Valve Seat Angles

The ***valve seat angle*** is the angle formed by the face or contact surface of the seat. Look at **Figure 10-14.** A 45° seat angle is commonly used on passenger car engines on both the intake and exhaust seats. However, some high-performance engines have a valve seat angle of 30° on the intake seats.

Interference Angle

An ***interference angle*** is a 1/2° to 1° angle difference between the valve seat face and the face of the valve. This is illustrated in **Figure 10-14B.** The interference angle reduces the amount of contact area between the seat and valve. This increases the pressure between the two faces and speeds valve seating (sealing) during engine break-in.

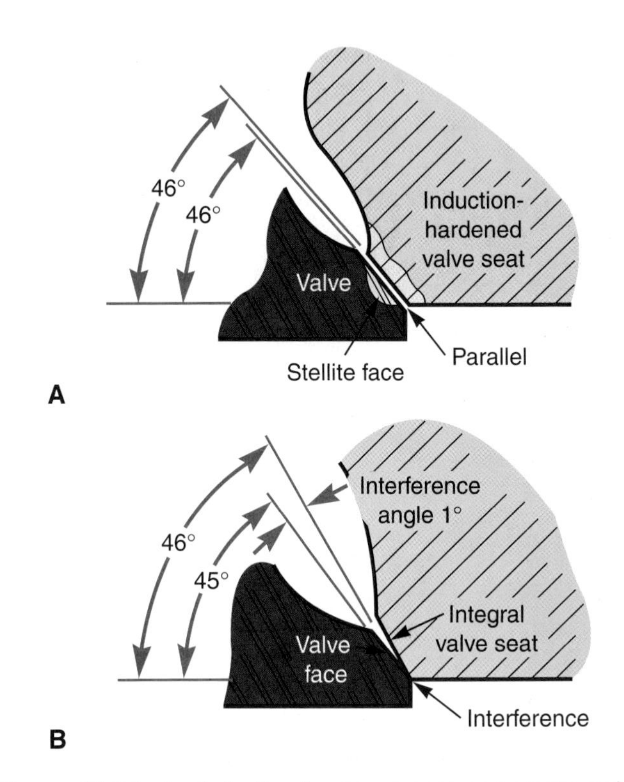

Figure 10-14. *A—Induction hardened seat and stellite valve face coating are used to increase service life. B—The interference angle is typically a 1/2° to 1° difference between the valve face angle and valve seat angle.*

Valve Construction

As you learned earlier, engine valves open and close to control flow into and out of the combustion chamber. The ***intake valve*** controls the flow of air-fuel mixture or just air through the cylinder head intake port. The ***exhaust valve*** controls the flow of burned gasses out of the combustion chamber. It is located over the cylinder head exhaust port. The exhaust valve is smaller than the intake valve. However, on a diesel engine, the intake and exhaust valves are much closer to the same size than on a gasoline engine.

Many engines have two valves per cylinder—one intake and one exhaust. However, today's high-performance engines often have four valves per cylinder, **Figure 10-15.** A few engines have three valves per cylinder—two intake and one exhaust, **Figure 10-16.**

Automotive engines commonly use ***poppet*** or ***mushroom valves,*** **Figure 10-17.** These terms are derived from the valve's shape and action. The valve looks like a mushroom and pops open.

Parts of Engine Valve

Almost every surface of a valve is machined. The stem must accurately fit the valve guide. The face must

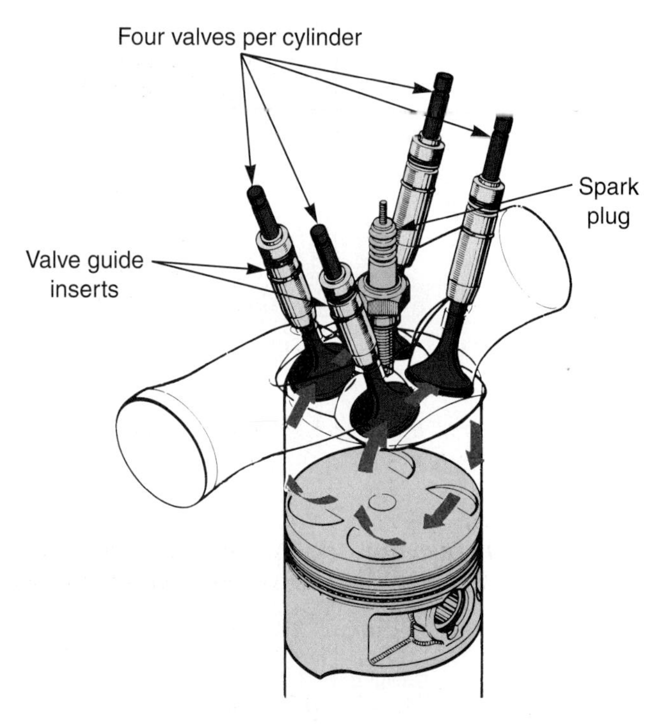

Figure 10-15. *This engine has four valves per cylinder. Note the valve guide inserts around the valve stems.*

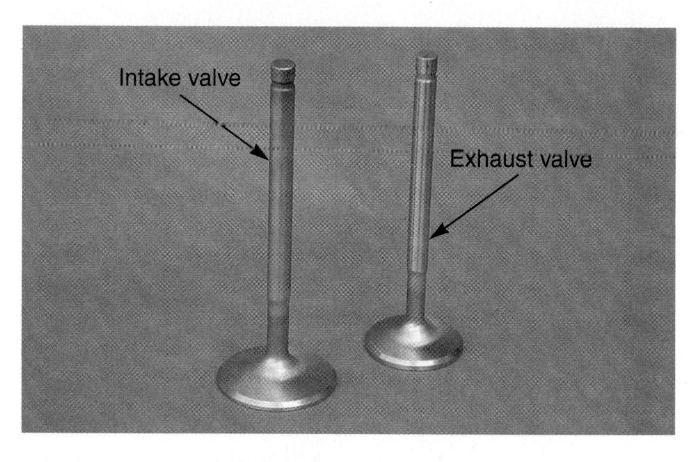

Figure 10-17. *Typical poppet or mushroom valves. The intake valve is normally larger than the exhaust valve.*

contact the seat perfectly. The margin must be thick enough to prevent valve burning. Grooves are accurately cut into the valve stem for the keepers. The basic parts of an engine valve include the following. Refer to **Figure 10-18.**

- ❑ Valve head.
- ❑ Valve face.
- ❑ Valve margin.
- ❑ Valve stem.
- ❑ Keeper or lock grooves.
- ❑ Stem tip.

The ***valve head*** is the large, disk-shaped surface exposed to the combustion chamber. The outside diameter of the valve head is the valve size.

The ***valve face*** is an angled, machined surface on the back of the valve head. This surface is what touches the valve seat in the cylinder head to form a seal.

The ***valve margin*** is the flat surface on the outer edge of the valve head. It is located between the valve head and face. The margin is needed to allow the valve to withstand the high temperatures of combustion. Without a margin, the valve head would melt and burn.

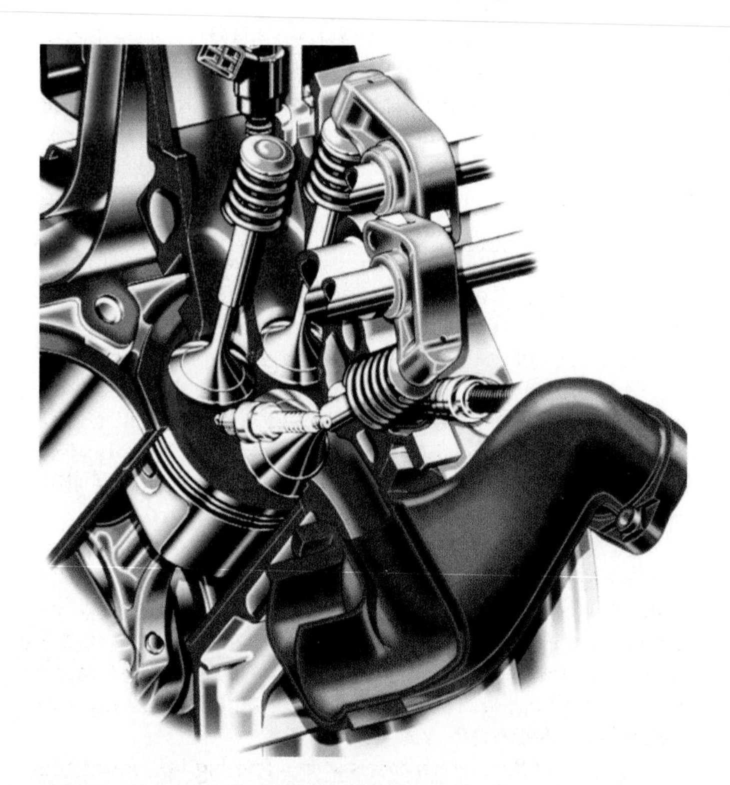

Figure 10-16. *This engine has three valves per cylinder—two intake and one exhaust. (Mercedes-Benz)*

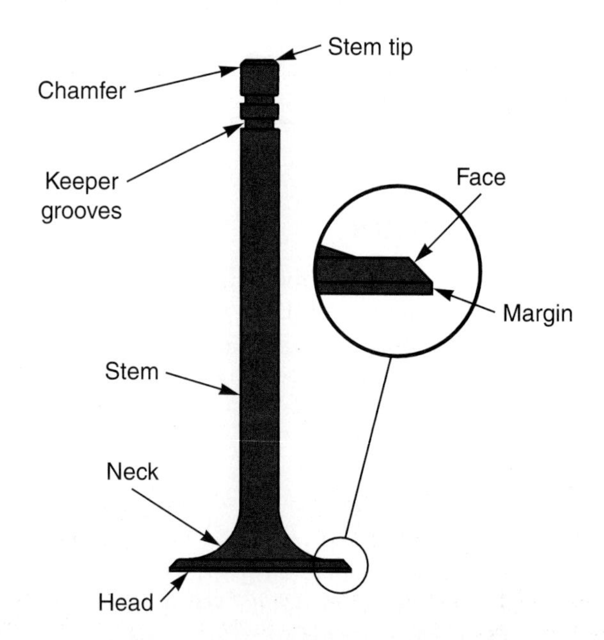

Figure 10-18. *Note the parts of a typical poppet valve. The valve size is determined by the diameter of the valve head.*

The ***valve stem*** is a long shaft extending out of the valve head. The stem is machined, polished, and sometimes chrome plated. It fits into the valve guide machined in the cylinder head.

Keeper grooves or ***lock grooves*** are machined into the top of the valve stem. They accept small keepers or locks that hold the spring retainer on the valve.

The ***stem tip*** is the upper end of the valve stem. It is ground perpendicular to the stem. A small chamfer is formed on the top edge of the stem tip. The tip may be coated with a hard alloy to resist wear from the rocker arm.

Valve Face Angle

The ***valve face angle*** is the angle formed between the valve face and valve head. Refer to **Figure 10-14.** Normally, the valve face angle is 45°, although some valves have a 30° angle.

Valve Stem-to-Guide Clearance

The ***valve stem-to-guide clearance*** is the small space between the valve stem and the valve guide. This clearance is what allows the valve to freely slide up and down in the valve guide. A small amount of engine oil can also enter this clearance to lubricate the valve stem.

Valve Shapes

Valve shape refers to the configuration of the valve head, **Figure 10-19.** The head of the valve can be recessed, flat, or oval. The shape of the valve head partially controls the flexibility of the valve. An engine that operates at high engine speeds needs a flexible valve. This is so that the valve does not bounce off of its seat when closing and the valve can conform to any irregularities in the seat.

Recessed- and tulip-headed valves are very flexible. These shapes also result in valves that are lighter than oval- or flat-headed valves because metal has been removed from the head. They are ideal for high-performance engines.

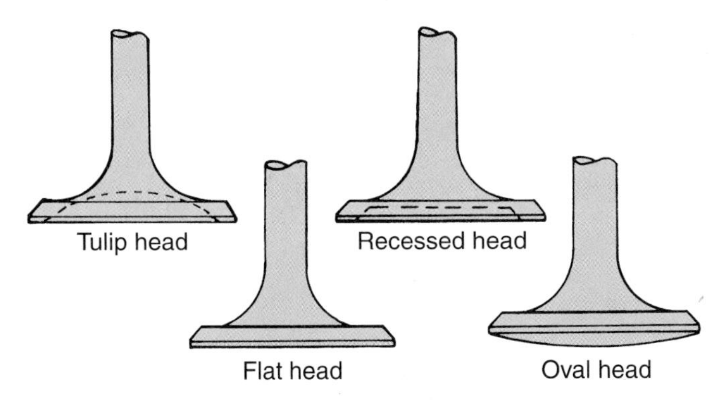

Figure 10-19. *Various valve shapes can be used. Tulip- or recessed-head valves are generally for high-performance engines. Flat- or oval-head valves are found in engines that operate at lower engine speeds. (TRW)*

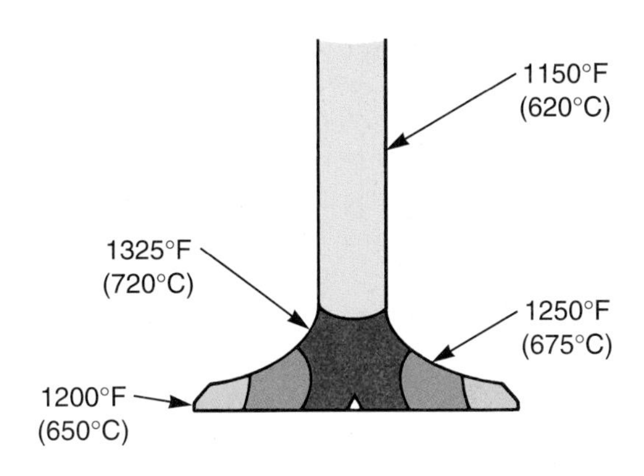

Figure 10-20. *Note the operating temperatures for various surfaces of an exhaust valve.*

Valve Operating Temperatures

Figure 10-20 illustrates the typical operating temperatures of an exhaust valve. The exhaust valve is exposed to higher temperatures than the intake valve. The intake valve has cooler outside air flowing over it. Hot combustion gasses blow over the exhaust valve. For the exhaust valve not to burn, it must transfer heat into the cylinder head.

Valve Material

Valves are usually made of steel or a steel alloy. However, there are valves made from other materials, that have coatings, or are filled with heat-absorbing materials. The next sections discuss some of the more common variations.

Stellite Valve

A ***stellite valve*** has a special, hard-metal coating on its face. A stellite coating is often used in engines designed to burn unleaded fuel. Refer back to **Figure 10-14.**

Sodium-Filled Valves

A ***sodium-filled valve*** is used when extra cooling and reduced weight are needed. The valve stem and valve head are hollow. Heat-absorbing sodium is placed in the cavity, **Figure 10-21.**

During engine operation, the sodium inside the valve melts. When the valve opens, the sodium splashes down into the valve head and collects heat. When the valve closes, the sodium splashes up into the valve stem. Heat is then transferred out of the sodium into the valve stem, to the valve guide, and to the engine coolant. In this way, the valve is cooled.

Sodium-filled valves are used in a few high-performance engines. They are very light and allow high engine speeds for prolonged periods without overheating the valve.

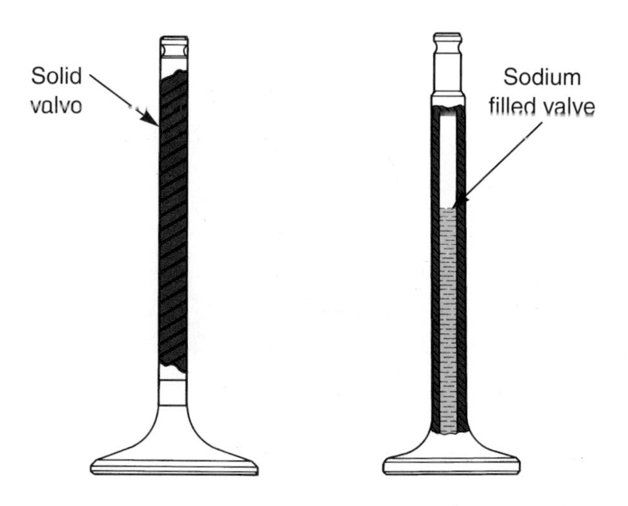

Figure 10-21. *A sodium-filled valve is hollow. It is lighter and runs cooler than solid stem valve. Sodium can cause severe burns; do not break open the valve. (Mercedes-Benz)*

Warning: Sodium is very reactive and can cause serious burns. Never break open a sodium-filled valve. If sodium is dropped into water, a reaction will occur that generates hydrogen gas and enough heat to ignite the hydrogen. In other words, an explosion will occur. When removing a damaged sodium-filled valve from an engine, use extreme caution. Wear gloves and eye protection!

Iconel Valves

Iconel valves are made of a hard, nonmagnetic, stainless steel alloy. Iconel exhaust valves are sometimes used so the valve can withstand extreme temperatures and friction. They are very expensive to manufacture and purchase.

Titanium Valves

Titanium valves are used on a few exotic engines to increase the engine speed capabilities. Titanium is nearly as light as aluminum, but almost as strong as steel. By decreasing the weight of the valve, the engine can rev to a higher speed without valve float. ***Valve float*** is when the valve spring is not able to fully close the valve before the camshaft begins to open the valve. The reduced weight of a titanium valve also decreases friction in the valve train because valve springs with lower tension can be used. Most titanium valves are manufactured by aftermarket companies and not available on production vehicles.

Valve Seal Construction

Valve seals prevent oil from entering the cylinder head ports through the clearance between the valve stems and valve guides. See **Figure 10-22.** The valve seals fit over the valve stems. Without valve seals, oil can be drawn into the engine cylinders and burned. Oil consumption and engine smoking could result. Valve seals come in two basic types—umbrella and O-ring. Both are commonly used on modern engines.

Umbrella Valve Seals

An ***umbrella valve seal*** is shaped like an inverted cup. **Figure 10-23A.** It can be made of neoprene, rubber, or plastic.

An umbrella valve seal slides down over the valve stem before the spring and retainer are installed. It covers the small clearance between the stem and guide. This keeps excess oil from splashing into the guide.

A positive lock umbrella oil seal snaps down over a groove cut into the top of the valve guide. A metal clip formed into the seal keeps the seal held down over the guide. See **Figure 10-24.**

A ***valve oil shedder*** is a variation of an umbrella type oil seal. It is made of hard plastic or nylon, **Figure 10-25.** The shedder simply keeps excess oil from splashing on the valve stem and flowing down through the guide.

O-Ring Valve Seals

An ***O-ring valve seal*** is a small, round seal that fits into an extra groove cut in the valve stem, **Figure 10-23B.** It seals the gap between the retainer and valve stem, not the guide and stem. This stops oil from flowing through the retainer, down the stem, and into the guide.

An O-ring valve seal fits onto the valve stem after the spring and retainer are installed. It is made of soft, synthetic

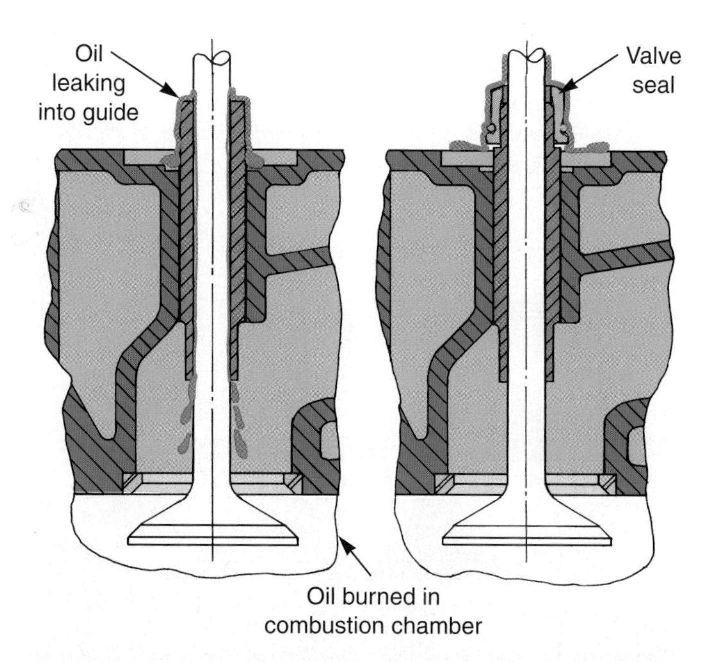

Figure 10-22. *Valve seals keep excess oil from entering the clearance between the valve guide and valve stem. (American Hammered Piston Rings)*

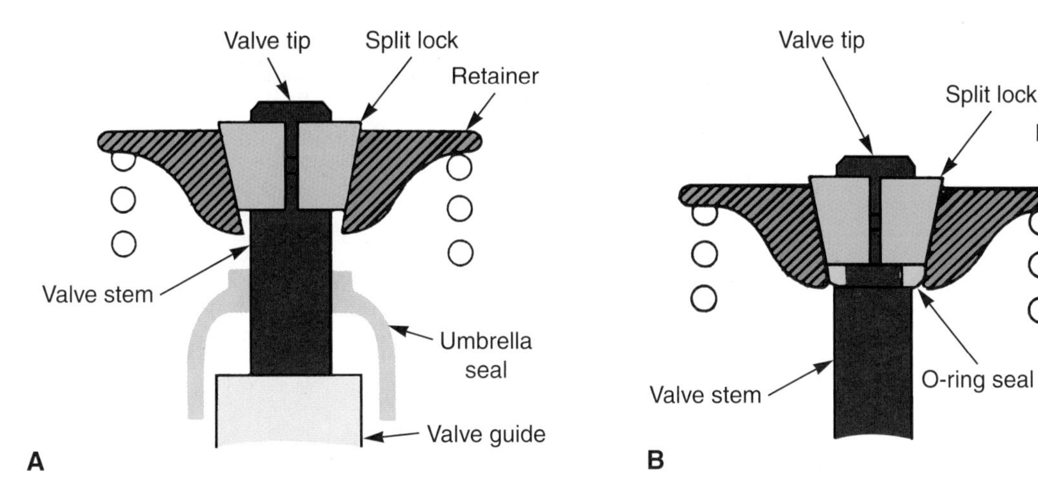

Figure 10-23. *A—An umbrella valve seal surrounds the opening of the valve guide. B—An O-ring valve seal keeps oil from flowing down through the retainer and into the valve guide. (DaimlerChrysler)*

rubber, which allows it to be stretched over the valve stem and into its groove.

Valve Stem Caps

Valve stem caps can be used on the tips of the valve stems, **Figure 10-26.** The cap is simply a hardened steel cup that installs over the valve stem. It is free to rotate on the tip to reduce friction and wear.

Valve stem caps are normally used to reduce rocker arm and stem tip wear. Sometimes, they are used to provide a means of adjusting valve-to-rocker arm clearance. The caps may be available in different thicknesses or can be ground to provide for adjustment.

Valve Spring Assembly Construction

The valve spring assembly is used to close the valve. It basically consists of a valve spring, retainer, and two keepers. Look at **Figure 10-27.**

Valve Spring Construction

Valve spring construction is basically the same for all engines. Spring steel is wound into a coil. However, the number and types of coils can vary. Extra coils can be used to increase the amount of pressure holding the valve closed. Multiple coils are needed for engines that operate at high engine speeds. Single coils are adequate for engines that operate at low engine speeds.

Valve Spring Specifications

Various specifications are given for valve springs. As you will learn in later chapters, these spring specifications are important. They affect valve action. For example, low spring tension can cause valve float and reduce engine performance. Specifications include:

- Tension.
- Free length.
- Open length.
- Installed height.

Valve spring tension refers to the stiffness of a valve spring. Spring tension is usually stated for both opened and

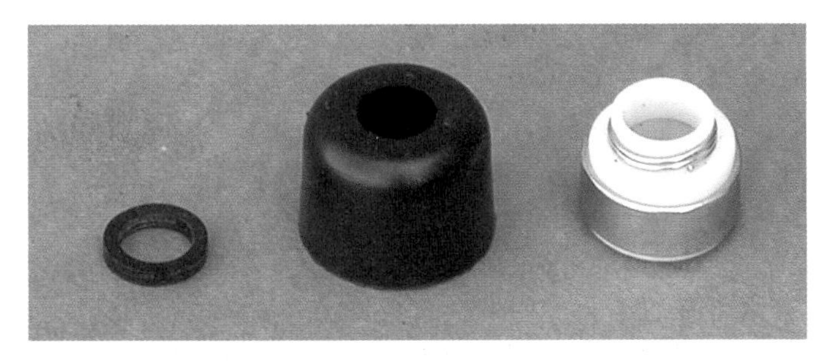

Figure 10-24. *These are three types of valve stem seals. From left to right, O-ring, umbrella, and positive lock umbrella.*

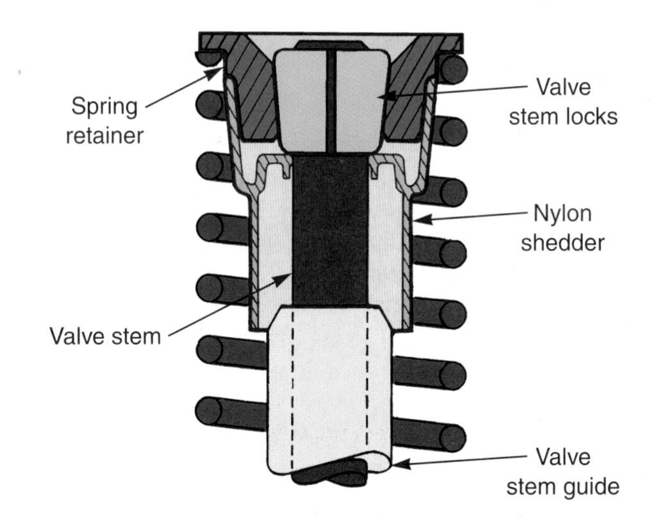

Figure 10-25. *A valve oil shedder is a large, nylon cup that keeps oil out of the valve guide. (General Motors)*

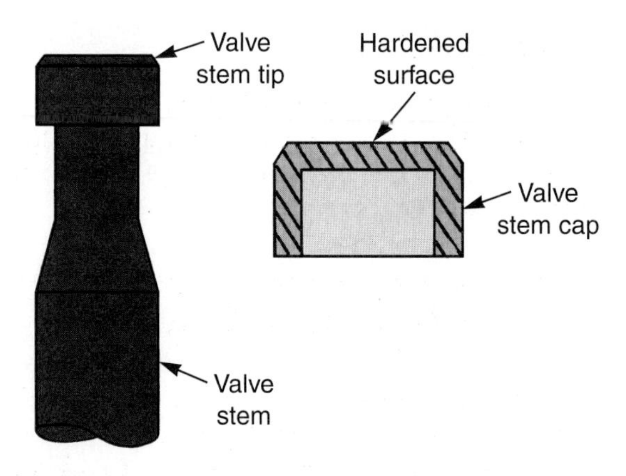

Figure 10-26. *A valve stem cap fits over the valve stem tip. It is used to reduce friction and, in some designs, provides for valve clearance (lash) adjustment. (Deere & Co.)*

closed valve positions. The tension specification is given in pounds or kilograms for specific compressed lengths.

Valve spring free length is the length (or height) of the valve spring when removed from the engine. ***Valve spring open length*** is the length (or height) of the valve spring when installed on the engine with that spring's valve fully open. It is measured from the bottom of the spring to the bottom of the spring retainer.

Valve spring installed height or ***closed length*** is the length (or height) of the valve spring when installed on the engine with the valve closed. It is measured from the bottom of the spring to the bottom of the spring retainer.

Valve Spring Shims

Valve spring shims are precise-thickness flat washers that fit under the valve springs to reduce the installed height of the valve spring. They can be used to increase or restore valve spring tension. When a shim is placed under a valve spring, it reduces the open and closed length (height) of the spring. This compresses the spring more to increase its closing pressure. See **Figure 10-28.**

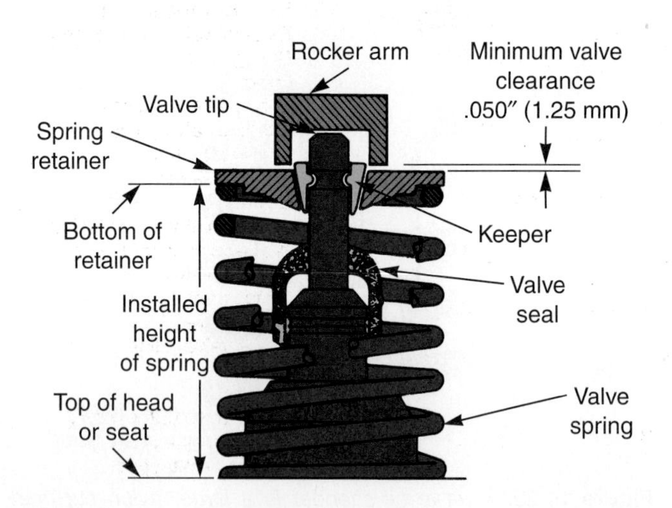

Figure 10-27. *A typical valve spring assembly. (DaimlerChrysler)*

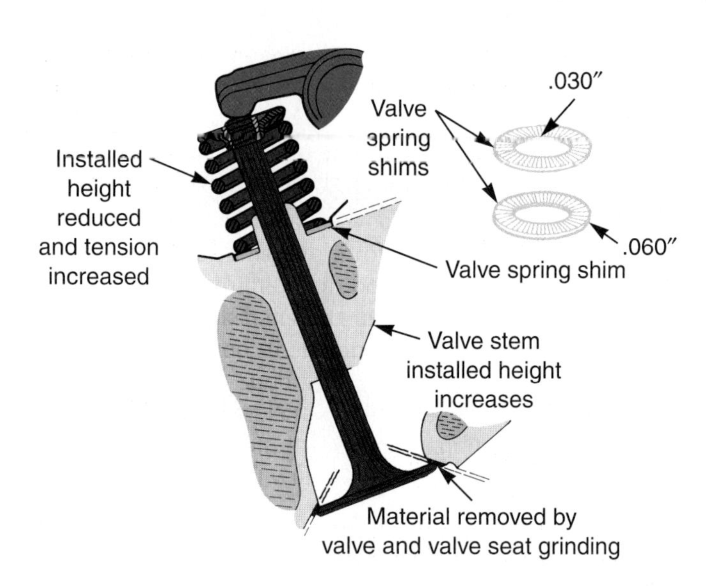

Figure 10-28. *Valve spring shims provide a means of restoring the correct spring tension. (Ford)*

A used valve spring may weaken and lose some of its tension. Also, valve and seat grinding during head reconditioning increase the spring installed height. Valve spring shims provide a means of restoring full spring tension and pressure without spring replacement.

Note: Selection and installation of valve spring shims are covered under cylinder head service.

Valve Rotators

Valve rotators are used on some engines to spin or turn the valves in their guides. Look at **Figure 10-29.** The

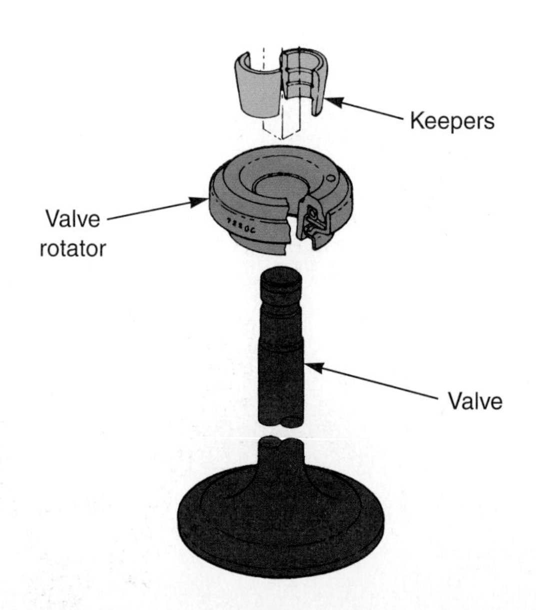

Figure 10-29. *Valve rotators spin the valve in the valve guide to reduce localized hot spots and carbon buildup. Rotators can fit under or over the valve spring. (Sealed Power Corporation)*

turning action helps prevent a carbon buildup on the valve. It also prevents a localized hot spot or wear pattern on the valve or seat.

A seat-type valve rotator is located under the valve spring. A retainer-type valve rotator is located on top of the valve spring. Rotators are commonly used on engine exhaust valves. The valve face on exhaust valves is exposed to more heat than the valve face on intake valves.

Valve Spring Retainer Construction

Valve spring retainers fit over the top of the valve springs, as shown in **Figure 10-27.** They have a machined lip so that they remain centered on the spring. Valve retainers are usually made of steel, but are sometimes made of lightweight alloys.

Valve Keeper Construction

Valve keepers fit between the valve retainers and valve stems to hold the springs on the valves. See **Figures 10-27** and **10-29.** The keepers are hardened steel to resist the pounding action of the valve's reciprocating motion. A valve keeper is usually two halves; both must be installed on each valve.

Valve Spring Seat

A ***valve spring seat*** is a cup-shaped washer installed between the cylinder head and bottom of the valve spring, **Figure 10-30.** It provides a pocket to hold the bottom of the valve spring squarely on the head and valve stem.

Valve Spring Shield

A ***valve spring shield*** is normally used in conjunction with an O-ring valve stem seal. The thin metal shield surrounds the top and upper sides of the spring and helps keep excess oil off of the valve stem.

Valve Train Construction

The ***valve train*** consists of the parts that control the opening and closing of the valves. **Figure 10-30** shows the valve train for an overhead camshaft engine. **Figure 10-31** illustrates the valve train for an overhead valve engine. Compare their similarities and differences. These are the two major types of valve train. Both are common in today's engines. Although the basic function of these parts is the same, their construction can vary. You must understand each type!

Camshaft Construction

A ***camshaft*** ensures that the engine valves are open or closed at the right time during each stroke, **Figure 10-32.** It can be powered by a chain, belt, or gears powered by the crankshaft. The camshaft turns at one-half of the crankshaft speed because one complete engine cycle is produced every two crankshaft rotations. The camshaft, or "cam," is usually made of cast iron or forged steel. The cam can be located in the block or in the cylinder head. Although most engines have only one camshaft, some have two or four camshafts. Refer to Chapter 6 *Engine Types and Classifications.*

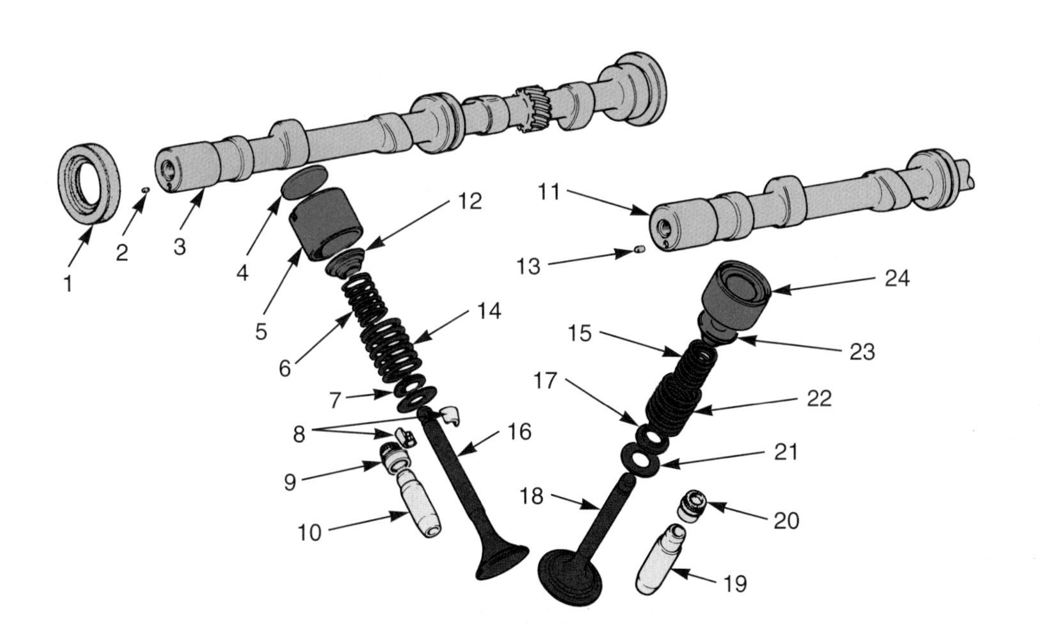

1. Camshaft seal.
2. Exhaust camshaft dowel.
3. Exhaust camshaft.
4. Tappet adjusting disc.
5. Exhaust valve tappet.
6. Exhaust valve inner spring.
7. Spring seat.
8. Locks.
9. Exhaust valve oil seal.
10. Exhaust valve guide.
11. Intake camshaft.
12. Spring retainer.
13. Dowel.
14. Exhaust valve outer spring.
15. Intake valve inner spring.
16. Exhaust valve.
17. Spring seat.
18. Intake valve.
19. Intake valve guide.
20. Oil seal.
21. Washer.
22. Intake valve outer spring.
23. Spring retainer.
24. Intake valve tappet.

Figure 10-30. *These are the basic parts of the valve train for an overhead camshaft engine. (Fiat)*

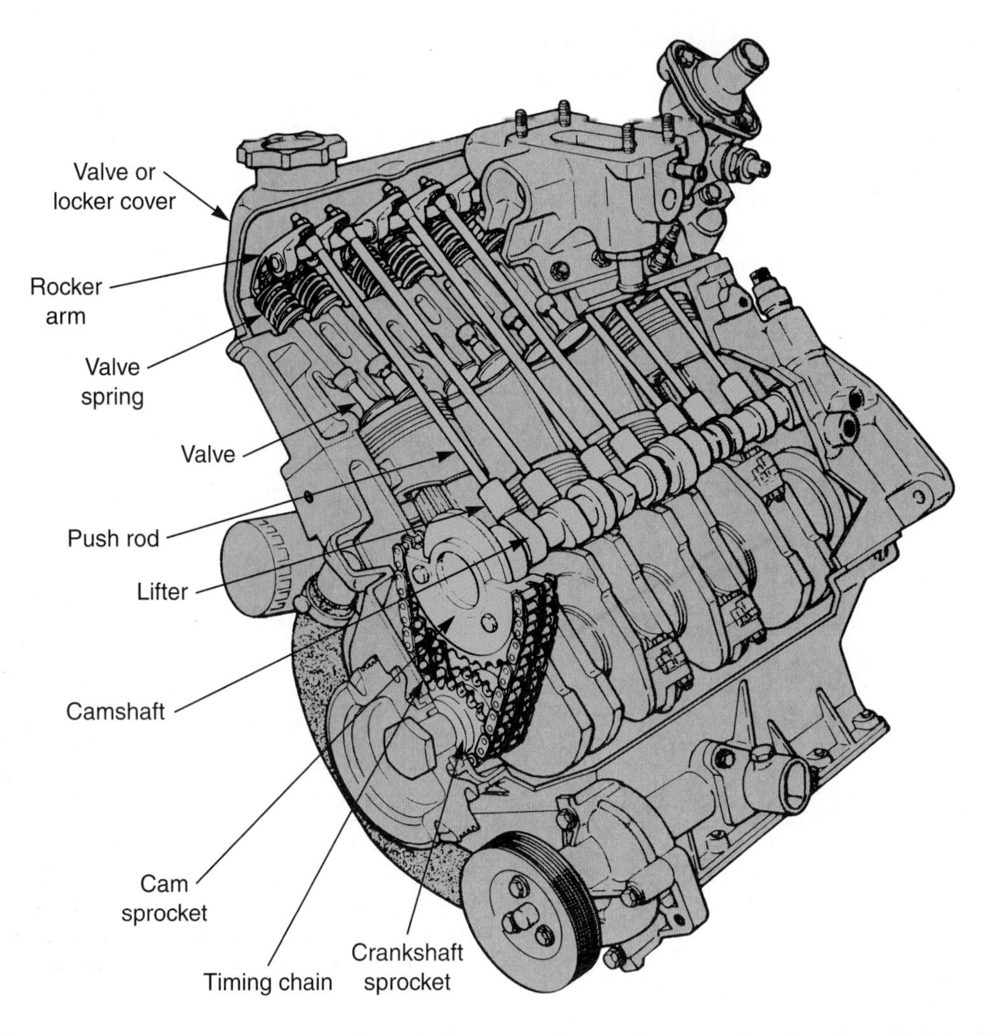

Figure 10-31. *These are the basic parts of the valve train for an overhead valve or push rod engine. (DaimlerChrysler)*

Camshaft Lobe Construction

The ***camshaft lobes*** are egg-shaped protrusions on the camshaft that are used to change the rotary motion of the camshaft into reciprocating motion of the valves. See **Figure 10-33.** They are precision-machined and polished. The lobes are also hardened to prevent rapid wear from operating the lifters. The camshaft-to-lifter contact point is one of the highest-friction points in an engine. As introduced in Chapter 8, the shape (profile) of the cam lobes affects:

- ❑ When each valve opens in relation to piston position.
- ❑ How long each valve stays open.
- ❑ How far each valve opens.

Figure 10-32. *The camshaft is used to open valves at the right time in relation to the engine crankshaft position. It rotates at one-half of the crankshaft speed.*

Figure 10-33. *This close-up of a camshaft shows its lobes. Note their shape.*

Usually, one cam lobe is provided for each engine valve. If the cylinder has two valves, there is an intake lobe and an exhaust lobe for that cylinder. A four-valve cylinder usually has two camshafts and two lobes per cylinder on each camshaft for a total of four lobes per cylinder.

Some camshafts are machined with dual cam lobes of different profiles. One cam lobe is designed for good low-speed efficiency. The other lobe profile provides power at high engine speeds. The ECM operates a solenoid valve that controls oil flow to shift the rocker arms from one lobe profile to the other. This is a form of variable valve timing.

Although the cam lobes are an integral part of most camshafts, a built-up camshaft consists of individual lobes that are mounted on a separate, hollow shaft. A built-up camshaft is light and extremely strong.

Camshaft Lift and Duration

Camshaft lift is the amount of valve train movement produced by the cam lobe. It partially determines how far the valve opens. Shown in **Figure 10-34,** camshaft lift is found by subtracting the base circle diameter from the overall height of the lobe. Lift is given in inches or millimeters. A typical camshaft lift is .450″ (11.5 mm).

Camshaft duration determines how long the valves stay open. The shape of the cam lobe nose and flank determine camshaft duration, **Figure 10-34.** For instance, a pointed cam lobe has a shorter duration compared to a rounded lobe, **Figure 10-35.** Camshaft duration is given in degrees of rotation. The larger the number of degrees, the longer the duration. For example, a camshaft might have a duration of 250°. This means that the valve is at least partially open for 250° of one camshaft rotation.

Valve Timing

Valve timing refers to when the valves open and close in relation to the position of the pistons in the cylinders. Valve timing is controlled by the camshaft and drive sprockets or gears. Timing marks are placed on the timing sprockets or gears. Aligning the marks ensures that valve timing is within specifications.

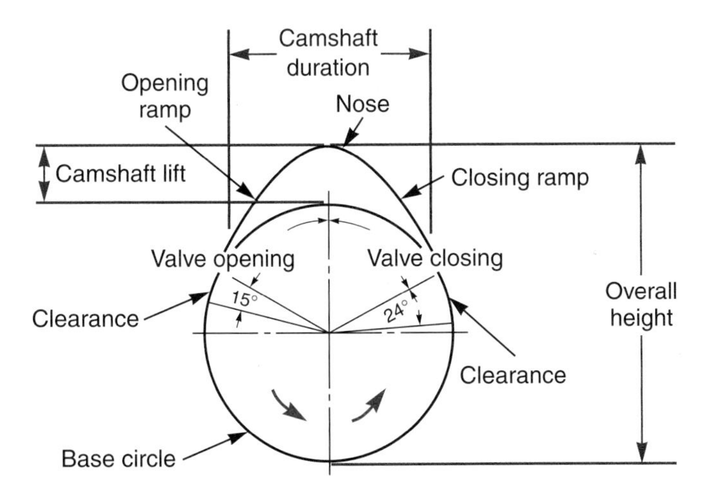

Figure 10-34. *Note the parts of a camshaft lobe. Also, note how the lift is determined. (DaimlerChrysler)*

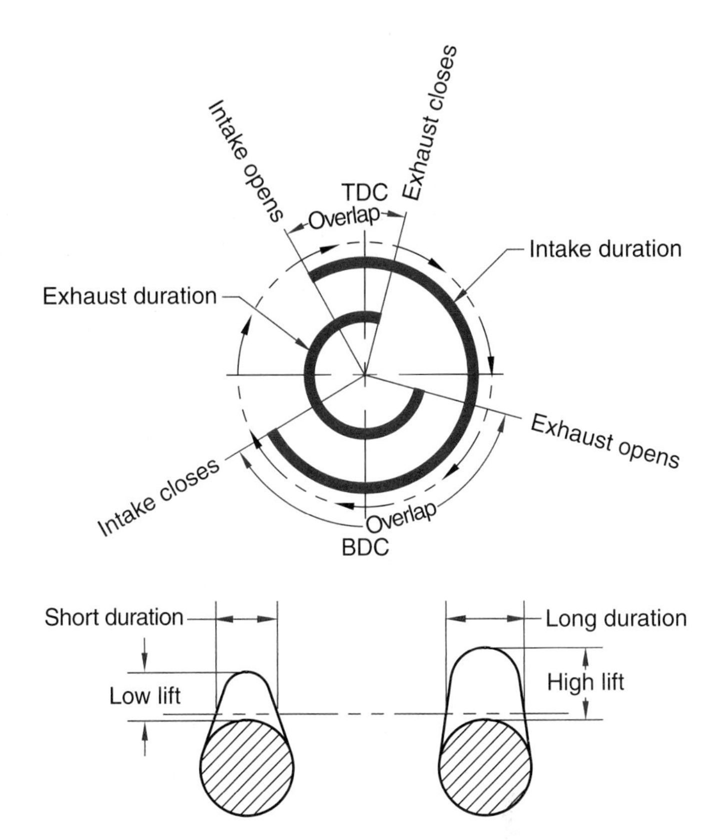

Figure 10-35. *The shape of the camshaft lobe determines when and for how long a valve is open. The diagram on top shows the action for a typical camshaft. Note the valve overlap. The lobe cross sections shown on the bottom demonstrates how the lobe profile affects lift and duration.*

Variable Valve Timing

Variable valve timing involves altering valve timing as engine speed changes. This is done to optimize engine power and efficiency at all operating speeds. Generally, the valves in these systems function with a normal duration at low engine speeds. As speed increases, however, the variable timing mechanism engages to hold the valves open longer, increasing volumetric efficiency. This allows the engine to vary when its valves open and close in relation to the engine operating parameters to improve performance and maintain acceptable emissions.

There are several methods of creating variable valve timing that have been explored. However, there are basically two common methods. One method has two different camshaft profiles for each valve. The second method uses a camshaft gear that can rotate on the camshaft.

In the system that works on two camshaft profiles, one profile is designed from optimal performance at low engine speeds. The second profile is designed for optimal performance at high engine speeds. At low engine speeds, the low speed cam lobe acts on the low speed follower to open the valve. See **Figure 10-36A.** The high speed cam lobe acts on the high speed follower, but the follower "freewheels" and does not act on any other component. However, once the engine reaches a certain speed, the computer opens a valve to allow oil pressure to push a pin through the low speed and high speed followers. Since the followers are now locked together, the high speed cam lobe controls the opening of the valve. See **Figure 10-36B.** An advantage of this variable valve timing design is that lift and duration can be changed along with the valve timing.

The system that uses a rotating camshaft gear cannot alter lift or duration, only timing. However, the system has the advantage of being somewhat less complex. The camshaft gear consists of two halves with a rotor contained inside. See **Figure 10-37.** One-half of the camshaft gear is attached to the rotor and camshaft. The other half is driven by a gear connected to the other camshaft or crankshaft. Oil pressure inside the gear locks the two halves together. By shifting oil pressure from one side of the rotor to the other, the valve timing is advanced or retarded. The computer controls the oil pressure inside the camshaft gear based on engine operating conditions. A variation of this system uses a shaft with helical splines instead of a rotor. As the camshaft gear slides along the splined shaft, the valve timing is adjusted.

Note: Some variable valve timing systems that have a rotating camshaft gear also have an additional system that allows the lift and duration to be altered.

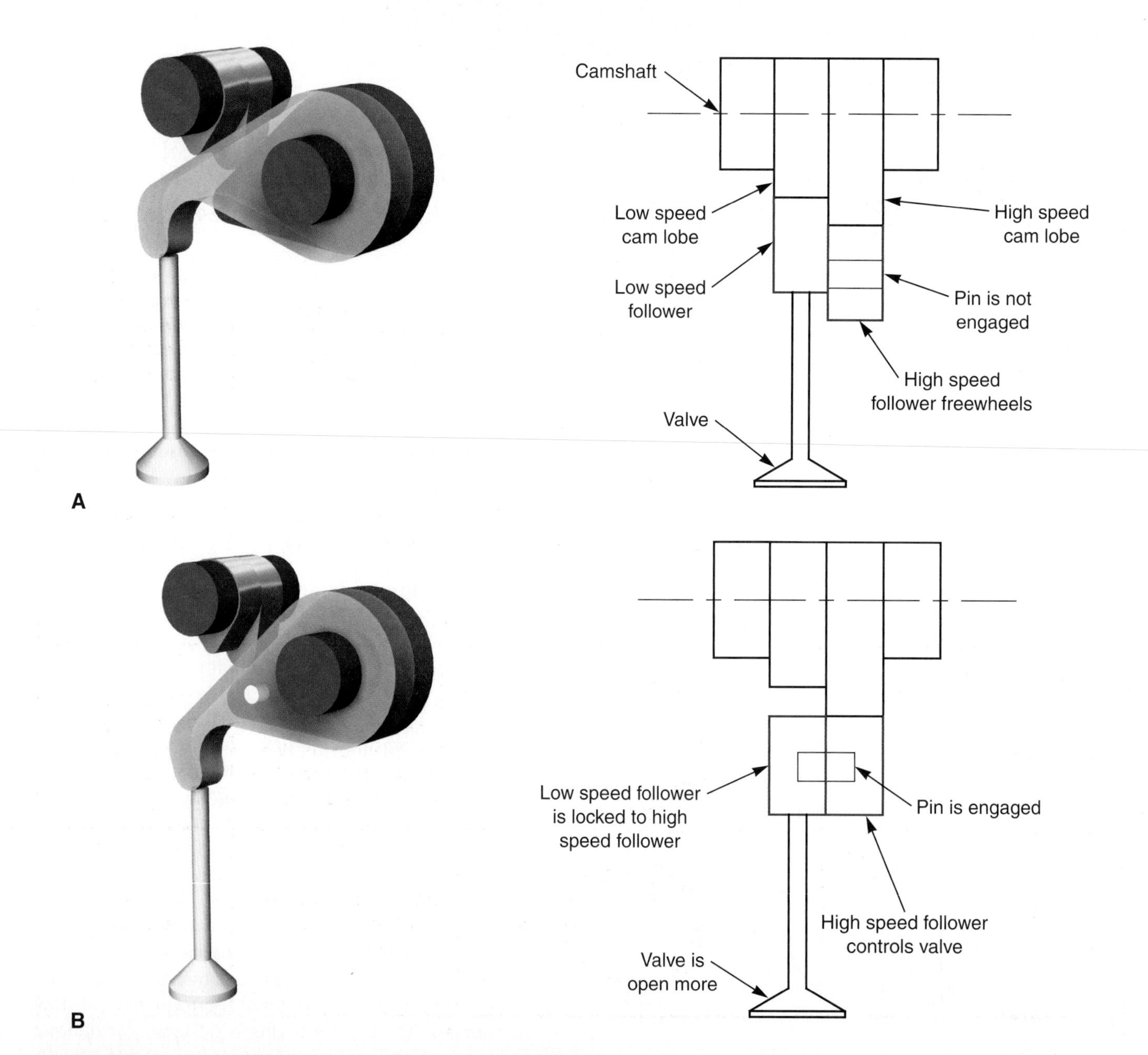

Figure 10-36. *A—At low engine speeds, the low speed cam lobe and follower control the valve action. B—At high engine speeds, the pin locks the low and high speed followers together and the high speed cam lobe controls the valve action.*

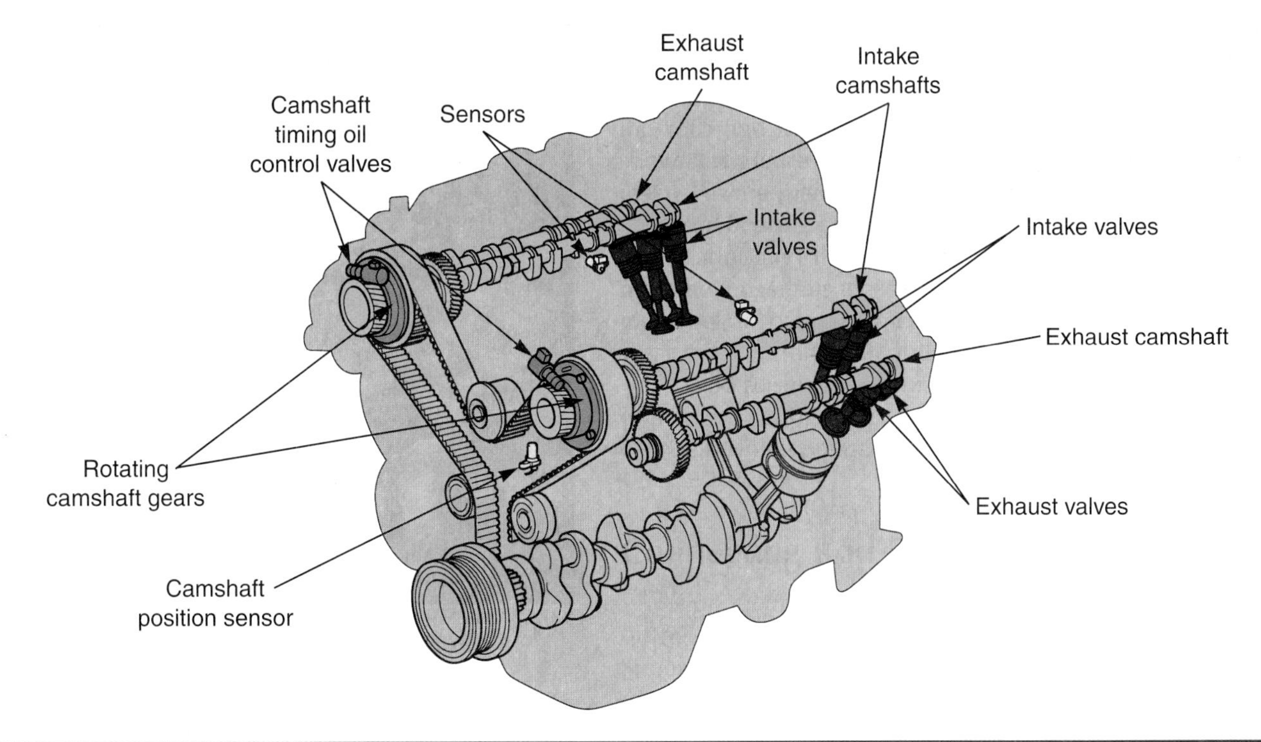

Operation State	Range	Valve Timing	Objective	Effect
In low to medium speed range with heavy load	4	TDC EX IN To advance side BDC	Advancing the intake valve close timing for volumetric efficiency improvement.	Improved torque in low to medium speed range.
In high speed range with heavy load	5	EX IN To retard side	Retarding the intake valve close timing for volumetric efficiency improvement.	Improved output.
At low temperatures	—	EX IN	Eliminating overlap to prevent blow back to the intake side leads to the lean burning condition, and stabilizes the idling speed at fast idling.	Stabilized fast idle rpm. Better fuel economy.
On starting/ stopping engine	—	EX IN	Eliminating overlap to minimize blow back to the intake side.	Improved startability.

B

Figure 10-37. *A—This DOHC engine has camshaft gears that rotate to alter valve timing. Note how the exhaust camshaft is driven by the intake camshaft. B—The operation ranges for a variable valve timing system that uses a rotating camshaft gear. (Lexus)*

Valve Overlap

Valve overlap is when both intake and exhaust valves in a cylinder are opened at the same time, **Figure 10-38.** Valve overlap is used to help scavenge burned gasses out of the cylinder. It also helps pull a larger air-fuel charge into the cylinder. Valve overlap is measured in degrees.

With both the intake and exhaust valves open at the same time, the inertia of the gasses through one cylinder head port and the cylinder act on the gasses in the other port. This results in slightly more flow into and out of the cylinder. Valve overlap helps engine breathing, especially at higher engine speeds.

Stock and High-Performance Camshafts

The term "stock camshaft" refers to the original camshaft design that is installed by the manufacturer. Usually, it has mild lift and duration to provide both good performance and low exhaust emissions.

An aftermarket high-performance or "race" camshaft has more lift and longer duration than a stock camshaft. Generally, more lift helps the engine breath at all engine speeds. Longer duration and more valve overlap generally increase maximum engine speed capabilities.

A high-performance camshaft usually decreases engine power, fuel economy, and exhaust cleanliness at normal engine speeds. If duration is extreme, the engine will idle very rough and produce very low vacuum at idle. Vacuum-dependent accessories, such as power brakes, may not have enough vacuum to properly function. The engine will be very inefficient from idle up to normal highway speeds. In fact, installing a high-performance camshaft without complementary engine modifications, such as exhaust headers or high-compression pistons, may actually reduce usable engine power. High-performance camshafts are designed to operate in a specific range of engine speeds. If the engine does not operate in that range, increased performance may not be achieved.

Note: The EPA does not allow engine modifications that increase exhaust pollution. By installing a camshaft with increased lift and/or duration, the engine will likely produce more pollution. Do not install parts that make a vehicle noncompliant with emissions regulations.

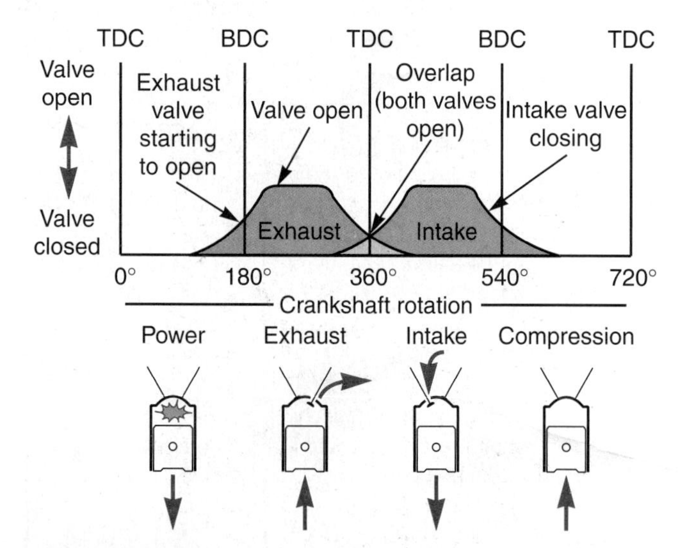

Figure 10-38. *Overlap is the period of time when intake and exhaust valves in a given cylinder are open at the same time. This graph shows valve action through complete cycle. (Volkswagen)*

Hydraulic and Mechanical Camshafts

A ***hydraulic camshaft*** is designed to be used with hydraulic lifters. Its cam lobes are shaped to initially open the valve more quickly. This type of lobe is called an accelerated cam lobe and makes the hydraulic lifters operate properly. Note that the term "hydraulic" refers to the operation of the lifter, not the camshaft.

A ***mechanical camshaft*** is made to work with solid lifters. The lobe is shaped to produce more constant opening of the valve. This type of lobe is called a constant-velocity lobe.

Caution: Do not use a hydraulic cam with solid lifters or a mechanical cam with hydraulic lifters. Doing so will result in increased valve train wear and noise.

Camshaft Thrust Plate

A ***camshaft thrust plate*** is used to limit the front-to-rear movement, or endplay, of the camshaft. The thrust plate is bolted to the front of the block or cylinder head. When the drive gear or sprocket is bolted in place, the thrust plate sets up a predetermined clearance.

Camshaft Bearings

Camshaft bearings are usually one-piece inserts pressed into the block or cylinder head, **Figure 10-39.** On OHC engines, the bearings may be two-piece inserts. The camshaft journals ride in the cam bearings. Cam bearings are usually constructed like engine main and connecting rod bearings, as described in Chapter 9.

For a cam-in-block engine, the camshaft journals normally get smaller toward the rear of the block. As a result, the set of press-in camshaft bearings have bearings of different diameters. Each bearing is marked or labeled and must be press fit into the correct bore. Also, the oil holes in the bearings must be aligned with the oil holes in the block.

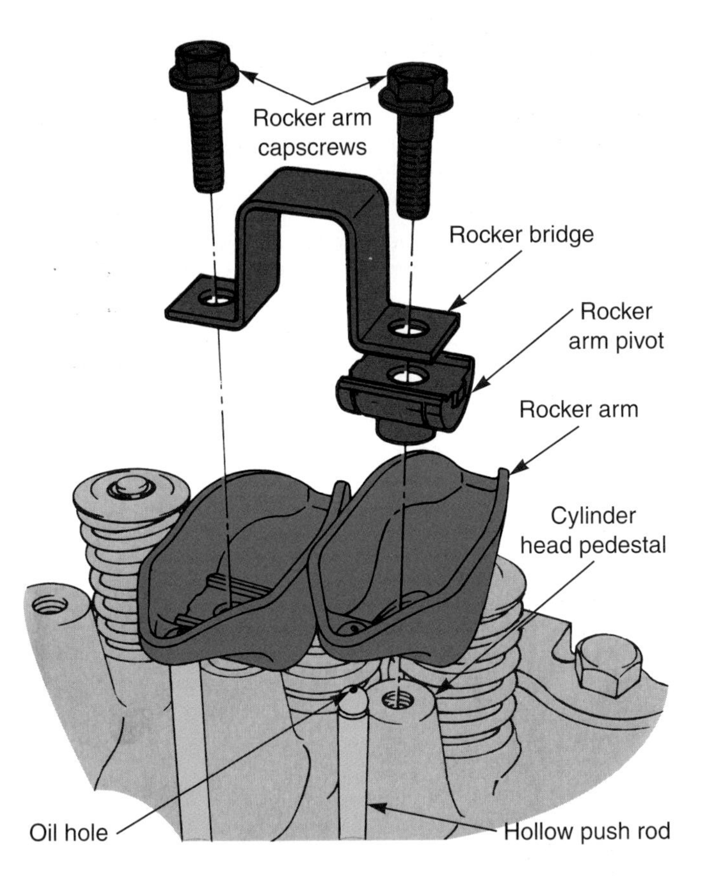

Figure 10-49. *This is a rocker arm pedestal. Note how the parts are assembled. (General Motors)*

the intake and .012" (0.3 mm) for the exhaust. Usually, the exhaust valve lash is slightly more than the intake valve lash. This is because the exhaust valve gets hotter and expands more during operation than the intake valve. Lash specifications will indicate if the engine is to be hot or cold when setting the lash.

Adjustable Rocker Arms

Adjustable rocker arms provide a means of changing valve lash. Either a screw is provided on the rocker arm or the rocker arm pivot point (the stud nut, for instance) can be screwed up or down.

Adjustable rocker arms *must* be used with mechanical lifters. Adjustable rockers are sometimes used with hydraulic lifters for a more accurate initial setting and for adjustment with severe part wear.

Nonadjustable Rocker Arms

Nonadjustable rocker arms provide no means of changing valve lash. They are only used with some hydraulic lifters. The rocker arm assembly is tightened to a specific torque. This presets the lifter plunger halfway in its travel. Then, during engine operation, the hydraulic lifter automatically maintains zero clearance.

Push rod lengths can be changed to adjust the valve lash on engines with nonadjustable rocker arms. Special push rod lengths can be purchased for unusual situations. For example, when the heads are milled, the rockers are actually closer to lifters than they were before. This must be corrected with new push rods.

Rocker Arm Ratio

Rocker arm ratio is a comparison of the length of each end of the rocker arm. It compares the length from the pivot point of the rocker on the valve and push rod sides. Look at **Figure 10-51.**

A typical rocker arm ratio might be 1.5:1. This means that camshaft lobe lift is multiplied by a factor of 1.5 by the action of the rocker arm. The valve opens a distance that is 1-1/2 more than the lift of the camshaft.

Rocker arm ratios vary from engine to engine. A larger ratio opens the valves wider for the same camshaft lift. However, in most cases the ratio cannot be changed without upsetting rocker geometry.

In some four-valve cylinder heads, the rocker arms open one of the intake valves more than the

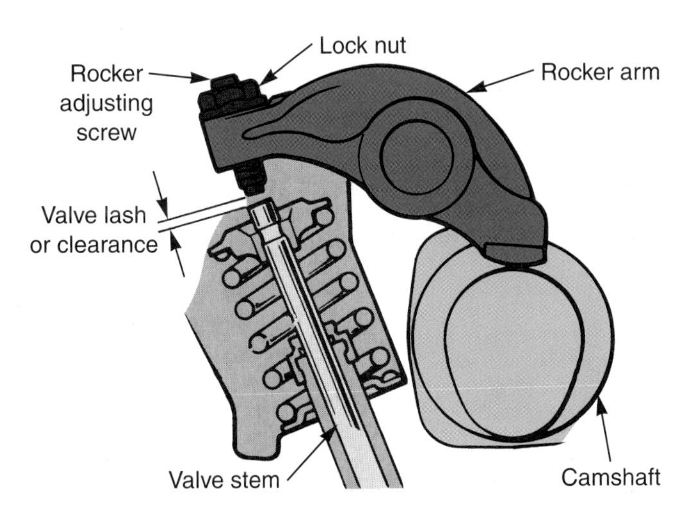

Figure 10-50. *Valve lash, or clearance, is the distance between the rocker arm tip and the valve stem tip. (General Motors)*

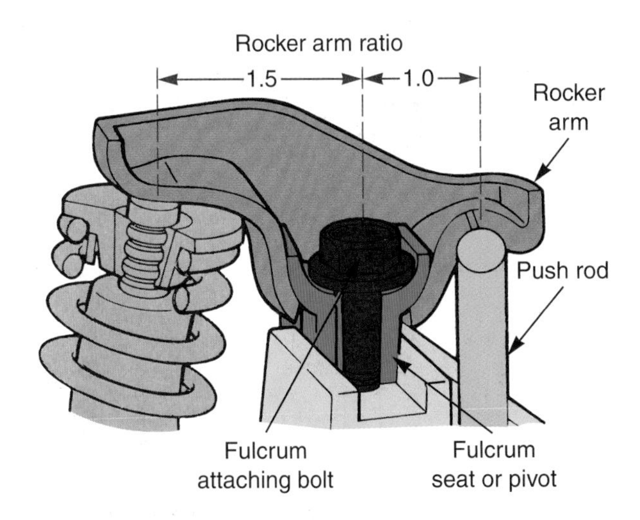

Figure 10-51. *Rocker arm ratio is a comparison of the distance from the center of the rocker pivot to the contact points on the push rod and valve stem tip. (Ford)*

other. This is done to produce a swirling action on the air-fuel mixture entering the combustion chamber. The mixing action helps reduce emissions at low engine speeds.

Valve and Camshaft Covers

Valve covers are sheet metal or plastic covers fastened to the top of the cylinder heads. The valve covers and valve cover gaskets prevent oil spraying from the valve train from leaking out of the engine. The cover is sealed by a gasket or sealant.

The term "valve cover" is used with overhead valve engines where the camshaft is mounted in the cylinder block. A ***camshaft cover*** is a lid over the top of the camshaft housing on overhead camshaft engines. It serves the same purpose as a valve cover on an OHV engine.

Some valve or camshaft covers are acoustically dampened. This means the part is designed to help reduce the transmission of sound. For example, by making the cover out of two sheets of steel formed around a center layer of sound-dampening material, the amount of valve train noise that can be heard is reduced.

Engine Gasket Construction

Engine gaskets prevent pressure, oil, coolant, and air leakage between engine components. They are very important to the performance and dependability of an engine. **Figure 10-52** shows major engine gaskets.

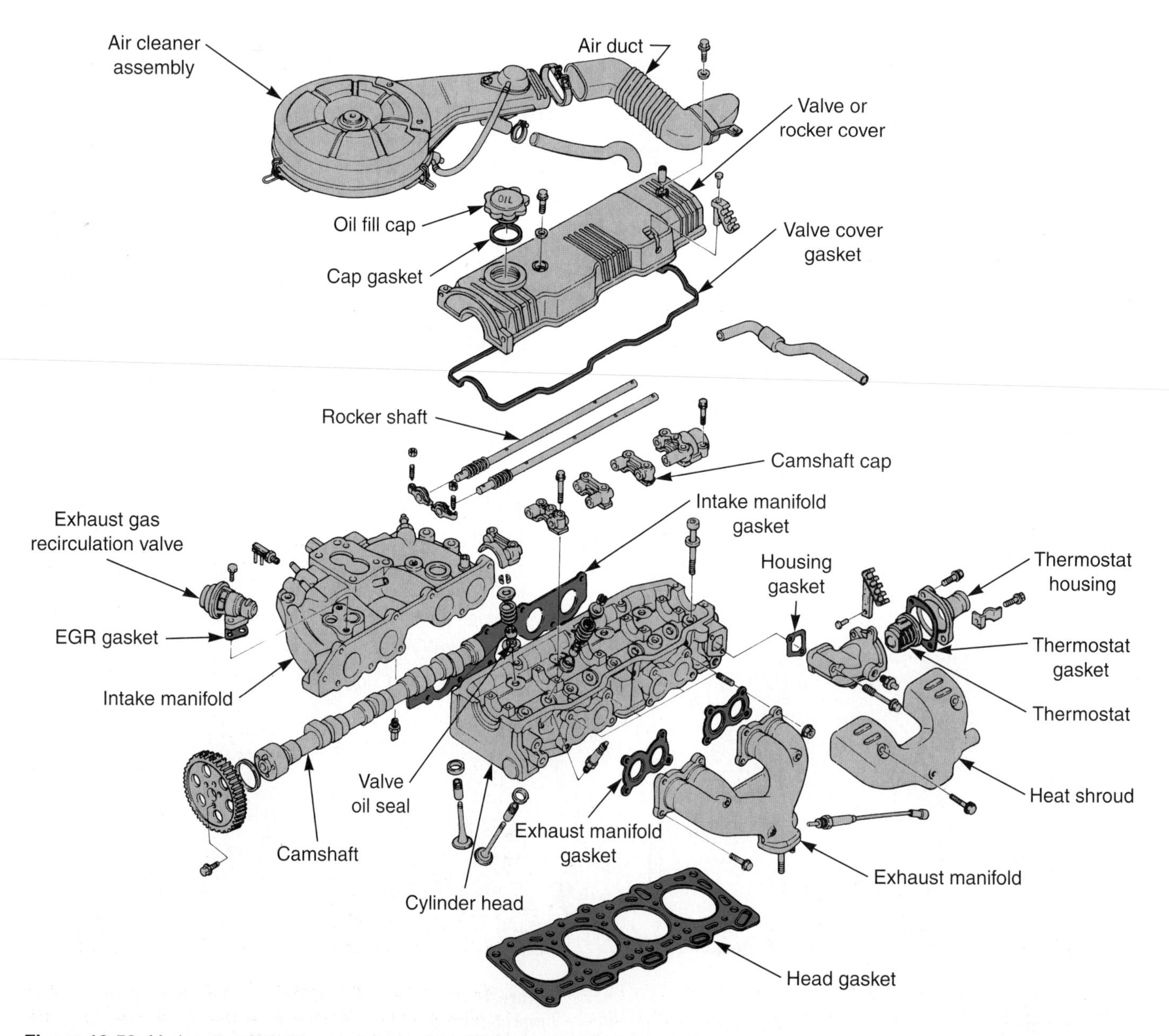

Figure 10-52. *Various gaskets are used on engines. They are usually named after the parts that they seal. Study the gasket names.*

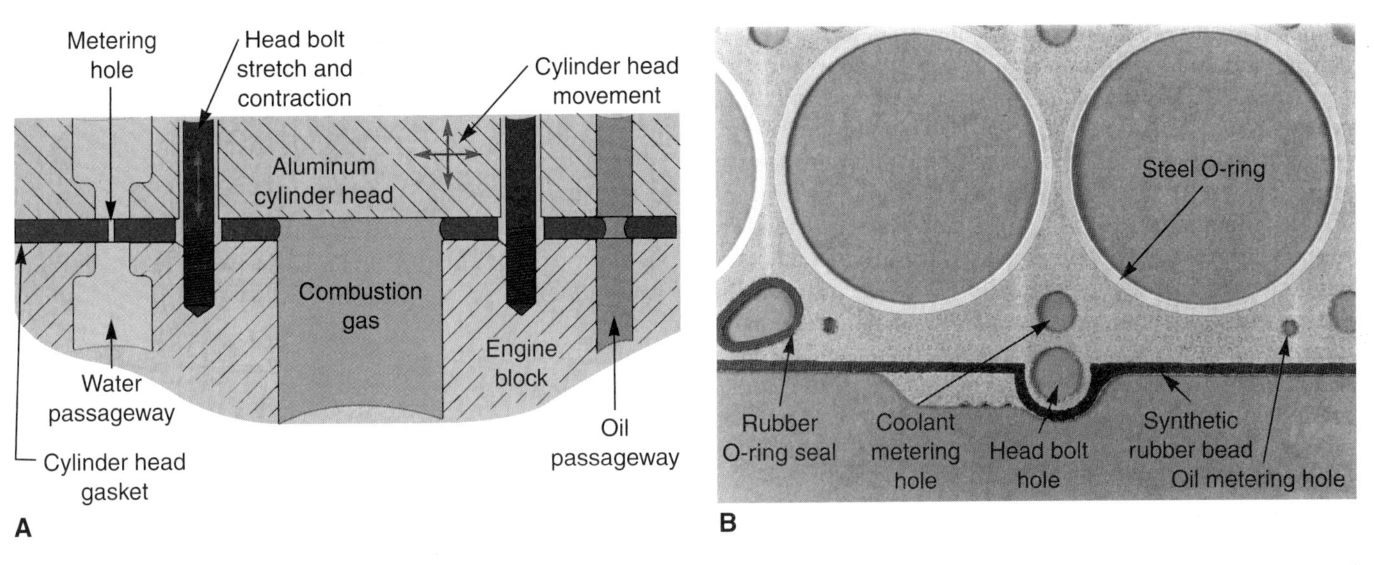

Figure 10-53. *A cylinder head gasket must operate under severe conditions. Combustion heat, pressure, oil, coolant, and head movement all try to break the seal. Modern head gaskets use advanced construction methods to prevent leakage. (Fel-Pro)*

Cylinder Head Gasket Construction

A ***cylinder head gasket*** must seal the pressure of combustion in the cylinder and prevent the flow of oil and coolant out of their respective passages. A head gasket must withstand tremendous pressure, chemical action, expansion and contraction, and severe change in temperature. See **Figure 10-53.**

Modern cylinder head gaskets are made of several materials. See **Figure 10-54.** A metal O-ring is normally

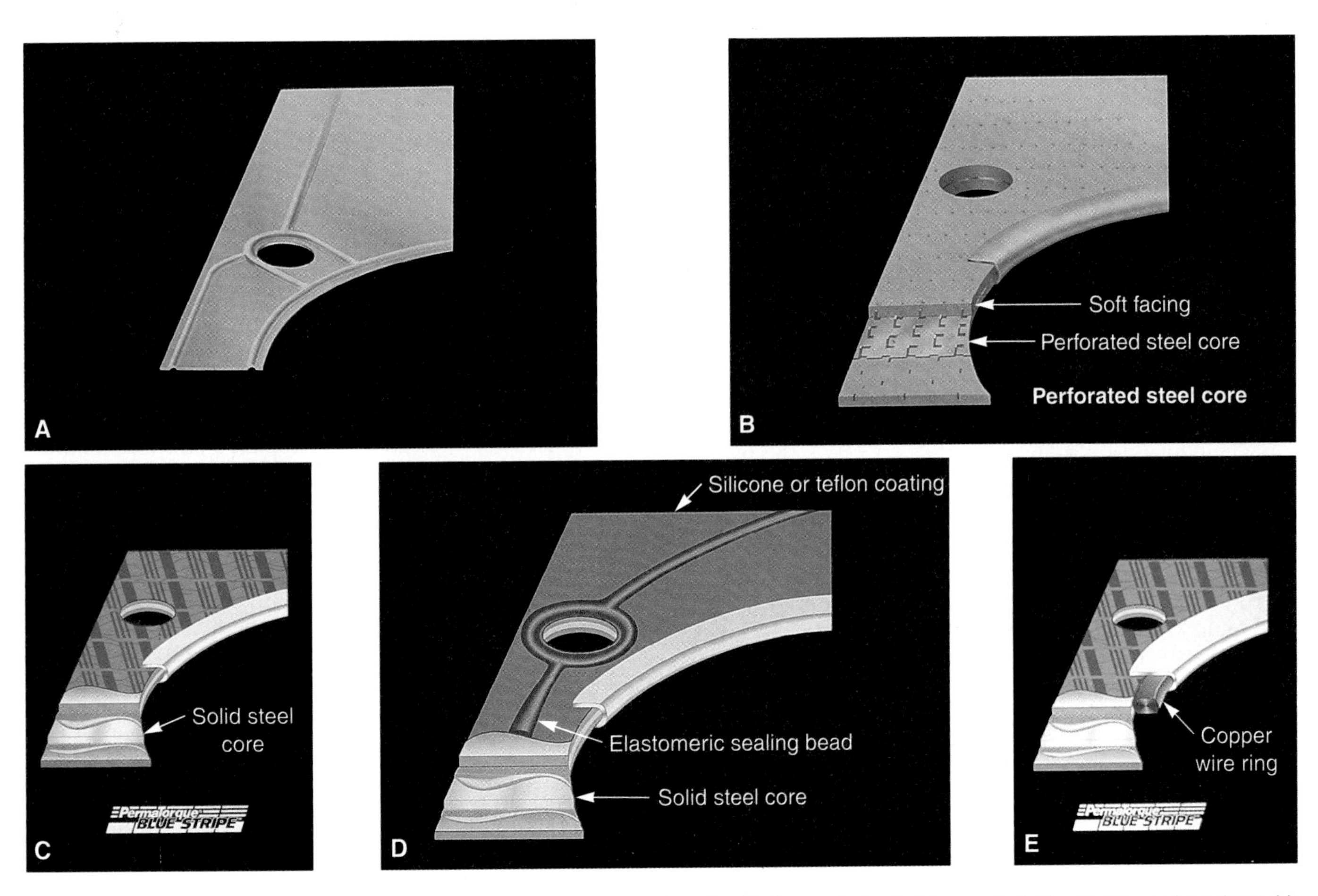

Figure 10-54. *Note the various types of head gasket construction. A—Embossed steel shim gasket. B—Multilayerd gasket with perforated steel core. C—Multilayered gasket with solid steel core. D—Multilayered gasket with elastomeric sealing bead. E—Multilayered gasket with copper ring and stainless steel armor around the combustion chamber. (Fel-Pro)*

formed around the cylinder opening. It is needed to withstand combustion heat and pressure. The body of the gasket is a composite material that can withstand the compression and relaxation.

When the engine warms, its parts expand and smash the head gasket tighter. Then, when the engine cools, the parts contract and less clamping pressure is produced on the head gasket. The head gasket must be resilient so that it can seal through thousands of these heating-cooling cycles.

Many modern head gaskets are coated with nonstick Teflon™. This is especially important when the cylinder head is aluminum and the block is cast iron. Aluminum expands more than iron and can produce a scrubbing action. The Teflon allows the surface of the head to move without gasket failure.

Many late-model vehicles have head gaskets with a bead of sealer formed around water and oil passages, as shown in **Figure 10-54D.** The bead is made from a rubber-like material and helps ensure that the gasket does not leak. Some head gaskets also have compression limiters, or metal rings, that surround bolt holes to help control gasket crush for better sealing.

High-quality head gaskets are permanent torque and do not have to be retorqued after a period of engine operation. They are resilient enough to maintain the initial torque. Many older steel shim gaskets or low-quality gaskets should be torqued a second time after engine operation.

Cylinder head gaskets for a diesel engine are typically available in different thicknesses. Since the piston in a diesel engine comes very close to the cylinder head at TDC, gasket thickness is critical. Also, a thinner head gasket increases the compression ratio. Notches or other markings on the gasket indicate the thickness. Be sure the correct thickness is used.

Cover and Housing Gasket Construction

Cover and housing gaskets is a broad category that includes the gaskets for the valve covers, oil pan, front cover, thermostat housing, intake and exhaust manifold gaskets, and so on. In **Figure 10-52,** note how the names of the gaskets relate to the parts that they seal. This makes it easy to remember gasket names. These gaskets are normally made of synthetic rubber, cork, or treated paper, or they may be a chemical gasket (RTV sealer). Their construction depends on their operating conditions.

When the engine cover or housing is made of thick metal, the gasket is made thinner or an anaerobic sealer may be used. However, with a thin, stamped cover, the gasket must be thicker or RTV (silicone sealer) is needed to conform to the irregular shape of the cover.

Summary

The engine top end refers to the parts that fasten above the engine short block. It controls the flow into and out of the engine cylinders. To properly service today's engines, you must understand engine design differences.

A bare cylinder head is the machined casting with no parts installed. Cylinder heads can be made of cast iron or aluminum.

Combustion chambers provide a place for the burning of the fuel mixture above the piston. The ports are passages in and out of the head for intake and exhaust. Valves open and close these ports.

Valve guides and valve seats can be integral or pressed into the head. Valve seats are usually induction hardened to improve wear resistance. Most valve seats have a 45° angle on their face. An interference angle of 1/2° to 1° is used between the valve and seat for better initial sealing.

The valve head is the large surface exposed to the combustion chamber. The face is the part that touches the seat. The margin is the lip between these two surfaces. Valve shape is determined by the shape of the valve head. A retainer and two keepers normally lock the valve spring onto the valve.

A sodium-filled valve is used in some high-performance engines. This type of valve is lighter and will transfer heat into the head quicker than a solid valve. A stellite valve has a hard coating on its face. This helps it resist heat and wear from the valve seat. Titanium and iconel valves are also used in limited applications.

Valve seals come in umbrella and O-ring designs. The umbrella seal shrouds the opening to the valve guide. The O-ring type seals the gap between the retainer and the valve stem. Both keep excess oil from flowing through the stem-to-guide clearance and into the head port.

Valve spring shims can be used to increase spring tension. This can help prevent valve float where the spring is too weak to close the valve at high engine speeds. Valve rotators can be used to turn the valve in its guide. This prevents localized hot spots and wear patterns that could reduce valve service life.

The camshaft is usually made of cast iron or forged steel. It has lobes that operate the valve train. Camshaft lift determines how wide the valves open. Camshaft duration controls how long the valves stay open.

Valve overlap is the period of time when both the intake and exhaust valves are open. It is used to help scavenge burned gasses out of the cylinder and pull more air-fuel mixture into the cylinder.

Valve lifters or tappets transfer the action of camshaft lobes to the rest of the valve train. A hydraulic lifter is for silent operation. It maintains zero valve clearance (lash). A mechanical camshaft is for severe or high-performance applications. It makes a clicking sound during operation. A cam follower is used in some overhead cam engines. Its function is the same as a lifter.

Push rods are hollow tubes that transfer the lifter motion to the rocker arms. Sometimes they carry oil to the rockers.

Rocker arms pivot to transfer push rod motion to the valves. They can be made of steel or cast iron. The rockers can be mounted on a steel shaft, studs, or bridge. Rocker arm ratio refers to the ratio between the length of each end of the rocker.

Valve lash or clearance is the space between the rocker arm tip and valve stem with the valve closed. It is needed with mechanical lifters to ensure complete closing of the valve.

Review Questions—Chapter 10

Please do not write in this text. Write your answers on a separate sheet of paper.

1. A(n) _____ is the machined head casting without any parts installed.
2. _____ are usually provided in the cylinder head and block for aligning the head and head gasket on the block.
3. On an OHC engine, what is the purpose of camshaft housings?
4. _____ valve guides are very common in aluminum heads.
5. In what situation are valve seat inserts normally used in cast iron heads?
6. What are *knurled valve guides?*
7. Valve seat inserts used in aluminum heads are usually made of _____.
8. _____-hardened valve seats are used to help improve wear resistance.
9. The most common valve seat angle is _____.
10. What is a *valve interference angle?*
11. The _____ valve is usually larger than the _____ valve.
12. What is a *valve margin* and why is it important?
13. Some valve stem tips are _____ coated to resist wear.
14. Why does a sodium-filled valve operate cooler than a solid valve?
15. How does an O-ring valve seal keep excess oil from entering the valve guide?
16. List and describe four valve spring specifications.
17. Valve spring _____ can be used to restore spring tension in used valve springs.
18. What is the difference between a valve retainer and a valve keeper?
19. Camshaft lobe profile affects which of the following?
    (A) When the valve opens.
    (B) How long the valve remains open.
    (C) How far the valve opens.
    (D) All of the above are correct.
20. Define *valve overlap.*

ASE-Type Questions—Chapter 10

1. Technician A says that a cylinder head for an OHV engine holds the valves and rocker arms. Technician B says that a cylinder head for an OHV engine holds the rocker arms, valves, and camshaft. Who is correct?
    (A) A only.
    (B) B only.
    (C) Both A and B.
    (D) Neither A nor B.
2. Technician A says that to grind the valve seat to form a proper interference angle, X should be between 45 1/2° and 46°. Technician B says that to grind the valve seat to form a proper interference angle, X should be between 1/2° and 1°. Who is correct?
    (A) A only.
    (B) B only.
    (C) Both A and B.
    (D) Neither A nor B.

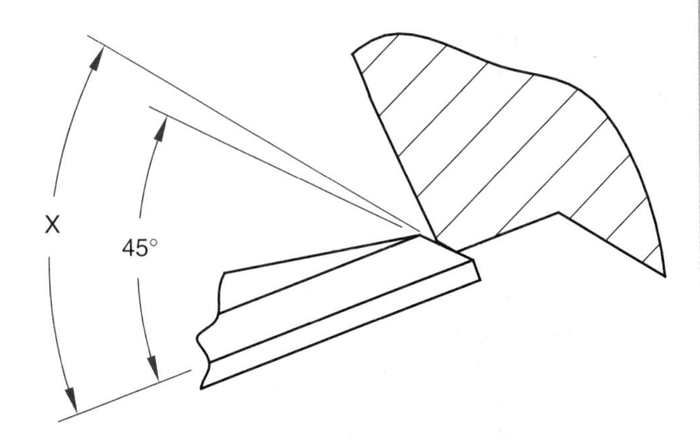

3. Technician A says that integral valve seats are common in cast iron cylinder heads. Technician B says that integral valve seats are common in both cast iron and aluminum cylinder heads. Who is correct?
    (A) A only.
    (B) B only.
    (C) Both A and B.
    (D) Neither A nor B.
4. All of the following are parts of a bare cylinder head *except:*
    (A) intake deck.
    (B) valve keepers.
    (C) coolant passages.
    (D) oil passages.

5. Technician A says that 30° is the most common valve seat angle. Technician B says that a valve seat angle of 45° is only used on some high-performance engines. Who is correct?
   (A) A only.
   (B) B only.
   (C) Both A and B.
   (D) Neither A nor B.
6. Technician A says that umbrella valve seals are used on modern automotive engines. Technician B says that O-ring valve seals are used on modern automotive engines. Who is correct?
   (A) A only.
   (B) B only.
   (C) Both A and B.
   (D) Neither A nor B.
7. Technician A says that valve spring shims can be used to increase or restore valve spring free length. Technician B says that valve spring shims can be used to increase valve spring tension. Who is correct?
   (A) A only.
   (B) B only.
   (C) Both A and B.
   (D) Neither A nor B.
8. All of the following are basic components of an overhead camshaft engine valve train *except:*
   (A) camshaft bearing.
   (B) intake valve tappet.
   (C) exhaust valve push rod.
   (D) intake valve spring.
9. Technician A says that camshaft lift determines how long the valves stay open. Technician B says that increasing camshaft lift and duration may cause low vacuum at idle. Who is correct?
   (A) A only.
   (B) B only.
   (C) Both A and B.
   (D) Neither A nor B.
10. Technician A says that the term *cam follower* refers to a valve tappet mounted in the cylinder head. Technician B says that a cam follower must be hydraulically operated. Who is correct?
    (A) A only.
    (B) B only.
    (C) Both A and B.
    (D) Neither A nor B.
11. Technician A says that valve lash is adjusted when the engine valve is open. Technician B says that valve lash may be adjusted when the engine is cold or hot, depending on specifications. Who is correct?
    (A) A only.
    (B) B only.
    (C) Both A and B.
    (D) Neither A nor B.
12. A customer complains of a rough idle, no power brakes at idle, decreased engine power at low speeds, and increased fuel consumption. The customer states that a high-performance camshaft has just been installed in the engine. Technician A says that the camshaft probably has too much lift and duration for the customer's driving conditions. Technician B says that there must be a vacuum leak and a fuel injection problem. Who is correct?
    (A) A only.
    (B) B only.
    (C) Both A and B.
    (D) Neither A nor B.

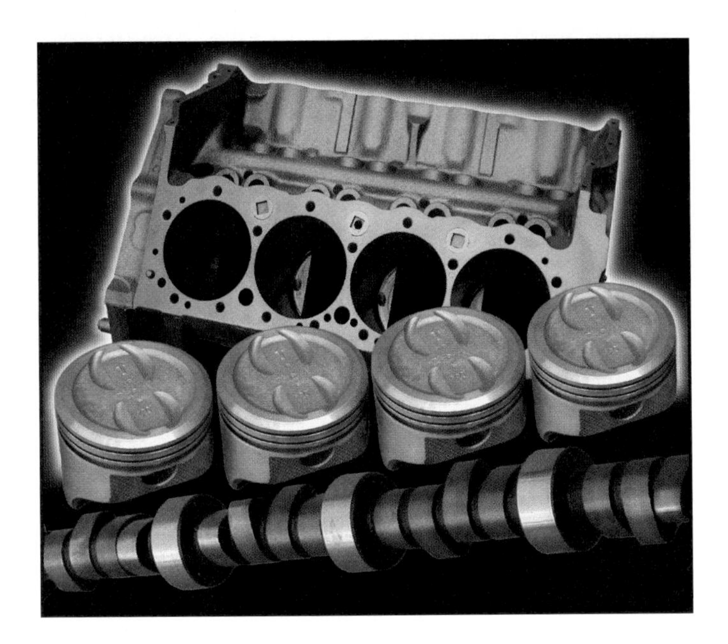

Chapter 11

Front End and Manifold Construction

After studying this chapter, you will be able to:

- ❑ Describe the three types of timing mechanism drives.
- ❑ Explain why the camshaft turns at one-half crankshaft speed.
- ❑ Explain a chain drive timing mechanism.
- ❑ Describe chain tensioners and chain guides.
- ❑ Explain the construction of a belt drive timing mechanism.
- ❑ Summarize the construction of a gear drive timing mechanism.
- ❑ Compare an engine front cover to a timing belt cover.
- ❑ Explain the purpose of crankshaft dampers.
- ❑ Describe the function of an intake manifold.
- ❑ Explain the function of an exhaust manifold.

Know These Terms

Belt drive timing mechanism
Chain drive timing mechanism
Chain guide
Chain tensioner
Dual plane intake manifold
Dual-mass vibration damper
Engine front cover
Engine front end
Exhaust manifold
Gear drive timing mechanism
Inertia ring
Intake manifold
Oil slinger
Sectional intake manifold
Single plane intake manifold
Split runner intake manifold
Timing belt
Timing belt cover
Timing belt sensor
Timing belt tensioner
Timing gear backlash
Timing marks
Timing mechanism
Torsional vibration
Tuned runner intake manifold
Variable induction system
Vibration damper

The ***engine front end*** consists of the parts that are attached to the front of the engine. These include the camshaft drive mechanism, front cover, water pump, auxiliary shafts, and so on. Engine front end construction has changed considerably from what it was twenty or thirty years ago. Today's engines frequently have overhead camshafts with a chain or belt drive. Some even provide variable camshaft timing. Auxiliary drive sprockets, chain tensioners, belt tensioners, chain guides, and balancer shaft drives can all be found on modern passenger vehicle engines.

In the past, overhead valve engines were the dominant design. In this design, the camshaft is located in the engine block. Now, most engines have overhead camshafts.

This chapter covers front end, vibration damper, intake manifold, and exhaust manifold construction. It will help you understand the many design variations found on today's engines and understand how the components operate and are constructed. As a result, you will be better prepared to diagnose, test, adjust, and repair these components.

Camshaft Drive Mechanism Construction

The ***timing mechanism,*** also called the ***camshaft drive mechanism,*** must turn the engine camshaft and keep it in time with the engine crankshaft and pistons. The camshaft drive ratio is two to one (2:1) because there is one complete engine cycle for every two crankshaft revolutions. As shown in **Figure 11-1,** the camshaft gear or sprocket turns at one-half of the engine (crankshaft) speed. The camshaft gear has twice as many teeth as the crankshaft gear and is twice the diameter.

Sometimes, the camshaft drive mechanism must also power other units, such as a balancer shaft or oil pump. The timing mechanism must provide smooth and dependable operation for all components. There are three basic types of camshaft drives, as shown in **Figure 11-2:**

- Chain drive.
- Belt drive.
- Gear drive.

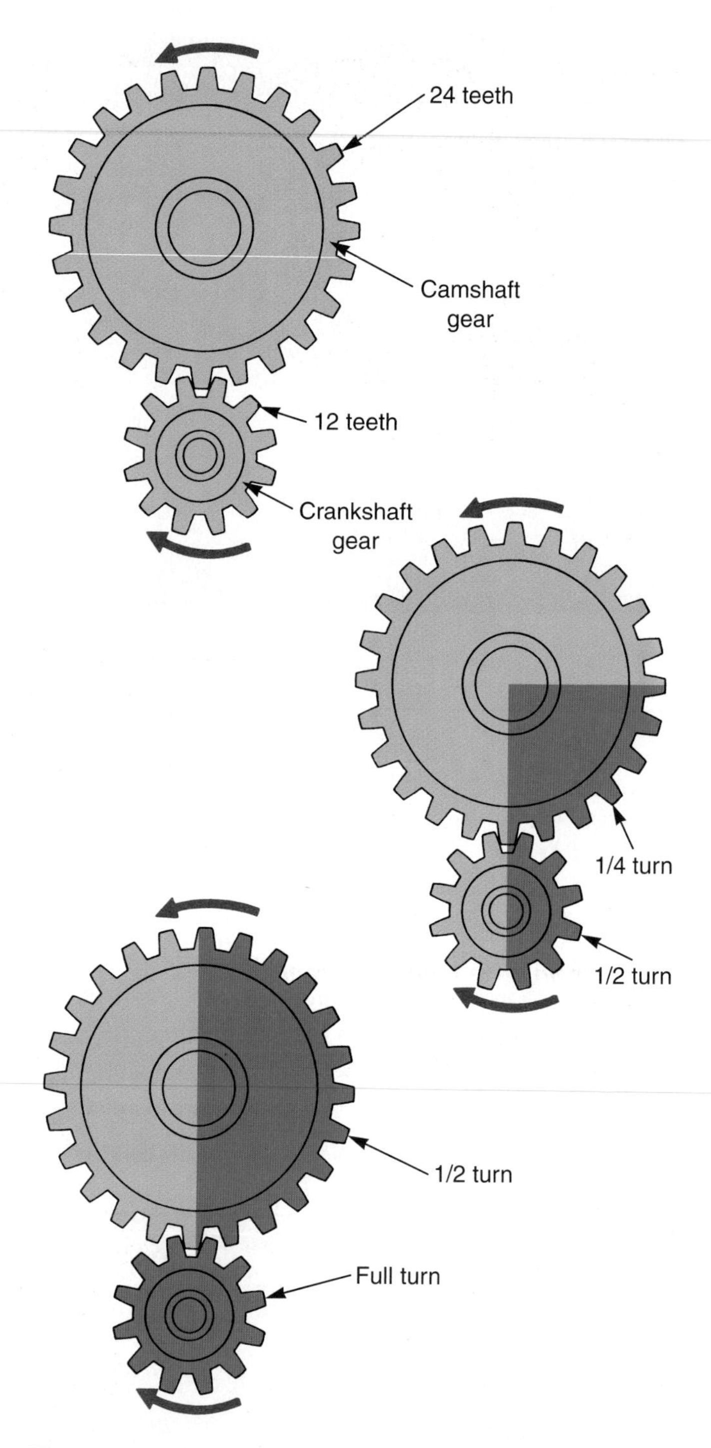

Figure 11-1. *The timing mechanism must turn the camshaft at one-half of the engine (crankshaft) speed. For every one revolution of the crankshaft gear or sprocket, the camshaft revolves one-half turn.*

Timing marks on the crankshaft and camshaft sprockets or gears indicate the proper orientation. The marks may be circles, indentations, or lines. Timing marks must be aligned when the Number 1 cylinder is at TDC for the camshaft to have the correct orientation to the crankshaft. See **Figure 11-3.** If this orientation is not correct, the valve timing will be off.

A ***crankshaft key*** is used to lock the crankshaft sprocket or gear to the snout of the crankshaft. This is shown in **Figure 11-4.** The key allows the sprocket to slide over the crankshaft but keeps it from turning on the snout. A ***camshaft key*** or dowel is used to align the camshaft sprocket or gear on the camshaft. These keys also ensure the timing mark on each gear is properly positioned in relation to the shaft.

Most camshaft and crankshaft position sensors do not need to be adjusted. However, depending on the design and manufacturer, the sensors may need to be adjusted, or indexed. This is done to prevent the sensor from contacting the rotating parts and to ensure the proper signal is transmitted to the computer. Refer to the service manual for specific details.

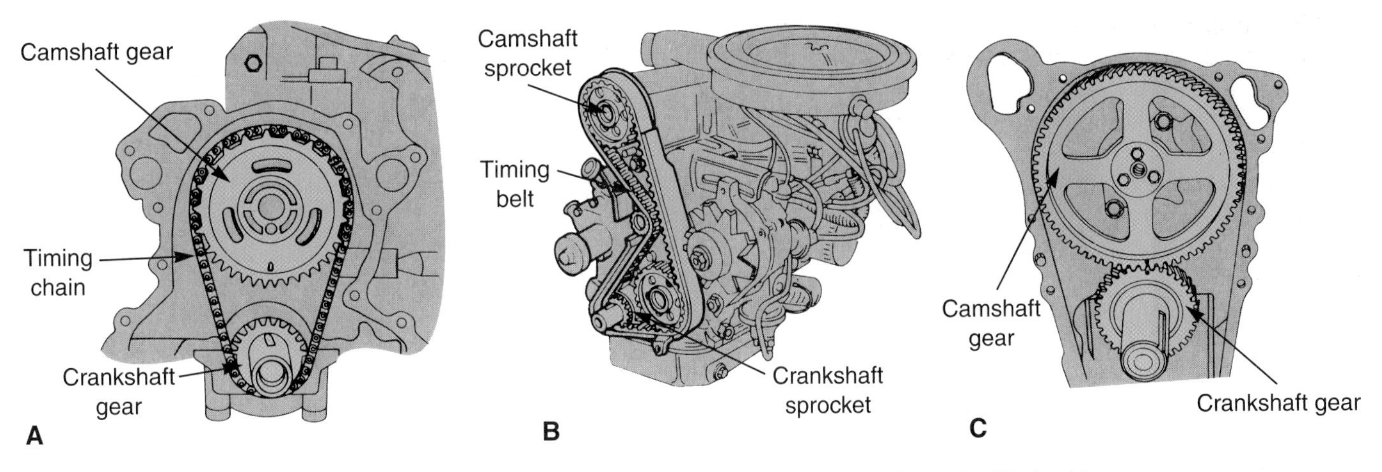

Figure 11-2. *These are the three basic types of camshaft drive found on today's engines. A—Chain drive. B—Belt drive. C—Gear drive. (Ford)*

Note: Some engines have a form of variable valve timing where the camshaft sprocket rotates on the camshaft based on engine operating conditions. By rotating the camshaft sprocket, the valve timing can be adjusted to match the engine operating conditions.

Chain Drive

A ***chain drive timing mechanism*** consists of a timing chain, crankshaft sprocket, and camshaft sprocket. This is the most common type of camshaft drive mechanism on cam-in-block engines. See **Figure 11-5.** It may, however, be used on OHC engines.

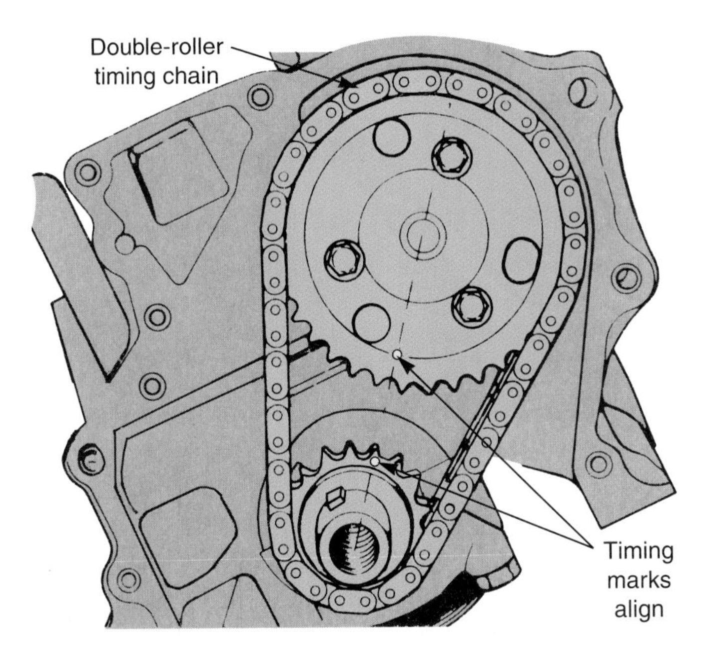

Figure 11-3. *Timing marks ensure correct orientation of the sprockets when the chain, belt, or set of gears is installed. (DaimlerChrysler)*

The crankshaft sprocket is mounted on the crankshaft snout. The camshaft sprocket is mounted to the camshaft. The camshaft sprocket may have metal, plastic, or composite teeth. The timing chain transfers power from the crankshaft sprocket to the camshaft sprocket. The timing chain can be a multiple-link (silent) or a double-roller type.

The basic construction of a chain drive is the timing chain and two sprockets. However, many designs have other components. These include:

- Chain tensioner.
- Chain guide.
- Auxiliary chain and sprockets.
- Oil slinger.

Chain Tensioner

A ***chain tensioner*** is to take up excess slack in the timing chain as the chain and sprockets wear. It is a spring-loaded or hydraulic device that pushes on the unloaded side of the chain to keep the chain tight. See **Figure 11-6.** A chain tensioner is commonly used on OHC engines, but can sometimes be found on cam-in-block engines.

There are several designs of chain tensioners. These include plunger, arm or lever, and oil-filled plunger, **Figure 11-7.** A plunger-type tensioner pushes straight out on the chain using spring tension. An arm- or lever-type tensioner is spring loaded and swings out to put pressure on the chain. An oil-filled-plunger-type tensioner is a hydraulic design that increases the pressure on the chain as engine speed and oil pressure increase.

The tensioner usually has a foot that makes contact with the chain. This foot is made of plastic, Teflon, or a composite fiber. These materials allow the foot to rub on the chain with a minimum amount of wear.

Chain Guide

A ***chain guide*** may be needed to prevent chain slap on long lengths of chain, **Figure 11-8.** Chain slap occurs when slack in the chain allows the chain to flap back and forth.

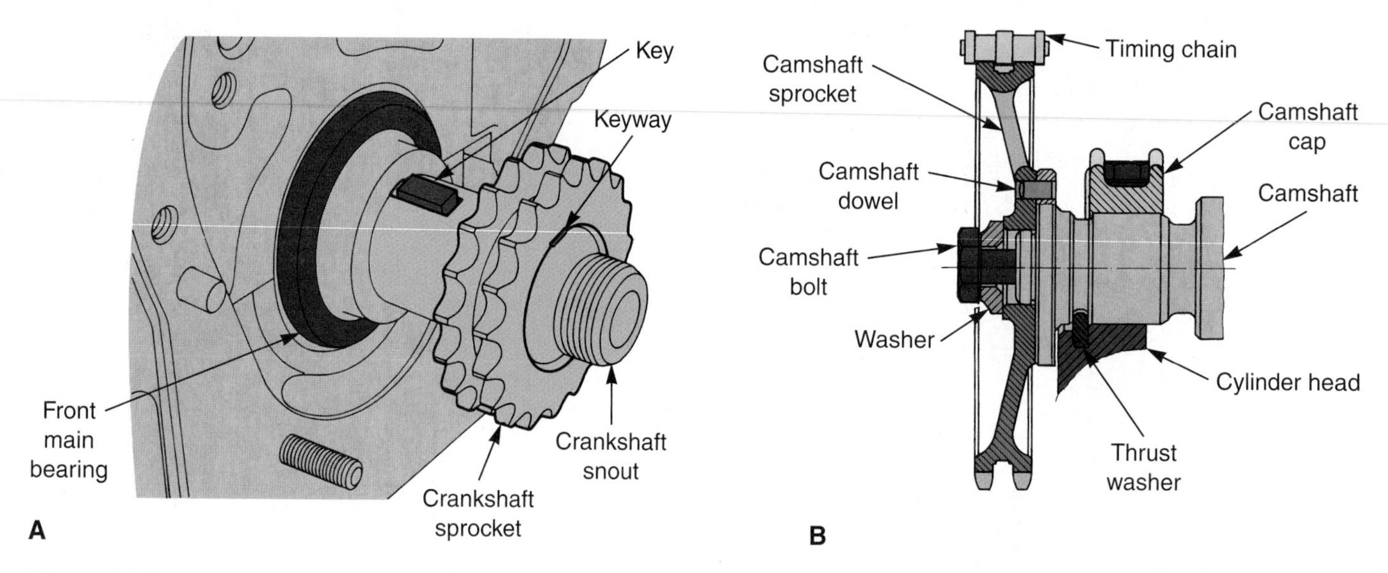

Figure 11-4. *A—A keyway is cut in the crankshaft snout and crankshaft sprocket or gear. A key then keeps the sprocket from spinning on the crankshaft. B—A camshaft sprocket is frequently locked to the camshaft with a steel dowel. A bolt holds the sprocket on the camshaft. (Peugeot, Mercedes-Benz)*

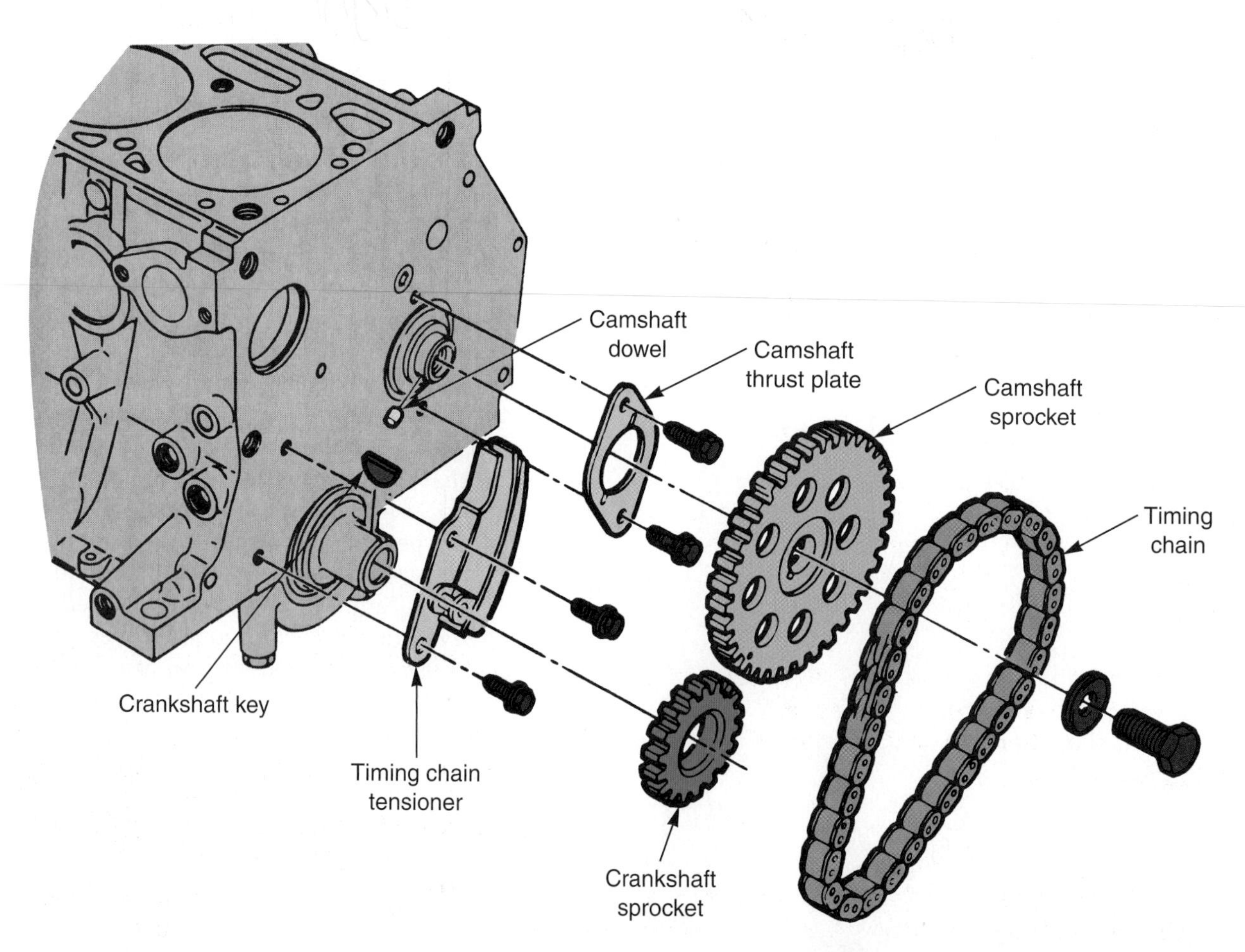

Figure 11-5. *The basic parts of a chain drive timing mechanism. This system also includes a timing chain tensioner. (Ford)*

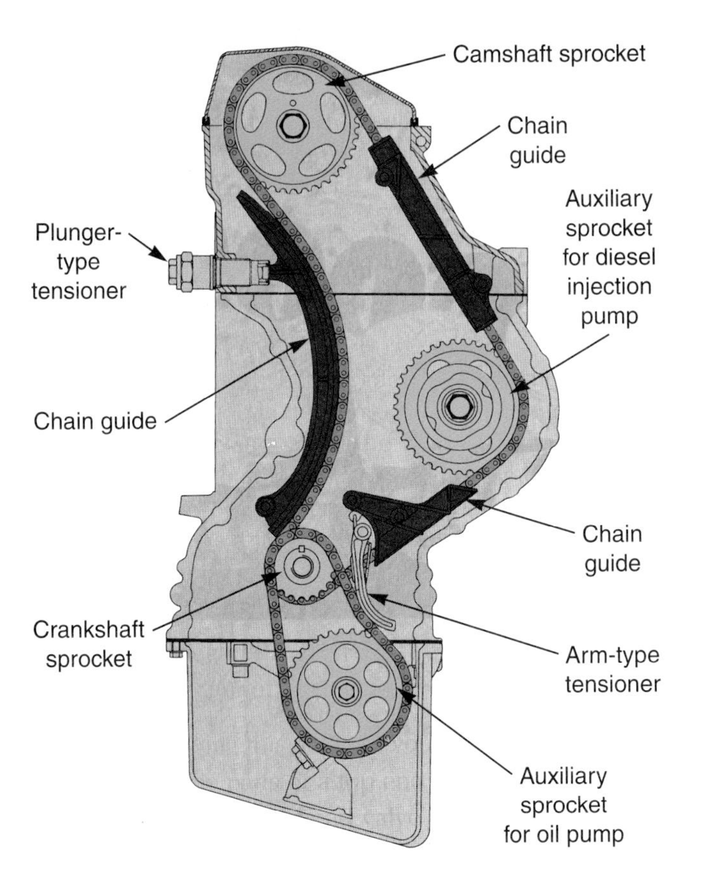

Figure 11-6. *This chain drive timing mechanism has two chain tensioners. (Mercedes-Benz)*

The guides have a metal body with a plastic, nylon, or Teflon face. This allows the chain to slide with minimum wear.

A chain guide is especially needed when there is a great distance between the chain sprockets. The guide pushes in slightly on the chain to keep it running smoothly. Slots in the bolt holes allow the guides to be correctly adjusted.

Auxiliary Chain and Sprockets

An auxiliary chain may be used to drive the engine oil pump, balancer shafts, and other units on the engine. An auxiliary chain is driven by an extra sprocket, usually placed in front of the crankshaft timing chain. Look at **Figure 11-9.** An auxiliary drive sprocket powers the accessory.

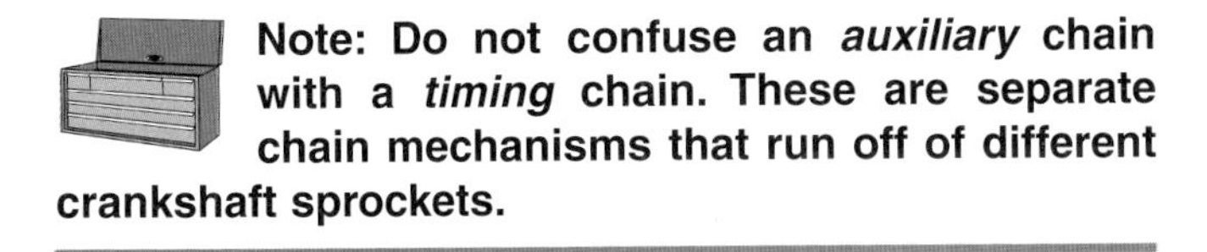

Note: Do not confuse an *auxiliary* chain with a *timing* chain. These are separate chain mechanisms that run off of different crankshaft sprockets.

Oil Slinger

An ***oil slinger*** helps spray oil onto the timing chain to prevent wear. It also helps prevent oil from leaking out of the front crankshaft oil seal. An oil slinger is a thin, washer-shaped, metal disk that fits in front of the crankshaft sprocket. See **Figure 11-10.**

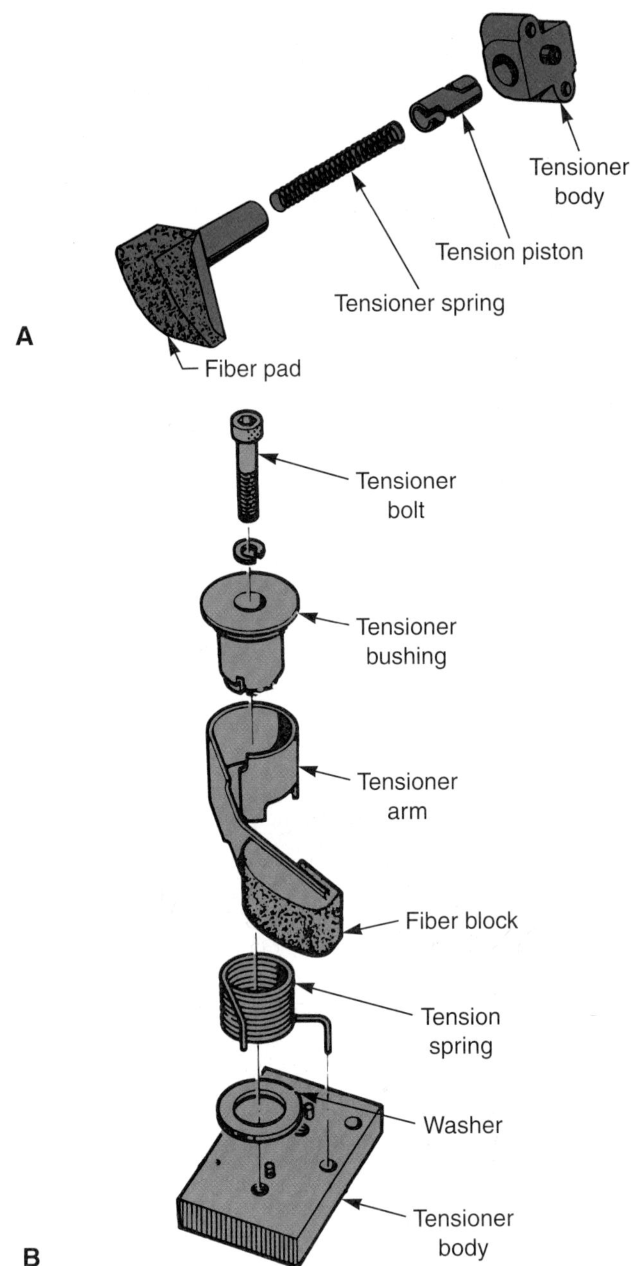

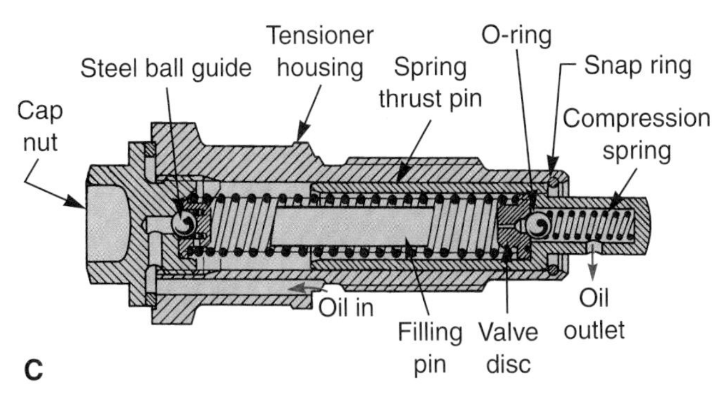

Figure 11-7. *Three types of chain tensioners. A—Plunger. B—Arm or lever. C—Oil-filled plunger.*

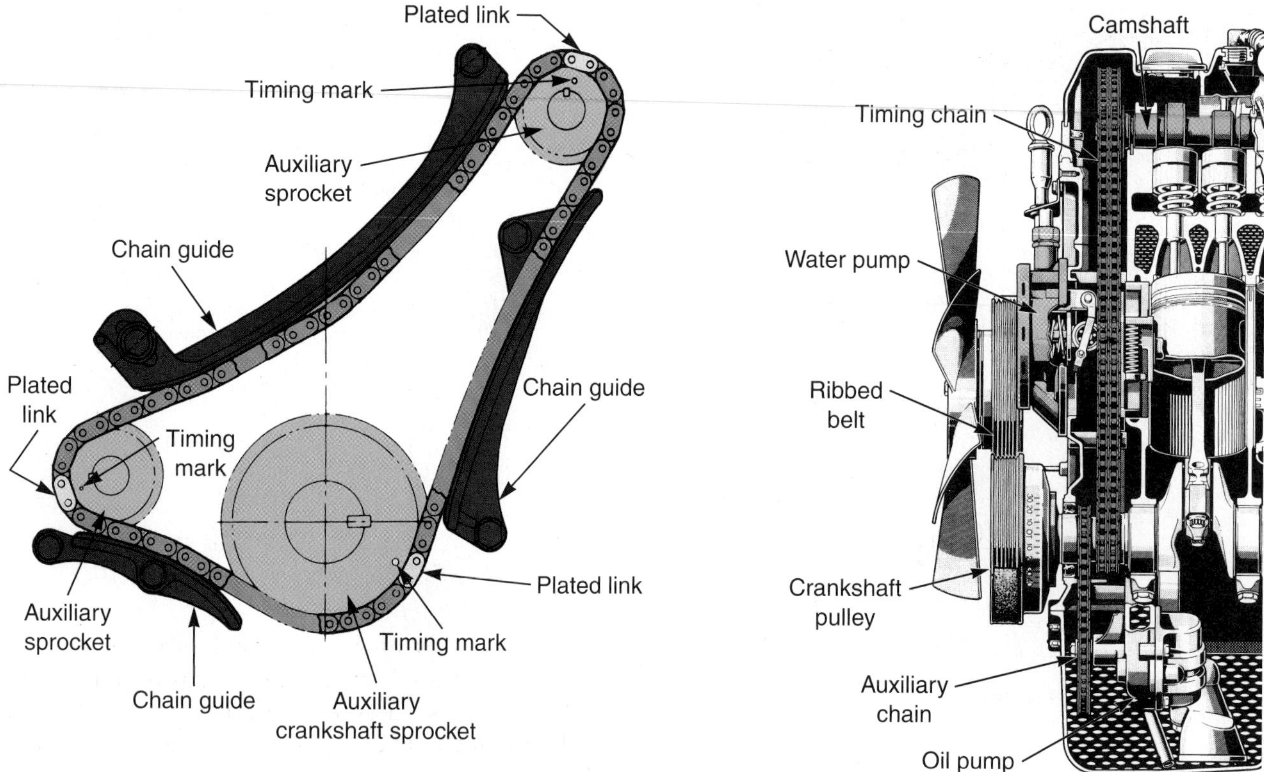

Figure 11-8. *Chain guides lightly touch the timing chain to keep it from slapping or flapping back and forth. This is an auxiliary chain that drives two balancer shafts, and it has three chain guides.*

Figure 11-9. *This side view of an engine shows how an auxiliary chain is used to drive the oil pump. (Mercedes-Benz)*

When the engine is running, oil squirts out of the front main bearing. Centrifugal force on the spinning slinger throws this oil outward, lubricating the timing chain or gears. Since the oil slinger is between the front main bearing and the front crankshaft oil seal, the amount of oil reaching the seal is minimized.

Belt Drive

A ***belt drive timing mechanism*** is commonly found on overhead camshaft (OHC) engines. See **Figure 11-11.** The basic parts of a belt drive system include a cogged timing belt, crankshaft sprocket, camshaft sprocket(s), and belt tensioner. This type of timing mechanism is an excellent way of driving the camshaft. It is very quiet and dependable. The belt runs very smoothly and without gear backlash or chain slack. However, the service life of a timing belt is much shorter than a timing chain or timing gears.

In an OHC engine, the crankshaft and camshaft are located relatively far away from each other, when compared to a cam-in-block engine. A chain drive system can

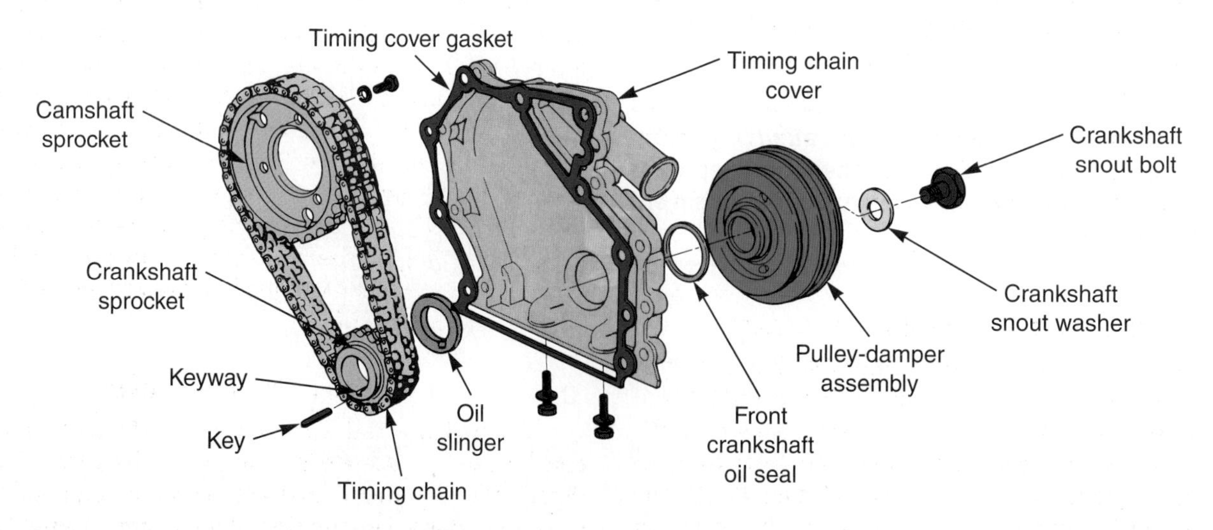

Figure 11-10. *The oil slinger fits on the crankshaft between the crankshaft sprocket and the front cover. (DaimlerChrysler)*

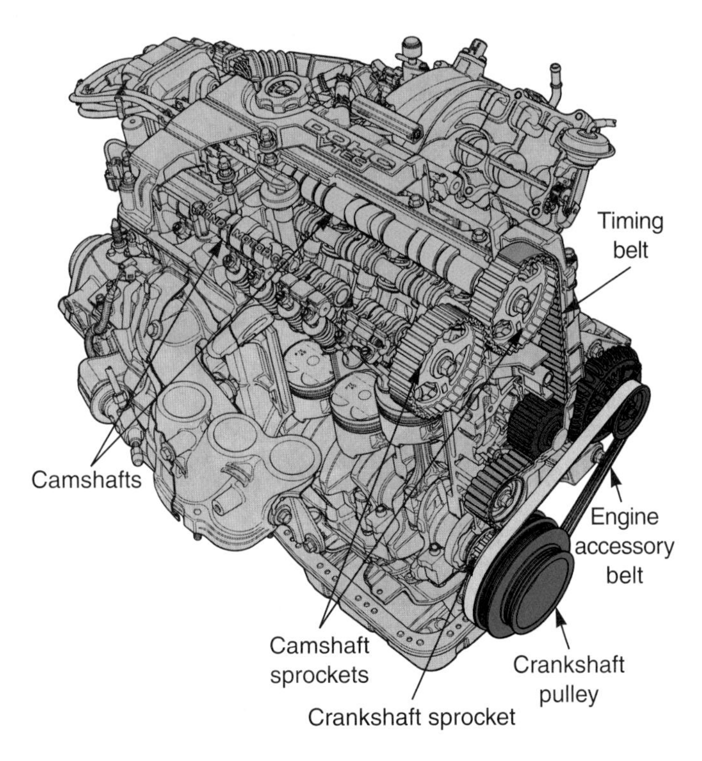

Figure 11-11. *A belt drive timing mechanism. Note the part names. This is a dual overhead camshaft (DOHC) engine. (Honda)*

be used, but the additional chain guides and a tensioner increase the friction in the system. Also, it is impractical to use a complex set of timing gears to transfer power to the camshaft in an OHC engine.

Timing Belt Construction

A ***timing belt*** is a cogged (square toothed) belt with internal reinforcing strands to limit stretch. Teeth are formed on the inside surface of the belt. They engage the teeth on the outside surface of the crankshaft and camshaft sprockets. This provides a positive, slip-free, low-vibration drive mechanism for the camshaft(s).

Some timing belts are made of fiberglass-reinforced nitril rubber for increased strength and durability. Some timing belts are designed to last the life of the engine.

Caution: Oil will damage most timing belts. Avoid spilling oil on a timing belt during an oil change. Also, warn the customer if you find a front oil seal leak.

Camshaft and Crankshaft Sprockets

The camshaft and crankshaft sprockets in a belt drive system perform the same function as the camshaft and crankshaft sprockets in a chain drive system. However, the sprockets in a belt drive system have square teeth to properly engage the cogged belt, as shown in **Figure 11-12.** The sprockets can be made of cast iron, steel, aluminum, or carbon fiber.

Figure 11-12. *Note the cogged belt and the square teeth in the sprockets. This is a DOHC engine.*

The camshaft sprocket in a belt drive system is aligned to the camshaft with a key or dowel pin. A bolt (or bolts) normally secures the sprocket to the front of the camshaft. An oil seal in the cylinder head keeps oil off the sprocket and belt.

A crankshaft sprocket in a belt drive system fits over the crankshaft snout and is keyed to the crankshaft. See **Figure 11-13.** The sprocket fits in front of the crankshaft oil seal. A large bolt screws into the crankshaft snout to hold the sprocket in place.

Timing Belt Tensioner

A ***timing belt tensioner*** is a wheel that keeps the timing belt tight on its sprockets, **Figure 11-14.** It performs the same function as a chain tensioner in a chain drive system. The tensioner pushes inward on the outer surface of the belt. This prevents the belt teeth from moving away from the sprockets and slipping on the sprocket teeth. It also keeps the belt from "flapping" possibly flying off the sprockets.

The belt tensioner is usually mounted on an adjustment bracket that allows it to be moved into or away from the belt. The wheel rides on an antifriction bearing that is filled with grease and permanently sealed at the factory.

Some belt tensioners use both spring and hydraulic pressure to maintain belt tightness. The spring tensioner keeps the belt tight when the engine is shut off and engine oil pressure drops. The hydraulic tensioner adjusts belt tension with engine speed. At higher engine speeds, belt tension is increased to keep the belt from "flapping," jumping teeth, and flying off.

Auxiliary Shaft Sprocket

An auxiliary shaft sprocket, or intermediate sprocket, can be used to operate the oil pump, balancer shaft, or other units requiring crankshaft power. It is an additional sprocket mounted to one side of the engine, **Figure 11-15.** The timing belt extends around this extra sprocket. The

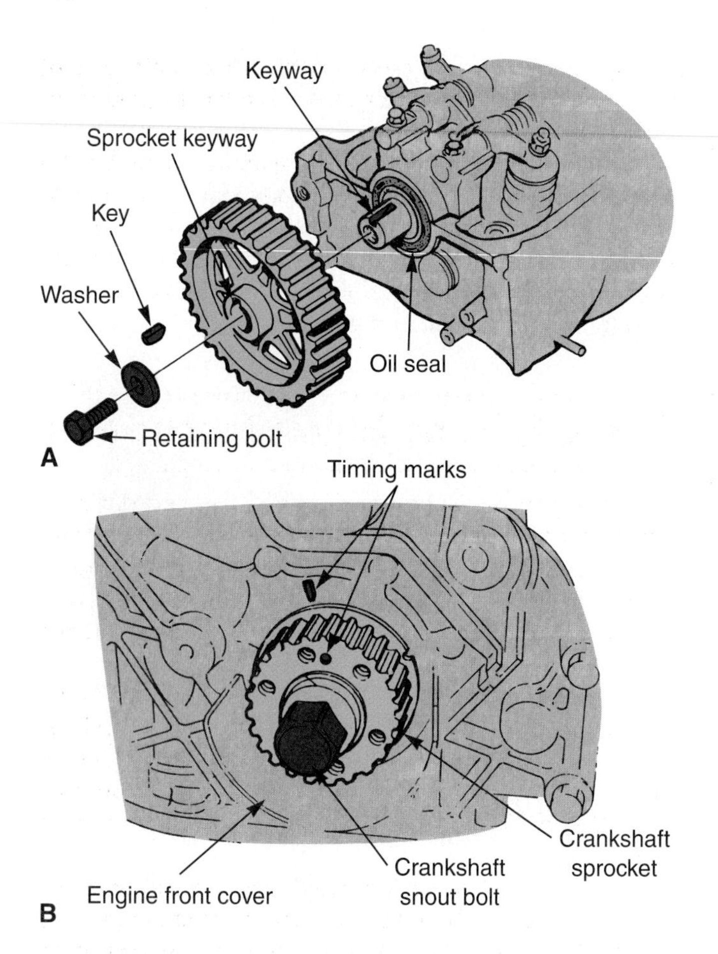

Figure 11-13. *A—The camshaft sprocket fits over a small snout on the camshaft in this design. A key locks the sprocket to the camshaft. B—The crankshaft sprocket is keyed to crankshaft. A large bolt holds the sprocket on the crankshaft snout. (Honda, Ford)*

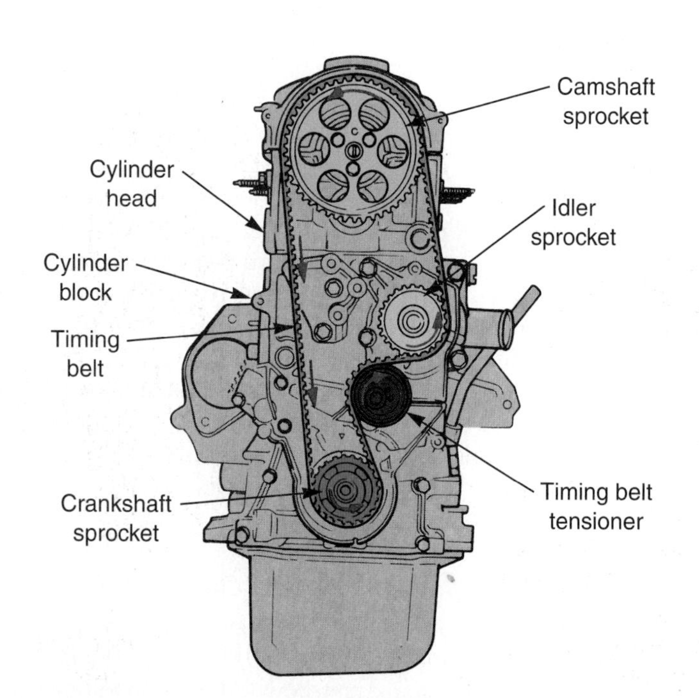

Figure 11-14. *A timing belt tensioner keeps the timing belt tight against the sprockets.*

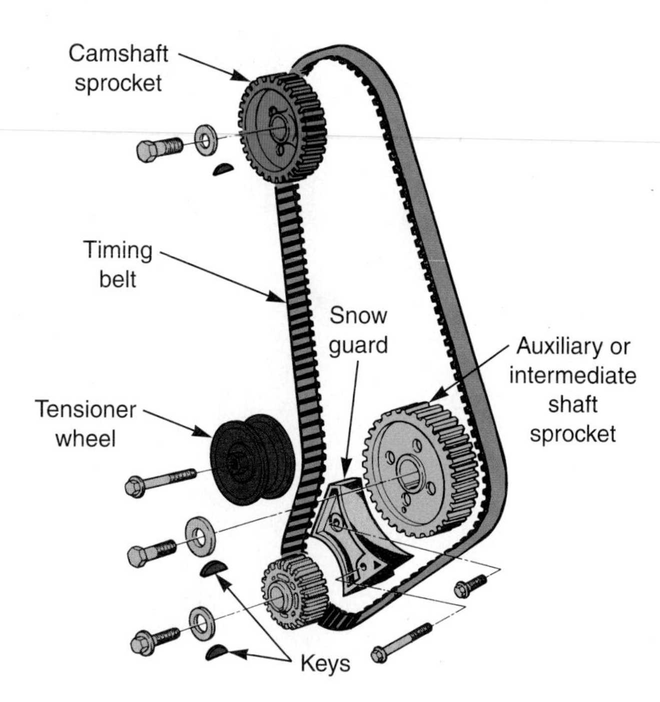

Figure 11-15. *An auxiliary shaft sprocket can be used to power an oil pump, balancer shaft, or other device. (DaimlerChrysler)*

diameter of the sprocket varies, depending on what device it is used to rotate. Its construction is similar to the camshaft and crankshaft sprockets.

Note: The auxiliary shaft in a chain drive system can be driven by the timing chain or a separate (auxiliary) chain. In a belt drive system, the timing belt drives the auxiliary shaft.

Timing Belt Sensor

A ***timing belt sensor*** detects excessive tensioner extension, which indicates excessive timing belt wear and stretch. Excessive stretch is due to possible belt failure. On interference engines, if the timing belt breaks or slips off its sprockets, the pistons will hit the valves. This may result in bent valves and piston damage.

When the sensor detects belt stretch, it signals the ECM. The ECM can then illuminate a dash light to warn the driver of the problem. This allows the belt to be replaced before it fails.

Some computer control systems can detect probable belt failure by using data from the crankshaft and camshaft position sensors. If the timing belt jumps a tooth or two, the ECM can warn the driver of the timing belt problem by illuminating the check engine light or malfunction indicator lamp (MIL) before major engine damage occurs.

Gear Drive

A ***gear drive timing mechanism*** consists of helical gears on the front of the engine that connect the crankshaft

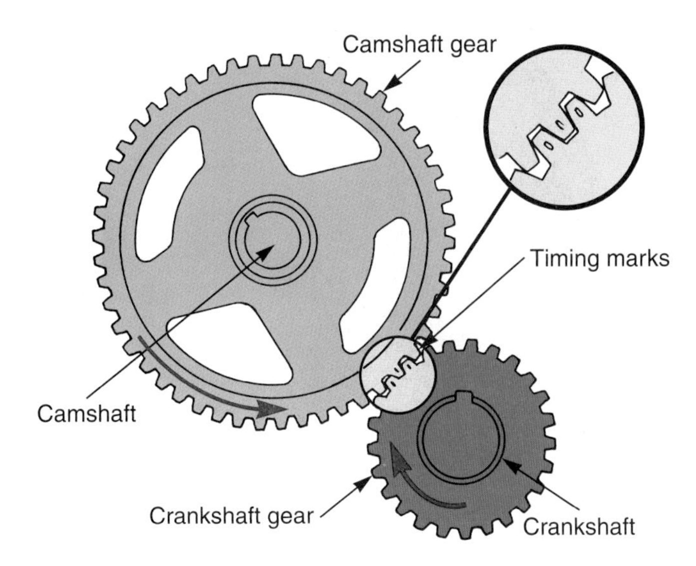

Figure 11-16. *A gear drive timing mechanism. Note the timing marks. (Deere & Co.)*

to the camshaft **Figure 11-16.** The crankshaft gear is keyed to the crankshaft snout. This gear turns the camshaft gear, which is connected to the end of the camshaft. A timing gear set is primarily used in cam-in-block engines where the crankshaft is close to the camshaft.

Timing gears are commonly used on heavy-duty applications, such as taxicabs, trucks, and some small diesel engines. They are very dependable and long lasting. However, they are noisier than a chain or belt drive. For this reason, their use is limited in passenger vehicles.

Timing gear backlash is the clearance between the teeth of the timing gears. A dial indicator is commonly used to measure timing gear backlash. Backlash is needed so that the gears can expand slightly when hot and still maintain clearance for oil. The gear teeth must be lubricated to prevent friction and wear.

Timing Belt and Engine Front Cover

An ***engine front cover,*** also called timing chain or timing gear cover, is a metal housing that bolts on the front of the engine. It encloses the timing chain or gears to keep oil from spraying out. It can be made of thin, stamped steel or cast aluminum. A gasket seals the joint between the cover and engine block.

As shown in **Figure 11-10,** the front cover holds the front crankshaft oil seal. Other engine parts can sometimes fasten to the front cover. For example, the fuel pump, water pump, oil pump, or other unit may be fastened to the front cover. When it supports other devices, the front cover is usually cast aluminum.

A ***timing belt cover*** is simply a sheet metal or, more commonly, plastic shroud around the camshaft drive belt, **Figure 11-17.** Its function is to protect the belt from external debris that could force the belt off its sprockets. It

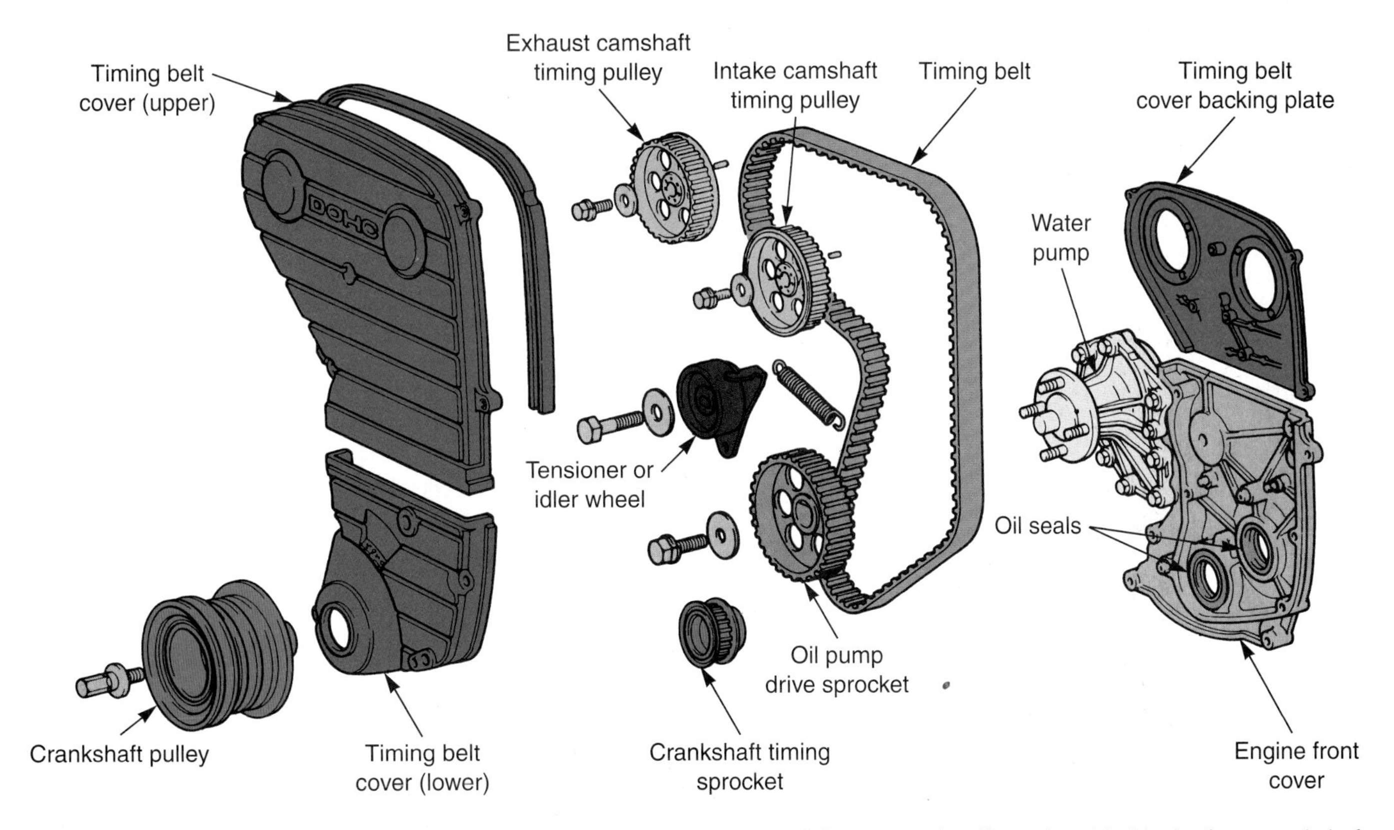

Figure 11-17. *A timing belt cover prevents debris from damaging the belt. A front cover is still used and holds the front crankshaft oil seal(s). (Toyota)*

also protects people from possible injury if the belt fails while the hood is open.

A timing belt cover should not be confused with an engine front cover. An engine front cover contains the crankshaft oil seal and keeps oil inside the engine. A timing chain is installed between the engine and front cover. A timing belt cover, on the other hand, fits over the timing belt, which is in front of the engine front cover. It does not provide any kind of pressure seal. **Figure 11-18** shows the difference between an engine front cover and a belt cover. Note that on an engine with a timing belt, there is still a front cover that contains the oil seal.

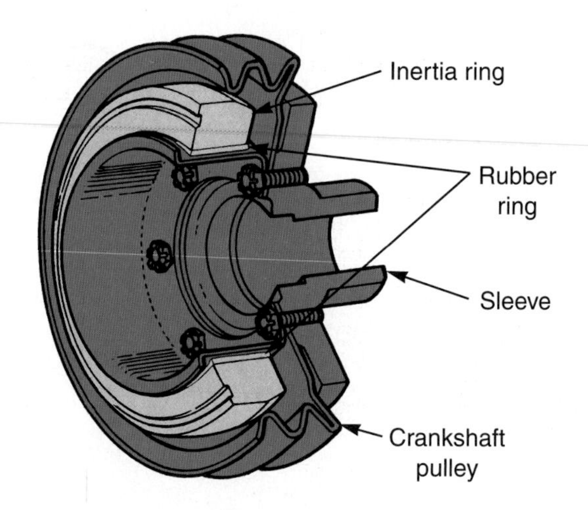

Figure 11-19. *Note the basic parts of a vibration damper. (DaimlerChrysler)*

Vibration Damper Construction

A ***vibration damper,*** also called a harmonic balancer or crankshaft damper, is a rotating mass that helps control torsional vibration in the engine. See **Figure 11-19.** It is a heavy steel ring mounted in rubber. The balancer is keyed to the crankshaft snout so it spins at the same speed as the crankshaft.

Torsional Vibration

Torsional vibration is a high-frequency vibration resulting from normal power pulses in the engine. Every time there is a power event, which starts with combustion in the cylinder, a brief pulse of power is created on the piston. This brief pulse quickly smoothes out. Each piston and rod assembly can exert over a ton of downward force on its journal. When the power pulse is transmitted to the journal, the crankshaft itself can actually twist about its centerline. This twist is called ***torsional movement*** because the motion is through an angle or radius. Then, as the power smoothes out, the crankshaft twists or snaps back in relation to its centerline. The result of this twisting and untwisting of the crankshaft is torsional vibration.

If torsional vibration is not controlled, serious engine damage can result. This is usually in the form of a broken crankshaft. High-horsepower engines are especially sensitive to torsional vibration.

Vibration Damper Construction

Figure 11-20 illustrates the basic construction of a typical vibration damper. Note how a rubber ring separates the outer ***inertia ring*** and the inner sleeve. The inertia and rubber rings set up a damping action on the crankshaft as it tries to twist and untwist. This damping minimizes torsional vibrations. A ***dual-mass vibration damper*** has one weight mounted on the outside of the pulley and another on the inside. The extra rubber-mounted weight helps reduce torsional vibration at high engine speeds.

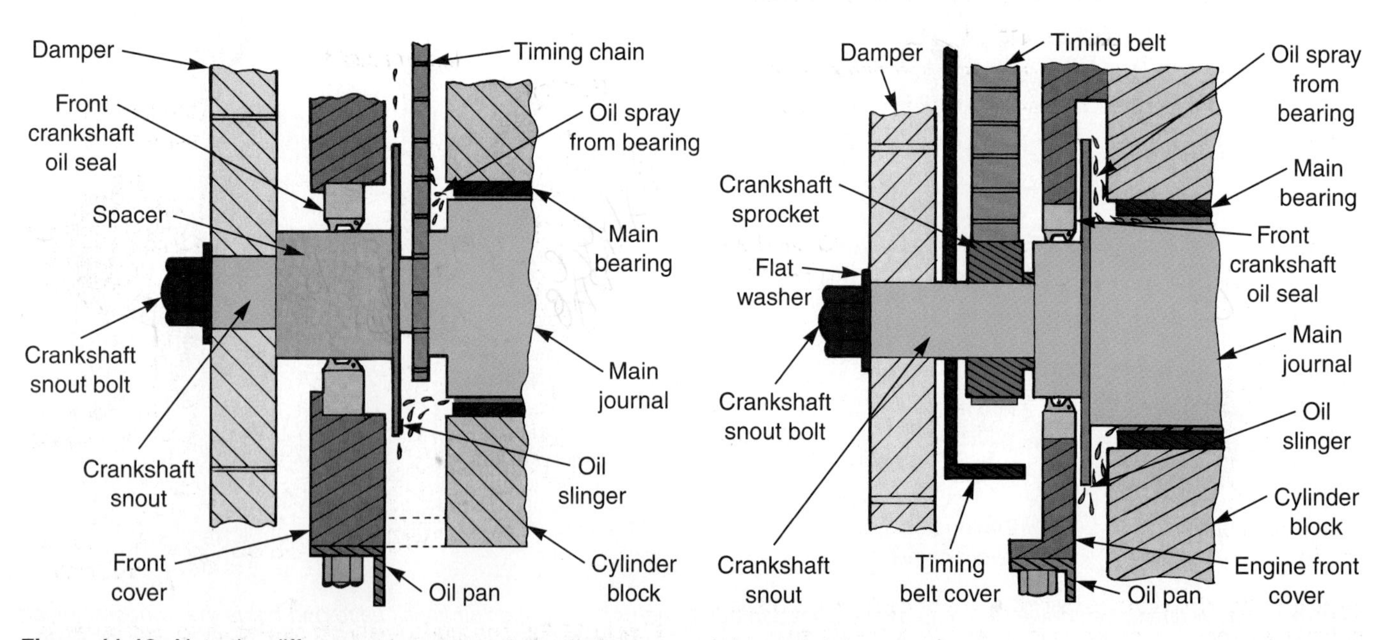

Figure 11-18. *Note the differences between a timing belt cover and the engine front cover.*

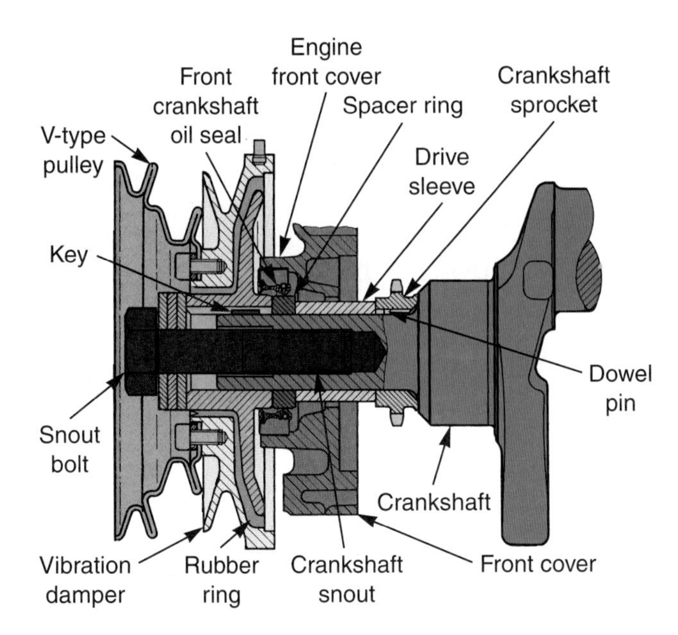

Figure 11-20. *This cutaway shows how a vibration damper is constructed. Note the location and construction of other parts as well. (Mercedes-Benz)*

Crankshaft Pulleys

Crankshaft pulleys are needed to operate the alternator, power steering pump, air conditioning compressor, air injection pump, and other devices. The pulleys can be formed as part of the vibration damper or they can be bolted to the damper, as shown in **Figure 11-21.**

A crankshaft pulley may be ribbed for a ribbed belt or V-type for use with V-belts. Crankshaft pulleys are usually made of stamped steel or cast aluminum. A hub on the center of the vibration damper centers the pulley on the crankshaft. Most crankshaft pulleys bolt into place. A few are machined as part of the vibration damper.

Intake Manifolds

The basic function of an ***intake manifold*** is to carry air to each cylinder. It is bolted over and covers the intake

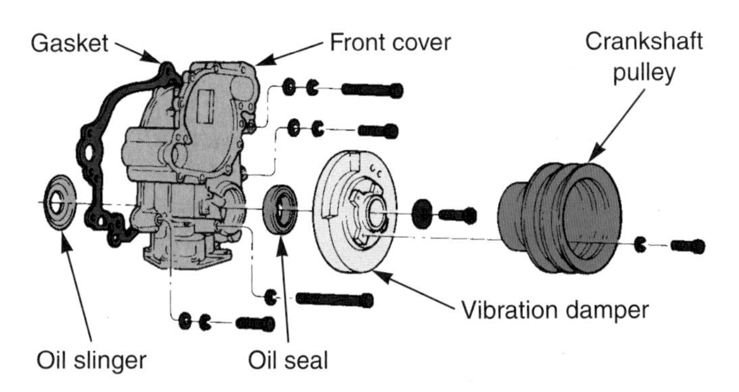

Figure 11-21. *Crankshaft pulleys can be bolted to the vibration damper, as shown here, or an integral part of the damper.*

ports on the cylinder head(s). See **Figure 11-22.** The intake manifold is usually made of lightweight aluminum, plastic, or a composite material. However, some older or heavy-duty manifolds are made of cast iron.

Lightweight materials are used for the intake manifold to help keep the weight of the engine down. This, in turn, helps improve vehicle fuel economy and acceleration. Also, the inside of a plastic intake manifold is smoother than metal, which improves airflow into the engine.

Note: In throttle body injection and older carburetion systems, the intake manifold must be designed to ensure an equal air-fuel charge reaches each cylinder.

Parts of an Intake Manifold

Figure 11-23 shows intake manifolds for inline and V-type engines. Note the differences. The manifold for the inline engine has the inlet positioned to one side and passages run out to meet the ports in the cylinder head. The manifold for a V-type engine has the inlet in the middle with passages feeding out each side for both cylinder heads.

The exact parts of an intake manifold vary from one design to another. However, most intake manifolds have:

- ❑ **Runners.** These are the passages in which the air travels to each cylinder, **Figure 11-24.**
- ❑ **Inlet or riser bore.** This is the opening for incoming air. The bottom of the throttle body bore matches up with the inlet.
- ❑ **Plenum.** This is a common chamber between the inlet and the runners. This chamber allows air pressure to equalize before airflow is split into the runners. Modern engines often have an intake manifold with a large plenum.

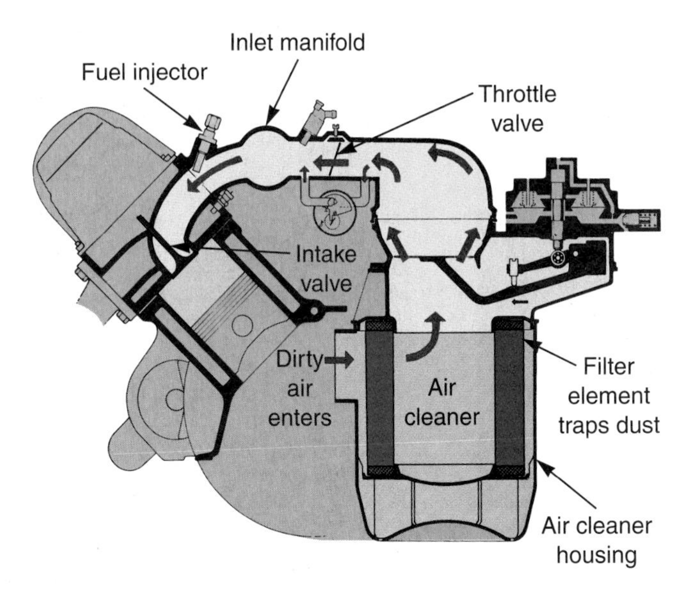

Figure 11-22. *The intake manifold carries air to each cylinder.*

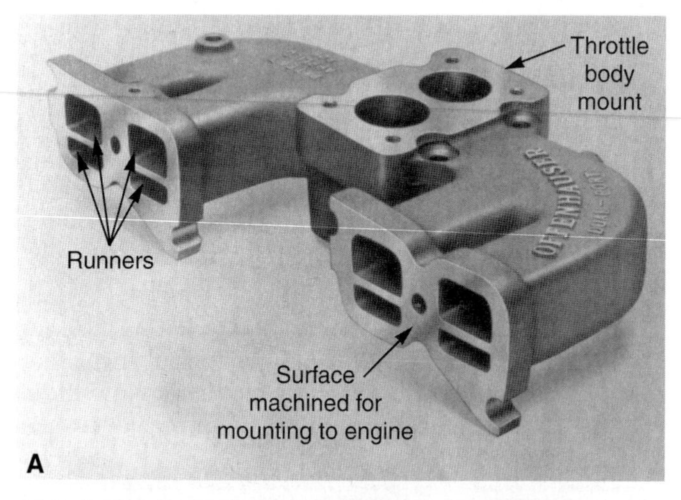

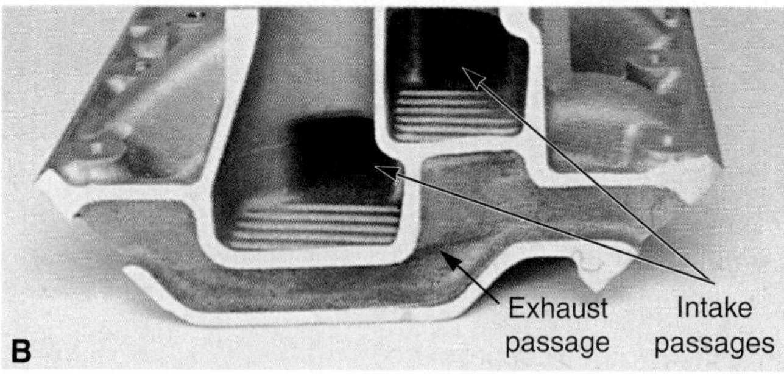

Figure 11-23. *Two basic intake manifold types. Note the parts of an intake manifold. A—Inline engine. B—V-type engine. (Offenhauser, Edelbrock)*

- ❑ **EGR passage.** This is an opening into the exhaust manifold that allows small amount of exhaust gasses to enter the plenum. This helps reduce NOx emissions.
- ❑ **Coolant passages.** Coolant circulates through these passages to remove excess heat from the intake manifold.

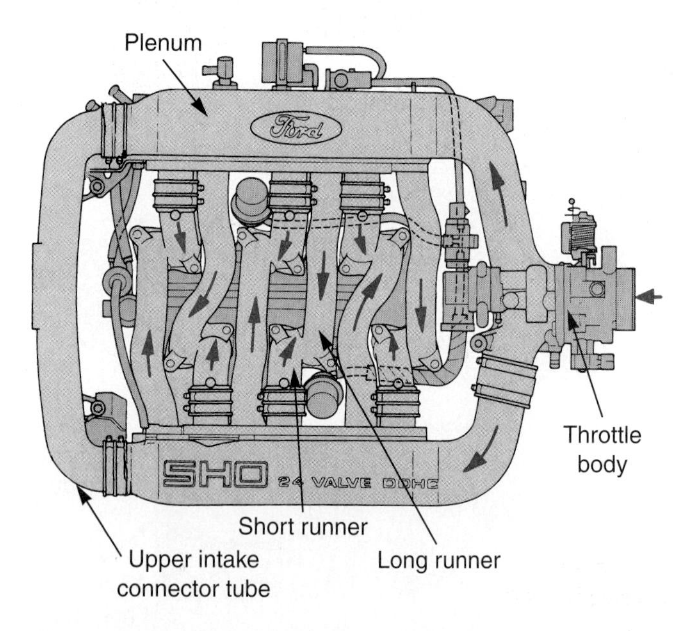

Figure 11-24. *This top view of an intake manifold shows how the runners feed out from the plenum. (Ford)*

- ❑ **Head mating surface.** This is a machined flange that mates to the cylinder head.
- ❑ **Other provisions.** Provisions for other parts may be included, such as openings for vacuum fittings, EGR valve, and fuel injectors.

Intake Manifold Types

Intake manifold types differ in plenum design, runner length, port shape, runner configuration, and method of construction. There are other considerations, but these are the most important.

Port Shape

An intake manifold can be classified by the shape of its ports. The three common port shapes are round (oval) port, D-port, and square (rectangular) port. These are pictured in **Figure 11-25.**

Split Runner Intake Manifold

A ***split runner intake manifold*** has two separate runners leading to each intake valve. Usually, the ECM operates a butterfly valve over the shorter of the two runners. At low engine speeds, air is forced into the cylinders through the long runner. When a specific engine speed is reached, the computer opens the butterfly valve over the short intake runner. This increases airflow into the engine for added power and torque.

Tuned Runner Intake Manifold

A ***tuned runner intake manifold,*** or ram intake manifold, has very long runners of equal length, **Figure 11-26.** The inertia of the air moving through the long runners tends to ram more charge into the combustion chambers. Even at low engine speeds, the large mass of air flowing through the runners wants to keep flowing even when the

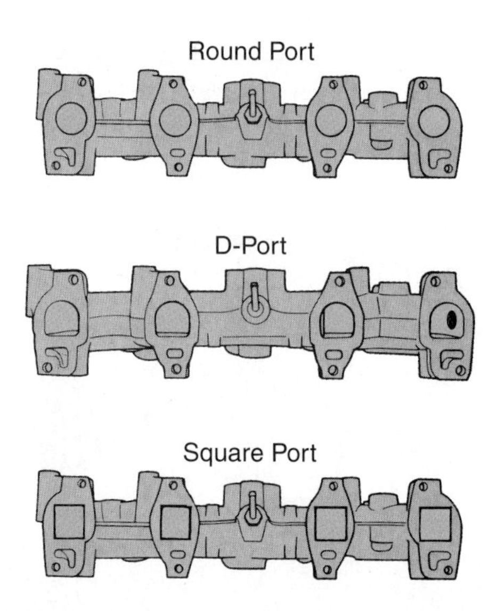

Figure 11-25. *Intake manifolds may be classified by the shape of their ports.*

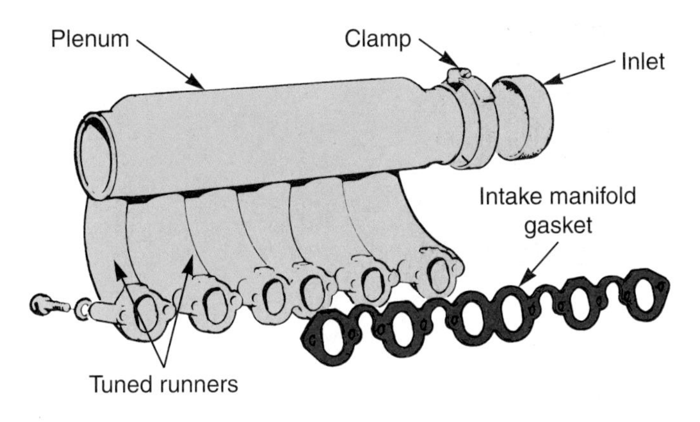

Figure 11-26. *A tuned runner intake manifold has long runners of equal length.*

cylinders are almost full. As a result, the inertia forces a little more air into the engine. The computer then increases the fuel from the injectors and horsepower is increased. This type of manifold is used on high-performance engines.

Sectional Intake Manifold

A ***sectional intake manifold*** is constructed in two or more pieces. When bolted together, the pieces form the intake manifold assembly, **Figure 11-27.** A sectional manifold allows for a more exotic shape for improved efficiency.

Single and Dual Plane Intake Manifold

In a ***single plane intake manifold,*** one plenum feeds all runners. This type is used primarily on high-performance

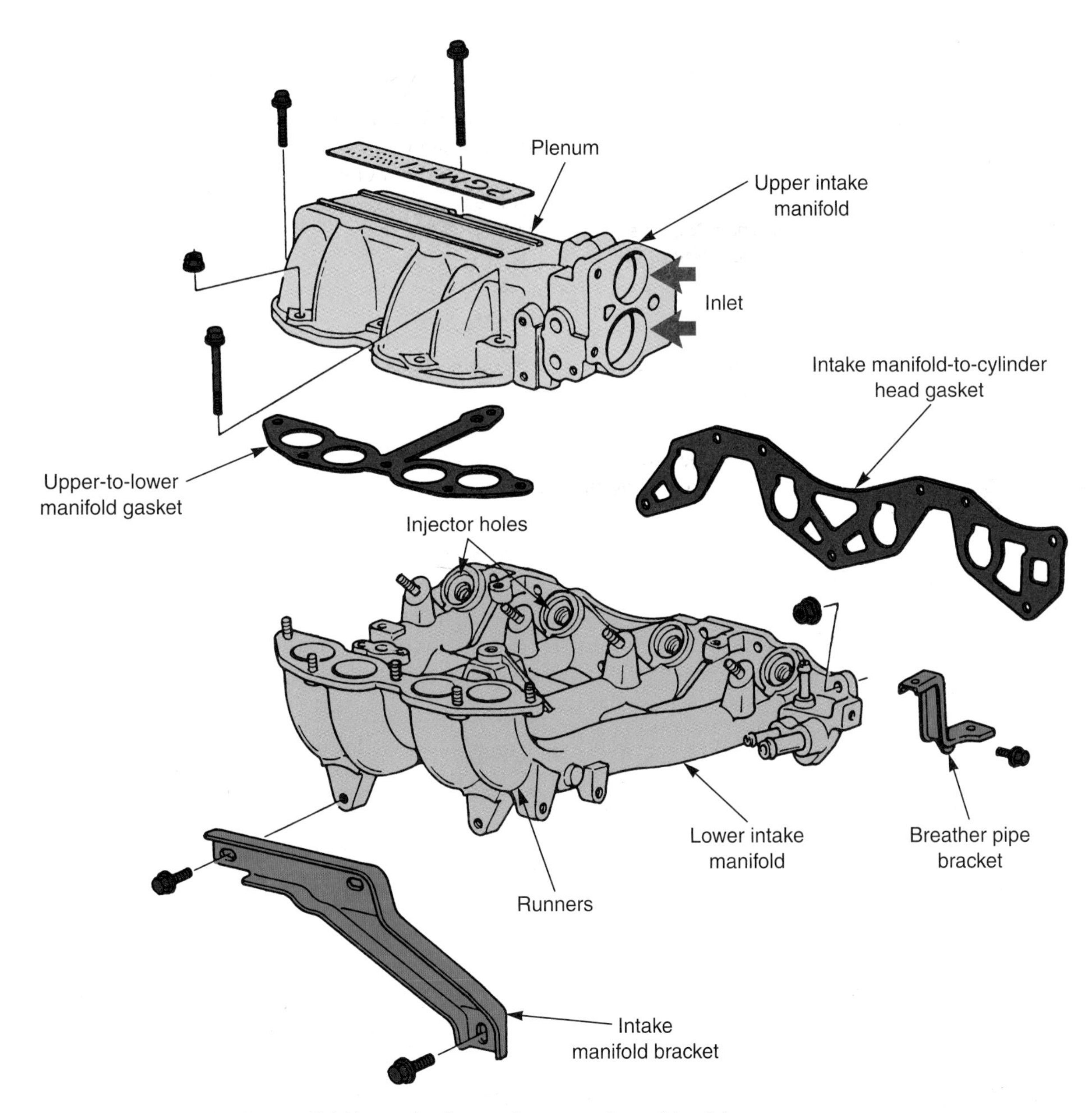

Figure 11-27. *A sectional intake manifold is made of more than one piece. (Honda)*

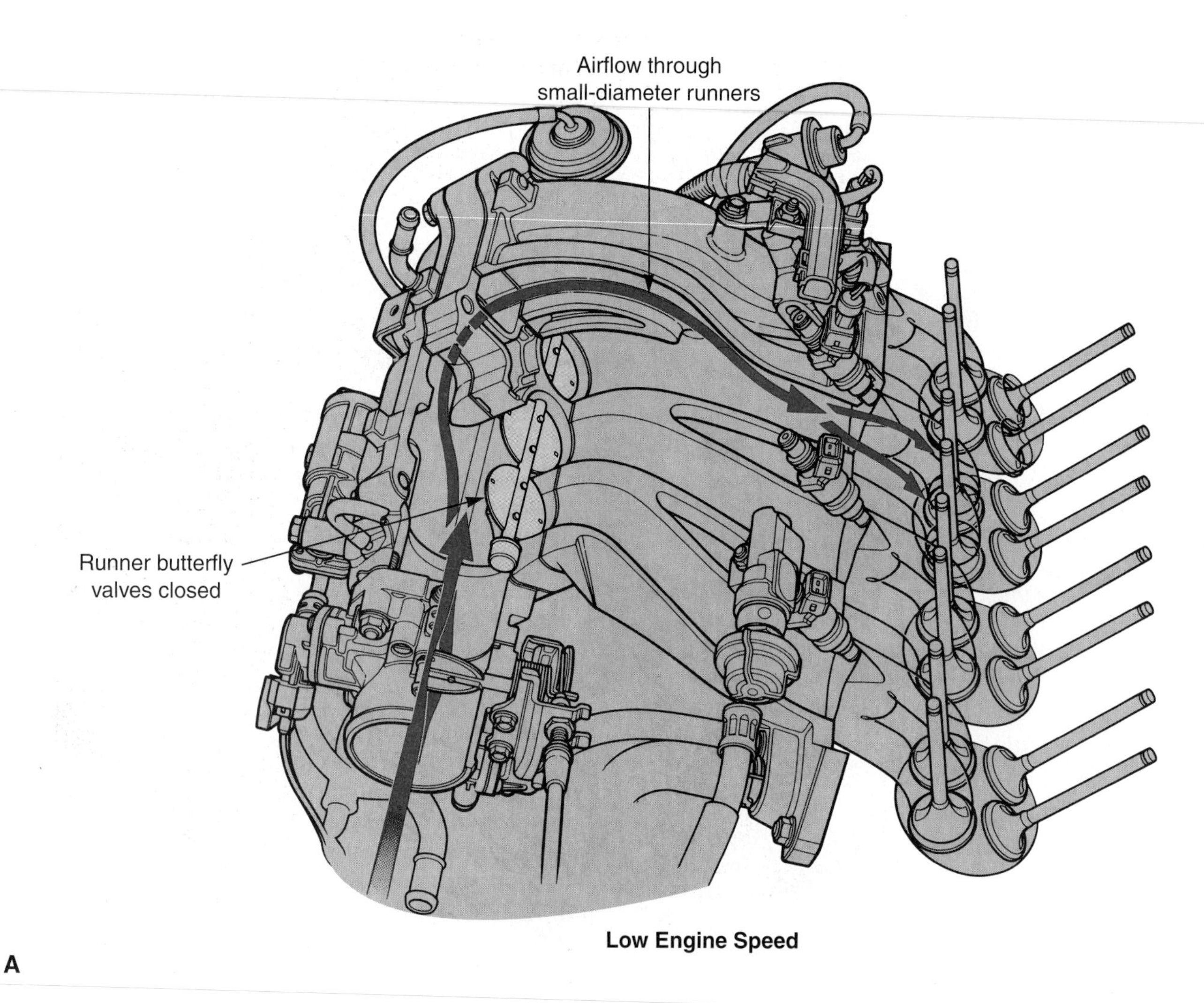

Figure 11-28. *A variable induction system. A—During low-speed operation, air is directed through the small-diameter runners. B—During high-speed operation, air can flow through both the small- and large-diameter runners.*

engines. It allows high airflow at higher engine speeds, thus increasing the engine's horsepower capability. However, a single plane manifold is not as efficient at lower engine speeds.

A ***dual plane intake manifold*** has a two-compartment plenum. This type is common on V-type engines with carburetors. It provides good fuel vaporization and distribution at low engine speeds. This makes it very efficient for everyday driving.

Variable Induction Systems

Some engines are equipped with a ***variable induction system*** to increase engine efficiency. Variable induction systems generally consist of two sets of intake runners. The primary runner has a smaller diameter than the secondary runner. Airflow through the runners is often controlled by butterfly valves. See **Figure 11-28.**

At low engine speeds, the butterfly valves in the secondary runners are closed. Consequently, only the primary runners provide airflow to the engine. This arrangement increases power and fuel economy at low engine speeds by improving fuel atomization and mixing. At higher engine speeds, the butterfly valves in the secondary runners are open. Both primary and secondary runners feed air into the combustion chambers. This improves volumetric efficiency and engine power.

Note: Intake manifold reed valves can be used in the intake runners to help prevent flow of the air-fuel mixture back into the intake manifold during valve overlap. They snap shut and only allow flow into the engine combustion chambers.

Exhaust Manifold Construction

The basic function of the ***exhaust manifold*** is to remove exhaust gasses from the cylinders. See **Figure 11-29.** During the exhaust stroke, hot gasses blow into this manifold

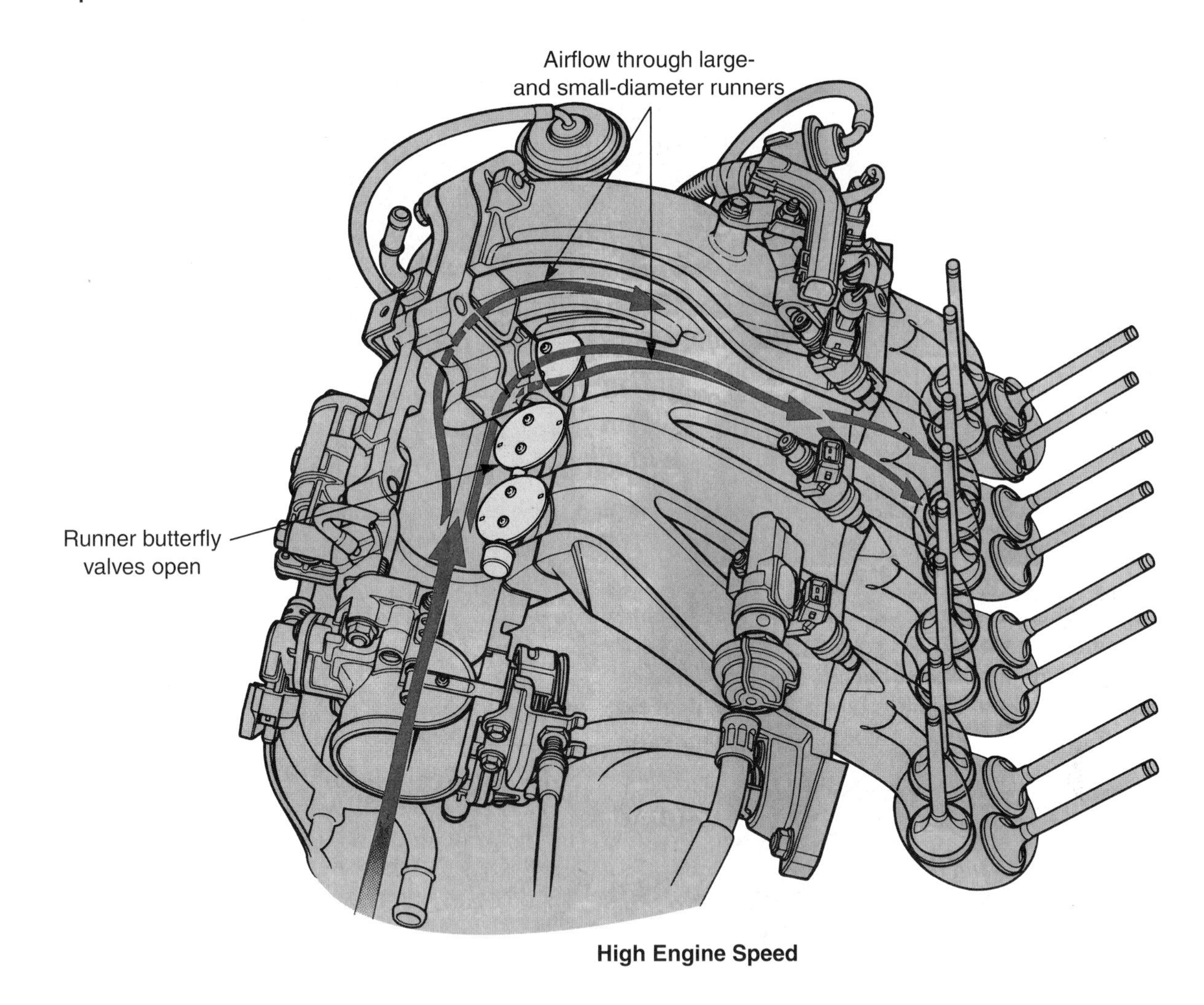

Figure 11-28. *Continued.*

before entering the rest of the exhaust system. An exhaust manifold is normally made of cast iron or lightweight steel tubing.

On naturally aspirated engines (not turbocharged), the goal of the exhaust manifold is to provide minimum exhaust backpressure. It must also be shaped to help scavenge gasses from the combustion chambers right before the exhaust valves close. To accomplish this, exhaust manifolds are often cast with smoothly curved runners to help keep the gasses flowing efficiently. Runners are the manifold pathways for the exhaust gasses.

The exhaust manifold for a turbocharged engine is designed to direct exhaust pressure into the turbocharger. The more efficiently pressure can be directed onto the turbo blades, the quicker boost pressure is developed and the less turbo lag results. Heat insulating materials can be used to help contain exhaust heat to improve thermal efficiency and to protect other parts in the engine compartment from damage. The exhaust manifold on a turbocharged engine may also have provisions for a waste gate. This device allows excess exhaust backpressure to be routed around the turbocharger to control boost.

Cast iron exhaust manifolds are used on engine applications where the manifold is exposed to very high operating temperatures. Heavy-duty engines or turbocharged engines are often equipped with cast iron exhaust manifolds.

Tubular exhaust manifolds, also called headers, are much lighter and more free flowing than cast iron exhaust manifolds. They can be shaped with very long, smooth-flowing runners to help direct exhaust gasses out of the cylinder head ports and into the exhaust system. They are often used on high-performance engines to improve power output at higher engine speeds.

When inspecting the exhaust system, look for signs of damage or leakage. The joints in the exhaust system may develop leaks over time. The muffler or catalytic converter may become clogged or restricted. Gaskets or the EGR valve may develop leaks. **Figure 11-30** shows a chart for diagnosing exhaust system problems.

Note: Exhaust manifolds are also discussed in Chapter 16, *Turbocharged and Supercharged Systems.*

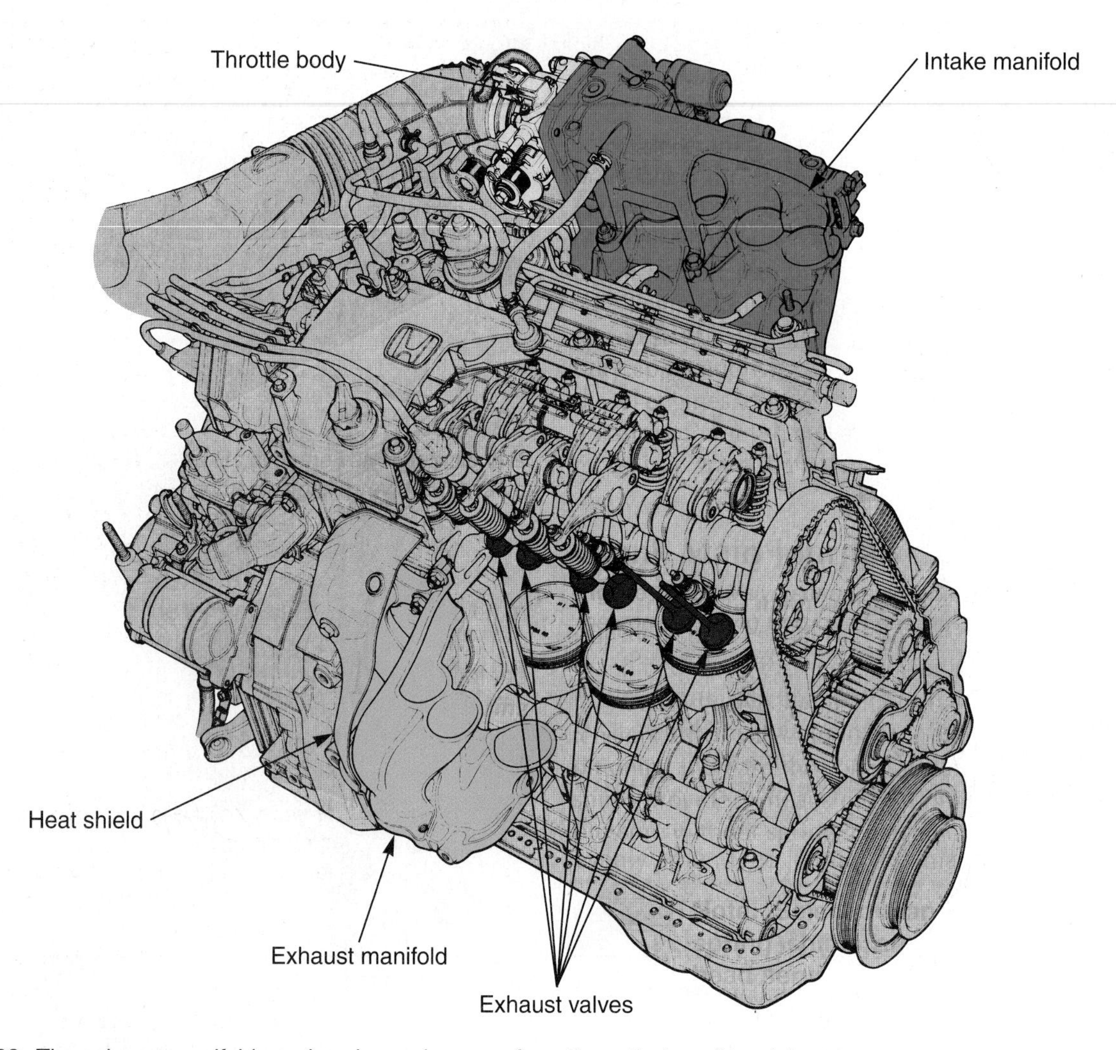

Figure 11-29. *The exhaust manifold receives burned gasses from the cylinders. (Honda)*

Summary

The engine front end consists of the parts that attach to the front of the engine, including the camshaft drive mechanism, front cover, water pump, auxiliary shafts, and so on. Various designs are now used to drive the timing mechanism and other accessory units.

A chain drive timing mechanism can be found on both cam-in-block and OHC engines. Either a double-roller or multiple-link chain can be used. The crankshaft sprocket is one-half as large as the camshaft sprocket. Timing marks on both sprockets are used when installing the chain to ensure proper orientation.

A chain tensioner is a spring-loaded device that presses on the chain to take up excess slack as the chain and sprockets wear out. A plastic, Teflon, or fiber block contacts the chain to reduce friction. A chain guide may be needed on long lengths of chain to prevent chain slap. An auxiliary chain may be used to drive the engine oil pump, balancer shafts, or other units. Extra auxiliary sprockets are also needed.

A belt drive timing mechanism is normally used on OHC engines because of the great distance between the crankshaft and camshaft. A timing belt is very quiet and smooth. The teeth of the cogged belt engage teeth on the crankshaft and camshaft sprockets. Timing marks on the sprockets ensure proper orientation when the belt is installed. A belt tensioner keeps the belt tight on its sprockets.

A gear drive timing mechanism is generally for heavy-duty applications. A gear drive is very dependable, but the gears are heavy and noisy. Timing marks on the gears ensure proper orientation when the gears are installed. Timing gear backlash is the small clearance between the gear teeth, which is needed to allow for heat expansion and lubricating oil.

The engine front cover surrounds the timing chain and sprockets or timing gears. It contains a crankshaft oil seal that keeps oil from leaking out around the crankshaft snout. The cover can be made of stamped steel or cast aluminum. A timing belt cover surrounds the belt and sprockets to keep debris away from the rubber belt.

A vibration damper is a heavy wheel mounted on rubber. It fits over the crankshaft snout. The damper is used to reduce torsional vibration produced by the twisting-untwisting action of the crankshaft due to power pulses. This protects the crankshaft from potential breakage.

Exhaust System Diagnosis		
Condition	**Possible cause**	**Correction**
Excessive exhaust noise (under hood).	1. Damaged exhaust manifold. 2. Manifold to cylinder head leak. 3. EGR valve leakage: (a) EGR valve to manifold gasket. (b) EGR valve to EGR tube gasket. (c) EGR tube to manifold tube nut. 4. Exhaust flex joint. (a) Spring height, installed not correct. (b) Exhaust sealing ring defective. 5. Pipe and shell noise from front exhaust pipe.	1. Replace manifold. 2. Tighten manifold and/or replace gasket. 3. (a) Tighten nuts or replace gasket. (b) Tighten nuts or replace gasket. (c) Tighten tube nut. 4. (a) Check spring height, both sides (specification is 32.5 mm, 1.28 inch) look for source of spring height variation if out of specification. (b) Inspect seal for damage on round spherical surface. If no damage is evident, check for exhaust obstruction causing high back pressure on heavy acceleration. 5. Characteristic of single wall pipes.
Excessive exhaust noise.	1. Leaks at pipe joints. 2. Burned or blown or rusted out muffler, tailpipe, or exhaust pipe. 3. Restriction in muffler or tailpipe. 4. Converter material in muffler.	1. Tighten clamps at leaking joints. 2. Replace muffler or muffler tailpipe or exhaust pipe. 3. Remove restriction, if possible or replace as necessary. 4. Replace muffler and converter assemblies. Check fuel injection and ignition systems for proper operation.
Exhaust system mechanical noise.	1. Improperly aligned exhaust system. 2. Support brackets loose, bent, or broken. 3. Incorrect muffler or pipes. 4. Baffle loose in muffler. 5. Manifold heat control valve rattles. 6. Worn engine mounts. 7. Damaged or defective catalytic converter.	1. Align system. 2. Tighten or replace brackets. 3. Install correct muffler or pipes. 4. Replace muffler. 5. Replace thermostatic spring. 6. Replace engine mounts. 7. Repair or replace converter.
Engine lacks power.	1. Clogged muffler. 2. Clogged or kinked exhaust or tailpipe. 3. Muffler or pipes too small for vehicle. 4. Catalytic converter clogged or crushed shut.	1. Replace muffler. 2. Replace pipe. 3. Install muffler and pipes of the correct size and type. 4. Replace with new converter.

Figure 10-30. *Use this chart to diagnosis problems with the exhaust system and determine repairs.*

Either V-type or a ribbed pulley is used on the crankshaft to operate the alternator, power steering pump, air conditioning compressor, and so on. The pulley may be bolted to the damper or can be an integral part of the damper.

There are several different designs for intake manifolds. Most designs have similar basic parts. These include a plenum, runners, intake, and coolant passages.

A tuned runner manifold has long, equal-length runners. The inertia of airflow forces extra air into the cylinders as the valves open at higher engine speeds. This increases engine power because of a supercharging type effect. A sectional intake manifold is made in two or more pieces. It allows a more complex shape. Other intake manifold designs exist, including variable induction systems.

Review Questions—Chapter 11

Please do not write in this text. Write your answers on a separate sheet of paper.

1. The engine front end includes:
   (A) Camshaft, valves, and timing mechanism.
   (B) Front cover, timing mechanism, and crankshaft.
   (C) Camshaft, front cover, and timing mechanism.
   (D) Front cover, timing mechanism, and water pump.
2. List three different camshaft drives.
3. The camshaft drive ratio is:
   (A) 3:1.
   (B) 1:1.
   (C) 2:1.
   (D) 4:1.
4. What is the purpose of *timing marks?*
5. The two types of timing chains are _____ and _____.
6. What is the purpose of a chain tensioner?
7. An auxiliary chain may be used to drive:
   (A) Balancer shafts.
   (B) Camshaft.
   (C) Alternator.
   (D) None of the above.
8. How does an *oil slinger* work?
9. A(n) _____ drive timing mechanism is usually used on heavy-duty, cam-in-block applications because of its dependability.
10. Define *timing gear backlash.*
11. What is the difference between an *engine front cover* and a *timing belt cover?*
12. _____ is a result of the twisting and untwisting of the crankshaft due to normal power pulses.
13. What is the basic function of an intake manifold?
14. List five basic parts of an engine intake manifold.
15. What is the basic function of an exhaust manifold?

ASE-Type Questions—Chapter 11

1. An engine's timing mechanism is sometimes used to power the _____.
   (A) alternator
   (B) oil pump
   (C) balance shaft(s)
   (D) Both B and C.
2. Technician A says that a timing belt is normally used in OHC engines. Technician B says that timing gears are primarily used in cam-in-block automotive engines. Who is correct?
   (A) A only.
   (B) B only.
   (C) Both A and B.
   (D) Neither A nor B.
3. Technician A says that the sprockets in a timing mechanism can be adjusted to compensate for timing chain or belt wear. Technician B says that a chain or belt tensioner is sometimes provided to compensate for timing chain or belt wear. Who is correct?
   (A) A only.
   (B) B only.
   (C) Both A and B.
   (D) Neither A nor B.
4. Technician A says that a timing belt cover holds the crankshaft oil seal. Technician B says that a timing belt cover contains oil and protects the timing belt mechanism. Who is correct?
   (A) A only.
   (B) B only.
   (C) Both A and B.
   (D) Neither A nor B.
5. Which of the following is another name for the vibration damper?
   (A) Intermediate pulley.
   (B) Harmonic balancer.
   (C) Crankshaft damper.
   (D) Both B and C.
6. Technician A says that the design of an engine's intake manifold can help create a ram effect. Technician B says that an engine's intake manifold carries air to the cylinders. Who is correct?
   (A) A only.
   (B) B only.
   (C) Both A and B.
   (D) Neither A nor B.
7. All of the following are basic parts of an engine's intake manifold *except:*
   (A) coolant passages.
   (B) riser bore.
   (C) cylinder mating surface.
   (D) runners.

8. Technician A says that the camshaft sprocket is twice the size of the crankshaft sprocket. Technician B says that the camshaft spins twice as fast as the crankshaft. Who is correct?
   (A) A only.
   (B) B only.
   (C) Both A and B.
   (D) Neither A nor B.
9. Technician A says that the vibration damper corrects for the twisting and untwisting of the crankshaft. Technician B says that the vibration damper corrects for torsional vibration in the crankshaft. Who is correct?
   (A) A only.
   (B) B only.
   (C) Both A and B.
   (D) Neither A nor B.
10. Technician A says that the basic function of the exhaust manifold is to remove exhaust gasses from the cylinders. Technician B says that exhaust gasses may be used to drive a turbocharger. Who is correct?
    (A) A only.
    (B) B only.
    (C) Both A and B.
    (D) Neither A nor B.

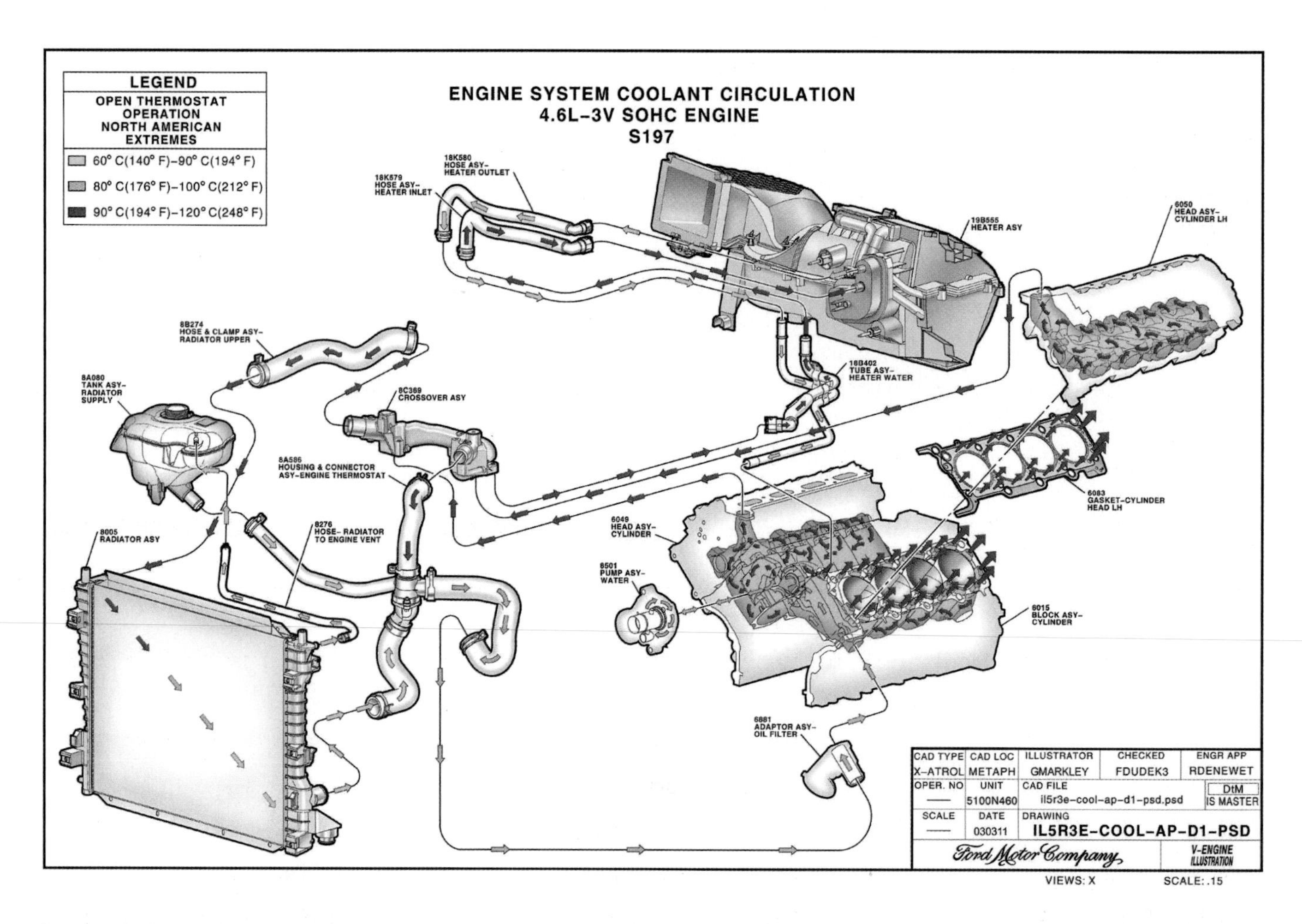

A properly functioning cooling system is an important part of the engine. Study this illustration of a cooling system. Then, read Chapter 12. (Ford)

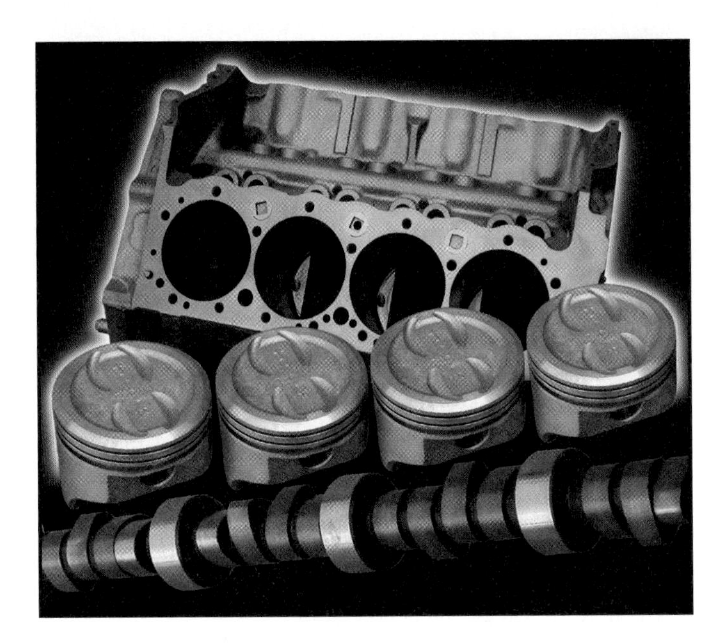

Chapter 12

Cooling System Operation and Service

After studying this chapter, you will be able to:

- ❑ Explain the operation of a cooling system.
- ❑ Compare cooling system types.
- ❑ Troubleshoot cooling system problems.
- ❑ Perform basic cooling system tests.
- ❑ Replace or repair faulty cooling system components.

Know These Terms

Air-cooled system
Antifreeze
Block heater
Chemical flushing
Closed cooling system
Coolant leaks
Coolant strength
Cooling system
Cooling system bleed screw
Cooling system fan
Cooling system flushing
Cooling system hydrometer
Cooling system pressure test
Conventional-flow cooling system
Electric engine fan
Engine coolant
Engine operating temperature
Engine overcooling
Engine overheating
Engine-powered fan
Fast flushing
Flex fan
Fluid coupling fan clutch
Heater hoses
Hot spot
Liquid-cooled system
Lower radiator hose
Open cooling system
Pressure tester
Radiator
Radiator cap
Radiator cap pressure test
Radiator hoses
Radiator shroud
Reverse flushing
Reverse-flow cooling system
Thermostat
Thermostatic fan clutch
Upper radiator hose
Water pump

The cooling system is vitally important to engine performance and service life. Without a cooling system, an engine will overheat and self-destruct in a matter of minutes. As an engine technician, you must be prepared to work on the cooling system. It is the source of many common causes of engine failure.

Cooling System Functions

A cooling system is shown in **Figure 12-1.** The engine's ***cooling system*** has four basic functions:

- Remove excess heat from the engine.
- Maintain a constant engine operating temperature.
- Quickly increase the temperature of a cold engine.
- Provide for heating of the vehicle's passenger compartment.

Removing Engine Heat

The burning of the air-fuel mixture inside the combustion chambers produces a tremendous amount of heat. Some of the heat from combustion produces expansion and pressure for piston movement. However, some of the heat of combustion flows out through the exhaust and into the parts of the engine. The flame temperature can reach 4500°F (2485°C). This is hot enough to melt metal parts. Without removal of this heat, the engine would be seriously damaged.

Maintain Operating Temperature

Engine operating temperature is the temperature of the engine coolant during normal engine operating conditions. When an engine warms to operating temperature, its parts expand. This ensures that all part clearances are correct. It also ensures proper combustion, emission output levels, and engine performance. Typically, an engine's operating temperature is between 180°F and 195°F (80°C and 90°C).

Reaching Temperature Quickly

An engine must warm up rapidly to prevent poor combustion, part wear, oil contamination, reduced fuel economy, and other problems. For instance, the aluminum pistons in a cold engine will not be expanded by heat to their correct size. This can cause too much clearance between the pistons and cylinder walls. The oil in a cold engine will be very thick. This can reduce lubrication protection and increase engine wear. The fuel mixture will also not vaporize and burn as efficiently in a cold engine as it does in a warn engine. Also, with a cold engine, too much combustion heat will conduct into the cold engine parts so less heat is available to produce pressure in the cylinder.

Heater Operation

A cooling system commonly circulates coolant to the vehicle's heater. The heater is used to warm the passenger

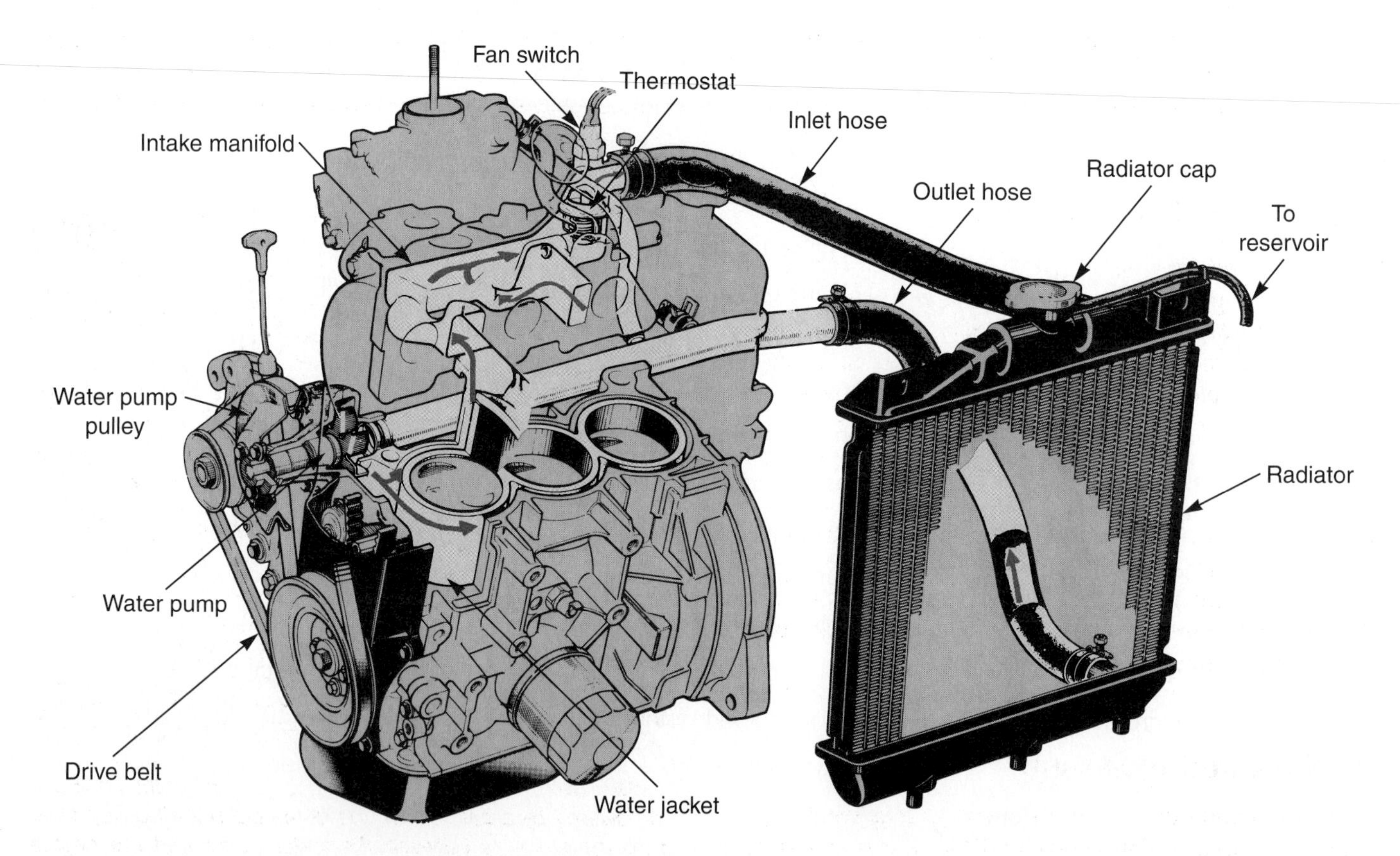

Figure 12-1. *Study the parts of a cooling system. (General Motors)*

compartment. Since the engine coolant is approximately 200°F (90°C), heat can be removed from the coolant and used to warm the passenger compartment in cold weather.

Cooling Systems Types

There are two major types of engine cooling systems: liquid and air. Modern automotive engines are liquid cooled. Air-cooled engines are no longer used in production cars and light trucks.

An ***air-cooled system*** uses cooling fins and outside airflow to remove excess heat from the engine. The cooling fins increase the surface area of the metal around the cylinder. This allows enough heat to be transferred from the engine to prevent damage.

Most air-cooled engines have plastic or sheet metal ducts and shrouds to route air over the cylinder fins. Thermostatically controlled flaps regulate airflow and engine operating temperature. This type of cooling system does not have a radiator, water pump, or coolant reservoir.

A ***liquid-cooled system*** circulates a coolant through internal passages cast in the engine called coolant passages, or "water jackets." The ***coolant*** is a solution of water and antifreeze. As the coolant circulates through the engine, it collects heat and carries it to a radiator. The radiator transfers the heat to the outside air. A liquid-cooled system has several advantages over an air-cooled system:

- More precise control of engine temperature.
- Less temperature variation inside the engine.
- Reduced exhaust emissions because of the better temperature control.
- Improved operation of the passenger compartment heater.

Closed Cooling System

A ***closed cooling system*** has an expansion tank or reservoir connected to the radiator by an overflow tube. The overflow tube is routed into the bottom of the reservoir. Pressure or vacuum acts on a valve in the radiator cap. Vacuum in the radiator will pull coolant out of the reservoir. Excess pressure in the radiator will push coolant into the reservoir. This keeps the cooling system filled at all times.

Warning: Some closed cooling system reservoirs are pressurized and have a cap similar to a radiator cap. Do *not* open these reservoirs when the cooling system is under pressure or severe burns could result.

Open Cooling System

The overflow tube in an ***open cooling system*** is not connected to a coolant reservoir. When excess pressure in the radiator pushes coolant out, the coolant leaks onto the ground. Also, there is no means of automatically adding coolant to the radiator when needed. Open cooling systems are no longer used on production vehicles.

Operation of a Liquid-Cooled System

A basic automotive liquid-cooled system consists of:

- Water pump.
- Radiator.
- Fan.
- Radiator hoses.
- Thermostat.

In a ***conventional-flow cooling system,*** coolant circulates through the engine block, travels to the head, and then returns to the radiator. Hot coolant from the engine enters the top of the radiator. The water pump draws the cooled liquid from the bottom of the radiator.

In a ***reverse-flow cooling system,*** however, coolant circulates through the head before moving through the block and into the radiator. This arrangement helps produce a more uniform temperature throughout the engine, especially around the hot exhaust valves. As a result, the engine is less likely to suffer spark knock or ping. A reverse-flow cooling system allows the use of higher compression ratios and greater spark advance to improve power and efficiency.

Water Pump

The ***water pump*** uses centrifugal force to circulate coolant through the engine coolant passages, hoses, and radiator. It is usually powered by a belt running off of the crankshaft pulley, **Figure 12-2.**

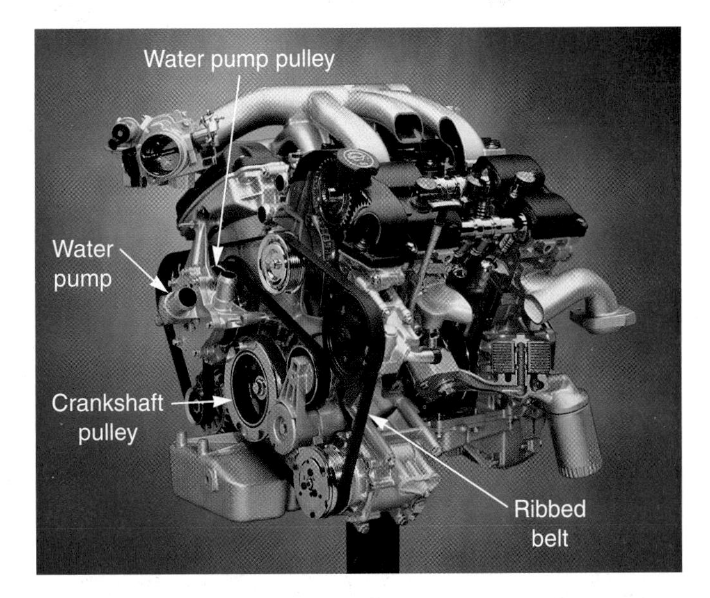

Figure 12-2. *This water pump bolts to the front of the engine. It is powered by a belt running off of the crankshaft pulley. Note how the pulley is between the water pump and the engine block. (Ford)*

The water pump is normally mounted on the front of the engine. A water pump gasket fits between the engine and pump housing to prevent coolant leakage. RTV sealer may be used instead of a gasket.

Radiator

The ***radiator*** transfers the heat collected by the coolant to the outside air. The radiator is normally mounted in front of the engine so cool, outside air can flow through it. Hot engine coolant is circulated through the radiator tanks and core tubes. Heat is transferred from the coolant to the core's tubes and fins. Since cooler air is flowing over the radiator fins, heat is transferred from the radiator's fins to the air. In this way, the temperature of the coolant is reduced before the coolant flows back into the engine.

Radiator Cap

The ***radiator cap*** performs several functions:

- Seals the top of the radiator filler neck.
- Allows the system to be pressurized.
- Relieves excess pressure and vacuum to protect against system damage.
- Allows coolant flow into and out of the coolant reservoir in a closed system.

The radiator cap locks onto the radiator filler neck. Rubber or metal seals make the cap-to-neck joint airtight. The radiator cap pressure valve has a spring-loaded disk that contacts the filler neck. The spring pushes the valve into the neck.

Many of the metal surfaces inside the coolant passages can be above 212°F (100°C). At sea level, water boils at 212°F (100°C). However, for every pound of pressure increase per square inch (psi increase), the boiling point goes up about 3°F (1.7°C). The radiator cap works on this principle. Typical radiator cap pressure is 12 psi to 16 psi (85 kPa to 110 kPa). This raises the boiling point of the engine coolant to about 250°F to 260°F (120°C to 130°C).

If the engine overheats and pressure exceeds the cap rating, the pressure valve opens, **Figure 12-3.** Excess pressure forces coolant out the overflow tube and into the reservoir. This prevents high pressure from rupturing the radiator, gaskets, seals, or hoses.

As the coolant temperature drops after engine operation, the pressure inside the radiator decreases to below atmospheric pressure. The radiator cap vacuum valve opens to allow reverse flow back into the radiator when pressure inside the radiator drops. It is a smaller valve located in the center of the bottom part of the cap. The cooling and contraction of the coolant and air in the system can decrease the coolant volume and pressure. Without the vacuum valve, the radiator hoses and radiator tanks could collapse when the engine cools.

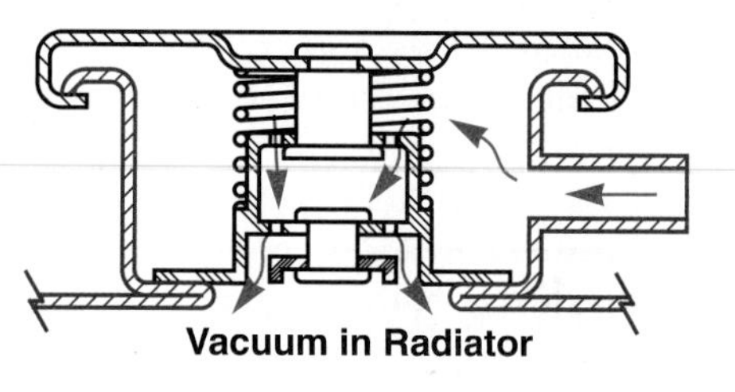

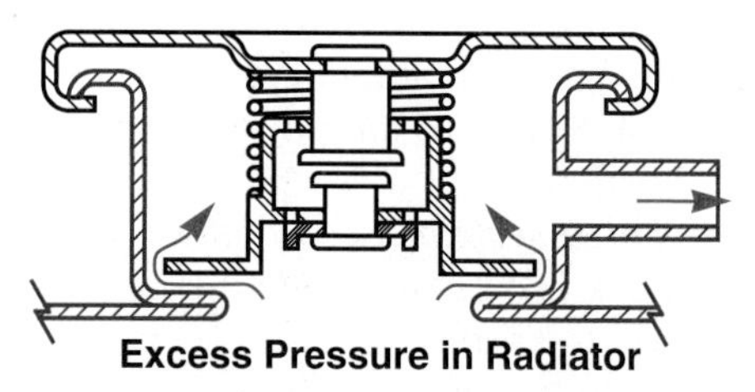

Figure 12-3. *When a vacuum is created inside the radiator, the vacuum valve in the radiator cap opens to allow fluid from the reservoir to enter the radiator (top). When there is excess pressure inside the radiator, the pressure valve opens to allow excess coolant into the reservoir (bottom).*

Radiator Shroud

A huge volume of air flows through the radiator. The ***radiator shroud*** keeps air from circulating between the back of the radiator and the front of the fan. Without a fan shroud, the engine could overheat. The shroud is typically attached to the rear of the radiator and surrounds the fan, **Figure 12-4.**

Cooling Fans

A ***cooling system fan*** pulls air through the core of the radiator and over the engine to help remove heat. It increases the volume of air flowing through the radiator, especially at low vehicle speeds. The fan may be powered by the engine or an electric motor.

Engine-Powered Fans

An ***engine-powered fan*** is bolted to the water pump hub and pulley. Sometimes, a spacer fits between the fan and pulley to move the fan closer to the radiator. This type of fan has generally been replaced by electric fans, which are discussed in the next section.

A ***flex fan*** has thin, flexible blades that alter airflow with engine speed. At low engine speeds, the fan blades remain curved to pull more air through the radiator. At higher engine speeds, the blades flex until they are almost straight. This reduces fan action and saves engine power.

A ***fluid coupling fan clutch*** is designed to slip at higher engine speeds to reduce fan action. The clutch is filled with a silicone-based oil. At a specific fan speed, there is enough load to make the clutch slip.

A ***thermostatic fan clutch*** has a temperature-sensitive, bimetallic spring that controls fan action. The spring controls oil flow in the fan clutch. When cold, the spring causes the clutch to slip, speeding engine warm-up.

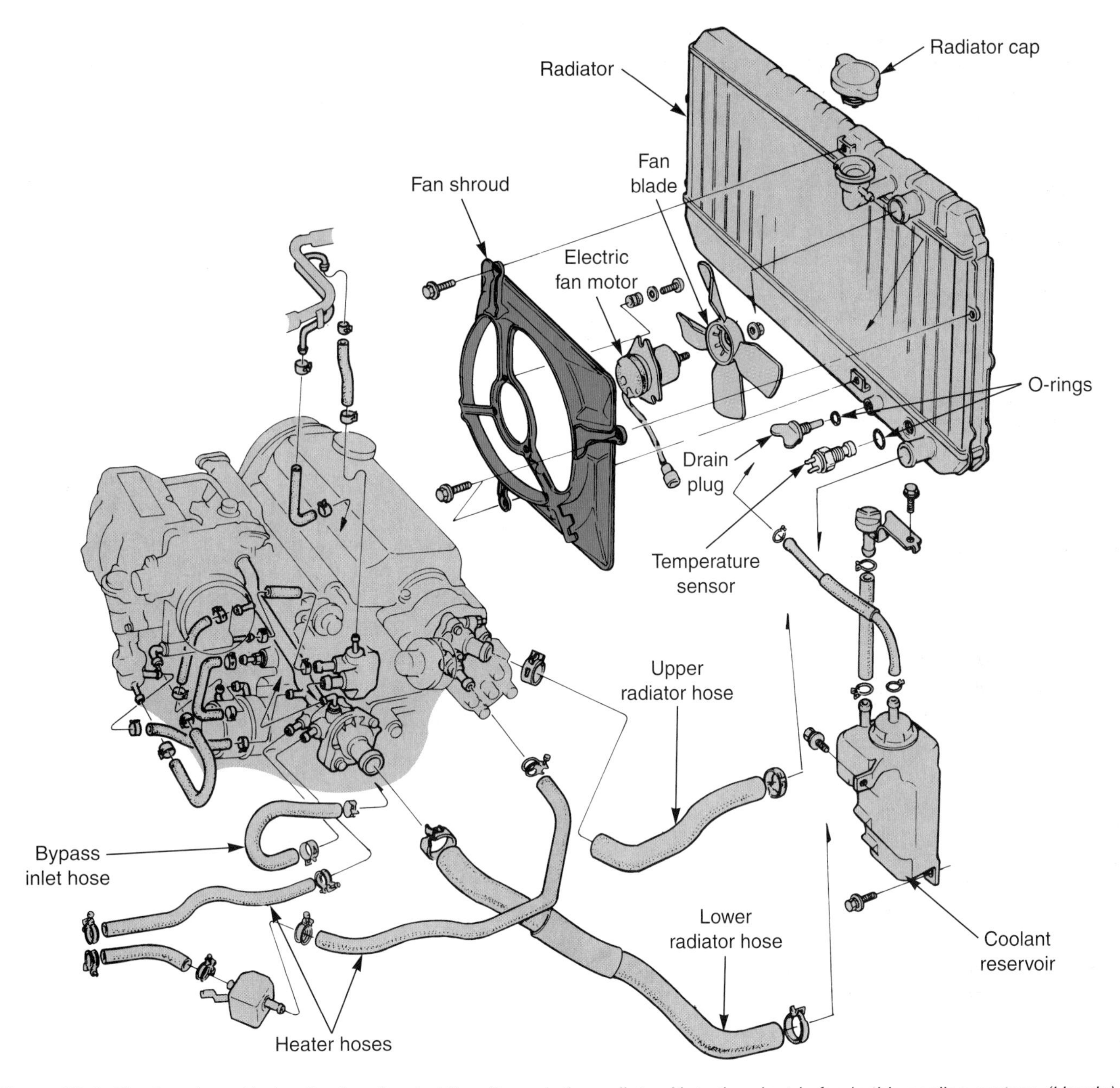

Figure 12-4. *The fan shroud helps the fan direct airflow through the radiator. Note the electric fan in this cooling system. (Honda)*

After reaching operating temperature, the spring moves and causes the clutch to lock up, providing forced-air circulation.

Electric Fans

An ***electric engine fan*** uses an electric motor and a thermostatic switch to provide cooling action. An electric fan is needed on transverse-mounted engines because the water pump is located away from the radiator. See **Figure 12-4.**

The fan motor is a small, dc motor mounted on a bracket secured to the radiator. A metal or plastic fan blade is mounted on the end of the motor shaft.

The fan switch is a temperature-sensitive switch that controls fan motor operation, **Figure 12-5.** When the engine is cold, the switch is open (off). This keeps the fan from spinning and speeds engine warm-up. After warm-up, the switch is closed (on) and the fan operates. Some electric fans are computer controlled.

Radiator and Heater Hoses

Radiator hoses carry coolant between the coolant passages in the engine and the radiator. The hoses are flexible to withstand the vibrating and rocking of the engine. The ***upper radiator hose*** is normally connected between the thermostat housing on the engine intake manifold or cylinder head and the radiator. The ***lower radiator hose*** is connected between the water pump inlet and the radiator. Hose clamps hold the radiator and heater hoses on their fittings.

The lower radiator hose frequently has a spring inside of it. This is designed to prevent collapsing of the hose. The

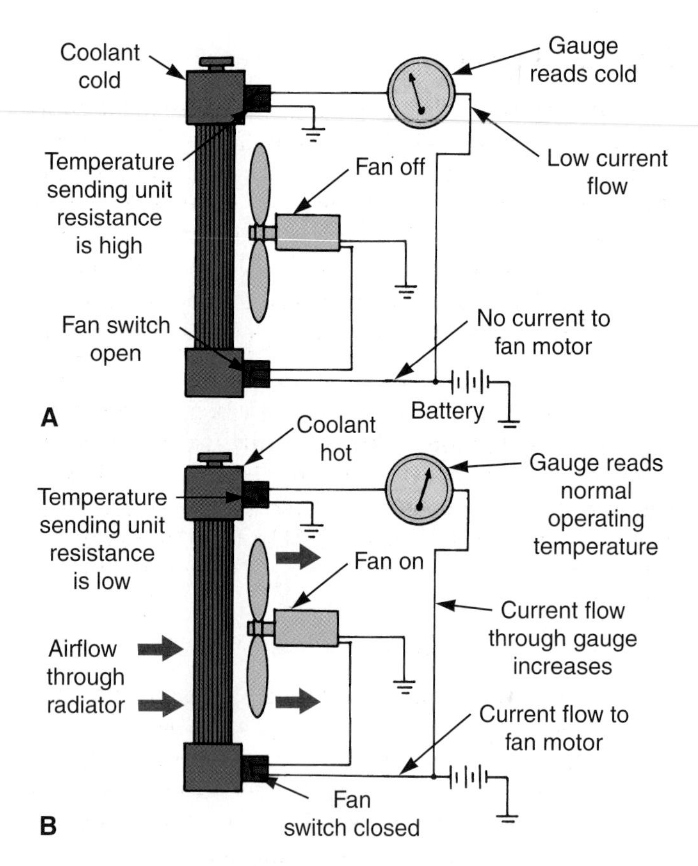

Figure 12-5. *Study how an electric cooling fan works. A—When the engine is cold, the fan switch is open; the fan stays off. B—When the engine is at operating temperature, the switch closes and the fan motor runs.*

lower hose is exposed to suction from the water pump. The spring ensures that the inner lining of the hose does not tear away or close up, stopping circulation.

Heater hoses are small-diameter hoses that carry coolant to the heater core. The heater core is a small, radiator-like device under the dash or on the firewall. By using a fan to blow air across the heater core, the passenger compartment can be heated in cold weather.

Thermostat

The ***thermostat*** senses engine temperature and controls coolant flow through the radiator, **Figure 12-6.** It reduces flow when the engine is cold and increases flow when the engine is hot. The thermostat normally fits under a thermostat housing that is located between the engine and one end of the upper radiator hose.

The thermostat has a wax-filled pellet contained within a cylinder and piston assembly. A spring holds the piston and valve in a normally closed position. When the coolant temperature increases, the coolant heats the thermostat and the pellet expands. This pushes the valve open. When the pellet and thermostat cool as the coolant temperature decreases, spring pressure closes the valve.

The thermostat temperature rating is stamped on the thermostat. This rating indicates the operating (opening) temperature of the thermostat. Normal ratings are between 180°F and 195°F (80°C and 90°C). High thermostat heat ranges are used in modern automobiles to help reduce exhaust emissions and increase combustion efficiency.

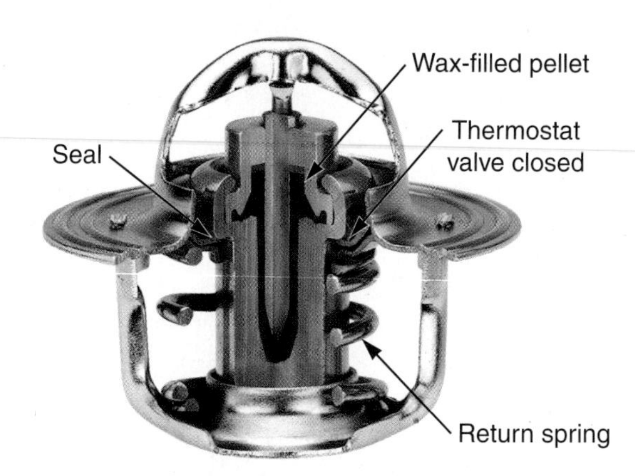

Figure 12-6. *A thermostat controls the flow of coolant through the engine and radiator. (Gates)*

A bypass valve permits coolant circulation through the engine when the thermostat is closed. If the coolant cannot circulate, hot spots will develop inside the engine.

Antifreeze

Engine coolant is a mixture of antifreeze and water. ***Antifreeze*** may have a base of ethylene glycol or propylene glycol. Antifreeze has several functions.

- ❑ Freeze protection.
- ❑ Rust prevention.
- ❑ Lubrication.
- ❑ Heat transfer.

Coolant freezing can cause serious cooling system or engine damage. Antifreeze keeps the coolant from freezing when the outside temperature is below 32°F or (0°C). Pure water freezes at 32°F (0°C). As ice forms, it expands. This expansion can produce tons of force. The water pump housing, cylinder head, engine block, radiator, or other parts could be cracked and ruined.

Antifreeze also prevents rust and corrosion inside the cooling system. It provides a protective film on part surfaces. The accumulation of rust and corrosion can reduce the cooling ability of the cooling system.

Antifreeze acts as a lubricant for the water pump. It increases the service life of the water pump bearings and seals.

Antifreeze conducts heat better than pure water. Therefore, a cooling system containing antifreeze and water will cool the engine better than if just water is used. For this reason, antifreeze is normally recommended in all seasons.

Engine Block Heater

A ***block heater*** may be used on an engine to aid engine starting and warm up in cold weather. It is simply a heating element mounted at some point in the cooling system, often in a coolant passage. Block heaters are commonly found on diesel engines.

Cooling System Troubleshooting and Service

A cooling system failure can lead to cracked cylinder blocks, warped cylinder heads, blown head gaskets, burned valves, melted pistons, and other expensive damage to the engine. It is important that you know how to correctly test, service, and repair this important engine system. An inspection will frequently let you find the source of the cooling system problem. Obvious troubles include:

- Coolant leaks.
- Loose or missing fan belts.
- Low coolant level.
- Leaves and debris covering the outside of the radiator.
- Coolant in the oil or oil in the coolant.
- Combustion leakage into coolant.

Cooling system problems can be grouped into three general categories: leaks, overheating, and overcooling.

Warning: Keep your hands and tools away from a spinning engine fan. Wear eye protection and stand behind, not over, the spinning fan blade. Then, if tools are dropped into the fan or a fan blade breaks, you are not likely to be injured by flying parts. Also, be aware that electric fans can turn on even when the ignition key is in the off position.

Coolant Leaks

Coolant leaks can occur almost anywhere in the cooling system, **Figure 12-7.** The leaking fluid will smell like antifreeze and have the same general color. Coolant leaks show up as wet, discolored areas in the engine compartment or on the ground. These areas are usually a darkened green, orange, or rust color. If not visible, the leak may be internal, such as from a cracked head or block or a blown head gasket. A low coolant level may also indicate a leak.

Warning: Never remove a radiator cap when the engine is hot. The release of pressure can make the coolant begin to boil and expand. Boiling coolant can cause severe burns! On a closed system, check the coolant level at the reservoir.

Engine Overheating

Engine overheating is a serious problem that can cause major engine damage. The driver may notice the engine temperature light glowing, temperature gauge reading high, or coolant boiling. In cases of extreme overheating, steam can blow out of the overflow reservoir. Common causes of engine overheating are:

- **Low coolant level.** A leak or lack of maintenance has allowed the coolant level to drop too low.
- **Rust or scale.** Mineral accumulations in the system have clogged the radiator core or built up in the coolant passages.
- **Stuck thermostat.** The thermostat fails to open normally, restricting coolant flow.

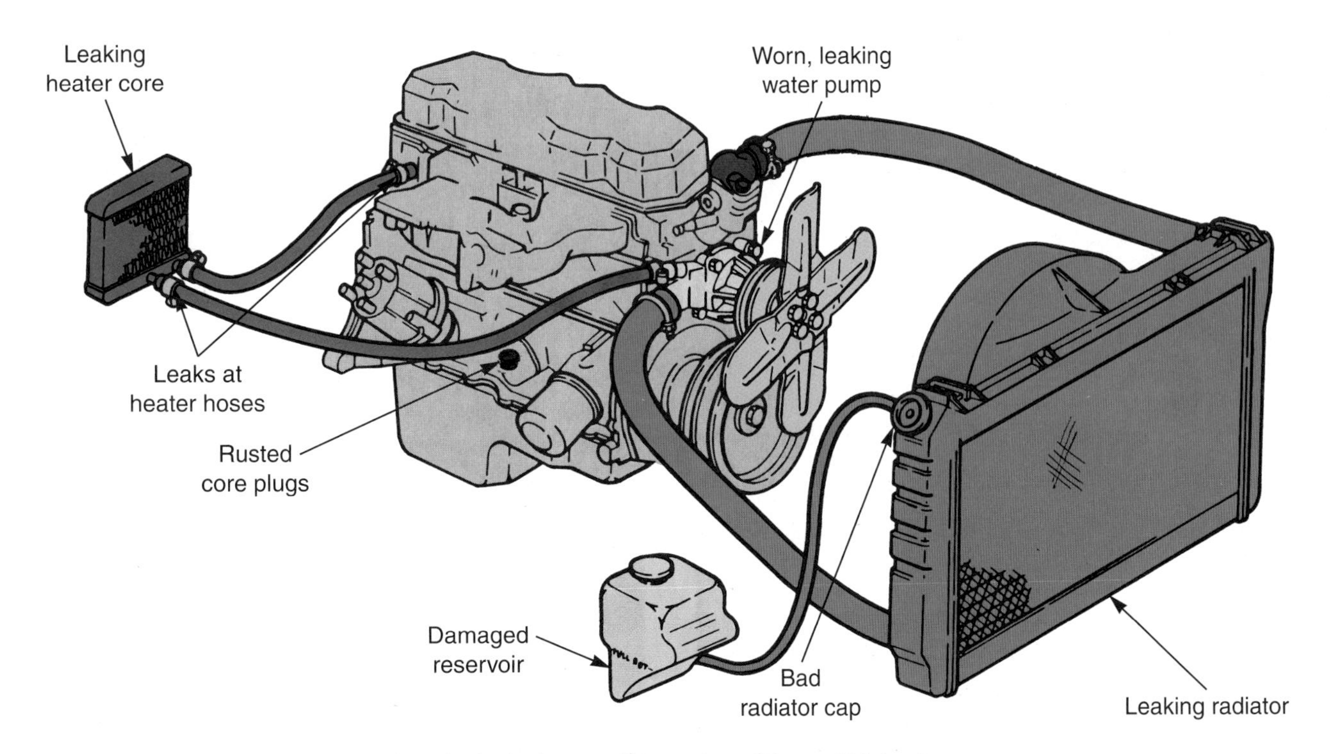

Figure 12-7. *These are common locations for leaks in a cooling system. (General Motors)*

- ❑ **Retarded ignition timing.** Late ignition timing transferring too much heat past the exhaust valves, through the ports, and into the exhaust manifold.
- ❑ **Loose fan belt.** The water pump drive belt slips under load and coolant circulation is reduced, **Figure 12-8.**
- ❑ **Bad water pump.** A leaking seal, broken pump shaft, or damaged impeller blade prevents normal pumping action.
- ❑ **Collapsed lower hose.** Suction from the water pump may collapse the lower hose if the spring is missing or the hose is badly deteriorated, **Figure 12-9.**
- ❑ **Missing fan shroud.** Air circulates between the fan and the back of the radiator, reducing airflow through the radiator.
- ❑ **Ice in coolant.** Coolant frozen from a lack of antifreeze (improper mixture) can block circulation and cause overheating.
- ❑ **Engine fan problems.** Problems with the fan clutch or electric fan can prevent adequate airflow through the radiator.

The electronic control module (ECM) in some vehicles helps protect the engine from overheating damage. The ECM monitors the coolant temperature sensor to detect overheating. If overheating is detected, the computer cuts off spark to one cylinder at a time in a predetermined sequence and retards ignition timing to reduce engine speed. Retarding the timing allows more hot gasses to be pulled into the exhaust system and away from the engine. The outside air pulled into the "dead cylinders" cools the engine and helps prevent overheating damage. However, this does not correct the problem. The vehicle must be serviced.

Figure 12-8. *A belt with improper tension may slip. On a serpentine system, a belt tensioner maintains proper tension. This belt tensioner has index marks to help determine how much the belt has stretched. The technician has drawn an arrow to indicate the position when the belt was new. As you can see, the belt has not stretched too much since it was installed.*

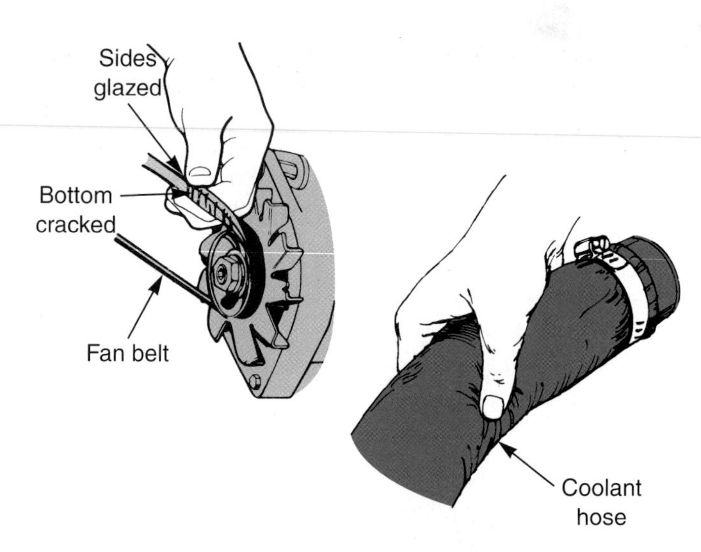

Figure 12-9. *Inspect the cooling system and check condition of belts and hoses. Check hoses for leaks, hardening, or softening. Also, check for loose or deteriorated belts. (Gates)*

Engine Overcooling

Engine overcooling causes slow engine warm-up, insufficient heat for the passenger compartment, and sluggish engine performance. Overcooling can result in increased part wear. Overcooling also reduces fuel economy because more combustion heat transfers into the metal engine parts. Overcooling may be caused by:

- ❑ **Stuck thermostat.** The thermostat sticks open, allowing too much coolant circulation.
- ❑ **Locked fan clutch.** The fan operates all of the time, causing excess airflow through the radiator.
- ❑ **Shorted fan switch.** An electric fan runs all the time, increasing warm-up time.

Engine Fan Service

A faulty engine fan can cause overheating, overcooling, vibration, and water pump wear or damage. Always check the fan for bent blades, cracks, and other problems. A flexible fan is especially prone to these problems. If any troubles are found, replace the fan.

Warning: A fan with cracked or bent blades is extremely dangerous. Broken blades can be thrown out with force, causing severe injury.

Testing a Fan Clutch

To test a thermostatic fan clutch, spin the fan with the engine off. The fan should turn without the fan shaft turning. Then, start the engine. When the engine warms up, the clutch should engage. Air should begin to flow

Hose Replacement

To remove a hose and install a new one:

1. Make sure the cooling system is not under pressure.
2. Carefully remove the radiator cap.
3. Place a drain pan under the radiator to collect the coolant.
4. Open the drain valve and allow the coolant to drain. Then, close the drain valve.
5. Loosen the hose clamps. Refer to **Figure 12-12A.**
6. Twist the hose while pulling it off of the fitting. If a new hose is to be installed, you may cut a slit in the end of the old hose to aid removal.
7. Clean the metal hose fittings.
8. Coat the metal hose fittings with a nonhardening sealer. Refer to **Figure 12-12B.**
9. Install the new hose.
10. Position the hose clamps so that they are fully over the metal hose fitting. Refer to **Figure 12-12C.**
11. Tighten the clamps.
12. Fill the cooling system with coolant.
13. Pressure test the system. Check all fittings for leaks.
14. Replace the radiator cap.

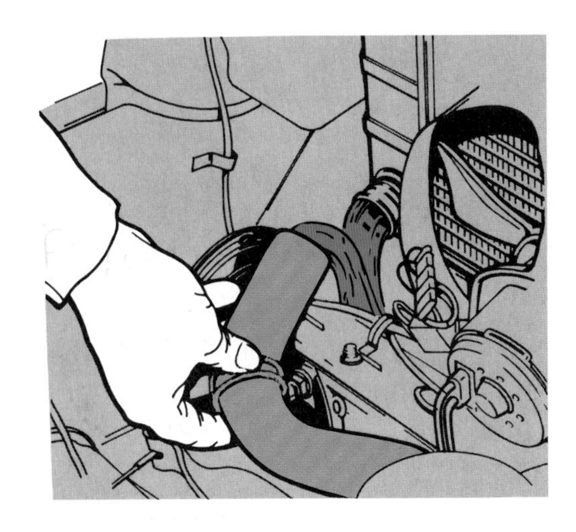

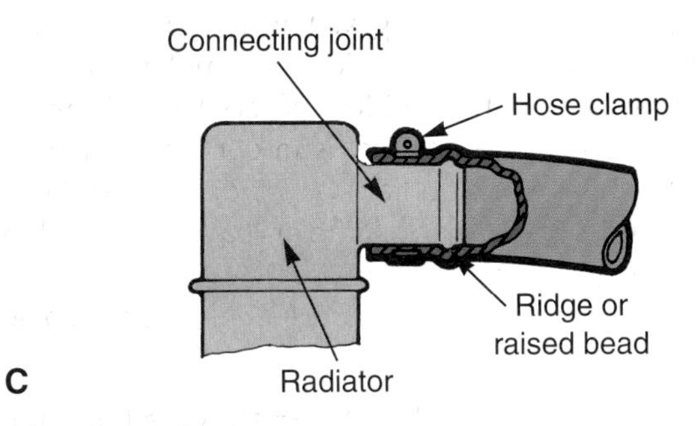

Figure 12-12. *Hose service. A—Loosen the hose clamps and twist the hose off. B—Clean the hose fittings and coat with a nonhardening sealer. C—Install the new hose. Properly position the hose clamps before tightening them. (Ford, DaimlerChrysler)*

Cooling System Pressure Test

A ***cooling system pressure test*** is used to quickly locate leaks in the cooling system. A ***pressure tester*** is a hand-operated air pump used to pressurize the cooling system for leak detection. The low air pressure forced into the system will cause coolant to pour or drip from any leak in the system.

1. Make sure the cooling system is not under pressure.
2. Carefully remove the radiator cap.
3. Install the pressure tester on the radiator filler neck, **Figure 12-13A.**
4. Using the hand pump, pressurize the system to the same pressure as the radiator cap rating, **Figure 12-13B.**
5. With pressure in the system, inspect all parts for leakage. Check all hose fittings, gaskets, and engine core (freeze) plugs. Also, check the water pump, radiator, and heater core.
6. If a leak is found, tighten, repair, or replace parts as needed.
7. Refill the cooling system with coolant as needed.
8. Replace the radiator cap.

Caution: Do not pump too much pressure into the cooling system or the radiator, a hose, or gaskets may be damaged. Never exceed the radiator cap pressure rating when pressure testing the system.

Figure 12-13. *Pressure testing a cooling system. A—Install the pressure tester on the radiator. B—Pump the tester handle until the pressure gauge reading equals the radiator cap pressure rating. Then, look for leaks.*

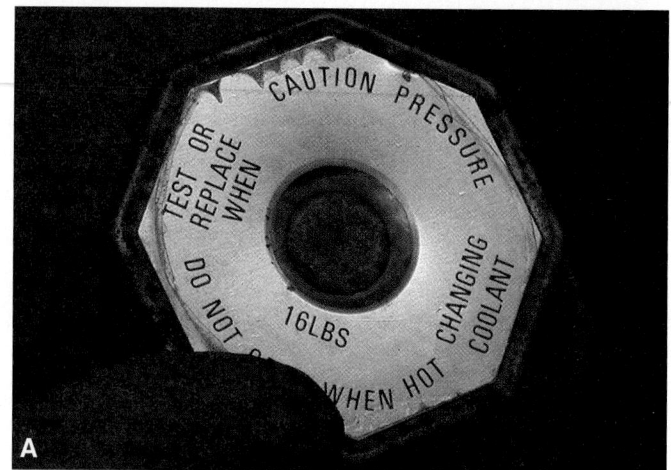

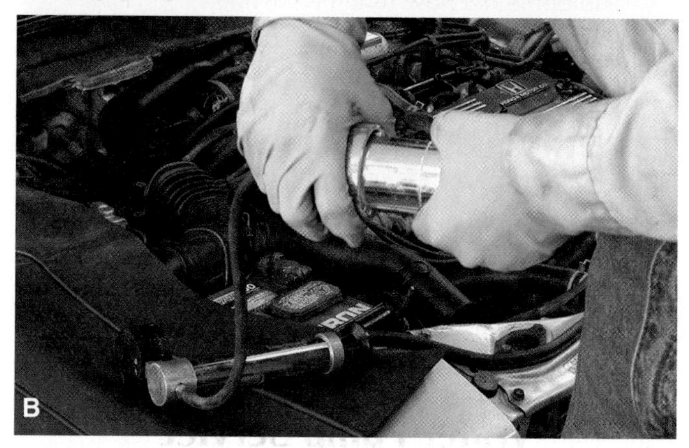

Figure 12-14. *Pressure testing the radiator cap. A—Inspect the cap for damage or seal deterioration. Also, note the pressure rating stamped on the cap. B—Install the adapter and radiator cap onto the pressure tester.*

Radiator and Pressure Cap Service

When overheating problems occur but the system is not leaking, check the radiator and pressure cap. Problems with these parts are common sources of overheating. The pressure cap may have bad seals, allowing pressure loss. The radiator may be clogged, not permitting adequate air or coolant flow.

Inspecting the Radiator and Pressure Cap

Inspect the outside of the radiator for leaves and road dirt. Also, make sure the radiator shroud is in place and unbroken. These troubles could limit air circulation through the core.

If needed, use a water hose to wash debris out of the fins. Spray water from the back to push debris out of the front of the radiator fins. You may also use compressed air, but make sure the pressure is low enough not to damage the radiator.

Inspect the radiator cap and filler neck. Check for cracks or tears in the cap seal. Check the filler neck sealing surfaces for nicks and dents. Replace the cap or have the neck repaired as needed.

Pressure Testing Radiator Cap

A ***radiator cap pressure test*** measures the cap opening pressure and checks the condition of the sealing washer. The cap is installed on a cooling system pressure tester. Refer to **Figure 12-14.**

1. Make sure the cooling system is not under pressure.
2. Carefully remove the radiator cap.
3. Install the radiator cap on the pressure tester using the adapter.
4. Pump the tester to pressurize the cap.
5. Watch the pressure gauge. The cap should release air at its rated pressure (pressure stamped on cap). It should also hold that pressure for at least one minute.
6. If the cap fails the test, replace it. Otherwise, reinstall the cap on the radiator.

Radiator Removal and Replacement or Repair

Radiator repairs are almost always done by a shop that specializes in radiator repair. Engine repair technicians rarely attempt radiator repairs. The radiator shop has the facilities to properly disassemble, clean, reassemble or repair, and pressure test a radiator. Many late-model vehicles have radiators with some plastic components. Special sealers, seals, and repair techniques are needed on these radiators. Refer to the service manual for details. The radiator in some late-model vehicles is designed to be replaced rather than rebuilt. The basic procedure for removing the radiator is:

1. Make sure the cooling system is not under pressure.
2. Carefully remove the radiator cap.
3. Drain all coolant.
4. Remove the radiator hoses.
5. Remove all fasteners holding the radiator to its supports.
6. Remove the radiator from the vehicle.

To install the new or repaired radiator, follow these steps in reverse order. Refer to the service manual for specific procedures for radiator removal and installation.

Water Pump Service

A bad water pump may leak coolant or fail to circulate enough coolant. Rust in the cooling system or lack of antifreeze in the coolant are common reasons for pump failure. These conditions could speed seal, shaft, and bearing wear.

Checking Water Pump

To check for a bad water pump seal, pressure test the system and watch for leakage at the pump. Coolant will leak out of the small drain hole at the bottom of the pump or at the end of the pump shaft. Replace a leaking water pump.

To check for worn water pump bearings, try to wiggle the fan or pump pulley up and down. If the water pump shaft can be deflected, the pump bearings are badly worn. A stethoscope can also be used to listen for worn, noisy water pump bearings. Replace the water pump if the bearings are bad.

To see if the water pump is properly circulating coolant, start and warm the engine to operating temperature. Then, shut off the engine. Squeeze the top radiator hose while someone restarts the engine. You should feel a pressure surge (hose swelling) if the pump is working. If not, a pump shaft or impeller problem is indicated, provided the thermostat is operating properly. You can also watch for coolant circulation in the radiator with the engine at operating temperature. Be sure there is no pressure in the cooling system before removing the radiator cap.

Water Pump Removal

The water pump on many late-model vehicles can be difficult to access and remove. For example, you may need to remove the radiator in some cases, but not in others. If in doubt, refer to the service manual for specific details. Keep all bolts, brackets, and parts organized to aid reassembly. To remove the water pump:

1. Make sure the cooling system is not under pressure.
2. Carefully remove the radiator cap.
3. Place a drain pan under the radiator.
4. Open the drain valve.
5. Allow the radiator to completely drain. Then, close the drain valve.
6. Remove the fan belt(s).
7. Unbolt all brackets and other components preventing pump removal. Often, the air conditioning compressor, power steering pump, or alternator must be moved out of the way to allow water pump removal.
8. Remove the radiator hose(s) from the water pump.
9. Remove the bolts holding the water pump to the engine.
10. Scrape off all old gasket or sealer material from the engine. The engine-to-pump mating surfaces must be perfectly clean to prevent coolant leakage when the new pump is installed. Be careful not to gouge or scratch the sealing surfaces, especially on aluminum parts.

Water Pump Installation

When installing the new water pump, a gasket must be installed between the pump and engine, **Figure 12-15.** You can, however, use an approved sealer in place of the gasket. If needed, refer to a service manual for details.

1. If using a water pump gasket, apply an approved adhesive to the water pump and stick the new gasket to the pump. To use sealer in place of a gasket, place a continuous bead of approved sealer about 1/8″ (3 mm) wide around the pump sealing surface.
2. Fit the pump onto the engine. Move it straight into place. Do not shift the gasket or break the bead of sealer.
3. Start all of the bolts by hand. Check that all bolt lengths are correct.
4. Tighten all of the fasteners to 1/2 of full torque in the specified torque sequence or a crossing pattern.
5. Tighten all of the fasteners to 3/4 of full torque in the specified torque sequence or a crossing pattern.
6. Tighten all of the fasteners to full torque in the specified torque sequence or a crossing pattern.

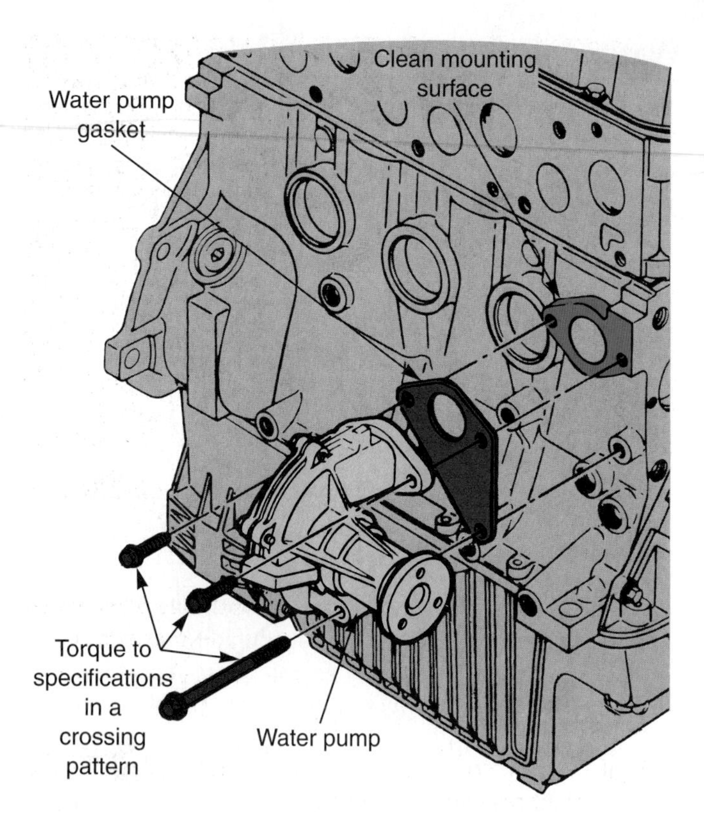

Figure 12-15. *Installing a water pump. (Ford)*

7. Tighten all of the fasteners to full torque one more time to ensure proper torque.
8. Reinstall the radiator hose(s) onto the water pump.
9. Attach any brackets or components that were removed to allow access to the water pump.
10. Reinstall the fan belt(s). Replace the belt(s) as needed.
11. Make sure the fan belt has the proper tension.
12. Fill the radiator with coolant.
13. Start the engine and look for leaks. Allow the engine to reach operating temperature.
14. Verify coolant flow through the radiator.
15. Install the radiator cap.

Thermostat Service

An improperly operating thermostat can cause either engine overheating or engine overcooling. If the thermostat is stuck shut, coolant will not circulate through the radiator. As a result, heat will not be removed from the engine and the engine can overheat in a matter of minutes. If a thermostat is stuck open, too much coolant may circulate through the radiator. As a result, the engine may not reach proper operating temperature. The engine may run poorly for extended periods in cold weather. Engine power, fuel economy, and driveability will be reduced.

Testing the Thermostat

To check the thermostat action, watch for coolant flow through the radiator.

1. Make sure the cooling system is not under pressure. The engine should also be cool.
2. Carefully remove the radiator cap.
3. Start the engine and observe the coolant in the radiator. There should be no flow.
4. Allow the engine to warm to operating temperature.
5. When the engine reaches operating temperature, coolant should begin to flow through the radiator.
6. If the thermostat allows flow when the engine is cold or it does not open at the correct temperature, the thermostat is defective and should be replaced.

In some instances, the thermostat may have to be removed from the engine for testing. Refer to the service manual for details. Place the thermostat and a thermometer in a container of water and heat the container on a hot plate or stove. Note the temperature at which the thermostat opens and compare that to specifications.

An infrared thermometer can also be used to check the action of the thermostat. When the engine is cool, the radiator side of the thermostat should not have much increase in temperature. However, the engine side of the thermostat should show an increasing temperature up to operating temperature. When the engine reaches operating temperature, both sides of the thermostat should show operating temperature. If the radiator side of the thermostat continues to show a lower temperature, the thermostat is defective and should be replaced.

Replacing the Thermostat

The thermostat is normally located under the thermostat housing. The upper radiator hose is usually connected to the thermostat housing. The thermostat housing may be located on the top or side of the engine.

1. Make sure the cooling system is not under pressure.
2. Carefully remove the radiator cap.
3. Place a drain pan under the radiator.
4. Open the drain valve and allow all coolant to drain. Close the drain valve.
5. Remove the upper radiator hose from the thermostat housing.
6. Remove the bolts holding the thermostat housing to the engine.
7. Tap the housing free with a rubber mallet. Lift the housing off of the engine and remove the thermostat.
8. Scrape all of the old gasket material off the thermostat housing and the sealing surface on the engine.

9. Make sure the thermostat housing is not warped. Place it on a flat surface and check for gaps between the housing and surface. If warped, file the surface flat.
10. Double-check that the temperature rating on the new thermostat is correct.
11. Place the new thermostat into the engine. Normally, the rod (pointed end) on the thermostat should face the radiator hose. The pellet chamber should face the inside of the engine.
12. Apply approved sealer to the mating surface on the engine, **Figure 12-16.** Place the gasket over the sealer.
13. Start the fasteners by hand.
14. Tighten the fasteners to 1/2 of the full torque. If there are more than three fasteners, use a crisscross pattern.
15. Tighten the fasteners to 3/4 of the full torque.
16. Tighten the fasteners to full torque.
17. Tighten the fasteners to full torque one more time to ensure proper torque.
18. Reinstall the upper radiator hose.
19. Fill the radiator with coolant.
20. Start the engine and check for leaks.
21. Verify no coolant flow in the radiator while the engine is cool.
22. Verify coolant flow through the radiator once the engine reaches operating temperature.
23. Reinstall the radiator cap.
24. Once pressure builds in the cooling system, check again for leaks.

Some thermostat housings have an O-ring in place of a gasket. In many cases, this O-ring can be reused if it is not damaged. If the O-ring shows any signs of damage, replace it. Some thermostat kits come with a new housing O-ring. Check the service manual for details on reusing the O-ring.

Note: Do not overtighten the thermostat housing bolts. The housing may become warped or cracked. Most housings are made of aluminum or "pot metal."

Cooling System Bleeder Screw

Many late-model vehicles have low hood lines. As a result, the top of the radiator is often lower than the highest coolant passage in the engine. Air tends to collect at the highest point in the cooling system, just as liquid tends to collect at the lowest point. Air can be trapped in the engine in pockets, which can result in hot spots. A ***hot spot*** is an area in the engine suffering from an abnormal buildup of heat. Hot spots can quickly cause engine overheating or

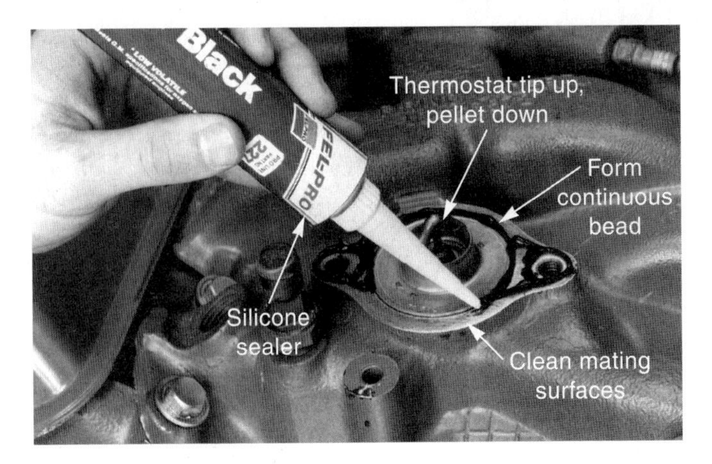

Figure 12-16. *When installing a thermostat, apply an approved sealer to the mating surface on the engine. (Fel-Pro)*

damage to the part near the hot spot. A ***cooling system bleed screw*** is sometimes provided to allow for removal of air.

1. Make sure the cooling system is not under pressure.
2. Carefully remove the radiator cap.
3. If the radiator is not full, add coolant. Also, fill the coolant reservoir, if needed.
4. Replace the radiator cap.
5. Start the engine and allow it to warm to full operating temperature.
6. Slowly open the bleed screw until air is released.
7. When coolant leaks from the bleed screw, all air has been bled from the system. Close the bleed screw.
8. Refill the coolant reservoir as needed.

Warning: *Never* fully remove a cooling system bleed screw, or any system connection, with the cooling system under pressure. Wear safety glasses and gloves when working with a cooling system under pressure.

Flushing a Cooling System

Cooling system flushing is a process of cleaning the cooling system. It involves running water or a cleaning chemical through the cooling system to wash out contaminants. Flushing should be done when rust or scale is found in the system.

Fast flushing is a common method of cleaning a cooling system because the thermostat does not have to be removed from the engine. A water hose is connected to an adapter fitting installed in a heater hose. The radiator cap is removed and the drain valve is opened. The water hose is turned on and water flows into the system, removing rust and loose scale.

Reverse flushing of a radiator requires a special adapter that is connected to the radiator outlet tank by a piece of hose, **Figure 12-17.** Another hose is attached to the inlet tank. Low-pressure compressed air is used to force water through the core in the opposite direction of normal flow. The engine block can be reverse flushed as well. However, the thermostat must be removed.

Chemical flushing is needed when scale buildup in the system is causing engine overheating. A chemical cleaner is added to the coolant. Then, the engine is operated for a specific amount of time to allow the chemical to act on the scale. Finally, the system is flushed with water to remove the chemical.

Warning: Always follow manufacturer's instructions when using a cooling system cleaning agent. The chemical may cause severe burns. Wear rubber gloves, an apron, and a face shield.

Summary

The cooling system must remove excess heat from the engine. It must also maintain a constant engine operating temperature, speed engine warm-up, and provide for heater operation.

The water pump circulates coolant through the coolant passages, hoses, and radiator. The water pump is driven by a belt running off of the engine crankshaft.

The radiator transfers heat from the coolant to the outside air. The radiator cap allows the system to be pressurized, which increases the boiling point of the coolant. Typical cap pressure is 12 psi to 16 psi (85 kPa to 110 kPa). A fan draws air through the radiator.

The thermostat controls coolant flow through the system. When the engine is below operating temperature, the thermostat blocks coolant circulation. This speeds warm-up of the engine. When the engine is at operating temperature, the thermostat opens to allow coolant flow.

Coolant is a mixture of water and antifreeze. It helps cool the engine better than water alone, prevents rust formation, and lubricates the water pump. A 50-50 mix of water and antifreeze is common.

Coolant should be changed at least every two years. After prolonged use, antifreeze turns corrosive and can accelerate rust formation in system. Coolant strength should be checked with a hydrometer. Freeze protection should be provided to about 10°F below the coldest possible temperature for the climate.

A cooling system pressure test is used to find leaks in the system. With the system pressurized, inspect the radiator, hoses, water pump, and other potential leakage points. The pressure tester can also be used to check the radiator cap opening pressure.

When installing a water pump, clean all mating surfaces. Use sealer or a gasket between the water pump and engine. Tighten bolts to specifications in a crisscross pattern.

Air in the cooling system can result in hot spots. Many late-model vehicles have a bleed screw in the cooling system. After filling the cooling system, bleed air using the bleed screw.

Coolant system flushing is needed to remove rust and scale. The system can be fast flushed, reverse flushed, or chemically flushed.

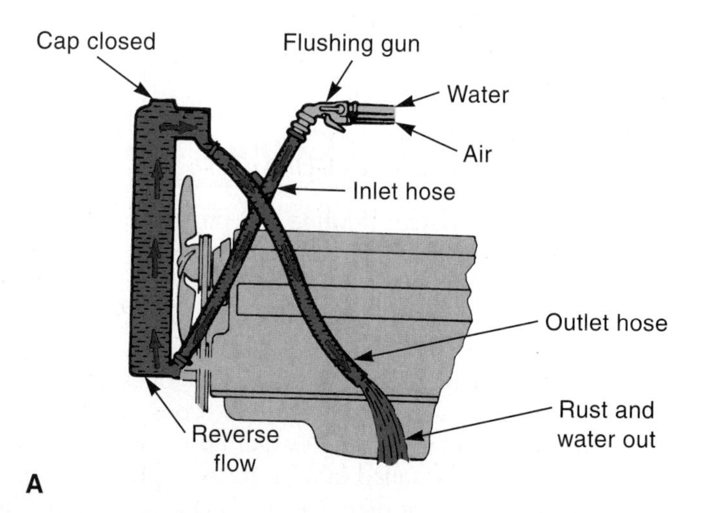

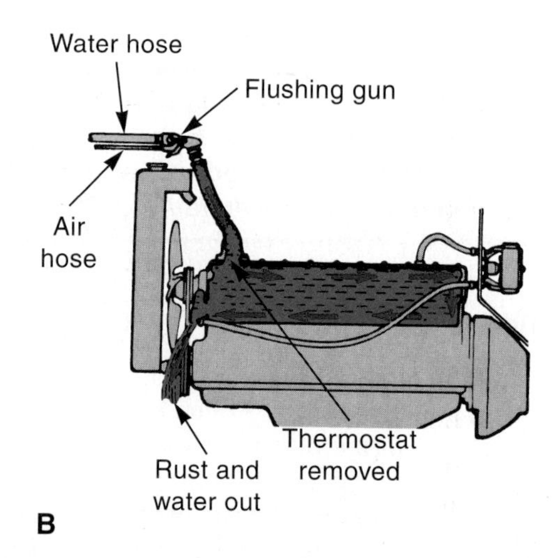

Figure 12-17. *A—Reverse flushing a radiator. B—Reverse flushing an engine block. (DaimlerChrysler)*

Review Questions—Chapter 12

Please do not write in this text. Write your answers on a separate sheet of paper.

1. List four functions of a cooling system.
2. _____-cooled engines are not commonly used as automotive engines.
3. The _____ circulates coolant through the engine.
4. List the five basic parts of an automotive cooling system.
5. This is *not* a purpose of the radiator cap:
   (A) Seal the top of the filler neck.
   (B) Relieve excess pressure in the cooling system.
   (C) Lower the boiling point of the coolant.
   (D) Relieve vacuum in the cooling system.
6. Explain the operation of the electric fan in a cooling system.
7. How can a missing lower hose spring cause engine overheating?
8. What are two ways of testing coolant?
9. What is the purpose of a cooling system pressure test?
10. How do you check for worn water pump bearings?
11. Briefly explain how to use an infrared thermometer to check the thermostat.
12. At what temperature should a thermostat be fully open?
13. What is the purpose of a cooling system bleed screw?
14. Describe when a cooling system should be flushed.
15. What is reverse flushing?

ASE-Type Questions—Chapter 12

1. A customer complains of engine overheating, but only at idle. The system is full of coolant and the belt has the proper tension. Technician A notices a missing fan shroud and says that could be the cause of overheating. Technician B says the thermostat stuck closed could cause the problem. Who is correct?
   (A) Technician A.
   (B) Technician B.
   (C) Both A and B.
   (D) Neither A not B.
2. An electric engine fan fails to turn on when the engine warms up. Technician A says the first step is to remove and test the fan motor. Technician B says simply apply battery voltage directly to the motor at the fan switch. Who is correct?
   (A) Technician A.
   (B) Technician B.
   (C) Both A and B.
   (D) Neither A nor B.
3. Technician A says that an automotive engine's operating temperature is normally between 220°F and 250°F (105°C and 120°C). Technician B says that an engine's operating temperature is normally between 180°F and 195°F (80°C and 90°C). Who is correct?
   (A) A only.
   (B) B only.
   (C) Both A and B.
   (D) Neither A nor B.
4. Technician A says that a liquid-cooled system is better than an air-cooled system because it improves the efficiency of the passenger compartment heater. Technician B says that a liquid-cooled system is better than an air-cooled system because engine temperature is more precisely controlled. Who is correct?
   (A) A only.
   (B) B only.
   (C) Both A and B.
   (D) Neither A nor B.
5. A closed cooling system has a(n) _____ to prevent excess coolant from draining onto the ground.
   (A) coolant reservoir
   (B) overflow tube
   (C) drain valve
   (D) None of the above.
6. All of the following will cause an automotive engine to overheat *except:*
   (A) retarded ignition timing.
   (B) a loose fan belt.
   (C) a locked fan clutch.
   (D) an inoperative electric cooling fan.
7. Technician A says that when the engine is cold, coolant should flow through the radiator. Technician B says that when the engine is hot, coolant should not flow through the radiator. Who is correct?
   (A) A only.
   (B) B only.
   (C) Both A and B.
   (D) Neither A nor B.

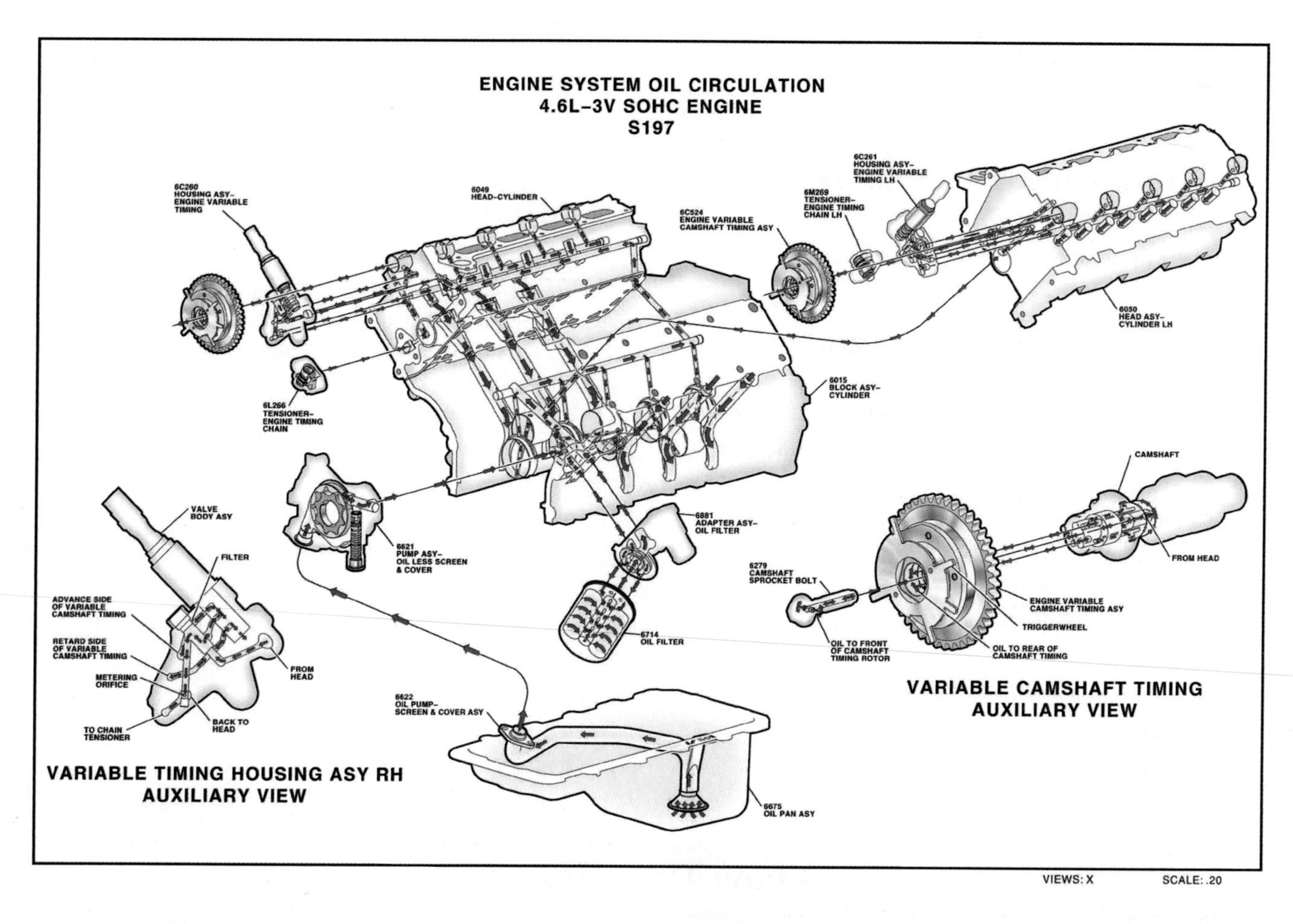

A properly functioning lubrication system is an important part of the engine. Study this illustration of a lubrication system. Then, read Chapter 13. (Ford)

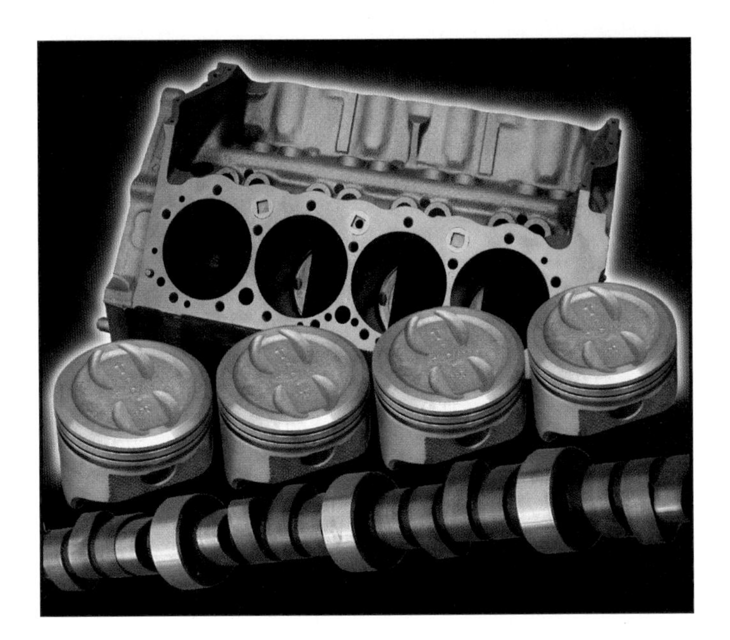

Chapter 13

Lubrication System Operation and Service

After studying this chapter, you will be able to:

- ❑ Explain the operation of an engine lubrication system.
- ❑ Describe engine oil ratings.
- ❑ Identify the types of oil pumps and explain their construction.
- ❑ Describe the types of oil filters and their operation.
- ❑ Explain the purpose and operation of a pressure-relief valve.
- ❑ Describe how low oil pressure warning circuits operate.
- ❑ Troubleshoot the lubrication system.
- ❑ Service the engine oil and replace the oil filter.
- ❑ Diagnose problems with the oil pump and make repairs.
- ❑ Service the pressure-relief valve.
- ❑ Locate engine oil leaks.
- ❑ Service the oil pan.
- ❑ Troubleshoot and repair the oil pressure indicator and gauge.

Know These Terms

API certification mark
Bypass lubrication system
Cartridge oil filter
Dry sump lubrication system
Element
Engine oil
Filter bypass valve
Full flow lubrication system
Gear oil pump
Lubrication system
Multiviscosity oil
Oil clearance
Oil cooler
Oil filter
Oil filter housing
Oil galleries
Oil level indicator
Oil pan
Oil pickup
Oil pickup screen
Oil pressure sending unit
Oil pressure test
Oil pump
Oil service rating
Oil viscosity
Pressure-fed oiling
Pressure-relief valve
Rotary oil pump
Spin-on oil filter
Splash oiling
Structural oil pan
Sump

This chapter summarizes the operation and construction of an engine lubrication system. Improper operation of the lubrication system can lead to engine wear and damage. As an engine repair technician, you must be able to troubleshoot and repair the lubrication system.

Lubrication System

An engine ***lubrication system*** has several functions. These include:

- Reducing friction and wear between moving parts.
- Helping remove heat from engine parts.
- Cleaning the inside of the engine by removing contaminants such as metal, dirt, and other particles.
- Cutting power loss and increasing fuel economy.
- Absorbing shocks between moving parts to quiet engine operation and increase engine life.

There are two methods for lubricating engine components: pressure-fed oiling and splash oiling. ***Pressure-fed oiling*** occurs when oil is forced through galleries and to the moving parts. For example, oil is forced through the crankshaft to the main and connecting rod bearings. ***Splash oiling*** occurs when oil is thrown from moving parts. For example, in an overhead valve engine, oil is forced (pressure-fed oiling) through hollow push rods and onto the rocker arms. As the rocker arms move, oil is thrown off (splashes) to lubricate all of the moving parts under the valve cover.

Engine Oil

Engine oil, also called motor oil, protects an engine from friction and the resulting heat. Without oil, an engine will self-destruct in a matter of minutes. Engine oil produces a lubricating film on the moving parts in an engine. A thin layer of oil separates engine parts to prevent metal-on-metal contact, **Figure 13-1.** Without the oil film, the parts would rub together, overheat, and wear rapidly.

Engine oil is commonly refined from crude oil or petroleum, but it may be synthetic. Synthetic oil is a highly processed blend of mineral, vegetable, and/or animal oils that contains molecularly altered compounds for improved lubricant performance. Synthetic oil normally provides longer service life, reduced friction, and increased fuel economy over conventional oil. However, synthetic oil is generally more expensive than conventional oil.

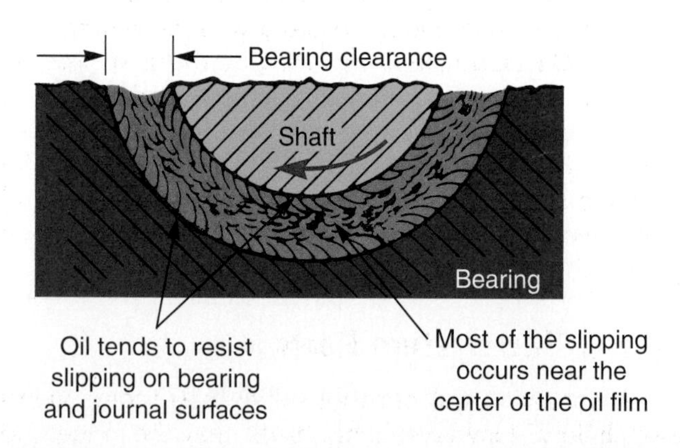

Figure 13-1. *An oil film separates parts to prevent friction and wear. (DaimlerChrysler)*

Note: Some engine manufacturers recommend synthetic oil in their engines. Always follow manufacturer recommendations.

Oil Clearance

Oil clearance, or bearing clearance, is the small space between moving engine parts for the lubricating oil film. The clearance allows oil to enter and prevent part contact, as shown in **Figure 13-1.** For example, a connecting rod bearing typically has a bearing or oil clearance of about .002″ (0.05 mm). This clearance results in a gap that is large enough to allow oil entry. However, it is also small enough to keep the parts from knocking or hammering together during engine operation.

Oil Viscosity

Oil viscosity is a measure of the oil's ability to flow. An oil's viscosity is often referred to as the oil's "weight." A high-viscosity oil is thick and resists flow, like honey. A low-viscosity oil is thin and flows easily, like water.

A viscosity numbering system is used to rate the thickness of engine oil. The higher the number, the thicker the oil. The lower the number, the thinner the oil. The oil's viscosity number is printed on the oil container.

The Society of Automotive Engineers (SAE) standardized the viscosity numbering system. For this reason, oil viscosity is written as SAE 30, SAE 40, etc., **Figure 13-2.** Engine oil viscosities for passenger cars and light trucks commonly range from thin SAE 5 to thick SAE 50. Vehicle manufacturers specify an SAE number or range of numbers for their engines.

Temperature Effects on Oil

Oil is thick and resists flow when it is cold. When heated, oil becomes thin and flows easily. This can pose a problem in the engine. The oil in a cold engine may be so thick that engine starting is difficult. This may increase starter drag. Also, the oil will not properly pump through the engine, resulting in poor lubrication. When the engine warms up, however, the oil film thins out and oil pressure may drop. If the oil becomes too hot and thin, the oil film can actually break down and parts may make contact.

Multiviscosity Oil

Multiviscosity oil, or multiweight oil, exhibits the operating characteristics of a thin, light oil when it is cold and a thick, heavy oil when it is hot. A multiviscosity oil can be numbered SAE 5W-30, 10W-40, 20W-50, and so on. The

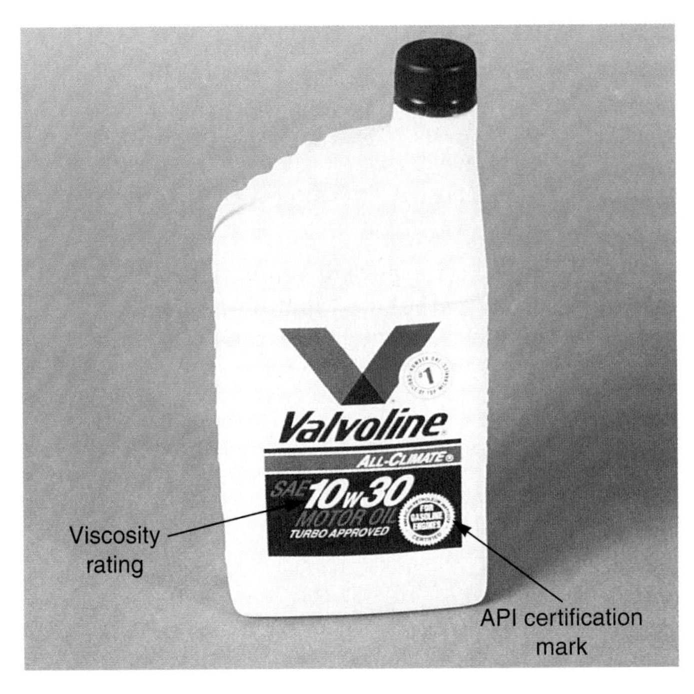

Figure 13-2. *A numbering system is used to indicate the oil viscosity. This is a multiviscosity oil.*

W stands for winter. The number before the W indicates the oil's viscosity when it is cold. The second number indicates the oil's viscosity when it is at the engine's operating temperature.

For example, a 5W-30 weight oil will have the flow properties of a thin 5 weight oil and flow easily when the engine is cold. It will then have the flow properties of a thick 30 weight oil when the engine warms to operating temperature. This will make the engine start more easily in cold weather, but will also provide adequate film strength (thickness) when the engine is at full operating temperature.

Selecting Oil Viscosity

Figure 13-3 is one automaker's chart showing recommended SAE viscosity number. Normally, always use the oil viscosity recommended by the manufacturer. However, oil with a higher-than-recommended viscosity may be beneficial in an old, high-mileage, worn engine. Thicker oil will tend to seal the rings and provide better bearing protection. It also may help reduce engine oil consumption and smoking.

Oil Service Rating

An ***oil service rating*** is a set of letters printed on the oil container to denote how well the oil will perform under operating conditions. This is a performance standard set by the American Petroleum Institute (API). The service rating categories are:

- **SL oil rating.** For use in all automotive gasoline engines. Introduced in July 2001.
- **SJ oil rating.** For use in 2001 and older automotive gasoline engines.
- **SH oil rating.** Obsolete; for use in 1996 and older automotive gasoline engines.

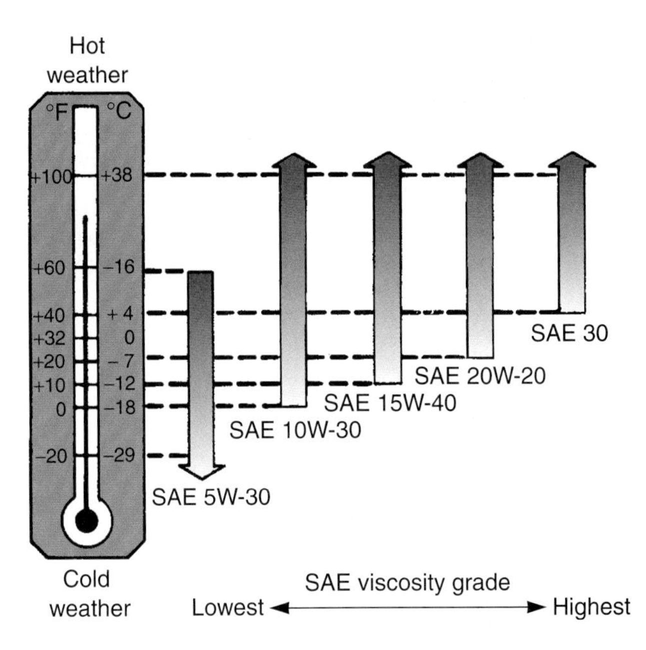

Figure 13-3. *This graph shows the recommended oil weight for various outside temperatures. (General Motors)*

- **SG oil rating.** Obsolete; for use in 1993 and older automotive gasoline engines.
- **SF oil rating.** Obsolete; for use in 1988 and older automotive gasoline engines.
- **SE oil rating.** Obsolete; for use in 1979 and older automotive gasoline engines.
- **SD oil rating.** Obsolete; for use in 1971 and older automotive gasoline engines.
- **SC oil rating.** Obsolete; for use in 1967 and older automotive gasoline engines.
- **SB oil rating.** Obsolete; use only when recommended for older engines.
- **SA oil rating.** Obsolete; use only when recommended for older engines.
- **CA through CI-4 oil ratings.** Oil recommended for diesel engines.

The owner's manual will give the service rating recommended for a specific vehicle. You can use a better (higher) service rating than recommended, but *never* a lower service rating! Doing so could void the warranty.

The ***API certification mark*** is a starburst symbol on the package that identifies the oil as intended for use in gasoline or diesel engines. See **Figure 13-4.** It also indicates that the oil meets the current requirements of the International Lubricant Standardization and Approval Committee (ILSAC). Most automakers recommend using engine oil brands that display the API certification mark.

Lubrication System Parts

A lubrication system consists of the basic components listed below. However, other parts may be added to increase the system efficiency.

- **Engine oil.** Lubricant for moving parts in the engine.

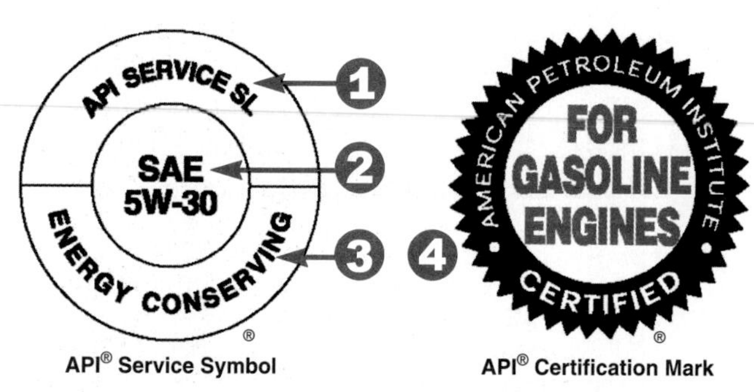

1. Performance Level:
Gasoline engine oil categories (for cars, vans, and light trucks with gasoline engines): Oils designed for gasoline-engine service fall under API's "S" (Service) categories. Look for current service categories SL and SJ. See reverse for descriptions of current and obsolete API service categories.

Diesel engine oil categories (for heavy-duty trucks and vehicles with diesel engines): Oils designed for diesel-engine service fall under API's "C" (Commercial) categories. Look for current categories CI-4, CH-4, CG-4, CF-4, CF-2, and CF.

2. Viscosity: The measure of an oil's thickness and ability to flow at certain temperatures.

3. Fuel Economy Rating: The "Energy Conserving" rating applies to oils intended for gasoline-engine cars, vans, and light trucks. Widespread use of "Energy Conserving" oils may result in an overall savings of fuel in the vehicle fleet as a whole.

4. API Certification Mark: An oil displaying this mark meets the current engine protection standard and fuel economy requirements of the International Lubricant Standardization and Approval Committee (ILSAC), a joint effort of U.S. and Japanese automobile manufacturers. Most automobile manufacturers recommend oils that carry the API Certification Mark.

Guide to SAE Grades of Motor Oil for Passenger Cars

Multigrade oils such as SAE 5W-30 and 10W-30 are widely used because, under all but extremely hot or cold conditions, they are thin enough to flow at low temperatures and thick enough to perform satisfactorily at high temperatures. **Note that vehicle requirements may vary. Follow your vehicle manufacturer's recommendations on SAE oil viscosity.**

If lowest expected outdoor temperature is	Typical SAE Viscosity Grades for Passenger Cars
0°C (32°F)	5W-20, 5W-30, 10W-30, 10W-40, 20W-50
−18°C (0°F)	5W-20, 5W-30, 10W-30, 10W-40
Below −18°C (0°F)	5W-20, 5W-30

Figure 13-4. *The API service symbol and certification mark identify the performance level, viscosity, and fuel economy rating. (Reproduced courtesy of the American Petroleum Institute from the 2002 API Motor Oil Guide)*

- **Oil pan.** Storage area for the engine oil.
- **Oil pump.** Forces oil through the oil galleries.
- **Oil filter.** Strains out impurities in the oil.
- **Oil galleries.** Oil passages through the inside of the engine.

Lubrication System Operation

With the engine running, the ***oil pump*** pulls oil out of the oil pan, **Figure 13-5.** A screen on the pickup tube removes large particles, such as bits of gasket material, from the oil before the oil enters the pump. The pump then pushes the oil through the oil filter and oil galleries. The ***oil filter*** cleans the oil by trapping small particles suspended in the lubricant. The filtered oil flows to the camshaft, crankshaft, lifters, rocker arms, and other moving parts. Oil finally drains back into the oil pan for recirculation, **Figure 13-6.**

When oil bleeds out of the engine bearings, which is normal, it sprays onto the outside of internal engine parts. For example, when oil sprays out of the connecting rod

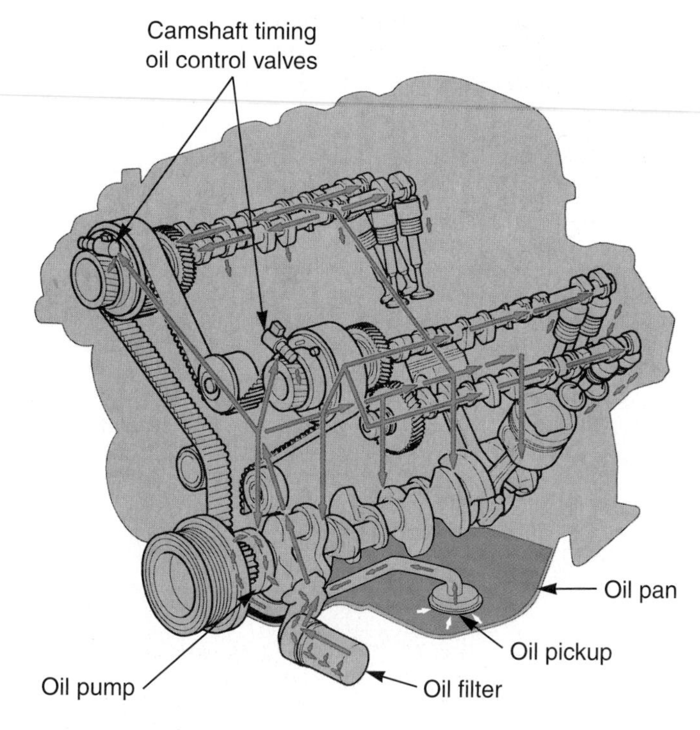

Figure 13-5. *Note how oil flows through the engine. (Lexus)*

bearings, it splashes onto the cylinder walls. This lubricates the pistons, piston rings, piston pins, and cylinders.

Oil Pumps

The oil pump is the heart of the engine lubrication system. It draws oil out of the pan and forces it through the engine filter and galleries to the engine bearings. There are two basic types of engine oil pumps: rotary and gear.

The oil pump is frequently driven by a gear on the engine camshaft or balancer shaft. It may also be driven by the timing belt or by a direct connection with the end of the crankshaft.

Rotary Oil Pumps

A ***rotary oil pump*** uses a set of star-shaped rotors in a housing to pressurize the motor oil, **Figure 13-7.** As the oil pump shaft turns, the inner rotor causes the outer rotor to spin. The eccentric action of the two rotors form pockets that change in size. A large pocket is formed on the inlet side of the pump. As the rotors turn, the oil-filled pocket becomes smaller as it nears the outlet of the pump. This squeezes the oil and makes it squirt out under pressure. As the pump spins, this action is repeated over and over to produce a relatively smooth flow of oil.

Gear Oil Pumps

A ***gear oil pump*** uses a set of gears to produce lubrication system pressure, **Figure 13-8.** One of the pump gears is

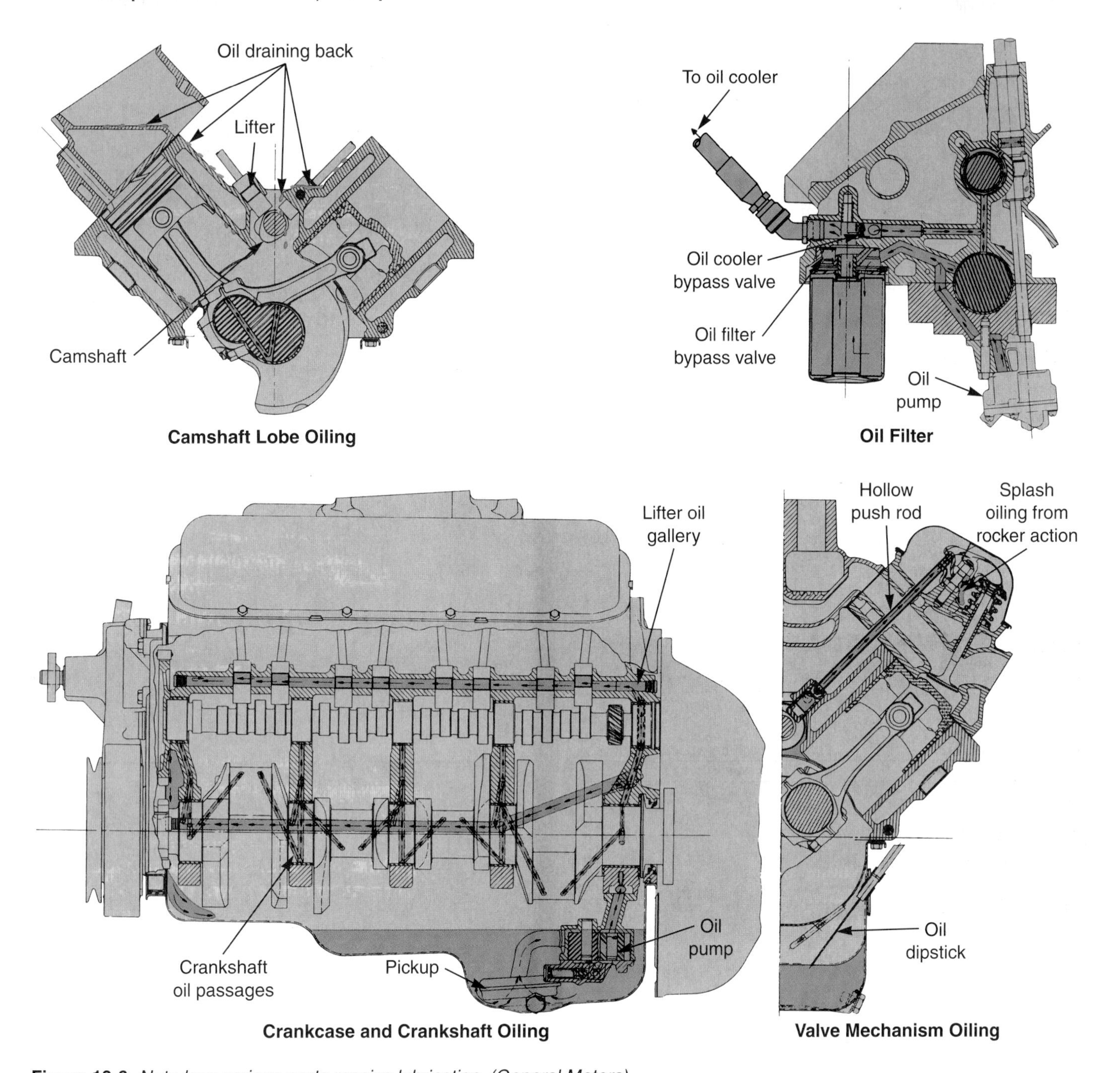

Figure 13-6. *Note how various parts receive lubrication. (General Motors)*

the drive gear. This gear turns the other gear, called the driven gear. Oil on the inlet side of the pump is caught in the gear teeth and carried around the outer wall inside the pump housing. When the oil reaches the outlet side of the pump, the gear teeth mesh and seal. Oil caught in each gear tooth is forced into the pocket at the pump outlet and pressure is formed. Oil flows out of the pump and to the engine.

Oil Pickup and Screen

The ***oil pickup*** is a tube and screen assembly extending from the oil pump to the bottom of the oil pan. One end of the pickup bolts or screws into the oil pump or to the engine block. The other end holds the pickup screen. The ***oil pickup screen*** prevents large particles from entering the pickup tube and oil pump, **Figure 13-9.** The screen is usually part of the pickup tube. Without the screen, the oil pump could be damaged by bits of seals, metal, and other large debris flushed into the pan.

Oil Filters

An oil filter removes small metal, carbon, rust, and dirt particles from the engine oil. It protects the moving

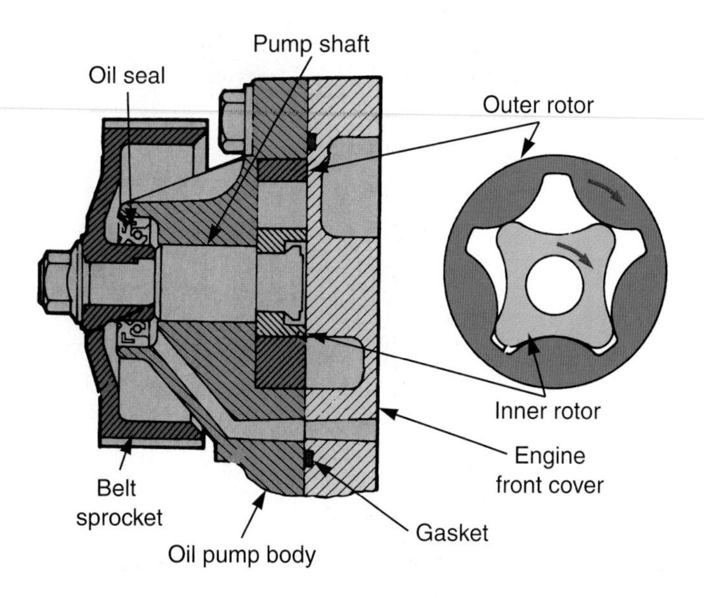

Figure 13-7. *This is a rotary oil pump that is driven by the timing belt. (DaimlerChrysler)*

engine parts from abrasive wear. The oil filter ***element*** is a paper, cotton, or synthetic filtering substance mounted inside the oil filter housing. It allows oil to flow, but blocks and traps small debris.

Most filters have a ***filter bypass valve*** to protect the engine from oil starvation if the filter element becomes clogged. The valve opens if too much pressure is formed in the filter. This allows unfiltered oil to flow to the engine bearings, preventing major part damage from lack of lubrication.

A ***full flow lubrication system*** forces all of the oil through the oil filter before the oil reaches the parts of the engine. It is the most common type. A ***bypass lubrication system*** does not filter all of the oil that enters the engine bearings. It filters the extra oil not needed by the bearings. Bypass lubrication systems are not very common.

Oil Filter Types

The two classifications of engine oil filters are spin-on and cartridge. The ***spin-on oil filter*** is a sealed unit having

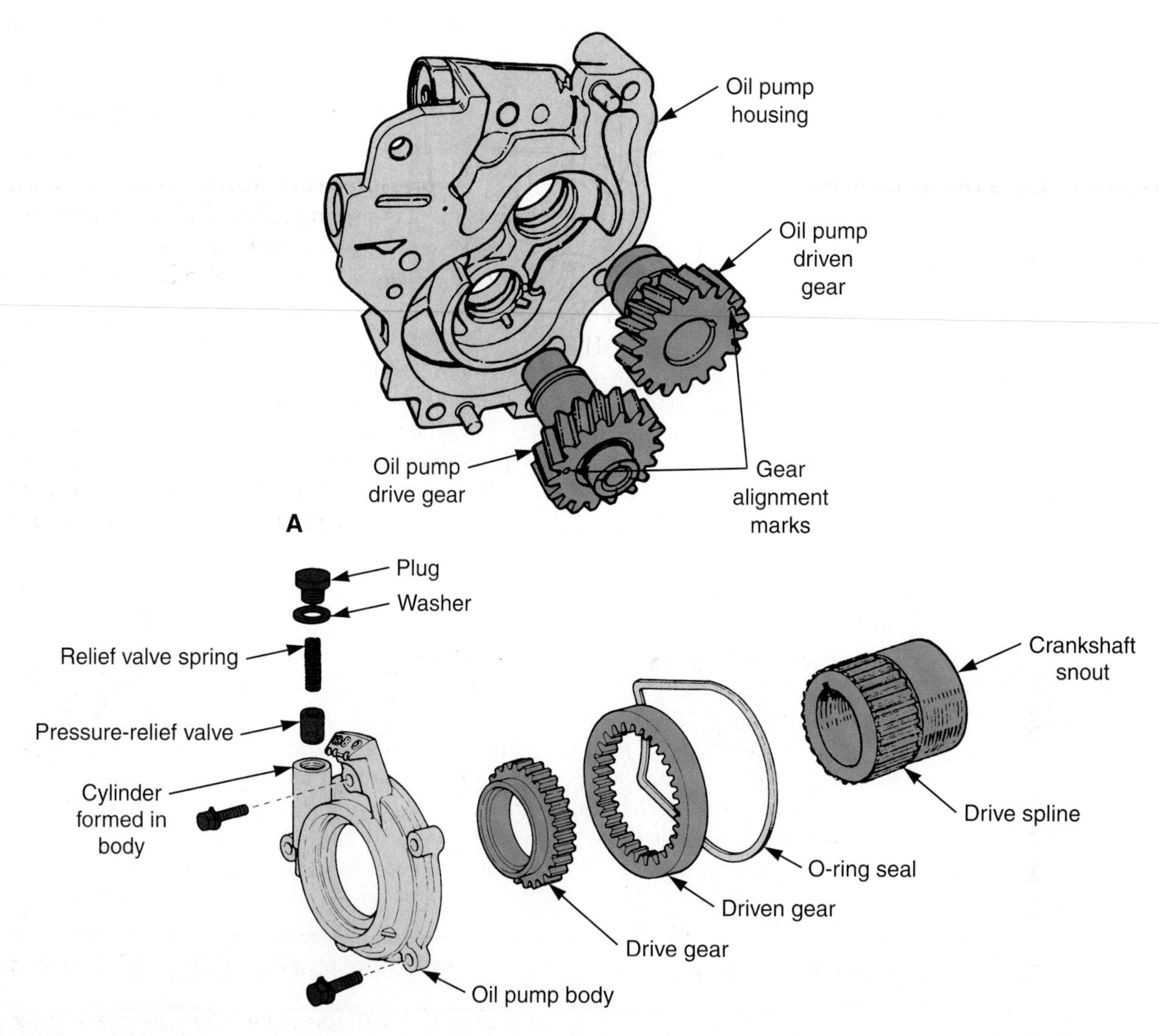

Figure 13-8. *Gear oil pumps. A—This oil pump mounts in the oil pan. B—This oil pump mounts on the crankshaft snout. (DaimlerChrysler, Toyota)*

Figure 13-9. *The oil pickup screen keeps large debris from entering the oil pump and being circulated in the engine. (Fel-Pro)*

the element permanently enclosed in the filter body, **Figure 13-10.** When it must be serviced, remove the old filter and simply screw a new filter into place. This is the most common type.

The ***cartridge oil filter*** has a separate element and canister, **Figure 13-11.** To service this type oil filter, remove the canister. Then, install a new element inside the existing canister. A cartridge oil filter is sometimes used on heavy-duty or diesel applications. Some gasoline-powered luxury cars also have this type of oil filter.

Oil Filter Housing

The ***oil filter housing*** is a metal part that bolts to the engine and provides a mount for the oil filter. The housing may also have a fitting for the oil pressure sending unit. A gasket normally fits between the engine and oil filter housing to prevent leakage. Sometimes, the pressure relief valve, filter bypass valve, or oil pump is inside this housing.

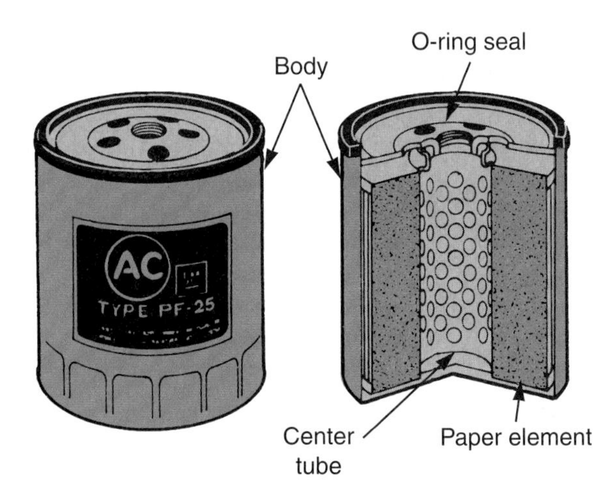

Figure 13-10. *A spin-on oil filter consists of a metal housing around a filter element. (AC-Delco)*

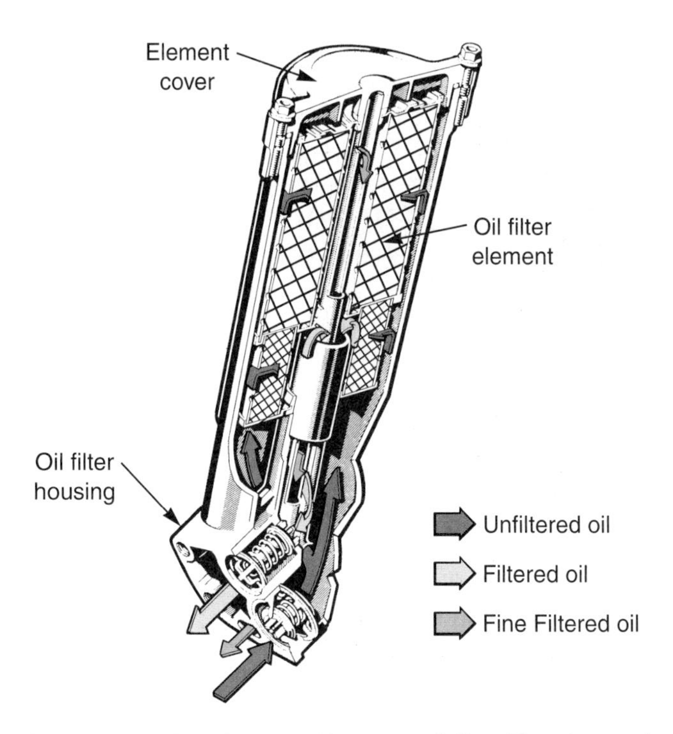

Figure 13-11. *This is a cartridge type oil filter. The element is removed from the housing to change the filter. (Mercedes-Benz)*

Pressure-Relief Valve

A ***pressure-relief valve*** limits the maximum oil pressure. It is a spring-loaded bypass valve in the oil pump, engine block, or oil filter housing. Under normal conditions, the spring holds the valve closed. All of the oil from the pump flows into the engine.

However, under conditions of abnormally high oil pressure, such as when the oil is cold and thick, the pressure-relief valve opens, **Figure 13-12.** Oil pressure pushes the small piston back in its cylinder by overcoming spring pressure. This allows some oil to bypass the main oil galleries and pour back into the oil pan. Most of the oil still flows to the bearings and a preset pressure is maintained.

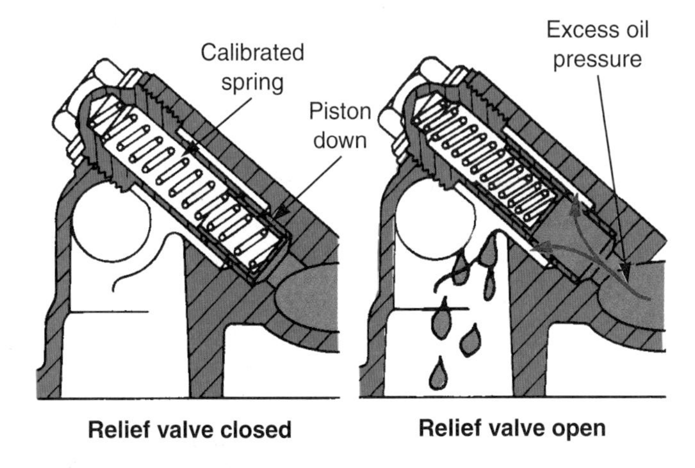

Figure 13-12. *The pressure-relief valve prevents excess engine oil pressure. When oil pressure is too high, the valve opens and bleeds some oil back into the pan. (DaimlerChrysler)*

Some pressure-relief valves are adjustable. By turning a bolt or screw or by changing spring shim thickness, the pressure setting can be altered.

Oil Pan and Sump

The ***oil pan*** holds an extra supply of oil for the lubrication system. It is normally made of sheet metal and bolted to the bottom of the engine block. The oil pan is fitted with a screw-in drain plug to allow for oil draining during oil changes. Baffles may be used to keep the oil from splashing around in the pan, **Figure 13-13.**

The ***sump*** is the lowest area in the oil pan where oil collects. As oil drains out of the engine, it fills the sump. Then the oil pump can pull oil out of the pan for recirculation.

A ***structural oil pan*** is designed to add strength to the engine bottom end and cylinder block. Most structural oil pans are made of lightweight, ribbed aluminum or composite construction. Some designs have the main cap bolts going through the oil pan.

A ***dry sump lubrication system*** uses two oil pumps and a separate oil tank. No oil is stored in the oil pan. The main oil pump pressurizes the lubrication system. The second oil pump, called the scavenging oil pump, pulls oil out of the oil pan.

A dry sump lubrication system is used on some high-performance engines to prevent the spinning engine crankshaft from stirring up and aerating the oil. ***Oil aeration*** results when air bubbles form in the oil and reduce its lubricating qualities. Because no oil remains in the pan to splash onto the crankshaft and pistons, it can increase engine power under extreme acceleration and cornering loads.

Oil Galleries

Oil galleries are small passages cast into or machined through the cylinder block and head for lubricating oil. They allow oil to flow to the engine bearings and other moving parts, **Figure 13-14.** The main oil galleries are large passages through the center of the block. They feed oil to the crankshaft bearings, camshaft bearings, cylinder heads, and valve train.

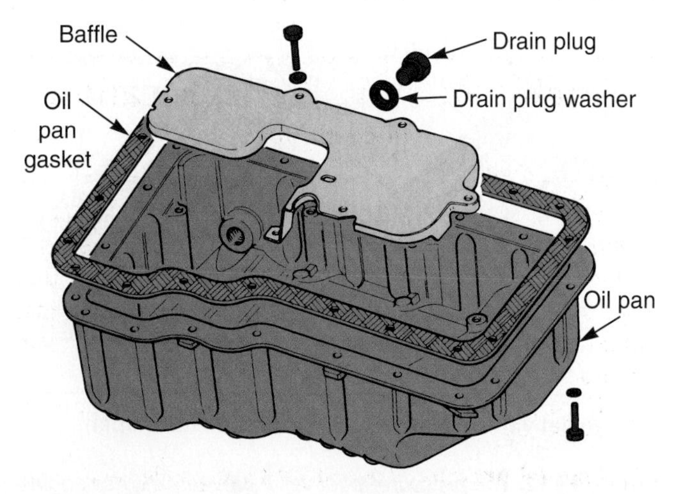

Figure 13-13. *The oil pan collects oil as it drains out of the block. Note the baffle.*

Oil Cooler

An ***oil cooler*** may be used to help lower and control the operating temperature of the engine oil. It is a radiator-like device connected to the lubrication system. One is shown in **Figure 13-14.** Oil is pumped through the cooler and back to the engine. Airflow through the cooler removes heat from the oil. Oil coolers are frequently used on turbocharged engines or heavy-duty applications.

Oil Level Indicator

An ***oil level indicator*** alerts the driver of a dangerously low oil level condition. As shown in **Figure 13-15,** one design uses a float mechanism located in the oil pan. The float rides on top of the engine oil in the pan. When the oil level drops too low, the float closes the circuit and current flows to the engine oil indicator light on the dash.

In a different design of oil level indicator, a special sensor is mounted inside the oil pan to detect low or dirty oil conditions. This sensing circuit uses a small light emitting diode (LED) and photo sensor assembly. If the oil drops below the sensor or if the oil is very opaque because it is dirty, a different amount of light is transferred across the sensor. The sensor output signal to the ECM changes. The ECM then illuminates the engine oil indicator light on the dash.

Oil Pressure Sending Unit

The ***oil pressure sending unit*** screws into one of the oil galleries to sense oil pressure. It is used with an indicator light or oil pressure gauge to warn the driver of low oil pressure. There are two types of oil pressure sending units—contact and resistance.

A contact-type oil pressure sending unit contains a diaphragm and electrical contacts. Oil pressure acts on the diaphragm. When oil pressure is normal, the diaphragm holds the contacts open. However, when oil pressure is too low, there is not enough pressure to hold the diaphragm off of the contacts. When the contacts are closed, the oil indicator light on the dash is illuminated. This warns the driver of low oil pressure.

A resistance-type oil sending unit is used with a dash-mounted oil pressure gauge. The resistance of the sending unit changes with a change in oil pressure. The reading on the oil pressure gauge varies accordingly. In some designs,

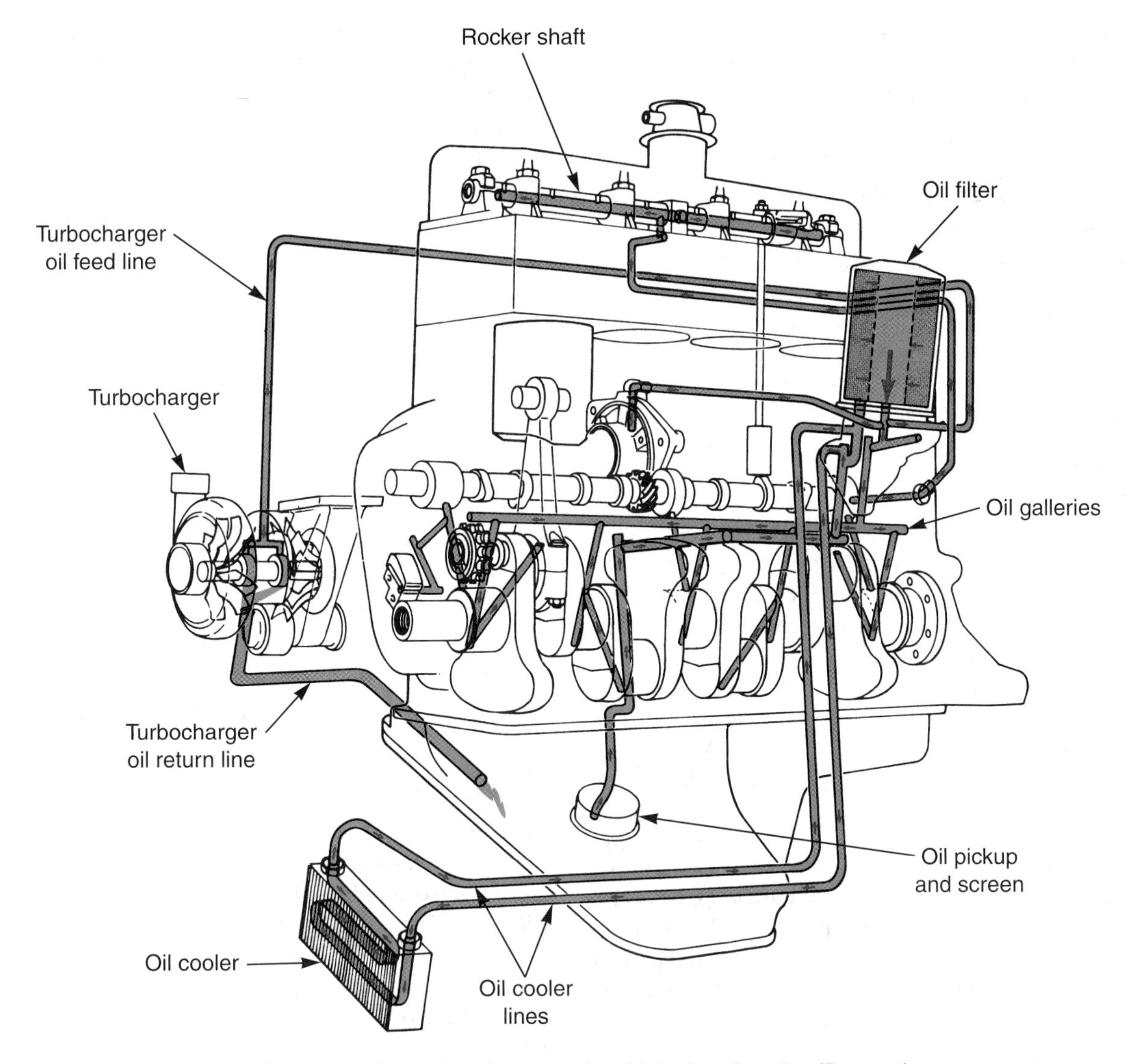

Figure 13-14. *Oil galleries pipe oil to the various points in the engine. Note the oil cooler. (Peugeot)*

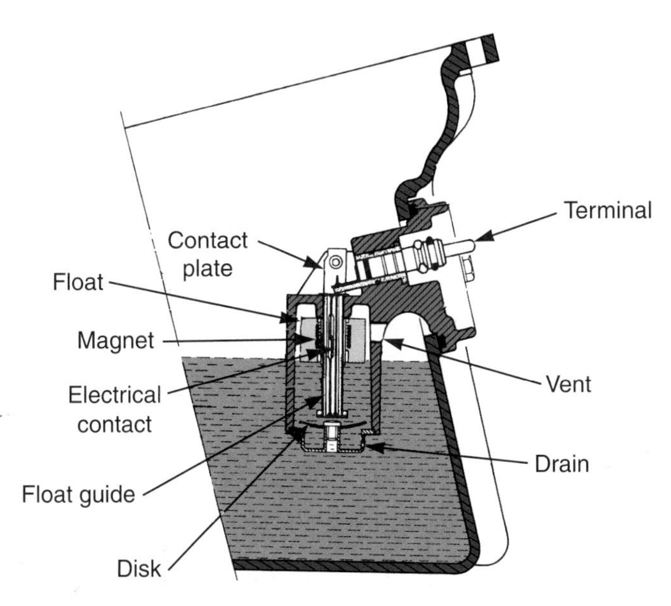

Figure 13-15. *An oil level indicator uses a simple float that rides on top of the oil to determine oil level. (Mercedes-Benz)*

an indicator light may be used in conjunction with the oil pressure gauge. The light is illuminated when oil pressure drops below a predetermined pressure.

Troubleshooting the Lubrication System

When diagnosing problems with the lubrication system, make a visual inspection of the engine for obvious problems. Check for oil leaks, disconnected sending unit wire, low oil level, damaged oil pan, or other trouble that would relate to the symptoms, **Figure 13-16.** Problems in the lubrication system are indicated by:

- ❑ **High oil consumption.** Oil must frequently be added to engine.
- ❑ **Low oil pressure.** The oil pressure gauge reads low, the indicator light glows, or abnormal engine noises can be heard.

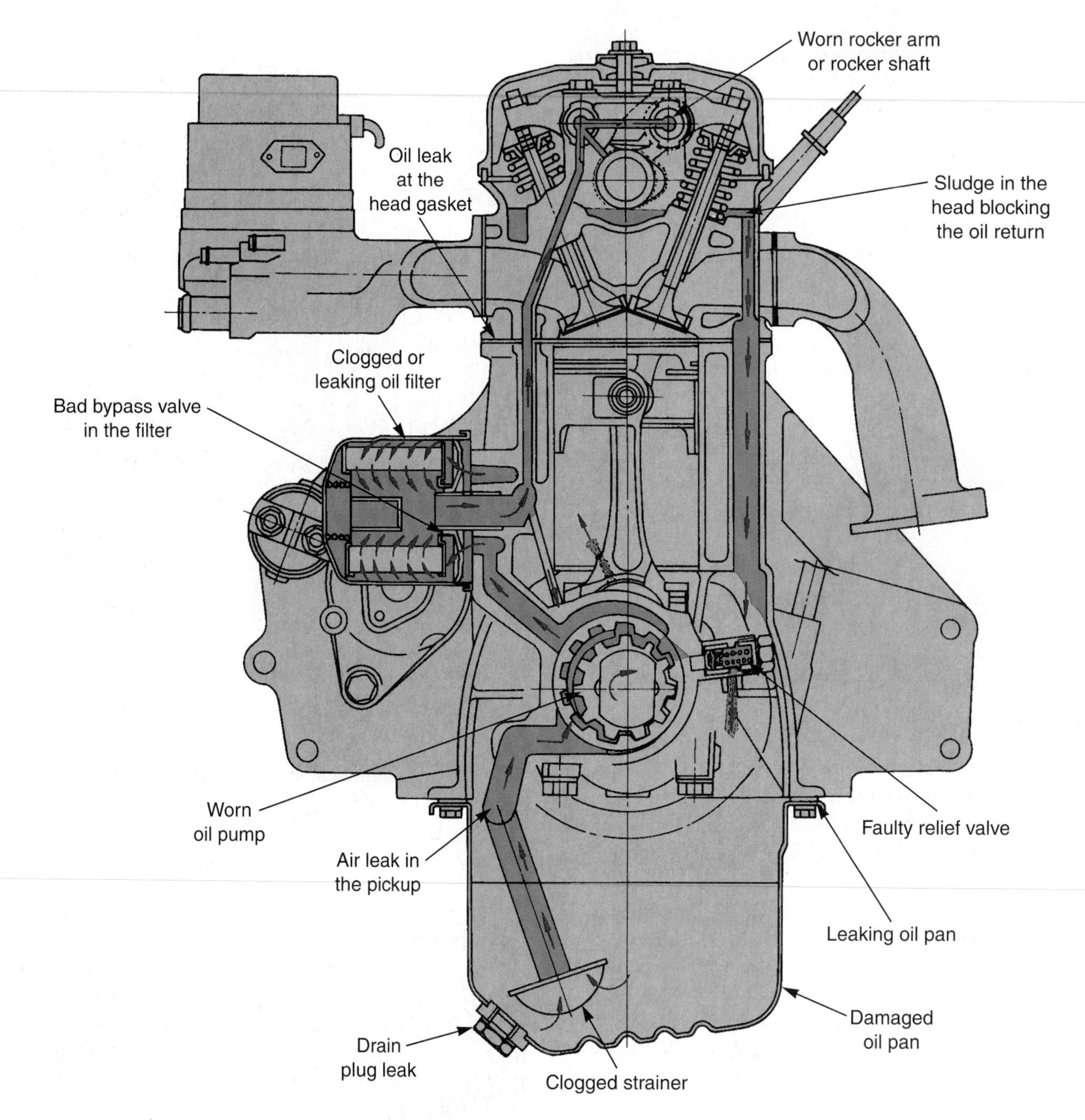

Figure 13-16. *Note some of the problems that can develop with an engine lubrication system.*

- ❑ **High oil pressure.** The oil pressure gauge reads high or the oil filter is swelled.
- ❑ **Defective indicator or gauge circuit.** Inaccurate operation of the indicator light or pressure gauge.

High Oil Consumption

High oil consumption is caused by external or internal oil leakage. External oil leakage is when the oil leaks out of the engine. Internal oil leakage occurs when oil leaks into the combustion chambers.

External oil leakage is easily detected as dark brown or black, oily wet areas on or around the engine. Oil may be found in small puddles under the car. Leaking gaskets or seals are usually the source of external engine oil leakage. To locate external oil leakage, you may need to raise the car on a lift and visually look for leaks under the engine. Trace the oil leakage to its highest point. The parts around the point of leakage will be washed clean by the constant dripping or flow of oil.

Internal oil leakage occurs when oil leaks into the combustion chamber. For example, if the piston rings and cylinders are badly worn, oil can enter the combustion chambers and be burned. Internal oil leakage is indicated by blue smoke coming out of the exhaust. Do not confuse black or white smoke with the blue smoke caused by engine oil. Black smoke is caused by excess fuel in the cylinder. White smoke is caused by coolant leakage into the cylinder.

Note: Engine oil consumption and smoking are covered fully in the chapter on engine mechanical problems.

Oil Pressure Problems

Low oil pressure is indicated when the oil indicator light glows, oil gauge reads low, or engine lifters or bearings rattle. Common causes of low oil pressure are:

- **Low oil level.** The oil not high enough in the pan to cover the oil pickup.
- **Worn connecting rod or main bearings.** The oil pump cannot provide enough oil volume to fill the bearing clearance.
- **Thin or diluted oil.** The viscosity of the oil is too low or gasoline is in the oil.
- **Weak or broken spring in the pressure-relief valve.** The spring allows the valve to open at too low of a pressure.
- **Cracked or loose oil pump pickup tube.** Air is being pulled into the oil pump instead of oil.
- **Worn oil pump.** Excess clearance between rotor or gears and housing prevents the pump from generating proper pressure.
- **Clogged oil pickup screen.** The amount of oil that can enter the pump is reduced.

High oil pressure is seldom a problem. The most frequent causes of high oil pressure are:

- **Pressure-relief valve stuck closed.** The valve does not open at the specified pressure.
- **High spring tension in the pressure-relief valve.** The wrong spring is installed or the spring has been improperly shimmed.
- **High oil viscosity.** Excessively thick oil or an oil additive that increases viscosity has been introduced into the engine.
- **Restricted oil gallery.** A defective block casting or debris in an oil passage is creating a restriction that increases oil pressure.

Defective Indicator or Gauge Circuit

A defective indicator or gauge circuit may appear to be a low or high oil pressure problem. The sending unit, circuit wiring, or gauge may be at fault. Inspect, test, remove, and replace the oil temperature and pressure sending units as described in the service manual.

Oil Pressure Test

An ***oil pressure test*** uses a gauge to measure the actual pressure in the engine. A pressure gauge is screwed into the hole for the oil pressure sending unit, **Figure 13-17.**

1. With the engine off, remove the oil pressure sending unit.
2. Install the oil pressure test gauge.
3. Start and warm the engine.

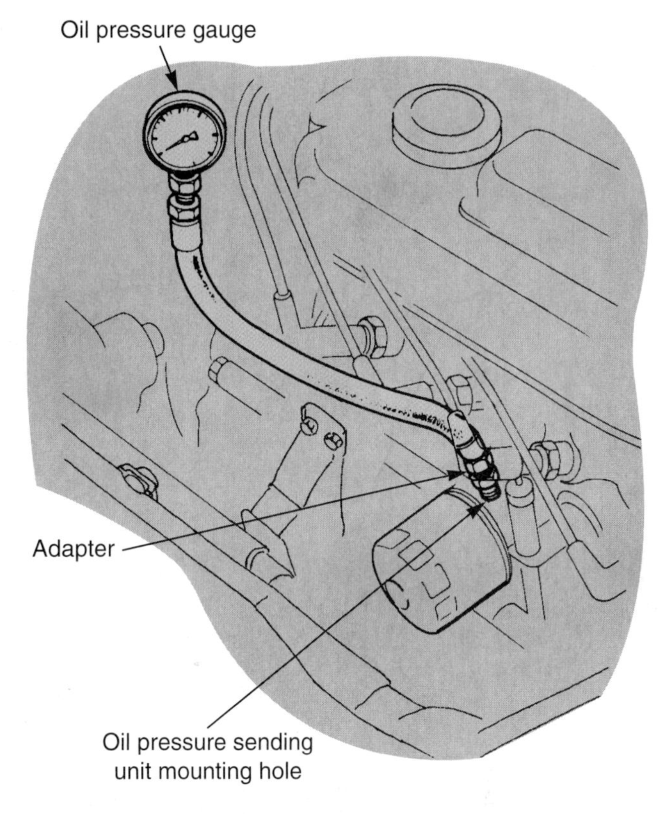

Figure 13-17. *Conducting an oil pressure test. (Honda)*

4. Run the engine at the rpm specified in the service manual.
5. Read the test gauge
6. Compare the reading to specifications.
7. If oil pressure is too low or high, make repairs as needed.

Depending on the number of miles on the engine and the type of engine, oil pressure should be at least 20 psi to 30 psi (140 kPa to 210 kPa) at idle and 40 psi to 60 psi (275 kPa to 410 kPa) at cruising speeds. Check the service manual for exact specifications.

Engine Oil and Filter Service

It is extremely critical that the engine's oil and oil filter are regularly serviced, **Figure 13-18.** Lack of maintenance can greatly shorten engine service life. Used oil will be contaminated with dirt, metal particles, carbon, gasoline, ash, acids, and other harmful substances. Some of the smallest particles and corrosive chemicals are not trapped in the oil filter. They will circulate through the engine, increasing part wear and corrosion.

Oil and Filter Change Intervals

Automakers give a maximum number of miles or kilometers a vehicle should be driven between oil

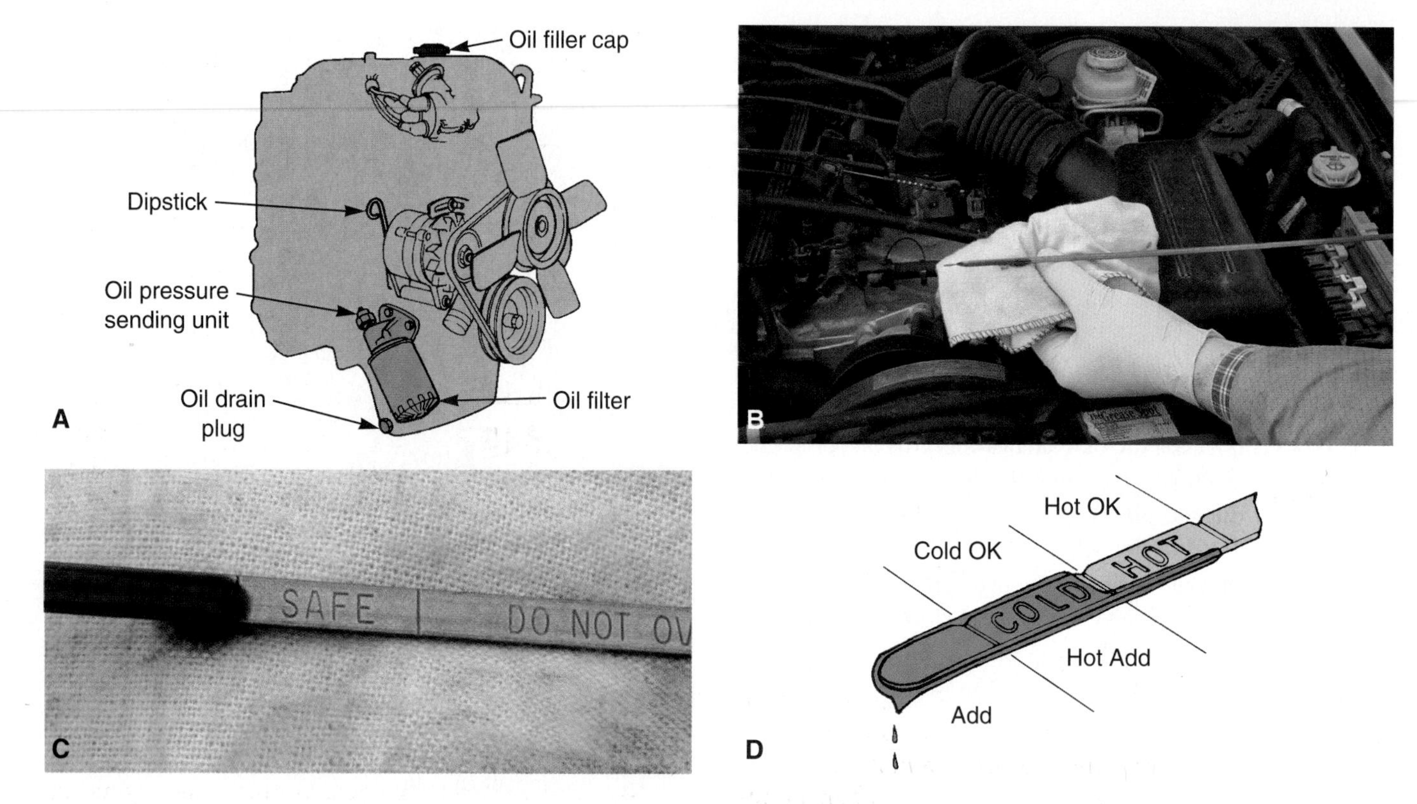

Figure 13-18. *Checking engine oil. A—Typical locations for the filler cap, dipstick, and oil filter. B—Wipe the dipstick off, reinsert it, and remove it to get an accurate oil level reading. C—Note the highest point of oil on the dipstick. D—When checking oil level, make sure you use the correct markings on the dipstick. (Mazda)*

changes. New vehicles can generally be driven about 6000 miles (10,000 km) between oil changes. If the oil is not changed at this interval, the warranty is voided.

Older vehicles should have their oil changed more frequently than new vehicles. An older, worn engine will contaminate the oil more quickly than a new engine. More combustion byproducts blow past the rings and enter the oil. Also, engine bearing clearances are larger and require more oil from the lubrication system.

If a vehicle is only driven for short periods, its oil should be changed more often than a vehicle that is driven for long periods. Since the engine may not be reaching full operating temperature, the oil can be contaminated with fuel, moisture, and other substances more quickly. Also, supercharged, turbocharged, and diesel engines usually require more frequent oil and filter service than naturally aspirated engines.

Changing the Engine Oil and Filter

A few rules to remember when changing engine oil and the oil filter are:

- ❑ Keep the car relatively level so all oil drains from the pan.
- ❑ Do not let hot oil pour out on your hand or arm!
- ❑ Check the condition of the drain plug threads and the O-ring washer. Replace them if needed.
- ❑ Do not overtighten the oil pan drain plug; the threads will strip very easily.
- ❑ Wipe oil on the filter O-ring before installation.
- ❑ Hand tighten the oil filter; do not use a wrench.
- ❑ Fill the engine with the correct amount and type of oil.
- ❑ Check for oil leaks with the engine running before releasing the vehicle to the customer.

To change the engine oil, warm the engine to full operating temperature. This will help suspend debris in the oil and make the oil drain more thoroughly. Then, turn off the engine and follow the basic steps below. Refer to **Figure 13-19.**

1. Use a lift or floor jack to raise the vehicle in level position. Place jack stands under the vehicle if a floor jack is used.
2. Place a drain pan under the oil pan drain plug.
3. Unscrew the drain plug and allow oil to pour into the pan.
4. When all oil is drained, replace the drain plug. Tighten to specifications.
5. Move the drain pan under the oil filter.
6. Use a filter wrench to remove the old filter. Turn counterclockwise.
7. Wipe clean oil on the gasket (O-ring) on the new filter.
8. Install the new oil filter by hand.
9. After the gasket (O-ring) makes contact, turn the filter 1/2 to a full turn more by hand.

10. Install the correct type and quantity of engine oil into the engine through the oil filler neck.
11. Start the engine and look for leaks.
12. Turn the engine off and check the oil level.

Engine Oil Disposal

Always send used engine oil to a recycling center. Recycling old oil not only saves our environment from possible pollution, it also helps save our natural resources. As an alternative, some shop heaters can burn used engine oil as a fuel.

Oil Pump Service

A bad oil pump will cause low or no oil pressure, which may result in severe engine damage. See **Figure 13-20.** When the pump's inner parts wear, the pump may have reduced output. The pump drive shaft can also strip, preventing pump operation. If tests point to a faulty oil pump, remove the pump and replace or rebuild it.

Most technicians install a new or factory-rebuilt pump when needed. It is usually too costly to completely rebuild an oil pump in the shop. However, you should have a general understanding of how to overhaul oil pumps. **Figure 13-21** summarizes how to inspect one type of oil pump.

The oil pump may be located inside the engine oil pan, on the front of the engine under the front cover, or on

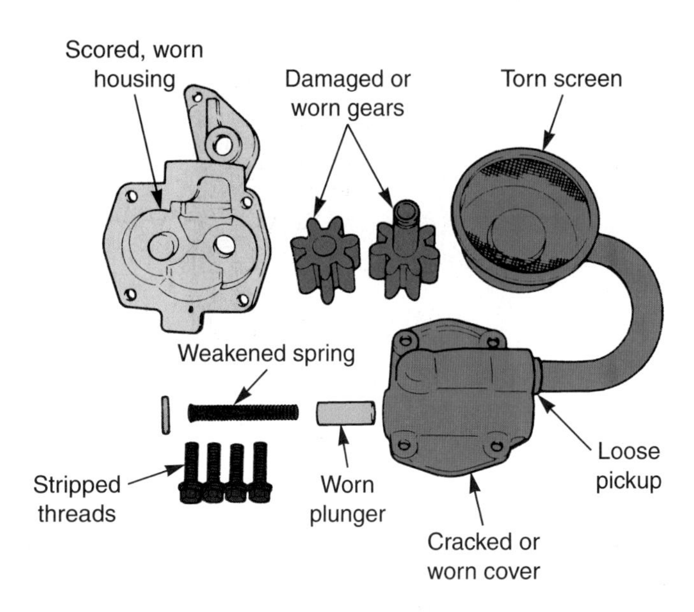

Figure 13-20. *Possible problems with the oil pump. (General Motors)*

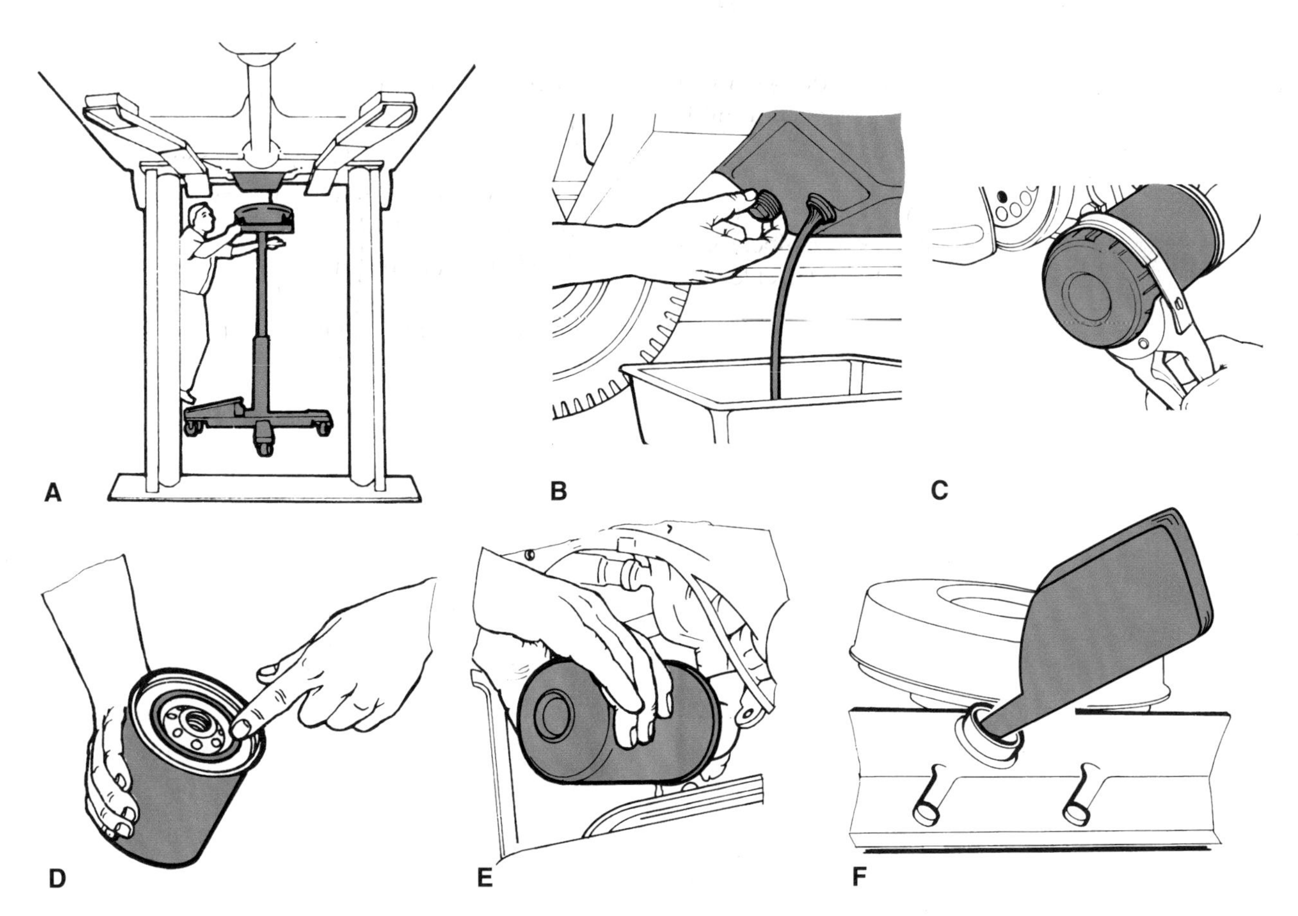

Figure 13-19. *Changing engine oil and filter. A—Lift the vehicle in a level position. B—Place a drain pan under the drain plug. Unscrew the drain plug and allow the oil to drain. C—Use a filter wrench to remove the old filter. D—Wipe clean oil on the new filter gasket (O-ring). E—Install the new oil filter by hand. F—Install the correct type and quantity of oil. (DaimlerChrysler)*

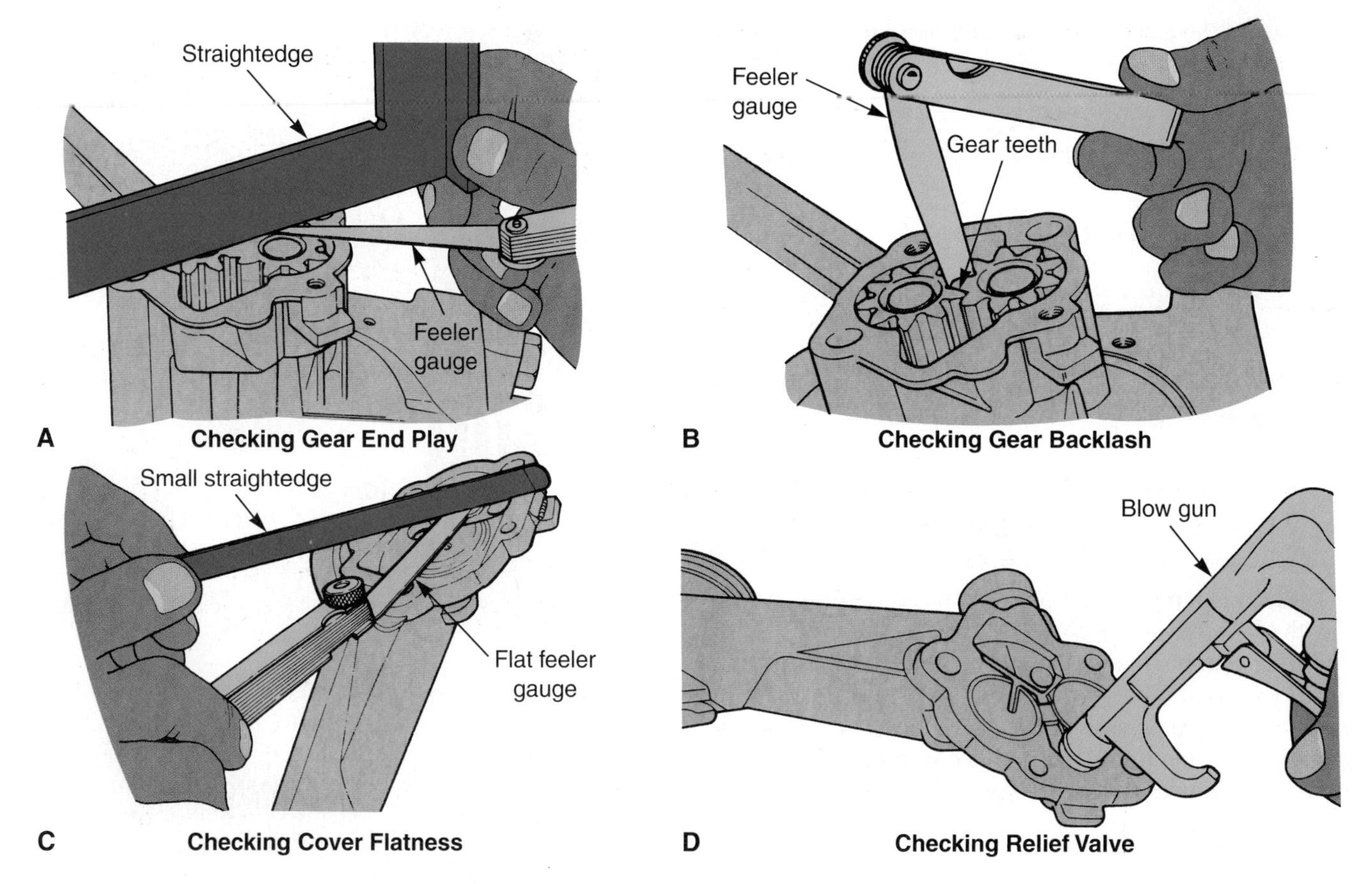

Figure 13-21. *Inspecting a gear-type oil pump. A—Use straightedge and flat feeler to check gear endplay. B—Checking gear backlash. C—Use a straightedge and feeler gauge to check the cover flatness. D—Checking the relief valve action with compressed air. (DaimlerChrysler)*

the side of the engine. Since removal procedures vary, refer to the service manual.

Before installation, prime the pump with engine oil. This can be done by pouring oil into the pumping cavity during assembly. You can also submerge the inlet into a container of oil and turn the pump by hand to draw oil into the pump. Priming will help protect the pump and ensure full engine oil pressure sooner on initial starting of the engine.

Install the pump in reverse order of removal. Double-check the position of any gaskets or O-rings. Torque all fasteners to specifications in the recommended torque sequence. If an oil pickup is used, make sure it is properly installed and tightened. **Figure 13-22** shows how to install a crankshaft-mounted oil pump.

Pressure-Relief Valve Service

A faulty pressure-relief valve can produce oil pressure problems. If symptoms point to the pressure-relief valve, it should be disassembled and serviced. It is also serviced during an engine overhaul. The valve may be located in the oil pump, filter housing, or engine block.

1. Remove the cup or cap holding the pressure-relief valve.
2. Slide the spring and piston out of their bore.
3. Measure spring free length and compare it to specifications. If the spring is too short or long, install a new spring.
4. If recommended, check the spring tension on a spring tester.
5. If recommended, shims can be used to increase the spring tension or adjust the opening pressure when needed.

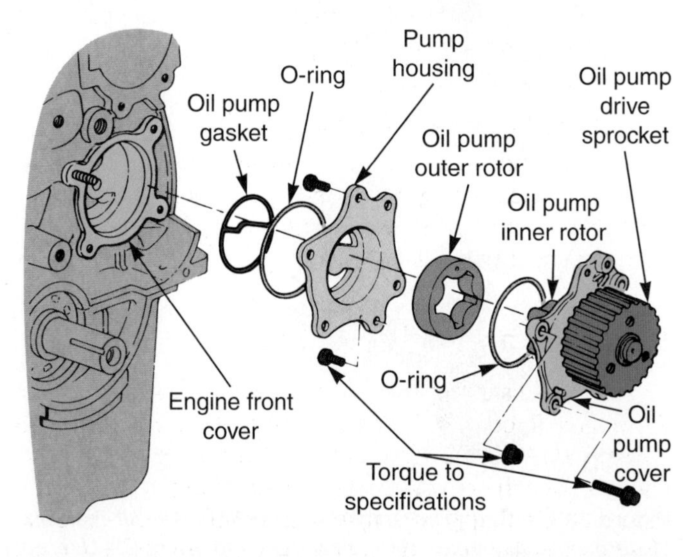

Figure 13-22. *Installing a front-mounted oil pump. (Honda)*

6. Use a micrometer and small hole gauge to check valve and valve bore wear.
7. Check the sides of the valve for scratches or scoring.
8. Slide the plunger in and out of its bore to check the action.
9. Replace parts as needed if any problems are found.

Oil Pan Service

An engine oil pan may need to be removed for various reasons:

- To service engine bearings or the oil pump.
- To repair damage to the pan, such as stripped drain plug threads in the pan.
- For pan replacement.
- During an engine overhaul.

Some engine oil pans can be removed with the engine in the vehicle. You may need to remove cross-members or other components and raise the engine off its motor mounts. In other cases, the engine must be lifted from the vehicle before the pan can be removed. Check the service manual for details.

Removing an Engine Oil Pan

To remove an engine oil pan, first drain the oil from the engine. Once the oil is drained, reinstall the drain plug.

1. Unscrew the bolts around the outside of the pan flange.
2. Tap on the pan lightly with a rubber hammer to free the pan from the cylinder block.
3. If the pan is stuck tight, carefully pry between the pan flange and the block. Do not bend the oil pan flange or leakage may result when reassembled.
4. Using a gasket scraper, remove all old gasket or silicone material from the pan and engine block. Do not gouge the sealing surfaces.
5. Check the inside of the pan for debris.

Debris indicates engine mechanical problems. Metal bits may come from bearing wear. Pieces of plastic may be from broken timing gear teeth. Rubber particles may come from damaged valve stem seals.

Installing an Engine Oil Pan

Before installing the engine oil pan, wash the pan thoroughly in cold soak cleaner. Check the drain plug hole for stripped threads. Also, lay the pan upside down on a flat workbench to check for a bent flange. If needed, straighten the flange with light hammer blows. A gasket or chemical sealer may be recommended for the oil pan, **Figure 13-23.**

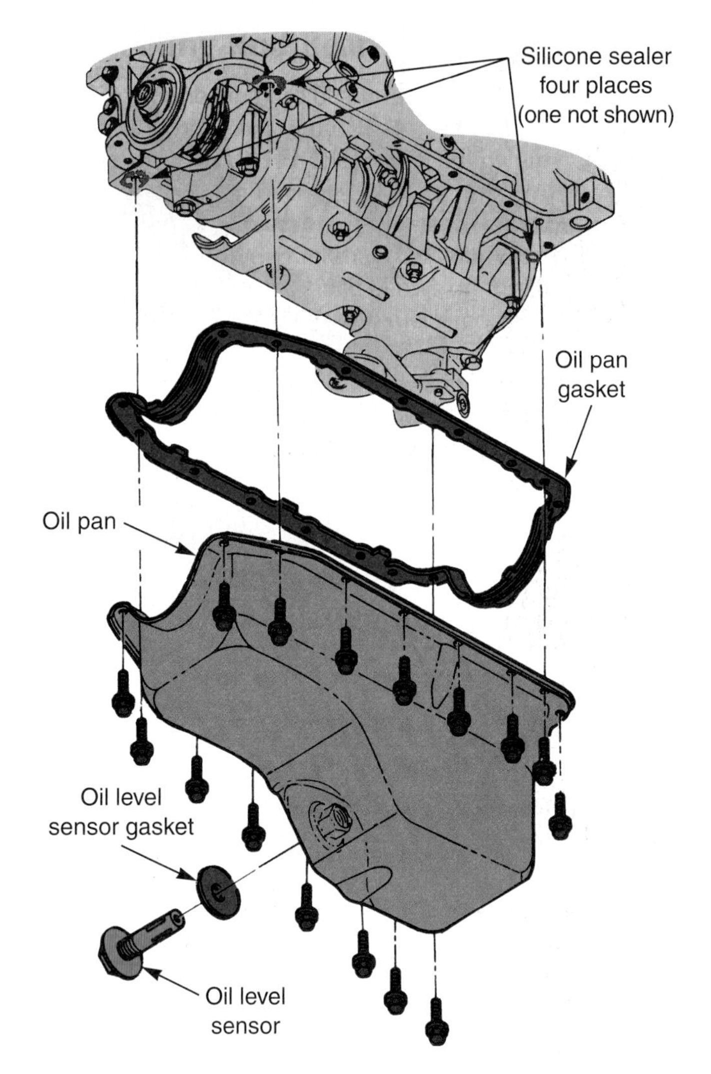

Figure 13-23. *Installing the oil pan. Use a new gasket or silicone sealer. (Ford)*

1. Use approved gasket adhesive to position the new gasket.
2. If rubber seals are used on each end, press them into their grooves.
3. Place silicone sealer where the gaskets meet the rubber seals, **Figure 13-24.**
4. Carefully fit the pan onto the block.
5. Start all of the pan bolts by hand.
6. Check that the gasket has not been shifted or the sealer smeared.
7. Then, tighten each bolt a little at a time in a crisscross pattern.
8. Torque the bolts to specifications.
9. Torque the drain plug to specifications.
10. Fill the engine with the correct type and quantity of oil.
11. Start the engine. Inspect for oil leaks.

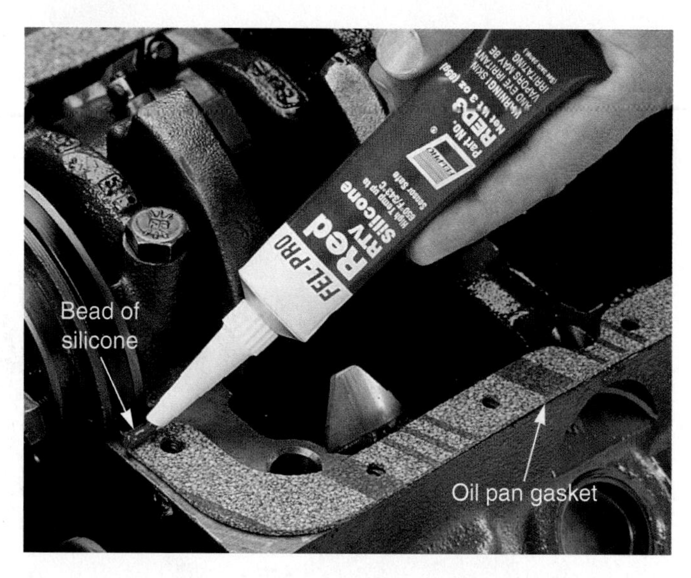

Figure 13-24. *Place a bead of sealer where the gasket meets a seal. (Fel-Pro)*

To use a chemical sealer, follow the same procedure. However, clean the pan and block mating surfaces with a suitable solvent. Then, in place of the gasket, run a continuous bead of sealer about 1/8″ (3 mm) wide around the pan or block flange, as recommended. Place extra sealer at part or gasket-seal joints.

Oil Pressure Indicator and Gauge Service

A bad oil pressure indicator or gauge may scare the customer into thinking there are major engine problems. The indicator light may flicker or stay on, pointing to a low pressure problem. Normally, the low oil pressure light will glow when oil pressure is below 5 psi to 10 psi (35 kPa to 70 kPa). The gauge can read low or high, also indicating a lubrication system problem.

Inspect the indicator or gauge circuit for problems. The wire going to the sending unit may be loose. In this case, the light may flicker or gauge needle may fluctuate. The sending unit wire may also be shorted to ground, in which case the light stays on or gauge always reads high.

Always check the service manual before testing an indicator or gauge circuit. Some automakers recommend a special gauge tester. This is especially important with some computer controlled systems. The tester places a specific resistance in the circuit to avoid circuit damage. Refer to the manual for instructions.

To check the action of the indicator or gauge, remove the wire from the sending unit, **Figure 13-25.** Ground the wire. The indicator light should glow or the oil pressure gauge should read minimum. If it does, the sending unit may be bad. If it does not, then the circuit, indicator, or gauge may be faulty.

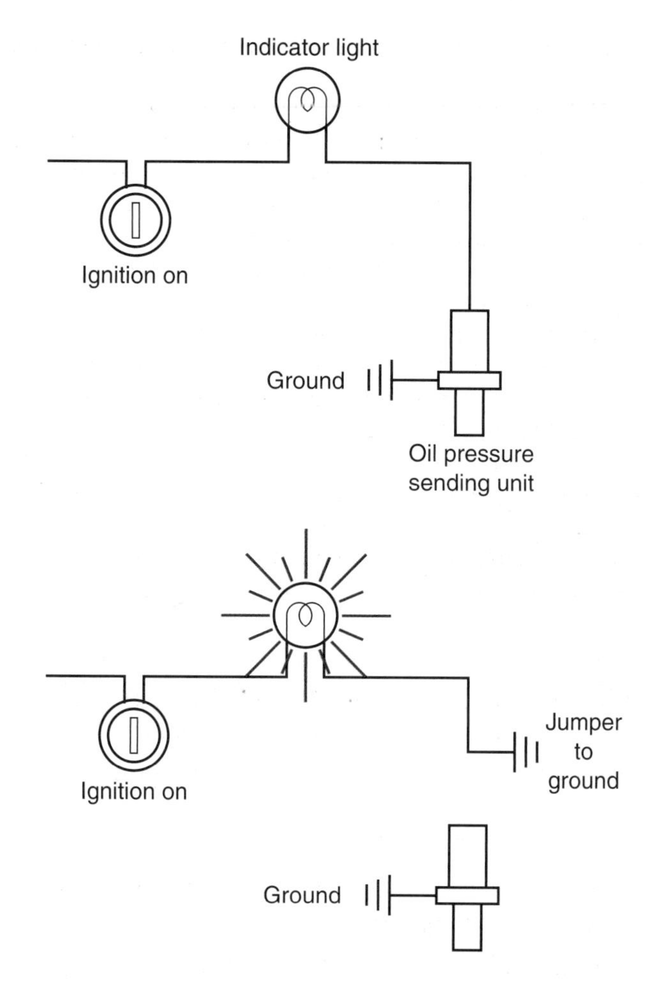

Figure 13-25. *Usually, you can check the oil pressure sending unit by grounding the wire going to the unit. When the lead is grounded, the indicator light should glow.*

Note: When testing the sending unit for an oil pressure gauge, do not hold the lead on ground too long or gauge damage could result. Also, some circuits could be damaged by grounding, so check the service manual for specific information.

Summary

The lubrication system uses the oil pump to force engine oil throughout the engine. An oil film separates moving parts to prevent wear.

Oil viscosity is a measure of the oil's ability to flow. A multiviscosity oil has the properties of a low-viscosity oil when cold and of a high-viscosity oil when hot. The SL service rating is the highest rating available. Never use a lower rating than what is recommended. Use the type and grade of oil recommended by the engine manufacturer.

Oil pumps can be rotary or gear types. Both use engine power to send oil through the oil galleries in the block. Oil is also forced through passages in the head for valve train lubrication.

An oil pressure sending unit screws into the engine and is exposed to engine oil pressure. A contact-type sending unit is used with an indicator light. A resistor-type sending unit is used with an oil pressure gauge.

An oil pressure test uses a gauge screwed into the oil pressure sending unit hole. This test determines the actual engine oil pressure.

The oil filter removes impurities from the oil. It can be a spin-on or cartridge type. The filter and engine oil should be changed at the recommended intervals. On some newer vehicles, this may be about 6000 miles (10,000 km).

The engine oil pan may have to be removed to replace the pan gasket, oil pump, or engine bearings. The oil pan can sometimes be removed without engine removal. Check in the service manual for details. Do not overtighten the oil drain plug; it will strip very easily.

Review Questions—Chapter 13

Please do not write in this text. Write your answers on a separate sheet of paper.

1. List five functions of the lubrication system.
2. The two methods of lubricating engine components are _____ oiling and _____ oiling.
3. Engine oil is refined from _____ or it may be _____.
4. Define *oil clearance.*
5. Briefly describe *oil viscosity* and how it is indicated on an oil container.
6. What effect(s) does temperature have on oil?
7. Briefly describe *multiviscosity oil* and how it is indicated on an oil container.
8. What is the *API certification mark* and what does it represent?
9. List the five basic components of a lubrication system.
10. What are the two types of oil pump?
11. What is the function of the *oil pickup screen?*
12. What is the purpose of the *filter bypass valve?*
13. A(n) _____ limits the maximum oil pressure.
14. What is the *sump?*
15. Define *oil galleries.*
16. Briefly describe the two types of oil sending units and how each type is used to indicate oil pressure.
17. Internal oil leakage shows up as _____ coming out the exhaust pipe.
18. A(n) _____ test is used to check the actual engine oil pressure.
19. List three reasons you may have to remove the engine oil pan.
20. When testing an oil pressure indicator light, _____ the wire to the sending unit and the light should glow.

ASE-Type Questions—Chapter 13

1. Technician A says that a 10W-30 engine oil has the properties of a 10 weight oil when it is hot. Technician B says that a low-viscosity engine oil is thin and flows easily. Who is correct?
   (A) A only.
   (B) B only.
   (C) Both A and B.
   (D) Neither A nor B.
2. All of the following are causes of low engine oil pressure *except:*
   (A) a clogged oil pump pickup screen.
   (B) worn connecting rod bearings.
   (C) restricted oil gallery.
   (D) worn main bearings.
3. When performing an oil pressure test, the pressure gauge should be attached to the _____.
   (A) oil filter fitting
   (B) oil pressure sending unit hole
   (C) oil pump
   (D) None of the above.
4. All of the following are basic rules to follow when changing the engine's oil and oil filter *except:*
   (A) apply oil to the oil filter gasket (O-ring) before installing filter.
   (B) do not overtighten oil pan drain plug.
   (C) use a filter wrench to tighten the oil filter.
   (D) keep the car relatively level when draining the oil.
5. Technician A says that before installing an oil pump, you should apply grease to the pickup screen. Technician B says that before installing an engine's oil pump, you should prime the pump with oil. Who is correct?
   (A) A only.
   (B) B only.
   (C) Both A and B.
   (D) Neither A nor B.

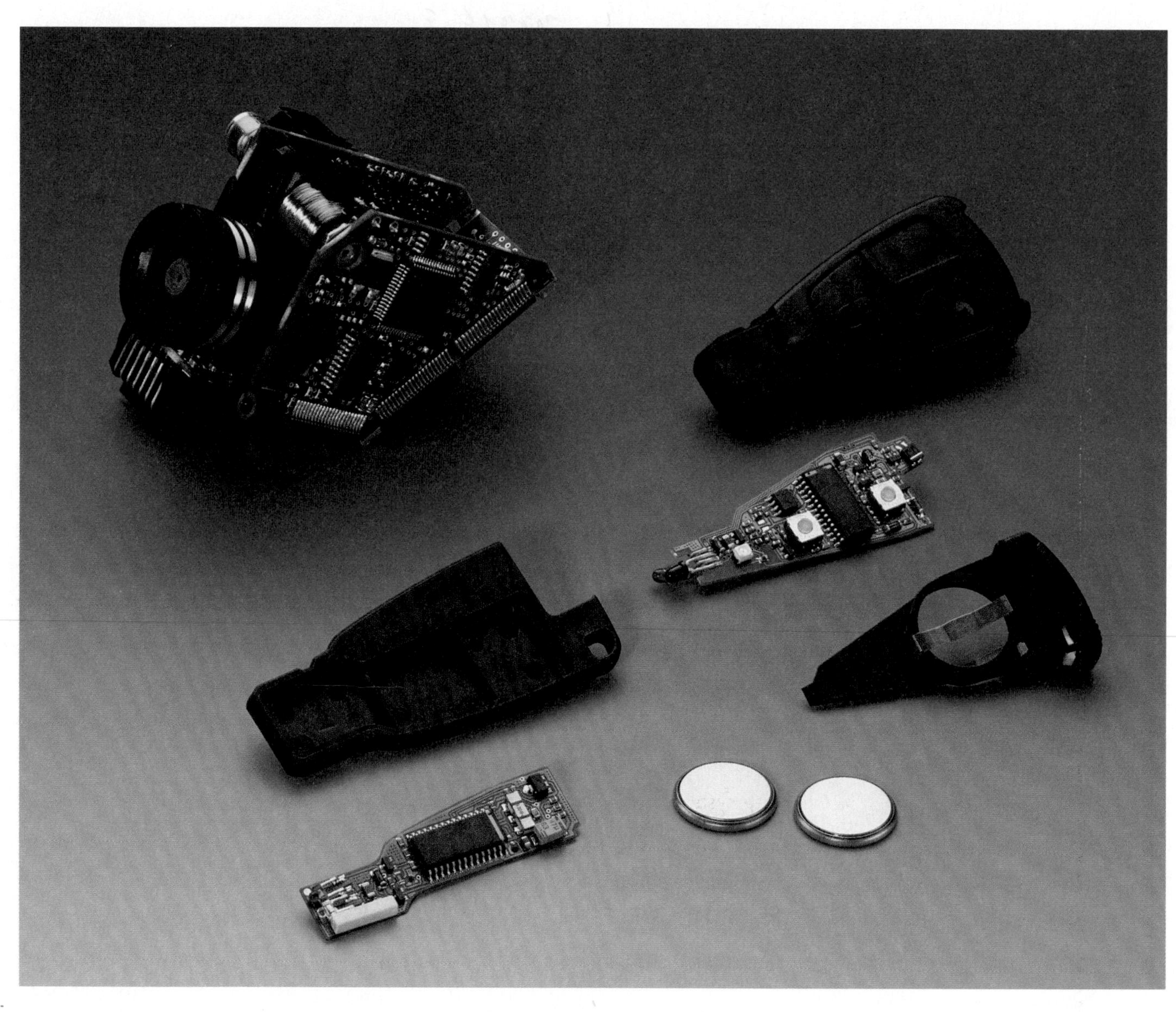

Some vehicles have electronic keys and ignition locks. (DaimlerChrysler)

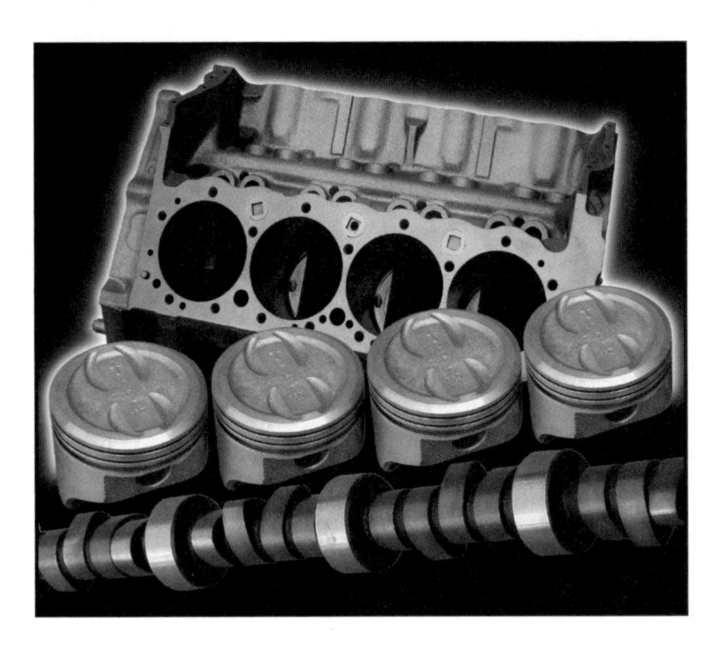

Chapter 14

Starting, Charging, and Ignition Systems

After studying this chapter, you will be able to:

- ❑ Summarize the operation of a starting system.
- ❑ Describe basic troubleshooting of the starting system.
- ❑ Explain how a charging system functions.
- ❑ Describe basic troubleshooting of the charging system.
- ❑ Explain the operation of an ignition system.
- ❑ Describe basic troubleshooting of the ignition system.

Know These Terms

Charging system	No-crank problem
Dead cylinder	Slow-crank problem
Ignition system	Spark intensity test
Jump starting	Starting system

As an engine technician, you must have a basic knowledge of the starting, charging, and ignition systems. You must be able to determine that a problem is an engine mechanical problem and not in one of these systems. This chapter covers the principles behind the starting, charging, and ignition system. Basic troubleshooting of these systems is also covered. This chapter will help in a review of the starting, charging, and ignition systems in preparation for taking the ASE certification tests.

Starting System Principles

The ***starting system*** rotates the crankshaft until combustion powers the engine. When the driver turns the ignition switch to the start position, a relay in the starter solenoid closes and current flows from the battery to the starter motor, **Figure 14-1.** A gear on the starter then engages the flywheel ring gear. As the starter turns, the flywheel turns and the engine rotates through its four-stroke cycle. The driver disengages the starter when combustion power takes over and the engine runs.

The starting system uses battery power and an electric motor to turn the engine crankshaft for engine starting. See **Figure 14-2.** The major parts of a starting system include:

- **Battery.** The source of stored energy for the starting system. It turns chemical energy into electrical energy.
- **Ignition switch.** Allows the driver to operate the starting system.
- **Solenoid.** A high-current relay (switch) for connecting the battery to the starter motor.
- **Starter motor.** A high-torque electric motor that engages and turns the engine flywheel.

Starting System Troubleshooting

A starting system is easy to troubleshoot. **Figure 14-3** shows the most common starting system troubles. If any of these parts have high resistance, low resistance, damage, or wear, the engine may not crank normally.

Common Starting System Problems

A no-start condition with normal cranking is usually *not* caused by the starting system. There may be trouble in the fuel or ignition systems or an engine mechanical problem.

A ***no-crank problem*** results when the engine crankshaft does not rotate properly with the ignition key in the start position. The most common causes are a dead battery, poor electrical connection, faulty system component, or engine mechanical problem.

Buzzing or clicking from the solenoid without cranking is commonly due to a discharged battery or poor battery cable connections. Low current flow is causing the solenoid plunger to rapidly kick in and out, making a chattering sound.

A single click without cranking may point to a bad starter motor, burned solenoid contacts, dead battery, or

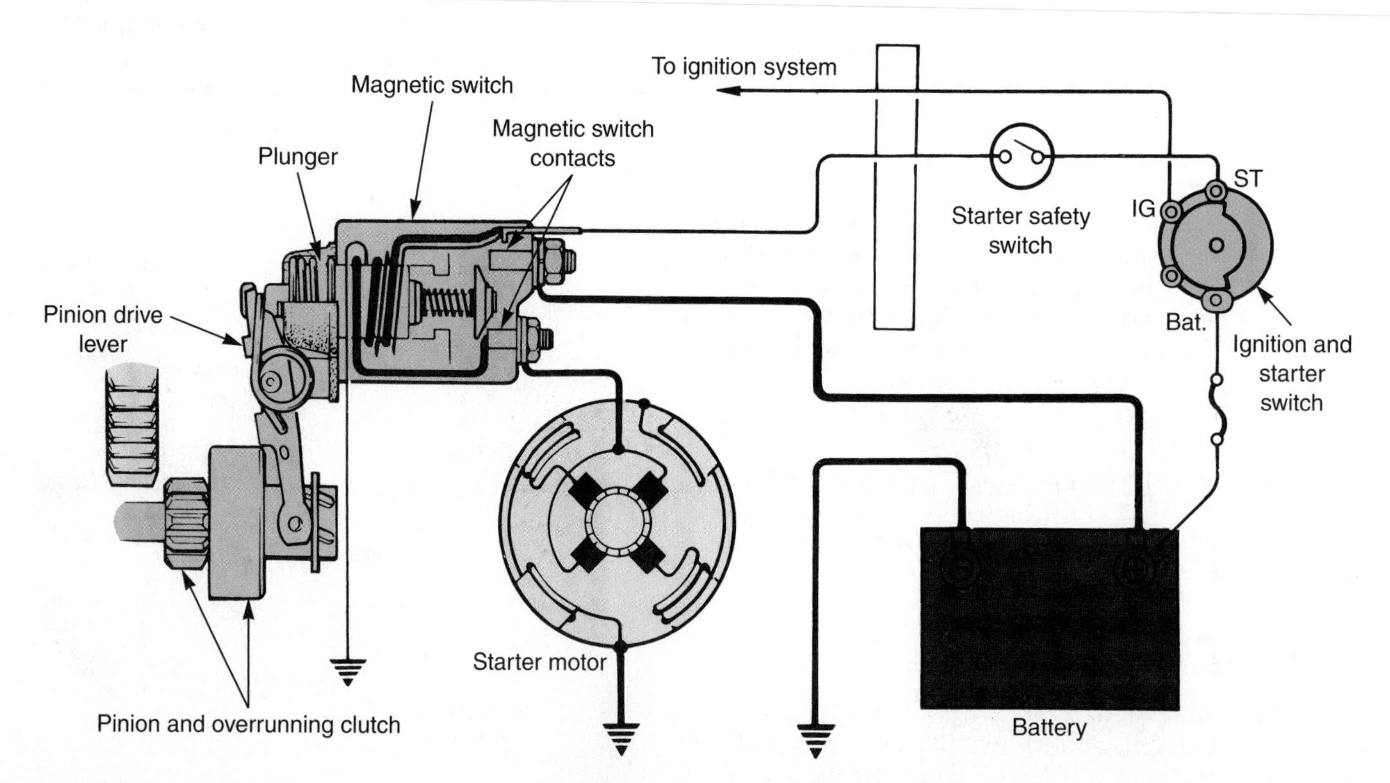

Figure 14-1. *This schematic shows how the solenoid completes the starter circuit and also engages the pinion gear to the flywheel ring gear. (General Motors)*

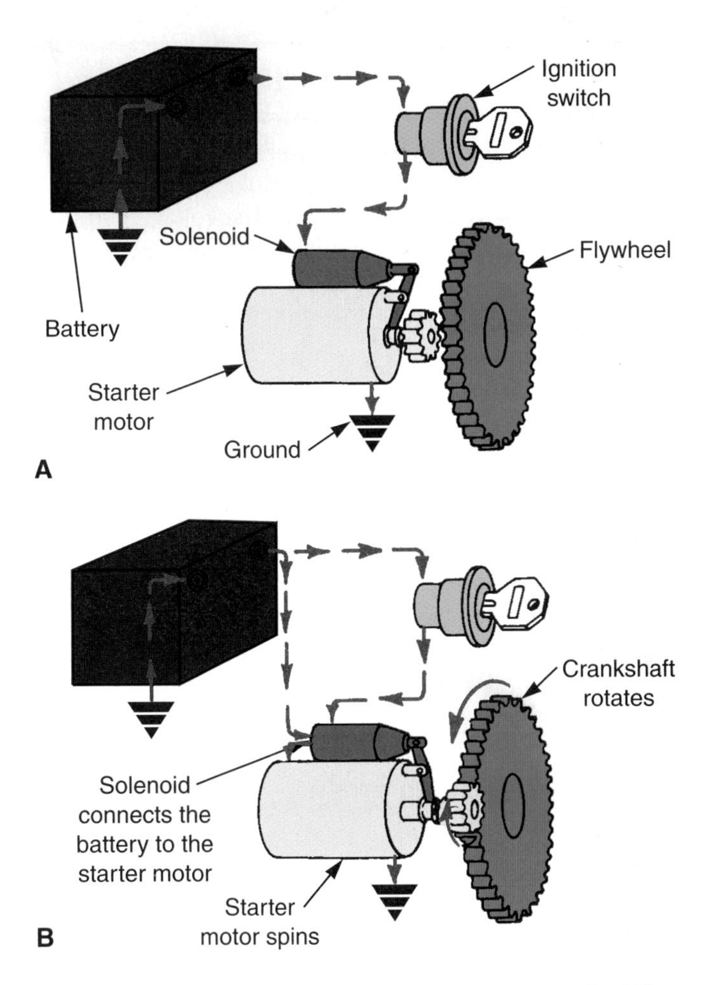

Figure 14-2. *A—The basic parts of a starting system. B—When the ignition key is turned to the start position, a small current activates the starter solenoid. Contacts in the solenoid close and conduct a larger current to turn the starter motor. (Deere & Co.)*

engine mechanical problem. The click is probably the solenoid closing or the pinion gear contacting the flywheel gear.

A humming sound after momentary engine cranking may be due to a bad starter overrunning clutch or pinion gear unit. Pinion gear wear can make the gear disengage from the flywheel gear too soon. This can allow the motor armature to spin rapidly, making a loud humming sound.

A metallic grinding noise may be caused by broken flywheel teeth or pinion gear teeth wear. The grinding may be the gears interfering with each other.

A ***slow-crank problem*** occurs when the engine crankshaft rotates at a lower-than-normal speed. It is usually caused by the same kind of problems producing a no-crank problem.

Check the Battery First

A dead or discharged battery is one of the most common reasons the starting system fails to properly crank the engine. A starter motor draws several times the amount of current as any other electrical component, over 200 amps. Make sure the battery is good and fully charged. A starter motor will not function without a fully charged and properly connected battery.

Battery Problems

A discharged battery, or dead battery, is a very common problem. The engine will usually fail to crank and start. Even though the lights and horn may work, there is not enough electricity in the battery to operate the starter motor. A discharged battery could be caused by:

- Defective battery due to sulfated plates.
- Charging system problem resulting in inadequate recharging.
- Starting system problem resulting in excessive current draw.
- Poor cable connections.
- Engine performance problem requiring excessive cranking time.
- Excessive current draw with the ignition key in the off position.

Warning: Wear eye protection when working around batteries. Batteries contain acid that could cause blindness. Even the film buildup on the outside of a battery can contain acid.

Checking Battery Voltage

The voltmeter function of a multimeter can be used to determine if the battery has a proper charge, **Figure 14-4.** A fully charged battery should have about 12.6 volts. If the battery does not have this voltage, charge the battery. If the battery does not take a charge, replace it.

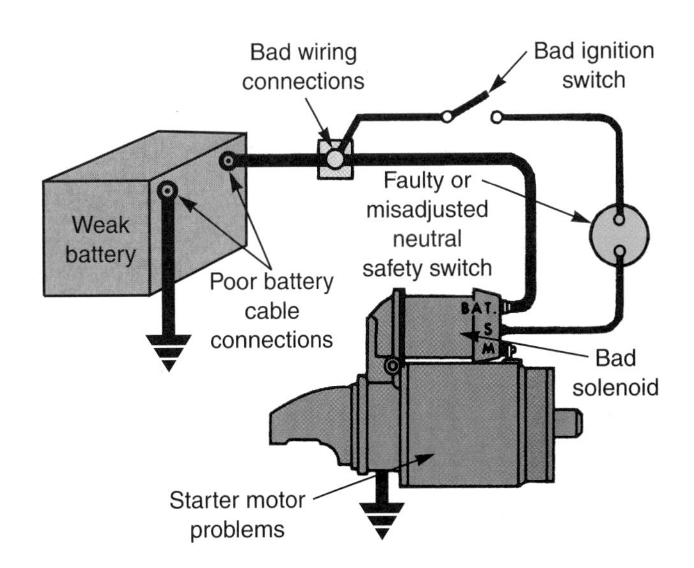

Figure 14-3. *These are the types of problems that you can find in the starting system. (Echlin)*

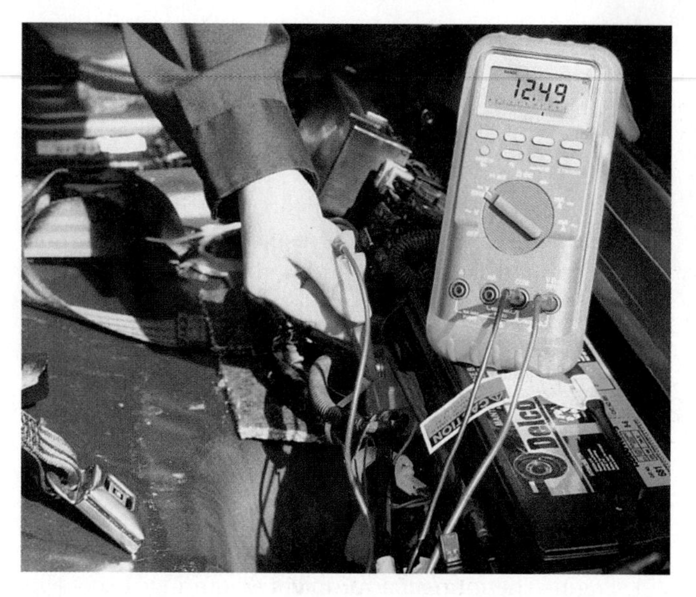

Figure 14-4. *Using a multimeter to check for battery charge. A fully charged battery should produce a reading of around 12.6 volts. (Fluke)*

Jump Starting

If the battery is discharged, the vehicle can be ***jump started*** by connecting another battery to the discharged battery. The two batteries are connected in parallel—positive to positive and negative to negative. Connect the red jumper cable to the positive terminal on both batteries. Then, connect the black jumper cable to any ground on both vehicles as shown in **Figure 14-5.** The last connection should be on the vehicle with the good battery at a location away from the battery to prevent a spark near the battery.

Warning: Do not short the jumper cables together or connect the batteries in series (negative to positive). This could cause serious damage to the charging or computer systems. Other electronic circuits, such as in the audio system, may also be damaged.

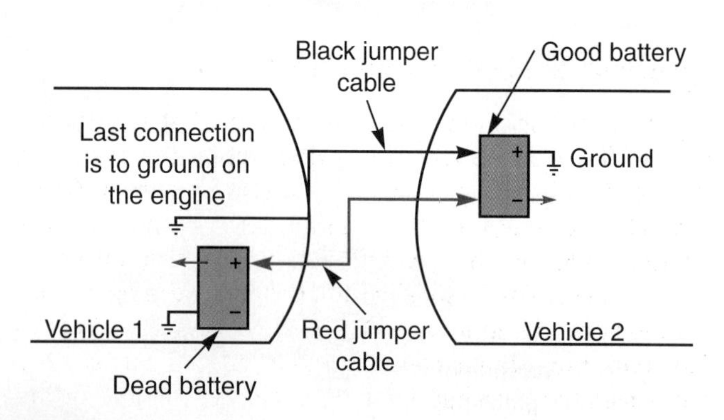

Figure 14-5. *Jump starting a vehicle with a dead battery. The last connection should be made away from the battery on any good ground.*

Charging System Principles

The ***charging system*** has several basic functions. It recharges the battery after engine cranking or use of electrical accessories with the engine off. The voltage output of the charging system is slightly higher than battery voltage. This allows the battery to be recharged. The charging system also supplies all of the vehicle's electricity when the engine is running. Finally, the charging system must change the current output to meet varying electrical loads.

The basic parts of a charging system are shown in **Figure 14-6.** They include:

- **Alternator.** An electric generator that uses mechanical energy from the engine to produce electrical energy.
- **Voltage regulator.** An electrical device for limiting the output voltage of the alternator. Charging voltage is between 13 and 15 volts.
- **Alternator drive (belt).** The system that connects the engine crankshaft pulley to the alternator pulley for driving the alternator.
- **Battery.** Provides electricity to initially excite (energize) the alternator. It also helps stabilize alternator output.

Charging System Troubleshooting

There are four common symptoms of trouble in the charging system:

- Dead battery, indicated by slow or no cranking.
- Overcharged battery.
- Abnormal noises in the alternator, such as grinding, squealing, or buzzing.
- Indicator light glows all the time or an incorrect gauge reading.

Even though a charging system only has two major parts (alternator and regulator), be careful when troubleshooting. Sometimes, a fault in another system, such as a bad starter, may appear to be a problem in the charging system.

Charging system tests should be done when symptoms point to low alternator voltage and current. These tests will quickly determine the operating condition of the charging system. There are four common tests made on a charging system:

- **Charging system output test.** Measures the current and voltage output of the charging system under a load.
- **Regulator voltage test.** Measures the charging system voltage under a low output, low load condition.
- **Regulator bypass test.** In this test, the voltage regulator is bypassed to determine if it is functioning properly.
- **Circuit resistance tests.** Measures the resistance in the insulated and ground circuits of the charging system.

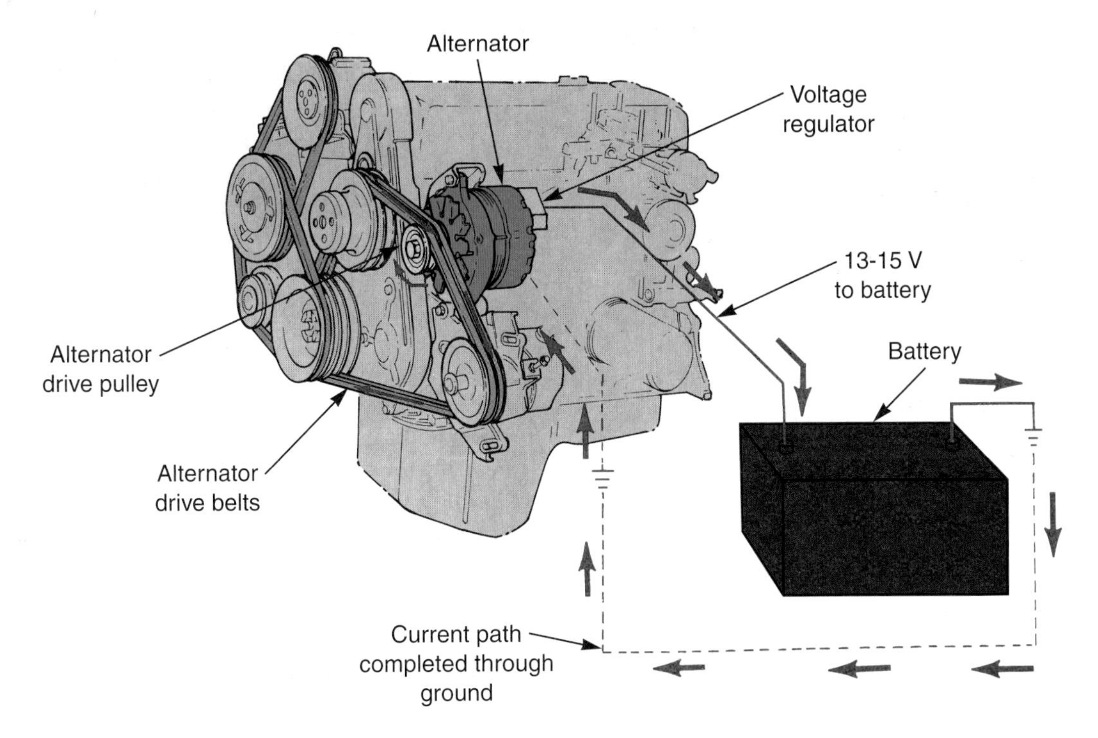

Figure 14-6. *The basic parts of a charging system. (Sun Electric)*

Ignition System Principles

The ***ignition system*** has several critical functions to perform. It produces a high voltage arc at the spark plug electrodes. The spark must be timed so that it occurs as the piston nears TDC on the compression stroke. The ignition system must also vary spark timing with engine speed, load, and other conditions. In addition, it must be capable of operating on various supply voltages (battery or alternator voltage). Finally, the ignition system provides a method of turning a gasoline engine on and off. Automotive gasoline engines have one of three basic types of ignition systems:

- **Direct ignition system.** A computer-controlled ignition coil is mounted on top of each spark plug, **Figure 14-7A.** The computer controls the timing of the spark.
- **Distributorless ignition system.** A computer operates multiple ignition coils that feed high voltage through plug wires, **Figure 14-7B.** The computer controls the timing of the spark.
- **Distributor ignition system.** A mechanical-electrical device sends high voltage generated by a single ignition coil through spark plug wires to each spark plug, **Figure 14-7C.** The timing of the spark is determined by the rotation of the rotor inside the distributor. The spark advance may be controlled by the ECM or through mechanical means.

Parts of Ignition System

The basic parts of an ignition system are illustrated in **Figure 14-8.** They include:

- **Battery.** The source of electrical energy for the ignition system when the engine is being started.
- **Ignition switch.** A switch for connecting and disconnecting electricity to the ignition system so the driver can turn the engine on and off.
- **Ignition coil.** A step-up transformer for changing battery voltage into a high voltage for supplying the spark plugs.
- **Switching mechanism.** A device for interrupting current flow through the ignition coil primary windings to induce a high voltage in the secondary windings.
- **Spark plugs.** Produces an electric arc across an air gap when supplied with high voltage.
- **Wiring.** Primary (low voltage) and secondary (high voltage) wires for connecting the components of the ignition system. Secondary wiring has much thicker insulation than primary wiring.

Ignition System Operation

With the ignition switch in the run or start position, current flows to the ignition system. When the switching device is closed (conducting current), current flows through and energizes the ignition coil. As the piston is nearing TDC on the compression stroke, the switching device opens. This induces a high voltage in the secondary windings of the ignition coil. The high voltage travels to the spark plug through the secondary wiring. An electric arc across the plug gap ignites the air-fuel mixture in the combustion chamber.

When the ignition key is turned to the off position, the primary circuit (battery-to-coil) is broken. Without current

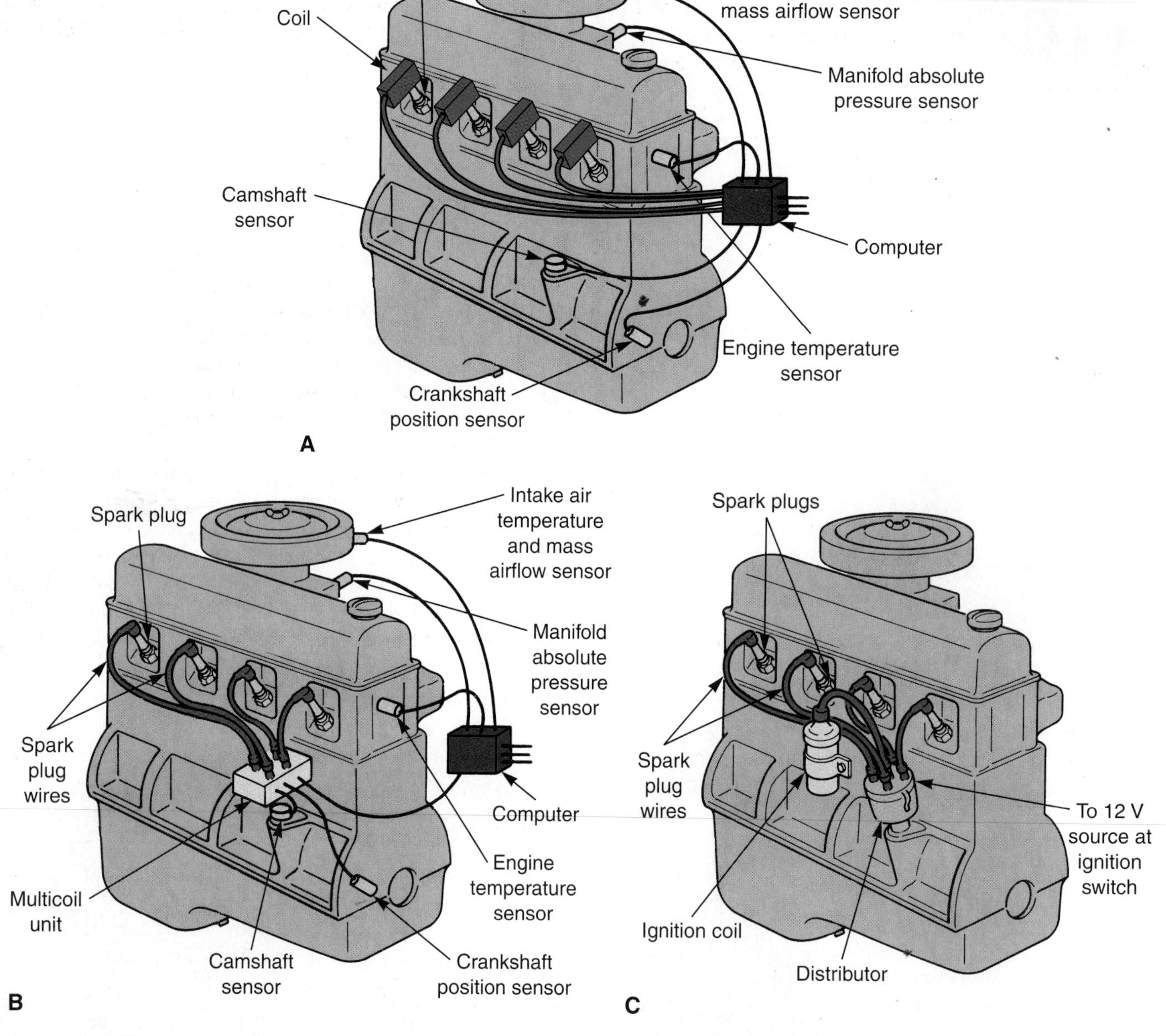

Figure 14-7. *The three basic types of ignition system. A—A direct ignition system has an ignition coil for each spark plug. A computer controls the spark timing. B—Distributorless ignition uses sensors, an electronic control unit, and a multiple-coil unit to fire the spark plugs. C—An older distributor ignition system. (Ford)*

to the ignition coil, sparks are not produced at the spark plugs and the engine stops running.

Ignition System Troubleshooting

Inspect the ignition system with and without the engine running. Look for obvious problems. **Figure 14-9** shows some typical problems:

- Poor electrical connections.
- Shorted, open, or crossed spark plug wires.
- Faulty sensor(s).
- Open or shorted coils.
- Bad electronic control module.
- Faulty ignition switch.
- Discharged battery.

If the problem cannot be located in the ignition or other system, it may be an engine mechanical problem. For example, a burned valve can prevent compression in a cylinder, which will prevent combustion.

Spark Test

A ***spark test,*** also called a ***spark intensity test,*** measures the brightness and duration of the electric arc (spark) produced by the ignition system. It is a quick way of checking the condition of the ignition system. A spark test is commonly used when an engine cranks but will not start.

A spark tester is a device with a very large air gap for checking ignition output voltage. It looks like a spark plug with a wide gap and a ground wire, **Figure 14-10.**

1. Remove the secondary wire or coil assembly from one of the spark plugs. Leave the spark plug in the engine.
2. Insert the spark tester into the wire or coil assembly.
3. Ground the tester on the engine.
4. Crank the engine.
5. Observe the spark at the tester air gap.

A strong spark indicates that the ignition coil, pickup coil, electronic control unit, and other parts are working. The engine no-start problem might be due to fouled spark plugs, a fuel system problem, or engine mechanical problems.

A weak spark or no spark shows that something is wrong in the ignition system. If spark is weak at all spark plug wires, the problem may be a bad ignition coil, rotor, or coil wire.

Caution: Only run the engine for a short time with a spark plug wire off. Unburned fuel from the dead cylinder can foul and ruin the catalytic converter.

Dead Cylinder

A ***dead cylinder*** is one in which combustion is not occurring. It could be due to ignition system troubles, engine mechanical problems, or problems in the fuel system. A very rough idle and a puffing noise in the engine exhaust may indicate a dead cylinder.

To check for a dead cylinder, start the engine. Then, pull off one spark plug wire at a time. Hold the wire next to an engine ground. Some engine analyzers automatically check for dead cylinders.

Engine speed should decrease and the engine should idle rough with the wire removed. If idle smoothness and engine speed stay the same with the plug wire off, that cylinder is dead (not producing power). Check for spark at the wire, the spark plug condition, and for low cylinder compression. Special test instruments may be needed to diagnose ignition system problems.

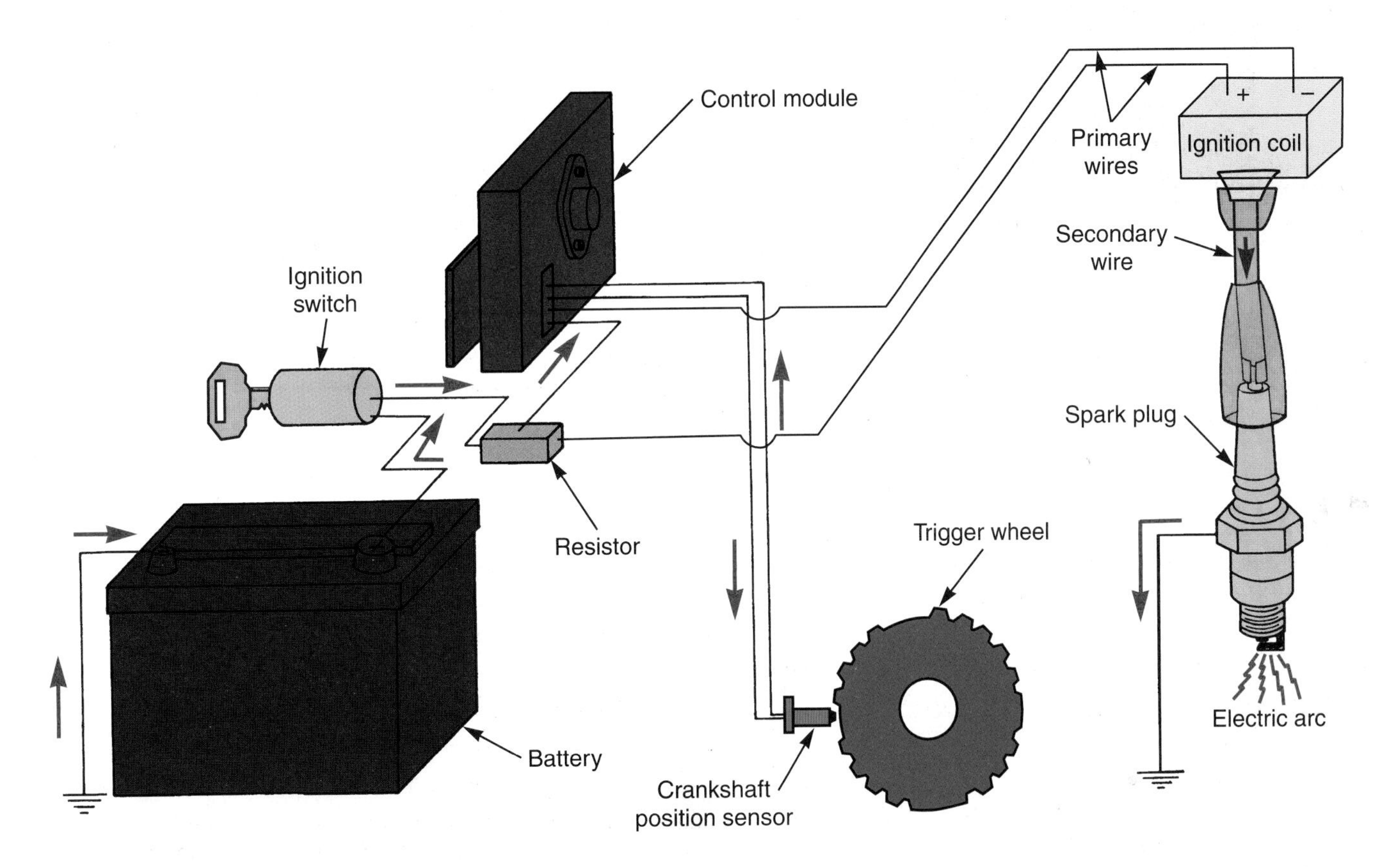

Figure 14-8. *The basic parts of an ignition system.*

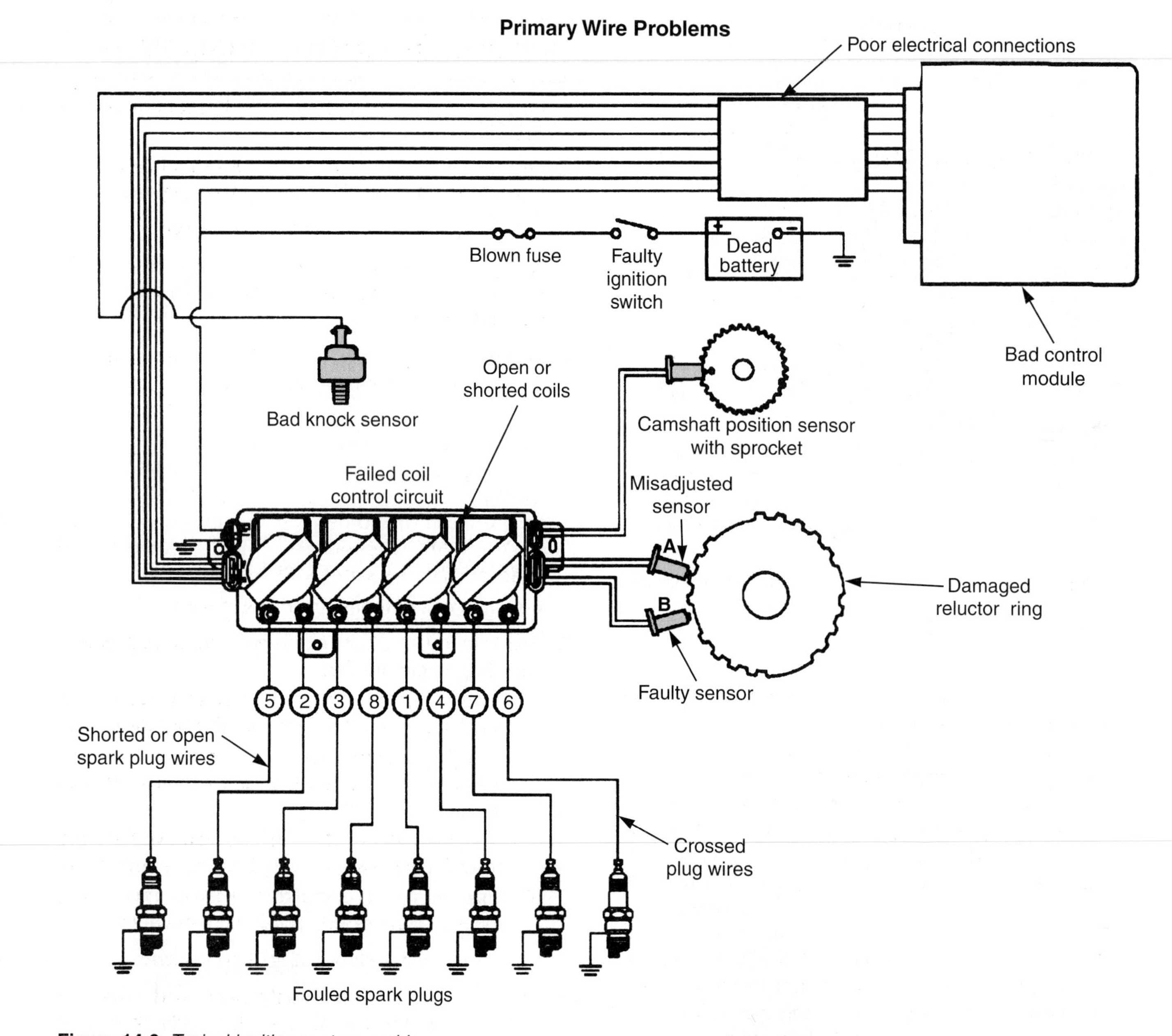

Figure 14-9. *Typical ignition system problems.*

Note: If the vehicle has adjustable ignition timing, verify and adjust the ignition timing as needed. Refer to the service manual for specific instructions.

Summary

The starting system rotates the crankshaft until combustion powers the engine. It draws the required electricity from the battery. The major parts of the starting system include the battery, ignition switch, solenoid, and starter motor.

Problems with the starting system are indicated by a no-crank or slow-crank condition. Always check the battery first. Make sure it is not discharged. Also, check cable connections. If the battery is discharged, you may need to jump start the vehicle.

The charging system recharges the battery and supplies electricity to the ignition system while the vehicle is running. A basic charging system consists of an alternator, voltage regulator, alternator drive (belt), and battery.

Four common symptoms of trouble in the charging system are a dead battery, overcharged battery, abnormal noises in the alternator, and an indicator light or gauge warning. There are four tests commonly conducted to locate a charging system problem. These are a charging

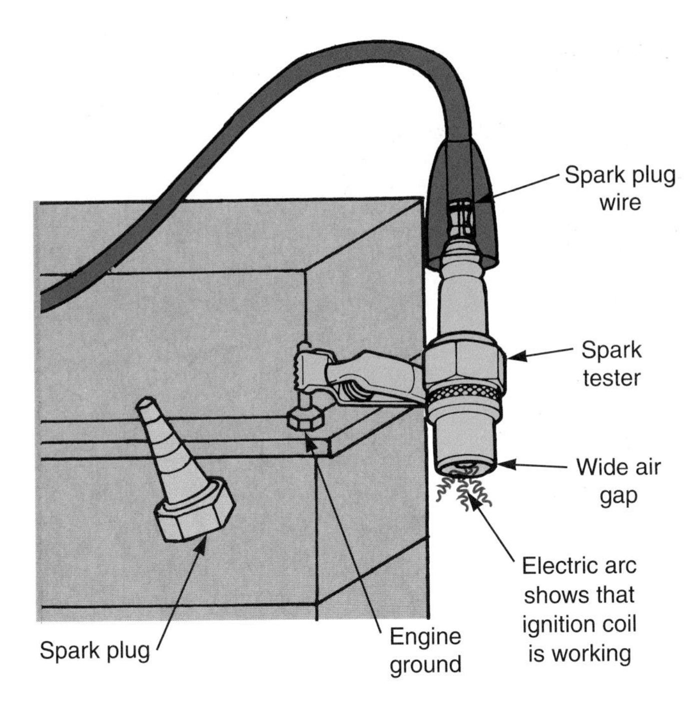

Figure 14-10. *Conducting a spark test to determine the condition of the ignition system.*

system output test, regulator voltage test, regulator bypass test, and circuit resistance tests.

The ignition system has several functions. It produces a high-voltage spark, times the spark with piston movement, and provides a method of turning the engine off. The ignition system may be direct ignition, distributorless ignition, or distributor ignition. The basic parts of an ignition system are a battery, ignition switch, ignition coil, switching mechanism, spark plugs, and wires.

There are several typical problems that may be found in an ignition system. Inspect the system with the engine running and off. A spark test is used to check the condition of the ignition system. Also, by removing spark plug wires one at a time with the engine on, a dead cylinder can be located. If the problem cannot be located in the ignition system, it may be an engine mechanical problem.

Review Questions—Chapter 14

Please do not write in this text. Write your answers on a separate sheet of paper.

1. The starting system uses _____ power and an electric motor to turn the engine crankshaft for engine starting.
2. List the four major parts of the starting system.
3. A(n) _____ problem results when the engine crankshaft does not rotate properly with the ignition key in the start position.
4. When checking battery voltage with the engine off, the multimeter reading should be about:
   (A) 14.0 volts.
   (B) 14.6 volts.
   (C) 12.6 volts.
   (D) 12.0 volts.
5. Why is the voltage produced by the charging system slightly higher than battery voltage?
6. List four parts of a basic charging system.
7. The four tests that are commonly performed on the charging system are the:
   (A) charging system output test, regulator current test, regulator bypass test, and circuit resistance tests.
   (B) charging system bypass test, regulator voltage test, regulator bypass test, and circuit resistance tests.
   (C) charging system output test, regulator voltage test, regulator bypass test, and circuit resistance tests.
   (D) charging system output test, regulator voltage test, regulator bypass test, and current draw tests.
8. What are the three types of ignition system?
9. A(n) _____ test is commonly performed when an engine cranks but will not start. It quickly checks the condition of the ignition system.
10. What is a *dead cylinder?*

ASE-Type Questions—Chapter 14

1. An automotive battery changes _____ energy into electrical energy.
   (A) heat
   (B) chemical
   (C) mechanical
   (D) All of the above.
2. An engine will not start but cranks normally. Technician A says that the problem is most likely in the starting system. Technician B says that an engine mechanical problem may be the cause. Who is correct?
   (A) A only.
   (B) B only.
   (C) Both A and B.
   (D) Neither A nor B.
3. An engine will not start and does not crank. Multiple buzzing or clicking sounds can be heard from the solenoid. Technician A says that the problem may be a discharged battery. Technician B says that low current flow is causing the solenoid plunger to rapidly kick in and out. Who is correct?
   (A) A only.
   (B) B only.
   (C) Both A and B.
   (D) Neither A nor B.
4. An engine will not start and does not crank. It is determined that the battery is discharged and the vehicle should be jump started. Technician A says that the two batteries should be connected in parallel. Technician B says that the last connection should be the black jumper cable to any ground on the vehicle with the good battery as long as the connection is away from the battery. Who is correct?
   (A) A only.
   (B) B only.
   (C) Both A and B.
   (D) Neither A nor B.
5. Technician A says that one of the functions of an automotive charging system is to supply all the vehicle's electricity when the engine is running. Technician B says that one of the functions of an automotive charging system is to provide voltage output slightly higher than battery voltage. Who is correct?
   (A) A only.
   (B) B only.
   (C) Both A and B.
   (D) Neither A nor B.
6. A charging system voltage regulator normally keeps alternator output at approximately _____.
   (A) 16–18 volts
   (B) 13–15 volts
   (C) 11–12 volts
   (D) 9–12 amps
7. Technician A says that a regulator bypass test can detect an alternator problem. Technician B says that a regulator bypass test can detect a faulty voltage regulator. Who is correct?
   (A) A only.
   (B) B only.
   (C) Both A and B.
   (D) Neither A nor B.
8. Technician A says that the primary wiring of an ignition system carries high voltage. Technician B says that a high voltage is induced in the secondary coil windings when the switching device opens. Who is correct?
   (A) A only.
   (B) B only.
   (C) Both A and B.
   (D) Neither A nor B.
9. Technician A says that one of the functions of an automotive ignition system is to time the spark so that it occurs as the piston nears TDC on the power stroke. Technician B says that one of the functions of an ignition system is to enable the engine to run on either battery or alternator voltage. Who is correct?
   (A) A only.
   (B) B only.
   (C) Both A and B.
   (D) Neither A nor B.
10. Technician A says that in a direct ignition system, a computer-controlled coil is mounted on top of each spark plug. Technician B says that in a distributorless ignition system, a computer operates multiple coils that feed high voltage through the plug wires. Who is correct?
    (A) A only.
    (B) B only.
    (C) Both A and B.
    (D) Neither A nor B.

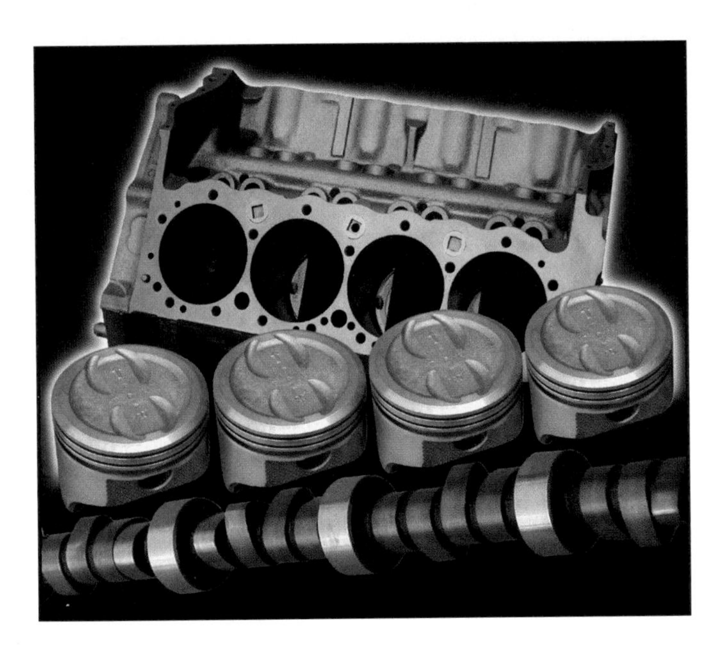

Chapter 15

Fuel and Emission Control Systems

After studying this chapter, you will be able to:

- ❑ Identify the three subsystems of a fuel system.
- ❑ Describe the function of the fuel supply system.
- ❑ Service fuel lines and hoses, fuel filters, and fuel pumps.
- ❑ Describe the function of the air supply system.
- ❑ Describe the function of the fuel metering system.
- ❑ List the basic components of a gasoline injection system.
- ❑ Identify the two types of gasoline injection.
- ❑ Explain on-board diagnostic systems.
- ❑ Use a noid light injector circuit tester.
- ❑ Describe the function of the emission control system.
- ❑ Identify the sources of vehicle emissions.
- ❑ List the four types of emissions produced by a vehicle.
- ❑ Describe engine modifications for emission control.
- ❑ List the systems used to control emissions.

Know These Terms

Air check valve
Air distribution manifold
Air injection system
Air supply system
Carbon monoxide (CO)
Catalyst
Catalytic converter
Charcoal canister
Data link connector (DLC)
Diagnostic trouble code (DTC)
Diverter valve
EGR valve
Emission control systems
Emissions
Evaporative emissions control (EVAP) system
Exhaust gas recirculation (EGR) system
Fuel filters
Fuel metering system
Fuel pump
Fuel supply system
Fuel tank pickup unit
Heated air inlet system
Hydrocarbons (HC)
Liquid-vapor separator
Multiport injection
Noid light
On-board diagnostics generation one (OBD I)
On-board diagnostics generation two (OBD II)
Oxides of nitrogen (NO_X)
Particulates
PCV valve
Positive crankcase ventilation (PCV)
Purge line
Purge valve
Rollover valve
Scan tool
Three-way converters
Throttle body injection
Vapor lock

As an engine technician, you must have a basic knowledge of the fuel and emission control systems. You must be able to quickly determine if problems are in either of these systems or the engine itself. This chapter covers the principles behind the fuel and emission control system. Basic troubleshooting of these systems is also covered.

Fuel System

Proper engine operation depends on an adequate supply of clean air and fuel. Without the correct mixture of fuel and air, an engine will perform poorly. This section of the chapter summarizes the most important information on air filters, fuel filters, and fuel pumps. It provides the information most commonly used by an engine technician.

A ***fuel system*** can be divided into three subsystems:

- Fuel supply system.
- Air supply system.
- Fuel metering system.

It is important that you understand the components included in each system when trying to diagnose and repair engine problems.

Fuel Supply System

The ***fuel supply system*** provides filtered fuel under pressure to the fuel metering system. It must also provide for storage of the fuel. See **Figure 15-1.**

Fuel Tanks

An automotive ***fuel tank*** must safely hold an adequate supply of fuel for prolonged engine operation. The size of a fuel tank partially determines a vehicle's driving range. ***Fuel tank capacity*** is the rating of how much fuel a tank can hold when full. An average fuel tank capacity is about 12 to 25 gallons (45 to 95 liters).

A ***fuel tank pickup unit*** extends down into the tank to draw out fuel. A coarse filter is usually placed on the end of the pickup tube to strain out larger debris. Sometimes, the pickup unit also operates the fuel gauge.

The fuel tank ***filler neck*** is the extension for filling the tank with fuel. The filler cap fits on the filler neck.

Fuel Lines and Hoses

Fuel lines and fuel hoses carry fuel from the tank to the engine. A main fuel line connects the tank to the engine. Fuel is pulled or pushed through this line to the engine, **Figure 15-2.**

Fuel lines are made from plastic, nylon, or double-wall steel tubing. Plastic and nylon tubing may be used in some areas of the fuel system to reduce the possibility of rust and corrosion. A fuel line must be strong enough to withstand the constant and severe vibrations produced by engine and vehicle operation.

Fuel hoses can be made of nylon or synthetic rubber covered with a plastic insulator. Hoses are used when movement between two or more parts can occur, such as between the body and the engine.

A fuel return system helps cool the fuel and prevent vapor lock. ***Vapor lock*** occurs when bubbles form in the overheated fuel and disrupt normal fuel flow. A second return fuel line is used to carry excess fuel back to the tank. This keeps cool fuel constantly circulating through the system.

Warning: Always keep a fire extinguisher handy when working on a fuel system. During a fire, a few seconds can mean the difference between life and death.

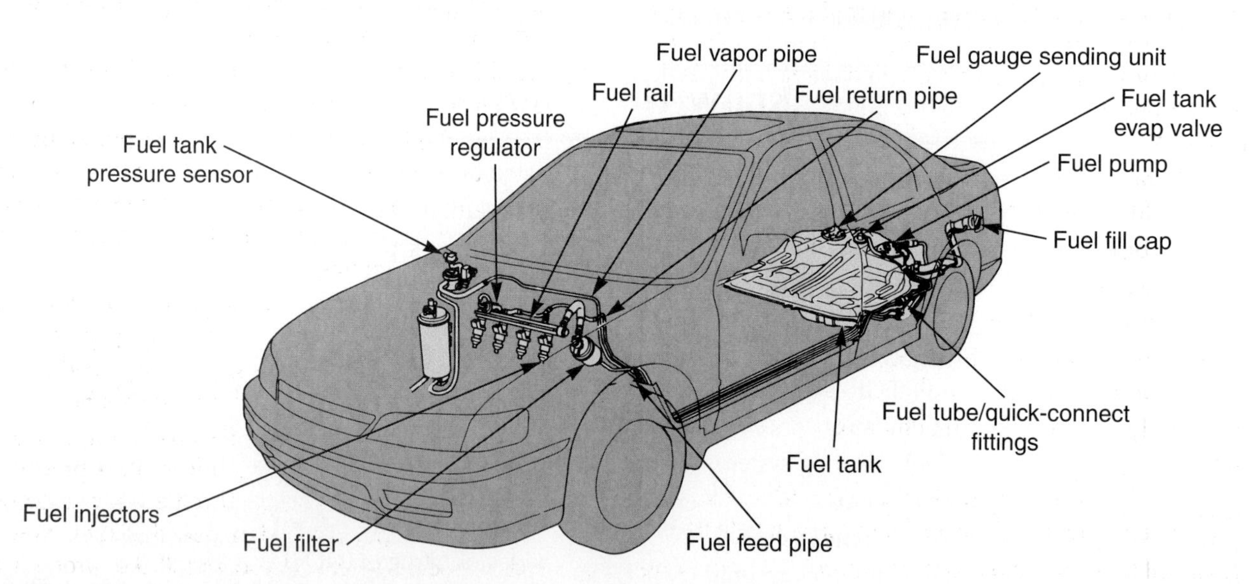

Figure 15-1. *Note the location of typical fuel supply system components. (Honda)*

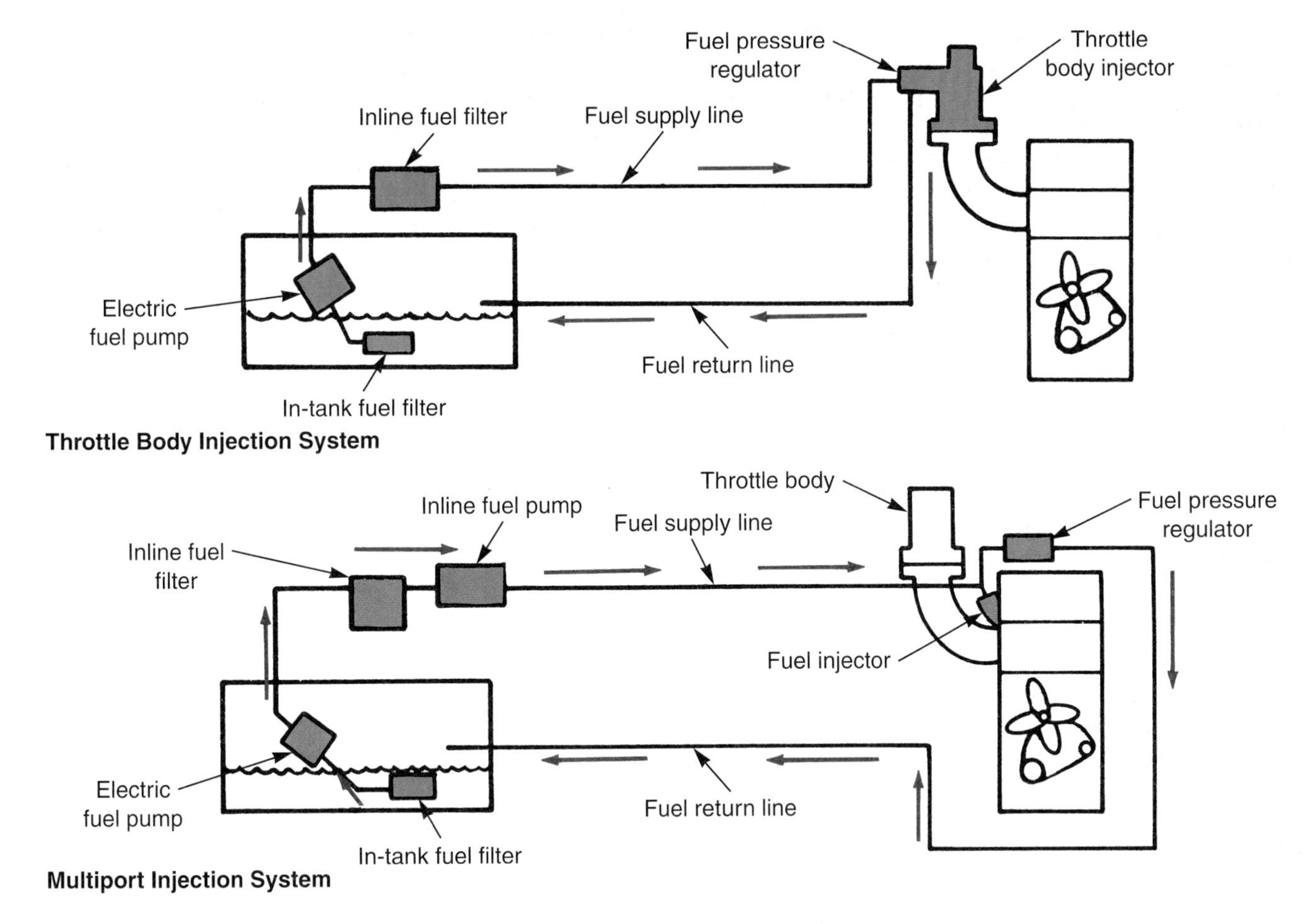

Figure 15-2. *Note the location of fuel lines, filters, pumps, and other parts of the fuel supply system.*

Fuel Line and Hose Service

Faulty fuel lines and hoses are a common source of fuel leaks. Fuel hoses can become hard and brittle after being exposed to engine heat and the elements. Engine oil can soften and swell fuel hoses.

Always inspect hoses closely and replace any in poor condition. Metal fuel lines seldom cause problems. However, they should be replaced when damaged or leaking. Remember these rules when working with fuel lines and hoses:

- ❑ Place a rag around the fuel line fitting during removal to keep fuel from spraying on you or the hot engine.
- ❑ Use a tubing (line) wrench on fuel system fittings.
- ❑ Only use approved tubing, such as double-wall steel, for fuel lines. Never use copper tubing.
- ❑ Make smooth bends when forming a new fuel line. Use a bending spring or tool.
- ❑ Form only double-lap or ISO flares on the ends of the fuel line.
- ❑ Reinstall fuel line hold-down clamps and brackets. If not properly supported, the fuel line can vibrate and fail.
- ❑ Route all fuel lines and hoses away from hot or moving parts. Double-check clearances after installation.
- ❑ Only use approved synthetic rubber hoses in a fuel system. If vacuum-type rubber hose is accidentally used, the fuel can chemically attack the hose. A dangerous leak could result.
- ❑ Make sure a fuel hose slides fully over its fitting before installing the clamps. Otherwise, pressure in the fuel system could force the hose off.
- ❑ Double-check all fittings for leaks. Start the engine and inspect the connections closely.
- ❑ Do not bend or mishandle plastic or nylon fuel lines. If the line becomes kinked, it will create a permanent restriction. If a plastic or nylon fuel line is damaged in any way, replace it. Do not attempt to repair a damaged plastic fuel line.
- ❑ Some fuel lines have a special, snap-type fitting called a ***push-on fitting.*** Do not pry the fitting apart or it will be damaged. In some cases, you may need a special tool to release the fuel line fitting for service.
- ❑ When servicing push-on fuel line fittings, you may also need to replace the O-ring seal to prevent fuel leakage. Make sure you purchase a seal of the correct material, shape, and size for the fuel line and make/model vehicle. If you install the wrong seal, a fuel leak and fire can result.

Warning: The fuel line may contain fuel under pressure as high as 50 psi (345 kPa) even when the ignition key is off. Always relieve fuel pressure before disconnecting any EFI fuel line. Loosen lines carefully until you verify the system is not under pressure. Most EFI systems have a special relief valve (test or service fitting). Or, the pressure regulator may allow pressure relief for bleeding pressure back to the fuel tank.

Fuel Filters

Fuel filters stop contaminants such as rust, water, corrosion, and dirt from entering the fuel metering system. A fuel filter, or strainer, is normally located on the fuel tank pickup tube. This is a coarse filter that blocks larger debris. A second fuel filter may be located in the main fuel line or inside the fuel pump to block smaller debris.

Fuel Filter Service

Fuel filter service involves periodic replacement or cleaning of system filters. It may also include locating clogged fuel filters. A clogged fuel filter can restrict the flow of fuel to the fuel metering system. Fuel filters can be located:

- In the fuel line before the fuel injectors.
- In the fuel line right after the fuel pump.
- Inside the fuel pump.
- Inside the injector body (strainer type).
- A fuel strainer is also located in the fuel tank on the end of the fuel pickup tube.

When in doubt, refer to the service manual for filter locations, service intervals, and related information.

Fuel Pump

A ***fuel pump*** produces fuel pressure and flow. Electric fuel pumps are commonly used on late-model vehicles. A small electric motor spins an impeller, which forces fuel through the system. The fuel pump operation is controlled by an on-board computer.

An electric fuel pump can be located inside the fuel tank as part of the fuel pickup unit, **Figure 15-3.** It can also be located in the fuel line between the tank and engine. Many in-tank fuel pumps also incorporate a sending unit for operating the fuel gauge in the dash, as shown in **Figure 15-3B.**

Note: Fuel pumps on older vehicles are mechanical and driven by the engine. This type of fuel pump is not used on modern, fuel-injected engines. A faulty mechanical fuel pump may leak fuel into the engine oil or onto the ground, or may fail to build fuel pressure.

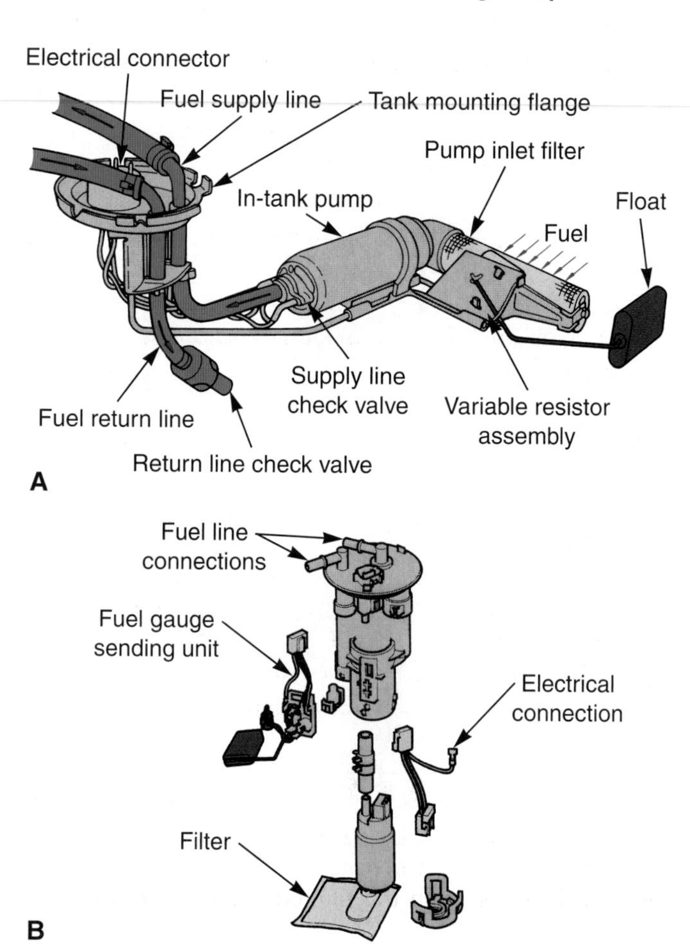

Figure 15-3. *A—This is an in-tank electric fuel pump. Note the fuel strainer, pump motor, fuel return, and float. B—Note the fuel gauge sending unit is part of this in-tank electric fuel pump. (DaimlerChrysler, Honda)*

Fuel Pump Service

Fuel pumps can fail after prolonged operation. Fuel pump problems usually show up as low fuel pressure, inadequate fuel flow, abnormal pump noise, or fuel leakage from the pump.

Low fuel pump pressure can be caused by worn motor brushes, worn motor bushings, clogged strainer, poor electrical connections, and similar troubles. Fuel pump leaks are caused by physical damage to the pump body or deterioration of the diaphragm or gaskets.

Most electric fuel pumps make a slight buzzing or humming sound when running. Only when the pump noise is abnormally loud should an electric fuel pump be considered faulty. A clogged tank strainer is a common cause of excess electric fuel pump noise. If an electric fuel pump is starving for fuel the pump will buzz loudly because it is pumping air and not fuel. Pump speed can increase because fuel is not entering the pump properly. Electric fuel pumps can also make noise after prolonged service when the motor bushing becomes worn.

If the pump is not pumping, check supply voltage to the pump. If there is voltage at the pump, check for a clogged fuel filter. If these are not the problems, the pump may need to be replaced.

Replacing a fuel pump is a relatively simple task, **Figure 15-4.** Refer to the service manual for specific instructions on pump replacement. Remember to:

- Relieve system fuel pressure.
- Place a rag around the pump fittings when loosening.
- Seal off the fuel lines when the pump is removed to prevent leakage and a potential fire.
- Make sure you have correct replacement pump. Match up the new pump with the old one.
- Tighten hose clamps, fittings, and fasteners properly.
- Check for fuel leaks before releasing the vehicle to the customer.

Air Supply System

The ***air supply system*** removes dust and dirt from the air entering engine intake manifold. An ***air filter*** traps the particles. The air filter fits inside a metal or plastic housing, **Figure 15-5.** Most air filters have a paper element. A few vehicles use a polyurethane foam element.

Fuel Metering System

The fuel supply system provides fuel to the fuel metering system. The ***fuel metering system*** controls amount of fuel that is mixed with the filtered air. The fuel metering system may be throttle body injection or multiport injection. The next section discusses the fuel metering system in detail.

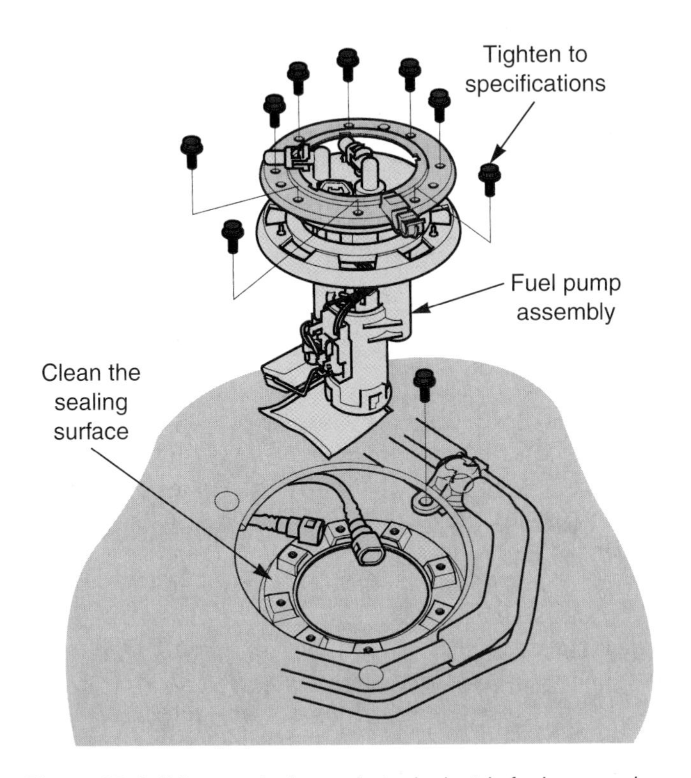

Figure 15-4. *When replacing an in-tank electric fuel pump, clean the sealing surface on the tank, install a new rubber seal (if needed), and torque the fasteners to specifications. Also, check that electrical connections and hoses are secure. (Honda)*

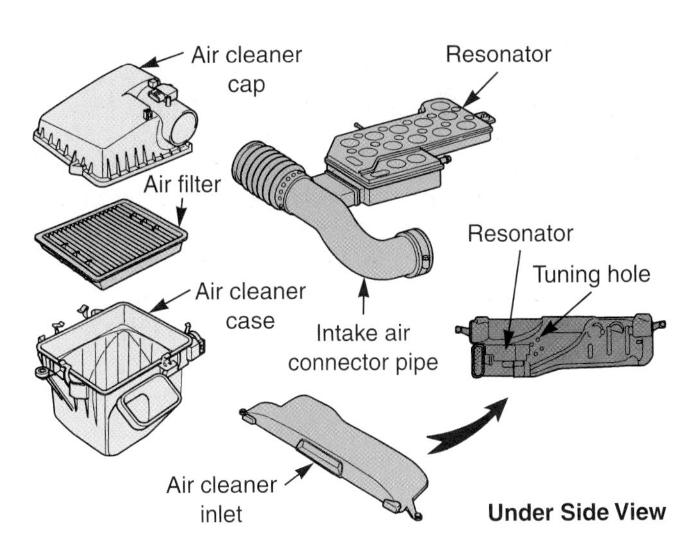

Figure 15-5. *The air filter mounts inside a metal or plastic housing. (Honda)*

Fuel Metering System

Modern gasoline injection systems use engine sensors, a computer, and solenoid-operated fuel injectors to meter the correct amount of fuel into the engine. Termed ***electronic fuel injection (EFI),*** these systems use electric and electronic devices to monitor and control engine operation. An electronic control module (ECM), or computer, receives electrical signals from the various sensors. The ECM processes this data and operates the injectors, ignition system, and other engine related devices.

Throttle Body and Multiport Injection

The two types of electronic fuel injection are multiport and throttle body injection, **Figure 15-6.** In ***throttle body injection,*** one or two injectors are mounted inside a throttle body assembly. Fuel is sprayed into one point or location at the center inlet of the engine intake manifold. In ***multiport injection,*** one injector is located just before each intake port. Fuel is injected into the engine in more than one location. This is the most common type and is often called ***port fuel injection.***

Gasoline Injection Components

The basic parts of a fuel injection system are:

- **Fuel rail.** This is a metal tube or casting that feeds fuel to the fuel pressure regulator and injectors, **Figure 15-7.** It usually has a service fitting.
- **Fuel pressure regulator.** A diaphragm-operated pressure-relief valve that maintains a constant pressure at the injectors, **Figure 15-8.**
- **Fuel return line.** A line and hose assembly that carries excess fuel back to the tank from the pressure regulator.

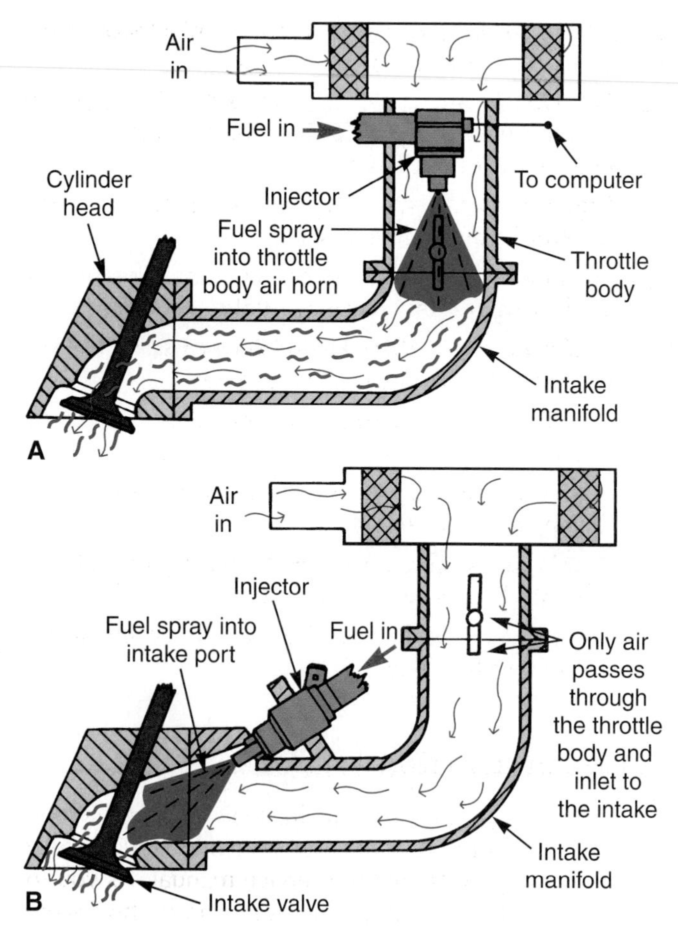

Figure 15-6. *A—In a throttle body injection system, the fuel injector is located in the throttle body. B—In a multiport injection system, an injector is located just before each intake valve.*

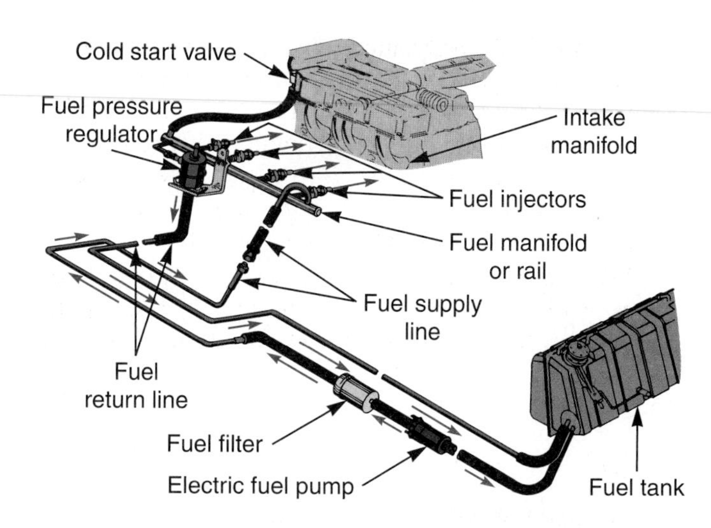

Figure 15-8. *A fuel pressure regulator is mounted on or next to the fuel rail in multiport fuel injection systems. It is mounted inside the throttle body in throttle body fuel injection systems.*

- **Injectors.** Solenoid-operated fuel valves that open when energized by the ECM.
- **Electronic control module (ECM).** A computer that processes sensor data to determine when and for how long injectors should be opened based on operating conditions.
- **Sensors.** Electronic devices capable of changing internal resistance or voltage output with a change in a condition, such as a change in temperature, pressure, or position.

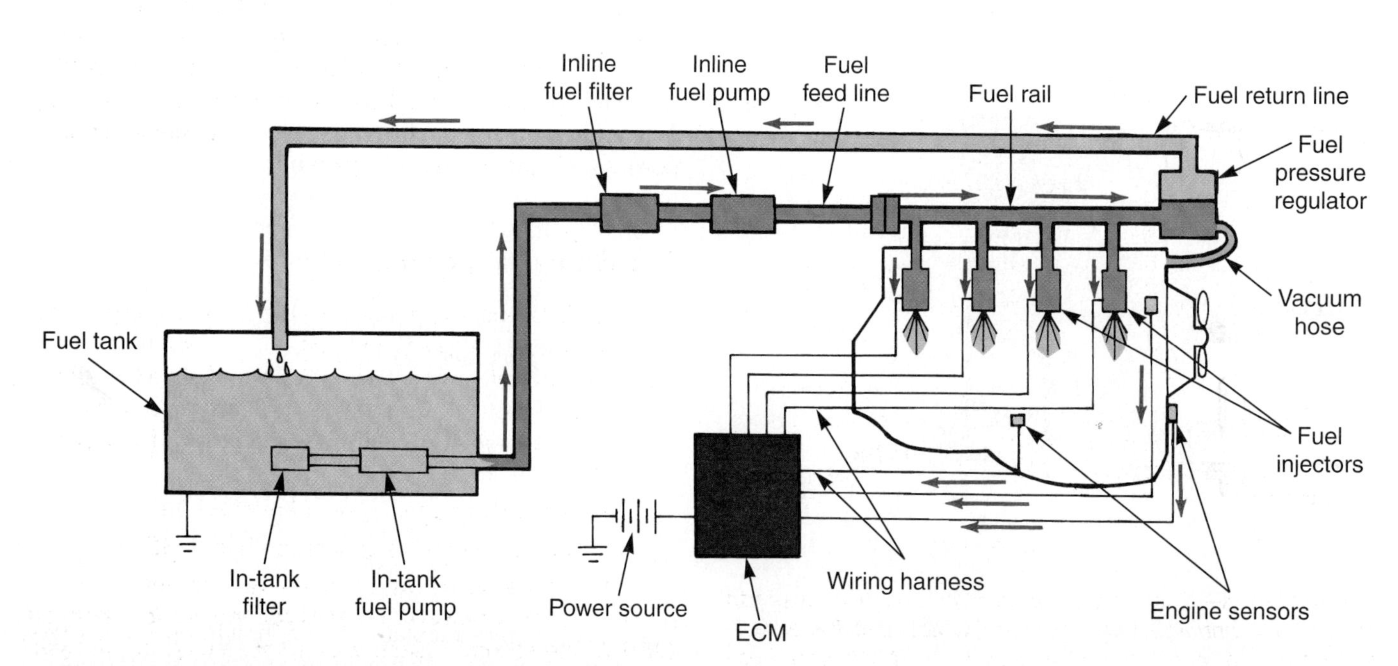

Figure 15-7. *Study the basic operation of a fuel injection system. A fuel pump forces fuel to the fuel rail, which supplies the fuel injectors. A pressure regulator limits fuel pressure and bleeds excess fuel into the return line. Fuel constantly circulates through the system. Engine sensors report engine conditions to the computer. The computer can then open injectors as needed.*

- **Throttle body.** A throttle body assembly for multiport injection is primarily used to control airflow into the engine. See **Figure 15-9A.** A throttle body assembly for throttle body injection controls airflow, but also contains the fuel injector(s) and pressure regulator. Look at **Figure 15-9B.** It should not be confused with a throttle body for multiport injection.
- **Idle speed motor.** Sometimes used on throttle body assemblies to control engine idle speed. The computer actuates the positioner to open or close the throttle plates.

Engine Sensors

An ***engine sensor*** is an electronic device that changes circuit resistance or voltage with a change in a condition, **Figure 15-10.** For example, the resistance of a temperature sensor may decrease as temperature increases. The computer can use the increased current flow through the sensor to calculate any needed change in injector opening. Typical sensors for an EFI system include:

- **Oxygen sensor.** Senses the amount of oxygen in engine exhaust, **Figure 15-11.** A high oxygen content indicates a lean mixture.
- **Engine coolant temperature sensor.** Senses the temperature of the engine coolant.
- **Mass airflow sensor.** Monitors the mass or volume of air flowing into the intake manifold.
- **Intake air temperature sensor.** Senses the temperature of the outside air entering the engine.
- **Throttle position sensor.** Monitors the position of the throttle plates.
- **Manifold absolute pressure sensor.** Senses vacuum in the engine intake manifold.
- **Crankshaft position sensor.** Monitors the rotation (speed) of the engine crankshaft.

Other sensors may also be used. Refer to the service manual for specific details as needed.

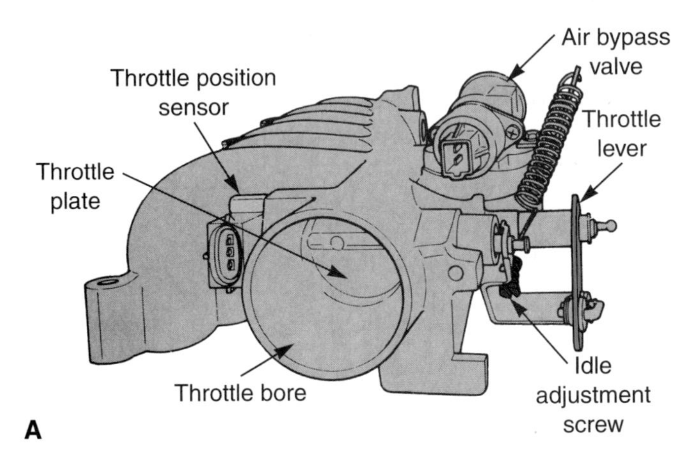

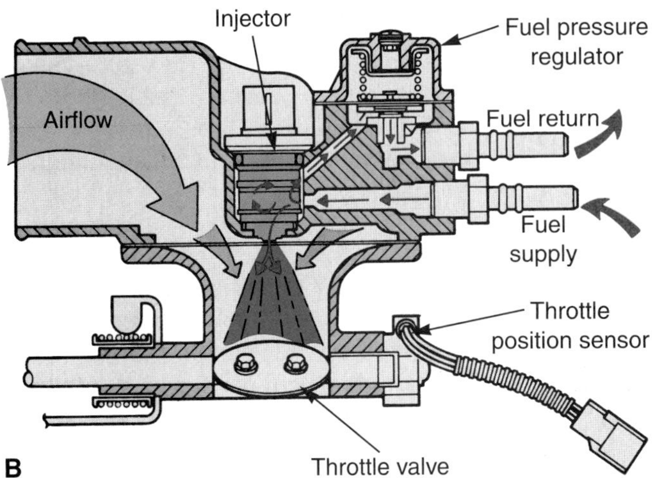

Figure 15-9. *A—This is a throttle body for a multiport injection system. It is primarily used to control airflow into the engine, which, in turn, controls engine speed. B—The throttle body assembly in a throttle body injection system controls air and fuel. The pressure regulator and injector are mounted inside the throttle body. (Ford)*

Gasoline Injection Service

To diagnose problems in the fuel injection system, you must use your knowledge of system operation, basic troubleshooting skills, and the service manual. As you try to locate problems, visualize the operation of the system. Relate the function of each component to the problem. This will let you eliminate several possible sources and concentrate on others.

Figure 15-12 shows several possible problems with the fuel injection system. Check the condition of all hoses, wires, and other parts. Look for fuel leaks, vacuum leaks, kinked lines, loose electrical connections, and other troubles. You may need to disconnect and check the terminals of the wiring harness. Inspect them for rust, corrosion, or burning. High resistance at terminal connections is a frequent cause of problems.

On-Board Diagnostics

The first thing a technician often does when diagnosing a problem in a computerized system is check for vehicle diagnostic trouble codes with a scan tool. Modern computer control systems test themselves and can indicate the location of a problem. There are two types of computer control systems in use today:

- On-board diagnostics generation one (OBD I).
- On-board diagnostics generation two (OBD II).

The first system has been in use for years and is referred to as ***on-board diagnostics generation one*** or ***OBD I.*** The second system was implemented on all new vehicles starting with the 1996 model year. This system is referred to as ***on-board diagnostics generation two*** or ***OBD II.*** It can detect sensor malfunctions before they

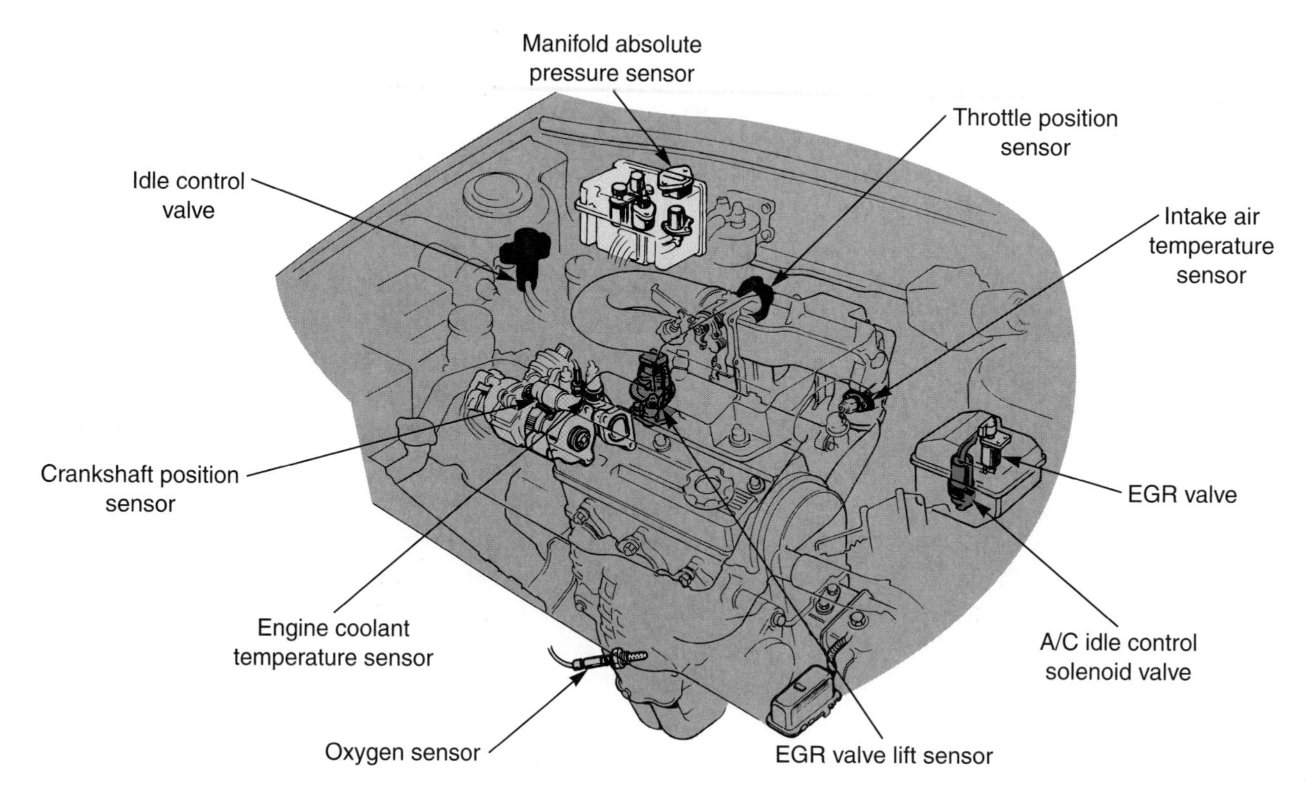

Figure 15-10. *There are many engine sensors on a modern vehicle. Common sensors are identified here. (Honda)*

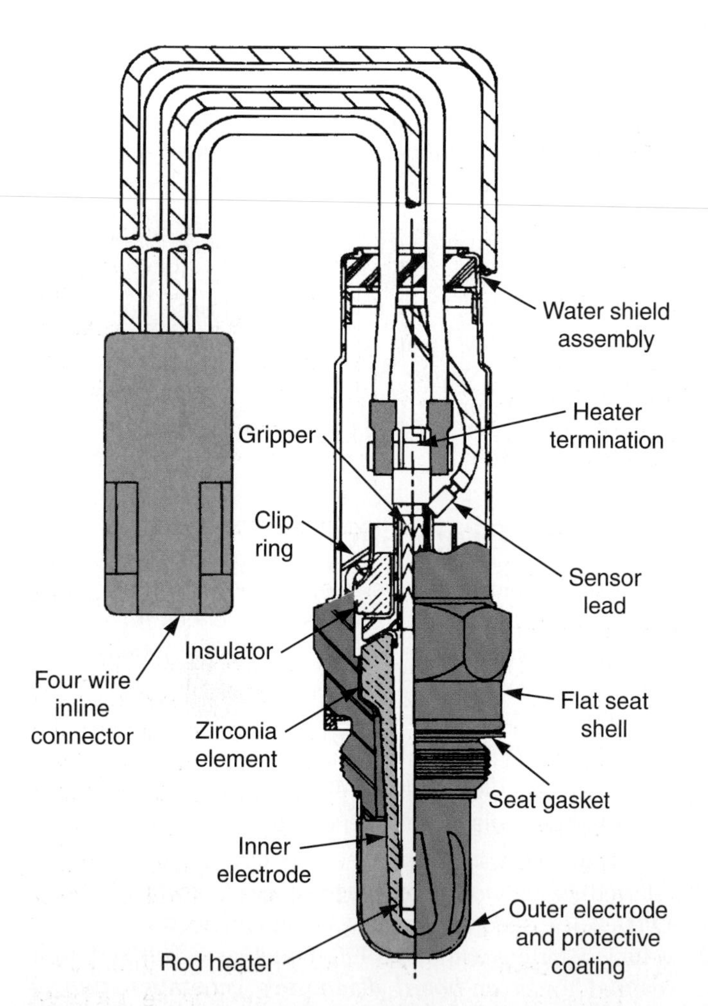

Figure 15-11. *An oxygen sensor monitors the air-fuel ratio by varying its output based on the amount of oxygen in engine exhaust.*

become noticeable to the driver or technician. OBD II is designed to help keep the vehicle running efficiently for at least 100,000 miles (160,000 km).

On-Board Diagnostics Generation Two (OBD II)

A poorly tuned or malfunctioning vehicle is a source of significant air pollution. It can produce several times the normal amount of emissions. For this reason, the California Air Resources Board (CARB) along with the Environmental Protection Agency (EPA) developed regulations that require on-board diagnostic systems to detect problems before they can result in the vehicle producing harmful exhaust emissions. These regulations also require vehicle manufacturers to standardize the performance monitoring systems on their cars and light trucks.

Today's on-board diagnostics will check the operation of almost every electrical and electronic part in every major vehicle system. A vehicle's engine control module can detect engine misfiring and air-fuel mixture problems. It monitors the operation of the fuel injectors, ignition coils, fuel pump, emissions system parts, and other major components that affect vehicle performance and emission control.

The on-board computers used in OBD II systems have greater processing speed, more memory, and more complex programming than computers used in OBD I systems. New vehicles now monitor more functions and can warn the driver and technician of more possible problems that affect driveability and emissions.

The OBD II diagnostic system can produce over 500 trouble codes related to engine performance. It checks the

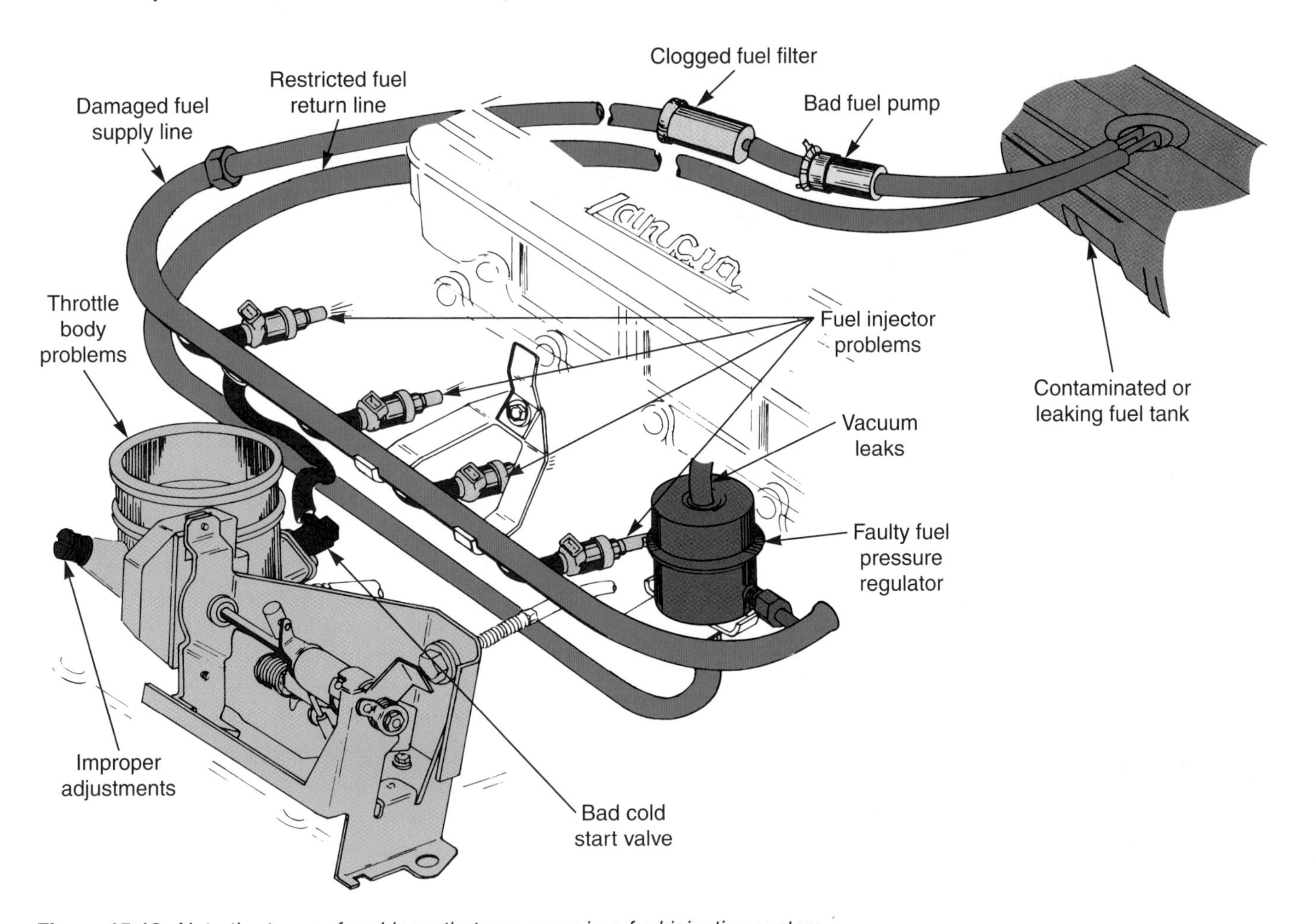

Figure 15-12. *Note the types of problems that can occur in a fuel injection system.*

operating parameters of switches, sensors, actuators, in-system components, their related wiring, and the computer itself. The computer actually scans its input and output circuits to detect an incorrect voltage, resistance, or current.

For example, with OBD II, engine misfires and fuel system malfunctions will cause the malfunction indicator light (MIL) in the dash to flash on and off. The misfire or poor mixture could damage the catalytic converter. The flashing MIL warns the driver that the vehicle could be damaged and should be serviced immediately.

OBD II systems have four levels of diagnostic trouble codes (DTC):

- Type A codes are emissions related. The MIL will illuminate or flash when present.
- Type B codes are also emissions related. However, the ECM will illuminate the MIL only when this type of code appears on two consecutive keystarts or "trips."
- Type C codes are non-emissions related. The ECM will not illuminate the MIL, however, it will store a DTC and illuminate a "service lamp" or the service message on vehicle equipped with a driver's information center.
- Type D codes are non-emissions related. The ECM will store a DTC, but will not illuminate any lamps.

Data link connections, trouble codes, sensor and output device terminology, and scan tool capabilities are also standardized with OBD II. In the past, a service facility needed to have data link connectors for each make of vehicle serviced. Sometimes, multiple connectors were needed on a single vehicle. To solve this problem, the federal government and the Society of Automotive Engineers (SAE) has set standards for all vehicle manufacturers to use.

Diagnostic Problem Indicators

While each manufacturer's systems have their differences, they are similar in many ways. Diagnostic systems indicate problems as follows:

- With OBD II systems, the ECM illuminates an MIL or check engine light on the dash. OBD II systems flash the MIL if a problem exists that could damage the vehicle's emission system. When the MIL light illuminates, it tells the driver and technician that a problem exists.
- With OBD I systems, the ECM displays a digital number code in the dash climate control or driver information center.
- On some older import vehicles, the ECM illuminates LEDs on the computer itself to indicate problem codes.

Diagnostic Trouble Codes

A ***diagnostic trouble code (DTC)*** is a digital signal produced and stored by the computer when an operating parameter is exceeded. An ***operating parameter*** is an acceptable minimum and maximum value. The parameter might be an acceptable voltage range from the oxygen sensor, a resistance range for a temperature sensor, current draw from a fuel injector coil winding, or an operational state from a monitored device. In any case, the computer stores the operating parameters for most inputs and outputs in permanent memory chips.

Note: Most computer system problems are not related to the computer, such as a loose electrical connection or mechanical problem. Only about 20% of all performance problems are caused by an actual fault in the computer or one of its sensors. For this reason, always check for the most common problems before testing more complex computer-controlled components.

Reading OBD I Codes

If the output is an OBD I digital display, simply read the number and compare it with the chart in the service manual. If the output is a code flashed from the dashboard MIL or ECM mounted LED, count how many flashes occur between each pause. Write down the number and compare it to the service manual. If an analog meter must be used on an early model vehicle, count the number of needle deflections between pauses. Write down the number and compare it to the chart in the service manual. Each system may vary. The service manual will give you instructions for the make and model of vehicle.

Data Link Connector (Diagnostic Connector)

The ***data link connector (DLC)*** is a multipin terminal used to link the scan tool to the computer. In the past, this connector was identified by a variety of names, including diagnostic connector and assembly line diagnostic link (ALDL). OBD I diagnostic connectors come in various shapes and sizes and have a varying number of pins or terminals. With OBD II, the DLC is a standardized, 16-pin connector. The female half of the connector is on the vehicle; the male half is on the scan tool cable. The diagnostic connector may be located:

- Under the left side of the dash, within arms reach when sitting in the driver's seat. This is the standard OBD II location.
- Near the firewall in the engine compartment, near or on the side of the fuse box, near the inner fender panel in the engine compartment, or under the center console on OBD I systems.

Using Scan Tools

A ***scan tool*** can be used to quickly check for problems in the engine and its support systems, transmission, suspension system, anti-lock brake system, and other vehicle systems. This has greatly simplified the troubleshooting of complex automotive systems. OBD II codes can only be read with a scan tool.

Modern scan tools provide prompts or step-by-step instructions in their display windows, **Figure 15-13.** The prompts tell you how to input specific vehicle information and run diagnostic tests. The scan tool may ask you to input VIN information. The VIN lets the scan tool know which engine, transmission, and options are installed on the vehicle. With some makes, however, the on-board computer contains this data and the scan tool will automatically download it. After the scan tool is set up, you can select which information you want displayed. Some of the information you can request includes:

- **Stored diagnostic trouble codes.** The trouble code numbers of any stored diagnostic trouble codes are displayed.
- **Fault description.** Explains what each stored diagnostic code means. This information is given with the trouble code number on most scan tools.
- **Datastream information.** Displays the operating values of all monitored circuits and sensors.
- **Run tests.** Performs sensor and actuator tests.
- **Oxygen sensor monitoring.** Performs detailed tests of the O_2 sensor signal.
- **Failure record.** Lists the number of times a particular trouble code has occurred by keystarts or warmups.

REVIEW CODES
** KEY ON, ENGINE OFF CODES-FIX FIRST **
54 AIR CHARGE TEMP SIGNAL HIGH CKT OPEN
** CONTINUOUS MEMORY CODES-FIX LAST **

A

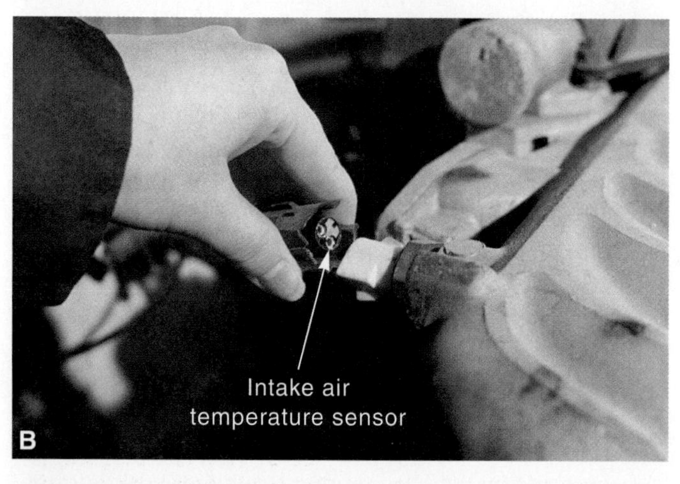

Figure 15-13. *Most technicians check for stored trouble codes using a scan tool. A—This scan tool readout shows a problem with the intake air temperature sensor. B—The technician would then check that sensor and its wiring for problems.*

- ❑ **Freeze frame.** Takes a snapshot of sensor and actuator values when a problem occurs.
- ❑ **Troubleshooting.** Provides help and instructions for diagnosing faults.

Caution: Make sure the scan tool is properly connected to the vehicle. Some technicians have mistakenly connected scan tools to the wrong connector (the tachometer connector, for example), which can damage the scan tool.

Interpreting Trouble Codes

Most technicians check the ECM for stored diagnostic trouble codes before performing tests on specific components. When verifying trouble code numbers, start from the lowest number and work up. For example, if codes 13 and 45 are displayed, correct trouble code 13 first and then 45. If multiple codes are stored, you can often eliminate one or more codes by using this method. Quite often, correcting one code will also fix the other code(s). For example, if a leaking fuel injector is repaired it can fix the code for a catalytic converter problem.

OBD I codes are two digit numbers, such as 32. These numbers correspond with a particular problem listed in the service manual. OBD II codes are alphanumeric with a four digit number and a letter designator. The letter appears first and indicates in which system the problem is located. The first number indicates whether the code is SAE or manufacturer specific. The second number indicates the function of the system where the fault is located. The last two numbers indicate the specific fault. **Figure 15-14** gives a breakdown of the OBD II diagnostic code.

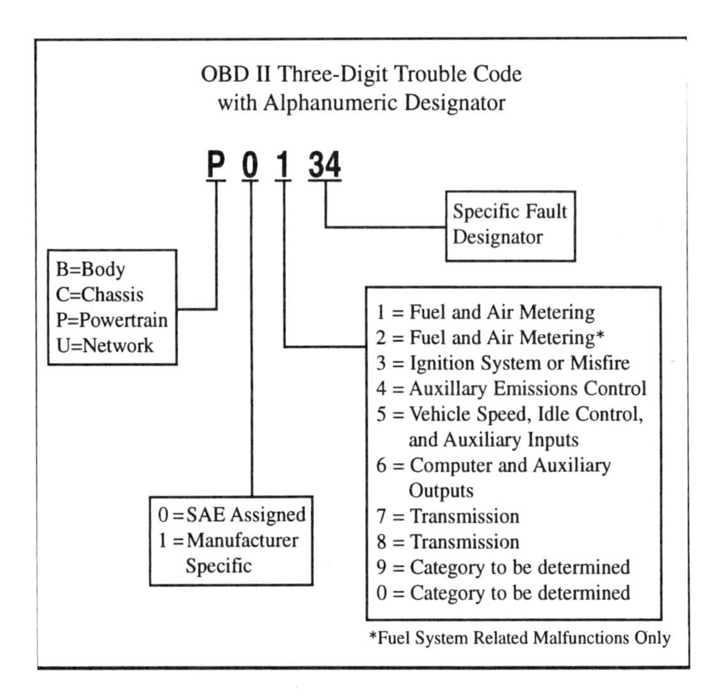

Figure 15-14. *OBD II systems produce a four digit code with a letter designator. The breakdown of this code is shown here.*

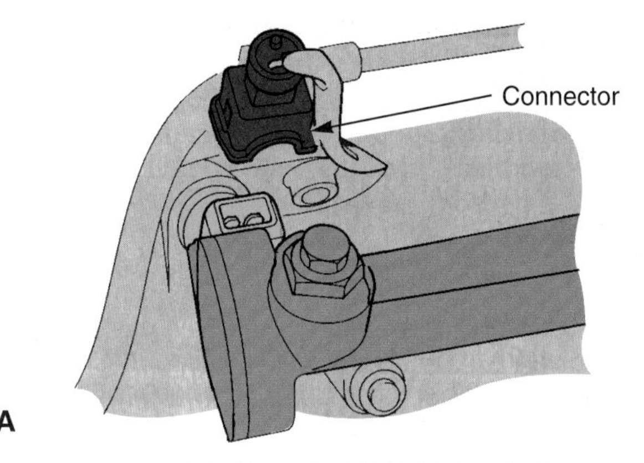

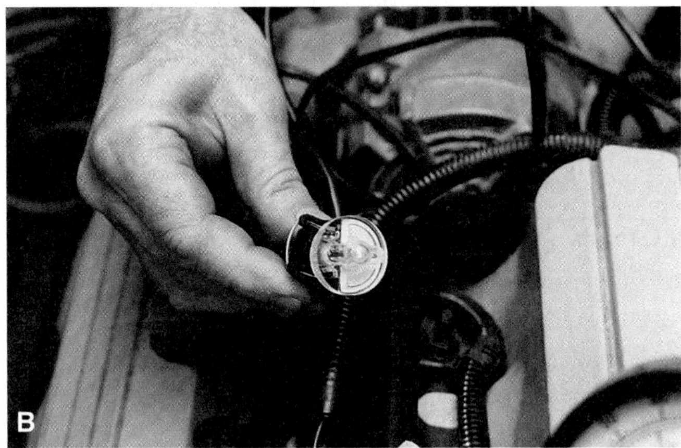

Figure 15-15. *Using a noid light. A—Disconnect the wiring harness from the injector without damaging the connector. B—Install the noid light in the harness. The light should flash with the engine running.*

Clearing Trouble Codes

To clear or remove the trouble codes from an OBD I computer system, you must usually disconnect the battery or the fuse to the computer for about 10 seconds. You must use a scan tool to clear codes on vehicles with OBD II systems.

After clearing the codes, check to see if the MIL comes back on after engine operation. If it does, this indicates there is still a problem in the system. Further tests are needed.

Noid Light Injector Circuit Testing

A ***noid light*** is a special test light for checking electronic fuel injector feed circuits. Different noid lights are made to fit the wiring harnesses on different vehicles. Usually the make of vehicle on which the noid light can be used is printed on the tool.

A stethoscope can be used to tell which injector is dead. A dead injector will not produce a clicking sound. To use a noid light:

1. Disconnect the wiring harness from the dead fuel injector, **Figure 15-15.** Most harness connectors are a positive-lock type to keep the wiring from vibrating loose. Make sure you release the connector properly to prevent part damage.

2. Fit the correct noid light into the injector harness connector.
3. Start the engine and check the noid light.
4. If the noid light flashes, power is reaching the injector from the control module or control unit. If the noid light does not flash, something is preventing current from reaching the noid light, and thus the injector. There may be an open in the wiring, bad connection, open injector resistors, or control module troubles.
5. Repeat the noid light test on any injector that is not operating.

Note: Refer to a wiring diagram when solving complex fuel injection electrical problems. The diagram shows all electrical connections and components that can upset the function of the injection system.

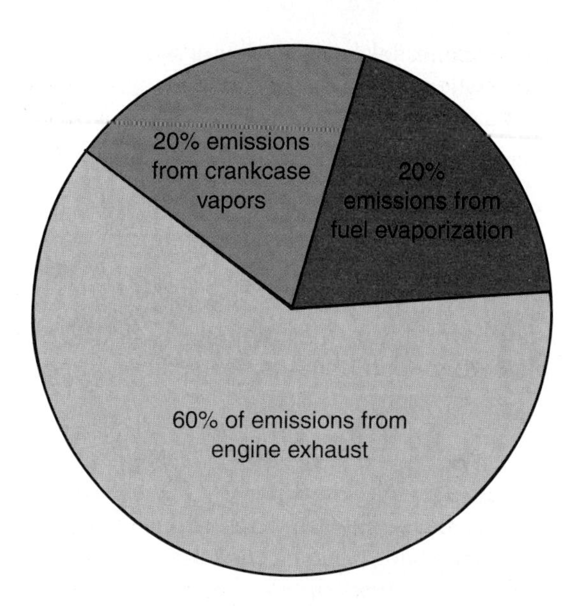

Figure 15-16. *These are the basic proportions of emissions produced by a vehicle. Note the high percentage from engine exhaust.*

Emission Control System

Emission control systems are used to reduce the amount of harmful chemicals released into the atmosphere from a vehicle. This section of the chapter introduces the terminology, parts, and systems that control emissions.

Emissions

Emissions are chemicals emitted into the environment from various sources:

- ❑ Volcanoes.
- ❑ Decaying animal and vegetable matter.
- ❑ Animal and human waste.
- ❑ Other natural sources.
- ❑ Vehicles.
- ❑ Lawnmowers.
- ❑ Power plants and factories.
- ❑ Other man-made sources.

Emissions from man-made sources must be controlled to maintain a balance in our environment. The four basic types of vehicle emissions are hydrocarbons, carbon monoxide, oxides of nitrogen, and particulates. **Figure 15-16** shows the makeup of vehicle emissions:

- ❑ Engine crankcase blowby fumes.
- ❑ Fuel vapors.
- ❑ Engine exhaust gasses.

Hydrocarbons (HC)

Hydrocarbons (HC) result from the release of unburned fuel into the atmosphere. Hydrocarbon emission can be caused by incomplete combustion or by fuel evaporation. For example, a rich mixture can result in increased HC levels because unburned fuel blows out of the combustion chamber and into the vehicle's exhaust system. Look at **Figure 15-17.** HC levels in the atmosphere can also be increased by fuel vapors escaping from a vehicle's fuel system. Hydrocarbon emissions can contribute to eye, throat, and lung irritation; other illnesses; and possibly cancer.

Carbon Monoxide (CO)

Carbon monoxide (CO) is an extremely toxic emission resulting from incomplete combustion. It is a colorless and odorless gas. CO poisoning can result in headaches, nausea, respiratory (breathing) problems, and death. CO prevents human blood cells from carrying oxygen to body tissues.

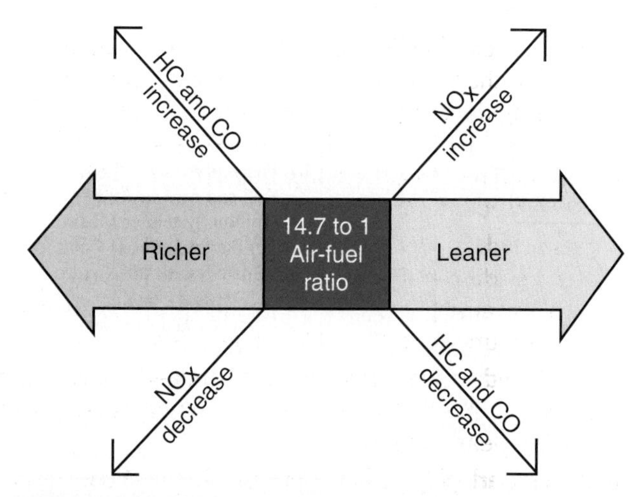

Figure 15-17. *Generally, a rich mixture increases HC and CO emissions while decreasing NO_X emissions. However, a lean mixture increases NO_X emissions while decreasing HC and CO emissions. (General Motors)*

Any factor that reduces the amount of oxygen present during combustion increases carbon monoxide emissions. For example, a rich air-fuel mixture increases CO levels. As the mixture is leaned, CO emissions are reduced, **Figure 15-17.**

Oxides of Nitrogen (NO_X)

Oxides of nitrogen (NO_X) are emissions produced by high temperatures during combustion. Air consists of about 80% nitrogen and 20% oxygen. With enough heat, above approximately 2500°F (1370°C), nitrogen and oxygen in the air-fuel mixture combine to form NO_X emissions.

Particulates

Particulates are solid particles of carbon soot and fuel additives that blow out of a vehicle's tailpipe. Carbon particles make up the largest percentage of these emissions. About 30% of all particulate emissions are heavy enough to settle out of the air in a relatively short period of time. The other 70%, however, can float in the air for an extended period of time. Particulate emissions are not usually a problem with gasoline engines.

Engine Modifications for Emission Control

The best way to reduce exhaust emissions is to burn all of the fuel inside the engine. For this reason, several engine modifications have been introduced to improve efficiency. Today's engines can have the following modifications to lower emissions:

- **Lower compression ratios.** Lower compression stroke pressure reduces combustion temperatures and NO_X emissions.
- **Leaner air-fuel mixtures.** In a lean mixture, more air is present to help all of the fuel burn. This lowers HC and CO emissions.
- **Heated air intake systems.** These systems speed engine warmup and permit the use of leaner mixtures during initial startup.
- **Smaller combustion chamber surface volumes.** A smaller chamber increases combustion efficiency by lowering the amount of heat dissipated out of the fuel mixture. Less combustion heat enters the cylinder head and more heat is left to burn the fuel. This results in reduced HC emissions.
- **Increased valve overlap.** A camshaft with more overlap dilutes the incoming air-fuel mixture with inert exhaust gasses. This reduces peak combustion temperatures and, thus, NO_X emissions.
- **Hardened valves and seats.** Hardened valves and seats are needed to prevent excessive wear when using unleaded fuels.
- **Wider spark plug gaps.** Wider gaps produce hotter sparks that can ignite hard-to-burn, lean air-fuel mixtures.
- **Reduced quench areas.** When the piston is too close to the cylinder head, there is a tendency to quench (put out) combustion, which results in increase emissions due to unburned fuel. Modern engines have cylinder heads and pistons designed to prevent high quench areas.
- **Higher operating temperatures.** If the metal parts in an engine are hotter, less combustion heat will dissipate out of the burning fuel and into the parts of the engine. This improves combustion and reduces HC and CO emissions. Higher-temperature thermostats are used to achieve higher operating temperatures.

Vehicle Emission Control Systems

Several systems are used to reduce the pollution produced by the engine and its fuel system. These include:

- PCV system.
- Heated air inlet system.
- Evaporation emissions control system.
- EGR system.
- Air injection system.
- Catalytic converter.

Variations of these systems and computer control are all used to make the modern engine very efficient.

Positive Crankcase Ventilation (PCV)

The ***positive crankcase ventilation (PCV)*** system keeps engine crankcase fumes out of the atmosphere. It uses engine vacuum to draw toxic blowby gasses into the intake manifold for reburning in the combustion chambers, **Figure 15-18.** Blowby gasses can cause:

- Air pollution.
- Corrosion of engine parts.
- Dilution of engine oil.
- Formation of sludge.

The PCV system is designed to prevent these problems. It helps keep the inside of the engine clean and also reduces air pollution.

A ***PCV valve*** is commonly used to control the flow of air through the crankcase ventilation system. It may be located in a rubber grommet in the valve cover or in a breather opening in the intake manifold.

The PCV valve changes airflow for idle, cruise, acceleration, wide open throttle, and engine-off conditions. This is shown in **Figure 15-19.** In case of an engine backfire, the PCV valve plunger is seated against the body of the valve. This keeps the backfire from entering the crankcase and igniting the fumes within.

Heated Air Inlet System

The ***heated air inlet system,*** also called a thermostatic air cleaner system, speeds engine warmup and keeps the temperature of the air entering the engine constant. A thermostatically controlled air cleaner maintains a constant temperature of air entering the engine, which improves combustion. An air temperature control valve is used to open and close the inlet to the air cleaner,

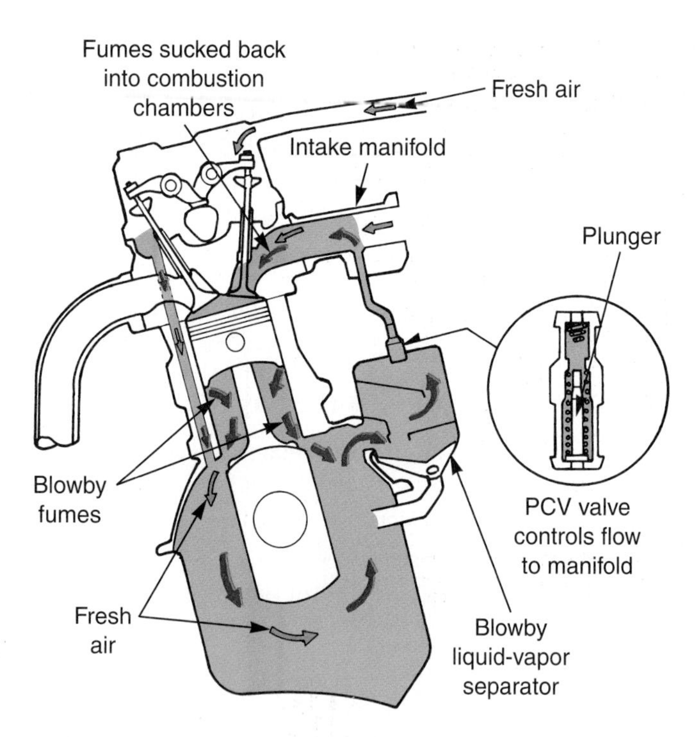

Figure 15-18. *The positive crankcase ventilation system uses engine vacuum to draw toxic crankcase fumes into the intake manifold. These fumes are then burned in the combustion chambers. The PCV valve controls flow so the mixture at idle is not too lean. (Honda)*

Engine not running or backfire
To intake manifold
PCV valve is closed
To valve cover

Normal operation
PCV valve is open, vacuum passage is large

Idling or decelerating
PCV valve is open, vacuum passage is small

Acceleration or high load
PCV valve fully open

Figure 15-19. *Study the operation of a PCV valve in different modes. (Toyota)*

Figure 15-20. By maintaining a more constant inlet air temperature, the fuel injection system can be calibrated leaner to reduce emissions.

Evaporative Emissions Control System

The ***evaporative emissions control (EVAP) system*** prevents vapors in the fuel system from entering the atmosphere, **Figure 15-21.** A fuel tank vent line carries vapors from the fuel tank to a charcoal canister in the engine compartment. The ***charcoal canister*** stores fuel vapors when the engine is not running. The metal or plastic canister is filled with activated charcoal granules. The charcoal is capable of absorbing fuel vapors. The top of the canister has fittings for the fuel tank vent line and the purge (cleaning) line. The bottom of the canister has an air filter that cleans the outside air as it enters the canister.

A ***liquid-vapor separator*** is frequently used to keep liquid fuel from entering the EVAP system. It is simply a metal tank located above the main fuel tank. Fuel vapor

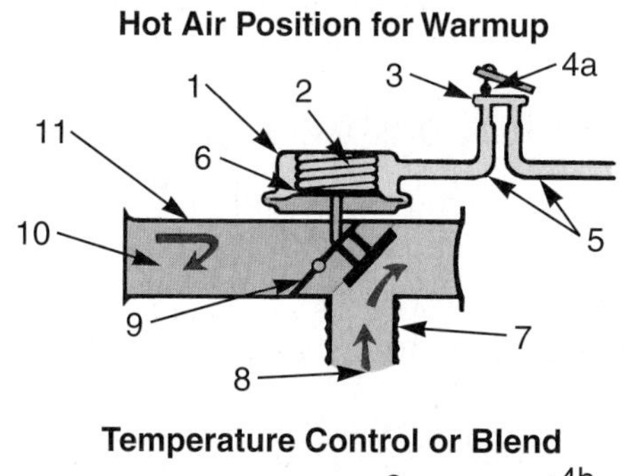

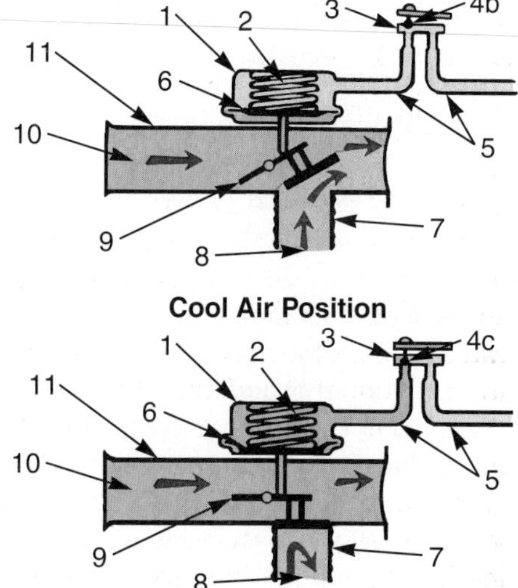

1 Vacuum diaphragm motor
2 Diaphragm spring
3 Temperature sensor
4a Air bleed valve—closed
4b Air bleed valve—partially open
4c Air bleed valve—open
5 Vacuum hoses
6 Diaphragm
7 Heat stove
8 Hot air (exhaust manifold)
9 Damper door
10 Outside inlet air
11 Snorkel

Figure 15-20. *A thermostatic air cleaner maintains a relatively constant air inlet temperature to improve combustion. (General Motors)*

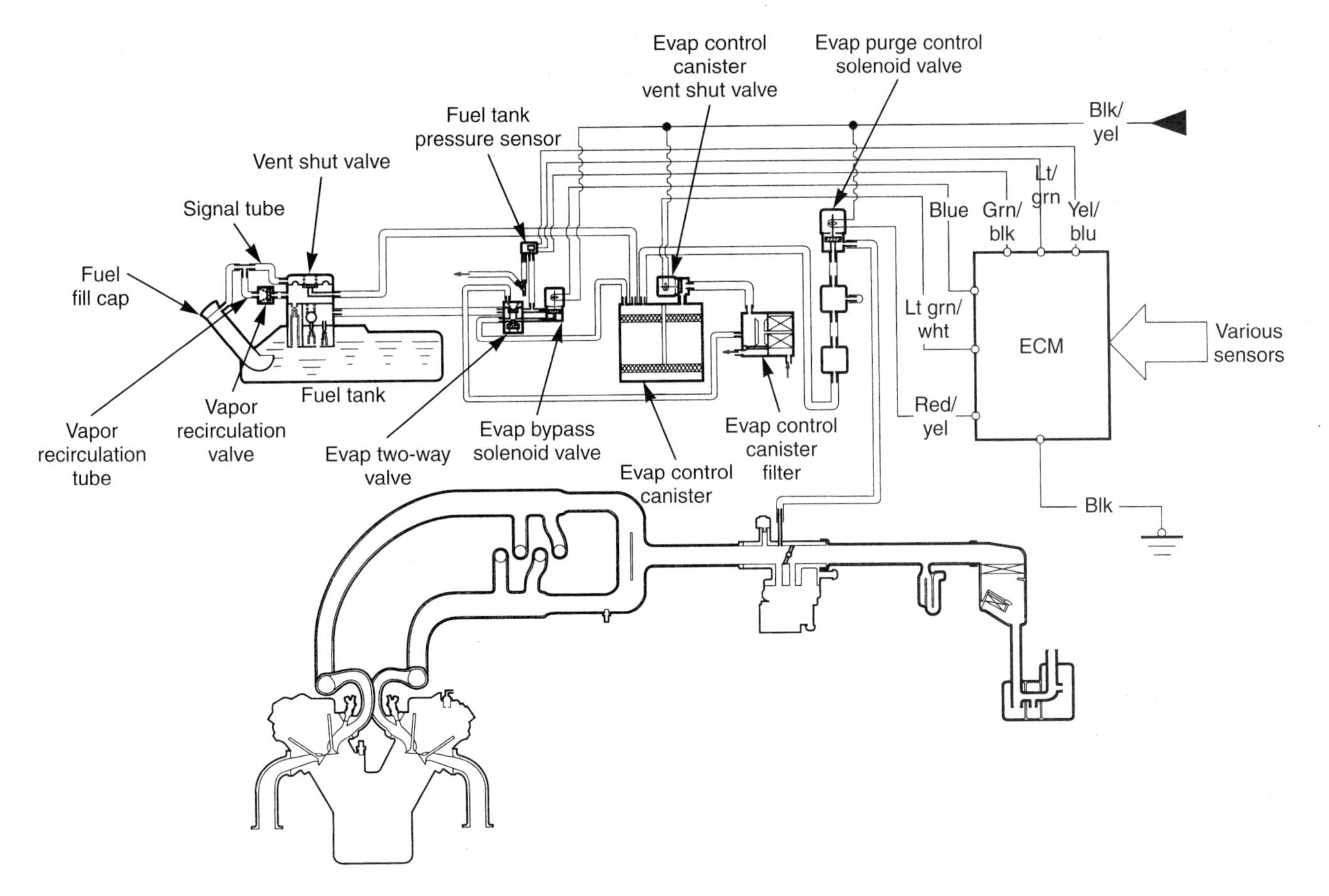

Figure 15-21. *The evaporative emissions control system collects and stores fuel vapors in a charcoal canister (EVAP control canister). When the engine is started, vapors are pulled out of the charcoal canister through the purge line and into the intake manifold. (Honda)*

condenses on the walls of the liquid-vapor separator. The liquid fuel then flows back into the fuel tank.

A ***rollover valve*** is sometimes used in the vent line from the fuel tank. It keeps liquid fuel from entering the vent line after an accident where the vehicle rolls upside-down.

A ***purge line*** is used for removing or cleaning the stored vapors out of the charcoal canister. It is connected to the canister and the engine intake manifold. When the engine is running, engine vacuum draws the vapors out of the canister and through the purge line.

A ***purge valve*** controls the flow of fuel vapor stored in the canister to the intake manifold. This vacuum- or computer-operated valve is located on the top of the canister or in the purge line. Purge valves generally allow flow when the engine reaches operating temperature and is operating above idle speed. This helps minimize emissions when the engine is cold and also prevents rough idle.

Exhaust Gas Recirculation System

The ***exhaust gas recirculation (EGR) system*** injects burned exhaust gasses into the engine intake manifold to lower the combustion temperature and reduce NO_X pollution. The ***EGR valve*** is usually bolted to the engine intake manifold or an adapter plate. It consists of a vacuum diaphragm, spring, plunger, exhaust gas valve, and a diaphragm housing. Exhaust gasses are routed through the cylinder head and manifold to the EGR valve.

At idle, the throttle plate in the throttle body is closed. This blocks engine vacuum so it cannot act on the EGR valve. The EGR spring holds the valve shut and exhaust gasses do not enter the intake manifold, **Figure 15-22A.** If the EGR valve opens at idle, the air-fuel mixture will be upset and the engine may stall.

When the throttle plate is opened to increase engine speed, engine vacuum is applied to the EGR diaphragm, which is pulled up. In turn, the diaphragm pulls the valve open. Engine exhaust can then enter the intake manifold and combustion chambers, **Figure 15-22B.**

Note: Some EGR valves are equipped with position and flow sensors that monitor system performance. Also, some EGR valves are completely electronic and do not use engine vacuum.

Air Injection System

An ***air injection system*** uses an air pump to force fresh air at low pressure into the exhaust ports of the engine, **Figure 15-23.** This helps burn any fuel present and reduce HC and CO emissions. The exhaust gasses leaving an engine can contain unburned and partially burned fuel. Oxygen in the air pumped in by the air injection systems causes this fuel to continue to burn.

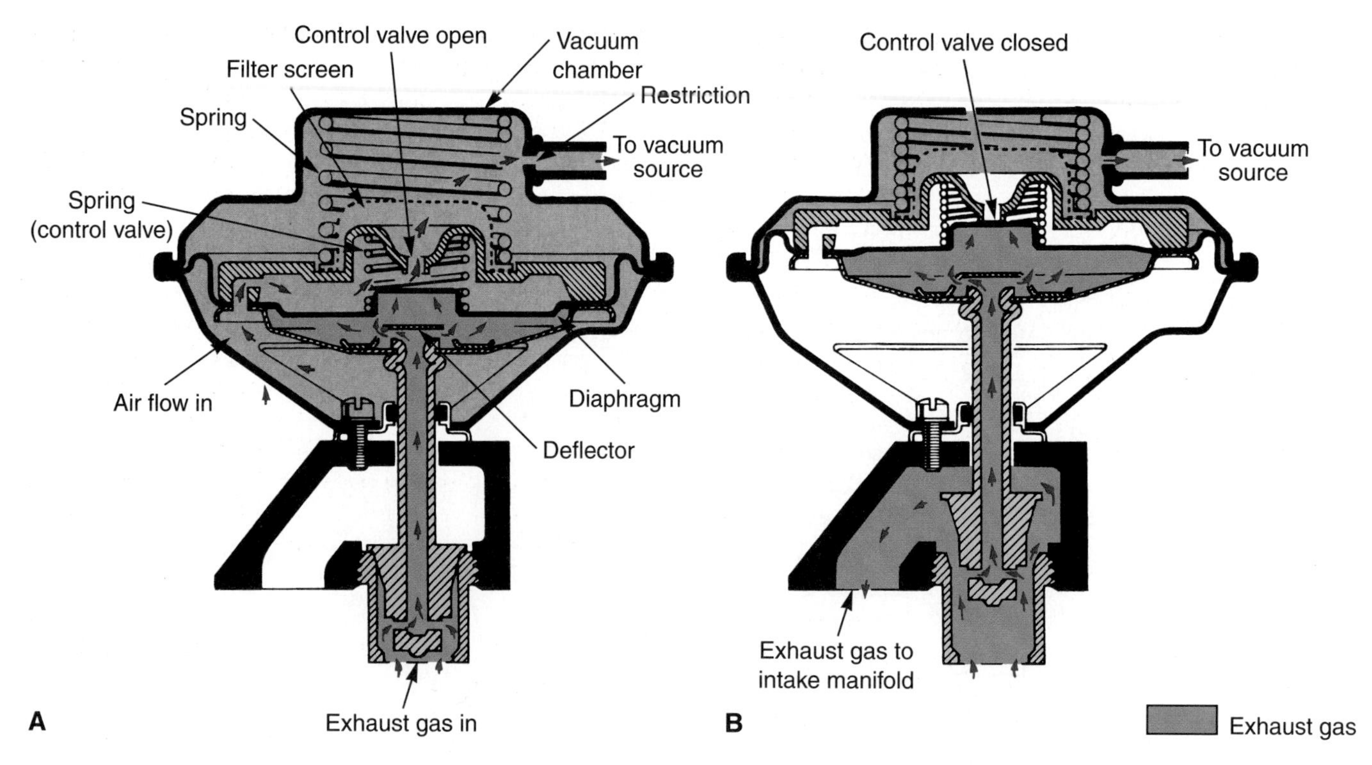

Figure 15-22. *EGR valve operation. A—When the engine is idling, the EGR valve is closed so the idle mixture is not too lean. B—As engine speed increases, intake vacuum is applied to the EGR valve. This pulls the diaphragm up and opens the valve so exhaust gasses can flow to the intake manifold and into the combustion chambers. (General Motors)*

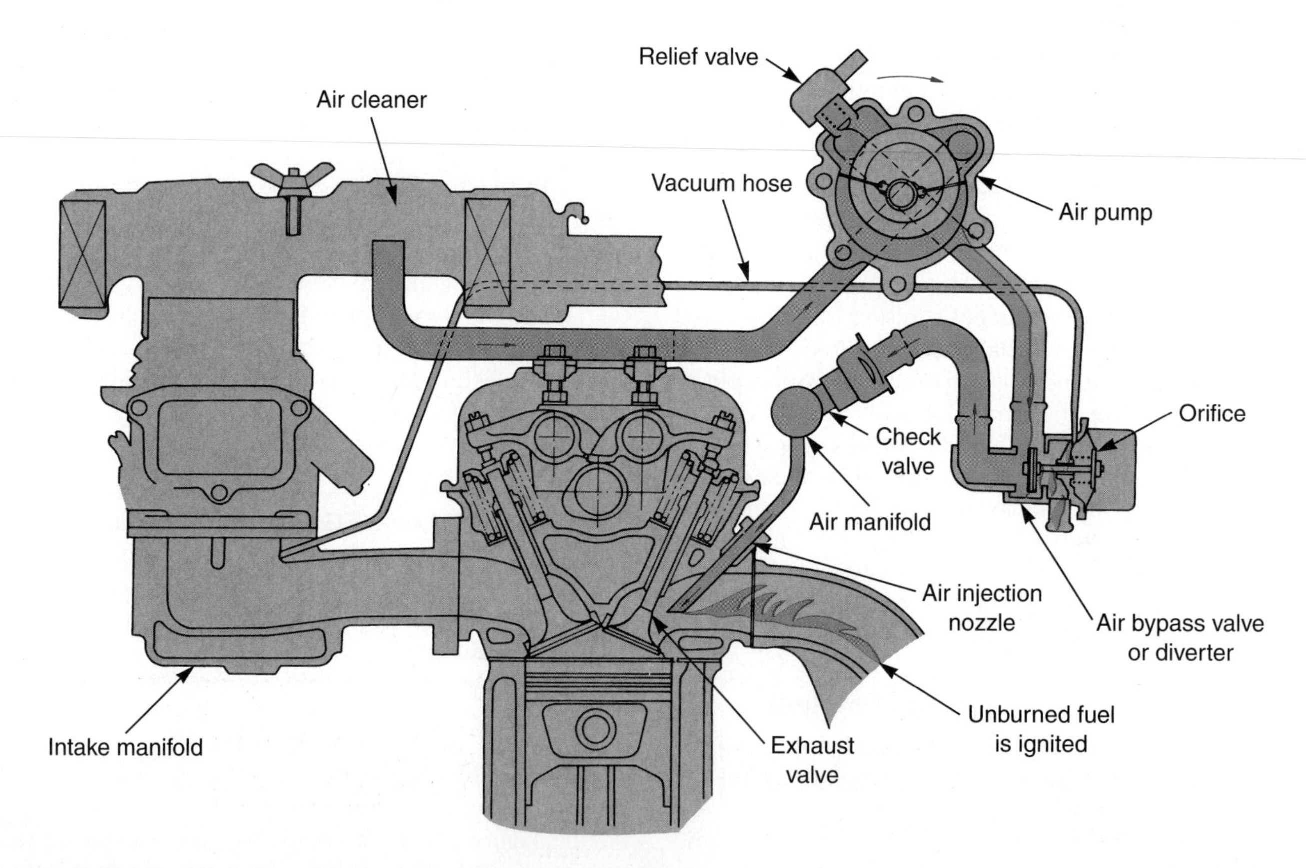

Figure 15-23. *An air injection system uses a low-pressure air pump to force fresh air into the exhaust ports or exhaust manifold. This makes any unburned fuel burn to reduce emissions. A diverter valve blocks air injection on deceleration. A check valve keeps exhaust from entering the air line. (General Motors)*

The ***air pump*** is usually belt driven. In some systems, the air pump is driven by a small electric motor. This reduces emissions even more because the air pump speed does not change with engine speed. A rubber hose or metal line connects the output of the pump to a diverter valve. The ***diverter valve*** keeps air from entering the exhaust system during deceleration. This prevents backfiring in the exhaust system. The diverter valve also limits the maximum system air pressure, when needed. It releases excess pressure through a silencer or muffler.

An ***air distribution manifold*** directs a stream of air toward each engine exhaust valve. Fittings on the air distribution manifold screw into threaded holes in the exhaust manifold or head.

An ***air check valve*** is usually located in the line between the diverter valve and the air distribution manifold. It keeps exhaust from entering the air injection system.

Pulse Air System

A ***pulse air system*** performs the same function as an air injection system. However, it uses the pressure pulses in the exhaust system to operate check valves. The check valves—also called aspirator valves, gulp valves, or reed valves—block airflow in one direction and allow airflow in the other direction. Pulse air systems are no longer used on modern vehicles.

Catalytic Converter

A ***catalytic converter*** is a thermal reactor for burning and chemically changing exhaust byproducts into harmless substances. It oxidizes (burns) the remaining HC and CO emissions that pass into the exhaust system. Extreme heat of approximately 1400°F (760°C) ignites these emissions and changes them into harmless carbon dioxide (CO_2) and water (H_2O), **Figure 15-24.** The catalytic operation starts treating emissions when the catalysts are above about 300°F (150°C).

A catalytic converter contains a catalyst, usually platinum, palladium, rhodium, or a mixture of these substances. A ***catalyst*** is any substance that speeds a chemical reaction without itself being changed. The catalysts platinum and palladium treat the HC and CO emissions. Rhodium acts on the NO_X emissions. Newer converters also contain a base metal called cerium to attract excess oxygen and release it into the exhaust.

A catalytic converter using a catalyst in the form of a ceramic block is often termed a ***monolithic catalytic converter.*** When small ceramic beads are used, it is called a ***pellet catalytic converter.*** Pellet converters are no longer used. A ***dual-bed catalytic converter*** contains two separate catalyst units enclosed in a single housing. A mixing chamber is provided between the two. Air is forced into the chamber to help burn the emissions.

Most modern catalytic converters are ***three-way converters.*** This type of converter can reduce HC, CO, and NO_X emissions. A three-way converter is usually coated with rhodium and platinum. **Figure 15-25** shows a catalytic converter system that uses three three-way converters and three oxygen sensors.

Diagnosing Problems Using an Exhaust Gas Analyzer

An exhaust gas analyzer is a testing instrument that measures the chemical content of engine exhaust gasses. Four- or five-gas analyzers are the most common type. The analyzer probe (sensor) is placed in the vehicle's tailpipe. With the engine running, the exhaust analyzer indicates the amount of pollutants and other gasses in the exhaust. The technician can use this information to determine the condition of the engine and other systems. An exhaust gas analyzer is an excellent diagnostic tool that will indicate:

- Engine mechanical problems.
- Vacuum leaks.
- PCV troubles.
- Fuel injection problems.
- Ignition system problems.
- Clogged air filter.
- Faulty air injection system.
- Evaporative control system problems.
- Computer control system troubles.
- Catalytic converter condition.

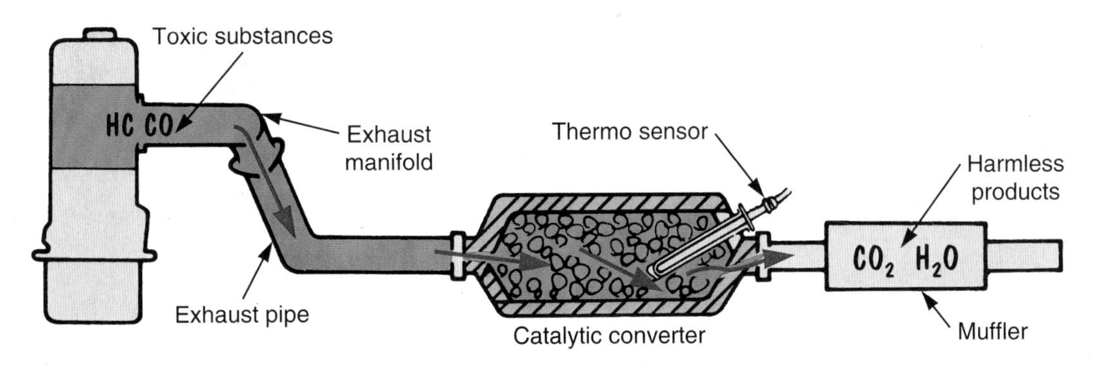

Figure 15-24. *A catalytic converter is a thermal reactor that burns any emissions entering the exhaust system. Ideally, harmless CO_2 and water come out of the exhaust pipe. (Toyota)*

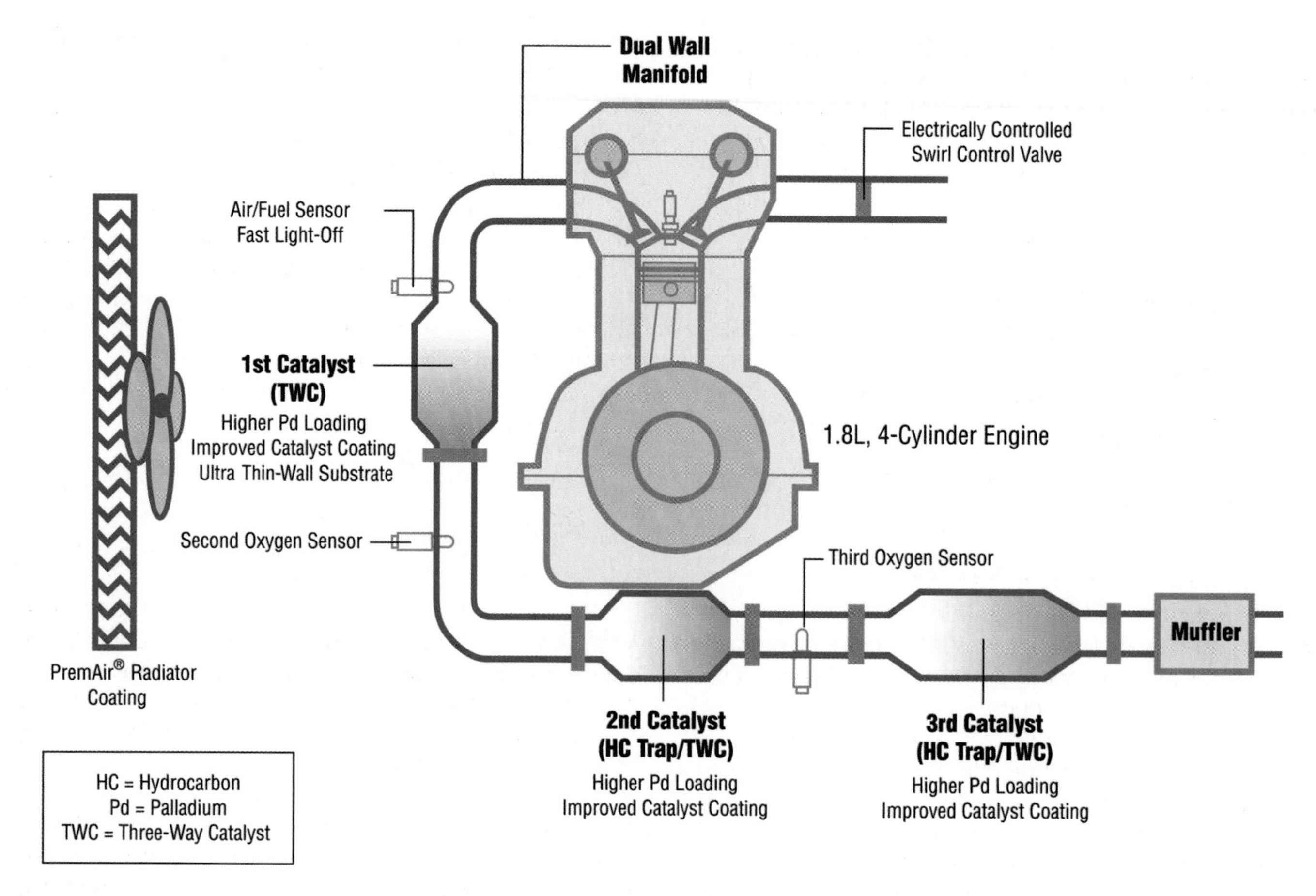

Figure 15-25. *This modern catalytic converter system has three three-way converters and three oxygen sensors to monitor and treat exhaust flow out of the engine. (Nissan)*

Drive the vehicle to bring it to operating temperature before testing. Never test a vehicle with a cold engine or inaccurate readings will result. Warm the analyzer as described by the manufacturer. Then, zero and calibrate the analyzer while sampling clean air.

Generally, take two readings, one with the vehicle at idle and a second at approximately 2500 rpm. Compare the analyzer readings with specifications. When testing some electronic (computer) fuel injection systems without a load, only readings at idle will be accurate. A dynamometer must be used to load the engine to simulate actual driving conditions.

HC Readings

An exhaust analyzer measures hydrocarbons (HC) in parts per million (ppm) by volume. An analyzer reading of 10 ppm means there are 10 parts of HC for every million parts of exhaust gas. If the exhaust analyzer reading is higher than specification, the vehicle's HC emissions (unburned fuel) are too high. An adjustment or repair is needed. Higher-than-normal HC readings can be caused by:

- **Rich or lean air-fuel mixture.** Leaking fuel injector, faulty pressure regulator, or improper fuel pressure.
- **Improper ignition timing.** Distributor, computer, or adjustment problem.
- **Engine problems.** Blowby, worn rings, burned valve, or blown head gasket.
- **Faulty emission control system.** PCV system problem, bad catalytic converter, faulty EGR valve, or evaporative control system problem.
- **Ignition system troubles.** Fouled spark plug, cracked distributor cap, or open spark plug wire.
- **Computer control system problems.** Defective input sensor, output actuator, or ECM.

Note: Always refer to the emission control sticker in the engine compartment or service manual for emission level specifications. Values vary from model year to model year.

CO Readings

An exhaust analyzer measures carbon monoxide (CO) in percentage by volume. A reading of 1% means that 1%

of the engine exhaust is CO. High CO is basically caused by incomplete burning and a lack of air (oxygen) during combustion. If the analyzer reading is higher than specifications, locate and correct the cause of the problem.

The CO reading is related to the air-fuel ratio. A low CO reading indicates a lean air-fuel mixture. A high CO reading indicates a rich mixture. Typical causes of high CO readings are:

- **Fuel system problems.** Bad injector, leaking fuel pressure regulator, restricted air cleaner, bad engine sensor, or computer control problems.
- **Emission control system troubles.** Almost any emission control system problem can upset CO readings.
- **Incorrect ignition timing.** Timing too advanced due to computer control system problem.

CO_2 Readings

An exhaust analyzer measures carbon dioxide (CO_2) in percent by volume. Typically, CO_2 readings should be above 8%. Carbon dioxide is a byproduct of combustion. CO_2 is not toxic at low levels.

Normally, oxygen and carbon monoxide levels are compared when evaluating the content of the engine exhaust. For example, if the percentage of CO_2 exceeds the percentage of oxygen, the air-fuel ratio is on the rich side of the stoichiometric (theoretically perfect) mixture. CO_2 is also a good indicator of possible dilution of the exhaust gas sample due to an exhaust leak.

O_2 Readings

An exhaust analyzer measures oxygen (O_2) in percentage by volume. Typically, O_2 readings should be between 0.1% and 7%. Oxygen is needed for the catalytic converter to burn HC and CO emissions. Without O_2 in the engine exhaust, exhaust emissions can pass through the converter and out of the vehicle's tailpipe.

The O_2 level in the exhaust sample is an accurate indicator of a vehicle's air-fuel mixture. When an engine is running lean, oxygen increases proportionately with the decrease in fuel. As the air-fuel mixture becomes lean enough to cause a lean misfire (engine miss), oxygen readings rise dramatically. The O_2 level is also a good indicator of a possible exhaust leak, which can dilute the exhaust sample.

NO_X Readings

Oxides of nitrogen (NO_X) are measured on a five-gas analyzer. High NO_X emissions can result from:

- **High combustion chamber temperatures.** Excessively high engine compression ratio, carbon deposits in the combustion chambers, low coolant level, blocked cooling passages, stuck thermostat, etc.
- **EGR system problems.** Burned gasses are not being injected into the intake manifold and combustion flame temperature is too high.

Summary

Proper engine operation depends on an adequate supply of clean air and fuel. Without the correct mixture of fuel and air, an engine will perform poorly. A fuel system can be divided into three subsystems: fuel supply system, air supply system, and fuel metering system.

The fuel supply system provides filtered fuel under pressure to the fuel metering system. The air supply system removes dust and dirt from the air entering engine intake manifold. The fuel metering system controls amount of fuel that is mixed with the filtered air.

The two types of electronic fuel injection (fuel metering) are multiport and throttle body injection. In multiport injection, one injector is located just before each intake port. Fuel is injected into the engine in more than one location. This is the most common type.

An engine sensor is an electronic device that changes circuit resistance or voltage with a change in a condition. The computer can use the increased current flow through the sensor to calculate any needed change in injector opening. There are several typical sensors used in an electronic fuel injection system.

The OBD II diagnostic system can produce over 500 trouble codes related to engine performance. It checks the operating parameters of switches, sensors, actuators, in-system components and their related wiring, and the computer itself. A diagnostic trouble code is a digital signal produced and stored by the computer when an operating parameter is exceeded.

An OBD II code is a four digit number with a letter designator. The letter designator indicates the system in which the problem is located. The first number indicates whether the code is SAE or manufacturer specific. The second number indicates the function of the system where the fault is located. The last two numbers indicate the specific fault.

Emission control systems are used to reduce the amount of harmful chemicals released into the atmosphere from a vehicle. The four basic types of vehicle emissions are hydrocarbons, carbon monoxide, oxides of nitrogen, and particulates. Vehicle emissions result from engine crankcase blowby fumes, fuel vapors, and engine exhaust gasses.

Several systems are used to reduce the pollution produced by the engine and its fuel system. The positive crankcase ventilation (PCV) system keeps engine crankcase fumes out of the atmosphere. The heated air inlet system speeds engine warmup and keeps the temperature of the air entering the engine constant. The evaporative emissions control (EVAP) system prevents vapors in the fuel system from entering the atmosphere. The exhaust gas recirculation (EGR) system injects burned exhaust gasses into the engine intake manifold to lower the combustion temperature and reduce NO_X pollution. An air injection system uses an air pump to force fresh air at low pressure into the exhaust ports of the engine. A catalytic converter is a thermal reactor for burning and chemically changing exhaust byproducts into harmless substances.

Review Questions—Chapter 15

Please do not write in this text. Write your answers on a separate sheet of paper.

1. The _____ system provides filtered fuel under pressure to the fuel metering system. It must also provide for storage of the fuel.
2. The three subsystems of the fuel system are the:
   (A) fuel supply system, fuel filtering system, and fuel metering system.
   (B) fuel filtering system, air supply system, and fuel metering system.
   (C) fuel supply system, air supply system, and fuel metering system.
   (D) fuel supply system, air supply system, and fuel filtering system.
3. What is the purpose of an *electric fuel pump?*
4. The _____ system removes dust and dirt from the air entering engine intake manifold.
5. The two types of electronic fuel injection are _____ and _____ injection.
6. Which type of fuel injection is the most common?
7. What is a *fuel rail?*
8. Describe the operation of a basic *engine sensor.*
9. *OBD II* is an acronym for _____.
10. What is a *data link connector?*
11. What is a *scan tool?*
12. List five types of information that can be requested from a scan tool.
13. An OBD II trouble code is an alphanumeric code consisting of _____.
14. For what is a *noid light* used?
15. What are the three sources of emissions from a vehicle?
16. List and describe the four types of vehicle emissions.
17. List six systems used to reduce emissions produced by the engine and its fuel system.
18. A PCV system uses vacuum to draw blowby gasses out of the _____ for reburning in the combustion chambers of the engine.
19. If HC and CO emissions are higher than normal, what does this indicate about the air-fuel mixture?
20. List four things that blowby gasses can cause.
21. What does the EVAP system do?
22. Which system injects burned exhaust gasses into the engine intake manifold to lower the combustion temperature and reduce NO_X pollution?
   (A) PCV system.
   (B) EVAP system.
   (C) EGR system.
   (D) computer system.
23. What is the purpose of the *catalytic converter?*
24. A(n) _____ is any substance that speeds a chemical reaction without itself being changed.
25. Most modern catalytic converters are _____ converters that can reduce HC, CO, and NO_X emissions.

ASE-Type Questions—Chapter 15

1. Technician A says that a vehicle's fuel return system is used to regulate fuel pressure in a gasoline injection system. Technician B says that a vehicle's fuel return system is used to cool the fuel and prevent vapor lock. Who is correct?
   (A) A only.
   (B) B only.
   (C) Both A and B.
   (D) Neither A nor B.
2. Technician A says that a fuel filter can be located inside the fuel pump. Technician B says that a fuel filter can be located inside an injector body. Who is correct?
   (A) A only.
   (B) B only.
   (C) Both A and B.
   (D) Neither A nor B.
3. Technician A says that a throttle body assembly for a multiport injection system is used to control airflow into the engine. Technician B says that a throttle body assembly for a throttle body injection system is used to control fuel flow into the engine. Who is correct?
   (A) A only.
   (B) B only.
   (C) Both A and B.
   (D) Neither A nor B.
4. All of the following are typical sensors used in a fuel injection system *except:*
   (A) knock sensor.
   (B) mass airflow sensor.
   (C) engine coolant temperature sensor.
   (D) oxygen sensor.

5. Technician A says that a throttle body for a throttle body injection system normally contains the system's fuel pressure regulator. Technician B says that the throttle body for a throttle body injection system normally contains the fuel injector(s). Who is correct?
   (A) A only.
   (B) B only.
   (C) Both A and B.
   (D) Neither A nor B.
6. Technician A says that to clear the trouble codes from most computer systems, you should disconnect the ECM from battery power. Technician B says that to clear the trouble codes from an OBD II computer system, you must use a scan tool. Who is correct?
   (A) A only.
   (B) B only.
   (C) Both A and B.
   (D) Neither A nor B.
7. Technician A says that hydrocarbon emission can be caused by incomplete burning of the fuel in the engine's combustion chambers. Technician B says that hydrocarbon emission can be caused by fuel evaporation. Who is correct?
   (A) A only.
   (B) B only.
   (C) Both A and B.
   (D) Neither A nor B.
8. Technician A says that wider spark plug gaps are used on today's vehicles to burn leaner fuel mixtures. Technician B says that today's automotive engines are equipped with hotter thermostats to reduce HC and CO emissions. Who is correct?
   (A) A only.
   (B) B only.
   (C) Both A and B.
   (D) Neither A nor B.
9. Technician A says that an engine's PCV valve can be located in a valve cover. Technician B says that an engine's PCV valve can be located in the intake manifold. Who is correct?
   (A) A only.
   (B) B only.
   (C) Both A and B.
   (D) Neither A nor B.
10. Which type of catalytic converter is most commonly used on modern automotive exhaust systems?
    (A) Pellet converter.
    (B) Three-way converter.
    (C) Four-way converter.
    (D) All of the above.

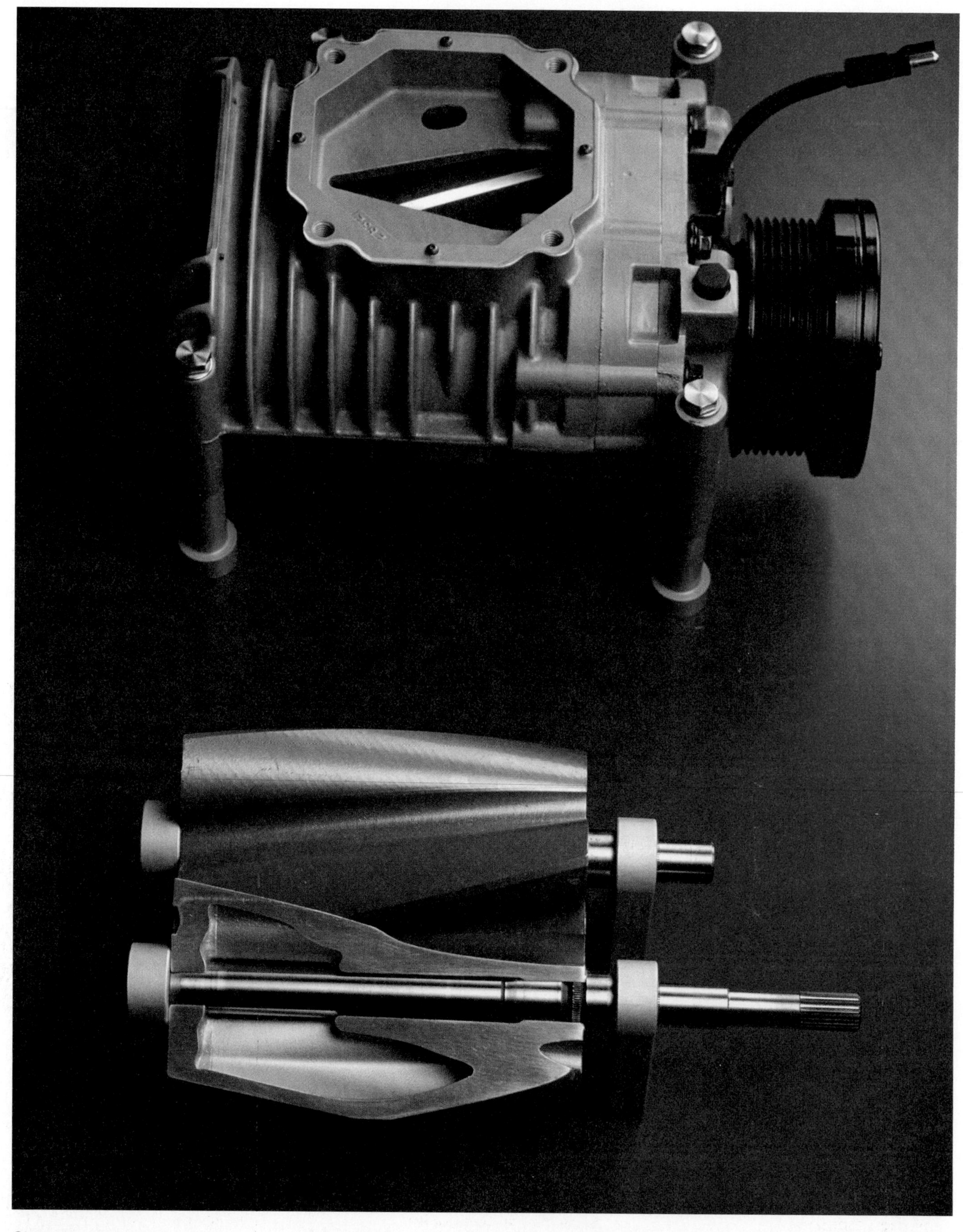

Superchargers can be used to increase power output from an engine. Note the supercharger vanes shown below the supercharger. (DaimlerChrysler)

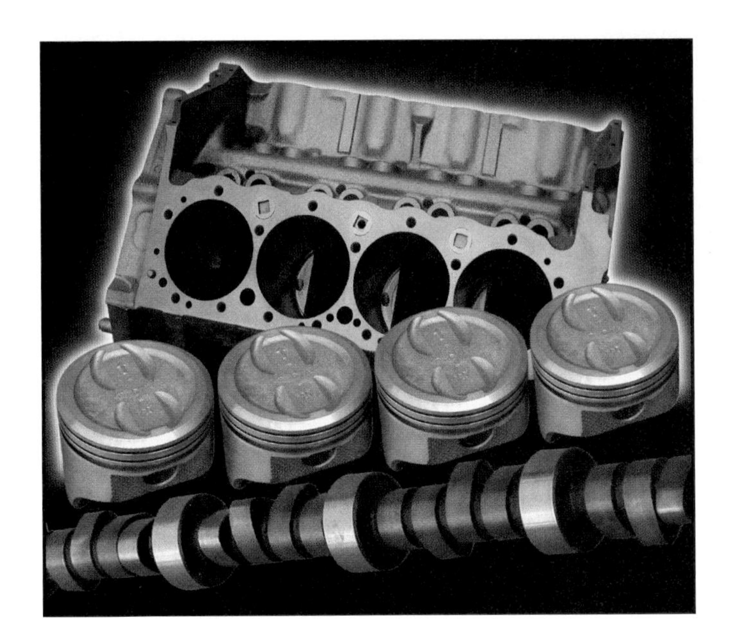

Chapter 16

Turbocharging and Supercharging Systems

After studying this chapter, you will be able to:

- ❑ Explain normal aspiration, turbocharging, and supercharging.
- ❑ Explain the construction of a turbocharger.
- ❑ Summarize turbocharging system operation.
- ❑ Describe the construction of a waste gate.
- ❑ Explain the operation of an intercooler.
- ❑ Diagnose common turbocharger problems.
- ❑ Replace a turbocharger and waste gate.
- ❑ Summarize supercharging system operation.
- ❑ Explain the construction of a supercharger.
- ❑ Diagnose common supercharger problems.
- ❑ Service a supercharger.

Know These Terms

Boost
Intercooler
Intercooler coolant
Intercooler lines
Intercooler pump
Intercooler radiator
Normal aspiration
Overboost
Sequential turbocharger system
Supercharger
Supercharger bypass valve
Supercharger drive pulley
Supercharger extension housing
Supercharger housing
Supercharger intercooler
Supercharger rotors
Supercharger timing gears
Turbocharger
Turbocharger intercooler
Turbo lag
Turbo timer
Volumetric efficiency
Waste gate

A turbocharger has the ability to increase the horsepower of an engine by as much as 50%. For this reason, they are commonly used on small displacement engines to increase performance. Turbochargers, as you will learn, can increase power under a load, yet have little adverse affect on fuel economy at cruising speeds.

Superchargers have been used on racing engines for many years. Manufacturers began installing superchargers on some production engines a few years ago. Several cars and trucks now have supercharging to help their engines develop over 400 horsepower.

To be a competent engine technician, you must have some knowledge of turbochargers and superchargers. Many of today's engines are equipped with these devices.

Normal Aspiration

Normal aspiration means the engine uses only outside air pressure (atmospheric pressure) to cause airflow into the combustion chambers. At sea level, atmospheric pressure is 14.7 psi (10.1 kPa). With only outside air pressure to carry oxygen into the engine cylinders, engine power is limited by the engine's volumetric efficiency.

Volumetric efficiency is a measure of how much air the engine can draw in on its intake strokes. A high volumetric efficiency means the engine "breathes" easily because of good intake port, valve, combustion chamber, and camshaft design. However, most engines do not have high volumetric efficiency. This is because of pumping losses due to restrictions to airflow through the intake manifold and cylinder head. As a result, they do not produce as much horsepower as theoretically possible for their size.

Turbocharging and supercharging pressurizes the incoming air. This has the effect of increasing the volumetric efficiency. Turbocharging and supercharging are used on many gasoline and diesel engines to improve volumetric efficiency and increase power output.

Turbocharging and Supercharging

Turbocharging and supercharging are both methods of increasing the intake manifold pressure. This forces more air into the cylinders than normal aspiration. As a result, power output from the engine is increased.

A ***turbocharger*** or ***turbo*** is a special fan assembly that uses engine exhaust gasses to turn the fan blades. A mechanical linkage with the engine is not needed. Look at **Figure 16-1A.**

A ***supercharger*** or ***blower*** is a special fan assembly that is driven by a belt on the engine. They are used on some passenger car engines and can be found on many racing engines. See **Figure 16-1B.**

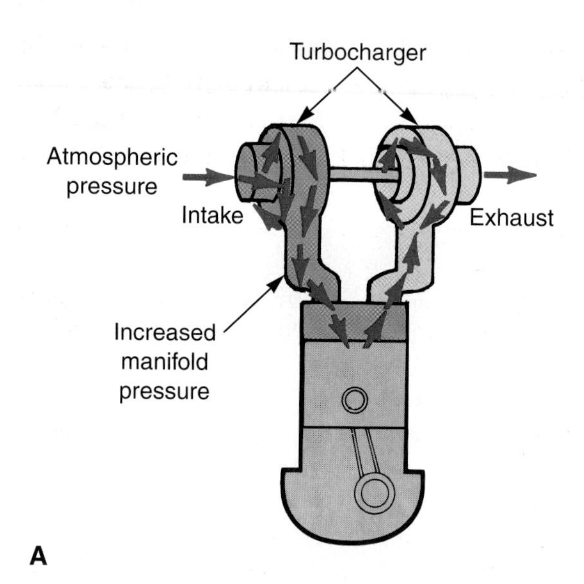

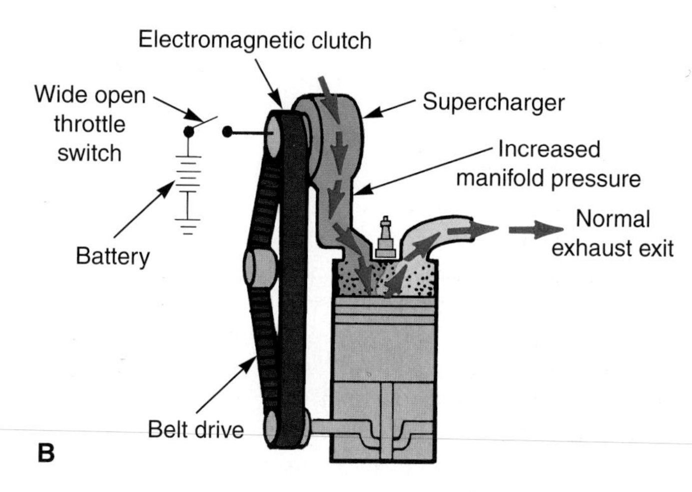

Figure 16-1. *A comparison of turbocharging and supercharging. A—A turbocharger is driven by exhaust gasses. It compresses the incoming air. B—A supercharger also compresses the incoming air. However, it is normally driven by an engine belt.*

Engine Modifications

Turbocharged and supercharged engines normally have several modifications to make them withstand the increased horsepower. A few of these modifications are:

- ❑ Lower compression ratio.
- ❑ Stronger rods, pistons, crankshaft, and bearings.
- ❑ Higher-volume oil pump.
- ❑ Increased oil capacity.
- ❑ Larger radiator.
- ❑ O-ring–type head gasket.
- ❑ Valves with increased heat resistance.

In addition, an oil cooler may be installed to help keep the engine oil cool. Oil helps cool the turbocharger. If the oil gets too hot, the turbocharger can be damaged.

Turbocharger Construction

Pictured in **Figure 16-2,** the major parts of a turbocharger are:

- **Turbine housing.** The enclosure that routes exhaust gasses over the turbine wheel.
- **Turbine wheel.** A fan driven by exhaust gasses. It turns the turbo shaft, which turns the compressor wheel.
- **Turbo shaft.** A steel shaft that connects the turbine and compressor wheels.
- **Compressor wheel.** A fan that forces air into engine intake manifold under pressure. It is driven by the turbo shaft.
- **Compressor housing.** The enclosure that surrounds the compressor wheel. Its shape helps pump air into the engine.
- **Bearing housing.** The enclosure around the turbo shaft that contains bearings, seals, and oil passages.

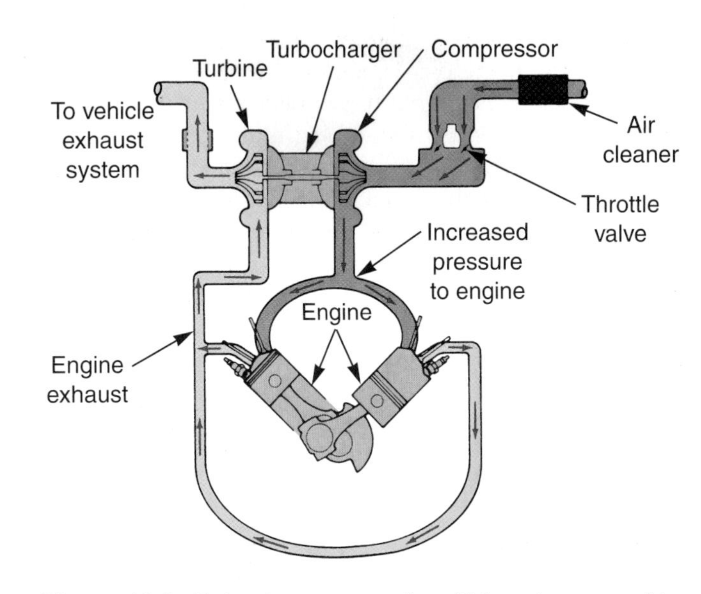

Figure 16-3. *Turbocharger operation. Exhaust gasses drive the turbocharger, which compresses the incoming air. (General Motors)*

Turbocharger Operation

During engine operation, hot exhaust gasses blow out of the open exhaust valves and into the exhaust manifold, **Figure 16-3.** The exhaust manifold and connecting tubing route these gasses into the turbine housing. As the gasses pass through the turbine housing, they strike the fins or blades on the turbine wheel. When engine load is high enough, there is enough exhaust gas flow to rapidly spin the turbine wheel.

Since the turbine wheel is connected to the compressor wheel by the turbo shaft, the compressor wheel rotates with the turbine. Compressor wheel rotation pulls air into the compressor housing. Centrifugal force throws the air outward. This causes air to flow out of the turbocharger and into the engine cylinder under pressure. With more air and fuel in the cylinder on the engine's intake stroke, more pressure and combustion force result during the engine's power stroke.

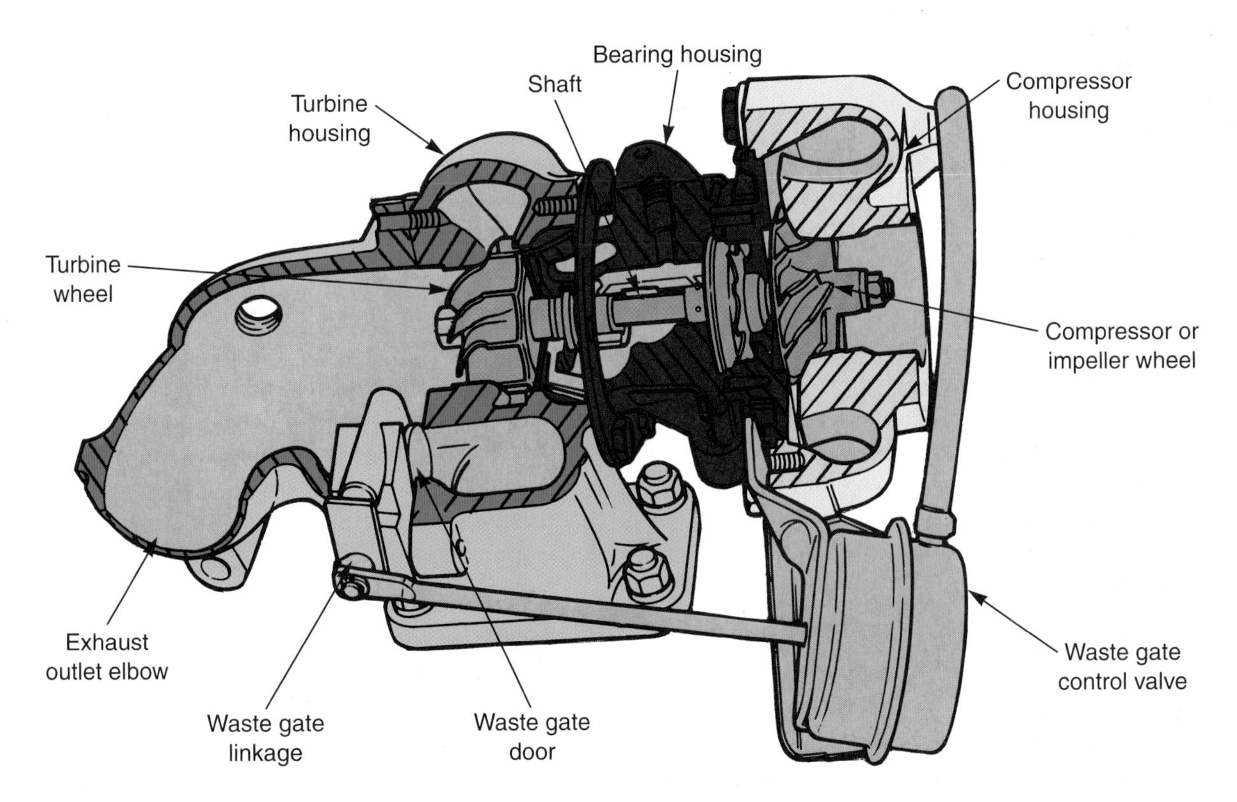

Figure 16-2. *The basic parts of a turbocharger. (DaimlerChrysler)*

Turbocharger Location

Turbochargers are normally bolted to the side, top, or front of the engine. Exhaust tubing routes engine exhaust gasses into the inlet of the turbine housing. Tubing or a hose also connects the outlet of the compressor housing to the engine intake manifold.

Theoretically, the turbocharger should be located as close to the engine exhaust manifold as possible. Then, a maximum amount of exhaust heat will enter the turbine housing. When the hot gasses blow onto the spinning turbine wheel, they are still burning and expanding to help rotate the turbine.

Turbo Lag

When the accelerator pedal is pressed down for rapid acceleration, the engine may lack power for a few seconds. ***Turbo lag*** is a short delay before the turbo develops sufficient boost. ***Boost*** is any pressure above atmospheric pressure in the intake manifold. Turbo lag is caused by the compressor and turbine wheels not spinning fast enough. It takes time for the exhaust gasses to bring the turbo up to operating speed.

Modern turbo systems suffer from very little turbo lag. Their turbine and compressor wheels are very light so that they can accelerate quickly. Some turbocharger impellers (wheels) are made of carbon fiber–reinforced plastic. This reduces impeller weight and the problem of turbo lag considerably.

Turbocharger Lubrication

Adequate lubrication is needed to protect the turbo shaft and bearings from damage. A turbocharger can operate at speeds up to 100,000 rpm. For this reason, the engine lubrication system forces engine oil into the turbo shaft bearings.

Oil passages are provided in the turbo housing and bearings, as shown in **Figure 16-4.** An oil supply line runs from the engine to the turbo. A drain line runs from the turbo to the engine. With the engine running, oil enters the turbo under pressure. The turbo shaft rides on a thin film of oil, avoiding metal-to-metal contact.

Sealing rings similar to piston rings are placed around the turbo shaft at each end. They prevent oil leakage into the compressor and turbine housings. A drain passage and drain line allow oil to return to the engine oil pan after passing through the turbo bearings.

Waste Gate

A ***waste gate*** limits the maximum boost pressure developed by the turbocharger. Without a waste gate, the turbo could produce too much pressure in the combustion chambers. This could lead to detonation and engine damage. A waste gate consists of:

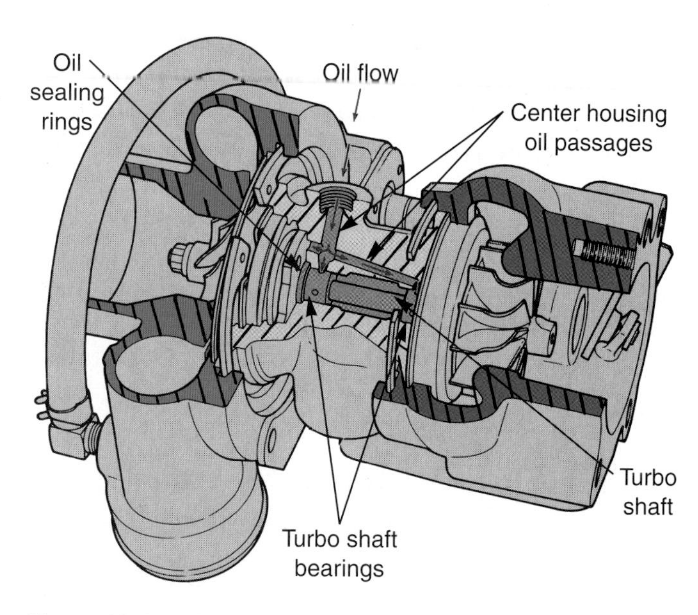

Figure 16-4. *Lubrication for the turbocharger is provided by the engine oil. Oil flow through a turbocharger is shown here. (DaimlerChrysler)*

- ❑ **Diaphragm.** A flexible membrane that reacts to different amounts of boost pressure.
- ❑ **Diaphragm spring.** A coil spring that holds the waste gate valve in the normally closed position.
- ❑ **Waste gate valve.** A poppet, flap, or butterfly valve that can open to bypass exhaust gasses away from the turbine wheel.
- ❑ **Housing.** An airtight, metal container that encloses parts and has a pressure hose fitting.
- ❑ **Pressure line.** The hose that connects the waste gate with a source of intake manifold pressure.

The operation of a turbocharger waste gate is shown in **Figure 16-5.** Under partial load, the waste gate is closed by the diaphragm spring. All exhaust gasses are directed against the turbine wheel blades. This ensures that there is adequate boost to increase engine power. The waste gate remains closed as long as boost pressure is not too high.

Under full load, boost increases as more exhaust gasses are produced and turn the turbine. When boost reaches a preset level (about 5 psi to 7 psi), pressure in the intake manifold or compressor housing acts on the diaphragm in the waste gate. It pushes the diaphragm down, which compresses the spring and forces the valve open. Some of the exhaust gasses are diverted from the turbine wheel blades. The turbo speed does not increase because less exhaust is left to spin the turbine. In this way, boost is limited.

Turbocharger Intercooler

As air is compressed by the turbocharger, the heat of compression increases the air temperature. By cooling the air entering the engine, engine power is increased because the air is more dense and contains more oxygen by volume. Cooling also reduces the tendency for engine detonation.

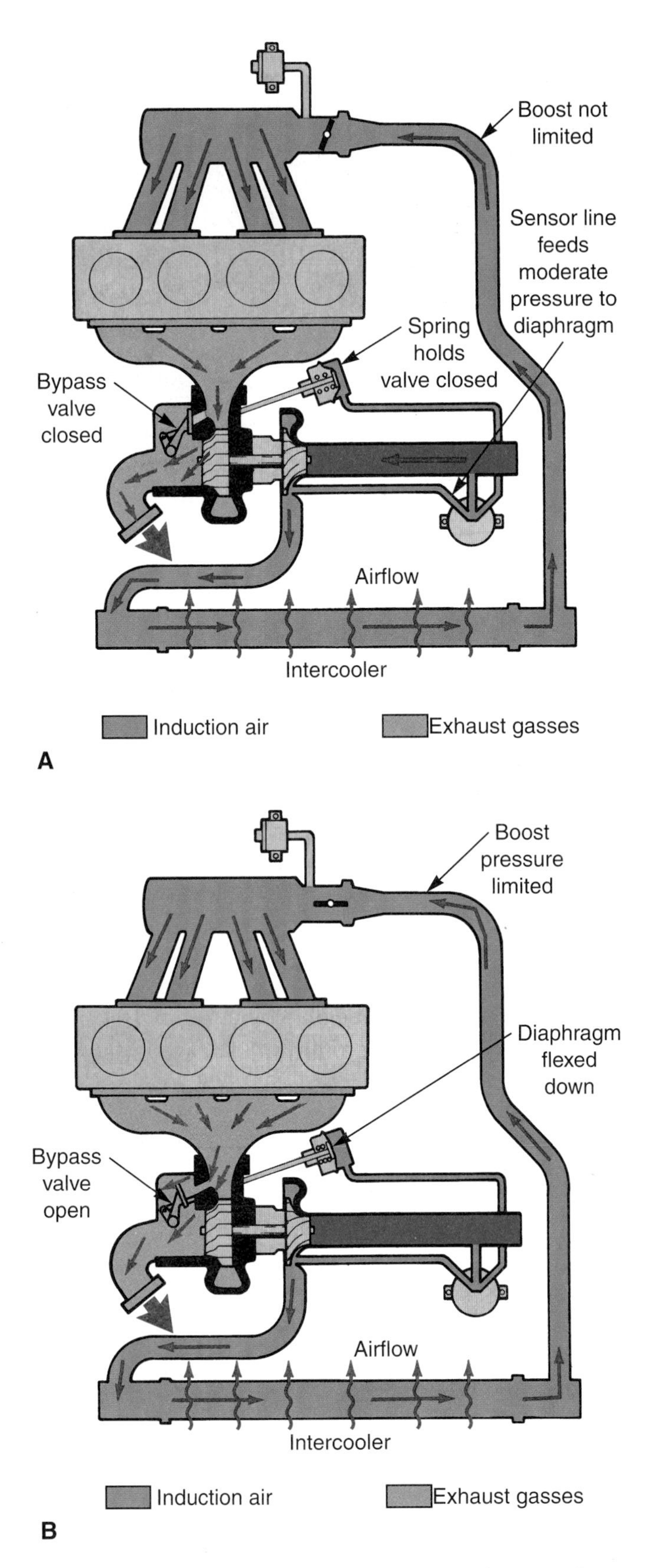

Figure 16-5. *Waste gate operation. A—As long as boost pressure is not too high, the waste gate remains closed and all exhaust gasses are directed against the turbine wheel blades. B—When boost reaches a preset level, the waste gate is opened and some of the exhaust gasses are diverted from the turbine wheel blades. (Saab)*

A ***turbocharger intercooler*** is an air-to-air or air-to-liquid heat exchanger that cools the air entering the engine. It is a radiator-like device mounted at the pressure outlet of the turbocharger. See **Figure 16-6.** Turbocharged engines often use an air-to-air intercooler. The intercooler is mounted near the front of the engine compartment so fresh outside air can flow through the intercooler fins.

In an air-to-air intercooler design, outside air flows over and cools the fins and tubes of the intercooler. The hot, compressed air from the turbocharger flows through the intercooler. Heat from the compressed air is transferred to the intercooler tubes and fins, and then to the outside air. An air-to-liquid design works in the same way except a coolant is used to cool the fins and tubes instead of outside air.

Computer-Controlled Turbocharging

In a computer-controlled turbocharging system, the electronic control module (ECM) operates the waste gate based on sensor signals, **Figure 16-7.** Several sensors, especially the knock sensor, speed sensor, and oxygen sensor, provide data to the computer. The ECM output is sent to a vacuum

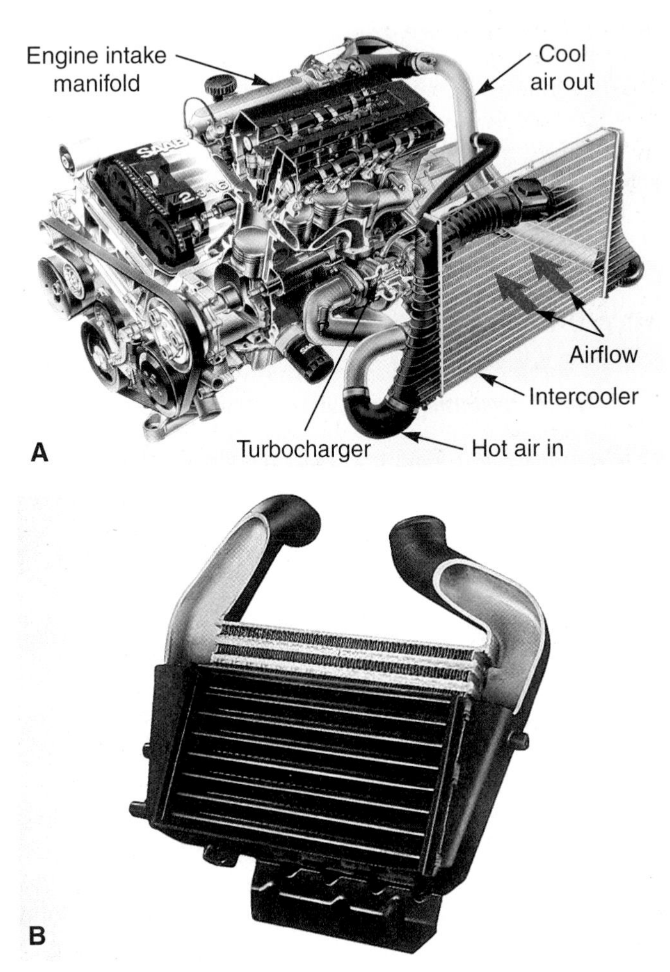

Figure 16-6. *A—An intercooler cools the air charge entering the engine for increased horsepower. B—This is an air-to-air intercooler unit used on a small-displacement, 4-cylinder engine. (Saab, Ford)*

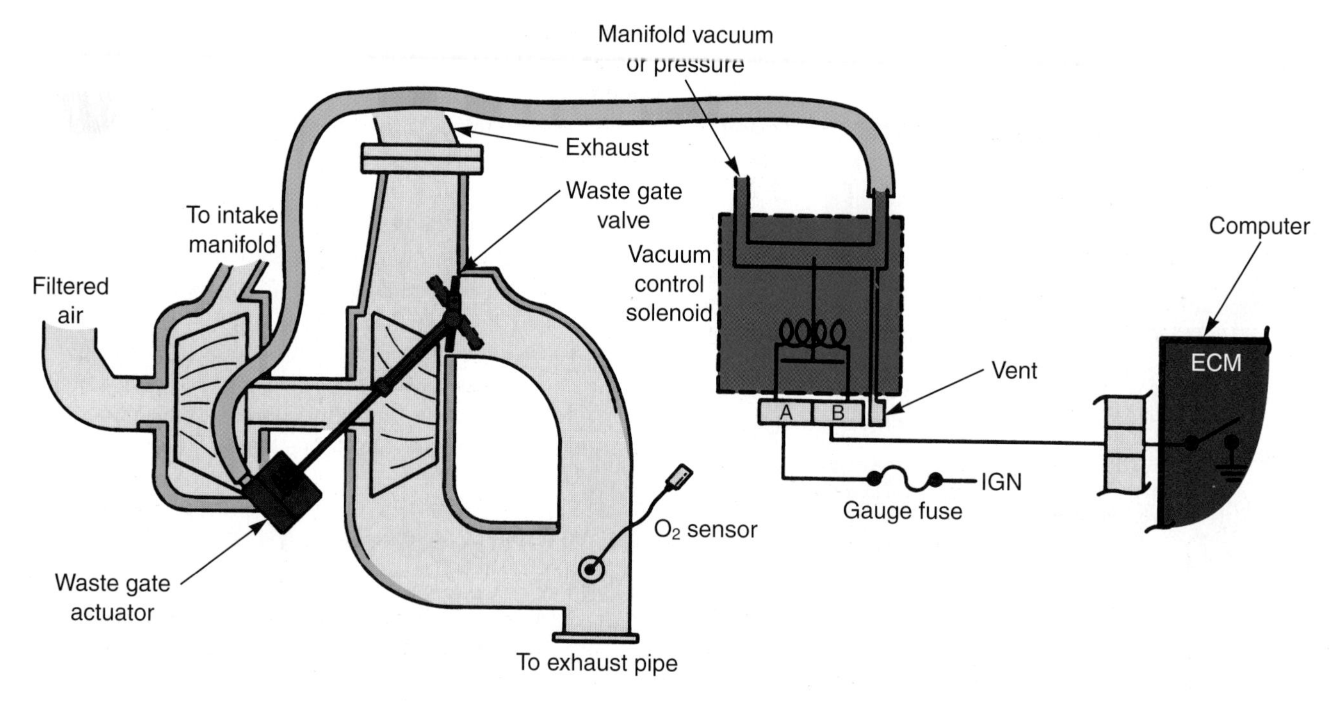

Figure 16-7. *A computer-controlled waste gate provides precise control of boost pressure based on sensor inputs. (General Motors)*

control solenoid. This solenoid valve either allows or blocks engine vacuum flow to the waste gate diaphragm.

For example, if the knock sensor detects pinging, the computer can shut off current to the vacuum solenoid. In turn, the solenoid allows vacuum to the waste gate diaphragm. The waste gate valve then opens to decrease boost. Decreasing boost can help prevent detonation and damage.

Sequential Turbocharger Systems

A ***sequential turbocharger system,*** also called a ***twin turbo system,*** uses two electronically controlled turbochargers that operate as needed. This system is sometimes used to help avoid turbo lag.

During low-engine-speed operation, only one turbocharger functions or provides boost pressure. Refer to **Figure 16-8.** An exhaust control valve and an intake control valve are closed, isolating the second turbocharger. An exhaust bypass valve is closed to divert exhaust gasses away from the turbine in the second turbocharger. All exhaust gasses are routed to the first turbocharger.

As the boost pressure reaches a predetermined level, the exhaust bypass valve opens to allow exhaust gasses to flow to the turbine of the second turbocharger. This helps smooth out the boost from the second turbocharger. As the engine speed increases, the intake and exhaust control valves are opened. This allows the second turbocharger to provide added boost. The waste gate limits the maximum boost of the system.

The size of the two turbochargers also helps reduce turbo lag. The first turbocharger is smaller than the second turbocharger. This allows the first turbocharger to spin up to speed very quickly for initial boost. Then, the second, larger turbocharger provides more boost at higher engine speeds.

Turbocharging System Service

Turbocharging system problems usually show up as:

- ❑ Lack of engine power, which may be inadequate boost pressure.
- ❑ Oil consumption, which may be leaking shaft seals.
- ❑ Vibration and noise, which may be due to damaged turbine or compressor wheels.
- ❑ Detonation, which may be too much boost pressure.

Refer to the service manual for a detailed troubleshooting chart. It will list the common troubles for the particular turbo system.

Increasing Turbocharger Life

The average service life of a turbocharger depends greatly on how well the driver follows manufacturer recommendations. With the high replacement cost of a turbo, it is wise to pass on to your customers some basic tips on driving and servicing a turbocharged engine:

- ❑ Change the engine oil and filter at the regular intervals specified by the manufacturer.
- ❑ Limit engine speeds until the engine and engine oil reach full operating temperature. An oil temperature gauge may help you determine when the oil reaches operating temperature.
- ❑ After driving the vehicle, allow the engine to idle for a few moments before shutting the engine off. The manufacturer may specify a cooldown time. This will let the turbo cool down from a potential internal temperature of about 1000°F (540°C). At idle, exhaust

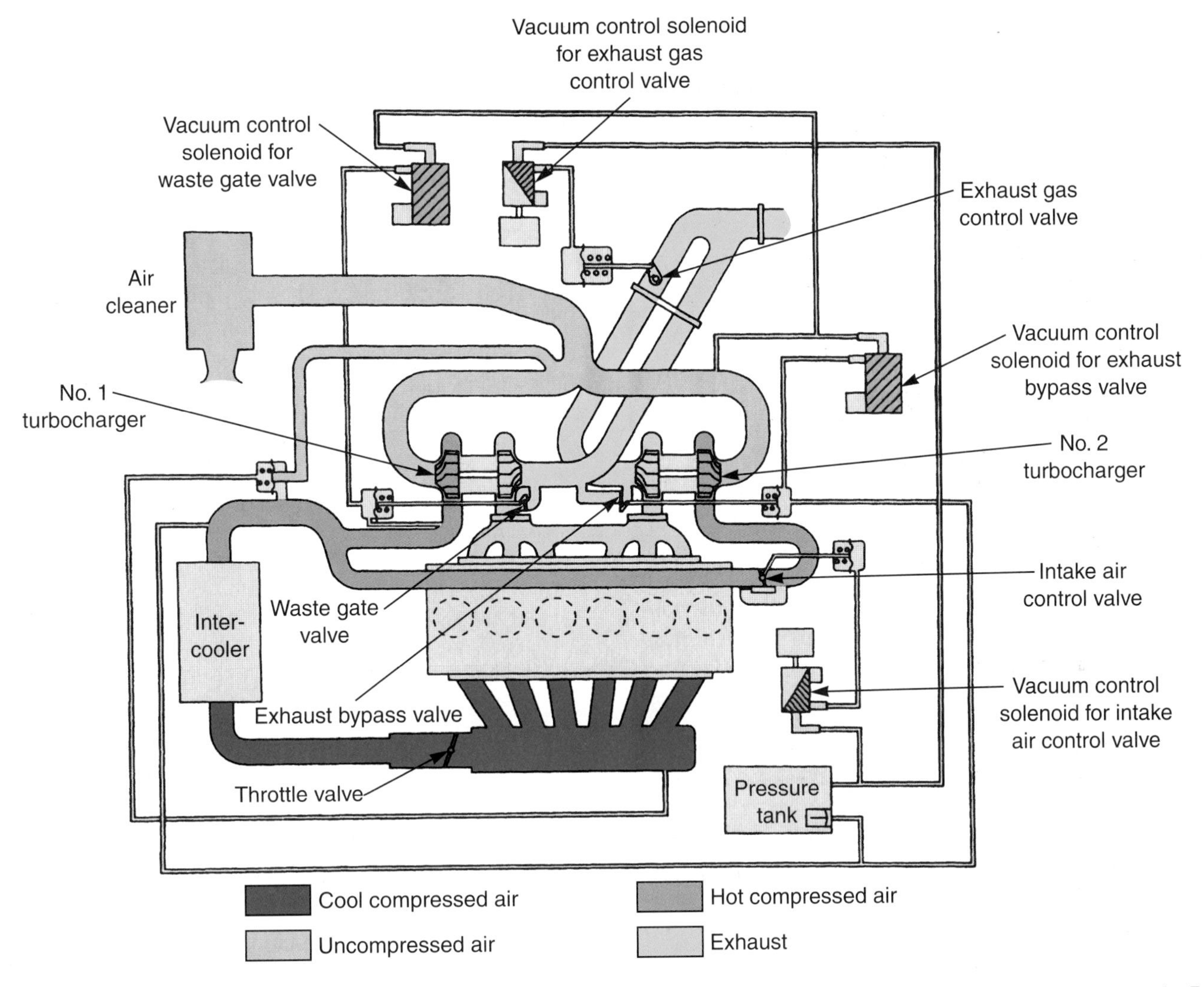

Figure 16-8. *A sequential turbocharger system contains two turbochargers. One turbocharger operates at low engine speeds. Both turbochargers operate at high engine speed. (Toyota)*

gasses are much cooler and will help prevent heat damage as the unit cools. Idling may also prevent oil starvation damage by allowing the turbo to slow down before the engine and oil supply stop.

A ***turbo timer*** is a device that keeps the engine running for a cooldown period after the ignition key has been shut off. It is an electronic circuit that includes the ignition and computer control circuits. When the driver shuts off the ignition, the turbo timer keeps the engine idling to provide time for the engine and turbo to cool off. Then, the circuit automatically shuts the engine off.

Checking the Turbocharging System

There are several checks that can be made to determine the condition of the turbocharging system. These include:

- ❑ Checking all vacuum lines to the waste gate and oil lines to the turbocharger for proper connection.
- ❑ Using regulated, low-pressure air to check for waste gate diaphragm leakage and operation.
- ❑ Using a dash gauge or a test gauge to measure boost pressure. If needed, connect the pressure gauge to an intake manifold fitting.
- ❑ Using a stethoscope to listen for bad turbocharger bearings.

Servicing a Turbocharger

Checking the condition of the inside of the turbocharger requires removing the turbocharger from the engine. Removal procedures may vary by manufacturer, so refer to the service manual for specific instructions. However, the basic steps are:

1. Remove the air cleaner.
2. Remove the turbocharger heat shield.
3. Remove the exhaust outlet pipe and exhaust down pipe.
4. Remove the boost control tube.
5. Remove the oil supply line.

6. Remove the throttle bracket.
7. Remove the waste gate actuator vacuum line.
8. Remove the EGR tube.
9. If necessary, remove the oil dipstick and tube.
10. Disconnect vacuum hoses as needed.
11. Remove the bolts attaching the turbocharger to the intake manifold. Also, remove any braces.
12. Remove the turbocharger from the engine.

After removing the unit, place it on a workbench. Before disassembling the unit, scribe an alignment mark on the housing. This will allow you to reassemble the unit in the same orientation. Then, open the housing and inspect the interior for oil contamination.

Also, check the turbine and compressor wheels. They should be clean and free of damage. Even the slightest imperfection could throw the wheels out of balance and cause severe vibration or wheel disintegration. Make sure the turbine assembly spins freely when rotated by hand. Also, make sure the assembly does not rub on the housing.

Figure 16-9 illustrates how to use a dial indicator to check for radial and axial wear on the turbo bearings. Generally, the radial wear should not exceed .003″ to .006″ (0.08 mm to 0.15 mm). Typical axial wear specifications are between .001″ and .003″ (0.03 mm and 0.08 mm).

Turbocharger Service

To check the turbocharger itself, shut the engine off and let it cool down. Remove the inlet ductwork to the turbocharger and look inside the air intake. If you find damage on the compressor wheel, such as parts that are bent, broken, gouged, marred, or missing, the turbo must be rebuilt or replaced. **Figure 16-10** shows an exploded view of a turbocharger.

Many technicians prefer to replace the entire center section and reuse the housing or replace the complete turbocharger to avoid repair problems. However, minor problems, such as a bad waste gate control diaphragm, leaking seals or hoses, or damaged housings, can be fixed in the shop. The following discussion provides an overview of service. Always refer to the service manual for the exact make and model of engine or turbocharger being serviced.

Push in and pull out on the turbocharger wheel as you spin it by hand. If you feel any bearing drag, binding, or rubbing, the turbocharger bearings and shaft journals are worn. A good turbocharger wheel should spin freely and easily. Turbocharger bearing failure is often the result of improper lubrication caused by oil breakdown or oil contamination. Compressor wheel damage is usually caused by foreign objects entering the turbocharger or worn bearings that allow wheel contact with the housing.

Turbocharger clearances, shaft endplay, and runout should be checked with a dial indicator to verify the condition of the unit. Always refer to the manufacturer's specifications, since clearances vary. Typically, endplay should be .001″ (0.03 mm) to .003″ (0.08 mm) and radial runout should be .003″ (0.08 mm) to .006″ (0.15 mm).

If the turbocharger oil return line becomes restricted or blocked, oil pressure can push oil past the oil seals and into the intake and exhaust tracts. A clogged turbocharger oil return line can cause blue engine smoke, fouled spark

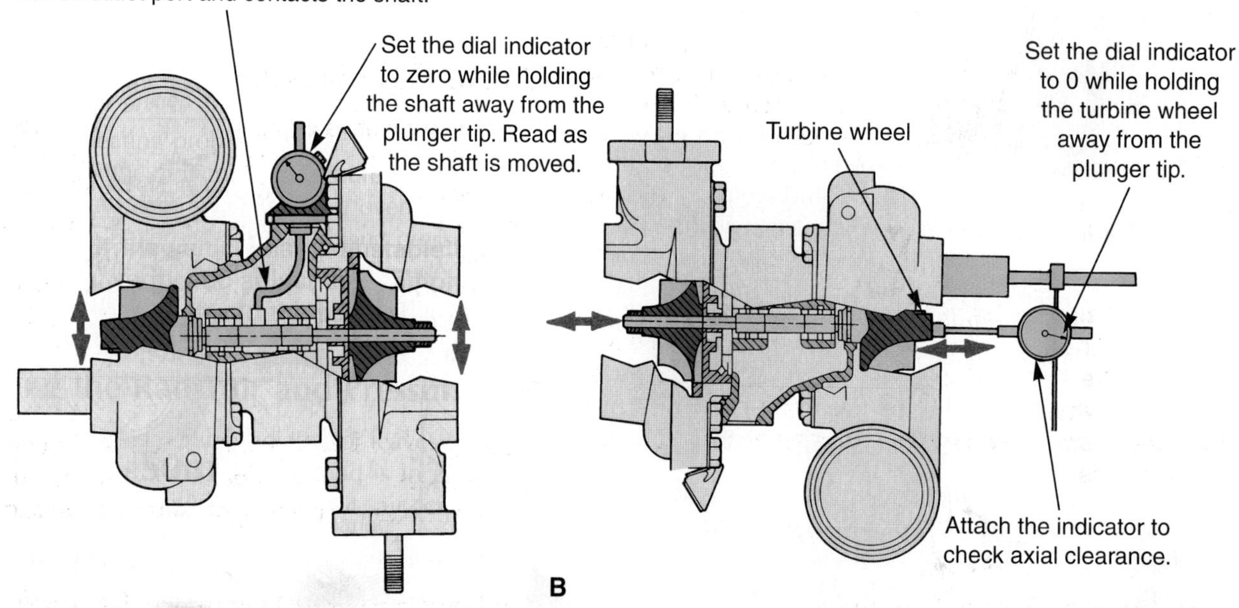

Figure 16-9. *Checking for turbo bearing wear. A—To measure bearing radial wear, mount a dial indicator as shown. Wiggle the wheels and shaft up and down while reading the dial indicator. B—To measure axial clearance, mount the dial indicator on the end of the turbo shaft. Slide the shaft back and forth in the housing while reading the dial indicator. (Ford)*

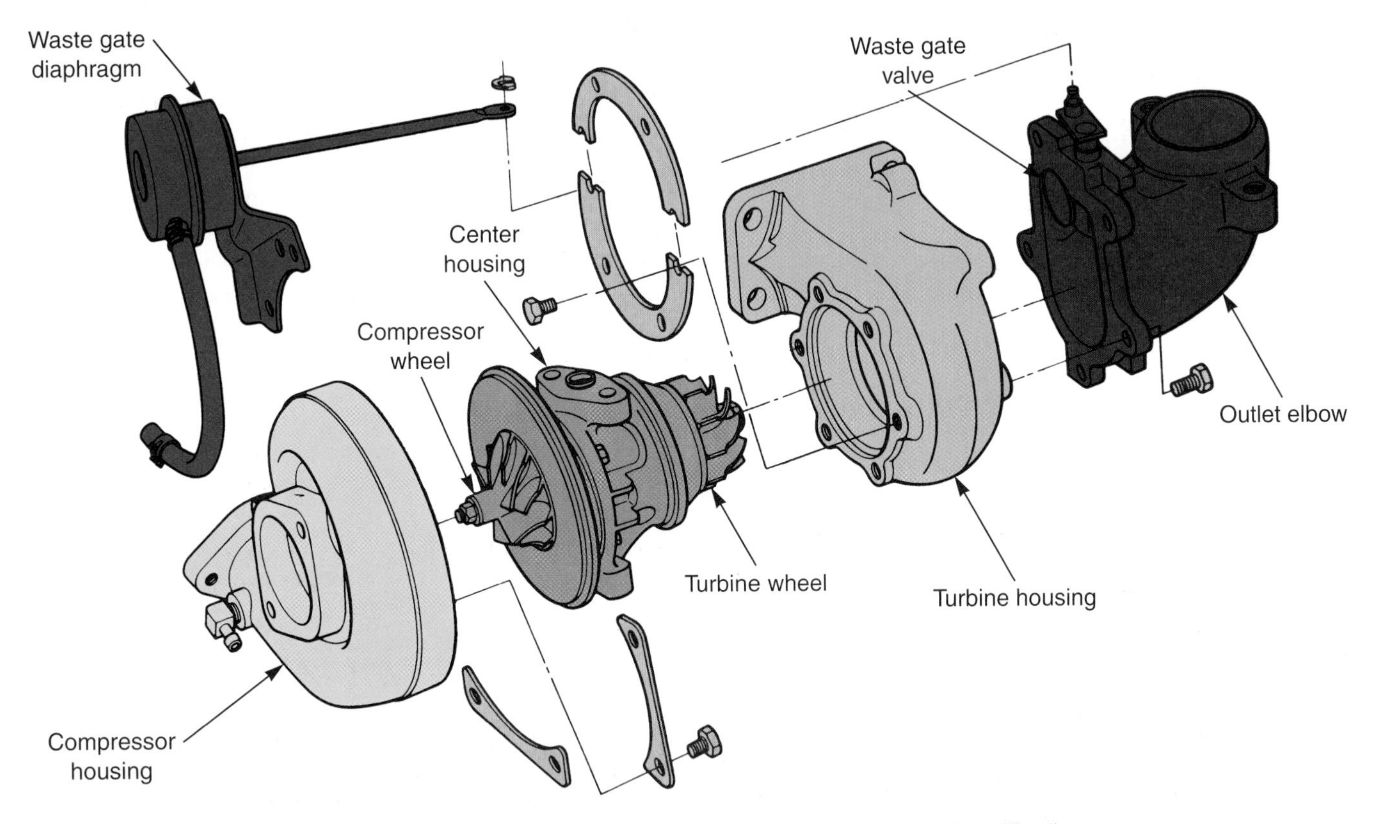

Figure 16-10. *An exploded view of a turbocharger showing how the internal parts fit together. (Ford)*

plugs, or a damaged catalytic converter. To check for blockage, remove the return line at the turbocharger and blow air back through the line. If you find blockage, it is best to replace the line with a new one instead of trying to remove the restriction.

When overhauling a turbocharger, it is very important that you do nothing to upset the balance of the turbocharger wheels. If you bend the turbo shaft loosening or even one blade slightly, the turbocharger will be out of balance and may self-destruct the first time it spins up to boost pressure.

When disassembling the turbocharger housings or wheels, mark their position. Use a scratch awl to make very small alignment marks. When loosening the wheel nut, be careful not to bend the turbo shaft. It is best to use a T-handle and socket to loosen the turbo shaft nuts without side-loading the shaft.

Caution: Never use a hard metal object or sandpaper to remove carbon deposits from the turbo wheels. If you gouge or remove metal, the wheel can vibrate and destroy the turbo. Use only a *soft* wire brush and solvent to clean the turbo wheels.

Installing a Turbocharger

Installing a turbocharger is basically the reverse of removing it. When installing a turbo, you should:

- ❑ Make sure the new turbo is the correct type.
- ❑ Use new gaskets and seals.
- ❑ Torque all fasteners to specifications.
- ❑ Change the engine oil and flush oil lines before starting engine, if needed.
- ❑ Check oil supply pressure in the feed line to the turbo.
- ❑ Check the waste gate for proper operation.

Waste Gate Service

An inoperative waste gate can cause either too much or too little boost pressure. However, always check other parts before diagnosing the waste gate. Check the knock sensor and the ignition timing. Make sure the vacuum and pressure lines are all connected properly.

Follow service manual instructions when testing a waste gate. **Figure 16-11** shows how to check the operation of a waste gate. If the waste gate is not functioning properly, it must be replaced.

Waste gate removal is simple. Remove the fasteners. Remove the lines and lift the unit off of the engine. Then, simply install the new waste gate. Many manuals recommend waste gate replacement rather than in-shop repairs.

Turbocharger Intercooler Service

Intercoolers are very dependable. However, they can be damaged or can deteriorate from extended service.

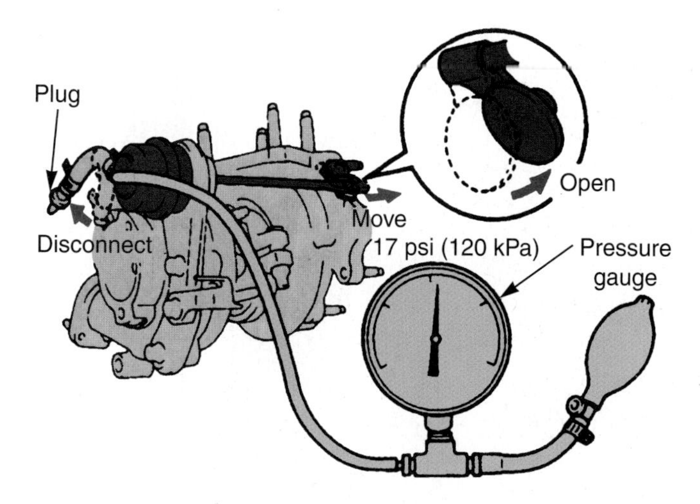

Figure 16-11. *To check the waste gate for proper operation, disconnect the pressure hose from the waste gate. Then, apply the specified pressure to the waste gate. The lever arm and valve should move. (Toyota)*

Intercoolers are relatively easy to service. **Figure 16-12** shows how to replace an air-to-air intercooler on a turbocharged engine. Simply remove covers to gain access to the intercooler. Then, remove hoses and mounting bolts.

Avoid hitting and damaging the intercooler during installation. They are often made of aluminum and can be dented and damaged easily.

Supercharger Construction

A supercharger for a passenger vehicle is a small unit mounted on the top of the engine intake manifold, **Figure 16-13.** It is usually belt driven off of the engine crankshaft pulley. Some superchargers have an electromagnetic clutch, similar to the clutch on an air conditioning compressor. This allows the supercharger to be turned on and off. However, most superchargers are direct drive without an electromagnetic clutch. A supercharger normally consists of:

- ❑ Housing.
- ❑ Rotors.
- ❑ Timing gears.
- ❑ Extension housing.
- ❑ Drive pulley.

The ***supercharger housing*** is the large aluminum enclosure for the rotors, rotor bearings, and drive gears. The inside of the housing is precision machined to form airflow chambers of equal size. The supercharger housing is normally bolted to the intake manifold. The throttle body is mounted at the inlet of the supercharger, as shown in **Figure 16-13.**

The ***supercharger rotors*** are meshing, twisted blades or fans that spin to produce pressure inside the supercharger housing. The rotors fit precisely in the supercharger housing and are supported by sealed ball or needle bearings. Refer to **Figure 16-14.**

Supercharger timing gears keep the two rotors timed and spinning at the same speed. The timing gears are attached to the rotors and are enclosed in the gear case, which is formed as part of the supercharger housing. One timing gear is driven by the drive pulley; the other is driven by the first timing gear. The rotor gears operate in a bath of oil. A small reservoir is formed into the front of the supercharger housing. As the timing gears spin, they are splash lubricated to prevent wear on the gear teeth.

The ***supercharger extension housing*** is bolted to the front of the supercharger housing. It encloses a shaft that connects one of the supercharger timing gears to the drive pulley. It also contains a large, sealed bearing for the drive pulley. See **Figure 16-14.**

The ***supercharger drive pulley*** is connected to one of the timing gears by the shaft, which is inside the supercharger extension housing. The drive pulley draws power from the crankshaft through a large serpentine belt.

Supercharger Bypass Valve

A ***supercharger bypass valve*** is used to improve engine efficiency by only providing maximum boost when needed. **Figure 16-15A** shows a schematic for a typical supercharger bypass circuit.

Under normal cruising conditions (low load), the bypass valve is held open to allow boost pressure to flow back up into the intake air. See **Figure 16-15B.** This reduces boost pressure and parasitic pumping losses to improve fuel economy.

When the driver presses on the accelerator pedal for full power and rapid acceleration, the bypass valve closes. All boost pressure flows into the engine. See **Figure 16-15C.**

Supercharger Operation

When the accelerator pedal is pushed to the floor for passing or getting up to highway speed, the supercharger spins faster to provide more boost. If equipped with an electromagnetic clutch, a switch sends power to the electromagnetic clutch on the supercharger pulley. This engages the supercharger and the rotors spin. Engine power is instantly increased.

Most superchargers used on production vehicles provide 4 psi to 10 psi (30 kPa to 70 kPa) of boost. Larger units that provide much more boost are used on racing engines and diesel trucks. The supercharger normally uses two rotors that compress air entering the engine. This pressurized air is fed into the intake manifold where it is mixed with fuel before being pulled into the combustion chamber.

Supercharging has the advantage of instant air pressure increase on demand. With a turbocharger, there is a slight delay in pressure and power increase (turbo lag) as the unit builds up speed. This delay does not occur with a supercharger since it is driven by the crankshaft.

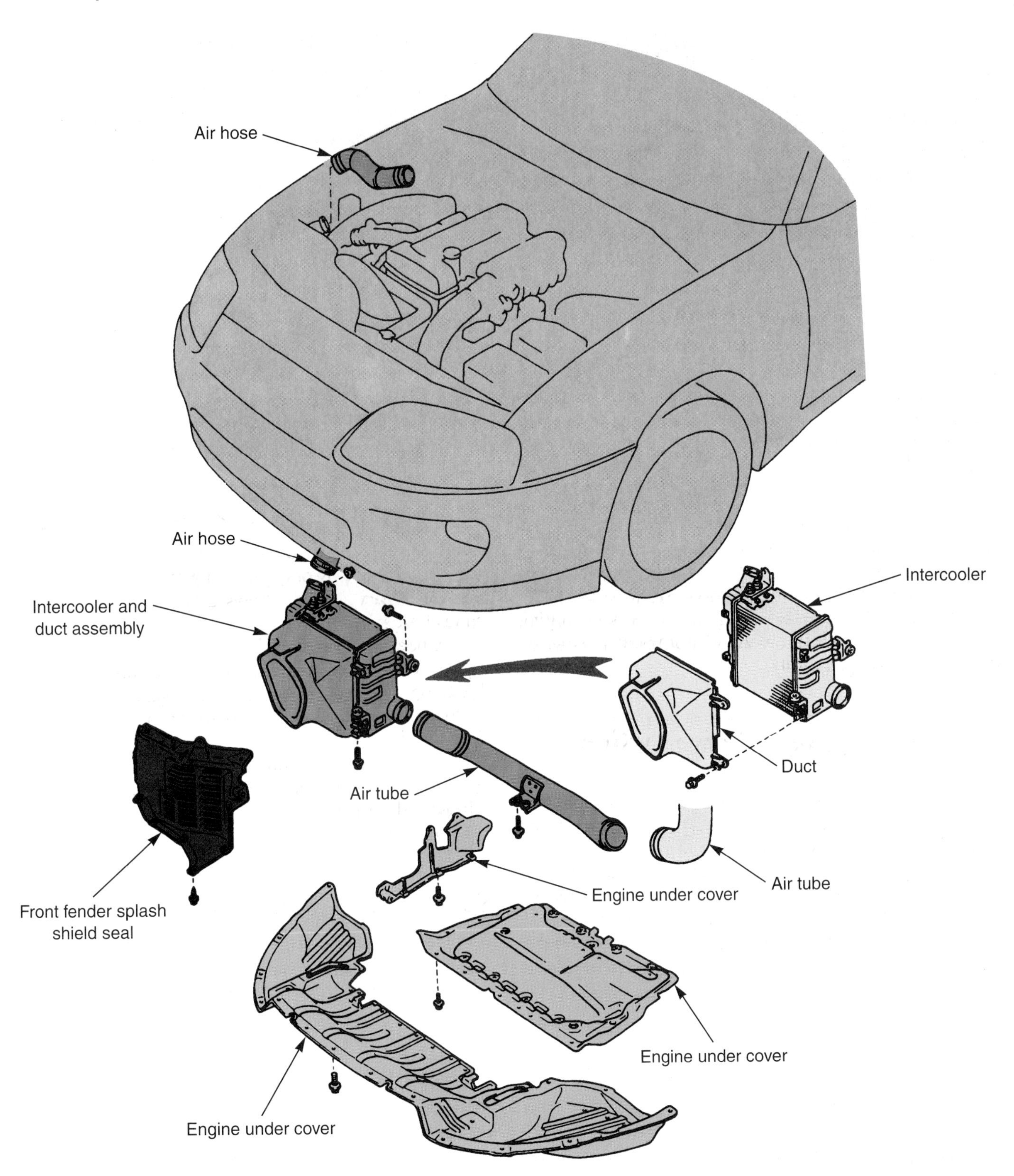

Figure 16-12. *This exploded view shows how a turbocharger intercooler is removed. (Toyota)*

Supercharger Intercooler

When the supercharger compresses the air and produces boost, the air is heated. This heat must be removed to avoid detonation. Detonation can damage pistons, crack cylinder heads, or blow out spark plugs. A ***supercharger intercooler*** removes heat from the compressed air before it enters the engine. See **Figure 16-16.** The intercooler provides the engine with a cooler, more dense charge of air than if the supercharged engine did not have an intercooler. A supercharger intercooler system consists of:

- Intercooler.
- Intercooler coolant.

Figure 16-13. *A supercharger is normally mounted on top of the intake manifold. Note the basic parts of a supercharger. (Land Rover)*

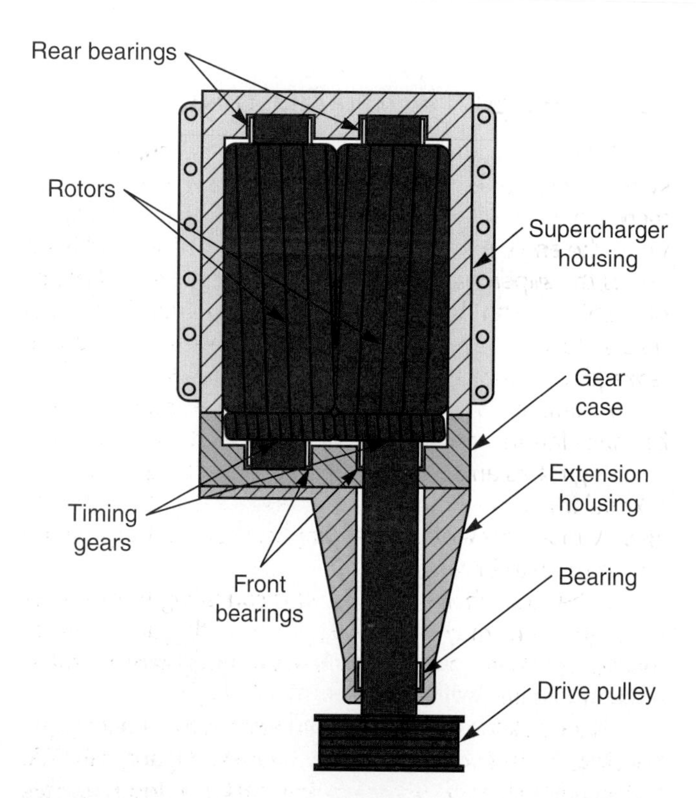

Figure 16-14. *This is a top cutaway view of a supercharger. Note the parts of the supercharger.*

- ❑ Intercooler lines.
- ❑ Intercooler pump.
- ❑ Intercooler radiator.

The ***intercooler*** is an air-to-liquid heat exchanger mounted between the supercharger and the engine. It consists of cooling tubes surrounded by cooling fins, similar to an engine radiator. The hot boost air leaving the supercharger flows through the intercooler fins and over the cooling tubes. Since the temperature of the coolant in the intercooler is lower than the hot boost air, heat is transferred from the air to the fins, cooling tubes, and into the coolant. The coolant carries this heat away from the intercooler.

The ***intercooler coolant*** is a mixture of 50% water and 50% antifreeze. A small reservoir is normally provided for servicing the coolant.

The ***intercooler radiator*** is a small heat exchanger for transferring heat from the intercooler coolant to the outside air. It is often mounted in the lower, front of the engine compartment. The intercooler radiator is positioned so that cool outside air can flow through the radiator fins.

Intercooler lines carry the coolant from the intercooler to the intercooler radiator and back to the intercooler. The lines can be metal or rubber. An ***intercooler pump*** forces the coolant through the system. This is a small electric water pump that keeps the coolant circulating through the cooling tubes and intercooler radiator.

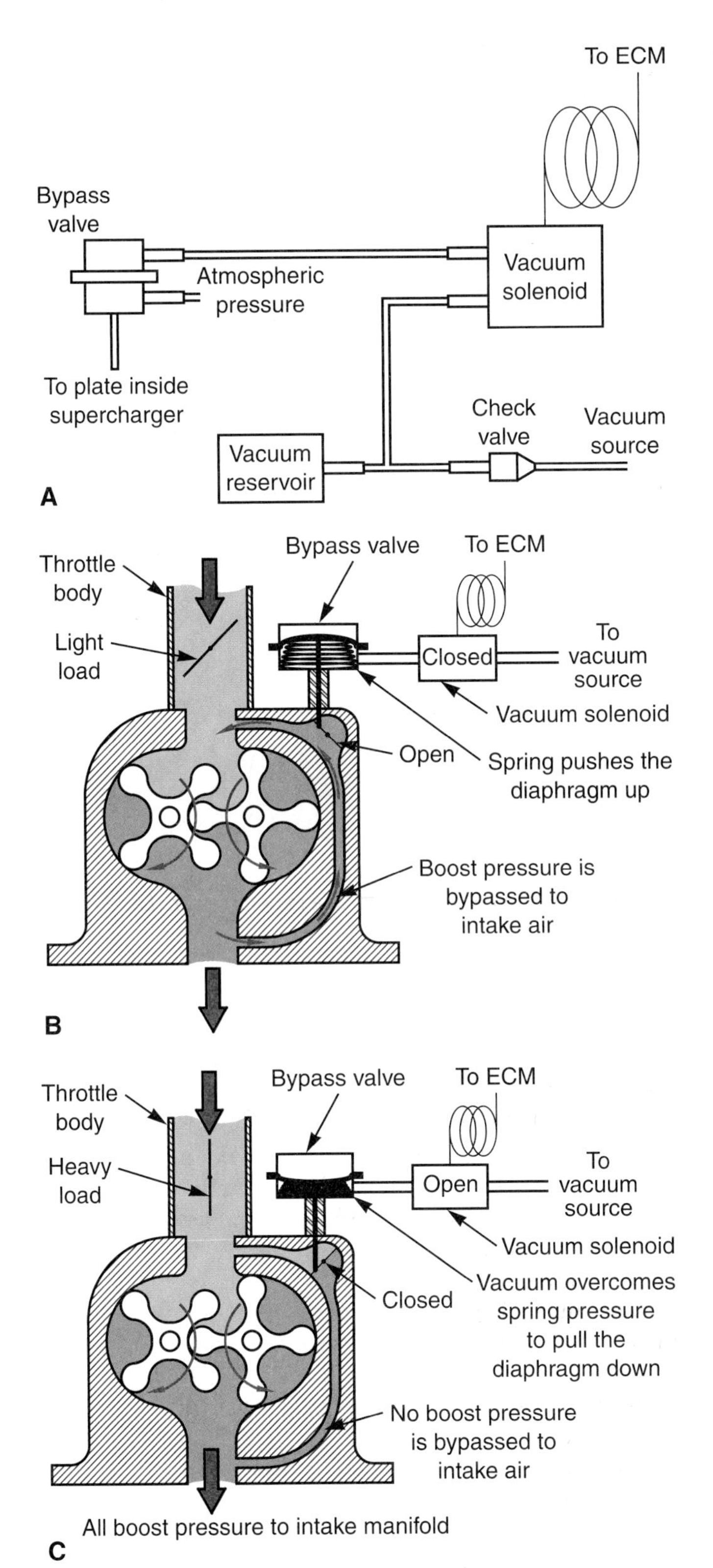

Figure 16-15. *A—This is a schematic for a typical supercharger bypass valve control system. B—When the engine is under full load, the ECM sends a control signal to open the vacuum solenoid. Vacuum pulls the diaphragm down, which closes the bypass valve. All supercharger boost pressure enters the intake manifold. C—When the engine is under light load, the ECM closes the vacuum solenoid. Spring pressure pushes the diaphragm up, which opens the bypass valve. Some of the supercharger boost pressure is returned to the inlet, reducing the amount of boost pressure that enters the intake manifold.*

Supercharger System Service

Many of the problems and tests associated with turbochargers relates to superchargers as well. Service may be required after extended use. Usually, a supercharger will fail because of worn bearings or damage to the impeller and housing.

Some supercharger noise or whine is normal. When a supercharged vehicle is accelerated, a loud whine or "scream" can usually be heard. This is the sound of the large supercharger rotors pulling air into the engine under full boost.

The most common supercharger service operations include replacement of the drive belt, checking supercharger oil, and supercharger replacement. Supercharger service can also involve tests and repairs to the intercooler, intercooler pump, bypass valve mechanism, and other support components.

Caution: Owners sometimes install a smaller drive pulley to increase rotor speed and boost pressure. This can result in supercharger or engine failure. Boost pressure that is higher than the factory specifications can cause severe detonation, which can result in cracked heads, broken piston crowns or skirts, spark plugs blown out of the head, and other major damage. Warn customers that modifications to a supercharged engine can void the manufacturer's warranty.

Supercharger System Diagnosis

When diagnosing problems with the supercharger system, inspect the unit for obvious problems first. Are there signs of drive belt slippage? Does the belt squeal when the engine is started? Are there any visible fluid leaks and is the supercharger oil reservoir full? Other symptoms of problems with the supercharger system include internal noises, low oil level in the reservoir, pressure or vacuum leakage, or out-of-specification boost pressure.

A rattling noise from the supercharger indicates rotor bearing failure and mechanical contact between the spinning rotors and the inside of the supercharger housing. A knocking noise indicates possible drive gear teeth breakage. A loud growling noise may indicate either bearing failure or teeth breakage.

If the supercharger drive belt is squealing from excess slippage, there may be internal drag or the unit may be locked up. This could indicate bearing seizure or other internal mechanical failure.

Place a stethoscope against the ends of the supercharger housing to isolate any bearing noises, **Figure 16-17A.** Failed supercharger bearings will make a loud rumble when the probe is placed over the bearings on the housing. If the unit is not too hot, you can also place your hand on

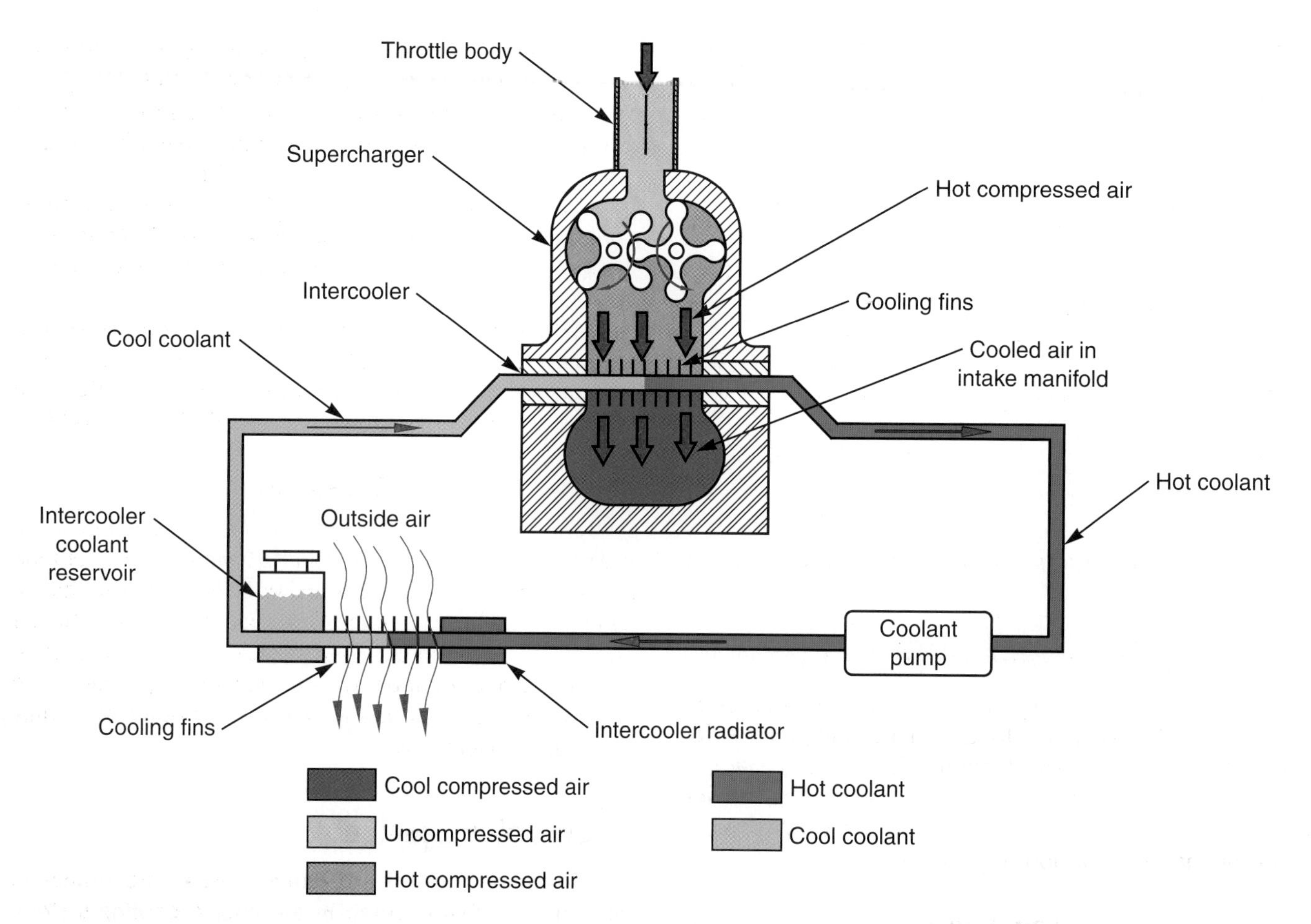

Figure 16-16. *Superchargers often have an air-to-liquid intercooler. The intercooler is a heat exchanger mounted on the outlet of the supercharger. The intercooler system contains a coolant that is circulated through the intercooler, where it picks up heat from the compressed air. The hot coolant is then circulated through the intercooler radiator, where heat is dissipated.*

the supercharger housing to feel for vibration caused by rotor-to-housing contact, **Figure 16-17B.**

A close visual inspection and partial disassembly are often needed to verify supercharger failures. Quite often, new seals or gaskets can be installed to correct pressure or vacuum leakage. However, if major mechanical problems are found in the supercharger, the unit should be replaced. For example, a supercharger explosion, resulting from a backfire or rotor failure, can crack the housing and allow oil leakage from the unit. In this case, the unit should be replaced.

Figure 16-17. *A—Use a stethoscope to check for unusual noises inside of the supercharger. B—If the supercharger is not hot, place your hand on the case to feel for vibration.*

Figure 16-18. *Using a pressure gauge to measure boost pressure.*

Measuring Supercharger Boost Pressure

Install a pressure gauge to measure actual boost pressure, **Figure 16-18.** Refer to the manufacturer's service information for the correct connection point. Compare actual boost pressure to specifications for the engine speed. If boost is low, check the pressure regulator circuit or for internal rotor or housing damage.

Low Boost

When a supercharged engine fails to generate proper boost pressure, it is often due to:

- ❑ A broken or slipping drive belt.
- ❑ An inoperative electromagnetic drive pulley mechanism.
- ❑ An inoperative pressure bypass valve system.
- ❑ Physical wear or damage inside the supercharger.
- ❑ A badly clogged air filter or an obstruction in the air intake.

Figure 16-19. *Use a hand-held vacuum pump to check for proper action of the bypass diaphragm.*

If there is low or no boost pressure, make sure the electromagnetic clutch is engaging if the vehicle is equipped with one. Also, make sure the supercharger is spinning. If the supercharger is spinning, check for loose or disconnected vacuum lines to the bypass valve.

To check the operation of the bypass valve, apply vacuum to the diaphragm. See **Figure 16-19.** Make sure the level or arm on the bypass valve moves as vacuum is applied.

If the bypass valve diaphragm is good but the bypass valve is not working, check the vacuum solenoid that controls vacuum to the bypass diaphragm. Make sure the ECM is sending a proper signal to the solenoid, **Figure 16-20.**

Overboost

If a supercharged engine has too much boost, or ***overboost,*** the engine may seem to have more power than normal, accelerate (rev up) too quickly, and detonate or knock under high-load conditions. Too much supercharger boost can be caused by a bypass valve stuck closed, often due to carbon buildup or a mechanical problem. An electronic control problem or engine modification may also lead to overboost.

Intercooler Service

Problems with the intercooler system are similar to the problems that develop in the engine cooling system. Some possible problems include an inoperative pump, deteriorated coolant, leaking intercooler radiator or intercooler, and leaking hoses or lines.

During service, check the level and condition of the coolant in the intercooler reservoir. This reservoir is usually located on the side of the intercooler radiator or engine compartment. See **Figure 16-21A.**

Figure 16-20. *Checking for a proper electrical signal to the vacuum solenoid. When the accelerator pedal is pushed to the floor, battery voltage should be present at the vacuum solenoid.*

Figure 16-21. *A—Check the level and condition of the intercooler coolant. B—Pressure testing the intercooler system.*

Warning: Never open a intercooler system that is under pressure. The operating temperature of intercooler coolant is hot enough to cause severe burns. Remove the intercooler radiator cap slowly to verify that the system is not under pressure.

If the intercooler coolant is low, there may be a system leak. Pressure test the system as you would the engine cooling system, **Figure 16-21B.** Refer to Chapter 12 for more information. A special adapter may be needed to install the pressure tester on the intercooler reservoir. Do not exceed the pressure rating given on the reservoir cap. If the pressure is exceeded, the intercooler, intercooler radiator, or intercooler coolant lines may be damaged.

With the intercooler system pressurized, check for leaks at the lines or hoses, pump, intercooler radiator, and intake manifold. If the pressure gauge "bleeds down" and the engine misses or white smoke comes out of the tailpipe, the intercooler unit is the likely source of the leak. The coolant is leaking into the intake manifold and being burned in the cylinders. In this case, the intercooler must be removed and repaired or replaced. Intercooler replacement is covered later in this chapter.

Supercharger Oil Service

To service the oil in the supercharger, remove the oil plug or dipstick, **Figure 16-22A.** With the vehicle level, the oil should be even with the threaded hole in the oil reservoir or the correct level on the dipstick, depending on the design. See **Figure 16-22B.** Refer to the service manual for the proper procedure for checking the oil level. If needed, use a flashlight to illuminate inside the hole. A small length of wire can also be used as a dipstick.

To replace the supercharger oil, use a hand pump as shown in **Figure 16-22C** to remove the existing oil. Fill the reservoir with the factory recommended oil. Then, replace the oil plug, **Figure 16-22D.**

Supercharger Belt Replacement

To replace a supercharger belt, first remove the belt cover, **Figure 16-23.** Usually, two or three small bolts secure the belt cover. To avoid injury, disconnect the battery to prevent accidental engine cranking.

A long wrench or special service tool is often needed to compress the tension spring so that the belt can be slid off of the supercharger drive pulley. This is shown in **Figure 16-24A.** With the belt off, you can rotate the supercharger pulley by hand to check for dry bearings, rotor-to-housing contact, or supercharger timing gear failure. See **Figure 16-24B.** Any roughness or binding when you turn the supercharger pulley by hand indicates mechanical problems.

Supercharger Clutch Service

Some superchargers are equipped with an electromagnetic clutch for engaging and disengaging the supercharger. See **Figure 16-25.** This clutch works just like an air conditioning compressor clutch. When current is sent to the clutch coil, the clutch plates are pulled together to lock the inner and outer hubs. The engine belt drives the outer hub. The inner hub is connected to the turbocharger shaft. In this way, the supercharger is engaged only when high boost is needed to increase engine power.

If the electromagnetic clutch fails to engage, first check for a bad clutch coil. Disconnect the clutch wiring harness and apply battery voltage to the coil using jumper wires. This should engage the clutch. A loud click should be heard as the clutch engages. You can also use a multimeter to check for an open in the coil windings. The coil should have a low resistance. Infinite resistance indicates an open in the windings. If the clutch engages when jumpered, check for the proper signal from the ECM.

Figure 16-22. *Servicing the supercharger oil. A—Remove the oil plug if the unit does not have a dipstick. B—Check the oil level in the oil plug hole or on the dipstick. C—Use a hand held pump to remove the old oil. Then, fill the supercharger with new oil. D—Replace the oil plug and tighten to specifications.*

Figure 16-23. *When servicing the supercharger belt, first remove all fasteners that secure the cover.*

Figure 16-24. *A—To release tension on the supercharger belt, use a long wrench or prybar. B—Once the belt is removed, turn the drive pulley by hand to check for binding.*

Supercharger Replacement

If supercharger removal is needed, follow the instructions given in the service manual. Procedures vary based on engine and supercharger design. Generally, first remove any parts that will interfere with supercharger removal. This includes the air plenum, drive belt, wires, hoses, lines, and throttle cable or linkage, **Figure 16-26.**

Next, remove the bolts that secure the supercharger to the engine, **Figure 16-27.** Free the gasket with light prying. Then, lift the unit off of the engine.

Superchargers are not normally serviced in the field and are simply replaced as a unit. However, seal kits are available that typically include the gaskets, oil seals, and special oil for the supercharger. These items are easily serviced.

To install the supercharger, use a new base gasket and apply sealer as recommended in the service manual, **Figure 16-28A.** Then, slowly lower the supercharger onto the engine without damaging the gasket or breaking the sealant bead. Start all bolts by hand. Then, use a torque wrench to tighten the bolts in the recommended torque sequence, **Figure 16-28B.** Reinstall any accessories and fill the supercharger with the correct oil.

Finally, reinstall the drive belt and related parts. Start and warm the engine. Test drive the vehicle to make sure you have corrected the problem. Make sure the engine has normal power, does not have abnormal noises, and does not smoke.

Intercooler Replacement

The intercooler on most supercharged V-6 and V-8 engines is mounted under the supercharger. Once the supercharger is removed, you can easily access the bolts securing the intercooler. Remove the bolts and separate the intercooler from the engine. On some designs, the intercooler is secured to top of the intake manifold. On other designs, it is secured to the bottom of the supercharger.

Once the intercooler is removed, inspect it for any signs of coolant leakage. If needed, send the intercooler to

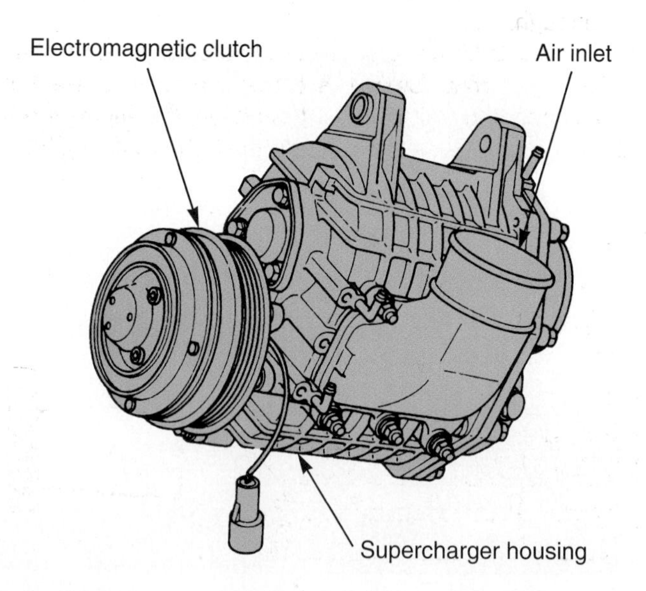

Figure 16-25. *Some superchargers are equipped with a electromagnetic clutch that allows the supercharger to be engaged and disengaged. (Toyota)*

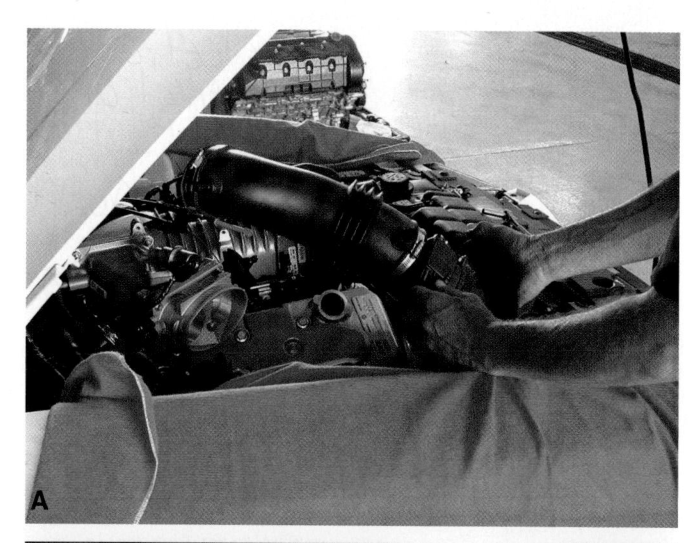

Figure 16-26. *To remove a supercharger, first disconnect all parts attached to the unit. A—Air plenum. B—Throttle cable. C—Wiring.*

Figure 16-27. *Remove all of the bolts that secure the supercharger to the engine.*

a radiator shop for pressure testing. A radiator shop can also flush the intercooler to remove scale or deposits in the cooling tubes.

When reinstalling the intercooler, carefully clean all gasket surfaces. Apply sealer as recommended and fit the new gaskets. Without hitting and damaging any of the fins, carefully lower the intercooler into place. Torque the intercooler fasteners to specifications in the specified pattern, **Figure 16-29.**

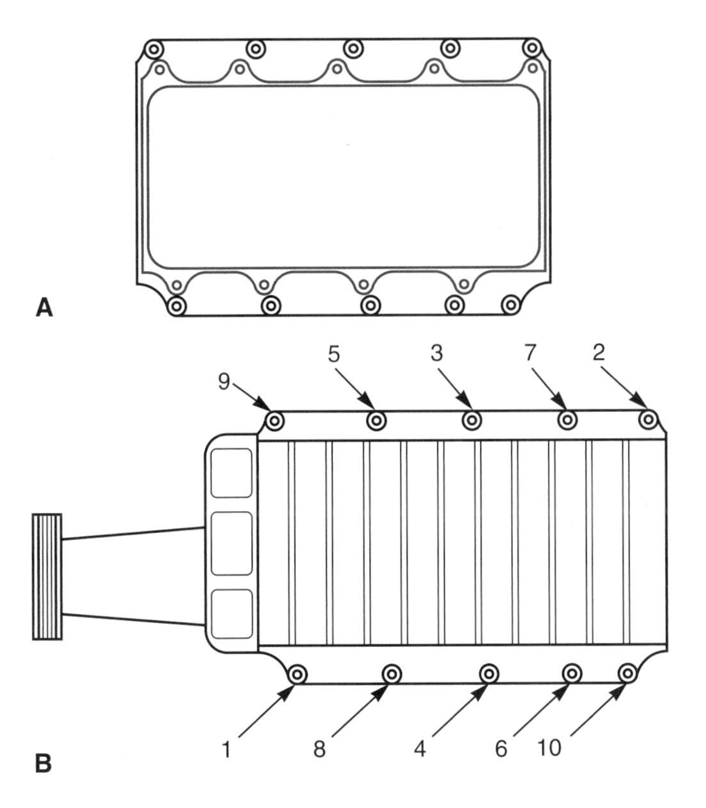

Figure 16-28. *Reinstalling a supercharger. A—Carefully clean the mating surfaces and install a new gasket. Use sealant as recommended. B—Torque the fasteners in the recommended sequence.*

Warning: Never open the intercooler cooling system when it is under pressure. The operating temperature of the intercooler coolant is hot enough to cause severe burns. Remove the intercooler radiator cap slowly to verify that the system is not under pressure.

Summary

Normal aspiration means the engine uses only outside air pressure to cause airflow into the combustion chambers. With only outside air pressure to carry oxygen into the engine cylinders, engine power is limited by the engine's volumetric efficiency. Turbocharging and supercharging pressurizes the incoming air. This has the effect of increasing the volumetric efficiency.

Turbocharged and supercharged engines normally have several modifications to make them withstand the increased horsepower. A few of these modifications include lower compression ratio; stronger rods, pistons, crankshaft, and bearings; higher-volume oil pump; larger radiator; O-ring–type head gasket; valves with increased heat resistance; and an oil cooler.

The major parts of a turbocharger are the turbine housing, turbine wheel, turbo shaft, compressor wheel, compressor housing, bearing housing. During engine operation, hot exhaust gasses spin the turbine wheel. The compressor wheel, which is connected to the turbine wheel by the turbo shaft, spins to compress the incoming air in the intake manifold.

Turbo lag is a short delay period before the turbo develops sufficient boost. Boost is any pressure above atmospheric pressure in the intake manifold. Turbo lag is caused by the compressor and turbine wheels not spinning fast enough.

Adequate lubrication is needed to protect the turbo shaft and bearings from damage. A turbocharger can operate at speeds up to 100,000 rpm. For this reason, the engine lubrication system forces engine oil into the turbo shaft bearings.

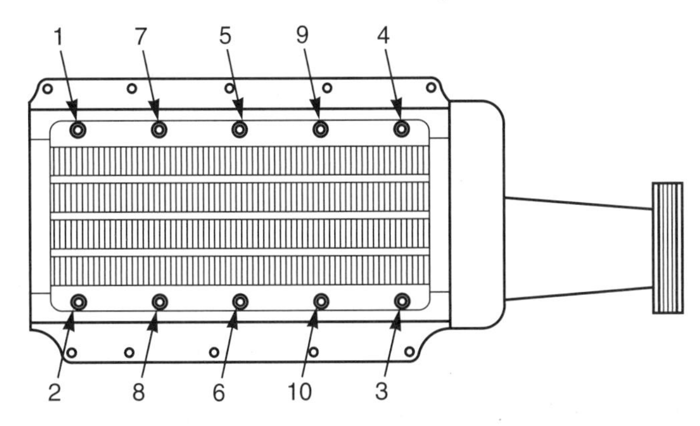

Figure 16-29. *When reinstalling an intercooler, torque the fasteners in the recommended sequence.*

A waste gate limits the maximum boost pressure developed by the turbocharger. Without a waste gate, the turbo could produce too much pressure in the combustion chambers.

A turbocharger intercooler is an air-to-air or air-to-liquid heat exchanger that cools the air entering the engine. It is a radiator-like device mounted at the pressure outlet of the turbocharger.

A supercharger normally consists of the housing, rotors, timing gears, extension housing, and drive pulley. A supercharger bypass valve may also be used to improve engine efficiency by only providing maximum boost when needed.

Most superchargers are direct drive. However, some have an electromagnetic clutch that allows the supercharger to be engaged and disengaged.

When the accelerator pedal is pushed to the floor for passing or getting up to highway speed, the supercharger spins faster to provide more boost. Engine power is instantly increased. Most superchargers used on production vehicles provide 4 psi to 10 psi (30 kPa to 70 kPa) of boost.

A supercharger intercooler removes heat from the compressed air before it enters the engine. A supercharger intercooler system consists of the intercooler, intercooler coolant, intercooler lines, intercooler pump, and intercooler radiator.

Supercharger service may be required after extended use. Usually, a supercharger will fail because of worn bearings or damage to the impeller and housing.

When diagnosing problems with the supercharger system, inspect the unit for obvious problems first. Some supercharger noise or whine is normal. A rattling noise from the supercharger indicates rotor bearing failure and mechanical contact between the spinning rotors and the inside of the supercharger housing. A knocking noise indicates possible drive gear teeth breakage.

When a supercharged engine fails to generate proper boost pressure, it is often due to a broken or slipping drive belt, an inoperative electromagnetic drive pulley mechanism, an inoperative pressure bypass valve system, physical wear or damage inside the supercharger, or a badly clogged air filter or an obstruction in the air intake. Overboost can be caused by a bypass valve stuck closed, electronic control problem, or engine modification.

Problems with the intercooler system are similar to the problems that develop in the engine cooling system. During service, check the level and condition of the coolant in the intercooler reservoir. Pressure test the system as you would the engine cooling system.

Superchargers are not normally serviced in the field and are simply replaced as a unit. However, seal kits are available that typically include the gaskets, oil seals, and special oil for the supercharger. The belt is also a normal wear item that requires periodic replacement.

Review Questions—Chapter 16

Please do not write in this text. Write your answers on a separate sheet of paper.

1. Define *normal aspiration.*
2. What is the basic purpose of turbocharging or supercharging?
3. List four engine modifications normally found on turbocharged or supercharged engines.
4. List the six parts of a turbocharger.
5. Describe the operation of a turbocharger.
6. _____ is the short delay before the turbo develops sufficient boost.
7. What is the function of a *waste gate?*
8. Why is an *intercooler* needed on turbocharged and supercharged engines?
9. Describe the basic operation of a *sequential turbocharger system.*
10. Minor problems, such as a bad waste gate control diaphragm, leaking seals or hoses, or damaged housings, can be done in the shop. Otherwise, the turbocharger must be _____.
11. List the five basic parts of a supercharger.
12. What is the purpose of a supercharger bypass valve?
13. A supercharger intercooler:
    (A) cools the air coming into the air inlet.
    (B) is typically an air-to-air heat exchanger.
    (C) cools the air coming out of the supercharger.
    (D) results in a cooler, less dense charge of air entering the intake manifold.
14. List four problems that may be indicated by low supercharger boost pressure.
15. List three problems that may result in overboost.

ASE-Type Questions—Chapter 16

1. A turbocharged vehicle has a severe detonation problem. Engine power is good and initial ignition timing is within specifications. Technician A says that the trouble could be in the knock sensor. Technician B says waste gate could be stuck closed. Who is correct?
   (A) Technician A.
   (B) Technician B.
   (C) Both A and B.
   (D) Neither A nor B.
2. Technician A says that a normally aspirated engine is equipped with a supercharger. Technician B says that a normally aspirated engine is equipped with a turbocharger. Who is correct?
   (A) A only.
   (B) B only.
   (C) Both A and B.
   (D) Neither A nor B.
3. All of the following are basic components of a turbocharger *except:*
   (A) bearing housing.
   (B) turbine wheel.
   (C) transfer pump.
   (D) turbo shaft.
4. Technician A says that with a supercharged engine, there is a slight delay in pressure and power increase because the unit must build up speed. Technician B says that a supercharger is driven by a belt connected to the crankshaft pulley. Who is correct?
   (A) A only.
   (B) B only.
   (C) Both A and B.
   (D) Neither A nor B.
5. Technician A says that volumetric efficiency is a measure of how much air an engine can draw in on its power strokes. Technician B says that a normally aspirated engine is limited by its volumetric efficiency. Who is correct?
   (A) A only.
   (B) B only.
   (C) Both A and B.
   (D) Neither A nor B.
6. Technician A says that boost refers to any pressure above atmospheric pressure in an engine's combustion chambers. Technician B says that boost refers to any pressure above atmospheric pressure in an engine's intake manifold. Who is correct?
   (A) A only.
   (B) B only.
   (C) Both A and B.
   (D) Neither A nor B.
7. A loud growling noise can be heard coming from the supercharger. Technician A says that the noise may indicate rotor bearing failure. Technician B says that the noise may indicate drive gear teeth breakage. Who is correct?
   (A) A only.
   (B) B only.
   (C) Both A and B.
   (D) Neither A nor B.
8. The boost on a supercharger is measured and it is found that the supercharger is producing no boost. Technician A says that the ECM may not be sending a signal to engage the electromagnetic clutch. Technician B says that the supercharger bypass valve may be stuck closed. Who is correct?
   (A) A only.
   (B) B only.
   (C) Both A and B.
   (D) Neither A nor B.
9. As a supercharger intercooler system is pressure tested, the engine begins to run rough and white smoke can be seen coming out of the tailpipe. Technician A says that the cylinder head may be cracked. Technician B says that the problem may be caused by bent cooling fins on the intercooler. Who is correct?
   (A) A only.
   (B) B only.
   (C) Both A and B.
   (D) Neither A nor B.
10. The electromagnetic clutch on a supercharger is not engaging. Technician A says to check for the proper signal from the ECM. Technician B says that the clutch coil should have high resistance in its windings. Who is correct?
   (A) A only.
   (B) B only.
   (C) Both A and B.
   (D) Neither A nor B.

An engine analyzer or scan tool is an important tool for diagnosing engine performance problems. (OTC Div. of SPX Corp.)

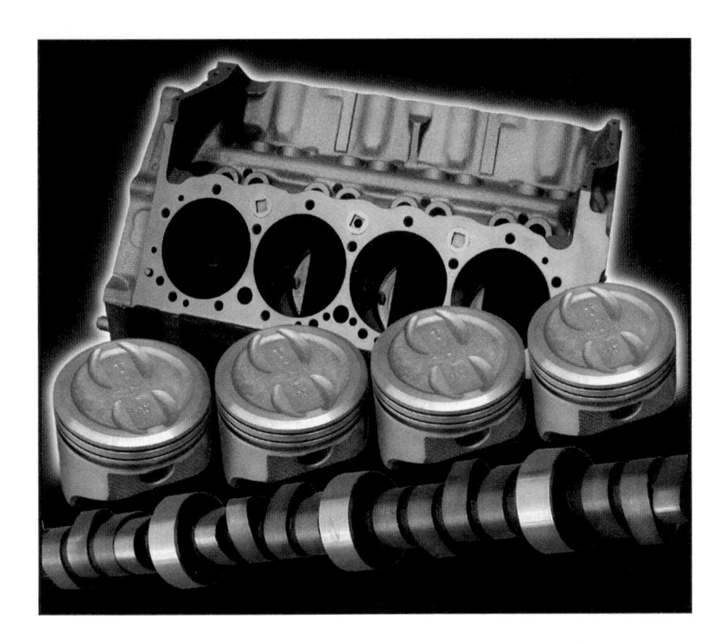

Chapter 17

Engine Performance Problems

After studying this chapter, you will be able to:

- ❑ Use a performance problem diagnosis chart.
- ❑ Read spark plugs.
- ❑ Identify and use various tools used to diagnose engine performance problems.
- ❑ Summarize typical engine performance problems and their causes.
- ❑ Use a systematic approach to diagnose engine performance problems.

Know These Terms

After-running
Backfiring
Dieseling
Digital multimeter
Engine analyzer
Engine performance problem
Fuel pressure gauge
Gas line freeze
Hard starting
Hesitation
Misfiring
No-start problem
Performance problem diagnosis chart
Pinging
Poor fuel economy
Reading spark plugs
Rough idle
Run-on
Scan tool
Sluggish engine
Stalling
Stumble
Surging
Systematic approach
Technical service bulletins (TSBs)
Test light
Vacuum gauge
Vapor lock

An ***engine performance problem*** is any trouble that lowers engine power, fuel economy, drivability (performance), or dependability. Performance problems can result from troubles in the fuel, ignition, emission control systems, or in the engine. This makes troubleshooting very challenging. You must use your knowledge of system and engine operation and basic testing methods to quickly locate and correct the source of engine performance problems.

This chapter summarizes the symptoms, causes, and corrective measures for typical performance problems. As a result, you will be better prepared to become employed as an engine technician.

Locating Engine Performance Problems

Use a systematic approach when trying to locate performance problems. Do not use "hit-and-miss" repairs. A ***systematic approach*** involves using your knowledge of auto technology and a logical process of elimination. Think of all of the possible systems and components that could upset engine operation. Then, one by one, mentally eliminate the parts that could *not* produce the symptoms. To troubleshoot properly, ask yourself these questions:

- ❑ What are the symptoms?
- ❑ Which system could be producing the symptoms?
- ❑ Where is the most logical place to start testing?

For example, suppose an engine with electronic fuel injection misses and emits black smoke after running for a few minutes. The black smoke would tell you that too much fuel is entering the engine. Since the engine does not run poorly when cold, the problem is related to engine temperature.

Through simple deduction, you might think of the engine temperature sensor. The air-fuel mixture is based on the operating temperature of the engine. Possibly, the temperature sensor is malfunctioning, which, in turn, causes an excessively rich air-fuel mixture to enter the engine. A very rich, cold start fuel charge could be flooding the fully warmed engine.

As you can see, logical thought will help you check the most likely problems. If your first idea is incorrect, rethink the problem and check the next most likely trouble source.

Performance Problem Diagnosis Charts

If you have trouble locating an engine performance problem, refer to a service manual ***performance problem diagnosis chart.*** As shown in **Figure 17-1,** this chart will list problems, possible causes, and corrections. The chart is written for the particular make and model of vehicle. First, locate the problem in the chart. Then, check the system(s) or part(s) identified as the possible cause. Once the problem is located, perform the repair specified.

Technical Service Bulletins

Technical service bulletins (TSBs) explain problems that frequently occur in one make or model of vehicle. They are published by the vehicle manufacturers and explain the symptoms of common problems and the tests and corrections for the problems. Auto manufacturers keep records of specific problems that have been fixed in the past. If a problem happens regularly, they describe and publish the problem in TSBs sent out to their dealerships. Always obtain and read TSBs. They give helpful information on fixing common problems with specific makes and models of cars and light trucks.

Problem	Possible cause	Remedy	Page
Rough idle, stalls or misses (con'd)	EGR valve faulty	Check valve	3–59
	Valve clearance incorrect	Adjust valves	2–7
	Compression low	Check compression	4–2
Engine hesitates/ poor acceleration	Spark plugs faulty	Inspect plugs	3–6
	High tension wires faulty	Inspect wires	3–6
	Vacuum leaks • PCV hoses • EGR hoses • Intake manifold • Air cleaner hoses	Repair as necessary	
	Incorrect ignition timing	Reset timing	2–26
	Air cleaner clogged	Check air cleaner	2–22
	Fuel system clogged	Check fuel system	5–2

Figure 17-1. *This is a partial page from a performance problem diagnosis chart. Note that the chart lists the problem, possible causes, remedy, and service manual page number for detailed coverage. (Toyota)*

Other Sources of Service Information

There are several other sources of service information that can be consulted when a problem is difficult to locate and correct. Trade magazines often discuss common and difficult-to-locate failures. Always read these types of magazines. On the Internet, use a search engine to find service information and review problems encountered by other technicians. You can type in information about the vehicle or problem and quickly search for additional information. Some vehicle manufacturers and part manufacturers provide service data on the Internet.

Reading Spark Plugs

Reading spark plugs involves inspecting the condition and color of the spark plug tips. This will show you whether combustion is normal or abnormal. An abnormal condition may be the result of a rich mixture, lean mixture, detonation, oil entering cylinder, and so on. As a result, you will have a better idea as to the condition of the engine and its related systems. See **Figure 17-2.**

To read plugs, remove each plug and place it in order on your fender cover or tool cart. Then, you will know which plug came from which cylinder. For example, if one plug is fouled with carbon and the others are clean, you would know that there is a problem in the cylinder from which the spark plug was removed.

Closely inspect the plugs to determine what is going on inside the cylinder or combustion chamber. A brown to gray-tan spark plug tip indicates normal combustion; combustion is clean and good. A black or wet plug indicates a dead cylinder or engine mechanical problem allowing oil to enter the cylinder. It could also indicate an extremely rich air-fuel mixture. A carbon-fouled plug may indicate a burned valve, bad plug wire, or broken piston ring. A compression test may be needed in that cylinder.

Vacuum Gauge

A ***vacuum gauge*** can be used to find some engine performance and mechanical troubles. The gauge is connected to a source of intake manifold vacuum. A normal reading should be between 18″ Hg and 22″ Hg. The needle on the vacuum gauge should be steady; the reading should not fluctuate.

To use a vacuum gauge to check the engine, connect the gauge to a vacuum fitting on the intake plenum or

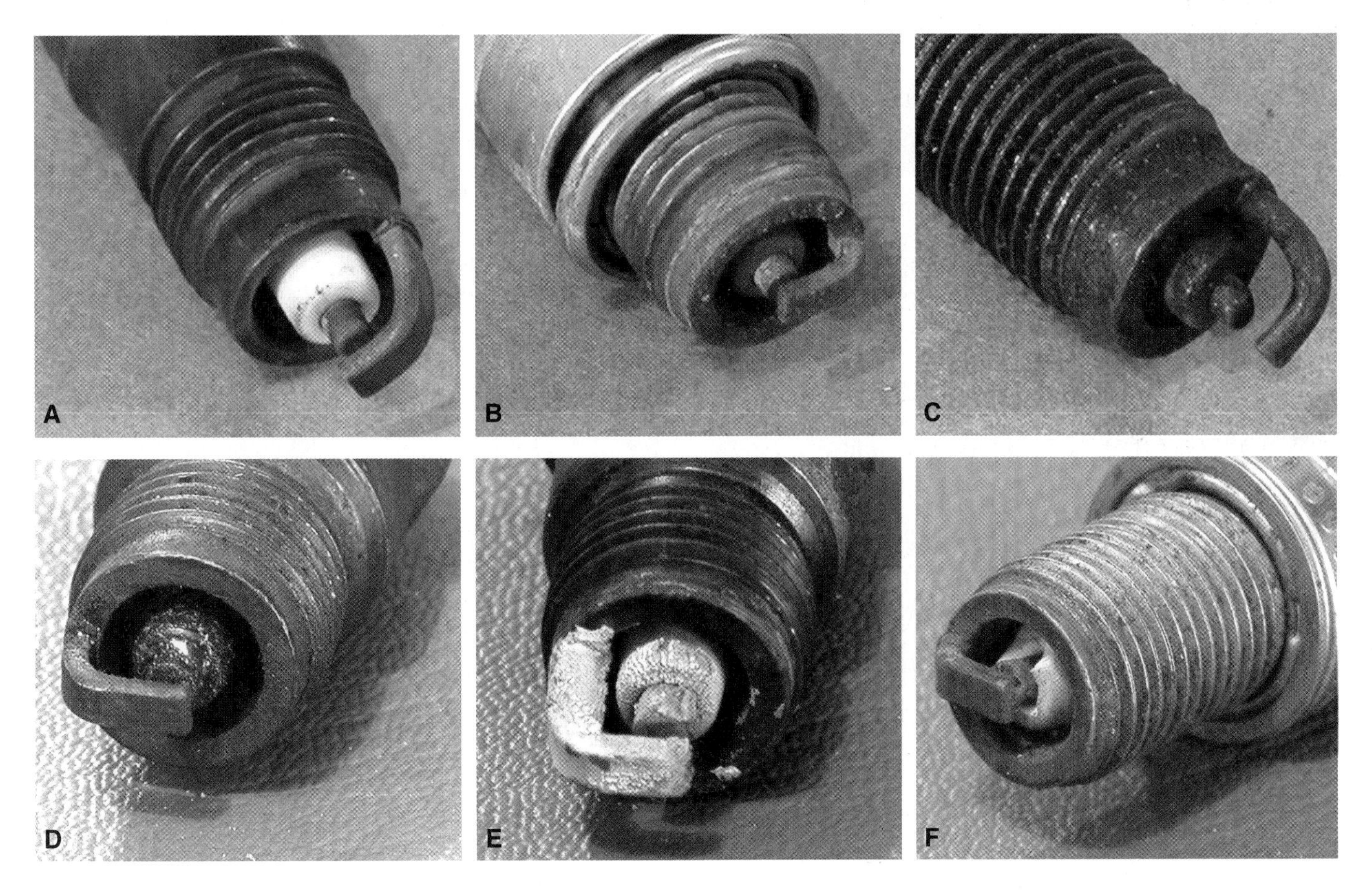

Figure 17-2. *Reading spark plugs will tell you about the condition of the engine and related systems. A—A normal burning plug should have a minimum of deposits and be gray-tan in color. B—This is a carbon-fouled plug that resulted from a prolonged rich air-fuel mixture. C—This high-mileage plug shows erosion at the gap. D—An oil-fouled plug is dark gray or black and shiny. E—An ash-fouled plug can be the result of burning low-quality gas. F—The physical damage on this spark plug indicates an internal engine failure. The piston, a broken valve, or foreign matter has struck the tip of the spark plug.*

manifold. Start the engine and note the reading on the gauge. Compare the gauge reading to specifications. **Figure 17-3** shows some sample vacuum gauge readings and the potential problem the readings indicate.

To help pinpoint problems that cannot be duplicated in the shop, technicians will often mount the vacuum gauge in the passenger compartment and run a long hose to the vacuum fitting on the engine. This allows the vehicle to be driven while checking for intermittent vacuum problems.

Test Light

A ***test light*** will quickly check for power in an electric circuit and can be used to find the source of several engine performance problems. For example, if an engine will not start and you discover no spark at a spark plug wire, use the test light to check for power at the coil. Connect the test light clip to a good frame ground. Then, touch the test light on the coil primary positive terminal. If the light glows, then you know that the ignition circuit before the coil (ignition switch, wiring, wiring connections, etc.) is functioning.

Caution: Do not connect a test light to certain electronic components. The light could draw too much current and ruin the electronic part. Refer to a service manual when in doubt.

Digital Multimeter

A ***digital multimeter*** is often used to check electronic circuits for normal voltage, current, or resistance, **Figure 17-4.** For example, if a scan tool detects a problem with a sensor circuit, you can check the wires to the sensor to see what voltage is present. If the voltage is too low or not present, you have found why the trouble code was set. Further testing is then needed to find any broken wire, loose connection, or ECM malfunction that could lower voltage to the sensor.

Scan Tools

A ***scan tool*** is a testing device to help you quickly read and interpret computer system trouble codes, **Figure 17-5.** The tool lead is connected to the vehicle's

Normal engine reading
Vacuum gauge should have reading of 18–22 inches of vacuum. The needle should remain steady.

Burned or leaky valves
Burned valve will cause pointer to drop every time burned valve opens.

Weak valve springs
Vacuum will be normal at idle but pointer will fluctuate excessively at higher speeds.

Worn valve guides
If pointer fluctuates excessively at idle but steadies at higher speeds, valves may be worn allowing air to upset fuel mixture.

Choked muffler
Vacuum will slowly drop to zero when engine speed is high.

Intake manifold air leak
If pointer is down 3 – 9 inches from normal at idle, throttle valve is not closing or intake gaskets are leaking.

Carburetor or fuel injection problem
A poor air-fuel mixture at idle can cause needle to slowly drift back and forth.

Sticking valves
A sticking valve will cause pointer to drop intermittently.

Figure 17-3. *Study these vacuum gauge readings and the conditions they indicate. (Sonco)*

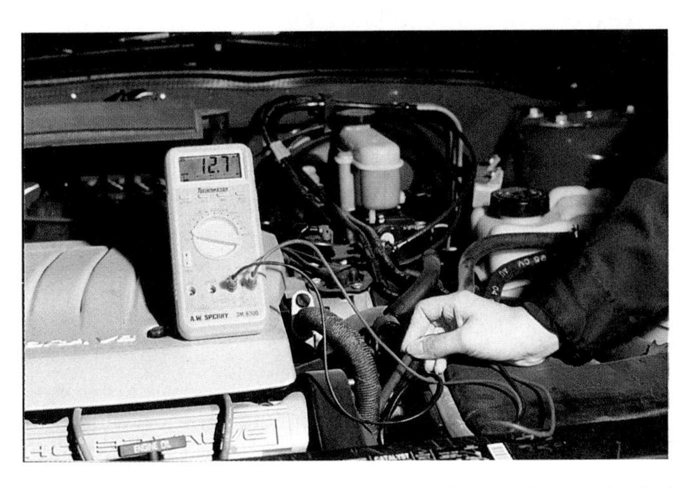

Figure 17-4. *A digital multimeter can be used to help find numerous engine performance problems.*

data link connector (DLC). This allows it to retrieve and display trouble codes. Scan tools will also let you test sensors, actuators, and the computer itself.

The scan tool cable should slide easily into the vehicle's data link connector. If not, something is wrong. Never force the two together or you can damage the pins on the tool cable or data link connector. You may need to use an adapter so the scan tool connector will fit the vehicle's DLC. If the scan tool is not powered through the DLC, connect the tool to battery voltage.

Caution: Make sure you are properly connecting the scan tool to the vehicle. Some technicians have mistakenly connected scan tools to the wrong connector, such as the tachometer connector, and the scan tool is damaged.

Engine Analyzer

An ***engine analyzer*** is a group of test instruments mounted in a roll-around cabinet, **Figure 17-6.** It is commonly used by the engine technician to find engine performance problems. As an engine technician in today's shop, you should know how to use an engine analyzer. The engine analyzer usually contains:

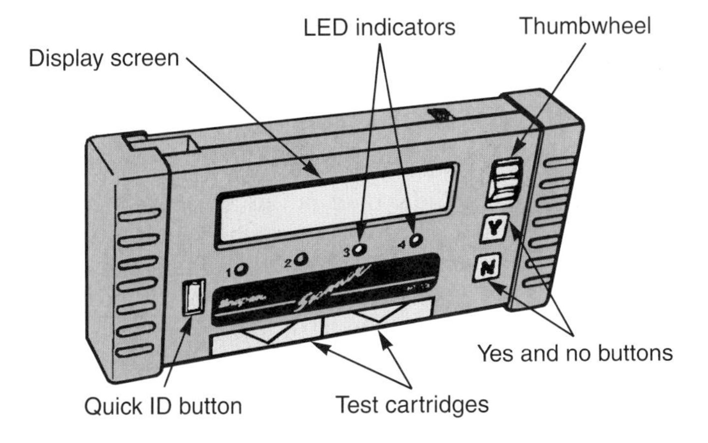

Figure 17-5. *A scan tool will communicate with the vehicle's on-board computer to help find the cause of engine performance problems. (Snap-on Tool Corp.)*

Figure 17-6. *An engine analyzer can be used to help find engine performance problems. (Snap-on Tool Corp.)*

- **Oscilloscope.** A voltmeter with a cathode ray tube (CRT) display for checking the operating condition of the ignition and other systems.
- **Tach-dwell.** A meter for checking engine speed, the ignition, and computer system.
- **Exhaust gas analyzer.** An instrument for measuring the chemical content of engine exhaust.
- **VOM.** A volt-ohm-milliammeter for measuring electrical values.
- **Compression gauge.** A pressure gauge for checking compression stroke pressure and engine condition.
- **Vacuum gauge.** A gauge for checking the condition of the engine and various vacuum-operated devices.
- **Cylinder balance tester.** A mechanism for electronically turning cylinders on and off (shorting) for finding a dead or weak cylinder.
- **Cranking current balance tester.** A device for checking engine compression by monitoring the amount of current needed to crank each cylinder through its compression stroke; low compression requires less current.
- **Timing light.** A strobe light for checking ignition timing.

Analyzer Connections

Analyzer connections differ with each type and model of analyzer. Nevertheless, most have the same general test connections. Special leads and hoses may be provided for measuring starting current, charging voltage, engine vacuum, fuel pump pressure, sensor signals, and exhaust gas content.

Most analyzers provide directions for connecting the test leads to the vehicle, **Figure 17-7.** If the analyzer does

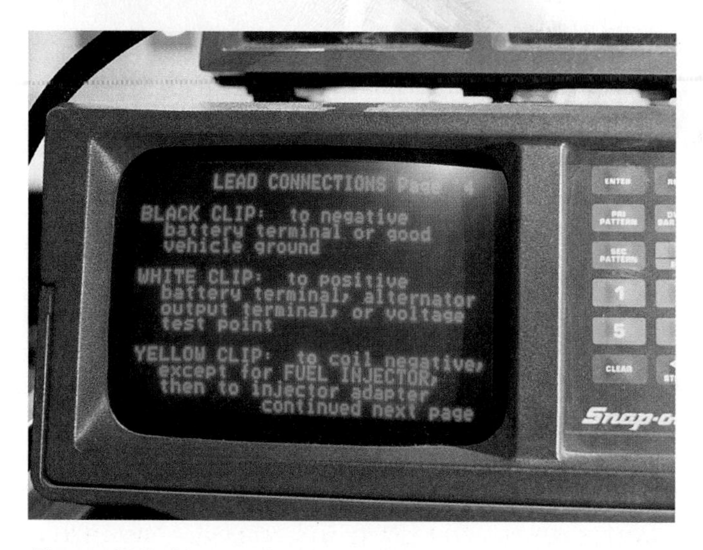

Figure 17-7. *Most analyzers provide on-screen instructions for connecting the various leads. (Snap-on Tool Corp.)*

not provide directions, the test leads should be connected as described in the user's manual.

Using an Analyzer

To use an analyzer, turn on the power to the machine. Then, set the controls and connect the test leads to the vehicle. If needed, read the operating manual for the analyzer.

Set the parking brake and start the engine. Many analyzer manufacturers recommend increasing the engine idle speed to around 1500 rpm during testing.

Caution: Before starting the engine, make sure all leads are away from hot or moving parts. The analyzer leads are very expensive and can easily be damaged by contact with a hot exhaust manifold, spinning pulley, or rotating fan.

Ignition Coil Output Test

An ***ignition coil output test*** measures the maximum available voltage produced by the ignition coil(s). A spark plug requires between 5 kilovolts (kV) and 20 kilovolts for operation. However, the ignition coil should have a higher reserve voltage. Without this extra voltage, the spark plugs could misfire when the engine is under load or at high speeds when voltage requirements are higher.

To perform the coil output test, set the analyzer controls and display to the highest kilovolt range. Increase the engine speed to between 1000 rpm and 1500 rpm. Then, using a pair of insulated pliers, disconnect a spark plug (secondary) wire. Hold the end of the plug wire away from ground to create a spark while watching the analyzer screen. A tall firing line should stand out from the others. Using the scale on the analyzer, read the voltage of this spike. This reading should match the specification for the ignition coil's maximum capacity. On coil pack and coil-over ignition systems, you will need to test each coil.

Modern ignition coils may produce 100,000 volts or more. Older ignition coil designs produce between 30,000 volts and 40,000 volts. If the ignition coil voltage is below specifications, do not condemn the coil until completing further tests. Low coil output may be due to low primary supply voltage, leaking secondary wires, or similar problems. Eliminate these potential problems before replacing the ignition coil.

Caution: Some electronic ignition systems may be damaged by disabling spark plugs while the engine is running. Be sure to check the manufacturer's recommendations before performing an ignition coil output test.

Load Test

A ***load test,*** also known as an ***acceleration test,*** measures the spark plug firing voltages when engine speed is rapidly increased. When an engine is accelerated, a higher voltage is needed to fire the spark plugs. A defective component may operate normally and produce a normal scope pattern at idle, but the component may not operate properly under load.

To perform a load test, set the analyzer on parade and increase the engine speed to between 1000 rpm and 1200 rpm. While watching the firing lines on the analyzer, quickly open and release the throttle valve. The firing voltage should increase, but it should not exceed certain limits. The upward movement of the firing lines during the load test should be the same on all lines. If any of the firing lines is high or low, a defect is present.

The highest firing line should not be more than 75% of maximum coil output. Typically, the voltage measured in this test should not exceed 20 kV for electronic ignition.

Cylinder Balance Test

A ***cylinder balance test,*** also known as a ***power balance test,*** measures the power output from each of the engine's cylinders. As each cylinder is shorted, the tachometer should register a drop in engine speed. During a cylinder balance test, the drop in engine speed should be the same for all cylinders (within 5%).

If a shorted cylinder does not produce an adequate drop in engine speed, the cylinder is not firing properly. If the drop in engine speed is below normal in one or more cylinders, a problem common to those cylinders is indicated. The cylinders may have low compression due to a burned valve, blown head gasket, or worn piston rings. The cylinders may be receiving a lean mixture due to a vacuum leak, faulty fuel injector, or computer malfunction. Other problems may be causing the low readings. Follow proper diagnostic procedures to determine the problem.

Figure 17-8. *A pressure gauge is needed to check for fuel pressure in a multiport fuel injection system.*

Caution: Never short cylinders for more than 15 seconds or damage to the catalytic converter may occur.

Cranking Balance Test

A ***cranking balance test*** is done to check the engine's mechanical condition. It can be used to isolate a cylinder with low compression due to a burned valve, worn piston rings, or other problems. The analyzer will show how much current is drawn by the starter motor as each cylinder goes through its compression stroke. Low current draw, indicated by a low display line, means that cylinder has low compression.

Fuel Pressure Gauge

A ***fuel pressure gauge*** is used to measure engine fuel pressure and can be used to find engine performance problems. See **Figure 17-8.** The fuel pressure gauge can be connected to the fuel line, usually at a test fitting on the fuel rail, to check the operation of the fuel pump, pressure regulator, fuel filters, and fuel lines. If pressure is within specifications, you have eliminated these parts as a source of the performance problem.

Fuel pump testing commonly involves measuring fuel pump pressure and volume. Exact procedures vary, depending on fuel system type. Refer to the service manual for specific testing methods.

Remember, there are several other problems that can produce symptoms like those caused by a faulty fuel pump. Before testing a fuel pump, check for:

- ❑ Restricted fuel filters.
- ❑ Smashed or kinked fuel lines or hoses.
- ❑ Air leaks in the vacuum side of the pump circuit.
- ❑ Injection system problems.
- ❑ Ignition system problems.
- ❑ Low engine compression.

The following procedure is used to measure fuel pump pressure. Remember, this is also checking the condition of the fuel pressure regulator.

1. Connect a pressure gauge to the output line of the fuel pump or to the appropriate test fitting on the fuel rail.
2. Activate the pump motor or start the engine.
3. Read the pressure gauge.
4. Compare the readings to specifications.

Fuel pressure on a gasoline injection system is usually 15 psi to 40 psi (100 kPa to 275 kPa). Use the factory values when determining fuel pump condition. Pressures vary from system to system.

If fuel pump pressure is not within specifications, check the pump volume and look for obstructions in the lines and filters before replacing the pump. Also, isolate the fuel pressure regulator from the pump. This can be done by pinching the fuel hose going to the fuel return line or by taking the regulator out of the system.

If needed, you can also connect the pressure gauge directly to the output of an in-tank electric fuel pump. This, in effect, removes the pressure regulator, feed lines, and other parts from the circuit to isolate the pump problem.

Typical Performance Problems

You must understand the most common engine performance problems. The following will help you use test equipment, troubleshooting charts, and a service manual during diagnosis.

No-Start Problem

A ***no-start problem*** occurs when the engine is turned over by the starter, but fails to fire and run on its own power. This is the most obvious and severe performance problem.

With a no-start problem, first check for spark. To check for spark, remove one spark plug wire or direct ignition coil from a spark plug. Install a spark tester or old spark plug on the wire end or into the coil terminal. Ground the spark tester or old spark plug to frame ground. A bright spark should jump the gap when the engine is cranked, **Figure 17-9.** If not, the problem is in the ignition system.

If you have spark, check for fuel pressure. With throttle body fuel injection, watch the injector outlet while cranking the engine. With multiport injection, install a pressure gauge on the fuel rail. Then, read the gauge with the ignition key in the run position. If the fuel pressure is low, then something is wrong with the fuel supply system.

Figure 17-9. *When diagnosing a no-start problem, use a spark tester to check for spark. When the engine is cranked, a strong spark should be produced. (Snap-on Tool Corp.)*

Gas line freeze results when moisture in the fuel turns to ice. The ice will block fuel filters and prevent engine operation. To correct gas line freeze, you may need to place the vehicle in a warm garage until the fuel is thawed.

If the fuel system has fuel pressure, check the injector harness for the presence of a pulse. Pull a harness connector loose and install a noid light in the harness connector. The noid light should flash when the engine is cranked. If the light does not flash, there may be a problem with the engine ECM, wiring harness, or engine speed sensor.

If you have both fuel and spark, check engine compression. A jumped timing chain or belt could be keeping the engine from starting. With a diesel, a slow cranking speed can prevent starting.

Hard Starting

Engine ***hard starting*** occurs when excessive cranking time is needed to start the engine. It is due to partial failure of a system. The engine coolant temperature sensor may be bad.

Vapor lock can cause hard starting, as well as engine stalling, lack of power, and no starting. ***Vapor lock*** occurs when the fuel is overheated, forming air bubbles that may reduce or block fuel flow. It is caused by too much engine heat transferring into the fuel.

Stalling

Engine ***stalling,*** also called ***dying,*** is a condition where the engine stops running. This may occur at idle, after cold starting, or after warmup. There are many problems that can cause stalling:

- ❑ Low idle speed.
- ❑ Injection system problem.
- ❑ Ignition system trouble.
- ❑ Severe vacuum leak.
- ❑ Inoperative thermostatic air cleaner.

A common cause of engine stalling is a faulty transmission torque converter lockup circuit. The solenoid or actuator that controls torque converter lockup can stick and keep oil flowing to the torque converter lockup clutches. The torque converter actuator keeps the clutches in the torque converter engaged and the engine stalls when the vehicle slows to a stop. To check for this condition, disconnect the wiring going to the torque converter lockup solenoid. If this corrects the stalling problem, replace the lockup solenoid.

Misfiring

Engine ***misfiring*** is a performance problem that occurs when the fuel mixture fails to ignite and burn properly. The engine may miss at idle, under acceleration, or at cruising speeds. The unburned fuel is pushed out of the engine and into the exhaust system. It can then damage the catalytic converter and pollute the environment.

If an engine only misses at idle, check the components that affect idle. With multiport fuel injection, an injector may not be opening or may be partially clogged or restricted. A fouled spark plug, open plug wire, cracked distributor cap, corroded terminals, faulty engine sensor, or vacuum leaks are a few other possible causes for a misfire at idle.

A misfire at higher engine speeds is often due to ignition systems troubles. Bad ignition coil(s), high resistance in the plug wires, wide spark plug gaps, or fouled spark plugs all may cause high-speed misfires. A contaminated oxygen sensor can make the fuel mixture too lean and lead to a high-speed misfire.

In a vehicle with on-board diagnostics generation two (OBD II), engine misfiring is monitored using the crankshaft position sensor. When a cylinder misfires, the crankshaft speed fluctuates slightly. The computer interprets this fluctuation in crankshaft speed as an indication of a poor running engine, or a misfire. A few OBD II systems detect misfire by electronically measuring the ionization at the spark plug electrodes. A misfire rate of less than 2% is acceptable because the catalytic converter can easily handle this amount of pollutants.

An OBD II scan tool can produce the following misfire data:

- ❑ **Misfire data values.** Indicates that something is causing an engine cylinder not to properly burn its air-fuel mixture. The misfire data can be recorded by the vehicle computer and stored in memory. The scan tool will retrieve this data to help you find problem sources.
- ❑ **Misfire history.** Indicates which cylinder was misfiring and how badly it has been misfiring.
- ❑ **Misfire passes.** Shows how many times the cylinder has *not* misfired.
- ❑ **Misfire failures.** Indicates how many misfire tests have been recorded.
- ❑ **Misfires per revolution status.** Shows accepted misfires (real misfires) and rejected misfires (false data due to a rough road or other cause).

- **Total misfires.** The average number of misfires recorded during the last 200 crankshaft revolutions.
- **Misfiring cylinder.** Shows the primary missing cylinder (cylinder with the most misses) and the secondary missing cylinder (cylinder with the second most misses) by cylinder number.
- **RPM at misfire.** Shows the engine speed (in rpm) at which the computer detected a cylinder misfire. This is handy for further diagnosis since you know the engine speed at which the problem occurs.
- **Load at misfire.** Gives the load at which the engine miss occurred. This is usually information gathered from the manifold pressure sensor, which measures engine load.

Other misfire data can also be produced depending on the vehicle make, model, and year. Refer to the service manual for more information.

Hesitation

A ***hesitation,*** also called a ***stumble,*** is a condition where the engine does not accelerate normally when the gas pedal is pressed. The engine may also stall before developing power.

Hesitation is usually caused by a temporary lean air-fuel mixture. The throttle position or manifold pressure sensor may be bad. Check the parts that aid engine acceleration.

Surging

Surging is a condition where engine power fluctuates up and down. When driving at a steady speed, the engine seems to speed up and slow down without movement of the gas pedal.

Surging is sometimes caused by an extremely lean fuel injection setting. Surging can also be caused by problems in the ignition or computer control systems.

Backfiring

Backfiring is caused by the air-fuel mixture igniting in the intake manifold or exhaust system. A loud bang or pop sound can be heard when the mixture ignites. A backfire can be caused by incorrect ignition timing, crossed spark plug wires, cracked distributor cap, exhaust system leakage, faulty air injection system, arcing across ignition coil terminals, or other system fault.

Dieseling

Dieseling, also called ***after-running*** or ***run-on,*** occurs when the engine fails to shut off. The engine keeps firing, coughing, and producing power. Dieseling is usually caused by a high idle speed, carbon buildup in the combustion chambers, low-octane fuel, or overheated engine.

Pinging

Pinging is a metallic tapping or light knocking sound, usually when the engine accelerates under load. Pinging is caused by abnormal combustion, either preignition or detonation. Pinging is normally caused by low-octane fuel, advanced ignition timing, carbon buildup in combustion chambers, or engine overheating.

Poor Fuel Economy

Poor fuel economy is a condition in which the engine consumes too much fuel. Fuel economy can be measured as the number of miles that can be driven on one gallon of fuel (kilometers per one liter).

Poor fuel economy can be caused by a wide range of problems. These include a rich air-fuel mixture, engine miss, incorrect ignition timing, or fuel leakage.

Lack of Engine Power

Lack of engine power, also termed a ***sluggish engine,*** causes the vehicle to accelerate slowly. When the gas pedal is pressed, the car does not gain speed properly. All cylinders are firing correctly and the engine runs smoothly, but normal power is not produced.

There are many troubles that can reduce engine power. These include fuel system problems, ignition system problems, emission control system problems, and engine mechanical problems.

Rough Idle

A ***rough idle*** is when the engine seems to vibrate on its mounts. In some cases, a popping noise can be heard at the tailpipe.

A vacuum leak is a common cause of rough idling. See **Figure 17-10.** For example, if a vacuum hose hardens and cracks, outside air will enter the engine intake manifold, bypassing the airflow sensor. This will cause an incorrect air-fuel mixture and prevent normal combustion.

Usually, a vacuum leak will produce a hissing sound. The engine roughness will smooth out when engine speed is increased.

A section of vacuum hose can be used to locate vacuum leaks. Place one end of the hose next to your ear. Move the other end around the engine. When the hiss becomes very loud, you have found the leak.

Summary

Use a systematic approach when trying to find the source of engine performance problems. Use your understanding of system operation and basic testing methods to check the most logical sources of trouble.

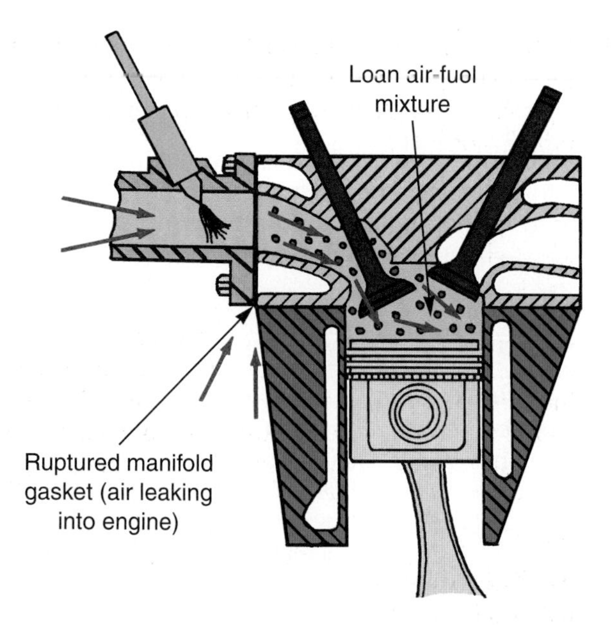

Figure 17-10. *A vacuum leak allows extra air to enter engine after the airflow sensor, which results in a lean mixture and rough idle.*

A service manual performance problem diagnosis chart should be used when a problem is difficult to locate and correct. TSBs can also provide valuable diagnostic information.

Reading spark plugs is a common method of finding the source of performance problems. The condition and color of the spark plug tips can indicate the air-fuel ratio, engine compression, condition of rings, and misfiring cylinders. A properly firing cylinder should have a plug tip that is gray-tan to brown in color.

A vacuum gauge is a handy tool for finding performance problems. A test light and multimeter are useful when diagnosing electrical troubles that affect engine performance. An engine analyzer is a group of test instruments in a single cabinet. A pressure gauge is commonly used to check fuel pressure.

There are several common engine performance problems. These include a no-start problem, hard start problem, stalling, misfiring, hesitation, surging, backfiring, dieseling, pinging, poor fuel economy, lack of engine power, and rough idle. Use a systematic approach when diagnosing the cause of these problems.

Review Questions—Chapter 17

Please do not write in this text. Write your answers on a separate sheet of paper.

1. An engine performance problem is any trouble that lowers engine _____, _____, _____, or _____.
2. What are the three questions you should ask yourself when troubleshooting engine performance problems?
3. Briefly explain how to use a *performance problem diagnosis chart.*
4. What does *TSB* stand for?
5. How do you read spark plugs?
6. List nine test instruments usually found on an engine analyzer.
7. When an engine cranks but fails to start, you should first check for _____.
8. This trouble will cause rough idle and make a hissing sound.
   (A) Hesitation.
   (B) Pressure leak.
   (C) Vacuum leak.
   (D) Dieseling.
9. A(n) _____ is caused by the air-fuel mixture igniting in the intake manifold or exhaust system.
10. Define *vapor lock.*

ASE-Type Questions—Chapter 17

1. A gasoline engine diesels when shut off. Technician A says that a high idle speed setting may be the cause. Technician B says that the fuel may have too high of an octane rating. Who is correct?
   (A) A only.
   (B) B only.
   (C) Both A and B.
   (D) Neither A nor B.
2. A spark plug is removed from a cylinder and read. It is found to be wet and black. Technician A says that this indicates an engine mechanical problem. Technician B says that this indicates an extremely rich air-fuel mixture. Who is correct?
   (A) A only.
   (B) B only.
   (C) Both A and B.
   (D) Neither A nor B.
3. Technician A says that a test light can be used to test all electronic engine components. Technician B says that a normal reading on a vacuum gauge should be between 18″ Hg and 22″ Hg. Who is correct?
   (A) A only.
   (B) B only.
   (C) Both A and B.
   (D) Neither A nor B.

4. An engine analyzer normally contains all of the following *except:*
   (A) compression gauge.
   (B) oscilloscope.
   (C) exhaust gas analyzer.
   (D) cylinder leak tester.

5. To test the operation of a multiport injection system's pressure regulator, the pressure gauge should be connected to the _____.
   (A) PCV valve vacuum port
   (B) fuel rail fitting
   (C) injector wiring harness
   (D) evaporative emissions purge valve

6. Technician A says that with a no-start engine problem, you should first check the engine's ignition system. Technician B says that gas line freeze can cause a no-start engine problem. Who is correct?
   (A) A only.
   (B) B only.
   (C) Both A and B.
   (D) Neither A nor B.

7. All of the following can cause an engine to stall *except:*
   (A) a severe vacuum leak.
   (B) an inoperative thermostatic air cleaner.
   (C) a lean air-fuel mixture.
   (D) low idle speed.

8. Technician A says that a cracked distributor cap may cause an engine to miss. Technician B says that a fuel injector malfunction may cause an engine to miss. Who is correct?
   (A) A only.
   (B) B only.
   (C) Both A and B.
   (D) Neither A nor B.

9. Technician A says that a noid light can be used to check the injector harness for a pulse when diagnosing a no-start problem. Technician B says that spark should be verified before using a noid light when diagnosing a no-start problem. Who is correct?
   (A) A only.
   (B) B only.
   (C) Both A and B.
   (D) Neither A nor B.

10. Technician A says that engine surging can be caused by a rich air-fuel mixture. Technician B says that hesitation may be caused by a faulty throttle position sensor. Who is correct?
    (A) A only.
    (B) B only.
    (C) Both A and B.
    (D) Neither A nor B.

Engine mechanical problems may be related to the valve train, timing mechanism, engine gaskets, cylinder head, cylinder block, pistons, connecting rods, crankshaft, or intake or exhaust manifolds. (Ford)

Chapter 18

Engine Mechanical Problems

After studying this chapter, you will be able to:

- ❑ Describe typical mechanical failures in an engine.
- ❑ Explain the causes of engine mechanical failures.
- ❑ Summarize the symptoms for common engine mechanical breakdowns.
- ❑ Diagnose engine mechanical problems.

Know These Terms

Bad valve stem seal
Bent connecting rod
Bent push rod
Bent valve
Blown head gasket
Broken connecting rod
Broken push rod
Burned piston
Burned valve
Camshaft bearing wear
Camshaft journal wear
Camshaft lobe wear
Camshaft thrust surface wear
Cracked block
Cracked cylinder head
Cracked valve
Crankcase warpage
Cylinder block problems
Engine smoking
Exhaust manifold problems
External lifter wear
Flywheel problems
Intake manifold problems
Internal lifter wear
Main bearing knock
Piston breakage
Piston knock
Piston pin knock
Piston slap
Rocker arm wear
Rocker shaft wear
Rocker stand wear
Spun rod bearing
Stuck valve
Timing mechanism problems
Valve guide wear
Valve stem breakage
Valve stem wear
Warped cylinder head
Worn main bearings

This chapter summarizes the causes, symptoms, and results of mechanical part failures in an engine. One of the most difficult aspects of being an engine technician is troubleshooting before engine teardown. You must be able to quickly and accurately locate mechanical problems and correct them so that the same problem does not happen again.

In previous chapters, you learned about engine operation, construction, and design. Now, you will learn about burned valves, blown head gaskets, and other similar troubles. This will help prepare you to diagnose engine problems and understand later chapters.

Valve Train Problems

The engine valve train is a common cause of mechanical problems in an engine. Valve train problems can cause engine noise, power loss, missing, rough idling, and even piston, cylinder, and head damage. Friction is high between valve train parts and most of the parts are lubricated by splash oiling. Wear can occur between parts after extended service or from lack of maintenance.

Camshaft Problems

There are several potential camshaft problems. These include lobe wear, journal wear, camshaft breakage, camshaft drive failure, and so on.

Camshaft lobe wear is a condition where friction from the lifter or follower has worn off some of the camshaft lobe. See **Figure 18-1.** This usually happens in a high-mileage engine or one that has not had oil changes at the recommended intervals.

Valve train clatter (noise), reduced valve lift, and poor engine performance will result from excess lobe wear. To check for major lobe wear, remove the valve cover and watch the rocker arms or valve springs. A badly worn lobe will not open the valve normally. To check for minor lobe wear, you must measure the rocker arm motion with a dial indicator.

Figure 18-1. *This camshaft lobe is badly worn and would not open its valve normally. This wear was due to extended service and lack of oil changes.*

Camshaft journal wear and ***camshaft bearing wear*** will increase part clearance and can reduce engine oil pressure. Again, this normally only happens after extended service or operating the engine with dirty oil.

Camshaft thrust surface wear can allow the camshaft to move too far forward and rearward in the engine. This can sometimes produce a light, slow knocking sound, timing chain or gear problems, and increased lifter and lobe wear.

A broken camshaft will prevent some of the valves from operating. The lobes on one end of the broken camshaft will not rotate and open their valves. Severe performance problems or valve damage can result. Removal of a valve cover will let you check valve train and camshaft action. Camshaft breakage is not very common, but should be considered if several cylinders are dead.

Hydraulic Lifter or Tappet Problems

Hydraulic lifter or tappet problems can produce valve train noise or a light tapping sound from inside the valve cover. A new lifter has a convex (crowned or humped) bottom, which makes it rotate in the lifter bore during engine operation. Look at **Figure 18-2.** High mileage can wear off this crown and speed camshaft lobe wear.

External lifter wear results in the bottom of the lifter becoming concaved (sunk in). This is due to friction between the lifter and camshaft lobe. This can increase the valve train clearance. Engine disassembly is needed to verify external lifter wear.

Internal lifter wear on a hydraulic lifter can prevent the lifter from pumping up to take up valve train clearance. This can cause the rocker arm to clatter as it strikes the valve stem tip. The noise will be near the top of the engine, as if a small ball peen hammer is hitting the engine. With

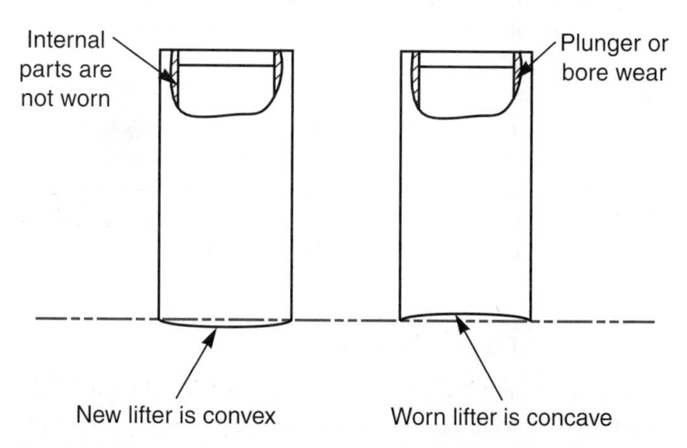

Figure 18-2. *A new lifter is crowned on the bottom. This helps rotate the lifter on its camshaft lobe to avoid wear. A worn lifter may be cupped on the bottom or its internal parts may be worn.*

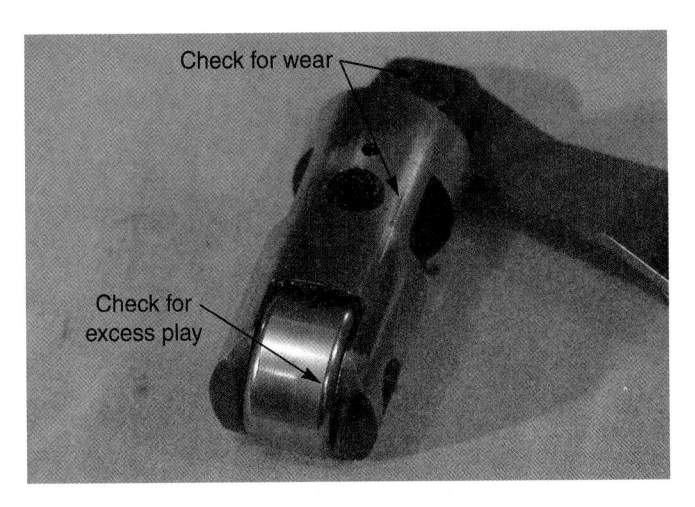

Figure 18-3. *A worn roller lifter may have a loose roller or failed needle bearings.*

the engine idling and the valve cover removed, push down on the push rod end of the rocker arm. This should increase the valve train noise. If not, the lifter for that valve is bad.

Low engine oil pressure can cause hydraulic lifter clatter. Check the oil level and oil pressure before condemning the lifter or lifters. Contaminated or dirty oil can also cause lifter noise.

Roller lifters also have needle bearings and a roller that can wear and fail. Lifter removal and close inspection is needed to find problems with the roller or needle bearings, **Figure 18-3.**

Push Rod Problems

A ***bent push rod*** is not straight, **Figure 18-4.** The damaged may be caused by a stuck valve, spring bind, spring breakage, piston hitting the valve, or other type of mechanical problem. If severely bent, the push rod can fall out of the rocker arm pocket. This will stop valve action and severe missing and tapping will result. Replace bent push rods; do *not* straighten them.

A ***broken push rod*** usually has the small ball on the end broken off. This will cause loud valve train noise and usually push rod disengagement from the rocker arm.

A ***clogged push rod*** has the internal oil hole restricted or blocked with hardened oil and sludge deposits. The tip of the push rod can also become mushroomed, which restricts oil flow. This will speed rocker arm wear because of reduced lubrication. Always check or clean the hole in push rods during an overhaul or valve job.

Rocker Arm Problems

Rocker shaft wear results from friction between the shaft and the rocker arm. This can happen because of inadequate lubrication, dirty oil, or high mileage. As shown in **Figure 18-5,** the clearance between the rocker arm and shaft will increase. This can cause valve train clatter. The bottom of the shaft is more prone to wear than the top of the shaft.

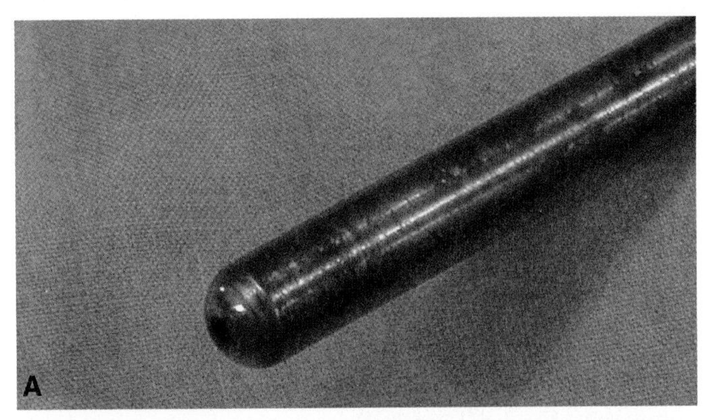

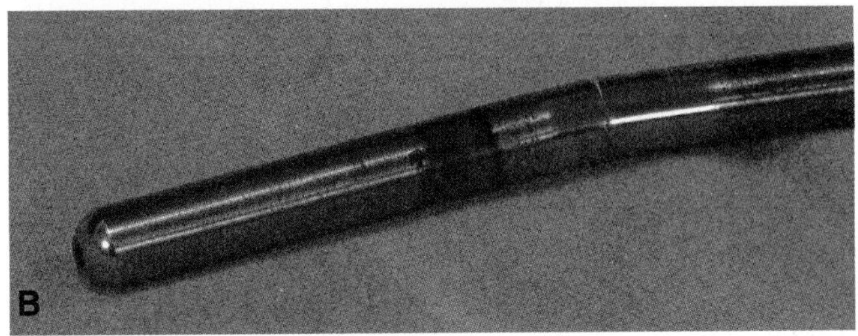

Figure 18-4. *A—Inspect the tip of a push rod for wear. Instead of being round, a worn push rod tip will be slightly pointed. B—This is a bent, cracked push rod. The failure resulted from over-revving the engine.*

Rocker arms or rocker shafts are prone to wear when engine is operated with dirty oil or a low oil level. Rocker arms or rocker shafts are the first parts to be starved for lubrication. Valve train noise can result.

Rocker arm wear can occur on the inside diameter of the rocker arm or where it contacts the valve or push rod, **Figure 18-6.** This can make the lifter plunger move too far out and result in a light valve train tapping noise. With an overhead camshaft engine, the rocker can wear at its contact point with the camshaft. Inspection is needed to verify rocker wear problems.

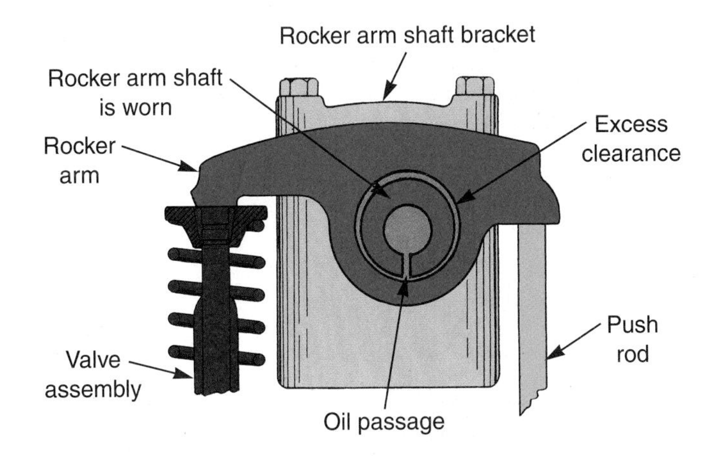

Figure 18-5. *A worn rocker arm shaft can result in excess clearance between the shaft and the rocker arm. This will allow the lifter plunger to extend too far up, resulting in valve train noise. (Perfect Circle)*

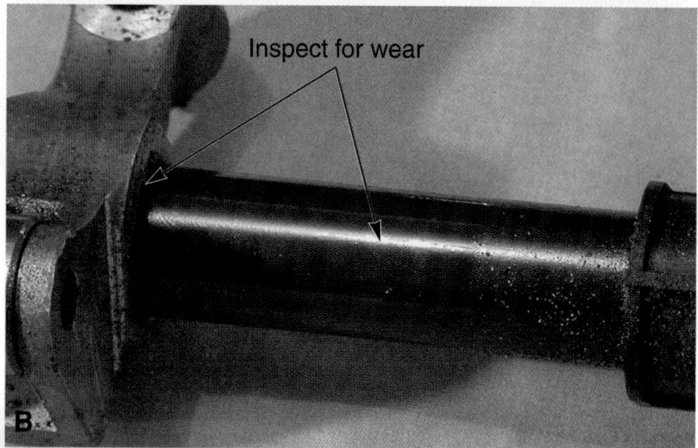

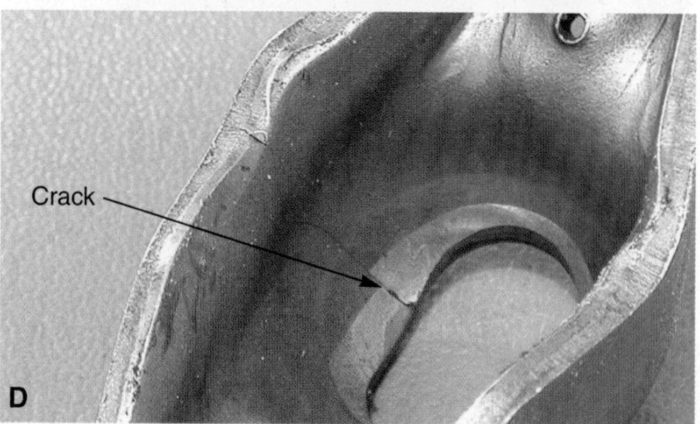

Figure 18-6. *A—This rocker arm rides on an aluminum pedestal, which will wear more quickly than steel. B—Inspect friction points on the rocker arm for wear. Slide the rocker arm over to check for wear on the rocker arm shaft. C—Rocker tips that ride on valve stems can also wear or fail. The swivel tip fell off of the rocker arm on the right. D—This rocker arm is cracked and must be replaced.*

Rocker stand wear is a common problem when the stand is made of aluminum. The stand will wear where it rubs on the rocker arm. This will cause valve train noise.

Rocker arm or stand breakage can occur if the valve strikes the piston from over-revving or if the valve is stuck in the head after the engine has been unused for a period of time and the valve stem becomes rusted. This will prevent valve action and cause a severe engine miss.

Valve Problems

Valves open and close millions of times during the life of an engine. They must do this while sealing the tremendous heat of combustion. After extended service, various valve problems can develop. See **Figure 18-7.** You must be able to diagnose and locate these problems efficiently.

Burned Valve

A ***burned valve*** has had a portion of the valve margin and face burned away. If the heat of combustion is too high, the heat can actually melt and burn away a portion of the face and margin, as shown in **Figure 18-7A.** Pressure can leak past the burned valve, reducing power in that cylinder. The engine will miss or run rough, especially at idle.

With a burned valve, you may be able to hear a puffing sound as pressure blows past the valve. There may be a popping or puffing sound at the throttle body (bad intake valve) or exhaust system tailpipe (burned exhaust valve). To fix a burned valve, you must remove the cylinder head from the engine.

Valve Face Erosion

Valve face erosion is the wearing away of the face where it contacts the seat in the cylinder head. Look at **Figure 18-7B.** This is a common problem when unleaded fuel is used in an older engine that was designed for leaded fuel.

Bent Valve

A ***bent valve*** is one in which the head of the valve is not perpendicular to the valve stem. Look at **Figure 18-7C.** It may result from over-revving the engine, which may allow the piston to hit the valve head. A bent valve may also result from a timing belt problem on an interference engine.

Cracked and Chipped Valves

A ***cracked valve*** often results from engine overheating. **Figure 18-7D** shows a cracked valve. A ***chipped valve*** is normally due to mechanical damage or a defective valve. **Figure 18-7E** shows a chipped valve, which was defective from the factory.

Sharp Margin

A ***sharp valve margin*** is caused by valve face wear due to friction or from metal removal from valve grinding. See

Figure 18-7F. A valve must have a margin between the valve head and face to withstand the high temperatures of engine operation. Normally, valve margin thickness is most important on the exhaust valves since they are exposed to higher operating temperatures than the intake valves.

Valve Stem Problems

Valve stem wear results from friction between the stem and the valve guide in the cylinder head. Normally, only a small amount of oil enters the guide, which makes stem wear common, **Figure 18-8A.** Valve stem problems are easy to overlook. Always inspect the stem, keeper grooves, and tip closely.

Severe stem or guide wear can produce a light rapping or tapping sound. The valve can flop sideways when acted on by the rocker arm or valve spring. To find a worn valve stem or guide, use a pry bar to pry sideways on the valve spring with the engine idling. This will tend to quiet or change the light tapping noise. Head removal and valve replacement are needed to correct worn valve guides and stems.

In addition to wear along the length of the valve stem, check the keeper grooves for wear or damage, **Figure 18-8B.** Also, if the chamfer on the stem tip is too small or missing, the tip is badly worn.

Valve stem breakage is a severe failure that usually occurs when the piston hits the valve head. This will bend and break off the valve head at the stem. A broken valve spring, weak valve spring, missing keepers, etc., can cause this problem.

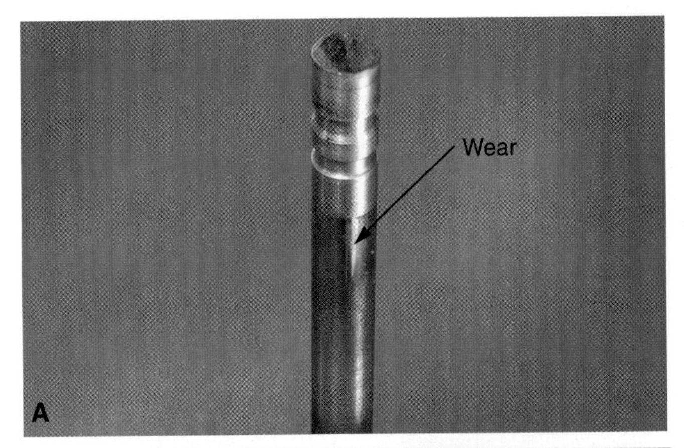

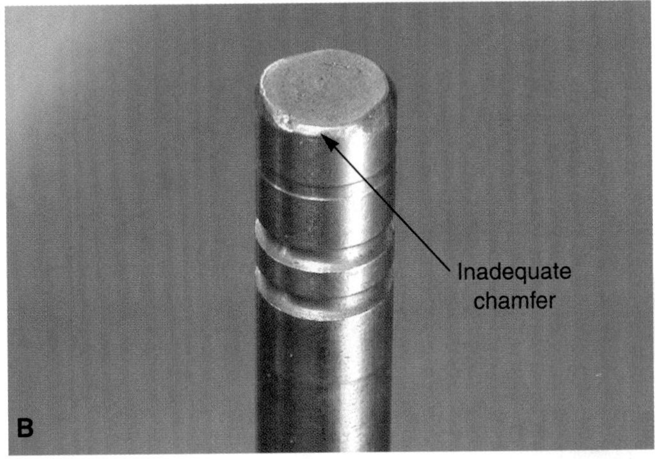

Figure 18-8. *A—This valve stem is worn. B—If the chamfer is too small or missing, the stem tip is badly worn.*

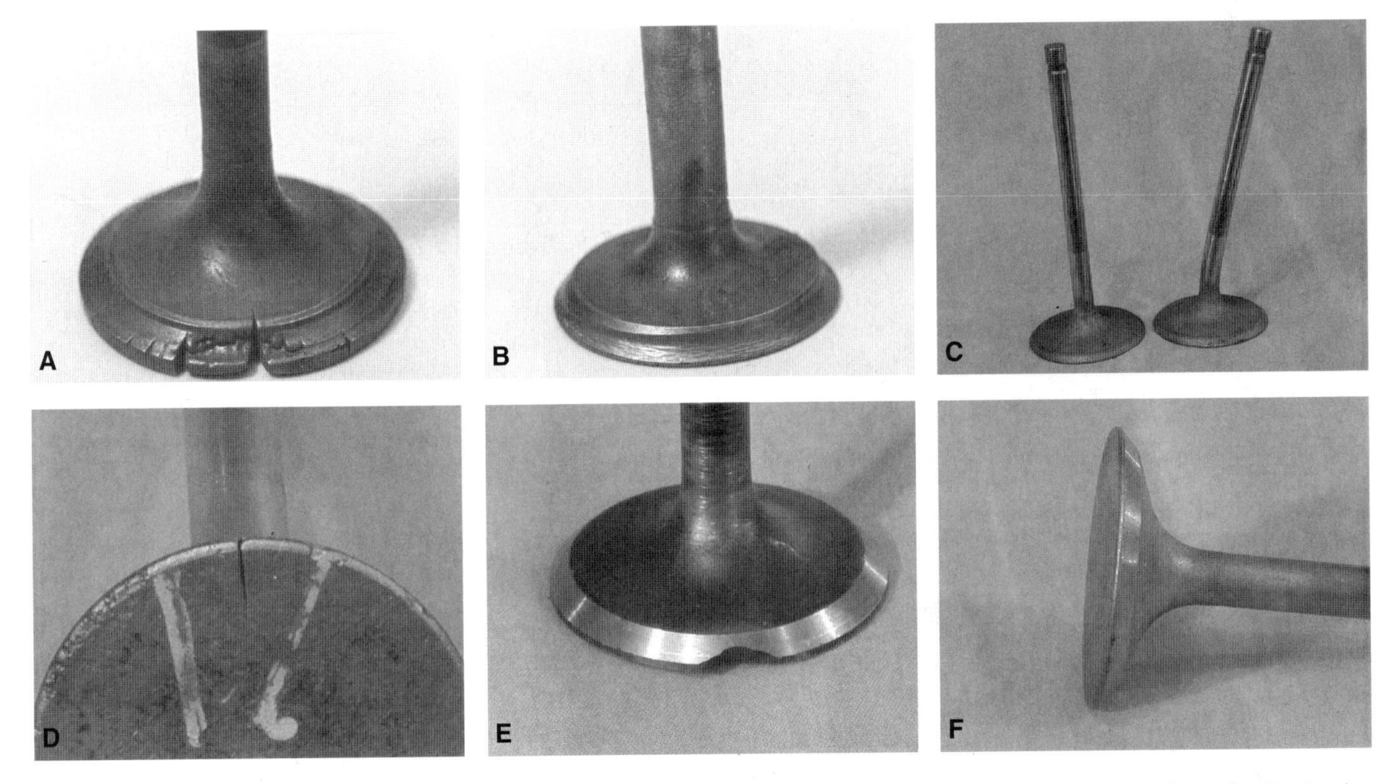

Figure 18-7. *Engine valves can suffer from wear and physical damage. A—Burned valve. B—Eroded valve face. C—Bent valve (right). D—Cracked valve. E—Chipped valve. F—Sharp valve margin (inadequate margin).*

When the valve head breaks off, it will usually be hit by the piston and forced up into the cylinder head. Piston and head damage normally results.

Valve stem scoring results when the stem has been overheated and scarred by severe friction. The rough surface of the scored valve stem will accelerate valve guide wear.

Valve Guide Problems

Valve guide wear produces the same symptoms as valve stem wear. The valve will be free to shift sideways in the guide, as shown in **Figure 18-9.** Oil can be drawn down through the excess clearance and burned in the combustion chamber. Worn valve guides will make an engine smoke on startup. A tapping noise can also result from valve guide wear.

A ***dropped valve guide*** is a valve guide that has broken free from its press-fit in the head. This will let the guide fall down around the head of the valve. It can wedge the valve open and cause severe part damage. This problem is more common on aluminum heads.

Bad Valve Stem Seals

A ***bad valve stem seal*** has hardened or broken from extended service or engine overheating. The synthetic rubber seals are no longer soft and pliable. With severe failure, the seal can disintegrate and fall down into the oil pan. Look at **Figure 18-10.**

Bad valve stem seals will make the engine emit blue smoke from the tailpipe, especially right after engine startup. Oil drips down past the seal and into the guides when the engine is not running. Then, on startup the oil is drawn into the cylinders and burned.

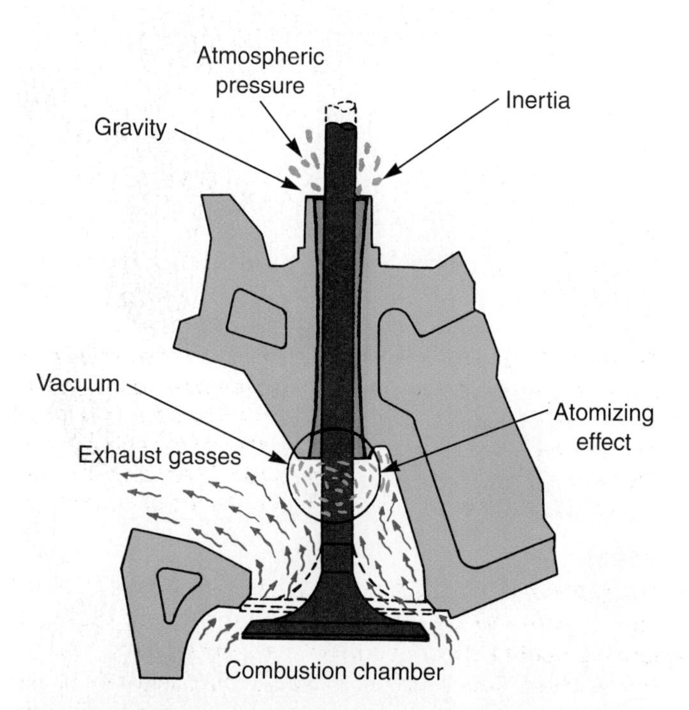

Figure 18-9. *Bad valve seals or worn valve guides will allow oil to flow down into the port. The oil is then pulled into the combustion chamber. A worn guide or stem can also allow the valve to flop sideways in the guide, resulting in a tapping noise.*

Figure 18-10. *Bad valve stem seals. A—Seals will harden after prolonged use. B—This seal has cracked. It will break up and fall down into the oil pan if not replaced. C—This seal has broken up and could clog oil passages, leading to major engine failures. (Fel-Pro)*

Valve stem seals can usually be replaced without removing the cylinder head. Air pressure is used to hold the valves closed. Then, a special tool is used to compress the valve spring. The keepers, springs, and seals can then be removed for service.

Stuck Valve

A ***stuck valve,*** or frozen valve, results when the valve stem rusts or corrodes and locks in the valve guide. This can happen when the engine sits in storage for an extended period.

A stuck valve can sometimes be fixed by squirting rust penetrant around the top of the valve guide. If this fails to correct the problem, remove the cylinder head for service.

If a valve is hit by a piston due to over-revving the engine or a weak or broken valve spring, the valve can stick and lock in its guide. The valve will hang open and fail to close with normal spring action.

Valve Spring Problems

Valve float is a condition where weakened valve springs or excessive engine speed cause the valves to remain partially open. This problem usually occurs at higher engine speeds. The engine may begin to miss, pop, or backfire as the valves float. Excess valve-to-rocker clearance may cause valve train noise.

Valve springs weaken after prolonged use. They lose some of their tension. As a result, the springs become too weak to close the valves properly. A broken valve spring will frequently let the valve hang partially open or even be hit by the piston head.

Inspection is often needed to verify a broken valve spring. The cylinder with a broken valve spring may pass a compression test.

Valve springs can be replaced without cylinder head removal. As with valve stem seal replacement, air pressure and a special tool will permit spring replacement.

Figure 18-11 shows how a broken valve spring resulted in a catastrophic engine failure. After the spring broke, the valve dropped down into the cylinder. The valve stem was then driven through the top of the piston.

Note: If a hydraulic lifter sticks in a position that is too high and the valve does not fully close, symptoms similar to those produced by valve float may be noticeable.

Figure 18-11. *A broken valve spring can lead to major damage. A—When the spring broke, the valve dropped down in the guide. B—Marks on the piston head show where the valve head touched the piston head. C—At the bottom of the piston travel, the valve slid completely out of the guide. Then, as the piston moved up, the valve was forced through the top of the piston.*

Timing Mechanism Problems

Timing mechanism problems can affect how the camshaft rotates in relation to the crankshaft to open the valves at the correct times. This involves the condition of the timing belt, timing chain, or timing gears.

Timing sprocket problems are common when the sprocket has plastic teeth. The plastic teeth can break off and cause timing chain slack, **Figure 18-12A.** The broken teeth can also allow the timing chain to jump over teeth, which upsets valve timing. Severe performance problems and even valve damage can result because the valves open at the wrong times.

Timing gear problems are not very common, but can happen after extended service or breakage of other parts. The most common problem is broken gear teeth, as shown in **Figure 18-12B.** Broken timing gear teeth can cause a loud knocking noise from inside the front cover. If enough teeth break off, the timing gears can jump and the valves will open incorrectly or the gears may lock up.

Timing chain problems can upset valve timing, which will reduce compression stroke pressure and engine power. Wear can cause slack between the crankshaft sprocket and camshaft sprocket. Then, the camshaft and valves will no longer be in time with the crankshaft and pistons.

To check for a worn timing chain, rotate the crankshaft back and forth while watching the rocker arms, camshaft, or distributor rotor. If you can turn the crankshaft several degrees without rocker arm or camshaft movement, the timing chain is worn and must be replaced. Excessive wear on the timing sprocket teeth also indicated a worn timing chain, **Figure 18-12C.** A compression test would show lowered compression in *all* cylinders. Incorrect valve timing and a worn chain affect all cylinders equally. Erratic timing mark motion when using a timing light during a tune-up could also be due to a worn timing chain.

A worn timing belt will usually break, jump off its sprockets, or skip over a few sprocket teeth. Severe performance problems or valve damage can result. With some engine designs, the pistons can move up into the open valves, bending or breaking the valves.

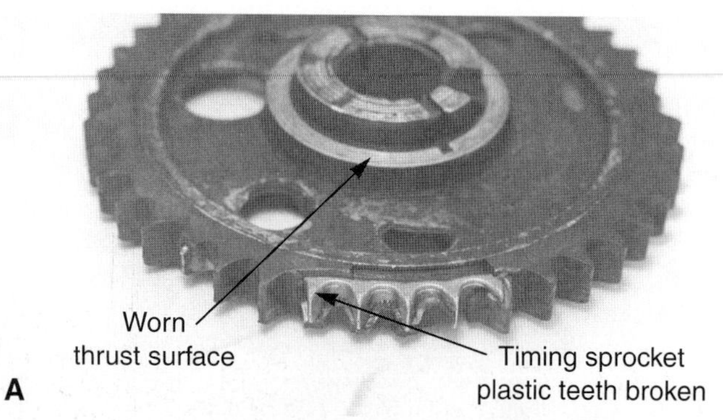

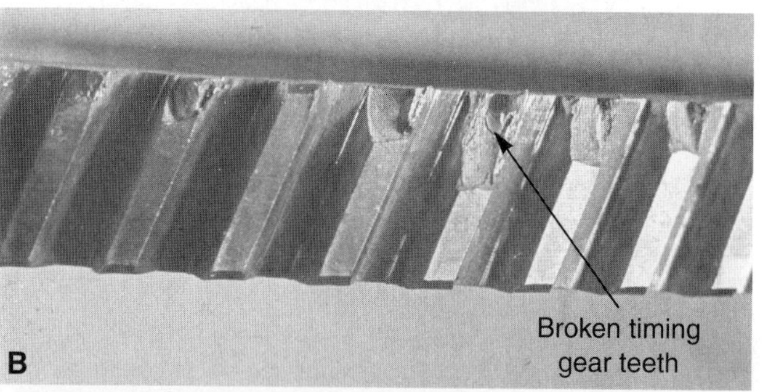

Figure 18-12. *Timing mechanism problems. A—The broken plastic teeth on this timing chain sprocket resulted in excess slack in the chain. B—This timing gear has broken teeth. It made a clunking sound with the engine running. C—Close inspection of sprocket teeth will often reveal teeth wear, which requires chain and sprocket replacement.*

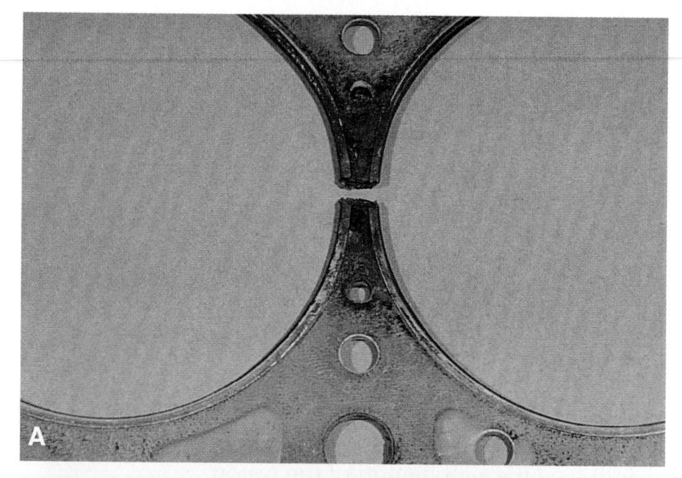

Figure 18-13. *Gasket problems are common. A—This is a blown head gasket that resulted from warped head. B—This leaking intake manifold gasket allowed oil to be drawn into the combustion chambers.*

A ***leaking exhaust manifold gasket*** is indicated by a clicking sound. As combustion gasses blow into the manifold and out of the bad gasket, an almost metallic-like rap is produced.

Part warpage is a condition in which the sealing surfaces on a part are not flat. It is a common cause of gasket failure, **Figure 18-14.** Always check for warpage when servicing a bad gasket.

Engine Gasket Problems

A ***blown head gasket*** can cause a wide range of problems, including overheating, missing, coolant or oil leakage, engine smoking, and burned mating surfaces on the cylinder head or block. Quite often, a blown head gasket will be indicated by a compression test. Two adjacent cylinders, usually the two center ones, will each have low pressure. A blown head gasket is shown in **Figure 18-13A.**

A ***leaking intake manifold gasket*** often results in a vacuum leak, which may result in rough idle. To check for an intake gasket leak, squirt oil along the edge of the gasket. The oil may temporarily seal the leak, in which case the intake gasket has ruptured. A low vacuum gauge reading can also indicate an intake manifold gasket leak. A leaking intake manifold gasket is shown in **Figure 18-13B.**

Cylinder Head Problems

A ***warped cylinder head*** has been overheated and is no longer true. Usually, the deck will warp and cause head gasket leakage. Coolant may leak into the engine oil, producing a milky substance. This is more of a problem with aluminum cylinder heads than with cast iron cylinder heads.

A ***cracked cylinder head*** may also result from engine overheating. Unequal heat expansion will break the head casting, usually near an exhaust port or on the deck between adjacent cylinders. See **Figure 18-15.** Engine coolant that is weak in antifreeze can also freeze in cold weather and crack a cylinder head.

A cracked head can leak coolant or combustion pressure. Coolant leakage can enter the combustion chamber or run down into the oil pan. Pressure leakage

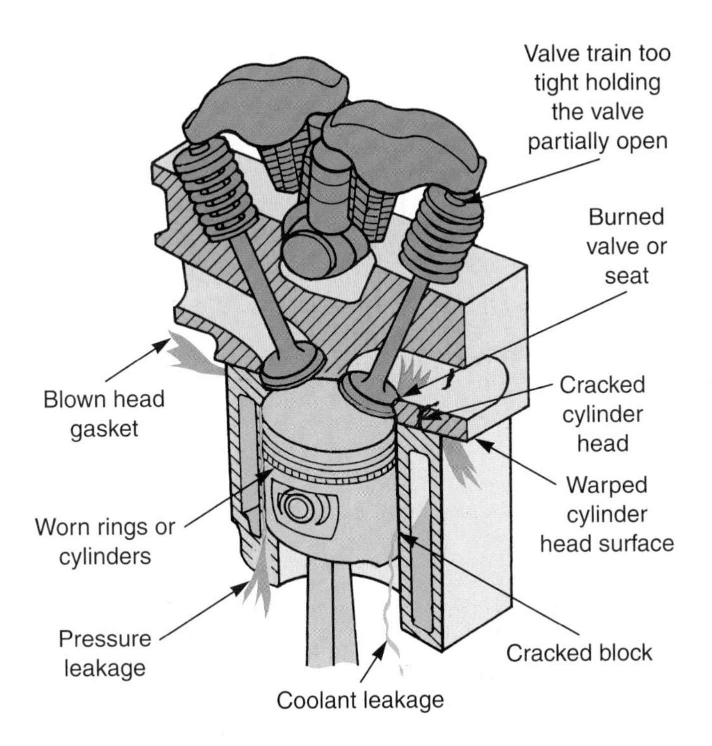

Figure 18-14. *Note some of the problems that can develop from part warpage.*

will allow combustion gasses to enter the cooling system. Coolant on a spark plug, milky oil, white exhaust smoke, and overheating are symptoms of a cracked head. A cracked head must be removed and welded or replaced.

Cylinder Block Problems

Cylinder block problems include cracks, warpage, worn cylinders, mineral deposits in coolant passages, misaligned main bores, and so on. Since the block is a large casting, it suffers from the same troubles as a cylinder head. Cylinder problems are major and require engine removal so the block can be bored or replaced.

Deck warpage resulting from overheating can cause head gasket leakage. This is not a very common problem.

Mineral deposits in the coolant passages can cause engine overheating. Lime can build up on the inner walls in the block. This can restrict coolant flow and cause localized hot spots, **Figure 18-16.**

Crankcase warpage will throw the main bearing bores out of alignment, as shown in **Figure 18-17.** This can make the crankshaft lock in the block during engine reassembly with new bearings. The block must be bored to correct this problem.

A ***cracked block*** can have fractures in the cylinder walls, deck surface, lifter gallery, and so on. This usually results from overheating, ice formation in cold weather, or breakage of pistons or connecting rods. **Figure 18-18** shows a badly damaged cylinder block. The connecting rod broke and knocked a large chunk of metal out of the cylinder. Sleeving is required to salvage the block.

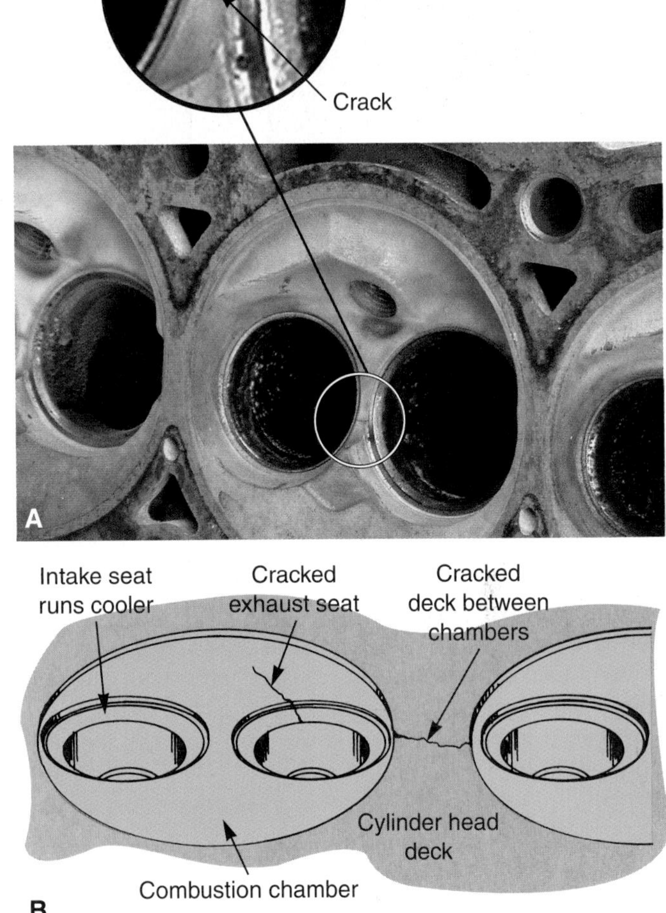

Figure 18-15. *A—This cylinder head has cracked between the ports. B—Cracks in a cylinder head are usually at the exhaust seat or between the combustion chambers on the deck surface.*

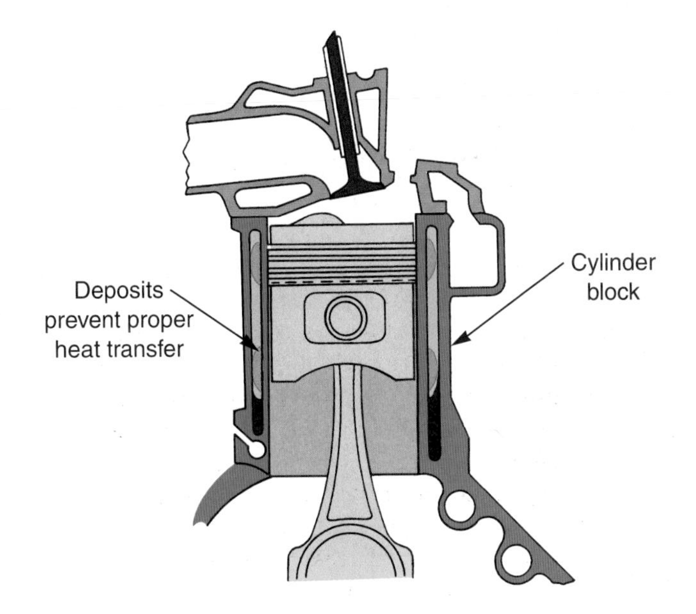

Figure 18-16. *Mineral deposits in coolant passages in the block can restrict coolant flow and cause overheating. (Deere & Co.)*

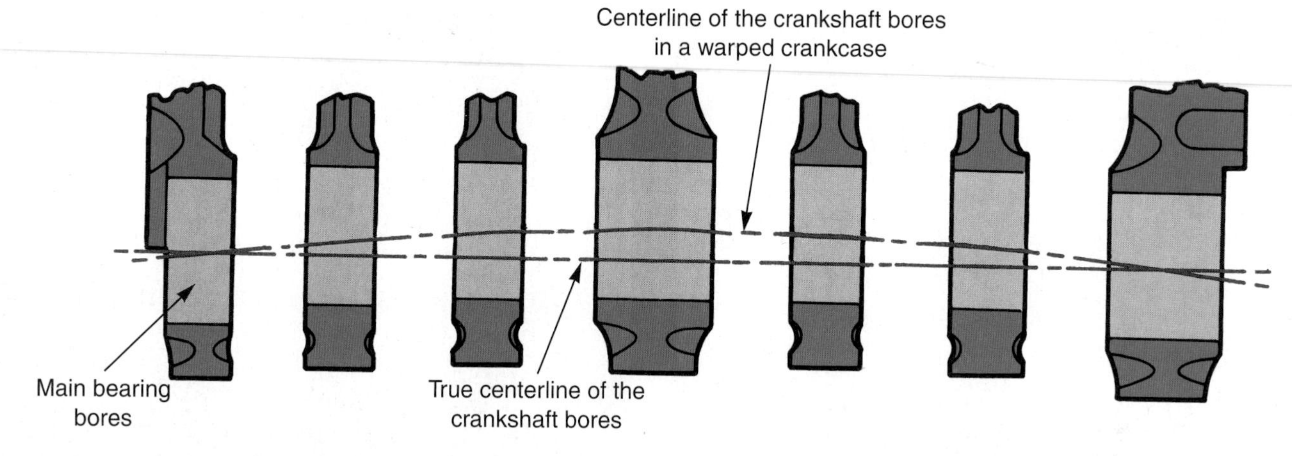

Figure 18-17. *Crankcase warpage can throw the main bearing bores out of alignment. Line boring is needed to fix the misalignment. (Federal Mogul)*

A ***worn cylinder block*** has a ring ridge in its cylinders. A ***ring ridge*** is a small lip formed where the rings do not wear the cylinder, **Figure 18-19.** Because there is more oil lubricating the bottom of the cylinder, the top of the cylinder will wear more. Since the piston rings do not rub at the very top of the cylinder, a ring ridge is formed. A ring ridge is an obvious sign of cylinder block wear. It must be cut out of the cylinder before piston removal.

As discussed in the previous chapter, you can read spark plugs to help diagnose an engine problem. For example, rings will not seal well on a bad cylinder and oil-fouled plugs can result.

Piston Problems

Piston problems are major and include skirt breakage, crown failure, worn ring grooves, and so on. Piston failure normally results in oil consumption, engine smoking, and sometimes knocking. **Figure 18-20** shows several worn or damaged pistons.

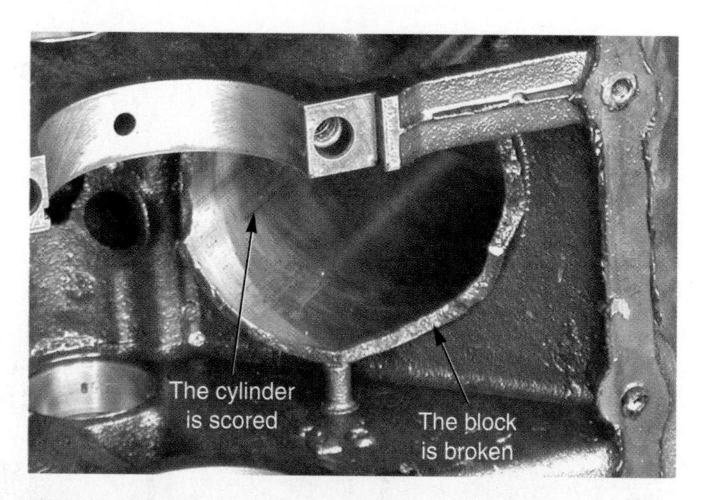

Figure 18-18. *This cylinder block was damaged when the connecting rod bolts broke off and the connecting rod was forced into the cylinder wall. Sleeving is needed to salvage the block.*

Figure 18-21 shows a piston that was damaged by a small bolt accidentally dropped into the intake manifold. The bolt eventually worked its way past the intake valve and into the cylinder. Once in the cylinder, it was hammered into the piston and cylinder head by piston movement. In addition to damaging the cylinder head and piston crown, the force of repeated impacts caused the ring grooves and rings to crack.

Piston Breakage

The most common type of ***piston breakage*** is a broken skirt. The piston then flops sideways in its cylinder and causes major cylinder damage. The affected cylinder will lose compression, burn oil, and make a knocking sound. Skirt breakage can result from too much piston-to-cylinder clearance, over-revving, or inadequate lubrication.

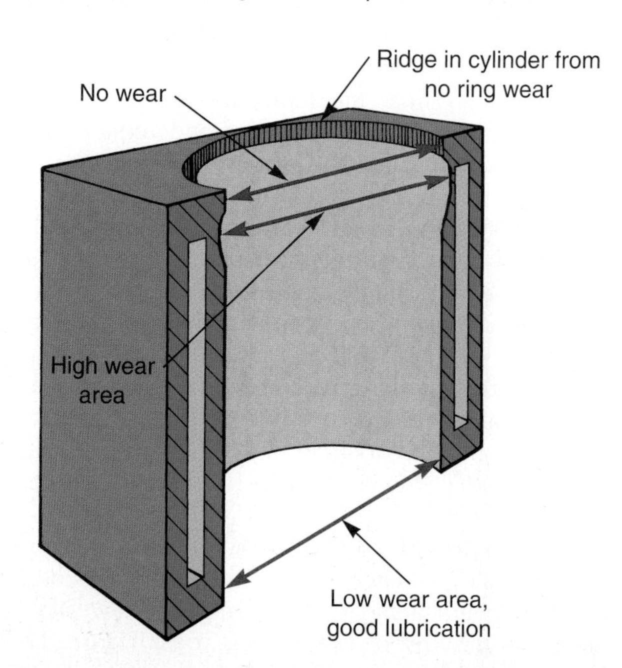

Figure 18-19. *There is more wear near the top of the cylinder and a ring ridge is formed where the piston rings do not rub on the cylinder wall.*

Figure 18-20. *A—This piston skirt shows normal wear for a high-mileage engine. B—This piston skirt suffered scoring when the engine overheated. C—Foreign matter is embedded in this piston skirt. When the engine was operated with low oil, the bearings eroded and that material was circulated through the engine. D—This skirt was damaged by broken connecting rod.*

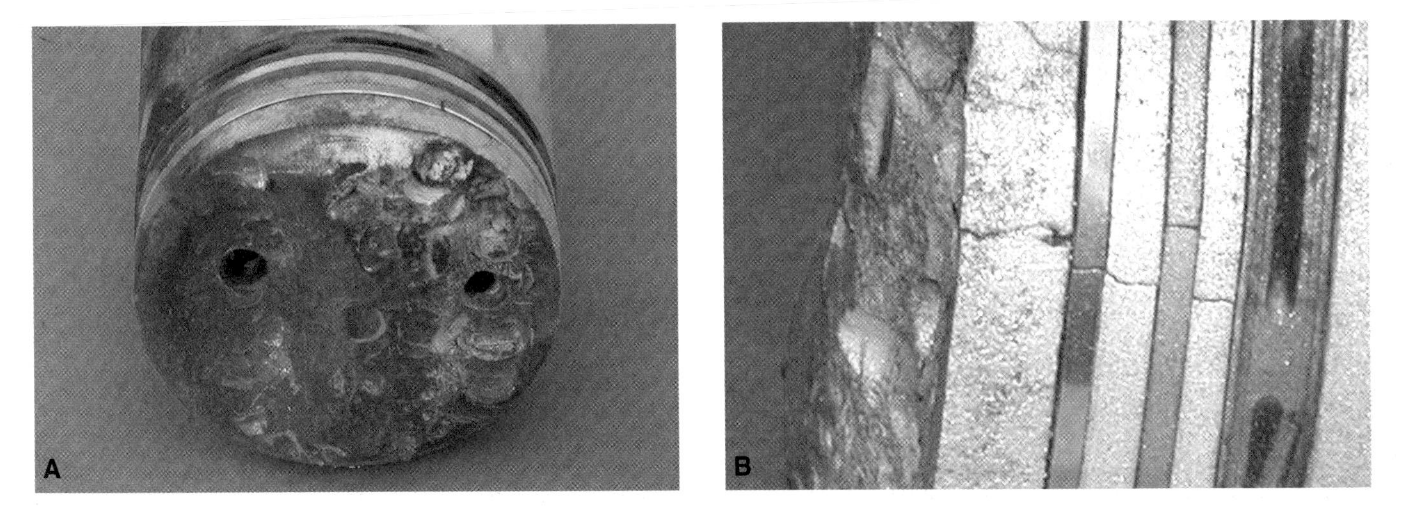

Figure 18-21. *A—A small bolt dropped into engine intake manifold caused this piston damage. B—As the bolt was repeatedly hammered into the piston head, the ring grooves and piston rings cracked.*

Piston Knock

Piston knock, or ***piston slap,*** is a loud metallic knocking sound produced when the piston flops back and forth inside its cylinder. It is caused by piston skirt or cylinder wear, excessive piston-to-cylinder clearance, or mechanical damage.

Piston slap is normally louder when the engine is cold. It tends to quiet down as the engine warms to operating temperature. As the aluminum piston heats up and expands, the clearance is reduced and the piston knock is reduced.

Piston Pin Knock

Piston pin knock occurs when too much clearance exists between the piston pin and piston pin bore or connecting rod bushing. Excessive clearance allows the pin to hammer against the rod or piston as the piston changes direction in the cylinder. Piston pin knock is usually indicated by a double knock. A double knock is two rapid knocks and then a short pause.

Burned Piston

A ***burned piston*** is often a result of preignition or detonation damage. Abnormal combustion, excessive pressure, and heat actually melt the piston and blow a hole in the crown or area around the ring grooves. The engine may smoke or knock, have excessive blowby, or present other symptoms.

Figure 18-22 shows a burned piston that was damaged by prolonged detonation. Heat melted the piston and pressure blew a hole along the side of the piston head and through the ring groove. The cylinder lost compression and blowby increased.

Figure 18-22. *This piston was burned from prolonged detonation. Heat melted the piston and pressure blew a hole along the side of piston head to the ring groove. This resulted in lower compression and increased blowby.*

A compression test or cylinder leakage test may indicate a burned piston. Engine disassembly is usually needed to verify the problem. However, if the cylinder wall is not damaged, the repair can be completed without removing the engine from the vehicle.

Melted aluminum deposits on a spark plug point to a melted or partially burned piston. Also, when the spark plug tip is damaged from detonation, you need to correct the cause of detonation and, in turn, possibly prevent piston damage.

Worn Piston Rings and Cylinders

Worn piston rings or ***worn cylinders*** result in blowby, blue-gray engine smoke, low engine power, spark plug fouling, and other problems caused by poor ring sealing. To check for ring and cylinder problems, increase engine speed while watching the tailpipe and valve cover breather opening. If blue-gray smoke pours out of the tailpipe with the engine under load, the oil rings and cylinders may need service. If excessive oil vapors and air blows out of the valve cover breather, blowby is entering the crankcase. Piston, compression ring, or cylinder problems are indicated. Worn ring grooves can also cause oil consumption.

If the engine cylinders are found to be worn after engine disassembly, the engine must be removed from the vehicle. The block must be sent to a machine shop for boring or sleeving. If just the rings are found to be worn, not the cylinders, the engine may be rebuilt while still installed in the vehicle. Check the service manual for details.

Connecting Rod Problems

A ***bent connecting rod*** is a condition where the rod is no longer straight. It results from detonation or excessive engine speed. As shown in **Figure 18-23,** a bent rod will load one side of the rod bearing and cause the piston to cock at an angle in the cylinder. A bent connecting rod is usually not discovered until engine disassembly. Uneven wear on the piston skirt or rod bearing is an indication of a bent connecting rod. Fixtures are available for checking rod straightness.

Figure 18-24 shows a connecting rod that was bent by hydraulic lock. The head gasket failed and one cylinder was filled with coolant. Since the engine was operating at high speed and the liquid coolant cannot be compressed, hydraulic lock occurred. The rod bent as the crankshaft rotated and attempted to force the piston assembly upward into the liquid on the compression stroke.

A ***broken connecting rod*** has suffered a complete failure of the I-beam or rod bolts and cap. See **Figure 18-25.** Rod breakage can result from excessive engine speed, running an engine after rod bearing failure, loss of lubrication, improper assembly, and so on.

When a connecting rod breaks, severe engine damage usually results. The spinning crankshaft can slam

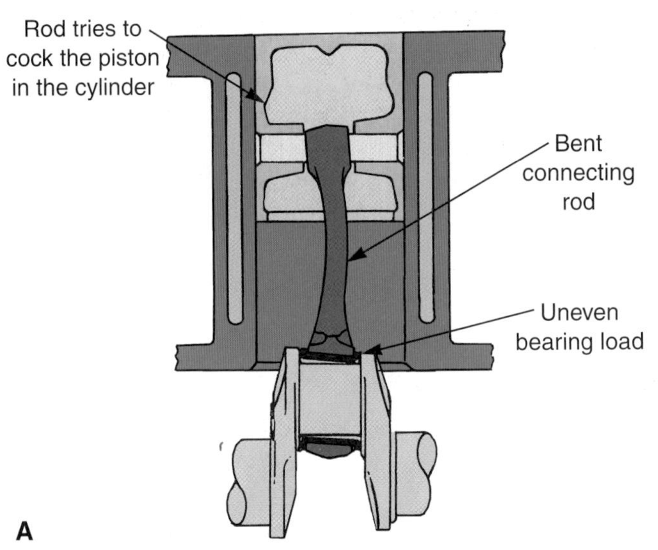

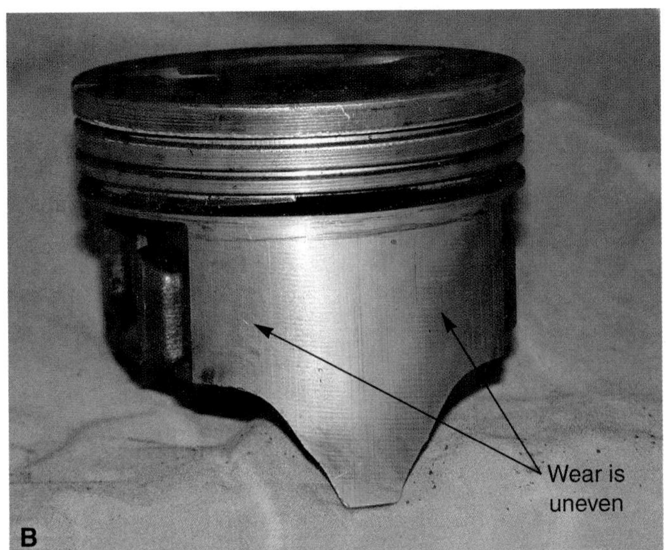

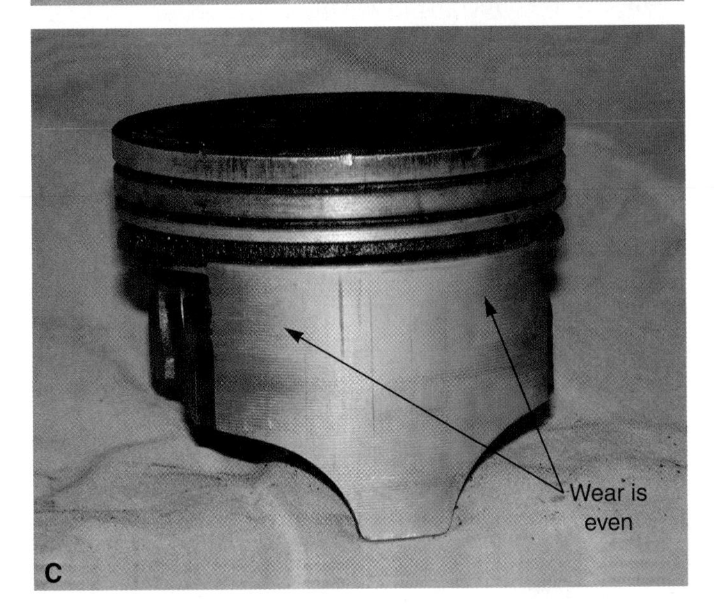

Figure 18-23. *A—A bent connecting rod loads one side of the piston skirt and rod bearing. B—Note the uneven piston wear pattern that resulted from a bent connecting rod. C—This is a normal wear pattern. Notice that the wear is even on both sides of the piston skirt.*

Figure 18-24. *This connecting rod was bent when the head gasket failed and a large amount of coolant flowed into the cylinder at a high engine speed. Since liquid cannot be compressed, the rod was badly bent on the compression stroke.*

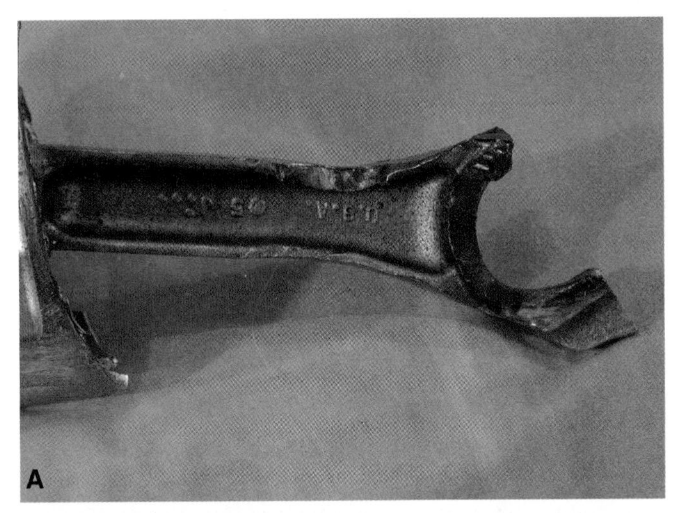

Figure 18-25. *Study major connecting rod failures. A—This connecting rod was improperly installed during an engine rebuild. During engine operation, the rod bolts snapped. The rod was then crushed into the block. B—This connecting rod snapped off near the piston pin when the engine was over-revved.*

into the disengaged rod, crushing the rod into the block and ruining the block. The piston can also fly upward and smash into the cylinder head, causing head and valve damage. Normally, complete engine disassembly, crankshaft turning, block boring, etc., are needed to repair a broken connecting rod.

Connecting rod bore problems result from extended service, rod bearing failure, oil pump problems, and so on. The big end of the rod can become out-of-round, in which case it can no longer properly support the rod bearing. Always inspect for rod bore problems during engine service. Marks on the bore can show rod bearing movement in the rod. This would tell you that the rod should be rebuilt by a machine shop or replaced.

Figure 18-26. *A bad rod bearing allowed this piston to hit the cylinder head, which produced a loud knocking noise. Note the shiny area that shows where the contact was being made.*

Crankshaft and Bearing Problems

The crankshaft and its bearings must withstand extreme loads and friction during engine operation. The slightest mistake during assembly, a lubrication problem, or driver abuse can lead to bearing or crankshaft troubles.

Rod Bearing Knock

Connecting rod bearing knock is caused by wear and excessive bearing-to-crankshaft clearance. It is indicated by a light, regular, rapping noise with the engine floating. ***Engine floating*** is a constant engine speed above idle. Rod bearing knock is loudest after engine warmup. Oil is thicker in a cold engine, which tends to cushion and quiet rod knock.

To locate a bad rod bearing, short out or disconnect each spark plug wire one at a time. The loose, knocking rod bearing may quiet down or change pitch when the corresponding spark plug is disabled.

With some engines, excessive rod bearing clearance can result in the piston head striking the cylinder head. This will make a loud knocking sound, similar to a failed piston. **Figure 18-26** shows a piston that has been hitting the cylinder head due to rod bearing failure. **Figure 18-27** shows how a badly worn rod bearing let the crankshaft counterweight rub on the bottom of a piston.

Spun Rod Bearing

A ***spun rod bearing*** is a bearing that has been hammered out flat and then rotated inside of the connecting rod bore. It will make a sound similar to a rod bearing knock, but much louder. When a rod bearing spins, it will normally damage the rod and crankshaft journal. A slight rod bearing knock often precedes total rod bearing failure.

Main Bearing Knock

Main bearing knock is similar to rod bearing knock, but the sound is slightly deeper or duller. It is usually more pronounced when the engine is pulling or lugging under a load. Bearing wear, and possibly journal wear, is allowing the crankshaft to move up and down inside the cylinder block.

Main bearing wear will usually reduce oil pressure significantly. To verify main bearing noise, remove the oil pan. Pressure test the lubrication system. Excessive oil flow out of one or more of the main bearings implies too much bearing clearance. If a pressure tester is not available, remove and inspect each of the main bearings. If a bearing is worn but the crankshaft is not, bearing insert replacement should correct the problem.

Figure 18-27. *The mark on the bottom of the piston boss is where the crankshaft counterweight was rubbing. This happened because of excess clearance due to a failed connecting rod bearing.*

Worn Main Bearings

Worn main bearings are too thin, which allows too much clearance between the crankshaft and bore. This excess clearance can allow the crankshaft to shift up and down in the block with load changes. A thin layer of metal has been worn off the bearing face, as shown in **Figure 18-28.**

Excess Crankshaft Endplay

Excess crankshaft endplay is caused by a worn main thrust bearing. Thrust bearing wear can produce a deep knock, usually when applying and releasing the clutch on a manual transmission. With an automatic transmission, the endplay problem may only show up as a single thud or knock on acceleration or deceleration.

With severe thrust bearing wear, you can sometimes verify bearing failure by watching the front crankshaft pulley with the engine running. When you accelerate and decelerate the engine, the crankshaft may "float" forward and rearward in the block. With engine off, you can also pry back and forth on the crankshaft damper with a large pry bar to verify excess main thrust bearing wear.

A knock occurring with clutch or torque converter action could also be caused by loose flywheel bolts or other drive train problems. Check out all possible causes before beginning repairs.

Flywheel Problems

Flywheel problems include loose bolts, failure of the ring gear teeth, friction surface scoring, warpage, and so on. A loose flywheel can make a knocking sound similar to main bearing knock. The teeth on the flywheel ring gear can wear or break off and cause starter engagement troubles. Keep these problems in mind during diagnosis.

Crankshaft Vibration Damper Problems

Crankshaft vibration damper problems include physical damage such as a deteriorated rubber ring, seal journal wear, slipped or sheared key, and so on. Engine vibration can result from a damaged vibration damper. Oil leakage can occur if the seal surface is worn or scored. If the outer metal ring slips on its rubber mount, the timing marks can be out of position. This can cause confusion when trying to set ignition timing. **Figure 18-29** illustrates a few of the problems you can encounter with a crankshaft vibration damper.

Intake and Exhaust Manifold Problems

Exhaust manifold problems are usually a cracked or warped manifold. See **Figure 18-30.** Engine exhaust heat can crack an exhaust manifold, allowing leakage and exhaust noise. Heat can also warp a manifold, allowing it to leak next to the cylinder head mating surface.

Intake manifold problems are not very common because the manifold is not exposed to as much heat as other parts. However, an intake manifold can sometimes warp and cause a vacuum leak. Also, passage plugs can also loosen or fall out, allowing vacuum or coolant leakage.

Mechanical Problem Diagnosis

If a technician does not know how to properly diagnose engine problems, a great deal of time, effort, and money will be wasted. An improperly trained technician may unknowingly rebuild an engine when a minor repair might have corrected the fault.

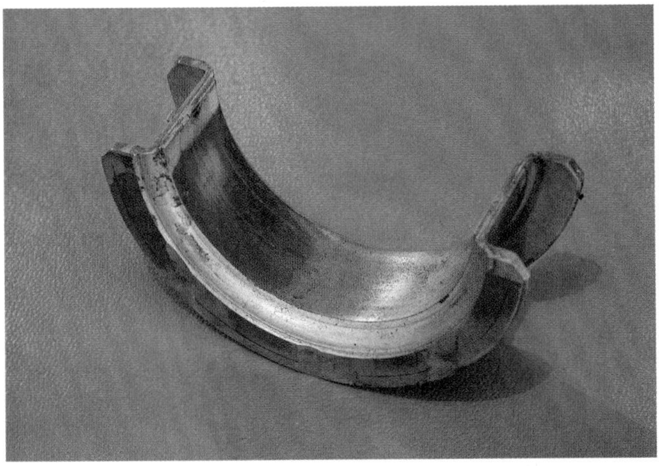

Figure 18-28. *Main bearing wear will lower oil pressure considerably and can cause knocking under load. This is a badly worn main thrust bearing.*

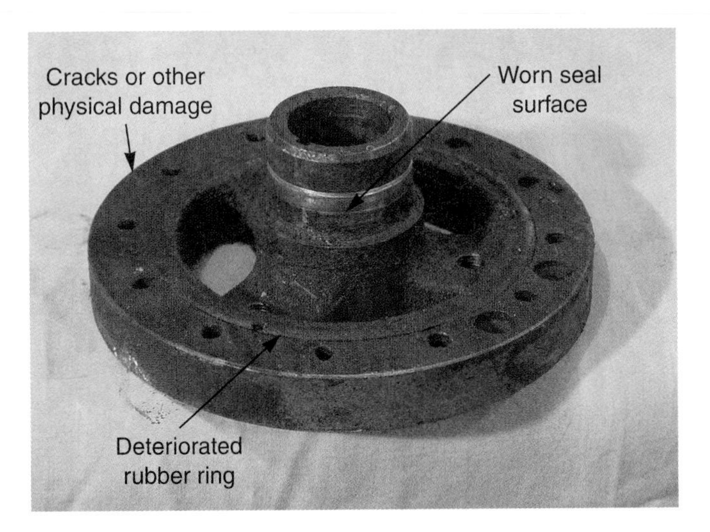

Figure 18-29. *Note the problems that can occur on a vibration damper. Keep these troubles in mind during engine service.*

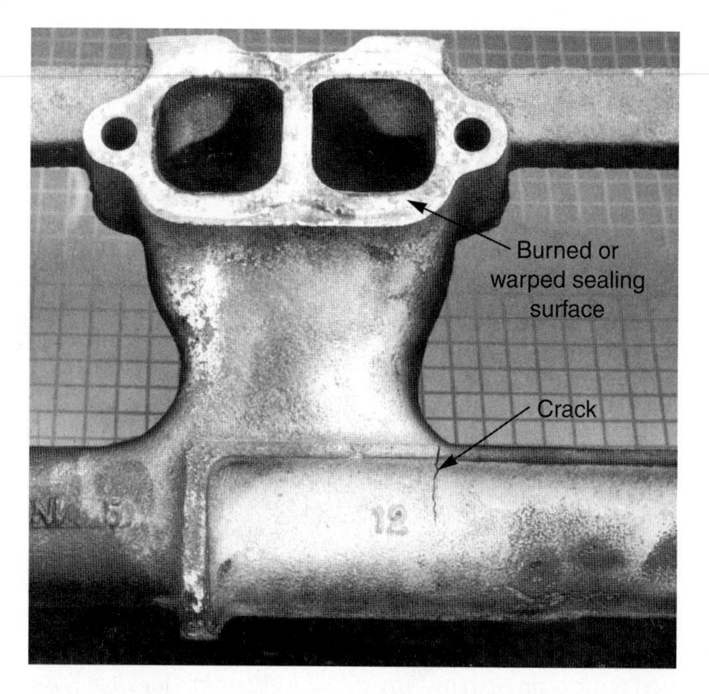

Figure 18-30. *A metallic, ticking sound indicates a cracked exhaust manifold. The sound is made as exhaust pressure blows out of the crack. This crack is on the back of the manifold and was difficult to locate. The leak was found by using a section of vacuum hose as a listening device. (Fel-Pro)*

For example, a worn or stretched timing chain can cause the engine valves to open and close at the wrong time in relation to piston movement. This could cause low compression pressure in all cylinders. The engine would have low power output and low compression readings. If not diagnosed correctly, the technician could overlook the actual trouble (worn timing chain) and complete more major repairs (replace piston rings or hone cylinders). A new timing chain would correct the problem, but the technician overhauls the engine instead. Unfortunately, the customer would have to pay for the technician's lack of training.

Engine Pre-Teardown Inspection

After gathering information from the customer or service writer, inspect the engine using your senses of sight, smell, hearing, and touch. First, look for external problems such as oil leaks, vacuum leaks, or part damage. If a fluid leak is found, smell the fluid to determine if it is oil, coolant, or other type of fluid. Listen for unusual noises that indicate part wear or damage. Increase engine speed while listening and watching for problems. The engine may run fine at idle, but act up at higher speeds.

Symptoms of Engine Mechanical Problems

Symptoms that result from engine mechanical problems include:

- ❑ **Excessive oil consumption.** Engine oil must be added frequently.
- ❑ **Crankcase blowby.** Combustion pressure blows past the piston rings, into the crankcase, and out of the breather.
- ❑ **Abnormal engine noises.** Knocking, tapping, hissing, or rumbling can be heard from the engine.
- ❑ **Engine smoking.** Blue-gray, black, or white smoke blows out of the tailpipe.
- ❑ **Poor engine performance.** Rough idle, slow acceleration, and engine vibrations indicate the engine is not performing properly.
- ❑ **Coolant in engine oil.** The engine oil has a white, milky appearance.
- ❑ **Engine locked up.** The crankshaft will not rotate.

With any of these symptoms, inspect and test the engine to determine the exact source of the problem. You must first find out which repair is needed. Then, determine whether the engine can be repaired in the vehicle or if it must be removed.

Symptom Diagnosis

Coolant in the oil is caused by a mechanical problem that allows engine coolant to leak into the engine crankcase. There may be a cracked block or head, leaking head gasket, leaking intake manifold gasket (V-type engine), or similar troubles.

Oil in the coolant is usually *not* an engine problem. It is caused by a leak in the radiator (transmission) oil cooler.

Oil-fouled spark plugs point to internal oil leakage into the engine combustion chambers. This may be caused by worn piston rings, worn or scored cylinder walls, or bad valve seals. You will need to perform additional tests to find the source of the problem.

External engine oil leaks occur when gaskets harden and crack, seals wear, fasteners work loose, or there is part damage (warped surfaces, cracked parts). To find external oil leaks, clean the affected area on the outside of the engine. Then, trace the leak upward to its source. Oil will usually flow down and to the rear of the engine because of the action of the cooling fan and airflow while driving.

External coolant leaks will show up as a puddle of coolant under the engine. Leaks can be caused by hose problems; rusted out core plugs; or warped, worn, or damaged parts. Use a cooling system pressure tester to locate external coolant leaks.

Engine blowby occurs when combustion pressure blows past the piston rings into the lower block and oil pan area of the engine. Pressure then flows up to the valve covers and out the breather. Excessive blowby will show up as smoke or oil around the breather. With the valve cover breather removed, oil vapors may blow out when engine speed is increased.

Engine vacuum leaks show up during inspection as a hissing sound, like air leaking out of a tire. Vacuum leaks are loudest at idle and temporarily quiet down as the engine is accelerated. This is because manifold vacuum drops with engine acceleration. Very rough engine idling usually accompanies a vacuum leak.

Engine Smoke

Engine smoking is normally noticed at the tailpipe when the engine speed is increased or decreased. The color of the smoke can be used to help diagnose the source of the problem. With a gasoline engine, the exhaust smoke may be:

- **Blue-gray smoke.** Indicates engine oil is entering combustion chambers; may be due to worn rings, worn cylinders, leaking valve stem seals, or other troubles.
- **Black smoke.** Caused by an extremely rich air-fuel mixture; not an engine mechanical problem.
- **White smoke.** May be due to internal coolant leak into the cylinders. May also be due to water condensation on cool day.

With a diesel engine, exhaust smoke may be:

- **Blue smoke.** Indicates engine oil is entering combustion chambers and being burned; may be due to ring, cylinder, or valve seal problems.
- **Black smoke.** Indicates an injection system problem or low compression is keeping fuel from burning.
- **White smoke.** Indicates unburned fuel, a cold engine, or coolant leaking into the combustion chambers.

Figure 18-31 shows engine exhaust smoke problems for an engine. Refer to the index for more information on exhaust smoke.

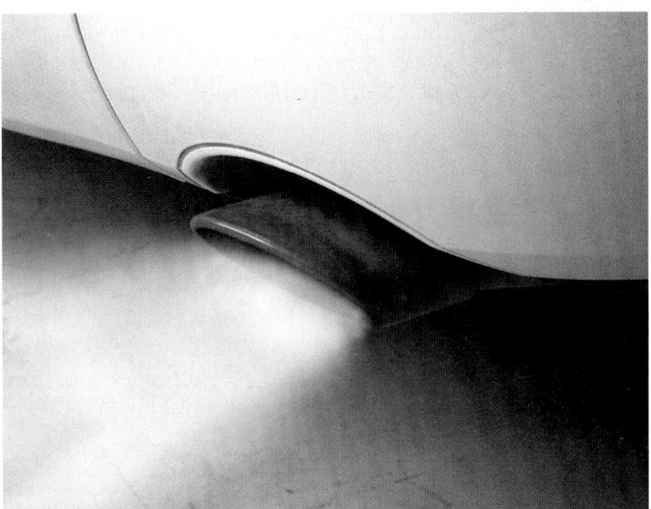

Figure 18-31. *The color of engine exhaust smoke is a good indicator of the type of engine problem. A—Blue-gray smoke indicates oil consumption. Oil is entering the cylinders and being burned. B—White smoke can be from coolant leakage into the combustion chambers or, in a diesel engine, unburned fuel. C—Black smoke indicates a rich air-fuel mixture and injection system trouble.*

Finding Abnormal Engine Noises

Abnormal engine noises include hisses, knocks, rattles, clunks, and popping. These noises may indicate part wear or damage. You must be able to quickly locate and interpret engine noises and determine what repairs are needed.

A stethoscope can be used to find internal sounds in parts. To use a stethoscope, place the headset earpiece in your ear. Then, touch the probe on different parts around the noise. When the sound becomes the loudest, you have touched near the part producing the abnormal noise. A long screwdriver can be used when a stethoscope is not available. Sounds will travel through the screwdriver, like with a stethoscope.

A section of hose can be used to locate vacuum leaks and air pressure leaks. Place one end of the hose next to your ear. Move the other end around the engine compartment. When the hiss is loudest, the hose is near the leaking part.

Warning: When using a listening device, keep it away from the spinning engine fan or belts. Severe injury could result if the stethoscope, screwdriver, or hose hits and is pulled into the fan or belts!

Figure 18-32. *This engine had burned valves and worn valve guides. The technician did not need to remove the engine from the vehicle. The cylinder heads were removed and sent to a machine shop for rebuilding. (Riverdale Auto Repair)*

Decide What Repair Is Needed

After performing all of the necessary inspections and tests, decide which part or parts must be repaired or replaced to correct the engine mechanical problem. Evaluate all data from your pre-teardown diagnosis. If you still cannot determine the exact problem, the engine may have to be partially disassembled for further inspection. Before you can properly repair engine problems, you must be able to:

- ❑ Explain the function of each engine part.
- ❑ Describe the construction of each engine part.
- ❑ Explain the cause of engine problems.
- ❑ Describe the symptoms of major engine problems.
- ❑ Select specific methods to pinpoint specific problems.
- ❑ Know which parts must be removed for certain repairs.
- ❑ Know whether the engine must be removed from the vehicle during the repair.
- ❑ Efficiently use service information to guide you through the repair process.

These skills can only be developed through study and work experience. For example, **Figure 18-32** shows an engine with its cylinder head removed. The engine had burned valves. Since tests indicated that the short block was in good condition, the engine could remain in the vehicle while the cylinder heads were removed for service. A valve job was performed to correct the problem and the heads were reinstalled.

Summary

The engine valve train is a common cause of mechanical problems in an engine. Valve train problems can produce light tapping noises, engine missing, power loss, and major engine damage.

A worn camshaft has lobes that are worn down by friction. This can prevent the valves from opening fully. Camshaft journal or bearing wear can reduce engine oil pressure.

A lifter problem will cause valve train clatter. External wear shows up as a concaved bottom on the lifter. Internal wear will make the lifter leak down too quickly.

A bent push rod results from a stuck valve, a valve hitting a piston, spring bind, spring breakage, and so on. It will make a tapping noise or can disengage from the rocker and prevent valve action. A push rod may also break or become clogged.

Rocker shaft wear results from friction between the shaft and the rocker arm. Rocker arm wear can occur on the inside diameter of the rocker arm or where it contacts the valve or push rod. Rocker stand wear occurs where the stand rubs on the rocker arm.

A burned valve has had some of its face and margin blown away. A loss of cylinder compression results. Valve face erosion is a wearing away of some of the valve face. A valve may also be bent, cracked, or chipped.

Worn valve guides can produce a tapping noise and oil consumption. Bad valve seals will also cause engine smoking, especially on initial engine startup.

Weak valve springs may cause valve float, a condition where the valves fail to close fully at high engine speed. A broken valve spring will cause valve train noise.

A worn timing chain will reduce engine compression in all cylinders because valve timing will be off. A worn timing belt can fly off its sprockets and result in valve damage.

A blown head gasket can cause overheating, missing, coolant or oil leakage, and so on. Intake manifold gaskets and exhaust manifold gaskets may leak. Gasket leakage usually occurs when one of the mating parts is warped.

Cylinder heads and blocks can be cracked and warped by engine overheating or ice formation in cold weather. A warped head would have to be milled.

A worn cylinder will have more wear at the top. As a result, a ring ridge is formed. A worn cylinder will cause oil consumption and smoking.

When a piston breaks, the break usually occurs in the skirt. Excess cylinder clearance lets the piston flop sideways, breaking off the bottom of the skirt. Smoking, knocking, and cylinder damage can result.

Piston knock or slap is a result of too much clearance. It is usually loudest when the piston is cold and quiets down as the engine warms.

A burned piston has a hole melted or blown in the head or land area. It is usually caused by detonation.

Worn piston rings cause engine oil consumption. Oil can be drawn past the rings and into the combustion chambers. Blue smoke is seen at the tailpipe. Smoking is more pronounced with the engine warm and under a load.

A connecting rod can bend or break. When a connecting rod breaks, it is usually hit by the spinning crankshaft and jammed into the side of the block. Parts are severely damaged.

Rod bearing knock results from too much crankshaft-to-bearing clearance. It is a light, regular rapping noise at a constant engine speed. Shorting out cylinders will affect the knock and help find the cylinder with the rod bearing knock.

A spun rod bearing has been hammered out and rotated inside the rod bore. Severe rod and crankshaft damage normally result. The excess clearance may also allow the piston to hit the cylinder head.

Main bearing knock is loudest when lugging the engine. The knock is deeper in pitch than a rod bearing knock. Worn main bearings cause main bearing knock and lower engine oil pressure considerably.

The symptoms of engine mechanical problems include oil consumption, blowby, abnormal noises, smoking, poor engine performance, coolant in the oil, and an engine that is locked up. Blue-gray smoke indicates oil is entering the combustion chambers. Black smoke is a sign that the air-fuel mixture is too rich. White engine smoke is from coolant leakage into the combustion chamber or from poor combustion in a diesel engine.

Review Questions—Chapter 18

Please do not write in this text. Write your answers on a separate sheet of paper.

1. A worn camshaft lobe results in reduced valve _____.
2. Worn camshaft bearings can reduce engine oil _____.
3. _____ lifter wear on a hydraulic lifter can prevent the lifter from pumping up to take up valve train clearance.
4. List four conditions that may result in a bent push rod.
5. Describe a *burned valve.*
6. When does valve stem breakage usually occur?
7. Define *valve float.*
8. How can you check for a worn timing chain?
9. This problem will lower compression, especially in two adjacent cylinders.
   (A) Burned valve.
   (B) Cracked head.
   (C) Leaking intake gasket.
   (D) Blown head gasket.
10. When a cylinder head cracks, where does the crack usually occur?
11. A ring ridge is a sign of a worn _____.
12. What condition will throw the main bearing bores out of alignment?
13. A burned piston is often the result of _____ or _____ damage.
14. What is *connecting rod bearing knock?*
15. Which of the following engine smoking conditions is not caused by an engine mechanical problem?
    (A) Blue-gray smoke.
    (B) White smoke.
    (C) Black smoke.
    (D) None of the above.

ASE-Type Questions—Chapter 18

1. An engine has valve train clatter. The engine has high mileage. Inspection found sludge or hardened oil in the engine. Technician A says to inspect components such as the rocker arms, push rods, and rocker stands for wear and replace them if worn. Technician B says you should check the hole in the center of the push rods for clogging. Who is correct?
   (A) Technician A.
   (B) Technician B.
   (C) Both A and B.
   (D) Neither A nor B.
2. An engine is misfiring and a puffing or popping noise can be heard at the throttle body. Technician A says there could be a leak at an intake valve caused by a bad spring or bent valve. Technician B says the engine could have a burned piston. Who is correct?
   (A) Technician A.
   (B) Technician B.
   (C) Both A and B.
   (D) Neither A nor B.
3. Technician A says that engine valve train problems can produce a rough idling condition. Technician B says that valve train malfunctions can cause engine cylinder damage. Who is correct?
   (A) A only.
   (B) B only.
   (C) Both A and B.
   (D) Neither A nor B.
4. All of the following are common symptoms caused by camshaft lobe wear *except:*
   (A) valve train noise.
   (B) excessive combustion pressure.
   (C) engine performance problems.
   (D) reduced valve lift.

5. Technician A says that a burned valve will cause an engine to run rough, especially at high speeds. Technician B says that a burned valve will cause an engine to run rough, especially at idle. Who is correct?
   (A) A only.
   (B) B only.
   (C) Both A and B.
   (D) Neither A nor B.

6. Technician A says that a crack in a cylinder head usually occurs near an intake port. Technician B says that a cylinder head crack usually results from overheating. Who is correct?
   (A) A only.
   (B) B only.
   (C) Both A and B.
   (D) Neither A nor B.

7. Technician A says that piston skirt breakage can result from over-revving an engine. Technician B says that piston skirt breakage can be caused by a malfunction in the engine's lubrication system. Who is correct?
   (A) A only.
   (B) B only.
   (C) Both A and B.
   (D) Neither A nor B.

8. Which of the following is a symptom of worn piston rings?
   (A) Blue-gray engine smoke.
   (B) Low engine power.
   (C) Fouled spark plugs.
   (D) All of the above.

9. Technician A says that a bent connecting rod is sometimes caused by engine detonation. Technician B says that a bent connecting rod is normally caused by using high-octane fuel. Who is correct?
   (A) A only.
   (B) B only.
   (C) Both A and B.
   (D) Neither A nor B.

10. Technician A says that black engine exhaust smoke is usually caused by a lean air-fuel mixture. Technician B says that black engine exhaust smoke is not caused by an engine mechanical problem. Who is correct?
    (A) A only.
    (B) B only.
    (C) Both A and B.
    (D) Neither A nor B.

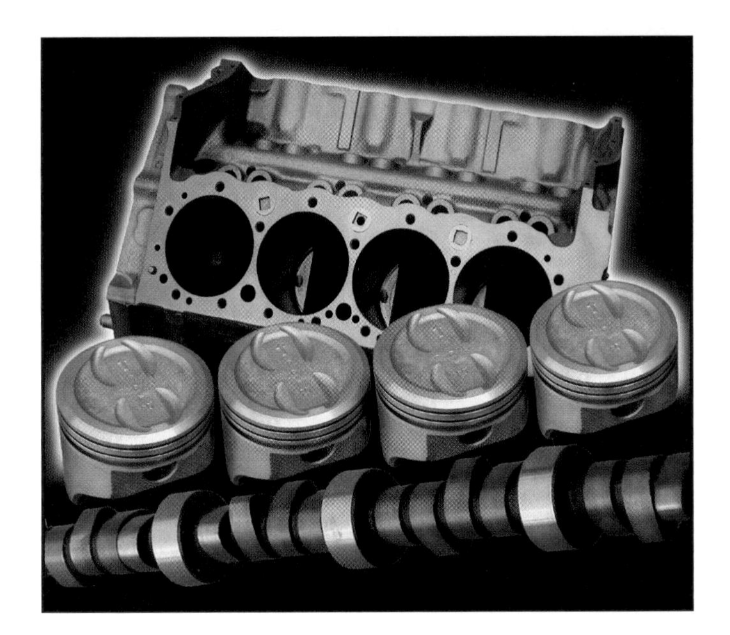

Chapter 19

Engine Problem Diagnosis and Tune-Up

After studying this chapter, you will be able to:

- Describe a minor tune-up and a major tune-up.
- List tune-up safety rules.
- List items to look for in a preliminary inspection.
- Perform a current starter draw test.
- Perform dry and wet compression tests and analyze the results.
- Perform a cylinder leakage test and analyze the results.
- Perform a vacuum test and analyze the results.
- Perform a cylinder balance test and analyze the results.
- Describe valve adjustment.
- List general tune-up rules.

Know These Terms

Compression test
Cranking balance test
Cylinder balance test
Cylinder leakage test
Cylinder pressure variations
Engine tune-up
Major tune-up
Minor tune-up
Vacuum test
Wet compression test

An engine tune-up is not only important to engine power and fuel economy, it is equally critical to air quality and the conservation of our natural resources. An untuned engine consumes a tremendous amount of fuel and produces high levels of air pollution. Both are harmful to our world as a whole.

In addition, many engine mechanical problems, such as worn piston rings, burned valves, or a slipped timing belt, are discovered when doing a tune-up. For example, a customer may bring a vehicle into the shop and indicate that the engine lacks power and seems to miss at idle. The customer asks for a tune-up thinking that will correct the problems. However, after you perform diagnostic tests, you find an engine mechanical failure. A tune-up would not correct the problems the customer indicated.

As an engine technician, you must be able to keep an engine running efficiently. This chapter summarizes what to do when initially checking out an engine to find and correct mechanical and performance problems. This chapter is an overview of subject matter presented in several other text chapters. For this reason, you may want to refer to the index to find additional information on specific subjects, such as injection system and ignition system service.

What Is a Tune-Up?

An ***engine tune-up*** is a maintenance operation that returns the engine to peak performance after part wear and deterioration. It involves the replacement of worn parts, basic service tasks, making adjustments, and, sometimes, minor repairs. A tune-up ensures that the ignition system, fuel system, emission control systems, and engine itself are within factory specifications.

Manufacturers normally recommend a tune-up after a specific period of engine operation. A tune-up is critical to the operation of an engine. It can affect:

- ❑ Engine power and acceleration.
- ❑ Fuel consumption.
- ❑ Exhaust pollution.
- ❑ Smoothness of engine operation.
- ❑ Ease of starting.
- ❑ Engine service life.

You must make sure you return every engine related system to peak operating condition during a tune-up. If you overlook just one problem, the engine will not perform properly and the customer will not be happy with your work.

The procedures for a tune-up vary from shop to shop. In one garage, a tune-up may include only the most routine tasks, such as visual inspection, changing filters, and changing spark plugs. In another shop, a tune-up may involve a long list of tests, adjustments, and repairs. Be sure to inform the customer what is covered by a tune-up performed by your shop.

Figure 19-1. *During an inspection, look for any troubles that could upset engine performance and dependability.*

Minor Tune-Up

A ***minor tune-up*** is preventative maintenance done when the engine is in good operating condition. For example, a new vehicle driven 20,000 miles to 30,000 miles (30,000 km to 40,000 km) may only require a minor tune-up. Most of the engine systems are in satisfactory condition with little or no wear. A minor tune-up typically involves these tasks:

- ❑ Visually inspect for obvious problems affecting engine performance, **Figure 19-1.** This includes loose wires, cracked vacuum hoses, oil leaks, coolant leaks, worn engine belts, and so on.
- ❑ Listen for abnormal engine noises, such as knocking, tapping, or hissing. A stethoscope or piece of vacuum hose will help you find the exact source of engine noises, **Figure 19-2.** The part emitting the loudest noise is near the source of the trouble.
- ❑ Check the condition of the engine oil and coolant. Pull out the dipstick and check to see if the oil is extremely dirty or milky, **Figure 19-3.** Milky oil indicates coolant in the oil. If the oil is dirty enough, it can trip a computer trouble code. Also, check the coolant to see if it looks rusty or smells acidic.
- ❑ Check for tripped trouble codes in the vehicle's computer.
- ❑ On older, high-mileage vehicles, replace the distributor cap, rotor, and spark plugs. Some new engines do not require plug replacement for 100,000 miles (160,000 km).
- ❑ Check and adjust the ignition timing, if adjustable.
- ❑ Replace air and fuel filters.
- ❑ Make fuel system adjustments, such as idle speed, if any adjustments are available.
- ❑ Clean the throttle body, idle air bypass, and EGR system.
- ❑ Test and service the emission control system. For example, replace the vapor canister filter.

Figure 19-2. *During an inspection, listen for abnormal engine noises. Use a listening device to pinpoint the trouble. A—By touching a stethoscope on the engine, you can find exactly where the noise is located. B—A section of vacuum hose or a stethoscope with the metal probe removed will help find vacuum and exhaust leaks.*

- ❑ Fill all fluid reservoirs to the correct level.
- ❑ Road test the vehicle.

Major Tune-Up

A ***major tune-up*** includes the steps for a minor tune-up plus diagnostic tests such as scope analysis, compression tests, or vacuum tests. These diagnostic tests help determine what should be done during the major tune-up. A major tune-up may also include fuel injection system cleaning or repairs, throttle body cleaning or rebuild, and other time-consuming jobs.

A major tune-up is more thorough than a minor tune-up. It is done when the ignition system, fuel system, emission control systems, or the engine itself is in poor condition. After prolonged use, parts can wear or deteriorate, requiring attention to restore them to proper operating condition. **Figure 19-4** gives a few problems that should be located and corrected during a major tune-up.

Tune-Up Safety

When doing a tune-up, remember these safety rules:

- ❑ Engage the emergency brake and block the wheels when the engine is to be running.
- ❑ Place an exhaust hose over the tailpipe when running the engine in an enclosed shop.
- ❑ Keep clothing, hands, tools, and equipment away from the spinning engine fan and belts.
- ❑ Disconnect the battery when recommended in the service manual. This will help avoid accidental engine cranking, electric fuel pump operation, or an electrical fire.
- ❑ Be careful not to touch the hot exhaust manifold when removing spark plugs.
- ❑ Keep test equipment leads away from the engine exhaust manifolds, fan, and belts.
- ❑ Wear eye protection when blowing debris from parts or when working near the spinning engine fan.
- ❑ Keep a fire extinguisher handy, especially when performing fuel system tests and repairs.
- ❑ With a diesel engine, disable the injection pump when removing an injection line. The system pressure is high enough to cause injury.

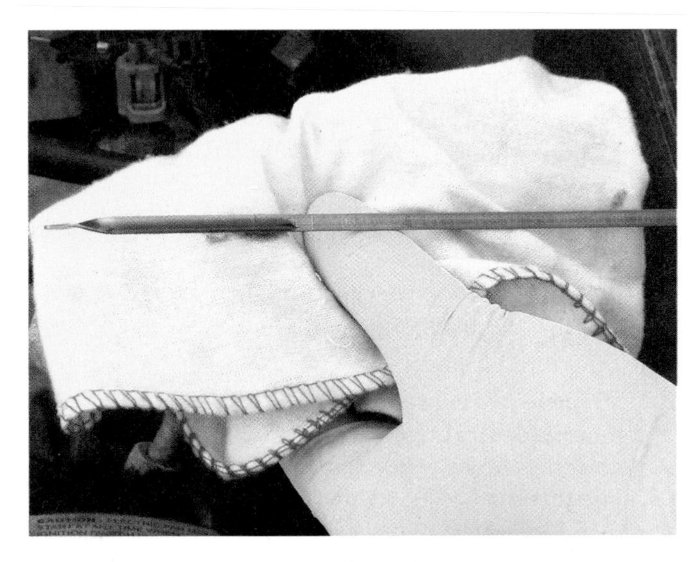

Figure 19-3. *Inspecting the oil on the engine dipstick is a good way to find indications of potential problems.*

Figure 19-4. *When inspecting the engine compartment, look for these kinds of problems. (Land Rover)*

Preliminary Inspection

To begin a tune-up, inspect the engine compartment. Try to find signs of trouble:

- **Battery problems.** Look for a dirty case top, corroded terminals, or physical damage.
- **Air cleaner problems.** Look for a clogged filter, inoperative air flap, or disconnected vacuum hoses or ducts.
- **Fuel system problems.** Check for leaks or a clogged fuel filter.
- **Belt troubles.** Inspect belts for fraying, cracks, slippage, or improper tension.
- **Deteriorated hoses.** Look for hardened or softened cooling system, fuel system, and vacuum hoses.
- **Poor electrical connections.** Check for loose or corroded connections, frayed wires, or burned insulation.

If any problems are found, correct them before beginning the tune-up. Many of these problems could affect engine performance.

Evaluating the Engine and Its Systems

During a tune-up, it is very important that you test and evaluate the condition of the engine and its systems. Various test instruments are used for this purpose. A compression gauge, leakdown tester, ignition scope, vacuum gauge, exhaust analyzer, timing light, and multimeter are a few of the test instruments used during a tune-up.

If your scan tool reading shows a problem with the O_2 sensors, remove and inspect the sensors. Reading oxygen sensors is similar to reading spark plugs. Note the color of its tip. This can give you information about the condition of the engine and its support systems. For example, a black tip indicates a rich air-fuel mixture caused by a fuel system problem, a leaky injector for example. See **Figure 19-5.** The new oxygen sensor must be identical to the existing one.

If the engine seems to be running too hot or cool, measure its actual operating temperature with a digital thermometer. Touch the test probe on the engine and take a reading. Refer to **Figure 19-6.** If the engine is running too cool, the thermostat may be stuck open. If running too hot,

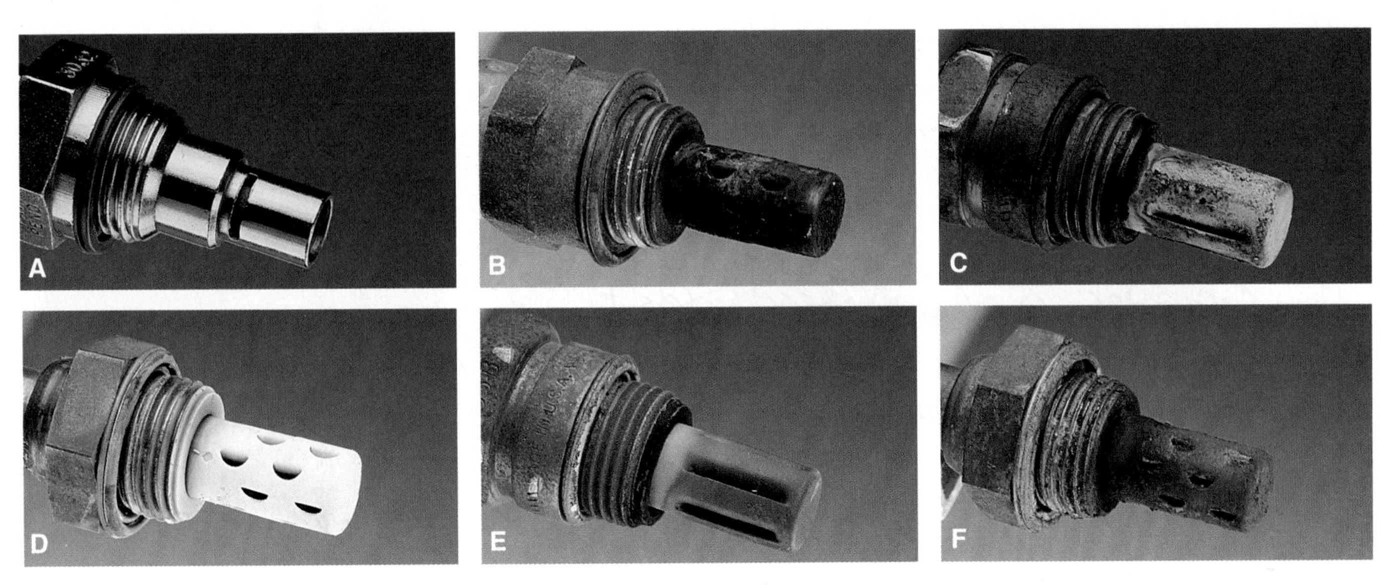

Figure 19-5. *After removing the oxygen sensor, note the color of its tip. This can give you information about the condition of the engine and its support systems. A—This is a new, high-quality, platinum oxygen sensor that is designed to provide long service life. B—A rich mixture caused by a fuel system problem has coated this sensor with black soot. C—Antifreeze contamination of this sensor was caused by a leaking head gasket or a crack in the engine block. D—Silicone contamination of this sensor resulted from the use of too much silicone sealer during an engine rebuild. E—Lead contamination of this sensor was caused by using leaded racing fuel or a lead-based octane additive. Check the fuel tank filler neck for modification. F—Oil contamination of this sensor was caused by worn piston rings and valve seals. (Tomco, Inc.)*

the thermostat could be stuck closed or the engine may have another trouble.

If a leaking head gasket, cracked head, etc., is allowing combustion gasses to blow into the engine coolant, bubbles can sometimes be seen down in the radiator neck. You can use an engine analyzer or handheld tester to check for a blown head gasket. Draw air from the radiator into the handheld combustion leak tester, **Figure 19-7A.** If there is a combustion leak into the coolant, the tester fluid changes color, usually to yellow. The exhaust gas analyzer will show the presence of hydrocarbon emissions, which indicates combustion gasses in the coolant. Look at **Figure 19-7B.** A bad head gasket, warped head, cracked head, or cracked cylinder block may be the problem if combustion gasses are found.

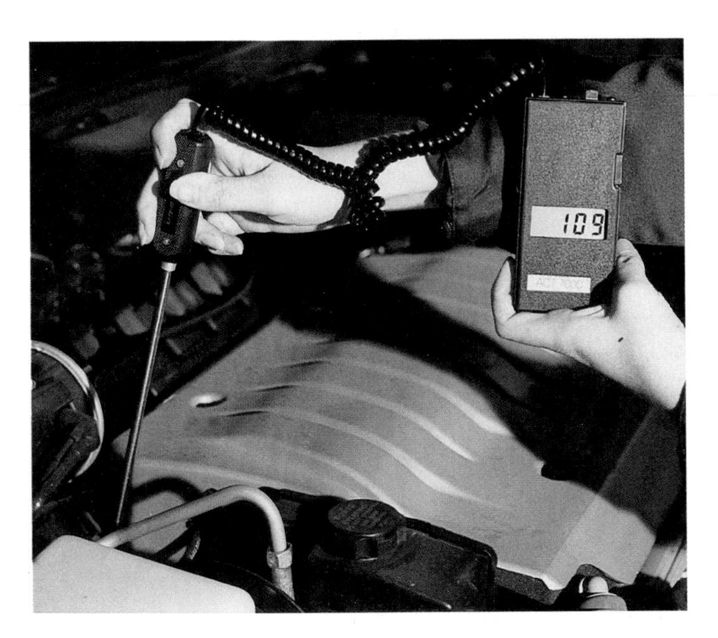

Figure 19-6. *A digital thermometer can be used to measure the actual operating temperature of the engine.*

To find hard-to-locate oil and coolant leaks, you can add a dye to the oil or coolant. One dye is designed to be added to the engine oil. Another type is for the coolant. After adding the dye, run the engine to full operating temperature. You can then move a black light around the engine to help find leaks. The black light and dye will make any leak glow so that it is easier to pinpoint. See **Figure 19-8.**

Cranking Balance Test

Modern engine analyzers can perform a ***cranking balance test*** to check the engine's mechanical condition. See **Figure 19-9.** This test is done to find cylinders with low compression. Low compression may be due to burned engine valves, worn piston rings, or other engine mechanical problems.

The engine analyzer monitors the amount of current drawn by the starter motor through a complete rotation of the crankshaft. As each cylinder goes through its compression stroke, a little more starter current draw is needed to rotate the crankshaft. In this way, an engine analyzer can find a cylinder with low compression. Lower peak current draw during any compression stroke shows low compression in that cylinder. A low current draw or low display line for any cylinder indicates the starter is not being loaded enough by the compression stroke.

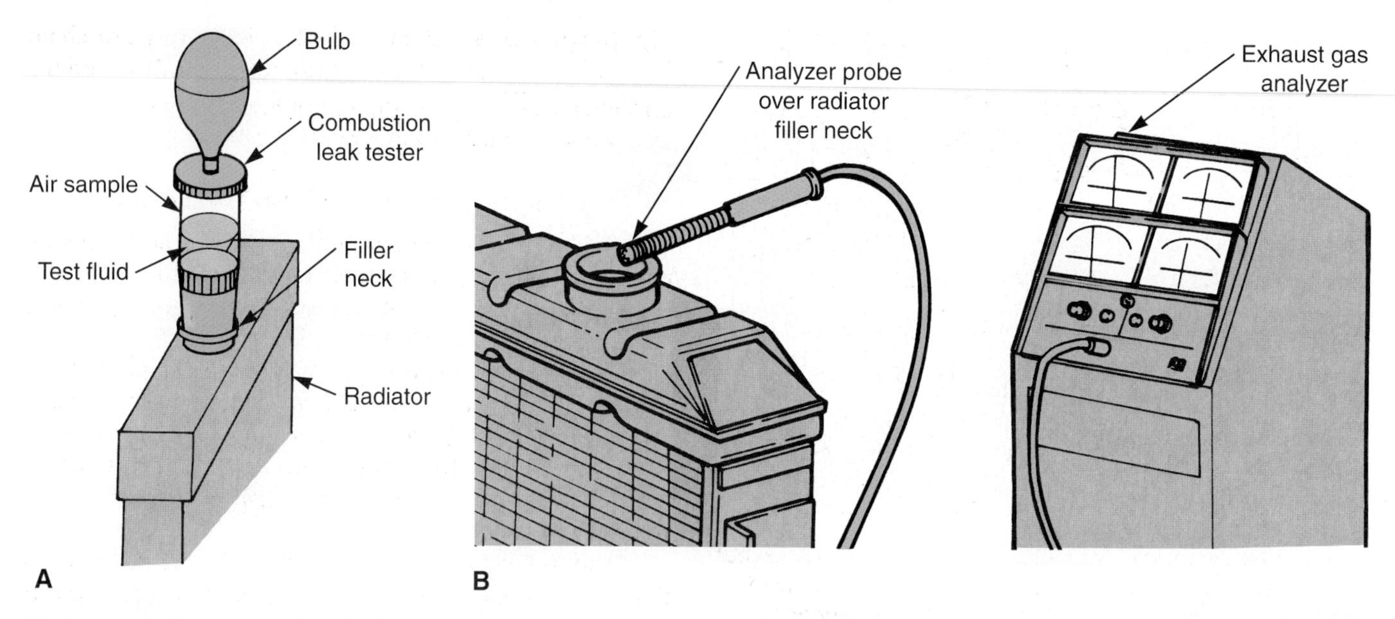

Figure 19-7. *Here are two ways to check for combustion gasses in the coolant, which may indicate a blown head gasket or cracked cylinder head or block. A—Draw air from the radiator into the handheld tester. If there is a combustion leak into the coolant, the tester fluid will change color. B—An exhaust gas analyzer can be used to check for combustion leakage into the coolant by detecting hydrocarbons.*

Figure 19-8. *Using dye and a black light to locate leaks. A—As you move the black light around the engine, it will make dye in the leaking fluid glow. B—Coolant seepage from a bad head gasket. C—Oil leaking from a bad oil pressure sending unit. D—This core plug is leaking coolant.*

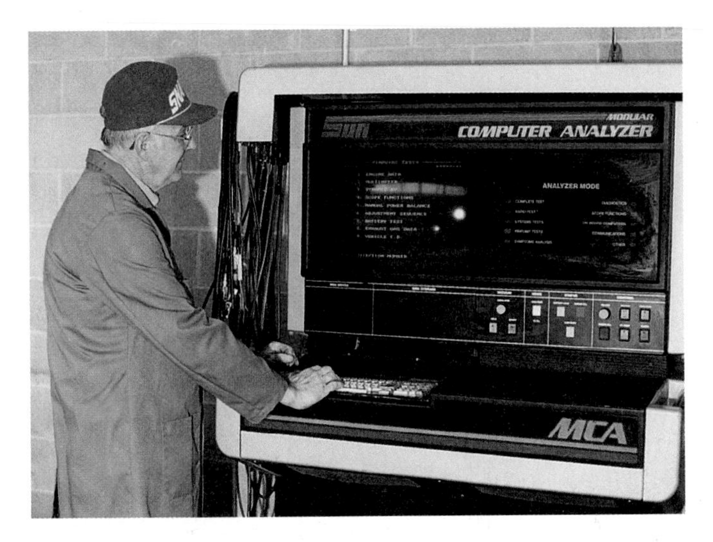

Figure 19-9. *An engine analyzer can be used to check out many engine problems. For example, most can be used to check engine compression by monitoring the current drawn by the starter while cranking the engine.*

Note: A compression test is often performed to verify low compression after a cranking balance test indicates trouble.

Compression Test

A ***compression test*** is used to determine compression stroke pressure. This test is a common method of determining engine mechanical condition. It should be done anytime symptoms point to cylinder pressure leakage. An extremely rough idle, popping noise from the air inlet or exhaust, excessive blue smoke, or blowby are all reasons to consider a compression test.

A compression gauge is used to measure compression stroke pressure during this test. If gauge pressure is lower than normal, pressure is leaking out of the engine combustion chamber.

A compression test is frequently made during a tune-up. It is impossible to tune an engine that is not mechanically sound. If the engine fails the compression test, mechanical repairs must be made *before* the tune-up. Low engine compression can be caused by:

- ❑ **Burned valve.** The valve face is damaged by combustion heat.
- ❑ **Burned valve seat.** The valve seat in the cylinder head is damaged by combustion.
- ❑ **Physical engine damage.** This may be a hole in a piston, broken valve, and so on.
- ❑ **Blown head gasket.** The head gasket has ruptured.
- ❑ **Worn rings or cylinders.** Part wear is preventing a good ring-to-cylinder seal.
- ❑ **Valve train troubles.** Valves have insufficient clearance, which keeps the valves from fully closing; a broken valve spring; and so on.
- ❑ **Jumped timing chain or belt.** A loose or worn chain or belt has jumped over teeth, upsetting valve timing.

For other, less-common sources of low compression, refer to a service manual troubleshooting chart.

Compression Test—Gasoline Engine

To do a compression test on a gasoline engine, remove all of the spark plugs so that the engine will crank easily. Block open the throttle plates. This will prevent a restriction of airflow into the engine.

Disable the ignition system to prevent sparks from arcing out of the disconnected spark plug wires or coils. Some electronic ignition systems may be damaged if operated with the spark plug wires disconnected. Usually, the primary wire going to the ignition coils can be removed to disable the system.

The fuel injection system should also be disabled so that fuel will not spray into the engine. You may need to unplug the wires to the injectors or remove the fuel pump fuse. Check the service manual for specific directions.

Screw the compression gauge into one of the spark plug holes. See **Figure 19-10.** Crank the engine and let the engine rotate for about six compression strokes. The needle on the compression gauge will move up on each compression stroke. Write down the gauge readings for each cylinder and compare the readings to specifications.

Compression Test—Diesel Engine

A diesel engine compression test is similar to a compression test for a gasoline engine. However, do not use a compression gauge intended for a gasoline engine. It can be damaged by the high compression stroke pressure. A diesel compression gauge must read up to approximately 600 psi (4200 kPa).

To perform a diesel compression test, remove either the injectors or the glow plugs. Refer to a service manual for instructions, but most suggest glow plug removal. Install the compression gauge in the recommended hole. Disconnect the fuel shut-off solenoid to disable the injection pump. Crank the engine about six compression strokes and note the highest reading on the gauge.

Note: During a compression test, the cranking speed must be up to specifications, about 200 rpm minimum. The engine should also be warm for accurate test results.

Compression Test Results

Gasoline engine compression readings should be around 125 psi to 175 psi (860 kPa to 1200 kPa) in each cylinder. Readings must be within about 10% to 15% of each other. Diesel engine compression readings will average 275 psi to 400 psi (1900 kPa to 2750 kPa), depending on the engine design and compression ratio.

Look for ***cylinder pressure variations*** during an engine compression check. See **Figure 19-11.** If some

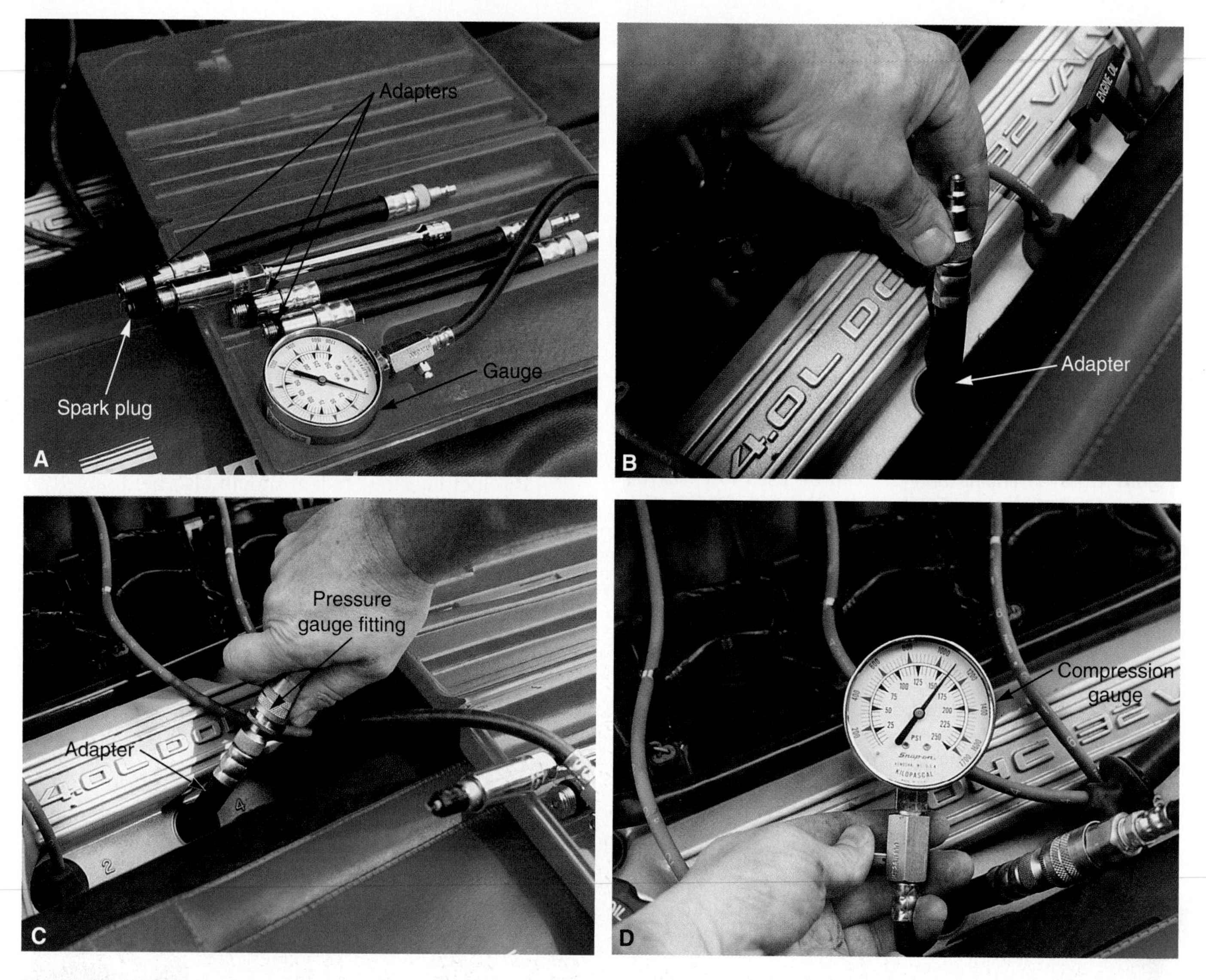

Figure 19-10. *Performing a compression test. A—Select the correct adapter for the spark plug threads. B—Screw the adapter into the spark plug hole and snug it down. C—Attach the compression gauge to the adapter using the quick-disconnect fitting. D—Crank the engine and note the gauge reading after about six compression strokes. (Snap-on Tool Corp.)*

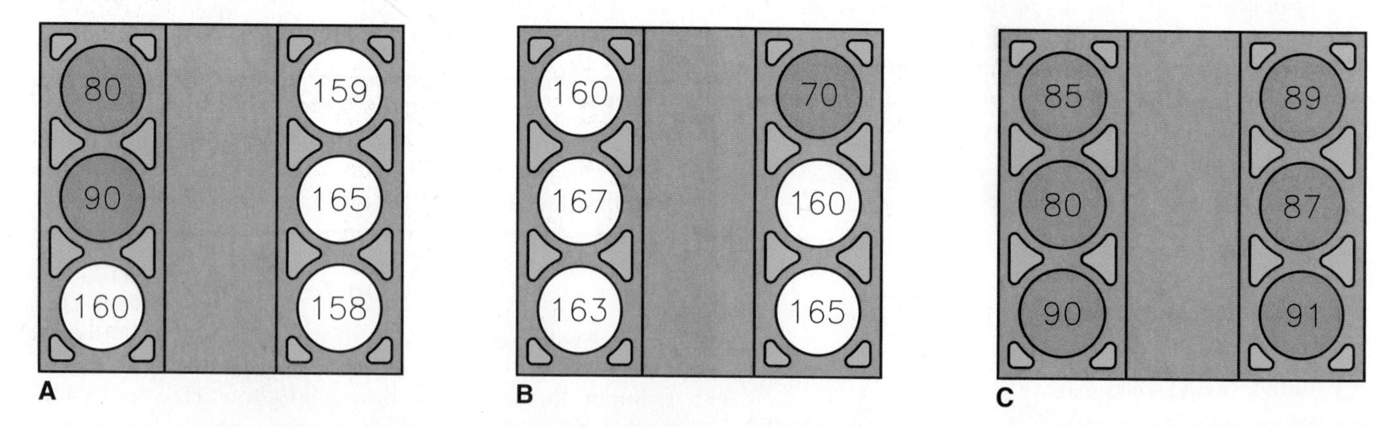

Figure 19-11. *Compression test results can help pinpoint possible causes of pressure losses from the cylinders. A—If two adjacent cylinders have low compression pressure, a blown head gasket or warped head may be the cause. B—If only one cylinder has low compression pressure, it is often due to a burned valve, valve train problem, or worn cylinder. C—If all cylinders have low compression pressure, the timing chain or belt may be worn or may have slipped.*

cylinders have normal pressure and one or two have low readings, engine performance will be reduced. The engine will have a rough idle and lack power.

If *all* of the cylinders have low pressure readings, the engine may run smoothly but lack power and gas mileage. If a camshaft is out of time with the crankshaft (due to a worn timing chain for example), compression pressure will be lowered in all cylinders.

If two adjacent cylinders have low pressure readings, it might point to a blown head gasket with the rupture between the two cylinders. A blown head gasket will sometimes produce a louder-than-normal puffing noise from the spark plug or glow plug holes with the compression gauge removed.

One end of an engine can sometimes run hotter and fail more often than middle cylinders due to coolant flow. Also, if the low cylinder is at the back of the engine, the camshaft may be broken.

Wet Compression Test

A ***wet compression test*** should be completed in any cylinder that has a compression test reading below specifications. See **Figure 19-12A.** The test will help you determine which engine parts are causing the problem.

To conduct the test, squirt a tablespoon of SAE 30 engine oil into the cylinder having the low pressure reading, **Figure 19-12B.** Do not squirt too much oil into the cylinder or a false reading will result. With excessive oil in the cylinder, compression readings will go up even if the piston rings and cylinders are not in good condition. Oil, like any liquid, will not compress. It will take up space in the cylinder, raising the compression ratio and gauge readings.

After placing a small amount of oil in the cylinder, install the compression gauge and recheck the cylinder pressure using the procedure for a standard (dry) compression test. If the compression gauge reading goes up with oil in the cylinder, the piston rings and cylinders may be worn and leaking pressure. The oil temporarily coated and sealed the bad compression rings to increase compression pressure. Look at **Figure 19-12C.**

If the low pressure reading stays about the same, then the engine valves or head gasket may be leaking. The oil will seal the rings, but not a burned valve or blown head gasket. In this way, a wet compression test will help isolate the cause of low compression.

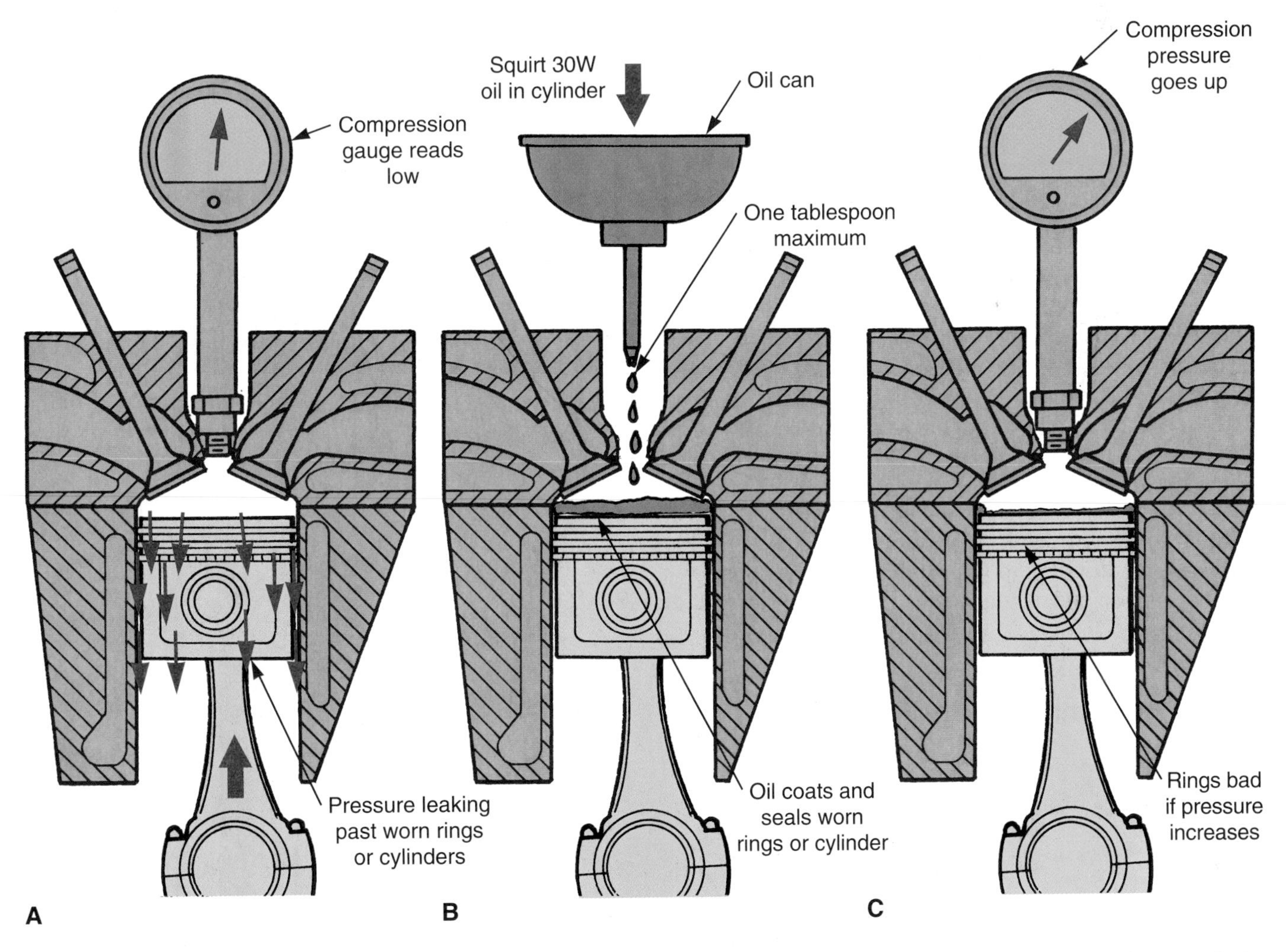

Figure 19-12. *A basic wet compression test. A—If a conventional, dry compression test indicates low compression, conduct a wet compression test. B—Squirt a tablespoon of oil into the cylinder with low compression. This will temporarily seal the rings. C—Measure compression pressure again. If the pressure reading goes up, that cylinder may have bad rings or a worn cylinder. If the pressure reading stays the same, burned valve or blown head gasket may be the problem.*

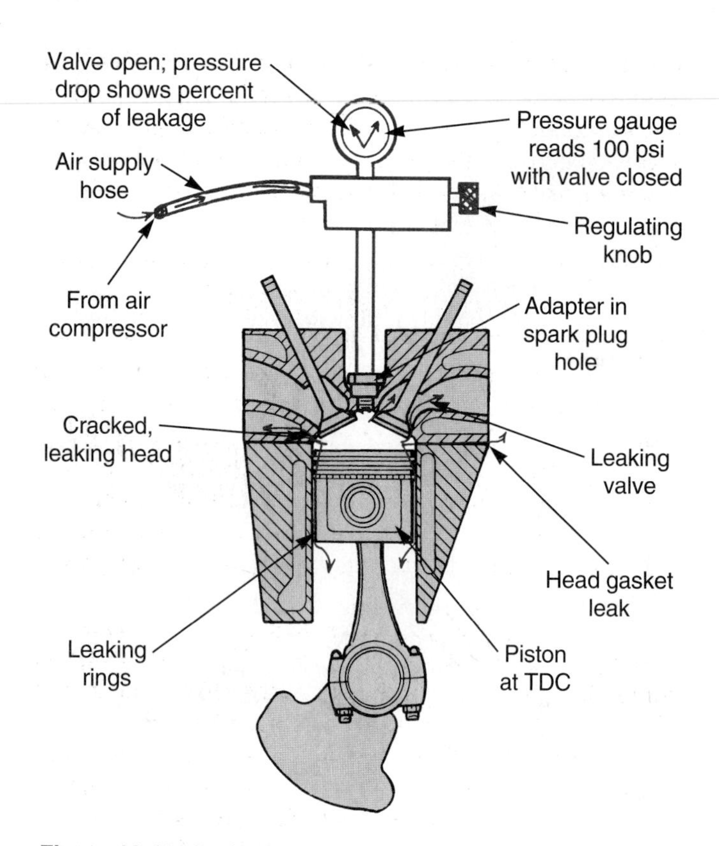

Figure 19-13. *Performing a cylinder leakage test.*

Caution: Some manufacturers warn against performing a wet compression test on a diesel. If too much oil is squirted into the cylinder, hydraulic lock and part damage could result because the oil cannot be compressed in the small cylinder volume.

Cylinder Leakage Test

A ***cylinder leakage test*** also measures air leakage out of the engine combustion chamber to find mechanical problems. It is an accurate way of finding the specific source of the trouble. Shop air is used to pressurize the cylinder and a pressure gauge is used to measure the actual percentage of cylinder leakage.

As shown in **Figure 19-13,** the gauge-valve unit is installed in a spark plug or glow plug hole. Shop air pressure is connected to the gauge-valve line. The crankshaft must be rotated so the cylinder being tested has its piston at TDC on the compression stroke (both valves closed).

First, adjust the air pressure so that the gauge reads zero leakage with the test hose disconnected or shut off. Then, allow air to enter the cylinder and the pressure gauge will drop. The amount of pressure drop depends on how well the rings, valves, head gasket, etc., seal pressure in the cylinder.

If the pressure gauge shows a 10% or less pressure drop, the engine is in fair condition. A higher percentage of pressure drop indicates excess leakage and a mechanical engine problem.

To find the source of leakage, listen for a hissing sound at the intake manifold inlet, exhaust pipe, and valve cover breather. Depending on where you hear leakage, the problems could include the following:

- **Hiss at intake manifold inlet.** A leaking intake valve, intake valve adjusted too tight, broken intake valve spring, or damaged intake valve.
- **Hiss at tailpipe.** A burned exhaust valve, exhaust valve adjusted too tight, broken exhaust valve spring, or bent exhaust valve.
- **Hiss at valve cover breather.** Worn piston rings, worn cylinder walls, blown head gasket, or burned piston.
- **Hiss around outside of head.** A blown head gasket, warped head, warped block deck, or cracked head or block.
- **Hiss from adjacent spark plug hole.** A blown head gasket, cracked head, or cracked block.

Note: During a cylinder leakdown test, make sure you have the piston at TDC on the compression stroke. If not, a valve could be open and cause false indications.

Vacuum Test

A vacuum gauge can be used to perform a ***vacuum test,*** which will measure intake manifold vacuum and indicate engine condition. The vacuum gauge is connected to an intake manifold vacuum fitting, as in **Figure 19-14.** Next, warm the engine. Then, run the engine at the correct idle speed and read the vacuum gauge. Vacuum should be steady and within specifications, about 15″ Hg to 17″ Hg (40 cm/Hg to 45 cm/Hg).

A burned valve, late ignition timing, worn timing chain or belt, restricted exhaust system, etc., will show up

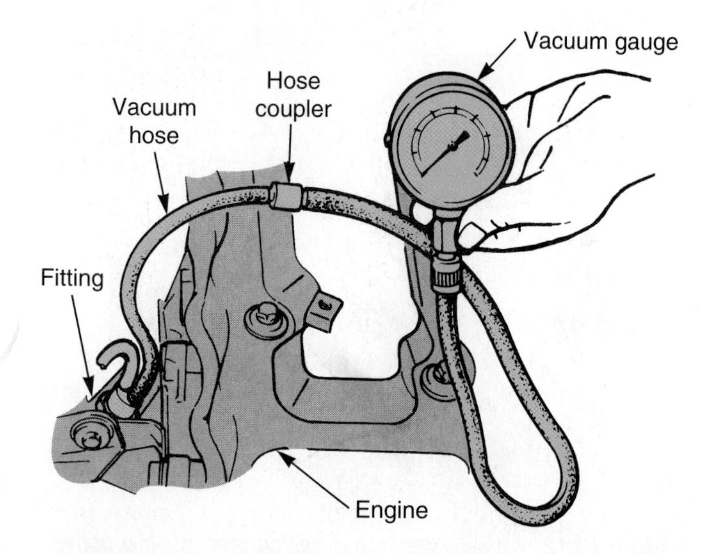

Figure 19-14. *A vacuum gauge can help find problems during a tune-up. At idle, engine vacuum should be steady and to specifications. (General Motors)*

as abnormal vacuum gauge readings. Refer to the service manual for more information.

Cylinder Balance Test

A ***cylinder balance test*** disables the spark plug in one cylinder at a time to find out if each cylinder is firing properly. For instance, if an engine has a dead miss at idle, you can pull one spark plug wire at a time. If engine speed drops and the engine idles more roughly, then that cylinder is firing normally. If a cylinder is disabled and idle speed and smoothness stay the same, that cylinder is not firing, or "dead." Something is preventing normal combustion.

A cylinder balance test can be performed with a tachometer, an engine analyzer, or by simply watching and listening as each wire is disconnected. **Figure 19-15** shows how a tachometer is used to complete a cylinder balance test. As you remove each wire, watch the tachometer display. The reading should drop equally as each cylinder is made inoperative. If the reading fails to drop enough on any cylinder, a compression test or leakage test may be done to find the source of the problem. A fouled spark plug or bad wire could be the cause, but the problem may be due to an engine mechanical problem.

On an engine analyzer or scan tool, there may be cylinder balance buttons that will short out cylinders electronically. Push a button and it will keep the coil from firing on that cylinder only. Some analyzers will short cylinders automatically and record the results on a paper printout. Follow the equipment operating instructions.

If the engine does not have spark plug wires (coil-over-plug), you can disconnect each coil's primary wire to disable the cylinder. If disconnecting a coil's source of voltage does not effect engine operation, that cylinder is dead and not producing power. Again, engine speed should drop and the engine should idle more roughly when a good cylinder is deactivated.

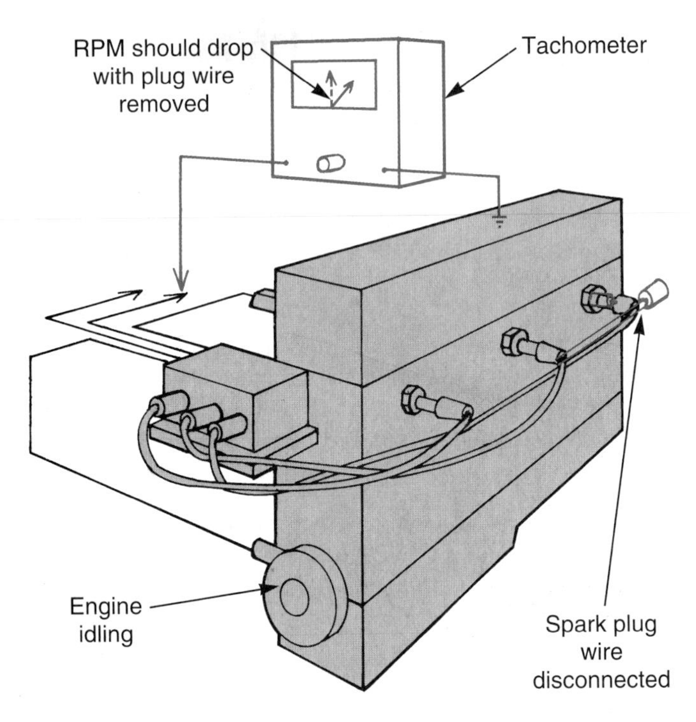

Figure 19-15. *Using a tachometer in performing a cylinder balance test. As good cylinders are disabled, the tachometer reading should go down. If not, the cylinder is dead.*

Valve Adjustment

When an engine has mechanical (solid) lifters, the valves require periodic adjustment. If a valve opens too much or not enough, it will upset the amount of air-fuel mixture pulled into the cylinder. Valve adjustment also affects valve lift and duration, which affects combustion and engine efficiency. Proper valve adjustment is important to the performance of the engine. It is sometimes done during a major tune-up.

On some engines, valve adjustment is done by turning rocker arm adjusting screws, **Figure 19-16A.** On some OHC engines, you must change shim thickness to alter valve lash (valve clearance), **Figure 19-16B.**

Basically, to adjust the valves, first remove the valve cover. Crank the engine until a camshaft lobe is pointing away from the rocker arm or until the valve is fully closed. Adjust the valve until a feeler gauge of the correct thickness fits over the valve or under the rocker arm. If clearance is incorrect, turn the adjusting screw or install a shim of different thickness. Repeat this procedure on all the valves.

Some engine manufacturers recommend valve adjustment while the engine is cold. Others require that the engine be warmed to full operating temperature before adjusting the valves. Always refer to the service manual for specific procedures and specifications.

Note: Usually the exhaust valve will require more lash, or clearance, than the intake valve. It runs hotter and will expand or lengthen more than the intake valve.

Tune-Up Parts Replacement

Depending on the age and condition of the engine, any number of parts may need replacement during a tune-up. With a new, late-model vehicle, you may only need to replace the spark plugs and filters. With an older, high-mileage vehicle, you may have to replace the spark plugs, injectors, distributor components, spark plug wires, or any other parts reducing engine efficiency. During a tune-up, you may have to replace or service:

- **Spark plugs.** Spark plugs are almost always serviced or replaced during tune-up, **Figure 19-17.** Remember, it is easy to strip out the plug threads in aluminum cylinder heads.

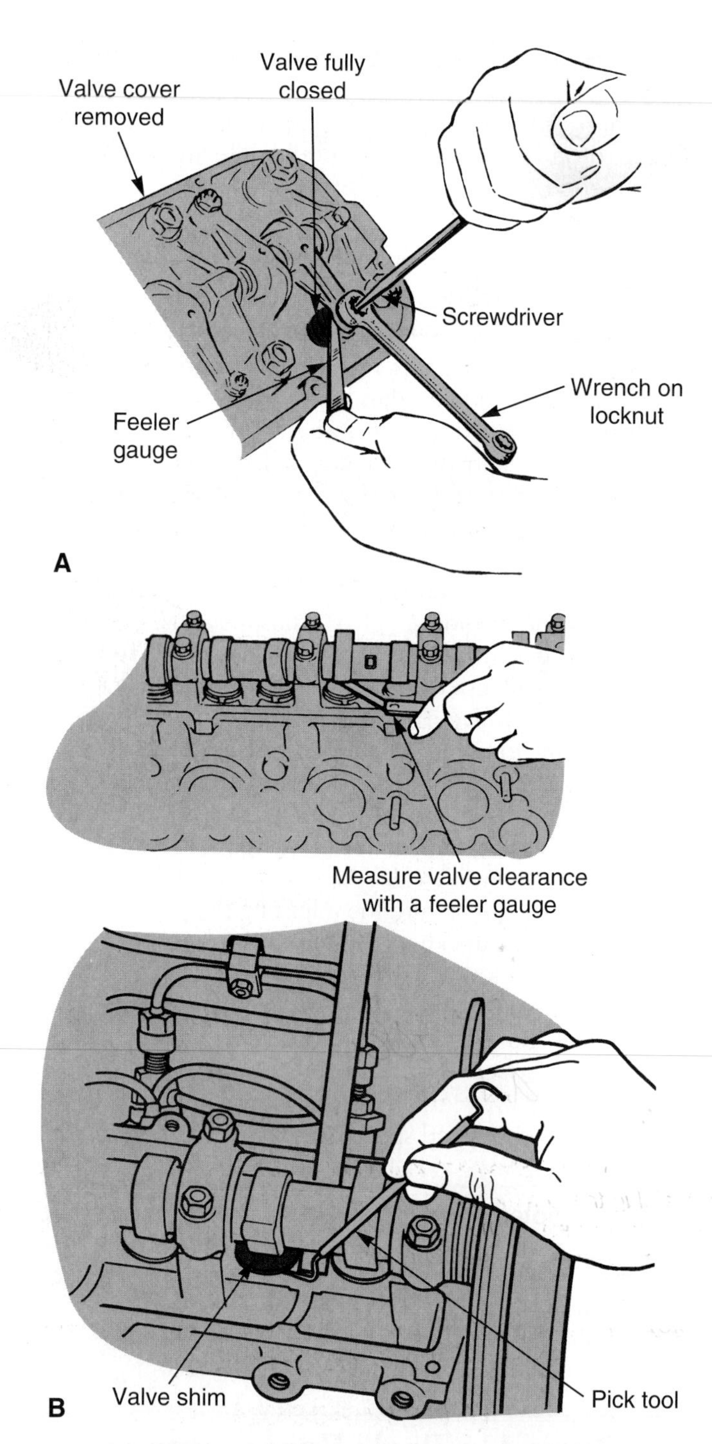

Figure 19-16. *Valve adjustment is sometimes needed during tune-up. A—Using an adjusting screw to change lash as needed. B—This OHC engine has valve adjusting shims. Install a shim of a different thickness to correct the lash. (Volvo, Ford)*

- **Spark plug wires.** Spark plug wires are a common source of engine performance problems. They can break internally, their insulation can leak high voltage to ground, current can be induced into another wire, or they can simply come off of the coil or spark plug. The wires should be checked for high resistance or electrical leakage through the insulation, **Figure 19-18.**
- **Distributor parts.** The cap and rotor may need to be replaced, the pickup coil air gap may need to be set, and the ignition timing may need to be adjusted.
- **Fuel system parts.** Check the idle speed, fast idle speed, and idle mixture and adjust if needed (if adjustment is possible). Filters should also be changed. Refer to Chapter 15 for details on fuel system service.
- **Emission control parts.** Change the canister filter, PCV valve, and so on. Refer to Chapter 15 for details on emission control system service.

Tune-Up Adjustments

The adjustments needed during a tune-up will vary with the make, model, and condition of the vehicle. A few of the most common tune-up adjustments include:

- Spark plug gap.
- Pickup coil air gap.
- Ignition timing (gasoline engine) or injection timing (diesel engine), **Figure 19-19.**
- Idle speed.

Other adjustments may also be needed. Follow the specific directions and specifications in the service manual. With late-model vehicles, the trend is less tune-up adjustments. The vehicle's computer is able to adjust idle speed, air-fuel mixture, and other variables affecting engine performance.

General Tune-Up Rules

Keep these general rules in mind when doing a tune-up:

- Gather information about the performance of the engine. Ask the customer about the vehicle. This may give you ideas about which components should be tested and replaced.
- Make sure the engine has warmed to full operating temperature. Usually, you cannot evaluate engine operation and make tune-up adjustments with a cold engine.
- Use professional, high-quality tools and equipment. Make sure the equipment is accurate and gives precise readings.
- Refer to the car service manual or emission sticker for specifications and procedures. Today's vehicles are so complex that the slightest mistake could ruin the tune-up.
- Use quality parts. Quality parts will ensure that your tune-up lasts a long time. Cheap, bargain parts are usually inferior and can fail quickly.
- Keep service records. You should write down all of the operations performed on the vehicle. This will

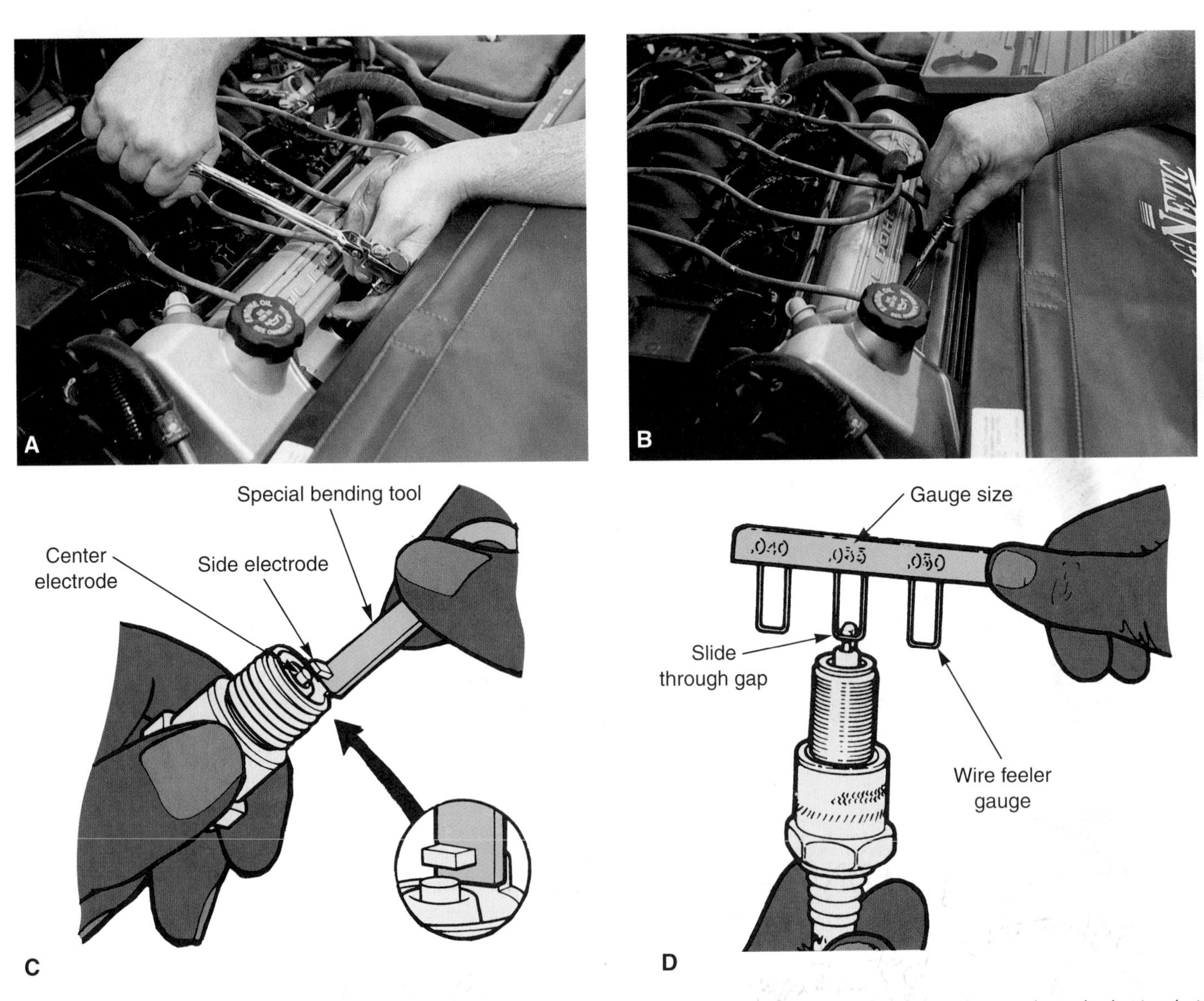

Figure 19-17. *Spark plugs are normally replaced during a tune-up. A—Use a swivel ratchet, long extension, and spark plug socket to remove each spark plug. B—When installing spark plugs, start the plug threads by hand to avoid crossthreading. Use a ratchet to snug down the plugs, but do not overtighten them. C—If the gap is not correct on the new plugs, use a bending tool to open or close the gap as needed. D—Use a wire feeler gauge to check the gap. (DaimlerChrysler, Snap-on Tool Corp.)*

give you and the customer a record for future reference. If a problem develops, you can check your records to help correct the trouble.

- ❑ Complete basic maintenance service. Although not part of a tune-up, lubricate door, hood, and trunk hinges, and trunk latches, during a tune-up. Also, check all fluid levels, belts, and hoses. This will build good customer relations and help ensure vehicle safety and dependability.
- ❑ Tune-up intervals are the recommended time spans in months or miles when an engine should be tuned up. They are given by the auto maker. One manufacturer may require tune-up operations every 36 months or 36,000 miles, while another may require more- or less-frequent service. Check the manual for details. The vehicle warranty may be void if tune-up intervals are missed.
- ❑ Always road test the vehicle when the tune-up is complete.

Diesel Engine Tune-Up (Maintenance)

Diesel engines do not require tune-ups like gasoline engines. A diesel does not have spark plugs to replace or an ignition system to fail. The diesel injection system is very dependable and only requires major service when problems develop.

A diesel engine tune-up, more accurately called ***diesel maintenance,*** typically involves the following. Refer to the service manual for more information on diesel engine maintenance.

- ❑ Replacing the air filter element.
- ❑ Cleaning, draining, or replacing fuel filters.
- ❑ Adjusting engine idle speed.
- ❑ Adjusting the throttle cable.

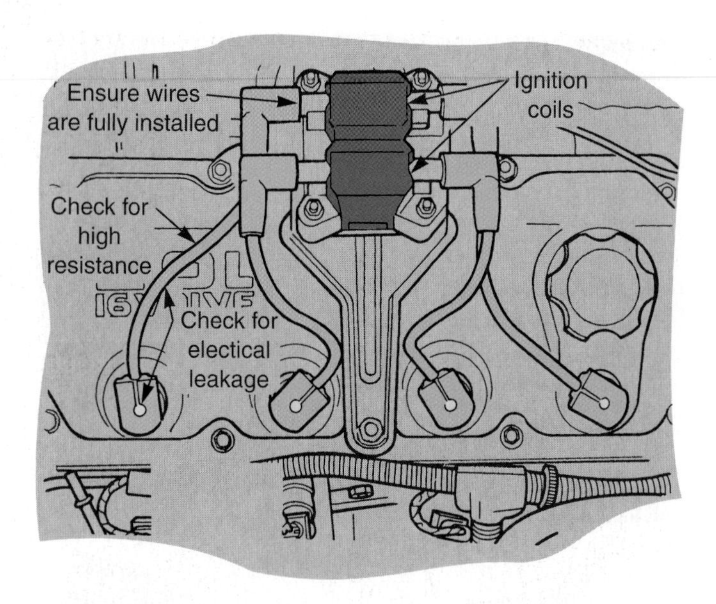

Figure 19-18. *Check for high resistance in the spark plug wires. Also, check for electrical leakage through the insulation. (DaimlerChrysler)*

- ❑ Inspecting the engine and related systems.
- ❑ Checking the injection timing.
- ❑ Changing the engine oil and oil filter.
- ❑ Periodic servicing of emission control systems.

Summary

An engine tune-up involves keeping all engine systems working at peak performance. It is a maintenance operation that must be done at regular intervals. Keep safety in mind during a tune-up; you will be working around a running engine.

A minor tune-up is preventative maintenance done on an engine in good condition. It involves spark plug replacement (as needed), filter service, and minor adjustments. A major tune-up can involve more time-consuming tasks, such as diagnostic tests and minor repairs.

Figure 19-19. *A timing light is used to check ignition timing on a gasoline engine. On a diesel engine, you need to check injection timing. (Snap-on Tool Corp.)*

Begin the tune-up with a preliminary inspection of the engine compartment. Look for battery problems, fuel system and ignition system troubles, bad hoses, loose wires, low fluid levels, and so on.

A compression test is done when engine missing or other symptoms point to mechanical problems. A compression test measures actual compression stroke pressure. If low, a mechanical trouble is allowing air to leak out of the engine combustion chamber through a burned valve, blown head gasket, or worn piston rings.

A cylinder leakage test is like a compression test, but uses shop air pressure to check for combustion chamber leakage. Air is pumped into the cylinder with the piston at TDC on the compression stroke. Any leak shows up as an excessive pressure drop on the gauge and a hissing sound.

A cylinder balance test shorts out spark plugs to see which the cylinders are firing. If a plug wire is disconnected, engine speed should drop. If engine idle speed stays the same, that cylinder is not producing power.

Valve adjustment is sometimes done during a tune-up when the engine has mechanical lifters or followers. Either adjustment screws or adjustment shims are provided to change valve lash (clearance).

During a tune-up, various parts may need replacement or adjustment. Usually, spark plugs are replaced. Sometimes, you must also test and replace spark plug wires, fuel filters, air filters, PCV valve, and other components. You must check out engine operation and refer to a shop manual and service records to determine what must be done during the tune-up.

Always check ignition timing and idle speeds during a tune-up. Road test the car after the tune-up to check your work.

Review Questions—Chapter 19

Please do not write in this text. Write your answers on a separate sheet of paper.

1. An engine tune-up can affect:
   (A) Engine power and acceleration.
   (B) Fuel economy.
   (C) Exhaust pollution.
   (D) All of the above.
   (E) None of the above.
2. An engine tune-up is a _____ operation that returns the engine to peak performance after part _____ and deterioration.
3. What is the difference between a *minor tune-up* and a *major tune-up?*
4. List six things to check for during a preliminary inspection.

5. A black tip on an oxygen sensor indicates _____.
6. How can a starter current draw test indicate low compression?
7. A(n) _____ test is used to determine actual compression stroke pressure.
8. List five problems that can cause *low engine compression.*
9. The results of a compression test should be within _____% of each other.
10. How do you perform a *wet compression test?*
11. Like a compression test, a(n) _____ test also measures air leakage out of the engine combustion chamber to find mechanical problems.
12. Engine vacuum at idle should be about _____.
13. What is the purpose of a *cylinder balance test?*
14. Which type of lifters require valve lash adjustment?
15. List three adjustments commonly made during a tune-up.

ASE-Type Questions—Chapter 19

1. A cylinder leakage test shows a 30% pressure drop in one cylinder. A hissing sound can be heard from the vehicle's tailpipe. Technician A says an intake valve could be damaged. Technician B says an exhaust valve could be burned. Who is correct?
   (A) Technician A.
   (B) Technician B.
   (C) Both A and B.
   (D) Neither A nor B.
2. A cylinder balance test shows that one cylinder is not causing a drop in engine speed. Technician A says that there could be pressure leakage and a mechanical problem. Technician B says that there could be a fouled spark plug or bad spark plug wire. Who is correct?
   (A) Technician A.
   (B) Technician B.
   (C) Both A and B.
   (D) Neither A nor B.
3. Technician A says that this illustration shows replacing a valve shim. Technician B says that the operation shown is done to adjust the valve train clearance. Who is correct?
   (A) A only.
   (B) B only.
   (C) Both A and B.
   (D) Neither A nor B.

(Ford)

4. Technician A says that a minor tune-up is performed when the engine is in poor running condition. Technician B says that a major tune-up includes all of the steps of a minor tune-up. Who is correct?
   (A) A only.
   (B) B only.
   (C) Both A and B.
   (D) Neither A nor B.
5. Technician A says that during a minor tune-up, the ignition timing should be checked and adjusted if needed (if adjustable). Technician B says that during a minor engine tune-up, the emission control system should be tested and serviced if needed. Who is correct?
   (A) A only.
   (B) B only.
   (C) Both A and B.
   (D) Neither A nor B.

6. Technician A says that you should check for fuel leaks when doing a major engine tune-up. Technician B says that checking for oil leaks should be performed during a minor tune-up. Who is correct?
   (A) A only.
   (B) B only.
   (C) Both A and B.
   (D) Neither A nor B.

7. Technician A says that a tune-up can affect the service life of an engine. Technician B says that a tune-up can affect fuel economy. Who is correct?
   (A) A only.
   (B) B only.
   (C) Both A and B.
   (D) Neither A nor B.

8. Technician A says that the illustration shows adjusting the valve train clearance. Technician B says that it shows adjusting valve lift. Who is correct?
   (A) A only.
   (B) B only.
   (C) Both A and B.
   (D) Neither A nor B.

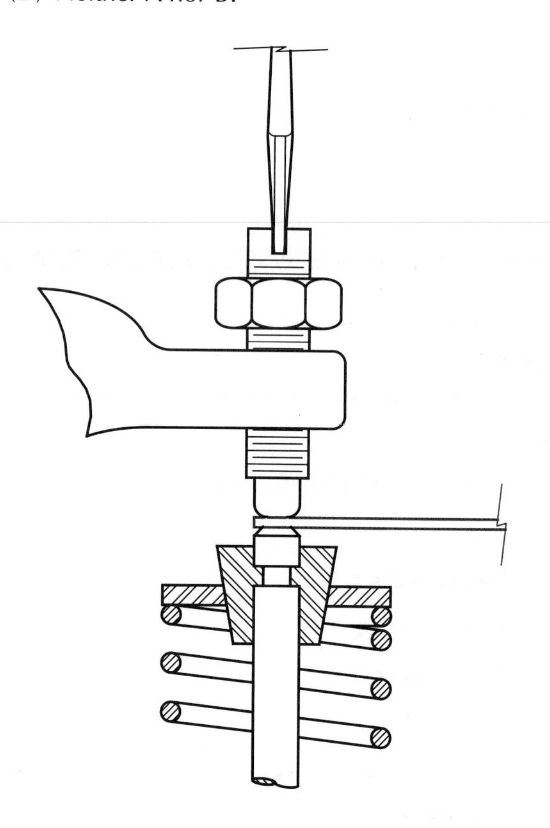

9. Low engine compression can be caused by all of the following except:
   (A) faulty spark plugs.
   (B) a burned valve seat.
   (C) a blown head gasket.
   (D) worn rings.

10. Technician A says that gasoline engine compression readings should be within approximately 10% to 15% of each other. Technician B says that a compression reading on a gasoline engine should be approximately 12.5 psi to 17.5 psi. Who is correct?
    (A) A only.
    (B) B only.
    (C) Both A and B.
    (D) Neither A nor B.

11. If a cylinder leakage test shows a 30% pressure drop in each cylinder, _____.
    (A) the leakage gauge is out of calibration.
    (B) the engine is in fair condition.
    (C) a sensor problem exists.
    (D) an engine mechanical problem exists.

12. Technician A says that a tachometer can be used to perform a cylinder balance test. Technician B says that an engine analyzer can be used to perform a cylinder balance test. Who is correct?
    (A) A only.
    (B) B only.
    (C) Both A and B.
    (D) Neither A nor B.

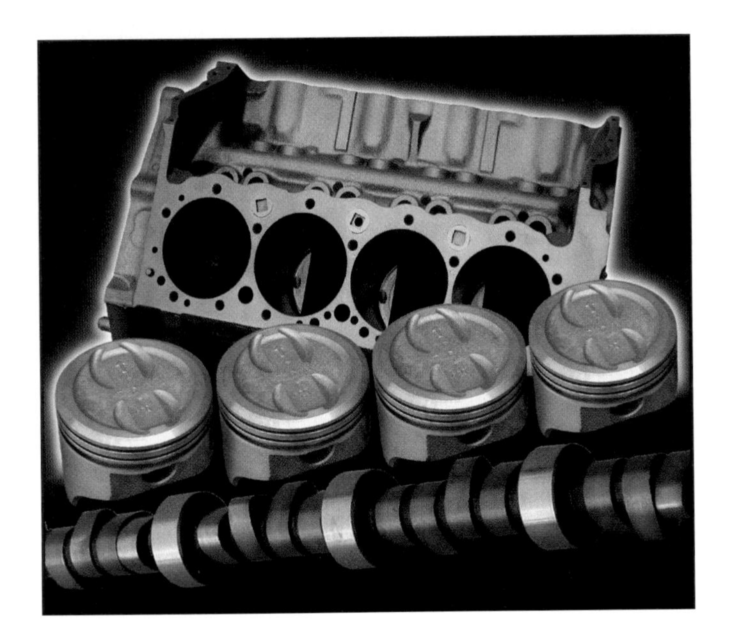

Chapter 20

Engine Removal, Disassembly, and Cleaning

After studying this chapter, you will be able to:

- ❑ Use a vehicle identification number (VIN).
- ❑ List engine repairs that require engine removal.
- ❑ List engine repairs that do not require engine removal.
- ❑ Summarize the steps in preparing for engine removal.
- ❑ Describe how to keep wires, lines, and parts organized during engine removal and teardown.
- ❑ Summarize how to attach and use engine lifting devices.
- ❑ Explain the major steps for engine disassembly.
- ❑ Describe the methods used to clean engine parts.

Know These Terms

Air blowgun
Casting numbers
Cold cleaning tank
Crane boom
Decarbonizing cleaner
Engine hoist
Engine overhaul
Engine rebuild
Engine stand
Heat-cleaning oven
Hot tank
In-vehicle engine overhaul
Lifting fixture
Media blasting
Mushrooming
Power brush
Scuff pad
Spray-on gasket solvent
Teardown
Thread chaser
Vehicle identification number (VIN)
Wire wheel

In previous chapters, you learned about engine design, part construction, performance problems, engine mechanical failures, and engine troubleshooting. This chapter expands your knowledge of engine service by giving the most important or typical steps for engine removal, disassembly (teardown), and cleaning. Keep in mind that exact steps for disassembly vary from engine to engine. However, there are many general procedures that apply to all makes and models of engines.

Engine Identification

The type of engine to be serviced must be identified when ordering parts. There are several ways of identifying an engine:

- Knowledge from work experience.
- Studying engine construction.
- Using the vehicle identification number (VIN).
- Casting numbers.
- Measurement after disassembly.

Identify the Engine Type

Simply looking at the engine, you can determine several things about it. You can tell how many cylinders it has. You can identify the cylinder arrangement, such as V-type, inline, or opposed. You can determine if the fuel system is multiport or throttle body injection. A longitudinal engine is mounted with the crankshaft in a line from front to rear. A transverse engine is mounted with the crankshaft in a line from side to side. Accessories such as air conditioning and power steering can be noted. Other facts, such as turbocharging or supercharging, may also be apparent. Using this type of information along with the year, make, model, and VIN of the vehicle, a parts person can find out what engine is installed in the vehicle and order the correct parts.

Vehicle Identification Number

The ***vehicle identification number (VIN)*** is a unique number identifying the vehicle and some of the equipment installed on the vehicle. Contained within the VIN may be codes for the engine type and size, transmission type, differential type, body style, optional equipment, and so on.

The VIN may be located in one or more of several different locations, **Figure 20-1.** The service manual will tell where to find the VIN. On late-model vehicles, the VIN is normally located in the top-left area of the dash and is visible through the windshield. See **Figure 20-2.** By comparing the letters and numbers in the VIN to service literature, you can find out what engine was installed in the vehicle at the factory. Refer to **Figure 20-3.**

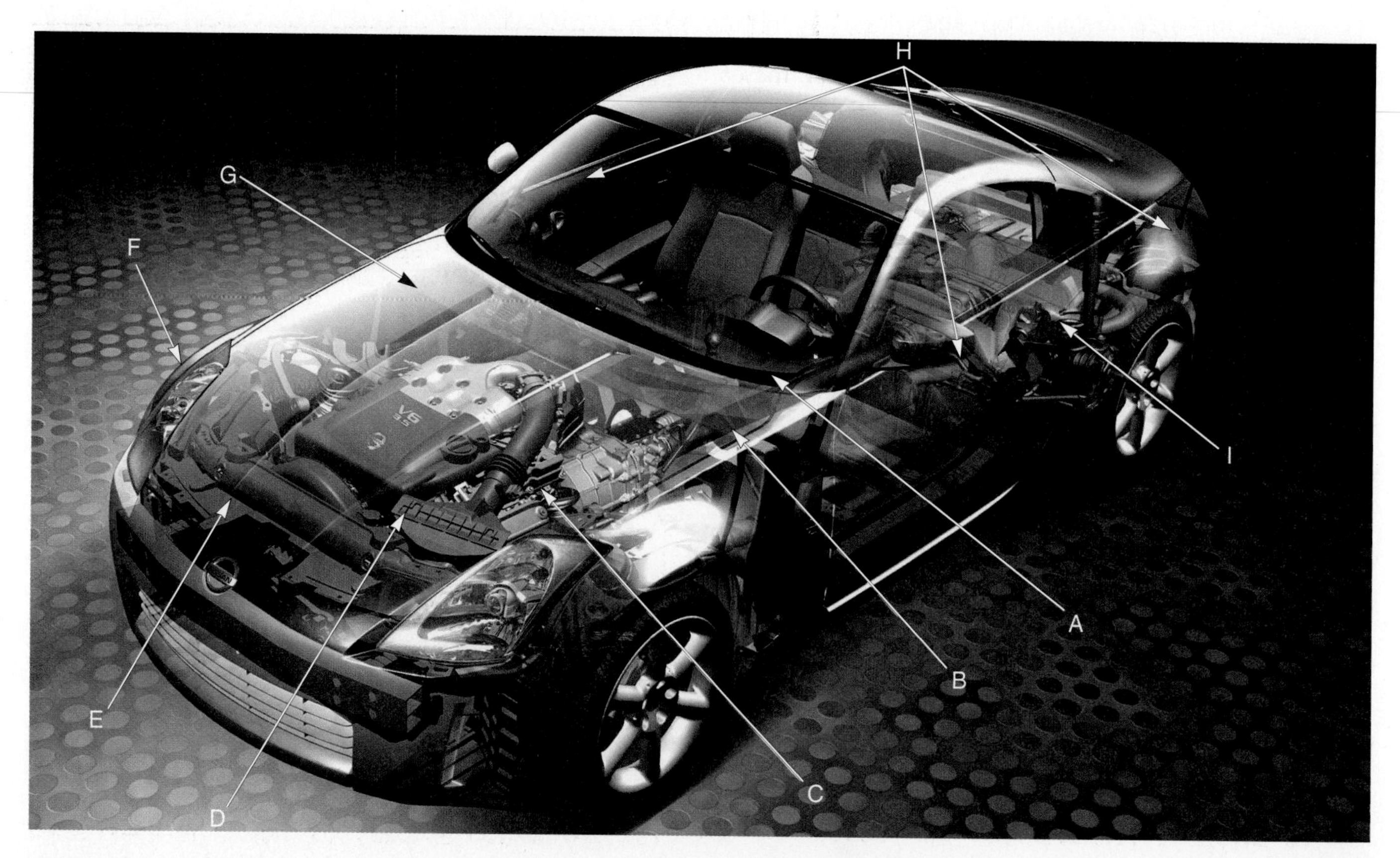

Figure 20-1. *The VIN may be located in one or more of these locations, depending on the make and model of vehicle. A—Left side of the dash. B—Upper-left corner of the firewall. C—Side of the engine. D—Front of the engine. E—Radiator support. F—Right-front wheelwell. G—Upper-right corner of the firewall. H—Body panels. I—Lockface on the left-hand door. (Nissan)*

Figure 20-2. *The VIN on most late-model vehicles can be found on the left, top of the dash and can be read through the windshield.*

Note: Before ordering parts, make sure the vehicle has its original engine. If the engine has been changed, the VIN may not match the engine and you could order the wrong parts.

Casting Numbers

Casting numbers are numbers formed into parts during manufacturing. They can be found on major components, such as the block, camshaft, and heads. These numbers can be used to find out information when ordering parts. Books or charts are available that decode casting numbers.

Part Measurement

Part measurement is another way of finding out which engine you are working on. You can measure cylinder bore, valve sizes, head port sizes, piston diameter, and so on. This will help you and the parts person determine the block, head, and engine type.

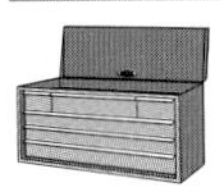

Note: For more information on engine identification, refer to the text chapters on engine construction and types. Knowing how to tell a "hemi head" from a "wedge head" or an OHC from a push rod engine is essential to the engine technician.

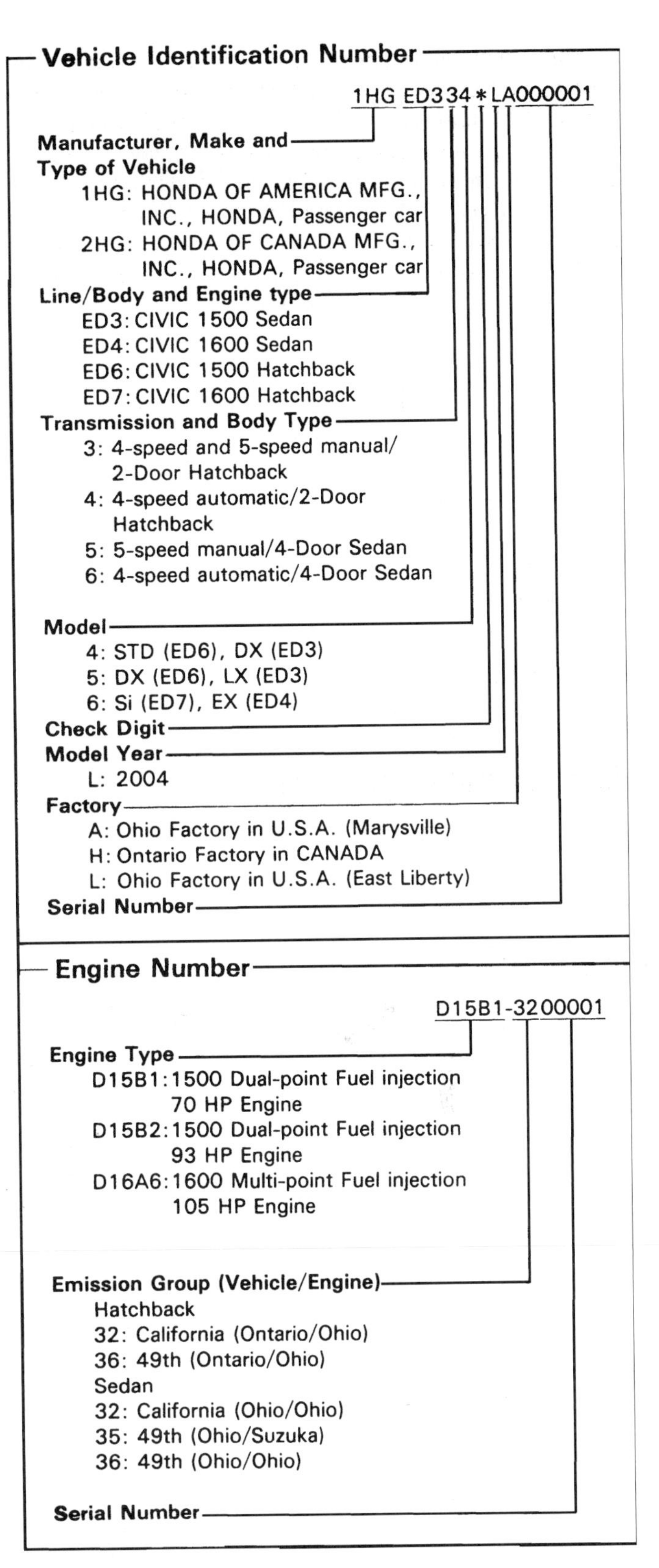

Figure 20-3. *Study how to read a VIN. (Honda)*

Is Engine Removal Necessary?

Many engine repairs can be made with the engine block in the chassis. Repairs limited to the cylinder head, valve train, and other external parts are commonly in-vehicle operations, **Figure 20-4.** Engines must be removed when the cylinder block or crankshaft is badly damaged.

Whether or not the engine must be removed for a repair often depends on the year, make, and model of the

Figure 20-4. *Many engine repairs can be done with the cylinder block still in the chassis. Here, the cylinder heads have been removed for service. (Riverdale Auto Repair)*

vehicle. For example, one make of vehicle might require engine removal simply to replace a damaged oil pan because access to parts is hampered by the body or chassis. On the other hand, a different make of vehicle may have unobstructed access to the oil pan, allowing for quick and easy in-vehicle oil pan removal. This design may also permit in-vehicle replacement of piston rings, oil pump, bearings, and other major components. Generally, full-size cars and light trucks provide easier access to, and removal of, external engine parts. Therefore, they are likely to allow more in-vehicle engine repairs.

When unsure if the engine must be removed for a repair, refer to the service manual. It provides directions for the vehicle and engine on which you are working.

A few of the engine repairs that can be done with the engine still installed in the chassis include:

- ❑ **Valve job.** This involves reconditioning the valve faces, seats, and other valve train or cylinder head parts. Normally, the intake manifold, exhaust manifold(s), and cylinder head(s) can be removed with the short block bolted in the chassis.
- ❑ **Valve seal and valve spring replacement.** These components can usually be replaced without removing the cylinder heads from the engine or the valves from the cylinder heads. Shop air pressure is injected into the cylinders to hold the valves closed. Then, a small valve spring compressor is used to compress the springs, allowing for removal and replacement of the keepers, retainers, springs, and seals.
- ❑ **Oil pump replacement.** The oil pump can often be replaced without removing the engine from the vehicle. However, if the oil pan cannot be removed by simply raising the engine off the motor mounts, engine removal is required.
- ❑ **Rod and main bearing replacement.** The rod and main bearings can sometimes be replaced with the block in the vehicle. After removing the oil pan, you can unbolt the rod and main caps to service the bearings. The top half of each main bearing insert can be rotated out of the block with a thin screwdriver or wooden tongue depressor. See **Figure 20-5A.** Pushing on the edge of the bearing rotates the upper bearing out. A special cotter pin–type tool that fits into a crankshaft oil hole can also be used. By turning the crankshaft, the tool rotates the bearing out, **Figure 20-5B.**
- ❑ **Rear main oil seal replacement.** This is an in-vehicle operation if the oil seal is a two-piece seal and the oil pan can be removed without removing the engine.

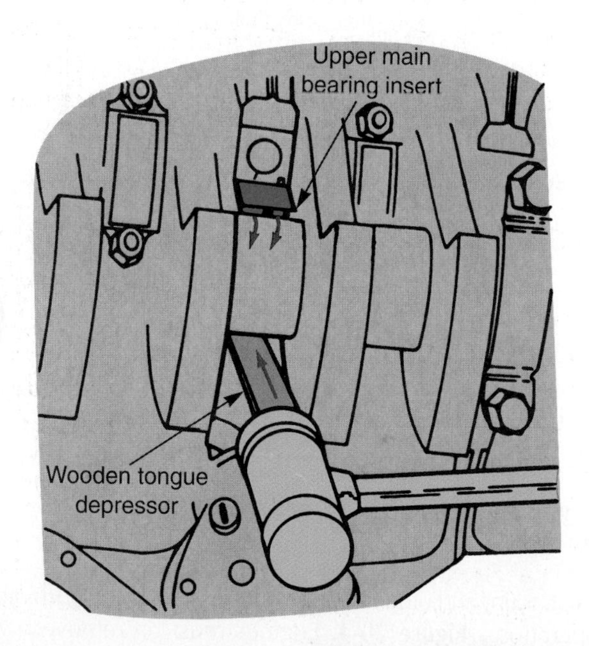

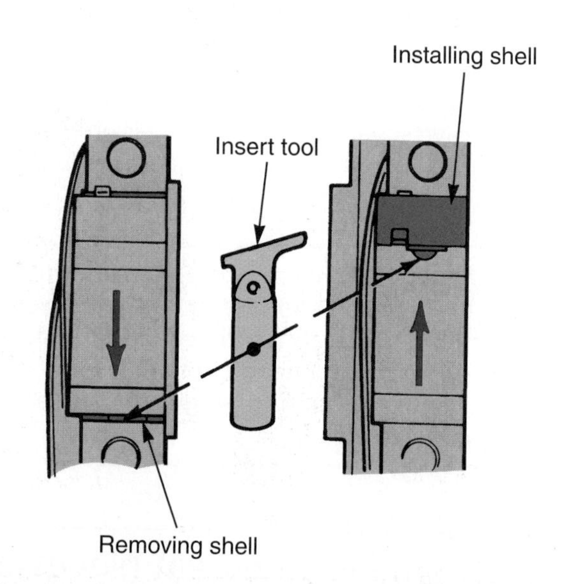

Figure 20-5. *It may be possible to replace rod and main bearings without removing the engine from the vehicle. A—Using wooden tongue depressor to rotate the old upper bearing insert out of the block. B—Using a special tool that fits into the oil hole in the crankshaft to remove the upper bearing insert. (Renault, General Motors)*

The top half of a two-piece neoprene oil seal can be pushed out of the block with a small screwdriver, **Figure 20-6.** The bottom half of a two-piece seal fits in the main cap.

- ❑ **Piston service.** If the block is in good condition, pistons can sometimes be serviced without removing the cylinder block from the vehicle. For example, a bad piston pin may make a double knock. Determine which one is knocking using a stethoscope. Then, pull the oil pan and the cylinder head on the correct side of the block. Remove the rod cap for that piston and push the piston out. Replace the piston and pin. Then, reassemble the engine. This would save time over pulling the engine out of the vehicle. However, it is normally more efficient to remove the engine from the vehicle for piston service. In-vehicle piston service is normally only appropriate on a low-mileage engine, like one that is under warranty.
- ❑ **External engine part service.** External engine parts include exhaust manifolds, intake manifolds, valve covers, and so on. These can normally be serviced without engine removal. When servicing exhaust manifolds, it may help to unbolt the motor mounts and lift the engine with a floor jack.

Seldom can any repairs be done to the block unless major parts are removed and the block is sent to a machine shop. Engine repairs that normally require engine removal include:

- ❑ **Crankshaft service.** If wear or damage to the crankshaft is excessive and the block must be machined, you need to pull the engine out of the vehicle. However, if only the crankshaft needs repair, you do not have to remove the cylinder heads, valve covers, rods, and other engine components, **Figure 20-7.** This saves hours of work!
- ❑ **Cylinder repair.** Wear or damage to the cylinder walls is typically repaired with the engine out of the vehicle.

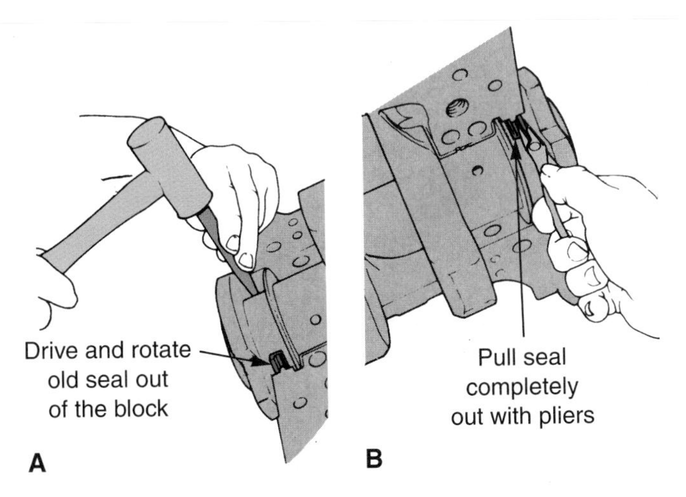

Figure 20-6. *A two-piece rear main oil seal can often be replaced without removing the engine from the vehicle. A—Driving the upper half of a neoprene seal out with a driver. B—Use pliers to pull the seal out of the block. (General Motors)*

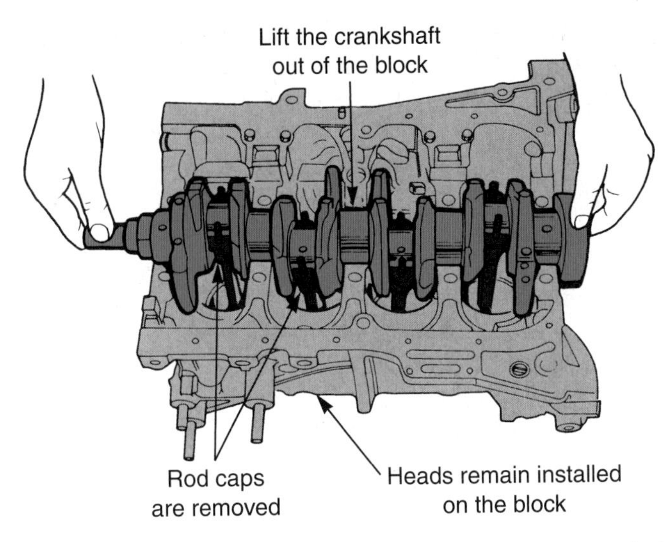

Figure 20-7. *It may be possible to service the crankshaft without removing the cylinder heads and other top end parts. However, if the rods, pistons, or cylinders are damaged, you will likely have to remove the engine. (Honda)*

Remember that there are exceptions to almost any rule! Machines are available for in-vehicle turning of crankshaft journals, but these machines are not common. In addition, some engines must be lifted halfway out of the vehicle just to change spark plugs. This makes it important for you to refer to the service manual when in doubt.

Engine Overhaul

The terms ***engine overhaul*** and ***engine rebuild*** are understood to mean a process that includes:

- ❑ Complete disassembly of the engine.
- ❑ Cleaning, inspection, and measurement of critical parts.
- ❑ Replacement or service of worn or damaged parts.
- ❑ Reassembly of the engine.

Parts Replaced or Reconditioned

Basically, the following parts are replaced or reconditioned during a major engine rebuild:

- ❑ Piston rings and worn or damaged pistons.
- ❑ Connecting rod, main (crankshaft), and camshaft bearings.
- ❑ Oil pump.
- ❑ Timing belt, chain, or gears.
- ❑ Core plugs.
- ❑ Camshaft(s) and worn lifters.
- ❑ Valve train components, especially worn valve springs.
- ❑ Valve guides and seats.
- ❑ All gaskets and seals.
- ❑ Other parts that are not within specifications.

In addition, as you will learn later, sometimes the block must be bored and oversize pistons installed. You may need to have the crankshaft turned and install undersize main and rod bearings. Camshaft bearings could be worn and lowering oil pressure, which means they need to be replaced.

In-Vehicle Engine Overhaul

An ***in-vehicle engine overhaul*** can only be done when cylinder and crankshaft wear are within specifications and the oil pan and heads can be easily removed with the engine installed in the vehicle. The general condition of the crankshaft can be determined by measuring oil pressure, listening for bearing knock, and/or inspection.

Generally, remove the top end from the engine. Check cylinder wear to make sure the cylinders will take new rings without boring. Remove the oil pan and rod caps. Drive the pistons out of the top of the block. Hone and clean the cylinders. Replace the main bearings. Then, reinstall the pistons and rods with new rings and rod bearings. Install the new oil pump, reconditioned heads, and other engine parts.

Note: An in-vehicle engine overhaul is only cost- and time-efficient when there is a large engine compartment and plenty of room to work under the engine. A light truck is a good example. In most cases, it is just as easy to remove the engine for major service.

Preparing for Engine Removal

To prepare for engine removal, most technicians disconnect everything under the engine first. This includes the motor mounts, bell housing bolts, torque converter bolts, exhaust system, and so on. See **Figure 20-8.** The parts under the engine are disconnected first so that oil, fuel, and coolant are not leaking out from the disassembled top end while working under the engine. Also, you can still turn the engine over using the starter to rotate the torque converter and remove its bolts. Next, disconnect the starter and drain the fluids. See **Figure 20-9.** Finally, disconnect everything on top of the engine that prevents engine removal. This includes the air cleaner, throttle linkage, wiring harnesses, and so on. See **Figure 20-10.**

Basically, use the following steps to get the engine ready for removal. Depending on whether the vehicle is rear- or front-wheel drive, you may need to modify this procedure.

1. Park the vehicle so there is plenty of space on both sides and in front of the vehicle.
2. Place the vehicle on jack stands. Disconnect or remove everything under the vehicle that prevents engine removal. The exhaust system can be rusted and the manifold or header pipes can be difficult to unbolt. Use rust penetrant and a six-point socket if the bolts or nuts are hard to remove.
3. Drain and collect the engine oil and coolant.
4. Lower the vehicle and place covers over the fenders to protect the paint.
5. Scribe lines around the hood hinges to aid realignment. Then, have someone help you remove the hood. Store the hood in a safe place where it cannot be scratched or damaged.
6. Disconnect the battery, negative terminal first, to prevent electrical shorts. Remove the battery if it is in the way.
7. Disconnect or remove the components on the top of the engine. Keep all fasteners in a separate container.
8. Unplug all electrical wires between the engine and chassis. If needed, use masking tape to label or identify the wires. This will simplify reconnection, **Figure 20-11.**
9. Remove all coolant, fuel, and vacuum hoses or lines that prevent engine removal, **Figure 20-12.** Label as needed, especially vacuum lines. On vehicles with an electric fuel pump, you may need to bleed off fuel pressure before disconnecting the fuel lines.
10. Do not disconnect any power steering or air conditioning lines unless absolutely needed. Usually, the power steering pump or air conditioning compressor can be unbolted and placed on one side of the engine compartment. If you *must* remove air conditioning lines, the refrigerant must be properly recovered.
11. Remove the radiator, fan, and other accessory units in front of the engine. Be careful not to hit or drop the radiator.
12. Remove any other part that prevents engine removal. Check behind the engine for hidden ground wires, dipstick tubes, and so on.

Keep fasteners organized in several different containers. For instance, keep all of the bolts and nuts from the front of the engine in one container. Keep engine top end and bottom end fasteners separate in two additional containers. This will speed reassembly.

Removing the Transmission with the Engine

Some front-wheel drive vehicles require that the engine and transaxle be removed as a single unit. Check the service manual for details.

You may want to drain the fluid from the transmission or transaxle if it is to be removed. On rear-wheel-drive vehicles, the drive shaft, transmission and clutch linkage, speedometer cable, rear motor mount, and other parts must also be removed. On a transaxle, the axle shafts must be disconnected.

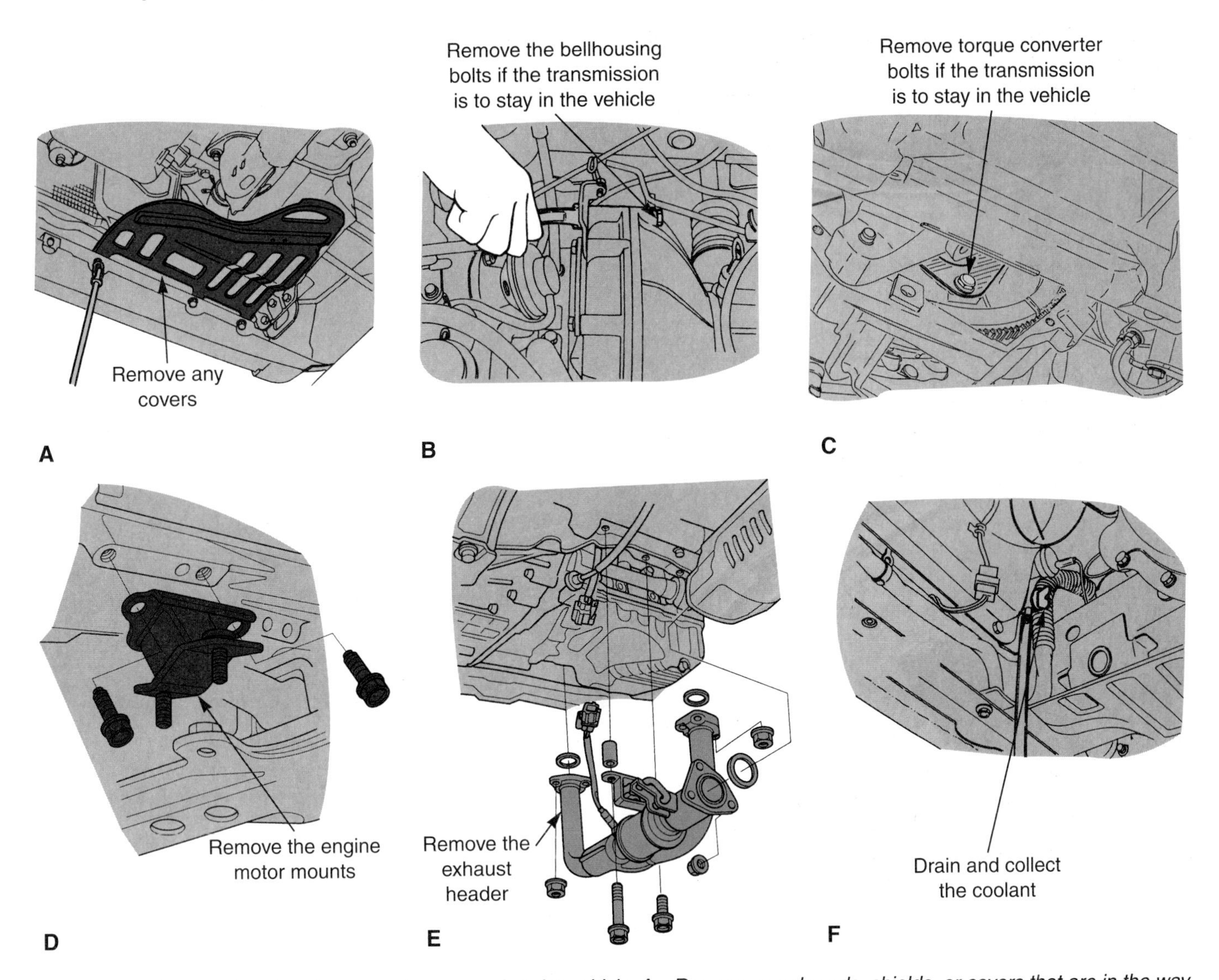

Figure 20-8. *Most technicians begin engine removal under vehicle. A—Remove any shrouds, shields, or covers that are in the way. B—Unbolt the bell housing if the transmission or transaxle is to remain in the vehicle. C—Unbolt the torque converter if the transmission is to stay in the vehicle. D—Unbolt the engine motor mounts. E—Unbolt and remove the exhaust manifold or header pipe. F—Drain the coolant and oil. (Subaru, Honda, Peugeot)*

Note: Keep your work area clean. Coolant and engine oil will usually drip onto the shop floor during engine removal. To prevent an accident, wipe up spills as soon as they occur. A clean work area also helps present a professional image to the customer.

Engine Removal

Before attempting removal of the engine, double-check everything to be sure nothing remains to be disconnected or removed. For example, check:

- ❑ Behind and under the engine for hidden wires or ground straps.
- ❑ That all bell housing bolts are removed.

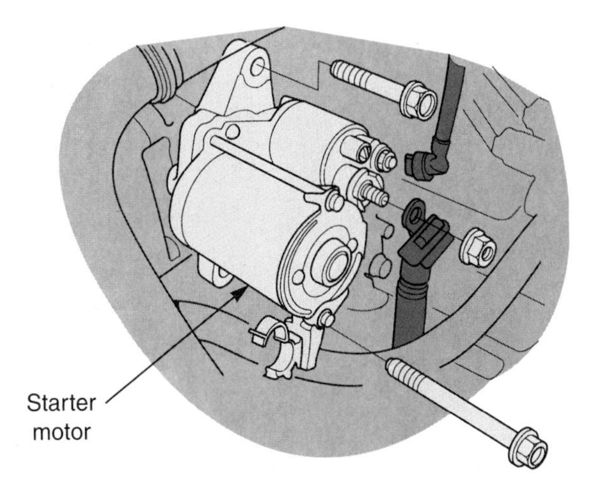

Figure 20-9. *Disconnect the starter. You may need to remove the starter to provide clearance for lifting the engine out of the vehicle. (Acura)*

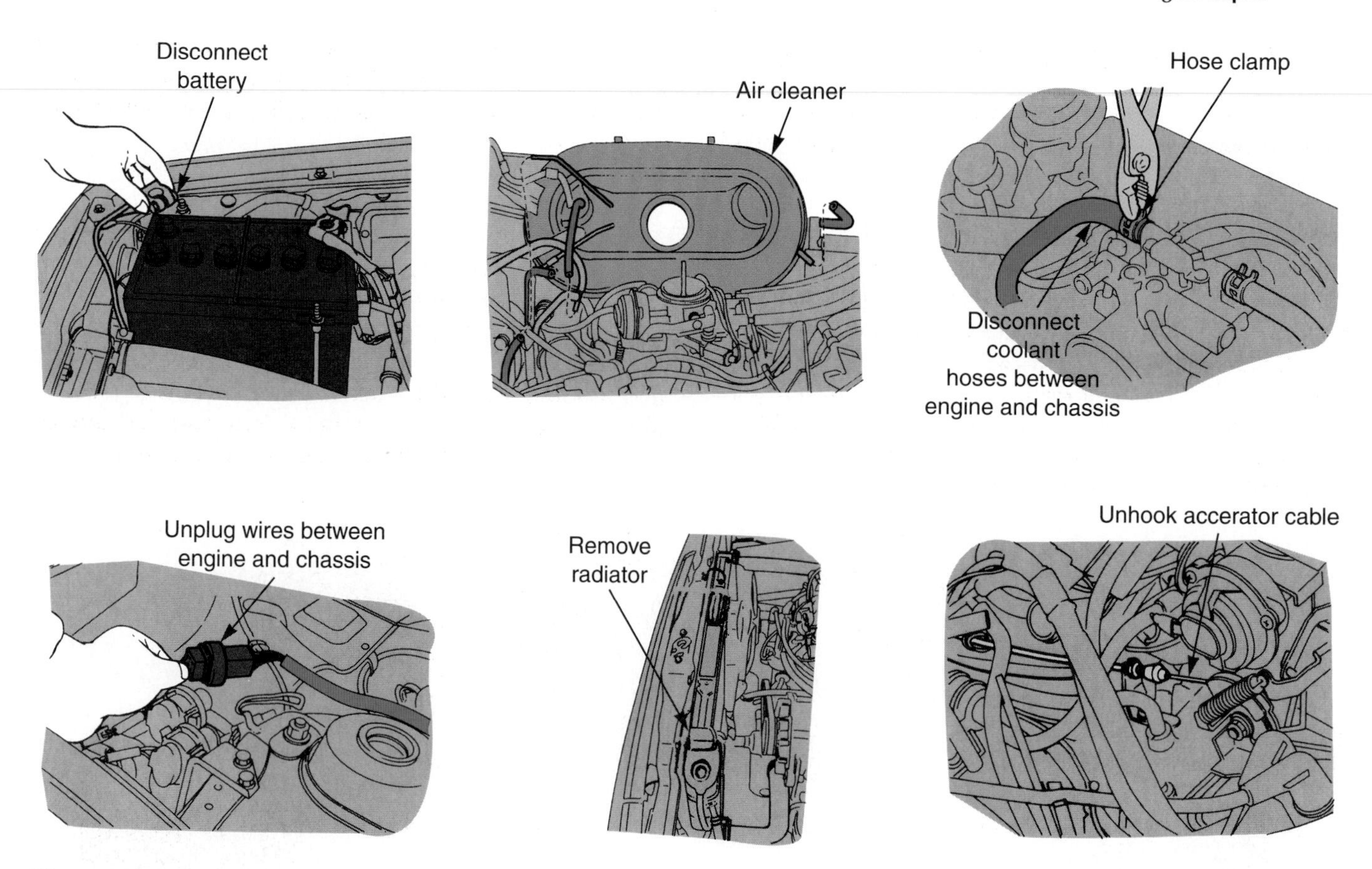

Figure 20-10. *These are some of the major components on top of engine compartment that must be disconnected. (Subaru)*

- ❑ The torque converter to see that all bolts are removed if the transmission or transaxle is to remain in the vehicle.
- ❑ All fuel lines to make sure they are disconnected and plugged.
- ❑ That motor mounts are unbolted.
- ❑ That a floor jack or holding bar is supporting the transmission.

Warning: An engine weighs several hundred pounds. Be very careful when lifting and moving an engine. It is very easy to get seriously injured. Be sure all necessary precautions are observed.

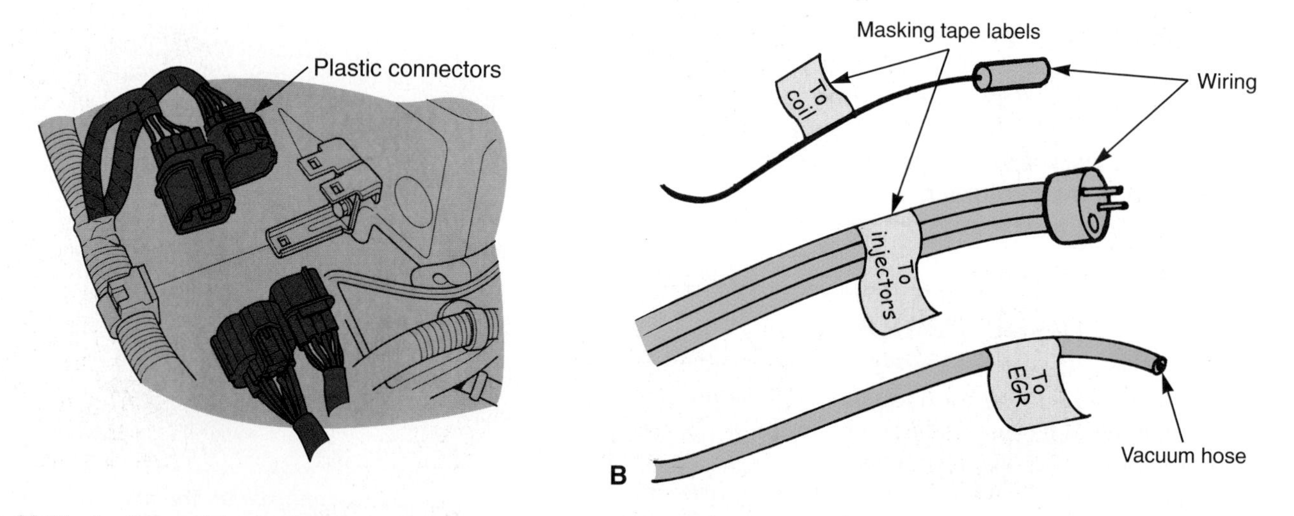

Figure 20-11. *A—Be careful not to break plastic connectors when disconnecting wire harnesses from the engine. B—Masking tape can be made into labels for hoses and wires. (Honda)*

Installing the Lifting Fixture or Chain

Position the ***lifting fixture*** or chain at the manufacturer's recommended lifting points on the engine, **Figure 20-13.** Sometimes, lift points are designed into the engine, often as brackets bolted to the block. Bolt the lifting fixture or chain to the engine. Use washers as needed to ensure the bolts do not pull out of the chain or lifting fixture.

Make sure the attaching bolts are the correct diameter for the hole in which they are installed. Also, the bolts must be fully seated and a proper length. They must thread into the hole a distance equal to at least 1-1/2 times the diameter of the bolt. This is illustrated in **Figure 20-14.**

Generally, position the fixture or chain so that it will raise the engine in a level manner, or slightly tilted down at the rear if needed. If one lifting point is at the right-front of the cylinder head, the other should be on the left-rear of the head. Use common sense and follow the manufacturer's instructions.

Caution: Avoid damaging the motor mounts during engine removal. Some motor mounts are liquid-filled units or computer-controlled, hydraulic units to reduce engine noise. The computer can alter the stiffness of the mounts by using solenoid valves that control hydraulic flow in the mounts. If you damage any of these mounts, it can be an expensive mistake.

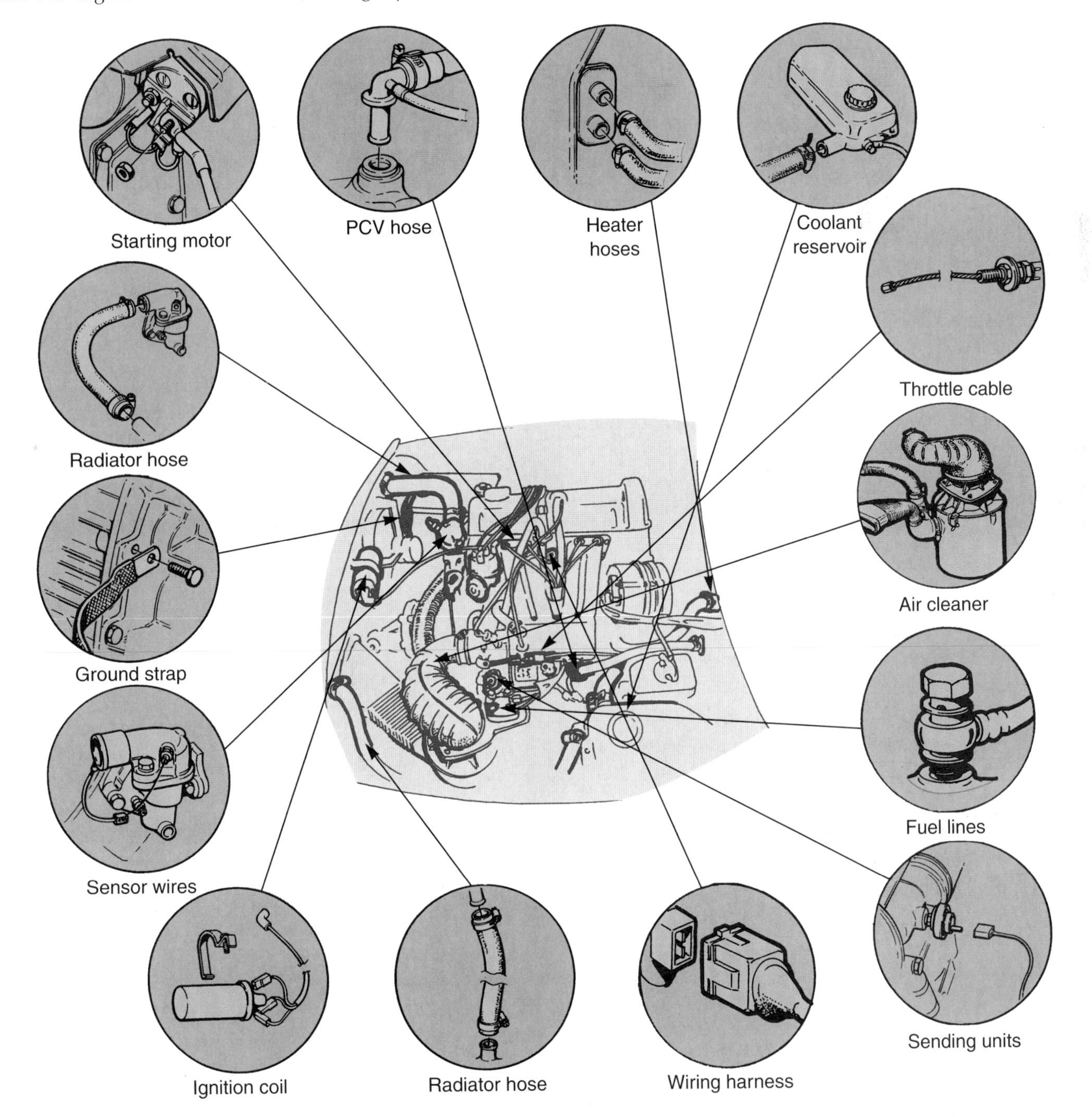

Figure 20-12. *Remove all components that will prevent removal of the engine. (Saab)*

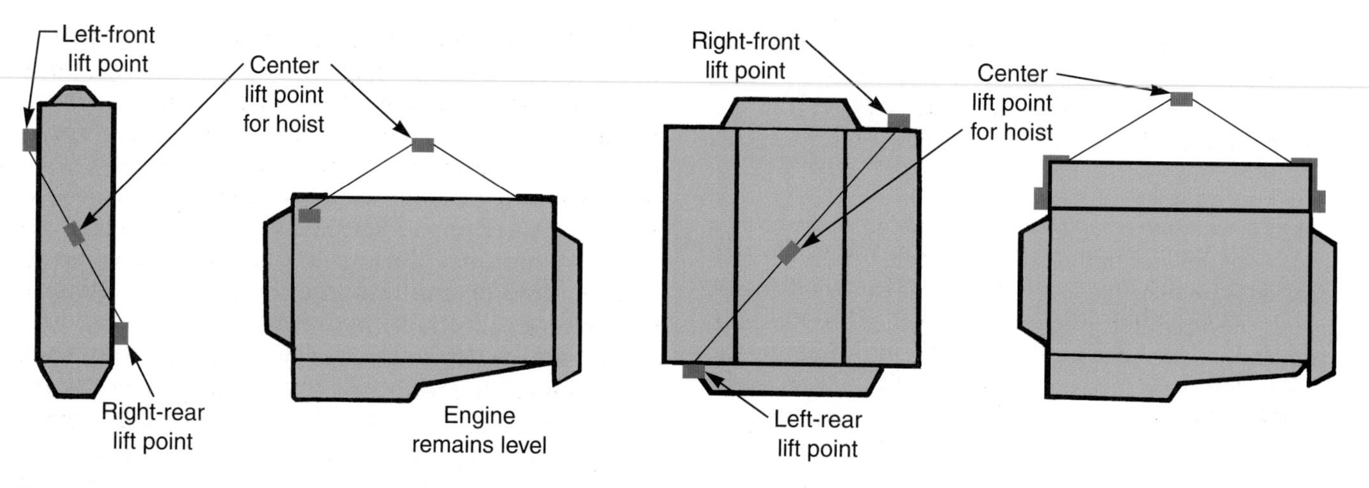

Figure 20-13. *Lift points should be on opposite sides and ends of the engine.*

Lifting the Engine out of the Vehicle

Make sure the ***crane boom*** or ***engine hoist*** is centered over the engine. Then, attach the crane or hoist to the lifting fixture or chain. Place a floor jack under the transmission if needed, **Figure 20-15.** If the transmission is to stay in the vehicle, support it. Do not let a transmission or transaxle hang unsupported after engine removal. This could damage the mounts, drive shaft, drive axles, and other parts.

1. Raise the engine slowly about an inch or two.
2. Check that everything is out of the way and disconnected.
3. Continue slowly raising the engine while pulling away from the transmission. This will separate the engine from the transmission or slide the transmission out from under the firewall. Do not let the engine bind or damage parts.
4. When the engine is high enough to clear the radiator support, roll the crane and engine straight out and away from the vehicle. With a stationary hoist, roll the vehicle out from under the engine.
5. As soon as you can, lower the engine to the ground or mount it on an engine stand.

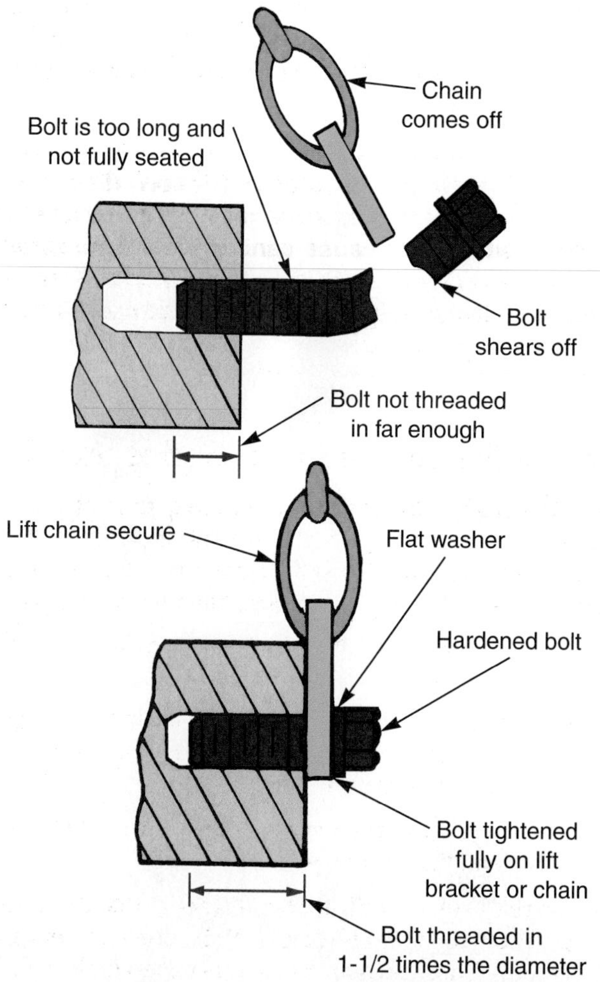

Chain
comes off
Bolt is too long and
not fully seated
Bolt
shears off
Bolt not threaded
in far enough
Lift chain secure
Flat washer
Hardened bolt
Bolt tightened
fully on lift
bracket or chain
Bolt threaded in
1-1/2 times the diameter

Figure 20-14. *Make sure the bolts used to fasten the lift chain to the engine are long enough and properly tightened. A—If the bolt extends too far from the hole, it can bend and break. B—A correctly installed bolt.*

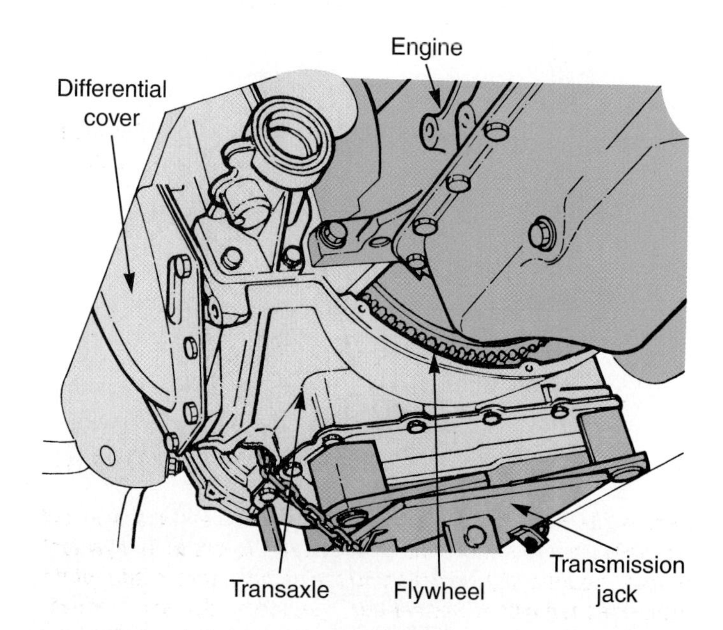

Figure 20-15. *If the transmission/transaxle is to stay in the vehicle, support it. Do not let it hang unsupported. (DaimlerChrysler)*

In some vehicles, the engine and transaxle are removed from underneath the vehicle. With some vans, the engine can be removed through the large door in the side of the body. Check the service manual for details on engine removal.

Warning: Never work on an engine that is held up by a crane or hoist. The engine could shift and fall, causing serious injury or damaging the engine! Also, keep your hands and feet out from under the engine when raising or lowering the crane or hoist.

Separating the Engine from the Transmission/Transaxle

If you pulled the engine and transmission/transaxle together, you will need to separate the transmission from the engine. Rest the assembly on the shop floor. Place blocks of wood under the engine, or leave the hoist and chain attached to the engine, to keep it from flipping over while sitting on the ground. Also, make sure the transmission/transaxle is sufficiently supported.

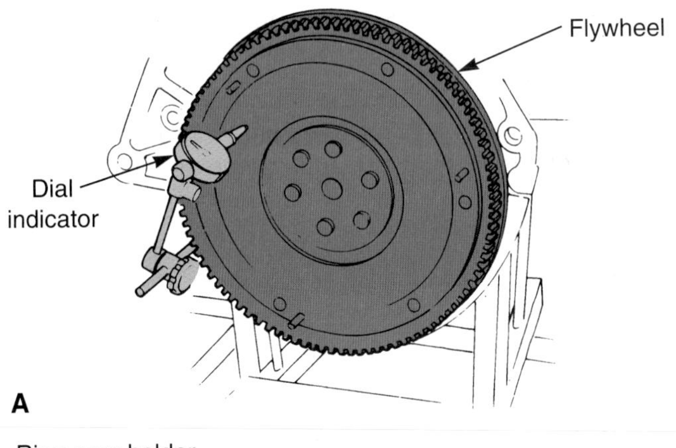

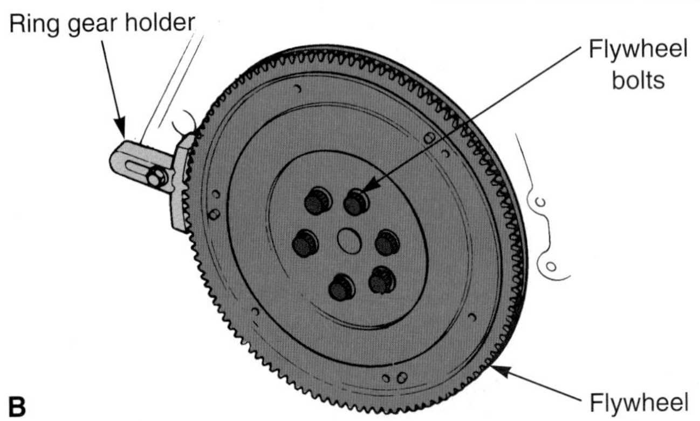

Figure 20-16. *Before mounting the engine on a stand, you will normally want to remove the flywheel. A—To check for flywheel runout, mount a dial indication on the friction surface and rotate the crankshaft with a breaker bar. The needle deflection equals the runout. B—To remove flywheel bolts, you may want to use holding fixture. (Honda)*

With an automatic transmission or transaxle, remove the bolts between the engine and bell housing. Check closely all the way around the engine because some can be hard to find. Remove the starter motor (if still attached) and any other parts preventing transmission removal. Some technicians prefer to leave the torque converter bolted to the flywheel. Others unbolt the converter so it will come out with the transmission. To separate the assembly, lift up on the transmission/transaxle, wiggle the unit, and gently pull it off of the engine.

With a manual transmission or transaxle, first unbolt the transmission. Then, remove the bell housing, clutch, and flywheel. Finally, separate the assembly by gently pulling the transmission/transaxle off of the engine.

Inspect the flywheel for problems. Check its friction surface for heat checking and cracks. Use a dial indicator to check the flywheel surface for runout, which indicates warpage, **Figure 20-16A.** Then, if needed, use a flywheel holding tool to keep the crankshaft from turning as you loosen and remove the flywheel bolts, **Figure 20-16B.**

You should also check the flywheel teeth for damage, as shown in **Figure 20-17.** If any of the teeth are broken or ground off, install a new ring gear on the flywheel. You do not want to rebuild the engine and have the starter motor grinding on a bad ring gear. Use heat from a torch to enlarge the ring gear so it can be driven off with a flat nose punch and hammer.

Warning: The clutch friction disc on a manual transmission may contain asbestos, which can cause cancer. Wear appropriate respiratory protection. Do not blow clutch dust with a blowgun. Always use appropriate dust containment and removal procedures.

Placing the Engine on an Engine Stand

Most ***engine stands*** have a removable, rotating plate to which the engine is bolted. It is usually easier to bolt the engine to this plate first while the plate is not attached to the stand. There are usually four adjustable arms on the plate.

1. Remove the engine mounting fixture (plate) from the engine stand.
2. Place the plate on the back of the engine and adjust the arms to align with the bell housing bolt holes. The arms should be evenly spaced.
3. Bolt the arms to the engine using Grade 8 bolts, but only hand tight at this point.
4. Move the plate and adjustable arms so that the plate's pivot point (collar) is roughly at the engine's center of gravity, which is usually just above the crankshaft.
5. Tighten the bolts in the adjustable arms.
6. Use the engine to raise the engine to approximately the height of the engine stand.

Figure 20-17. *A—This flywheel surface is badly worn and has tiny surface checks or cracks. B—This ring gear is badly worn and must be replaced. It has been removed from the flywheel.*

7. Move the stand into position by the mounting fixture.
8. Slowly raise or lower the engine until the collar on the mounting plate aligns with the pivot hole in the stand.
9. Insert the collar into the hole and lock it into place; most stands have a locking pin or bolt(s).
10. Slowly lower the crane boom until there is no tension on the lifting chain. See **Figure 20-18.**
11. Once you are certain the engine is securely attached to the engine stand, you can remove the lifting chain.

Warning: Be very careful when releasing the tension on the lifting chain. If the engine is not secure or not balanced and centered, the engine and stand may shift, flip, or fall causing severe injury.

Engine Disassembly

With the engine bolted to an engine stand, you are ready to begin disassembly, or teardown. ***Teardown*** is the basic disassembly of the engine. Engine teardown methods vary somewhat from engine to engine. However, general procedures are similar and apply to all engines. The following sections provide a general guide.

Go slowly and inspect each part for signs of trouble as it is removed. Look for wear, cracks, damage, seal leakage, or gasket leakage. Remember, if you overlook even one problem, the rebuilt engine may fail in service. All of your work could be for nothing, and your employer may have to assume the responsibility of fixing the vehicle again.

Engine Front End Disassembly

The engine front end should be disassembled first, especially with OHC engines. Engine front end disassembly is simple if a few basic rules and service manual instructions are followed.

1. Remove the water pump and any other parts in front of the engine front cover. If the engine has a timing belt, remove the timing belt cover. Mark the direction of rotation on the belt. Then, loosen the tensioner and slip off the belt.
2. The vibration damper is commonly press-fit onto the crankshaft. A vibration damper puller ("wheel puller") is normally needed to remove the vibration damper.
3. If specified by the service manual, remove the front oil seal at this point. **Figure 20-19A** shows how to use a puller to remove the seal. However, the seal is often removed from the front cover after the front cover is removed from the engine.
4. Unbolt and remove the engine front cover. If prying is necessary, do it lightly while tapping with a rubber or plastic hammer. Do not bend the cover or mar the mating surfaces.

Figure 20-18. *This technician has checked to make sure the engine is secured to the stand. The lifting chain can now be removed.*

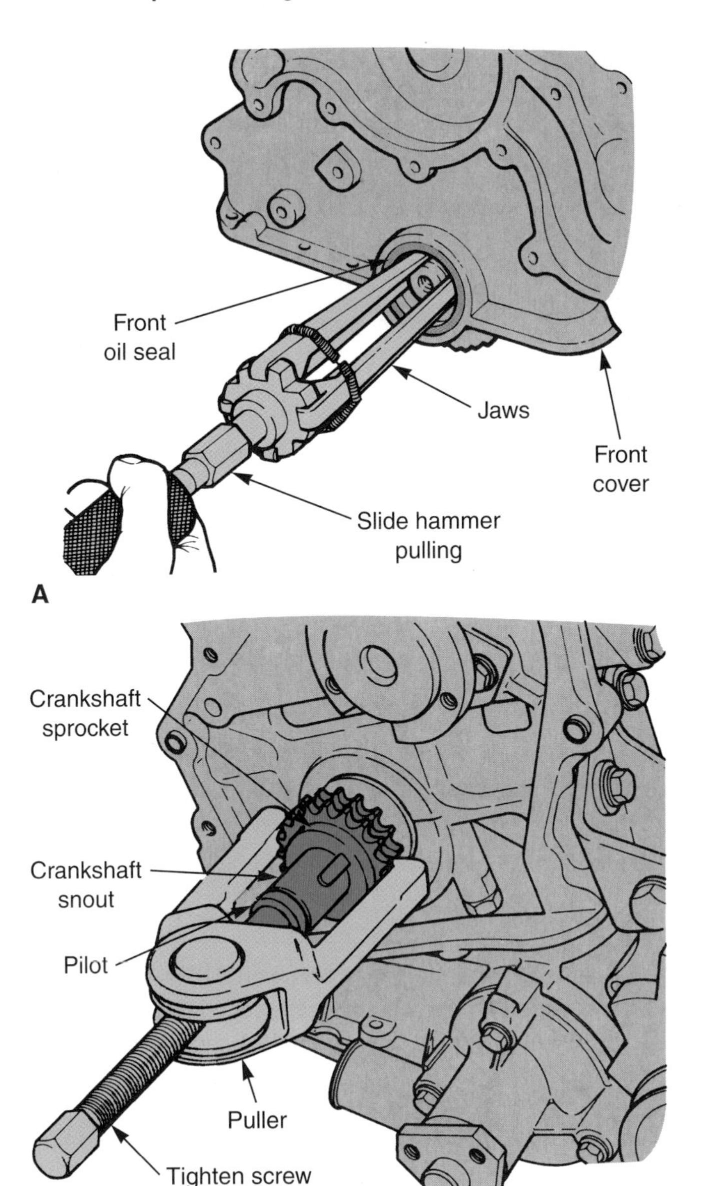

Figure 20-19. *A—A slide hammer puller is being used to remove the front oil seal from the front cover. B—A puller is being used to remove the timing sprocket from the crankshaft snout. (Ford, DaimlerChrysler)*

5. Remove the oil slinger and timing mechanism. Usually, the timing gears or sprockets will slide off with light taps from a brass or plastic hammer. If not, use a gear puller, **Figure 20-19B.**
6. If the oil pump or other components are in the front cover, remove them and inspect for wear.

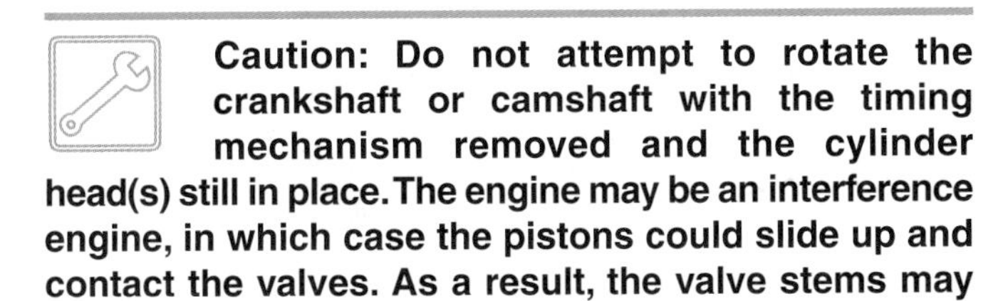

Caution: Do not attempt to rotate the crankshaft or camshaft with the timing mechanism removed and the cylinder head(s) still in place. The engine may be an interference engine, in which case the pistons could slide up and contact the valves. As a result, the valve stems may be bent or the pistons damaged.

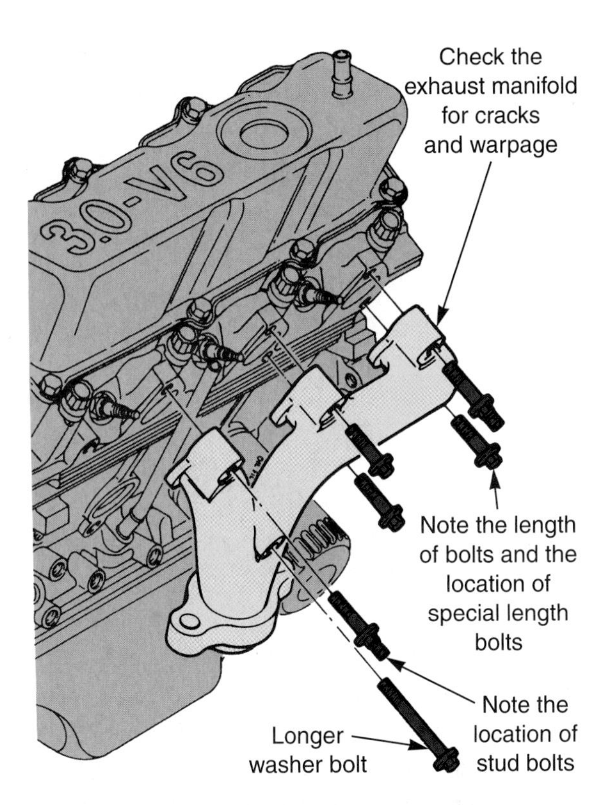

Figure 20-20. *When removing the exhaust manifold, note odd bolt lengths and the location of stud bolts. They must be reinstalled in the same holes. (Ford)*

Engine Top End Disassembly

The engine top end generally includes the valve train, cylinder head, and related components. The top end is normally disassembled second, after the front end.

1. Remove all external engine parts, such as the fuel rail, throttle body unit, spark plug wires, coil(s), and so on. Take off all parts that could be damaged or that prevent removal of the cylinder head(s).
2. Remove the exhaust manifold(s). Note the location of any stud bolts. They must be replaced in the same hole, **Figure 20-20.** Brackets, such as the alternator bracket, or other devices mount to these studs.
3. Unbolt the valve or camshaft cover(s), **Figure 20-21.** If light prying is needed, be careful not to damage the mating surfaces. Light taps with a rubber mallet may also free a cover. Be careful not to crack or damage the cover. Take extra care if the cover is plastic.
4. Remove the rocker arm assemblies. If the engine has rocker shafts, loosen the fasteners using the sequence shown in **Figure 20-22.** This will prevent the end fasteners from being bent by valve spring tension.
5. If the lifters, push rods, and rocker arms are to be reused, keep them in exact order by using an organizing tray. Wear patterns and select-fit parts require that most components be installed in their original locations.

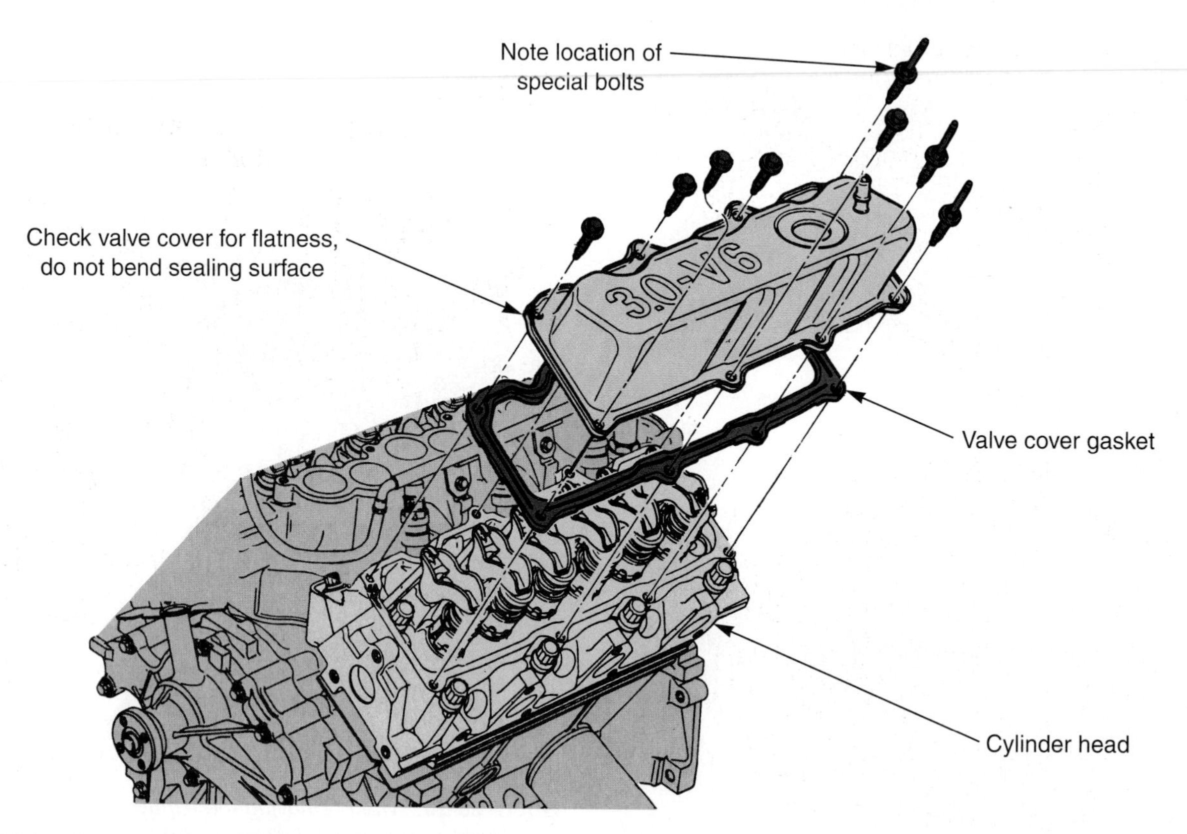

Figure 20-21. *Do not bend the valve covers during removal. A rubber mallet will sometimes free the cover from the head. Light prying under the cover may be needed, but do not bend the lip. (Ford)*

6. Remove the intake manifold. See **Figure 20-23.** With V-type, push rod engines, you may need to remove the valve train components before the intake manifold. In some designs, the push rods pass through the bottom of the intake.
7. Loosen and remove the cylinder head bolts, **Figure 20-24.** Most technicians use an air impact or a breaker bar and six-point socket to loosen the bolts. On a V-type or opposed engine, use number or letter punches to mark the heads as right and left.
8. Frequently, the cylinder head is stuck to the block. To free the head, insert a breaker bar in one of the ports and pry up. Avoid prying on mating surfaces! Some heads have a unmachined notch on the deck that allow for prying without damaging the sealing surfaces.

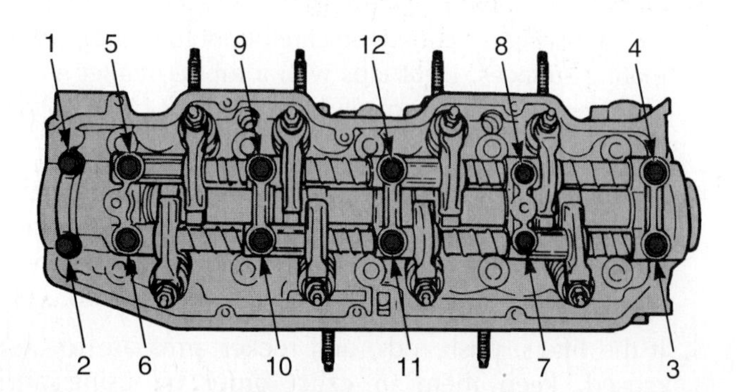

Figure 20-22. *When unbolting rocker shafts, use this sequence. If you start on one end and work across, spring tension could bend the rocker shafts or break the caps. (General Motors)*

9. Inspect the head gasket and deck surfaces for signs of leakage. Also, look for oil in the combustion chambers. This may indicate a seal or ring problem.

Caution: Do not unbolt an aluminum cylinder head when it is hot. This could cause head warpage. Allow the head to cool before loosening the bolts.

Cylinder Head Disassembly

Disassembly procedures for cylinder heads can vary, especially between overhead valve and overhead camshaft engines. However, keep the following rules in mind during cylinder head teardown:

- ❑ Inspect for signs of trouble before disassembly. Look for oil-fouled combustion chambers, which may indicate bad rings, valve seals, or valve guides. Inspect for a blown head gasket, signs of coolant leakage, rust, and cracks. Close inspection may find other troubles.
- ❑ Use a valve spring compressor to remove the valve keepers, retainers, and valve springs. Before compressing the springs, strike the valve retainers, not the valve tips, with a brass hammer. This will free the valve retainers from the keepers and allow the compressor to squeeze the springs.

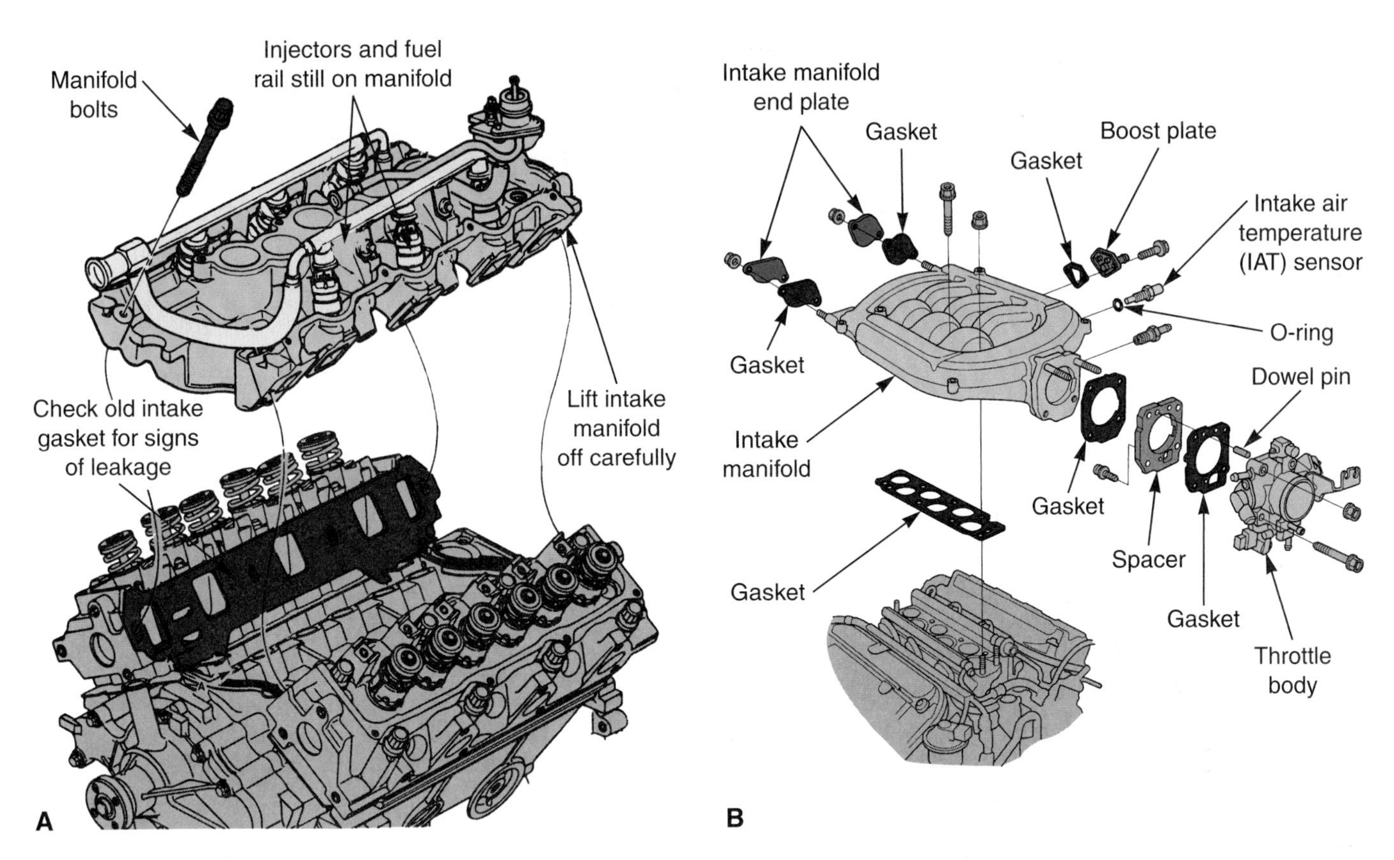

Figure 20-23. *A—The intake manifold can be removed with injectors and fuel rail attached. B—This two-piece intake manifold requires you to remove the upper plenum to gain access to the lower half of the intake manifold and the injectors. (Ford, Honda)*

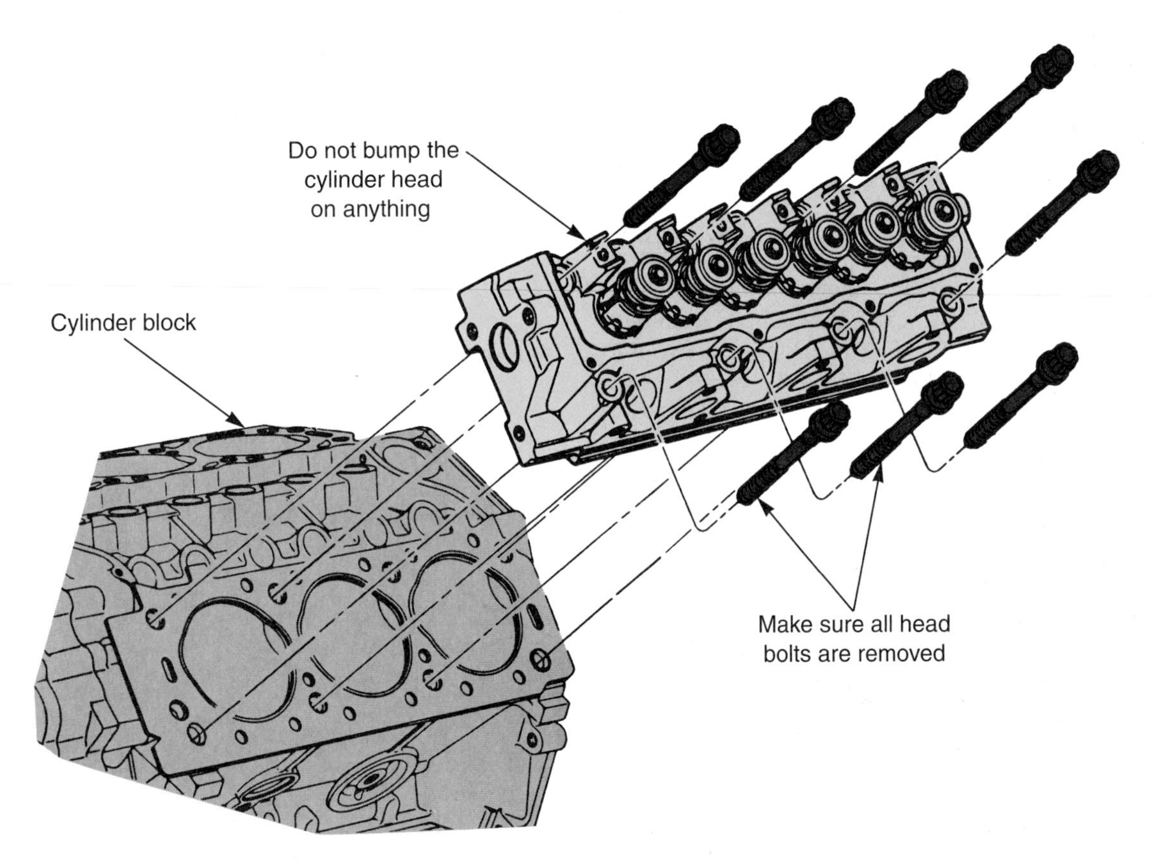

Figure 20-24. *First, remove the cylinder head bolts. Free the head by inserting a breaker bar into a port and prying up. (Ford)*

- A special valve spring compressor is needed for some overhead camshaft cylinder heads. It reaches down inside the tappet bore to compress the springs.
- Before removing the valves, check their tips for ***mushrooming.*** This is a condition where the tips have been hammered larger by the action of the rocker arms. If mushrooming exists, file a chamfer on the valve tips. Never try to drive mushroomed valves out of the head. This can score or split the valve guides or crack the head.
- Remove any other parts that prevent head service.
- Keep all parts organized. During reassembly, it is best to return reused parts to their same locations in the engine. Parts can be select-fit from the factory and their sizes may be matched for more accurate clearances.

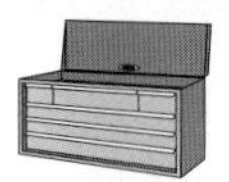

Note: Detailed information on cylinder head disassembly and rebuilding is given in Chapter 22.

Engine Bottom End Disassembly

After top end and front end disassembly, you are ready to take the bottom end apart. The bottom end includes the pistons, connecting rods, crankshaft, and related bearings. In addition, some modern engines are equipped with balance shafts that are driven by the crankshaft. These shafts are normally located in the block or below the crankshaft, depending on the engine design. Balance shafts are normally removed during engine disassembly.

1. Inspect the cylinders for signs of excess wear. Use your fingernail to feel for a lip or ridge at the top of the cylinder wall, **Figure 20-25A.** This ridge is formed because the piston rings do not wear the cylinder wall to the top of the cylinder.
2. Use a ridge reamer to cut away the ridge, **Figure 20-25B.** Use a wrench to rotate the reamer and cut away the metal lip, **Figure 20-25C.** Cut until the lip is flush with the rest of the cylinder wall. This prevents damage to the piston during removal.
3. Use compressed air to blow the metal shavings out of the cylinder after ridge reaming to prevent cylinder or piston scoring.
4. Unbolt and remove the oil pan and oil pump, **Figure 20-26.** Inspect the bottom of the pan for debris. Metal chips or plastic bits may help you diagnose engine problems. Aluminum chips may be from broken piston skirts, black rubber bits from deteriorated valve seals, plastic or fiber from broken timing gear teeth, or babbit from failed engine bearings.
5. Unbolt one of the connecting rod caps. Then, use a wooden hammer handle or wooden rod to tap the piston and rod out of the cylinder, **Figure 20-27.** Be careful not to drop the piston and rod assembly.

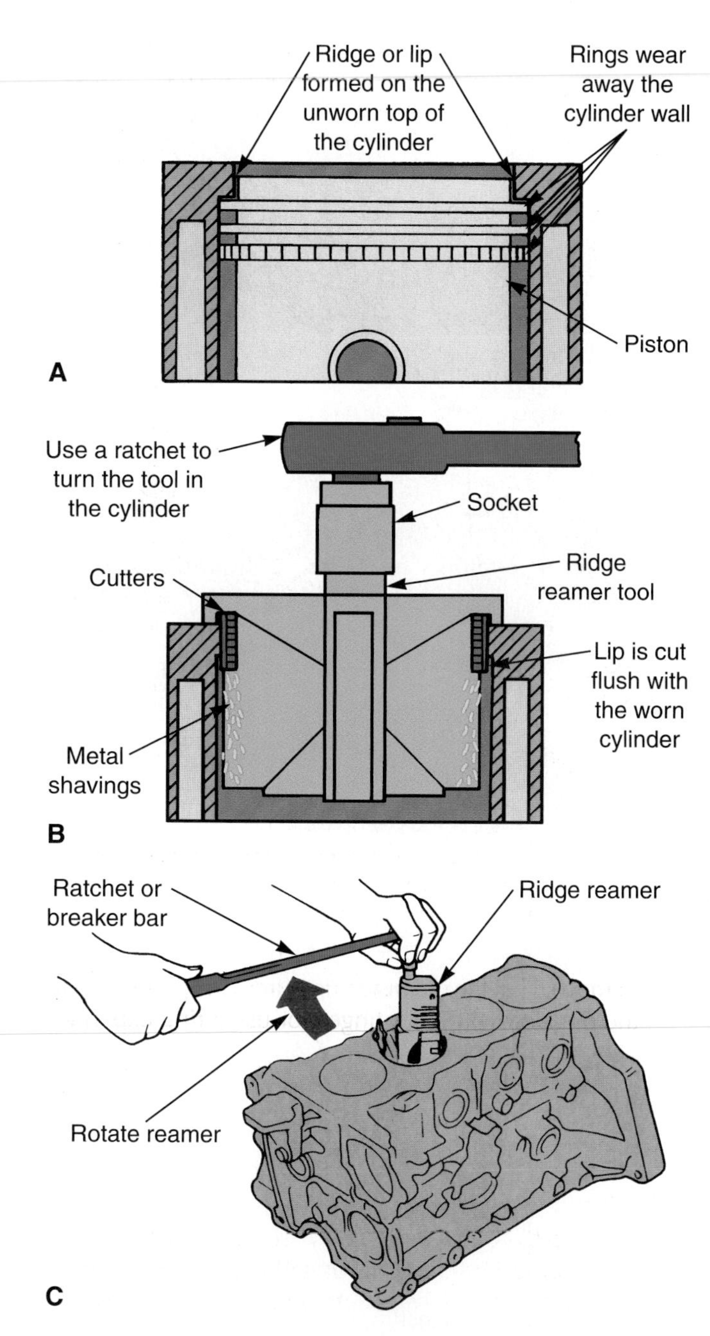

Figure 20-25. *A—A ridge may be formed at the top of the cylinder where the rings do not wear. B—A ridge reamer is used to remove the ridge. C—Turn the reamer until the ridge is removed flush with the cylinder wall.*

6. As soon as the piston is out, replace the rod cap. Also, check the piston head and connecting rod for identification markings. The piston will usually have an arrow pointing to the front of the engine. The connecting rod and rod cap should have numbers matching the cylinder number, **Figure 20-28.** If not, use a punch set to mark the rod and cap. If needed, mark the piston heads with arrows or numbers.
7. Remove the other piston and rod assemblies in the same way, one at a time. Reinstall each cap on its rod. Mark the rods, caps, and pistons as needed.

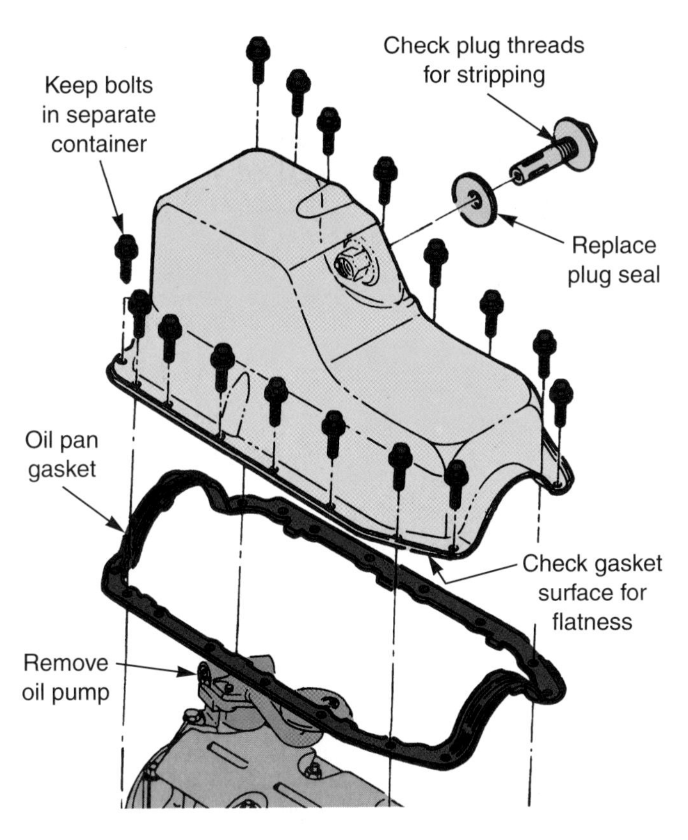

Figure 20-26. *Turn the engine over on the engine stand and remove the oil pan and oil pump. (Ford)*

8. Remove all of the old piston rings from the pistons. Spiral the rings off with your fingers or use a ring expander.
9. Before removing the main bearing caps, check to see that they are numbered. Normally, numbers and arrows are cast on each cap. Number one cap is at the front of the engine. If they are not marked, use a punch set to mark the main caps.
10. Remove the main cap bolts. To remove each cap, pull the main cap bolts out about halfway and wiggle them back and forth while pulling up on the main cap.

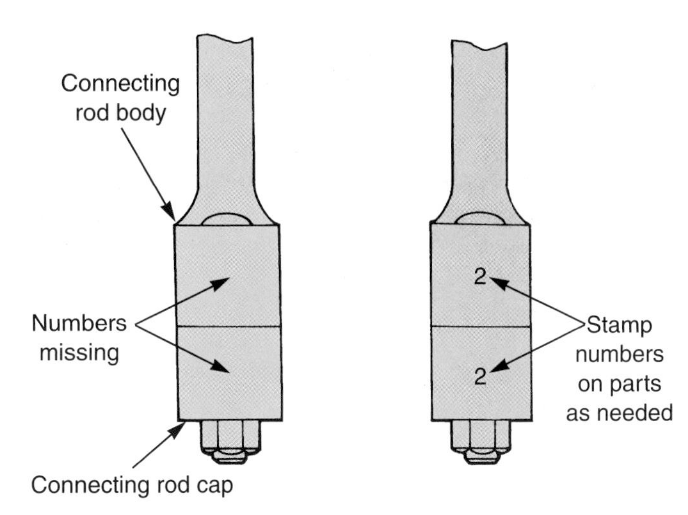

Figure 20-28. *As each piston/rod assembly is removed, make sure the rod is marked on the body and cap. If not, use a number punch set to mark each rod body and cap.*

11. Lift the crankshaft carefully out of the block, **Figure 20-29.** It is heavy; be careful not to hit and nick the journals. Place the crankshaft in a location where it cannot be damaged.
12. To remove the balance shaft, begin by removing any retainers holding the shaft in the engine. Then, slide the shaft out. In some cases, it may be necessary to use a slide hammer to remove the shaft.
13. If the engine is old or it is going to be hot tanked, pry out the core plugs from the head and block. Core plugs rust out and leak after prolonged service. Since the engine is disassembled, it is a good time to replace them.
14. If the block must be sent to a machine shop for boring and hot tank cleaning, make sure all external hardware is removed. This includes rubber motor mounts, oil and coolant temperature sending units, sensors, actuators, and so on.

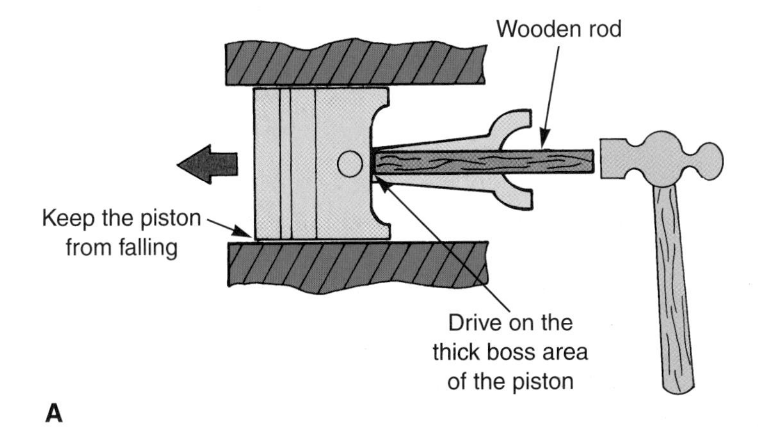

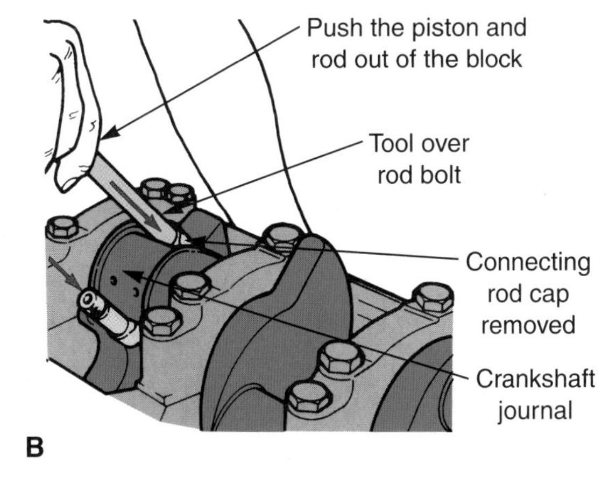

Figure 20-27. *A—Wooden rod or hammer handle is one way to drive out pistons. Place the end of the handle on the piston boss. B—This technician is using a threaded rod that fits over a rod bolt to hand-force the piston out of the block.*

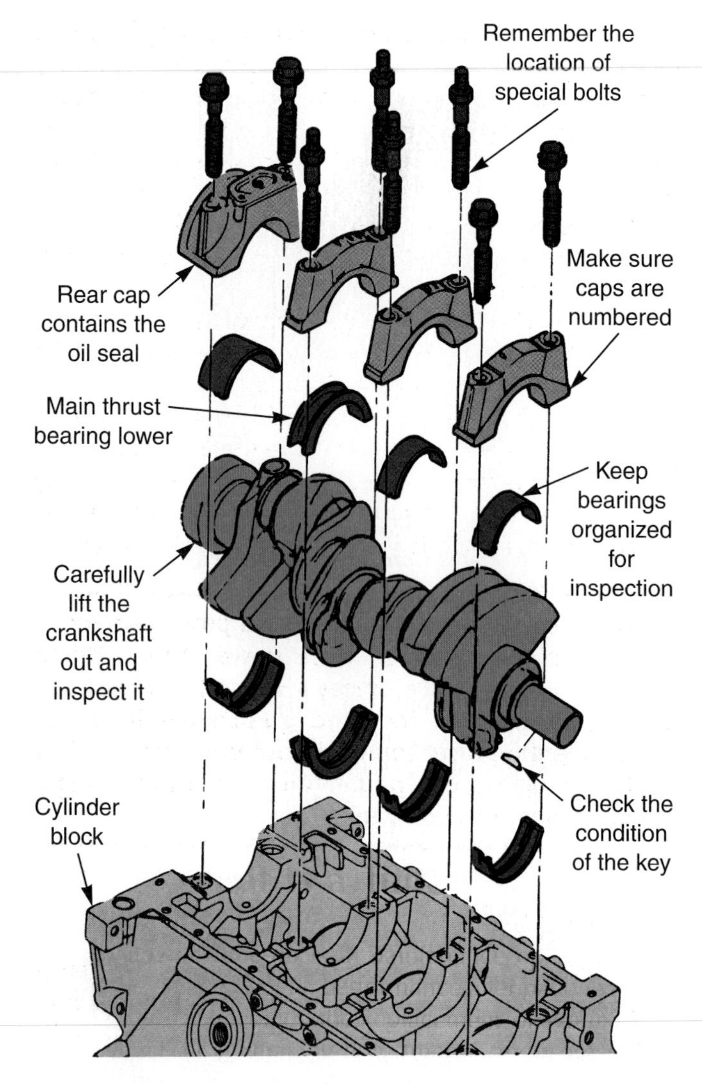

Figure 20-29. *Make sure main caps are numbered before removal. Then, carefully lift the crankshaft out of the block without hitting the journals. (Ford)*

Caution: Do not do anything to dent or mar the connecting rods. The I-beam portion of the rod is especially sensitive to damage. For example, never clamp the rod in unprotected vise jaws. The dent or indentation in the metal could cause a stress point. This can lead to cracking and breakage of the rod during engine operation. When marking the rods, only mark on the side of the cap and rod body.

Cleaning Engine Parts

Parts can be very dirty. For example, valves and combustion chambers can be coated with carbon; the valve covers, lifter valley, and bottom of the intake manifold can have sludge buildup; and most parts will have an oil film. See **Figure 20-30.**

After you have removed all of the parts from the engine block and cylinder head(s), everything should be cleaned. Different cleaning techniques are needed depending on part construction and type of material.

Closer part inspection can be done during and after part cleaning. Some problems are difficult to see when a part is covered with oil, grease, gasket material, or carbon deposits.

Scrape Off Old Gaskets and Hard Deposits

Begin cleaning the engine by scraping off all old gasket material and hard deposits, **Figure 20-31.** Scrape off the gaskets for the valve covers, head, front cover, intake manifold, oil pan, and other components. Also, scrape off as much sludge and carbon as you can.

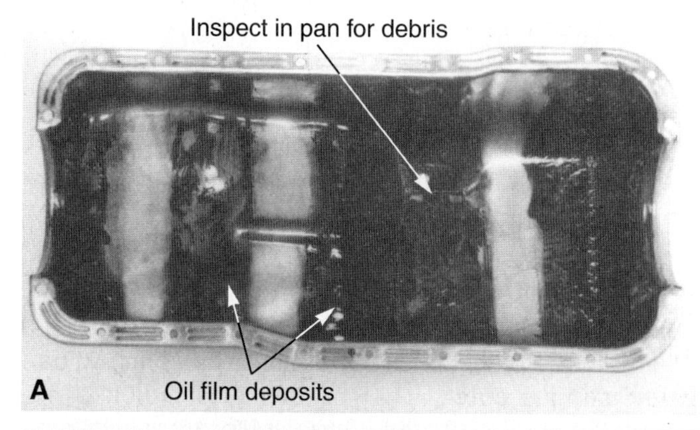

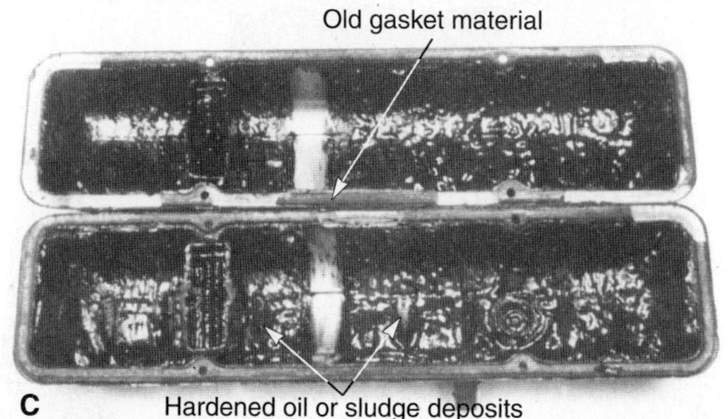

Figure 20-30. *Three common types of engine deposits. A—Light oil film. B—Carbon deposits. C—Sludge. (Texaco)*

Figure 20-31. *This technician is scraping old gasket material from the deck of a cast iron block.*

When cleaning aluminum parts, use a dull scraper and work carefully. See **Figure 20-32.** The soft material is easily nicked. This could lead to gasket leakage or failure when the engine is returned to service.

Warning: When using a gasket scraper, push away from your body; do not pull toward your body. Even a dull scraper can cause serious cuts.

Power Cleaning Tools

A ***scuff pad*** mounted in an air drill or similar tool can cut through and lift off gasket material and silicone. Yet, the pad is too soft to easily damage metal surfaces. Move the tool back and forth along the gasket material until the material is removed.

Figure 20-32. *A dull scraper should be used on aluminum parts. A sharp scraper may dig into the aluminum and damage the part. (Fel-Pro)*

A ***power brush,*** or stiff rotary brush, mounted in a drill can be used to remove hard carbon deposits. It is especially handy inside hard-to-reach areas; combustion chambers, for example, **Figure 20-33A.** Move the tool over the carbon deposit until the deposit is removed. Often, it may be easiest to start at the edge of the deposit, rather than in the middle. A small power brush can be used to clean out valve guides, **Figure 20-33B.** However, use a power brush carefully on aluminum parts.

A ***wire wheel*** is also used to remove carbon deposits, usually off of valves. However, other small parts can be cleaned with a wire wheel, such as brackets. Use an appropriate coarseness of wheel. For small or delicate parts, you may need to use a fine wheel. Heavier parts may be able to withstand an aggressive, coarse wheel. This may speed up the cleaning process. Large parts, such as the cylinder head, are too large to hold against the wheel.

Warning: Always wear eye protection when cleaning parts with power tools. When using a wire wheel mounted in a grinder, make sure the tool rest and guards are in place. Never operate power tools without the proper safety devices in place. Wear a full-face shield when using a bench grinder or wire wheel.

Cleaning Solvents

After scraping off the gaskets, various chemical cleaners (cleaning solvents) are used to remove hard-to-reach or stubborn deposits. For example, sometimes gaskets can be

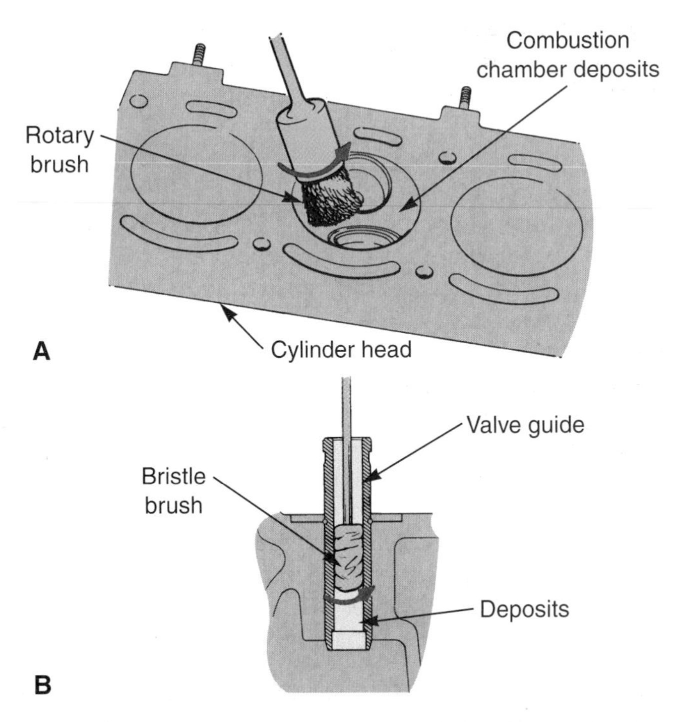

Figure 20-33. *A—A power brush is handy for cleaning hard-to-reach areas, such as combustion chambers. B—A small power brush can be used to clean out valve guides. (Toyota)*

Figure 20-34. *Sometimes gaskets can become very hard to remove. Chemical sprays can dissolve gaskets. The softened gaskets can then be easily scraped off. (Fel-Pro)*

very difficult to scrape off. A ***spray-on gasket solvent*** can be used to soften the gasket, **Figure 20-34.** After spraying, let the chemical work for a few minutes. Then, scrape off the softened gasket with a scraper.

Warning: Never use gasoline as a cleaning solvent. The slightest spark or flame could ignite the fumes, causing a deadly fire!

Cold Tank

Most auto shops have a ***cold cleaning tank*** or machine for removing oil and grease from parts, **Figure 20-35.** It will not remove hard carbon or mineral deposits. It has a pump and filter that circulates clean, mild naphtha solvent out of a spout. To clean parts, place them in the work area of the tank. Then, direct the stream of solvent onto the part while scrubbing the part with a soft bristle brush.

Decarbonizing cleaner ("carb cleaner dip tank") is another type of "cold soaking." This cleaner is a very powerful chemical that can remove carbon, paint, gum, and most other deposits. It is, however, an expensive solvent and not commonly used unless necessary. Parts are typically placed in a basket that is lowered into a tank containing the cleaner. The parts are allowed to soak in the cleaner for a couple of hours. Then, the parts are removed and rinsed.

Cold solvent is cool to the touch and not as flammable as gasoline. However, it is a fraction of crude oil and can be harmful if proper safety precautions are not followed.

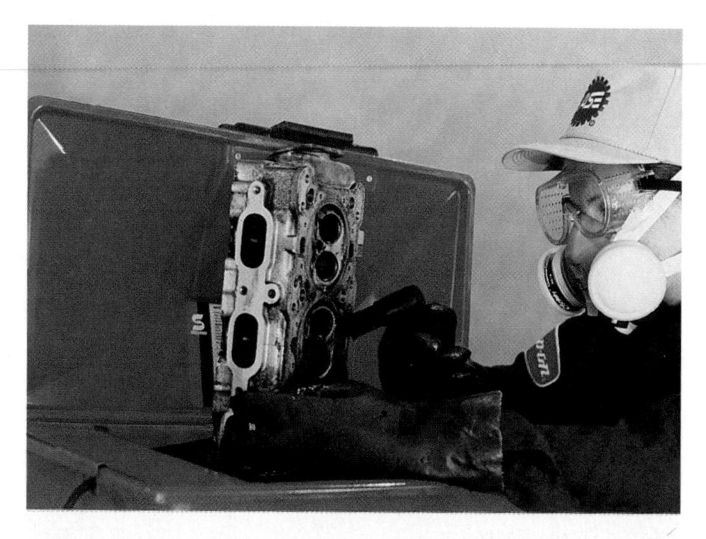

Figure 20-35. *A cold soak tank is for removing oil film and grease. If the parts are covered with sludge, scrape off the sludge before cleaning the part in a cold soak tank.*

When using a cold solvent tank, wear a cartridge respirator, rubber gloves, and a rubber or plastic apron. Part cleaning solvents can be harmful if absorbed into your skin, inhaled, or ignited.

Hot Tank

A ***hot tank*** is used to aggressively remove mineral deposits, hard carbon deposits, oil, grease, and even paint. See **Figure 20-36.** In a complete engine rebuild, the cylinder head(s) and block are usually hot tanked, or "boiled out." Other parts, such as valve covers or oil pans,

Figure 20-36. *Most automotive machine shops have a hot tank for cleaning blocks, heads, and other cast iron or steel parts.*

may also be hot tanked. Most automotive machine shops have a hot tank.

Aluminum engine parts can be etched and damaged by the powerful acidic solvents in some hot tanks. Hot solvent tanks can be filled with chemicals of different strengths, or caustic levels. Stronger hot tank solvents can etch aluminum engine parts. When hot tanking aluminum heads, blocks, manifolds, and similar castings, make sure the solvent is formulated for use on aluminum.

Caution: Before hot tank or oven cleaning, all non-metal parts (sensors, harness clips, hoses, etc.) must be removed to avoid being damaged. However, these parts can often remain in place when cleaning by hand in a cold solvent tank.

Heat Cleaning Ovens

Automotive machine shops may use a heat-cleaning oven in place of a hot tank to clean engine parts. A ***heat cleaning oven,*** also termed a burn-off oven, is designed to remove organic materials such as oil, grease, paint, varnish, and plastic from metal engine parts without the use of chemicals. Superheated air is circulated over the parts to burn off the organic material. The part temperature is raised just high enough to remove deposits without affecting the metal in the part.

Heat cleaning ovens eliminate the disposal and liability problems associated with acid or chemical cleaning. Modern ovens are designed to operate while producing minimum emissions.

When placing parts in a heat-cleaning oven, stack and space the parts so that air can circulate between them. Do not stack large castings flat on top of each other or surfaces may not get hot enough to be cleaned. Make sure the oven temperature and timer are set correctly for the type of metal being cleaned. Most parts can be cleaned in a couple of hours. Refer to the equipment manufacturer's instructions for details.

After heat cleaning in an oven, engine parts are often shot peened or media blasted to remove ash and soot. The engine parts can then be washed in a cold solvent tank and blown dry with an air nozzle.

Warning: When removing parts from an oven, wear leather gloves and a shop apron. Be careful not to get burned on the hot parts.

Media Blasting

Media blasting can remove paint, carbon, and other dry deposits from parts. Cylinder heads, intake manifolds, valve covers, etc., can be cleaned and readied for painting with media blasting. However, the parts must be free of oil and sludge to prevent clogging the machine.

Be careful when media blasting aluminum parts. Only use enough blast action to clean off deposits. Too much blasting can erode and etch the aluminum and ruin the part.

Carbon coated piston heads are very hard to clean. Media blasting can quickly remove these deposits, but careful use is required. Only clean the heads of the piston, not the skirt. The media can quickly erode the skirt and ruin the piston.

Note: Make sure you are using the correct media for the application. Various types of media are available for different applications.

Cleaning Threads

Most technicians clean, or "chase," threads during a major engine rebuild. Rust and scale can collect inside threaded holes. This can prevent the bolts from threading in smoothly. In turn, the tightening of the bolt can be altered and the parts may not be properly clamped together. Leakage or part failure may result.

A ***thread chaser,*** not a tap, should be run down through all critical bolt holes to clean the threads. Use a speed handle to lightly run the chaser down through bolt holes. Do not tighten the chaser, just run it down lightly to the bottom of each hole. This is shown in **Figure 20-37.**

Air Blowgun

Using an ***air blowgun*** is normally the last step in cleaning parts. Blow off small bits of dirt, solvent, water, and other debris. Be sure to blow out bolt holes, coolant passages, and oil passages. When done with this step, there should be no dirt, debris, or deposits on or in the engine and its parts.

Warning: Use extreme care when using an air gun. Wear eye protection. Never aim the gun at your body or other individuals. If air is forced through your skin and into your blood stream, a blood clot may result that could cause death!

Summary

The vehicle identification number (VIN) provides valuable information. From it, you can determine which engine was installed in the vehicle at the factory. Other information, such as transmission type and body style, can also be determined from the VIN.

Figure 20-37. *During a major engine rebuild, run a thread chaser through all threaded bolt holes. A—Hand start the chaser and use speed handle to "chase" the threads. B—Spin the chaser lightly down into the hole, but do not tighten it in the bottom of a hole.*

Many engine repairs can be done with the engine block still bolted to the chassis. This includes front end and top end service. Valve jobs, camshaft service, and valve train repairs can all be done without removing the engine block from the vehicle.

Crankshaft and cylinder block service requires engine removal. In some cases, however, you can service pistons and rings by removing the heads and oil pan. The engine can be rebuilt in-vehicle. However, it is often just as easy to remove the engine for major service.

An engine overhaul involves cleaning, inspecting, measuring, and replacing parts as needed. When part measurements are not within specifications, the parts must be repaired or replaced. Usually, piston rings, bearings, seals, gaskets, core plugs, timing mechanism, oil pump, and any other worn parts are replaced during an overhaul.

During engine removal, you should try to keep everything organized. Label wires and hoses. Place fasteners in separate containers. Use one container for bolts off the front of the engine; another for lower end fasteners, and so on. This will save time and eliminate confusion during reassembly.

When attaching the lift chain, make sure it is positioned to lift the engine level or at the specified angle. Also, bolts should be the correct diameter and fully threaded into holes a distance equal to at least 1-1/2 times their diameter.

When raising the engine out of the vehicle, double-check that everything is disconnected. Keep hands out from under the engine. Only raise the engine high enough to clear the radiator support. As soon as the engine is out, lower it to the ground. Never work on an engine held in the air with a crane or hoist. Mount the engine on a stand.

Generally, start with the engine front end when disassembling the engine. Then, remove the top end: intake, exhaust manifolds, valve train, and heads. Finally, remove the bottom end.

Pullers are needed to remove some vibration dampers, seals, and timing mechanisms. A ridge reamer is needed to cut out any lip formed at the top of worn cylinders before piston removal. Check for numbers on the rods as they are removed. Number them as needed. This also applies to block main caps and camshaft caps.

After disassembly, clean all parts using approved methods. Inspect parts closely for problems before and after cleaning. Look for signs of leakage, cracks, stripped threads, and so on.

Scrapers are used to remove old gaskets and sludge. A power brush will clean off carbon in combustion chambers, from valve guides, and on valves. Cold soak cleaner is for removing an oil film. Decarbonizing cleaner and hot tank cleaning will remove more stubborn deposits. Do not place aluminum parts in a hot tank because they will be ruined. When all done cleaning, use a blowgun to blow all solvent, water, and remaining debris from the parts.

Review Questions—Chapter 20

Please do not write in this text. Write your answers on a separate sheet of paper.

1. Which of the following may be coded in the VIN?
   (A) Engine type.
   (B) Body style.
   (C) Transmission type.
   (D) All of the above.
2. Define *engine overhaul.*
3. Define *casting number.*
4. Engine repairs limited to the _____, valve train, and other external parts are commonly in-vehicle operations.
5. When preparing to remove an engine from a vehicle, why are the components underneath the engine usually disconnected first?
6. Why should you label wires and vacuum hoses when they are disconnected?
7. Do not disconnect _____ or _____ lines unless absolutely necessary.
8. When attaching a lifting chain to an engine, the bolts should extend into the bolt hole a distance equal to at least _____ times the diameter of the bolt.
9. Define *teardown.*
10. A(n) _____ is sometimes needed to remove a vibration damper, seal, or crankshaft sprocket.
11. Why should most parts be returned to their original locations in an engine?
12. Why should you never drive out a valve that has a mushroomed tip?
13. A(n) _____ is needed to cut out a lip at the top of a worn cylinder to prevent piston damage when the piston is removed.
14. Which parts are typically cleaned in a hot tank?
15. When cleaning threads, use a _____, not a tap.

ASE-Type Questions—Chapter 20

1. Tests find a burned engine valve. The engine is not smoking and has good oil pressure. Technician A says that this should be an in-vehicle operation; the block can remain bolted to the chassis. Technician B says that the engine must be removed from the vehicle to service the valves. Who is correct?
   (A) A only.
   (B) B only.
   (C) Both A and B.
   (D) Neither A nor B.
2. Technician A says that the VIN can be used to identify the engine type. Technician B says that a VIN can be used to identify the transmission type. Who is correct?
   (A) A only.
   (B) B only.
   (C) Both A and B.
   (D) Neither A nor B.
3. All of the following operations can be performed with the engine still installed in the vehicle *except:*
   (A) valve job.
   (B) intake manifold replacement.
   (C) hot tank cleaning of the block.
   (D) valve seal removal and replacement.
4. Technician A says that the VIN may be located on the front of the engine. Technician B says that the VIN may be located on the upper-left corner of the firewall. Who is correct?
   (A) A only.
   (B) B only.
   (C) Both A and B.
   (D) Neither A nor B.
5. Technician A says that an engine's timing belt should be replaced during a major engine rebuild. Technician B says that an engine's cylinder head(s) should be inspected and reconditioned as needed during a major engine rebuild. Who is correct?
   (A) A only.
   (B) B only.
   (C) Both A and B.
   (D) Neither A nor B.
6. Technician A says that when preparing to remove the engine from a vehicle, you do not need to disconnect the vehicle's battery. Technician B says that when preparing to remove the engine from a vehicle, you should remove the automobile's radiator. Who is correct?
   (A) A only.
   (B) B only.
   (C) Both A and B.
   (D) Neither A nor B.
7. Technician A says that when removing the engine on certain front-wheel-drive vehicles, you must remove the engine and transaxle as a unit. Technician B says that on some vehicles, the engine is removed from below by raising the vehicle. Who is correct?
   (A) A only.
   (B) B only.
   (C) Both A and B.
   (D) Neither A nor B.

8. Technician A says that the illustration shows a boring bar in an engine cylinder. Technician B says that the use of a ridge reaming tool is being illustrated. Who is correct?
   (A) A only.
   (B) B only.
   (C) Both A and B.
   (D) Neither A nor B.

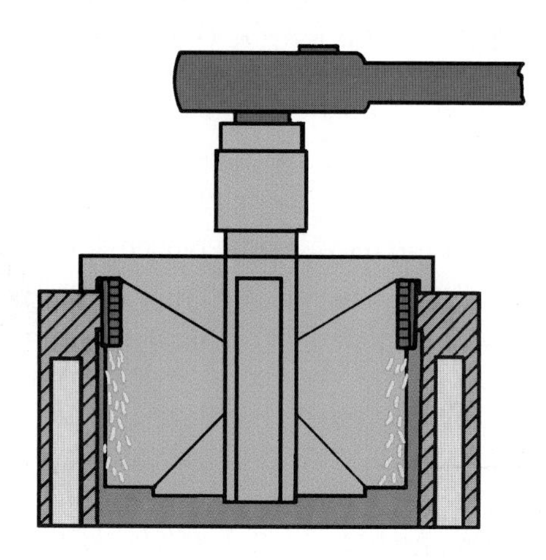

9. Teardown on an OHC engine commonly begins with _____ disassembly.
   (A) front end
   (B) top end
   (C) bottom end
   (D) Any of the above.

10. An engine block has deep ridges in its cylinders. Technician A says that a brush hone should be used to recondition the cylinders. Technician B says that the block should be sent to a machine shop to have the cylinders bored. Who is correct?
   (A) A only.
   (B) B only.
   (C) Both A and B.
   (D) Neither A nor B.

11. Technician A says that aluminum engine components can be cleaned in a cold solvent tank. Technician B says that aluminum engine components can be corroded if cleaned in a hot tank. Who is correct?
   (A) A only.
   (B) B only.
   (C) Both A and B.
   (D) Neither A nor B.

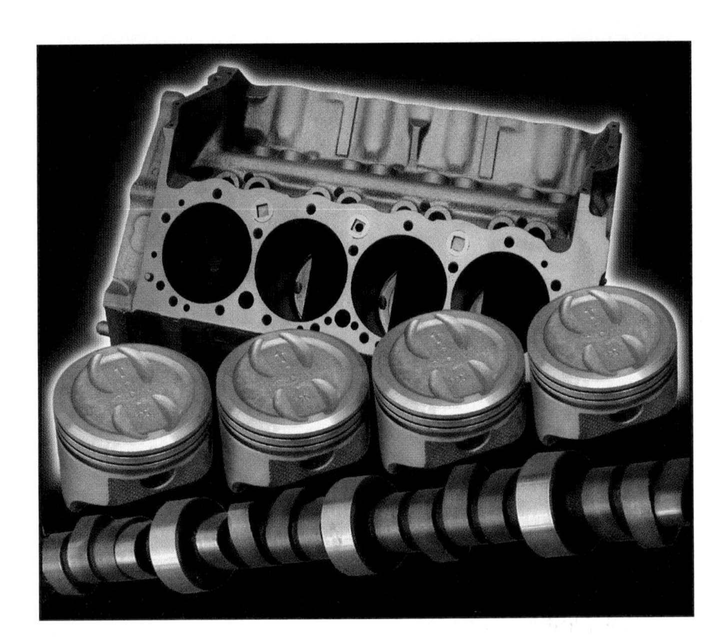

Chapter 21

Short Block Rebuilding and Machining

After studying this chapter, you will be able to:

- ❑ Locate cracks in the cylinder block and crankshaft.
- ❑ Check for main bearing bore wear.
- ❑ Measure deck warpage.
- ❑ Check for cylinder taper and out-of-round.
- ❑ Determine piston-to-cylinder clearance.
- ❑ Hone a cylinder.
- ❑ Describe the steps in boring a cylinder oversize.
- ❑ Summarize camshaft bearing installation.
- ❑ Install core plugs.
- ❑ Measure crankshaft journal taper, out-of-round, and cheek width.
- ❑ Describe crankshaft turning.
- ❑ Identify undersize bearings.
- ❑ Measure bearing clearance.
- ❑ Properly assemble a short block.
- ❑ Describe the steps in engine balancing.

Know These Terms

Block boring
Boring bar
Connecting rod side clearance
Crankshaft endplay
Crankshaft service
Crankshaft turning
Cylinder honing
Cylinder out-of-round
Cylinder sleeving
Cylinder taper
Deck warpage
Deglazing
Dye penetrant
Engine balancing
Fluorescent penetrant
Free-floating piston pin
Line boring
Line honing
Magnafluxing
Overbore limit
Piston clearance
Piston ring gap
Piston ring side clearance
Piston wear
Piston-to-cylinder clearance
Plateau honing
Power hone
Press-fit piston pin
Ring expander
Ring spacers
Torque-to-yield specifications

You learned about engine problems, diagnosis, teardown, and part cleaning in earlier chapters. This chapter summarizes the complete overhaul of an engine short block. You will learn how to prep a block; inspect and measure part dimensions; and install the main bearings, crankshaft, pistons, rings, rods, piston pins, rod bearings, and camshaft bearings.

Remember, the slightest mistake during short block reassembly can result in engine failure. One improperly tightened rod bolt, a mixed up rod cap, one wrong part clearance, or other seemingly minor mistake can be catastrophic. The engine may experience a major mechanical failure as soon as you start and run the engine. One minor mistake can cost you or your shop thousands of dollars when you have to rebuild the engine again for free.

If needed, refer to earlier chapters. These chapters explained engine removal, disassembly, and part cleaning. These operations must be understood before you can properly reassemble an engine short block.

Cylinder Block Service

Cylinder block service typically includes:

- ❑ Inspection for cracks, excess wear, scored cylinders, and other block damage.
- ❑ Checking for main bore and deck warpage; having the block machined if needed.
- ❑ Measuring cylinder wear; having the block bored or sleeved if cylinders are worn beyond specifications.
- ❑ Honing or deglazing cylinders to prepare cylinder walls for new rings.
- ❑ Installing camshaft and balancer shaft bearings.
- ❑ Final cleaning of the block before assembly.

As you will learn, most engine repair tasks can be done in the average automotive repair shop. However, milling and boring operations usually require the help of a machine shop.

Inspecting the Cylinder Block

Begin short block service by closely inspecting the block for problems. Look for the kind of troubles illustrated in **Figure 21-1.** Inspect the cylinder block closely to make sure it can be reused and has no major problems. In particular, check the cylinder walls for scoring, heavy wear, cracks, and so on. Check the top of each cylinder for wear. If there is a large ridge at the top of a cylinder, cylinder wall wear is severe and block boring, sleeving, or replacement is required.

Using a shop light, closely inspect the surface of each cylinder wall. Rub your fingernail around in the cylinder. This will help you feel and locate problems. Also, inspect the crankshaft main bores for signs of bearing movement, a spun bearing, or other damage in the crankcase area.

Finding Cracks in the Block

If symptoms such as coolant in the oil or overheating point to possible cracks in the block, you should magnaflux a cast iron block or use dye penetrant on an aluminum block. ***Magnafluxing*** involves using a magnet and a ferrous (iron) powder to highlight cracks and pores in cast iron blocks. **Figure 21-2** shows a large magnafluxing machine found in some machine shops. A large electromagnet surrounds the part with a magnetic field to aid crack detection. Smaller, hand-held units are also available.

1. Place the magnet over the area to be tested.
2. Spread the powder over the block.
3. A break, void, or crack in the block metal sets up a magnetic field that pulls the powder to the crack or flaw. This will make the powder collect in the flaw. The light-colored powder will highlight the crack or pore so it is easy to detect, **Figure 21-3.**

Dye penetrant is commonly used to find cracks and casting imperfections in aluminum cylinder blocks. Since aluminum is a nonferrous metal, magnafluxing cannot be used.

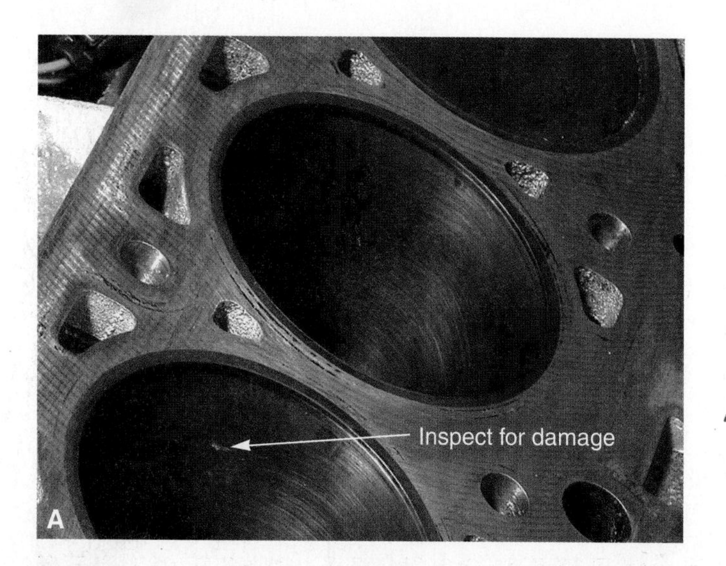

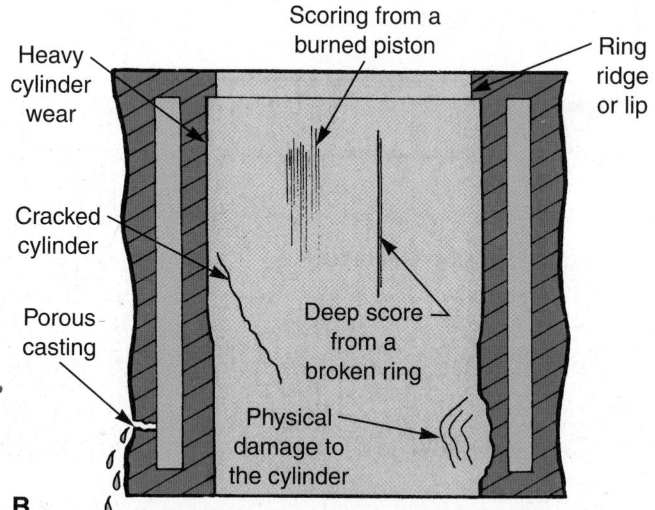

Figure 21-1. *A—Use your fingernail to feel for cracks and scoring. B—Closely inspect the cylinder walls for these kinds of trouble.*

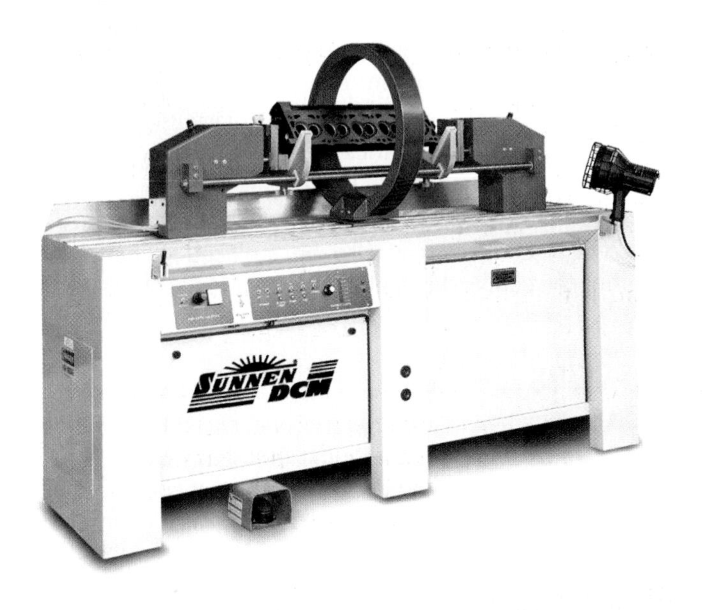

Figure 21-2. *This is a magnetic particle inspection, or mangaflux, machine. Parts are mounted in the machine and a large electromagnet ring is moved over them as iron powder is spread onto the parts. The powder is attracted to, and highlights, any crack. (Sunnen)*

1. Spray the special dye over the block.
2. The dye will collect in any crack.
3. Spread or spray the powder-like developer over the penetrant.
4. The developer makes the penetrant turn red if it has collected in any crack or casting flaw(s).

Fluorescent penetrant is similar to dye penetrant, but it requires an ultraviolet (black) light to illuminate the penetrant in the cracks. When the chemical is sprayed over the part, it collects in any crack. When the part is placed under a black light, the chemical concentrated in the crack glows, **Figure 21-4.**

Pressure testing involves filling the coolant or oil passages in the block with air pressure to find cracks or flaws. All openings to passages in the cylinder block are sealed using rubber expander plugs or gaskets. Sheets of rubber are bolted over the deck surface with metal plates. This makes the inside of the block airtight. An air hose is connected to the coolant or oil passages and the block is lowered into a large tank of water. Air bubbles will be released from any crack or pore indicating where the problem is located.

Repairing Block Cracks

Block repairs are usually done by a specialized machine shop that has the equipment and skills needed to

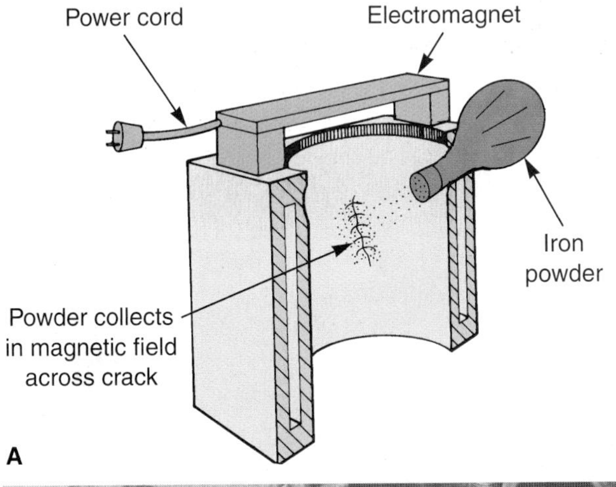

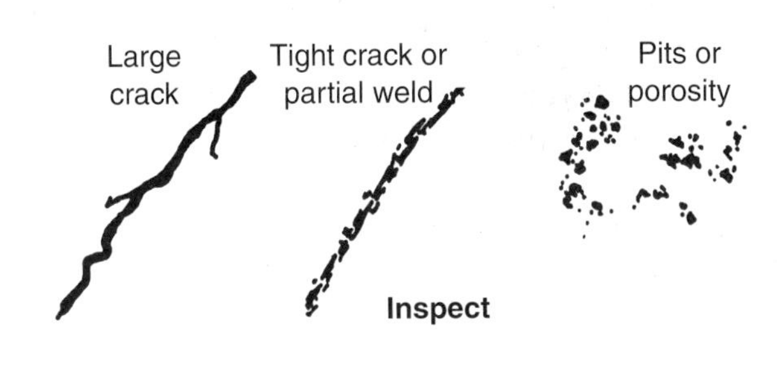

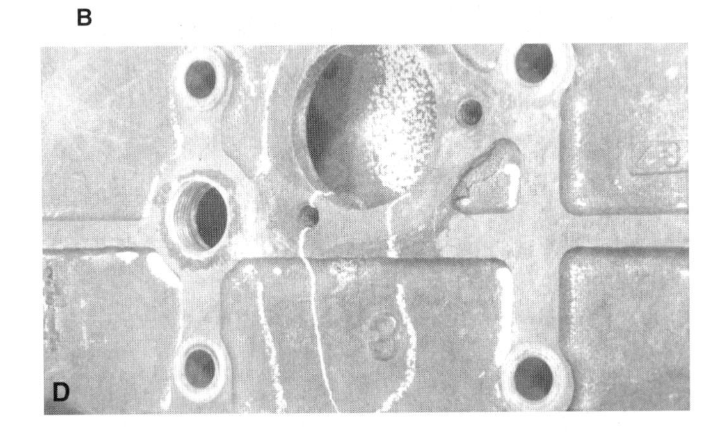

Figure 21-3. *A—Magnafluxing equipment is used to locate a crack in a cylinder. B—The pattern of iron particles indicates the type of problem. C—A cracked block caused by part breakage. D—Cracks in a block caused by freezing of weak coolant. (K-line, Magnaflux Corp.)*

Figure 21-4. *Fluorescent penetrant can be applied to parts to find cracks. The chemical collects in cracks, which show up when the part is placed under black light. (Magnaflux Corp.)*

weld large castings. Cracks and pores in a cylinder block can be corrected by:

- Installing a sleeve in a cylinder if the crack is in the cylinder wall area.
- Welding the crack or pore. Cast iron is welded with a Ni (high nickel) rod after heating in a furnace. An aluminum block is welded with a GMAW welder and aluminum rod.
- Plugging, which involves drilling series of holes along crack and threading in repair plugs.
- Using special epoxy or metallic plastic to seal pores in certain areas of block, **Figure 21-5.** Epoxy should *not* be used on cracks, however.

If the cracks and pores cannot be repaired, the block must be replaced. The block may also be replaced if the cost of repairing the crack is higher than that of a replacement block.

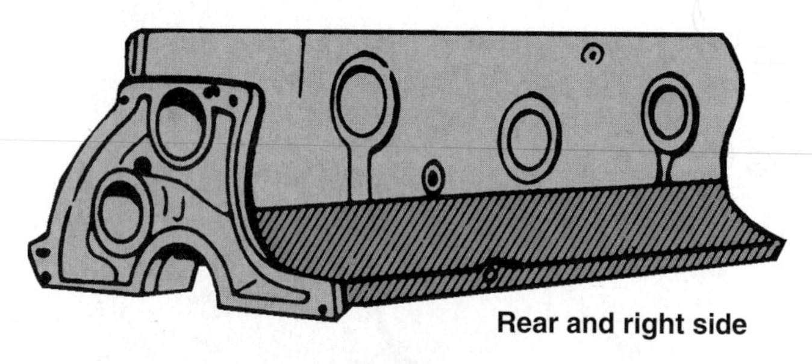

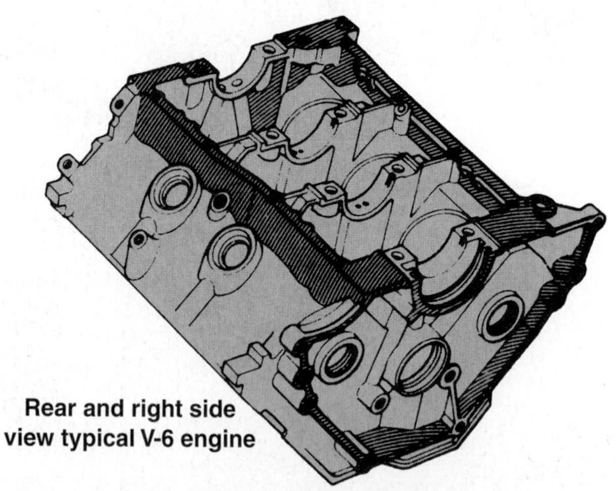

Figure 21-5. *Casting flaws or pores can sometimes be repaired with special epoxy or plastic. This is a service manual illustration from one manufacturer that shows the areas of the block that can be sealed with epoxy. (Ford)*

Checking Main Bearing Bores

After overheating or repeated heating and cooling, the main bearing bores in the cylinder block can warp or twist. This will affect main bearing insert alignment and crankshaft fit in the block. Under severe cases of main bearing bore misalignment, the crankshaft can lock up when the main caps are torqued with the new bearing installed. Always check for bore misalignment after main bearing failure.

1. Lay a straightedge on the main bores with the main bearing inserts removed, **Figure 21-6.**

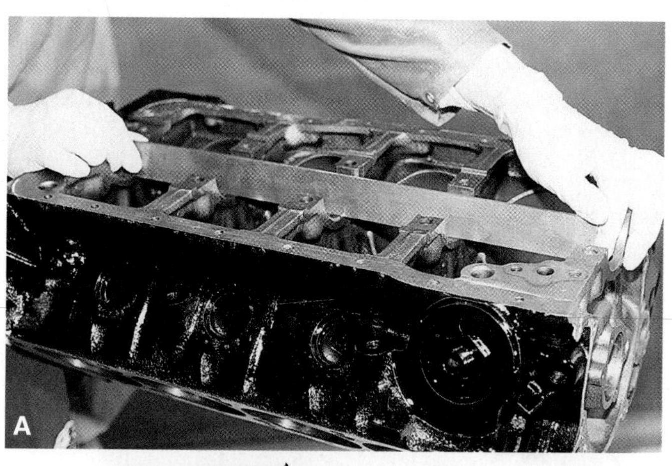

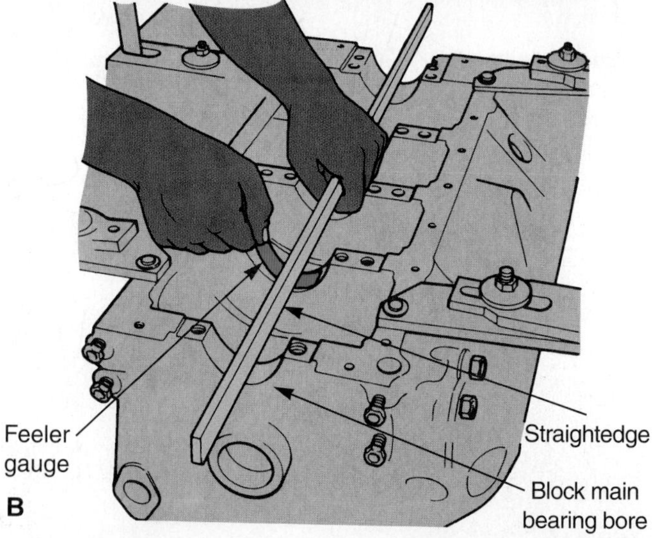

Figure 21-6. *A— To check main bearing bore alignment, place a straightedge into the main bore of the cylinder block. B—Try to fit a .0015″ feeler gauge between the straightedge and block bore. If it fits, the block should be line bored by a machine shop. (Federal Mogul)*

2. Try to slide a .0015" (0.038 mm) feeler gauge, or use the factory recommended size, under the straightedge.
3. If the blade fits between the straightedge and the main bore, the main bearing bores are misaligned.
4. Check for warpage at the bottom of each main bore and on each side near the main cap parting surface.

Line Boring the Block

A machine shop will have a boring or honing bar for machining the bore back into alignment. Block ***line boring,*** using a machine tool cutter, or ***line honing,*** using rigid stones, can be used to straighten or true misaligned main bearing bores. A block should be line bored when misalignment is beyond factory specifications or a main bearing has "spun" and damaged the block. **Figure 21-7** shows the major steps a machine shop uses to line bore a cylinder block main bore.

1. Machine the main cap mounting surfaces. When the caps are installed on the block, the main bore will be smaller in diameter and slightly out of round. This allows the bore to be machined back to the original diameter.
2. Mount the cylinder block on the line boring machine so that its main bore is aligned with the boring bar.
3. Install the caps on the block and torque the fasteners to specifications.
4. Adjust the cutter bit or hone on the line boring machine to the correct bore diameter for the engine.
5. Engage the large drive head, which spins the boring bar or hone. Move the bar or hone through the main bores.
6. Verify the bore diameter with an inside micrometer or a telescoping gauge and outside micrometer.

Measuring Cylinder Block Deck Warpage

Cylinder block deck warpage often results from engine overheating, especially with thin-wall blocks. Aluminum blocks are especially prone to warpage. ***Deck warpage*** is measured with a straightedge and feeler gauge

Figure 21-7. *Note the basic procedure for line boring a cylinder block. A—The cylinder block is mounted in the line boring machine. B—Main caps are installed and the cap bolts are torqued to specifications. C—The large drive head spins the boring bar as the bar passes through the main bores. D—The cutter bit must be adjusted to the correct bore diameter. (Sunnen)*

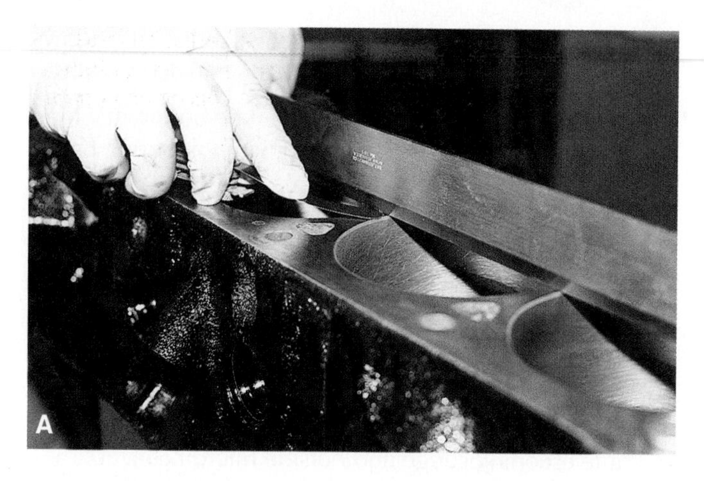

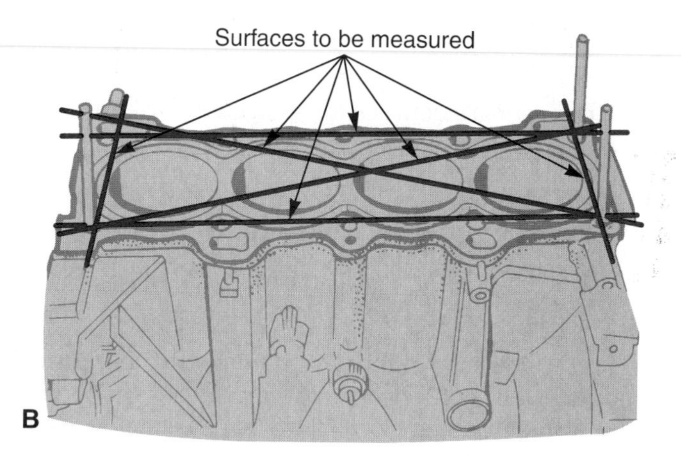

Figure 21-8. *A—To check for deck warpage, place a straightedge on the deck and try to slide a feeler gauge between block and straightedge. B—Check for warpage at these locations. (Honda)*

on the head gasket sealing surface(s) of the block. It should be checked, particularly when the old head gasket was blown and leaking.

1. Lay a straightedge on the clean block deck surface, **Figure 21-8.**
2. Try to slip feeler gauges of different thickness between the block and straightedge.
3. The thickest blade that fits shows the deck warpage.
4. Check in different locations on the cylinder block deck.

If beyond specifications, generally about .003″ to .005″ (0.08 mm to 0.13 mm), replace the block or send it to a machine shop for decking. See **Figure 21-9.** ***Decking*** a block involves surface milling the cylinder head mounting surfaces until they are parallel and equidistant from the main bore. This is also called "squaring" the block.

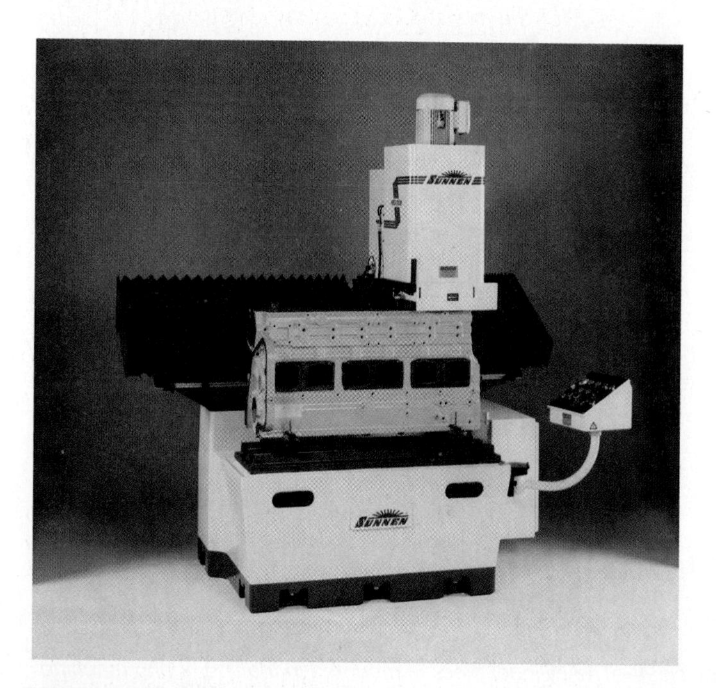

Figure 21-9. *This is a machine shop milling machine being used to true the block deck. (Sunnen)*

Measuring Cylinder Diameter

Before measuring cylinder wear, you should check the cylinder bore diameter. It is possible that the block was bored previously and the cylinders are not the original-specification diameter. To check cylinder diameter:

1. Check the service manual for the standard (original) bore specification.
2. Use a dial bore gauge, inside micrometer, or telescoping gauge and outside micrometer to measure the cylinder diameter at its widest point near the top of the cylinder.
3. Measure the cylinder diameter at the bottom of the cylinder in an unworn section.
4. Compare your measurements to specifications.

As an example, you measure the cylinder bore near the bottom in an area of the cylinder where the rings do not wear and determine the diameter to be 4.010″. However, the standard specification calls for a 4.000″ cylinder diameter. From this, you can determine that the cylinder has been bored .010″ oversize. This would allow you to order the right size piston rings, know correct overbore limits, etc.

Cylinder Wear

A worn cylinder is not round, or "true." It is wider across its thrust surface. In addition, it is normally larger in diameter near the top than at the bottom. This is illustrated in **Figure 21-10.**

Cylinder taper is the difference in the diameter from the top to the bottom of the cylinder measured on the piston ring contact surface of the cylinder wall. It is caused by less oil lubricating the top of the cylinder wall. More oil splashes off of the crankshaft and onto the lower area of the cylinder wall. As a result, the top of the cylinder wears more than the bottom, producing a taper, **Figure 21-10A.**

Cylinder out-of-round is a difference in cylinder wall diameter when measured front-to-rear and side-to-side in

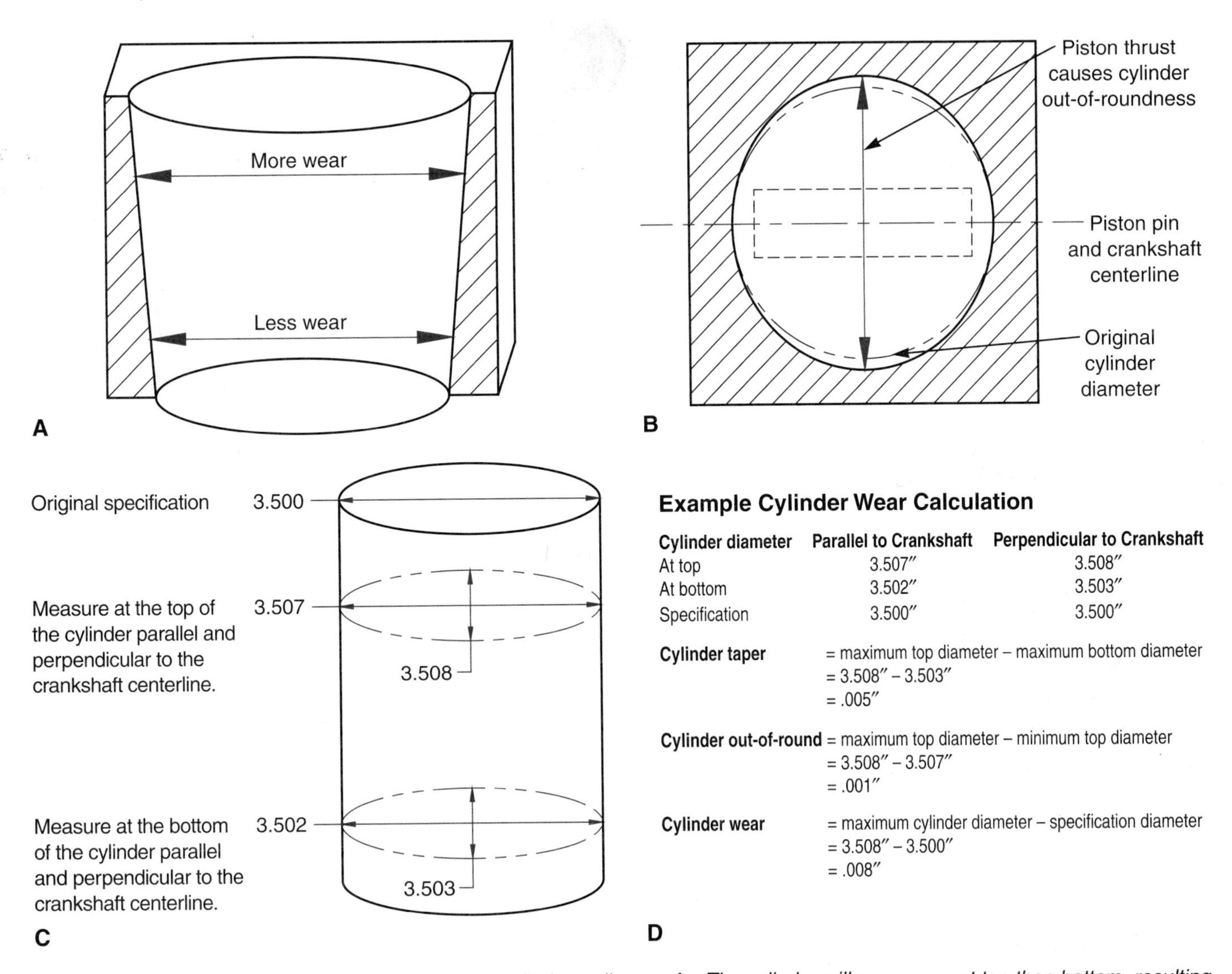

Cylinder diameter	Parallel to Crankshaft	Perpendicular to Crankshaft
At top	3.507″	3.508″
At bottom	3.502″	3.503″
Specification	3.500″	3.500″

Cylinder taper = maximum top diameter – maximum bottom diameter
= 3.508″ – 3.503″
= .005″

Cylinder out-of-round = maximum top diameter – minimum top diameter
= 3.508″ – 3.507″
= .001″

Cylinder wear = maximum cylinder diameter – specification diameter
= 3.508″ – 3.500″
= .008″

Figure 21-10. *Study the basic steps for measuring cylinder wall wear. A—The cylinder will wear more at top than bottom, resulting in taper. B—The cylinder will also wear more on the piston thrust surface, resulting in out-of-round. C—Measure the cylinder in these locations and write down your readings. D—Calculation of cylinder wear based on the measurements shown in C.*

the block. Piston thrust action normally makes the cylinders wear more at a right angle to the centerline of the crankshaft and piston pin, **Figure 21-10B.**

If new rings are installed in an out-of-round cylinder, blowby will result. There will be a gap where the round ring does not contact the worn cylinder. Pressure and oil can leak past this gap resulting in oil consumption and blue exhaust smoke.

In a tapered cylinder, the piston rings expand and contract every time the piston slides up and down in the cylinder. At high engine speeds, the rings may fail to stay in contact with the cylinder. A loss of power, blowby, and oil consumption can result. New rings installed in a tapered cylinder will quickly wear or break. Cylinder taper must be removed during an engine rebuild.

Measuring Cylinder Wear

If the cylinder is not badly scratched, scored, or otherwise damaged, you must measure cylinder wear to ensure that the new rings will seal properly, **Figure 21-10C.** New, round rings cannot seal in worn, out-of-round or tapered cylinders. Also, cylinder and piston diameter measurements will let you determine piston-to-cylinder clearance.

Piston-to-cylinder clearance can be found by measuring cylinder diameter and piston diameter. The difference between these two measurements is the piston clearance. See **Figure 21-11.**

A dial bore gauge can be used to quickly and precisely measure cylinder taper and out-of-round. Look at **Figure 21-12.**

1. Use an outside micrometer to adjust the dial bore gauge to the specification diameter.
2. Slide the bore gauge up and down in the cylinder.
3. The maximum dial indicator movement equals the cylinder taper. If the top of the cylinder is worn beyond specifications, no other measurements need to be taken and the cylinder must be machined.
4. To determine out-of-round, check the indicator movement with the gauge measuring parallel and perpendicular to the crankshaft centerline.

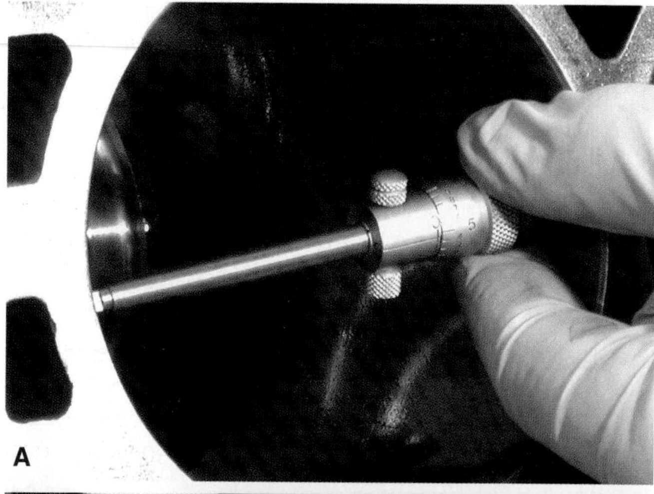

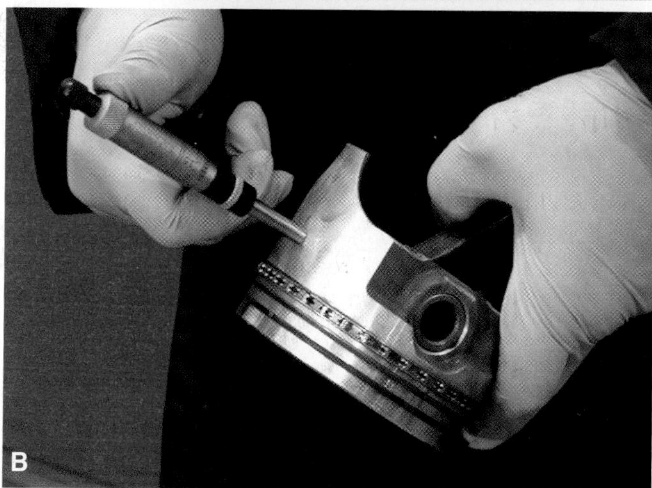

Figure 21-11. *Determining piston clearance. A—Use an inside micrometer to measure the cylinder diameter. B—Use an outside micrometer to measure across the piston skirt to find the piston diameter. Subtract this from the cylinder diameter to find the piston clearance.*

A large inside micrometer, or a telescoping gauge and outside micrometer, can also be used to measure cylinder dimensions. This method is slightly more time-consuming, however. Refer to Chapter 2 for more information on measuring tools.

Note: Write down your cylinder measurements carefully, as was shown in *Figure 21-10.* The slightest mistake in math could be critical to an engine rebuild.

Cylinder Wear Limits

Generally, cylinder boring, sleeving, or block replacement is needed when:

- Cylinder out-of-round is more than .005″ (0.13 mm).
- Cylinder taper is more than .008″ (0.20 mm).
- The maximum cylinder diameter stated by the engine manufacture has been exceeded. This limit is typically .010″ (0.25 mm) over the bore specification.

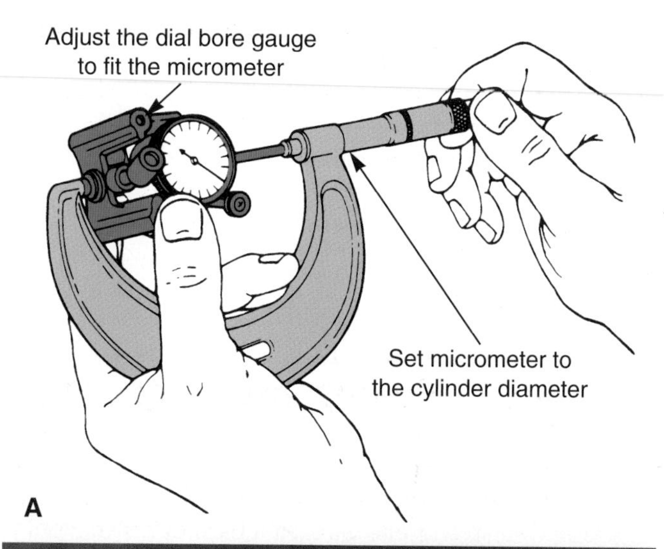

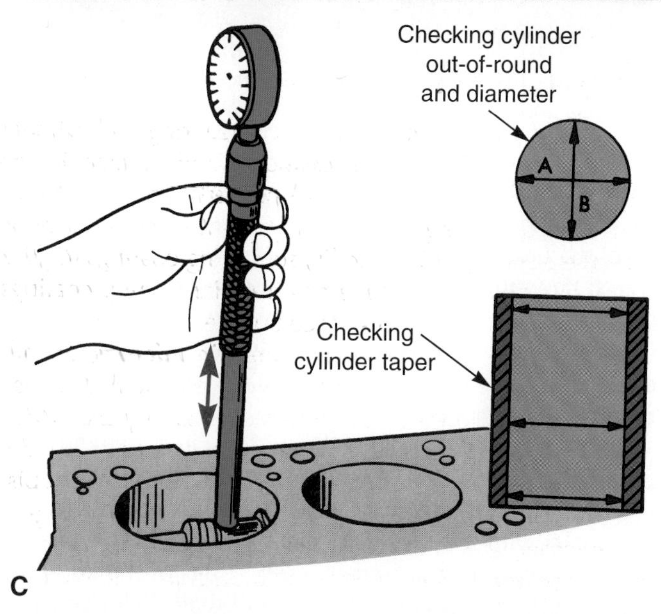

Figure 21-12. *Using a dial bore gauge to check cylinder wear. A—Use a micrometer to adjust the dial bore gauge to fit inside the cylinder. B—Slide the dial bore gauge up and down in the cylinder to read taper. C—Rotate the gauge around the cylinder to read out-of-round. Cylinder wear is the difference between new specification and the actual readings. Note the locations for cylinder wear measurement. (General Motors)*

Always refer to the specifications for the engine on which you are working and follow those values.

If the cylinders are worn beyond the maximum allowable limit, you will have to bore, sleeve, or replace the block. If bored oversize, all cylinders must be bored and oversize pistons will be needed for all cylinders. Alternately, a sleeve may be installed in a damaged or badly worn cylinder that is bored to accept the sleeve.

Cylinder Honing and Deglazing

Cylinder honing, also called ***deglazing,*** is used to remove the smooth surface and provide a lightly roughened or textured surface on the walls of a used cylinder. Honing must also be done to smooth the rough cylinder wall surface after cylinder boring. Most ring manufacturers recommend deglazing or honing before new ring installation.

The pockets or valleys in the hone marks or scratches hold oil to prevent ring scoring. The peaks or high points in the honed metal surface contact the piston rings. They quickly wear down from ring friction so that the ring can match the shape of the cylinder wall.

A cylinder hone produces a precisely textured, cross-hatched pattern on the cylinder wall to aid ring seating and sealing. Tiny scratches resulting from the action of the hone cause the initial ring and cylinder wall break-in wear. This makes the ring fit in the cylinder after only a few minutes of engine operation.

There are three basic types of hone: brush, flex, and rigid. Refer to Chapter 2 for an explanation of the types of hones.

Hone Grit Selection

If the cylinder surface is too coarse, ring and cylinder scoring may result. If the cylinder wall surface is too smooth, the rings may not seal properly and lead to oil consumption. Therefore, it is important to select a hone with the correct grit. Generally, softer ring coatings require a coarser stone and surface texture. Harder ring coatings require a finer stone and surface texture.

Surface smoothness is measured in ***microns,*** or millionths of an inch. Special testers are available that can be passed over a surface to measure root mean square (RMS), which is a measurement of surface smoothness. Cylinder smoothness or RMS should be matched to the type of piston ring. Stone coarseness is measured in grit. The higher the number, the finer the stone. Always use the honing stone recommended by the manufacturer. Remember these rules:

- ❑ Use 220 grit stones to produce a textured 25 to 30 RMS surface texture for cast iron rings operating on cast iron cylinder walls.
- ❑ Use 280 grit stones to produce a medium 20 to 25 RMS surface for chrome rings (hard) operating on cast iron cylinder walls.
- ❑ Use 320 grit stones to produce a very fine 10 to 15 RMS surface for moly rings (very hard) operating on cast iron cylinder walls.
- ❑ Use 150 grit, aluminum oxide stones to lightly hone and clean up nickel-ceramic cylinder walls, or cylinder walls with a similar coating. Finish hone the coated cylinder wall with 500 grit, diamond stones. Refer to Chapter 9 for information on piston rings for use on coated cylinder walls.

Caution: If the incorrect honing procedure or piston ring type is used in an engine, the new piston rings may not seal properly. Oil consumption, blowby, and engine smoking can result.

Using a Power Hone

A ***power hone*** is a machine shop tool that sizes and finishes the cylinder walls in an engine block. A rigid power hone can be used instead of a boring bar to remove cylinder taper and out-of-round. It can remove smaller amounts of metal to clean up a worn or slightly scored cylinder wall. This makes the cylinder equal in size at all points. See **Figure 21-13.**

Modern power hones are computer controlled. The machinist can enter specifications for hone speed, feed, diameter, and so on. Some machines have a display that shows the amount of taper, out-of-round, and diameter of cylinder. This allows the machinist to know how much more material must be removed to recondition the cylinder.

Typically, power honing is used when less than .030″ (0.76 mm) of material must be removed from the cylinder walls. If more material must be removed, as when sleeving a cylinder, the cylinder is bored instead of power honed. Generally, standard rings and pistons can be used when a power hone is used to enlarge a cylinder by a maximum of .008″ (0.20 mm). Most new standard piston ring sets will seal in a cylinder that is .010″ oversize. This will save you from having to bore the cylinder .010″ larger and install oversize pistons and rings.

Plateau honing involves using extra fine stones to final size the cylinder dimensions and final smooth the surface finish. This step removes the "peaks" produced by the previous honing steps. It also produces the desired crosshatch pattern in the cylinder walls to speed piston ring seating and sealing during engine break-in.

A ***deck plate*** is a rigid metal cap bolted onto the block deck surface when power honing or boring. When torqued down, the deck plate stresses the block as if a cylinder head is installed. If you bore or power hone a cylinder without a deck plate, the cylinders can become out-of-round or tapered when the cylinder head is torqued in place. A deck plate avoids this potential problem when rebuilding an engine.

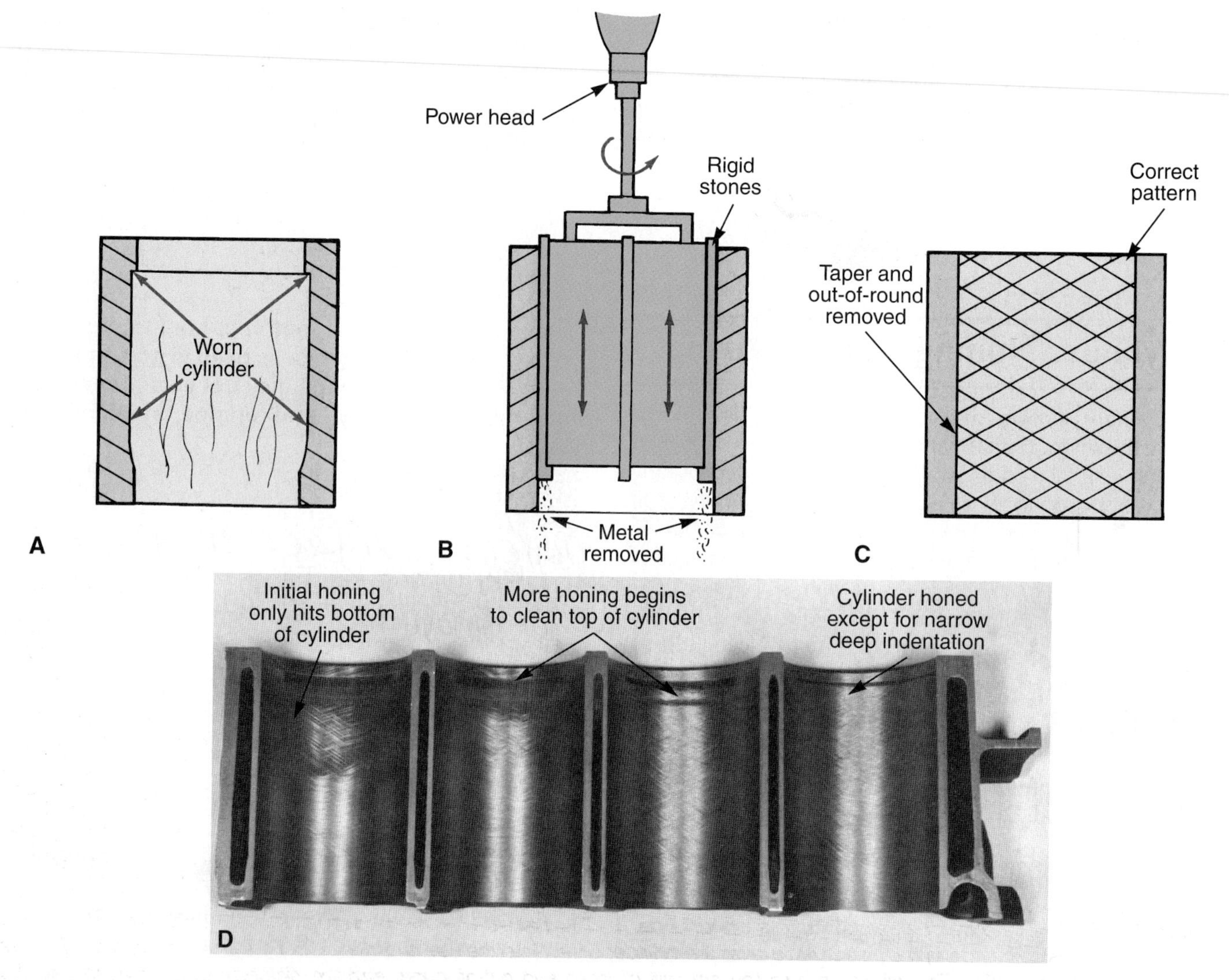

Figure 21-13. *A—This cylinder has minor scratches and a small ring ridge. B—The power hone spins rigid stones inside the cylinder. C—After honing, taper and out-of-round are removed and the cylinder wall has the correct crosshatch pattern. D—Note the stages of power honing. The small, dark ring at the top of the finished cylinder is only .001" deep and will not affect ring sealing. (Sunnen)*

Figure 21-14 shows the major steps that a machinist uses when power honing a cylinder block. Basically, a series of rotating stones (a hone) is passed through the cylinder block bores until the final diameter and finish are achieved. Generally, the first hone used is a very coarse, 80 grit, diamond stone set. These coarse stones can rapidly remove large amounts of metal. The second power hone used is a 180 grit, semi-finish stone set. This provides initial smoothing. Finally, a 220 grit, finish stone set is used. This smoothes the cylinder walls even more and also final sizes the cylinder diameter.

Note: For new rings to seal, the most critical cylinder characteristic is roundness. The next most critical characteristic is taper. Round rings must have a round cylinder to prevent blowby. Rings can still seal a slightly oversize cylinder wall as long as the cylinder is round and untapered.

Using a Flex Hone

To hone a cylinder using a flex hone, follow the equipment manufacturer's instructions. Install the hone in a large, low-speed electric drill, **Figure 21-15.** Compress the stones (squeeze inward) and slide them into the cylinder. Be careful not to scratch the cylinder. Turn on the electric drill and move the spinning hone up and down in the cylinder.

Move the hone up and down in the cylinder fast enough to produce a 50° to 60° crosshatch pattern. This is illustrated in **Figure 21-16.** Moving the hone up and down faster or slower will change the pattern. It is usually acceptable if a small band in the cylinder under the ring ridge fails to polish up when honing. This will have little effect on ring sealing.

If the cylinder has not been bored or power-honed true, hone more at the bottom of the cylinder. This will remove taper and enlarge the bottom slightly to make the cylinder equal in diameter at the top and bottom. Hone at the bottom for a few seconds and then move up and down to produce the 50° to 60° crosshatch.

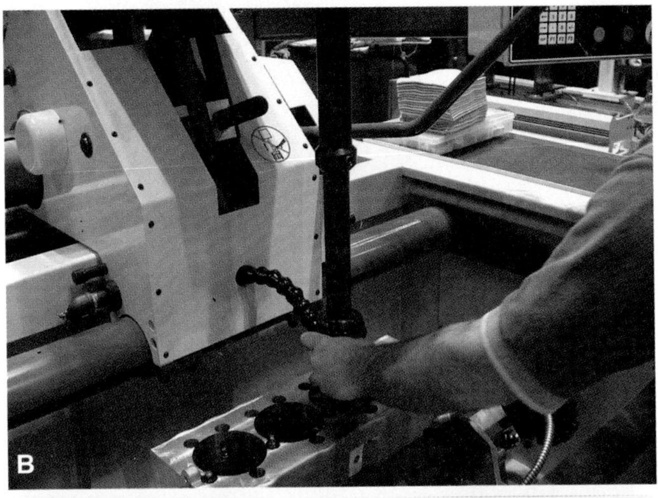

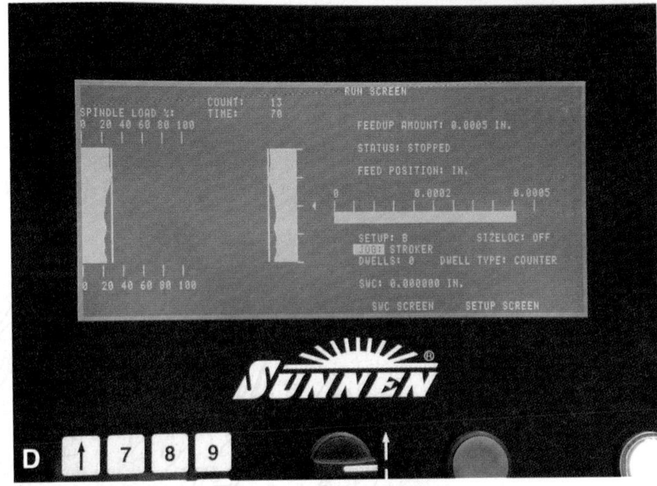

Figure 21-14. *Study the major steps for using a power hone. A—The first set of stones is very coarse for rapid removal of metal. B—The machinist is centering the power hone over the cylinder bore. Note the use of a deck plate on top of the engine. C—The power hone in action. D—This display shows the amount of taper and out-of-round, and the diameter of the cylinder, so the machinist knows how much more material must be removed. (Sunnen)*

Warning: Make sure you do not pull the hone too far out of the cylinder while the hone is spinning. The hone could break and bits of stone may fly out.

Block Boring

Block boring is needed to correct excess cylinder wall wear, scoring, or scratching. Boring will true the cylinder walls so the rings and pistons fit properly. It is also done when installing a repair sleeve, **Figure 21-17.** Boring oversize and installing new pistons and rings will make the engine like new. The new pistons and rings operate on a freshly machined cylinder surface providing excellent ring sealing and service life.

The trend in industry is to use a honing machine to bore engine block cylinders when sleeving is not required;

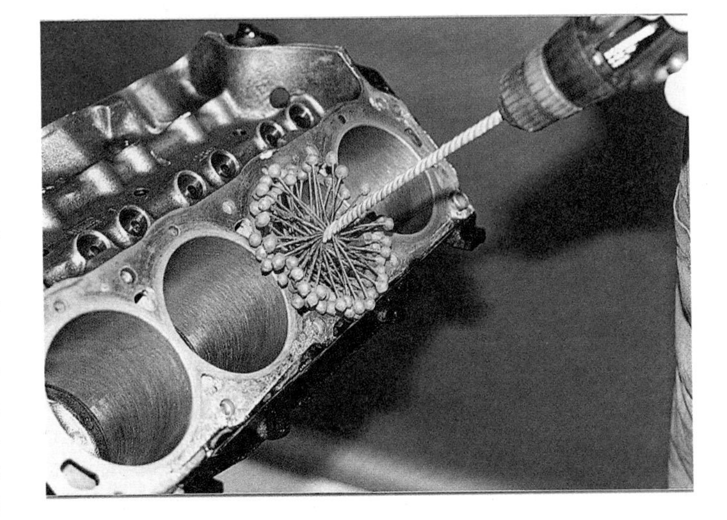

Figure 21-15. *A brush hone is used to provide the final finish bore after power honing or flex honing. Move the drill up and down fast enough to produce the crosshatch pattern. Be careful not to pull the spinning hone too far out of the cylinder.*

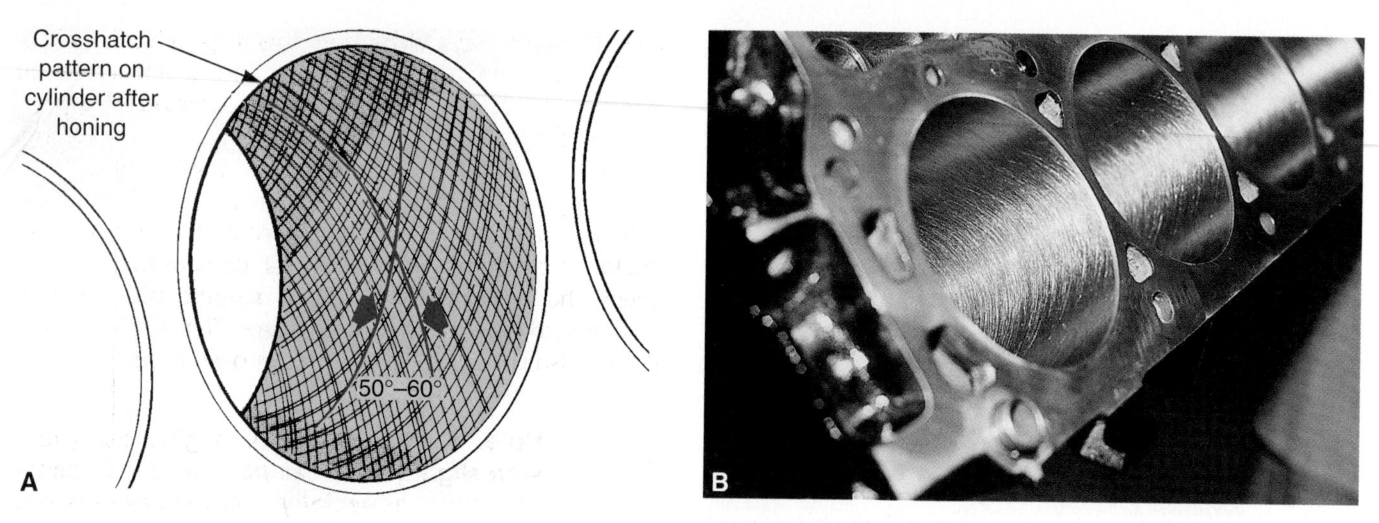

Figure 21-16. *A—When honing, try to produce a 50° to 60° crosshatch pattern. B—The crosshatch pattern is clearly visible on these finished cylinders.*

Figure 21-17. *Study the major steps for boring a block. A—The cylinder block must first be mounted and squared up on the boring bar. B—When turned on and engaged, the boring bar rotates while being automatically fed down into the cylinder. C—The machinist is adjusting the cutter to make a second cut. D—The second cut is being made. Notice the metal shavings generated as metal is removed. (Van Senus Auto)*

in other words, when the cylinder can be overbored within factory bore limits. However, when more material must be removed to sleeve cylinders, a boring bar is used instead of a honing machine. The cutter bit in a boring bar can cut more deeply and quickly than honing stones.

Boring is needed when your measurements show cylinder wear is beyond specifications. If power honing will not clean up the cylinder, more material must be removed. This process is called boring and can be done with a boring bar or a power hone. If taper is more than .012″ and out-of-round is more than .005″, boring is also needed. After boring, oversize pistons and rings must be installed. Unless a sleeve is to be installed, *all* cylinders are bored to the same oversize.

A ***boring bar*** is a machine shop tool that will cut a thin layer of metal out of the cylinders. The cylinder block must first be mounted, squared on the boring bar, and centered on the table. This will allow the boring bar to pass down through existing bores properly. When turned on and engaged, the boring bar rotates while being automatically fed down into the cylinder. The cutter tip on the boring bar removes metal from the cylinder wall. This will correct cylinder wear, out-of-round, or taper problems or allow for the installation of a repair sleeve. The maximum limit on each pass is generally .060″. Multiple passes are needed for larger oversizes, such as when a sleeve is being installed.

Caution: Always use a reputable machine shop. Your rebuild work is only as good as the machine work done on the parts. Always verify measurements on any machined part before assembly.

Overbore Sizes

Normally, a block is bored in increments of .010″ (0.25 mm). This depends on whether piston oversizes are available in standard or metric sizes. You must make sure what size of oversize pistons are available and overbore the block accordingly. Many shops order the oversize pistons, measure them, and then bore the block to perfectly match the size of the new oversize pistons. This ensures precise piston-to-cylinder clearance.

Depending on the amount of wear, the block cylinders may be bored .010″, .020″, .030″, .040″, .050″, or .060″ oversize; larger on thick-wall castings. Most shops bore cylinders either .030″ or .060″ oversize because these piston oversizes are the most readily available.

When oversize pistons come in metric dimensions, the cylinders need to be bored either 0.25 mm, 0.50 mm, 0.75 mm, 1.0 mm, or 1.5 mm oversize. These overbore sizes match the sizes of metric pistons available to replace the stock diameter pistons.

Overbore Limit

The ***overbore limit*** is the maximum amount that a block should be bored oversize without resulting in cylinder walls that are too thin. It is a specification given by the engine manufacturer. In the past, blocks could be bored as much as .120″ (1.5 mm) oversize. However, many of the lightweight, thin-wall castings found in today's engines can only be bored .020″ (0.25 mm) oversize.

The overbore limit depends on the thickness of the cylinder walls. If too thin after boring, the cylinder wall could crack from the pressure of combustion. Always check the overbore limit of the specific block before recommending an overbore diameter. The machine shop should also be able to determine the overbore limit.

Note: Boring increases engine compression slightly. This can be critical with some engines, especially high-compression, turbocharged, and supercharged engines.

Oversize Pistons and Rings

Oversize pistons and rings are required to fit a cylinder block that has had its cylinders bored oversize. The pistons must be purchased to match cylinder oversize. If the block is bored .030″ oversize, then you would have to order .030″ oversize pistons and rings. The proper clearances are manufactured into the oversize pistons.

Caution: Make sure the correct piston oversize is available before having the block bored. If .020″ oversize pistons are not available, you do not want to bore the block oversize to that diameter. Many shops will not bore a block until after obtaining the new, oversize pistons. This ensures that the new pistons will properly fit the new cylinder bore dimensions. Special pistons can be ordered and machined, but these can cost several times as much as off-the-shelf pistons.

Cylinder Sleeving

Cylinder sleeving involves machining or boring a cylinder oversize and pressing in a cylinder liner. Sleeving is needed, for example, after part breakage has severely gouged, nicked, chipped, or cracked the cylinder wall. If the damage is too deep to be cleaned up with power honing or boring, sleeving is a way to salvage the block.

Sleeving also allows a cylinder to be restored to its original diameter. If the existing piston is not damaged, it can be reused. For example, if only one piston and cylinder are damaged in a low-mileage block, all of the other pistons and cylinders may be good and usable. In this case, sleeving would salvage the block and save the customer money on the repair. Sleeving can also be used if one or more cylinders require boring but the customer wants to retain the stock cylinder diameter.

Sleeve installation is a machine shop operation. If you have an aluminum block with cast iron sleeves, sleeve replacement is simple. The machine shop can drive out the old sleeve and press in a new one. Boring is not needed. This can be done for a badly damaged cylinder, for example when a rod breaks and hits the cylinder wall.

Sleeving a cast iron block is more complex. The old cylinder must be bored oversize to a diameter slightly smaller than the outside diameter of the sleeve. The sleeve is then pressed into the block. Finally, the sleeve is bored to the correct cylinder diameter using a power hone or boring bar.

Sleeve protrusion or ***liner protrusion*** is the distance that the sleeve sticks up above the block deck. Sleeve protrusion is needed to seal combustion pressure at the head gasket. If the sleeve is even or slightly below the deck surface, the hot combustion flame can blow past the cylinder sleeve into the coolant or oil, between cylinders, or out of the engine. A few thousands of an inch (hundredths of a millimeter) sleeve protrusion will ensure the combustion chamber seals properly while still allowing the compressible head gasket to seal coolant and oil passages. **Figure 21-18** shows how to check sleeve protrusion with a dial indicator.

Cleaning Cylinder Walls

After boring, sleeving, and honing, it is very important to clean the engine to remove all bits of stone and metal from inside the engine. If not removed, this material, or grit, will act like grinding compound circulating through the engine. It can wear bearings, rings, and other vital engine parts.

A pressure washer is the most efficient way of cleaning a cylinder block. The high-pressure soap and water will blow all grit out of the cylinders and crankcase. After pressure washing, blow the block dry and wipe the cylinders down with motor oil. See **Figure 21-19.**

If a pressure washer is not available, hand wash the cylinders with warm, soapy water. See **Figure 21-20.** A soft-bristle brush, not a wire brush, will quickly loosen grit inside the hone marks. A clean rag and the detergent will then lift grit out of the hone marks. Keep rinsing the rag out and washing the cylinders until the rag comes out clean. After the rag comes out clean, wash the cylinders one more time. Then, blow the block dry.

Next, place new motor oil on a clean shop towel. Wipe the cylinder down thoroughly with the oil-soaked towel. The heavy oil will pick up any remaining grit embedded in the cylinder honing marks. Refer to **Figure 21-21.** Wipe the cylinders down until the rag comes out clean.

After cleaning, recheck the cylinder for scoring or scratches. If light honing did not remove all of the vertical scratches in the cylinder, cylinder boring or sleeving may be needed.

Check all threads in the block for damage and cleanliness. Clean all threads. Repair any damaged threads. Thread repair and cleaning are discussed in Chapter 3 and Chapter 20.

Caution: Before engine reassembly, double-check to make sure the block is in an acceptable condition. Remember, the block is the foundation of the engine rebuild. If the block has problems, you are building on an unstable foundation that will fail.

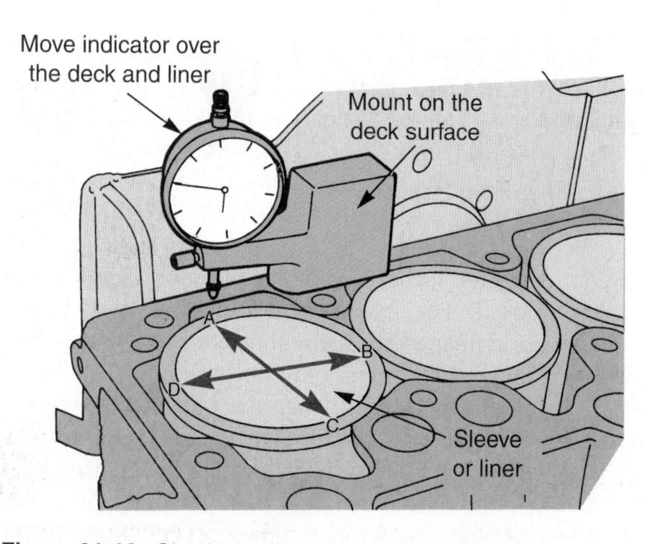

Figure 21-18. *Checking sleeve protrusion. (Renault)*

Figure 21-19. *Blow the block dry after washing.*

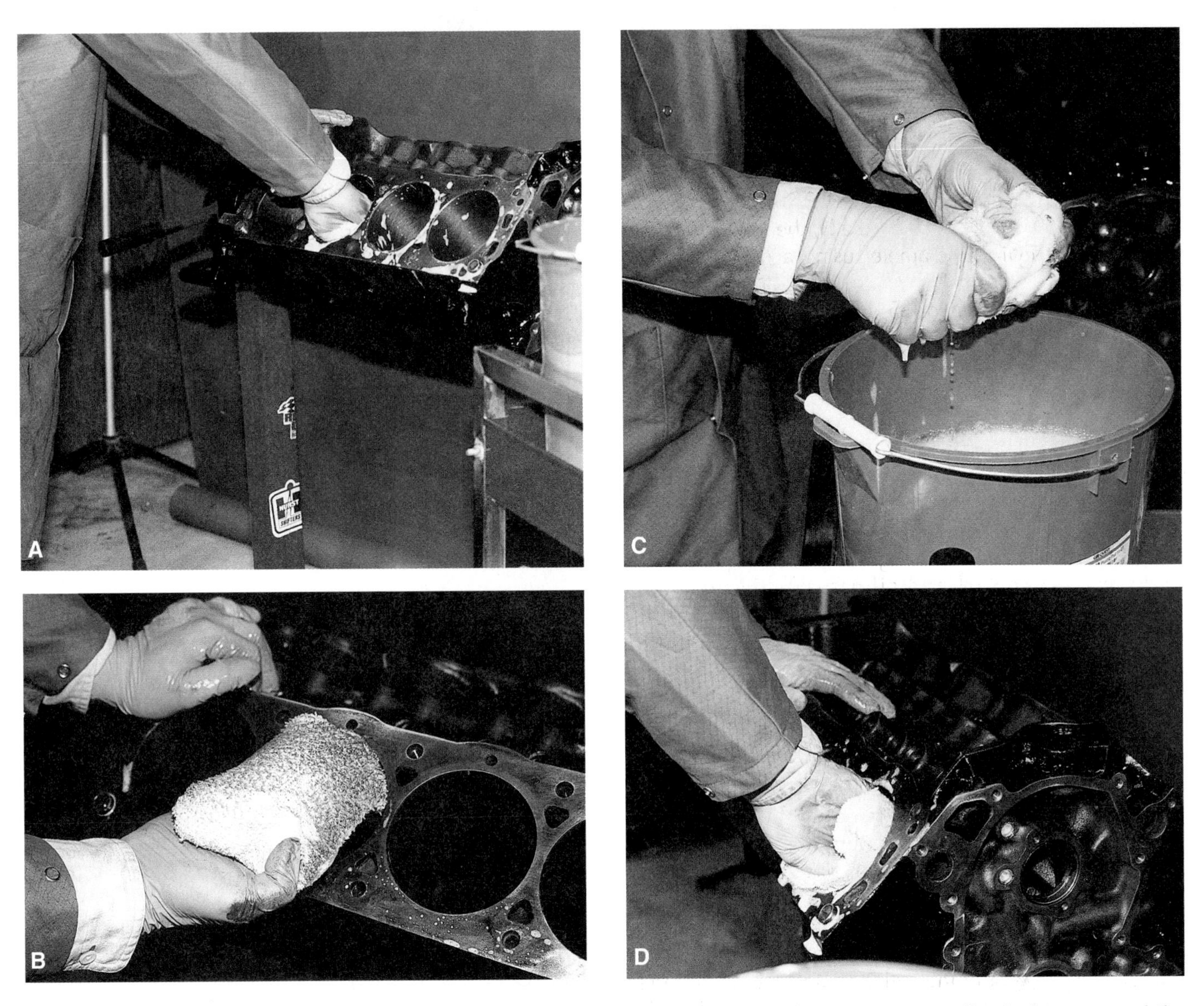

Figure 21-20. *A—If a pressure washer is not available, wash the cylinders with warm, soapy water. B—A clean rag and the detergent will lift grit out of the hone marks. C—Keep rinsing the rag out and washing the cylinders until the rag comes out clean. D—After the rag comes out clean, wash the cylinders one more time.*

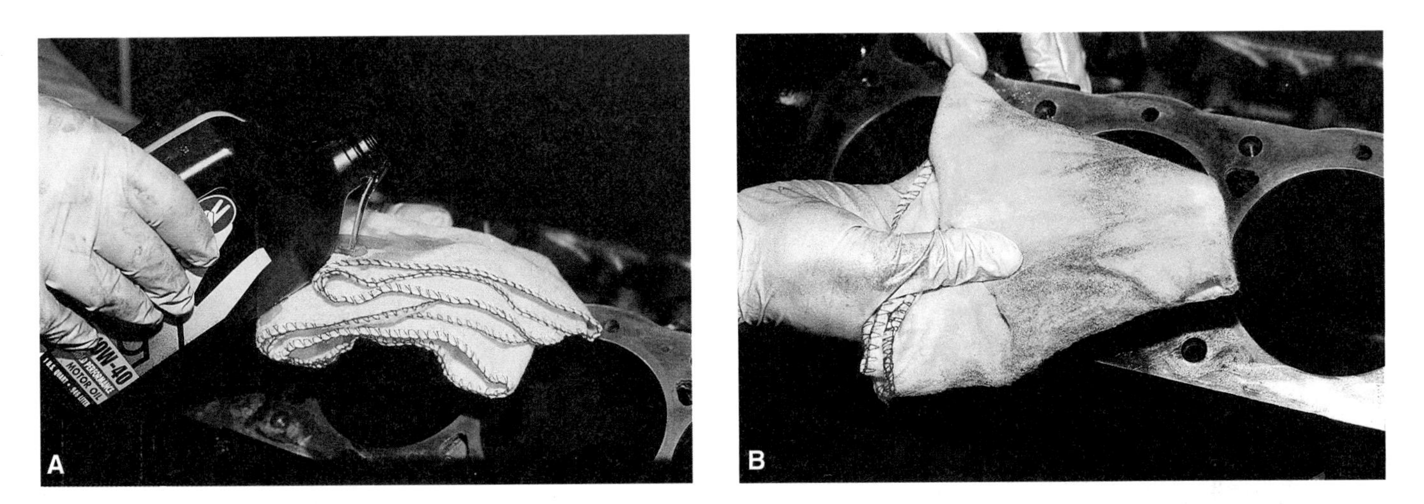

Figure 21-21. *A—After washing and drying the block, pour clean motor oil on a folded shop towel. B—Wipe the cylinders with the oil-soaked rag. Note how the oil pulls more grit out of the hone marks.*

Camshaft and Balancer Bearing Installation

New camshaft and balancer shaft bearings should be installed during major engine service. Worn camshaft bearings are common and can lower engine oil pressure. The camshaft bearings on an OHC engine may be split bearings, similar to a main or connecting rod bearing. However, in cam-in-block engines, the camshaft bearings are a one-piece design and pressed into the block.

Many technicians have the machine shop install camshaft bearings. However, some technicians have the special drivers needed to press the camshaft bearings into and out of the block.

Figure 21-22 summarizes how camshaft or balancer shaft bearings are removed and installed in an OHV engine. A driver is used to force the bearings out of their bores. The new bearings are then forced into place using a similar bearing driver.

It is critical for the oil holes in the block to line up with the oil holes in the bearings. If the holes do not align, severe valve train wear and damage can result from lack of lubrication.

Note: Bearing service for an overhead camshaft is discussed in Chapter 22. Installation of an overhead camshaft is discussed in Chapter 24.

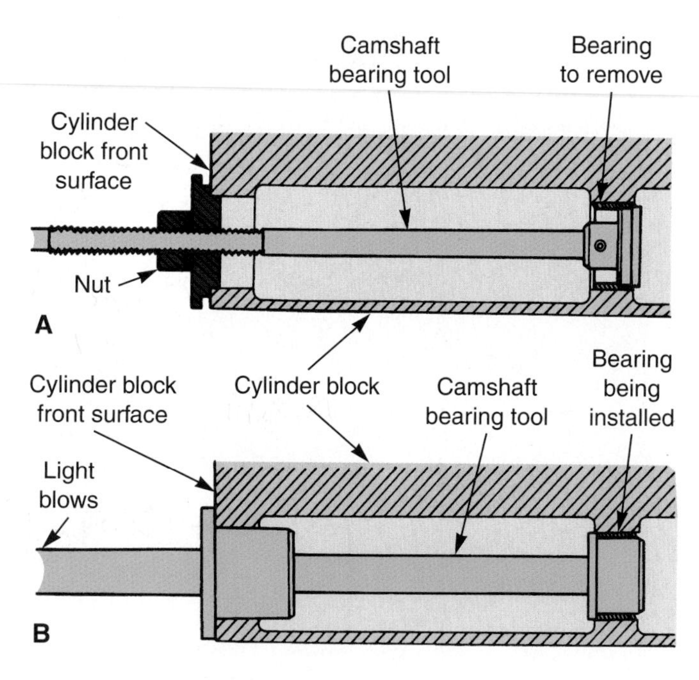

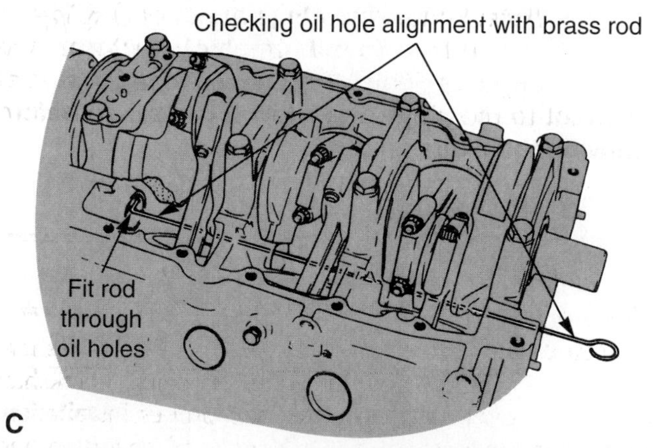

Figure 21-22. *A—The proper driver is being used to pull the old camshaft bearings out of the block. B—The proper driver is being used to push the new camshaft bearings into place. C—Make sure the oil holes in the camshaft bearings align with the oil holes in the block or severe valve train damage can result. D—Here, you can see the oil hole in the bearing aligns with the oil hole in the block.*

Core Plug Installation

Core plugs are installed over water jackets and oil gallery openings in the outside of the cylinder block. Unless the engine is a relatively new, low-mileage engine, new core plugs are installed during a major engine rebuild.

The old core plugs must be popped out and new ones driven in. A full-shank screwdriver or pointed pry bar can be driven through the center of the core plug. The old plug can then be pried out. Be careful not to scratch or damage the plug sealing surface in the block with your prying tool.

Before installing a new core plug, sand the block core plug opening with sandpaper or emery cloth to remove rust or debris from the sealing surface. Apply recommended sealer to the outside of the core plug and drive it in. Use a correct diameter driver and light hammer blows. Some technicians like to "stake" oil core plugs in place. Use a dull chisel or punch to place a couple indentations in the block so that the core plug cannot be forced out by oil or coolant pressure. This is shown in **Figure 21-23.**

Figure 21-23. *The machine shop has staked the core plug openings so that the plugs cannot be forced out by oil or coolant pressure.*

Note: Some core plugs are threaded. These are often found in high-performance engines. You may need to apply rust penetrant to the plugs and allow it to soak in before removing the plugs.

Crankshaft Service

Crankshaft service involves cleaning the crankshaft, measuring crankshaft journal wear, checking crankshaft straightness, checking for cracks, and proper installation. Quite often, the crankshaft will be in good condition and can be reinstalled without machining.

Crankshaft journals can be worn or damaged in service. In a high-mileage engine, normal wear can cause the journal dimensions to be out of specifications. If a bad oil pump has not been providing normal engine oil pressure or if the customer has not changed the oil at regular intervals, crankshaft journal wear can be high even in a low-mileage engine. Unusually high crankshaft wear can also be caused by operating the engine at high engine speeds or otherwise "abused." Connecting rod or rod bearing failure can also badly damage crankshaft journals.

If you take measurements and find the crankshaft journals are worn or damaged beyond specifications, the crankshaft must be rebuilt or replaced. Crankshaft rebuilding involves welding material to the journals so they are larger than the correct size and then turning (machining) them to fit the rod and main bearing inserts.

Before inspection, make sure the crankshaft is clean. Use compressed air to blow out all oil passages. Then, look at each connecting rod and main journal surface closely. Look for scratching, scoring, and any signs of wear. The slightest nick or groove is very serious. Fine crocus cloth can be used to polish off minor burrs or marks on journals. Polish around the journal, not across it.

Measuring Journal Taper, Out-of-Round, and Cheek Width

If one side of a crankshaft journal is worn more than the other, the journal is tapered. To measure journal taper, use an outside micrometer, as shown in **Figure 21-24.** Measure both ends of each journal, **Figure 21-25.** Any difference between the two ends indicates taper. Taper beyond recommended limits must be fixed by having the crankshaft turned.

When you measure for journal taper, also measure journal out-of-round. Out-of-round occurs when the journal is worn unevenly around its circumference. As shown in **Figure 21-25,** measure the journal diameter at two points around its circumference. The difference between the diameters indicates out-of-round. If not within specification limits, send the crankshaft to a machine shop for turning or repair.

Generally, crankshaft journal taper and journal out-of-round must not exceed .0005" to .001" (0.013 mm to 0.025 mm). However, always follow manufacturer specifications. If worn more, you should have the crankshaft turned undersize. Check service manual specifications for the exact make and model engine. Also, check to see what type and size of oversize bearings are available.

Use a telescoping gauge and micrometer to measure the crankshaft cheek width. Later, after measuring the rod width(s), you can subtract the rod width(s) from the crankshaft cheek width to determine rod side clearance.

Checking for a Cracked Crankshaft

A cracked crankshaft can result from part breakage, vibration damper failure, flywheel failure, or driver abuse. Cracks can be hard to find on a crankshaft because they may be very small. However, if cracks are not located and the crankshaft is reinstalled, it may break during service, as shown in **Figure 21-26.**

Magnafluxing is used to locate cracks in a crankshaft. This is basically the same as locating cracks in a cast iron cylinder block, as discussed earlier. Sprinkle the metal powder over the crankshaft. The magnetic field will make the powder collect in any crack. Cracks usually form in a crankshaft on the side of the journals. If the crankshaft shows even the smallest crack, replace the crankshaft.

Checking Crankshaft Straightness

A bent crankshaft can ruin new main bearings or cause the engine to lock up when the main caps are tightened. To measure crankshaft straightness, mount the crankshaft on V-blocks or in a lathe, **Figure 21-27.** The crankshaft can also be placed in the block main bearings with all main caps installed except the center main cap. Then, mount a dial indicator against the center main journal.

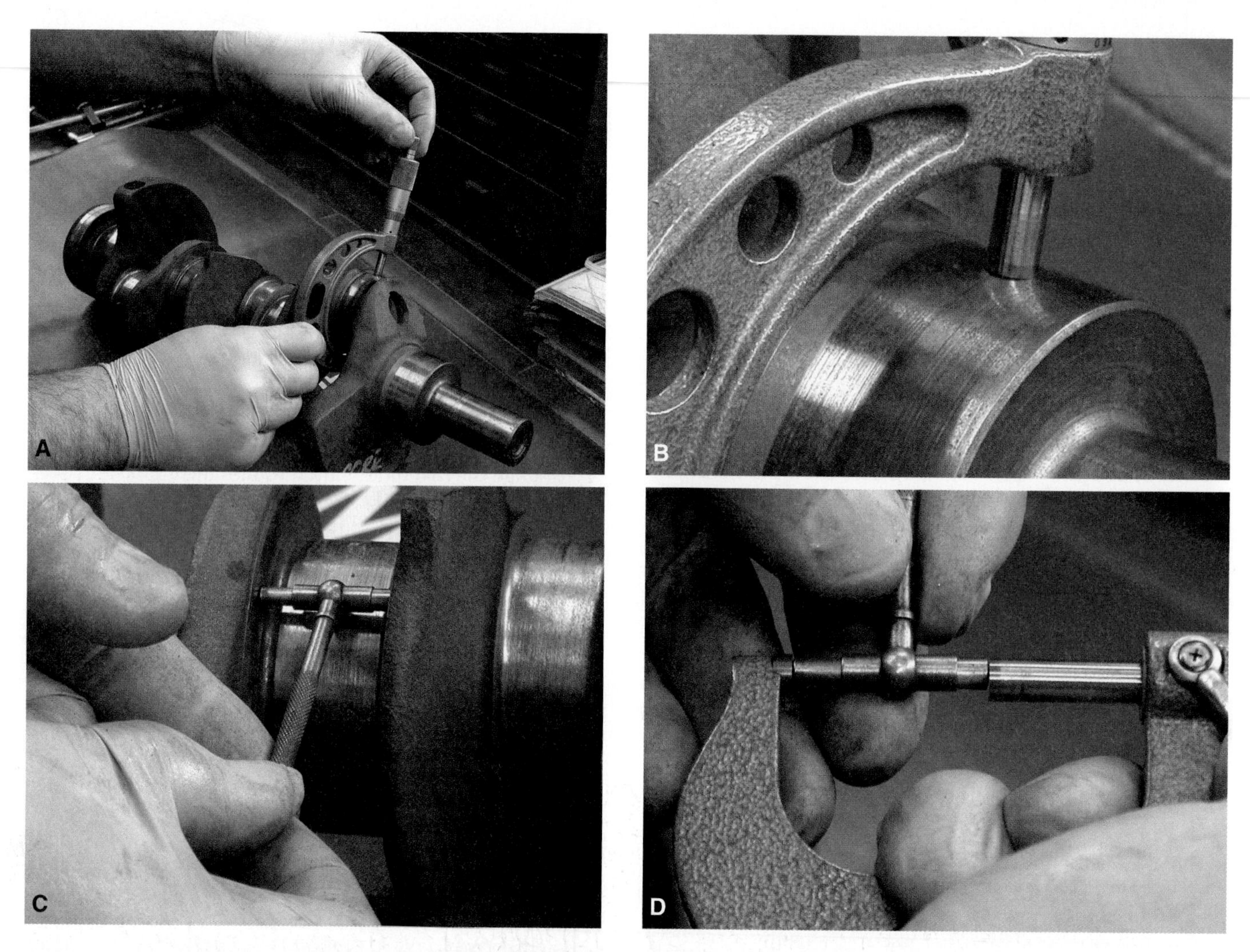

Figure 21-24. *Measuring a crankshaft. A—Measure every journal to make sure none are worn more than others. B—Take measurements at different locations on each journal to check for taper or out-of-round. C—Use a telescoping gauge to measure crankshaft cheek width. D—Use a micrometer to measure the telescoping gauge. By measuring the crankshaft cheek width and connecting rod width, you can calculate rod side clearance.*

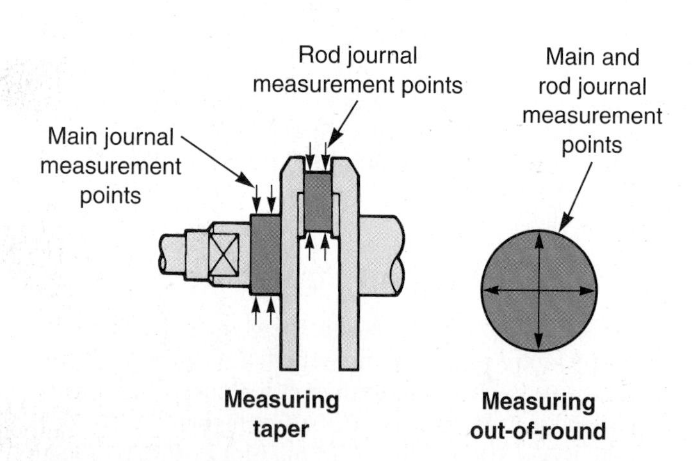

Figure 21-25. *Measure crankshaft and main journals at these locations. This will check for journal taper and out-of-round. (General Motors)*

Figure 21-26. *This cracked crankshaft was reassembled without checking for cracks. A catastrophic engine failure resulted. When the crankshaft broke, the main bearing next to the fracture also failed.*

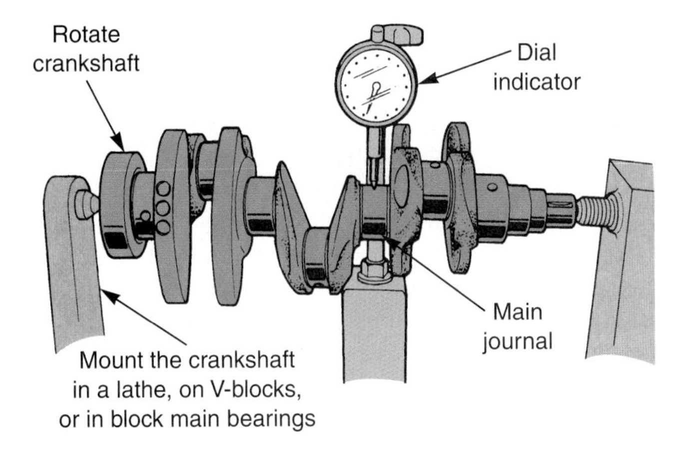

Figure 21-27. *Checking crankshaft runout. (General Motors)*

Slowly turn the crankshaft while watching the indicator. Indicator movement equals the crankshaft runout, or bend. If runout is not within specification limits, about .001″ (0.025 mm), replace the crankshaft or have it straightened and turned by a machine shop.

Crankshaft Turning

Crankshaft turning involves grinding the rod and main journals smaller in diameter to fix journal wear or damage. The worn crankshaft can also be repaired by welding additional material onto the journals and then grinding the journals to original specifications. Both of these operations are done by a machine shop.

To turn a crankshaft, it is mounted in a crankshaft grinding machine, **Figure 21-28.** This machine has a large-diameter grinding stone that can resurface the crankshaft main and rod journals to precise diameters.

1. Mount the crankshaft so that its centerline aligns with the machine's chucks. Use dial indicators to make sure the crankshaft is centered in the chucks.
2. Engage the machine, which rotates the crankshaft and the grinding wheel.
3. Slowly move the grinding wheel the proper amount into and across each journal.
4. Turn off the machine. Use a micrometer to check the journal diameters.

Figure 21-28. *Note the major steps for grinding a crankshaft. A—The crankshaft is mounted in the grinding machine. B—The crankshaft spins as the grinding wheel is slowly moved into and across each journal. C—As material is removed, a micrometer is used to check the journal diameter. D—This close-up shows the grinding wheel in action.*

5. Continue grinding as needed until the crankshaft journal is turned to the correct diameter.

Undersize bearings must be used with a crankshaft that has been turned. Since the journals have a smaller diameter, the bearings must be thicker to provide the correct bearing-to-journal fit.

Undersize Bearing Markings

Always look at the old crankshaft bearings to determine whether they are undersize. Undersize bearings may be used on rebuilt engines or even in a few new engines.

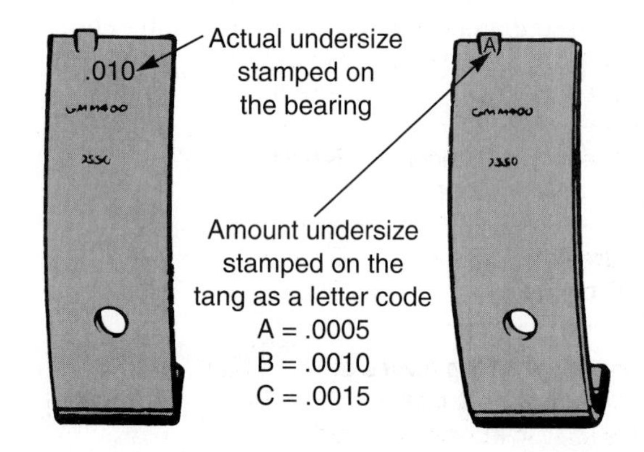

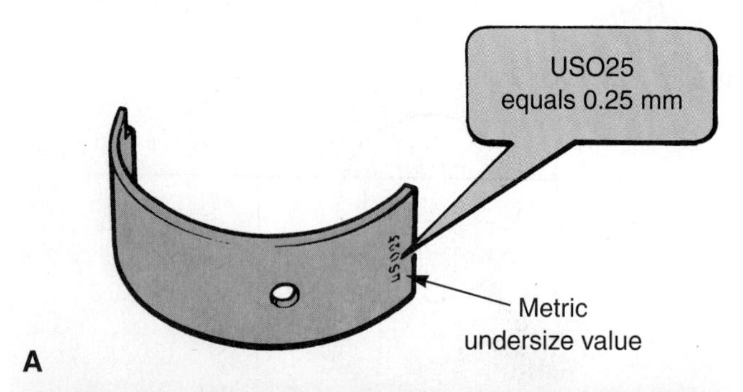

Figure 21-29. *A—Bearing undersize is marked on the back of the bearing. B—This close-up shows the bearing undersize marking on the back of a main bearing. This bearing is undersize by .020″. (General Motors)*

If the crankshaft is not turned and the connecting rod and main bores are not bored, standard size bearings are used when the crankshaft is reinstalled.

Inspect the back of the old bearings for an undersize number, as in **Figure 21-29.** The number stamped on the bearing back denotes bearing undersize. For example, .010 stamped on the bearing indicates that the bearing is .010″ undersize. A letter code can also be used to give bearing undersize. Do not confuse metric undersize markings with US customary undersize markings.

Some engines that have metric dimensions have letters or bars stamped on the block and crankshaft to denote bearing sizes, **Figure 21-30.** You can use these markings and the service manual chart to find which oversize or undersize bearings are in the engine. If the markings on the block are dirty and cannot be read, clean them with a solvent or detergent. Do not damage the markings with a wire brush, scuff wheel, or scraper or you may not be able to read them.

Cleaning the Crankshaft

The crankshaft must be perfectly clean before installation into the block. This is especially critical if the crankshaft has been ground. Debris such as metal shavings and stone grit may be inside the oil passages. Also, metal bits from failed bearings can collect inside the crankshaft. If not removed, this material can flow out and onto the new bearings during engine operation, where it will act as an abrasive. Bearing damage and failure can result.

Debris can collect inside the hollow crankshaft. The inside of the crankshaft must be cleaned to prevent bearing damage on engine operation. Remove any core or gallery plugs in the crankshaft, as shown in **Figure 21-31.**

Wash the crankshaft carefully in solvent. Then, use compressed air to blow out all passages and dry off the journals. Reinstall the core or gallery plugs. Finally, wipe the journals down with an oil-soaked rag.

Caution: Freshly overhauled engines have been ruined when technicians failed to remove crankshaft core or gallery plugs and clean the oil passages inside the crankshaft.

Installing the Rear Main Oil Seal

The ***rear main oil seal*** prevents oil from leaking out of the engine past the crankshaft spinning in the rear main bearing. There are two basic types of rear main oil seals: two-piece synthetic rubber (neoprene) seal and one-piece synthetic rubber seal. Each requires a different installation technique.

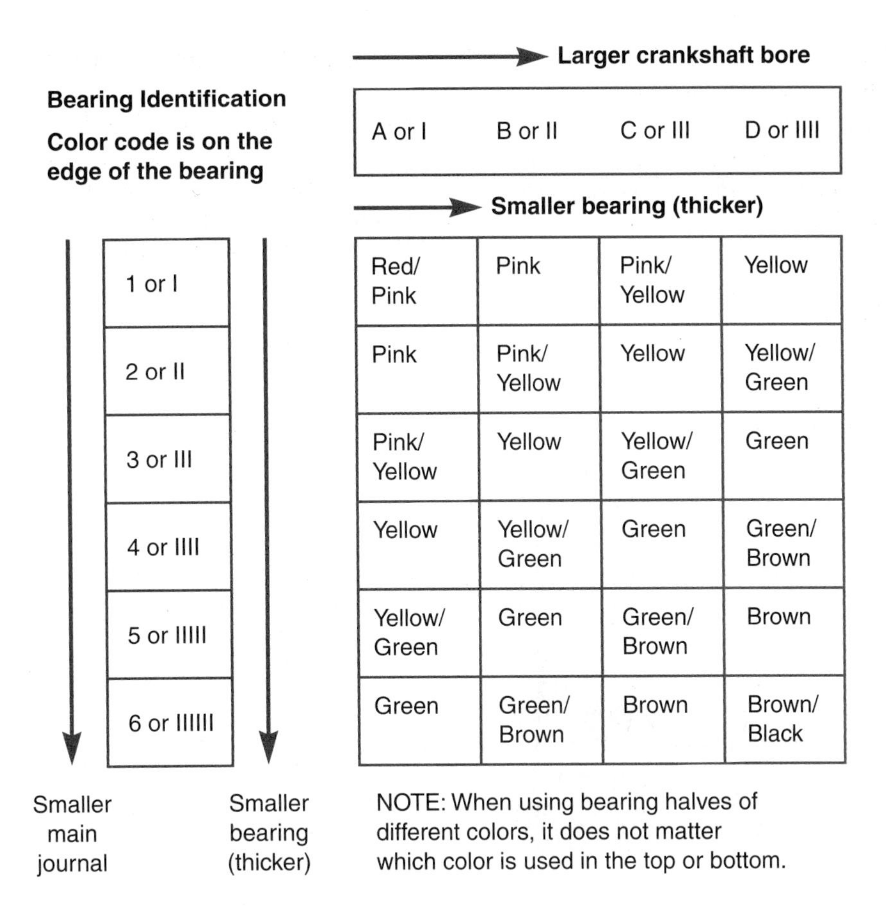

	A or I	B or II	C or III	D or IIII
1 or I	Red/ Pink	Pink	Pink/ Yellow	Yellow
2 or II	Pink	Pink/ Yellow	Yellow	Yellow/ Green
3 or III	Pink/ Yellow	Yellow	Yellow/ Green	Green
4 or IIII	Yellow	Yellow/ Green	Green	Green/ Brown
5 or IIIII	Yellow/ Green	Green	Green/ Brown	Brown
6 or IIIIII	Green	Green/ Brown	Brown	Brown/ Black

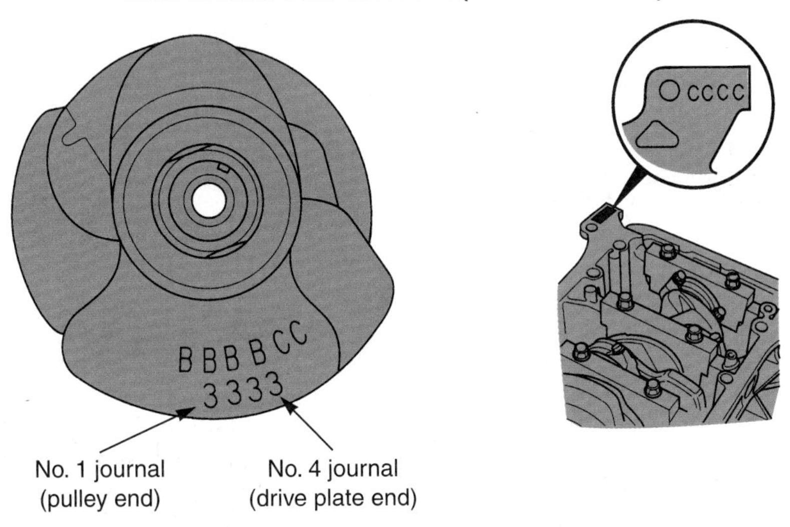

Figure 21-30. *This engine has letters and numbers (or bars) to denote bearing sizes. The crankshaft and cylinder block are both marked. (Honda)*

Two-Piece Seal

A two-piece neoprene rear seal is very easy to remove and install. See **Figure 21-32.** Use a small screwdriver to remove the old seal from the cap and block; do not scratch the surfaces. The sealing lip on the rear main seal must point toward the inside of the engine. Look at the new seal closely to identify the sealing lip and make sure it faces the inside of the block when installed. If installed backward, oil will pour out of the seal when the engine is first started. Lubricate the sealing lip with motor oil, but keep the back of the seal clean and dry. Some manufacturers recommend sealant on the outside of the rear main seals. Finally, press each half of the seal into place in the block and rear main cap.

If additional side seals are provided for the cap, follow the instructions with the gasket set. Sealer may be recommended for the main cap side seals.

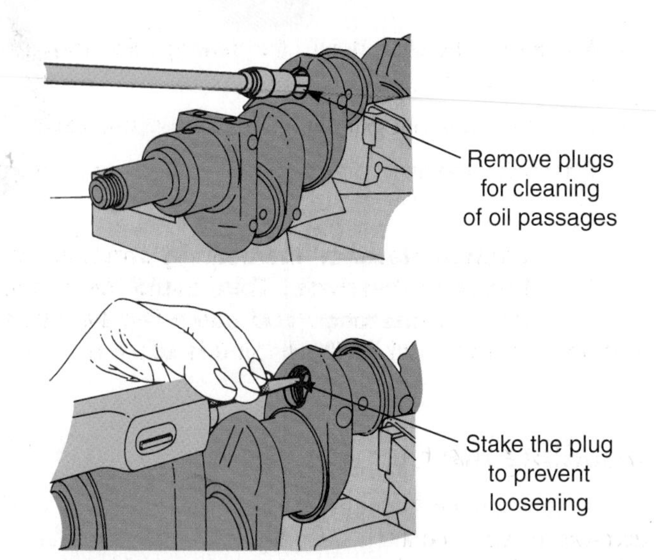

Figure 21-31. *After grinding, make sure you remove any plugs and clean the inside of the crankshaft.*

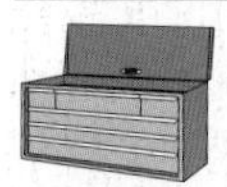

Note: Silicone sealer or anaerobic sealer is commonly recommended on the rear main cap to prevent oil leakage. The sealer keeps oil from seeping between the main cap and block mating surfaces. Check the manual to see which type of sealant to use.

One-Piece Seal

A one-piece neoprene seal is usually removed and installed after the rear main cap has been bolted to the block. See **Figure 21-33.** Pry out the old seal without scratching the block. Lubricate the inside diameter of the seal with an engine assembly lube. If recommended, coat the outside diameter with an approved sealer.

A special seal-installing tool is required to install a one-piece seal. Use the correct diameter driving tool to force the new seal into position in the block rear main bore. After installation, make sure the seal is square and undamaged. Also double-check that the seal is properly oriented with the seal lip facing inward.

Note: A rope-type seal was used as a rear main seal in the past. New neoprene seals should be installed in place of the rope-type seals.

Main Bearing and Crankshaft Installation

With the block and crankshaft prepped, you are ready to install the main bearings and crankshaft. Make sure you have the correct size main bearings to go with the crankshaft.

Installing the Main Bearings

Before installing the main bearings and crankshaft, carefully clean the cylinder block main bores, main cap mounting surfaces, and crankshaft journals one last time. See **Figure 21-34.** Use compressed air to blow out the bolt holes, oil holes and bearing mounting surfaces. Then, using solvent and a clean, dry rag, wipe the main bores free of any dust or oil. Also clean the crankshaft in a similar manner.

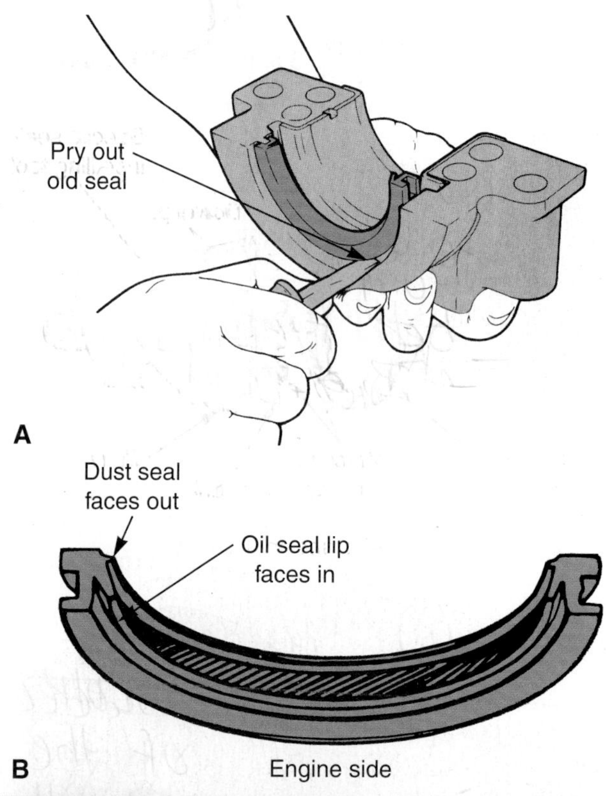

Figure 21-32. *Removing and installing a two-piece neoprene rear main oil seal. A—Use small screwdriver to pop out the old seal. B—Look at the new seal closely to identify the sealing lip. It must face the inside of the block when installed. C—Once the seal is installed, verify that the sealing lip points to the inside of the engine. (General Motors)*

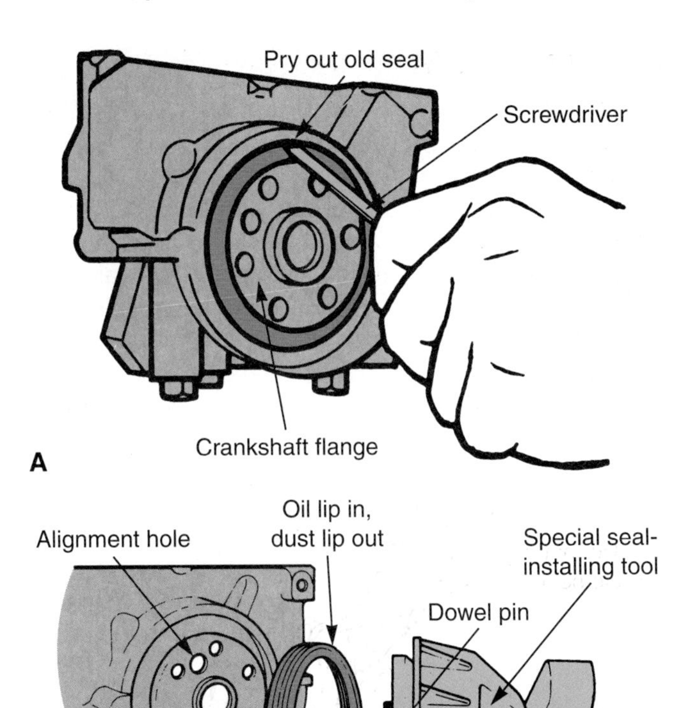

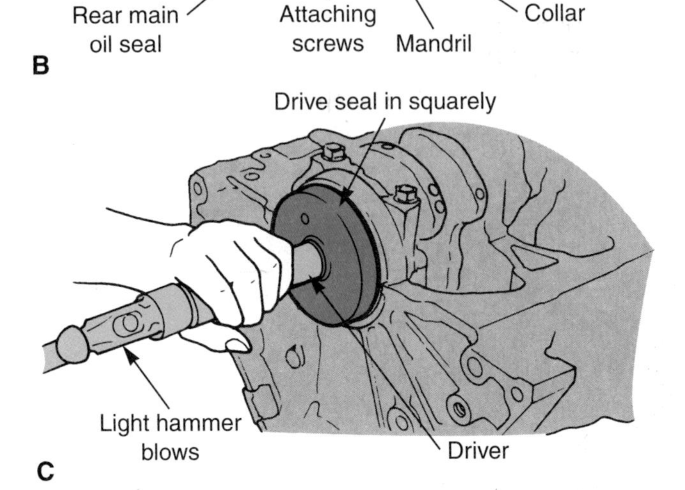

Figure 21-33. *Removing and installing a one-piece neoprene rear main oil seal. A—Pry out the old seal without scratching the surfaces. B—This special seal installing tool is tightened with a wrench. C—This oil seal is installed with a driver. Light hammer blows are used to push the seal squarely into place. (General Motors)*

1. Verify the bearing sizes marked on the box with that stamped on the back of the bearings.
2. With everything clean and dry, press the new main bearings down into place without touching the front of the bearing inserts, **Figure 21-35.**
3. The top of each bearing half should be almost flush with block on both sides.
4. The bearing tab fits into a notch in the main bore.
5. Make sure the oil holes in the bearings line up with oil holes in the cylinder block.
6. Snap the other bearing halves into the main caps.
7. Check that the main thrust bearing is located properly.

Caution: Never oil the bearing bores or the backs of bearings. This could decrease bearing clearance and make the bearings spin in their bores when the engine is started.

Installing the Crankshaft

With the main bearings and rear main seal installed (two-piece type), coat the bearing faces and rear seal lip with assembly lube, heavy engine oil, or white grease, **Figure 21-36A.** Spread the lube over the entire surface of the bearing half, **Figure 21-36B.** Then, carefully lower the crankshaft straight down into the main bearings, **Figure 21-36C.** Be careful not to damage the bearings or journals by bumping or hitting them.

Checking Main Bearing Clearance

Plastigage™ is used to check the oil clearance between the crankshaft journal and main bearing. Refer to **Figure 21-37.** Place a small bead of Plastigage on the crankshaft. Be careful not to deform the bead.

Install and torque the main bearing cap. Then, remove the main cap. Compare the smashed Plastigage to the paper scale. If clearance is not correct, check bearing sizes and crankshaft journal measurements.

An average main bearing clearance is .001″ to .002″ (0.025 mm to 0.05 mm). However, check the service manual for the exact specification. Some technicians only check one main bearing clearance; others check all of the bearing clearances.

Torquing the Main Caps

Once the main bearing clearance has been verified, coat the face of each bearing insert in the main caps with assembly lube. Then, place each main cap into the block. Check main cap numbers and arrows to be sure they are positioned correctly. Each cap must be facing the right direction and in its original location.

Lubricate the bolt threads with the factory-recommended type of lubricant, **Figure 21-38.** The viscosity of the thread lubricant can affect bolt torque.

Never squirt oil down into the bolt holes. This may cause hydraulic lock and prevent normal bolt tightening or cause the block to crack. Lubricate bolt threads, not hole threads. Then, tighten the bolts hand tight. Lightly tap each cap with a hammer to make sure the cap is fully seated.

When installing the rear main cap, double-check the rear main oil seal. Make sure the sealing lip is pointing

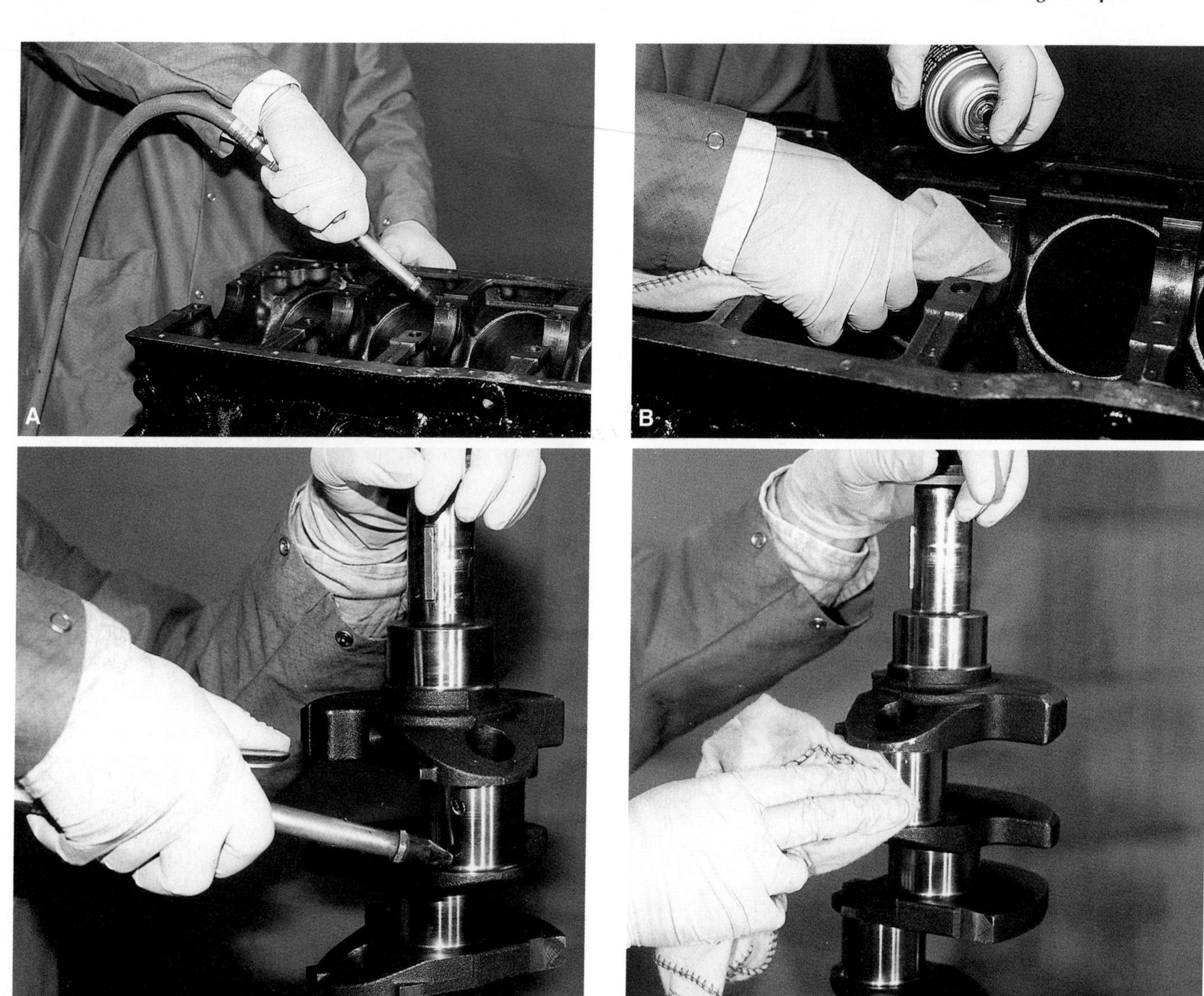

Figure 21-34. *Before installing the main bearings and crankshaft, carefully clean the main bores, main cap mounting surfaces, and crankshaft journals. A—Use compressed air to blow the parts clean. B—Wipe off the main bores with solvent and a clean shop towel. C—Use compressed air to blow out any oil or bolt holes. D—Using a clean shop towel and solvent, wipe the crankshaft journals.*

Figure 21-35. *With everything clean and dry, press the new main bearings down into place. The top of each bearing should be almost flush with the cap mounting surface on both sides.*

toward the inside of the engine, **Figure 21-39A.** Remember to place sealer on the rear cap if specified, **Figure 21-39B.**

Look up the torque specifications for the main bearing cap bolts in the service manual. Next, make sure all caps are fully seated into the block. Tap on each with a soft-faced mallet. Then, use a torque wrench to tighten the bolts to about one-half of the full torque. Use an even pull on the wrench. Torque each cap in the sequence recommended in the service manual, **Figure 21-40.** This will pull each cap down squarely. Next, tighten each bolt to three-fourths of the full torque. Finally, tighten each bolt to the full torque. Repeat the torque sequence at full torque at least one more time to ensure proper torque and that no bolts have been skipped.

After torquing all of the main cap bolts, make sure that the crankshaft will turn freely. Install the front snout bolt

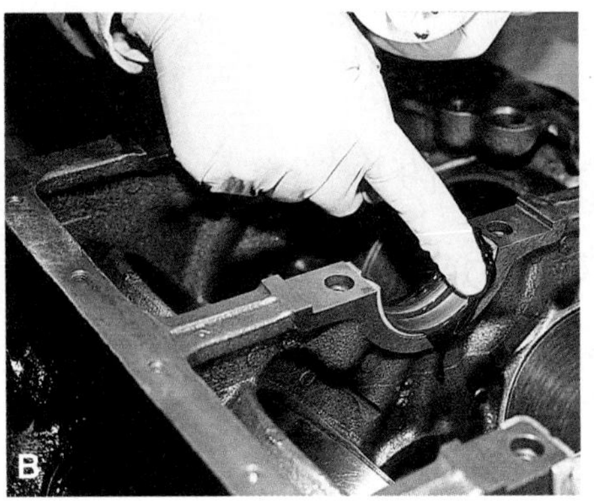

Figure 21-36. *A—Apply bearing assembly lube to all of the main bearing inserts. B—Spread bearing lube over surface of main bearings. C—Slowly lower the crankshaft straight down into the main bearings without damaging the bearings or journals.*

and turn the crankshaft by hand with a wrench or breaker bar. See **Figure 21-41.** If the crankshaft does not rotate easily, something is wrong. Remove the main caps and check the bearing sizes, look for debris in the bearings, and verify measurements and clearances.

Torque-to-Yield

Most late-model engines have ***torque-to-yield specifications.*** To properly tighten bolts with a torque-to-yield specification, first tighten each bolt to a specified torque,

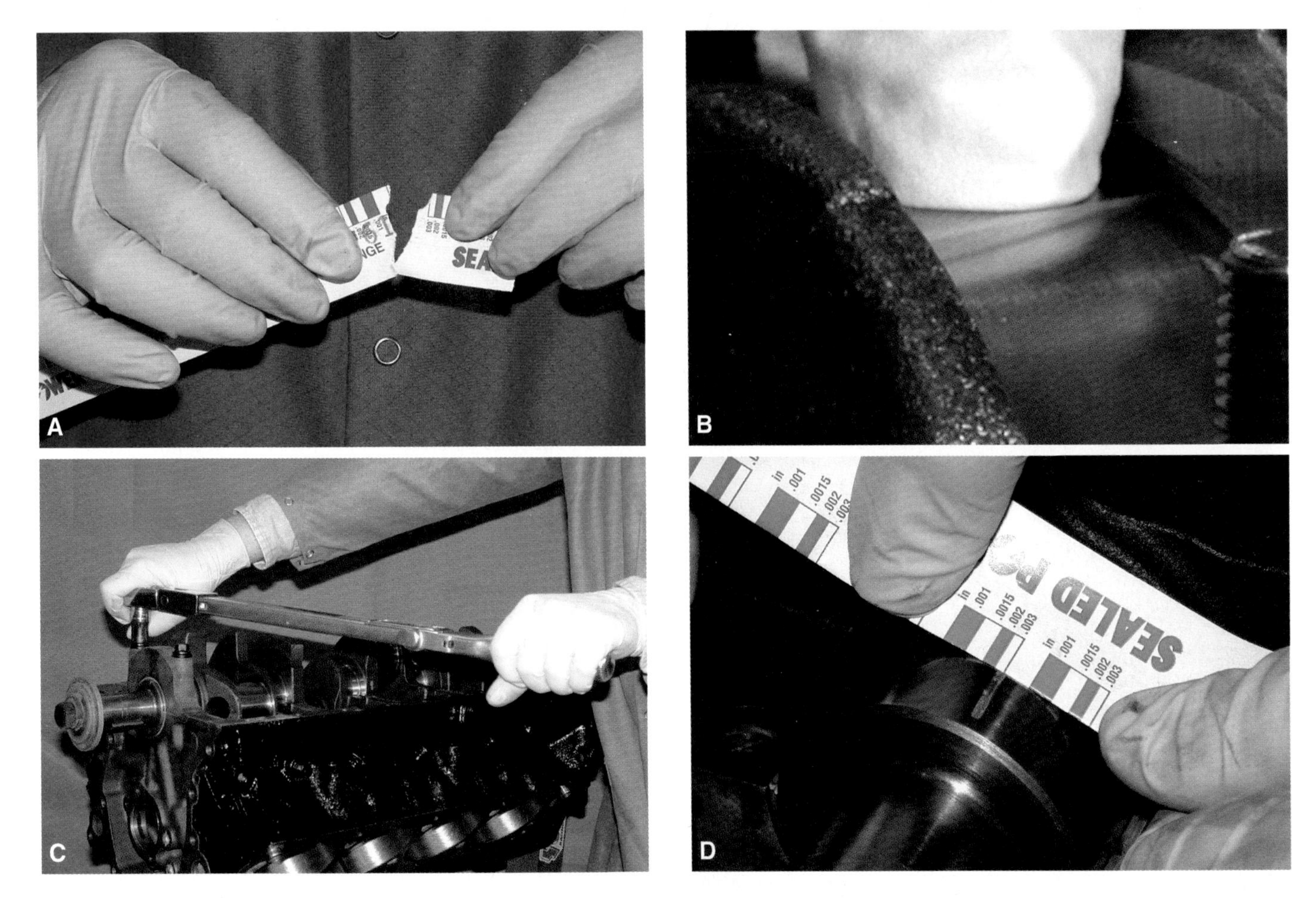

Figure 21-37. *Using Plastigage. A—Tear off short piece of the Plastigage; it is inside the paper wrapper. Be careful not to squeeze the Plastigage inside the wrapper. B—Remove the Plastigage from the paper and place it across the crankshaft journal. C—Install and torque the cap and bearing down onto the Plastigage. D—Remove the cap and compare the paper scale with the width of the crushed Plastigage.*

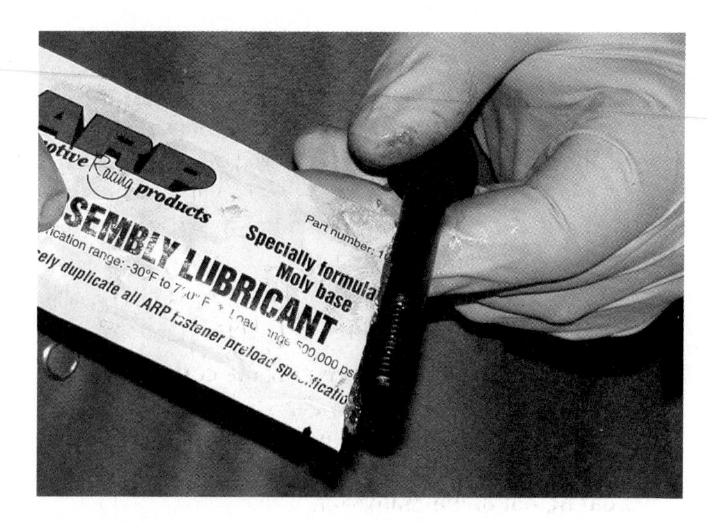

Figure 21-38. *Before installing main bearing bolts, lubricate the bolt threads with the factory-recommended lubricant.*

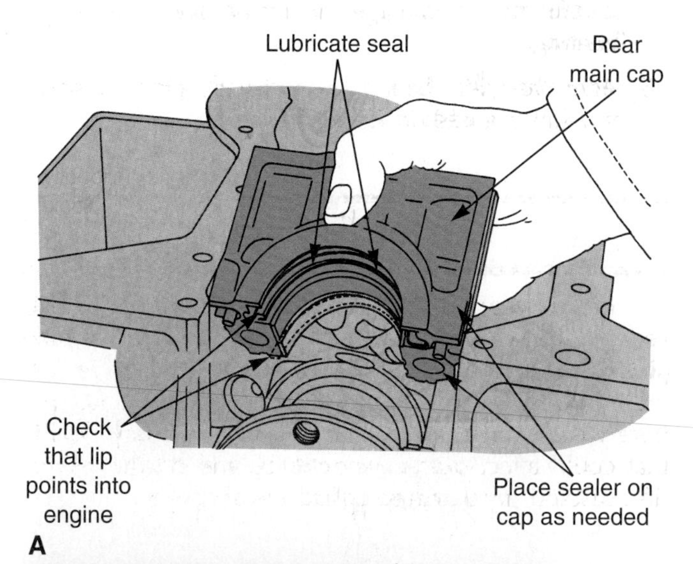

Figure 21-39. *A—When installing the rear main cap, double-check that the rear main oil seal is installed in the correct orientation. B—With a two-piece rear oil seal, the mating surfaces between the rear main cap and block usually have sealer applied to them. (Peugeot)*

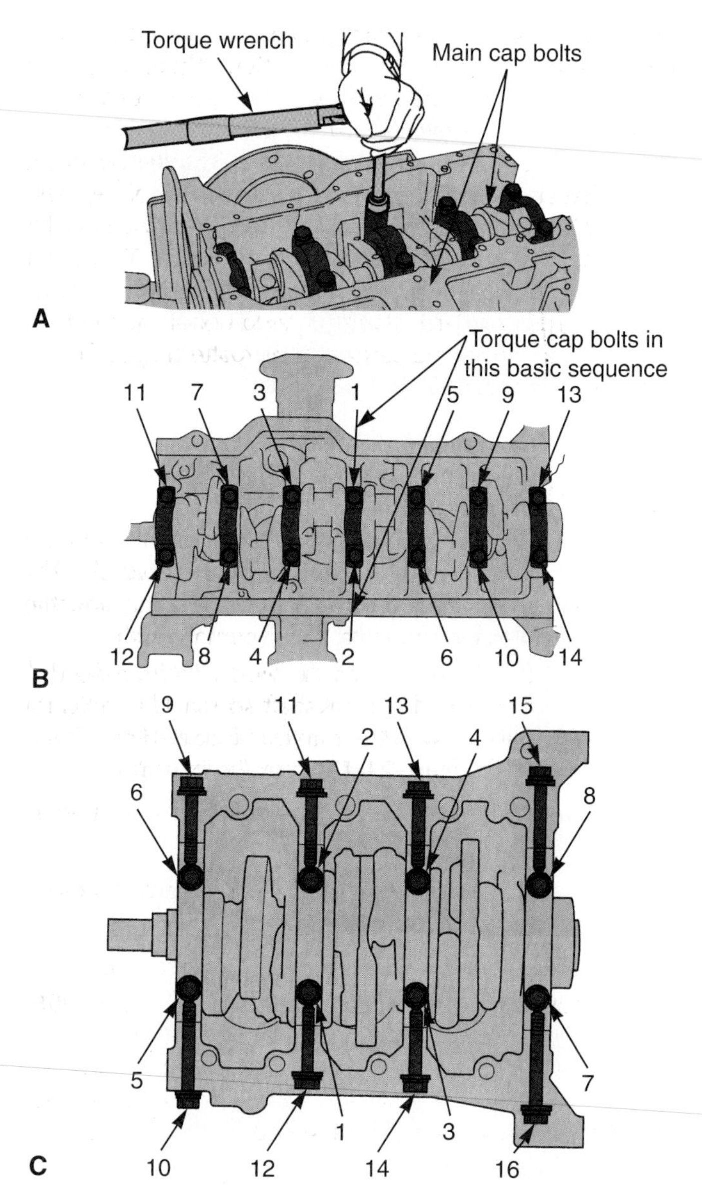

Figure 21-40. *Torquing main cap bolts. A—Torque each bolt a little at a time up to the full torque specification. B—Torque each cap in sequence as shown in the service manual. C—Note the torque sequence recommended for this engine with cross-bolt main caps. (Ford, Honda)*

Figure 21-41. *After torquing all of the main caps, make sure the crankshaft turns freely.*

Figure 21-42A. For tightening main caps, use the procedure described in the previous section. Then, you must turn the bolt a specified number of degrees beyond the torque point. This stretches and preloads the fastener.

A dial gauge, or degree wheel, is installed between the socket and wrench to measure the number of degrees, **Figure 21-42B.** The small arm on the degree tool must be placed against the cap or other stationary part. Then, you must zero the degree gauge by turning the dial to read zero. Finally, turn the bolt an additional number of degrees. Use the torque sequence or pattern specified in the service manual.

Checking Crankshaft Endplay

Crankshaft endplay is the amount of front-to-rear movement of the crankshaft in the block. It is controlled by the clearance between the main thrust bearing and the crankshaft cheeks, thrust surface, or journal width.

1. Mount a dial indicator on the block. Position the dial indicator against the crankshaft so that the indicator stem is parallel with the crankshaft centerline. This is illustrated in **Figure 21-43.** Zero the indicator.
2. Pry back and forth on a counterweight with a small pry bar.
3. The maximum movement of the indicator needle is the crankshaft endplay.
4. Compare your measurements to specifications.

If not within specifications, usually about .006" (0.15 mm), check the main thrust bearing and the width of the crankshaft journal thrust surfaces. You may have the wrong main thrust bearing or the crankshaft thrust surface may have been incorrectly machined.

Replacing the Pilot Bearing

If the vehicle has a manual transmission or transaxle, you should replace the pilot bearing in the rear of the crankshaft when rebuilding an engine. This bearing supports the end of the transmission input shaft. If worn, clutch and transmission problems can develop.

1. To remove the pilot bearing, fill the cavity in the crankshaft and pilot bearing with heavy grease.
2. Use an old input shaft or driver of the correct diameter to drive out the old pilot bearing. As the driver is forced into the grease, the grease will transfer the force to the back of the pilot bearing. This pushes the bearing out of the crankshaft.
3. Select the proper driver to install the new pilot bearing, as shown in **Figure 21-44.**
4. Drive the pilot squarely into the crankshaft. Be careful not to damage the inside diameter of the bearing.
5. After the pilot bearing is installed, place a small amount of grease inside the cavity.

Piston Service

Pistons are primarily made of aluminum, which is very prone to wear and damage. It is very critical that each piston be thoroughly checked. Look for cracked or collapsed skirts, worn ring grooves, cracked ring lands, pin bore wear, or other problems. You must find any trouble that could affect piston performance and engine service life. Discard any damaged or badly worn pistons.

Figure 21-42. *Tightening torque-to-yield fasteners. A—First, torque the main cap bolts to factory specifications. Use the recommended torque sequence. B—Next, install the degree wheel onto the wrench. Position the stop against a stationary part and zero the degree wheel. Tighten the bolt the specified number of additional degrees.*

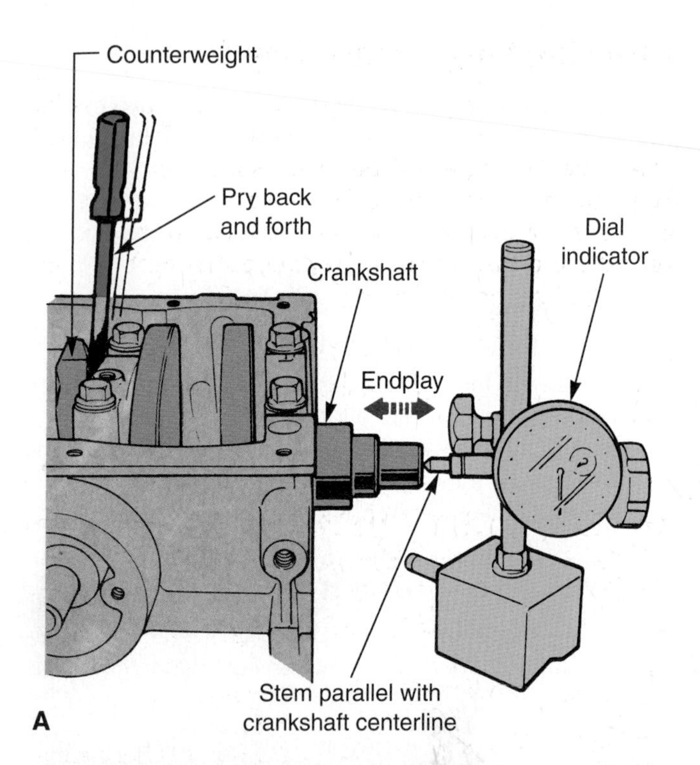

Figure 21-43. *Checking crankshaft endplay. A—Make sure the indicator plunger stem is parallel to the crankshaft centerline. B—Pry back and forth on a crankshaft counterweight while reading the indicator. (General Motors)*

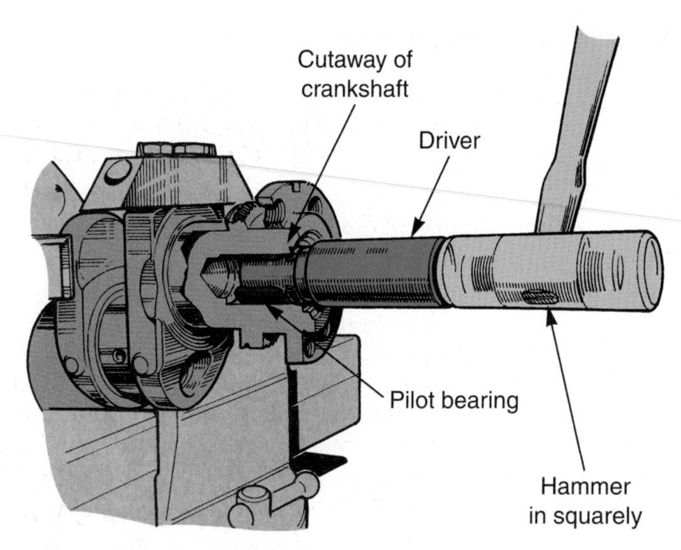

Figure 21-44. *Installing a new pilot bearing. (Renault)*

Measuring Piston Wear

Piston wear is the amount of decrease in the piston diameter due to friction. Piston size (diameter) is usually measured on the skirts, just below the piston pin hole, as in **Figure 21-45.** Measure perpendicular to the piston pin. A large outside micrometer is used to measure the piston diameter. Adjust the micrometer for a slight drag as it is pulled over the piston. Subtract the reading from the specification for piston diameter to find the piston wear. If the wear is more than that allowed by specifications, usually .005″ (0.13 mm), replace or knurl the piston(s).

Most engine manufacturers specify that measurements are to be made just below the piston pin, perpendicular to the pin. However, some engine manufacturers require measurement even (aligned) with the pin hole. Other manufacturers require measurement lower on the skirt. If you measure in the wrong location, your piston diameter reading will be inaccurate. If in doubt, check the service manual for directions on properly measuring piston wear.

Cleaning Piston Ring Grooves

Use a ring groove cleaner to remove any carbon deposits from the piston grooves. This is detailed in the previous chapter. All hard deposits must be removed from inside the piston ring grooves. If not removed, the carbon deposits could force the new rings out against the cylinder walls when the piston heats up and expands. This could score the rings and cylinder walls. It could also cause enough friction to heat the rings enough so they melt and weld to the cylinder wall, causing the engine to seize (lock up).

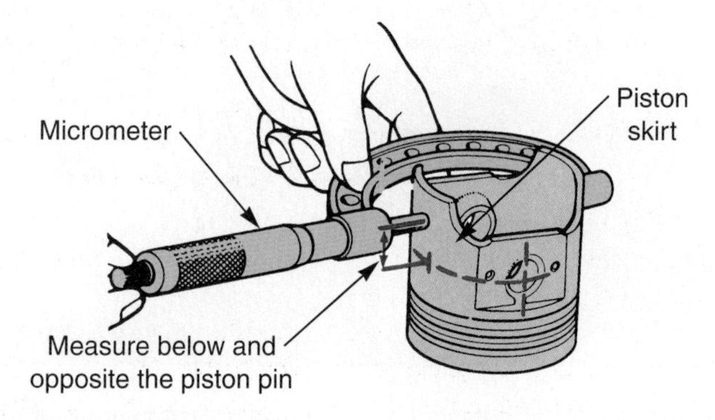

Figure 21-45. *Piston wear is usually measured on the piston skirt just below the piston pin. (Toyota)*

Knurling a Piston

Piston knurling can be used to slightly increase the diameter of the skirt, up to .005″ (0.13 mm) oversize. Knurling makes grooves in the skirts and the metal next to the grooves is pushed up. This increases the piston diameter and also provides grooves for motor oil to reduce friction.

Measuring Piston Clearance

Piston clearance is the distance between the side of the piston and the cylinder wall. To find piston clearance, subtract the piston diameter measurement from the cylinder diameter measurement. The difference is the piston clearance. Piston clearance is usually .001″ to .003″ (0.025 mm to 0.08 mm). Since specifications vary, always refer to the service manual.

Another way of measuring piston clearance is with a feeler gauge strip. A long, flat feeler gauge strip is placed on the piston skirt. Then, the gauge and piston are pushed into a cylinder. A spring scale is used to pull the feeler gauge strip out of the cylinder. When the spring scale reading equals specifications, the size of the feeler gauge equals piston clearance.

When piston-to-cylinder clearance is excessive, it must be corrected. This can be done by one of the following methods.

- Install a new standard size piston, providing cylinder is not worn beyond specifications.
- Bore all of the cylinders and purchase oversize pistons.
- Sleeve the cylinder.
- Knurl the piston.

Measuring Piston Ring Gap

Piston ring gap is the clearance between the ends of a ring when it is installed in the cylinder. This gap is very important. If the gap is too small, the ring could lock up or score the cylinder when it heats up and expands. If the ring gap is too large, ring tension against the cylinder wall may be low, causing oil blowby.

1. Compress a compression ring and place it in a cylinder (not installed on a piston).
2. Push the ring to the bottom of normal ring travel with the head of a piston. This will square the ring in the cylinder and locate it at the smallest cylinder diameter, **Figure 21-46.**
3. Measure the ring gap with a flat feeler gauge.
4. Compare your measurement to specifications.

If not correct, usually between .010″ to .020″ (0.25 mm to 0.51 mm), you may have the wrong piston ring set or the cylinder dimensions may be off.

Some manufacturers allow ring filing to increase the piston ring gap. ***Ring filing*** involves using a special, thin grinding wheel to remove metal from the ends of a ring. However, some manufacturers do not allow this practice. Check the service manual before filing a ring.

Checking Ring Groove Depth

To check ***ring groove depth,*** fit a new compression ring into the piston groove, as shown in **Figure 21-47.** Make sure the ring will fit below the surface of the piston. This helps ensure you have the correct ring set. If the ring groove is not deep enough, the rings may be forced out against the cylinder walls when the piston heats up and expands. Part damage could result.

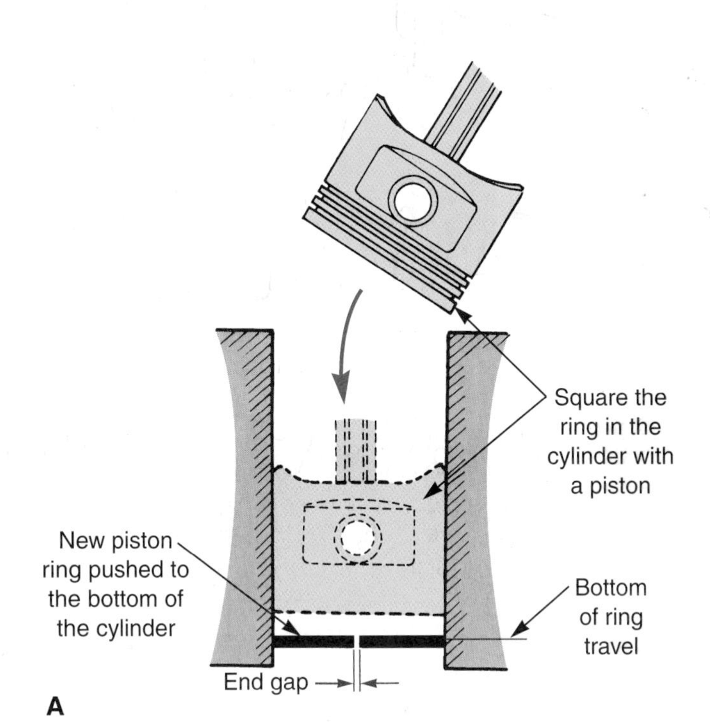

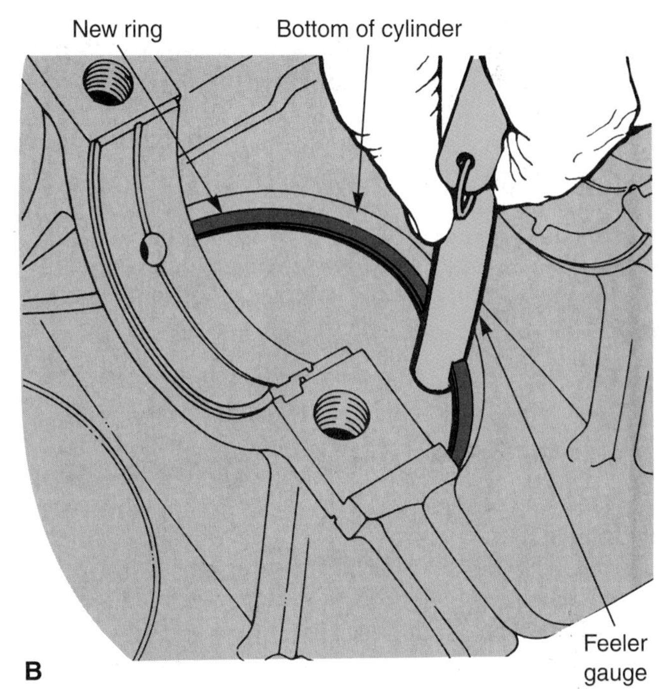

Figure 21-46. *Checking ring gap. A—Use a piston to push the new ring to the bottom of the ring travel in the cylinder. B—Use a feeler gauge to check ring gap. (Toyota, DaimlerChrysler)*

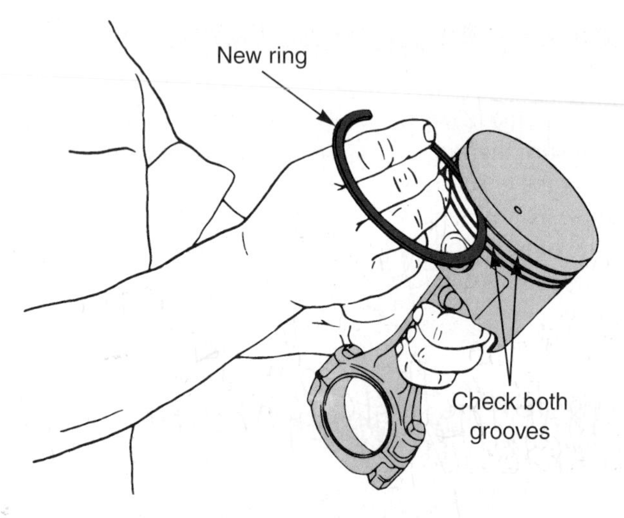

Figure 21-47. *To check ring groove depth, install the ring into the groove and make sure the ring goes below the surface of the piston. (General Motors)*

Piston Pin Service

Depending on the type and make of engine, the piston pin may either be free floating or press fit. A ***free-floating piston pin*** will turn in both the rod and piston. A ***press-fit piston pin*** is force fitted into either the rod or piston. Other setups have been manufactured, but are not common.

Piston Pin Clearance

During piston and rod service, check the pin clearance on both free-floating and press-fit pins. To check for excessive piston pin clearance, lightly clamp the connecting rod I-beam in a vise. Use vise caps to protect the rod. Try to move the piston up and down and sideways against normal pin movement, **Figure 21-48A.**

If play can be detected, the pin, rod bushing, or piston pin bore is worn. After the pin is removed, a small telescoping gauge and outside micrometer should be used to measure the piston pin bore and piston pin. Then, you can determine exact part wear and clearances.

Free-Floating Pin Service

To remove a free-floating piston pin from the piston, use snap ring pliers to compress and lift out the snap rings on each end of the pin, **Figure 21-48B.** Then, push the pin out of the piston by hand. A brass drift and light hammer blows may be needed. Make sure the piston and rod are marked and kept organized, **Figure 21-48C.**

To reassemble, insert the connecting rod into the piston. Make sure the piston is facing in the right direction in relation to the connecting rod. Normally, a piston will have some form of marking on its head indicating the front of the engine. The connecting rod may have one edge of the big end bore chamfered. This faces the outside of the journal on V-type engines. The rod may also have an oil spray hole or rod numbers that must face in only one direction. If needed, check the vehicle's shop manual for directions.

After the rod and piston are assembled in the correct orientation, insert the piston pin. Then, install the snap rings to secure the pin. Double-check that the snap rings are fully seated in their grooves.

Pressed-In Pin Service

A hydraulic press is normally used to remove a pressed-in piston pin from the piston and rod, as shown in **Figure 21-49.** Mount the piston in a proper holding fixture to avoid part damage. Select a driver that is slightly smaller than piston pin bore. With the piston assembly mounted and the driver installed, slowly lower the ram onto the piston assembly. The piston pin should slide out of the rod and piston bore.

You can use a hydraulic press in a similar manner to reassemble the pin. You can also use heat to fit the pin in the piston. Basically, the rod small end is heated to make the pin bore slightly larger. Then, the cool pin can be easily inserted. Before inserting the pin, make sure the rod and

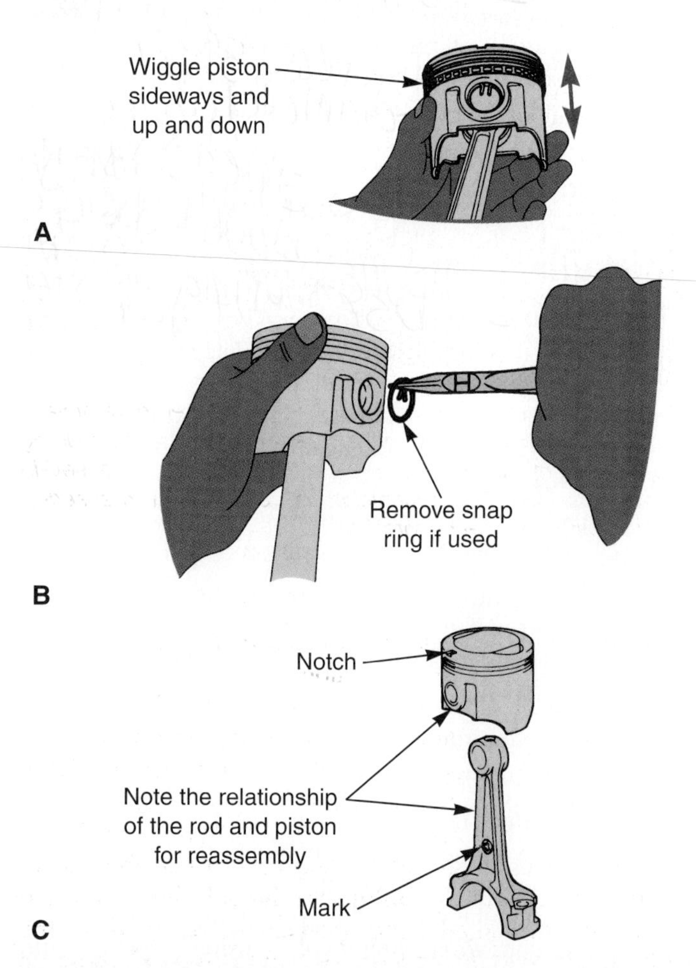

Figure 21-48. *A—Checking for a loose piston pin. B—Removing a free-floating piston pin. C—Make note of how the piston is installed on the rod. (Toyota)*

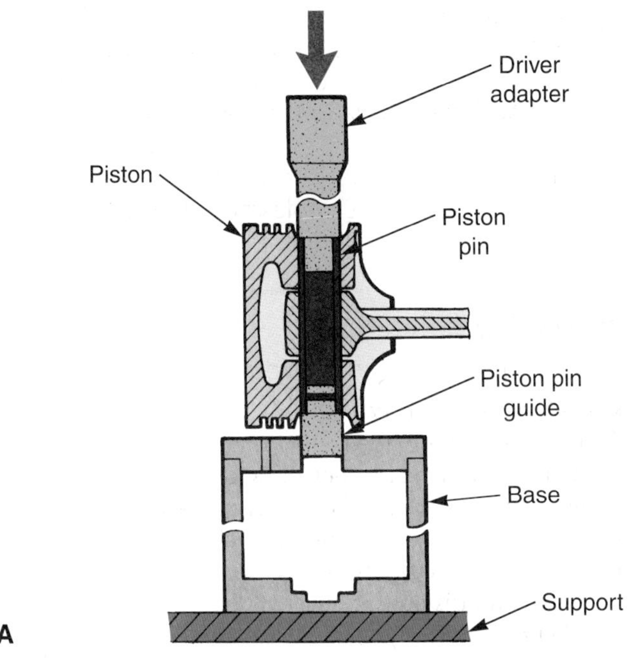

Figure 21-49. *A hydraulic press is normally used to remove a pressed-in piston pin. A—This cutaway shows how a holding fixture and hydraulic press are used to drive out a pressed-in piston pin. B—This technician is removing a pressed-in piston pin. (General Motors)*

piston are assembled in the correct orientation, as described in the previous section.

Figure 21-50 shows how to use an electric rod heater to properly assemble a pressed-in piston pin. Turn on the electric heating element and allow it to warm up. Place the rod small end into the heating element. Allow enough time to heat the rod and enlarge the pin bore. Then, remove the rod from the heater, insert it into the piston, and quickly push the piston pin through the piston and rod. Use the small hand-operated driver to push the piston pin into the piston and rod. The rod must be centered on the pin before it cools. When the rod cools and shrinks, it will lock the pin in place. This, in effect, creates a press fit.

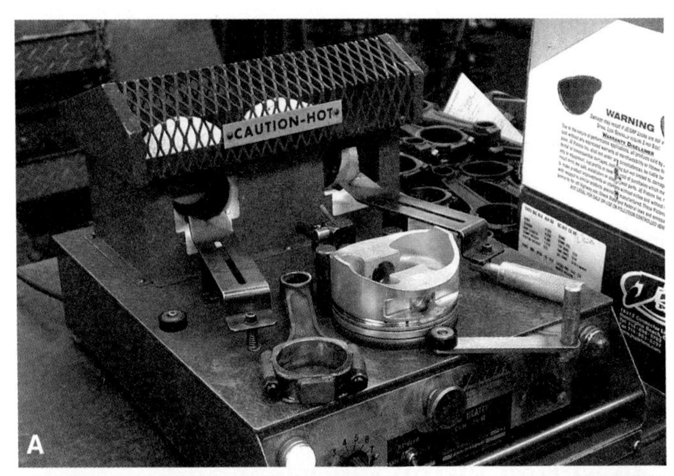

Figure 21-50. *Study the major steps for using a rod heater to press fit a piston pin. A—Turn on the heating element and allow it to warm up. B—Place the rod small end into the heating element. D—When the rod small end is heated enough, fit it into the piston and use the hand-operated driver to push the piston pin into the piston and rod.*

Warning: When using a hydraulic press, wear eye protection and make sure the piston is mounted properly. Also, when using a rod heater, take the necessary precautions to avoid burns.

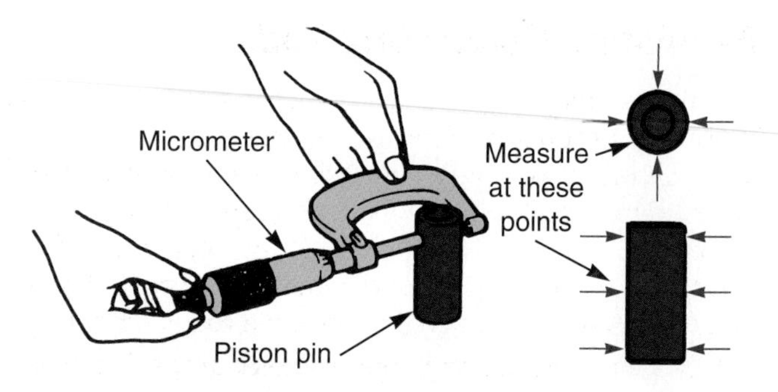

Figure 21-51. *Measure the piston pin in the locations shown to determine wear.*

Measuring Piston Pin Wear and Clearance

To determine piston pin wear, measure the diameter of the pin with an outside micrometer. Measure in the middle and at both ends of the pin (floating types only). Look at **Figure 21-51.** Then, compare the diameter to specifications. The difference between the measured diameter and the specification is the amount of wear. If worn too much, replace the piston pins.

To find piston pin clearance with a floating pin, you must measure the diameter of the pin bore in the piston, **Figure 21-52.** By subtracting pin diameter from piston pin hole diameter, you can calculate piston pin clearance.

A typical piston pin clearance is only .0001″ to .0003″ (0.003 mm to 0.008 mm). This specification is in ten-thousandths of an inch, not thousandths (thousandths of a millimeter, not hundredths). Keep this in mind while calculating piston pin wear, piston bore wear, and rod small end wear; measure accurately.

When the pin is worn, it should be replaced. If the pin bore is larger than specifications, replace the piston or rod.

Figure 21-52. *Use a telescoping gauge and micrometer to measure the diameter of the piston pin bore. Then, subtract the piston pin diameter from the bore diameter to calculate piston pin clearance.*

The pin bore may also be reamed larger. Oversize piston pins can then be used. Pin bore reaming is usually done by a machine shop.

Note: Oversize piston pins may be available (.0015″ to .003″ or metric oversize). They can be used in reamed pistons and rods to correct wear.

Connecting Rod Service

Connecting rods are subjected to tremendous force during engine operation. As a result, they can wear, bend, or even break. The old piston and bearing inserts will indicate the condition of the connecting rod. If any piston or bearing wear abnormalities are found, there may be a problem with the connecting rod.

For example, if one side of a bearing is worn, that connecting rod may be bent. If the back of a bearing insert has marks on it, that rod big end may be distorted, allowing the bearing insert to shift inside the rod.

Rod Small End Service

Measure the rod small end with a telescoping gauge and a micrometer, **Figure 21-53A.** Make at least two measurements at 90° to each other. Then, compare the measurement to specifications. If worn beyond specifications, have a machine shop replace the rod bushing, **Figures 21-53B** and **21-53C.** The bushing will need to be reamed so the pin is fitted in the rod for proper clearance.

Rod Big End Service

To check the connecting rod big end for problems, remove the bearing inserts and bolt the rod cap to the rod. Torque the fastener to specifications without bending or damaging the rod. Then, use a telescoping gauge and micrometer to measure the rod bore diameter, **Figure 21-54.** Take two measurements on each edge at 90° to each other.

The difference in diameter between bore edges is rod big end taper. Any difference in the cross diameters on a single edge is rod big end out-of-round. If taper or out-of-round are greater than specifications, have a machine shop rebuild the rod or purchase a new rod.

Checking Rod Straightness

To determine if a rod is bent, a special rod alignment fixture is needed. See **Figure 21-55.** It can be used to check whether the rod small end and big end are perfectly parallel. Use the operating instructions provided with the particular fixture or send the rods to a machine shop. Replace any bent rods.

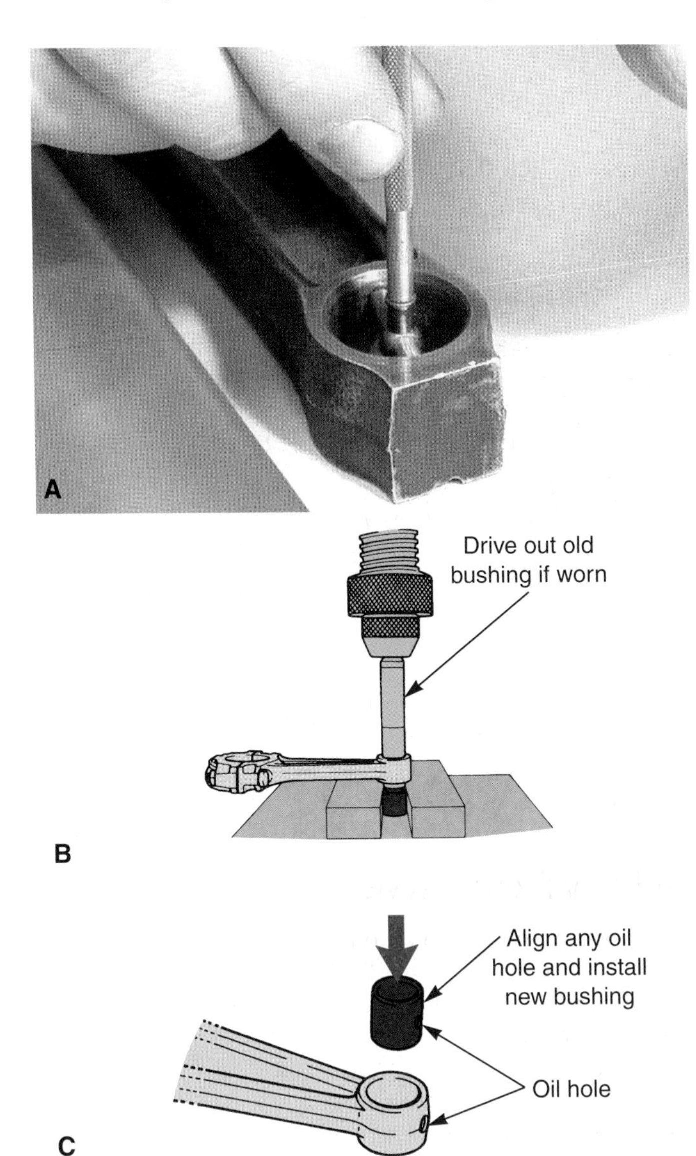

Figure 21-53. *A—Use a telescoping gauge and micrometer to measure the rod small end diameter, which can be used to calculate wear. B—Using a press to push out the old bushing. C—When driving in the new bushing, make sure any oil holes align. (Toyota)*

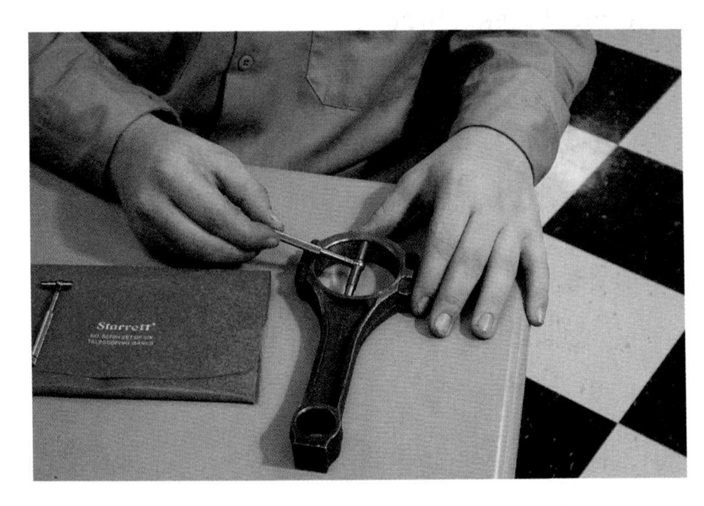

Figure 21-54. *Use a telescoping gauge and micrometer to measure the diameter of the rod big end bore.*

Rebuilding Connecting Rods

Connecting rods can be rebuilt by most machine shops. This should be done when inspection or measurement shows major problems. Sometimes, only one rod needs to be rebuilt, as when a bearing spins and damages the rod big end bore. In a high-mileage engine, all of the connecting rods may require rebuilding. The following operations are done when a connecting rod is rebuilt.

- ❑ Machine the big end bore diameter to specifications.
- ❑ Replace the small end bushing.
- ❑ Check rod bore alignment and correct if needed.
- ❑ Install new rod bolts, if needed.

Installing Piston Rings

With the cylinders, pistons, and connecting rods all in proper working condition and the piston, rod, and piston pin assembled, you are ready to install the new piston rings

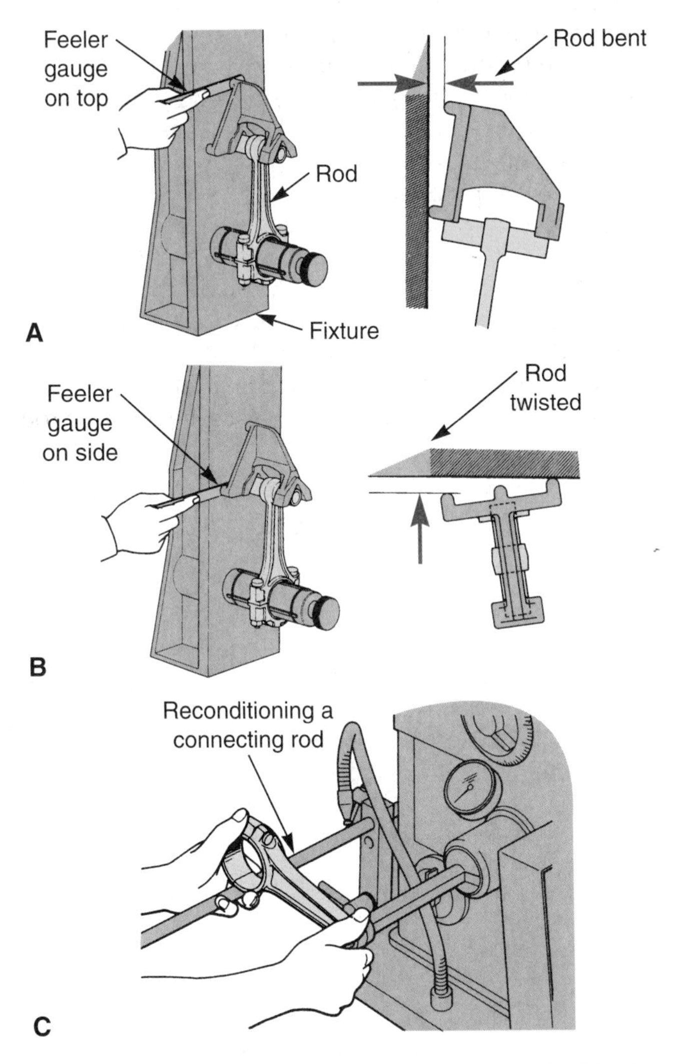

Figure 21-55. *Using a special rod alignment fixture. A—Checking for a bent rod. B—Checking for rod twist. C—Honing the rod small end bore. (Toyota)*

on the pistons. Make sure you checked ring gap in the block as explained earlier.

Installing Oil Rings

Install the steel oil ring first, which consists of two rails and one expander-spacer. To hold the piston, lightly clamp the rod in a vise. Use wooden blocks or brass or lead vise caps to prevent rod damage. The bottom of the piston should touch the top of the vise to keep the piston from swiveling.

1. Wrap the expander-spacer around its groove. Butt the ends together.
2. Insert one end of one oil ring rail in the groove under the expander-spacer.
3. In a spiral manner, carefully work the ring into the groove around the piston, **Figure 21-56A.**
4. Insert one end of the second oil ring rail in the ring groove above the expander-spacer.
5. Work the ring into the groove by feeding it down into the spacer and piston ring groove, **Figure 21-56B.**
6. Double-check that the expander-spacer is not overlapped. Its ends *must* butt together, **Figure 21-56C.**
7. Make sure the oil ring assembly will rotate on the piston. There should be a moderate drag as the oil ring assembly is turned.
8. Rotate one of the oil ring rails so that its ring gap is almost aligned with the end of the piston pin. Rotate the other rail so that its gap is at the opposite end of the pin. If the oil ring gaps align, oil can blow through the gaps.

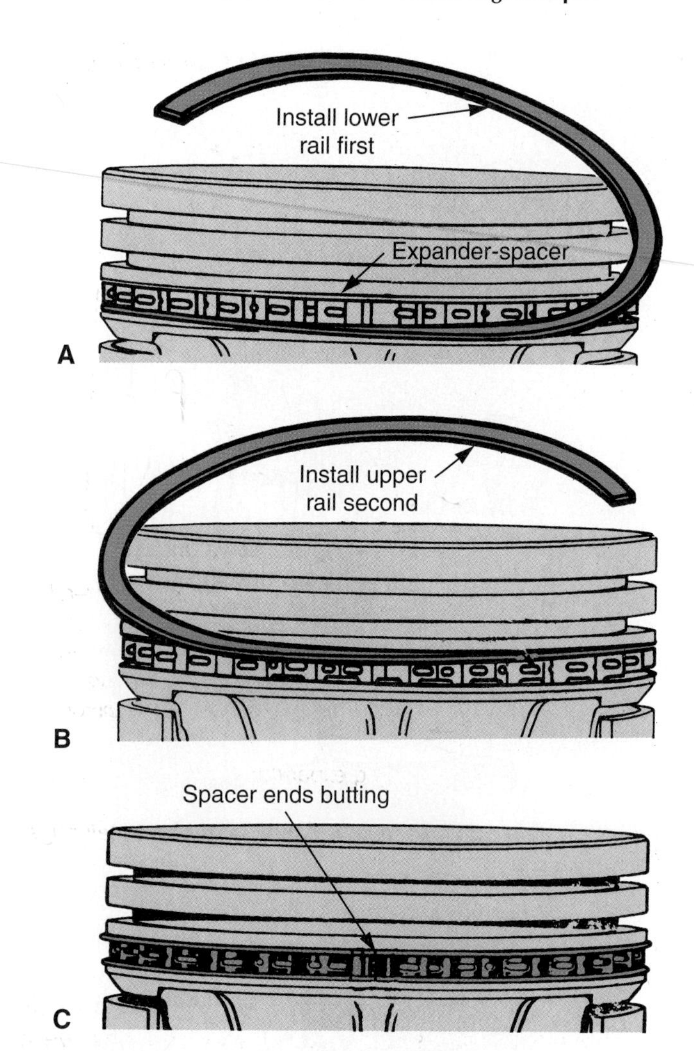

Figure 21-56. *Installing the oil ring onto the piston. A—Install the expander-spacer; make sure the ends are butted. Then, install the lower rail. B—Install the top rail. C—Double-check that expander-spacer ends are not overlapped.*

Installing Compression Rings

Read the instructions supplied with the new piston ring set. There are usually hints on proper ring installation. The shape of compression rings can vary. The shape is designed to help hold the ring against the cylinder wall for a proper seal.

Most compression rings have a top and a bottom. If installed upside down, ring failure or pressure or oil leakage may result. Ring markings are usually provided to show how compression rings should be installed. The markings show the top of each ring and which ring goes into the top or second piston groove. Refer to **Figure 21-57.**

Using a piston ring expander, slip the compression rings into their grooves. A ***ring expander*** is a special tool for spreading and installing rings. See **Figure 21-58.** If a ring expander is not available, carefully use your fingers. Without overexpanding the rings, carefully insert each compression ring into the ring grooves.

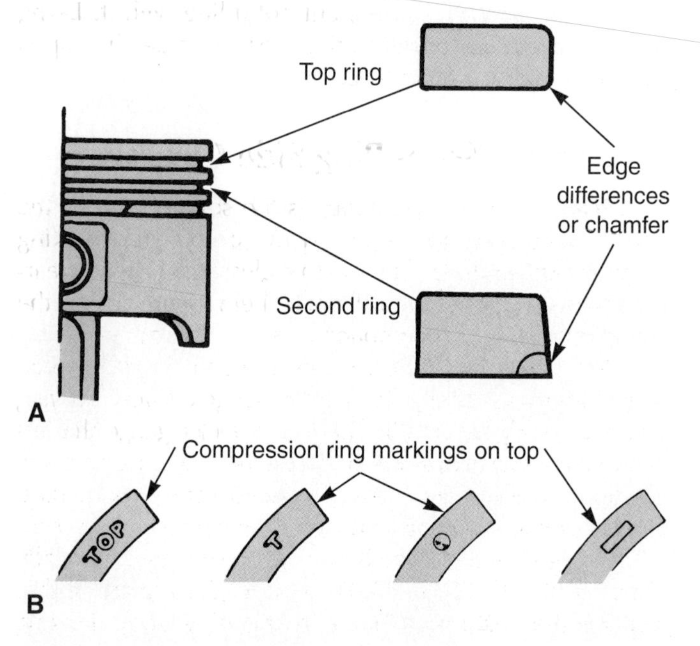

Figure 21-57. *A—Examples of different compression ring shapes. B—Marks indicate the top of the compression ring, when needed. (Honda, Ford)*

Caution: Compression rings are made of very brittle cast iron. They will easily break if expanded or twisted too much.

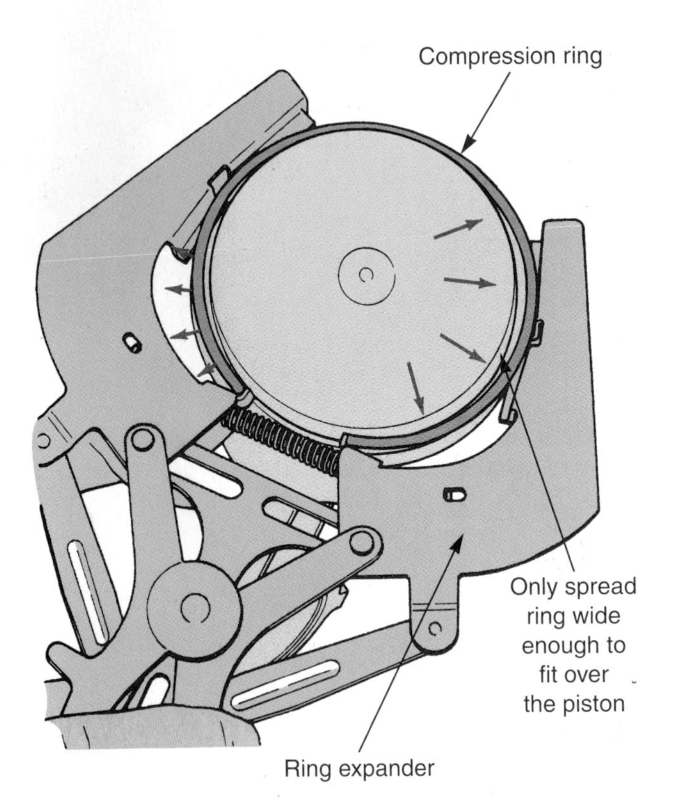

Figure 21-58. *If available, use a ring expander to install rings. (DaimlerChrysler)*

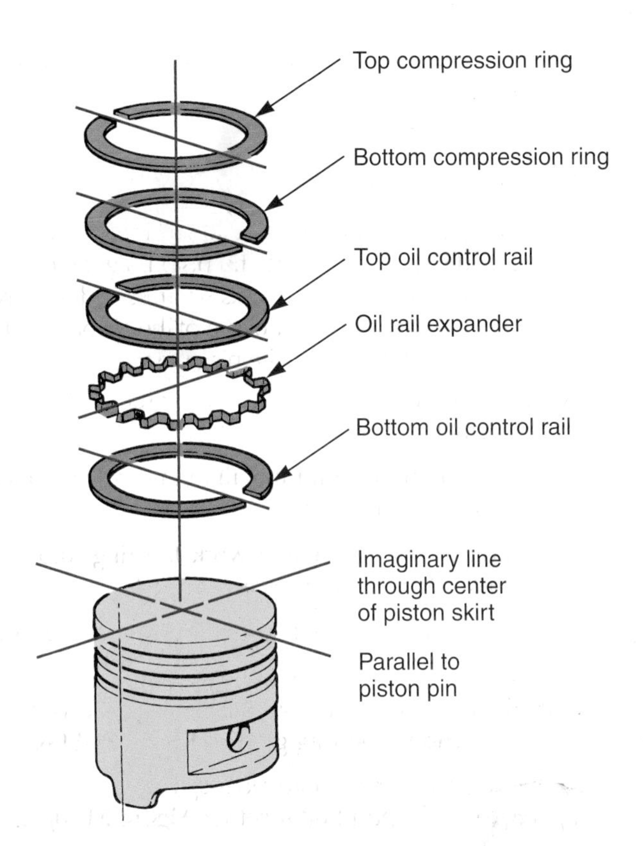

Figure 21-59. *Space piston ring gaps away from each other as shown. Compression ring gaps should be spaced 180° apart to minimize blowby through the gaps. (DaimlerChrysler)*

Piston Ring Gap Spacing

A specific piston ring gap spacing is normally recommended. This is to reduce blowby and ring wear. **Figure 21-59** shows a typical method.

Try to position each gap directly opposite the one next to it. This provides maximum distance between each gap for minimum pressure leakage. The gaps should also be close to the piston pin hole, but not inline with it. Being next to the pin hole reduces ring wear because the gap is not on a major thrust surface.

Measuring Piston Ring Side Clearance

Piston ring side clearance is the space between the side of a compression ring and the piston groove. Ring groove wear tends to increase this clearance. If this clearance is too large, the ring will not be held square against the cylinder wall. Oil consumption and smoking can result.

To measure ring side clearance, install the new piston ring in its groove. Then, slide a feeler gauge between the ring and groove, **Figure 21-60.** The largest feeler gauge that fits between the ring and groove indicates the ring side clearance. The top ring groove is usually checked because it is exposed to more combustion heat and wear than the second groove.

If ring side clearance is beyond specifications, usually about .002″ to .005″ (0.05 mm to 0.13 mm), either replace the pistons or have a machine shop fit ring spacers into the grooves. ***Ring spacers*** are thin, steel rings that fit below the compression rings. The piston groove is machined wider to accept the spacer. This will restore ring side clearance to within limits. **Figure 21-61** shows a tool for cutting ring grooves wider to accept ring spacers.

Installing the Piston and Rod Assembly

To install the piston and rod assembly into the engine, first fit the rod bearing inserts into the connecting rods,

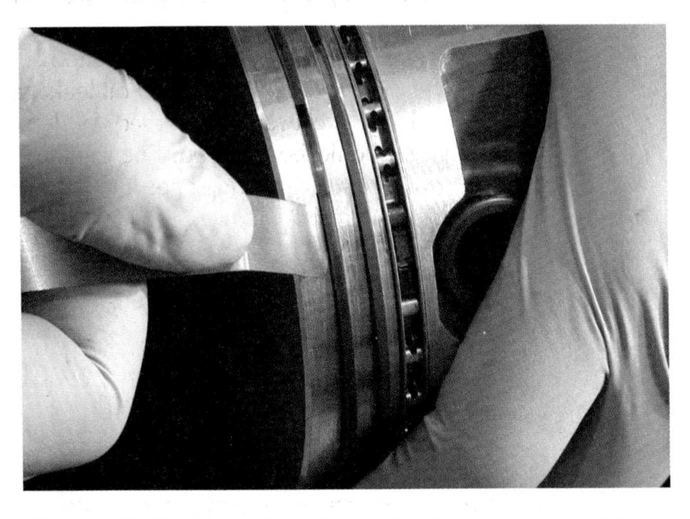

Figure 21-60. *Measuring ring side clearance. The thickest feeler gauge that fits between the ring and groove indicates the clearance.*

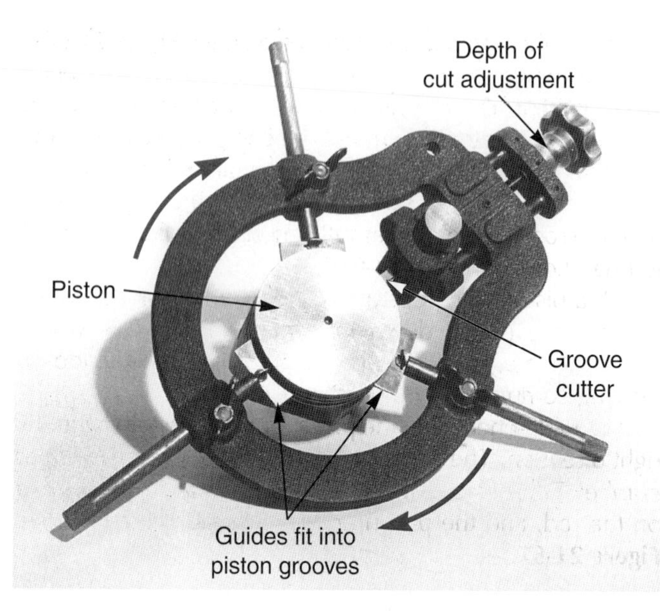

Figure 21-61. *This tool can be used to cut ring grooves wider. The tool is rotated around the piston. (K-Line Tools)*

Figure 21-62A. Fit the matching inserts into the correct caps. The rod bores and backs of the bearing inserts must be perfectly clean and dry (*not* oiled). Make sure the tang on each bearing insert fits down into the notch in the rod and rod cap, as shown in **Figure 21-62B.**

You should also double-check the connecting rod bolts. Make sure the threads are in perfect shape, **Figure 21-63.** Also, make sure the bolts are fully seated in the rod body, if applicable.

To prevent part damage during initial engine start-up, you must lubricate all moving parts on the piston and rod assemblies. Apply heavy oil or assembly lube to the piston rings, piston skirts, piston pin, and faces of the rod bearings. This is illustrated in **Figure 21-64.** Oil is applied on the piston rings and skirts because they can heat up quickly on initial startup if not prelubricated. Some technicians like to dip the head of the piston and new rings into a container of clean engine oil. Double-check that the ring gaps are still spaced properly. Make sure the face of the rod bearing is coated with assembly lube.

Clamp a ring compressor around the rings, as in **Figure 21-65.** While tightening the compressor, hold the compressor square on the piston. The small dents or indentations around the edge of the compressor should face the bottom of the piston. The piston and rod assembly is now ready to install into the block.

Protecting the Crankshaft

Slide special rod bolt covers over the connecting rod bolts. Plastic or rubber hoses can be used if covers are not available. This will prevent the rod bolts from scratching the crankshaft journals.

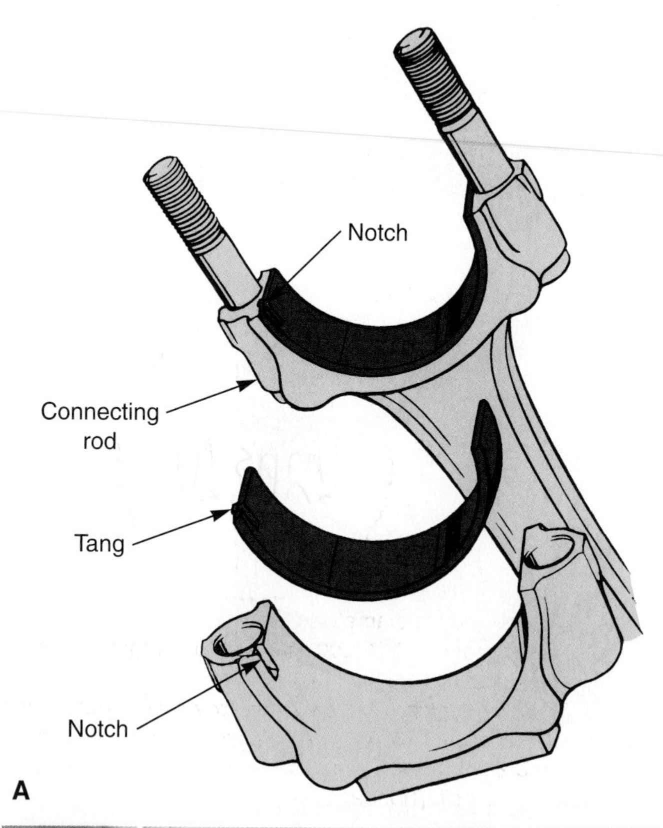

Figure 21-62. *A—Fit clean, dry rod bearings into the connecting rod body and cap. Make sure bearing is square in the bore. B—Make sure the tang on the bearing insert is fully down into its notch in the rod and cap. (General Motors)*

Installing Piston and Rod in Block

Turn the crankshaft until the corresponding rod journal is at or close to BDC. Double-check the markings on rod and piston. Make sure the rod is facing the right direction and that the cap number matches the rod number. Also, check that the piston notch or arrow is facing the front of the engine. For example, the number one rod normally goes in the number one cylinder with the piston marking to the front. Check the service manual if in doubt.

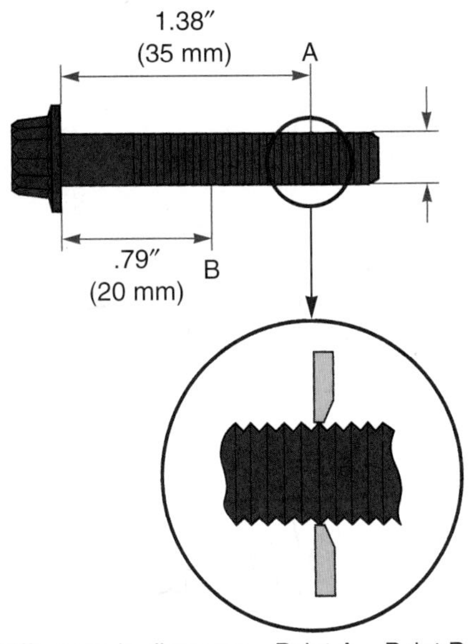

Figure 21-63. *Checking rod bolts for excessive stretch and distortion. Use a micrometer to measure the bolt diameter at Points A and B. If the difference is greater than the specification, replace the bolt. (Honda)*

Place the piston and rod into its cylinder, **Figure 21-66A.** The ring compressor should be square against the block deck. While guiding the rod bolts over the crankshaft with one hand, tap the piston down into the engine, **Figure 21-66B.** A soft wooden or plastic hammer handle will not mark or damage the head of the piston. Keep tapping on the piston until the rod bearing makes full contact with the crankshaft journal. Look at **Figure 21-66C.**

If a piston ring pops out of the compressor, do not try to force the piston down into the cylinder. This would break or damage the piston rings or piston. Instead, loosen the ring compressor and start over.

Double-check that the piston and rod are facing in the right direction. The rod number should match the cylinder number. The number on the cap should match the number on the rod, and the piston arrow must face correctly. See **Figure 21-67.**

Checking Rod Bearing Clearance

To measure connecting rod bearing clearance, use Plastigage. Place a bead of Plastigage across the crankshaft rod journal, **Figure 21-68A.** Assemble the rod cap and bolts. When installing a connecting rod cap, make sure the rod and cap numbers are the same. If the rod is numbered with

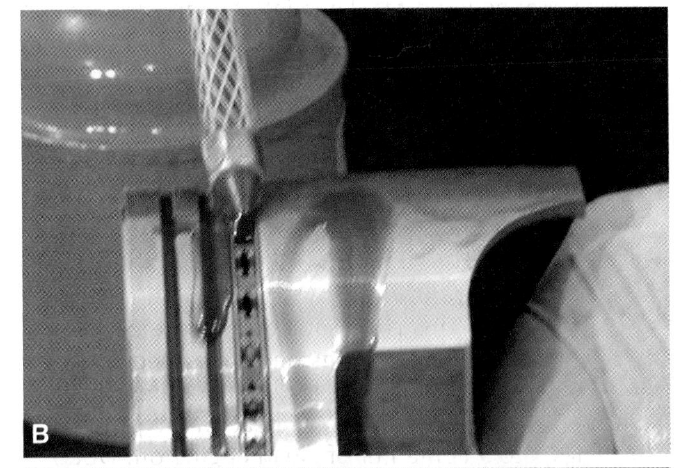

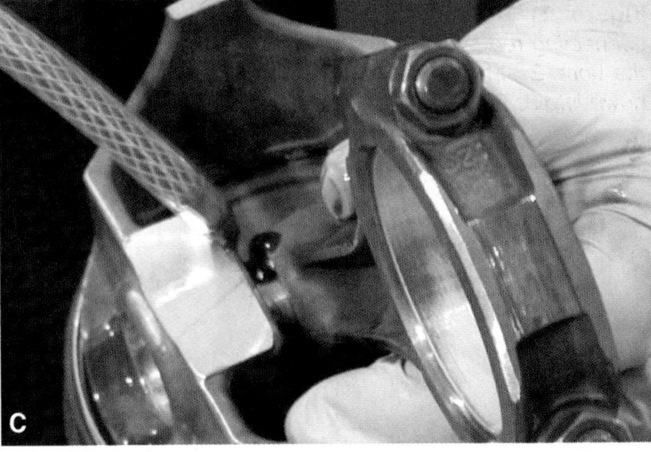

Figure 21-64. *Lubricate all moving parts to prevent part damage during initial engine start up. A—Piston skirts. B—Piston rings. C—Piston pin. Slide the rod back and forth in the piston as you lubricate the pin. D—Rod bearing faces.*

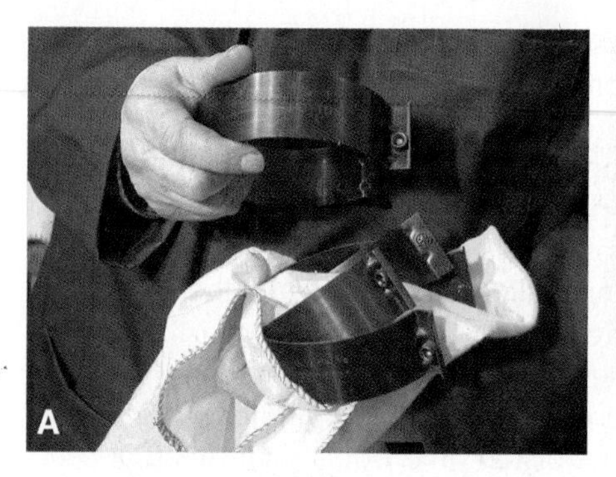

Figure 21-65. *A—Select the correct diameter ring compressor to fit over the piston and rings. B—Make sure the arrow on the ring compressor faces down. C—Tighten the ring compressor enough to compress the rings.*

a 5, then the cap should also have 5 stamped on it. Mixing up rod caps can damage the bearings or crankshaft. Then, torque the rod cap bolts, as described in the next section.

Next, remove the cap and compare the smashed Plastigage to the paper scale. The width of the smashed Plastigage determines the bearing clearance, **Figure 21-68B.** Typical rod bearing clearance is about .0015″ (0.038 mm). This would equal a 1.5 reading on the Plastigage paper scale.

Torquing Connecting Rods

It is very important for you to properly torque each rod nut or bolt to specifications. Connecting rod bolts are the most-stressed fasteners in an engine. At high engine speeds, rod bolts must withstand thousands of pounds of force as the rod changes direction. If a rod bolt or nut is overtightened, the rod bolt could snap off during engine operation. If a rod is under-tightened, the bolts could stretch under load allowing the bearing to spin or hammer against the crankshaft. Severe block, crankshaft, piston, and cylinder head damage could result in either case.

After checking the bearing clearance, lubricate the rod bearing, as shown in **Figure 21-69A.** Also, lubricate the crankshaft journal, **Figure 21-69B.** Next, install the rod cap, **Figure 21-69C.** Make sure the cap numbers match the rod number. Also, make sure the tangs on the bearing face in the right direction.

If the bolts are used, place locking compound on the threads. If nuts are used, make sure they face in the right direction. The shiny side of reused rod nuts must face down. Then, tighten the bolts or nuts hand tight. Make sure the cap is properly seated on the rod and journal.

Using a torque wrench, tighten each connecting rod fastener a little at a time. First, tighten all rod fasteners to about 1/2 of the specified torque. Then, tighten all rod fasteners to about 3/4 of the specified torque. Next, fully torque all connecting rod fasteners. Finally, retorque all rod fasteners at full torque, **Figure 21-70A.** This procedure will pull the rod cap down squarely and securely on the connecting rod body.

The engine in many late-model vehicles have torque-to-yield specifications for the rod bolts. This process is the same as discussed earlier in this chapter. First, tighten the rod fastener to the specified torque. Then, install the degree gauge on the wrench and tighten each rod fastener an additional number of degrees. See **Figure 21-70B.**

Note: Always refer to the service manual when assembling an engine. Torque specifications, clearances, torque-to-yield, and other procedures can vary with engine design.

Checking Rod Side Clearance

Connecting rod side clearance is the distance between the side of the connecting rod and the cheek of the crankshaft journal or side of the other rod. To measure rod side clearance, insert different size feeler gauge blades into the gap between the rod and the crankshaft, as in **Figure 21-71.** The largest feeler blade that slides between the rod indicates side clearance.

Compare your measurements to specifications. Connecting rod side clearance specifications vary from .005″ to .020″ (0.13 mm to 0.51 mm). If not within specifications, the crankshaft thrust journal or connecting rod width is incorrect. The engine must be disassembled and the crankshaft replaced or welded and reground to correct the rod side clearance problem.

Engine Balancing

Engine balancing, or ***crankshaft balancing,*** is needed when the weight of the pistons, connecting rods, or crankshaft is altered by installing new parts of a different weight. For example, if new, oversize pistons are installed and they weigh more than the old, standard pistons, engine (crankshaft) balancing may be required. If the engine is not balanced, severe engine vibration would result and the engine may be damaged. Most large, automotive machine shops have engine balancing equipment.

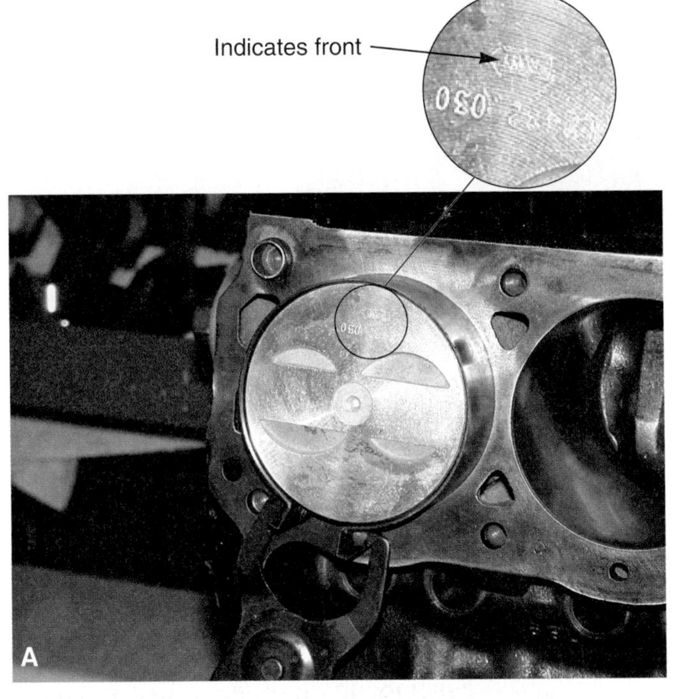

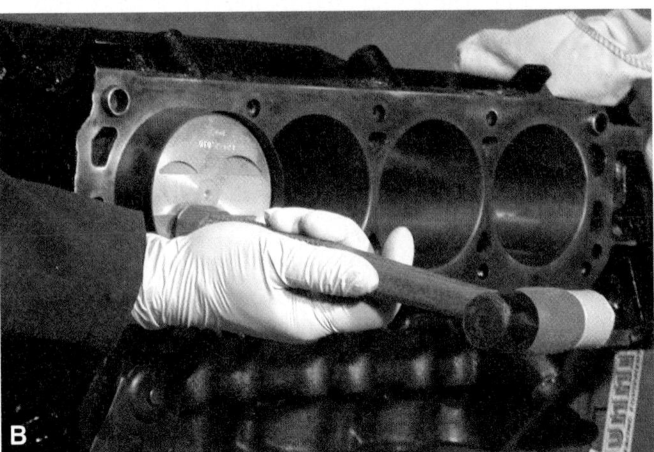

Figure 21-66. *Installing the piston and rod assembly. A—Slowly slide the piston and rod down into the block. The arrow on the piston must face the front of the engine. B—Use a hammer handle to tap the piston and rod down into the block. C—While tapping the piston down, guide the connecting rod over the crankshaft journal.*

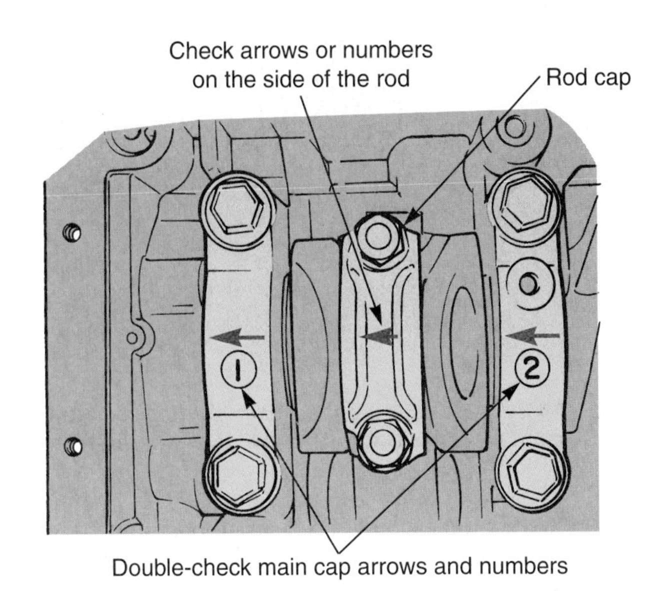

Figure 21-67. *Check that the rod caps and main caps are installed properly. The numbers must match and arrows must point in the correct direction. (General Motors)*

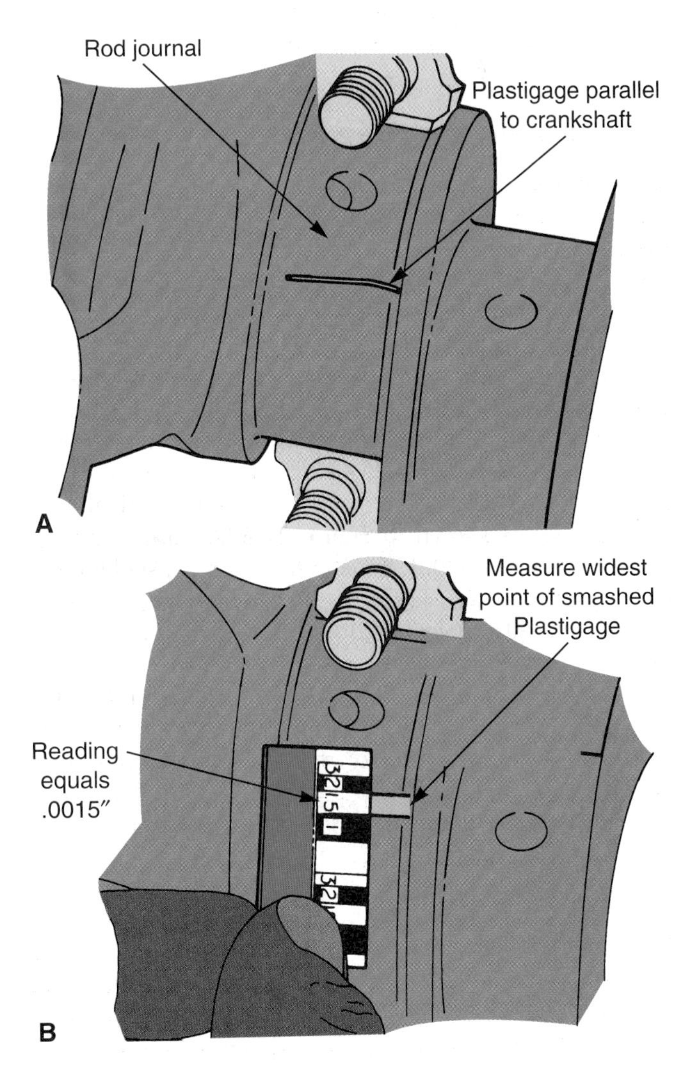

Figure 21-68. *Checking rod bearing clearance. A—Place a piece of Plastigage on the journal. Install the cap and torque the bolts. B—Remove the cap. Compare the paper scale with the smashed Plastigage. (General Motors)*

Figure 21-69. *Installing a rod cap. A—Apply heavy oil to the face of the rod bearing insert. B—Lubricate the crankshaft journal. C—Fit the rod cap down into place on the rod body.*

The first step in balancing an engine is to weigh each piston and record its weight. See **Figure 21-72A.** All pistons should weigh the same. Material must be removed from heavier pistons to make all pistons the same weight. If needed, metal is machined off of the bottom of the piston pin boss.

Next, weigh each connecting rod assembly and record its weight. The weight of each rod assembly must be the same. All rod big ends and small ends should also be equal in weight. See **Figure 21-72B.** You can machine pads on each end of the rods to equalize their weights as needed.

The weight of the piston and rod assembly is called the ***reciprocating weight.*** To replicate this weight as the crankshaft is being balanced, ***bob weights*** are installed on the crankshaft. The calculated reciprocating weight must account for *all* weight, including bearings, rings, and oil. See **Figure 21-72C.**

After you have calculated the correct bob weights, assemble the bob weights. See **Figure 21-72D.** Double-check the bob weights on a scale. Then, mount the bob weights on the crankshaft. Also, mount the vibration damper and flywheel on the crankshaft

Next, mount the crankshaft assembly in the engine balancing machine. Using the balancing machine controls, spin the crankshaft at the speed required. The balancing machine readout indicates where and how much weight needs to be removed from or added to the crankshaft counterweights. Metal is added to a counterweight by welding and removed by drilling.

Figure 21-70. *A—Tighten the rod bolts to factory torque specifications. B—If the fastener has a torque-to-yield specification, use a degree wheel to tighten it the specified number of degrees after the fastener is torqued.*

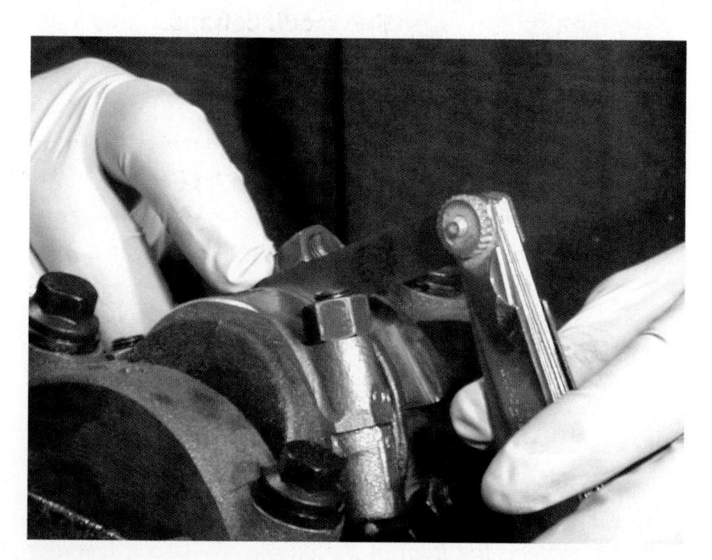

Figure 21-71. *To measure connecting rod side clearance, use a flat feeler gauge. The largest blade that fits indicates the clearance.*

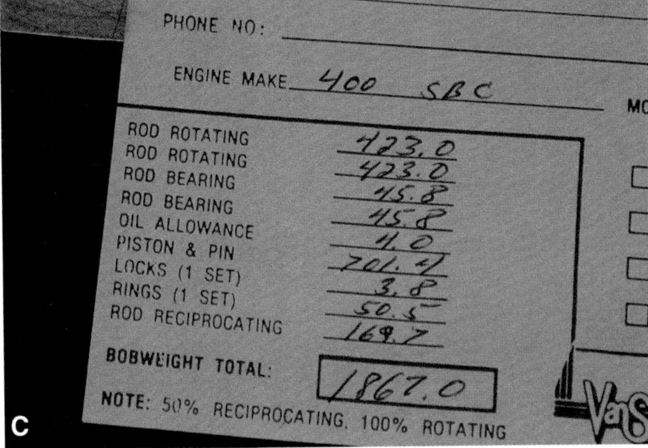

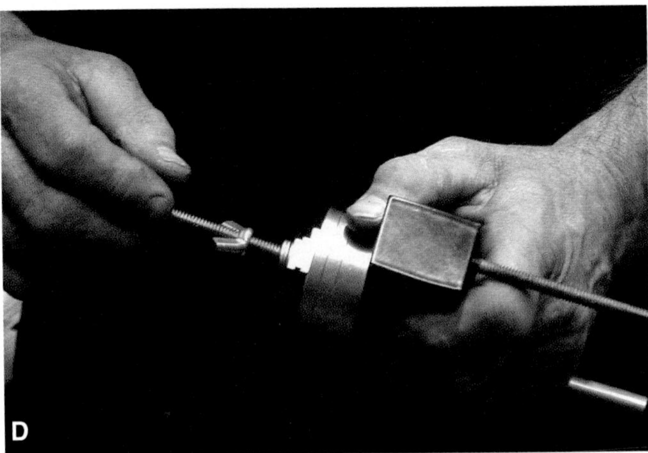

Figure 21-72. *Study the major steps for balancing an engine. A—Weigh the new piston and record its weight. B—Weigh the rod and record your readings. This device measures the big end and small end separately. C—Calculate the reciprocating weight and calculate the needed bob weight. D—Assemble the bob weights to match the calculated weight.*

Note: Proper engine balance is very critical with today's small-displacement, high-rpm engines. Keep engine balancing in mind when making major engine modifications.

Balancer Shaft Service

Balancer shaft service involves the same basic procedures as crankshaft service. You must measure shaft runout, journal wear, and bearing clearance. Also, inspect the shaft for nicks, burrs, and wear. Often, special tools are needed to service the balancer shaft. Check the service manual for specific service procedures.

Use a dial indicator to check for excessive runout, which indicates a bent balancer shaft. See **Figure 21-73.** Mount the shaft on V-blocks. Place the indicator tip on the journals. Rotate the shaft and read the indicator. Replace the balancer shaft if the runout exceeds specifications. The balancer shaft and bearings should be replaced as a set if damaged or worn.

Use a micrometer to determine balancer shaft wear. Measure the diameter of the balancer shaft. Then, subtract the diameter from the specification for original diameter. The difference is the wear. Compare the wear to the amount allowed by specifications.

Also, inspect the balancer shaft bearings for wear. A hole gauge is often used to measure the balancer shaft bearing diameters.

Summary

Inspect the cylinder block closely for problems before assembly. Magnafluxing is a quick way of finding cracks in cast iron blocks. Dye or fluorescent penetrant will find cracks in aluminum blocks. Cracked blocks can sometimes be repaired by welding, special epoxy, plugging, and sleeving.

Main bore alignment and deck warpage are checked with a straightedge and feeler gauge. Machining is needed to correct misalignment or warpage of the block.

Cylinder wear is best checked with a dial bore gauge. However, an inside micrometer or a telescoping gauge and

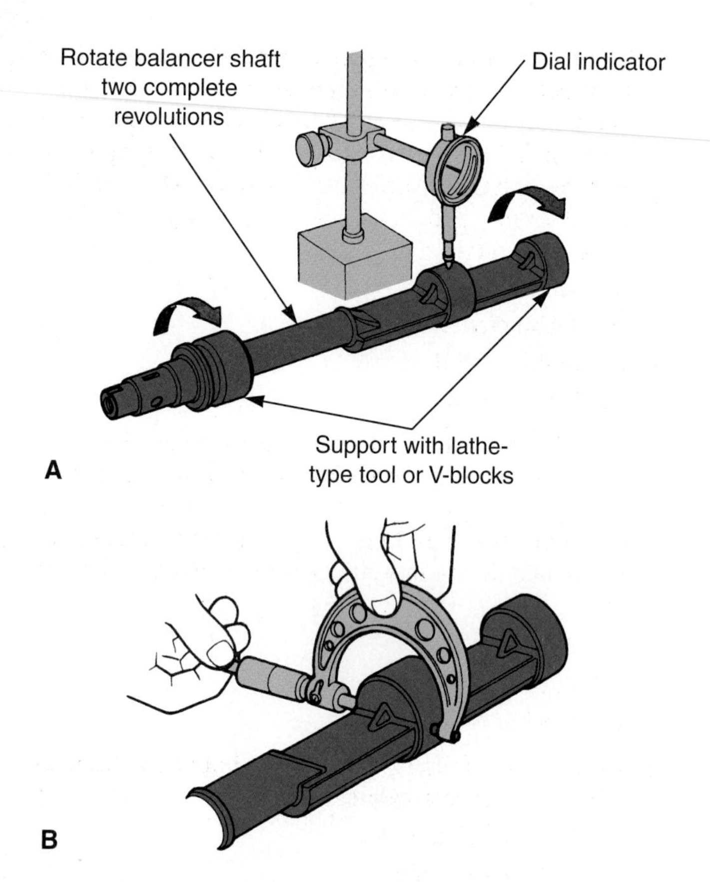

Figure 21-73. *A—Measuring balancer shaft runout. B—Measuring the diameter of the balancer shaft journals.*

outside micrometer can also be used. Check for cylinder out-of-round, taper, and diameter.

Cylinder honing is needed on used cylinders or cylinders that have been bored oversize. Honing produces the correct surface texture for proper ring break-in. Power honing will help remove cylinder taper and out-of-round. Flex or brush honing simply prepares the cylinder surface. A finer grit stone is needed when hard chrome or moly piston rings will be installed. Do not pull a spinning hone too far out of the cylinder or tool breakage can result.

A block should be bored when cylinder wear is beyond specifications. A machine shop can bore the block cylinders oversize. Then, oversize pistons must be installed. The overbore limit depends on the manufacturer; check specifications. Cylinder sleeving can be done to repair a badly worn or damaged cylinder wall. The sleeve can be installed so that the same size piston will fit.

Always clean the block carefully after boring or honing. You must remove all grit so that it cannot circulate through the engine during operation. Use a pressure washer or hand wash the block with soap and water. Wipe the block down with an oil-soaked rag to remove the last of the grit.

Measure the crankshaft journals for wear. Check for general diameter, taper, and out-of-round. Also, check the crankshaft for cracks. If worn too much, have a machine shop grind the crankshaft journals undersize. Then, undersize bearings must be installed. Bearing undersize is usually stamped on the back of the bearing inserts.

Make sure you install the rear main oil seal correctly. The sealing lip must face inside the engine. Use the proper driver to install a one-piece rear seal to prevent seal damage.

Never oil the back of engine bearings. They should be installed clean and dry. Check oil hole alignment. Coat the bearing faces with heavy engine oil or assembly lube. Check bearing clearance with Plastigage. Torque main and rod bearing caps to specifications. Start out at about 1/2 torque and work up to full torque. Recheck final torque at least once. If the specification is torque-to-yield, turn each fastener the specified number of degrees past the torque specification.

Clean piston ring grooves of carbon with a ring groove cleaner or broken ring. Measure the piston diameter with an outside micrometer. Usually, measure just below the piston pin on the skirt. Piston wear is found by subtracting the measured diameter from the specification for diameter. Piston clearance is found by subtracting the piston diameter from the cylinder diameter. Clearance must not be too large or piston slap can result.

Measure piston pins and their connecting rod bores. Make sure the dimensions are within specifications. Replace parts or have the rods rebuilt if needed.

Check the piston ring gap by sliding the ring to the bottom of the cylinder bore with a piston. Then, use a feeler gauge to check the end gap. If the gap is too small, file the ring ends until within specifications, if allowed by the manufacturer.

Install the oil ring on the piston first. Make sure the expander-spacer ends butt together and do not overlap. Use your fingers or a ring expander to install the compression rings. Space the end gaps away from each other and slightly to one side of the piston pin ends.

Ring side clearance is checked between the compression ring and its groove. Slide the correct size feeler gauge in next to the ring. If the groove is too wide, have the grooves machined and install spacers or replace the pistons.

Use a ring compressor to install the piston and rod into the block. Cover the rod bolts to protect the crankshaft. Then, tap the piston down into the block with a hammer handle. Torque rod and main cap bolts properly. Check rod side clearance with a feeler gauge.

Engine balancing may be needed if components are altered or changed. The reciprocating weight is the weight of the piston, rod, bearings, and oil. The counterweight on the crankshaft must balance this weight to prevent engine vibration.

Review Questions—Chapter 21

Please do not write in this text. Write your answers on a separate sheet of paper.

1. When _____, a ferrous powder and magnet are used to find cracks in iron and steel parts.
2. Explain three ways to repair cracks in blocks.

3. How do you check main bore alignment?
4. How do you measure cylinder taper and out-of-round?
5. Generally, a block should be bored if cylinder out-of-round is more than _____ inch or taper is more than _____ inch.
6. The _____ is the maximum amount that a block should be bored oversize.
7. Define *plateau honing.*
8. What is the purpose of deglazing a cylinder?
9. What purpose does a deck plate serve?
10. How do you measure crankshaft main and rod journal wear?
11. Cylinders are normally bored oversize in increments of _____.
12. Define *cylinder sleeving.*
13. Why is it important to clean the engine after honing or boring?
14. Check main bearing clearance with _____.
    (A) a flat feeler gauge
    (B) a dial indicator
    (C) Plastigage
    (D) a micrometer
15. Which direction must the lip on the rear main oil seal point?
16. Describe *torque-to-yield.*
17. How do you measure piston wear?
18. To find piston clearance, subtract the _____ from the _____.
19. Why are the piston ring gaps spaced apart from each other?
20. How do you check rod bearing clearance?

ASE-Type Questions—Chapter 21

1. A cast iron, V-6 block is almost new and has very low mileage. However, a piston skirt broke off and scored one cylinder. All of the other pistons and cylinders are in good condition. Technician A says to bore the block oversize and install six new, oversize pistons to correct the cylinder damage. Technician B says to sleeve the damaged cylinder and buy a new piston. Who is correct?
    (A) A only.
    (B) B only.
    (C) Both A and B.
    (D) Neither A nor B.
2. Technician A says you should torque each main bolt to about 1/2 torque in sequence, then to 3/4 torque, and finally to full torque. Then, torque each bolt once more at full torque. Technician B says to torque the first rod bolt to full torque and then the other rod bolt to full torque, and then check torque one more time. Who is correct?
    (A) Technician A.
    (B) Technician B.
    (C) Both A and B.
    (D) Neither A nor B.
3. Technician A says that checking for main bore and deck warpage is a typical procedure performed during short block service. Technician B says that cylinder head milling is a common procedure performed during short block service. Who is correct?
    (A) A only.
    (B) B only.
    (C) Both A and B.
    (D) Neither A nor B.
4. Which of the following problems should you look for when inspecting a cylinder block?
    (A) Cylinder ring ridge.
    (B) Cylinder scoring.
    (C) Cracked cylinder
    (D) All of the above.
5. Technician A says that block deck warpage is measured with a straightedge and feeler gauge. Technician B says that cylinder wear is measured with an inside micrometer and dial indicator. Who is correct?
    (A) A only.
    (B) B only.
    (C) Both A and B.
    (D) Neither A nor B.
6. Technician A says that a short block should be line bored when misalignment is over approximately .0015″ (0.038 mm). Technician B says that a short block should be line bored when misalignment is over approximately .0001″ (0.003 mm). Who is correct?
    (A) A only.
    (B) B only.
    (C) Both A and B.
    (D) Neither A nor B.
7. Cylinder taper is the difference in cylinder diameter when measured _____.
    (A) from front-to-rear and side-to-side
    (B) at the top and bottom
    (C) at the ridge
    (D) on the deck

8. All of the following could cause the condition shown here *except:*
   (A) the crankshaft was improperly ground.
   (B) the piston skirt is cracked.
   (C) the connecting rod is bent.
   (D) the connecting rod was installed incorrectly.

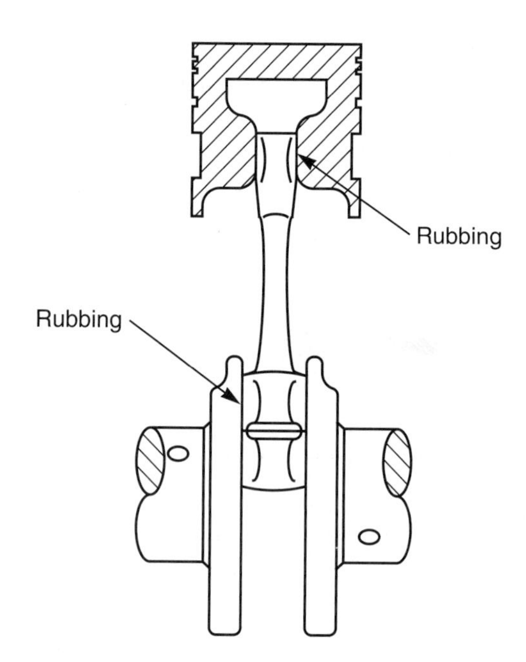

9. Technician A says that .030″ (0.75 mm) is a common oversize to which machine shops bore cylinders. Technician B says that .060″ (1.5 mm) is a common oversize to which machine shops bore cylinders. Who is correct?
   (A) A only.
   (B) B only.
   (C) Both A and B.
   (D) Neither A nor B.

10. Technician A says that block decking reduces engine compression. Technician B says that block boring increases engine compression. Who is correct?
   (A) A only.
   (B) B only.
   (C) Both A and B.
   (D) Neither A nor B.

11. Technician A says that the illustration shows how to check for main bore misalignment. Technician B says that the illustration shows how to use Plastigage™ to check for bearing clearance. Who is correct?
   (A) A only.
   (B) B only.
   (C) Both A and B.
   (D) Neither A nor B.

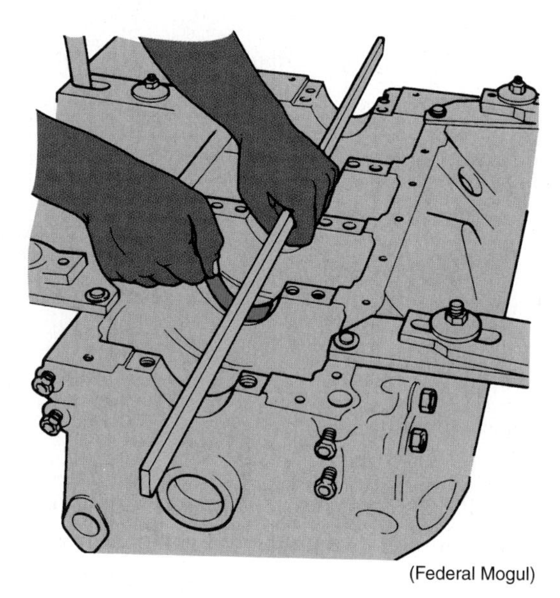

(Federal Mogul)

12. Technician A says that an engine's compression rings are usually made of steel. Technician B says that an engine's oil ring is normally made of cast iron. Who is correct?
   (A) A only.
   (B) B only.
   (C) Both A and B.
   (D) Neither A nor B.

Chapter 22

Top End Rebuilding and Machining

After studying this chapter, you will be able to:

- ❑ Check a cylinder head for cracks, warpage, and other problems.
- ❑ Summarize cylinder head machining operations.
- ❑ Measure valve guide and valve stem wear.
- ❑ Knurl, ream, or replace valve guides.
- ❑ Replace valve seats.
- ❑ Recondition valve seats.
- ❑ Recondition valves.
- ❑ Test, shim, and install valve springs.
- ❑ Properly install valve stem seals.
- ❑ Reassemble a head.
- ❑ Measure camshaft lobe and journal wear, camshaft end play, and camshaft bearing clearance.
- ❑ Service rocker arms and rocker arm shafts.
- ❑ Service push rods and lifters.

Know These Terms

Camshaft bearing clearance
Camshaft bearing wear
Camshaft endplay
Camshaft grinding
Camshaft journal wear
Camshaft lobe wear
Camshaft service
Camshaft straightness
Concentricity
Cracked cylinder head
Cylinder head milling
Dye penetrant
Fluorescent penetrant
Head pressure testing machine
Head warpage
Interference angle
Leak-down rate
Lifter service
Magnafluxing
Pilot
Prussian blue
Push rod service
Rocker assembly service
Seat narrowing
Seat width
Spring bind
Staking
Three-angle valve
Valve grinding
Valve guide wear
Valve retrusion
Valve seat reconditioning
Valve spring free height
Valve spring installed height
Valve spring squareness
Valve spring tension
Valve springs shims

The cylinder head has to seal the top of the block and control flow into and out of the cylinders. With the tremendous heat and pressure of combustion, this is a difficult task. After prolonged service, cylinder heads can warp, valves can burn, valve train parts can wear, and engine performance can suffer.

This chapter outlines the steps for reconditioning a cylinder head and its related parts. You will learn how to return these parts to specifications for maximum engine power, smoothness, and economy. **Figure 22-1** shows some of the things you will learn to check for and correct in this chapter.

Cylinder Head Problems

With the cylinder head disassembled, inspect it closely for problems. Mount the head on a stand and check for cracks, burned seats, burned deck surface between combustion chambers, cracked valve guides, and other problems. See **Figure 22-2.** Keep in mind that if any trouble is not found, your repair could fail in service.

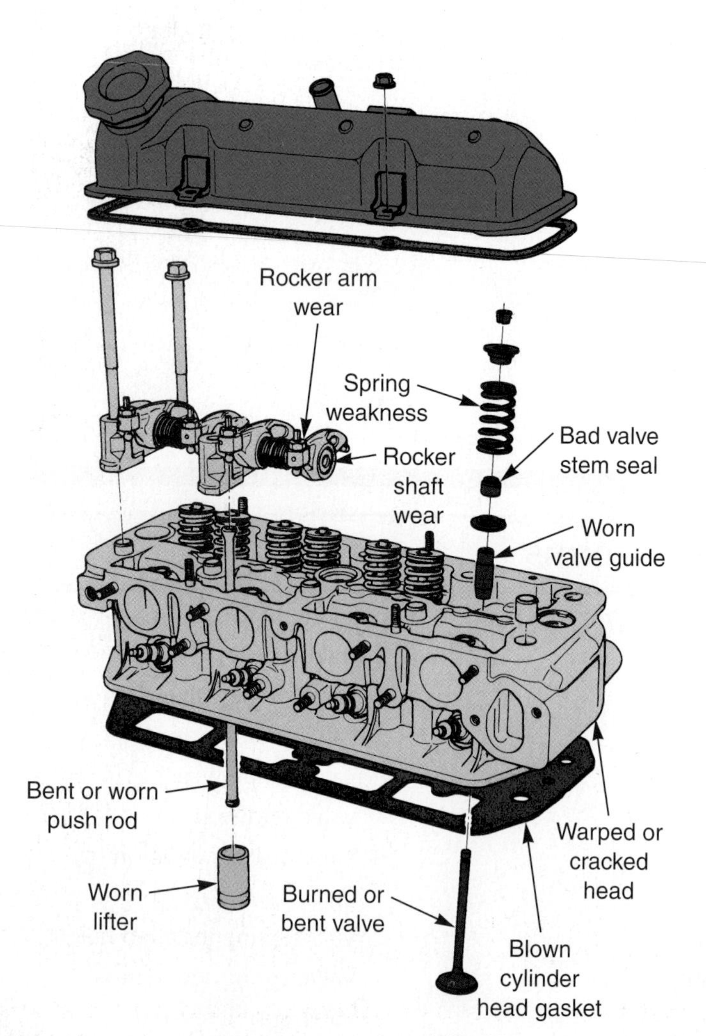

Figure 22-1. *These are some of the parts you will learn to service in this chapter. (DaimlerChrysler)*

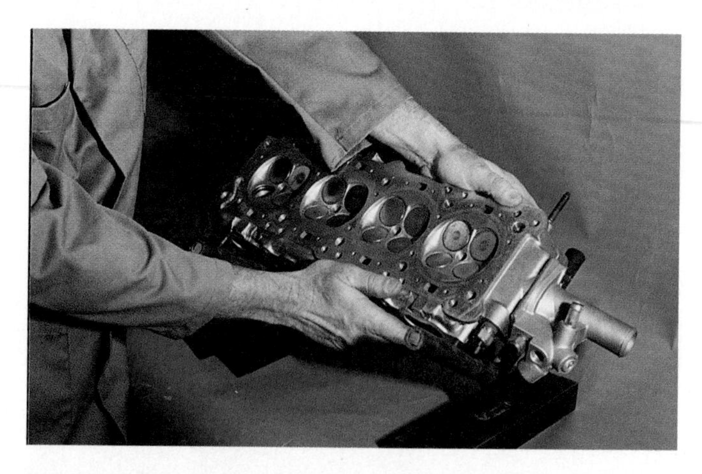

Figure 22-2. *Cylinder head stands are handy and will allow you to position the head in different orientations while working.*

Checking for Cracked Cylinder Head

A ***cracked cylinder head*** can result from engine overheating or physical damage from other broken parts. Usually, the crack happens at an exhaust seat or between chambers, allowing coolant or pressure leakage. This can lead to further engine overheating, coolant in the oil, and other symptoms, as described in earlier chapters. Checking for cracks in a cylinder head is basically the same as checking for cracks in a cylinder block, as detailed in the previous chapter.

Magnafluxing is a common way of finding cracks in cast iron cylinder heads. Pictured in **Figure 22-3**, an electromagnet is placed on the head where a crack might be located. Cracks commonly occur around the exhaust seats and between the cylinders. Then, a squeeze bulb is used to spread fine iron powder over the head. A crack will set up a disruption in the magnetic field that then attracts the powder. The powder collects in the crack and makes it visible.

Magnafluxing will not work with nonferrous materials, such as aluminum. ***Dye penetrant*** or ***fluorescent penetrant*** can be used to find cracks in aluminum cylinder heads. You must follow manufacturer instructions since procedures vary. Refer to **Figure 22-4.**

Some larger machine shops have a cylinder head pressure testing machine. A ***head pressure testing machine*** seals all oil and coolant passages and then pressurizes the head to check for leakage. Basically, a rubber gasket is clamped down on the deck surface to cover and seal passages through the cylinder head. Then, air is forced into the water and oil jackets in the head. The assembly is lowered into a vat of water. Any leakage, such as from cracks or casting flaws, will be evident by air bubbles in the tank of water. Study **Figure 22-5.**

Repairing Cracked Heads

When a cylinder head is cracked, it can either be welded, plugged, or replaced. Plugging involves drilling a series of holes and inserting metal plugs to fix the crack. Replacement is usually the most cost-effective fix. With expensive heads, however, welding may be desirable.

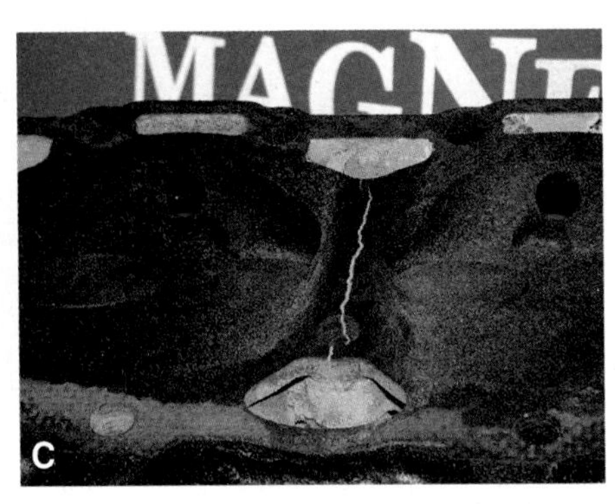

Figure 22-3. *Magnafluxing a head. A—The electromagnet is placed over the combustion chambers and then energized. B—Iron particles are spread over the head using a squeeze bulb. C—Any crack will be highlighted by the iron particles.*

The welding of cast iron heads normally requires a specialty shop. The head should be heated in a furnace. Then, an Ni (high nickel) welding rod is used to repair the crack. Aluminum heads can be welded using a TIG welder. Preheating in a furnace is usually not needed with aluminum.

Caution: Crack repair in heads and blocks should be done by an expert. Special welding skills are essential.

Figure 22-4. *Using dye penetrant to find cracks in an aluminum head. A—Spray dye onto the head surface and allow it to soak in for a few seconds. B—Wipe off excess dye penetrant with a clean shop towel. C—Spray developer onto the head. D—The developer makes any dye that has collected in any cracks, like this one around the spark plug hole, show up as a bright red (or other color). (Van Senus Auto)*

Figure 22-5. *Pressure testing a head to check for cracks. A—The head is mounted in the machine and ready for testing. All oil and coolant passages are sealed. B—These are rubber gaskets used to seal the head. C—A metal flange is placed over the rubber gasket so that the clamp pressure will seal all oil and coolant passages. D—After the head is lowered into the vat of water, air pressure is injected into the head. (Sunnen)*

Measuring Head Warpage

A head that is warped has actually become slightly curved from overheating. ***Head warpage*** is more common with modern, thin-cast heads, especially aluminum heads. A warped head can lead to head gasket failure because of uneven clamping pressure on the gasket. This can cause coolant leakage into the cylinder and oil, compression loss, and other troubles. Before checking for warpage, all of the gasket material must first be cleaned off of the head deck surface, **Figure 22-6.**

1. Peel off the old gasket by hand. A gasket scraper or putty knife can be used.
2. Use a single-edge razor blade to remove the last bits of the gasket. Be careful not to scratch the surface of an aluminum head.
3. A good way to final clean the head deck surface is by using a hand sanding board with 180 to 220 grit sandpaper. Block sanding the head surface with a medium grit sandpaper will remove gasket residue and level any minor surface imperfections.
4. Lay a straightedge across the cylinder head deck.
5. Try to slip different feeler gauge blade thicknesses under the straightedge, **Figure 22-7.**
6. The thickest blade that fits under the straightedge indicates the head warpage.
7. Check warpage in different positions across the head surfaces. See **Figure 22-8.**

The most common location for warpage is between the two center combustion chambers. Generally, head warpage should not exceed about .003″ to .005″ (0.08 mm to 0.13 mm). You can also check the exhaust and intake manifold mounting surfaces for warpage in this manner.

On the cylinder head from an overhead camshaft engine, you should also use the straightedge and feeler gauge to check the camshaft bore, **Figure 22-9.** Place the straightedge in the bore and try to slip feeler gauge blades

Figure 22-6. *A—Before checking the head for warpage, its surfaces must be cleaned. B—A good way to final clean the head deck is by hand sanding using a board and 180 to 220 grit sandpaper.*

under it. If warped beyond specifications, over approximately .001″ (0.03 mm), have a machine shop line bore the camshaft bores.

Cylinder Head Milling

Cylinder head milling is a machine shop operation where a thin layer of metal is cut or machined off the head gasket (deck) surface of the cylinder head. It is done to correct head warpage. The major steps for milling a cylinder head are shown in **Figure 22-10.**

1. Mount the head in a machine tool fixture so that the fixture sits squarely on the cylinder head deck surface.
2. Place the fixture-head assembly on the mill table. By first mounting the head in a fixture, it is easier and quicker to get the head squared up on the mill table.
3. Clamp the cylinder head to the mill table.
4. Remove the fixture.
5. The head should be level and square on the mill table. Use a dial indicator to double-check that the head is properly mounted.
6. Adjust the milling machine for the depth of cut.
7. Turn the mill on and engage the feed. The cutter is automatically fed across the cylinder head surface.

Only remove enough material to correct warpage. Milling increases the engine's compression ratio and can affect valve train geometry. It can also affect alignment of the manifold on V-type engines. On a V-type engine, the intake manifold must also be milled if an excess amount of metal is removed from the cylinder head.

With a diesel engine, cylinder head milling is very critical. A diesel engine compression ratio is so high, any milling can reduce clearance volume and increase compression pressures beyond acceptable limits. Some

Figure 22-7. *Checking for cylinder head warpage. A—The technician has placed a straightedge on the deck. B—Then, a feeler gauge is slid under straightedge. The thickness of the largest gauge equals the amount of warpage.*

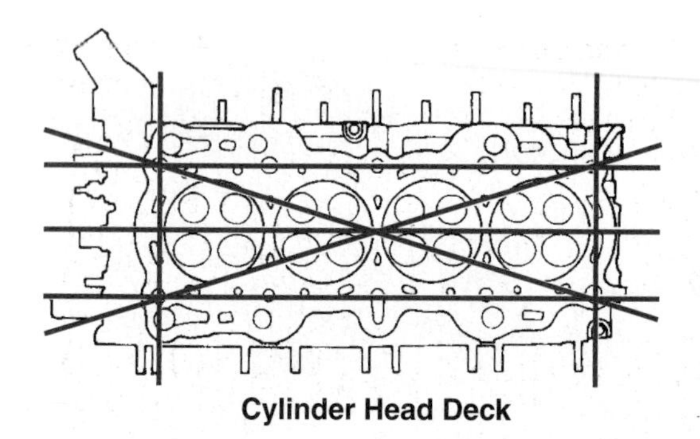

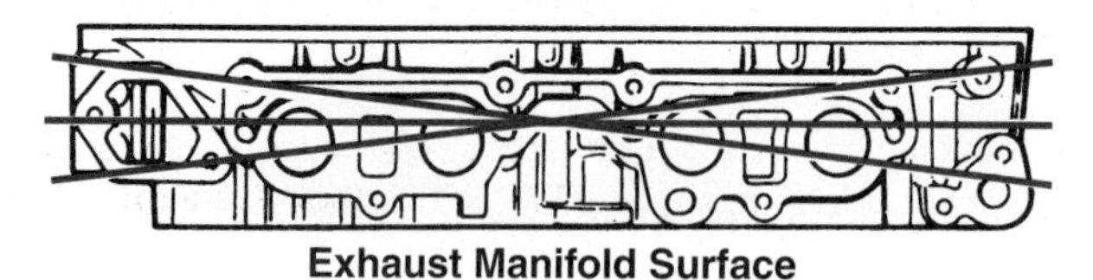

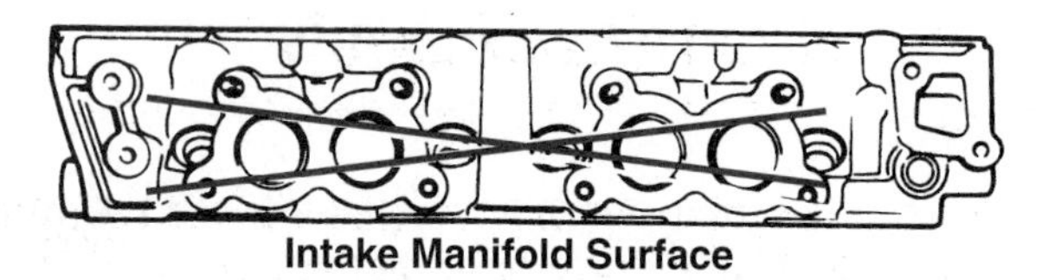

Figure 22-8. *Check the head for warpage at these locations. (Honda, General Motors)*

diesel engine cylinder heads are case hardened and cannot be milled. They must be replaced when warped. Check manufacturer directions when in doubt.

Diesel Precombustion Chamber Service

Automotive diesel engines have precombustion chambers in the cylinder head. These are small cups pressed into the cylinder head The tips of the diesel injectors and glow plugs extend into the precombustion chambers. After prolonged use, the precombustion chambers may require removal for cleaning or replacement.

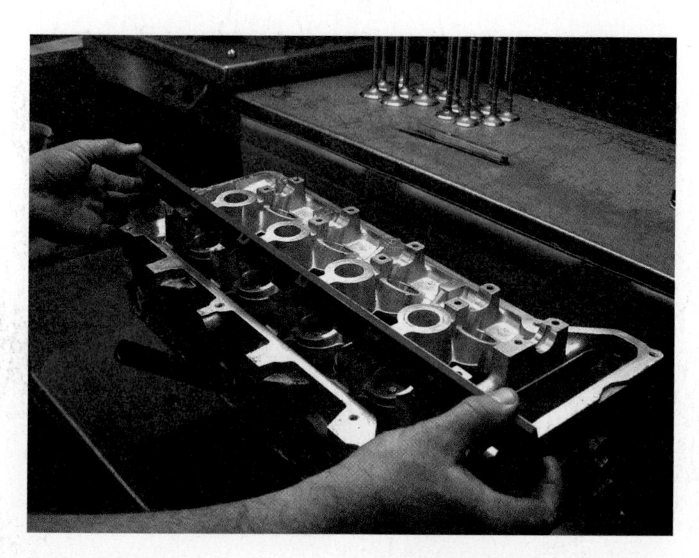

Figure 22-9. *A technician is checking the camshaft bore for warpage.*

Figure 22-10. *Note the major steps for milling a cylinder head. A—The head is first mounted in a fixture, which helps in mounting the head square on the mill table. B—A dial indicator can be used to double-check that the head is mounted properly on mill. C—The head is being machined.*

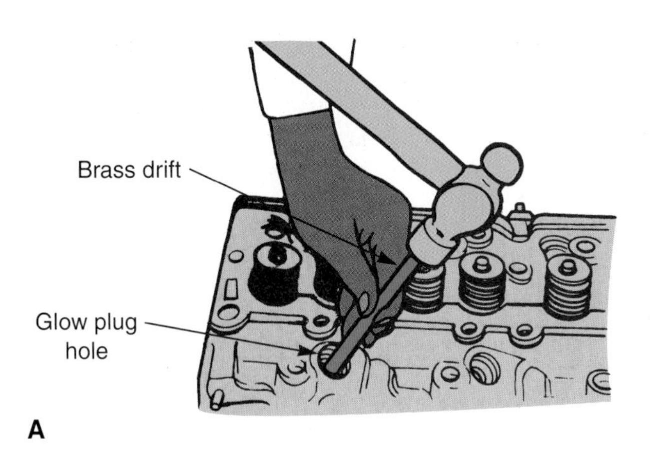

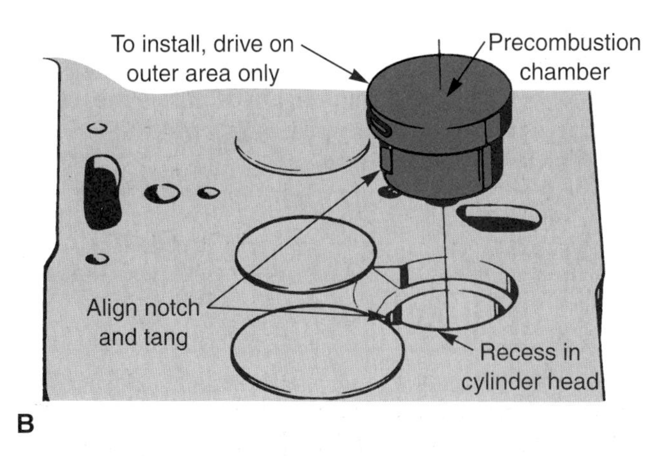

Figure 22-11. *Servicing a diesel precombustion chamber. A—Drive out the existing precombustion chamber through the glow plug or injector hole. B—Align the precombustion chamber and drive it in without damaging it. (General Motors, Fel-Pro)*

Precombustion Chamber Removal

A brass drift and hammer are commonly used to remove a precombustion chamber, **Figure 22-11A.** The drift may be inserted through the glow plug or the injector hole in the head. Light blows with the hammer will drive out the chamber. Compare the dimensions of the new prechamber to those on the old one.

Precombustion Chamber Installation

When installing a precombustion chamber, be careful not to damage the chamber. Use a special driver or brass hammer to tap the cup back into the head, **Figure 22-11B.** Hammer only on the outer edge of the chamber. The center area of the precombustion chamber may be damaged by hammering.

Make sure the precombustion chamber is perfectly flush with the cylinder head deck. If it is not, head gasket leakage can result.

Head Disassembly

Cylinder head disassembly was explained in Chapter 20 and is shown in **Figure 22-12.** Basically, you must:

1. Hit the spring retainers with a plastic hammer to free the keepers from the valve stems.
2. Compress the springs with a valve spring compressor.
3. Use your fingers or a small magnet to lift the keepers out of the valve stem grooves.
4. Release the compressor and remove the spring assembly from the valve.
5. Repeat this procedure on the other valve assemblies.

Keep all parts organized so that they can be reinstalled in the same locations. Valve stems may be select-fit from the factory.

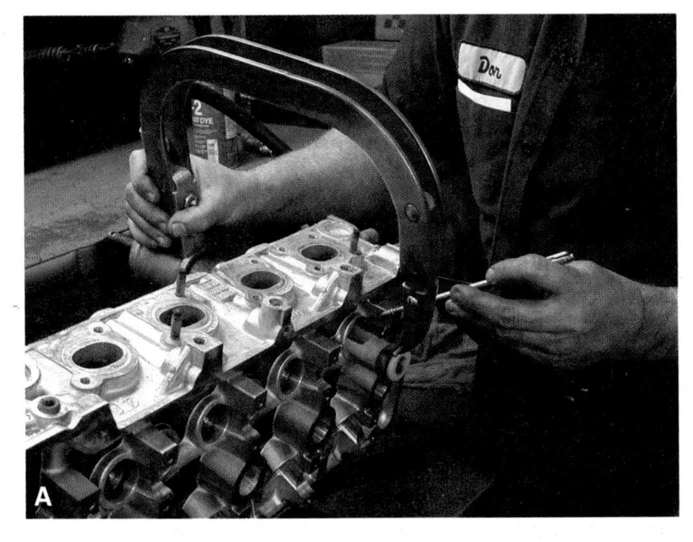

Figure 22-12. *Disassembling a cylinder head. A—Fit the valve spring compressor over the valve head and valve retainer. A—Once the valve spring is compressed, a small magnet is handy for removing the valve keepers.*

Warning: Wear eye protection when using a valve spring compressor. If it slips off the retainer, spring pressure could make parts fly off and injure your eyes.

Valve Guide Service

Excess ***valve guide wear*** allows the valve stem to move sideways in its guide during operation. It is a common problem that can lead to oil consumption, a light knocking or tapping sound, burned valves, or valve breakage. Oil consumption occurs when oil leaks past the valve seal and through the guide. A burned valve occurs because of the poor valve-to-seat seal. **Figure 22-13** shows the types of measurements you must make on the valve and valve guide.

Measuring Valve Guide and Stem Wear

To check for valve guide wear, slide the valve into its guide. Pull it open about one-half inch. Then, try to wiggle the valve sideways. If the valve head visibly moves sideways in any direction, the guide or stem is worn.

Figure 22-14 shows how a small hole gauge and outside micrometer are used to measure guide and valve stem wear. **Figure 22-15** shows how a dial indicator is used to measure valve stem clearance at the stem. If valve stem wear and clearance are not within specifications, part replacement or repair is needed.

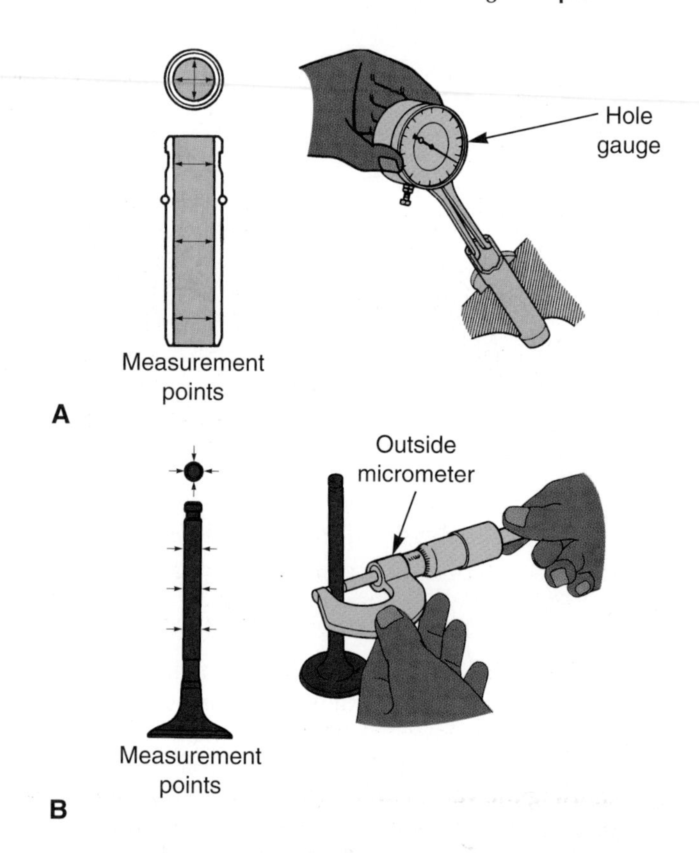

Figure 22-14. *A—Use a hole gauge to measure in these locations. B—Measure the valve stem at these points. Subtract the valve stem diameter from the valve guide diameter to calculate the clearance. (Toyota)*

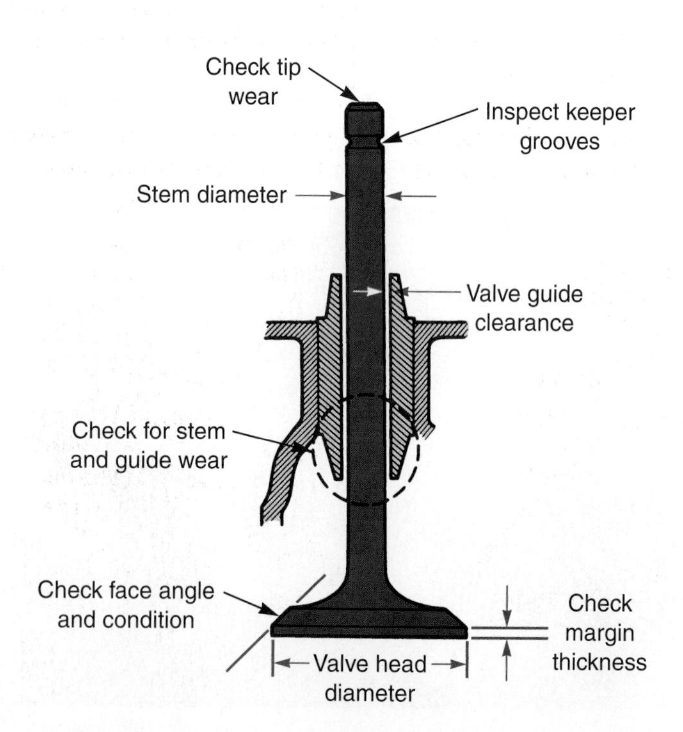

Figure 22-13. *These are the kinds of problems you must look for when servicing valves and valve guides. (DaimlerChrysler)*

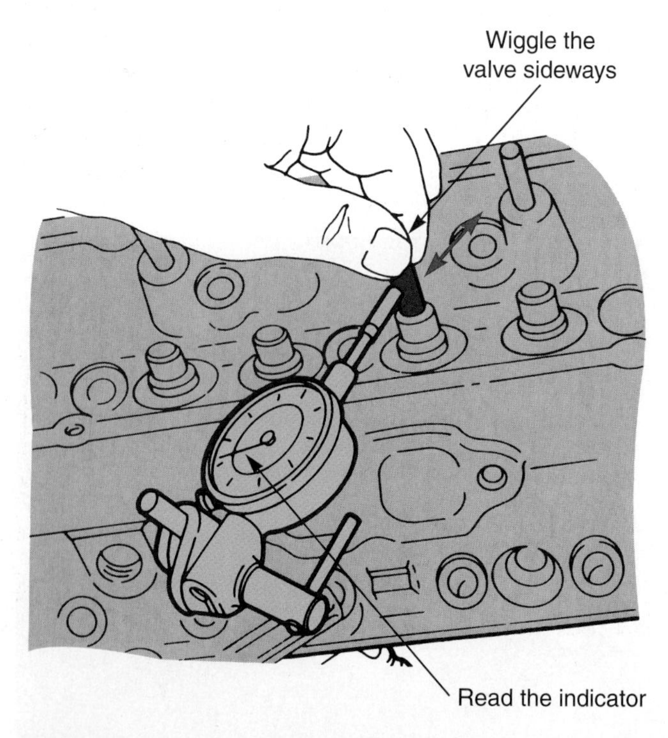

Figure 22-15. *Valve guide wear and clearance can also be measured with dial indicator at the stem or valve head. (General Motors)*

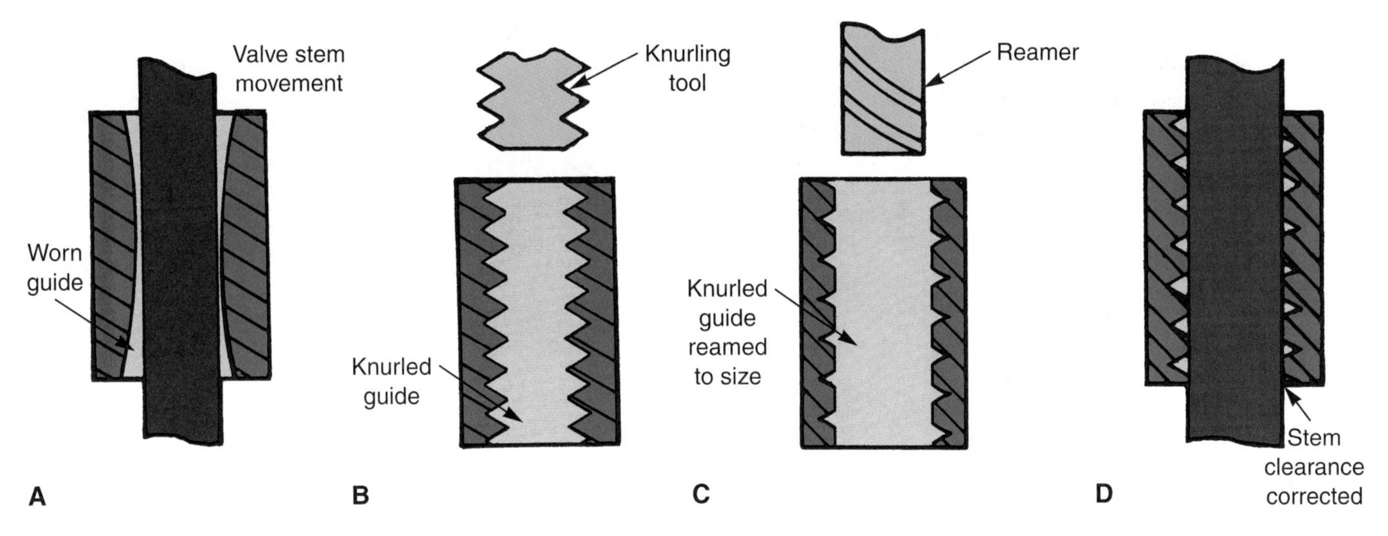

Figure 22-16. *A—The old valve guide is worn and the valve is free to flop sideways. B—A knurling tool is run down through the valve guide, pushing metal outward. C—The valve guide is reamed to the original diameter. D—The valve stem now fits correctly in the valve guide.*

Repairing Valve Guide Wear

There are four common methods used to repair worn valve guides. These include:

- ❑ **Knurling the valve guide.** A machine shop tool is used to press indentations into the guide to reduce its inside diameter, **Figure 22-16.** A knurled guide must be reamed.
- ❑ **Reaming the valve guide.** The guide is reamed to larger diameter and new valves with oversize stems installed. See **Figure 22-16.**
- ❑ **Installing a valve guide insert.** The old guide is pressed or machined out and a new guide is pressed into the head. **Figure 22-17** shows the basic steps for guide replacement with factory pressed-in valve guides. If possible, heat the head in hot water. This will expand the valve guide bore and help free the valve guide. Then, drive the valve guide out from the combustion chamber side of the head. Finally, drive in the new valve guide.
- ❑ **Installing a bronze guide liner.** The old guide is reamed and a thin guide liner is installed.

Most machine shops have a valve guide machine to speed the rebuilding of cylinder head valve guides. This machine is basically a large, precision drill press that can accurately ream the valve guide holes to a larger diameter.

Figure 22-18 shows how to mount and square a cylinder head on a valve guide machine. Place the head on the movable table and initially square it by sight. An electronic gauge can then be inserted into a valve guide to help in final squaring of the cylinder head deck on the machine. The electronic readout shows if the head is square on the machine or needs to be moved. A bubble gauge can also be used to square the head on the machine. If the machine is level, the head must be level. The bubble must be centered inside the level lines on the gauge in both directions.

If the guides are an integral part of the head (not pressed in), they must be machined oversize to accept new valve guide inserts. **Figure 22-19** summarizes installing a

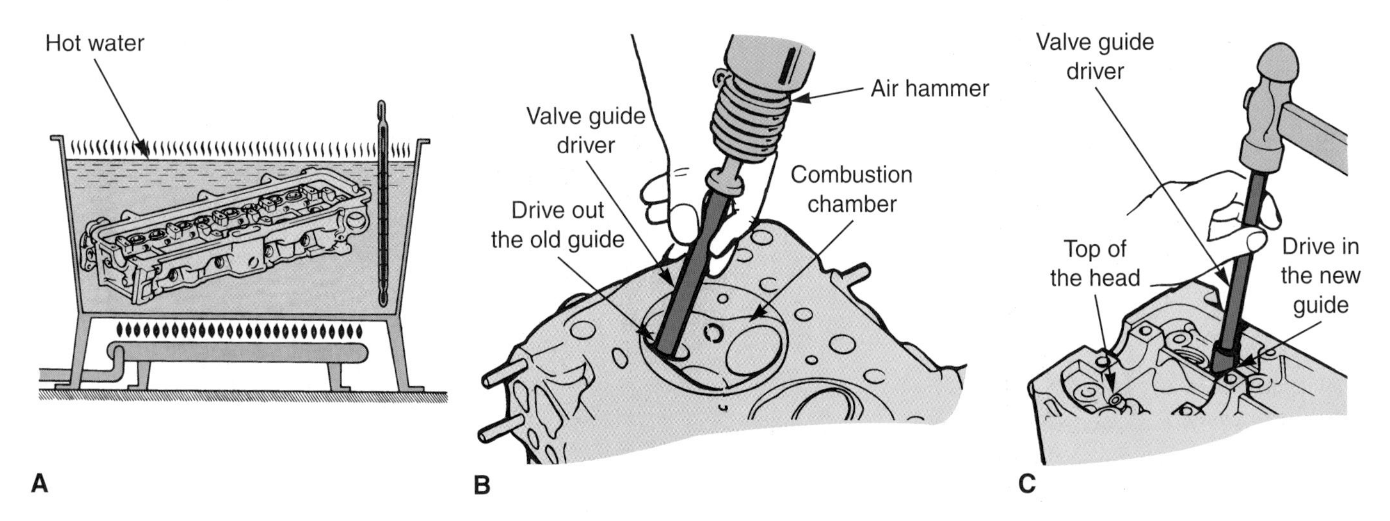

Figure 22-17. *This aluminum head has pressed-in valve guide inserts. A—Heat the head in a hot water bath. B—Drive out the old guide from the combustion chamber side of the head. C—Drive in the new insert from the top of the head. (Toyota)*

Figure 22-18. *Mounting a cylinder head in a valve guide machine. A—Place the head on the movable table and square it by sight. B—Use an electronic gauge inserted into a valve guide to square the head on the machine. C—A bubble level can also be used to square the head on the machine. The bubble must be inside level lines. (Sunnen)*

valve guide insert in place of an integral valve guide using a valve guide machine.

1. Select the correct diameter drill bit and reamer specified for the valve guide insert.
2. Install the core drill into the machine. This is used to rough machine the bore to within about .015″ (0.38 mm) of the size needed for the insert.
3. Center the core drill over the integral guide.
4. Engage the machine head and lower the drill through the guide.
5. Disengage the machine head. Install the correct reamer.
6. Engage the machine head and ream the guide to the final size. The reamer also creates a smooth surface for the press fit of the new guide insert.
7. Repeat this operation on other valve guides that are worn and must be replaced.
8. Use an air hammer and driver to force the new guide insert(s) into the cylinder head.

Valve Seat Reconditioning

The valve seats are also exposed to tremendous heat, pressure, and wear. ***Valve seat reconditioning*** involves using a grinding stone or carbide cutter to resurface the cylinder head valve seats.

Checking Valve Seats

Inspect the valve seats closely for problems such as burning, cracks, excess width, and retrusion. If the seat is badly worn or burned, too much grinding may be needed to recondition the seat.

Valve retrusion occurs when the valve seat and valve face wear, resulting in the head of the valve sinking into the head. See **Figure 22-20.** Valve seat retrusion results from heat erosion or seat grinding that causes the valve head to sink into the combustion chamber. The valve stem also protrudes too far out of the other side of the head. Retrusion causes lowered compression, reduced flow through the port, incorrect valve train geometry, and decreased performance. Seat replacement is needed to correct retrusion.

Valve Seat Replacement

Normally, valve seats can be ground or cut and returned to service. Valve seat replacement is only needed when seat wear or damage is severe.

Most technicians send the cylinder head to an automotive machine shop for seat replacement. Most service

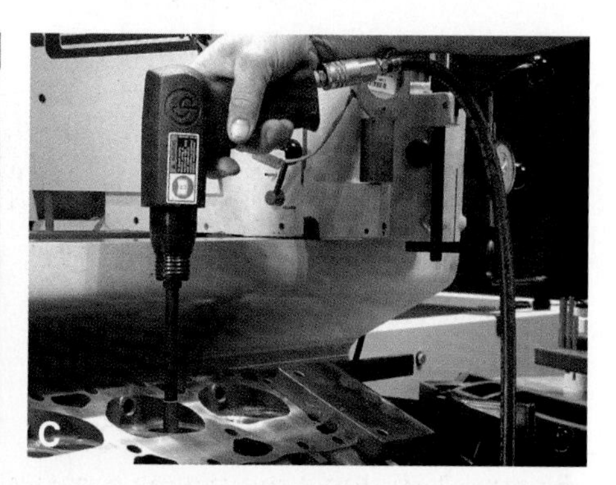

Figure 22-19. *Installing a valve guide insert. A—Select the correct size of core drill and install it into the machine. B—After using the core drill, use a reamer to machine the bore to the final size and smoothness. C—Use an air hammer and driver to force the new guide insert into the cylinder head. (Sunnen)*

shops do not have the specialized tools required for seat removal and installation, especially with integral seats. A new valve seat insert should have an interference fit of around .002″ to .004″ (0.05 mm to 0.10 mm). This locks the insert in the head.

Pressed-in valve seat replacement is not as difficult. Machining is often not required. The old seat can easily be removed and a new one installed. **Figure 22-21** shows the tools recommended by one manufacturer for pressed-in valve seat service. Another way to remove a pressed-in seat is to split the old seat with a sharp chisel. Then, pry out the seat with a hook-nose chisel. Extreme care must be taken not to damage the cylinder head.

To install a seat, some machinists chill the seat in dry ice to shrink it; others heat the head in a furnace to expand the valve seat bore. Either method eases seat installation, yet the seat is locked in the head after the parts reach room temperature. Use a driving tool to force the seat into the recess in the head. Seat installation tools vary. Follow equipment directions.

Staking the seat involves placing small dents in the cylinder head next to the seat, **Figure 22-22.** The stakes push part of the metal in the head over the seat. This forms a lip that keeps the seat in place during engine operation. Top cutting may also be needed to make the top of the seat flush with the surface of the combustion chamber.

Valve Seat Resurfacing

After new seat installation or when the old valve seats are in serviceable condition, grind or cut the face of the valve seats. The equipment needed to grind valve seats is shown in **Figure 22-23.**

Selecting Stone and Pilot

The ***seat stone*** is used to cut a fresh surface on the face of the valve seat. It is a grinding stone mounted on a threaded drive attachment. To find the correct-size stone for the valve seat, adjust dividers to the largest diameter of the seat, **Figure 22-24.** Then, compare the dividers to different stones until you find a stone that matches. The

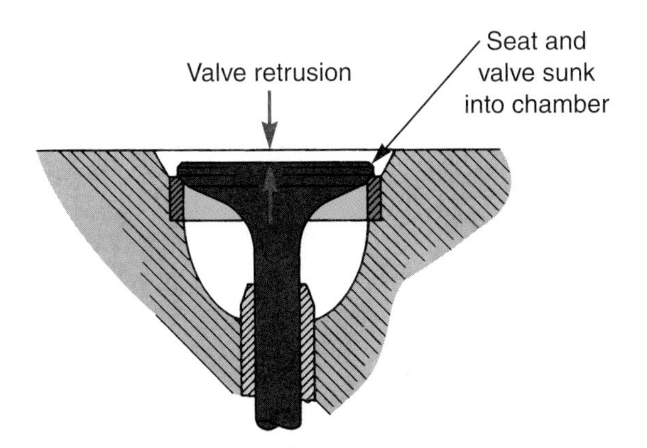

Figure 22-20. *Valve retrusion is a condition where the valve sinks into the combustion chamber due to wear on the valve seat and valve face. The stem also moves further through the head. (Ford)*

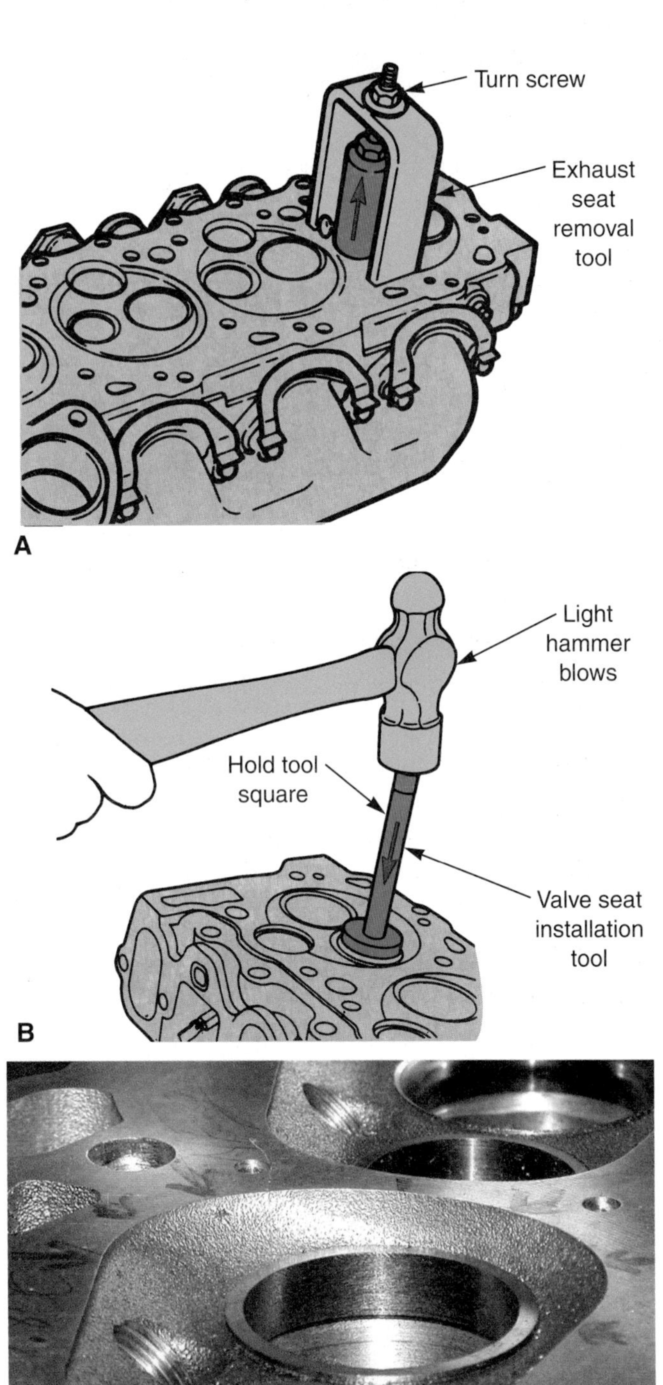

Figure 22-21. *Removing and installing a pressed-in valve seat. A—Use a puller to remove the old, damaged seat. B—A driver and hammer are used to force the new seat into the head. C—The new seat must be ground for the correct valve angle and contact. (Ford)*

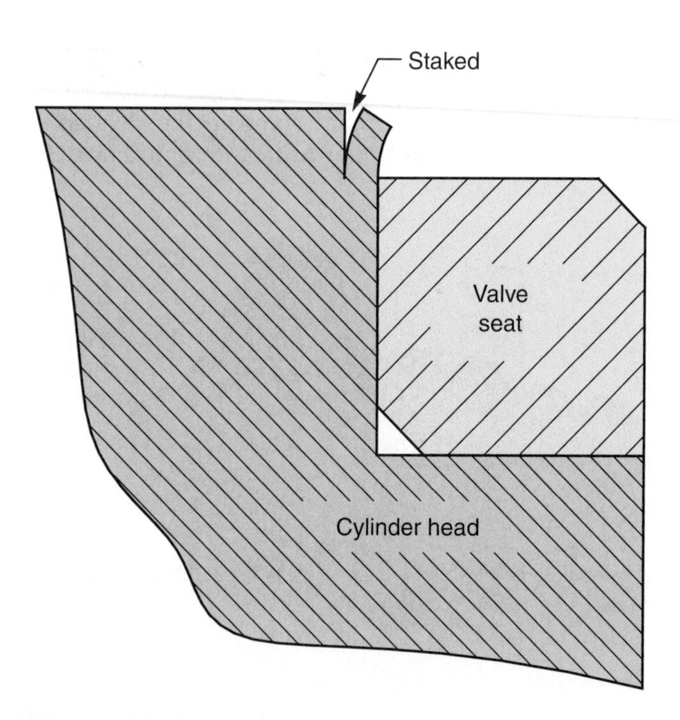

Figure 22-22. *Staking forms a lip that keeps the seat in place during engine operation. (TRW)*

Figure 22-23. *This is the equipment needed to grind valve seats. (Sioux Tools)*

stone should be slightly larger than the seat, but it must not hit the side of the combustion chamber. The stone must also have the correct angle, usually 45°, but sometimes 30°.

The ***pilot*** is a steel shaft that is installed in the valve guide to support the grinding stone. Since valve guide diameters vary, you must find the correct size pilot. Find one that slides down into the guide far enough, yet fits the guide snugly.

Dressing Stone

A ***stone dressing tool*** is used to resurface the face of the seat stone. See **Figure 22-25.** The angle of the stone dressing tool must be set at the angle to which the seat will be ground. If you are going to do a three-angle valve job, you will need three stones, one dressed at 30°, one at 60°, and one at 45° (typically).

1. Mount the correct size stone on its pilot and place it on the stone dressing tool.
2. Lubricate the pilot with a drop of oil. Adjust the diamond cutter so it almost touches the stone. Make sure it will clear the stone slightly before spinning the stone with the power head.
3. Verify that everything is adjusted properly.
4. Spin the stone with the power head.
5. Slowly feed the diamond into the stone until it just touches the stone. The diamond cutter can be ruined if it is adjusted too far into the stone. The stone can grind metal away from the diamond and the diamond can fall out.

Figure 22-24. *Determining the correct stone size. A—Adjust dividers to the diameter of the valve seat. B—Compare the dividers to stones to find the correct size.*

Figure 22-25. *Dressing a valve seat grinding stone. A—Place the stone and arbor on the dressing tool. Spin the stone with the power head. B—Slowly feed the diamond dressing tool into the stone.*

6. Move the lever up and down to true the entire face of the stone.
7. Only remove as much of the stone as needed to provide a proper grinding surface. This will resurface the stone and ready it for use on the cylinder head.

Warning: When dressing a seat stone, keep your hands, loose clothing, jewelry, and long hair away from the spinning stone.

Grinding Valve Seats

A T-handle tool is usually provided for installing the pilot. Push down and twist as you fit the pilot into the guide. If the pilot is an expanding type, slide the pilot down into the guide and then tighten its adjusting bolt lightly until the pilot is locked into the guide. See **Figure 22-26A.** The pilot must not be loose in the guide.

Screw the stone onto its drive head and check its fit on the seat. If the stone is too small, the seat can be ruined by even momentary grinding. After you are sure the stone is the correct size and dressed properly, you are ready to grind the valve seat.

1. Place a drop of oil on the pilot.
2. Slide the stone down over the pilot and onto the seat, **Figure 22-26B.**
3. Install the power head onto the stone driver.
4. Use one hand to support the power head and the other to operate the trigger, **Figure 22-26C.**
5. Engage the power head and grind the seat for a second or two. Do not push down; just let the weight of the tool do the work. Simply hold the tool square and steady.
6. Remove the stone and inspect the seat.
7. If the seat is shiny all the way around, no more grinding is needed, **Figure 22-26D.**

Grind as little as possible while achieving acceptable results. The more you grind, the more the valve will sink into the head, which in turn lowers compression and affects valve train geometry.

Work by type of valve, not by combustion chamber. Grind all of the intake or exhaust seats first. Then, install the correct size stone for the other seats and grind those. This will keep you from repeatedly changing the stone. Dress the stone every couple of seats or as needed.

Cutting Valve Seats

The term *cutting* valve seats is used when a carbide cutter, not an abrasive stone, is used to recondition the cylinder head seats. **Figure 22-27** shows a carbide valve seat cutting tool. Note how the small carbide blades fit into the tool. This particular tool has a 15° angle on the top and 45° angle on the bottom.

To use the cutter, select the correct size and angle cutter. Make sure the blades are installed properly. Install the pilot in the head and slide the cutter over the pilot. Then, use the handle to rotate the cutter on the seat. Only cut long enough to clean up all dark areas on the seat. The cutter may also be rotated by a low-speed drill, but refer to the operating instructions supplied with the cutter. Repeat on the other seats as needed.

Checking Seat Concentricity and Width

After grinding or cutting the seats, most technicians check ***concentricity*** or seat runout. To do this, a special runout gauge is mounted on the seat, **Figure 22-28A.** Then, the gauge is rotated around the seat face. The indicator shows the seat runout. If not within specifications, the seat was installed or ground incorrectly.

Figure 22-26. *Grinding valve seats. A—Select the right size pilot and push it into the valve guide. B—Fit the stone and arbor assembly over the pilot. C—Spin the stone and arbor with the power head. D—Inspect each seat to make sure all pits and dark areas have been removed.*

Grinding or cutting widens the seat. You must check the ***seat width.*** This is the area of the valve seat that actually makes contact with the valve face. Measure the seat width as shown in **Figure 22-28B.** If wider than specifications, you will have to narrow the seat face to reposition the seat contact point.

Valve Seat Narrowing

After prolonged service, the seat can wear away from heat and friction created by the valve action. Then, after grinding, the seat can be too wide or contact the valve too far out on the face. **Figure 22-29** shows typical seat specifications for both intake and exhaust valve seats. Note how the exhaust seat is wider so it can dissipate heat better.

Seat narrowing, also called seat positioning, is needed so that the seat contacts the valve face correctly. For example, if the seat angle is 45°, use 15° or 30° and 60° stones to move the seat and narrow it. Grinding the outer and inner seat surface can be done to narrow and move the seat. This is sometimes called a ***three-angle valve*** job because three different stone angles are used. It is sometimes done to increase engine power. A rounded or curved seat can be cut by some cutters. The radius allows more air-fuel flow past the opened valve. This is commonly done to racing engines to gain a few horsepower.

Figure 22-30 shows how a carbide cutter is used to position and narrow a valve seat. Estimating the seat width and location before cutting will help you know how much to cut with each angle. This principle also applies to using cutting stones to do three-angle valve job. First, use a 60° cutter to remove material from the inside diameter of the seat. Then, use a 15° cutter to remove metal from the outside diameter of the seat as needed. Finally, use the correct cutter angle (45° for example) to resurface the face of the seat.

Moving the Seat Contact Point

If the valve seat does not touch the valve face properly because of the wrong width or location on the valve, regrind the seat using different stone angles. Just as when narrowing the seat, use 15° or 30° and 60° stones.

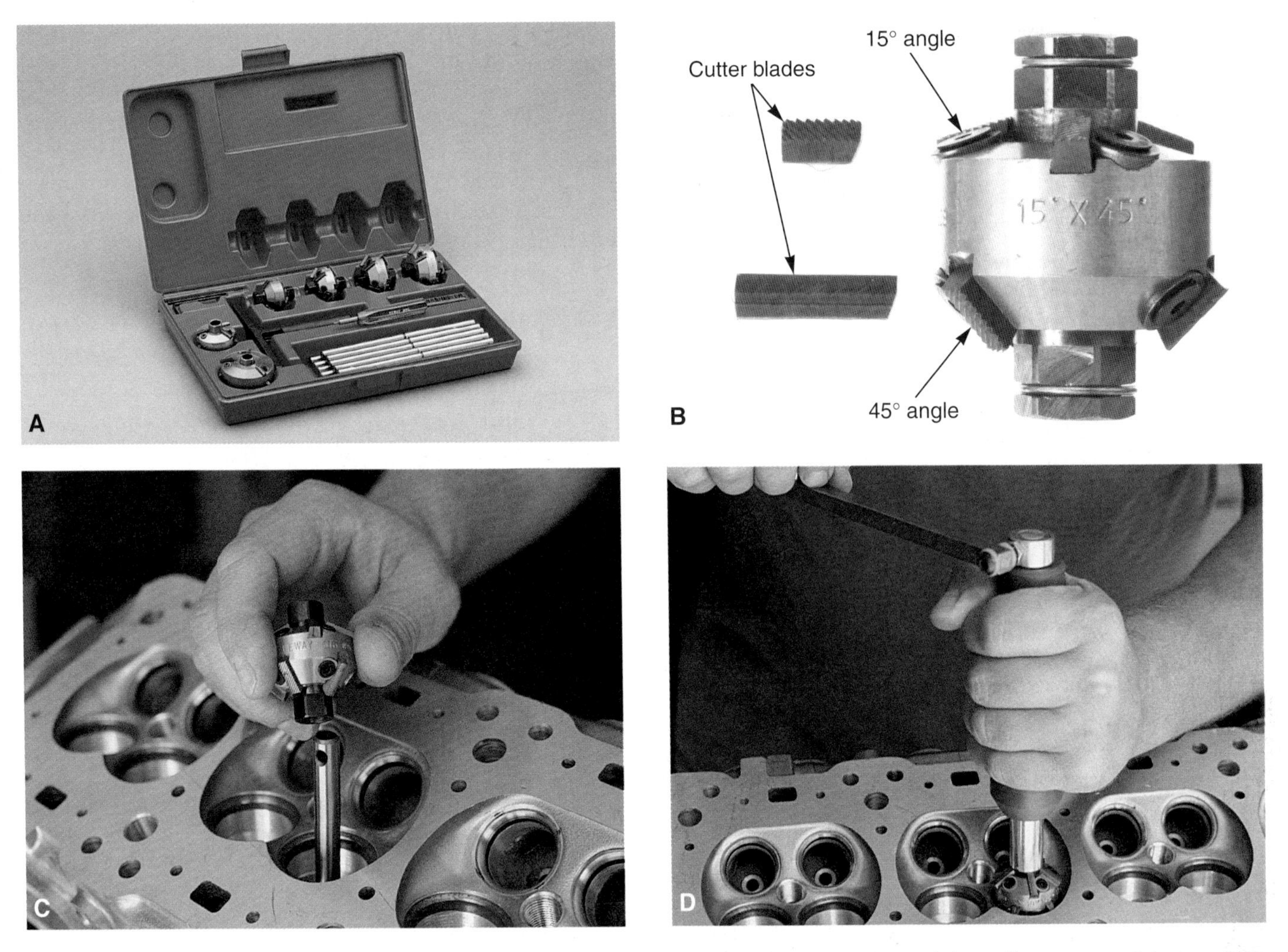

Figure 22-27. *A—A carbide valve seat cutter kit. B—The cutters are made of very hard metal that will remove metal accurately. C—Select the correct size pilot. Also, select the correct size and angle of cutter. D—Rotate the cutter against the seat by hand. (Neway)*

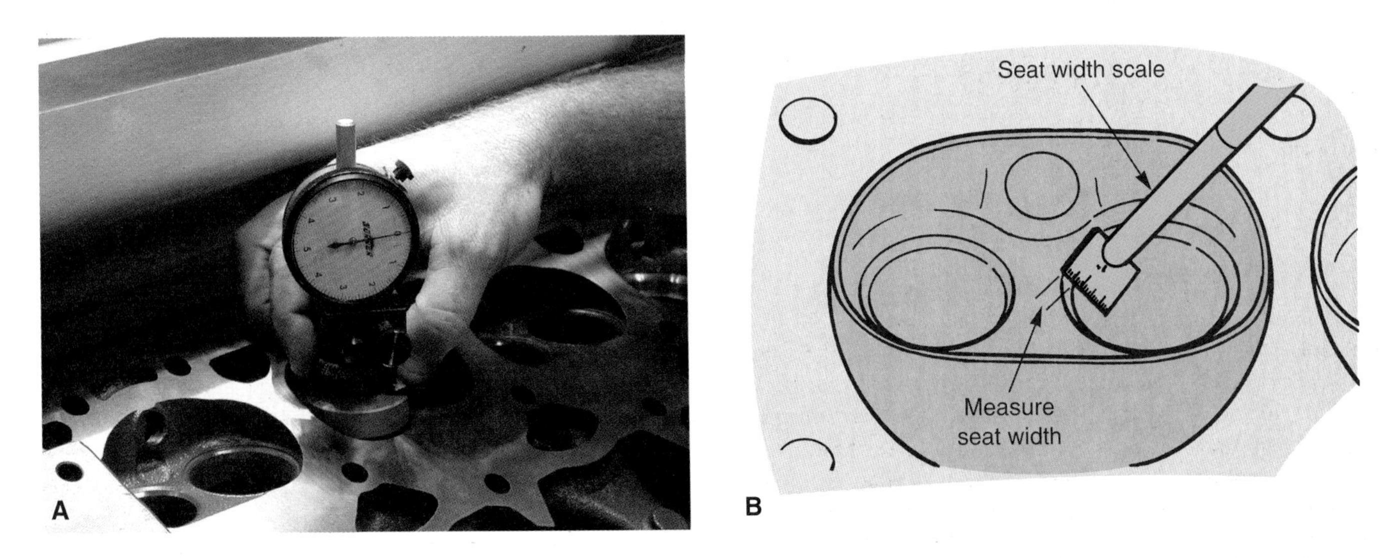

Figure 22-28. *After grinding or cutting, check concentricity and seat width. A—Measure concentricity or runout as shown. B—Use a small scale to check the seat width. (Ford)*

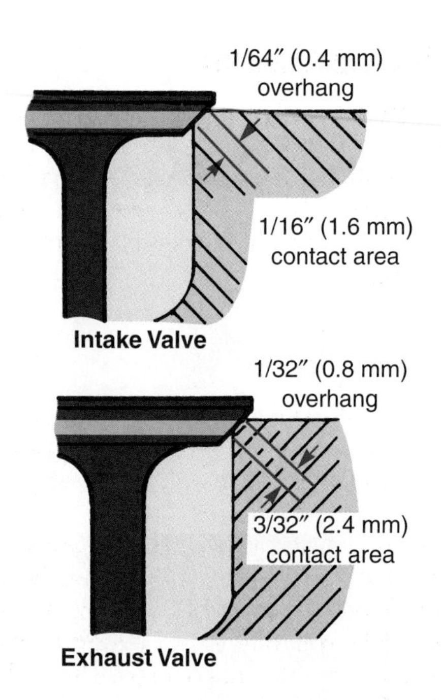

Figure 22-29. *These are typical specifications for intake and exhaust valves. Always use manufacturer specifications, when available. (TRW)*

To move the seat *in* and narrow it, grind the valve seat with a 15° or 30° stone. Use a 30° stone with a 45° seat and 15° stone with a 30° seat angle. This will remove metal from around the top of the seat. The seat face (valve contact surface) will move closer to the valve stem.

To move the seat *out* and narrow it, grind the valve seat with a 60° stone. This will remove metal from the inner edge of the seat. The seat contact point will move toward the margin, or outer edge of the valve.

Valve Seat Machine

A ***valve seat machine*** is a piece of equipment designed for valve guide and valve seat service. It is a specialized milling machine designed to service automotive cylinder heads. The procedures for using any valve seat machine is similar. Refer to **Figure 22-31.**

1. Mount and square the head on the table. If the machine table is level, you can use a small bubble level to square the head on the machine.
2. Install the correct size pilot into the valve guide in the cylinder head.

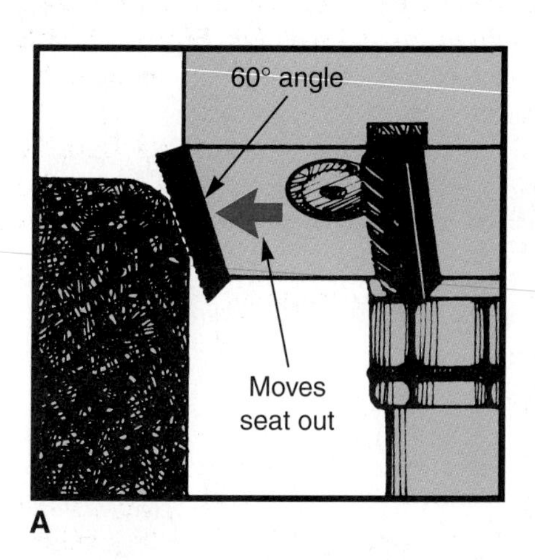

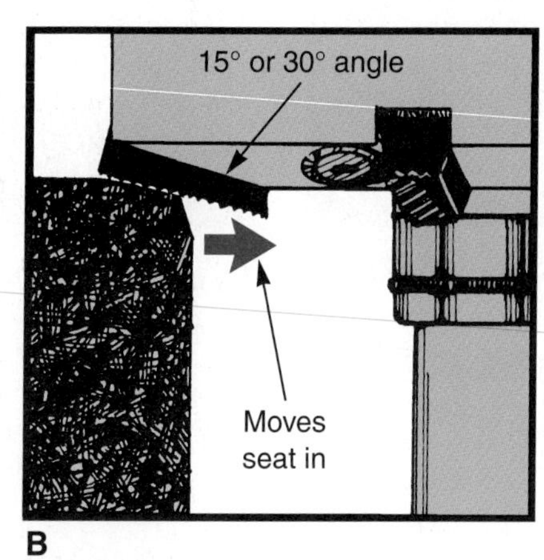

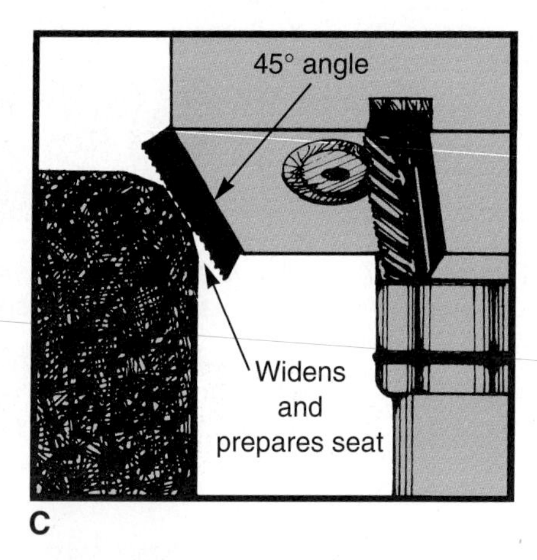

Figure 22-30. *Using a carbide cutter to do a three-angle valve job. A—The first cut cleans and reconditions the area below the seat. B—The second cut cleans and reconditions the area above the seat. C—The final cut creates the proper angle. (Neway)*

Figure 22-31. *Using a valve seat machine. A—Fit the cutter down over pilot to check its size. B—Lower the power head and fit it onto the cutter. C—Using the correct cutting speed and feed, rotate the cutter head on the valve seat. (Sunnen)*

3. Slide the damper spring over the pilot.
4. Select the correct cutter.
5. Fit the cutter down over the pilot to check its size.
6. Lower the power head and fit it onto the cutter.
7. Using the correct cutting speed and feed, cut the valve seat.
8. Use a dial indicator to check seat concentricity or roundness.

Valve Grinding

Valve grinding creates a fresh, smooth surface on the valve face and stem tip. Valve faces suffer from burning, pitting, and wear caused by opening and closing millions of times during the service life of an engine. Valve stem tips wear because of friction from the rocker arms or cam follower. **Figure 22-32** shows valve problems.

Before grinding, inspect each valve face for burning and each stem tip for wear. Replace any valves that have badly burned or worn faces. New and reused valves must be ground.

Warning: Wear a face shield when grinding valves. The stone could shatter, throwing debris into your face.

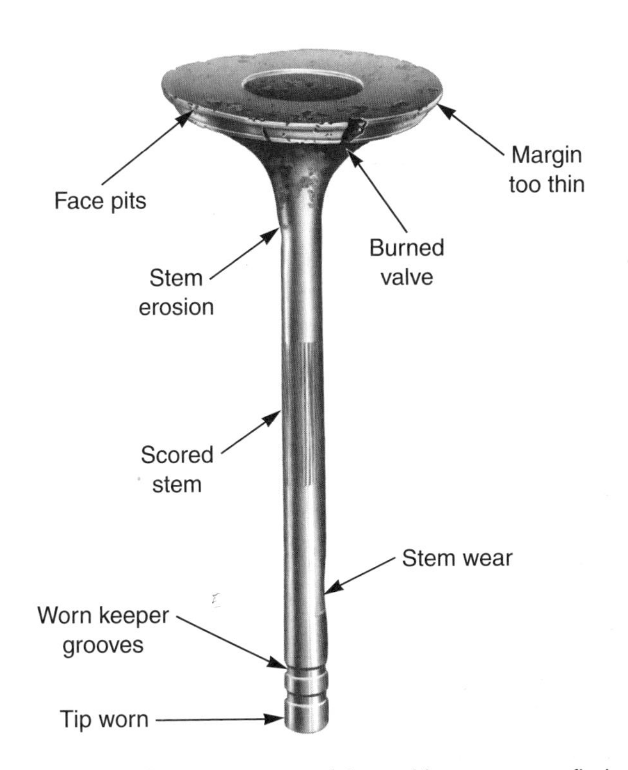

Figure 22-32. *These are some of the problems you may find on used valves. (Sioux Tools)*

Valve Grinding Machine

A ***valve grinding machine*** will resurface the valve faces and stem tips, **Figure 22-33.** Although there are some variations in design, most valve grind machines are basically the same. They use a grinding stone and precision chuck to remove a thin layer of metal from the valve face and tip.

Valve Grinder Setup

Dress the stone with a diamond cutter to true the stone surface before grinding the valves. A diamond tipped cutting attachment is provided with the machine. Follow the equipment manufacturer's instructions.

Set the chuck angle by rotating the chuck assembly on the valve grinding machine. A degree scale or digital readout is provided so that the cutting angle can be set precisely. Normally, you must loosen a locknut and swivel the chuck assembly to the desired angle. This is illustrated in **Figure 22-34A.**

An ***interference angle*** is normally a 1° difference between the valve face angle and the valve seat angle. It is set on the valve grinding machine. If the valve seat angle is 45°, the chuck is set to 44°. This produces a 1° interference angle between the valve face and seat. See **Figure 22-34B.** The break-in and sealing time of the valve is reduced by using an interference angle.

Chuck the valve in the valve grinding machine by inserting the valve stem into the chuck. Make sure the chuck grasps the machined surface of the stem near the valve head. The chuck must not clamp onto an unmachined surface of the stem or runout will result.

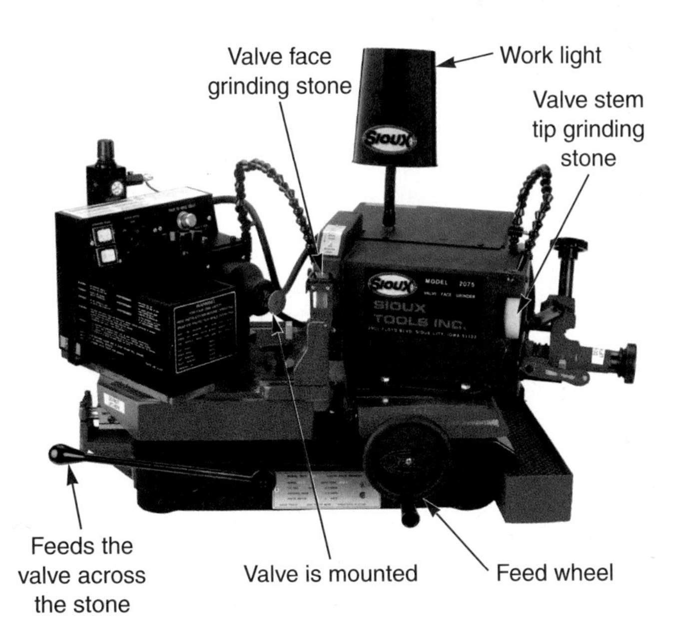

Figure 22-33. *A valve grinding machine is used to resurface valve faces and stem tips. Note the names of the machine parts. (Sioux Tools)*

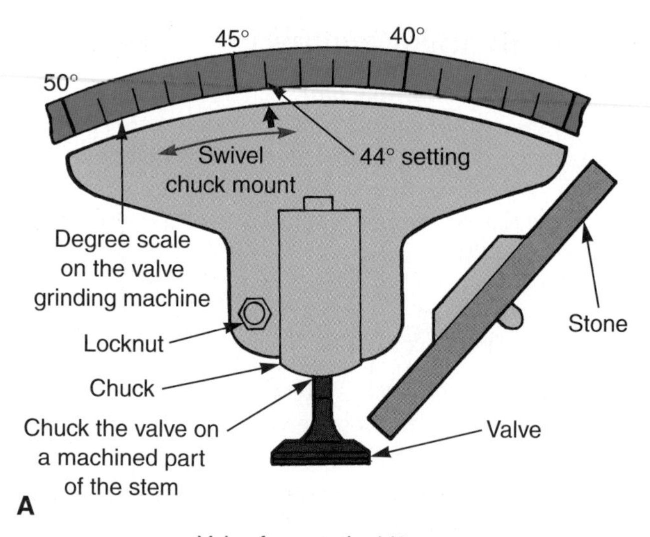

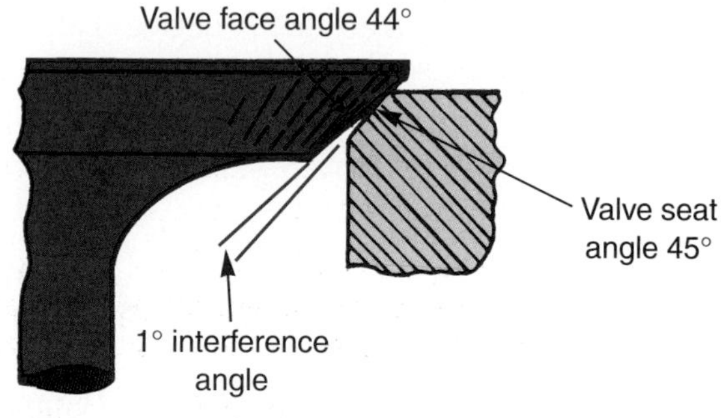

Figure 22-34. *A—Adjusting the valve angle on a valve grinding machine. B—An interference angle speeds the initial valve sealing and prevents rapid widening of the valve face in service.*

Warning: Be very careful when using a diamond tool to dress a stone. Wear eye protection and feed the diamond into the stone slowly. If fed in too fast, tool or stone breakage may result.

Grinding Valve Faces

Once the machine is properly set up and the valve properly mounted, turn on the valve grind machine and cooling fluid. Slowly feed the valve face into the stone. While feeding, slowly move the valve back and forth in front of the stone. Use the full face of the stone, but do not let the valve face move out of contact with the stone while cutting. Look at **Figure 22-35.** Only grind the valve long enough to clean up its face. When the full face looks shiny with no darkened pits, shut off the machine and inspect the valve face more closely. Carefully look for pits or grooves.

Repeat the grinding and inspecting operation on the other valves. Return each ground valve to its place in an organizing tray. Used valves should be returned to the same valve guide in the cylinder head. The stems may have been select fit.

If not noticed during the initial inspection, a burned valve will show up when excess grinding is needed to clean up the valve face. A normal amount of grinding will not remove a deep pit or groove in the valve face, **Figure 22-36A.** Replace the valve if it is burned.

A bent valve will show up when the valve head wobbles as it turns in the chuck. If the valve is chucked improperly, it can appear like a bent valve. Shut off the machine and find the cause, **Figure 22-36B.** V-blocks and a dial indicator can be used to check the valve to see if it is bent.

By grinding metal from the valve face, the valve stem will extend through the head more. This will affect spring tension and rocker arm geometry. Grind the face of each valve as little as possible.

A sharp valve margin indicates excess valve face removal and requires valve replacement. Look at **Figure 22-37.** Refer to the specification for minimum valve margin thickness. If the margin is too thin, the valve can burn when returned to service. It may not be thick enough to dissipate heat fast enough. The head of the valve can actually begin to melt, burn, and blow out the exhaust port.

Figure 22-35. *Grinding a valve face. A—Install the valve in the chuck. B—As the stone starts to grind, move the valve back and forth over the stone. Keep the valve in contact with the stone. (Sunnen)*

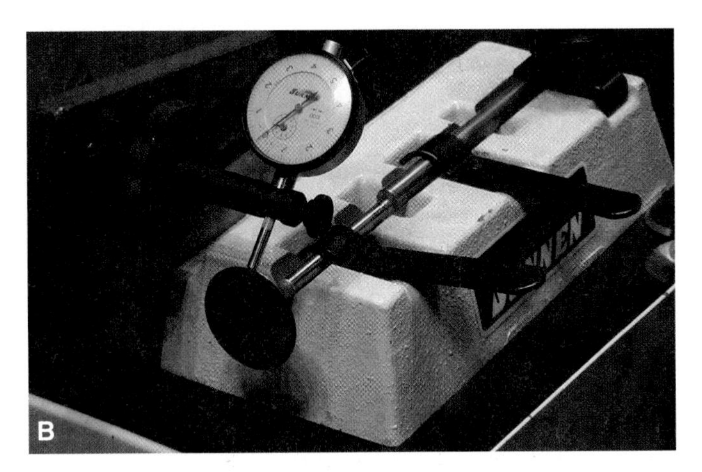

Figure 22-36. *A—This valve must be replaced. The erosion could not be removed without excessive grinding. B—Mount the valve in V-blocks and use a dial indicator to check valve runout before grinding the valve. (Sunnen)*

Grinding the Valve Stem Tip

After grinding valve faces, you should also true the valve stem tips. This is done so the valve tip is square to the valve stem. Also, this corrects for the extra distance that the valve stem extends through the head after valve grinding. Proper valve train geometry is restored. However, if too much material has been worn off of a tip, the valve should be replaced. Check the service manual for specifications.

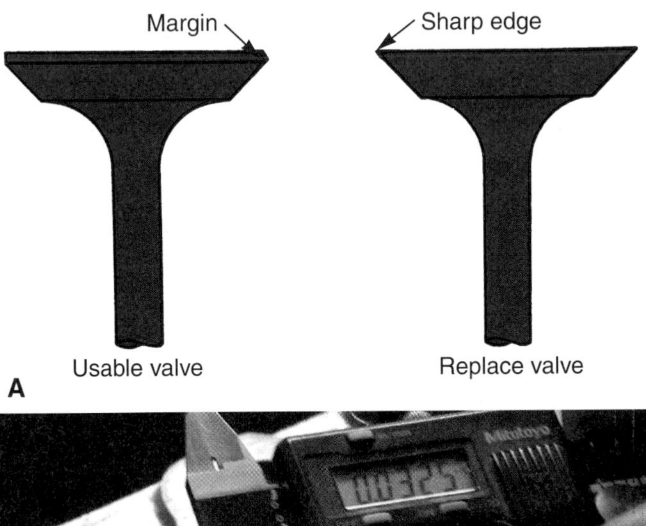

Figure 22-37. *A—After valve grinding, check the margin. B—Calipers can be used to measure the valve margin. (TRW)*

The second stone on a valve grind machine, which is usually on the right-hand side of the machine, is provided for truing and chamfering valve stem tips. **Figure 22-38** shows how to grind and chamfer valve stems.

1. Mount the valve in the V-block chuck that is perpendicular to the stone. Before turning the machine on, make sure the stem is close to the stone, but not too deep.
2. Use the thumbwheel to slowly feed the stem tip into the stone. The thumbwheel has graduations in thousands of an inch or hundredths of a millimeter to keep track of how far you feed the tip into the stone.
3. Remove approximately the same amount of material from the tip as you removed from the valve face. Generally, cut the same amount of metal off of the face and stem of every valve. This will help keep valve train geometry and engine compression ratio correct. Grind as little material as needed to clean up and true the surface of the tip.
4. If you need to chamfer the tip, mount the valve in the V-block chuck that is at an angle to the stone. See **Figure 22-39.** Then, slowly feed the tip into the stone as you rotate the valve.

Caution: Many valve stems are hardened. Too much grinding will cut through the hardened layer of metal. Since the softer metal is exposed to wear, the tip will rapidly wear when the valve is returned to service.

Checking Valve-to-Seat Contact

After the valve is ground, you need to check for proper valve-to-seat contact. To do so, use valve lapping compound or Prussian blue. Some manufacturers do not recommend valve lapping, so refer to a service manual for details.

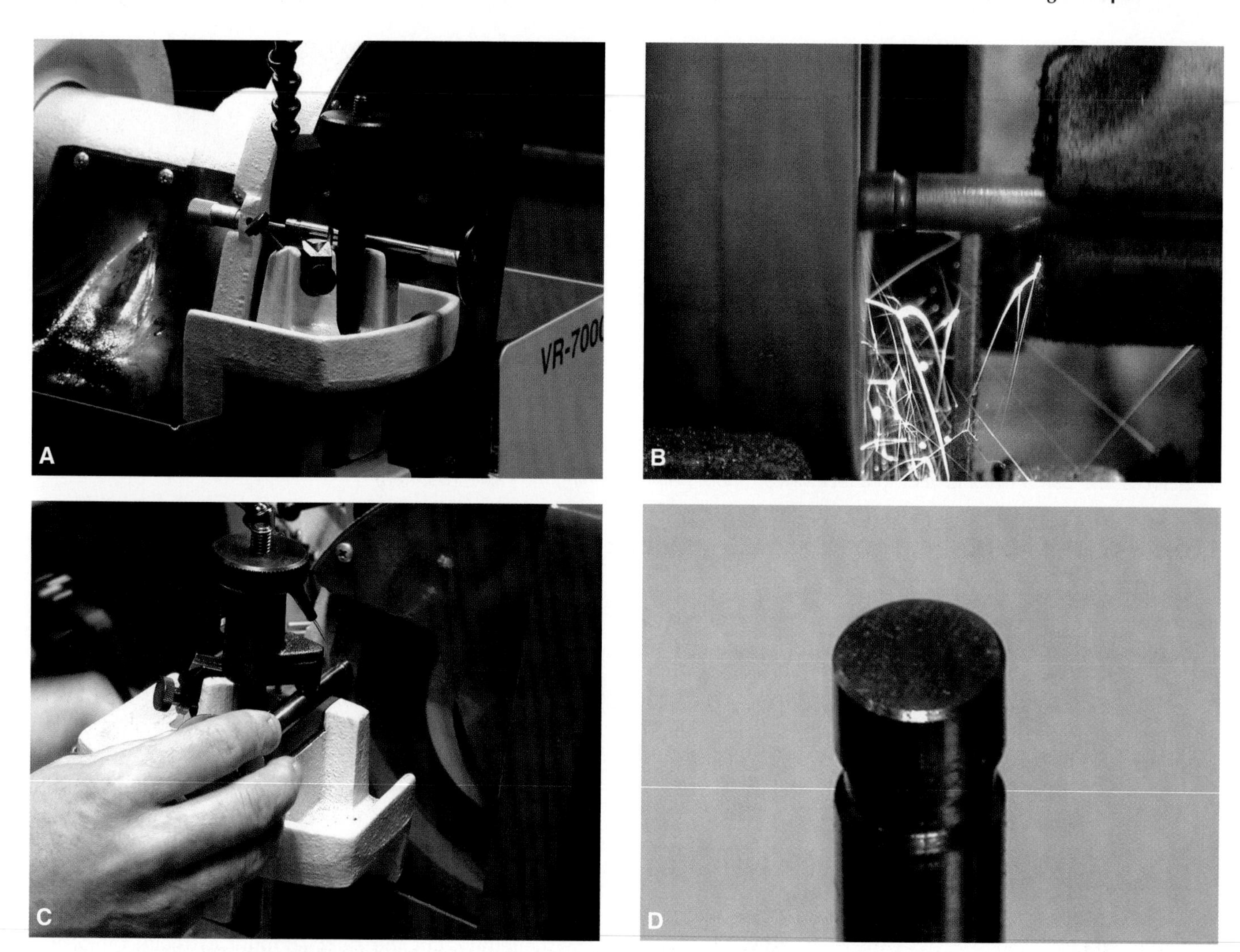

Figure 22-38. *Grinding a valve stem tip. A—Mount the valve in the V-block chuck that is perpendicular to the stone. B—Move the tip into the stone. C—Grind as little material as needed to clean up and true the surface of the tip. D—Note how the valve stem tip has been resurfaced flat and a small chamfer placed on the tip.*

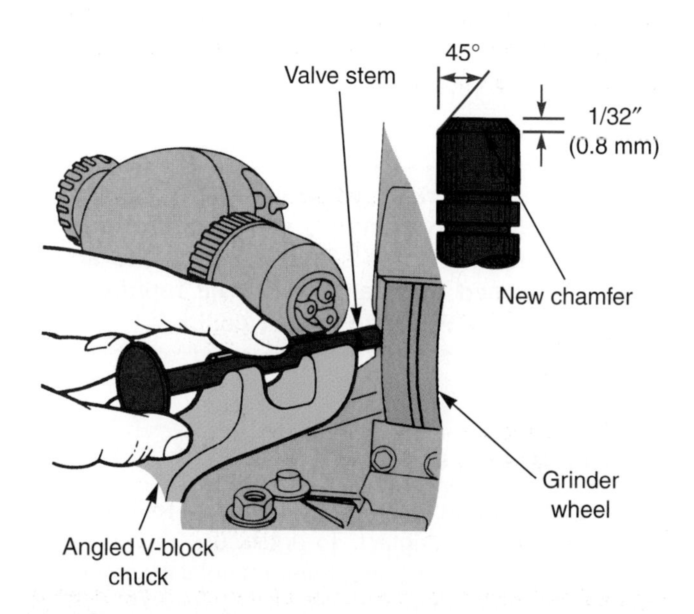

Figure 22-39. *Placing a chamfer on the valve stem tip. (Ford)*

Valve lapping

Valve lapping compound is an abrasive paste material. To use lapping compound to check the valve-to-seat contact, proceed as follows.

1. Place a small amount of the compound on the valve face. Spread the compound around the valve face.
2. Place the valve into place in the head.
3. Use a lapping stick to spin the valve on the seat. A ***lapping stick*** is wooden rod with a suction cup on the end. To turn the stick, rub the palms of your hands together on each side of the stick.
4. Rotate the valve back and forth while picking it up and turning it to a new position on the seat, **Figure 22-40A.**
5. Remove the valve and check the contact point, **Figure 22-40B.**

A dull gray stripe around the seat and face is where the valve is making contact with the seat. This will help you determine how to move the valve seat contact point, if needed.

Figure 22-40. *Lapping is an easy way to check seat location on the valve. A—Rotate the valve back and forth while picking it up and turning it to new position on seat. B—Remove the valve and look for the dull stripe on the valve and seat. This shows how the seat touches the valve face. (Van Senus Auto)*

Caution: Make sure you clean all of the valve lapping compound off of the valve, seat, and cylinder head. The compound can cause rapid part wear if left in the engine during service.

Prussian blue

Prussian blue is a metal stain often used to layout metal parts. It can be used to check the valve-to-seat contact.

1. Wipe the dye over the full surface of the valve face.
2. Allow the dye to dry.
3. Insert valve into its correct valve guide in the head.
4. By hand, snap the valve closed on the seat several times.

The contact with the seat will wipe off some of the Prussian blue, making the valve face shiny at the point of contact. This will show you where the valve seat and valve face touch. Refer to **Figure 22-41.**

Contact patterns

An intake valve should generally have a valve-to-seat contact width of about 1/16″ (1.5 mm). An exhaust valve should have a slightly larger valve-to-seat contact width of about 3/32″ (2.5 mm). This helps dissipate heat. Check the manual for exact specifications. **Figure 22-42** shows good and bad seat contact patterns.

Valve Spring Service

Valve springs tend to weaken, lose tension, or even break after extended service. Always test valve springs to make sure they can be reused. If the engine has over 100,000 miles (160,000 km) on it, install new valve springs.

Inspecting Valve Springs

Valve spring free height can be measured with a combination square, caliper, or valve spring tester. Simply measure the length of each spring in a normal, uncompressed condition. Compare the measurement to specifications. If the spring is too long or too short, replace it, **Figure 22-43A.**

Valve spring squareness is easily checked with a combination square. Place the spring next to the square on a flat work surface, **Figure 22-43B.** Rotate the spring while checking for a gap between the side of the spring and the square. Replace the spring if it is not square.

Valve spring tension or pressure is measured on a spring tester. Compress the spring to installed height or the height given in the specification. Then, read the scale, **Figure 22-44.** The spring pressure must be within specifications. If too low, the spring has weakened and needs replacement or shimming. If too high, high-performance springs may have been installed, which may lead to premature valve train wear.

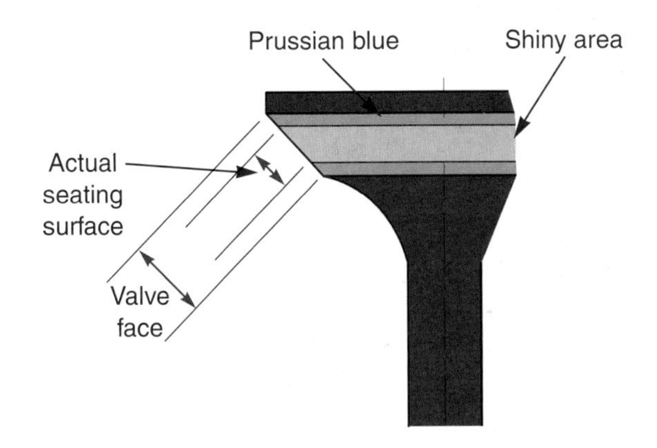

Figure 22-41. *When using Prussian blue to check the valve-to-seat contact, the shiny area shows the point of contact. (Honda)*

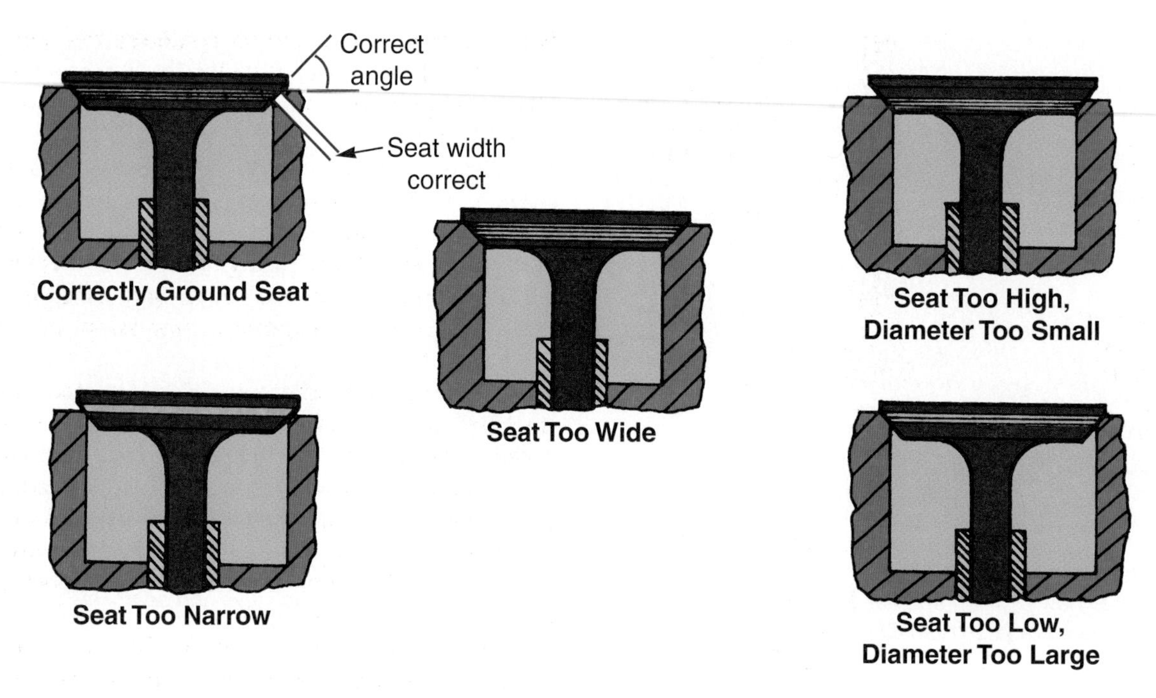

Figure 22-42. *Examples of a good and several bad seat contact patterns. (Ford)*

Shimming Valve Springs

When valves and seats are ground, the valve stems stick through the cylinder head more. This is illustrated in **Figure 22-45A.** This increases valve spring installed height and reduces spring pressure. ***Valve spring installed height*** is the distance from the top to the bottom of the valve spring when the spring is installed on the cylinder head. To measure valve spring installed height:

1. Inspect the valve spring retainers and locks for wear. Also, inspect the grooves on the valve stem for wear.
2. Install the valve, retainer, and keepers as shown in **Figure 22-45B.**
3. Measure the height from the top of the head to the bottom of the retainer. This distance is equal to the valve spring installed height.
4. Compare the measured height to the specifications. If the height is greater than the specification, spring tension will be lower than normal and valve float may result.

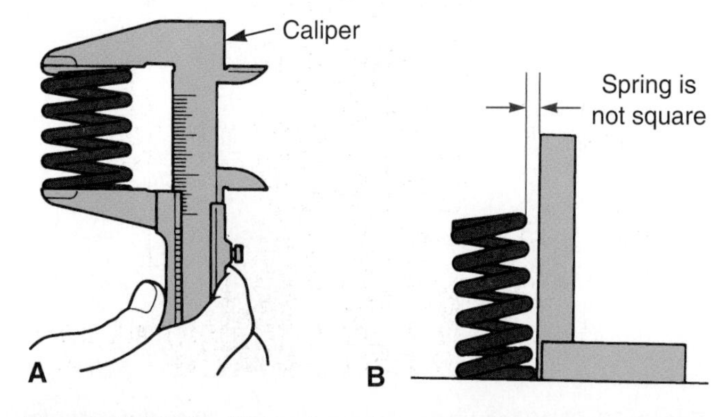

Figure 22-43. *A—Measuring valve spring free height or length. B—Making sure a valve spring is square. (General Motors)*

Figure 22-44. *Using a valve spring tester to measure valve spring tension. A—Place the spring on the tester and pull the handle to compress the spring. B—With the spring compressed to the specification height, read the pressure scale. Compare the reading to specifications.*

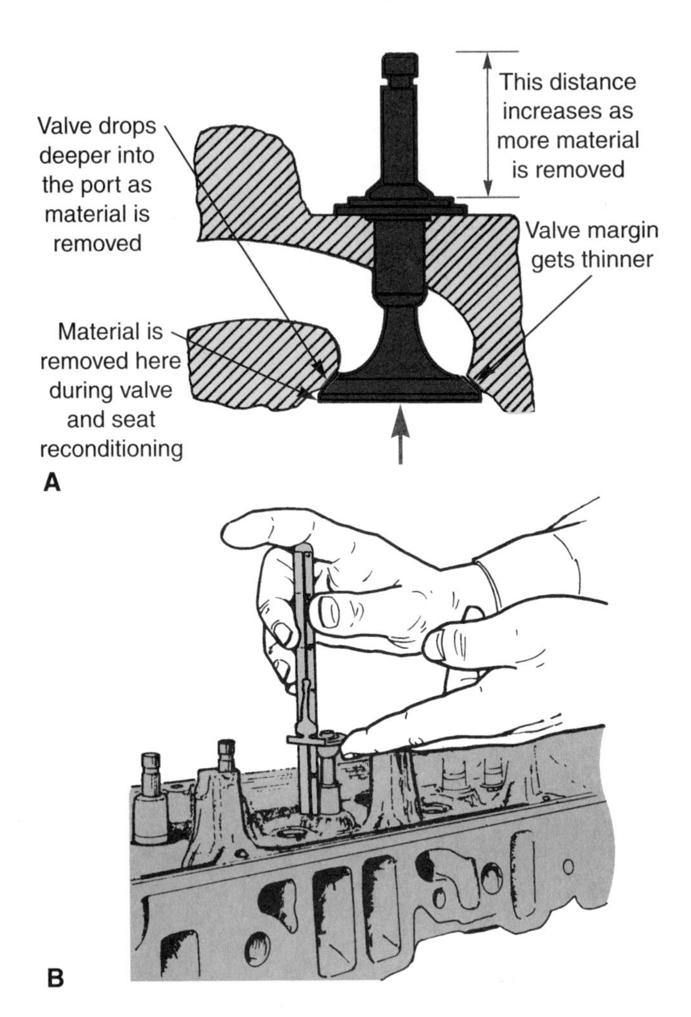

Figure 22-45. *A—When the valve and valve seat are ground, the valve stem will extend farther through the head. B—After grinding valves and valve seats, install the valve, retainer, and keepers as shown. Then, measure the valve spring installed height to determine if valve spring shims are needed. (General Motors)*

Valve springs shims are used to maintain the correct spring tension when the spring is installed on the cylinder head. If the valve spring installed height is greater than specifications, subtract the stock spring height from the measured height. The result is the total shim thickness needed to return the spring pressure to normal. Place the shim(s) under the valve spring, as shown in **Figure 22-46.** The serrated side of the shim(s) should face away from the spring.

A valve spring tester is used to determine the needed shim thickness if the spring pressure is lower than specifications. Place the valve spring in the compressor and continue as follows.

1. Press down on the lever until the pressure scale equals the spring pressure specification.
2. Read the height scale to find how far the spring has been compressed.
3. Subtract this measurement from the specification for installed spring height to determine the thickness of shim needed.

For example, if you had to compress the spring .010″ more than the installed spring height specification to attain the spring pressure specification, the shim needs to be .010″ thick.

Generally, never shim valve springs over .060″ (1.5 mm). Adding more shim thickness than this can cause the spring to bind and lead to valve train part breakage. Look at **Figure 22-47.** ***Spring bind*** refers to when a valve spring is compressed too much and its coils hit each other, stopping more spring compression and the valve opening action.

Valve spring bind can be caused by too much shim thickness under a spring, too much valve lift, incorrect rocker arm ratio, or the incorrect springs. When a valve spring binds, it will no longer compress. Instead, it acts as a solid piece of steel. When the camshaft tries to open the valve, push rods can be bent or rocker arms broken because of spring bind. Always visually inspect for spring bind during engine reassembly.

While measuring spring height, you should also check that the valve retainers do not hit and bind on the top of the valve guides. There should be at least 1/16″ to 1/8″ (1.5 mm to 3 mm) clearance between the retainer and guide at full valve lift. This is only a problem if you installed new valve guides or different valve spring retainers with different-than-stock dimensions.

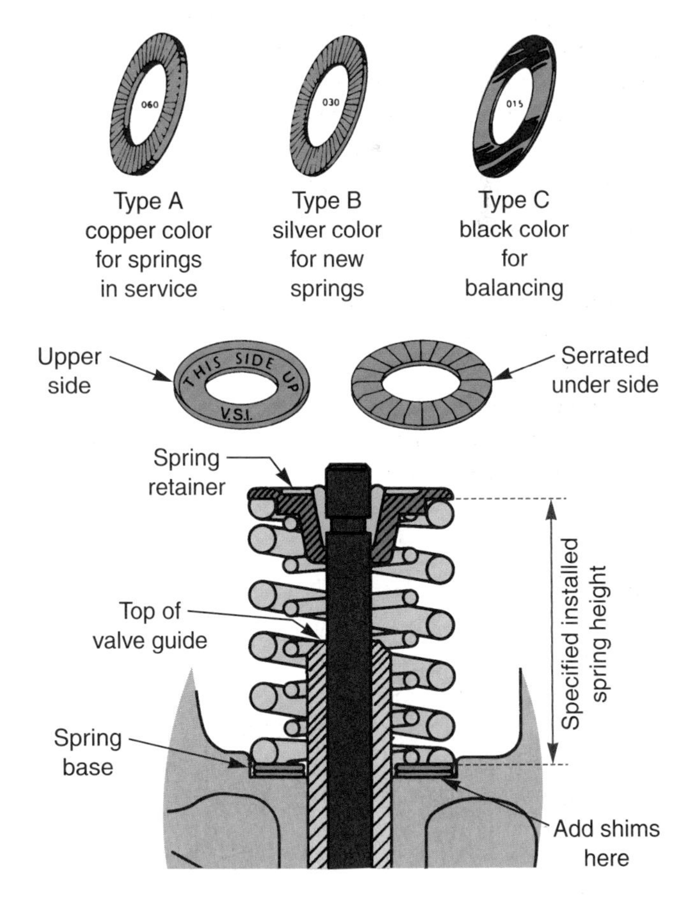

Figure 22-46. *Shims are placed between the valve spring and the head. The serrated side should face down. (K-Line Tools, TRW)*

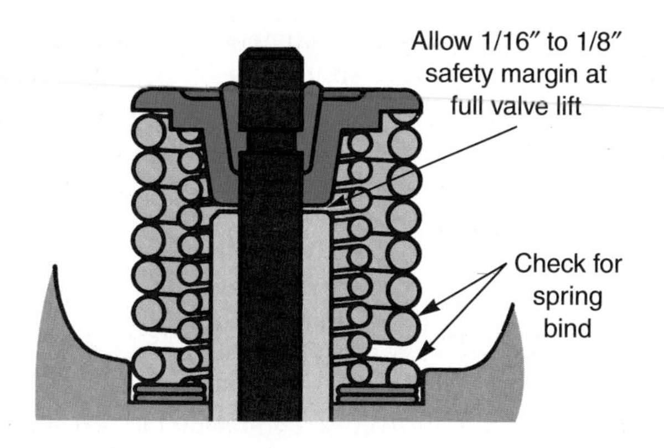

Figure 22-47. *Spring bind is a result of too much camshaft lobe lift or shims that are too thick. Also, a new guide must be short enough to clear the spring retainer. (TRW)*

Cylinder Head Reassembly

With all parts cleaned, inspected, measured, and machined, you are ready to assemble the cylinder head. First, lubricate the valve stems with motor oil or assembly lube. Then, slide the valves into their guides. Continue as described in the next sections.

Installing Valve Seals

To install an umbrella valve seal, simply slide the seal over the valve stem. This is done before the spring is installed. Refer to **Figure 22-48A.**

When installing O-ring valve seals, compress the valve spring *before* fitting the seal on the valve stem, **Figure 22-48B.** If you install the seal first, it will be cut, split, or pushed out of its groove. Engine oil consumption and smoking will result.

In some OHC applications, the valve seal is part of the spring seat. To install these integral seats/seals, fit them down into their pockets in the cylinder head, **Figure 22-48C.**

With a positive-lock-type valve seal, the top of the valve guide has been machined so that the seal locks onto the guide. First, a protective cap is placed over the valve stem. Then, an installation tool is used to force or snap the valve seal into position.

Installing Valve Springs

To reinstall the valve springs, place the valve spring over the valve stem and onto its seat. Some springs have a top and bottom so make sure the spring is not upside-down. Next, place the spring retainer over the spring and against the valve stem. Then, position the valve spring compressor over the spring retainer and valve head, **Figure 22-49A.**

Hold the compressor square on the valve and compress the spring. Then, fit the retainer and keepers into their grooves on the valve stem. Be careful; keepers are very small and can be easily dropped.

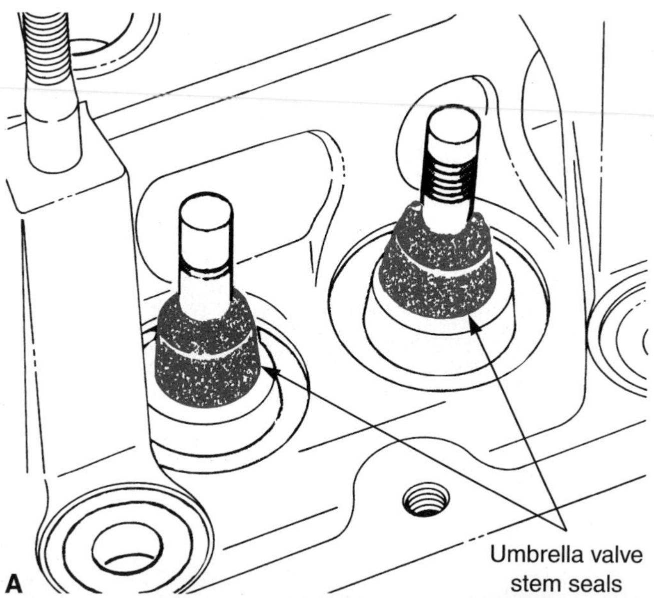

Figure 22-48. *A—Slide an umbrella-type valve seal over the valve before installing the spring. B—When installing an O-ring valve seal, compress the spring and fit the seal into the groove in the valve stem. C—Some OHC engines have valve seals that are integral to the valve spring seat. These seats/seals must be installed in the bottom of the spring pocket before installing the springs. (DaimlerChrysler, Fel-Pro)*

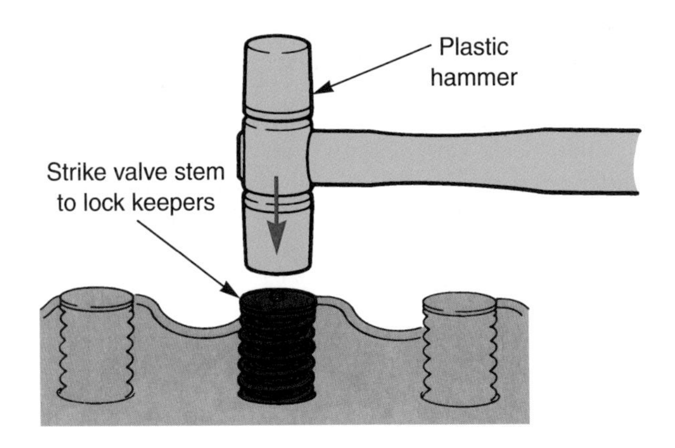

Figure 22-49. *Tap valve stems with a plastic mallet to fully seat the keepers in their grooves. (Toyota)*

Once the keepers are in their grooves, release the valve spring compressor. Check that the keepers stayed in place. Tap on the valve stems with a plastic or brass hammer to seat the keepers in their grooves, **Figure 22-49.** To avoid severe engine damage, it is critical that keepers cannot pop out from between the valve stem and the retainer.

Some OHC engines require a special valve spring compressor, as shown in **Figure 22-50A.** It is needed to reach down into the valve spring pockets. If working on a overhead camshaft head with deep pockets for the springs, use an adapter on the spring compressor. You can reach through the adapter to fit the small valve keepers down into their groove in the valve stem. A small pick-type tool will help work the keepers down into place.

With some overhead camshaft engines, you cannot directly hit the stem with a mallet to seat the keepers. Instead, use flat nose punch and hammer to hit each valve stem tip. Hit straight down on the tips to seat the keepers. **Figure 22-50B** shows how to seat the keepers in an OHC head with the springs recessed down into hard-to-reach pockets.

Warning: Wear safety glasses when using a valve spring compressor.

Checking for Valve Leakage

Placing a liquid in the cylinder head ports is a good way to double-check for valve leakage after head reconditioning. Lay the cylinder head on its side with the ports up. If intake and exhaust ports are on opposite sides of the head, you will need to test one side and then the other. Pour or spray clean solvent or water into the ports. See **Figure 22-51.** Then, shake the head to splash the liquid around the valve heads.

As you splash liquid around in the cylinder head, watch for leakage at the valve heads. If solvent or water drips from around a valve and its seat, that valve is leaking. When solvent is used, a small amount of *seepage* is acceptable. Solvent is normally thinner than water and can seep through smaller clearances. This solvent *seepage* is not a concern. The valve will wear and seal this seepage after only a few minutes of engine operation. However, the solvent must not *leak* or *pour* from the valve-to-seat contact point. If the valve leaks badly, remove the valve from the head and find the problem.

Rocker Arm Stud Service

Rocker arm studs need to be replaced if their threads are damaged or if pressed-in rocker arms pull out of the head. Some rocker arm studs are press-fit in the head; others screw into the head. **Figure 22-52** summarizes how to replace pressed-in cylinder head rocker arm studs.

1. Use a puller to remove the old stud. Unscrew threaded rocker arm studs.

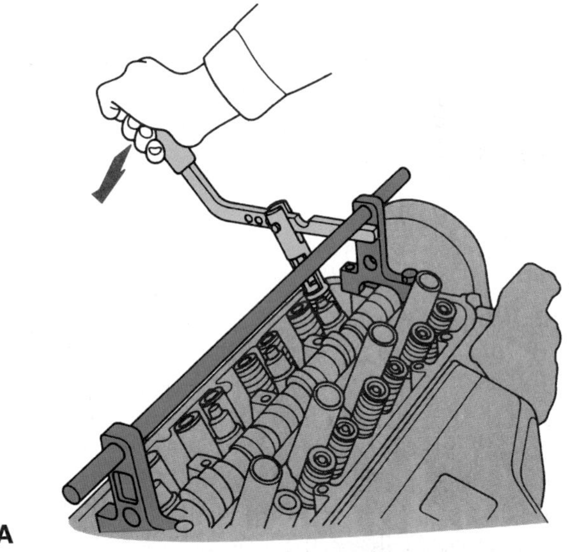

Figure 22-50. *A—In this application, a special valve spring compressor is needed. Note how it fastens to the camshaft cap studs. B—A flat nose punch and hammer can be used to tap the valve stem tips when they are recessed into a pocket. (Honda)*

2. You may need to ream the hole larger. For example, if the rocker arm stud had pulled out during service, the hole may be damaged. If the hole is reamed oversize, an oversize rocker arm stud can be installed.
3. Finally, use an appropriate driver to press-in a new rocker arm stud.

To replace threaded rocker arm studs:

1. Use a deep socket to unscrew the rocker arm stud.
2. Verify that the threads in the cylinder head are not damaged. Make repairs as needed.
3. Compare the old rocker arm stud to the new one to make sure they are identical.
4. Thread the new rocker arm stud into the hole.
5. Use a torque wrench to tighten the rocker arm stud to specifications.

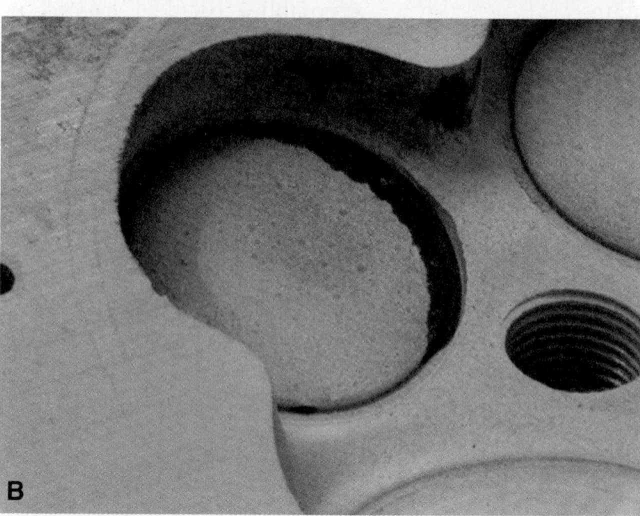

Figure 22-51. *Double-checking for valve leakage. A—Pour or spray clean solvent or water into the ports. B—If water or solvent pours out around the valve head and seat, something is wrong.*

In-Vehicle Valve Seal Service

Valve seals and springs can be serviced without cylinder head removal. This will allow in-vehicle seal or spring replacement.

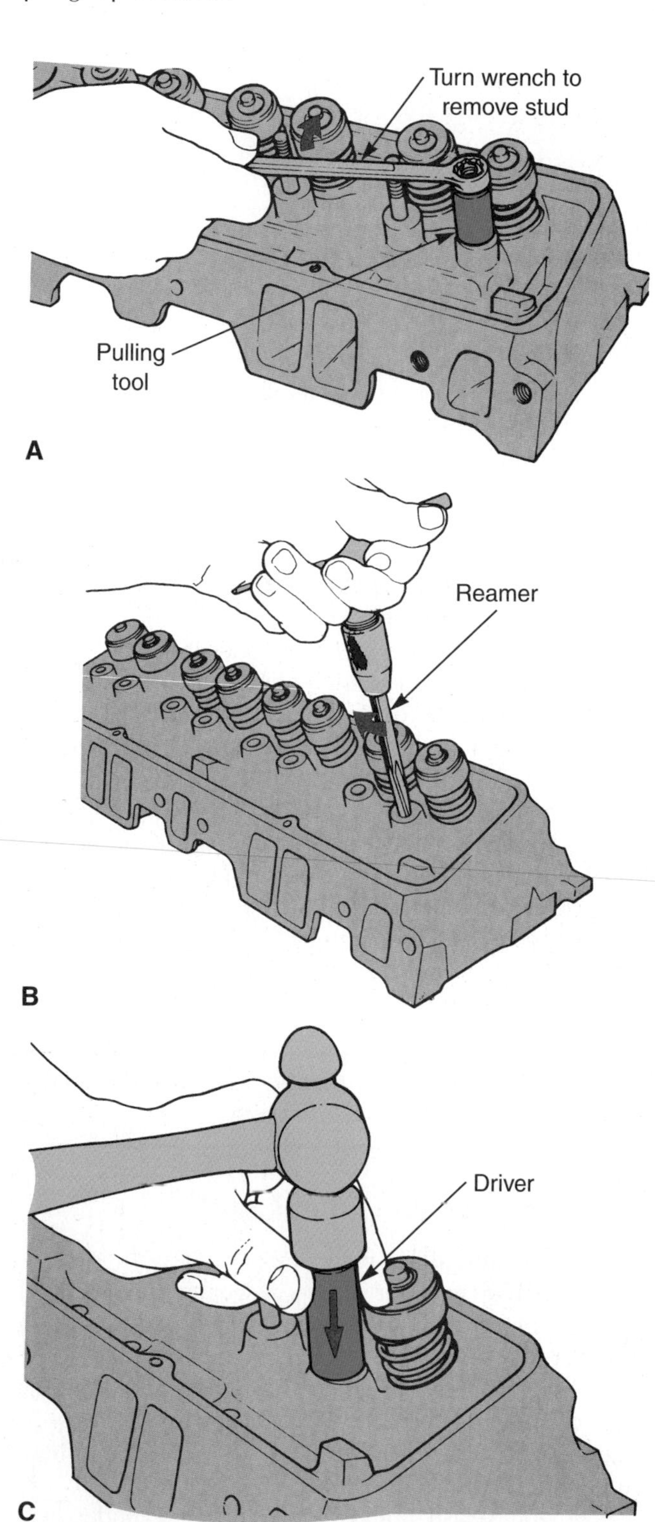

Figure 22-52. *Rocker arm stud replacement. A—Use a puller to remove the old, pressed-in stud. B—If needed, ream the hole oversize for an oversize stud. C—Drive the new stud in place. (General Motors)*

1. Remove the spark plugs, valve covers, and rocker arms.
2. Rotate the crankshaft until all valves in the cylinder to be worked on are closed.
3. Use a shop air hose and special fitting to inject air into the cylinder, **Figure 22-53A.** The air pressure will hold the valves in that cylinder up against their seats.
4. Only work on one spring at a time. Use a special pry bar- or screw-type compressor to compress the spring, **Figure 22-53B.**
5. Remove the keepers and spring.
6. Install a new valve seal on the valve.
7. Reassemble the valve spring, retainer, keepers, seal (if O-ring type), and any needed shims onto the valve.
8. Repeat for the other valve(s) in the cylinder.
9. Repeat for other cylinders as needed.
10. Reassemble the rocker arms, valve covers, and spark plugs.

Figure 22-53. *Valve seals and broken springs can be serviced without cylinder head removal. A—Use shop air to hold the valves against their seats. B—Use a special spring compressor to remove the keepers and springs for that cylinder.*

Camshaft Service

Camshaft service involves measuring camshaft lobe and journal wear. It also includes measuring camshaft endplay and camshaft bearing clearance.

Measuring Camshaft Wear

Camshaft lobe wear can be measured with a dial indicator with the camshaft installed in the engine. Mount the indicator so its stem is parallel to a push rod and contacts the rocker arm or end of the push rod. Then, zero the gauge and rotate the camshaft. The difference between the maximum and minimum indicator values is the valve lift. If the camshaft is out of the engine, use a micrometer to measure all lobes.

Subtract the measured lift from the specification for lift to determine wear. If wear on any lobe is out of specifications, over about .005″ (0.13 mm), replace or grind the camshaft.

Camshaft straightness is checked with the camshaft out of the engine. Mount the camshaft in V-blocks. Then, mount a dial indicator as pictured in **Figure 22-54** and rotate the camshaft. If the dial indicator reads more than specifications, usually about .002″ (0.05 mm), the camshaft is bent or warped and must be replaced.

Camshaft journal wear is measured with an outside micrometer, as in **Figure 22-55.** If worn more than specifications, approximately .002″ (0.05 mm), the camshaft is usually replaced. Worn camshaft journals lower engine oil pressure considerably.

Figure 22-56 shows a technician using a telescoping gauge to check the dimensions of camshaft bearings. By subtracting the camshaft journal diameter from the camshaft bearing diameter, you can determine the bearing clearance. Normally, each camshaft journal is a different diameter. Check the service manual for specifications for each camshaft journal and bearing.

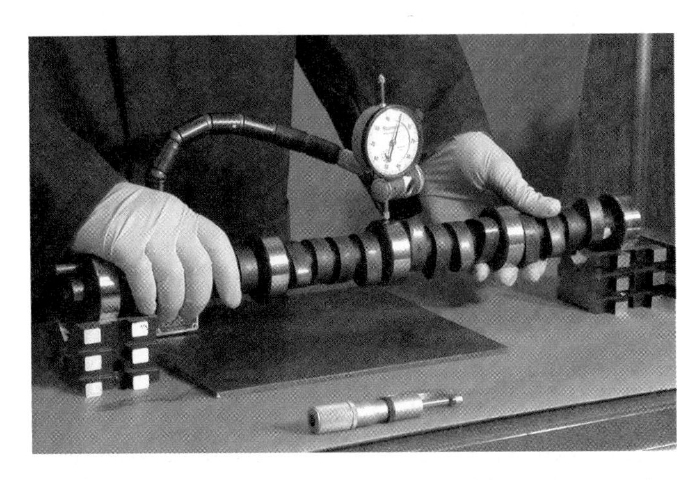

Figure 22-54. *Checking camshaft straightness by measuring runout.*

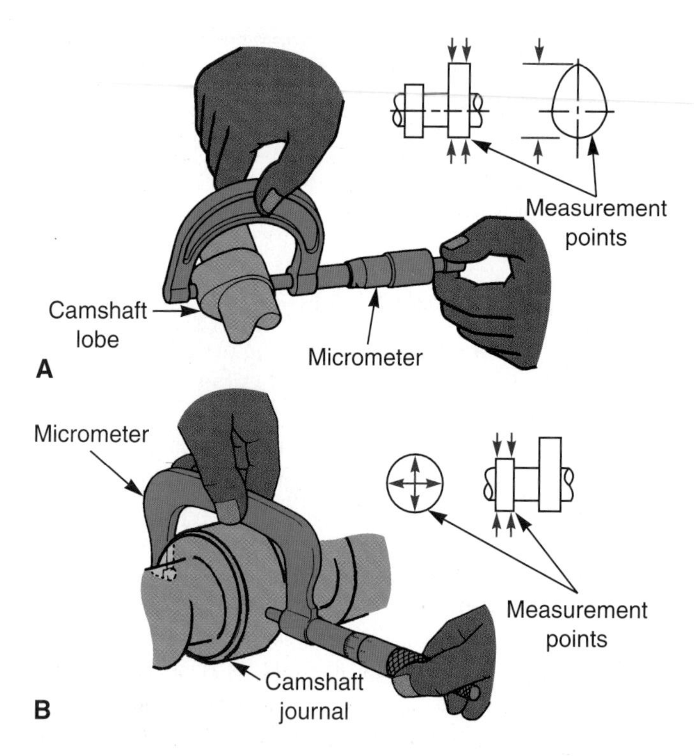

Figure 22-55. *A—Measuring camshaft lobe wear. B—Measuring camshaft journal wear. (Toyota)*

Camshaft Grinding

Camshaft grinding is done to restore camshaft lobe and journal dimensions to specifications. A camshaft grinding machine regrinds *all* of the camshaft lobes. If a journal is badly damaged, it can be welded and then ground down to size.

A ground camshaft may be less expensive than a new camshaft. So long as the lobes are hardened properly, it can give good service life. Always compare the price of a new camshaft with that of reconditioning a camshaft. Sometimes, a new camshaft may be more cost effective when "down time" is considered or even less expensive off the shelf.

Figure 22-56. *Measuring the camshaft bearing diameters. Use a micrometer to measure the telescoping gauge.*

Installing an Overhead Valve Camshaft

Before installing the camshaft, lubricate the camshaft journals and camshaft bearings with assembly lube. See **Figure 22-57A.** Since the camshaft lobes endure very high friction on engine startup, thick assembly lube is very important. It will protect the camshaft lobes and lifters from rapid wear during initial engine start-up. To install an overhead valve (OHV) camshaft into the cylinder block, use the following procedure.

1. Slowly slide the camshaft into the camshaft bearings. Do not hit or nick the camshaft bearings.
2. Slowly rotate the camshaft to help feed it through each bearing, **Figure 22-57B.**
3. Make sure you support the camshaft as it is inserted.
4. To help feed the very end of the camshaft into the engine, a threaded handle or bolt should be installed into the camshaft. This will allow you to lift and guide the camshaft completely into the camshaft bearing bores. You must lift up the rear of the camshaft to get it to slide into its bearing.

Figure 22-57. *Installing a camshaft in an OHV engine. A—Apply thick assembly lube on the camshaft journals and lobes. B—Carefully slide the camshaft into the camshaft bearings.*

5. Install the camshaft retainer plate. Make sure it is facing in the right direction. Sometimes the retainer plate has oil grooves that must face toward the inside of the engine, **Figure 22-58.**
6. Torque the bolt(s) to specifications.

Note: The inspecting, removing, and installing of camshaft bearings for an overhead valve engine is discussed in the previous chapter.

Camshaft Endplay

Camshaft endplay is the front-to-rear movement of the camshaft in its bores. It is commonly measured with a dial indicator. Install a bolt in the end of the camshaft. Then, mount the dial indicator so the stem is parallel with the camshaft centerline and contacts the bolt. See **Figure 22-59.** Zero the indicator. Then, pry back and forth on the bolt in the end of the camshaft. The maximum indicator needle movement equals the camshaft endplay.

If endplay is not within specifications, replace or repair parts as needed. The flange on the camshaft may be worn or the thrust surface on the head or block may be damaged.

OHC Bearing Service

Camshaft bearings in the cylinder head of an overhead camshaft engine are serviced in much the same way as the camshaft bearings in cam-in-block engines. You must check the bearings for wear, damage, and clearance. Worn camshaft bearings can reduce engine oil pressure.

Figure 22-58. *Installing the camshaft thrust plate. Make sure the face that has oil grooves (if so equipped) points toward the engine.*

Figure 22-59. *Measuring camshaft endplay.*

Camshaft bearing wear is checked by measuring the bearing's inside diameter. Use a dial bore gauge or telescoping gauge and micrometer, **Figure 22-60A.** Subtract the specification diameter from the measured diameter to determine wear. If wear is more than specifications, about .002″ (0.05 mm), replace the bearings.

Most technicians replace the camshaft bearings during major engine service regardless of wear. The bearings are not very expensive and are important to engine service life. Cylinder head camshaft bearings can be a one- or two-piece design. If a one-piece design, similar to cam-in-block camshaft bearings, you must drive them in and out with a bearing driver. This is often done by a machine shop.

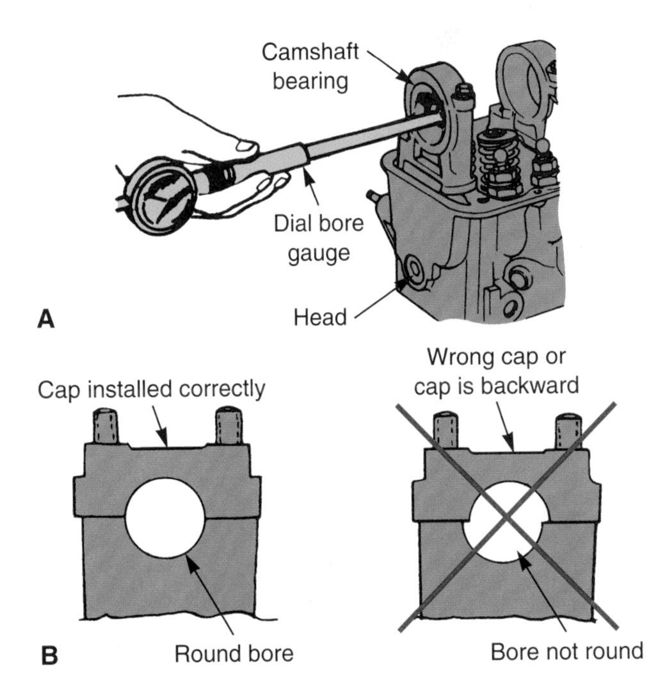

Figure 22-60. *A—Measuring camshaft bearing wear with a dial bore gauge. B—Make sure camshaft caps are correctly installed. (VW, Nissan)*

If a two-piece design, they simply snap into place like rod or main bearings. The back of each bearing should be clean and dry. Always make sure the camshaft bearing caps are installed in their correct locations in the proper orientation. If mixed-up or turned backward, the camshaft bearing bore will not be round. This is shown in **Figure 22-60B.**

When installing any camshaft bearing, make sure the oil holes in the bearing align with the oil holes in the cylinder head and cap (if applicable). Do not dent or mar the bearing surfaces.

With two-piece camshaft bearings, you can measure ***camshaft bearing clearance*** with Plastigage. Use the same basic procedure discussed in the previous chapter for determining main bearing or connecting rod bearing clearance. To find the bearing clearance on a one-piece bearing, subtract the measured camshaft journal diameter from the measured camshaft bearing diameter to calculate clearance.

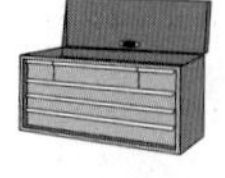

Note: Installation of an overhead camshaft is covered in Chapter 24. On some OHC engines, the cylinder head must be installed before the camshaft.

Rocker Assembly Service

Rocker assembly service involves cleaning, inspection, and measurement of the rocker arms, rocker arm shafts, and related components. A typical assembly is shown in **Figure 22-61.**

Inspect all wear points closely. Check the rocker shafts for signs of heavy wear or scoring. Check for bushing wear or wear on the external surfaces where the rocker arm makes contact with the valve, lifter, push rod, or camshaft lobe. Also, check the adjuster threads to make sure they are not damaged.

A dial bore gauge is needed to measure rocker arm bushing or inside diameter wear, **Figure 22-62.** A telescoping gauge and micrometer can also be used. Measure for out-of-round. If more than specifications, replace parts or install new rocker arm bushings, if applicable.

Use an outside micrometer to check the rocker arm shaft for wear. Measure the diameter of the rocker arm shaft at each location where the rocker arm contacts the shaft. If the diameter is not within specifications at any point, replace the shaft. Also, check the surface of the rocker arm shaft. It should be smooth without any nicks, gouges, or scoring.

During rocker arm reassembly, make sure the rocker shaft oil holes point down. The bottom of the shaft is the load-bearing side of the shaft. Oil is needed to reduce wear at this area of high friction. If the oil holes are installed up, the rocker arm shaft and rocker arms will wear rapidly due to lack of lubrication.

With a ball- or stand-type valve mechanism, inspect the ball, pivot, or stand for wear. If the rocker stand or pivot is aluminum, inspect it closely. Aluminum is soft and prone to wear. Frequently, replacement is required. Replace any rocker part worn beyond specifications.

Push Rod Service

Push rod service involves checking for tip wear and straightness. When a push rod tip is worn, the center where the push rod fits into the rocker arm tends to become pointed. A push rod does not wear as much at the lifter end of the push rod because it is submersed in oil.

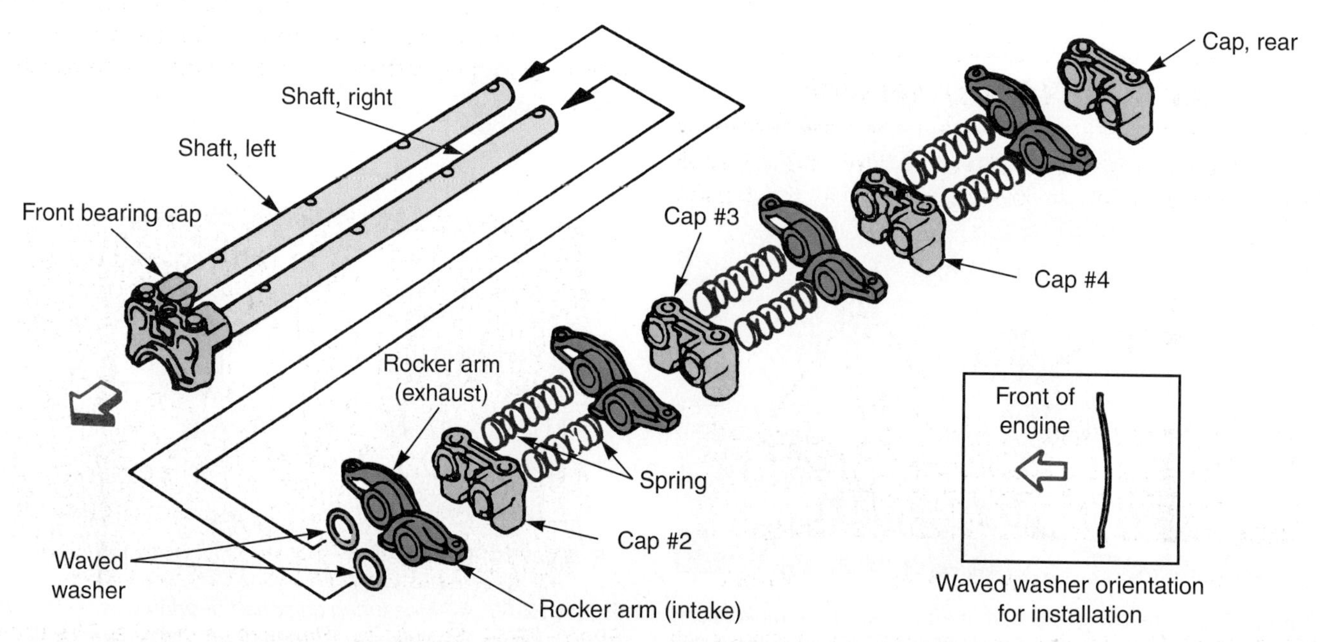

Figure 22-61. *This is a typical rocker arm shaft assembly. (DaimlerChrysler)*

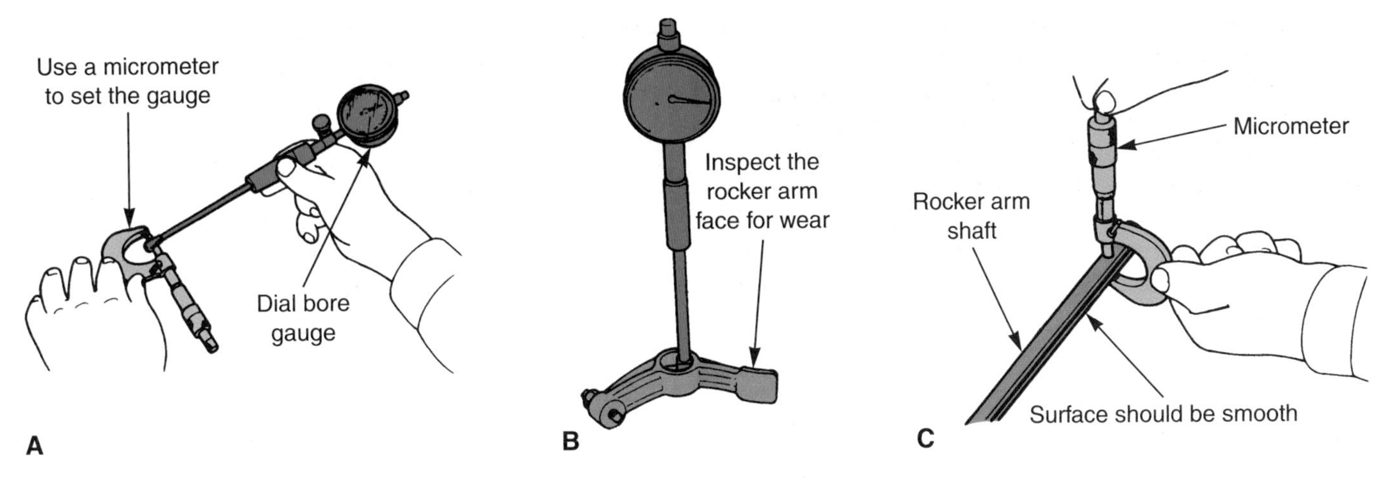

Figure 22-62. *Measuring rocker arm bushing wear. A—Adjust the dial bore gauge to the correct size for the rocker arm bushing. B—Insert the dial bore gauge into the rocker arm bushing and measure wear and out-of-round. C—Measure rocker arm shaft wear with a micrometer. (Honda)*

You can usually check push rod straightness by rolling the push rods on a smooth, flat work surface. A more accurate, but more time-consuming, method is to mount the push rod in V-blocks and measure runout with a dial indicator. This is shown in **Figure 22-63.** A bent push rod can result from a valve hitting a piston, stuck valve, spring bind, and other similar problems.

Push rod straightness can also be checked by using a drill press. Chuck the push rod in the drill press, turn the drill press on, and watch for push rod movement. If the push rod wobbles in the drill press, it is bent and must be replaced.

Replace any bent push rod. Do not attempt to straighten a push rod. Doing so will decrease the push rod's strength.

Most push rods are hollow to feed oil up to the rocker arms. If the push rod is hollow, look down through the push rods to make sure the hole is not clogged with debris. Run wire through the push rod if needed to clean it out.

Lifter (Tappet) Service

The contact surface between a lifter and camshaft lobe is one of points in an engine that has the highest friction and wear. Hydraulic lifters can also wear internally, causing valve train noise. ***Lifter service*** involves inspecting, testing, and, if needed, rebuilding.

Some lifters can be oversize from the factory. This makes it important for you to keep all lifters in order so they can be reinstalled in the same location in the engine. **Figure 22-64** shows how the lifter bore is marked, denoting an oversize lifter and lifter bore.

Inspecting Lifters

Inspect the bottoms of the lifters for wear. A reusable lifter has a slight hump or convex shape on the bottom. See **Figure 22-65.** Worn lifters are flat or concave on the bottom. Replace the lifters if the bottoms are worn.

Never install used lifters on a new camshaft. The wear on the old lifters, even if acceptable, will cause rapid wear of the camshaft lobes and additional wear on the lifters. Normally, install new lifters when a camshaft is replaced.

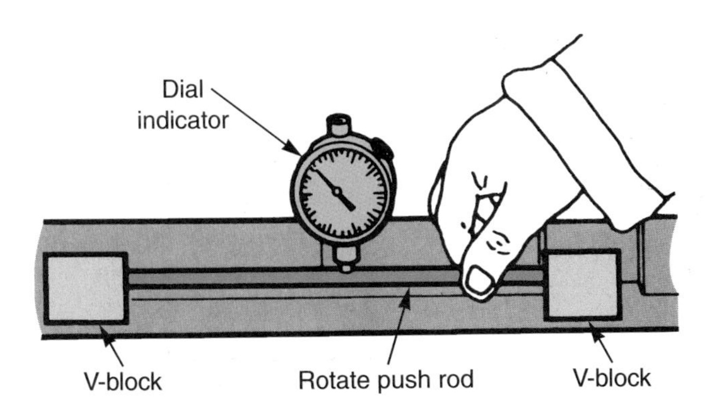

Figure 22-63. *Use V-blocks and a dial indicator to check push rods for straightness. (Ford)*

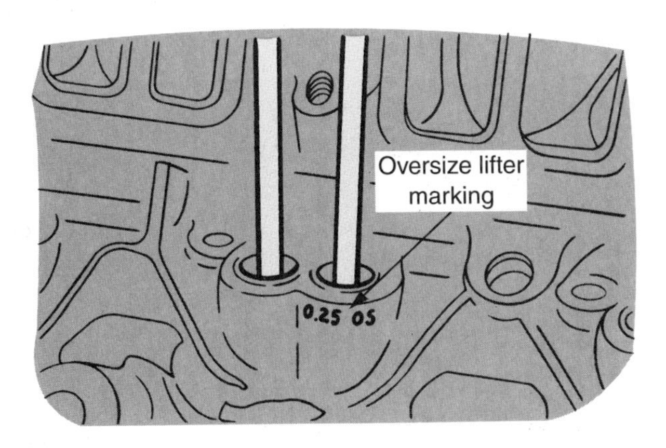

Figure 22-64. *Sometimes, lifters can be oversize. This should be marked on the block.*

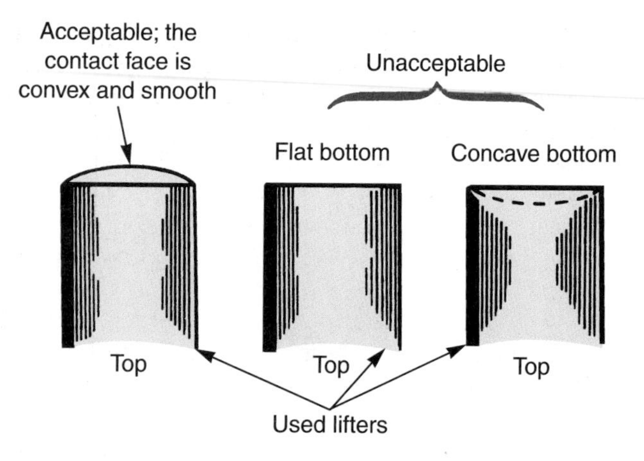

Figure 22-65. *A lifter that can be reused has a slight crown or convex shape on the bottom. If not, the lifter is worn and should be replaced. (Ford)*

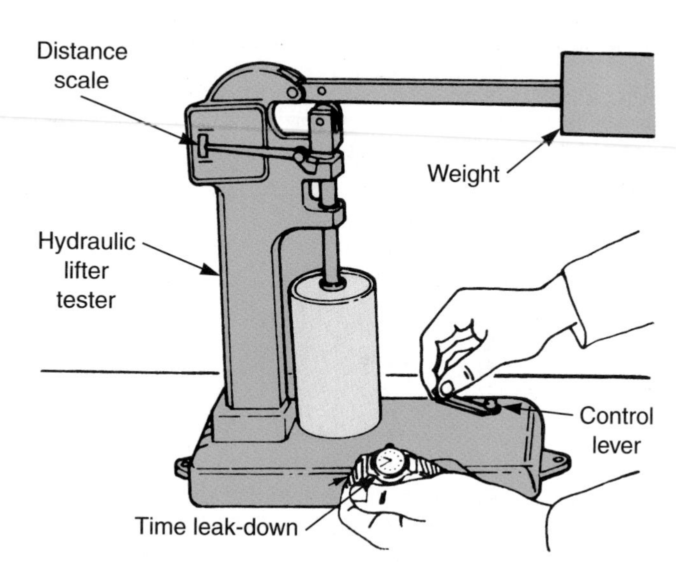

Figure 22-66. *Performing a leak-down test on a hydraulic lifter. (General Motors)*

Testing Lifter Leak-Down Rate

Hydraulic lifter ***leak-down rate*** is the amount of time it takes the pressure inside the lifter to bleed out. It is measured using a lifter tester, which pushes the lifter plunger to the bottom of its stroke under controlled conditions, **Figure 22-66.** The lifter leak-down rate is timed and compared to specifications.

Generally, fill the tester with a special test fluid. Place the lifter in the tester. Then, follow the tester's instructions to determine the lifter leak-down rate. If leak-down is too fast or slow, replace or rebuild the lifter.

Rebuilding Hydraulic Lifters

Rebuilding hydraulic lifters typically involves complete disassembly, cleaning, and measurement of lifter components. All worn or scored parts must be replaced following manufacturer instructions. Many shops do not rebuild hydraulic lifters. If the lifters are defective, new ones are installed. Replacing lifters usually saves time and money over rebuilding. However, on some OHC lifters or followers it may be more cost effective to rebuild rather than replace the units.

Bleeding Cam Followers

Many technicians will bleed down used cam followers before engine assembly. Hydraulic cam followers may retain oil after the engine is off. If the engine is rebuilt and the cam followers are not bled down, starting problems can result. The oil in the followers will hold the valves open and prevent a normal compression stroke. This will not allow engine starting until an extended period of engine cranking. The engine will not start until valve spring pressure pushes the extra oil out of the cam follower.

To avoid an engine starting problem after major repairs, you should bleed down hydraulic cam followers as shown in **Figure 22-67.** This will also allow you to check for follower wear.

1. Pull the plunger out of the cam follower with pliers.
2. Use a pin punch to depress the check valve and relieve the oil pressure.
3. Inspect the follower for wear.
4. Push the plunger back into the follower.

The cam follower is now bled down and ready for reassembly.

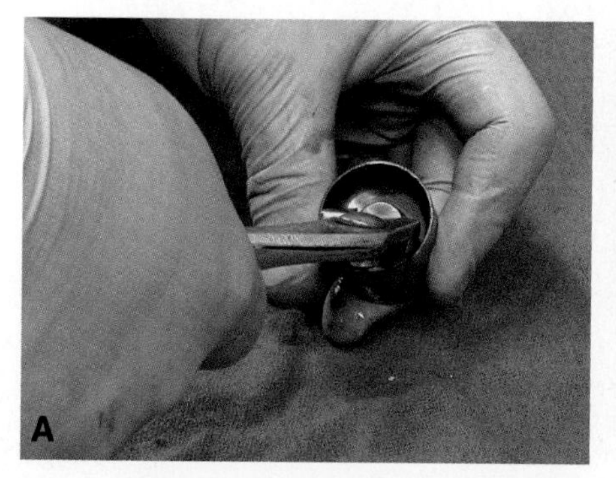

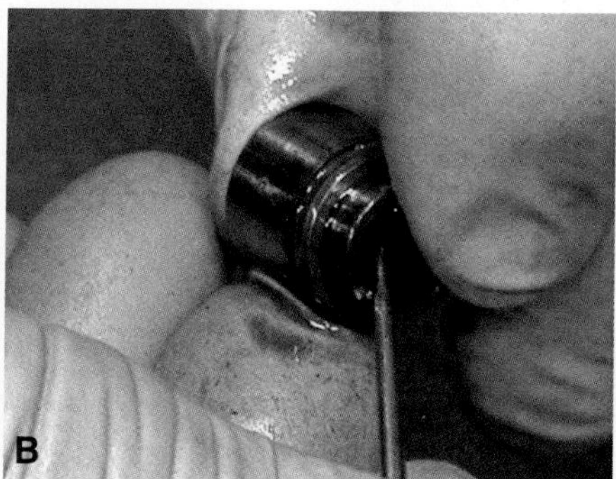

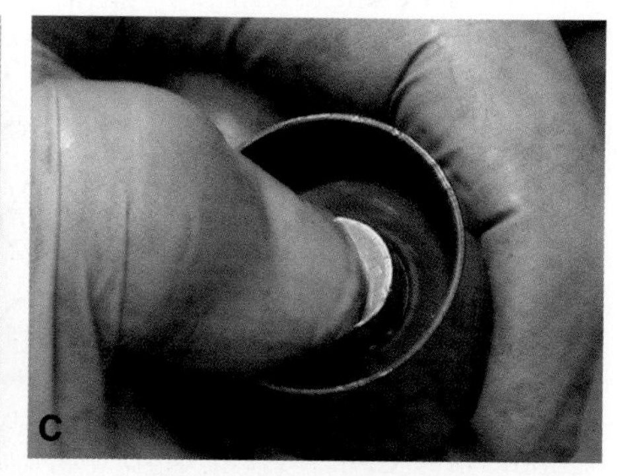

Figure 22-67. *Bleeding down a cam follower. A—Pull the plunger out of the follower with pliers. B—Use a pin punch to depress the check valve and relieve the oil pressure. C—Push the plunger back into the follower.*

More Information

The next two chapters provide more information relating to cylinder head service. These chapters discuss torquing cylinder heads and manifolds, installing timing covers, timing belts, valve covers, front covers, and so on. The next chapter details timing mechanism service. Chapter 24 discusses engine reassembly.

Summary

Cylinder heads can suffer from a wide range of problems, including warpage, cracks, burned seats, burned valves, worn valve guides, and so on. Magnafluxing is commonly used to find cracks in cast iron heads. Dye penetrant is used to find cracks in aluminum heads. Cracks can sometimes be repaired by welding. If badly damaged, it might be better to install a good used or new head.

Head warpage is checked with a straightedge and feeler gauge. Check at different angles across the head. Milling is used to correct head warpage. Do not mill too much off of a head or it can affect the compression ratio and, on V-type engines, the fit of the intake manifold.

Valve guide wear is common and can be checked with a dial indicator or small telescoping gauge. Valve guide wear can be corrected by knurling, reaming oversize, installing a bronze liner, or installing a new guide insert.

Valve seat reconditioning involves grinding or cutting a new face on the seat. Valve retrusion is a problem caused by wear or grinding when the valve sinks into the combustion chamber. Valve seat replacement is commonly done by a machine shop.

Valve seats can be resurfaced with a grinding stone or carbide cutter. Seats are usually 45°, but a few high-performance heads have a 30° angle seat. Only grind as little as needed off of the seat. You may have to move the seat in or out so it touches the valve properly. Use 15° or 30° and 60° stones or cutters to reposition the valve-to-seat contact point.

Valve lapping compound or Prussian blue can be used to check seat location on the valve. An intake valve should have a seat contact width of about 1/16″ (1.6 mm). The exhaust valve needs a little more contact area, 3/32″ (2.4 mm) being typical. Make sure you clean off all lapping compound to avoid part wear.

Valves are reconditioned on a valve grinding machine. The valve faces and stem tips are ground smooth and true. Remove as little as possible off of the valves. Wear eye protection when grinding and follow equipment operating instructions.

Valve springs should be replaced or checked closely for problems. Check spring squareness, spring tension, and spring free length. Valve spring shimming may be needed to produce the correct spring pressure.

Use a valve spring tester to measure spring pressure. Measure the spring installed height. Use spring shims to reduce installed height to equal the tester value and meet spring pressure specifications.

Install valve seals carefully. Umbrella seals simply slide onto the valve stems before spring installation. However, you must compress the valve springs before installing O-ring type valve seals. This will prevent O-ring seal damage.

Check the camshaft closely for wear or damage. Measure lobe height, journal diameter, and runout. Replace or have a machine shop grind the camshaft if needed.

Inspect all valve train parts closely. Replace any components showing wear or damage. Measure lifter leak-down rate, camshaft bearing wear, push rod straightness, and so on.

Review Questions—Chapter 22

Please do not write in this text. Write your answers on a separate sheet of paper.

1. List two ways to locate cracks in a head.
2. To measure _____, lay a straightedge across the cylinder head deck and slide feeler gauges under it.
3. _____ is a machine shop operation that removes a thin layer of metal from the head deck surface.
4. Explain four ways of correcting valve guide wear.
5. Valve seat retrusion can cause lowered _____, reduced _____, incorrect valve train _____, and decreased engine _____.
6. A seat grinding stone should be slightly _____ in diameter than the valve seat.
7. What is a stone dressing tool?
8. How does valve seat *cutting* differ from valve seat *grinding?*
9. The most common valve seat angle is _____ degrees, but in some applications the seat is ground at _____ degrees.
10. Briefly describe how to move the valve seat contact point.
11. What are typical valve-to-seat contact widths for intake and exhaust valves?
12. Define *interference angle.*
13. If a valve seat is ground at a 45° angle, what is the setting on the valve grinding machine in order to create an interference angle?
14. A(n) _____ indicates too much valve face removal and requires valve replacement.
15. Why is the valve stem tip ground?

16. If the valve spring installed height is greater than specifications, how do you determine the proper valve spring shim thickness?
17. Define *spring bind.*
18. After the cylinder head is reassembled, how can you check for valve leakage?
19. How do you measure camshaft straightness?
20. What is the purpose of camshaft grinding?
21. How do you measure camshaft bearing wear?
22. How do you measure the camshaft bearing clearance on a two-piece camshaft bearing in an OHC engine?
23. When installing a rocker arm shaft, the oil holes should face _____.
24. How can a drill press be used to check for push rod straightness?
25. Why should you bleed down cam followers before engine reassembly?

ASE-Type Questions—Chapter 22

1. An engine has coolant in the oil, which is apparent from the milk-like substance in the crankcase. Technician A says the engine could have a warped head and blown head gasket. Technician B says that the engine could have a cracked head or block. Who is correct?
   (A) A only.
   (B) B only.
   (C) Both A and B.
   (D) Neither A nor B.
2. Technician A says that umbrella and O-ring valve seals are installed after compressing the valve spring. Technician B says to install umbrella or O-ring valve seals and then compress the valve spring. Who is correct?
   (A) A only.
   (B) B only.
   (C) Both A and B.
   (D) Neither A nor B.
3. Technician A says that magnafluxing should be used to detect cracks in cast iron cylinder heads. Technician B says that dye penetrant should be used to find cracks in aluminum cylinder heads. Who is correct?
   (A) A only.
   (B) B only.
   (C) Both A and B.
   (D) Neither A nor B.
4. Technician A says that some cracked cylinder heads can be welded. Technician B says that some cracked cylinder heads can be plugged. Who is correct?
   (A) A only.
   (B) B only.
   (C) Both A and B.
   (D) Neither A nor B.
5. Technician A says that grinding material at Points 1 and 2 will reduce valve spring tension. Technician B says that adding the correct thickness of shim at Point 3 will restore spring tension specifications. Who is correct?
   (A) A only.
   (B) B only.
   (C) Both A and B.
   (D) Neither A nor B.

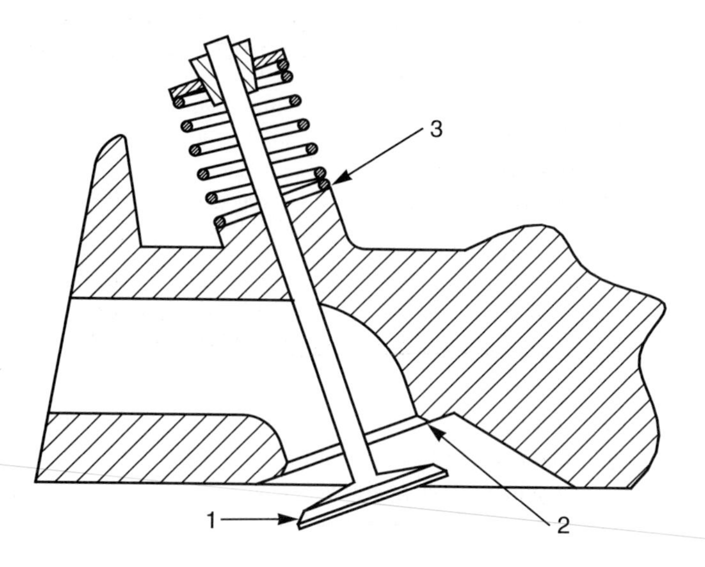

6. Technician A says that valve guide wear can cause engine coolant leakage. Technician B says that valve guide wear can cause engine oil consumption. Who is correct?
   (A) A only.
   (B) B only.
   (C) Both A and B.
   (D) Neither A nor B.
7. Technician A says that knurling is a common method used to repair worn valve guides. Technician B says that reaming is a common method used to repair worn valve guides. Who is correct?
   (A) A only.
   (B) B only.
   (C) Both A and B.
   (D) Neither A nor B.

8. Valve seat replacement should be performed _____.
   (A) anytime the cylinder head is removed from the engine
   (B) only when seat wear is severe
   (C) when replacing the valves
   (D) when replacing the valve springs
9. Technician A says that an interference angle is normally a 4° difference between the valve face angle and the valve seat angle. Technician B says that an interference angle reduces the break-in and sealing time of the valve. Who is correct?
   (A) A only.
   (B) B only.
   (C) Both A and B.
   (D) Neither A nor B.
10. Technician A says that you should never shim a valve spring over .120″. Technician B says that you should never shim a valve spring over .145″. Who is correct?
   (A) A only.
   (B) B only.
   (C) Both A and B.
   (D) Neither A nor B.
11. Technician A says that camshaft lobe wear should be measured with a telescoping gauge. Technician B says that camshaft lobe wear can be measured with the camshaft installed in the engine. Who is correct?
   (A) A only.
   (B) B only.
   (C) Both A and B.
   (D) Neither A nor B.
12. Technician A says that a bent push rod can result from a valve hitting a piston. Technician B says that a bent push rod can result from a stuck valve. Who is correct?
   (A) A only.
   (B) B only.
   (C) Both A and B.
   (D) Neither A nor B.
13. The 45° contact area should be moved out and narrowed by using a _____ stone.
   (A) 5°
   (B) 30°
   (C) 45°
   (D) 60°

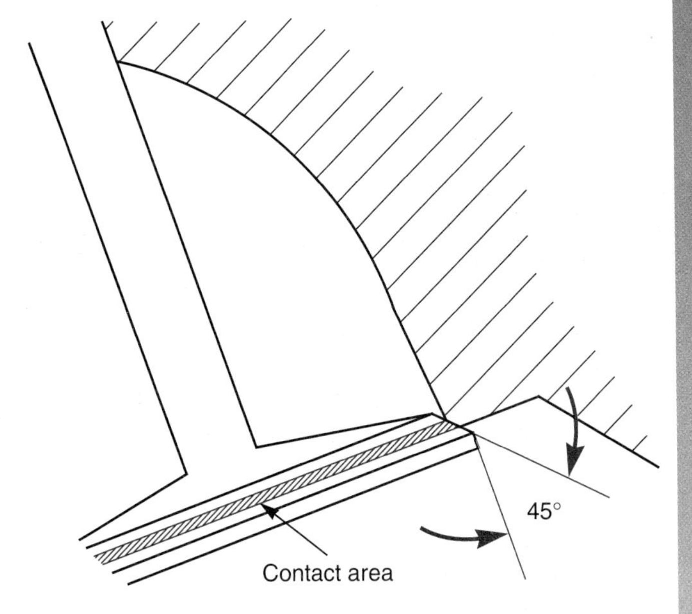

Front end service includes maintenance on the timing chain, belt, or gears and the front cover. (Ford)

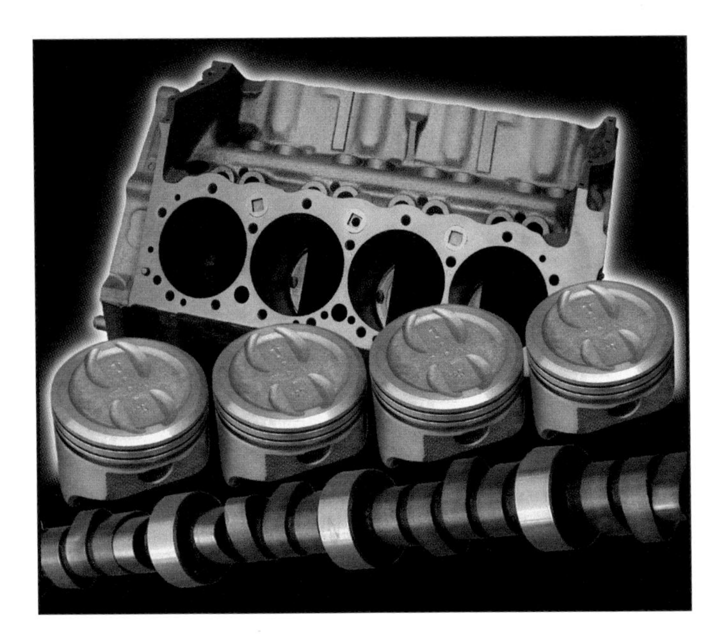

Chapter 23

Front End Service

After studying this chapter, you will be able to:

- ❑ Check a timing chain for wear.
- ❑ Install a timing chain assembly.
- ❑ Install a timing sprocket and belt.
- ❑ Service timing chain tensioners.
- ❑ Adjust timing belt tension.
- ❑ Install a timing belt cover.
- ❑ Install timing gears.
- ❑ Check timing gear runout and backlash.
- ❑ Properly torque timing gear or sprocket fasteners.
- ❑ Replace a front cover oil seal.
- ❑ Install an engine front cover.

Know These Terms

Chain tensioner

Crankshaft front seal

Engine front cover

Oil slinger

Timing belt

Timing belt cover

Timing belt tension

Timing chain

Timing gear backlash

Timing gear runout

Timing gears

Timing marks

Today's engines, as you have learned, are much more complex than the ones used in passenger vehicles just a few years ago. V-8 engines with two or four overhead camshafts and variable valve timing are now built by several manufacturers. Long timing chains or timing belts are needed to smoothly and dependably rotate the camshafts with the crankshaft. Conventional cam-in-block engines with short timing chains or timing gears are also manufactured. This makes the service of an engine front end more challenging than ever before. An engine's front end includes the timing mechanism, front cover, and related seals and parts. See **Figure 23-1.**

This chapter summarizes the most critical information concerning the service and repair of timing chains, timing belts, and timing gears. Timing mechanism service is critical. If a mistake is made, severe engine damage or performance problems may result. Improper crankshaft-to-camshaft timing can cause the engine pistons to slam into the opened valves. Valves can be bent or snapped off at their heads. Pistons and cylinder heads can also suffer severe damage when a valve is broken off by incorrect valve timing.

Note: Earlier chapters covered information useful in understanding the material in this chapter. For instance, Chapter 20 explained engine disassembly, including front end disassembly. Chapter 11 explained the types and construction of camshaft drives.

In-Vehicle Timing Mechanism Service

Often, a timing chain, belt, or gears can be serviced without removing the engine from the vehicle. Usually, the parts on the front of the engine and the timing cover can be removed without complete engine disassembly. In some OHC engines, the camshaft(s) must be secured before timing mechanism removal. If this is not done on these engines, the camshaft(s) will slip out of time and can cause engine damage. Refer to the service manual for specific directions as needed.

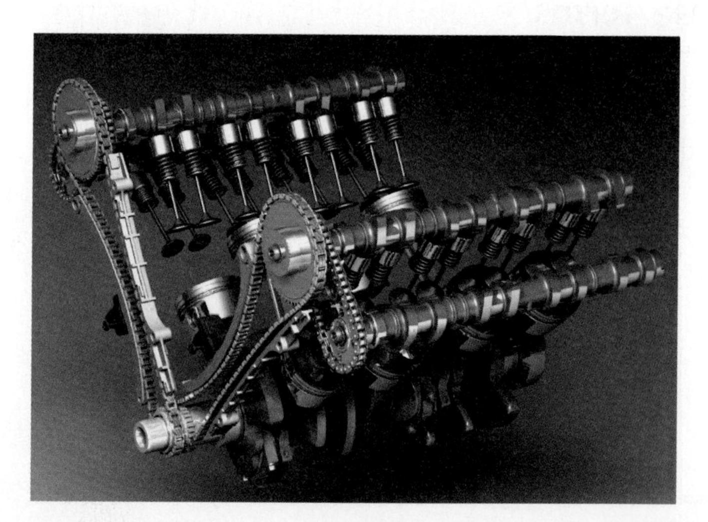

Figure 23-1. *The front end or timing mechanism on today's engines can be complex. You must be knowledgeable of proper repair methods to successfully service modern engines. (Jaguar)*

Timing Chain Service

Timing chains transfer rotation from the crankshaft to the camshaft. In some designs, they also power other devices. Timing chains can be found in both overhead valve and overhead camshaft engines. OHC timing chains often have a complex set of guides, chain adjusters, and idler gears to maintain chain tension and prevent chain slap.

Timing Chain Problems

Timing chains and their drive sprockets suffer from wear after prolonged service. The chain links can wear and stretch. The sprocket teeth can also wear down. This can cause excess chain slack or the chain can be loose on its sprockets. A badly worn timing chain can have so much slack that it hits the side of the front cover and produces a loud noise.

Timing chain wear and slack can also allow the camshaft to go out of time with the crankshaft. As a result, valve timing can be thrown off. The valves can open too late during the four-stroke cycle. This can reduce compression, power, fuel economy, and exhaust cleanliness.

Inspecting a Timing Chain

You should inspect a timing chain before removal. **Figure 23-2** shows one method of checking OHC timing chain wear. Use a wrench to rotate the camshaft sprocket back and forth. Measure the distance the camshaft sprocket turns before the crankshaft begins to turn. Compare your measurement to specifications.

Another method to check wear is to measure the maximum deflection of the chain. This is also shown in **Figure 23-2.** Generally, if you can deflect one side of a timing chain more than about 1/2″ (13 mm), replace the chain and sprockets.

You can also check timing chain wear on an OHV engine without removing the timing cover. First, remove the valve cover. Then, place a large wrench on the crankshaft snout bolt. Rotate the crankshaft one way and then the other while watching the valve train. The valves should operate with little crankshaft movement. If you can turn the crankshaft more than about 5° to 10° without valve motion, the timing chain is worn and loose.

If the timing chain is worn more than specifications, it should be replaced. Many manufacturers recommend replacement of the chain and both sprockets as a set.

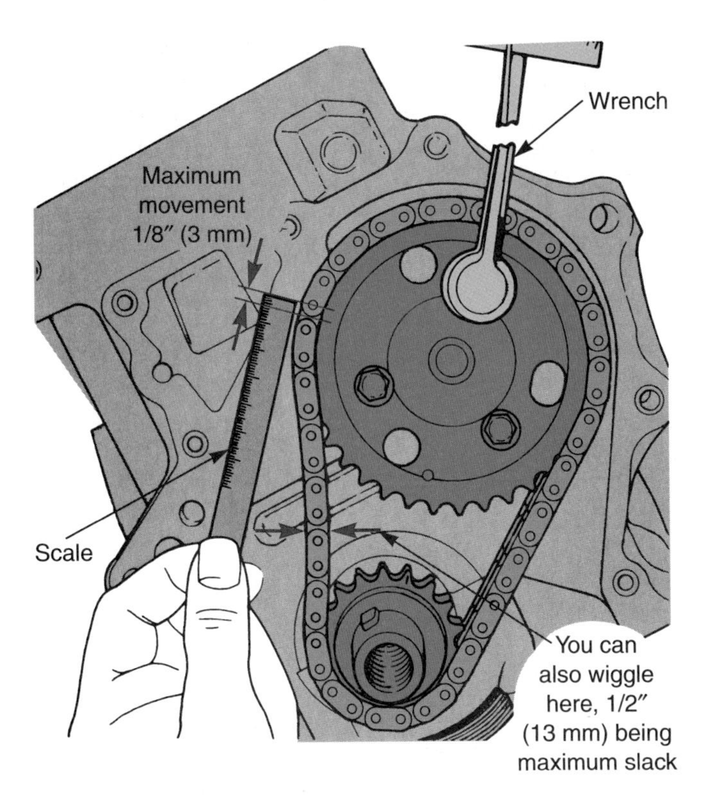

Figure 23-2. *Timing chain wear can be checked by rotating the crankshaft back and forth. If sprocket rotation is excessive, the chain and sprockets should be replaced. You can also check for wear by measuring the deflection of the chain. (DaimlerChrysler)*

Others allow you to reuse steel sprockets if they are not excessively worn.

Some camshaft sprockets have plastic teeth. The plastic teeth are prone to wear and failure. Plastic or fiber timing sprockets should be replaced whenever they are removed or disassembled for service.

If the timing chain has a tensioner, check it for wear. The fiber rubbing block can wear down and its spring can loose tension. Replace all parts as needed.

Installing a Timing Chain

Before installing a timing chain, make sure the crankshaft key is the correct size, unworn, and properly installed in its groove. Refer to **Figure 23-3.** Also, check that the camshaft is installed properly. If needed, measure camshaft endplay and check any fasteners holding a thrust plate on the front of the engine, as described in the previous chapter.

You must align ***timing marks*** on the crankshaft sprocket and camshaft sprocket, **Figure 23-4.** Timing marks may be dots, lines, circles, or other shapes indented or cast into the timing sprockets. Sometimes, a keyway or dowel can be used as a timing mark. If needed, refer to the service manual. These sprocket markings must be pointing in the correct direction to time the camshaft and valves with the crankshaft and pistons.

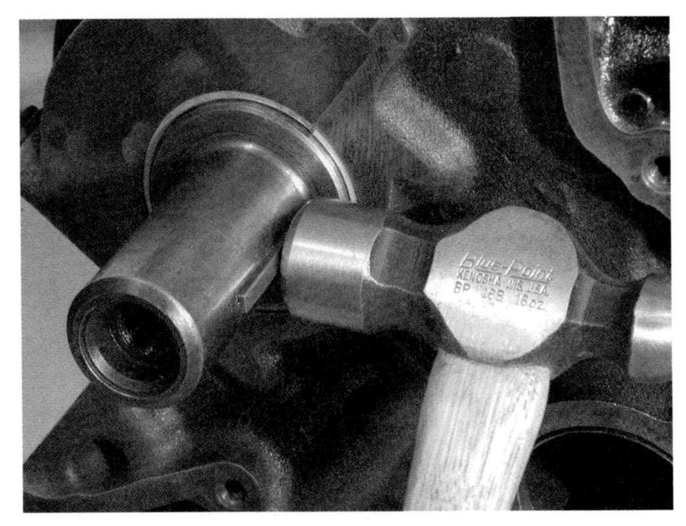

Figure 23-3. *Make sure the crankshaft key is undamaged and properly installed in the crankshaft snout. Light hammer taps can be used to insert the key squarely into its groove. Some engines have two crankshaft keys.*

1. Install the sprockets without the chain, **Figure 23-5.**
2. Turn the camshaft and crankshaft so that their timing marks align.
3. Remove the sprockets without turning the camshaft or crankshaft.
4. Fit the timing chain over the sprockets. The timing marks should align.
5. Fit the sprockets and chain onto the engine with the camshaft and crankshaft in correct timing, **Figure 23-6.**
6. If so equipped, mount the fuel pump eccentric on the camshaft sprocket. Usually, the camshaft dowel fits through a hole in the eccentric and the camshaft sprocket bolt passes through the eccentric center.

Figure 23-4. *The timing marks must align when the chain and sprockets are installed, as shown here. Refer to the service manual for the engine on which you are working.*

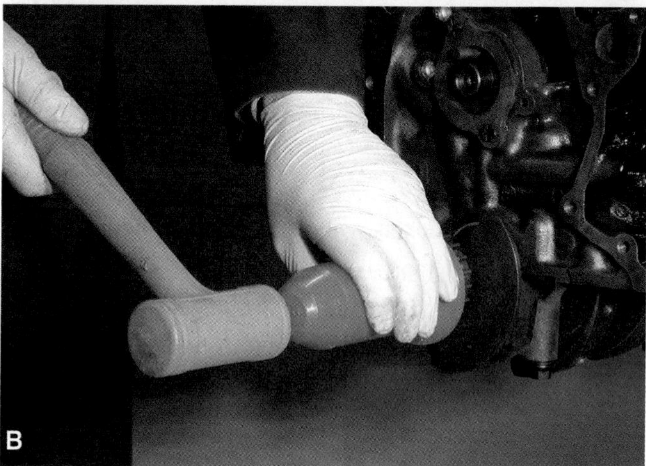

Figure 23-5. *A—A driver is the best way to seat a crankshaft or camshaft sprocket. B—Fit the driver over the sprocket and hit the driver with a hammer until the sprocket is fully seated.*

7. Torque the camshaft sprocket fastener(s) to specifications. See **Figure 23-7.** Many engine builders like to place a drop of thread lock on the camshaft sprocket bolt(s). This will help keep the bolt(s) from loosening during engine operation.
8. Double-check the alignment by turning the crankshaft by hand until the timing marks realign.

Usually, tightening the fastener(s) or hand pressure will slide the sprockets over the camshaft and crankshaft. If not, use a large socket or driver and a hammer to exert equal pressure in the center. This will drive the sprockets on squarely. Do not hammer on the outside of the sprockets to force them in place. Blows on the outside of the sprocket may result in misalignment.

Figure 23-8 shows the timing mechanism for one design of overhead camshaft engine. Note that the chain has special plated or color-coded links. These must be aligned with marks on the chain sprockets. The number one piston must be at TDC on the compression stroke with these marks aligned.

Figure 23-6. *With the camshaft and crankshaft in the correct positions, fit the chain and sprockets into place. The timing marks on the sprockets should align.*

Figure 23-7. *Torque the camshaft sprocket fastener(s) to specifications.*

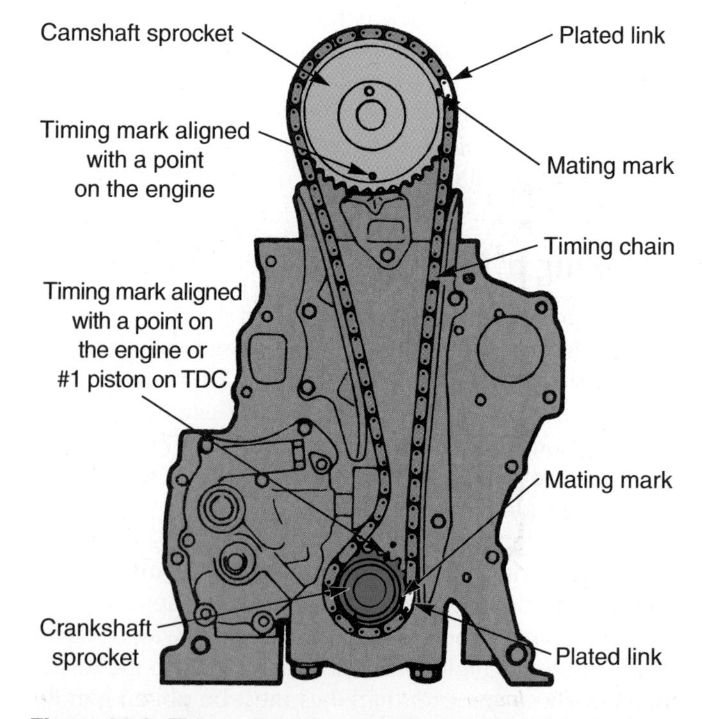

Figure 23-8. *This engine has special chain links that should be aligned with timing marks on the sprockets. (DaimlerChrysler)*

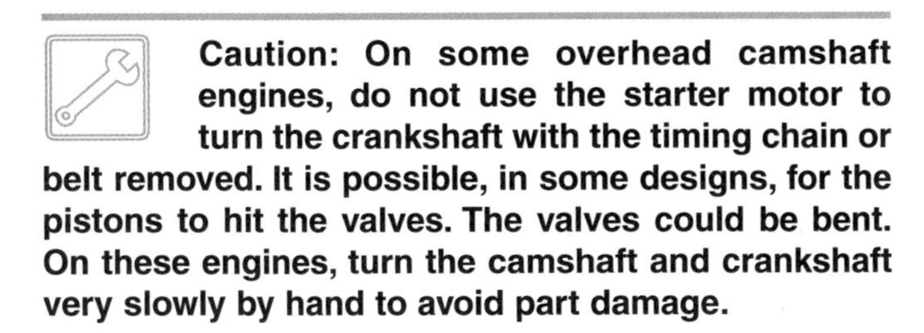

Caution: On some overhead camshaft engines, do not use the starter motor to turn the crankshaft with the timing chain or belt removed. It is possible, in some designs, for the pistons to hit the valves. The valves could be bent. On these engines, turn the camshaft and crankshaft very slowly by hand to avoid part damage.

Installing a Chain Tensioner

A ***chain tensioner*** is used in most OHC timing chain mechanisms. Inspect it closely, **Figure 23-9.** The fiber rubbing block can wear off. Replace it if needed.

Fit the tensioner onto the engine. Then, pull on the chain to make sure the tensioner moves properly. Some plunger-type tensioners can bind if worn and extended too much.

Installing an Oil Slinger

The ***oil slinger*** fits in front of the crankshaft sprocket. Look at **Figure 23-10.** It slings oil out of the chain and also prevents front seal leakage. This is usually a trouble-free part. However, if bent or damaged, it should be straightened or replaced.

Make sure the slinger faces the correct direction. If installed backward, it can rub on the front cover. This mistake would require front cover removal, a time-consuming procedure. Then, slide the slinger over the crankshaft snout.

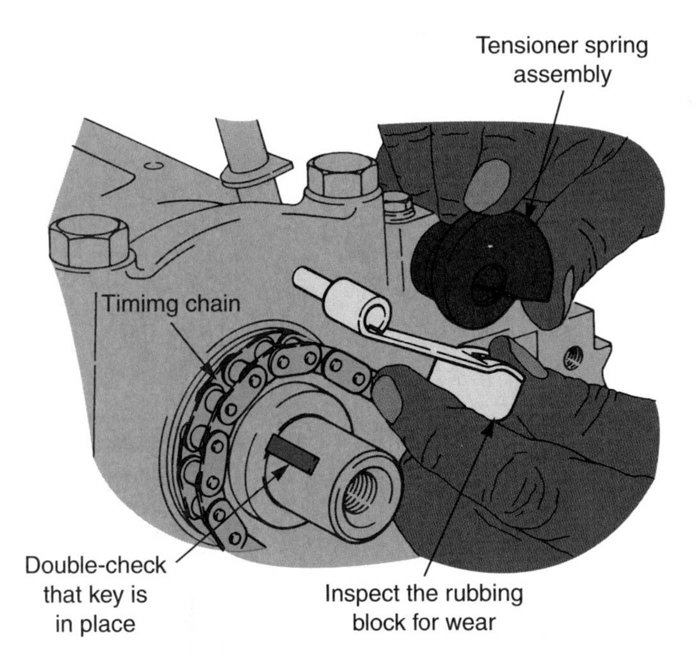

Figure 23-9. *Inspect the chain tensioner closely before installation. The fiber rubbing block can wear. Refer to wear specifications. (Ford)*

Figure 23-10. *Make sure the oil slinger is installed in the correct orientation.*

Timing Belt Service

Many OHC engines use a synthetic rubber ***timing belt*** to operate the engine camshaft. The cogged (square-toothed) belt provides an accurate, quiet, light, and dependable means of turning the camshaft, **Figure 23-11.** Timing belt service is very important. If the timing belt breaks or is improperly installed, engine valves, pistons, and other parts could be damaged.

Inspecting the Timing Belt Mechanism

Most manufacturers recommend belt replacement about every 50,000 miles (80,500 km). The timing belts on some late-model engines are designed to last the life of the engine without adjustment or replacement. However, the timing belt should be replaced if oil or coolant has contaminated the belt. Refer to the owner's manual or service manual for belt service intervals if in doubt.

The timing belt is normally replaced during major engine repairs. Since timing belt removal can be very time-consuming, most technicians replace the belt whenever it is removed, regardless of how long the belt has been in service. Timing belts are relatively inexpensive.

During timing belt service, inspect not only the belt, but the sprockets, tensioner, and other parts for problems. Inspect the timing sprockets for rounded teeth or physical damage. Spin the tensioner wheel by hand to check for a bad bearing. See **Figure 23-12.** The wheel should spin freely and quietly. Replace the tensioner bearing if you can feel any roughness. Since timing belt tensioners are sealed, many technicians like to replace a tensioner whenever the front end is disassembled, especially on high-mileage engines.

Look for oil leaks that will allow oil onto the timing belt. These must be fixed. Oil will cause rapid failure of the timing belt.

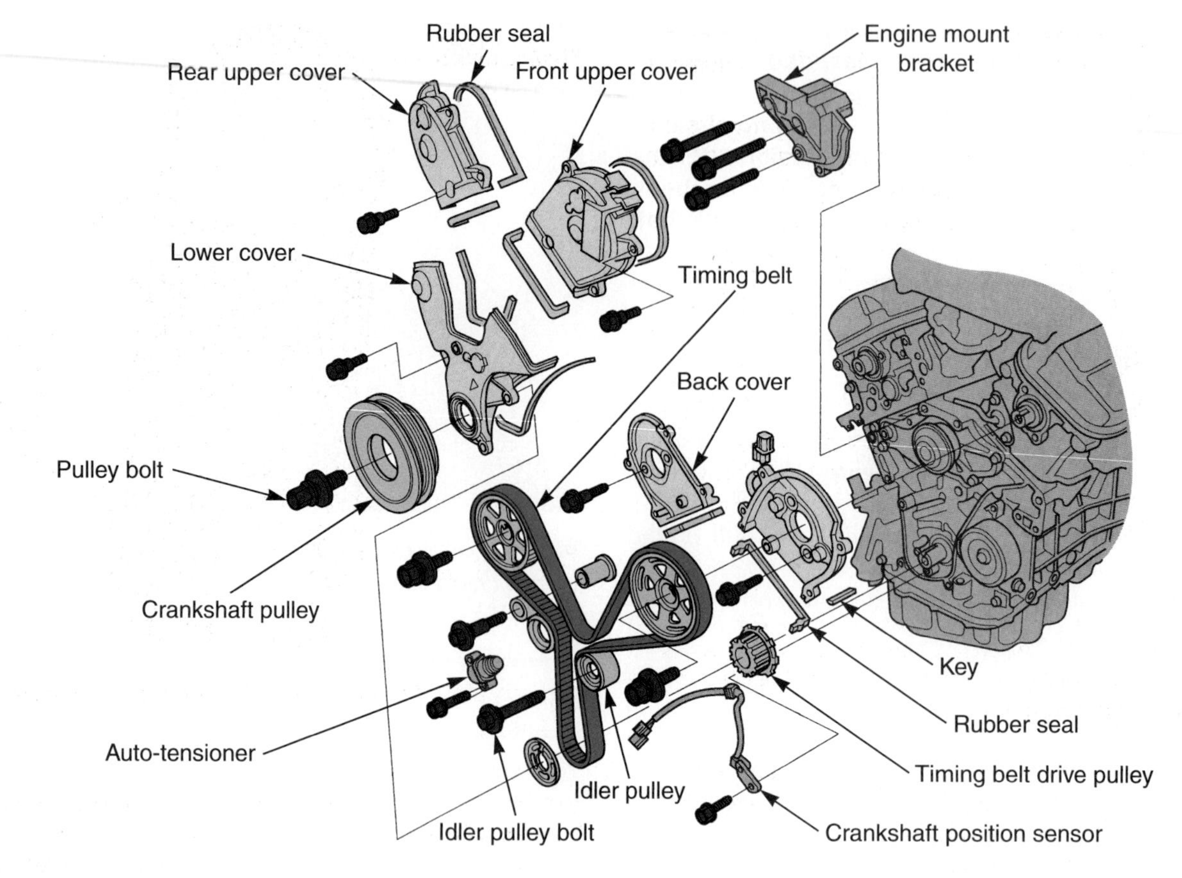

Figure 23-11. *This is an exploded view of the front end on an OHC engine that has a timing belt. (Honda)*

Caution: The lubrication in the belt tensioner bearings can dry out and the bearing can lock up. This can cause timing belt failure and severe engine damage. Never install a worn tensioner that has dry bearings.

Installing Timing Belt Sprockets

The camshaft sprocket usually fits over a small steel dowel in the camshaft, as in **Figure 23-13.** The sprocket is secured by one or more bolts.

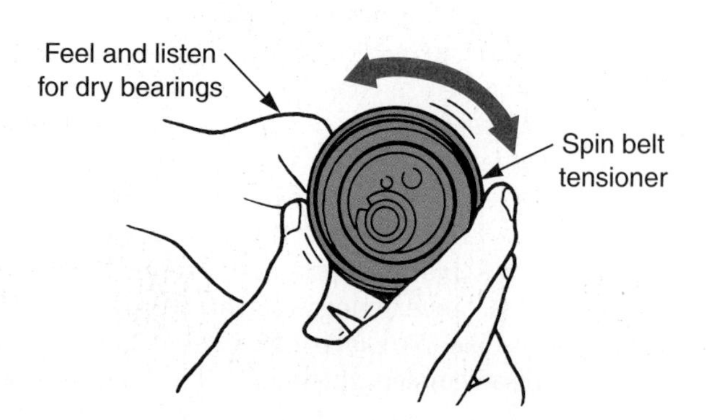

Figure 23-12. *Spin the timing belt tensioner by hand to check for dry or rough bearings. Replace the tensioner if it does not spin freely. (Mazda)*

1. Install the camshaft sprocket.
2. Torque the camshaft sprocket fastener(s) to specifications. Usually, a flat is provided on the camshaft so you can hold the camshaft while tightening the bolt(s), **Figure 23-14.** Make sure you do not damage a camshaft lobe or the sprocket teeth when trying to hold the camshaft! Grasp an unmachined surface if needed.

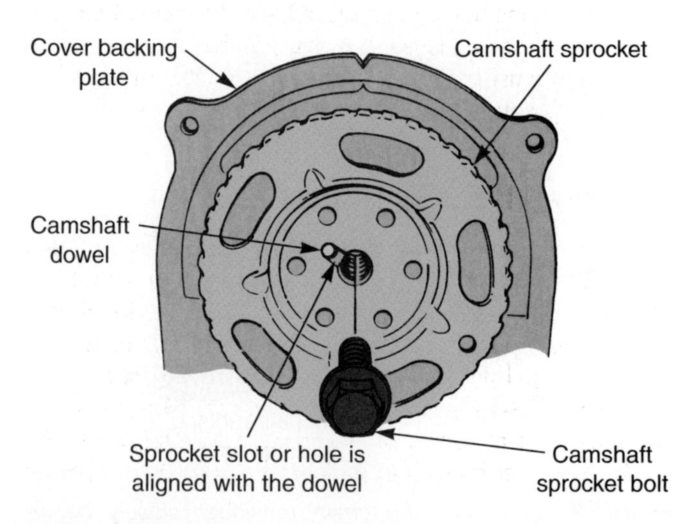

Figure 23-13. *A small dowel pin is often used to position the timing belt sprocket.*

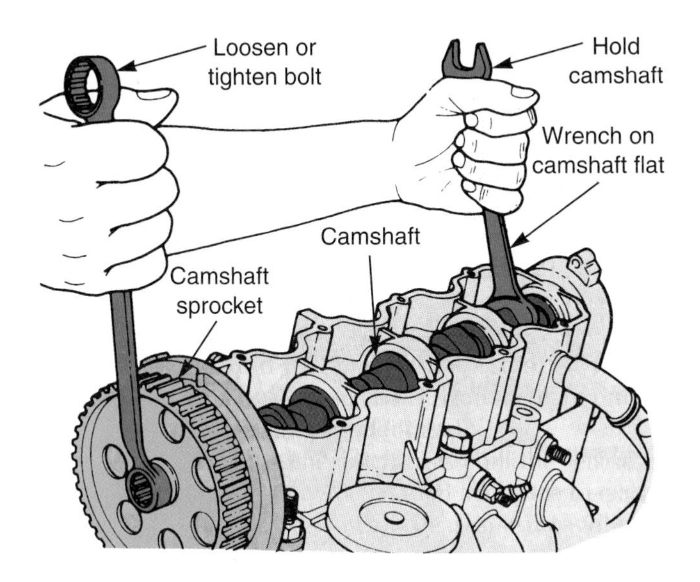

Figure 23-14. *A flat surface is frequently ground on the camshaft so a wrench can be used to hold the camshaft stationary when removing or installing the camshaft sprocket. With some engine designs, a special camshaft holding device is needed. (General Motors)*

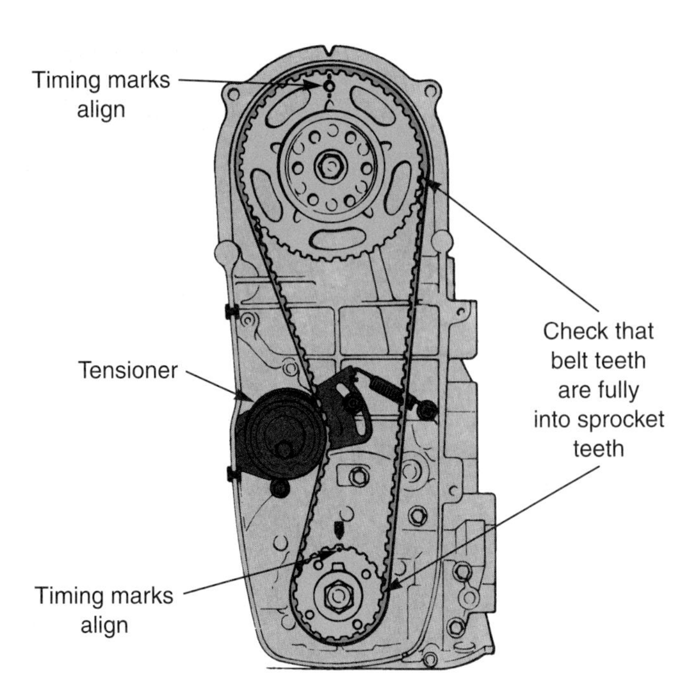

Figure 23-15. *This is a typical OHC engine with the timing belt installed.*

3. Fit the crankshaft sprocket into position, as was described earlier for a timing chain.
4. Torque the crankshaft snout bolt to specifications.
5. Turn the sprockets until the timing marks align or until number one piston is at TDC on the compression stroke.

Installing the Timing Belt

Since exact belt installation procedures vary, refer to the service manual. It will explain special methods for aligning timing marks, positioning the camshaft and crankshaft, and installing the belt tensioner. **Figure 23-15** shows the front view of an OHC engine with the timing belt in place. Note how the timing marks align.

If the timing belt is to be reused, the direction of rotation should have been marked on the belt before removal. A reused belt must be installed so that it rotates in the same direction as before. This will prevent premature failure.

Adjusting Belt Tension

It is important that ***timing belt tension*** is adjusted properly. If too loose, the belt could flap or vibrate in service and fly off its sprockets. If too tight, the timing belt could snap. Both could lead to severe valve damage. The service manual will detail how to properly adjust tension.

Figure 23-16 shows a belt tensioner. Note how you can loosen the attaching bolts and swivel the tensioner. Pry on the tensioner bracket to preload the belt. Do not pry on the wheel.

Figure 23-17 shows a special tool used to correctly set belt tension. The adjusting tool is turned with a wrench until the belt gauge reads within specifications. Tighten the tensioner bracket bolts to secure the adjustment.

Figure 23-18 shows how one manufacturer recommends using a battery hold-down bolt to pretension the timing belt. You must turn the prong on the battery hold-down bolt with your fingers to hold the tensioner in place while working.

There is a simple way to check timing belt tension if a gauge or tool is not available. Adjust the belt tension until moderate finger and thumb pressure is needed to twist the belt about one-quarter turn. This should provide good belt service.

After installing and tensioning the belt, rotate the engine by hand. Use a wrench on the crankshaft snout bolt. Turn the crank in the direction of normal rotation until the timing marks come around and realign. Then, double-check alignment of the timing marks.

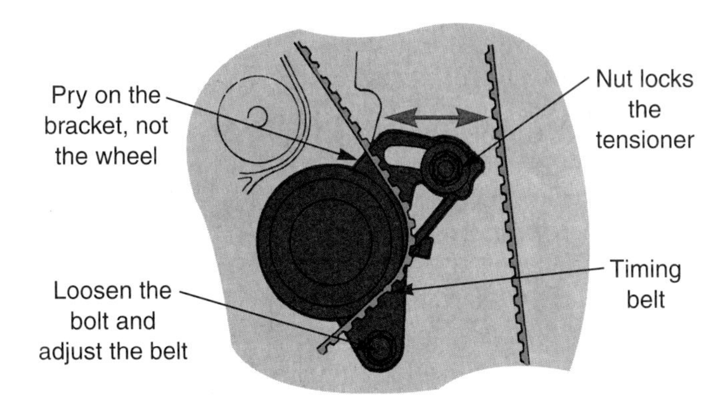

Figure 23-16. *This timing belt adjuster has a slot that lets you swivel adjuster to the side. Pry on the adjuster and tighten the fastener. Then, check for proper timing belt tension.*

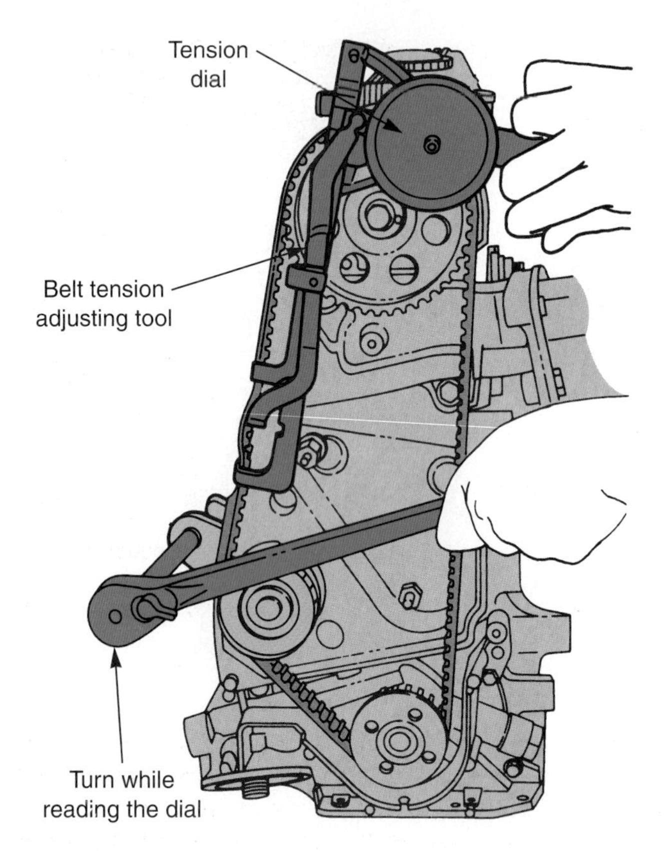

Figure 23-17. *This is a special tool for adjusting belt tension. It measures the deflection on a straight section of the belt. If it is not available, adjust the belt tension until a moderate finger pressure is needed to twist the belt one-quarter turn. (General Motors)*

Installing the Timing Belt Cover

The ***timing belt cover*** is simply a plastic or sheet metal shroud around the timing belt. It keeps debris off the belt and protects your hands when working in the engine compartment. Unlike the front cover, the timing cover does *not* contain an oil seal, **Figure 23-19.**

To install the cover, simply fit it into place on the front of the engine. If rubber washers are used, make sure they are installed. They prevent cover vibration, noise, and cracking of the cover. Check that all fasteners are in place and that the belt and sprockets do not rub on the timing belt cover.

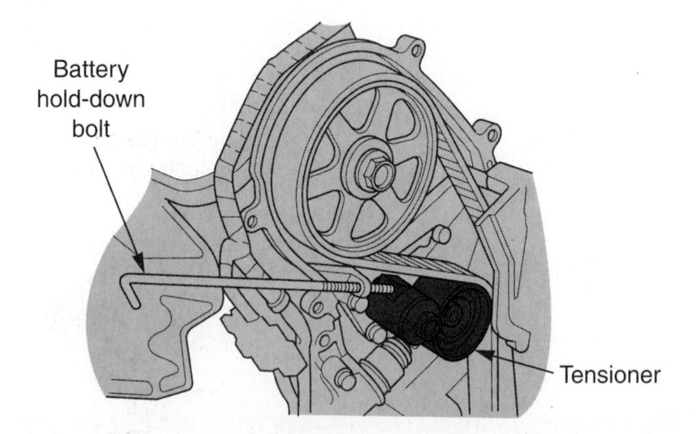

Figure 23-18. *This manufacturer recommends using a battery hold-down bolt to pretension the timing belt. (Honda)*

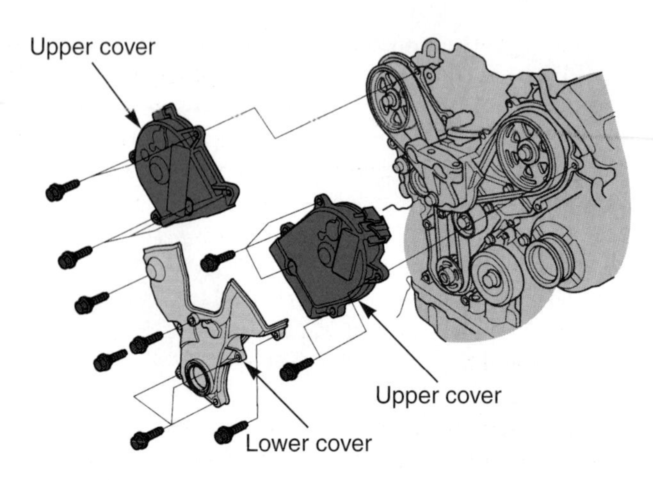

Figure 23-19. *When installing the timing belt cover, make sure you use the correct bolts and that all rubber grommets or spacers are in place. (Honda)*

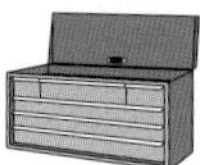

Note: Front cover installation is discussed later in this chapter. The timing belt cover is installed after the front cover.

Timing Gear Service

Timing gears are normally more dependable than timing chains. See **Figure 23-20.** They will provide thousands of miles of trouble-free engine operation. However, after prolonged use, timing gear teeth can wear or become chipped and damaged, requiring replacement.

Timing gear service is similar to timing chain service. Inspect the old timing gears carefully. Look for any signs of wear or other problems. Replace the gears as a set if needed.

Installing Timing Gears

Timing gears are usually press-fit on the crankshaft and camshaft. A wheel puller is normally needed to remove the crankshaft gear.

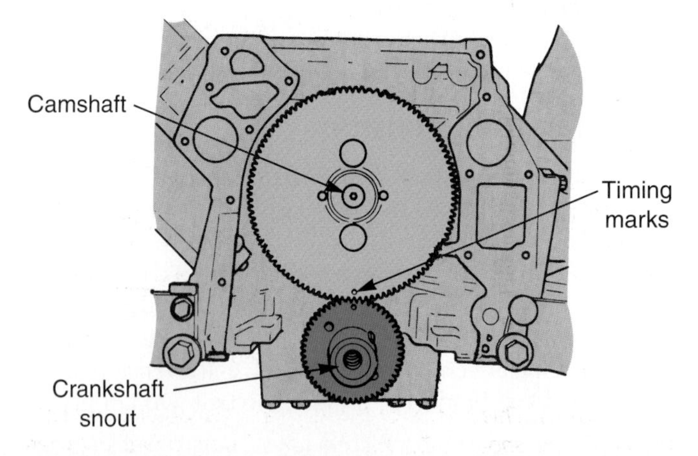

Figure 23-20. *Timing gears are very dependable. However, after prolonged service the teeth can wear or break. (Ford)*

Timing gears can usually be installed with light blows from a brass hammer. Make sure the keys and keyways are aligned. Also, make sure all timing marks are aligned. Tap in a circular motion around the gear to force it squarely into position. Do not hit and damage the gear teeth. A press may be needed to install the camshaft gear onto the camshaft.

Figure 23-21 shows how a press is used to service the camshaft gear on the camshaft. With this engine design, the camshaft must be out of the engine for timing gear replacement. A clearance must be maintained between the thrust plate and the camshaft cheek. This clearance allows for proper camshaft endplay.

Double-check the alignment of the timing marks. The timing marks must be positioned properly to time the camshaft with the crankshaft.

Measuring Timing Gear Runout

Timing gear runout or wobble is measured with a dial indicator. To measure timing gear runout:

1. Position the indicator stand on the engine block. Place the indicator stem on the outer edge of the camshaft timing gear. The stem should be parallel with the camshaft centerline, as shown in **Figure 23-22.**
2. Turn the engine crankshaft while noting indicator needle movement.
3. The indicator reading equals the runout.
4. If runout is greater than specifications, remove the timing gears and check for problems. The gear may not be fully seated or it may be machined improperly. Also, check the camshaft for runout or straightness.

Measuring Timing Gear Backlash

Timing gear backlash is the amount of clearance between the timing gear teeth. Too little clearance can make the gears lock up when expanded by engine heat. Too much clearance can reduce engine power and increase gear noise. Backlash can also be used to determine timing gear teeth wear. Use a dial indicator to measure timing gear backlash.

1. Set the indicator stand on the engine. Look at **Figure 23-23.**
2. Locate the indicator stem on one of the camshaft gear teeth. The stem must be parallel with gear tooth travel.
3. Move the camshaft gear one way and then the other, without turning the crankshaft.
4. Observe the indicator needle. The needle travel is equal to the backlash.

Refer to the service manual for backlash specifications. If timing gear backlash is greater than specifications, the gears are worn. They should be replaced.

If a dial indicator is not available, backlash can be measured with a feeler gauge. The size of the blade that fits between the gear teeth equals the backlash.

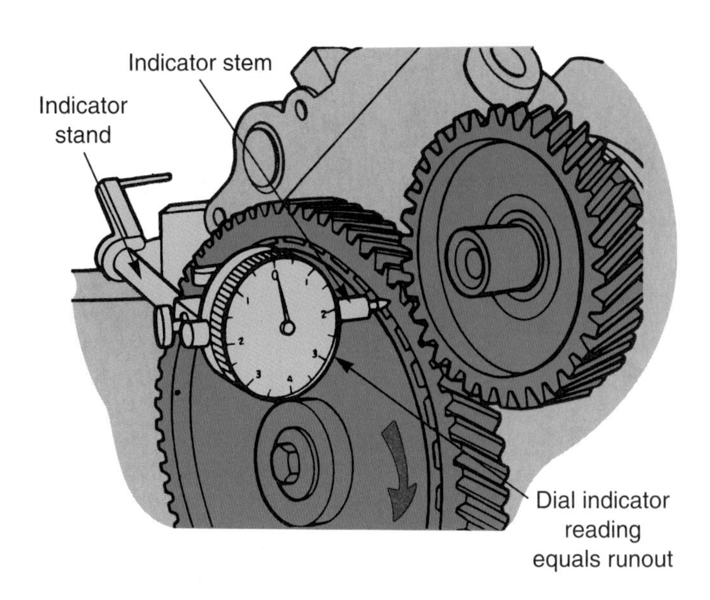

Figure 23-22. *Measuring timing gear runout or wobble. (Ford)*

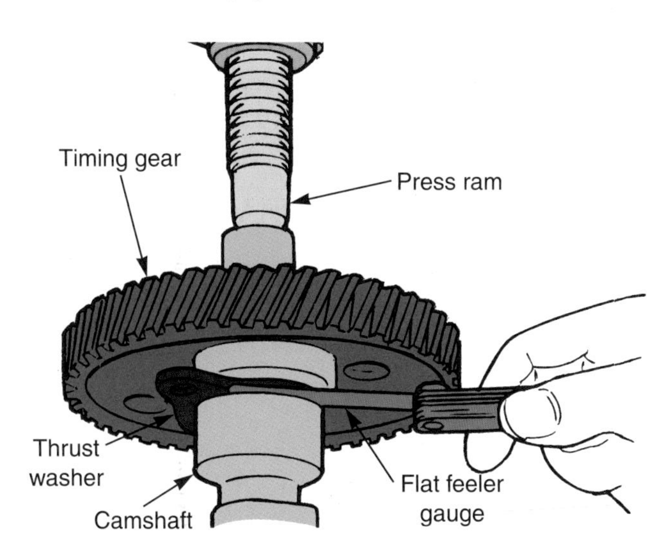

Figure 23-21. *This particular timing gear must be pressed onto the camshaft snout. Clearance must be maintained between the thrust plate and the camshaft cheek to allow proper camshaft endplay. (Ford)*

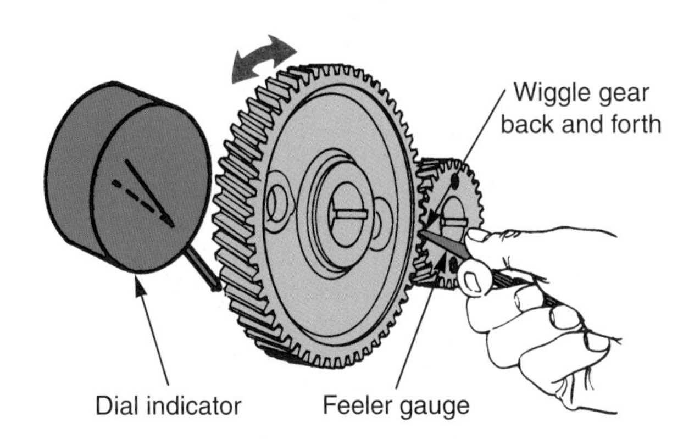

Figure 23-23. *Measuring timing gear backlash.*

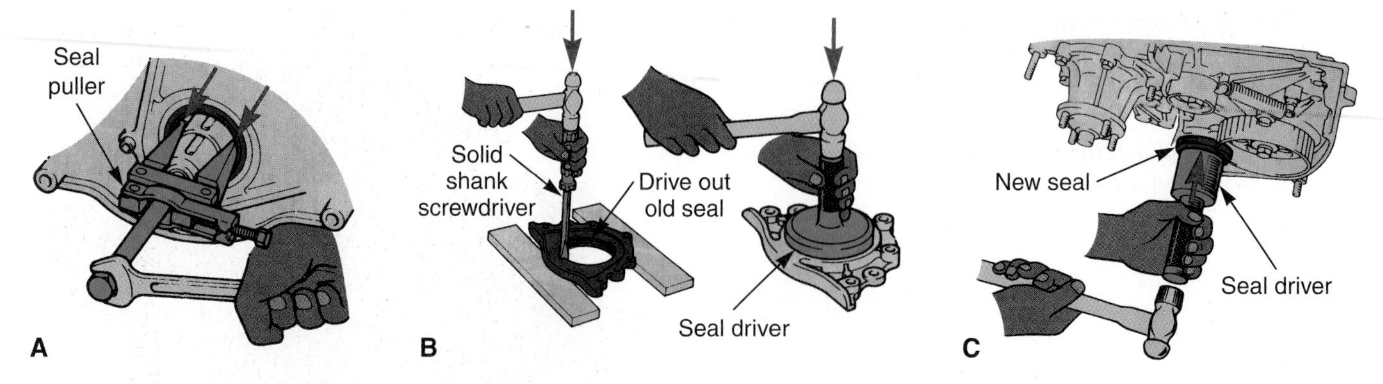

Figure 23-24. *A—Sometimes the front crankshaft seal can be replaced without removing the front cover. A special puller is used to remove the seal. B—On most engines, the front cover must be removed for front crankshaft seal service. The oil seal is driven out from the back. Then, a seal driver is used to install the new seal. C—Driving in a new front crankshaft seal with the front cover installed on the engine. (Toyota)*

Engine Front Cover Service

The ***engine front cover,*** also called the timing cover, holds the front oil seal and encloses the timing gears or chain. It can be made of thin stamped, sheet metal or cast aluminum.

Scrape off all gasket material. Then, install the new oil seal as described in the previous section. Make sure the cover sealing surface is true. Lay it on a flat work surface and check for gaps under the cover. Straighten the cover if needed.

Front Seal Service

The ***crankshaft front seal*** keeps engine oil from leaking out from between the crankshaft snout and the engine front cover. Replace the front seal whenever it is leaking or the front cover is removed.

The front seal can be replaced in the vehicle with only partial engine disassembly. Typically, you must remove the radiator and other accessory units on the front of the engine. Use a puller to remove the vibration damper.

Sometimes, the front crankshaft seal can be replaced without front cover removal. Shown in **Figure 23-24A,** a special seal puller may be used to remove the old seal. Some crankshaft front seals are pressed-in from the rear of the timing cover. In these cases, timing cover removal is needed for seal replacement. After the cover is removed, drive the old seal out, as shown in **Figure 23-24B.**

Always compare the old seal with the new one. The outside and inside diameters of the two seals must be the same. You may also be able to match part numbers stamped on the metal flange of the seals.

Before installing a front seal, coat the outside diameter of the seal with nonhardening sealer. This will prevent oil seepage between the seal body and the front cover. Coat the rubber sealing lip with engine oil. This will lubricate the seal during initial engine startup. Then, use a seal driver to squarely seat the new seal into its bore, **Figures 23-24B** and **23-24C.** If the front cover has been removed, install it as described in the next section.

When available, a seal driver should be used to install the front seal. However, if you do not have one, a block of wood can be used. Drive the seal in squarely without damaging the metal housing of the seal.

Some camshafts also have a front seal. It seals the front camshaft bearing of an OHC engine. **Figure 23-25** shows a technician installing a camshaft front seal in a cylinder head. The outside diameter of this type of seal should be coated with nonhardening sealer. The seal housing surface should be dry. Place engine oil or grease on the inside diameter of the seal and on the camshaft. Then, drive the seal in.

Sealing the Front Cover

A front-end gasket set often includes the gaskets, oil seals, sealer, and other parts needed to replace a timing cover with the oil pan on the engine. Either a conventional gasket or chemical sealant can be used during front cover installation. Usually, both a gasket and sealer are used. Make sure you use a bead of sealer where two gaskets or parts come together. This is a common leakage point. Also,

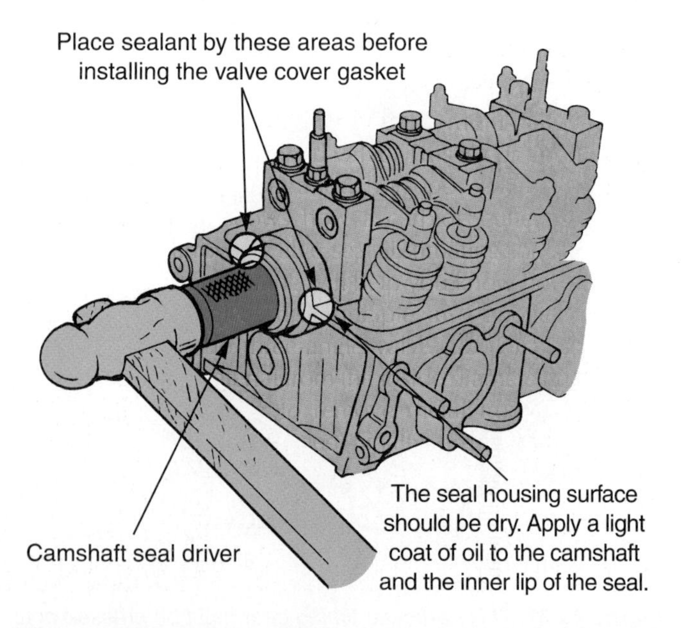

Figure 23-25. *With an OHC engine, a front seal may be mounted in the cylinder head. A seal driver is the best way to install the seal without damaging it. (Honda)*

check that all holes in the gasket align with the corresponding holes on the engine. **Figure 23-26** shows the major steps for installing a front cover on a cam-in-block engine.

1. Apply sealer to the front cover gasket. This holds the gasket in place during assembly and helps keep it from leaking during service.
2. When servicing a front cover with the engine in the vehicle and the oil pan is in place, you must often cut off the existing oil pan gasket flush with the front of the block.
3. Cut a piece from a new oil pan gasket to replace the piece you removed.
4. Place sealer at the joint between the existing and new oil pan gaskets. This will help seal between the new piece of gasket and the existing gasket.

Figure 23-27 shows how anaerobic and silicone sealers are placed on an aluminum front cover. Anaerobic sealer is used on the sealing surface between two thick or solid pieces. RTV or silicone sealer is recommended where a flexible part, such as the oil pan, mates with the front cover. Use a small bead of sealer. Make sure you have a continuous bead to prevent oil or coolant leakage. Circle all holes that carry oil or coolant.

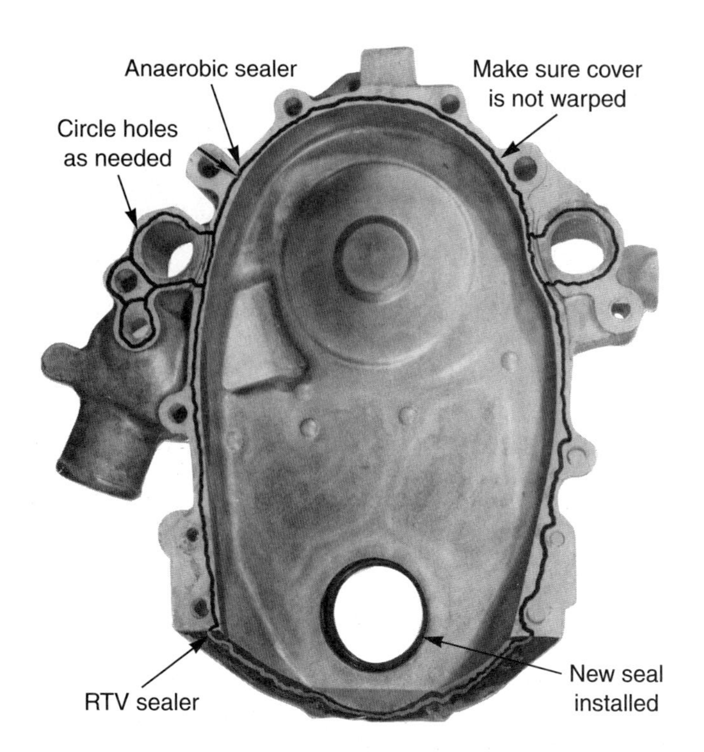

Figure 23-27. *On this cast aluminum front cover, anaerobic sealer is used where two thick pieces mate. RTV or silicone sealer is recommended where the flexible oil pan mates with the front cover. (General Motors)*

Installing the Front Cover

Before installing the front cover, check that all fasteners under the cover are properly torqued. Check the installation of the chain tensioner, if used. Double-check timing mark alignment.

When installing the front cover, do not shift the gasket or smear the sealer. If the oil pan is on the engine, you may need to loosen the front oil pan bolts. Pry down on the pan lightly so the front cover will fit into place. You may need to use a seal centering tool that fits over the crankshaft snout. It positions or centers the front cover seal correctly around the crankshaft.

1. Check bolt lengths and locations.
2. Place a small amount of nonhardening sealer on any bolt entering a coolant or oil passage. The service manual will indicate which bolts require sealer, if in doubt.
3. Carefully fit the front cover and gasket onto the engine, **Figure 23-28.** For in-vehicle service, make sure the oil pan gasket is in place.
4. Install the front cover bolts.
5. Tighten the bolts to specifications in gradual steps following a crisscross pattern.
6. If used, remove the seal alignment tool.
7. Double-check that all bolts are installed properly. It is easy to miss a hidden hole under the lower section of the front cover.
8. Install the front oil pan bolts, if needed.
9. Assemble all other external components, such as the water pump, brackets, vibration damper, and so on.

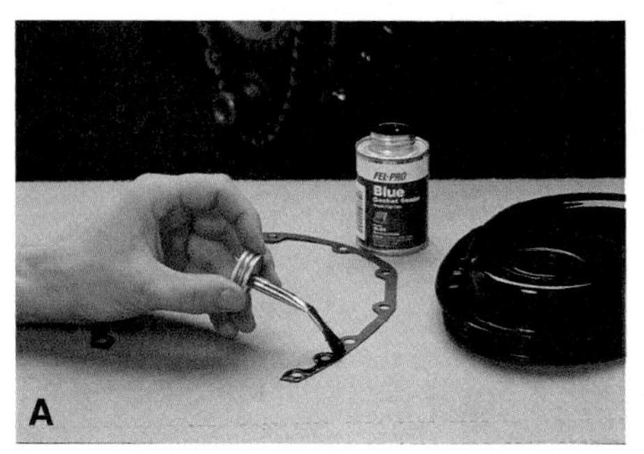

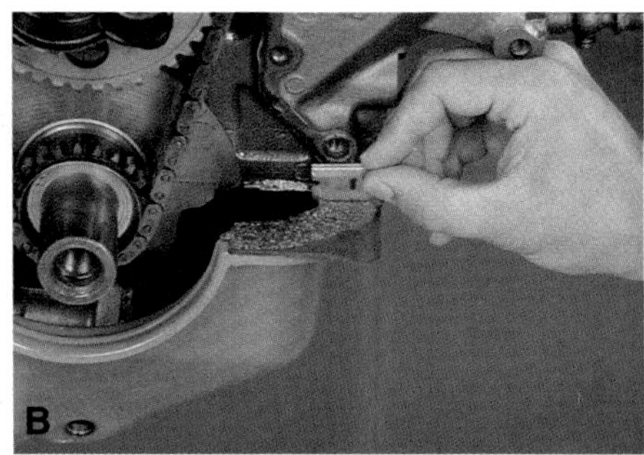

Figure 23-26. *Installing the front cover on an OHV engine. A—Apply sealer to the front cover gasket. B—If the oil pan is in place, you may need to cut off the existing gasket flush with the front of the block. C—Place sealer at the joint between existing and new oil pan gasket. (Fel-Pro)*

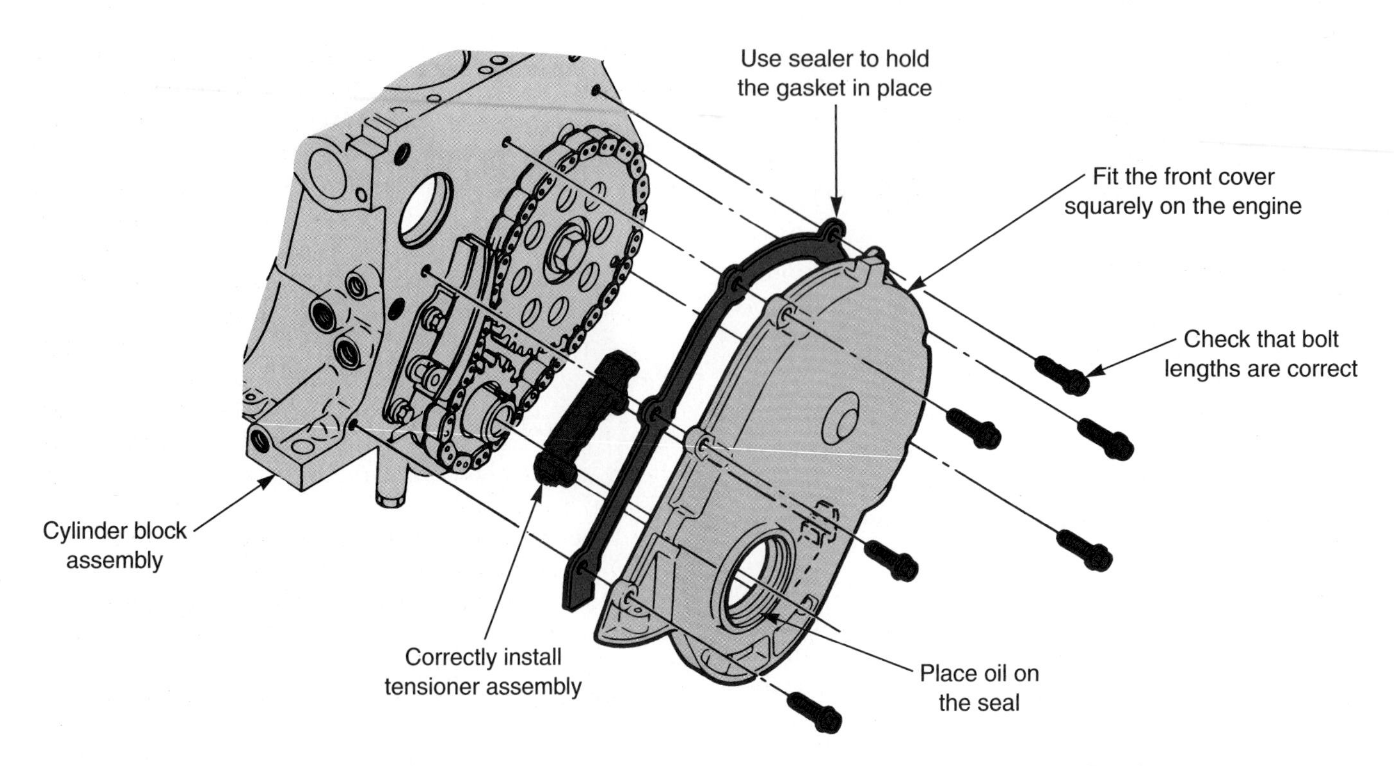

Figure 23-28. *Carefully position the front cover onto the engine without moving or shifting the gasket. (Ford)*

Vibration Damper Installation

Before installing the vibration damper, check that the crankshaft snout key is properly installed and tight in its groove. Then, place lubricant on the damper seal surface, **Figure 23-29A.** Fit the damper onto the crankshaft without pushing the key out of its groove, **Figure 23-29B.** Install the crankshaft snout bolt and washer.

Usually, the crankshaft snout bolt can be tightened to force the damper into position. If light hammering is necessary, tap in the center of the damper. Never hammer on its outer ring. The outer ring is mounted on rubber. Hammering on the outer ring could shift the ring or damage the rubber and ruin the damper. Sometimes, a special driver is needed to install the damper, **Figure 23-30.**

Make sure the damper seats squarely on the crankshaft snout. If it becomes misaligned during assembly, stop and align the damper to prevent part damage. Once the damper is fully seated, torque the crankshaft snout bolt, if so equipped.

Install any other parts, such as the crankshaft pulley, timing pointer, brackets. **Figure 23-31** shows a special holding tool recommended by one manufacturer to hold the pulley while torquing the pulley nut.

Summary

Timing chain wear can allow the camshaft to go out of time with the crankshaft. This can reduce engine power

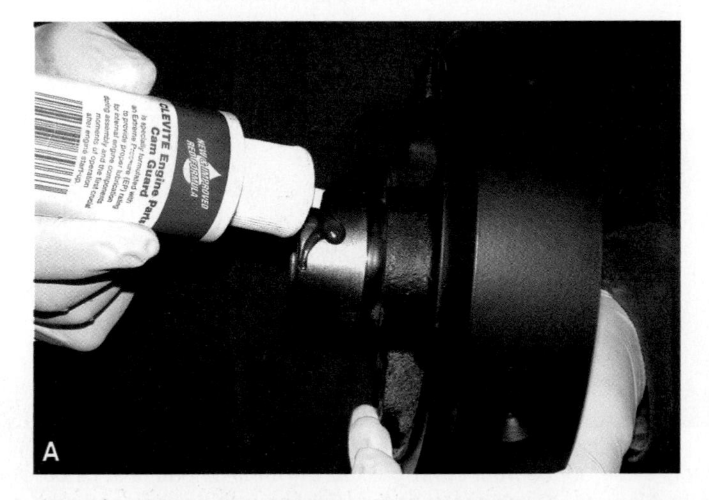

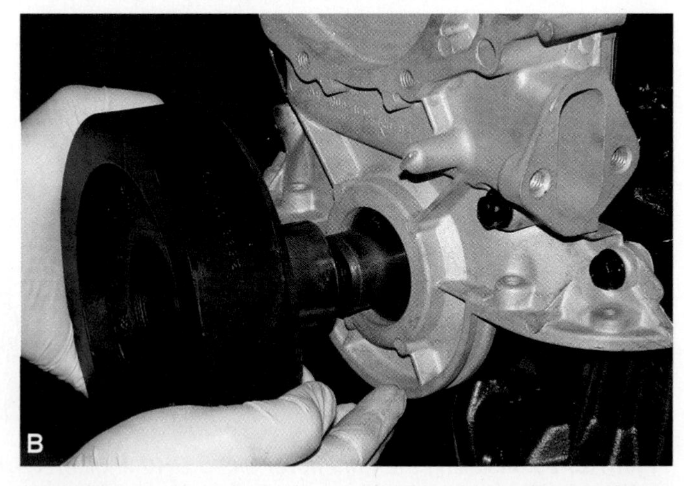

Figure 23-29. *A—Apply assembly lube to the seal journal on the vibration damper. B—Align the slot on the damper with the crankshaft snout key and slide the damper into place.*

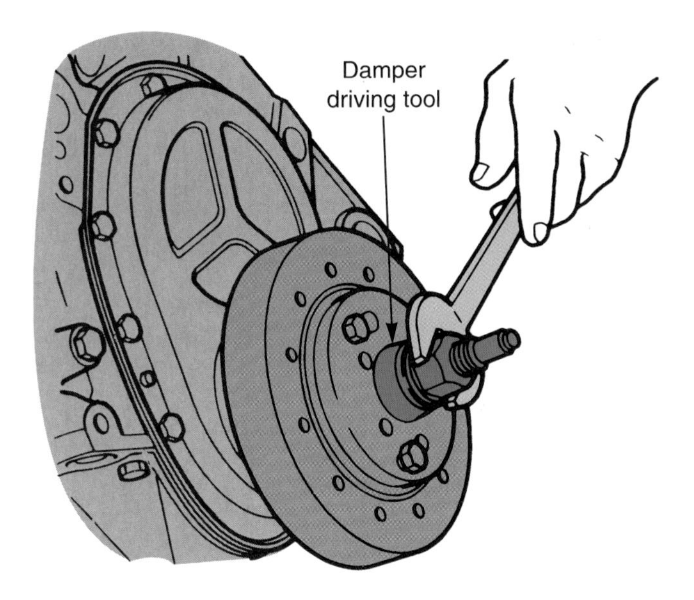

Figure 23-30. *A special tool is required to install this press-fit damper. (General Motors)*

and increase emissions. Check chain slack by turning the crankshaft back and forth. If it can be turned excessively without valve train movement, the chain is worn.

Timing sprockets with plastic gear teeth are a common source of problems. The plastic teeth can wear or break off. A bad chain tensioner can also cause excess chain slack.

To begin timing chain installation, check any keys or dowel. Turn the crankshaft and camshaft to align the timing marks. The timing marks on the sprockets must align. Timing marks can be small dents, circles, or other figures on the sprockets. The service manual will usually illustrate them. Fit the chain and sprockets in place together.

Do not hammer on the outer edges of the sprockets to force them into place. Use light blows in the center of the sprockets.

Do not crank an engine with the starter motor with the timing chain, belt, or gears removed. The pistons could slide up and bend the valves.

An oil slinger fits in front of the crankshaft sprocket to prevent oil leakage and to lubricate the timing chain or gears. Make sure it is not installed backwards.

Inspect a timing belt for signs of deterioration. Most manufacturers recommend timing belt replacement every

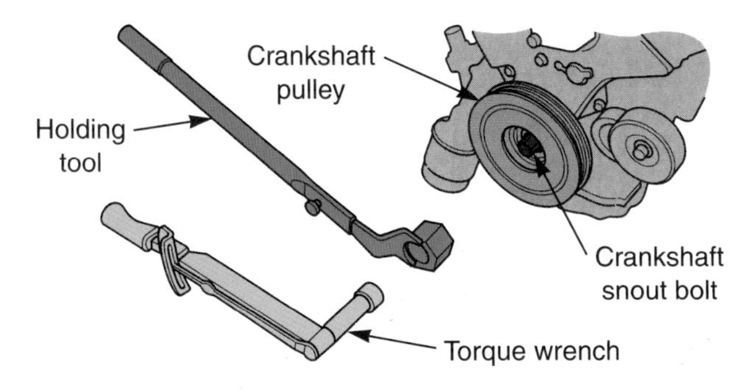

Figure 23-31. *A special tool is recommended to hold the vibration damper while torquing the crankshaft snout bolt.*

50,000 miles or 80,500 km. However, most mechanics replace the belt whenever it is removed for service.

Install the timing belt as you would a timing chain. Align the timing marks. The number one piston should usually be at TDC on the compression stroke when the marks align. If the belt is reused, it should be installed to rotate in the same direction as before.

Adjust timing belt tension to prevent belt breakage or looseness. Special belt tension gauges are available. However, you can adjust the belt until moderate finger pressure is needed to twist the belt about one-quarter turn.

Timing gears are very dependable. However, gear teeth can break and wear after extended service. During timing gear service, measure gear backlash and runout with a dial indicator.

When installing a new front oil seal, drive out the old seal without damaging the front cover. Coat the outside diameter of the new seal with nonhardening sealer. Wipe engine oil on the inside of the seal. Drive the new seal in squarely.

An engine front cover can be installed using a conventional gasket, chemical sealants, or both. Use extra silicone where two seals come together. Torque all fasteners to specifications.

Review Questions—Chapter 23

Please do not write in this text. Write your answers on a separate sheet of paper.

1. Generally, if you can deflect a timing chain more than _____ the chain should be replaced.
2. How can you check timing chain wear on an OHV engine without removing the timing cover?
3. _____ must align to time the camshaft and valves with the crankshaft and pistons.
4. Explain how to fit the timing chain and sprockets onto the engine.
5. Which part of a chain tensioner should be inspected for wear?
6. Why should you never crank an engine with the starting motor when the timing chain or belt is removed?
7. This part keeps oil from spraying out the front seal and helps lubricate the timing chain or gears.
   (A) Tensioner.
   (B) Shroud.
   (C) Slinger.
   (D) Front cover.
8. A timing belt is normally replaced at least every _____ miles.
9. If you do not have a belt tension gauge, how can you adjust belt tension properly?

10. What is the difference between a *timing belt cover* and an *engine front cover?*
11. This is measured with a dial indicator with the indicator plunger resting on one of the timing gear teeth.
    (A) Timing gear runout.
    (B) Timing gear diameter.
    (C) Timing gear out-of-round.
    (D) Timing gear backlash.
12. What is the purpose of the *crankshaft front seal?*
13. What type of sealer should be used on the outside diameter of the crankshaft front seal?
14. What is the purpose of the *engine front cover?*
15. Why should you never hammer on the outer ring of the vibration damper?

ASE-Type Questions—Chapter 23

1. An engine is being serviced. Worn camshaft lobes are found and the camshaft is being replaced. The car has 35,000 miles (56,300 km) on it. Technician A says that the timing belt should be replaced. Technician B says that the belt is good for 50,000 miles (80,500 km) and it should be reused. Who is correct?
    (A) A only.
    (B) B only.
    (C) Both A and B.
    (D) Neither A nor B.
2. Technician A says that most engines must be removed from the vehicle in order to replace the timing chain. Technician B says that normally, a timing chain can be replaced without removing the engine from the vehicle. Who is correct?
    (A) A only.
    (B) B only.
    (C) Both A and B.
    (D) Neither A nor B.
3. Technician A says that timing chain wear can reduce engine compression. Technician B says that timing chain wear can affect exhaust cleanliness. Who is correct?
    (A) A only.
    (B) B only.
    (C) Both A and B.
    (D) Neither A nor B.
4. Most manufacturers recommend replacing an engine's timing belt about every _____.
    (A) 50,000 miles (80,500 km).
    (B) 70,000 miles (110,000 km)
    (C) 100,000 miles (160,000 km)
    (D) 200,000 miles (320,000 km)
5. Technician A says that the camshaft sprocket mark should normally be aligned with a mark on the side of the engine. Technician B says that the camshaft sprocket mark should normally be aligned with a mark on the front of the engine. Who is correct?
    (A) A only.
    (B) B only.
    (C) Both A and B.
    (D) Neither A nor B.
6. Technician A says that a loose timing belt could lead to severe valve damage. Technician B says that if a timing belt is too tight, severe valve damage can occur. Who is correct?
    (A) A only.
    (B) B only.
    (C) Both A and B.
    (D) Neither A nor B.
7. Technician A says that the function of a timing belt cover is to protect the oil seal and belt from debris. Technician B says that the function of a timing belt cover is to keep debris off the belt and protect your hands when working in the engine compartment. Who is correct?
    (A) A only.
    (B) B only.
    (C) Both A and B.
    (D) Neither A nor B.
8. Timing gear runout should be measured with a _____.
    (A) dial indicator
    (B) inside micrometer
    (C) outside micrometer
    (D) feeler gauge
9. Technician A says that timing gear backlash is the amount of clearance between the crankshaft snout and timing gear. Technician B says that timing gear backlash is an indication of the wear on the timing gear teeth. Who is correct?
    (A) A only.
    (B) B only.
    (C) Both A and B.
    (D) Neither A nor B.
10. Technician A says that normally the snout bolt can be tightened to force the vibration damper into position. Technician B says that you can hammer on the outer edge of the vibration damper to force it into place on the crankshaft. Who is correct?
    (A) A only.
    (B) B only.
    (C) Both A and B.
    (D) Neither A nor B.

Chapter 24

Engine Reassembly, Installation, Startup, and Break-In

After studying this chapter, you will be able to:

- ❑ List the general engine reassembly rules.
- ❑ Install an oil pump and oil pan.
- ❑ Install a cylinder head assembly.
- ❑ Install lifters.
- ❑ Install exhaust and intake manifolds.
- ❑ Install a camshaft housing cover.
- ❑ Install an overhead camshaft.
- ❑ Adjust valve train clearance.
- ❑ Install valve and camshaft covers.
- ❑ Paint the engine.
- ❑ Complete final engine assembly.
- ❑ Install the engine into the vehicle.
- ❑ Conduct proper engine startup.
- ❑ Perform proper engine break-in.

Know These Terms

Camshaft cover
Camshaft housing
End seals
Exhaust manifold
Exhaust manifold warpage
Head gasket
Intake manifold
Intake manifold warpage
Nonadjustable rocker arms
Oil pan
Oil pump
Torque-to-yield bolts
Valve cover
Valve train clearance adjustment
Valve train noise
Zero valve lash

This chapter summarizes the basic methods for installing the major assemblies of an engine. Previous chapters explained how to assemble the short block and cylinder head. This chapter builds on this knowledge by explaining how to install the oil pan, cylinder head(s), manifolds, and various covers on the short block, **Figure 24-1.** You will also learn about valve train clearance adjustment, engine installation, startup, and break-in.

General Engine Assembly Rules

There are several basic rules you must remember when assembling and installing an engine. These include:

- ❑ All parts and tools should be perfectly clean and organized. Remember, the slightest bit of dirt, metal, plastic, gasket material, etc., may cause major engine damage.
- ❑ Parts that might be select fit should be reinstalled in the same location. For example, if the pistons on a low mileage engine are going to be reused, reinstall them in the same cylinder bores from which they were removed. Piston and cylinder diameters will vary slightly.
- ❑ Lubricate parts as needed. All moving parts should be coated with the appropriate lubricant: assembly lube, engine oil, camshaft lube, or white grease. This will provide protection during initial engine startup.
- ❑ Make sure bolts and nuts are correct. It is best to use the same fastener as supplied by the manufacturer. Some engines have both conventional and metric fasteners. Do not mix them or thread damage will result. Also, check bolt lengths. A bolt that is too long could damage parts by bottoming out. A bolt that is too short could strip and fail in service. Some bolts must be replaced with new ones every time they are removed from the engine, so check with the service manual when in doubt. Connecting rod and head bolts may require replacement.

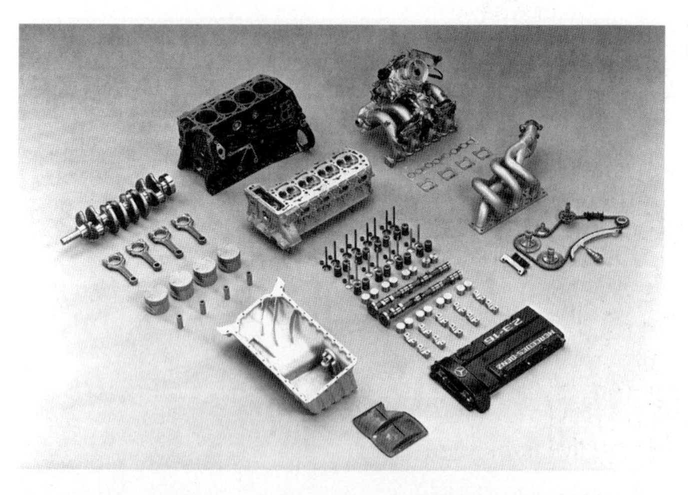

Figure 24-1. *Before starting engine reassembly, all components should be perfectly clean and organized. (Mercedes-Benz)*

- ❑ Keep components covered until you are ready to assemble them. For example, keep a cloth over the short block while it is sitting on the engine stand. This will help keep dust and debris out of the cylinders and off of other areas in the block. Keep the intake manifold inlet covered, especially when the manifold is installed on the engine. Anything dropped into the manifold could cause severe damage to the pistons, valves, cylinder head, and cylinder block.
- ❑ Torque all components properly. Do not try to save time by using an impact wrench on critical fasteners such as connecting rod bolts, main cap bolts, head bolts, manifold bolts, and oil pump bolts. Modern engines are cast with very thin walls. Improper torque can cause fluid leakage, part warpage, or even part breakage.
- ❑ Inspect everything one last time as you install engine components. Look for signs of trouble, such as unusual wear, cracks, or scoring. You do not want to find something wrong *after* the engine is assembled and installed in the vehicle.
- ❑ Do not get too excited and rush. It is very easy to become excited about finishing an engine overhaul and wanting to see the engine "fire up" and run. You must have control and work at a steady pace.
- ❑ Avoid talking to anyone while working on an engine. If distracted, it is very easy to forget to do something. Forgetting to torque even one bolt could result in major engine damage or a serious leak, at your expense.
- ❑ Refer to a service manual if in doubt about any service task. The manual will answer any questions you might have concerning an assembly method, torque value, valve train clearance specification, and so on.

Note: Lubricant on threads will affect bolt torque. Only use the lubricant recommended by the manufacturer.

Engine Reassembly

This section describes the typical steps for engine reassembly. The methods given in this chapter are general and apply to most makes and models. However, always follow the more specific procedures given in the service manual. The manual provides information that might be unique to the particular engine.

The short block should be placed on an engine stand. You should have already cleaned all parts, as described in Chapter 20. The cylinder head should be reconditioned, as explained in Chapter 22. The short block should also be assembled, as discussed in Chapter 21. All connecting rods should be torqued and all surfaces perfectly clean.

The front cover and timing mechanism should also be installed, as discussed in Chapter 23. Refer to these chapters for more information as needed.

Note: In some designs, the oil pump is driven by the timing belt. In these applications, the timing belt and timing cover should not be installed until the oil pump is installed.

Installing the Oil Pump

With the front cover and timing mechanism in place, you can usually begin engine final assembly by installing the engine ***oil pump.*** As explained in Chapter 23, make sure the used pump is in good condition. Many technicians prefer to use a new or factory-rebuilt oil pump. The oil pump is very important to engine service life and is usually not expensive.

When possible, prime the oil pump. This can be done by pouring oil into the pumping cavity during assembly. You can also submerge the inlet into a container of oil and turn the pump by hand to draw oil into the pump. Priming will help protect the pump and ensure full engine oil pressure sooner on initial starting of the engine.

If the oil pump is mounted on the front cover, position the new gaskets, seals, rotors, and the other parts in the engine. Torque all pump fasteners to specifications. Turn the pump by hand to check for smoothness. Make sure the pump is well oiled, **Figure 24-2A.**

If the oil pump is installed on the bottom of the engine and extends into the oil pan, fit the oil pump drive into

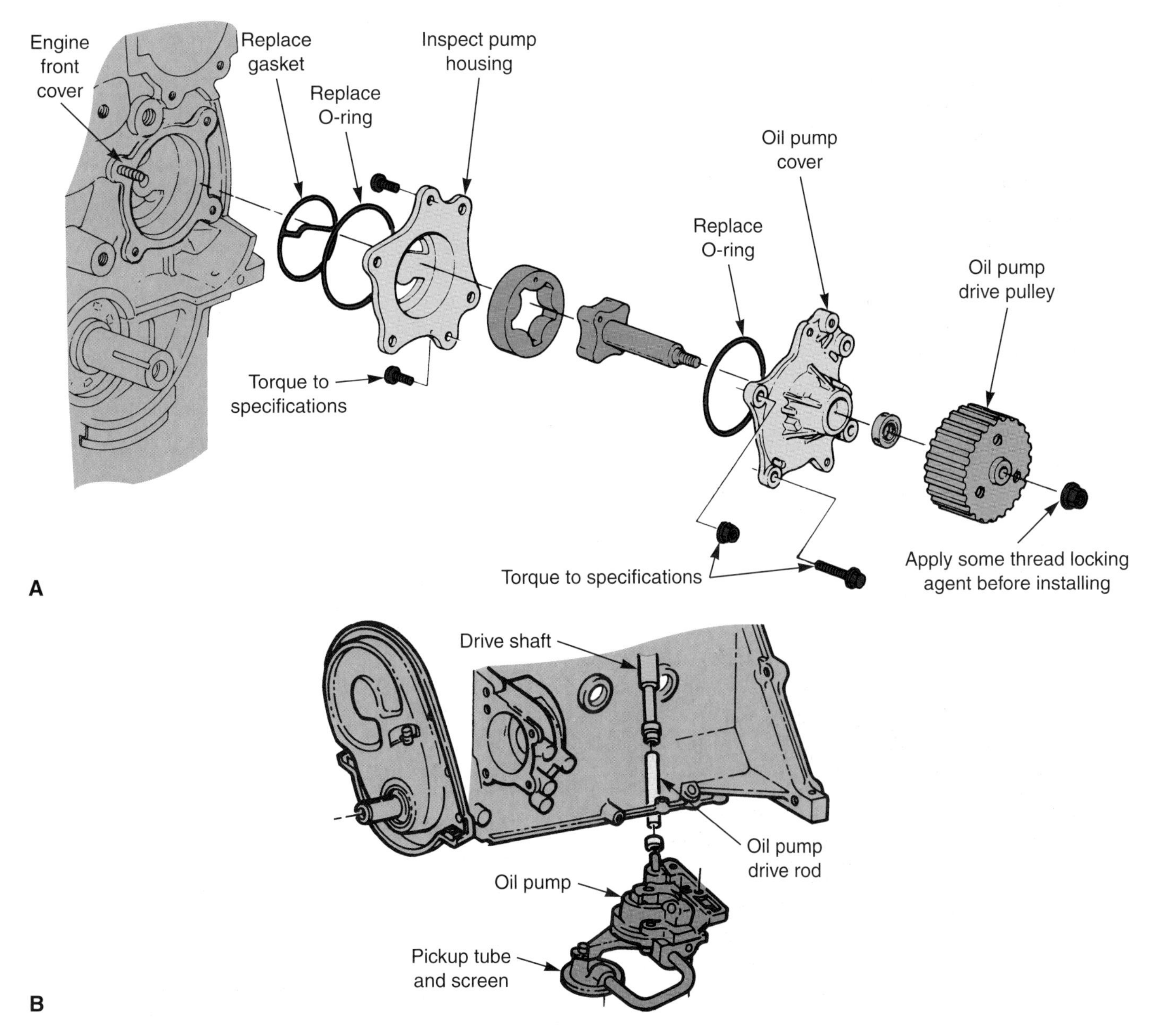

Figure 24-2. *A—Use new gaskets and seals when installing an oil pump that is mounted on the front cover. B—When installing an oil pump in the oil pan, make sure the drive rod is in the hole in the block and the pickup tube is secure.*

position. This is shown in **Figure 24-2B.** Make sure the oil pump drive rod or shaft is located in its guide hole in the block. As you fit the oil pump to the block, the drive rod must fit into the pump and the hole in the block.

Check that the oil pump gasket holes align properly. Avoid using sealer on the oil pump gasket. A gob of loose sealer could lock up the oil pump. The oil pump gasket is usually installed clean and dry. Start the pump bolts by hand and then torque them to specifications.

If support brackets are provided on the oil pump, make sure they are installed properly. They prevent pump vibration that could break the pump housing or cause bolt failure. The oil pump pickup tube and screen must also be in good condition and properly installed.

After pump installation, double-check everything. The pump gasket should be in place. The pump drive rod must be located in the block correctly. The pickup tube must not leak at the pump or block.

Note: Oil pump mounting bolts must be properly tightened. If these bolts loosen or fail in service, the engine will lose oil pressure and major bearing damage can result. It might be wise to use thread locking compound on the oil pump bolt threads to prevent the bolts from loosening due to engine vibration.

Installing the Oil Pan

With the pump and front cover in place, you are ready to install the ***oil pan.*** Lay the pan face down on a flat work surface. If the pan flange is bent or untrue, straighten it as shown in **Figure 24-3.** A bent flange can cause oil pan gasket leakage and the return of the customer with a valid complaint.

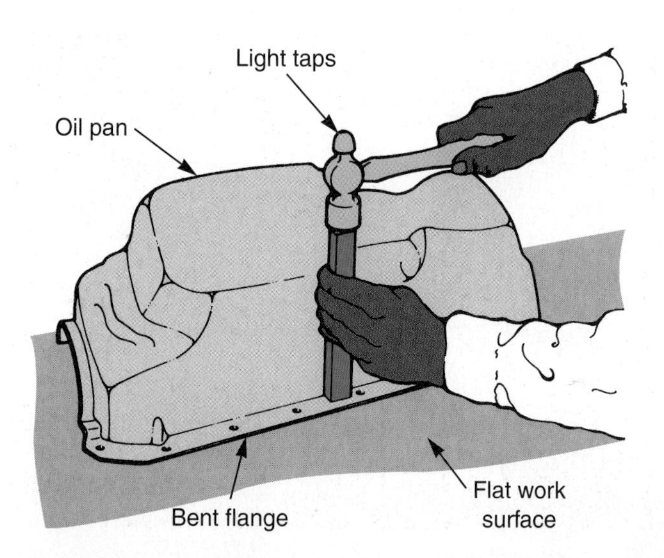

Figure 24-3. *Lay the oil pan on a flat workbench to check for a bent flange. If needed, the flange on a sheet metal oil pan can be straightened. Also, check the condition of the oil drain plug and the threads in the pan. (Ford)*

Turn the pan upright. Lay the oil pan gasket in place on the pan. Check that all bolt holes align and that you have the correct gasket.

You can either adhere the gasket to the pan or to the engine block. If the engine is installed in the car, it might be easier to glue the gasket to the oil pan. If you are working on an engine stand, it might be better to bond the gasket to the block, **Figure 24-4.** Then, there is less chance of the gasket falling or shifting during assembly.

Sometimes, the end seals for the oil pan have to be worked down into grooves in the oil pan or in the engine front cover and block. If this is the case, the pan gasket should be installed next to or under the rubber end seals. RTV sealer should be used to help prevent leakage where the gasket and end seals overlap.

Allow the gasket sealer to set up slightly before installing the pan. This will help keep bolt holes aligned. Fit the pan into place on the engine block.

Start all bolts by hand before tightening any of the fasteners. See **Figure 24-5.** If you tighten any of the bolts before others are started, the gasket can compress and block the other bolt holes. With all bolts started, tighten the oil pan bolts to specifications. Use a crisscross pattern, or other specified pattern, to ensure even gasket compression. Do not overtighten the bolts. Overtightening is a common mistake that will lead to oil pan leakage.

Installing Accessory Units

With the engine upside down on the stand, install any components that are easily accessible. For example, you might bolt on the engine motor mounts, any sending units, oil filter bracket, and so on. These parts are easy to access with the engine on the stand after oil pan installation.

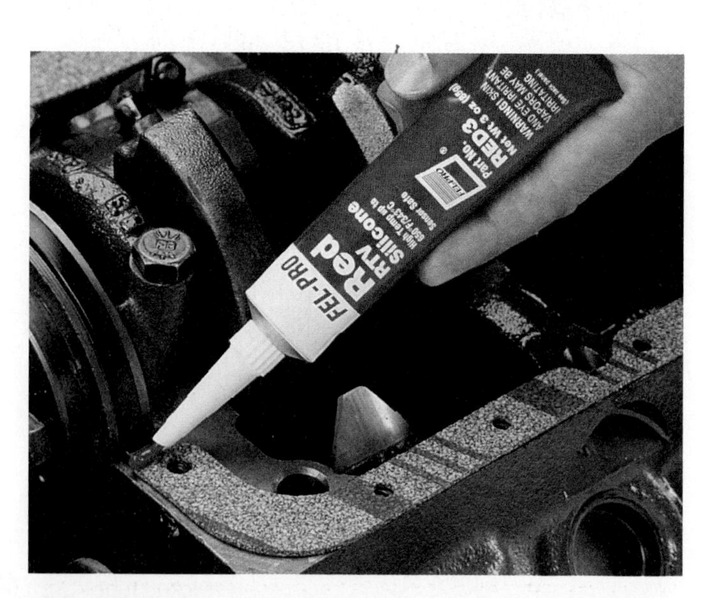

Figure 24-4. *When installing an oil pan, sealer is commonly used to hold the oil pan gasket in alignment during pan installation. The sealer also helps avoid oil pan leaks. To avoid leakage where a gasket and seal come together, place an extra dab of sealer at that point. (Fel-Pro)*

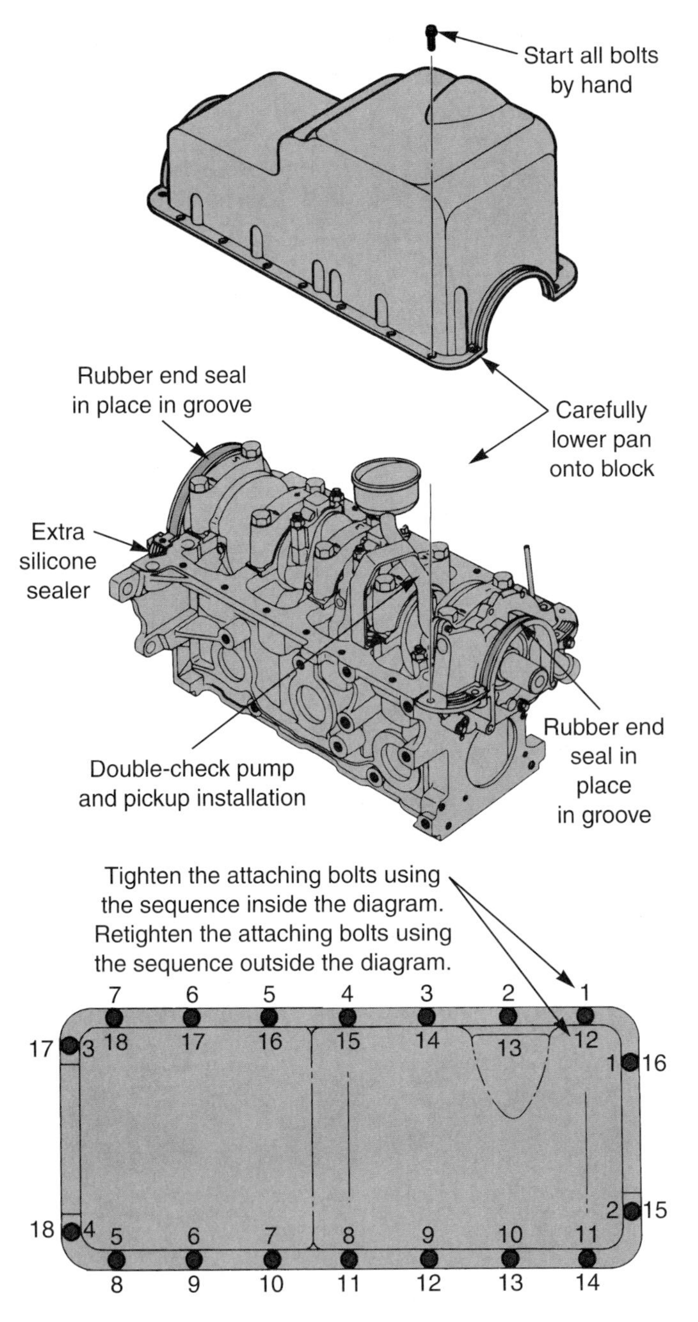

Figure 24-5. *When installing the oil pan, start all bolts by hand before tightening any bolts. Torque the bolts to specifications in the recommended sequence. (Ford)*

When installing the water pump, check the fit of the new gasket. Then, use gasket adhesive to hold the gasket on the pump. If a chemical gasket or RTV is used, form a continuous bead and circle all bolt holes and coolant passages. Fit the pump on the engine and start all bolts by hand. Check bolt lengths. Place sealer on threads of any bolts that extend into coolant and oil passages in the block.

Cylinder Head Installation

Installation methods are very critical when bolting the cylinder head to the block. The engines in late-model vehicles have thin-wall, cast iron or aluminum heads that can leak, warp, or crack if installed improperly. The engines in late-model vehicles also operate at higher temperatures than in the past, which can increase the strain on the seal between the head and block. Great care must be taken to ensure proper installation techniques for cylinder heads.

Installing the Head Gasket and Cylinder Head

Before installing the head gasket(s) and cylinder head(s), double-check that the pistons are installed in the right direction. Check the arrows or other markings on the pistons to make sure they are facing correctly.

A ***head gasket*** can usually only be installed one way. If installed backward, the gasket may fit, but coolant or oil passages may be blocked. This mistake could cause oil starvation to the valve train or engine overheating.

Cylinder head gasket markings are normally provided to show the front or top of a head gasket, **Figure 24-6.** The

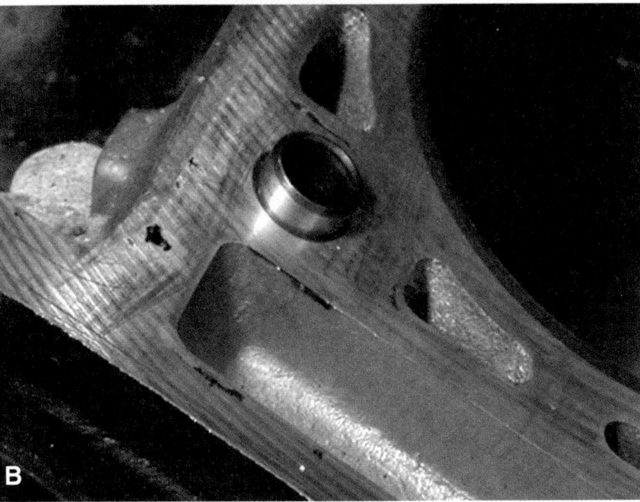

Figure 24-6. *A—A word, arrow, or notch indicates the front or top of the head gasket. These markings must be oriented correctly when the gasket is installed. B—Dowels are often used to hold the head gasket in alignment as the head is installed.*

head gasket may be marked "top" or "front," or it may have a line to show the installation direction. Metal dowels are frequently provided to hold the head gasket on the block during assembly. If dowels are not provided, use two long stud bolts to align the head gasket and head during assembly.

Most modern head gaskets are a Teflon-coated, permanent-torque type. Permanent torque means that they do not need retorquing after engine operation. These head gaskets should be installed clean and dry. Sealer is not recommended. However, some steel-shim or low-quality head gaskets may require retorquing and sealer. Refer to the gasket manufacturer's instructions when in doubt.

1. Turn the engine upright on the engine stand.
2. Make sure the block deck has been cleaned and all old gasket material is removed.
3. Check that the dowels for holding the head gasket in alignment are in place.
4. Find the markings on the head gasket that show top and front.
5. Place the head gasket onto the block clean and dry. Do not use sealer unless specified.
6. Make sure the head gasket is over the dowels. If dowels are not provided on the block deck, install stud bolts in the block. See **Figure 24-7.**

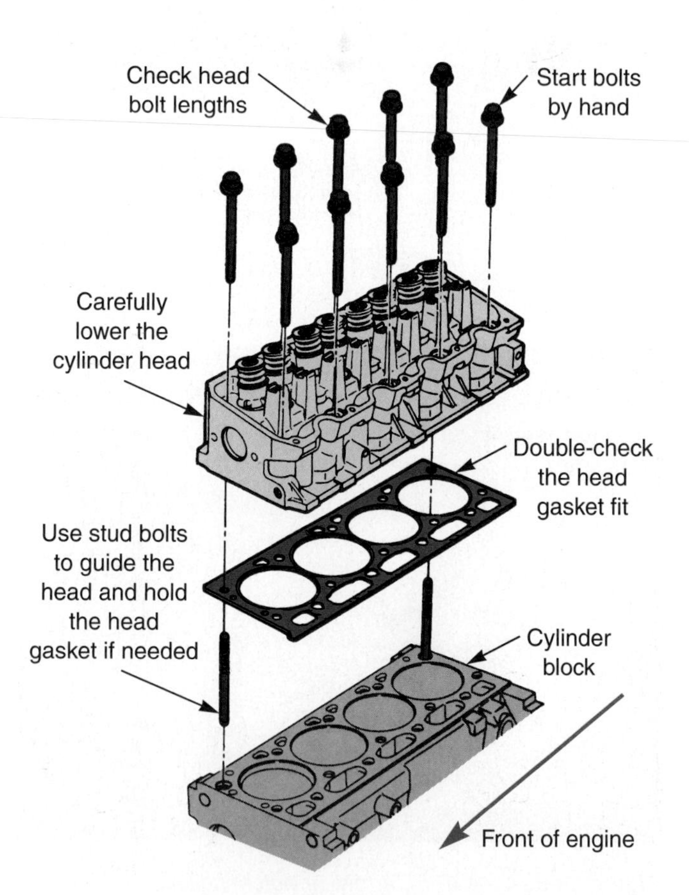

Figure 24-7. *If dowels are not provided, use two long stud bolts to align the head gasket and cylinder head during assembly. (Ford)*

7. Carefully place the cylinder head over the head gasket and block. Do not bump and damage the gasket.
8. Make sure the head is over the dowels or stud bolts.

Diesel Cylinder Head Installation

With diesel engines, head gasket thickness and bore size are very critical. Head gaskets are provided in different thicknesses to allow for cylinder head milling or varying the block deck height. Gasket thickness may be denoted with a color code, series of notches, holes, or other marking system.

When a diesel engine is bored oversize, it also requires a special gasket. You must request an overbore gasket from the parts supplier. A standard bore gasket will usually stick out into the cylinder, causing problems.

When buying a diesel engine head gasket, make sure you have the right one. Gasket thickness and bore size must be correct for the engine being repaired. Refer to the manual for details.

Threaded Fasteners

Some engines are assembled with ***torque-to-yield bolts.*** This type of bolt must be replaced when reassembling the engine. Refer to the service manual to find out if the head bolts are torque-to-yield and must be replaced.

Some engines have head bolts of different lengths. Make sure you have not accidentally installed a head bolt in the wrong hole. This could cause severe part damage or engine failure. Double-check that the head bolts are sticking out the same amount before threading them into their holes.

Some head bolts may extend into a coolant or oil passage. Place a nonhardening sealer on the threads of these bolts, **Figure 24-8.** The service manual will show which

Figure 24-8. *Head bolts that extend into an oil or coolant passage should have a nonhardening sealer applied to the threads. Refer to factory service literature for details about sealing head bolt threads. (Fel-Pro)*

bolts must be sealed. If sealer is not applied to the threads on these fasteners, oil or coolant may leak past the threads.

Make sure all bolt holes in the block are clean. Use compressed air to blow them out if needed. Place a drop of oil on the head bolt threads that do not have sealer on them. This will ensure an accurate torque when tightening the head bolts. To prevent hydraulic lock, do not squirt oil into the bolt holes. Hydraulic lock could prevent proper torquing or lead to part damage. To ensure accurate bolt torque, use only the lubricant recommended by the manufacturer. Using the wrong thread lubricant can greatly affect bolt tightening and clamping pressure.

Installing and Torquing Head Bolts

After checking head bolt lengths and applying sealer to the threads that need it, you are ready to install the bolts. Place the bolts in the correct holes and start threading them by hand.

1. Use a speed handle or ratchet to tighten the head bolts about finger tight. Use a crisscross pattern or the proper torque sequence, **Figure 24-9.** Generally, tighten the bolts in the middle first and work your way outward. The service manual will give the best sequence.
2. If stud bolts were installed to align the gasket and head, remove them. Install the correct bolts in their place.
3. Using a torque wrench, tighten the head bolts to 1/2 of the full torque specification in the proper sequence. For example, if the head bolt torque is 100 ft-lb (130 N·m), tighten all of the bolts to 50 ft-lb (70 N·m). A slow, steady pull on the wrench will produce the most accurate bolt torque. Refer to **Figure 24-10.**
4. Tighten the bolts to 3/4 of the full torque specification in the proper sequence. If the head bolt torque is 100 ft-lb (130 N·m), tighten the bolts to 75 ft-lb (100 N·m).
5. Tighten the bolts to the full torque specification in the proper sequence.
6. Tighten each bolt to full torque again to ensure proper torque.

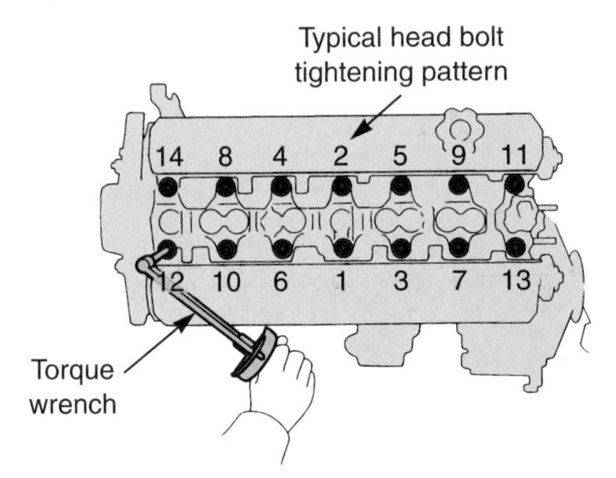

Figure 24-9. *When tightening head bolts, generally start in the middle and work your way to the outer bolts. Always follow the torque sequence given in the service manual. (Toyota)*

Figure 24-10. *Torque the head bolts to specifications using an even pull on the torque wrench. (Snap-on Tool Corp.)*

Lifter Installation

If you are working on a V-type engine, you will need to install the lifters before the intake manifold. Place heavy assembly lube on the bottom and sides of each lifter. Then, fit each lifter down into its bores in the block valley. Look at **Figure 24-11.**

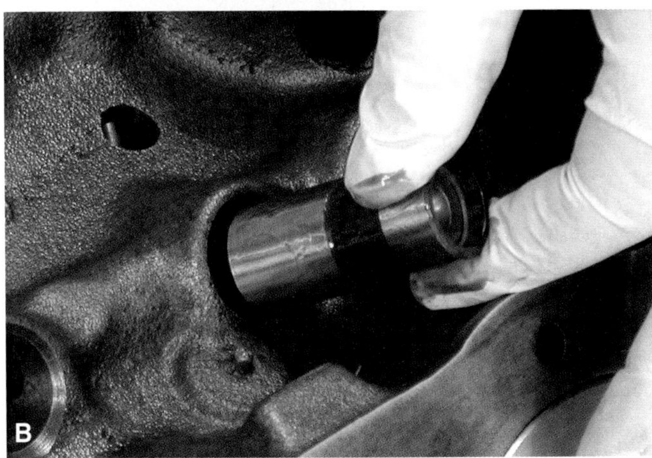

Figure 24-11. *A—Lubricate the bottom and sides of each lifter with thick assembly lube. B—Place each lifter down into its bore. Reused lifters must be installed in their original location.*

If the lifters and camshaft are being reused, make sure each lifter is placed in the same hole from which it was removed. This will match wear patterns on the camshaft and lifters to prevent premature wear. The lifters might have also been select fit from the factory.

Intake Manifold Installation

The installation of an engine intake manifold is critical. Some manifolds not only seal vacuum, they carry coolant and hot exhaust gasses. If incorrectly installed, the manifold can allow a vacuum or coolant leak. Partial engine teardown might be needed to correct the problem.

Inspecting the Intake Manifold

Intake manifold warpage is a measure of the flatness of the manifold's sealing surfaces. Use a straightedge and flat feeler gauge to check the intake manifold sealing surfaces for warpage.

1. Lay the edge over the port openings.
2. Try to slide feeler gauge blades of different thicknesses under the straightedge. Measure at different points on the manifold.
3. The thickest blade that easily slides under the straightedge indicates the intake manifold warpage.
4. If warpage is more than specifications, the intake manifold sealing surfaces must be machined or a new manifold installed.

Also, inspect the intake manifold for cracks, burned areas, and other problems. Closely inspect the bottom of the manifold and near the heat passage from the cylinder head. Hot exhaust gasses for the EGR or for manifold warming can burn and crack the manifold. Also, check the head mating surface near the heat passage. A ruptured gasket could have allowed the hot gasses to burn the mating surface.

Installing the Intake Manifold Gasket and Intake Manifold

Place the new ***intake manifold gasket*** over the head to check its fit. Check all openings in the gasket and head to make sure they align. Any misalignment of ports or passages could cause a vacuum leak, coolant leakage, or even serious engine overheating damage.

If the intake manifold gasket is a metal-valley-tray type, use RTV or silicone sealer. The sealer should circle all ports and coolant passages. The thin metal gasket can leak easily; sealer provides added protection against leakage. However, before applying sealer, make sure the manifold is ready to be installed.

If the intake manifold gasket is a multipiece, fiber type, you do not have to circle every intake port. Simply use

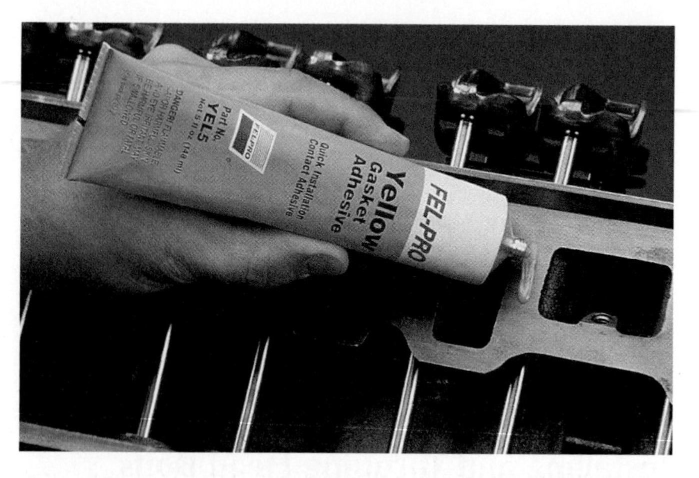

Figure 24-12. *Gasket adhesive is commonly used around ports and coolant passages to help hold the intake manifold gaskets in place during intake manifold installation. (Fel-Pro)*

gasket adhesive to hold the gasket in place, **Figure 24-12.** However, some technicians prefer to circle coolant passages with sealer for added protection.

With the head sections of the intake manifold gasket in place, install the ***end seals.*** Adhesive can be used to hold the seals in place if needed, **Figure 24-13A.** If large

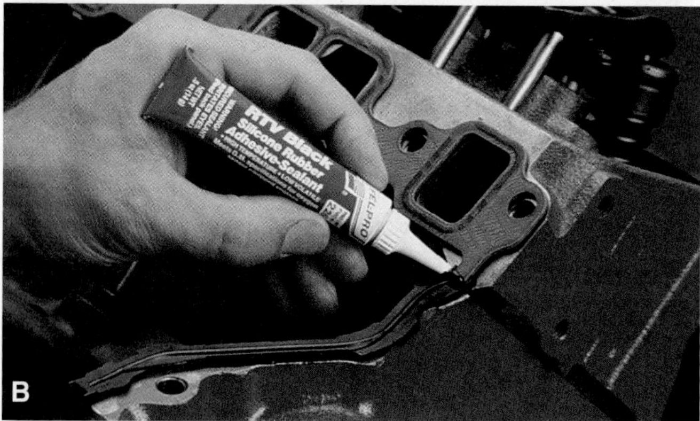

Figure 24-13. *A—Applying adhesive to hold the end seals in place during intake manifold assembly. B—Additional sealer should be placed where the end seals and gaskets come together. This is a common point for oil leakage on V-type engines. (Fel-Pro)*

barbs are provided on the seals and they fit properly down into the block, no adhesive is needed. Place a small bead of silicone sealer where the end seals and main intake gaskets overlap, **Figure 24-13B.** This will keep engine oil from weeping out during engine operation.

Some late-model vehicles have plastic intake manifolds. Depending on the design, these manifolds may have a separate rubber seal for each intake port in place of a traditional gasket. First, remove the old rubber seals, **Figure 24-14A.** Then, clean all mating surfaces and the seal grooves, **Figure 24-14B.** Finally, press the seals into their grooves, **Figure 24-14C.** Make sure the seals are properly installed so they do not shift or fall out while lowering the manifold onto the engine.

Carefully lower the intake manifold onto the engine. Be careful not to bump and shift the gasket or smear the sealer. In some applications, you might want to use guide studs so the manifold lowers straight down into place.

Note: You should be ready to install the intake manifold after the intake manifold gasket is in place. If you wait too long, the sealer can completely cure and affect the manifold sealing. The sealer or adhesive can be allowed to cure slightly, but should not completely dry or cure.

Torquing the Intake Manifold

After the intake manifold and gasket are in place, start all intake manifold bolts by hand. Otherwise, you will not be able to shift the manifold and start the remaining bolts.

1. With all bolts threaded in a few turns, run them down lightly.
2. Using a torque wrench, torque each bolt to 1/2 of the full torque. Use a crisscross pattern or the specified torque sequence to draw the manifold down evenly.
3. Tighten each bolt to 3/4 of full torque.
4. Tighten each bolt to full torque.
5. Tighten each bolt at full torque one more time to ensure proper torque.

Figure 24-15 shows a multipiece intake manifold. Use the same basic procedure described above to install the base of the manifold. Tighten the bolts in a crisscross pattern to the torque specifications. Tighten them to 1/2, then 3/4, and finally full torque twice. Then, do the same for the center and upper sections of the intake manifold.

The upper section of the intake manifold, or upper plenum, may be constructed of plastic. With plastic engine parts, like intake plenums, brass thread inserts are usually pressed or molded into the plastic. This allows parts such as sensors and vacuum fittings to be bolted to the plastic part without thread failure.

Check the Thermostat

When doing major engine service, you should check the engine thermostat. Make sure one is installed and that it operates properly. Also, check that the thermostat heat range matches the service manual specification. Most technicians replace the thermostat as a routine part of an engine rebuild.

When installing the thermostat housing, make sure the thermostat is centered under the housing. Also, make sure the new gasket or O-ring is in place. Then, tighten each bolt a little at a time until the full torque specification is reached, **Figure 24-16.** Remember that the housing will warp or crack if the bolts are tightened too much or improperly.

Exhaust Manifold Installation

Exhaust manifolds route extremely hot gasses into the vehicle's exhaust system. As a result, the mainfolds can suffer from cracks, warpage, rust, and other problems. Always inspect an exhaust manifold closely before installation. Look for cracks, burned areas, eroded mating surfaces, and severe rust damage. To check for ***exhaust manifold warpage,*** use a straightedge and flat feeler gauge, **Figure 24-17.** Use the same procedure described earlier for checking intake manifold warpage. If warped slightly, a

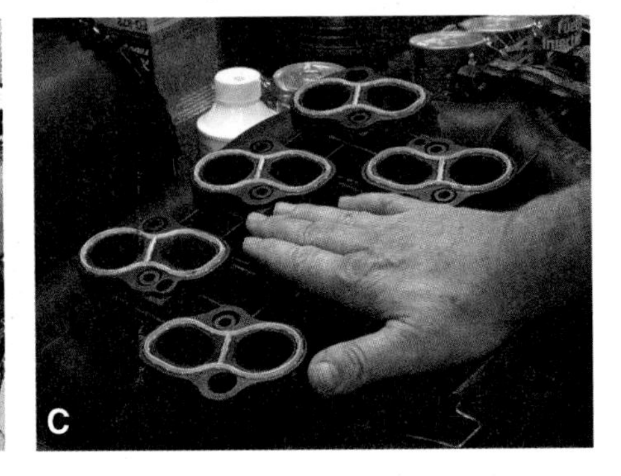

Figure 24-14. *Installing rubber seals on a plastic intake manifold. A—Remove the old seals. B—Clean all mating surfaces and the seal grooves. C—Install the new seals. Make sure they are fully seated in their grooves.*

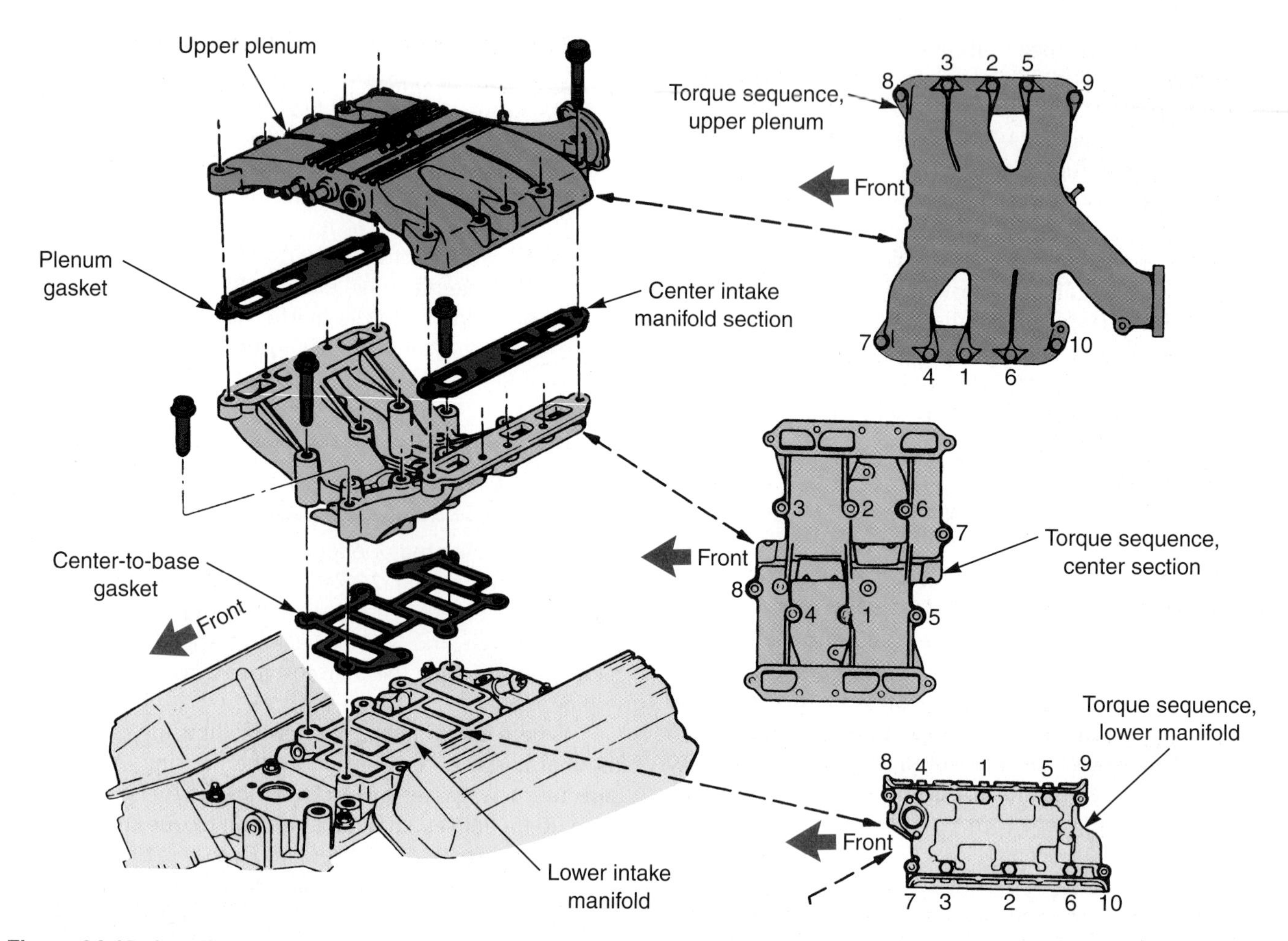

Figure 24-15. *Installing a multipiece intake manifold. Install the lower manifold first, followed by the center section, and finally the upper plenum. (General Motors)*

machine shop can usually true the surface. Replace the manifold if it is excessively warped.

A leaking exhaust manifold will be indicated by a clicking sound as the exhaust gasses blow out between the cylinder head and manifold. Most used exhaust manifolds should have a gasket installed between the manifold and head to prevent leakage, **Figure 24-18.** Some exhaust manifold gaskets have slots. The slotted gasket allows you to start the manifold bolts, holding the exhaust manifold in place, and then drop the gasket into place.

Position the gasket on the head. Then, slide the exhaust manifold into position. Stud bolts can be helpful in guiding the manifold and gasket in place and holding them while the bolts are started.

Start all bolts by hand. Then, turn the bolts down until snug. Use a torque wrench to tighten the exhaust manifold bolts to specifications. Use a crisscross pattern to pull the manifold down evenly and to compress the gasket (if used), **Figure 24-19.** First, torque each bolt to 1/2 of full torque, then 3/4 of full torque, and finally to full torque twice.

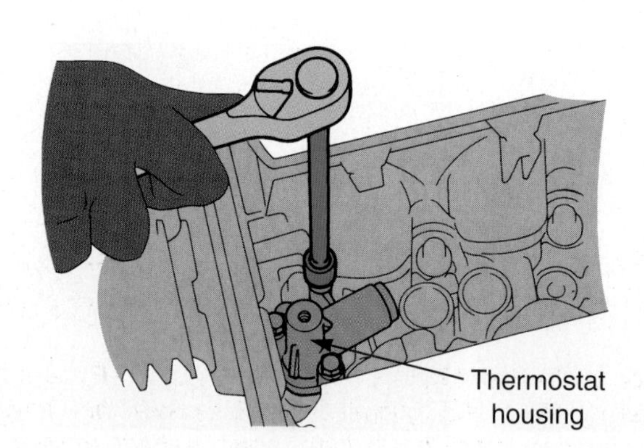

Figure 24-16. *Tighten the thermostat housing bolts evenly to prevent warpage or breakage.*

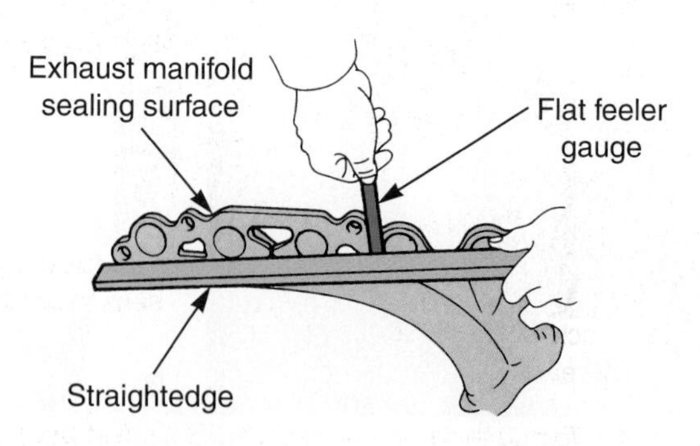

Figure 24-17. *Checking the exhaust manifold for warpage. Also, check it for cracks.*

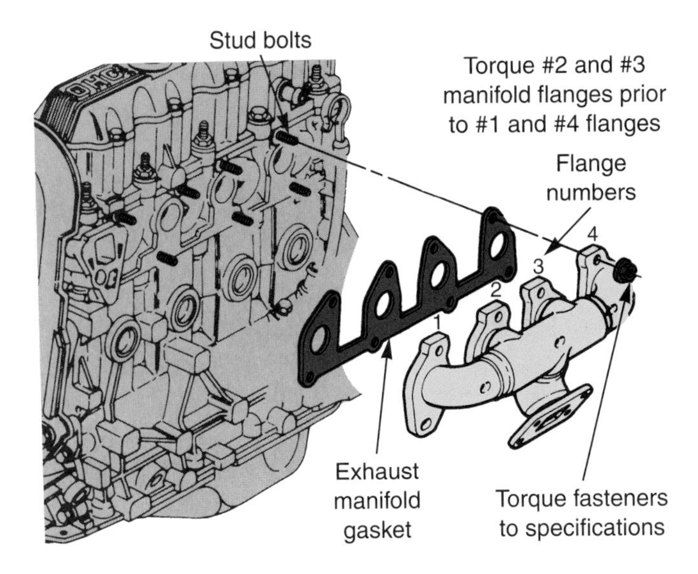

Figure 24-18. *An exhaust manifold gasket is usually installed between the cylinder head and exhaust manifold. (General Motors)*

Note: When the exhaust manifold is installed, it is a good idea to check the condition of the oxygen sensor. Replace the sensor if it looks contaminated or if it has been in service for a long time.

Camshaft Housing Installation

Some OHC engines have a ***camshaft housing*** that mounts on top of the cylinder head. See **Figure 24-20.** The camshaft housing holds the cam followers and camshaft(s).

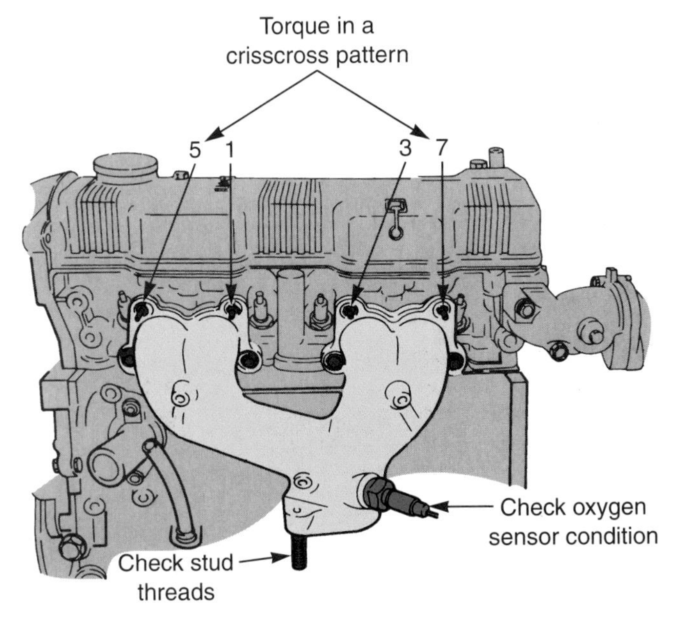

Figure 24-19. *Torque exhaust manifold bolts in a crisscross pattern to pull the manifold down evenly. This is a good time to check the oxygen sensor and replace it if needed. (General Motors)*

1. Place a gasket on the top of the cylinder head.
2. Make sure everything aligns properly and you have the correct gasket.
3. Lower the camshaft housing onto the gasket and head.
4. Start all bolts by hand.
5. Tighten the bolts to specifications using the recommended torque sequence. Tighten in steps.

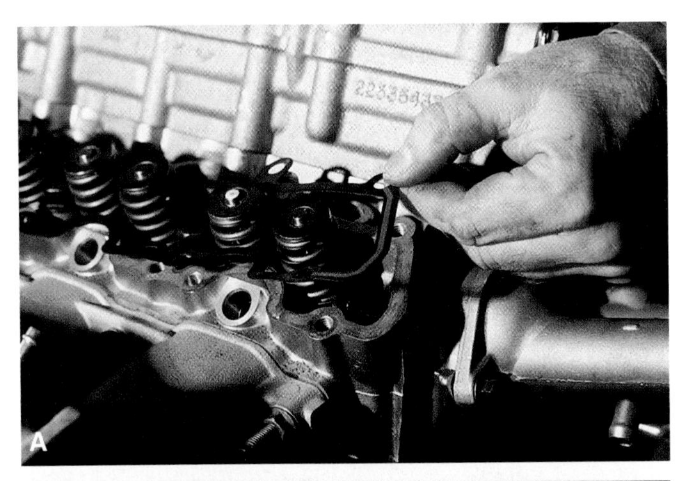

Figure 24-20. *Installing a camshaft housing. A—Place a new gasket on top of the cylinder head. B—Lower the camshaft housing down onto the cylinder head without moving the gasket. C—Place the cam followers down into their pockets. Reused followers must go back into their same pocket.*

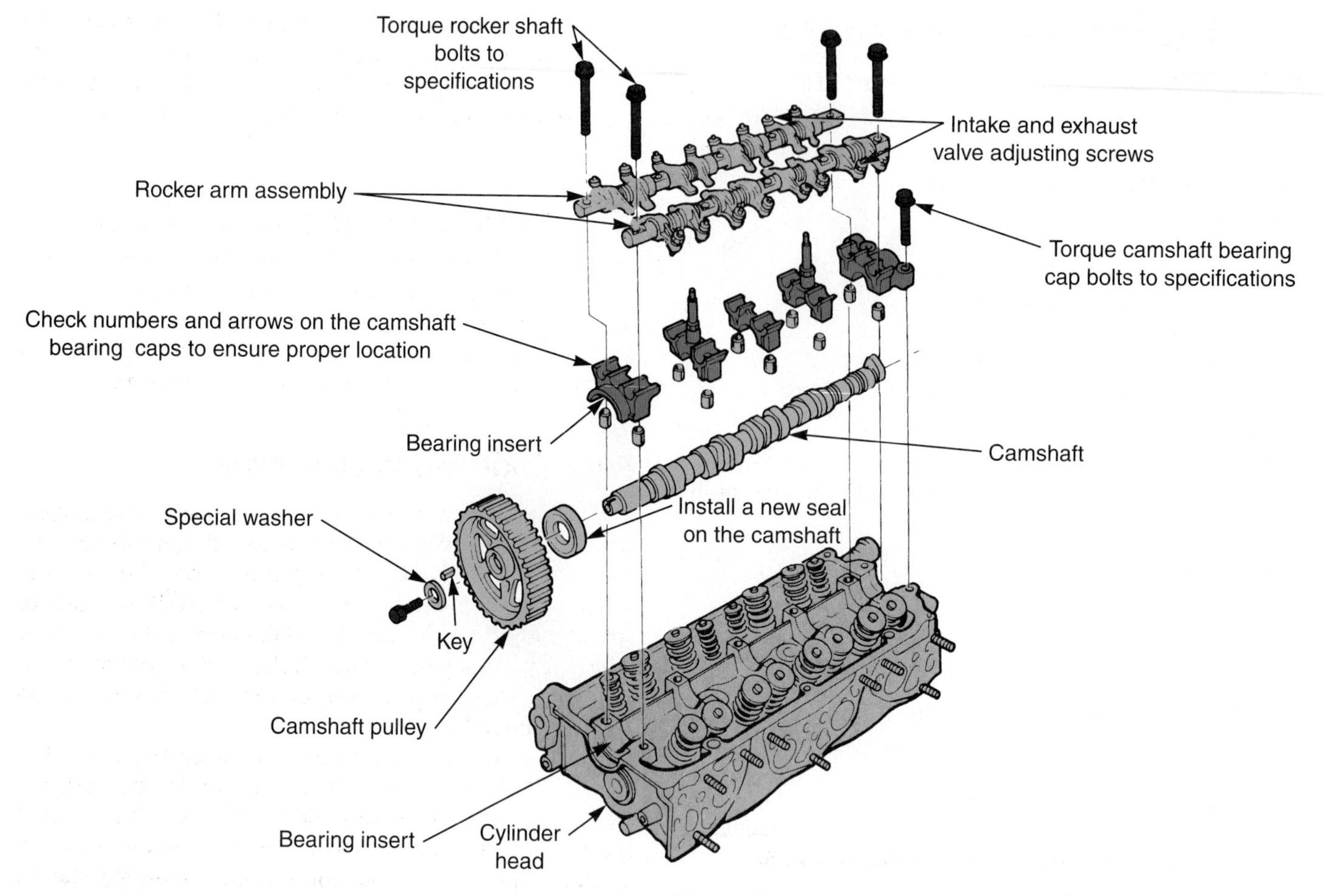

Figure 24-21. *This engine has two-piece camshaft bearings. Camshaft bearing caps secure the camshaft bearings and the camshaft to the cylinder head. (Honda)*

6. Place the cam followers into their pockets in the camshaft housing.
7. Install the camshaft(s), as described in the next section.

Overhead Camshaft Installation

The camshaft bearings in an overhead camshaft engine may be a one- or two-piece design. The one-piece bearings are pressed into the cylinder head. The two-piece bearings snap into the head and camshaft bearing caps, **Figure 24-21.**

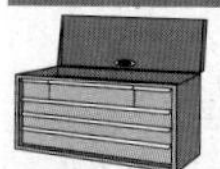

Note: Servicing overhead camshaft bearings is discussed in Chapter 22.

Installing a Camshaft with One-Piece Bearings

Make sure the camshaft bearings have been inspected and replaced if needed. White grease or thick assembly lube is commonly recommended to reduce friction and wear during initial startup.

1. Lubricate the camshaft journals and lobes.
2. Slowly slide the camshaft into the camshaft bearings. Do not hit or nick the camshaft bearings.
3. Slowly rotate the camshaft to help feed it through each bearing, **Figure 24-22.**
4. Make sure you support the camshaft as it is inserted.
5. Install the timing sprocket and camshaft seal if applicable. Install any remaining parts as needed, such as the retainer plate, oil seal, and thrust bearing.

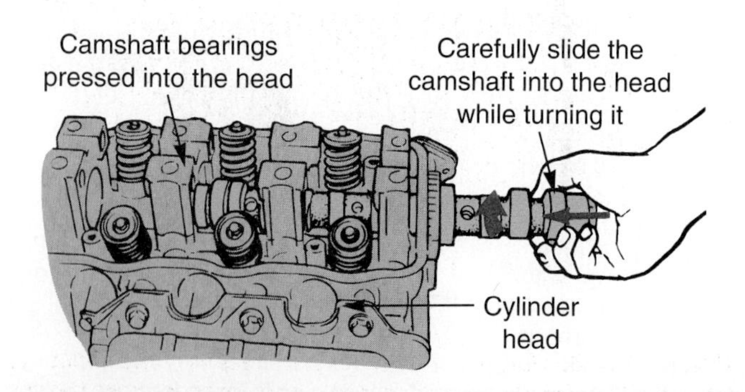

Figure 24-22. *Installing a camshaft into one-piece camshaft bearings. (General Motors)*

Installing a Camshaft with Two-Piece Bearings

Some OHC engines have two-piece bearings. Camshaft bearing caps hold the upper cam bearing inserts and bolt over the camshaft journals. Just like block main caps, camshaft bearing caps are numbered and must be installed in their original locations. If you mix up caps or install them backwards, the camshaft can lock in the cylinder head. The bearings and camshaft can be severely damaged on engine starting.

Before installing the camshaft, make sure the camshaft bearings are properly installed in the head and bearing caps. Also, make sure the proper bearing cap locations are identified.

1. Apply assembly lube or engine oil to the face of the bearings in the head.
2. Carefully lower the camshaft into the bearings. Make sure the front of the camshaft is at the front of the engine.
3. Apply assembly lube or oil to the camshaft journals.
4. Position the bearing caps on the head.
5. Install the bearing cap bolts hand tight.
6. Torque the bearing cap bolts to 1/2 of the full torque specification. Use the specified torque sequence or the general sequence shown in **Figure 24-23.**
7. Torque the bearing cap bolts to 3/4 of the full torque specification in sequence.
8. Torque the bearing cap bolts to full torque specification in sequence.
9. Torque the bearing cap bolts to full torque one more time to verify proper torque.
10. Apply assembly lube to the camshaft lobes. Coat the entire surface of each lobe.
11. Make sure you can still turn the camshaft after torquing the camshaft bearing cap bolts. If not, something is wrong. Check bearing size, camshaft straightness, cap locations, and so on. Correct any problems.

Assembling the Valve Train

Procedures for installing valve train components will vary by engine design. Refer to the service manual for specific instructions. However, the following sections provide overviews of the most common procedures.

Push Rods and Rocker Arms

If you are working on an overhead valve engine (cam-in-block), the push rods may pass through the bottom of the intake manifold or through the cylinder heads. If the push rods extend through the intake manifold, you cannot install the push rods until the intake manifold is installed. If the push rods pass through holes in the cylinder head, you can install the push rods and rocker arms before the intake manifold.

Make sure that the lower end of each push rod is installed fully into its lifter. It is very easy for the push rod to slide out of the lifter. Also, be careful if you have to pull a push rod back out of a lifter. Oil between the lifter and push rod may result in suction that can cause the lifter to be pulled out of its bore. You might then have to remove the intake manifold to put the lifter back into place.

Sometimes the intake and exhaust rocker arms are different. Usually, an E for exhaust or an I for intake will be stamped on the rocker arms if they are not the same. Refer to **Figure 24-24.**

If a rocker arm shaft is used, fit the assembly on the cylinder head. Start the bolts by hand. Then, torque the

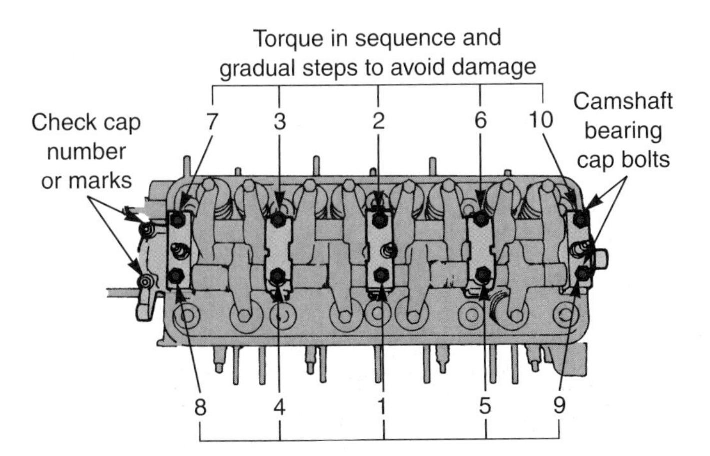

Figure 24-23. *Torque the camshaft bearing cap bolts to specifications in the proper torque sequence of this general sequence. Make sure you can still turn the camshaft after tightening the bolts. (Honda)*

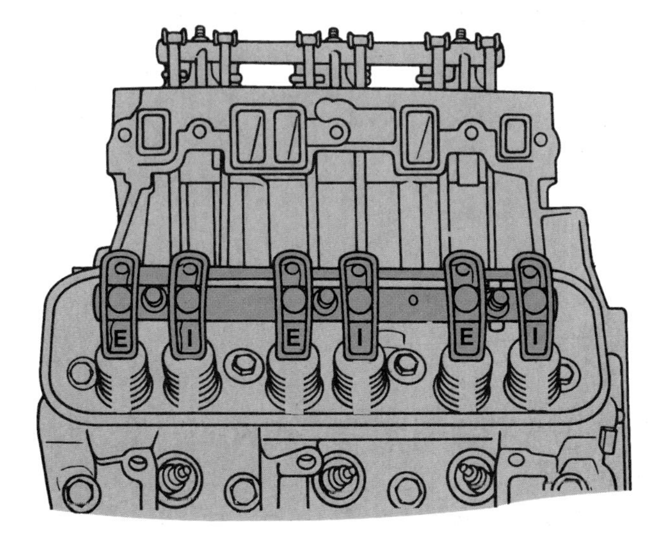

Figure 24-24. *Some rocker arms are marked for intake (I) and exhaust (E) and must be installed on the correct valve. (General Motors)*

bolts to specifications in the sequence given in the service manual. Tighten in steps from 1/2 full torque, to 3/4 full torque, and then full torque. Tighten at full torque a second time to ensure proper torque.

If nonadjustable, individual rocker arm stands or studs are used, place the rocker arm over the stand or stud. Then, start the bolt or nut by hand. Torque the bolt or nut to specifications. Tighten in steps from 1/2 full torque, to 3/4 full torque, and then full torque. Tighten at full torque a second time to ensure proper torque.

With adjustable rocker arms, start the bolt or nut by hand. However, do not tighten the bolt or nut. You will have to adjust the valve train clearance, as described in the next section. You have to adjust the valve train clearance if solid lifters are used as well.

Always make sure all parts are unworn and in usable condition. Inspect the push rod tips and make sure no push rods are bent. Check the rocker arm friction surfaces for wear. Replace any parts worn or damaged.

Adjusting Valve Train Clearance

After installation of the valve train components, you may have to adjust the clearance in the valve train. ***Valve train clearance adjustment,*** also called ***valve lash adjustment*** or ***valve adjustment,*** is done by turning the rocker arm fastener or adjusting screw, installing OHC follower shims, or using longer push rods.

Valve train clearance adjustment is critical to engine performance and service life. If the valves are too loose, valve train noise, sometimes termed lifter clatter, will result. ***Valve train noise*** is a light tapping or clicking sound produced from under the valve or camshaft cover. If the valves are too tight, burned valves can result. The valves may not close tightly and hot exhaust gasses can erode away metal from the valve face and valve seat.

Valve Train Clearance Adjustment with Nonadjustable Rocker Arms

Nonadjustable rocker arms are used on some push rod engines with hydraulic lifters. The hydraulic lifters automatically compensate for changes in valve train clearance to maintaining a zero valve lash. ***Zero valve lash*** is no clearance in the valve train and it provides quiet operation. The hydraulic lifters can adjust valve train clearance as parts wear, with changes in temperature (part contraction or expansion), and with changes in oil thickness.

If adjustment is needed because of valve grinding, head milling, or other modifications, different push rod lengths can sometimes be purchased for use with nonadjustable rocker arms. Refer to the service manual or a part supplier for details. Normally, torquing the rocker arm nut or bolt to specifications properly adjusts the valve train clearance.

Caution: Overtightening the fastener on a nonadjustable rocker arm can cause engine missing and rocker stud damage. Torque the fastener to specifications.

Valve Train Clearance Adjustment with Hydraulic Lifters

With some OHV engine designs, you must adjust the rocker arms to center the hydraulic lifter plungers. Centering the lifter plungers in their bores allows the lifters to automatically adjust to take up or allow more valve train clearance. Some manuals recommend adjustment with the engine off. However, many technicians adjust hydraulic lifters with the engine running.

To adjust the valve train clearance in a valve train with hydraulic lifters with the engine off, remove the valve cover. Turn the crankshaft until the first lifter is on the camshaft base circle, not on the lobe. The valve must be fully closed.

1. Loosen the rocker arm fastener until you can wiggle the push rod up and down.
2. Slowly tighten the rocker arm fastener until all play is out of the valve train (cannot wiggle the push rod).
3. Tighten the rocker arm fastener about one more turn to center the lifter plunger. Refer to the service manual for details because this can vary with engine design.
4. Repeat this procedure on the other lifters.

To adjust the valve train clearance in a valve train with hydraulic lifters with the engine running, first remove the valve covers. Then, install oil deflecting rocker arm clips. These clips will prevent oil spray off of the rocker arms.

1. Start and warm the engine to operating temperature.
2. Tighten all of the rocker arm fasteners until the valve train is quiet.
3. On the first rocker arm, loosen the fastener until valve train noise appears.
4. Tighten the rocker arm fastener slowly until the valve train noise disappears. At this point, there is zero valve lash.
5. To center the lifter plunger, tighten the rocker arm fastener about one-half to one turn more. Tighten the fastener slowly to give the lifter time to leak down and prevent engine missing or stalling.
6. Repeat the adjustment on the other rockers.

Other adjustment methods may be recommended. Check the service literature for more detailed information.

Valve Train Clearance Adjustment with Mechanical Lifters

Mechanical lifters are used on some heavy-duty engines, such as those in taxi cabs, pickup trucks, and diesel engines, and on some high-performance engines. Since mechanical lifters, also called solid lifters, are not self-adjusting, you must periodically adjust the valve train clearance to maintain zero valve lash. A valve train that contains mechanical lifters makes a clattering or pecking sound during normal engine operation.

Check the service manual for valve train clearance adjustment intervals and specifications. A typical valve train clearance, or valve lash, specification is approximately .014″ (0.35 mm) for the intake valves and .016″ (0.40 mm) for the exhaust valves.

To adjust the valve train clearance in a valve train with mechanical lifters, first remove the valve covers. Then, crank the engine until the first piston is at TDC on its compression stroke. On the compression stroke, you will be able to feel air blow out of the spark plug hole with the spark plug removed. This positions the intake and exhaust lifters for that cylinder on the camshaft base circle; intake and exhaust valves are fully closed. The valve train clearance can be adjusted for both intake and exhaust valves in that cylinder.

1. Select the correct size of flat feeler gauge.
2. Slide the feeler gauge between the rocker arm and the valve stem, as shown in **Figure 24-25.**
3. When properly adjusted, the feeler gauge should slide between the valve and rocker arm with a slight drag.
4. If the feeler gauge does not fit, loosen the rocker arm fastener or adjusting screw until the feeler gauge fits.
5. If the feeler gauge fits without a drag, tighten the rocker arm fastener or adjusting screw.
6. Repeat this procedure on the other lifters.

To tighten or loosen a rocker arm adjusting screw, you will normally have to loosen a locknut and then turn the adjusting screw. Refer to **Figure 24-26.** After making the adjustment, tighten the locknut and recheck the clearance.

Figure 24-25. *When adjusting valve train clearance on an engine with mechanical (solid) lifters, tighten the rocker arm fastener until there is a slight drag on the feeler gauge.*

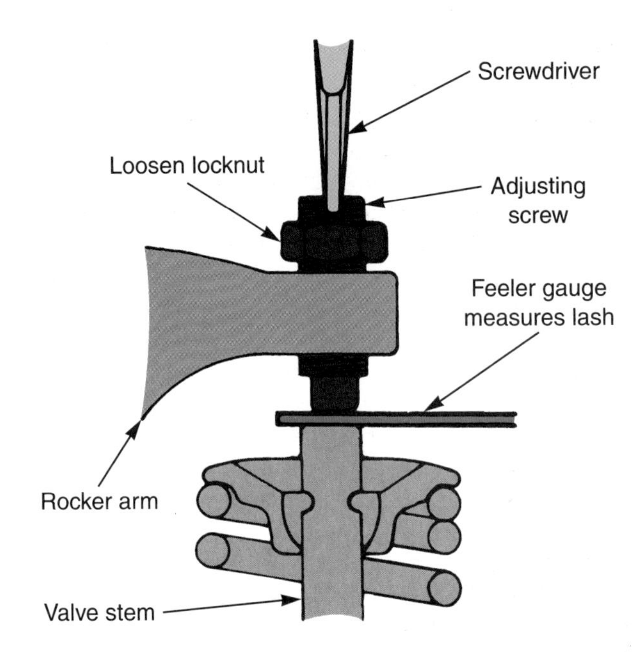

Figure 24-26. *To tighten or loosen a rocker arm adjusting screw, you will normally have to loosen a locknut and then turn the adjusting screw. (General Motors)*

Note: In some cases, valve train clearance is adjusted with the engine cold. In other cases, the adjustment must be made with the engine at operating temperature. Check the engine manufacturers recommendations. A change in temperature will cause part expansion or contraction. This, in turn, causes a change in valve train clearance.

OHC Valve Train Clearance Adjustment

There are many methods of adjusting the valve train clearance on OHC engines. Sometimes the adjustment is made like the adjustment for a valve train with mechanical lifters in a push rod engine. The rocker arm adjuster is turned until the correct size feeler gauge fits between the rocker or camshaft lobe and the valve stem or cam follower.

Some OHC engines have hydraulic cam followers. These followers are self-adjusting, like hydraulic lifters. Valve train clearance adjustment may not be necessary on these engines.

Valve adjusting shims may also be used on OHC engines. By installing different shims, the clearance between the camshaft lobe and the valve stem or follower can be adjusted, **Figure 24-27.** To calculate the proper shim thickness, first measure the clearance between the camshaft lobe and shim with a feeler gauge. Compare the measurement to the clearance specifications. For example, if the engine requires .015″ clearance and the measured clearance is .020″, the shim must be .005″ thicker than the existing one (.020″ – .015″ = .005″). Use a micrometer to measure the thickness of the existing shim and add .005″ to determine the correct thickness of the new shim.

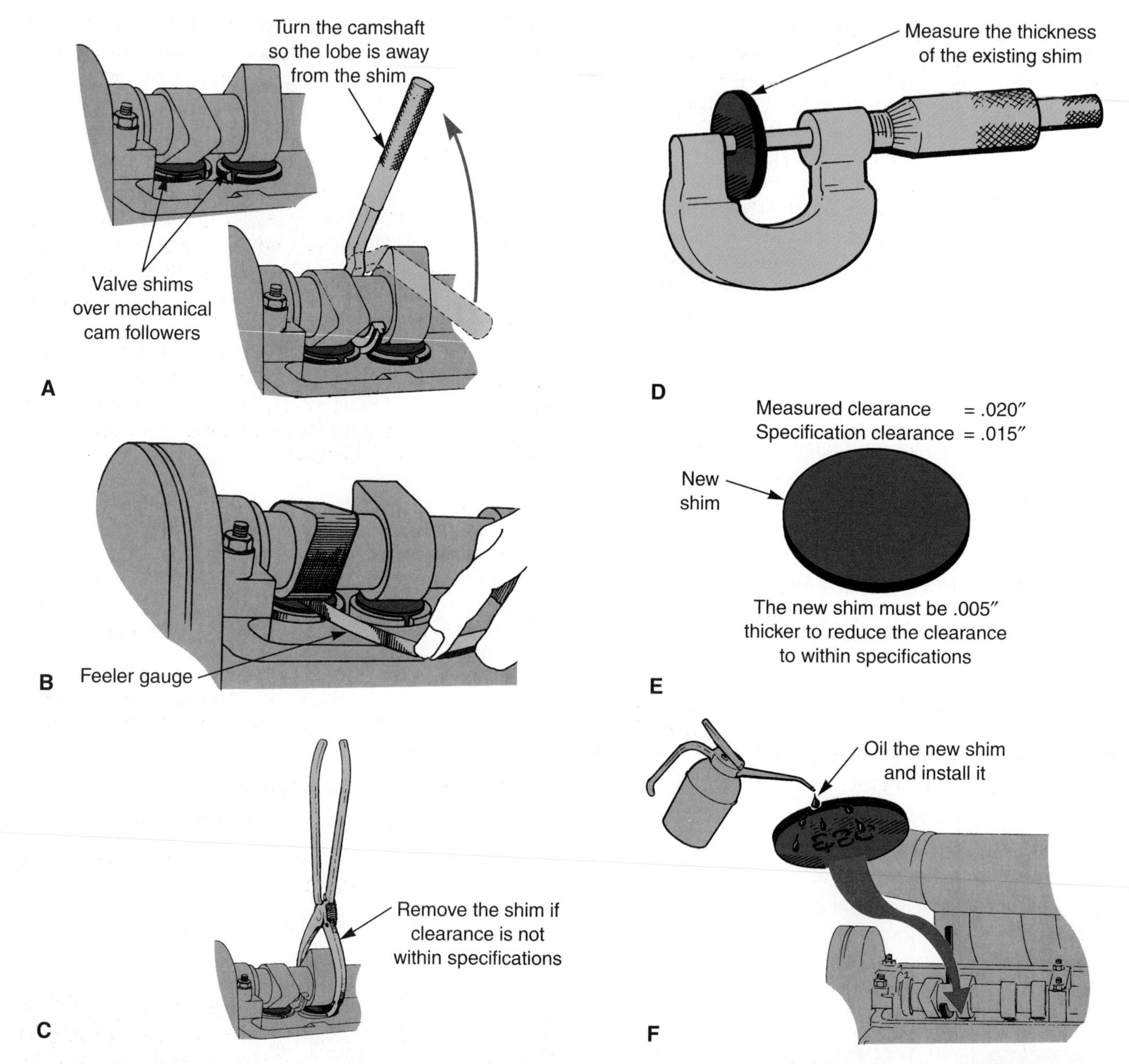

Figure 24-27. *Study the basic steps for adjusting valve train clearance with shims on an OHC engine. A—Position the camshaft. B—Measure the clearance with a feeler gauge. C—If the clearance is not within specifications, remove the existing shim. D—After measuring the thickness of the existing shim, calculate the required shim thickness. E—Obtain a new shim that will result in the correct clearance. F—Oil and install the new shim. (Volvo)*

Some OHC engines have an adjusting screw in the cam followers. Turning the screw changes the valve train clearance. Refer to the service manual for detailed directions. Exact procedures for valve train clearance adjustment vary with engine design.

Figure 24-28 shows how one engine manufacturer recommends that you check the action of the hydraulic followers in a variable valve timing engine. Push down on each rocker arm (primary, mid, and secondary) to check that each moves properly. If the rocker arms move properly, the hydraulic follower for the variable valve timing is functioning normally.

Camshaft and Valve Cover Installation

A ***valve cover*** is a sheet metal or plastic cover fastened to the top of the cylinder heads. The valve covers and valve cover gaskets prevent oil spraying from the valve train from leaking out of engine. A ***camshaft cover*** serves the same purpose as a valve cover, but is used on an OHC engine. It fits over the camshaft housing of an OHC engine. Most camshaft covers are made of cast aluminum.

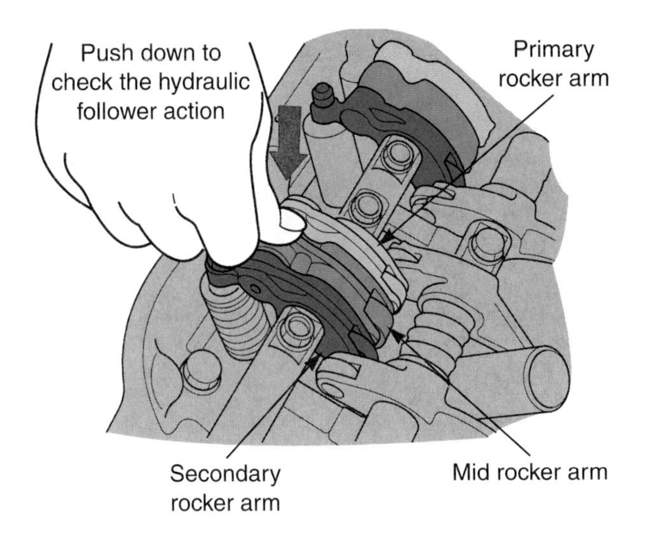

Figure 24-28. *This service manual illustration shows how to check the action of hydraulic followers on a particular variable valve timing system. (Honda)*

Camshaft Cover Installation

Camshaft covers are often installed with a gasket or seal between the cover and cylinder head or camshaft housing. In some designs, O-rings are also placed around the spark plug holes. When installing a camshaft cover, make sure all gaskets and seals are placed properly. If any fall out during cover installation, you will have a major oil leak and engine repair failure. See **Figure 24-29.**

Valve Cover Installation

If not installed properly, a valve cover, also termed rocker cover, can leak oil very easily. Some valve covers use a cork or synthetic rubber gasket. Valve covers on a few late-model vehicles are sealed at the factory with silicone sealer. When reinstalling the valve cover, either sealer or gaskets may be used, depending on cover design and accessibility of the cover.

Checking the Valve Cover Sealing Surface

Before installing a valve cover, make sure the cover flange is not bent. Lay the cover on a flat workbench. View between the gasket surface and the workbench to detect gaps. A straightedge can also be used to verify that the flange is not bent, **Figure 24-30.** Gaps are caused by dents, bends, or warpage.

If the valve cover is made from sheet metal, the flange can be straightened with taps from a small, ball peen hammer. A cast aluminum valve cover can sometimes be sent to a machine shop for resurfacing. A warped plastic valve cover must be replaced.

Installing the Valve Cover Gasket

To install a valve cover gasket:

1. Place a very light coating of approved adhesive around the edge of the valve cover, **Figure 24-31.** This is mainly needed to hold the gasket in place during assembly.
2. Fit the gasket on the cover and align the bolt holes. Inserting a couple of bolts through the holes can help in alignment.
3. Let the adhesive cure slightly.
4. Make sure the mating surface on the cylinder is clean.
5. Carefully place the cover and gasket onto the cylinder head. See **Figure 24-32.**
6. Hand start all of the valve cover bolts.

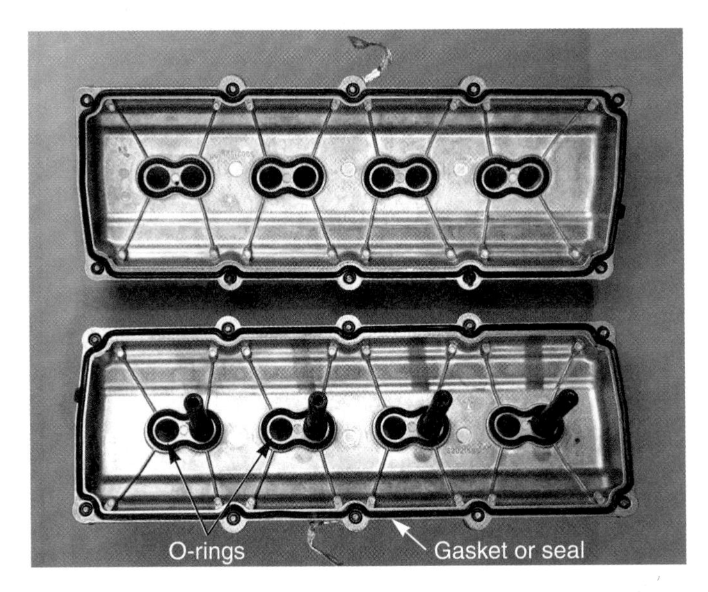

Figure 24-29. *Note the O-rings that must be installed around the spark plugs in this camshaft cover design. (DaimlerChrysler)*

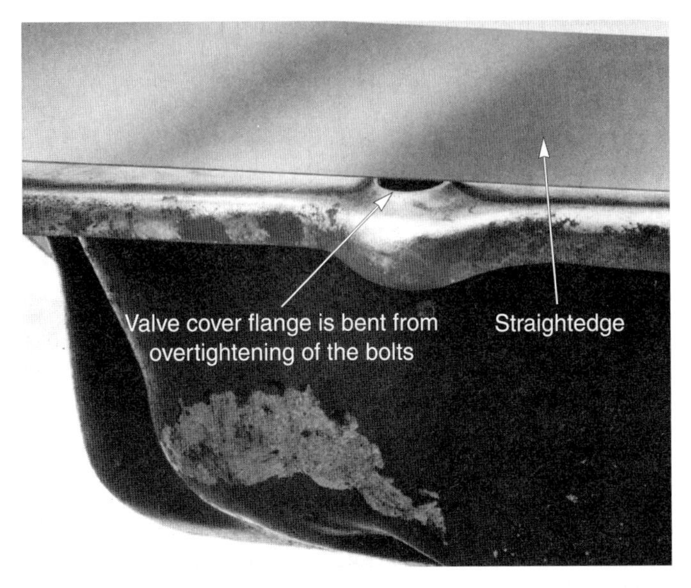

Figure 24-30. *Check the valve cover closely for a bent or warped flange. Here, note how a bolt hole is dented from overtightening. This could cause oil leakage. Use a hammer to straighten the flange. (Fel-Pro)*

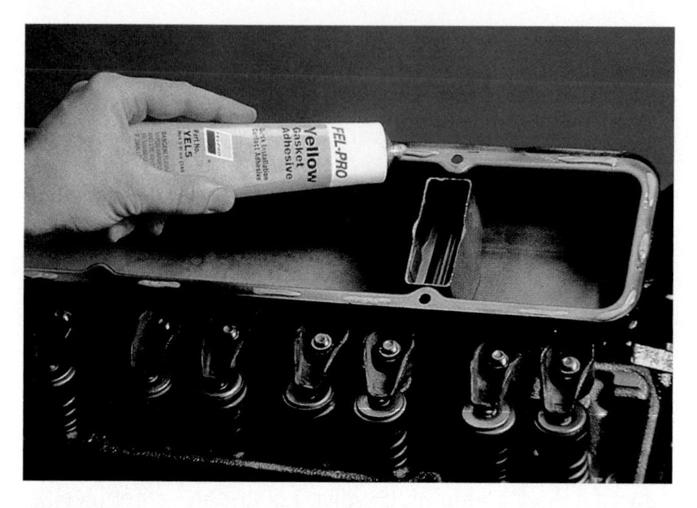

Figure 24-31. *Adhere the new gasket to the valve cover using an approved adhesive. (Fel-Pro)*

7. Torque the valve cover bolts to specifications. Use the recommended torque sequence or a crisscross pattern.

Note: A very common mistake is to overtighten valve cover bolts. Overtightening can smash and split the gasket or bend the valve cover. Oil leakage will result. Torque the bolts to the specification, which is generally just enough to lightly compress the gasket.

Installing a Valve Cover with Silicone Sealer

To use silicone or RTV sealer on the valve cover in place of a gasket:

1. Double-check that the valve cover and cylinder head mating surfaces are perfectly clean. Sealer will not bond to and seal dirty, oily surfaces.
2. Apply a continuous bead of sealer all the way around the valve cover sealing surface. Typically, the bead should be about 3/16″ (1.5 mm) wide.
3. Carefully lower the valve cover onto the head. Do not wait for the sealer to cure.
4. Make sure the bead of silicone sealer is not smeared or broken as the valve cover is lowered.
5. Hand start all of the valve cover bolts.
6. Torque the fasteners to specifications. Use the recommended torque sequence or a crisscross pattern.

Painting the Engine

Most cast iron cylinder blocks, heads, and intake manifolds must be painted to prevent rusting. Steel oil pans and valve covers may also require painting to restore engine cosmetics. An engine is usually painted after

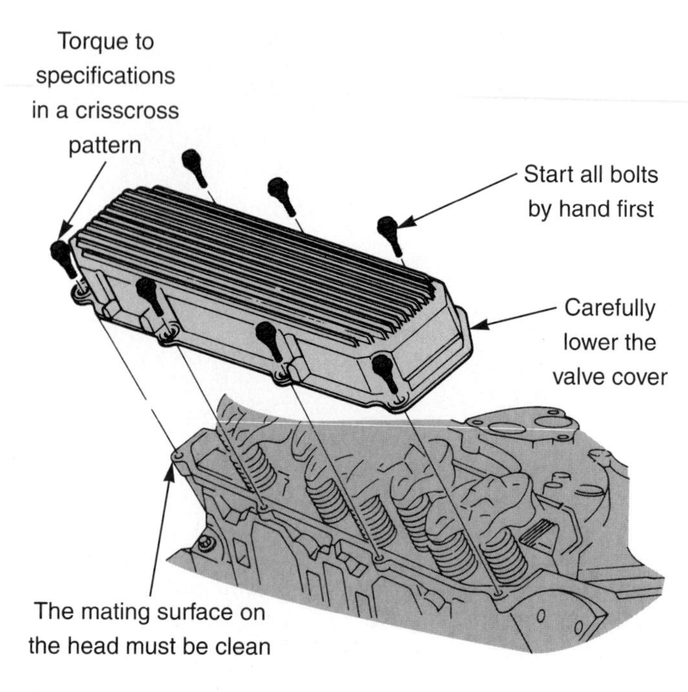

Figure 24-32. *Do not bump or shift the gasket while lowering the valve cover onto the engine. Start all bolts by hand before tightening any bolts. (General Motors)*

reassembly, but before installing parts that must not be painted. The engine should be primed with an epoxy primer before painting. The primer will bond to the bare metal and provide a good base for the paint.

1. Clean all surfaces to be painted with wax and grease remover or clean solvent. All oil and grease must be removed for proper paint adhesion.
2. Use a clean, dry shop towel to wipe off the cleaning agent before it dries.
3. Use masking tape or plastic plugs to mask any part or opening not to be painted, such as oil holes, coolant passages, and exhaust ports. Petroleum jelly can be placed on gasket mating surfaces, for example, to prevent paint adhesion.
4. Wipe off all surfaces to be painted with a tack rag while blowing the engine surfaces off with compressed air, **Figure 24-33A.**
5. Mix the primer following label directions. With epoxy primer and paint, you will have to mix in a hardening agent for proper drying.
6. Filter the primer into the spray gun cup, **Figure 24-33B.**
7. Spray a medium-wet coat of primer onto all surfaces to be painted, **Figure 24-33C.**
8. Allow the primer to cure properly before applying paint.
9. Thoroughly clean the spray gun.
10. Mix the paint following label directions.
11. Filter the paint into the clean spray gun.

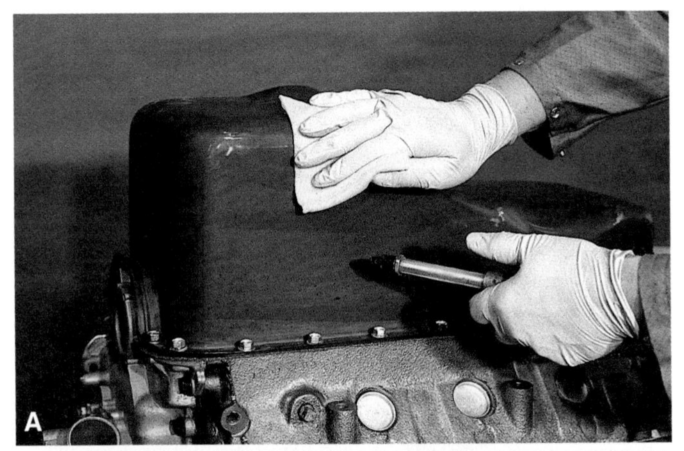

Figure 24-33. *Painting a rebuilt engine. A—After cleaning with solvent, wipe off the engine surfaces with a tack rag while blowing off with compressed air. B—After mixing the primer following label directions, filter it into the spray gun cup. C—Prime all bare metal surfaces with a medium-wet coat. D—While holding spray gun about 10″ to 12″ from surface, spray on a medium-wet coat of paint.*

12. Apply a medium-wet, thin coat of paint over the primer, **Figure 24-33D.** Hold the spray gun about 10″ to 12″ from the surface being painted.
13. Apply a second coat of paint to provide full coverage.
14. Thoroughly clean the spray gun.
15. Allow the paint to cure properly before further work on the engine.

Warning: When painting, use an approved respirator, wear safety glasses, and work in an area that has proper ventilation. Inhaled paint and fumes can cause health problems. Dispose of paint and solvents in an approved manner. Follow all related EPA and OSHA regulations.

Final Engine Assembly

Before the engine is ready to be installed in the vehicle, final engine assembly must be completed. This involves installing the flywheel and clutch or torque converter. Final assembly also involves attaching external components, such as brackets or sensors.

Note: Before installing the flywheel and clutch or torque converter, the engine must be removed from the engine stand. Support the engine with an engine lift or hoist. Refer to Chapter 20 for instructions on installing a lift chain and using an engine lift. Then, unbolt the engine from the stand. Lower the engine to the floor, but maintain tension on the lifting chain so the engine does not shift or fall over as you work on it.

Installing External Engine Parts

Before fitting the engine back into the vehicle, install all external parts that will not get in the way of engine installation. This includes engine sensors, actuators, brackets, engine mounts, and sometimes the oil filter and other parts. It will save time if many of these parts are installed now rather than waiting until the engine is back in the vehicle.

Installing the Flywheel

Before installing the flywheel, check its ring gear teeth. Make sure none of the teeth are worn or broken off.

Ring gear teeth damage will cause problems with starter motor engagement and operation. If needed, replace the ring gear or flywheel.

Automatic transmission and transaxle flywheels do not typically have a removable ring gear. However, most manual transmission and transaxle flywheels have a removable ring gear. To replace the ring gear on a manual transmission or transaxle flywheel, use a chisel to split and remove the damaged or worn ring gear, **Figure 24-34A.** Evenly heat the new ring gear with a torch to expand it. See **Figure 24-34B.** Use a hammer to tap the heat-expanded ring gear over the flywheel lip until it is seated on the flywheel. Do not strike the gear teeth with the hammer. Tap on the inner diameter of the ring gear.

Once the ring gear is replaced, if needed, install the flywheel on the crankshaft. Do not forget to use any spacer plate that fits between the engine and transmission or transaxle. The flywheel and crankshaft flange normally have offset bolt holes so the flywheel will only go on in one position. Turn the flywheel on the crankshaft until all bolt holes align. Install and torque the flywheel bolts to specifications.

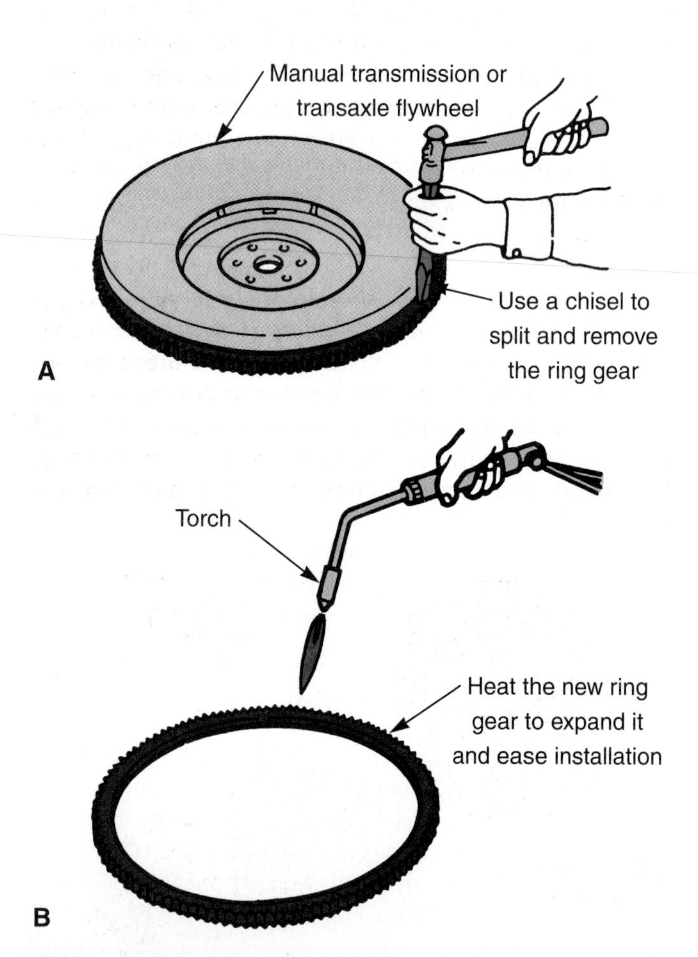

Figure 24-34. *Replacing a manual transmission or transaxle flywheel ring gear. A—Use a chisel to break and remove the existing ring gear. B—Heat the new ring gear to expand it. Drive the hot ring gear over the flywheel until the gear is fully seated. (General Motors)*

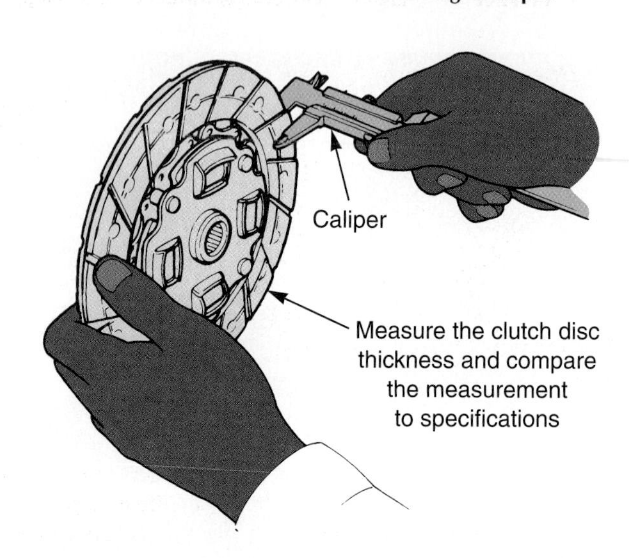

Figure 24-35. *Inspect the clutch disc closely. Measure the thickness. If the disc is worn too thin, it must be replaced. (Honda)*

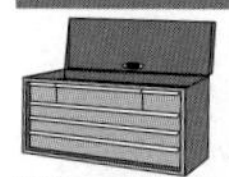

Note: Inspection of the flywheel is discussed in Chapter 20.

Installing the Clutch

With a manual transmission or transaxle, you will have to check and install the clutch before engine installation. Most technicians prefer to use a new pilot bearing, clutch disc, and throw-out bearing. These parts wear and fail in relatively few miles. Some technicians even like to replace the pressure plate. Obtain customer approval before replacing these parts. In any case, check these parts closely before reuse. Be sure to wear appropriate respiratory protection when working with clutches.

Figure 24-35 shows how to measure clutch disc wear. If the disc is thinner than specifications, it should be replaced. Also, replace the disc if it is contaminated with oil or shows signs of slippage and overheating.

Check the pressure plate friction surface for heat checking, signs of overheating, worn release arms, weakened springs, and so on. The throw-out bearing must turn freely with no signs of bearing roughness.

To install the clutch, fit the disc and pressure plate onto the flywheel. Insert an alignment tool or old transmission input shaft through the clutch disc and into the pilot bearing, **Figure 24-36.** This will hold the clutch and center it while torquing the pressure plate bolts. Use the proper torque sequence specified in the service manual.

Most technicians install a new throw-out bearing whenever the clutch is disassembled. Fit the throw-out bearing onto its release arm and make sure it is secure, **Figure 24-37.** The throw-out bearing must stay in place when bolting the transmission or transaxle to the engine.

Bolt the bell housing to the rear of the engine block. Torque the bell housing fasteners to specifications. Attach any other components to the bell housing, such as the clutch bracket.

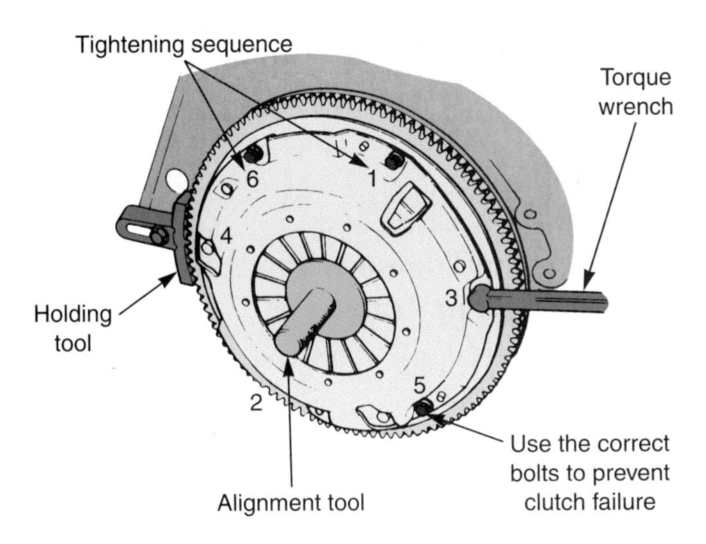

Figure 24-36. *Place an alignment tool in the pilot bearing to align the clutch assembly on the flywheel. Torque the factory bolts to specifications using the proper torque sequence.*

Caution: Make sure you use the factory bolts to install a pressure plate. They have a high tensile strength and are usually not threaded near the head. If you use incorrect bolts, they can break and lead to serious part damage!

Installing the Torque Converter

When installing the torque converter, make sure it is completely seated over its shaft. Wiggle the converter up and down while pushing in. Look down the side of the transmission or transaxle to make sure the converter is not sticking out too far.

If you accidentally bolt the transmission or transaxle to the engine and the converter is not completely over its shaft, severe transmission or transaxle damage can occur. The front pump, housing, or converter could be ruined—a time-consuming and costly mistake.

Installing the Transmission or Transaxle

If the transmission or transaxle is out of the vehicle, you can usually install the engine and transmission or transaxle into the vehicle as a single unit. To do so, the transmission or transaxle must be attached to the engine. Refer to the service manual for specific details. However, the general procedure is discussed in this section.

Make sure the engine lifting chain has tension on it from the engine lift. This will ensure the engine will not fall over. Next, using an appropriate lifting device, lift the transmission or transaxle to the level of the engine.

On a manual transmission or transaxle, slide its input shaft through the clutch. You might have to wiggle the unit while pushing in. Make sure to hold the transmission or transaxle perfectly straight with the centerline of the crankshaft or it will not slide fully into the pilot bearing. Once fully seated against the clutch housing, install the transmission or transaxle bolts to attach the unit to the bell housing. See **Figure 24-38.** Torque the fasteners to specifications using the appropriate torque sequence.

On an automatic transmission or transaxle, slide the transmission forward until the torque converter meets the flywheel. If the torque converter has a drain plug, rotate the torque converter so the plug is down. The hole in the flywheel for the drain plug must also be rotated down. When the torque converter is bolted to the flywheel, the converter drain plug will stick through this hole in the flywheel. This will also help you align everything as you bolt the automatic transmission or transaxle to the engine. Install the transmission or transaxle bolts to attach the unit to the back of the engine. Torque the fasteners to specifications using the appropriate sequence. Then, align the bolt holes in the torque converter with those in the flywheel. Install the torque converter bolts through the flywheel.

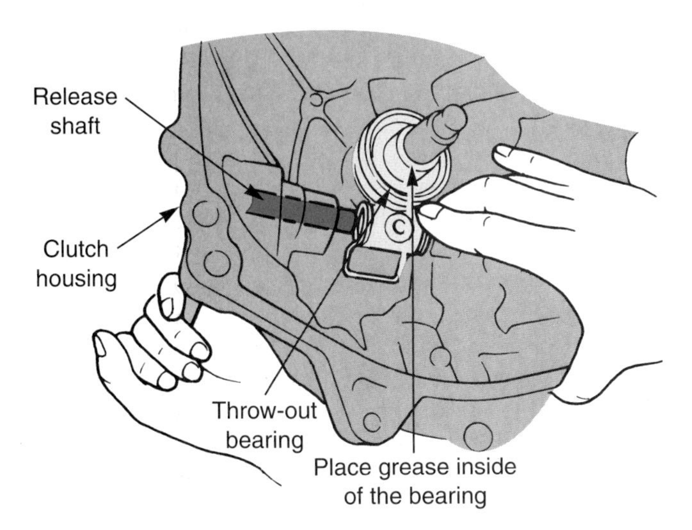

Figure 24-37. *Carefully position the throw-out bearing. It must stay in place when bolting the transmission or transaxle to the engine. (Honda)*

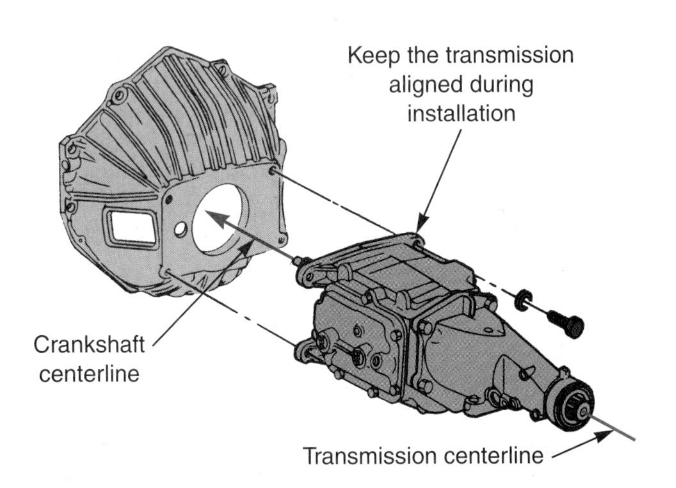

Figure 24-38. *When bolting the transmission or transaxle to the engine, the transmission centerline and the engine (crankshaft) centerline must be in perfect alignment. (General Motors)*

Figure 24-39. *Before installing the engine, place all parts that you will need near your work area. However, keep walkways free to prevent tripping and other accidents.*

Torque the fasteners to specifications using the appropriate torque sequence.

Caution: Do not use the transmission or transaxle bolts to pull the unit up to the bell housing. The unit should slide freely up against the bell housing. If you use the bolts to pull them together, the input shaft can smash into the side of the pilot bearing and cause part damage.

Installing the Engine in the Vehicle

Before installing the engine, double-check that all small components have been reinstalled. For example, check that all core and oil plugs have been installed in the front, sides, and rear of the cylinder block and heads. If you fail to install the rear camshaft bore plug, for example, and try to start the engine, the engine will lack oil pressure and oil will pour out of the back of the engine. The engine will need to be removed from the vehicle to correct this mistake.

When preparing to install the engine, make sure all needed parts are close by. See **Figure 24-39.** This can save time looking for brackets, bolts, cables, etc., when you are installing the engine.

Engine Installation Rules

There are a few basic safety rules to keep in mind when installing an engine. Some of these are listed below. Always observe accepted safety practices and use common sense.

- ❑ Keep your hands and feet out from under the engine.
- ❑ Never work on an engine that is supported only by the hoist or lift.
- ❑ Lower the engine slowly. Check for any obstructions as you lower.
- ❑ Never use the bell housing bolts to force the engine against the transmission or transaxle or part damage may result.

Installing the Engine without the Transmission or Transaxle

Installing the engine in the vehicle is basically the reverse of removing it.

1. Slowly lower the engine into the vehicle while watching for clearance all around the engine compartment. Position the engine so that its crankshaft centerline aligns with the transmission input shaft centerline.
2. Push the engine back and align the engine dowel pins with the holes in the transmission. Use a large bar to shift the engine.
3. As soon as the dowel pins slide fully into their holes, install an engine-to-transmission bolt, but do not tighten it. Start another bolt on the other side of the engine.
4. Check that the torque converter is properly lined up with the holes in the flywheel, **Figure 24-40.** Turn the converter if needed. Never use the bell housing bolts to force the engine against the transmission or transaxle or part damage can result. The two should slide together freely.
5. Finish installing the other components: motor mounts, oil filter, fuel lines, wiring, vacuum hoses, throttle cable or linkage, battery cable ground, fan belts, front drive axles, and so on. Refer to **Figures 24-41** through **24-47.** Never reroute wires on computer-controlled vehicles. Keep all wires in their original locations and reinstall them in their original positions. If you reroute spark plug wires or some high-current-carrying wires, current can be induced into a low-current sensor or computer feedback wires, which could upset computer system operation.
6. Fill the engine with motor oil, radiator with coolant, and transmission or transaxle with fluid, **Figure 24-48.**
7. Replace and align the hood. Visually double-check everything.

Installing the Engine with the Transmission or Transaxle

If the engine and transmission or transaxle are to be installed together as an assembly, first mate the engine to the transmission or transaxle.

1. Secure the engine with an engine crane and lifting fixture. You might want to place the engine on large block of wood to raise it a little off of the floor.

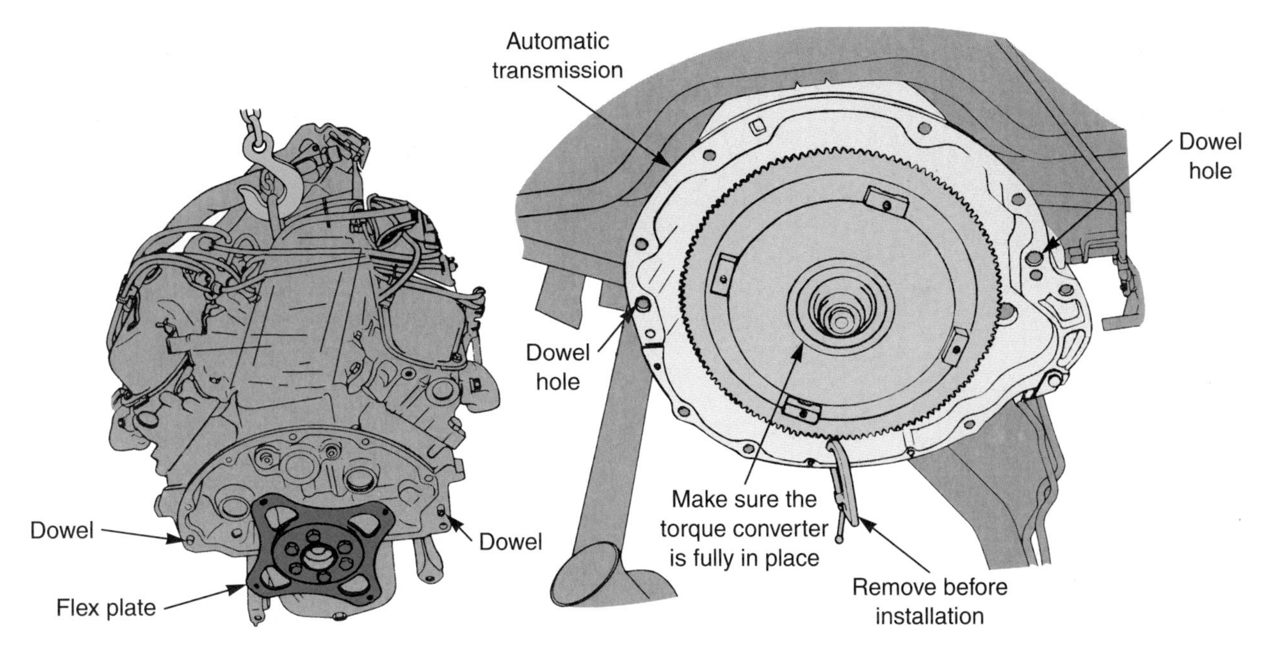

Figure 24-40. *With an automatic transmission or transaxle, make sure the torque converter is fully installed over its shaft. (DaimlerChrysler)*

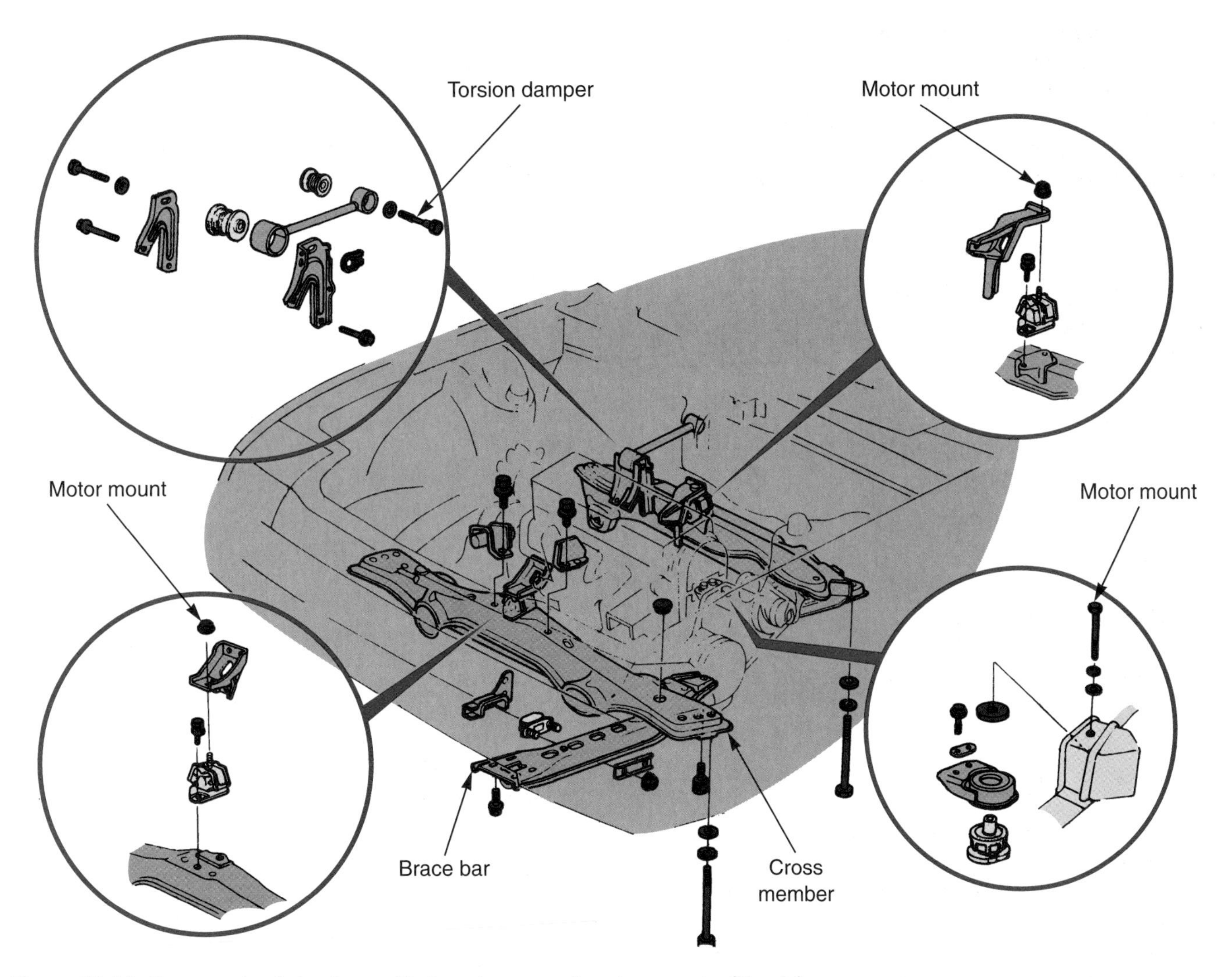

Figure 24-41. *Once engine is in place, attach and secure all motor mounts. (Honda)*

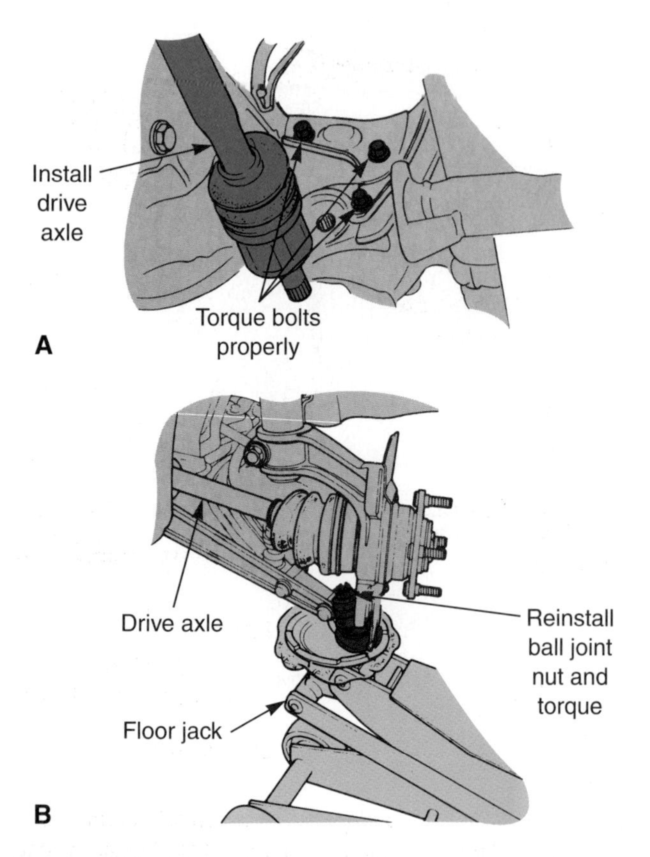

Figure 24-42. *With a front-wheel-drive car, you will have to install the front drive axles as you install the engine and transaxle. Refer to service manual for particular car. A—Install the drive axle. B—Reassemble the ball joint and related parts.*

2. Secure the transmission or transaxle to another crane or a floor jack. Position it at the rear of the engine.
3. Raise the engine off the floor only if needed to provide working room to mate the transmission or transaxle to the engine.
4. Raise the transmission or transaxle until its centerline is perfectly aligned with the crankshaft centerline.

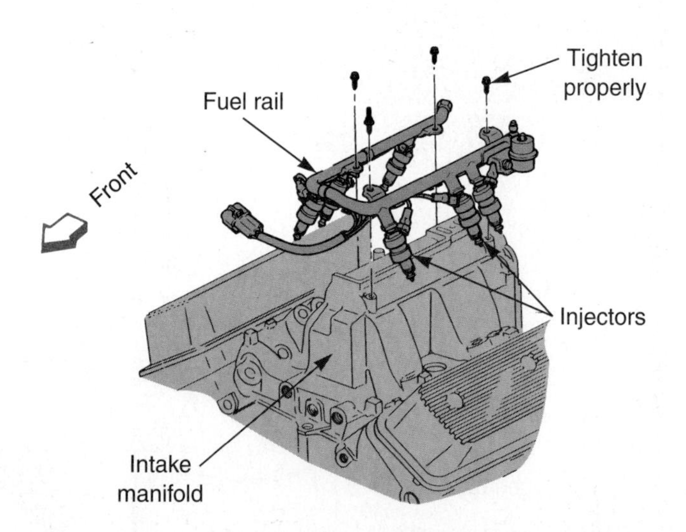

Figure 24-43. *Fuel injectors are usually installed after the engine is installed in the vehicle. This prevents damage that may occur during installation. (General Motors)*

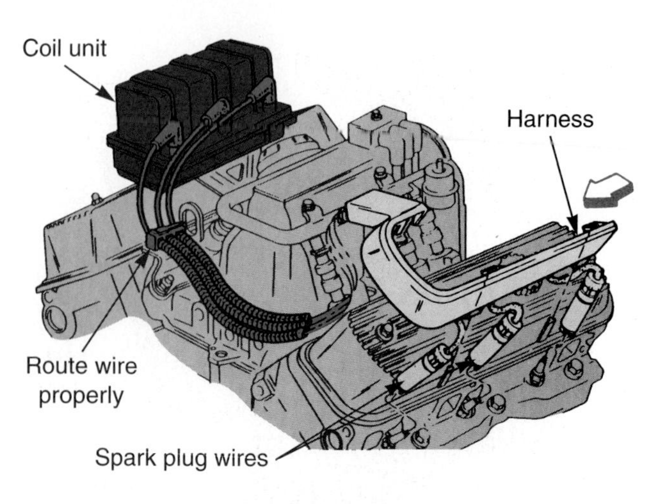

Figure 24-44. *Route all ignition system wires away from hot or moving parts. Double-check that wires feed to the correct spark plugs. (General Motors)*

5. Have some one hold the engine as you slide the transmission or transaxle into the engine. The dowel pins on the rear of the engine should smoothly engage the holes in the bell housing.
6. Double-check that the torque converter is fully installed over its input shaft. If the torque converter has partially slid out, the engine and transmission or transaxle will not fully mate. Make sure any torque converter stud bolts feed through the holes in the flex plate or flywheel.
7. Wiggle the transmission or transaxle sideways while pushing it against the engine. After the dowels have fully seated in the bell housing, start the bell housing bolts.

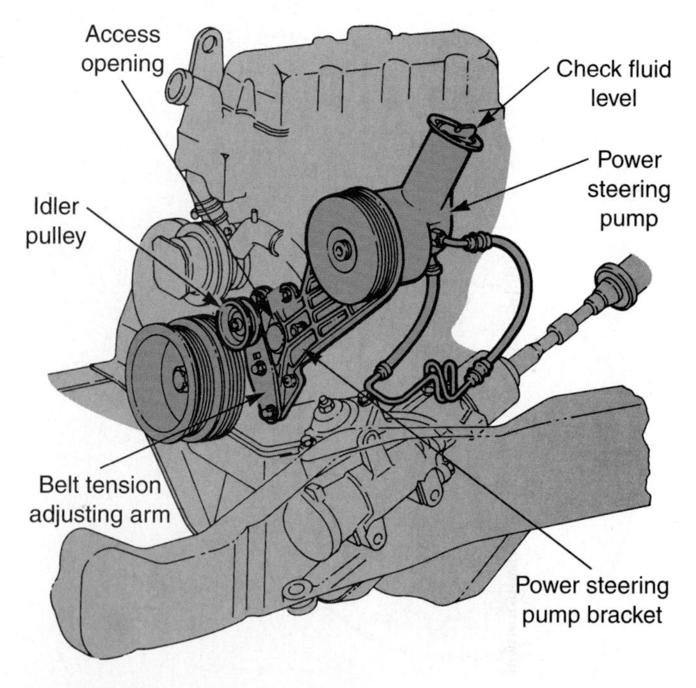

Figure 24-45. *Install all accessory units in correct sequence. Sometimes, one bracket will block access to other bolts. (Ford)*

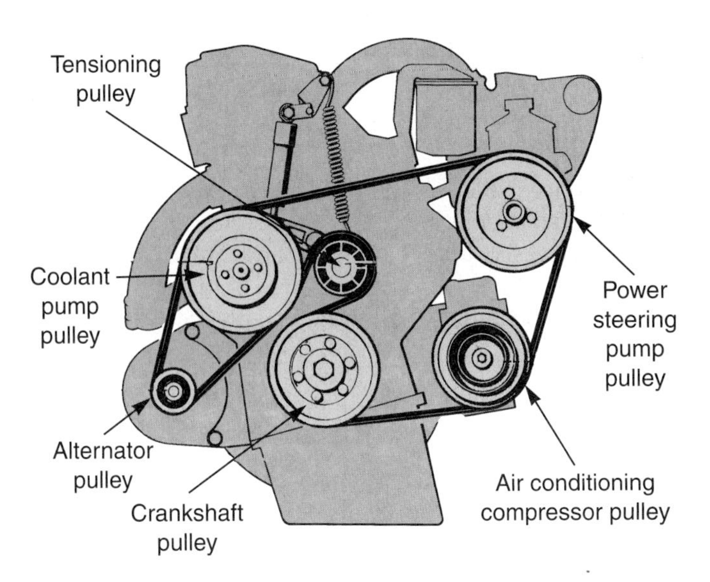

Figure 24-46. *Make sure belts are installed properly and in good condition. (Mercedes-Benz)*

8. Tighten all bell housing bolts using the proper torque and sequence.
9. Remove the jack or crane from the transmission or transaxle. Lower the engine and transmission or transaxle assembly so it is resting on the floor.

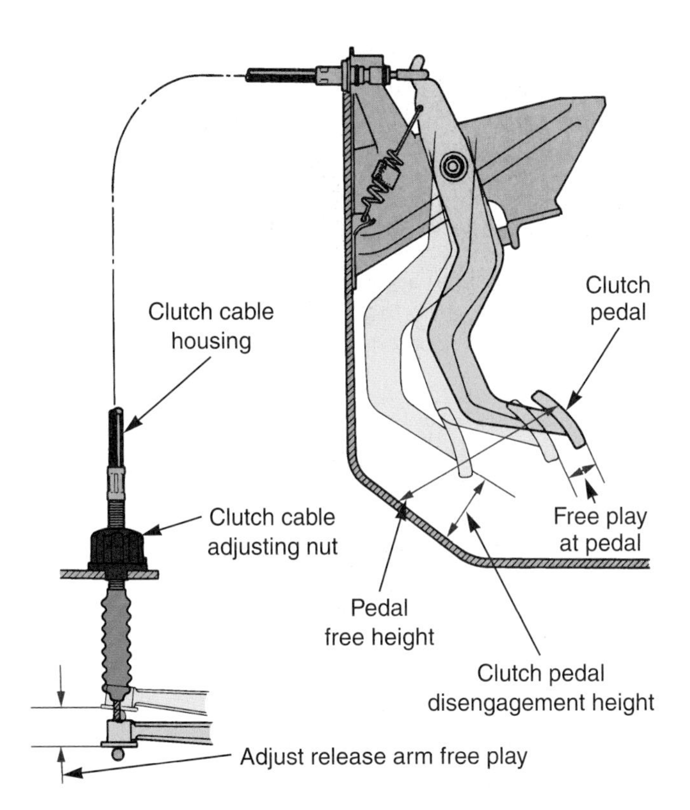

Figure 24-47. *The manual clutch or automatic transmission/ transaxle linkage may require adjustment. Follow service manual directions. (Honda)*

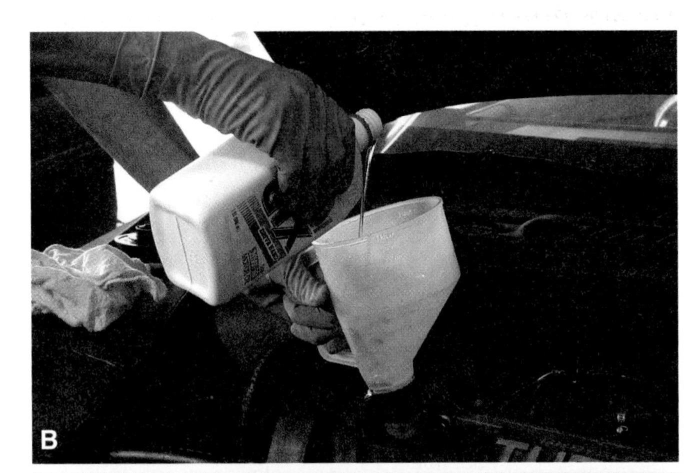

Figure 24-48. *Preparing for engine startup. A—Clean the battery cable connections and reinstall the cables. B—Fill the engine with the proper amount of the recommended motor oil. C—Fill the radiator and reservoir with coolant. D—Add fluid to the transmission or transaxle, if needed.*

10. Install all parts onto the engine and transmission or transaxle that will not prevent engine installation in the vehicle, such as the transmission dipstick, brackets, sensors, and so on.
11. Reposition the lifting fixture so that the engine and transmission or transaxle assembly can be raised level.
12. Raise the engine and transmission or transaxle assembly so it is just off the floor. Roll the crane in front of the engine compartment.
13. Check the engine bay to make sure all wires, hoses, brackets, and other parts are out of the way. They should be lying on the firewall or fender aprons to prevent damage during engine installation.
14. Raise the crane just high enough so the assembly clears the radiator support. Then, roll the crane back so the assembly is over the engine compartment. Tilt the assembly as needed so the rear of the transmission clears the firewall. Keep moving the assembly rearward as you lower it so the engine rests on the motor mounts.
15. While still supporting the engine and transmission or transaxle assembly with the crane, install any cross member that had to be removed. Also, install the motor mount bolts.
16. Once the engine is secured in the vehicle, lower the crane and remove the lifting fixture.
17. Install all remaining parts, such as wiring, hoses, brackets, throttle cable, etc., onto the engine.
18. Fill the engine with motor oil, radiator with coolant, and transmission or transaxle with fluid.
19. Replace and align the hood. Visually double-check everything.

Installing the Engine from Under the Vehicle

Many front-wheel-drive vehicles require that the engine and transaxle be removed and installed from underneath the engine compartment because the assembly will not clear the unibody or frame. Procedures vary from one vehicle to another, so be sure to consult the service manual for exact instructions. The basic procedure for installing an engine and transaxle assembly from the bottom is:

1. Secure the engine with an engine crane and lifting fixture.
2. Secure the transaxle to another crane or a floor jack. Position it at the rear of the engine.
3. Raise the engine off of the floor only if needed to provide working room to mate the transaxle to the engine.
4. Raise the transaxle until its centerline is perfectly aligned with the crankshaft centerline.
5. Have some one steady the engine as you slide the transaxle into the engine. The dowel pins on the rear of the engine should smoothly engage the holes in the bell housing.
6. Double-check that the torque converter is fully installed over its input shaft. If the torque converter has partially slid out, the engine and transaxle will not fully mate. Make sure any torque converter stud bolts feed through the holes in the flex plate or flywheel.
7. Wiggle the transaxle sideways while pushing it against the engine. After the dowels have fully seated in the bell housing, start the bell housing bolts.
8. Tighten all bell housing bolts using the proper torque and sequence.
9. If the suspension cradle (front suspension components) was removed with the engine and transaxle assembly, place the cradle on jack stands or wooden blocks. See **Figure 24-49.** Then, raise the engine and transaxle assembly so it clears the suspension cradle. Position the engine and transaxle assembly over the suspension cradle, lower the assembly, and secure the suspension to the assembly.
10. Raise the engine and transaxle assembly and secure it to the engine removal/installation jack. Once the assembly is secured to the jack, lower the crane and remove the engine lifting fixture.
11. Raise the vehicle on a lift.
12. Position the engine removal/installation jack under the vehicle.
13. Slowly raise the jack making sure the engine and transaxle assembly is not hitting on anything on the chassis.

Figure 24-49. *When installing the engine from under the vehicle, you will have to reinstall the front drive axles, exhaust downpipe, engine cradle, and related parts that were disassembled during engine removal.*

14. By hand, start the bolts that secure the suspension components to the chassis. Once all bolts have been started, tighten them to the proper specification.
15. Reinstall the front drive axles, exhaust downpipe, and other parts that are attached between the engine and transaxle assembly and the chassis. Also, install any brackets, cross members, or motor mounts between the engine and transaxle assembly and the chassis.
16. Reinstall all other parts underneath the engine and transaxle that were removed during engine removal.
17. Reinstall all parts in the top of the engine compartment, such as wires, throttle cable, hoses, air conditioning, transaxle cooler lines, dipstick, etc., that were removed or disconnected during engine removal.
18. Once the engine is secured in the vehicle, lower the crane and remove the lifting fixture.
19. Fill the engine with motor oil, radiator with coolant, and transmission or transaxle with fluid.
20. Replace and align the hood. Visually double-check everything.

Note: If you do not have access to an engine removal/installation jack for installing an engine from below, you can use a standard transmission jack and use the vehicle lift to lower the vehicle down onto the engine and transaxle assembly.

Engine Startup

1. Start and fast idle the engine until warm. Watch the engine oil light or gauge to make sure you have good oil pressure. Also, watch the engine temperature for signs of overheating.
2. After the engine "fires" and warms up, check the dash oil pressure warning light or gauge immediately. Make sure the engine has oil pressure.
3. If the engine will not start, check the ignition timing and spark intensity, spark plug wire routing, primary wire routing to ignition coil(s) and engine sensors, fuel supply to the injection system, and valve train adjustment.
4. Let the engine run at a fast idle. Watch for leaks under the engine and make sure the engine does not overheat.

Engine Break-In

Engine break-in is done mainly to seat and seal new piston rings. It also aids the initial wearing-in of other components under controlled conditions.

1. After warm-up at a fast idle, most mechanics road test the car. At the same time, they use moderate acceleration and deceleration for engine break-in.
2. Generally, accelerate the vehicle to about 40 mph (65 km/h). Then, release the gas pedal fully and let the vehicle coast down to about 20 mph (30 km/h).
3. Do this several times while carefully watching engine temperature and oil pressure.

Do not allow the engine to overheat during break-in; ring and cylinder scoring may result.

Caution: When road testing and breaking in an engine, drive the vehicle on a road that is free of traffic. Do not exceed posted speed limits or normal safe driving standards.

Customer Delivery

Inform the vehicle owner of the following rules concerning the operation of a freshly overhauled engine:

- ❑ Avoid prolonged highway driving during the first 100 to 200 miles (150 to 300 km). This will prevent ring friction from overheating the rings and cylinders, possibly causing damage.
- ❑ Do not worry about oil consumption until after about 2000 miles (3,200 km) of engine operation. It will take this long for full ring seating.
- ❑ Check engine oil and other fluid levels frequently.
- ❑ Change the engine oil and filter after approximately 2000 miles (3,200 km) of driving. This will remove any particles trapped in the oil from the engine.
- ❑ Inform the customer of any problems not corrected by the engine repair or overhaul. For example, if the radiator is in poor condition, tell the customer about the consequence of not correcting the problem. If the customer refuses the repair, have them sign a release form, which will protect you if the unserviced part fails.

Summary

When assembling an engine, all parts must be perfectly clean. Parts should be reinstalled in their same locations when possible. Lubricate moving parts. Make sure bolt lengths are correct. Keep components covered until assembly. Torque all fasteners properly. Inspect everything as it is installed.

When installing an oil pump, make sure all parts are in perfect condition. Check the fit of the gasket. If an oil pump drive rod is used, make sure the rod fits up into its hole in the block and into the pump. Torque the pump bolts to specifications.

When installing the oil pan, straighten its flange if bent. Adhere the gasket to the pan or block. Use silicone sealer where the end seals overlap the pan gasket.

When installing a head gasket, check its fit on the block. Make sure it is facing in the right direction. Markings are normally provided showing the top or front of a head gasket. If dowels are not provided in the engine, install long stud bolts to hold the gasket and guide the head.

Lower the cylinder head onto the block without hitting the head gasket. Start all of the bolts by hand and turn them down lightly. Use new bolts if they are torque-to-yield fasteners. Use a torque wrench to tighten the head bolts in an approved sequence. Tighten the bolts in steps and at full torque at least twice.

Before intake manifold installation, check its mating surfaces for warpage with a straightedge. Also, check the fit of the gasket. Adhere the gasket in place, if needed. With a metal-type intake gasket, circle all ports and coolant passages with silicone sealer. If the gasket is a multipiece, fiber type, extra sealer is not needed. However, place a bead of sealer where the gasket and end seals overlap.

Place the intake manifold into position without shifting the gasket or end seals. Start all bolts by hand and then torque them to specifications using a crisscross sequence. Go over all of the bolts several times.

Check the exhaust manifold for damage or warpage before installation. Check bolt lengths and torque the manifold bolts properly.

Lubricate lifters and fit them into their bores. Then, install the push rods and rocker arms. With an overhead camshaft engine, position the cam followers.

Valve train clearance adjustment methods vary. Many engines do not require valve train clearance adjustment. Simply torquing the rocker arm nut centers the lifter plunger and provides for quiet engine operation. Adjustable rockers must be turned to center the lifter plunger.

To adjust valve train clearance in an engine with hydraulic lifters with the engine off, position the camshaft lobe away from the lifter or rocker. Turn the nut until all play is out of the valve train. Then, turn the rocker nut about one more turn. To adjust the valve train clearance with the engine running, back off the rocker nut until the valve train clatters. Then, tighten the nut slowly until the noise stops. One more turn should center the hydraulic lifter.

To adjust the valve train clearance on an engine with mechanical lifters, the engine must be off. Turn the camshaft lobe away from the lifter or rocker. Slide the correct size feeler gauge between the rocker arm and valve stem tip. Turn the adjusting nut until the feeler drags slightly when pulled back and forth. Tighten the locknut and recheck the adjustment.

Valve adjusting shims are used on some OHC engines. You must obtain the correct shim thickness to provide proper valve train clearance. An adjustment screw is provided on hydraulic tappets in some OHC cylinder heads. Turning the screw changes the valve train clearance.

Check the valve cover flange for straightness before installation. Adhere the gasket to the cover or use a continuous bead of silicone in place of the gasket. All surfaces should be perfectly clean. Carefully lower the cover into place. Start all bolts by hand and then tighten them in a crisscross sequence. Do not overtighten valve cover bolts or you can bend the cover or split the gasket.

Engine installation is basically the reverse of engine removal. Keep the engine crankshaft and transmission or transaxle input shaft in perfect alignment during installation. Do not use the bell housing bolts to force the transmission against the engine. Make sure all external parts are installed, including motor mounts, sending units, sensors, flywheel, clutch, and so on.

Fill the engine and radiator before starting the engine. Start and fast idle the engine. Check oil pressure and do not let the engine overheat. Watch for leaks. Verify ignition timing and idle speed.

Test drive the vehicle. Accelerate and decelerate several times to help seat the piston rings. Advise the vehicle owner about driving methods after an engine overhaul.

Review Questions—Chapter 24

Please do not write in this text. Write your answers on a separate sheet of paper.

1. List five general rules to remember during engine assembly.
2. What is the purpose of priming the oil pump?
3. Most modern head gaskets are:
   (A) Integral.
   (B) Steel-shim type.
   (C) Teflon-coated, permanent-torque type.
   (D) Silicone or RTV.
4. If dowels are not provided in the block for alignment of the head gasket and cylinder head, how should you provide for alignment?
5. Existing head bolts that are torque-to-yield fasteners must be _____ when the engine is reassembled.
6. If a bolt will extend into an oil or coolant passage, what must be done when installing the bolt?
7. The lifters are installed before the push rods. Also, if you are working on a V-type engine, you will need to install the lifters before the _____.
8. Briefly describe how to install an intake manifold.
9. When doing major engine service, make sure the _____ is installed, that it operates properly, and that its heat range matches the service manual specification to help ensure proper cooling system operation.
10. To check the exhaust manifold for _____, use a straightedge and a flat feeler gauge.

11. Some overhead camshaft engines have a(n) _____ that holds the cam followers and camshaft(s).
12. With nonadjustable rocker arms, you must:
    (A) Run the rocker nuts down with an impact wrench.
    (B) Grind material off of valve stems for adjustment.
    (C) Torque the rocker nuts or bolts to specifications.
    (D) None of the above are correct.
13. The valve train clearance on an engine with _____ lifters can be adjusted with the engine off or with the engine running.
14. Briefly describe how valve adjusting shims are used to adjust the clearance between the camshaft lobe and the valve stem or follower.
15. On a freshly rebuilt engine, oil consumption during the first _____ miles is not unusual.

ASE-Type Questions—Chapter 24

1. When assembling and installing an engine, you should _____.
   (A) tighten head bolts as tight as possible
   (B) reuse torque-to-yield bolts
   (C) coat all moving parts with motor oil or white grease
   (D) install the oil pump dry
2. Technician A says that you should use thread locking compound when installing oil pump mounting bolts. Technician B says that the oil pump should be primed before installation. Who is correct?
   (A) A only.
   (B) B only.
   (C) Both A and B.
   (D) Neither A nor B.
3. Technician A says that the oil pan gasket can be adhered to the engine block mating surface. Technician B says that the oil pan gasket can be adhered to the oil pan. Who is correct?
   (A) A only.
   (B) B only.
   (C) Both A and B.
   (D) Neither A nor B.
4. Technician A says that new head bolts should always be used when installing an engine's cylinder head. Technician B says that torque-to-yield head bolts can be reused during cylinder head installation. Who is correct?
   (A) A only.
   (B) B only.
   (C) Both A and B.
   (D) Neither A nor B.
5. Technician A says that when torquing head bolts, you should start at one end of the cylinder head and work inward. Technician B says that torque-to-yield head bolts are first torqued, then tightened an additional number of degrees. Who is correct?
   (A) A only.
   (B) B only.
   (C) Both A and B.
   (D) Neither A nor B.
6. Technician A says that an improperly installed intake manifold may result in a coolant leak. Technician B says that an improperly installed intake manifold may be indicated by a clicking sound. Who is correct?
   (A) A only.
   (B) B only.
   (C) Both A and B.
   (D) Neither A nor B.
7. You should use a(n) _____ to check an intake manifold sealing surface for warpage.
   (A) dial indicator and flat feeler gauge
   (B) straightedge and flat feeler gauge
   (C) outside caliper and straightedge
   (D) dial indicator and straightedge
8. Technician A says that the camshaft bearing caps in an OHC engine must be installed in their original location. Technician B says that the camshaft bearings in an OHC engine may be a pressed-in, two-piece design. Who is correct?
   (A) A only.
   (B) B only.
   (C) Both A and B.
   (D) Neither A nor B.
9. Technician A says that the valve train clearance in an engine with hydraulic lifters can be adjusted with the engine running. Technician B says that the valve train clearance in an engine with hydraulic lifters can be adjusted with the engine off. Who is correct?
   (A) A only.
   (B) B only.
   (C) Both A and B.
   (D) Neither A nor B.
10. After an engine has been overhauled, prolonged highway driving should be avoided during the first _____.
    (A) 500 miles (800 km)
    (B) 2000 miles (3200 km)
    (C) 1000 miles (1600 km)
    (D) 100 miles (150 km)

ASE certification can show your employer and customers that you are fully qualified to work on a system of a vehicle or an engine. These technicians are employed in a dealership service center. (Ford)

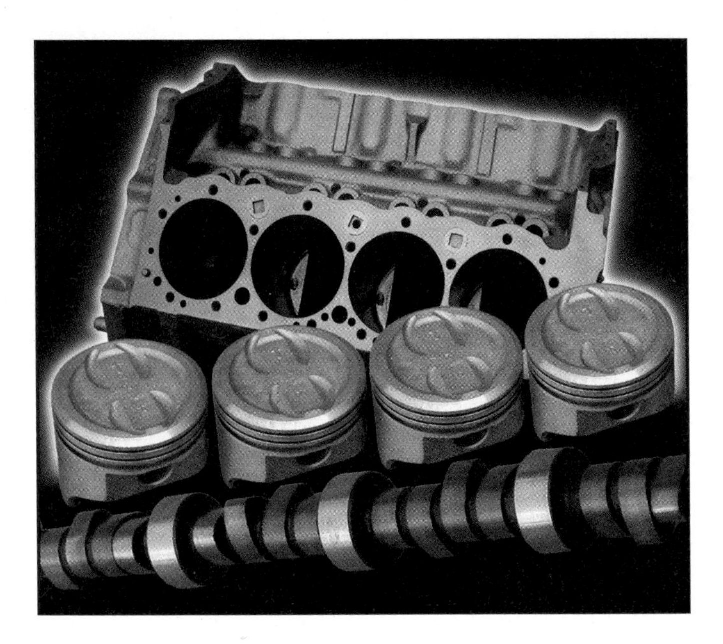

Chapter 25

ASE Engine Certification

After studying this chapter, you will be able to:

- Summarize the ASE testing program.
- Describe why ASE certification can be helpful to both the technician and shop owner.
- Explain the engine repair test (A1) content.
- Explain the engine performance test (A8) content.
- Identify the tests in the engine machinist series.
- Summarize the content of the engine machinist test series.
- Summarize tips for successfully taking an ASE test.

Know These Terms

ASE

ASE Automobile Technician

ASE Master Automobile Technician

Automotive technician

Certified

Engine machinist test series

Engine performance test

Engine repair test

This chapter provides an overview of the ASE testing program. It explains the tests available in engine service and gives hints on how to take them. This textbook has covered information that will be helpful when taking the ASE engine tests.

Remember! Always try to learn more about automotive technologies. Engines, and vehicles in general, are changing every year at a rapid pace. If you do not continue to study and read technical publications, you will fall behind in your knowledge. As a result, your ability to diagnose and repair engines will suffer.

With the precision and complexity of today's cars, the person who services and repairs cars and light trucks is called an ***automotive technician.*** The term *technician* implies the high degree of skill required to service and repair modern vehicles. The "shade tree mechanic" is not qualified to service and repair modern automobile designs.

ASE

The National Institute for Automotive Service Excellence, or ***ASE,*** is a nonprofit, nonaffiliated (no ties to industry) organization formed to help ensure the highest standards in automotive service. ASE directs an organized program of self-improvement under the guidance of a 40-member board of directors. These members represent all aspects of the automotive industry—educators, repair shop owners, consumer groups, government agencies, aftermarket parts companies, and auto manufacturers. This broad group of experts guides the ASE testing program and helps it stay in touch with the needs of the industry.

Voluntary Certification

Technicians take the ASE certification tests to show their employer and customers that they are fully qualified to work on a system of a vehicle or an engine. ASE tests are voluntary; they do not have to be taken. However, some countries and a few states have made technician certification a requirement.

ASE does not *license* technicians. When a technician passes an ASE test, the technician will be ***certified*** by ASE, with the appropriate work experience. ASE states that about 420,000 technicians are ASE certified. About 100,000 technicians take ASE tests each year, **Figure 25-1.** After certification, technicians must be retested and recertified every five years to maintain their credentials.

Benefits of ASE Certification

ASE certification serves as good public relations, showing everyone that you are well trained to work on today's complex vehicles. When you pass an ASE test, you will be given a shoulder patch for your work uniform. The patch has the ASE Blue Seal insignia and identifies you as an Automotive Technician or a Master Automobile Technician.

Figure 25-1. *Thousands of technicians take ASE certification tests every year. (ASE)*

Certification also demonstrates to employers that you are someone special that has taken extra effort to prove your value as a technician. It should lead to more rapid advancement and more income as customers indicate their preferences for a certified technician.

Test Categories

For car and light truck technicians, there are eight test categories. **Figure 25-2** gives a breakdown of what each automobile service test covers.

- ❑ Engine Repair (A1)
- ❑ Automatic Transmission/Transaxle (A2)
- ❑ Manual Drive Train and Axles (A3)
- ❑ Suspension and Steering (A4)
- ❑ Brakes (A5)
- ❑ Electrical/Electronic Systems (A6)
- ❑ Heating and Air Conditioning (A7)
- ❑ Engine Performance (A8)

You can take any one or all of these tests. However, only four tests (200 questions maximum) should be taken

A1 Engine Repair Test
60 Questions

General engine diagnosis	17 questions
Cylinder head and valve train diagnosis/repair	14 questions
Engine block diagnosis/repair	14 questions
Lubrication and cooling systems diagnosis/repair	8 questions
Fuel, electrical, ignition, and exhaust systems inspection/service	7 questions

A2 Automatic Transmission/Transaxle
50 Questions

General transmission/transaxle diagnosis	25 questions
Transmission/transaxle maintenance and adjustment	4 questions
In-vehicle transmission/transaxle repair	8 questions
Off-vehicle transmission/transaxle repair	13 questions

A3 Manual Drive Train and Axles
40 Questions

Clutch diagnosis/repair	6 questions
Transmission diagnosis/repair	6 questions
Transaxle diagnosis/repair	8 questions
Drive shaft/half shaft and universal joint/CV joint diagnosis/repair	6 questions
Rear axle diagnosis and repair	7 questions
Four-wheel drive/all-wheel drive component diagnosis/repair	7 questions

A4 Suspension and Steering
40 Questions

Steering systems diagnosis/repair	10 questions
Suspension systems diagnosis/repair	11 questions
Related suspension and steering service	2 questions
Wheel alignment diagnosis/ adjustment/repair	12 questions
Wheel and tire diagnosis/repair	5 questions

A5 Brakes
50 Questions

Hydraulic system diagnosis/repair	14 questions
Drum brake diagnosis/repair	6 questions
Disc brake diagnosis/repair	12 questions
Power assist units diagnosis/repair	4 questions
Misc. systems diagnosis/repair	7 questions
Antilock brake systems diagnosis/repair	7 questions

A6 Electrical/Electronic Systems
50 Questions

General electrical/electronic system diagnosis	13 questions
Battery diagnosis/service	4 questions
Starting system diagnosis/repair	5 questions
Charging system diagnosis/repair	5 questions
Lighting systems diagnosis/repair	6 questions
Gauges, warning devices, and driver information systems diagnosis/repair	6 questions
Horn and wiper/washer diagnosis/repair	3 questions
Accessories diagnosis/repair	8 questions

A7 Heating and Air Conditioning
50 Questions

A/C system service/diagnosis/repair	12 questions
Refrigeration system component diagnosis/repair	10 questions
Heating and engine cooling systems diagnosis/repair	5 questions
Operating systems and related controls diagnosis/repair	17 questions
Refrigerant recover, recycling, handling, and retrofit	6 questions

A8 Engine Performance
60 Questions

General engine diagnosis	10 questions
Ignition system diagnosis/repair	10 questions
Fuel, air induction, and exhaust systems diagnosis/repair	11 questions
Emission control systems diagnosis/repair	9 questions
Computerized engine controls diagnosis/repair	16 questions
Engine electrical systems diagnosis/repair	4 questions

Note: Each test may contain additional questions for statistical research. These questions are not identified, but do not affect the scoring of the test.

Figure 25-2. *These are the certification test categories for car and light truck technicians. If you pass a test and have two years of related work experience, you will be certified as an ASE Automobile Technician. If you pass all of them and have the work experience, you will be certified as an ASE Master Automobile Technician. (ASE)*

Figure 25-3. *When you pass any ASE automobile test, you are certified as an ASE Automobile Technician in that area if you have at least two years of experience.*

in any one testing session. For every test you pass, you will be certified as an ***ASE Automobile Technician*** for that area, **Figure 25-3.** If you pass all eight tests, you will be certified as an ***ASE Master Automobile Technician.***

There are ASE tests in other areas as well. There are seven tests in medium/heavy-duty truck repair, five tests in collision repair and refinishing, and three tests in the engine machining area.

Engine Repair (A1) Test

The ***engine repair test*** has questions relating to the service of the valve train, cylinder head, and block assemblies. The questions primarily cover how to do repairs or find mechanical problems. Questions are also given on lubricating, cooling, ignition, fuel, exhaust, and electrical system service.

If you plan on taking this test, review the textbook chapters that explain these topics. In particular, study the service chapters in this book, which includes Chapters 12 through 24.

Engine Performance (A8) Test

The ***engine performance test*** has questions dealing with the diagnosis of the engine and engine-related systems. Troubleshooting, tune-up problems are the general thrust of this test. Two sample questions are shown in **Figure 25-4.** Try to answer them before reading the answer in the caption. These questions are similar to the types of questions you will find on the engine performance test.

If you plan on taking this test, review the textbook chapters that focus on diagnosis. Concentrate on Chapters 12 through 19.

Engine Machinist Test Series

After completing this course, you may wish to pursue a career in automotive machining. This is discussed in more detail in the next chapter. Certification through the ASE ***engine machinist test series*** is a way in which engine machinists or advanced engine repair technicians can prove their skills to their customers and employers, **Figure 25-5.** The engine machinist test series contains three tests:

- Cylinder Head Specialist (M1)
- Cylinder Block Specialist (M2)
- Assembly Specialist (M3)

See **Figure 25-6** for a breakdown of the questions. You can certify as a gasoline or diesel specialist, or both. Each test has a group of questions that relates to both gasoline and diesel engines. This portion of the test is about 80% of the total questions. Then, you choose a group of questions related to either gasoline or diesel engines. You can choose to answer all questions and receive both gasoline and diesel certifications from the same test.

1. A compression test shows that one cylinder is too low. A leakage test on that cylinder shows that there is too much leakage. During the test, air could be heard coming out of the tailpipe. Which of these could be the cause?
   (A) Broken piston rings.
   (B) A bad head gasket.
   (C) A bad exhaust gasket.
   (D) An exhaust valve not seating.

2. After the compression readings shown below were taken, a wet compression test was performed. The second set of readings was almost the same as the first. Technician A says that a burned valve could cause these readings. Technician B says that a broken piston ring could cause these readings. Who is correct?
   (A) A only.
   (B) B only.
   (C) Both A and B.
   (D) Neither A nor B.

   140
   135
   5
   140
   Compression specifications 140 psi

Figure 25-4. *These are sample questions for the ASE engine performance test (A8). Try to answer them. The answers are 1. (D) and 2. (A). (ASE)*

Figure 25-5. *When you pass the engine machinist tests, you are certified as an ASE Engine Machinist if you have at least two years of experience.*

Who Can Take ASE Tests?

To take ASE certification tests and receive certification, you must either have two years of on-the-job experience or one year of approved educational credit and one year of work experience. However, you may take the tests even if you do not have the required two years experience. You will be sent a score for the test, but not certification credentials. After you have gained the mandatory experience, you can notify ASE and they will mail you a certificate.

You will be granted credit for formal training by one, or a combination, of the following types of schooling:

- ❑ High school training for three full years in automotives may be substituted for one year of work experience.
- ❑ Post-high school training for two years in a public or private facility can be substituted for one year of work experience.
- ❑ Two months of short training courses can be substituted for one month of work experience.
- ❑ Three years of an apprenticeship program, where you work under an experienced technician as a form of training, can be substituted for two years of work experience.

To have schooling substituted for work experience, you must send a copy of your transcripts (list of courses taken), a statement of training, or certificate to verify your training or apprenticeship. Each should give your length of training and subject area. This should accompany your registration form and fee payment.

Test Locations and Dates

The ASE tests are administers twice a year in over 750 locations across the country. The test sites are usually community colleges or high schools. In addition, ASE is developing computer-based testing (CBT) in a pilot program at 200 sites. Tests are given in May and November of each year. Contact ASE for more specific dates and locations for the tests.

M1 Cylinder Head Specialist	
55 Questions	
Cylinder head disassembly and cleaning	9 questions
Cylinder head crack repair	5 questions
Cylinder head inspection and machining	31 questions
Cylinder head assembly	10 questions
M2 Cylinder Block Specialist	
60 Questions	
Cylinder block disassembly and cleaning	4 questions
Cylinder block crack repair	3 questions
Cylinder block machining	24 questions
Crankshaft inspection and machining	12 questions
Connecting rods and pistons inspection and machining	10 questions
Balancing	3 questions
Cylinder block preparation	4 questions
M3 Assembly Specialist	
60 Questions	
Engine disassembly, inspection, and cleaning	10 questions
Engine preparation	9 questions
Short block assembly	15 questions
Long block assembly	18 questions
Final assembly	8 questions

Note: Each test may contain additional questions for statistical research. These questions are not identified, but do not affect the scoring of the test.

Figure 25-6. *These are the certification test categories for engine machinists. If you pass a test and have two years of related work experience, you will be certified as an ASE Engine Machinist. If you pass all three tests and have the work experience, you will be certified as an ASE Master Engine Machinist. (ASE)*

Test Results

The results of your test will be mailed to your home. Only you will find out how you did on the tests. You can then inform your employer, if you like.

Test scores will be mailed out a few weeks after you have completed the test. If you pass a test, you can consider taking more tests. If you fail, you will know that more study is needed before retaking the test.

Test Taking Techniques

Each test consists of multiple choice questions. You must carefully read the question and evaluate it. Then, read through the possible correct answers. Finally, select the *most-correct* response. Sometimes more than one response is correct, however, one answer will be *more* correct than the others. **Figure 25-4** gives two example questions.

You will not be required to recall exact specifications unless they are general and apply to most makes and models of cars. For example, compression test pressure specifications are typically about the same with all gasoline engines, and so are many engine clearances. This type of general information might be needed to answer some questions.

Test Taking Tips

The following are a few tips that might help you pass ASE certification tests.

- Read the statements or questions slowly. You might want to read them twice to make sure you fully understand the question.
- Analyze the statement or question. Look for hints that make some of the possible answers wrong.
- Analyze the questions as if you are the technician trying to fix the vehicle. Think of all possible situations and use common sense to pick the most correct response.
- When two technicians give statements concerning a problem, try to decide if either one is incorrect. If both are valid statements about a situation, choose the answer that indicates both technicians are correct. If only one is correct or neither is correct, mark the answer accordingly. This is one of the most difficult types of questions.
- If the statement only gives limited information, make sure you do not pick one answer as correct simply because it describes a more common condition. If the statement does not let you conclude that one answer is better than another, both answers are equally correct.
- Your first thought about which answer is correct is usually the correct response. If you think about a question too much, you will usually read something into the question that is not there. Read the question carefully and make a decision.
- Do not waste time on any one question. Make sure you have time to answer all of the questions on the test.
- Visualize how you would perform a test or repair when trying to answer a question. This will help you solve the problem more accurately.

Types of ASE Questions

ASE tests are designed to measure your knowledge of three things:

- The operation of various automotive systems and components.
- The diagnosis and testing of various automotive systems and components.
- The repair of automotive systems and components.

All test questions are multiple choice and contain four possible answers. Each chapter in this text contains ASE-type questions at the end of the chapter. Additional sample questions are given in the next sections. The answer to each question is explained in detail.

One-Part Questions

In a one-part question, you must choose the best answer out of all of the possibilities.

1. Which of the following components ignites the fuel in a gasoline engine?
   (A) Injector.
   (B) Valve.
   (C) Spark plug.
   (D) Glow plug.

The spark plug produces the electric arc to start the fuel burning. Therefore, the correct answer is (C). The injector simply sprays fuel into the engine. The valve allows the air-fuel mixture to flow into the engine. The glow plug is used only on diesel engines to warm the combustion chamber to aid combustion.

Two-Part Questions

Two-part questions require you to read two statements and decide if they are true. Both statements can be true or both can be false. In some cases, only one of the statements is true.

1. Technician A says that new head bolts should always be used when installing an engine's cylinder head. Technician B says that torque-to-yield head bolts can be reused during cylinder head installation. Who is correct?
   (A) A only.
   (B) B only.
   (C) Both A and B.
   (D) Neither A nor B.

In this question, the statement made by Technician A is wrong. In many cases, the head bolts can be reused. The statement made by Technician B is also wrong. Torque-to-yield head bolts *must* be replaced. If they are reused, the bolts may fail in service. Therefore, the correct answer is (D).

Negative Questions

Some questions require you to identify the *incorrect* answer from the list of possibilities. A negative question usually contains the word *except.*

1. An engine contains all of the following bearings *except:*
   (A) connecting rod bearings.
   (B) main crankshaft bearings.
   (C) camshaft bearings.
   (D) reverse idler bearings.

The correct answer is (D). While you may not know that reverse idler bearings are used in transmissions, you should know that engines do not contain these bearings.

A variation of the negative question contains the word *least.*

1. An engine is experiencing detonation. Which of the following is *least* likely to be the cause?
   (A) Inoperative waste gate.
   (B) Low coolant level.
   (C) Fouled spark plug.
   (D) Faulty fuel injector.

A waste gate is used on a turbocharged engine to limit boost pressure. Too much boost pressure can cause detonation. An inoperative waste gate may allow excess boost pressure and lead to detonation. Therefore, answer (A) is a possible cause. A low coolant level can make the engine run hot. Higher-than-normal operating temperatures can lead to detonation. Therefore, answer (B) is a possible cause. A faulty fuel injector may not be injecting enough fuel, resulting in a lean air-fuel mixture. A lean mixture can lead to detonation. Therefore, answer (D) is a possible cause. However, a fouled spark plug cannot lead to detonation. Therefore, answer (C) is *not* a possible cause and is the correct answer.

Completion Questions

Some test questions are simply sentences that must be completed. One of the four possible answers correctly completes the sentence.

1. A torque wrench is used to measure:
   (A) twisting force on fasteners.
   (B) shear applied to fasteners.
   (C) horsepower applied to fasteners.
   (D) camshaft degree angles.

Torque is a twisting force. Therefore, a torque wrench measures the twisting force on fasteners. Answer (A) is correct.

More Information

For more information on ASE certification tests, write for a registration booklet from:

ASE Registration Booklet
101 Blue Seal Drive, SE
Leesburg, VA 20175

The bulletin will give test locations, testing dates, costs, sample questions, and other useful information. You can also visit the ASE website at www.ase.com for more information. Many of the registration forms and catalogs are available for download.

Summary

ASE is a nonprofit, nonaffiliated organization. Its goal is to help ensure the highest standards in automotive service.

The ASE tests are voluntary. They will help show your employer and customers that you are fully competent to work on cars and light trucks. For car and light truck technicians, there are eight tests. Two tests related to engines—engine repair test (A1) and engine performance test (A8). There are also three tests used to certify automotive machinists.

Anybody can take the certification tests. However, you will only be certified if you have two years of related work experience. A master technician has passed all eight certification tests.

When taking the test, read the questions and statements slowly. Try to imagine you are really working on a vehicle when reading the question. Then, eliminate the incorrect responses. Usually, your first choice for a correct answer is the best answer. If you think too much about an answer, you might pick the second-most-correct response. Do not waste too much time on any question or you may not have time to answer all of the questions.

Write to ASE for a registration booklet. It will give test dates, locations, costs, and so on. Visit their website at www.ase.com for more information.

Review Questions—Chapter 25

Please do not write in this text. Write your answers on a separate sheet of paper.

1. The National Institute for Automotive Service Excellence, also known as _____, is a nonprofit, nonaffiliated organization formed to help ensure the highest standards in automotive service.
2. ASE certifies technicians, it does not _____ technicians.

3. List the eight certification tests available for car and light truck technicians.
4. The questions in the A1 test primarily cover:
   (A) Cooling, ignition, and mechanical problems.
   (B) Diagnosing and repairing mechanical problems.
   (C) Troubleshooting and tune-up problems.
   (D) Lubrication, exhaust, and electrical problems.
5. List the three tests in the engine machinist series.
6. In addition to passing an ASE test, a technician must have at least _____ years of related work experience.
7. The questions on the ASE tests are multiple choice and contain _____ possible answers.
8. What are the four types of questions that may be found on an ASE test?
9. When answering a test question, choose the _____ answer.
10. What is the ASE website address?

ASE-Type Questions—Chapter 25

1. Technician A says that some of the members of the ASE board of directors are repair shop owners. Technician B says that some of members of the ASE board of directors represent government agencies. Who is correct?
   (A) A only.
   (B) B only.
   (C) Both A and B.
   (D) Neither A nor B.
2. Technician A says that questions relating to exhaust system service are part of the ASE engine repair certification test. Technician B says that questions relating to cooling system service are part of the ASE engine repair certification test. Who is correct?
   (A) A only.
   (B) B only.
   (C) Both A and B.
   (D) Neither A nor B.
3. Technician A says that in order to take ASE certification tests, you must have two years of on-the-job experience. Technician B says that in order to receive ASE certification, you must have two years of on-the-job experience. Who is correct?
   (A) A only.
   (B) B only.
   (C) Both A and B.
   (D) Neither A nor B.
4. Technician A says that ASE has an engine machinist series of certification tests. Technician B says that an ASE Master Automobile Technician has passed all eight automotive tests (A1 through A8). Who is correct?
   (A) A only.
   (B) B only.
   (C) Both A and B.
   (D) Neither A nor B.
5. Technician A says that each ASE engine test consist of multiple choice questions with five possible answers each. Technician B says that each ASE engine test consist of a combination of multiple choice, true-false, and short answer questions. Who is correct?
   (A) A only.
   (B) B only.
   (C) Both A and B.
   (D) Neither A nor B.

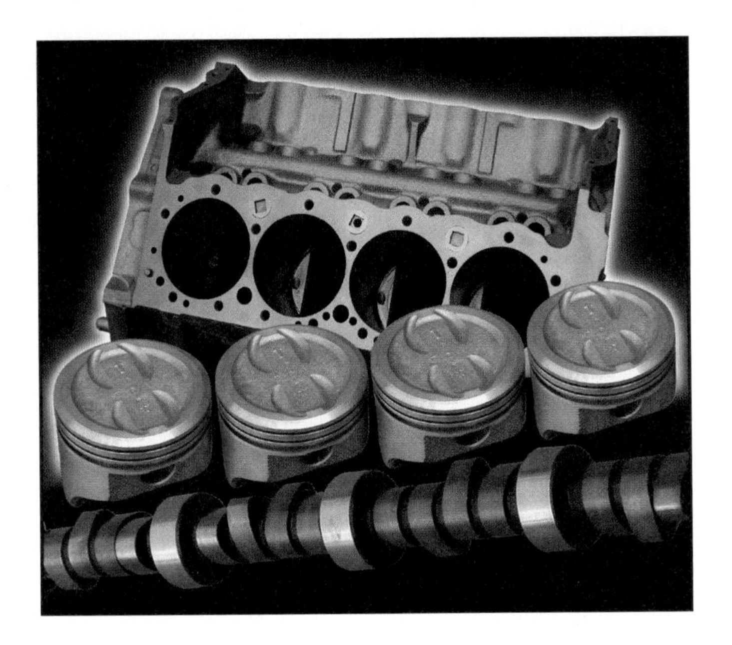

Chapter 26

Career Success

After studying this chapter, you will be able to:

- ❑ Discuss how dependability is important to career success.
- ❑ Describe how a team effort helps all technicians.
- ❑ Explain why the continuing study of automotive technology is essential to job success.
- ❑ Describe how both work quality and work speed are equally important to job success.
- ❑ Summarize the different systems used to pay technicians.
- ❑ Explain the types of repair facilities.
- ❑ Summarize how you can advance in the automotive field.

Know These Terms

Comebacks	Parts department
Commission	Private garage
Commission plus parts	Service station
Continuing education	Service writer
Dealerships	Shop supervisor
Dependability	Team effort
Engine machine shop	Warranty work
Entrepreneur	Work clothes
Flat rate manual	Work quality
Hourly rate	Work speed

This chapter briefly covers factors that can determine your degree of success when employed as an automotive technician. There is more to being a good technician than just being able to work on engines. There are many simple things that can help or hurt you as an engine technician.

Dependability

Dependability is critical to career success. The shop suffers if you fail to report to work on time, leave early, or take too many days off. The shop owner or service writer is responsible for scheduling repair work based on a certain number of technicians. If one or more technicians do not report to work, this schedule is useless and customers may have to wait for their vehicles longer than planned.

If customers are unhappy about being inconvenienced, they will not return to the shop. The technicians will have less work and the shop will not be as profitable. Everyone can suffer when a technician is constantly absent or late for work. The number one reason technicians are fired is lack of dependability.

Team Effort

Technicians should consider all employees in a shop as a team. In a ***team effort,*** all employees should work together to make the service facility a safe, enjoyable, and efficient operation.

If technicians do not get along and do not cooperate, morale and work output suffer. Keep in mind, a shop is competing with other shops for business. Technicians should never compete with each other.

Many times, tasks are difficult to complete by yourself. For example, you might need help lowering a heavy intake manifold onto an engine or replacing a hood. If you have helped other technicians with similar tasks, they will be glad to help you.

Also, if you do not have a vehicle to work on, offer to assist another technician who is completing a job. He or she should be willing to return the favor at a later date. As a result, you will make more money and so will the other employees.

Constantly Learn

To be successful in engine repair, you must constantly learn about new engine technologies and engine repair techniques. Engine technology is changing at a frantic pace. If you think you know everything, you are very quickly going to be left behind. ***Continuing education*** is a way to stay abreast of new and changing engine technologies.

To be a wise technician, you must always try to learn something new. Engine repair is a very complex and challenging area of employment. You must always try to read service manuals, service bulletins, automotive magazines, and other literature. Try to learn something new each day.

Work Quality

A primary concern of any engine technician, shop owner, or customer is ***work quality. Comebacks*** are when a repair must be done over because it failed to correct the problem. Comebacks are an indication of poor work quality. Repairs must be done correctly the first time following factory specifications.

It is tempting for a technician to try to work too fast in an attempt to make more money. The excess speed usually backfires and costs everyone money—the technician, customer, and shop owner. When a technician works too quickly, comebacks increase. More customers return unsatisfied with the repairs done on their vehicle. Then, the technician must do the repair again, without charging the customer. The technician loses money on lost work time, the shop owner loses money on wasted stall time, and the owner of the vehicle loses because they had to return to the shop.

Work Speed

While working too fast may lead to poor work quality, ***work speed*** is still very important to the engine technician. For those working on a percentage basis, income will depend on the amount of work completed each week. For those on an hourly basis, employers may not give pay raises because the technician is not making the shop enough money.

To increase work speed without affecting work quality, constantly think about new methods to increase work efficiency. Consider a new tool or technique that will save time without lowering the quality of the repair. This kind of mental attitude will help you improve your skills. As a result, you will become more productive each day.

Earnings

There are three basic methods shops use to pay their technicians. These are hourly rate, commission, and commission plus parts.

Hourly Rate

The ***hourly rate*** is simple; a stated amount of pay is given for each hour worked. This pay method is found in every type of shop. It may be desirable when you do not like the pressure of producing a quantity of work each week. However, you may make less money per year being paid an hourly rate than with other pay systems.

Commission

When you earn a ***commission,*** you obtain a set percent of the flat rate labor charged to the customers. For example, with a 50% commission at $60 an hour shop rate, you would get $30 an hour for work completed.

If you are fairly quick and can match or exceed the flat rate manual, you can earn a good wage. The ***flat rate manual*** is a book that states how long each repair should take. However, if you are slow or if there is not enough work to keep you busy, your income could be as low or lower than a technician paid an hourly rate.

Commission Plus Parts

Some technicians earn ***commission plus parts.*** This is a variation of the commission system. In this pay system, a technician being paid on commission also receives a small commission for the parts installed during a repair.

Types of Shops

There are several different types of repair shops in which you may find employment. Each type of shop has advantages and disadvantages.

Dealerships

Large auto ***dealerships*** usually pay well and have a large volume of work to be done, **Figure 26-1.** Generally, dealerships also offer good benefits. In addition, since a dealership is aligned with a vehicle manufacturer, specialized training may be available for the types of vehicles on which you will be working. If you are fairly experienced and fast, a large dealership may be a good place to seek employment.

However, dealerships are required to perform warranty work. ***Warranty work*** is repairs done on vehicles or engines that are still under warranty. The rate of pay is considerably lower under warranty.

Figure 26-1. *Working as an automotive technician in a dealership can be a rewarding career. (ASE)*

Private Garages

A ***private garage*** can also be an excellent place to work. The pay is normally good and private garages do not perform warranty work. The shop owner controls the work atmosphere and conditions for the technicians. If you want to work in a private garage, locate a reputable shop that has plenty of volume to keep you busy. ASE certification should be encouraged by the shop. Inquire about insurance and other benefits; many private garages offer these.

Service Stations

A ***service station*** having a repair area (not a "gas station") may be an excellent place to work, especially when you are a beginner. Much of the work will be quick service repairs that only take an hour or two. You will replace belts, hoses, water pumps, thermostats, tires, and other items. Many service stations offer a parts percentage incentive, which can increase your earnings. You can learn much about engine repair in a service station.

Specialized Engine Machine Shops

An ***engine machine shop*** specializes in the rebuilding of engines. Many machine shops will rebuild or overhaul engines. The machine shop has all of the specialized equipment needed to bore blocks, power hone cylinders, resurface decks, magnaflux components, and so on. In most instances, the engine must be removed from the vehicle and taken to the engine shop. Some machine shops will disassemble the engine; others will not.

Private garages, service stations, and other types of facilities may use the service of a specialized engine shop. These businesses make money on the installation and removal of the engine. The engine machine shop makes money on the rebuild.

Work Clothes

Work clothes are a vital part of a technician's preparation for work. They should be comfortable, well fitting, attractive, and clean. To the public, your work clothes reflect your mechanical abilities and work attitudes. If you look dirty and sloppy, customers will suspect that your work may be less than desirable. You must always project a professional image!

Many shops provide a standard set of work clothes for their technicians. The technicians usually pay a small fee for purchasing or renting the work clothes and having them cleaned.

Shop Supervisor and Service Writer

Always maintain good working relations with everybody in the shop, including the ***shop supervisor*** and ***service writer.*** In larger shops, these individuals assign jobs to technicians, **Figure 26-2.** If you do not treat them with respect, you may be assigned more than your share of less-desirable or less-profitable jobs.

Parts Department

As just mentioned, keep good relations with everyone in the shop. This includes the workers in the ***parts department.*** You must depend on parts people to get parts quickly and correctly for engine repairs. This also applies to the people working at a parts house. If you do not give them respect, do not expect them to go out of their way to help you obtain a hard-to-get part.

Advancement in Automotive Technology

Keep in mind the many opportunities in automotive technology. A few of these include:

- Automotive instructor.
- Service manager.
- Service writer.
- Shop supervisor.
- Technical representative for an automaker.
- Technical representative for an aftermarket parts company.
- Aftermarket sales representative.
- Automotive engineer.
- Shop owner.

Some of these positions will require you to take special training. You may have to attend a college or university or take specialized courses. You may also want ASE master technician certification, **Figure 26-3.** However, a basic knowledge of automotive technology will give you an edge over others seeking the same kind of position.

The automotive repair and manufacturing field is one of the largest employment areas in the nation. Your chances for succeeding in this field depend primarily on your initiative and willingness to learn.

Entrepreneurship

An ***entrepreneur*** is someone who starts a business. This might be a muffler shop, tune-up shop, parts house, or similar facility. For example, the technician in **Figure 26-4** owns a shop that specializes in diagnosing and correcting driveablity problems.

To be a good entrepreneur, you must be able to organize all aspects of the business. This includes bookkeeping, payroll, facility planning, and hiring. After gaining experience in the automotive field, you might want to consider starting your own business.

Summary

Being an engine technician offers many challenges and rewards. Your success will depend on your ability to develop competent skills, cooperate with employers and fellow workers, and keep up-to-date with developments in automotive technology.

Dependability is critical to career success. The shop suffers if you fail to report to work on time, leave early, or take too many days off. As an engine technician, you will be expected to work efficiently and rapidly while producing quality work. All employees are part of a team.

Figure 26-2. *A service writer often assigns work to the technicians after gathering information from the customer. (ASE)*

Figure 26-3. *An ASE Master Automobile Technician has a vast knowledge of cars and light trucks. (ASE)*

Figure 26-4. *This entrepreneur started a business that specializes in diagnosing and correcting driveablility problems. (ASE)*

There are three basic ways in which a technician may be paid. Depending on the shop or employer, you may be paid an hourly rate, straight commission on the dollar amount of your work, or commission plus a percentage of the parts you install.

There are several types of service facilities in which you may find employment as a technician. These include dealerships, private garages, service stations, and specialty engine machine shops.

Your work clothes are an important part of your work preparation. They should be comfortable, well fitting, attractive, and clean. Many shops provide a standard set of work clothes for their technicians.

Always maintain good working relations with all employees. This includes the shop supervisor and service writer. These individuals assign work to the technicians. Also, maintain a professional relationship with the parts department.

There are many opportunities available in the automotive field outside of technician. Some careers may require additional education or work experience. You may even choose to become an entrepreneur and start your own business.

Review Questions—Chapter 26

Please do not write in this text. Write your answers on a separate sheet of paper.

1. Briefly explain why *dependability* is important to career success.
2. What is meant by the term *team effort?*
3. Why is *continuing education* important to the technician?
4. Why is work quality equally important as work speed?
5. Define *comebacks.*
6. Explain three ways in which you can be paid as an engine technician.
7. This would be used to figure your pay for a certain repair.
   (A) Service manual.
   (B) Service bulletin.
   (C) Commission manual.
   (D) Flat rate manual.
8. Technicians may take engines to a(n) _____ for major rebuilding.
9. List eight positions of advancement in automotive technology.
10. What is an *entrepreneur?*

ASE-Type Questions—Chapter 26

1. The number one reason a technician is usually fired is due to lack of _____.
   (A) mechanical skills
   (B) dependability
   (C) education
   (D) productivity
2. Which of the following determine(s) the success of a technician?
   (A) Work quality.
   (B) Work speed.
   (C) Method of earnings.
   (D) All of the above.
3. Technician A says that some shops pay their technicians on an hourly rate basis. Technician B says that some shops pay their technicians on a commission basis. Who is correct?
   (A) A only.
   (B) B only.
   (C) Both A and B.
   (D) Neither A nor B.
4. Which of the following is typical of the work normally performed at a service station?
   (A) Water pump replacement.
   (B) Serpentine belt replacement.
   (C) Thermostat replacement.
   (D) All of the above.
5. Technician A says that most specialized machine shops perform engine removal and installation when completing an engine rebuild. Technician B says that an entrepreneur is someone who has started their own business. Who is correct?
   (A) A only.
   (B) B only.
   (C) Both A and B.
   (D) Neither A nor B.

OBD II Trouble Codes

These are the generic SAE codes generated by OBD II diagnostic systems. Some late model OBD I ECMs will provide these codes to some scan tools along with the normal two digit code. Manufacturer specific codes (codes that begin with a P1 alphanumeric designator) should be looked up in the service manual.

P01XX Fuel and Air Metering

Note: For systems with single O_2 sensors, use codes for Bank 1 sensor. Bank 1 contains cylinder #1. Sensor 1 is closest to the engine.

P0100 Mass or Volume Airflow Circuit Malfunction
P0101 Mass or Volume Airflow Circuit Range/Performance Problem
P0102 Mass or Volume Airflow Circuit Low Input
P0103 Mass or Volume Airflow Circuit High Input
P0104 Mass or Volume Airflow Circuit Intermittent
P0105 Manifold Absolute Pressure/Barometric Pressure Circuit Malfunction
P0106 Manifold Absolute Pressure/Barometric Pressure Circuit Range/Performance Problem
P0107 Manifold Absolute Pressure/Barometric Pressure Circuit Low Input
P0108 Manifold Absolute Pressure/Barometric Pressure Circuit High Input
P0109 Manifold Absolute Pressure/Barometric Pressure Circuit Intermittent
P0110 Intake Air Temperature Circuit Malfunction
P0111 Intake Air Temperature Circuit Range/Performance Problem
P0112 Intake Air Temperature Circuit Low Input
P0113 Intake Air Temperature Circuit High Input
P0114 Intake Air Temperature Circuit Low Input
P0115 Engine Coolant Temperature Circuit Malfunction
P0116 Engine Coolant Temperature Circuit Range/Performance Problem
P0117 Engine Coolant Temperature Circuit Low Input
P0118 Engine Coolant Temperature Circuit High Input
P0119 Engine Coolant Temperature Circuit Intermittent
P0120 Throttle/Pedal Position Sensor/Switch A Circuit Malfunction
P0121 Throttle/Pedal Position Sensor/Switch A Circuit Range/Performance Problem
P0122 Throttle/Pedal Position Sensor/Switch A Circuit Low Input
P0123 Throttle/Pedal Position Sensor/Switch A Circuit High Input
P0124 Throttle/Pedal Position Sensor/Switch A Circuit Intermittent
P0125 Insufficient Coolant Temperature for Closed Loop Fuel Control
P0126 Insufficient Coolant Temperature for Stable Operation
P0130 Oxygen Sensor Circuit Manfunction (Bank 1 Sensor 1)
P0131 Oxygen Sensor Circuit Low Voltage (Bank 1 Sensor 1)
P0132 Oxygen Sensor Circuit High Voltage (Bank 1 Sensor 1)
P0133 Oxygen Sensor Circuit Slow Response (Bank 1 Sensor 1)
P0134 Oxygen Sensor Circuit No Activity Detected (Bank 1 Sensor 1)
P0135 Oxygen Sensor Heater Circuit Manfunction (Bank 1 Sensor 1)
P0136 Oxygen Sensor Circuit Manfunction (Bank 1 Sensor 2)
P0137 Oxygen Sensor Circuit Low Voltage (Bank 1 Sensor 2)
P0138 Oxygen Sensor Circuit High Voltage (Bank 1 Sensor 2)
P0139 Oxygen Sensor Circuit Slow Response (Bank 1 Sensor 2)
P0140 Oxygen Sensor Circuit No Activity Detected (Bank 1 Sensor 2)
P0141 Oxygen Sensor Heater Circuit Malfunction (Bank 1 Sensor 2)
P0142 Oxygen Sensor Circuit Malfunction (Bank 1 Sensor 3)
P0143 Oxygen Sensor Circuit Low Voltage (Bank 1 Sensor 3)
P0144 Oxygen Sensor Circuit High Voltage (Bank 1 Sensor 3)
P0145 Oxygen Sensor Circuit Slow Response (Bank 1 Sensor 3)
P0146 Oxygen Sensor Circuit No Activity Detected (Bank 1 Sensor 3)
P0147 Oxygen Sensor Heater Circuit Malfunction (Bank 1 Sensor 3)
P0150 Oxygen Sensor Circuit Malfunction (Bank 2 Sensor 1)
P0151 Oxygen Sensor Circuit Low Voltage (Bank 2 Sensor 1)

P0152 Oxygen Sensor Circuit High Voltage (Bank 2 Sensor 1)
P0153 Oxygen Sensor Circuit Slow Response (Bank 2 Sensor 1)
P0154 Oxygen Sensor Circuit No Activity Detected (Bank 2 Sensor 1)
P0155 Oxygen Sensor Heater Circuit Malfunction (Bank 2 Sensor 1)
P0156 Oxygen Sensor Circuit Malfunction (Bank 2 Sensor 1)
P0157 Oxygen Sensor Circuit Low Voltage (Bank 2 Sensor 2)
P0158 Oxygen Sensor Circuit High Voltage (Bank 2 Sensor 2)
P0159 Oxygen Sensor Circuit Slow Response (Bank 2 Sensor 2)
P0160 Oxygen Sensor Circuit No Activity Detected (Bank 2 Sensor 1)
P0161 Oxygen Sensor Heater Circuit Malfunction (Bank 2 Sensor 2)
P0162 Oxygen Sensor Circuit Malfunction (Bank 2 Sensor 2)
P0163 Oxygen Sensor Circuit Low Voltage (Bank 2 Sensor 3)
P0164 Oxygen Sensor Circuit High Voltage (Bank 2 Sensor 3)
P0165 Oxygen Sensor Circuit Slow Response (Bank 2 Sensor 3)
P0166 Oxygen Sensor Circuit No Activity Detected (Bank 2 Sensor 3)
P0167 Oxygen Sensor Heater Circuit Malfunction (Bank 2 Sensor 3)
P0170 Fuel Trim Malfunction (Bank 1)
P0171 System too Lean (Bank 1)
P0172 System too Rich (Bank 1)
P0173 Fuel Trim Malfunction (Bank 2)
P0174 System too Lean (Bank 2)
P0175 System too Rich (Bank 2)
P0176 Fuel Composition Sensor Circuit Malfunction
P0177 Fuel Composition Sensor Circuit Range/Performance
P0178 Fuel Composition Sensor Circuit Low Input
P0179 Fuel Composition Sensor Circuit High Input
P0180 Fuel Temperature Sensor A Circuit Malfunction
P0181 Fuel Temperature Sensor A Circuit Range/Performance
P0182 Fuel Temperature Sensor A Circuit Low Input
P0183 Fuel Temperature Sensor A Circuit High Input
P0184 Fuel Temperature Sensor A Circuit Intermittent
P0185 Fuel Temperature Sensor B Circuit Malfunction
P0186 Fuel Temperature Sensor B Circuit Range/Performance
P0187 Fuel Temperature Sensor B Circuit Low Input
P0188 Fuel Temperature Sensor B Circuit High Input
P0189 Fuel Temperature Sensor B Circuit Intermittent
P0190 Fuel Rail Pressure Sensor Circuit Malfunction
P0191 Fuel Rail Pressure Sensor Circuit Range/Performance
P0192 Fuel Rail Pressure Sensor Circuit Low Input
P0193 Fuel Rail Pressure Sensor Circuit High Input
P0194 Fuel Rail Pressure Sensor Circuit Intermittent
P0195 Engine Oil Temperature Sensor Malfunction
P0196 Engine Oil Temperature Sensor Range/Performance
P0197 Engine Oil Temperature Sensor Low
P0198 Engine Oil Temperature Sensor High
P0199 Engine Oil Temperature Sensor Intermittent

P02XX Fuel and Air Metering

P0200 Injector Circuit Malfunction
P0201 Injector Circuit Malfunction–Cylinder 1
P0202 Injector Circuit Malfunction–Cylinder 2
P0203 Injector Circuit Malfunction–Cylinder 3
P0204 Injector Circuit Malfunction–Cylinder 4
P0205 Injector Circuit Malfunction–Cylinder 5
P0206 Injector Circuit Malfunction–Cylinder 6
P0207 Injector Circuit Malfunction–Cylinder 7
P0208 Injector Circuit Malfunction–Cylinder 8
P0209 Injector Circuit Malfunction–Cylinder 9
P0210 Injector Circuit Malfunction–Cylinder 10
P0211 Injector Circuit Malfunction–Cylinder 11
P0212 Injector Circuit Malfunction–Cylinder 12
P0213 Cold Start Injector 1 Malfunction
P0214 Cold Start Injector 2 Malfunction
P0215 Engine Shutoff Solenoid Malfunction
P0216 Injection Timing Control Circuit Malfunction
P0217 Engine Over Temperature Condition
P0218 Transmission Over Temperature Condition
P0219 Engine Overspeed Condition
P0220 Throttle/Pedal Position Sensor/Switch B Circuit Malfunction
P0221 Throttle/Pedal Position Sensor/Switch B Circuit Range/Performance Problem
P0222 Throttle/Pedal Position Sensor/Switch B Circuit Low Input
P0223 Throttle/Pedal Position Sensor/Switch B Circuit High Input
P0224 Throttle/Pedal Position Sensor/Switch B Circuit Intermittent
P0225 Throttle/Pedal Position Sensor/Switch C Circuit Malfunction
P0226 Throttle/Pedal Position Sensor/Switch C Circuit Range/Performance Problem
P0227 Throttle/Pedal Position Sensor/Switch C Circuit Low Input
P0228 Throttle/Pedal Position Sensor/Switch C Circuit High Input
P0229 Throttle/Pedal Position Sensor/Switch C Circuit Intermittent
P0230 Fuel Pump Primary Circuit Malfunction
P0231 Fuel Pump Secondary Circuit Low
P0232 Fuel Pump Secondary Circuit High
P0233 Fuel Pump Secondary Circuit Intermittent
P0235 Turbocharger Boost Sensor A Circuit Malfunction
P0236 Turbocharger Boost Sensor A Circuit Range/Performance

P0237 Turbocharger Boost Sensor A Circuit Low

P0238 Turbocharger Boost Sensor A Circuit High

P0239 Turbocharger Boost Sensor B Circuit Malfunction

P0240 Turbocharger Boost Sensor B Circuit Malfunction Range/Performance

P0241 Turbocharger Boost Sensor B Circuit Low

P0242 Turbocharger Boost Sensor B Circuit High

P0243 Turbocharger Wastegate Solenoid A Malfunction

P0244 Turbocharger Wastegate Solenoid A Range/Performance

P0245 Turbocharger Wastegate Solenoid A Low

P0246 Turbocharger Wastegate Solenoid A High

P0247 Turbocharger Wastegate Solenoid B Malfunction

P0248 Turbocharger Wastegate Solenoid B Range/Performance

P0249 Turbocharger Wastegate Solenoid B Low

P0250 Turbocharger Wastegate Solenoid B High

P0251 Injection Pump A Rotor/Cam Malfunction

P0252 Injection Pump A Rotor/Cam Range/Performance

P0253 Injection Pump A Rotor/Cam Low

P0254 Injection Pump A Rotor/Cam High

P0255 Injection Pump A Rotor/Cam Intermittent

P0256 Injection Pump B Rotor/Cam Malfunction

P0257 Injection Pump B Rotor/Cam Range/Performance

P0258 Injection Pump B Rotor/Cam Low

P0259 Injection Pump B Rotor/Cam High

P0260 Injection Pump B Rotor/Cam Intermittent

P0261 Cylinder 1 Injector Circuit Low

P0262 Cylinder 1 Injector Circuit High

P0263 Cylinder 1 Contribution/Balance Fault

P0264 Cylinder 2 Injector Circuit Low

P0265 Cylinder 2 Injector Circuit High

P0266 Cylinder 2 Contribution/Balance Fault

P0267 Cylinder 3 Injector Circuit Low

P0268 Cylinder 3 Injector Circuit High

P0269 Cylinder 3 Contribution/Balance Fault

P0270 Cylinder 4 Injector Circuit Low

P0271 Cylinder 4 Injector Circuit High

P0272 Cylinder 4 Contribution/Balance Fault

P0273 Cylinder 5 Injector Circuit Low

P0274 Cylinder 5 Injector Circuit High

P0275 Cylinder 5 Contribution/Balance Fault

P0276 Cylinder 6 Injector Circuit Low

P0277 Cylinder 6 Injector Circuit High

P0278 Cylinder 6 Contribution/Balance Fault

P0279 Cylinder 7 Injector Circuit Low

P0280 Cylinder 7 Injector Circuit High

P0281 Cylinder 7 Contribution/Balance Fault

P0282 Cylinder 8 Injector Circuit Low

P0283 Cylinder 8 Injector Circuit High

P0284 Cylinder 8 Contribution/Balance Fault

P0285 Cylinder 9 Injector Circuit Low

P0286 Cylinder 9 Injector Circuit High

P0287 Cylinder 9 Contribution/Balance Fault

P0288 Cylinder 10 Injector Circuit Low

P0288 Cylinder 10 Injector Circuit High

P0290 Cylinder 10 Contribution/Balance Fault

P0291 Cylinder 11 Injector Circuit Low

P0292 Cylinder 11 Injector Circuit High

P0293 Cylinder 11 Contribution/Balance Fault

P0294 Cylinder 12 Injector Circuit Low

P0295 Cylinder 12 Injector Circuit High

P0296 Cylinder 12 Contribution/Balance Fault

P03XX Ignition System or Misfire

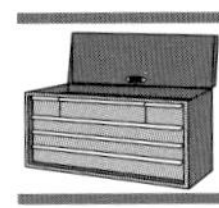

Note: Bank 1 contains cylinder #1.

P0300 Random/Multiple Cylinder Misfire Detected

P0301 Cylinder 1 Misfire Detected

P0302 Cylinder 2 Misfire Detected

P0303 Cylinder 3 Misfire Detected

P0304 Cylinder 4 Misfire Detected

P0305 Cylinder 5 Misfire Detected

P0306 Cylinder 6 Misfire Detected

P0307 Cylinder 7 Misfire Detected

P0308 Cylinder 8 Misfire Detected

P0309 Cylinder 9 Misfire Detected

P0310 Cylinder 10 Misfire Detected

P0311 Cylinder 11 Misfire Detected

P0312 Cylinder 12 Misfire Detected

P0320 Ignition/Distributor Engine Speed Input Circuit Malfunction

P0321 Ignition/Distributor Engine Speed Input Circuit Range/Performance

P0322 Ignition/Distributor Engine Speed Input Circuit No Signal

P0323 Ignition/Distributor Engine Speed Input Circuit Intermittent

P0325 Knock Sensor 1 Circuit Malfunction (Bank 1 or Single Sensor)

P0326 Knock Sensor 1 Circuit Range/Performance (Bank 1 or Single Sensor)

P0327 Knock Sensor 1 Circuit Low Input (Bank 1 or Single Sensor)

P0328 Knock Sensor 1 Circuit High Input (Bank 1 or Single Sensor)

P0329 Knock Sensor 1 Circuit Input Intermittent (Bank 1 or Single Sensor)

P0330 Knock Sensor 2 Circuit Malfunction (Bank 2)

P0331 Knock Sensor 2 Circuit Range/Performance (Bank 2)

P0332 Knock Sensor 2 Circuit Low Input (Bank 2)

P0333 Knock Sensor 2 Circuit High Input (Bank 2)

P0334 Knock Sensor 2 Circuit Input Intermittent (Bank 2)

P0335 Crankshaft Position Sensor A Circuit Malfunction

P0336 Crankshaft Position Sensor A Circuit Range/Performance

P0337 Crankshaft Position Sensor A Circuit Low Input

P0338 Crankshaft Position Sensor A Circuit High Input

P0339 Crankshaft Position Sensor A Circuit Intermittent

P0340 Camshaft Position Sensor Circuit Malfunction

P0341 Camshaft Position Sensor Circuit Range/Performance

P0342 Camshaft Position Sensor Circuit Low Input
P0343 Camshaft Position Sensor Circuit High Input
P0344 Camshaft Position Sensor Circuit Intermittent
P0350 Ignition Coil Primary/Secondary Circuit Malfunction
P0351 Ignition Coil A Primary/Secondary Circuit Malfunction
P0352 Ignition Coil B Primary/Secondary Circuit Malfunction
P0353 Ignition Coil C Primary/Secondary Circuit Malfunction
P0354 Ignition Coil D Primary/Secondary Circuit Malfunction
P0355 Ignition Coil E Primary/Secondary Circuit Malfunction
P0356 Ignition Coil F Primary/Secondary Circuit Malfunction
P0357 Ignition Coil G Primary/Secondary Circuit Malfunction
P0358 Ignition Coil H Primary/Secondary Circuit Malfunction
P0359 Ignition Coil I Primary/Secondary Circuit Malfunction
P0360 Ignition Coil J Primary/Secondary Circuit Malfunction
P0361 Ignition Coil K Primary/Secondary Circuit Malfunction
P0362 Ignition Coil L Primary/Secondary Circuit Malfunction
P0370 Timing Reference High Resolution Signal A Malfunction
P0371 Timing Reference High Resolution Signal A too Many Pulses
P0372 Timing Reference High Resolution Signal A too Few Pulses
P0373 Timing Reference High Resolution Signal Intermittent/Erratic Pulses
P0374 Timing Reference High Resolution Signal A No Pulse
P0375 Timing Reference High Resolution Signal B Malfunction
P0376 Timing Reference High Resolution Signal B too Many Pulses
P0377 Timing Reference High Resolution Signal B too Few Pulses
P0378 Timing Reference High Resolution Signal B Intermittent/Erratic Pulses
P0379 Timing Reference High Resolution Signal B No Pulse
P0380 Glow Plug/Heater Circuit Malfunction
P0381 Glow Plug/Heater Indicator Circuit Malfunction
P0385 Crankshaft Position Sensor B Circuit Malfunction
P0386 Crankshaft Position Sensor B Circuit Range/Performance
P0387 Crankshaft Position Sensor B Circuit Low Input
P0388 Crankshaft Position Sensor B Circuit High Input
P0389 Crankshaft Position Sensor B Circuit Intermittent

P04XX Auxiliary Emission Controls

Note: Bank 1 contains cylinder #1.

P0400 Exhaust Gas Recirculation Flow Malfunction
P0401 Exhaust Gas Recirculation Flow Insufficient Detected
P0402 Exhaust Gas Recirculation Flow Excessive Detected
P0403 Exhaust Gas Recirculation Circuit Malfunction
P0404 Exhaust Gas Recirculation Circuit Range/Performance
P0405 Exhaust Gas Recirculation Sensor A Circuit Low
P0406 Exhaust Gas Recirculation Sensor A Circuit High
P0407 Exhaust Gas Recirculation Sensor B Circuit Low
P0408 Exhaust Gas Recirculation Sensor B Circuit High
P0410 Secondary Air Injection System Malfunction
P0411 Secondary Air Injection System Incorrect Flow Detected
P0412 Secondary Air Injection System Switching Valve A Circuit Malfunction
P0413 Secondary Air Injection System Switching Valve A Circuit Open
P0414 Secondary Air Injection System Switching Valve A Circuit Shorted
P0415 Secondary Air Injection System Switching Valve B Circuit Malfunction
P0416 Secondary Air Injection System Switching Valve B Circuit Open
P0417 Secondary Air Injection System Switching Valve B Circuit Shorted
P0420 Catalyst System Efficiency Below Threshold (Bank 1)
P0421 Warm Up Catalyst Efficiency Below Threshold (Bank 1)
P0422 Main Catalyst Efficiency Below Threshold (Bank 1)
P0423 Heated Catalyst Efficiency Below Threshold (Bank 1)
P0424 Heated Catalyst Temperature Below Threshold (Bank 1)
P0430 Catalyst System Efficiency Below Threshold (Bank 2)
P0431 Warm Up Catalyst Efficiency Below Threshold (Bank 2)
P0432 Main Catalyst Efficiency Below Threshold (Bank 2)
P0433 Heated Catalyst Efficiency Below Threshold (Bank 2)
P0434 Heated Catalyst Temperature Below Threshold (Bank 2)
P0440 Evaporative Emission Control System Malfunction
P0441 Evaporative Emission Control System Incorrect Purge Flow
P0442 Evaporative Emission Control System Leak Detected (Small Leak)
P0443 Evaporative Emission Control System Purge Control Valve Circuit Malfunction

P0444 Evaporative Emission Control System Purge Control Valve Circuit Open
P0445 Evaporative Emission Control System Purge Control Valve Circuit Shorted
P0450 Evaporative Emission Control System Pressure Sensor Malfunction
P0451 Evaporative Emission Control System Pressure Sensor Range/Performance
P0452 Evaporative Emission Control System Pressure Sensor Low Input
P0453 Evaporative Emission Control System Pressure Sensor High Input
P0454 Evaporative Emission Control System Pressure Sensor Intermittent
P0455 Evaporative Emission Control System Leak Detected (Gross Leak)
P0460 Fuel Level Sensor Circuit Malfunction
P0461 Fuel Level Sensor Circuit Range/Performance
P0462 Fuel Level Sensor Circuit Low Input
P0463 Fuel Level Sensor Circuit High Input
P0464 Fuel Level Sensor Circuit Intermittent
P0465 Purge Flow Sensor Circuit Malfunction
P0466 Purge Flow Sensor Circuit Range/Performance
P0467 Purge Flow Sensor Circuit Low Input
P0468 Purge Flow Sensor Circuit High Input
P0469 Purge Flow Sensor Circuit Intermittent
P0470 Exhaust Pressure Sensor Malfunction
P0471 Exhaust Pressure Sensor Range/Performance
P0472 Exhaust Pressure Sensor Low
P0473 Exhaust Pressure Sensor High
P0474 Exhaust Pressure Sensor Intermittent
P0475 Exhaust Pressure Control Valve Malfunction
P0476 Exhaust Pressure Control Valve Range/Performance
P0477 Exhaust Pressure Control Valve Low
P0478 Exhaust Pressure Control Valve High
P0479 Exhaust Pressure Control Valve Intermittent

P05XX Vehicle Speed, Idle Control, and Auxiliary Inputs

P0500 Vehicle Speed Sensor Malfunction
P0501 Vehicle Speed Sensor Range/Performance
P0502 Vehicle Speed Sensor Circuit Low Input
P0503 Vehicle Speed Sensor Intermittent/Erratic/High
P0505 Idle Control System Malfunction
P0506 Idle Control System RPM Lower Than Expected
P0507 Idle Control System RPM Higher Than Expected
P0510 Closed Throttle Position Switch Malfunction
P0530 A/C Refrigerant Pressure Sensor Circuit Malfunction
P0531 A/C Refrigerant Pressure Sensor Circuit Range/Performance
P0532 A/C Refrigerant Pressure Sensor Circuit Low Input
P0533 A/C Refrigerant Pressure Sensor Circuit High Input
P0534 Air Conditioner Refrigerant Charge Loss
P0550 Power Steering Pressure Sensor Circuit Malfunction
P0551 Power Steering Pressure Sensor Circuit Range/Performance
P0552 Power Steering Pressure Sensor Circuit Low Input
P0553 Power Steering Pressure Sensor Circuit High Input
P0554 Power Steering Pressure Sensor Circuit Intermittent
P0560 System Voltage Malfunction
P0561 System Voltage Unstable
P0562 System Voltage Low
P0563 System Voltage High
P0565 Cruise Control On Signal Malfunction
P0566 Cruise Control Off Signal Malfunction
P0567 Cruise Control Resume Signal Malfunction
P0568 Cruise Control Set Signal Malfunction
P0569 Cruise Control Coast Signal Malfunction
P0570 Cruise Control Acceleration Signal Malfunction
P0571 Cruise Control/Brake Switch A Circuit Malfunction
P0572 Cruise Control/Brake Switch A Circuit Low
P0573 Cruise Control/Brake Switch A Circuit High
P0574 through **P0580** Reserved for Cruise Control System Codes

P06XX Computer and Auxiliary Outputs

P0600 Serial Communication Link Modification
P0601 Internal Control Module Memory Check Sum Error
P0602 Control Module Programming Error
P0603 Internal Control Module Keep Alive Memory (KAM) Error
P0604 Internal Control Module Random Access Memory (RAM) Error
P0605 Internal Control Module Read Only Memory (ROM) Error
P0606 PCM Processor Fault

P07XX Transmission

P0700 Transmission Control System Malfunction
P0701 Transmission Control System Range/Performance
P0702 Transmission Control System Electrical
P0703 Torque Converter/Brake Switch B Circuit Malfunction
P0704 Clutch Switch Input Circuit Malfunction
P0705 Transmission Range Sensor Circuit Malfunction (PRNDL Input)
P0706 Transmission Range Sensor Circuit Range/Performance
P0707 Transmission Range Sensor Circuit Low Input
P0708 Transmission Range Sensor Circuit High Input
P0709 Transmission Range Sensor Circuit Intermittent
P0710 Transmission Fluid Temperature Sensor Circuit Malfunction
P0711 Transmission Fluid Temperature Sensor Circuit Range/Performance
P0712 Transmission Fluid Temperature Sensor Low Input
P0713 Transmission Fluid Temperature Sensor Circuit High Input
P0714 Transmission Fluid Temperature Sensor Circuit Intermittent
P0715 Input/Turbine Speed Sensor Circuit Malfunction
P0716 Input/Turbine Speed Sensor Circuit Range/Performance

P0717 Input/Turbine Speed Sensor Circuit No Signal
P0718 Input/Turbine Speed Sensor Circuit Intermittent
P0719 Torque Converter/Brake Switch B Circuit Low
P0720 Output Speed Sensor Circuit Malfunction
P0721 Output Speed Sensor Circuit Range/Performance
P0722 Output Speed Sensor Circuit No Signal
P0723 Output Speed Sensor Circuit Intermittent
P0724 Torque Converter/Brake Switch B Circuit High
P0725 Engine Speed Input Circuit Malfunction
P0726 Engine Speed Input Circuit Range/Performance
P0727 Engine Speed Input Circuit No Signal
P0728 Engine Speed Input Circuit Intermittent
P0730 Incorrect Gear Ratio
P0731 Gear 1 Incorrect Ratio
P0732 Gear 2 Incorrect Ratio
P0733 Gear 3 Incorrect Ratio
P0734 Gear 4 Incorrect Ratio
P0735 Gear 5 Incorrect Ratio
P0736 Reverse Incorrect Ratio
P0740 Torque Converter Clutch Circuit Malfunction
P0741 Torque Converter Clutch Circuit Performance or Stuck Off
P0742 Torque Converter Clutch Circuit Stuck On
P0743 Torque Converter Clutch Circuit Electrical
P0744 Torque Converter Clutch Circuit Intermittent
P0745 Pressure Control Solenoid Malfunction
P0746 Pressure Control Solenoid Performance or Stuck Off
P0747 Pressure Control Solenoid Stuck On
P0748 Pressure Control Solenoid Electrical
P0749 Pressure Control Solenoid Intermittent
P0750 Shift Solenoid A Malfunction
P0751 Shift Solenoid A Performance or Stuck Off
P0752 Shift Solenoid A Stuck On
P0753 Shift Solenoid A Electrical
P0754 Shift Solenoid A Intermittent
P0755 Shift Solenoid B Malfunction
P0756 Shift Solenoid B Performance or Stuck Off
P0757 Shift Solenoid B Stuck On
P0758 Shift Solenoid B Electrical
P0759 Shift Solenoid B Intermittent
P0760 Shift Solenoid C Malfunction
P0761 Shift Solenoid C Performance or Stuck Off
P0762 Shift Solenoid C Stuck On
P0763 Shift Solenoid C Electrical
P0764 Shift Solenoid C Intermittent
P0765 Shift Solenoid D Malfunction
P0766 Shift Solenoid D Performance or Stuck Off
P0767 Shift Solenoid D Stuck On
P0768 Shift Solenoid D Electrical
P0769 Shift Solenoid D Intermittent
P0770 Shift Solenoid E Malfunction
P0771 Shift Solenoid E Performance or Stuck Off
P0772 Shift Solenoid E Stuck On
P0773 Shift Solenoid E Electrical
P0774 Shift Solenoid E Intermittent
P0780 Shift Malfunction
P0781 1-2 Shift Malfunction
P0782 2-3 Shift Malfunction
P0783 3-4 Shift Malfunction
P0784 4-5 Shift Malfunction
P0785 Shift/Timing Solenoid Malfunction
P0786 Shift/Timing Range/Performance
P0787 Shift/Timing Solenoid Low
P0788 Shift/Timing Solenoid High
P0789 Shift/Timing Solenoid Intermittent
P0790 Normal/Performance Switch Circuit Malfunction

Useful Tables

CONVERSION CHART

METRIC/US CUSTOMARY UNIT EQUIVALENTS

Multiply:	by:	to get: / Multiply:	by:	to get:
ACCELERATION				
feet/sec^2	x 0.3048	= meters/sec^2 (m/s^2)	x 3.281	= feet/sec^2
inches/sec^2	x 0.0254	= meters/sec^2 (m/s^2)	x 39.37	= inches/sec^2
ENERGY OR WORK (watt-second = joule = newton·meter)				
foot–pounds	x 1.3558	= joules (J)	x 0.7376	= foot–pounds
calories	x 4.187	= joules (J)	x 0.2388	= calories
Btu	x 1055	= joules (J)	x 0.000948	= Btu
watt–hours	x 3600	= joules (J)	x 0.0002778	= watt–hours
kilowatt–hours	x 3.600	= megajoules (MJ)	x 0.2778	= kilowatt-hrs
FUEL ECONOMY AND FUEL CONSUMPTION				
miles/gal	x 0.42514	= kilometers/liter (km/L)	x 2.3522	= miles/gal
Note: 235.2/(mi/gal) = liters/100km; 235.2/(liters/100 km) = mi/gal				
LIGHT				
foot-candles	x 10.76	= lumens/meter^2 (lm/m^2)	x 0.0929	= foot-candles
PRESSURE OR STRESS (newton/sq meter = pascal)				
inches Hg(60 °F)	x 3.377	= kilopascals (kPa)	x 0.2961	= inches Hg
pounds/sq in	x 6.895	= kilopascals (kPa)	x 0.145	= pounds/sq in
inches H_2O(60 °F)	x 0.2488	= kilopascals (kPa)	x 4.0193	= inches H_2O
bars	x 100	= kilopascals (kPa)	x 0.01	= bars
pounds/sq ft	x 47.88	= pascals (Pa)	x 0.02088	= pounds/sq ft
POWER				
horsepower	x 0.746	= kilowatts (kW)	x 1.34	= horsepower
ft–lbf/min	x 0.0226	= watts (W)	x 44.25	= ft–lbf/min
TORQUE				
pounds–inches	x 0.11298	= newton–meters (N·m)	x 8.851	= pound–inches
pound–feet	x 1.3558	= newton–meters (N·m)	x 0.7376	= pound–feet
VELOCITY				
miles/hour	x 1.6093	= kilometers/hour (km/h)	x 0.6214	= miles/hour
feet/sec	x 0.3048	= meters/sec (m/s)	x 3.281	= feet/sec
kilometers/hr	x 0.27778	= meters/sec (m/s)	x 3.600	= kilometers/hr
miles/hour	x 0.4470	= meters/sec (m/s)	x 2.237	= miles/hour

COMMON METRIC PREFIXES

mega	(M)	= 1 000 000	or 10^6	centi	(c)	= 0.01	or 10^{-2}
kilo	(k)	= 1 000	or 10^3	milli	(m)	= 0.001	or 10^{-3}
hecto	(h)	= 100	or 10^2	micro	(µ)	= 0.000 001	or 10^{-6}

METRIC/US CUSTOMARY UNIT EQUIVALENTS

Multiply:	by:	to get: / Multiply:	by:	to get:
LINEAR				
inches	x 25.4	= millimeters (mm)	x 0.03937	= inches
feet	x 0.3048	= meters (m)	x 3.281	= feet
yards	x 0.9144	= meters (m)	x 1.0936	= yards
miles	x 1.6093	= kilometers (km)	x 0.6214	= miles
inches	x 2.54	= centimeters (cm)	x 0.3937	= inches
microinches	x 0.0254	= micrometers (µm)	x 39.37	= microinches
AREA				
inches^2	x 645.16	= millimeters^2 (mm^2)	x 0.00155	= inches^2
inches^2	x 6.452	= centimeters^2 (cm^2)	x 0.155	= inches^2
feet^2	x 0.0929	= meters^2 (m^2)	x 10.764	= feet^2
yards^2	x 0.8361	= meters^2(m^2)	x 1.196	= yards^2
acres^2	x 0.4047	= hectares (10^{4}m^2) (ha)	x 2.471	= acres
miles^2	x 2.590	= kilometers^2 (km^2)	x 0.3861	= miles^2
VOLUME				
inches^3	x 16387	= millimeters^3 (mm^3)	x 0.000061	= inches^3
inches^3	x 16.387	= centimeters^3 (cm^3)	x 0.06102	= inches^3
inches^3	x 0.01639	= liters (l)	x 61.024	= inches^3
quarts	x 0.94635	= liters (l)	x 1.0567	= quarts
gallons	x 3.7854	= liters (l)	x 0.2642	= gallons
feet^3	x 28.317	= liters (l)	x 0.03531	= feet^3
feet^3	x 0.02832	= meters^3 (m^3)	x 35.315	= feet^3
fluid oz	x 29.57	= milliliters (ml)	x 0.03381	= fluid oz
yards^3	x 0.7646	= meters^3 (m^3)	x 1.3080	= yards^3
teaspoons	x 4.929	= milliliters (ml)	x 0.2029	= teaspoons
cups	x 0.2366	= liters (l)	x 4.227	= cups
MASS				
ounces (av)	x 28.35	= grams (g)	x 0.03527	= ounces (av)
pounds (av)	x 0.4536	= kilograms (kg)	x 2.2046	= pounds (av)
tons (2000 lb)	x 907.18	= kilograms (kg)	x 0.001102	= tons (2000 lb)
tons (2000 lb)	x 0.90718	= metric tons (t)	x 1.1023	= tons (2000 lb)
FORCE				
ounces—f (av)	x 0.278	= newtons (N)	x 3.597	= ounces—f (av)
pounds—f (av)	x 4.448	= newtons (N)	x 0.2248	= pounds—f (av)
kilograms—f	x 9.807	= newtons (N)	x 0.10197	= kilograms—f

TEMPERATURE

32 98.6 212

°F -40 0 40 80 120 160 200 240 280 320 °F

°C -40 -20 0 20 40 60 80 100 120 140 160 °C

°Celsius = 0.556 (°F – 32)

°F = (1.8 °C) + 32

TAP/DRILL CHART

COARSE STANDARD THREAD (UNC) Formerly US Standard Thread					FINE STANDARD THREAD (UNF) Formerly SAE Thread				
Sizes	Threads Per Inch	Outside Diameter at Screw	Tap Drill Sizes	Decimal Equivalent of Drill	Sizes	Threads Per Inch	Outside Diameter at Screw	Tap Drill Sizes	Decimal Equivalent of Drill
1	64	.073	53	0.0595	0	80	.060	3/64	0.0469
2	56	.086	50	0.0700	1	72	.073	53	0.0595
3	48	.099	47	0.0785	2	64	.086	50	0.0700
4	40	.112	43	0.0890	3	56	.099	45	0.0820
5	40	.125	38	0.1015	4	48	.112	42	0.0935
6	32	.138	36	0.1065	5	44	.125	37	0.1040
8	32	.164	29	0.1360	6	40	.138	33	0.1130
10	24	.190	25	0.1495	8	36	.164	29	0.1360
12	24	.216	16	0.1770	10	32	.190	21	0.1590
1/4	20	.250	7	0.2010	12	28	.216	14	0.1820
5/16	18	.3125	F	0.2570	1/4	28	.250	3	0.2130
3/8	16	.375	5/16	0.3125	5/16	24	.3125	I	0.2720
7/16	14	.4375	U	0.3680	3/8	24	.375	0	0.3320
1/2	13	.500	27/64	0.4219	7/16	20	.4375	25/64	0.3906
9/16	12	.5625	31/64	0.4843	1/2	20	.500	29/64	0.4531
5/8	11	.625	17/32	0.5312	9/16	18	.5625	0.5062	0.5062
3/4	10	.750	21/32	0.6562	5/8	18	.625	0.5687	0.5687
7/8	9	.875	49/64	0.7656	3/4	16	.750	11/16	0.6875
1	8	1.000	7/8	0.875	7/8	14	.875	0.8020	0.8020
1 1/8	7	1.125	63/64	0.9843	1	14	1.000	0.9274	0.9274
1 1/4	7	1.250	1 7/64	1.1093	1 1/8	12	1.125	1 3/64	1.0468
					1 1/4	12	1.250	1 11/64	1.1718

BOLT TORQUING CHART

METRIC STANDARD

Grade of Bolt		5D	8G	10K	12K	
Min. Tensile Strength		71,160 psi	113,800 psi	142,200 psi	170,679 psi	
Grade Markings on Head		5D	8G	10K	12K	Size of Socket or Wrench Opening
Metric		Foot-Pounds				Metric
Bolt Dia.	US Dec Equiv.					Bolt Head
6 mm	.2362	5	6	8	10	10 mm
8 mm	.3150	10	16	22	27	14 mm
10 mm	.3937	19	31	40	49	17 mm
12 mm	.4720	34	54	70	86	19 mm
14 mm	.5512	55	89	117	137	22 mm
16 mm	.6299	83	132	175	208	24 mm
18 mm	.709	111	182	236	283	27 mm
22 mm	.8661	182	284	394	464	32 mm

SAE STANDARD

Grade of Bolt	SAE 1 & 2	SAE 5	SAE 6	SAE 8		
Min. Tensile Strength	64,000 psi	105,000 psi	133,000 psi	150,000 psi		
Markings on Head					Size of Socket or Wrench Opening	
US Standard	Foot-Pounds				US Regular	
Bolt Dia.					Bolt Head	Nut
1/4	5	7	10	10.5	3/8	7/16
5/16	9	14	19	22	1/2	9/16
3/8	15	25	34	37	9/16	5/8
7/16	24	40	55	60	5/8	3/4
1/2	37	60	85	92	3/4	13/16
9/16	53	88	120	132	7/8	7/8
5/8	74	120	167	180	15/16	1
3/4	120	200	280	296	1-1/8	1-1/8

DECIMAL CONVERSION CHART

FRACTION	INCHES	mm
1/64	.01563	0.397
1/32	.03125	0.794
3/64	.04688	1.191
1/16	.6250	1.588
5/64	.07813	1.984
3/32	.09375	2.381
7/64	.10938	2.778
1/8	.12500	3.175
9/64	.14063	3.572
5/32	.15625	3.969
11/64	.17188	4.366
3/16	.18750	4.763
13/64	.20313	5.159
7/32	.21875	5.556
15/64	.23438	5.953
1/4	.25000	6.350
17/64	.26563	6.747
9/32	.28125	7.144
19/64	.29688	7.541
5/16	.31250	7.938
21/64	.32813	8.334
11/32	.34375	8.731
23/64	.35938	9.128
3/8	.37500	9.525
25/64	.39063	9.922
13/32	.40625	10.319
27/64	.42188	10.716
7/16	.43750	11.113
29/64	.45313	11.509
15/32	.46875	11.906
31/64	.48438	12.303
1/2	.50000	12.700

FRACTION	INCHES	mm
33/64	.51563	13.097
17/32	.53125	13.494
35/64	.54688	13.891
9/16	.56250	14.288
37/64	.57813	14.684
19/32	.59375	15.081
39/64	.60938	15.478
5/8	.62500	15.875
41/64	.64063	16.272
21/32	.65625	16.669
43/64	.67188	17.066
11/16	.68750	17.463
45/64	.70313	17.859
23/32	.71875	18.256
47/64	.73438	18.653
3/4	.75000	19.050
49/64	.76563	19.447
25/32	.78125	19.844
51/64	.79688	20.241
13/16	.81250	20.638
53/64	.82813	21.034
27/32	.84375	21.431
55/64	.85938	21.828
7/8	.87500	22.225
57/64	.89063	22.622
29/32	.90625	23.019
59/64	.92188	23.416
15/16	.93750	23.813
61/64	.95313	24.209
31/32	.96875	24.606
63/64	.98438	25.003
1	1.00000	25.400

Internet Resources

The following are Internet addresses for some of the companies and organization that may help you in your study of automotive technology.

www.3m.com 3M WorldWide
www.acdelco.com ACDelco
www.airtexproducts.com Airtex Automotive Division
www.alfaromeo.com Alfa Romeo, Inc.
www.ambac.net AMBAC International (formerly American Bosch)
www.honda.com American Honda Motor Co.
www.applied-power.com Applied Power, Inc.
www.armstrongtools.com Armstrong Bros. Tool Co.
www.astonmartin.com Aston Martin Lagonda, Inc.
www.bendix.com Bendix
www.binks.com Binks
www.blackanddecker.com Black & Decker, Inc.
www.bmwusa.com BMW of North America, Inc.
www.borg-warner.com Borg-Warner Corp.
www.boschtools.com Bosch Power Tools
www.buick.com Buick Motor Division
www.cadillac.com Cadillac Motor Division
www.caterpillar.com Caterpillar, Inc.
www.federal-mogul.com/champion Champion Spark Plugs
www.chevrolet.com Chevrolet Motor Division
www.chrysler.com Chrysler Division
www.crcindustries.com CRC Industries, Inc.
www.cummins.com Cummins Engine Co., Inc.
www.daimlerchrysler.com/index_e.htm DaimlerChrysler
www.dake-div-jsjcorp.com Dake
www.dana.com Dana Corp.
www.dayco.com Dayco Corp.
www.deere.com Deere & Co.
www.delcoremy.com Delco-Remy
www.detroitdiesel.com Detroit Diesel
www.ethyl.com Ethyl Corp.
www.exxon.com Exxon
www.federal-mogul.com Federal Mogul Corp.
www.federal-mogul.com/felpro Fel-Pro
www.fluke.com Fluke Corporation
www.ford.com Ford Motor Company
www.fram.com Fram Filters
www.gates.com Gates Rubber Co.
www.generaltire.com General Tire and Rubber Company
www.gm.com General Motors Corporation
www.goodallmfg.com Goodall Manufacturing, LLC
www.gunk.com Gunk Chemical Division
www.hastingsmfg.com Hastings Mfg. Co.
www.helicoil.com Heli-Coil Products
www.holley.com Holley High Performance Products, Inc.
www.hunter.com Hunter Engineering Co.
www.irtools.com Ingersoll-Rand Co.
www.isuzu.co.jp Isuzu
www.jaguar.com Jaguar
www.kwik-way.com Kwik-Way Products
www.lexus.com Lexus
www.lislecorp.com Lisle Corp.
www.mactools.com Mac Tools, Inc.
www.mazdausa.com Mazda Motors of America, Inc.
www.mbusa.com Mercedes-Benz of North America, Inc.
www.mitsucars.com Mitsubishi Motor Sales of America
www.mobil.com Mobil Oil Corp.
www.federal-mogul.com/moog Moog Automotive
www.motorola.com Motorola
www.napaonline.com NAPA
www.asecert.org National Institute for Automotive Service Excellence (ASE)
www.nissanusa.com Nissan Motor Corp.
www.oldsmobile.com Oldsmobile Division of GM
www.otctools.com OTC Tools & Equipment
www.parker.com Parker Hannifin Corp.
www.pontiac.com Pontiac Motor Division
www.pureoil.com Purolator Filter Division
www.quakerstate.com Quaker State Corp.
www.renault.com Renault
www.rolls-royce.com Rolls-Royce, Inc.
www.saabusa.com Saab
www.sellstrom.com Sellstrom Mfg. Co.
www.shell.com Shell Oil Co.
www.simpsonelectric.com Simpson Electric Co.
www.siouxtools.com Sioux Tools, Inc.
www.snapon.com Snap-on Tools Corp.
www.stewartwarner.com Stewart-Warner
www.subaru.com Subaru of America
www.sunnen.com Sunnen Product Co.
www.texaco.com Texaco
www.goodyear.com The Goodyear Tire and Rubber Company
www.starrett.com The L.S. Starrett Co.
www.tif.com TIF Instruments
www.tomco-inc.com Tomco (TI) Inc.
www.toyota.com Toyota
www.trw.com TRW Inc.
www.citroen.com/site/htm/en Citroen
www.uniroyal.com Uniroyal, Inc.
www.valvoline.com Valvoline Oil Co.
www.vw.com Volkswagen of America, Inc.
www.volvocars.com Volvo Cars of North America, LLC
www.waukeshaengine.com Waukesha Engine Division, Dresser Industries, Inc.

Acknowledgments

The production of a textbook of this type would not be possible without assistance from the automotive industry. The author would like to thank the following companies for their assistance in the preparation of *Auto Engine Repair.*

Acura
AC-Delco
American Hammered Piston Rings
American Motors
American Petroleum Institute
ASE
Audi
Belden
Cadillac
Caterpillar
Champion Spark Plugs
Clayton
Clevite
DaimlerChrysler
Dana
Deere & Co.
Echlin
Edelbrock
Ethyl Corporation
Federal Mogul
Fel-Pro
Fiat
Florida Voc. Ed.
Fluke
Ford
Gates
General Motors†
General Tool Corp.
Heyco
Honda
Jaguar
K-D Tools
Klein Tools
K-Line Tools
Land Rover
Lexus
Lisle Tools
Magnaflux Corp.
Mazda
Mercedes-Benz
Mitchell Manuals
Motorola
MSD Ignition Systems
NAPA
Neway
Nissan
Offenhauser
OTC Div. of SPX Corp.
Parker
Perfect Circle
Peugeot
Plew Tools
Renault
Riverdale Auto Repair
Robert Bosch
Saab
Saturn
Sealed Power
Sioux Tools
Snap-on Tool Corp.
Sonco
Starret
Storm Vulcan
Subaru
Sun Electric
Sunnen
Texaco
Tomco, Inc.
Toyota
TRW
Van Senus Auto
Volkswagen
Volvo

†Portions of materials contained herein have been reprinted with permission of General Motors Corporation, Service Technology Group.

Glossary

A

Abnormal engine noises: Include hisses, knocks, rattles, clunks, and popping. These noises may indicate part wear or damage.

Acceleration test: Measures the spark plug firing voltages when engine speed is rapidly increased.

Actuator subsystem: Based on signals from the control module, this system moves parts, opens injectors, closes the throttle, turns on the fuel pump, and performs other tasks needed to increase the overall efficiency of the vehicle.

Adjustable rocker arms: Provide a means of changing valve lash; must be used with mechanical lifters.

Advance: The ignition timing causes the spark to happen earlier in the compression stroke.

After-running: The engine will not shut off, but keeps firing, coughing, and producing power. Also called *dieseling* or *run-on.*

Air blowgun: Used to blow off small bits of dirt, solvent, water, and other debris and to blow out bolt holes, coolant passages, and oil passages.

Air check valve: Keeps exhaust from entering the air injection system. Usually located in the line between the diverter valve and the air distribution manifold.

Air compressor: The source of compressed air for an automotive service facility. An air compressor normally has an electric motor that spins an air pump, which forces air into a large, metal storage tank.

Air-cooled system: Uses cooling fins and outside airflow to remove excess heat from the engine. Circulates air over cooling fins on the cylinders and cylinder heads. This removes heat from the cylinders and heads to prevent overheating.

Air distribution manifold: Directs a stream of air toward each engine exhaust valve.

Air filter: Part of the air supply system that prevents particles in the air from entering the engine intake manifold.

Air-fuel ratio: A comparison of the amount of fuel and air entering an engine.

Air injection system: Uses an air pump to force fresh air at low pressure into the exhaust ports of the engine to help burn any unburned fuel that exits the combustion chamber.

Air jet combustion chamber: A small, extra valve allows a stream of air to enter the combustion chamber to aid swirl and combustion efficiency.

Air pump: Forces fresh air at low pressure into the exhaust ports of the engine. The pump is usually belt driven, but in some systems may be driven by a small electric motor.

Air supply system: Supplies air to the engine and prevents dust and dirt from the air entering engine intake manifold.

Air tools: Tools that use air pressure for operation.

Alternating current (ac): Current that changes from positive polarity to negative polarity. The current flows back and forth through the circuit.

Anaerobic sealer: Sealer that cures in the absence of air and is designed for tightly fitting, thick parts.

Antifreeze: Provides freeze protection, rust prevention, lubrication, and heat transfer. May have a base of ethylene glycol or propylene glycol.

Antifriction bearing: A bearing that uses ball, roller, or needle bearings between two moving surfaces.

API certification mark: A starburst symbol on an oil package that identifies oil as intended for use in gasoline or diesel engines.

ASE: National Institute for Automotive Service Excellence. A nonprofit, nonaffiliated (no ties to industry) organization formed to help ensure the highest standards in automotive service.

Asphyxiation: A condition in which the body has too little oxygen or too much carbon dioxide.

Atmospheric pressure: Air pressure produced by the weight of the air above the surface of the earth.

Atomization: Breaking of fuel into tiny droplets in order to increase the rate of combustion.

Atoms: The building blocks that make up everything in our universe. An atom is made up of tiny particles called protons, neutrons, and electrons.

B

Backfiring: Caused by the air-fuel mixture igniting in the intake manifold or exhaust system. A loud bang or pop sound can be heard when the mixture ignites.

Balancer shaft: An extra shaft with counterweights that are positioned to counteract the non-power-producing strokes. Used to help smooth the operation of the engine. Also called a *silent shaft.*

Bare block: The cylinder block without any parts installed.

Bare cylinder head: A machined casting without any parts installed.

Battery: Stores chemical energy that can be changed into electrical energy.

Bearing: A component used to reduce friction and absorb shock between other components.

Bearing clearance: See *oil clearance.*

Bearing conformability: The bearing's ability to move, shift, or adjust to imperfections in the journal surface.

Bearing corrosion resistance: The bearing's ability to resist corrosion from acids, water, and other impurities in the engine oil.

Bearing crush: Used to help prevent the bearing from spinning inside its bore during engine operation.

Bearing embedability: The bearing's ability to absorb dirt, metal, or other hard particles.

Bearing load strength: The bearing's ability to withstand pounding and crushing during engine operation.

Bearing spread: Used on split-type engine bearings to hold the bearing in place during assembly. The distance across the parting line of the bearing is slightly wider than the bearing bore. This causes the bearing insert to stick in its bore when pushed into place.

Belt drive timing mechanism: A cogged rubber belt is used to turn the camshaft. Basic parts include a cogged timing belt, crankshaft sprocket, camshaft sprocket(s), and belt tensioner. Commonly found on overhead camshaft (OHC) engines.

Bent connecting rod: The rod is no longer straight. This condition results from detonation or excessive engine speed.

Bent push rod: A push rod that is not straight. Damage that may be caused by a stuck valve, spring bind, spring breakage, piston hitting the valve, or other type of mechanical problem.

Bent valve: The head of the valve is not perpendicular to the valve stem.

Black light: Ultraviolet light used with a dye to locate engine leaks.

Block boring: Corrects excess cylinder wall wear, scoring, or scratching. Boring will true the cylinder walls so the rings and pistons fit properly. It is also done when installing a repair sleeve.

Block heater: A heating element mounted at some point in the cooling system, often in a coolant passage. It aids engine starting and warm up in cold weather.

Blowby: Combustion pressure leaking past the piston rings into the engine crankcase.

Blower: A special fan assembly that is driven by a belt on the engine.

Blowgun: See *air blowgun.*

Blown head gasket: A condition where the head gasket is ruptured. Can cause a wide range of problems, including overheating, missing, coolant or oil leakage, engine smoking, and burned mating surfaces on the cylinder head or block.

Bob weights: Weights installed on the crankshaft for crankshaft balancing to replicate the reciprocating weight.

Boost: Any pressure above atmospheric pressure in the intake manifold.

Bore/stroke ratio: The relationship between the bore and the stroke. It is calculated by dividing the bore by the stroke.

Boring bar: A machine shop tool that will cut a thin layer of metal out of a worn engine cylinder to make it larger in diameter.

Bottom dead center (BDC): The point of travel where the piston is at its lowest point in the cylinder.

Bottom end: The cylinder block with all of its internal parts installed.

Brake horsepower (bhp): The usable horsepower at the engine crankshaft.

Broken connecting rod: A complete failure of the I-beam or rod bolts and cap.

Broken push rod: Usually has the small ball on the end broken off. This will cause loud valve train noise and usually push rod disengagement from the rocker arm.

Brush hone: Hone with small beads of abrasive material on the ends of spring-steel wires. It is used to final deglaze cylinders before assembly.

Burette: A graduated cylinder with a valve for metering out liquid. Used to measure combustion chamber volume.

Burned piston: Abnormal combustion, excessive pressure, and heat melt the piston and blow a hole in the crown or area around the ring grooves.

Burned valve: A portion of the valve margin and face is burned away.

Bypass lubrication system: Does not filter all of the oil that enters the engine bearings. It filters the extra oil not needed by the bearings.

C

Cam bearing driver: Used to force cam bearings into or out of the cylinder block.

Cam follower: A valve tappet mounted in the cylinder head. The camshaft lobe acts on the follower, which in turn pushes down to open the valve.

Cam ground piston: Piston that is machined slightly out-of-round to compensate for different rates of piston expansion (enlargement) due to differences in the material thickness.

Cam-in-block engine: Uses push rods to transfer camshaft motion to the rocker arms and valves. The term *overhead valve (OHV)* is sometimes used to refer to this design.

Camshaft: Opens the valves and allows the valve springs to close the valves at the right time during each stroke.

Camshaft bearing: Inserts pressed into the block or cylinder head; may be two-piece type in some overhead camshaft engines. The camshaft journals ride in the cam bearings.

Camshaft bearing clearance: The oil gap between the camshaft bearing and the camshaft journal.

Camshaft bore: A series of holes machined through the center of the cylinder block or cylinder head to receive the camshaft bearings.

Camshaft cover: A lid over the top of the camshaft housing on overhead camshaft engines. It serves the same purpose as a valve cover on an OHV engine.

Camshaft drive mechanism: Turns the engine camshaft and keeps it in time with the engine crankshaft and pistons. Also called the *timing mechanism.*

Camshaft duration: The amount of time the valves stay open. It is determined by the shape of the cam lobe nose and flank.

Camshaft endplay: The front-to-rear movement of the camshaft in its bores.

Camshaft grinder: A machine shop tool that can be used to resurface the cam lobes and bearing journals on a worn camshaft.

Camshaft housing: Mounts on top of the cylinder head and holds the cam followers and camshaft(s).

Camshaft journal wear: Reduction in the size of the camshaft journal that results in increased part clearance and may reduce engine oil pressure.

Camshaft key: Used to align the camshaft sprocket or gear on the camshaft.

Camshaft lift: The amount of valve train movement produced by the cam lobe.

Camshaft lobes: Egg-shaped protrusions on the camshaft that are used to change the rotary motion of the camshaft into reciprocating motion of the valves.

Camshaft lobe wear: Reduced lift or duration resulting from friction wearing off some of the camshaft lobe.

Camshaft service: Measuring camshaft lobe and journal wear, as well as measuring camshaft endplay and camshaft bearing clearance.

Camshaft straightness: A measure of how bent or warped the camshaft is.

Camshaft thrust plate: Used to limit the front-to-rear movement, or endplay, of the camshaft.

Camshaft thrust surface wear: Allows the camshaft to move too far forward and rearward in the engine. This can sometimes produce a light, slow knocking sound, timing chain or gear problems, and increased lifter and lobe wear.

Camshaft timing: Ensures that the valves properly open and close in relation to the crankshaft. A belt, chain, or set of gears is used to turn the camshaft at one-half of the crankshaft speed and keep the camshaft in time with the crankshaft.

Cap screws: Bolts threaded into a part without the use of a nut.
Carbon monoxide (CO): An extremely toxic emission resulting from incomplete combustion. It is a colorless and odorless gas.
Cartridge oil filter: Filter with a separate element and canister.
Casting numbers: Numbers formed into parts during manufacturing. They can be found on major components, such as the block, camshaft, and heads.
Casting plugs: See *core plugs.*
Cast iron crankshaft: Made by pouring molten iron into a mold. It is a common type for light- and medium-duty applications.
Cast iron exhaust manifolds: Made by pouring molten iron into a mold. Used on engine applications where the manifold is exposed to very high operating temperatures.
Catalyst: Any substance that speeds a chemical reaction without itself being changed.
Catalytic converter: Device that chemically converts byproducts of combustion into harmless substances. It oxidizes (burns) the remaining HC and CO emissions that pass into the exhaust system.
Chain drive timing mechanism: Consists of a timing chain, crankshaft sprocket, and camshaft sprocket.
Chain guide: Prevents chain slap on long lengths of chain. The guides have a metal body with a plastic, nylon, or Teflon face, allowing the chain to slide with minimum wear.
Chain tensioner: Takes up excess slack in the timing chain as the chain and sprockets wear. It is a spring-loaded or hydraulic device that pushes on the unloaded side of the chain to keep the chain tight.
Charcoal canister: A metal or plastic canister filled with activated charcoal granules that absorb fuel vapors. Stores fuel vapors when the engine is not running.
Chemical flushing: A chemical cleaner is added to the coolant to remove scale buildup.
Chipped valve: Valve in which part of it has broken away. Normally due to mechanical damage or a defective valve.
Clogged push rod: The internal oil hole is restricted or blocked with hardened oil and sludge deposits. The tip of the push rod can also become mushroomed, which restricts oil flow.
Closed cooling system: An expansion tank or reservoir is connected to the radiator by an overflow tube. Vacuum in the radiator will pull coolant out of the reservoir. Excess pressure in the radiator will push coolant into the reservoir.
Closed length: The length (or height) of the valve spring when it is installed on the engine with the valve closed.
Clutch: Allows the driver to engage or disengage the engine power from the drive train. It is mounted onto the engine flywheel between the engine and transmission or transaxle.
Clutch alignment tool: Tool is used to hold the manual clutch disc onto the flywheel at the exact center of the flywheel when installing a transmission/clutch assembly onto the rear of the engine.
Cogged belt: A belt with teeth or ribs that run from side to side, which fit in matching ribs in the belt sprockets. Commonly used to drive the engine camshaft and auxiliary shaft.
Cold cleaning tank: Tank used for removing oil and grease from parts. Also called a *cold solvent tank.*
Combination square: A sliding square mounted on a steel rule.
Combustion: A chemical reaction that releases the energy stored in the fuel as heat.
Combustion chambers: Cavities or pockets formed in the cylinder head deck surface above the cylinders. They are located directly over the pistons.
Comebacks: When a repair must be done over because it failed to correct the problem.
Commission: The technician obtains a set percent of the flat rate labor charged to the customers.
Commission plus parts: A technician being paid on commission also receives a small commission for the parts installed during a repair.
Component location diagram: Illustrates where each electrical or electronic part is positioned in the engine compartment.
Compression gauge: Used to measure the amount of pressure in a cylinder during the engine compression stroke.
Compression ignition: Uses the heat generated by the high compression pressure to heat the air and ignite the fuel.
Compression pressure: The amount of pressure produced in the engine cylinder on the compression stroke.
Compression rings: Seal the clearance between the block and piston to contain the pressure formed in the combustion chamber.
Compression stroke: Squeezes the air-fuel mixture to make it more combustible. The piston slides up and compresses the mixture into the small area in the combustion chamber.
Compression test: Test used to determine compression stroke pressure. It should be done anytime symptoms point to cylinder pressure leakage.
Concentricity: A measure of how closely two circular cross sections share a common axis.
Connecting rod: Rod that transfers the force of the piston to the crankshaft.
Connecting rod bearing knock: Caused by wear and excessive bearing-to-crankshaft clearance. It is indicated by a light, regular, rapping noise with the engine floating.
Connecting rod bore problems: The big end of the rod is out-of-round, in which case it can no longer properly support the rod bearing.
Connecting rod cap: Cap that is bolted to the bottom of the connecting rod. It can be removed for disassembly of the engine.
Connecting rod side clearance: The distance between the side of the connecting rod and the cheek of the crankshaft journal or side of the other rod.
Control module: Controls the ignition system, fuel system, transmission or transaxle, emission control systems, and other systems.
Control subsystem: Looks at the inputs from the sensors and determines what actions need to take place.
Conventional-flow cooling system: Coolant circulates through the engine block, travels to the head, and then returns to the radiator.
Coolant: A solution of water and antifreeze.
Coolant passages: Water jackets formed in the cylinder head that allow coolant to circulate around the combustion chambers. This removes excess heat from the burning fuel.
Coolant strength: A measure of the concentration of antifreeze compared to the concentration of water.
Cooling system: Carries the heat of combustion and friction away from the engine. Basically consists of a radiator, water pump, fan, thermostat, water jackets, and connecting hoses.
Cooling system bleed screw: Provided to allow for removal of air.
Cooling system fan: Pulls air through the core of the radiator and over the engine to help remove heat. It increases the volume of air flowing through the radiator, especially at low vehicle speeds.
Cooling system flushing: A process of cleaning the cooling system by running water or a cleaning chemical through the cooling system to wash out contaminants.
Cooling system hydrometer: Device used to measure the freezing point of the cooling system antifreeze solution.

Cooling system pressure test: Test used to quickly locate leaks in the cooling system.

Core plugs: Small metal plugs pressed or screwed into holes in a cylinder block. Also called *casting plugs.*

Corrosive hazard: A material or waste that dissolves metals and other materials or burns human skin.

Cotter pin: A safety device often used in conjunction with a slotted nut. It fits through a hole in the bolt or part, keeping the slotted nut from turning and possibly coming off.

Counterweights: Heavy lobes on the crankshaft to help prevent engine vibration.

Cracked block: Fractures in the cylinder walls, deck surface, lifter gallery, and so on. This usually results from overheating, ice formation in cold weather, or breakage of pistons or connecting rods.

Cracked cylinder head: Results from engine overheating or physical damage from other broken parts. Usually, the crack happens at an exhaust seat or between chambers, allowing coolant or pressure leakage.

Cracked valve: Results from overheating.

Crankcase: The lower area of the block in which the crankshaft spins.

Crankcase warpage: Throws the main bearing bores out of alignment. This can make the crankshaft lock in the block during engine reassembly with new bearings.

Cranking balance test: Test performed by an engine analyzer to find cylinders with low compression due to a burned valve, worn piston rings, or other problems.

Crankshaft: Converts the reciprocating (up and down) movement of the connecting rod and piston into rotary (spinning) motion.

Crankshaft balancing: Needed when the weight of the pistons, connecting rods, or crankshaft is altered by installing new parts of a different weight.

Crankshaft endplay: The amount of front-to-rear movement of the crankshaft in the block. It is controlled by the clearance between the main thrust bearing and the crankshaft cheeks, thrust surface, or journal width.

Crankshaft flange: A round disk with threaded holes to hold the flywheel bolts.

Crankshaft front seal: Keeps engine oil from leaking out from between the crankshaft snout and the engine front cover.

Crankshaft grinder: A large machine tool that rotates the crankshaft against a spinning stone, grinding the crank journals to a smaller diameter.

Crankshaft key: Used to lock the crankshaft sprocket or gear to the snout of the crankshaft.

Crankshaft oil seals: Keep oil from leaking out of the front and rear of the engine.

Crankshaft rod journals: Precision-machined and polished surfaces for the connecting rod bearings. The connecting rods bolt around these journals. Also termed *crankpins.*

Crankshaft service: Cleaning the crankshaft, measuring crankshaft journal wear, checking crankshaft straightness, checking for cracks, and proper installation.

Crankshaft snout: A shaft machined on the front of the crankshaft to hold the damper, camshaft drive sprocket or gear, and crank pulleys.

Crankshaft thrust surfaces: The area on each side of the main and rod journals where the sides of the grinding stone rub on the crankshaft during its initial machining. Also called cheeks.

Crankshaft turning: Grinding the rod and main journals smaller in diameter to fix journal wear or damage.

Crankshaft welder: Used to weld damaged journals during crankshaft rebuilding.

Crossflow combustion chamber: Has the intake ports on one side of the head and the exhaust ports on the other side.

Crown: The top of the piston. Also called *piston head.*

Crude oil: Oil taken directly from the ground. It is a mixture of various elements, including carbon, hydrogen, sulfur, nitrogen, and oxygen.

Cylinder: Large holes machined into the cylinder block for the pistons.

Cylinder arrangement: The position of the cylinders in the engine block in relation to the crankshaft.

Cylinder balance test: Measures the power output from each of the engine's cylinders.

Cylinder block: Supports or holds all of the other engine components. Various surfaces and holes are machined in the block for attaching these other parts.

Cylinder block deck: A flat, machined surface for the cylinder head.

Cylinder block problems: Include cracks, warpage, worn cylinders, mineral deposits in coolant passages, and misaligned main bores.

Cylinder bore: The diameter of the engine cylinder.

Cylinder deactivation engine (CDE): Uses solenoid-operated rocker arms to alter the number of engine cylinders that function during engine operation, reducing displacement of the engine.

Cylinder head: Bolted to the top of the block deck to enclose the top of the cylinders and form the top of the combustion chamber.

Cylinder head assembly: The bare head with the valves, springs, retainers, and related parts installed.

Cylinder head gasket: Seals the pressure of combustion in the cylinder and prevents the flow of oil and coolant out of their respective passages.

Cylinder head milling: A machine shop operation where a thin layer of metal is cut or machined off the head gasket surface (deck) of the cylinder head. It is done to correct head warpage.

Cylinder head stand: Used to hold a cylinder head while servicing the valves and valves seats.

Cylinder hone: Tool used to deglaze a cylinder wall during an engine rebuild.

Cylinder honing: Used to remove the smooth surface and provide a lightly roughened or textured surface on the walls of a used cylinder.

Cylinder leakage test: Measures air leakage out of the engine combustion chamber to find mechanical problems.

Cylinder numbers: Numbers that identify the cylinders, pistons, and connecting rods of the engine.

Cylinder out-of-round: A difference in cylinder diameter when measured front-to-rear and side-to-side in the block.

Cylinder sleeving: Machining or boring a cylinder oversize and pressing in a cylinder liner.

Cylinder taper: The difference in the diameter from the top to the bottom of the cylinder measured on the piston ring contact surface of the cylinder.

D

Data link connector (DLC): A multipin terminal used to link the scan tool to the computer.

Dead cylinder: A cylinder in which combustion is not occurring.

Decarbonizing cleaner: A very powerful chemical that can remove carbon, paint, gum, and most other deposits.

Deck: A flat surface machined on the top of the block for the cylinder head.

Deck plate: A rigid metal cap bolted onto the block deck surface when power honing or boring.

Deck warpage: Often results from engine overheating, especially with thin-wall or aluminum blocks. It is measured with a straightedge and feeler gauge on the head gasket sealing surface(s) of the block.

Deglazing: See *cylinder honing.*

Degree wheel: Used to measure how much the crankshaft has been rotated when performing camshaft degreeing.

Depth micrometer: Precisely measures the depth of an opening.

Detonation: Abnormal combustion where the last unburned portion of the fuel mixture almost explodes in the combustion chamber.

Diagnostic trouble code (DTC): A digital signal produced and stored by the computer when an operating parameter is exceeded.

Dial bore gauge: A tool often used for measuring cylinder wear in an engine block. It is a variation of a dial indicator and is slid up and down in the cylinder bore.

Dial indicator: Used to measure part movement in thousandths of an inch or hundredths of a millimeter. It is frequently used to check gear teeth backlash (clearance), shaft end play, cam lobe lift, and other similar kinds of part movements.

Die: A tool for cutting external threads, such as on rods, bolts, shafts, and pins.

Die grinder: A hand-held tool with a shaft that spins at high speed.

Diesel engine: Engine that burns diesel oil (fuel), which is a thicker fraction of crude oil. Diesel fuel is injected directly into the engine combustion chamber or a precombustion chamber.

Dieseling: The engine will not shut off after the ignition key switch is turned off. It keeps firing, coughing, and producing power. Also called *after-running* or *run-on.*

Diesel injection: System that forces fuel directly into the engine's combustion chamber.

Diesel injector: A spring-loaded valve that is normally closed and blocks fuel flow. When the injection pump forces fuel into the injector under high pressure, the injector opens and fuel is sprayed directly into the combustion chamber or a precombustion chamber.

Digital caliper: A measuring tool with an accuracy of typically .001″ (0.01 mm). It is a Vernier caliper with a digital indicator.

Digital multimeter: Meter used to check electronic circuits for normal voltage, current, or resistance.

Direct current (dc): Current that is always the same polarity, either positive or negative. The current only flows in one direction.

Dished piston: Piston with a concave crown to lower compression.

Displacement: The size of an engine; determined by cylinder diameter, the amount of piston travel per stroke, and the number of cylinders.

Distributor wrench: A long-handled, box-end wrench for removing fasteners that are under obstructions.

Diverter valve: Valve that keeps air from entering the exhaust system during deceleration and limits the maximum system air pressure, when needed.

Domed piston: Piston with a head that is curved upward. This type is normally used with a hemispherical cylinder head.

Dowel holes: Dowels in the block fit into these holes. This holds the head gasket and cylinder head in alignment when installing the head on the block.

Drive train: Uses power from the engine to turn the vehicle's drive wheels.

Drive wheel horsepower: An engine horsepower rating that takes into account the friction losses required to operate the transmission/transaxle, drive axles or drive shaft joints, differential, axle bearings, and other drive train parts that consume energy.

Dropped valve guide: A valve guide that has broken free from its press-fit in the head.

Dry sleeve: Relatively thin sleeve that is not exposed to engine coolant.

Dry sump lubrication system: Uses two oil pumps and a separate oil tank. The main oil pump pressurizes the lubrication system. The second oil pump, called the scavenging oil pump, pulls oil out of the oil pan.

Dual-bed catalytic converter: A catalytic converter having two separate catalyst units enclosed in a single housing.

Dual-mass vibration damper: Has one weight mounted on the outside of the pulley and another on the inside. The extra rubber-mounted weight helps reduce torsional vibration at high engine speeds.

Dual overhead camshaft (DOHC): Engine that has two camshafts per cylinder head. One camshaft operates the intake valves and the other camshaft operates the exhaust valves.

Dual plane intake manifold: Manifold with a two-compartment plenum. It provides good fuel vaporization and distribution at low engine speeds.

Duration: Period of time a spark is held or sustained.

Dye penetrant testing: A special dye is sprayed over a nonmagnetic engine part. The dye collects in any cracks that are in the part and appears as a dark line.

Dying: See *stalling.*

E

EGR valve: Controls the amount of exhaust gas fed into the intake manifold. Consists of a vacuum diaphragm, spring, plunger, exhaust gas valve, and a diaphragm housing.

Electrical components: Devices that use electricity and moving parts to accomplish a task.

Electrical fires: Fires that occur when a current-carrying wire shorts to ground. This causes unlimited current flow, which in turn causes the wire to heat up, melt its insulation, and burn.

Electrical symbols: Wiring diagram symbols that represent the electrical components in a circuit.

Electric engine fan: Uses an electric motor and a thermostatic switch to provide cooling action.

Electrolysis: An electric current is sent through water to release the oxygen and hydrogen.

Electronic control module (ECM): A computer that uses very complex electronic circuits to perform various functions.

Electronic devices: Devices that use semiconductors instead of moving parts.

Electronic fuel injection (EFI): Gasoline injection system that uses engine sensors, a computer, and solenoid-operated fuel injectors to meter the correct amount of fuel into the engine.

Electronic fuel injection tester: Used to check the operation of electronic fuel injection systems.

Electronic noise detector: Amplifies sounds emitted from engine parts.

Element: A paper, cotton, or synthetic filtering substance mounted inside the oil filter housing. It allows oil to flow, but blocks and traps small debris.

Emission control systems: Reduce the amount of harmful chemicals and compounds (emissions) that enter the atmosphere from a vehicle.

Emissions: Chemicals emitted into the environment from the engine.

End gas: Unburned portion of the air-fuel mixture.

Engine: The source of power for moving the vehicle and operating the other systems.

Engine analyzer: A group of test instruments mounted in a roll-around cabinet. It usually contains an oscilloscope, tach-dwell, exhaust gas analyzer, VOM, compression gauge, vacuum gauge, cylinder balance tester, cranking current balance tester, and timing light.

Engine balancer: Used to spin the engine crankshaft to determine if weight must be added to or removed from its counterweights.

Engine block: The framework or "backbone" of an engine to which many of the other components are fastened.

Engine compression ratio: Comparison of cylinder volumes with the piston at the extreme top and bottom of its travel.

Engine coolant: A mixture of antifreeze and water.

Engine crane: Used to remove (raise) an engine from or install (lower) an engine into a vehicle. It has a hydraulic hand jack for raising and a pressure release valve for lowering.

Engine dynamometer (dyno): Used to measure brake horsepower.

Engine efficiency: The ratio of power produced by the engine and the power supplied to the engine (energy content of the fuel).

Engine floating: A constant engine speed above idle.

Engine flywheel: A very heavy, round disk mounted to the back of the crankshaft to help keep the crankshaft spinning between power strokes and smooth engine operation.

Engine front cover: Holds the front oil seal and encloses the timing gears or chain. Also called the *timing cover.*

Engine front end: The parts that are attached to the front of the engine, including the camshaft drive mechanism, front cover, water pump, and auxiliary shafts.

Engine gaskets: Prevent pressure, oil, coolant, and air leakage between engine components.

Engine machine shop: Shop that specializes in the rebuilding of engines.

Engine oil: Oil that protects an engine from friction and the resulting heat. Also called *motor oil.*

Engine operating temperature: The temperature of the engine coolant during normal engine operating conditions.

Engine overcooling: To much heat is removed from the engine.

Engine overhaul: A process that includes complete disassembly of the engine; cleaning, inspection, and measurement of critical parts; replacement or service of worn or damaged parts; and reassembly of the engine. Also termed engine rebuild.

Engine overheating: Not enough heat is being removed from the engine.

Engine performance problem: Any trouble that lowers engine power, fuel economy, drivability (performance), or dependability.

Engine-powered fan: Fan that is bolted to the water pump hub and pulley. This type of fan has generally been replaced by electric fans.

Engine prelubricator: Used to fill the oil passages in an engine with oil before starting the engine and to find worn engine bearings and other lubrication system problems during pre-teardown diagnosis.

Engine rebuild: See *engine overhaul.*

Engine sensor: An electronic device that changes circuit resistance or voltage with a change in a condition.

Engine smoking: A condition where smoke is expelled from the exhaust system. The color of smoke can be used to diagnose engine problems.

Engine stand: A metal framework on small casters or wheels. It is used to hold an engine for disassembly and reassembly.

Engine top end: The parts that fasten above the engine's short block. They include the cylinder heads, valves, camshaft, rocker arms and other related valve train components.

Engine torque: A rating of the turning force at the engine crankshaft.

Engine tune-up: A maintenance operation that returns the engine to peak performance after part wear and deterioration. It involves the replacement of worn parts, basic service tasks, making adjustments, and, sometimes, minor repairs.

Engine valves: Valves that control the flow into and out of the engine cylinder or combustion chamber.

Evaporative emissions control (EVAP) system: Prevents vapors in the fuel system from entering the atmosphere through the use of a charcoal canister.

Excess crankshaft endplay: Too much front-to-back movement of the camshaft.

Exhaust deck: A machined surface on the head for bolting on the exhaust manifold.

Exhaust gas analyzer: Measures the chemical content of the engine's exhaust.

Exhaust gas recirculation (EGR) system: Injects burned exhaust gasses into the engine intake manifold to lower the combustion temperature and reduce NO_X pollution.

Exhaust manifold: Carries burned gasses from the cylinder head exhaust port to the other parts of the exhaust system.

Exhaust manifold warpage: A condition in which the deck on the exhaust manifold is no longer flat.

Exhaust ports: Passages leading from the combustion chambers to the exhaust manifold to carry burned gasses out of the cylinders.

Exhaust stroke: The burned gasses are pushed out of the cylinder and into the vehicle's exhaust system.

Exhaust valve: Controls the flow of burned gasses out of the combustion chamber. It is located over the cylinder head exhaust port.

Explosion: A violent expansion of gasses due to rapid combustion.

External lifter wear: Results in the bottom of the lifter becoming concaved (sunk in). This is due to friction between the lifter and camshaft lobe.

Externally balanced: Extra weight is added to certain points on the flywheel and balancer to prevent engine vibration.

F

Fast flushing: A water hose is connected to an adapter fitting installed in a heater hose. The radiator cap is removed and the drain valve is opened. The water hose is turned on and water flows into the system, removing rust and loose scale.

Feeler gauge: Gauge with precision ground, steel blades of various thicknesses, normally used to measure small distances between parallel surfaces.

Fiber-reinforced piston: Piston made from a special alloy reinforced with fibers to increase piston dependability under severe operating conditions.

Filler neck: The extension for filling the fuel tank with fuel.

Filter bypass valve: Protects the engine from oil starvation if the oil filter element becomes clogged.

Finger pickup tool: Used to retrieve dropped parts or tools.
Fire: The result of heat, fuel, and oxygen in the correct proportions to start combustion.
Firing order: The sequence in which combustion occurs in the engine. The order in which cylinders fire.
Flaring tool: Tool used to form a flare on the end of a metal line.
Flat head engine: See *L-head engine.*
Flat rate manual: A book that states how long each repair should take.
Flat top piston: Piston with a crown that is almost flat and parallel to the block deck.
Flex fan: Fan with thin, flexible blades that alter airflow with engine speed.
Flex hone: Hone with hard, flat, abrasive stones affixed to spring-loaded, movable arms. It is used when cylinder wear is minor and only a small amount of cylinder honing is needed.
Floating piston pin: Pin that is secured by snap rings and is free to "float" or rotate in both the piston pin bore and the connecting rod small end.
Fluid coupling fan clutch: Fan clutch designed to slip at higher engine speeds to reduce fan action.
Fluorescent penetrant: Similar to dye penetrant, but requires an ultraviolet (black) light to illuminate the penetrant in the cracks.
Flywheel: A large, steel disk mounted on the rear flange of the crankshaft.
Flywheel lock: Tool that keeps the engine crankshaft from turning during engine disassembly or reassembly.
Flywheel problems: Include loose bolts, failure of the ring gear teeth, friction surface scoring, and warpage.
Flywheel rotating tool: A special pry bar with fingers for grasping the teeth on the flywheel ring gear.
Forged steel crankshaft: Made by hammering hot steel into a mold using tons of force. It is used in high-performance applications.
Form-in-place gasket: A special sealer used in place of a conventional fiber or rubber gasket.
Four-stroke cycle: Every two up and two down strokes of the piston results in one power-producing cycle. Two complete revolutions of the crankshaft are needed to complete one four-stroke cycle.
Four-valve combustion chamber: Has two intake valves and two exhaust valves for each engine cylinder.
Fractionating tower: Device in which different hydrocarbons, or fractions, contained in crude oil are separated by weight.
Frame ground circuit: Uses the vehicle's metal unibody or frame as a conductor to complete an electrical circuit.
Friction bearing: A plain bearing that has a smooth surface of soft metal, such as copper or babbit, on which a part slides or rotates.
Frictional horsepower (fhp): The power needed to overcome engine friction and a measure of the resistance to movement between engine parts.
Fuel cell: Used to combine or alter chemicals to produce water and electrons (electrical energy).
Fuel filter: Stops contaminants such as rust, corrosion, and dirt from entering the fuel metering system. It is normally located on the fuel tank pickup tube.
Fuel hoses: Hoses made of nylon or synthetic rubber covered with a plastic insulator.
Fuel injected gasoline engine: Sprays fuel into airstream in the intake manifold, either at the port or in the throttle body.
Fuel injector: An electrically-operated fuel valve.
Fuel lines: Plastic, nylon, or double-wall steel tubing.
Fuel metering system: Controls amount of fuel that is mixed with the filtered air.
Fuel pressure gauge: Measures engine fuel pressure and can be used to find engine performance problems.
Fuel pump: Produces fuel pressure and flow.
Fuel supply system: Provides filtered fuel under pressure to the fuel metering system. It also provides for storage of the fuel.
Fuel system: Meters the right amount of fuel into the engine for efficient combustion under different conditions. Includes the fuel supply system, air supply system, and fuel metering system.
Fuel tank: Safely holds an adequate supply of fuel for prolonged engine operation.
Fuel tank pickup unit: Extends down into the tank to draw out fuel.
Full flow lubrication system: Forces all of the oil through the oil filter before the oil reaches the parts of the engine.
Fuse: Protects a circuit against damage caused by excessive current draw (overcurrent), such as that created by a short circuit.

G

Gas line freeze: Moisture in the fuel turns to ice and blocks fuel flow, thus preventing engine operation.
Gasket: A soft, flexible material placed between parts to fill small gaps, scratches, or other imperfections in the mating part surfaces.
Gasoline engine: Engine that burns gasoline.
Gasoline injection: System that uses fuel pump pressure to spray fuel into the engine intake manifold, usually near the cylinder head's intake port.
Gear drive timing mechanism: Consists of helical gears on the front of the engine that connect the crankshaft to the camshaft.
Gear oil pump: Uses a set of gears to produce lubrication system pressure.
Glow plug: An electric heating element that warms the air in the combustion chamber.
Grade markings: Markings on the bolt head that specify the strength of the bolt.
Grinding stone dresser: A tool used to resurface the grinding stones in the valve seat grinder.
Gross horsepower (ghp): The engine power available with only basic accessories installed.
Ground wires: Wires that connect electrical components to the chassis or ground of the vehicle.

H

Hand brushes: Brushes used with solvent to remove oil and grease.
Hard starting: Excessive cranking time is needed to start the engine.
Hardening sealer: Sealer used on permanent assemblies and for filling uneven surfaces.
Hazardous waste: A solid, liquid, or gas that can harm people or the environment.
Head bolts: Secure the head to the block.
Head gasket: Creates a seal between the block and head surfaces to prevent oil, coolant, and combustion-chamber pressure leakage.
Head pressure testing machine: Seals all oil and coolant passages and then pressurizes the head to check for leakage.

Head warpage: A condition in which the head deck is no longer flat.

Heat-cleaning oven: Oven designed to remove organic materials such as oil, grease, paint, varnish, and plastic from metal engine parts without the use of chemicals.

Heated air inlet system: Speeds engine warmup and keeps the temperature of the air entering the engine constant. Also called a *thermostatic air cleaner system.*

Heater hoses: Small-diameter hoses that carry coolant to the heater core.

Hemispherical combustion chamber: Shaped like a dome when viewed from the side. The valves are canted (tilted) on each side of the chamber. The spark plug is located near the center.

Hesitation: The engine does not accelerate normally when the gas pedal is pressed, or stalls before developing power. Also called *stumble.*

High-pressure washer: Used to remove heavy deposits of dirt, grease, and oil from the outside of engines.

High-tension cable: See *secondary wire.*

Hole gauge: Measures very small holes in parts where an internal micrometer cannot be used.

Horsepower: Unit of power that originated as the average strength of one horse.

Hose clamps: Used to secure hoses tightly to their fittings.

Hot spot: An area in the engine suffering from an abnormal buildup of heat.

Hot tank: Tank that uses heat and a very powerful cleaning agent to clean engine parts. It is used to aggressively remove mineral deposits, hard carbon deposits, oil, grease, and paint.

Hybrid power source: Uses two different methods of propulsion, usually a gasoline engine and a large electric motor.

Hydraulic camshaft: Camshaft designed to be used with hydraulic lifters. Its cam lobes are shaped to initially open the valve more quickly.

Hydraulic jacks: Used to lift the vehicle, engine, and transmission during engine removal or installation, or when replacing motor mounts, exhaust manifolds, oil pans, etc.

Hydraulic lifter: Type of lifter that is filled with engine oil and can take up all of the clearance in the valve train when the engine is running.

Hydraulic press: Produces tons of force for assembly or disassembly of pressed-together components.

Hydrocarbons (HC): Vehicle emissions resulting from the release of unburned fuel into the atmosphere.

I

Iconel valves: Valves made of a hard, nonmagnetic, stainless steel alloy that can withstand extreme temperatures and friction.

Ignitable hazard: A material that will easily ignite and burn.

Ignition coil: Steps up battery voltage to make the electricity jump the spark plug gap.

Ignition system: Ignites and burns the air-fuel mixture. Consists of a spark plug, plug wire, ignition coil, switching device, and power source.

Ignition timing: The relation of spark to valve opening and piston position.

I-head engine: Engine in which intake and exhaust valves are in the cylinder head. Also called an overhead valve engine (OHV).

Indicated horsepower (ihp): The amount of power formed in the engine combustion chambers.

Induction-hardened valve seats: An electric-heating operation that makes the surface of the metal much harder and more resistant to wear.

Inertia ring: Part of a vibration damper that sets up a damping action on the crankshaft as it tries to twist and untwist.

Injection pump: A high-pressure, mechanical pump. It is powered by the engine and forces fuel to the diesel injector under very-high pressure.

Inline crankshaft: Only has one connecting rod fastened to each rod journal.

Inline engines: Engines with cylinders positioned one after the other in a straight line.

Inside micrometer: Tool used for measuring internal measurements in large holes, cylinders, or other part openings.

Intake deck: A machined surface on the head for bolting on the intake manifold.

Intake manifold: Carries the air-fuel mixture into the cylinder head intake ports.

Intake manifold gasket: Seals the joint between the intake manifold and the head.

Intake manifold warpage: A measure of the flatness of the manifold's sealing surfaces.

Intake ports: Large passages through the cylinder head from the intake manifold runners to the combustion chambers through which the air-fuel mixture enters the cylinders.

Intake stroke: Draws air and fuel into the combustion chamber.

Intake valve: Controls the flow of air-fuel mixture or just air through the cylinder head intake port.

Integral cylinder: Cylinder that is part of the block.

Integral valve guides: Valve guides that are made as part of the cylinder head casting.

Integral valve seat: Valve seat made from the head casting.

Intensity: Energy of the spark to start and maintain the combustion of a lean fuel mixture.

Intercooler: An air-to-liquid heat exchanger mounted between the supercharger and the engine. It consists of cooling tubes surrounded by cooling fins, similar to an engine radiator.

Intercooler lines: Metal or rubber lines that carry the coolant from the intercooler to the intercooler radiator and back to the intercooler.

Intercooler pump: A small electric water pump that keeps the coolant circulating through the cooling tubes and intercooler radiator.

Intercooler radiator: A small heat exchanger for transferring heat from the intercooler coolant to the outside air.

Interference angle: A 1/2° to 1° angle difference between the valve seat face and the face of the valve.

Internal combustion engine: Engine that burns fuel inside of itself.

Internally balanced crankshaft: Uses the counterweights on the crankshaft to offset the weight of the rod and piston assemblies.

In-vehicle engine overhaul: Overhaul that can only be done when cylinder and crankshaft wear are within specifications and the oil pan and heads can be easily removed with the engine installed in the vehicle.

J

Jumper cables: Used to start a vehicle with a dead (discharged) battery.

Jumper wire: Used to test switches, relays, solenoids, wires, and other nonresistive components.

Jump starting: Connecting another battery to the discharged battery with jumper cables.

K

Keeper grooves: Grooves machined into the top of the valve stem to accept small keepers or locks that hold the spring retainer on the valve.

Key: Fits into a groove cut into a shaft and part.

Keyway: Groove into which a key fits.

Knock sensing system: A knock sensor and the control module (computer) retard ignition timing or limit boost pressure to prevent detonation and preignition.

Knock sensor: Acts as a microphone to "listen" for the sound of engine knock. When the sensor "hears" engine knock, an electrical pulse is sent to the control module.

Knurled valve guides: Valve guides that have spiral grooves pressed into the inside diameter of the guide.

L

Lapping stick: A wooden rod with a suction cup on the end used to spin the valve on the seat when checking valve-to-seat contact.

Leak down rate: The amount of time it takes the pressure inside the lifter to bleed out.

Lean fuel mixture: More than 14.7 parts of air for every one part of fuel.

L-head engine: Engine in which both the intake and exhaust valves are located in the block. Also called a *flat head engine.*

Lifter bores: Precision-machined and honed holes in the cylinder block to accept tappets or lifters.

Lifter service: Inspecting, testing, and, if needed, rebuilding the lifter.

Lifter valley: The area in the block for the lifters and push rods.

Line bending tool: Tool that creates bends in a metal line without pinching or collapsing the line.

Line boring: Straightening or truing misaligned main bearing bores using a machine tool cutter.

Line hone: Hone used to repair minor damage and straighten the main bearing bores of a cylinder block.

Liner protrusion: The distance that the sleeve sticks up above the block deck. Needed to seal combustion pressure at the head gasket.

Liquid-cooled system: Circulates a coolant through internal passages cast in the engine called coolant passages, or "water jackets." As the coolant circulates through the engine, it collects heat and carries it to a radiator. The radiator transfers the heat to the outside air.

Liquid-vapor separator: A metal tank located above the main fuel tank. It is used to keep liquid fuel from entering the EVAP system.

Load test: Measures the spark plug firing voltages when engine speed is rapidly increased.

Load tester: Used to check the condition of a vehicle's battery, charging system, and starting system. It can also perform other electrical tests.

Long block: The short block with the completed heads installed.

Lower radiator hose: Connected between the water pump inlet and the radiator.

Lubrication system: Circulates engine oil to high-friction points in the engine. Basically consists of an oil pump, oil pickup, oil pan, and oil galleries.

M

Machine screws: Similar to bolts, but with screwdriver-type heads.

Magnafluxing: Using a magnet and a ferrous (iron) powder to highlight cracks and pores in cast iron blocks and heads.

Magnetic pickup tool: Attracts and holds ferrous objects so they can be picked up.

Mag-tach timing meter: A magnetically triggered tachometer for measuring engine speed (rpm) and for adjusting diesel injection timing.

Main bearing clearance: The space between the crankshaft main journal and the main bearing insert.

Main bearing knock: Similar to rod bearing knock, but the sound is slightly deeper or duller. Bearing wear, and possibly journal wear, is allowing the crankshaft to move up and down inside the cylinder block.

Main bearings: Bearings that provide an operating surface for the crankshaft.

Main bore: A series of holes machined from the front to the rear of the cylinder block for the crankshaft main bearings.

Main caps: Fastened to the bottom of the cylinder block, the main caps form one-half of the main bore. They hold the crankshaft and main bearings in place.

Main journals: Precision-machined and polished surfaces that ride on the main bearings. They are machined along the centerline of the crankshaft.

Major tune-up: Includes the steps for a minor tune-up plus diagnostic tests such as scope analysis, compression tests, or vacuum tests.

Material Safety Data Sheet (MSDS): Document that lists all of the known dangers and treatment procedures for a specific chemical.

Mechanical camshaft: Camshaft made to work with solid lifters. The lobe is shaped to produce more constant opening of the valve.

Mechanical efficiency: Comparison of brake horsepower (bhp) to indicated horsepower (ihp). It is a measurement of mechanical friction.

Mechanical lifter: Transfers camshaft lobe action to the push rod. It is not self-adjusting, requires periodic setting, and is noisy during engine operation.

Media blaster: A blast cabinet fitted with long rubber gloves allows you to blast rusted surfaces while viewing through a glass window. Shop air pressure is used to direct a steam of air and media, such as sand, iron slag, or other abrasive material, into the part, scrubbing the surfaces clean.

Media blasting: Removes paint, carbon, and other dry deposits from parts.

Metal lines: Double-wall steel tubing used as fuel lines and brake lines.

Micrometer: Tool capable of making engine measurements accurate to one one-thousandth of an inch (.001") or one one-hundredth of a millimeter (0.01 mm).

Microns: Millionths of an inch.

Miller-cycle engine: Engine that uses a modified four-stroke cycle. It is designed with a shorter compression stroke and a longer power stroke to increase efficiency.

Milling machine: Machine used to resurface cylinder heads and blocks.

Mineral deposits: Deposits in the coolant passages that can cause engine overheating.

Minor tune-up: Preventative maintenance done when the engine is in good operating condition.

Mirror probe: Used to inspect for problems in an area that is blocked from view.

Misfiring: Occurs when the fuel mixture fails to ignite and burn properly.

Mixture jet combustion chamber: Uses an extra passage running to the intake valve port to aid swirl in the port and combustion chamber to help burning.

Monolithic catalytic converter: A catalytic converter using a catalyst in the form of a ceramic block.

Motor oil: See *engine oil.*

Multimeter: Used to measure voltage, current, and resistance.

Multiport injection: Electronic fuel injection system in which one injector is located just before each intake port. Also called *port fuel injection.*

Multiviscosity oil: Exhibits the operating characteristics of a thin, light oil when it is cold and a thick, heavy oil when it is hot. Also called *multiweight oil.*

Mushrooming: A condition where valve tips have been hammered larger by the action of the rocker arms.

Mushroom valves: Engine valve that looks like a mushroom and pops open.

N

Net horsepower: The maximum power developed when an engine is loaded down with all of its accessories.

No-crank problem: The engine crankshaft does not rotate properly with the ignition key in the start position.

Noid light: A special test light for checking electronic fuel injector feed circuits.

Nonadjustable rocker arms: Provide no means of changing valve lash.

Noncrossflow combustion chamber: An older design that has been phased out for the more efficient crossflow combustion chamber.

Nonhardening sealer: Sealer used for semipermanent assemblies.

Normal combustion: Occurs in a gasoline engine when a single flame spreads evenly and uniformly away from the spark plug electrodes.

Normally aspirated engine: Uses atmospheric pressure to force air into the engine.

No-start problem: The engine is turned over by the starter, but fails to fire and run on its own power.

Number punch set: Used to stamp identifying numbers onto parts for reference during engine reassembly.

O

Octane rating: Gasoline rating that indicates its ability to resist ping, or knock.

Oil aeration: Air bubbles form in the oil and reduce its lubricating qualities.

Oil clearance: The small space between moving engine parts for the lubricating oil film. Also called *bearing clearance.*

Oil-cooled piston: Piston that uses oil pressure to direct a stream of oil through and onto the piston to help cool the piston head.

Oil cooler: A radiator-like device connected to the lubrication system used to help lower and control the operating temperature of the engine oil.

Oil deflecting rocker arm clips: Small metal clips used to prevent oil from squirting out of the engine rocker arms when operating the engine with the valve covers removed.

Oil filter: Cleans the oil by trapping small particles suspended in the lubricant.

Oil filter housing: A metal part that bolts to the engine and provides a mount for the oil filter.

Oil filter wrench: Wrench designed to grasp and hold the outside diameter of an engine oil filter, allowing the oil filter to be turned for removal.

Oil galleries: Small passages that lead to the crankshaft bearings, camshaft bearings, and valve train components for lubricating oil.

Oil level indicator: Alerts the driver of a dangerously low oil level condition.

Oil pan: Holds an extra supply of oil for the lubrication system.

Oil passages: Holes in the cylinder head so oil can enter from the block.

Oil pickup: A tube and screen assembly extending from the oil pump to the bottom of the oil pan.

Oil pickup screen: Prevents large particles from entering the pickup tube and oil pump. Usually part of the pickup tube.

Oil pressure gauge: Gauge designed to screw into an oil plug hole for measuring engine oil pressure.

Oil pressure sending unit: Screws into one of the oil galleries to sense oil pressure.

Oil pressure test: A gauge is used to measure the actual pressure in the engine.

Oil pump: Pulls oil out of the oil pan and pushes it through the oil filter and oil galleries.

Oil ring: Ring that fits into the lowest groove in the piston to scrape excess oil from the cylinder wall.

Oil service rating: A set of letters printed on the oil container to denote how well the oil will perform under operating conditions.

Oil slinger: Thin, washer-shaped, metal disk that fits in front of the crankshaft sprocket to help spray oil onto the timing chain to prevent wear. Also helps prevent oil from leaking out of the front crankshaft oil seal.

Oil spurt hole: Part of a connecting rod that provides added lubrication for the cylinder walls, piston pin, and other surrounding parts.

Oil viscosity: A measure of the oil's ability to flow.

On-board diagnostics generation one (OBD I): Computer control system that has been in use for years.

On-board diagnostics generation two (OBD II): Computer control system implemented on all new vehicles starting with the 1996 model year. It can detect sensor malfunctions before they become noticeable to the driver or technician.

Open cooling system: Overflow tube is not connected to a coolant reservoir. When excess pressure in the radiator pushes coolant out, the coolant leaks onto the ground. There is no means of automatically adding coolant to the radiator when needed.

Opposed engine: Engine with two banks of cylinders that lay flat or horizontal on each side of the crankshaft. Also termed *pancake engine.*

O-ring seal: A stationary type seal that fits into a groove between two parts. When the parts are assembled, the synthetic rubber seal is partially compressed and forms a leakproof joint.

O-ring valve seal: A small, round seal that fits into an extra groove cut in the valve stem. It seals the gap between the retainer and valve stem.

Outside micrometer: Tool used for measuring external dimensions, diameters, or thicknesses.

Overboost: Too much supercharger boost.

Overbore limit: The maximum amount that a block should be bored oversize without resulting in cylinder walls that are too thin.

Overhead valve engine (OHV): See *cam-in-block engine.*

Overhead camshaft engine (OHC): Engine in which the camshaft is located in the cylinder head.

Oversize bearing: Bearing designed to be used in a bore that has been machined to a larger diameter.

Oversquare engine: Engine with a bore dimension that is larger than the stroke dimension.

Oxides of nitrogen (NO_X): Emissions produced by high temperatures during combustion.

P

Pancake combustion chamber: Older, less common combustion chamber design with valve heads almost parallel with the top of the piston.

Pancake engine: See *opposed engine.*

Particulates: Solid particles of carbon soot and fuel additives that blow out of a vehicle's tailpipe.

PCV valve: Commonly used to control the flow of air through the crankcase ventilation system.

Pellet catalytic converter: A catalytic converter using a catalyst in the form of small ceramic beads.

Pent-roof combustion chamber: Similar to the hemispherical combustion chamber, but it has flat, angled surfaces rather than a domed surface.

Performance problem diagnosis chart: Chart that lists engine performance problems, possible causes, and corrections.

Petroleum: Oil taken directly from the ground. It is a mixture of various elements, including carbon, hydrogen, sulfur, nitrogen, and oxygen.

Pilot: A steel shaft that is installed in the valve guide to support the grinding stone.

Pilot bearing: Mounted in the rear of the crankshaft to support the manual transmission input shaft.

Pin: Used to secure a gear or pulley to its shaft.

Pin hole: Machined through the pin boss for the piston pin.

Pinging: See *preignition.*

Piston: Converts the pressure of combustion into movement.

Piston boss: A reinforced area around the piston pin hole that supports the piston pin under severe loads.

Piston breakage: Part of the piston is cracked or broken, most commonly the skirt.

Piston clearance: The distance between the side of the piston and the cylinder wall.

Piston displacement: The volume the piston displaces (moves) from BDC to TDC. It is determined by multiplying the cylinder diameter by the piston stroke.

Piston-guided connecting rod: Has thrust surfaces formed next to the piston pin to limit axial movement of the rod small end.

Piston head: See *crown.*

Piston head coating: The substance covering a piston head; designed for improved durability and power production.

Piston knock: A loud metallic knocking sound produced when the piston flops back and forth inside its cylinder. Caused by piston skirt or cylinder wear, excessive piston-to-cylinder clearance, or mechanical damage.

Piston pin: Pin that allows the connecting rod to swing back and forth inside the piston.

Piston pin knock: Excessive clearance exists between the piston pin and piston pin bore or connecting rod bushing allowing the pin to hammer against the rod or piston as the piston changes direction in the cylinder.

Piston pin offset: Locates the piston pin hole slightly to one side of the piston centerline in order to quiet the piston operation.

Piston protrusion: A comparison of the piston head location at TDC in relation to the block deck.

Piston ring compressor: Tool used to squeeze the piston rings into their grooves so that the piston assembly can be installed into its cylinder.

Piston ring expander: Tool used to remove and install piston rings.

Piston ring gap: The clearance between the ends of a piston ring when it is installed in cylinder.

Piston ring grooves: Slots machined in the piston for the piston rings.

Piston rings: Rings that fit into grooves machined into the sides of the piston to keep combustion pressure from entering the crankcase and engine oil from entering the combustion chamber.

Piston ring shape: The cross-section of a piston ring.

Piston ring side clearance: The space between the side of a compression ring and the piston groove.

Piston shape: Generally refers to the shape of the piston head or crown.

Piston skirt: The side of the piston below the last ring. Prevents the piston from tipping in its cylinder.

Piston slap: A loud metallic knocking sound produced when the piston flops back and forth inside its cylinder. Caused by piston skirt or cylinder wear, excessive piston-to-cylinder clearance, or mechanical damage.

Piston stroke: The distance that the piston moves from top dead center (TDC) to bottom dead center (BDC).

Piston taper: The top of the piston is machined slightly smaller than the bottom, making the piston almost equal in size at the top and bottom when in operation (hot).

Piston-to-cylinder clearance: The distance between the edge of the piston and the cylinder wall.

Piston wear: The amount of decrease in the piston diameter due to friction.

Plain main bearing: A split-type bearing that limits up and down movement of the crankshaft in the block.

Plastigage: A special measuring device normally used to check clearances between internal surfaces. Commonly used to measure clearances in engine connecting rod bearings, crankshaft main bearings, and oil pumps.

Plateau honing: Using extra fine stones to final size the cylinder dimensions and final smooth the surface finish.

Pollutants: Undesirable compounds produced by combustion.

Poor fuel economy: A condition in which the engine consumes too much fuel.

Poppet: Engine valve that looks like a mushroom and pops open.

Pop-up piston: Piston that has a head that is curved upward. This type is normally used with a hemispherical cylinder head.

Port fuel injection: See *multiport injection.*

Ports: Passages in the cylinder head that allow air-fuel mixture to enter the combustion chamber and exhaust gasses to flow out of the engine.

Positive crankcase ventilation (PCV) system: Pulls fumes from the engine crankcase into the intake manifold so they can be burned before entering the atmosphere.

Postignition: Surface ignition that occurs after the spark plug fires.

Power: Work done over a period of time.

Power balance test: Measures the power output from each of the engine's cylinders.

Power brush: Brush mounted in a drill and used to remove hard carbon deposits.

Power hone: A machine shop tool that sizes and finishes the cylinder walls in an engine block.

Power stroke: The air-fuel mixture is ignited and burned to produce gas expansion, pressure, and a powerful downward piston movement.

Preignition: A hot spot in the combustion chamber ignites the air-fuel mixture before the spark plug fires. Also known as *pinging.*

Pressed-in valve guides: Sleeves made of iron, steel, or bronze that are force-fitted into the cylinder head.

Pressed-in valve seat: A separate, machined valve seat force-fitted into the cylinder head.

Press-fit piston pin: Piston pin that is force-fitted into either the rod or piston.

Pressure-fed oiling: Oil is forced through galleries and to the moving parts.

Pressure gauge: Gauge used to check various pressures, such as fuel pump pressure, pressure regulator pressure, engine oil pressure, or turbo boost pressure.

Pressure regulator: Used to set a specific pressure in the shop's compressed-air system.

Pressure-relief valve: A spring-loaded bypass valve in the oil pump, engine block, or oil filter housing that limits the maximum oil pressure.

Pressure tester: A hand-operated air pump used to pressurize the cooling system for leak detection.

Pressure testing: Filling the coolant or oil passages in the block with air pressure to find cracks or flaws.

Primary wire: Small wire that carries battery or alternator voltage. It normally has plastic insulation to prevent shorting.

Prony brake: Older device used to measure brake horsepower.

Prussian blue: A metal stain often used to layout metal parts. It can be used to check the valve-to-seat contact.

Puller: Tool that uses a screw action to force a hub, damper, pulley, gear, etc., off of its shaft.

Pulse air system: Performs the same function as *an air injection system;* however, it uses the pressure pulses in the exhaust system to operate check valves.

Purge line: Removes or cleans the stored vapors out of the charcoal canister. It is connected to the canister and the engine intake manifold.

Purge valve: Vacuum- or computer-operated valve that controls the flow of fuel vapor stored in the charcoal canister to the intake manifold.

Push-on fitting: Special, snap-type fitting on a fuel line.

Push rod: Metal tube with specially formed ends that transfers motion between the lifters and the rocker arms in cam-in-block engines.

Push rod guide plates: Used to hold the push rods in alignment with the rocker arms.

Q

Quick seal rings: Piston rings with soft coatings to help the ring wear in quickly and form a good seal.

R

Radiator: Transfers heat from the coolant to the outside air.

Radiator cap: Seals the top of the radiator filler neck, allows the system to be pressurized, relieves excess pressure and vacuum to protect against system damage, and allows coolant flow into and out of the coolant reservoir in a closed system.

Radiator cap pressure test: Measures the cap opening pressure and checks the condition of the sealing washer.

Radiator hoses: Flexible hoses that carry coolant between the coolant passages in the engine and the radiator.

Radiator shroud: Keeps air from circulating between the back of the radiator and the front of the fan; typically attached to the rear of the radiator and surrounds the fan.

Reactivity hazard: Anything that reacts violently or releases poisonous gasses when in contact with other materials. Also, materials that generate toxic mists, fumes, vapors, and flammable gasses.

Reading spark plugs: Inspecting the condition and color of spark plug tips.

Rear main oil seal: Seal that prevents oil from leaking out of the engine past the crankshaft spinning in the rear main bearing.

Reciprocating weight: The weight of the piston and rod assembly.

Refining: Heating and breaking down crude oil into different compounds, such as gasoline, diesel fuel, and liquefied petroleum gas.

Regenerative braking: Uses the energy generated by slowing the vehicle to generate electricity, which is stored in the batteries.

Remote starter switch: Used to turn the engine over without turning the ignition key to the start position.

Respirators: Filter masks that should be worn when working around any kind of airborne impurities.

Retarding: The ignition timing causes the spark to happen later in the compression stroke.

Reverse-flow cooling system: Coolant circulates through the head before moving through the block and into the radiator.

Reverse flushing: Low-pressure compressed air forces water through the core in the opposite direction of normal flow.

Ribbed belt: A belt that has a flat or rectangular cross section with small ridges formed around its inside diameter.

Rich fuel mixture: Less than 14.7 parts of air for every one part of fuel.

Ridge reamer: Tool with small cutter blades for removing a cylinder ridge.

Rigid hone: Hone with hard, flat, abrasive stones attached to stationary, but adjustable, arms. Used when the worn cylinders do not require boring, but a large amount of honing is needed to true and texture the cylinder walls in the block.

Ring expander: A special tool for spreading and installing rings.

Ring filing: Using a special, thin grinding wheel to remove metal from the ends of a ring.

Ring gap: The split or space between the ends of a piston ring when installed in the cylinder.

Ring groove cleaner: Tool used to remove carbon deposits from inside of piston ring grooves.

Ring lands: The areas between and above the ring grooves.

Ring radial wall thickness: The distance from the face of the ring to its inner wall.

Ring ridge: A lip formed where the rings do not wear the cylinder.

Ring spacers: Thin, steel rings that fit below the compression rings to restore ring side clearance to within limits.

Ring width: The distance from the top to the bottom of a piston ring.

Rocker arm: Transfers movement of the push rod to the valve.

Rocker arm pedestal: A variation of a rocker arm stud. Raised pads are formed in the cylinder head. An aluminum pivot and bridge bolt to the pedestal to support the rocker arm.

Rocker arm ratio: A comparison of the length of each end of the rocker arm.

Rocker arm shaft: Long, steel tube that allows the rocker arms to pivot up and down.
Rocker arm stud: Threaded stud in the head that supports the rocker arm.
Rocker assembly service: Cleaning, inspection, and measurement of the rocker arms, rocker arm shafts, and related components.
Rocker-mounted tappet: Has the hydraulic lifter installed inside the end of the rocker arm.
Rod bearing clearance: Small space between the rod bearing and crankshaft rod journals, which allows oil to enter the bearing.
Rod rebuilding machine: Used to re-machine engine connecting rod bores during repairs.
Roll hardened journal: Journal that is specially treated during manufacturing to make it harder and more wear resistant.
Roller lifter: Hydraulic or mechanical lifter with a small roller that operates on the camshaft lobe.
Rollover valve: Keeps liquid fuel from entering the vent line after an accident where the vehicle rolls upside-down.
Rotary engine: Uses a triangular rotor instead of conventional pistons. Also called a *Wankel engine.*
Rotary oil pump: Uses a set of star-shaped rotors in a housing to pressurize the motor oil.
Rough idle: The engine seems to vibrate on its mounts. In some cases, a popping noise can be heard at the tailpipe.
RTV sealer: Sealer that cures (dries) from moisture in the air. Used to form a gasket on thin, flexible flanges.
Run-on: The engine will not shut off, but keeps firing, coughing, and producing power. Also called *after-running* or *dieseling.*

S

Scan tool: Device that can interface with the on-board computer to check for problems in vehicle systems and display diagnostic trouble codes.
Scavenging: A condition in which the exhaust and intake valves are open so the incoming air-fuel charge helps push out exhaust gasses.
Scrapers: Commonly used by an engine technician to remove old gasket material and oil buildups.
Scuff pad: Pad mounted in an air drill or similar tool that can cut through and lift off gasket material and silicone, yet is too soft to easily damage metal surfaces. Also called a scuff wheel.
Seal: Used to prevent leakage between a stationary part and a moving part. Normally made of synthetic rubber molded onto a metal body.
Seat narrowing: Grinding the outer or inner valve seat surface to narrow and move the seat. Also called seat positioning.
Seat stone: A grinding stone mounted on a threaded drive attachment used to cut a fresh surface on the face of the valve seat.
Seat width: The area of the valve seat that actually makes contact with the valve face.
Secondary wire: A wire with extra thick insulation that carries high voltage from the ignition coil to the spark plug. Also called high-tension cable.
Sectional intake manifold: Manifold that is constructed in two or more pieces.
Self-powered test light: Test light that contains batteries. Used to check for circuit continuity in a circuit with its power source removed.
Sensor: A device that can change its electrical signal based on a change in a condition.
Sensor sockets: Sockets with a special shape or opening in the side to allow service of engine and other sensors.
Sequential turbocharger system: Uses two electronically controlled turbochargers that operate as needed. Also called *twin turbo system.*
Series circuit: Only has one path for current flow.
Series-parallel circuit: A circuit consisting of at least one series circuit and one parallel circuit.
Serpentine belt: A single belt that "snakes" around many pulleys.
Setscrew: A fastener designed to, typically, lock a pulley or gear onto a shaft.
Sharp valve margin: The valve margin is too small.
Short block: Cylinder block with all internal parts installed.
Short circuit: A defective wire or component touches ground and causes excess current flow.
Siamese cylinders: Cylinders connected with cast metal ribs without a coolant passage between the cylinders.
Siamese ports: One large port divides into two smaller ports right before the engine valve.
Silent shaft: See *balancer shaft.*
Single overhead camshaft (SOHC): Engine that has one camshaft per cylinder head.
Single plane intake manifold: Manifold in which one plenum feeds all runners. Allows high airflow at higher engine speeds.
Slant engines: Inline engines with cylinders that are angled to vertical.
Sleeve: Thick-wall, tube-shaped inserts that fit into a cylinder block and form the cylinder walls.
Sleeve protrusion: The distance that a cylinder sleeve sticks up above the block deck.
Sleeveless aluminum block: Aluminum cylinder block with a special alloy coating bonded onto the cylinder walls to improve piston ring and cylinder wall sealing and wear resistance.
Slipper skirt: Produced when the portion of the piston skirt below the piston pin ends are removed to provide clearance between the piston and crankshaft counterweights.
Slow-crank problem: The engine crankshaft rotates at a lower-than-normal speed.
Sluggish engine: Lack of engine power that causes the vehicle to accelerate slowly.
Snap ring: Fits into a groove in a part and commonly holds shafts, bearings, gears, pins, and other similar components in place.
Snap-ring pliers: Pliers with specially shaped, usually pointed tips used to expand or contract snap rings for removal and installation.
Sodium-filled valve: Heat-absorbing sodium is contained in the hollow valve stem and valve head. Used when extra cooling and reduced weight are needed.
Solenoid: An electromagnet with a sliding core.
Solid lifter: Transfers camshaft lobe action to the push rod. It is not self-adjusting, requires periodic setting, and is noisy during engine operation.
Spark ignition: Uses an electric ignition and a spark plug to start the combustion of the fuel.
Spark intensity test: Measures the brightness and duration of the electric arc (spark) produced by the ignition system.
Spark knock: A form of preignition caused when the electric arc at the spark plug happens too early in the compression stroke.
Spark plug: The "match" that starts the air-fuel mixture burning in the combustion chamber.
Spark plug cleaner: Uses shop air pressure and blast media (sand-like particles) to remove carbon deposits from a spark plug.
Spark test: See *spark intensity test.*
Spark tester: Used to check the basic operation of the engine

ignition system.

Spin-on oil filter: A sealed unit having the element permanently enclosed in the filter body.

Splash oiling: Oil is thrown from moving parts.

Splayed rod journal: Has two separate rod journals machined offset from each other on the same crankpin.

Split runner intake manifold: Manifold with two separate runners leading to each intake valve.

Spray-on gasket solvent: Solvent used to soften a gasket so that it can be scraped off more easily.

Spring bind: A valve spring is compressed too much and its coils hit each other, stopping more spring compression and the valve opening action.

Spun rod bearing: A bearing that has been hammered out flat and then rotated inside of the connecting rod bore.

Square engine: Engine with equal bore and stroke dimensions.

Squish area: When the piston reaches top dead center (TDC), it comes very close to the bottom of the cylinder head. This squeezes the air-fuel mixture in that area and forces it out into the main part of the chamber.

Staking: Placing small dents in the cylinder head next to the valve seat to keep the seat in place during engine operation.

Stalling: A condition in which engine stops running. Also called *dying.*

Standard-size bearing: Bearing that has the original dimensions specified by the engine manufacturer.

Starter motor: An electric motor used to spin the flywheel, and thus the crankshaft, until the engine starts and runs.

Starting system: Uses a battery, ignition switch, high-current relay, and electric motor to rotate the crankshaft until the engine can begin running on its own power.

Stellite valve: Valve with a special, hard-metal coating on its face.

Stem tip: The upper end of the valve stem.

Stethoscope: A listening device that amplifies the sound generated by internal parts.

Stoichiometric fuel mixture: A chemically correct air-fuel ratio (14.7 parts air to 1 part fuel).

Straightedge: Normally used with a flat feeler gauge to check or measure the warpage (flatness) of engine parts.

Straight skirt: Piston skirt that is flat across the bottom.

Stratified charge combustion chamber: Uses a flame in a small combustion prechamber to ignite and burn the fuel in the main, large combustion chamber.

Structural oil pan: Designed to add strength to the engine bottom end and cylinder block.

Stuck valve: Condition in which a valve stem rusts or corrodes and locks in the valve guide.

Stumble: The engine does not accelerate normally when the gas pedal is pressed, or stalls before developing power. Also called *hesitation.*

Sump: The lowest area in the oil pan where oil collects.

Supercharger: A special fan assembly driven by a belt on the engine that compresses intake air.

Supercharger bypass valve: Used to improve engine efficiency by only providing maximum boost when needed.

Supercharger drive pulley: Draws power from the crankshaft through a large serpentine belt to drive the supercharger.

Supercharger extension housing: Bolted to the front of the supercharger housing, it encloses a shaft that connects one of the supercharger timing gears to the drive pulley.

Supercharger housing: The large aluminum enclosure for the rotors, rotor bearings, and drive gears.

Supercharger intercooler: Removes heat from the compressed air before it enters the engine.

Supercharger rotors: Meshing, twisted blades or fans that spin to produce pressure inside the supercharger housing.

Supercharger timing gears: Gears that keep the two rotors timed and spinning at the same speed.

Surface ignition: A surface in the combustion chamber is heated to a temperature that is high enough to ignite the air-fuel mixture.

Surging: When driving at a steady speed, engine power fluctuates up and down.

Swirl combustion chamber: The shape of the intake and exhaust ports and the combustion chamber roof create a swirling effect to help mix the air-fuel mixture.

Systematic approach: Using knowledge of auto technology and a logical process of elimination to diagnose engine performance problems.

T

Tach-dwell meter: A tachometer and a dwell meter combined into one instrument. The tachometer is for measuring engine speed (rpm). The dwell meter measures in degrees.

Tap: A tool for cutting internal threads.

Tappets: Ride on the camshaft lobes and transfer motion to the rest of the valve train.

Taxable horsepower (thp): A general rating of engine size used to find the tax placed on a vehicle.

Teardown: The basic disassembly of the engine.

Technical service bulletins (TSBs): Bulletins published by vehicle manufacturers to explain the symptoms of common problems and the tests and corrections for problems that frequently occur in one make or model of a vehicle.

Telescoping gauge: Used in conjunction with a micrometer to measure internal part bores or openings.

Temperature probes: Probes used to measure the temperature of various components, such as radiator temperature, exhaust manifold temperature, air inlet temperature, and catalytic converter temperature.

Test light: Used to check a circuit for power or voltage, or to quickly check for circuitry continuity.

Thermal efficiency: Comparison of the amount of energy contained in the fuel used to the amount of energy produced by the engine.

Thermostat: A temperature-sensing valve that controls the flow of coolant through the radiator.

Thermostatic air cleaner system: See *heated air inlet system.*

Thermostatic fan clutch: An engine fan that contains a temperature-sensitive, bimetallic spring to control fan action.

Thread chaser: A tool for cleaning up damaged threads.

Thread pitch gauge: Gauge used to determine thread series.

Thread repair insert: Repairs damaged internal threads and allows the use of the original size fastener.

Three-angle valve job: Three different stone angles are used to grind the outer and inner seat surface in order to narrow and move the seat.

Three-way converter: A catalytic converter, usually coated with rhodium and platinum, to reduce HC, CO, and NO_X emissions.

Throttle body injection: Electronic fuel injection system in which one or two injectors are mounted inside a throttle body assembly.

Throttle valve: Connected to the accelerator pedal, it controls airflow, engine speed, and engine power.

Thrust main bearing: Crankshaft bearing needed to limit crankshaft endplay.

Thrust washers: Serve the same purpose as a conventional main thrust bearing, but not made as part of the main bearing.

Timing belt: Cogged (square-toothed) belt that provides an accurate, quiet, light, and dependable means of turning the camshaft.

Timing belt cover: A plastic or sheet metal shroud around the timing belt to protect the belt from external debris.

Timing belt sensor: Detects excessive tensioner extension, which indicates excessive timing belt wear and stretch.

Timing belt tensioner: A wheel that keeps the timing belt tight on its sprockets.

Timing chain: Transfers rotation from the crankshaft to the camshaft and may power other devices.

Timing cover: See *engine front cover.*

Timing gear backlash: The amount of clearance between the timing gear teeth.

Timing marks: Dots, lines, circles, or other shapes indented or cast into the timing sprockets, engine block, and cylinder head and used to ensure valve timing.

Timing mechanism: See *camshaft drive mechanism.*

Top dead center (TDC): The point of travel where the piston is at its highest point in the cylinder.

Torque sequence: Pattern of tightening fasteners that ensures parts are clamped together evenly.

Torque specifications: Values given by the vehicle manufacturer to which fasteners should be tightened.

Torque wrench: Applies a measured amount of turning force to a fastener (bolt or nut).

Torque-to-yield: A bolt tightening method in which the bolt is tightened until it is deformed (stretched).

Torquing: The process of properly tightening a fastener.

Torsional vibration: A high-frequency vibration resulting from normal power pulses in the engine.

Toxicity hazard: Materials like lead, cadmium, chromium, arsenic, and other heavy metals that can pollute and make water and soil harmful. Also, used motor oil, solvents, and other chemicals in the auto shop.

Tubing cutter: Tool used to cut a metal line.

Tubing equipment: Equipment used to make new lines running to an engine, such as tube cutters, flaring tools, and reamers.

Tubular exhaust manifolds: Exhaust manifolds shaped with very long, smooth-flowing runners. Also called headers.

Tuned runner intake manifold: Manifold with very long runners of equal length. Also called a ram intake manifold.

Turbo lag: The short delay before a turbocharger develops sufficient boost.

Turbo timer: A device that keeps the engine running for a cooldown period after the ignition key has been shut off.

Turbocharger: A special fan assembly driven by engine exhaust gasses that compresses intake air.

Turbocharger intercooler: Removes heat from the compressed air before it enters the engine.

Turbulence: Swirl of air and fuel in the combustion chamber.

Twin turbo system: See *sequential turbocharger system.*

Two-piece piston: Piston constructed in two parts—the crown and the skirt.

Two-valve combustion chamber: Has one intake valve and one exhaust valve per cylinder.

U

Umbrella valve seal: Seal shaped like an inverted cup. It covers the small clearance between the stem and guide, keeping excess oil from splashing into the guide.

Undersize bearing: Bearing designed to be used on a journal that has been machined to a smaller diameter.

Undersquare engine: Engine with a stroke dimension that is larger than the bore dimension.

V

Vacuum: An area of pressure below atmospheric pressure, or negative pressure.

Vacuum gauge: Gauge commonly used to measure negative pressure or vacuum (suction). It reads in inches of mercury (in/Hg) or centimeters of mercury (cm/Hg).

Vacuum test: Using a vacuum gauge to measure intake manifold vacuum and indicate engine condition.

Valve adjusting wrench: A special tool for turning the rocker arm adjusting screws.

Valve adjustment: See *valve train clearance adjustment.*

Valve covers: Sheet metal or plastic covers fastened to the top of the cylinder heads to prevent oil spraying from the valve train from leaking out of engine.

Valve duration: How long the engine valves stay open.

Valve face: The angled part of the valve that contacts the valve seat.

Valve face angle: The angle formed between the valve face and valve head.

Valve face erosion: The wearing away of the valve face where it contacts the seat in the cylinder head.

Valve float: When the valve spring is not able to fully close the valve before the camshaft begins to open the valve.

Valve grinding: Creates a fresh, smooth surface on the valve face and stem tip.

Valve grinding machine: Machine used to resurface valve faces and valve stem tips.

Valve guide inserts: Sleeves made of iron, steel, or bronze that are force-fitted into the cylinder head and in which the valve stem rides.

Valve guides: Small holes machined through the top of the head into the intake and exhaust ports in which the engine valve stems slide up and down.

Valve head: The large, disk-shaped surface exposed to the combustion chamber.

Valve keepers: Fit between the valve retainers and valve stems to hold the springs on the valves.

Valve lapping compound: An abrasive paste material used to check the valve-to-seat contact.

Valve lash: The space between the rocker arm tip or cam follower and the valve stem tip with the valve closed. Also called valve clearance.

Valve lash adjustment: See *valve train clearance adjustment.*

Valve lift: How far the engine valves open.

Valve lifters: Ride on the camshaft lobes and transfer motion to the rest of the valve train.

Valve margin: The flat surface on the outer edge of the valve head.

Valve oil shedder: A variation of an umbrella type oil seal.

Valve overlap: When the intake and exhaust valves in a cylinder are opened at the same time, helping to scavenge burned gasses out of the cylinder.

Valve relief piston: Piston with small indentations either cast or machined into the crown to provide ample piston-to-valve clearance.

Valve retrusion: The head of the valve sinking into the cylinder head.

Valve rotators: Used on some engines to spin or turn the valves in their guides in order to help prevent carbon buildup on the valve.
Valve seals: Prevent oil from entering the cylinder head ports through the clearance between the valve stems and valve guides.
Valve seat angle: The angle formed by the face or contact surface of the seat.
Valve seat grinder: Tool used to resurface the valve seats in a cylinder head.
Valve seat machine: A specialized milling machine designed for valve guide and valve seat service.
Valve seat reconditioning: Using a grinding stone or carbide cutter to resurface the cylinder head valve seats.
Valve seats: Round, machined surfaces in the port openings to the combustion chambers.
Valve shape: The configuration of the valve head.
Valve spring: Keeps the valve in a normally closed position.
Valve spring compressor: Clamp-like device that squeezes (compresses) the valve spring so that the valve keepers can be removed or installed.
Valve spring free height: The length of the spring in a normal, uncompressed condition.
Valve spring installed height: The length (or height) of the valve spring when installed on the engine with the valve closed.
Valve spring open length: The length (or height) of the valve spring when installed on the engine with that spring's valve fully open.
Valve spring seat: A cup-shaped washer installed between the cylinder head and bottom of the valve spring to hold the bottom of the valve spring squarely on the head and valve stem.
Valve spring shield: Thin metal shield that surrounds the top and upper sides of the spring and helps keep excess oil off of the valve stem.
Valve spring shims: Precise-thickness flat washers that fit under the valve springs to reduce the installed height of the valve spring and maintain the correct spring tension.
Valve spring squareness: Measure of the valve spring's centerline to its seating surface.
Valve spring tension: The stiffness of a valve spring.
Valve spring tester: Used to measure the tension of valve springs.
Valve stem: A long shaft extending out of the valve head.
Valve stem cap: A hardened steel cup that installs over the valve stem. It is free to rotate on the tip to reduce friction and wear.
Valve stem-to-guide clearance: The small space between the valve stem and the valve guide.
Valve timing: Determines when the valves open in relation to the piston position.
Valve train: Operates the engine valves, timing valve opening and closing to produce the four-stroke cycle.
Valve train clearance adjustment: Done by turning the rocker arm fastener or adjusting screw, installing OHC follower shims, or using longer push rods.
Valve train noise: A light tapping or clicking sound produced from under the valve or camshaft cover.
Vapor lock: Fuel is overheated, forming air bubbles that may reduce or block fuel flow.
Variable backpressure exhaust engine (VBEE): Uses different paths from the engine exhaust manifolds to improve engine efficiency.
Variable compression piston: An experimental, two-piece piston that has a variable height controlled by engine oil pressure.
Variable compression ratio engine (VCRE): Engine that can alter the volume of its combustion chambers, and thus its compression ratio, for improved operating efficiency.
Variable intake engine (VIE): Adjusts its intake manifold runner length to match engine speeds.
Variable valve timing engine (VVTE): Engine that can alter valve opening and closing independent of crankshaft rotation.
V-belt: A belt with a cross section that looks like the letter V.
V-blocks: Metal stands for holding parts, usually a camshaft, for measuring straightness.
Vehicle identification number (VIN): A unique number identifying the vehicle and some of the equipment installed on the vehicle.
Vibration damper: A rotating mass that helps control torsional vibration in the engine. Also called a harmonic balancer or crankshaft damper.
Volt-ohm-milliammeter (VOM): Used to measure voltage, current, and resistance.
Volumetric efficiency: A measure of how much air the engine can draw in on its intake strokes.
V-type crankshaft: Has two connecting rods bolted to the same journal.
V-type engine: Engine with two banks (sets) of cylinders that lay at an angle from vertical on each side of the crankshaft.

W

Wankel engine: See *rotary engine.*
Warped cylinder head: Cylinder head that is no longer true (flat).
Warranty work: Repairs done on vehicles or engines that are still under warranty.
Waste gate: Limits the maximum boost pressure developed by the turbocharger.
Water jackets: See *coolant passages.*
Water pump: Uses centrifugal force to circulate coolant through the engine coolant passages, hoses, and radiator.
Wedge combustion chamber: A combustion chamber that is shaped like a triangle or wedge when viewed from the side.
Wet compression test: Compression test in which a small amount of oil is placed in the cylinder to determine if bad rings are causing low compression.
Wet sleeve: Cylinder sleeve that is exposed to engine coolant.
Wild knock: A pinging sound that can last for a few moments or come and go for no apparent reason.
Wire feeler gauge: Gauge with precision-sized wires labeled by diameter or thickness, normally used to measure slightly larger spaces or gaps than a flat feeler gauge.
Work: When force causes movement.
Wrist pin: Pin that fits through the hole in the piston and the small end of the connecting rod, allowing the piston to swing on the connecting rod.
W-type engine: Similar to a V-type engine in that it has two banks of cylinders, but the cylinder bores in each bank are staggered.

Z

Zero valve lash: No clearance in the valve train; provides quiet operation.

Index

A

B

C

D

E

F

G

H

I

J

K

L

M

N

O

P

Q

R

S

T

U

V

W

Z